36 $\displaystyle\int \frac{du}{u^2\sqrt{a^2 - u^2}} = -\frac{1}{a^2 u}\sqrt{a^2 - u^2} + C$

37 $\displaystyle\int (a^2 - u^2)^{3/2}\, du = -\frac{u}{8}(2u^2 - 5a^2)\sqrt{a^2 - u^2} + \frac{3a^4}{8}\sin^{-1}\frac{u}{a} + C$

38 $\displaystyle\int \frac{du}{(a^2 - u^2)^{3/2}} = \frac{u}{a^2\sqrt{a^2 - u^2}} + C$

Forms Involving $\sqrt{u^2 - a^2}$

39 $\displaystyle\int \sqrt{u^2 - a^2}\, du = \frac{u}{2}\sqrt{u^2 - a^2} - \frac{a^2}{2}\ln\left|u + \sqrt{u^2 - a^2}\right| + C$

40 $\displaystyle\int u^2\sqrt{u^2 - a^2}\, du = \frac{u}{8}(2u^2 - a^2)\sqrt{u^2 - a^2} - \frac{a^4}{8}\ln\left|u + \sqrt{u^2 - a^2}\right| + C$

41 $\displaystyle\int \frac{\sqrt{u^2 - a^2}}{u}\, du = \sqrt{u^2 - a^2} - a\cos^{-1}\frac{a}{u} + C$

42 $\displaystyle\int \frac{\sqrt{u^2 - a^2}}{u^2}\, du = -\frac{\sqrt{u^2 - a^2}}{u} + \ln\left|u + \sqrt{u^2 - a^2}\right| + C$

43 $\displaystyle\int \frac{du}{\sqrt{u^2 - a^2}} = \ln\left|u + \sqrt{u^2 - a^2}\right| + C$

44 $\displaystyle\int \frac{u^2\, du}{\sqrt{u^2 - a^2}} = \frac{u}{2}\sqrt{u^2 - a^2} + \frac{a^2}{2}\ln\left|u + \sqrt{u^2 - a^2}\right| + C$

45 $\displaystyle\int \frac{du}{u^2\sqrt{u^2 - a^2}} = \frac{\sqrt{u^2 - a^2}}{a^2 u} + C$

46 $\displaystyle\int \frac{du}{(u^2 - a^2)^{3/2}} = -\frac{u}{a^2\sqrt{u^2 - a^2}} + C$

Forms Involving $a + bu$

47 $\displaystyle\int \frac{u\, du}{a + bu} = \frac{1}{b^2}(a + bu - a\ln|a + bu|) + C$

48 $\displaystyle\int \frac{u^2\, du}{a + bu} = \frac{1}{2b^3}\left[(a + bu)^2 - 4a(a + bu) + 2a^2\ln|a + bu|\right] + C$

49 $\displaystyle\int \frac{du}{u(a + bu)} = \frac{1}{a}\ln\left|\frac{u}{a + bu}\right| + C$

50 $\displaystyle\int \frac{du}{u^2(a + bu)} = -\frac{1}{au} + \frac{b}{a^2}\ln\left|\frac{a + bu}{u}\right| + C$

51 $\displaystyle\int \frac{u\, du}{(a + bu)^2} = \frac{a}{b^2(a + bu)} + \frac{1}{b^2}\ln|a + bu| + C$

52 $\displaystyle\int \frac{du}{u(a + bu)^2} = \frac{1}{a(a + bu)} - \frac{1}{a^2}\ln\left|\frac{a + bu}{u}\right| + C$

53 $\displaystyle\int \frac{u^2\, du}{(a + bu)^2} = \frac{1}{b^3}\left(a + bu - \frac{a^2}{a + bu} - 2a\ln|a + bu|\right) + C$

54 $\displaystyle\int u\sqrt{a + bu}\, du = \frac{2}{15b^2}(3bu - 2a)(a + bu)^{3/2} + C$

55 $\displaystyle\int \frac{u\, du}{\sqrt{a + bu}} = \frac{2}{3b^2}(bu - 2a)\sqrt{a + bu} + C$

56 $\displaystyle\int \frac{u^2 du}{\sqrt{a + bu}} = \frac{2}{15b^3}(8a^2 + 3b^2u^2 - 4abu)\sqrt{a + bu} + C$

57 $\displaystyle\int \frac{du}{u\sqrt{a + bu}} = \frac{1}{\sqrt{a}}\ln\left|\frac{\sqrt{a + bu} - \sqrt{a}}{\sqrt{a + bu} + \sqrt{a}}\right| + C, \quad \text{if } a > 0$

$\qquad\qquad = \frac{2}{\sqrt{-a}}\tan^{-1}\sqrt{\frac{a + bu}{-a}} + C, \quad \text{if } a < 0$

58 $\displaystyle\int \frac{\sqrt{a + bu}}{u}\, du = 2\sqrt{a + bu} + a\int \frac{du}{u\sqrt{a + bu}}$

59 $\displaystyle\int \frac{\sqrt{a + bu}}{u^2}\, du = -\frac{\sqrt{a + bu}}{u} + \frac{b}{2}\int \frac{du}{u\sqrt{a + bu}}$

60 $\displaystyle\int u^n\sqrt{a + bu}\, du = \frac{2u^n(a + bu)^{3/2}}{b(2n + 3)} - \frac{2na}{b(2n + 3)}\int u^{n-1}\sqrt{a + bu}\, du$

61 $\displaystyle\int \frac{u^n\, du}{\sqrt{a + bu}} = \frac{2u^n\sqrt{a + bu}}{b(2n + 1)} - \frac{2na}{b(2n + 1)}\int \frac{u^{n-1}\, du}{\sqrt{a + bu}}$

62 $\displaystyle\int \frac{du}{u^n\sqrt{a + bu}} = -\frac{\sqrt{a + bu}}{a(n - 1)u^{n-1}} - \frac{b(2n - 3)}{2a(n - 1)}\int \frac{du}{u^{n-1}\sqrt{a + bu}}$

Calculus

WITH ANALYTIC GEOMETRY

SECOND EDITION

The Prindle, Weber & Schmidt Series in Mathematics

Calculus

WITH ANALYTIC GEOMETRY

SECOND EDITION

DENNIS G. ZILL

LOYOLA MARYMOUNT UNIVERSITY

PWS-KENT PUBLISHING COMPANY

BOSTON

PWS–KENT
Publishing Company

20 Park Plaza
Boston, Massachusetts 02116

PWS-KENT Publishing Company is a division of Wadsworth, Inc.

Printed in the United States of America

88 89 90 91 92 — 9 8 7 6 5 4 3 2

Library of Congress Cataloging-in-Publication Data

Zill, Dennis G., 1940–
    Calculus with analytic geometry/Dennis G. Zill,
   —2nd ed.
     p. cm.
    Includes index.
    ISBN 0-534-91620-1
    1. Calculus.   2. Geometry, Analytic.   I. Title.
QA303.Z53 1988   87-25392
515'.15—dc19   CIP

Sponsoring Editor: David Geggis
Production Coordinator: Helen Walden
Production: Miller/Scheier Associates
Copy Editor: Carol Dondrea
Interior Design: Sara Waller/Helen Walden
Cover Design: Helen Walden
Art Coordinator: Bernie Scheier
Interior Illustrations: J&R Art Services
Typesetting: Jonathan Peck Typographers
Cover Printing: New England Book Components
Printing and Binding: R.R. Donnelley & Sons

Photographs appearing with biographical boxes courtesy of The Bettmann Archive and Culver Pictures Inc.

Cover photo: Gregory W.P. Boquist and Jennifer Spalt for Technology International Corporation. This delayed exposure shows an experimental rocket changing course in mid-flight.

Preface

This book is intended for a three-semester or four-quarter course in calculus for students of the sciences, engineering, mathematics, or business. It reflects several of my more strongly held, though hardly original, viewpoints: that a calculus text should consider all of the trigonometric functions as early as possible; that it should motivate and explain the idea of a limit in a manner that is as simple as possible; and that there should be an abundance of examples and problems, and an emphasis on applications.

There appear to be at least three schools of thought regarding the calculus of the trigonometric functions. In one opinion, the study of these functions should be postponed until the second semester or second quarter of the course; in another, only the derivatives and integrals of the sine and cosine functions should be introduced in the beginning chapters of the text. The third opinion holds with the early development of the calculus of all six trigonometric functions. Because I feel that students of mathematics will benefit from working with the trigonometric functions throughout the entire course in calculus rather than in just two thirds of the course, because a substantial number of students encounter these functions early on in courses in science and engineering, because the Chain Rule of differentiation can be illustrated in its full power, because there is a wider variety of applications, and simply because I find calculus to be intrinsically more interesting, I subscribe to the third opinion. The six trigonometric functions are reviewed in Chapter 1, and their derivatives are considered in Chapter 3. The integrals of these functions are discussed in Chapter 4.

Many features of the first edition have been retained.

───── Pedagogical Features ─────

- Sections marked by the symbol [O] are optional and may be skipped if the instructor so chooses.
- All examples are set off by three-sided boxes to delineate them clearly from the general discussion and are numbered for easy reference.
- Important formulas and lists are emphasized by the use of a second color. Definitions and theorems are emphasized by means of colored shading. Important topics in the textual discussion are designated by means of marginal side heads.
- The end of a thorem/proof is indicated by a solid colored ■.
- Each chapter begins with its own table of contents and introduction to the material covered in that chapter and ends with Chapter Review Exercises, which include true/false and fill-in-the blank questions.
- Some of the longer sections (such as partial fractions), although unified by subject matter, are partitioned into subsections to accommodate their coverage in several class periods if desired. To facilitate homework assignments, the exercise sets at the end of these sections are appropriately marked to correspond to these subsections.
- Many sections conclude with brief informal discussions labeled Remarks. These discussions include possible applications of the mathematics just covered, a little bit of history, a review of some basic mathematics pertinent to that chapter, a few words of caution to the student about typical errors to watch out for, or cautions about possible misinterpretations or unwarranted generalizations of definitions and theorems.
- Illustrations are used generously throughout the text (there are over 1200).
- Most exercise sets are massive and include the usual abundance of drill problems along with Miscellaneous Problems, which are either more challenging in nature or expound on material not formally presented, and Calculator/Computer Problems, which are intended to be done on a calculator or computer.
- Answers (including graphs) to odd-numbered problems are included in the text.
- For ease of reference, the right-hand edges of pages in the answer section are marked with color tabs. The page for the appropriate answers is given at the beginning of each exercise set.

───── Supplements ─────

For the Instructor

Complete Solutions Manual by Warren J. Wright, Loyola Marymount University; *Even Numbered Answers*; *Computerized Test Generator* (for the IBM-PC and compatibles); *PWS-KENT GradeDisk*.

For the Student

Student Supplement, Volumes I and II, also by Warren Wright, contains solutions for every third exercise; *True BASIC Calculus Software* by True BASIC, Inc.; *Calculus and the Computer* by Sheldon Gordon, Suffolk Community College.

Improvements in the Second Edition

• Almost every exercise set has been expanded (there are now about 6500 problems). In some cases the added problems are routine, others demand insight into the interpretation of graphs, and others demand greater analytical skills.

• A great effort was made to rid the manuscript of those annoying "math error/typo bugs" that like to infest first editions.

• There are many additional examples, figures, and computer graphics. My philosophy is to supply a figure whenever possible and to encourage the student to start the analysis of a problem by first drawing a picture.

• More applications have been added throughout the text.

• There is an increased emphasis on the use of the computer in the second edition. BASIC computer programs have been added in the sections on Newton's method, approximate integration, and Reimann sums.

• The biographical sketches of mathematicians who were influential in the development of the calculus have been reorganized. The number of these notes has also been increased.

• Several users of the first edition thought that the Chapter Tests were too long, and hence not reflective of the true nature of a test. Consequently, they have been expanded and renamed Chapter Review Exercises.

• Certain topics that were introduced in the exercise sets in the first edition, such as direction cosines, surfaces of revolution, and the root test, are now considered in the body of the text.

• The odd function and even function integration rules and a discussion of integration of piecewise continuous functions have been added to Section 5.7.

• A discussion of cycloidal curves has been added to Section 13.1.

• A section on vectors in two dimensions has been added to Chapter 14.

• Tree diagrams have been introduced as a mnemonic for the Chain Rule for partial derivatives.

• A thorough discussion of the motion of weight on spring has been added to Chapter 19.

Contents

There is a great emphasis in Chapter 1 on understanding the key concept of a function. In preparation for later applications of the derivative, the exercises include many problems on setting up a function by interpreting word descriptions in terms of symbols.

Chapter 2 is all about limits. It is my belief that calculus is more tractable for the student, especially for those who have had no prior exposure to it, when the somewhat obscure, albeit precise, epsilon-delta formulation of the limit is downplayed in favor of a geometric or "intuitive" presentation. I de-emphasize the often unrewarding chase after the unknown "delta."

Students can gain a tremendous amount of insight into the nature of a limit by examining graphs and doing numerical calculations. The epsilon-delta definition of a limit is presented in such a manner that it can be skipped if desired. I must say, however, that in my own classes, I still expect a student to know and to be able to interpret graphically this formal definition of a limit. I want him or her to know that intuition, graphs, and calculation may be convincing, but often have, so to speak, their own limits. Epsilon-delta proofs are not presented in the text proper; for those desiring them, proofs of some of the basic limit theorems are given in Appendix II.

The derivative is motivated in Chapter 3 by both the tangent line problem and the velocity problem.

In Chapter 4, I have introduced rates and rectilinear motion before the discussion of related rates. In the section on related rates I encourage the student to translate the words of a problem into symbols and figures while bearing in mind several components in its solution: What rates are *given*? What rates do we *want*? and, What mathematics do we *know* relating the variables used in the problem?

Chapter 5, on the integral, begins with the concept of the antiderivative or indefinite integral. The substitution method for finding indefinite integrals of powers of functions is the subject of the second section. Summation notation, summation formulas, regular partitions, and finding areas under graphs by computing limits of sums are considered prior to the definition of the definite integral.

Chapter 6 is concerned with the applications of the integral. These applications are presented before the techniques of integration are considered. I feel that the student should be exposed to both the applications of the derivative and the applications of the integral in the first semester. Since Chapter 6 is quite extensive the instructor has latitude to choose those topics that are appropriate to his or her course or program. A section devoted to a review of applications such as arc length problems, liquid pressure, work, pump problems, separation of variables, and centroids, follows the examination of the techniques of integration in Chapter 9. Of course "standard" applications of the integral such as area, volumes of solids of revolution, and so on, are not simply presented and forgotten but appear in the exercise sets whenever possible.

The notion of an inverse function is purposely delayed until Chapter 7 in order that this concept be fresh in a student's mind when dealing with the inverse trigonometric functions in this chapter and the exponential function and inverse hyperbolic functions in the next.

The natural logarithm is defined in Chapter 8 by means of an integral. The student is informed that there is a difference of opinion as to how the natural logarithm should be defined and why, in the author's view, the integral is the preferred definition. However, the alternative approach to the natural logarithm as the inverse of an exponential function is discussed in a separate section. Also, an entire section is devoted to logarithmic differentiation. Given the importance of the hyperbolic functions in subsequent courses in applied mathematics, their treatment is not slighted in this chapter.

The techniques of integration are considered in Chapter 9. Although a section of this chapter is devoted to the use of integral tables, each example in this section is worked in at least one alternative manner. The point is made to the reader that a problem may often be solved in a faster manner through a little thoughtful analysis than by spending untold minutes perusing through unfamiliar formulas in a table.

Chapter 12, on analytic geometry, includes a section on polar equations of conic sections.

Chapter 13, on parametric equations, now has a separate section devoted to the tangent line and arc length problems.

Chapter 14 now starts with a discussion of free vectors and vectors in two dimensions.

Tree diagrams have been added to Section 16.6 as an aid in remembering the Chain Rule for partial derivatives. Optimization problems for functions of three variables subject to two constraints have been added to the discussion of Lagrange Multipliers in Section 16.10.

In Chapters 17 and 18 the similarities between the five steps leading to the definition of the definite integral and the steps leading to the definitions of double, triple, line, and surface integrals are emphasized by summarizing these steps in boxes.

The text ends with Chapter 19, which is an introduction to some of the standard topics in differential equations. Since, in all likelihood, this chapter will not be covered in a traditional three-semester or four-quarter sequence, I have chosen to introduce the important concept of separation of variables in Chapter 8. Exact differential equations are considered in Chapter 16 following the discussion of the differential of a function of two variables.

A review of basic mathematics, such as the laws of exponents, the Binomial Theorem, determinants, Cramer's Rule, and complex numbers can be found in Appendix I.

Acknowledgments

A text often has its inception in the classroom where, after a not quite happy experience with a particular text, the instructor muses "Why doesn't someone write . . . ?" So it was with this book. When I was given the opportunity to write a text on calculus I was faced with the pleasant prospect of being able to codify my partialities, and even sentiments, which have been carefully gathered and honed over many years of college teaching. But I think it is safe to say that any text written at a level below the upper strata of graduate school, when finally published, is a compromise and a synthesis of the ideas from the author, reviewers, and editors.

Of course, the refinements that make up a revision of a text are influenced by even more individuals. Accordingly, I would like to recognize and express my appreciation to the following persons:

the production and editorial staff and PWS-KENT Publishing Company for another job well done,

my fellow faculty members who, after teaching parts of the first edition, were never bashful in offering suggestions,

my former students, who at times actually read the text and told me about it, those users of the first edition who took the time to write or call me with either a critique or a word of encouragement,

Roy Myers, Penn State University, New Kensington, for the new computer graphics on page 780, as well as David Smith and Christopher Morgan for those retained from the first edition,

and the reviewers of this current revision, who contributed many ideas for its improvement:

Salvatore Anastasio, SUNY, New Paltz

Thomas Bengston, Penn State University, Delaware County

Dietrich Burbulla, University of Toronto

Maurice Chabot, University of Southern Maine

Hugh Easler, College of William & Mary

Jane Edgar, Brevard Community College

Shirley Goldman, University of California at Davis

Walter Gruber, Mercy College of Detroit

Bernard Harvey, California State University, Long Beach

Martin Kotler, Pace University

Timothy Loughlin, New York Institute of Technology

William Mastrocola, Colgate University

Susan Prazak, College of Charleston

Susan Richman, Penn State University, Harrisburg

Rod Ross, University of Toronto

Margaret Suchow, Adirondack Community College

John Suvak, Memorial University of Newfoundland

George Szoke, University of Akron

In conclusion, I would like to thank Sheldon R. Gordon for the BASIC computer programs that appear in this edition. These programs are slight modifications of programs that appear in his text *Calculus and the Computer*.

Dennis G. Zill
Los Angeles

Contents

7

Inverse Trigonometric Functions   388

8

Logarithmic and Exponential Functions   415

9

Techniques of Integration   481

10

Indeterminate Forms and Improper Integrals   525

11

Sequences and Series   549

12

Analytic Geometry in the Plane   611

APPENDICES   A1

1

Functions

The word *calculus* is a diminutive form of the Latin word *calx*, which means stone. In ancient civilizations small stones or pebbles were often used as a means for reckoning. Consequently, the word *calculus* can refer to any systematic method of computation. However, over the last several hundred years a definition of *calculus* has evolved to mean that branch of mathematics concerned with the calculation and application of entities known as derivatives and integrals. Thus, the subject known as calculus has been divided into two rather broad but related areas known as *differential calculus* and *integral calculus*.

Before launching into the study of differential calculus, we are going to examine some basic mathematics. Undoubtedly, in previous courses you have encountered many of the topics that we shall consider in this first chapter: real numbers, inequalities, graphs, slope, lines, circles, functions, and trigonometry. However, do not let a sense of familiarity with these topics lead to crippling complacency. Most of the topics covered in this initial discussion will be used in subsequent chapters with little explanation or fanfare. Learn them (or, as the case may be, relearn them) well.

In conjunction with Chapter 1 you are also encouraged to review further by reading Appendix I.

1.1 The Real Numbers

Real numbers are classified as either **rational** or **irrational**. A rational number can be expressed as a quotient a/b, where a and b are integers and $b \neq 0$. A number that is *not* rational is said to be irrational. For example, $\frac{5}{4}$, -6, $\frac{22}{7}$, $\sqrt{4} = 2$, and $1.32 = \frac{132}{100}$ are rational numbers, whereas $\sqrt{2}$, $\sqrt{3}/2$, and π are irrational. The sum, difference, and product of two real numbers is a real number. The quotient of two real numbers is a real number provided the divisor is not zero.

The set of real numbers is denoted by the symbol R; the symbols Q and H are commonly used to denote the set of rational numbers and the set of irrational numbers, respectively. In terms of the union of two sets we have $R = Q \cup H$. The fact that Q and H have no common elements is summarized using the intersection of two sets: $Q \cap H = \varnothing$, where $\varnothing$ is the empty set.

Number Line

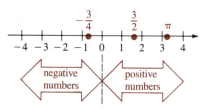

Figure 1.1

The set of real numbers can be put into a one-to-one correspondence with the points on a horizontal line, which is called the **number line** or **real line**. A number a associated with a point P on the number line is called the **coordinate** of P. The point chosen to represent 0 is called the **origin**. As shown in Figure 1.1, **positive numbers** are placed to the right of the origin and **negative numbers** are placed to the left of the origin. The number 0 is neither positive nor negative. The arrowhead on the number line points in the **positive direction**. As a rule the words "point a" and "number a" are used interchangeably with "point P with coordinate a."

Inequalities

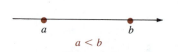

Figure 1.2

The set R of real numbers is an **ordered** set. A real number a is **less than** a real number b, written $a < b$, if the difference $b - a$ is positive. For example, $-2 < 5$ since $5 - (-2) = 7$ is a positive number. On the number line $a < b$ means that the number a lies to the left of the number b. See Figure 1.2. The statement "a is less than b" is equivalent to saying "b is greater than a" and is written $b > a$. Thus, a number a is positive if $a > 0$ and negative if $a < 0$. Expressions such as $a < b$ or $b > a$ are called **inequalities**.

Properties of Inequalities

The following is a list of some **properties of inequalities**.

> *(i)*   If $a < b$, then $a + c < b + c$ for any real number c.
>
> *(ii)* If $a < b$ and $c > 0$, then $ac < bc$.
>
> *(iii)* If $a < b$ and $c < 0$, then $ac > bc$.
>
> *(iv)* If $a < b$ and $b < c$, then $a < c$.

Property *(iii)* indicates that if an inequality is multiplied by a negative number, then the sense of the inequality is *reversed*.

The symbolism $a \leq b$ is read "a is **less than or equal** to b" and means $a < b$ or $a = b$. The preceding four properties hold when $<$ and $>$ are replaced by $\leq$ and $\geq$, respectively. If $a \geq 0$, the number a is said to be **nonnegative**.

Example 1

Solve the inequality $2x < 5x + 9$.

Solution   First we use (i) to add $-5x$ to both sides of the inequality:

$$2x + (-5x) < 5x + 9 + (-5x)$$
$$-3x < 9$$

Then by multiplying both sides of the last inequality by $-\frac{1}{3}$, property (iii) implies

$$x > -3$$

Intervals   If $a < b$, the set of real numbers x that are *simultaneously* less than b and greater than a is written $\{x \mid a < x < b\}$ or simply as $a < x < b$. This set is called an **open interval** and is denoted by (a, b). The numbers a and b are the **endpoints** of the interval.

Various kinds of intervals and their **graphs** on the number line are summarized in the following table.

Name	*Symbol*	*Definition*	*Graph*
Open interval	(a, b)	$\{x \mid a < x < b\}$	
Closed interval	$[a, b]$	$\{x \mid a \le x \le b\}$	
Half-open intervals	$(a, b]$	$\{x \mid a < x \le b\}$	
	$[a, b)$	$\{x \mid a \le x < b\}$	
Infinite intervals	(a, ∞)	$\{x \mid x > a\}$	
	$[a, \infty)$	$\{x \mid x \ge a\}$	
	$(-\infty, b)$	$\{x \mid x < b\}$	
	$(-\infty, b]$	$\{x \mid x \le b\}$	
	$(-\infty, \infty)$	$\{x \mid -\infty < x < \infty\}$	

The symbols ∞ and $-\infty$ are read "infinity" and "negative infinity," respectively. These symbols do not represent real numbers. Their use is simply shorthand for writing an *unbounded* interval.

Example 2

(*a*)   The set $\{x \mid -3 < x < 8\}$ is $(-3, 8)$ in interval notation.

(*b*)   The set $\{x \mid 5 < x \le 6\}$ is $(5, 6]$ in interval notation.

(*c*)   The set $\{x \mid x \le -1\}$ is $(-\infty, -1]$ in interval notation.

An interval that is a subset of another interval I is said to be a **subinterval** of I. For example, $[1, 2]$, $[2, \frac{5}{2}]$, and $(3, 6)$ are each subintervals of $[1, 6]$.

_____ **Example 3** _____

Solve the inequality $4 < 2x - 3 < 8$.

Solution From the properties of inequalities it follows that

$$4 < 2x - 3 < 8$$
$$4 + 3 < 2x < 8 + 3 \qquad [\text{by } (i)]$$
$$\frac{7}{2} < x < \frac{11}{2} \qquad [\text{by } (ii)]$$

In interval notation the solution is $(\frac{7}{2}, \frac{11}{2})$.

The inequalities considered in Examples 1 and 3 are **first-degree** or **linear** inequalities. In the next example we shall solve a **second-degree** or **quadratic** inequality.

_____ **Example 4** _____

Solve the inequality $(x + 3)(x - 4) < 0$.

Solution We begin by putting the two roots, -3 and 4, of the quadratic equation $(x + 3)(x - 4) = 0$ on the number line. By ascertaining the algebraic sign of each factor on the subintervals $(-\infty, -3)$, $(-3, 4)$, and $(4, \infty)$ of $(-\infty, \infty)$, we can then determine the algebraic sign of the product. For example, by replacing x by *any* number in $(-\infty, -3)$ we see that both factors $x + 3$ and $x - 4$ are negative, and so the product $(x + 3)(x - 4)$ must be positive. In this case, the original inequality is not satisfied. Inspection of Figure 1.3 indicates that the solution of the problem is $(-3, 4)$.

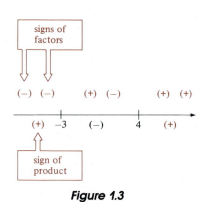

Figure 1.3

Absolute Value If a is a real number then its **absolute value** is

$$|a| = \begin{cases} a, & a \geq 0 \\ -a, & a < 0 \end{cases}$$

On the number line, $|a|$ is the distance between the origin and the number a. In general, the **distance between two numbers** a and b is either $b - a$ if $a < b$ or $a - b$ if $b < a$. In terms of absolute value the distance between a and b is $|b - a|$. See Figure 1.4. Note that $|b - a| = |a - b|$.

Figure 1.4

Example 5

(a) $\quad |-5| = -(-5) = 5$

(b) $\quad |2x - 1| = \begin{cases} 2x - 1 & \text{if } 2x - 1 \geq 0 \\ -(2x - 1) & \text{if } 2x - 1 < 0 \end{cases}$

That is,

$$|2x - 1| = \begin{cases} 2x - 1 & \text{if } x \geq \dfrac{1}{2} \\ -2x + 1 & \text{if } x < \dfrac{1}{2} \end{cases}$$

Example 6

The distance between -5 and 7 is the same as the distance between 7 and -5:

$$|7 - (-5)| = |12| = 12 \text{ units}$$
$$|-5 - 7| = |-12| = 12 \text{ units}$$

Absolute Values and Inequalities Since $|x|$ gives the distance between a number x and the origin, the solution of the inequality $|x| < b$, $b > 0$ is the set of real numbers x that are *less* than b units from the origin. As shown in Figure 1.5,

$$|x| < b \quad \text{if and only if} \quad -b < x < b \qquad (1.1)$$

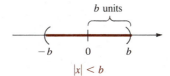

b units

$$|x| < b$$

Figure 1.5

On the other hand, the solution of the inequality $|x| > b$ is the set of real numbers that are *greater* than b units from the origin. Consequently,

$$|x| > b \quad \text{if and only if} \quad x > b \quad \text{or} \quad x < -b \qquad (1.2)$$

See Figure 1.6. The results in (1.1) and (1.2) hold when $<$ and $>$ are replaced with $\leq$ and $\geq$, respectively, and when x is replaced by $x - a$. For example, $|x - a| \leq b$, $b > 0$ represents the set of real numbers x such that the distance from x to a is less than or equal to b units. As illustrated in Figure 1.7, $|x - a| \leq b$ if and only if x is a number in the closed interval $[a - b, a + b]$.

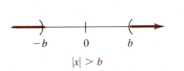

$$|x| > b$$

Figure 1.6

b units

$$|x - a| \leq b$$

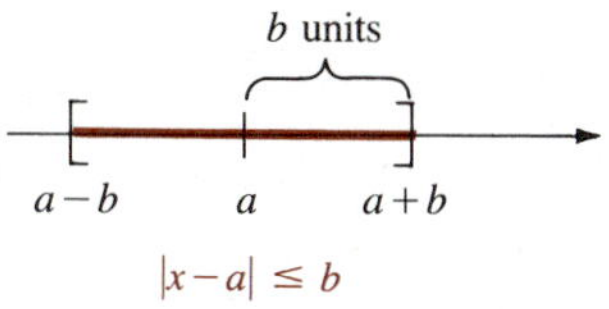

Figure 1.7

Example 7

Solve the inequality $|x - 1| < 3$.

Solution Using (1.1), first we rewrite the inequality as

$$-3 < x - 1 < 3$$

Therefore,
$$-3 + 1 < x - 1 + 1 < 3 + 1$$
$$-2 < x < 4$$

The solution is the open interval $(-2, 4)$.

Example 8

Solve the inequality $|x| > 2$.

Solution From (1.2) we have immediately,

$$x > 2 \quad \text{or} \quad x < -2$$

The solution is a union of infinite intervals: $(-\infty, -2) \cup (2, \infty)$.

Example 9

Solve the inequality $0 < |x - 2| \leq 7$.

Solution The given inequality means $0 < |x - 2|$ *and* $|x - 2| \leq 7$. In the first case, $0 < |x - 2|$ is true for any real number except $x = 2$. In the second case, we have

$$-7 \leq x - 2 \leq 7$$
$$-5 \leq x \leq 9$$

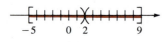

Figure 1.8

or $[-5, 9]$. The set of real numbers x that satisfies both inequalities consists of all numbers in $[-5, 9]$ except 2. Written as a union of intervals, the solution is $[-5, 2) \cup (2, 9]$. See Figure 1.8.

Triangle Inequality In conclusion, the proof of the so-called **triangle inequality**

$$|a + b| \leq |a| + |b| \tag{1.3}$$

is left as an exercise.

Remarks

(*i*) Every real number has a nonterminating decimal representation. The rational numbers are characterized as *repeating* decimals; for example, by long division we see

$$\frac{7}{11} = 0.636363\ldots \quad \text{and} \quad \frac{3}{4} = 0.750000\ldots$$

repeats

Hence, irrational numbers are the *nonrepeating* decimal numbers. For example,

$$\pi = 3.141592\ldots \quad \text{and} \quad \sqrt{3} = 1.732050\ldots$$

nonrepeating decimal

(*ii*) In Example 7 there is no way of writing the solution of the inequality $|x| > 2$ as a single interval. The statement that $x > 2$ or $x < -2$ is *not* equivalent to $2 < x < -2$. There are no real numbers that are simultaneously greater than positive 2 and less than negative 2.

(*iii*) Before moving on, a brief word about the square root of a nonnegative number x is in order:

The square root of $x \geq 0$, written $\sqrt{x}$, is always nonnegative.

For example, $\sqrt{64} = 8$, *not* ± 8. In fact, the square root is related to the absolute value:

$$\sqrt{x^2} = |x|$$

Thus, $\sqrt{x^2} = x$ if $x \geq 0$ and $\sqrt{x^2} = -x$ if $x < 0$. For example,

$$\sqrt{5^2} = 5 \quad \text{since} \quad 5 > 0 \text{ and } \sqrt{(-5)^2} = -(-5) = 5 \quad \text{since} \quad -5 < 0.$$

Exercises 1.1

Answers to odd-numbered problems begin on page A-16.

In Problems 1–4 write the given inequality using interval notation.

1. $-4 \leq x < 20$

2. $x \geq 5$

3. $x < -2$

4. $\frac{1}{2} < x < \frac{7}{4}$

In Problems 5–8 write the given interval as an inequality.

5. $(\frac{3}{2}, 6)$

6. $[-1, 5]$

7. $[20, \infty)$

8. $(-\infty, -7)$

In Problems 9–12 write the interval for the given graph.

9.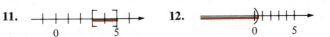

10.

11.

12.

In Problems 13–28 solve the given inequality. Write the solution in interval notation.

13. $3x < -9$

14. $-2x > 8$

15. $4x + 1 > 10$

16. $-\frac{1}{2}x + 6 \leq 0$

17. $4x \geq 5x - 7$

18. $x + 12 \leq 5x$

19. $-4 < 1 - x \leq 3$

20. $1 \leq \dfrac{2x + 14}{3} < 2$

21. $x \leq 3x + 2 \leq x + 6$

22. $10 - x < 4x \leq 25 - x$

23. $(x - 1)(x - 9) < 0$

24. $x^2 + 2x - 3 \geq 0$

25. $10 - 3x^2 \geq 13x$

26. $-x^2 < 6x$

27. $4x^2 - 4x + 1 \geq 0$

28. $x^2 > 25$

In Problems 29–32 write the given expression without absolute value symbols.

29. $|4 - a|$, $4 - a$ is a negative number

30. $|-6a|$, a is a positive number

31. $|a + 10|$, a is greater than or equal to -10

32. $|a^2 - 1|$, a is a number in $(-1, 1)$

In Problems 33–36 solve for x.

33. $|4x| = 36$ **34.** $|-2x| = 16$

35. $|3 - 5x| = 22$ **36.** $|12 - \frac{1}{2}x| = x$

In Problems 37–46 solve the given inequality. Write the solution in interval notation.

37. $|x| < 4$ **38.** $|-\frac{1}{3}x| \leq 3$

39. $|1 - 2x| \leq 1$ **40.** $|5 + 4x| < 17$

41. $\left|\dfrac{x + 3}{-2}\right| < 1$ **42.** $0 < |x + 1| \leq 5$

43. $|x| > 6$ **44.** $|4 - x| > 0$

45. $|5 - 2x| > 7$ **46.** $|x + 9| \geq 8$

47. If $1/x < 4$, does it follow that $x > \frac{1}{4}$?

48. If $x^2 < 6x$, does it follow that $x < 6$?

In Problems 49–52 express the given statement as an inequality involving an absolute value. Express the solution of the given statement using interval notation.

49. The set of real numbers less than 4 units from 9

50. The set of real numbers less than or equal to $\frac{1}{2}$ unit from 1.5

51. The set of real numbers greater than 2 units from 2

52. The set of real numbers greater than -1 and less than 7

53. When equipment is depreciated linearly and loses all its initial worth of A dollars over a period of n years, its value V in x years $(0 \leq x \leq n)$ is given by $V = A(1 - x/n)$. If a computer costs \$100,000 initially and is depreciated over 20 years, determine the values of x such that $30,000 \leq V \leq 80,000$.

54. According to one theory, the most beneficial effect of exercise such as jogging is obtained when the pulse rate is maintained within a certain interval. The endpoints of the interval are obtained by multiplying the number $(220 - \text{age})$ by 0.70 and 0.85. Determine this pulse rate interval for a 30-year-old jogger. For a 40-year-old jogger.

Calculator Problems

In Problems 55–58 replace the comma between the given pair of real numbers with one of the symbols $<$, $>$, or $=$. Use a calculator.

55. $\pi, \dfrac{22}{7}$ **56.** $\dfrac{\pi}{2}, 1.5$

57. $\dfrac{180}{\pi}, 57.29$ **58.** $\dfrac{1}{\pi}, \dfrac{1}{3}$

Miscellaneous Problems

The **midpoint** of an interval with endpoints a and b is the number $(a + b)/2$. In Problems 59–62 use this information to find an inequality $|x - c| < d$ whose solution is the given interval.

59. $(0, 8)$ **60.** $(-1, 6)$

61. $(-3, 4)$ **62.** $(-10, -2)$

63. Use $-|a| \leq a \leq |a|$ and $-|b| \leq b \leq |b|$ to prove the triangle inequality (1.3). (*Hint:* Add the inequalities.)

1.2 The Cartesian Plane

It is virtually impossible to pick up a text, journal, or news magazine without encountering some sort of graphical display of data, such as illustrated in Figure 1.9, in a *coordinate plane* formed by the intersection of two perpendicular number lines. In mathematics such a coordinate plane is called a **Cartesian plane*** (the footnote is on page 9).

     In the following general discussion, we shall assume that the same scale has been used to mark off each number line. The point of intersection of these number lines, corresponding to the number 0 on both lines, is called the **origin** and is denoted by O. The horizontal number line is called the **x-axis** and the vertical line is called the **y-axis**. Numbers to the right of the origin on the x-axis

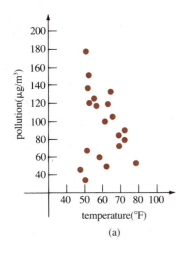

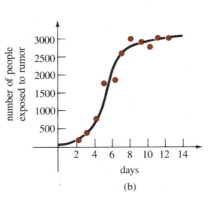

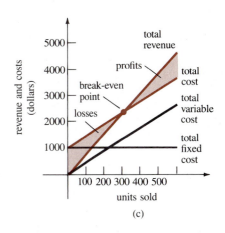

Figure 1.9

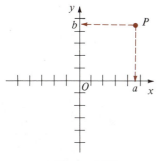

Figure 1.10

are positive; numbers to the left of O are negative. On the y-axis numbers above the origin are positive; numbers below O are negative. Because the horizontal and vertical number lines are labeled with the letters x and y, a Cartesian plane is often simply called an **xy-plane**.

If P denotes a point in a Cartesian plane, we can draw perpendicular lines from P to both the x- and y-axes. As Figure 1.10 shows, this determines a number a on the x-axis and a number b on the y-axis. Conversely, we see that specified numbers a and b on the x- and y-axes determine a unique point P in the plane. In this manner a one-to-one correspondence between points in a Cartesian plane and ordered pairs of real numbers (a, b)[†] is established. We call a the **x-coordinate** or **abscissa** of the point P; b is called the **y-coordinate** or **ordinate** of the point. The axes are also called **coordinate axes** and P is said to have **coordinates** (a, b).

René Descartes

* **René Descartes (1596–1650)**   The Cartesian plane was named in honor of the French nobleman, lawyer, scientist, engineer, mathematician, gentleman soldier, playboy, philosopher, and prominent architect of the so-called Age of Reason, René Descartes. In mathematics Descartes is credited with being one of the principal "inventors" of the field of analytic geometry, which deals with the correspondence between equations in two variables and curves in a coordinate plane. Legend has it that the fundamental concepts of analytic geometry came to Descartes while he was lying in bed, observing a fly crawling on the ceiling. Although his philosophical treatises *Rules for the Direction of the Mind*, *Discourse on Method*, and *Meditations* are still of some significance, Descartes probably achieved everlasting fame by penning the dictum: *Cogito ergo sum*—"I think therefore I am." Always of fragile health, Descartes died in 1650 at the age of 54 of influenza while serving as a philosophy tutor to Queen Christina of Sweden.

[†]This is the same notation used for an open interval. The reader should be able to tell from the context of the discussion whether this symbol refers to a point or an interval.

Quadrants   The coordinate axes divide the Cartesian plane into four regions known as **quadrants**. Algebraic signs of the *x*-coordinate and *y*-coordinate of any point (a, b), located in each of the four quadrants, are indicated in Figure 1.11(a). Points on a coordinate axis, such as $(2, 0)$ and $(0, -3)$ in Figure 1.11(b) are not considered to be in any quadrant. This method of describing points in a plane is called a **rectangular** or **Cartesian coordinate system**. In this system, two points (a, b) and (c, d) are **equal** if and only if $a = c$ and $b = d$.

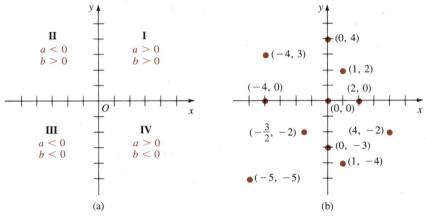

Figure 1.11

Distance Formula   We can obtain the distance $d(P_1, P_2)$ between two points $P_1(x_1, y_1)$ and $P_2(x_2, y_2)$ from the Pythagorean Theorem. As shown in Figure 1.12, the three points P_1, P_2, and P_3 form a right triangle with a hypotenuse of length d and sides of lengths $|x_2 - x_1|$ and $|y_2 - y_1|$. Thus,

$$d^2 = |x_2 - x_1|^2 + |y_2 - y_1|^2$$

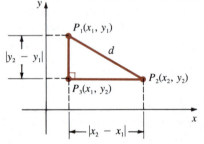

Figure 1.12

yields the **distance formula**

$$d(P_1, P_2) = \sqrt{(x_2 - x_1)^2 + (y_2 - y_1)^2} \qquad (1.4)$$

Example 1

Find the distance between the points $(-2, 3)$ and $(4, 5)$.

Solution   By identifying P_1 as $(-2, 3)$ and P_2 as $(4, 5)$ we obtain from (1.4)

$$d(P_1, P_2) = \sqrt{(4 - (-2))^2 + (5 - 3)^2} = \sqrt{6^2 + 2^2} = \sqrt{40} = 2\sqrt{10}$$

Since $(x_2 - x_1)^2 = (x_1 - x_2)^2$ and $(y_2 - y_1)^2 = (y_1 - y_2)^2$, it does not matter which points are designated P_1 and P_2. In other words, $d(P_1, P_2) = d(P_2, P_1)$.

Midpoint of a Line Segment

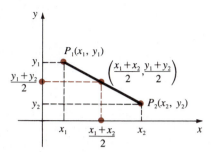

Figure 1.13

Throughout the study of calculus we shall use the distance formula many times in proofs and definitions. As a first application, we shall prove that the coordinates of the **midpoint of a line segment** from a point $P_1(x_1, y_1)$ to a point $P_2(x_2, y_2)$ are

$$\left(\frac{x_1 + x_2}{2}, \frac{y_1 + y_2}{2}\right) \tag{1.5}$$

See Figure 1.13. If M denotes a point on the segment P_1P_2 with coordinates given in (1.5), then M is the midpoint of P_1P_2 provided

$$d(P_1, M) = d(M, P_2) \quad \text{and} \quad d(P_1, P_2) = d(P_1, M) + d(M, P_2)$$

From (1.4) and a little algebra we find

$$d(P_1, M) = \sqrt{\left(\frac{x_1 + x_2}{2} - x_1\right)^2 + \left(\frac{y_1 + y_2}{2} - y_1\right)^2}$$

$$= \frac{1}{2}\sqrt{(x_2 - x_1)^2 + (y_2 - y_1)^2} = \frac{1}{2}d(P_1, P_2) \tag{1.6}$$

$$d(M, P_2) = \sqrt{\left(x_2 - \frac{x_1 + x_2}{2}\right)^2 + \left(y_2 - \frac{y_1 + y_2}{2}\right)^2}$$

$$= \frac{1}{2}\sqrt{(x_2 - x_1)^2 + (y_2 - y_1)^2} = \frac{1}{2}d(P_1, P_2) \tag{1.7}$$

Evidently $d(P_1, M) = d(M, P_2)$. Moreover, adding (1.6) and (1.7) gives $d(P_1, P_2)$, as was to be shown.

Example 2

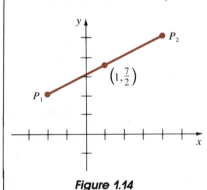

Figure 1.14

Find the coordinates of the midpoint of the line segment from $P_1(-2, 2)$ to $P_2(4, 5)$.

Solution   From (1.5) we have

$$x = \frac{-2 + 4}{2} = \frac{2}{2} = 1 \quad \text{and} \quad y = \frac{2 + 5}{2} = \frac{7}{2}$$

The midpoint $(1, \frac{7}{2})$ is shown in Figure 1.14.

Graphs   A **graph** is any set of points (x, y) in the Cartesian plane. A graph can be an infinite set of points such as the points on the line segment P_1P_2 in Figure 1.14 or simply a finite set of points as in Figure 1.9(a). The **graph of an equation** is the set of points (x, y) in the Cartesian plane that are solutions of the equation. An ordered pair (x, y) is a **solution** of an equation if substitution of x and y into the equation reduces it to an identity.

Example 3

The point $(-2, 2)$ is on the graph of $y = 1 - \frac{1}{8}x^3$ since

$$2 = 1 - \frac{1}{8}(-2)^3 \quad \text{is equivalent to} \quad 2 = 1 + \frac{8}{8} \quad \text{or} \quad 2 = 2$$

Point Plotting One way of sketching the graph of an equation, often done in elementary courses, is to **plot points** and then connect these points with a smooth curve. To obtain points on the graph we assign values to either x or y and then solve the equation for the corresponding values of y or x. Of course, we need to plot enough points until the shape, or pattern, of the graph is evident.

Example 4

Graph $y = 1 - \frac{1}{8}x^3$.

Solution By assigning values of x we find the values of y given in the accompanying table. It seems reasonable, by connecting the points shown in Figure 1.15(a) with a smooth curve, that the graph of the equation is that given in Figure 1.15(b).

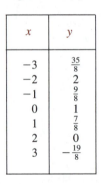

x	y
-3	$\frac{35}{8}$
-2	2
-1	$\frac{9}{8}$
0	1
1	$\frac{7}{8}$
2	0
3	$-\frac{19}{8}$

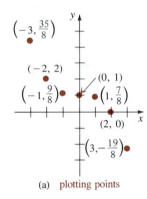

(a) plotting points

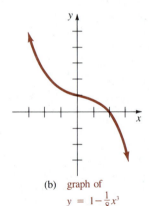

(b) graph of $y = 1 - \frac{1}{8}x^3$

Figure 1.15

Symmetry Before plotting points, you can determine whether the graph of an equation possesses **symmetry**. Figure 1.16 shows that a graph is

(*i*) **symmetric with respect to the y-axis** if, whenever (x, y) is a point on the graph, $(-x, y)$ is also a point on the graph;

(*ii*) **symmetric with respect to the x-axis** if, whenever (x, y) is a point on the graph, $(x, -y)$ is also a point on the graph; and

(*iii*) **symmetric with respect to the origin** if, whenever (x, y) is a point on the graph, $(-x, -y)$ is also a point oh the graph.

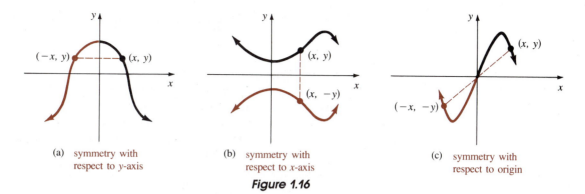

(a) symmetry with
respect to y-axis

(b) symmetry with
respect to x-axis

(c) symmetry with
respect to origin

Figure 1.16

Tests for Symmetry   For an equation, (*i*), (*ii*), and (*iii*) yield the following three **tests for symmetry**. A graph of an equation is symmetric with respect to

> (*i*)   the **y-axis** if replacing x by $-x$ results in an equivalent equation;
>
> (*ii*) the **x-axis** if replacing y by $-y$ results in an equivalent equation; and
>
> (*iii*) the **origin** if replacing x by $-x$ and y by $-y$ results in an equivalent equation.

Example 5

Determine whether the graph of $x = |y| - 2$ possesses any symmetry.

Solution

Test (*i*): $-x = |y| - 2$ or $x = -|y| + 2$ is not equivalent to the original equation.

Test (*ii*): $x = |-y| - 2$ is equivalent to the original equation since $|-y| = |y|$.

Test (*iii*): $-x = |-y| - 2$ or $x = -|y| + 2$ is not equivalent to the original equation.

We conclude that the graph of $x = |y| - 2$ is symmetric with respect to the x-axis.

Inspection of Figure 1.15(b) shows pictorially that the graph of $y = 1 - \frac{1}{8}x^3$ has none of the symmetries we are considering.* You are encouraged to verify that each of the three tests for symmetry fails to yield an equivalent equation.

Detecting symmetry before plotting points can often save time and effort. For example, if the graph of an equation is shown to be symmetric with respect to the y-axis, then it is sufficient to plot points with x-coordinates that satisfy $x \geq 0$. As suggested in Figure 1.16(a), we can find points in the second and third quadrants by taking the mirror images, through the y-axis, of the points in the first and fourth quadrants.

---

*We are confining our attention to symmetry with respect to the two coordinate axes and the origin. A graph could, of course, possess other types of symmetries.

Example 6

Graph $x = |y| - 2$.

Solution   The entries in the accompanying table were obtained by assigning values to y and solving for x. We have taken $y \geq 0$ since we saw in Example 5 that the graph of the equation is symmetric with respect to the x-axis. In Figure 1.17(a) color is used to indicate points on the graph gained by symmetry. The graph of the equation, which seems to consist of two straight lines, is given in Figure 1.17(b).

x	y
2	4
1	3
0	2
−1	1
−2	0

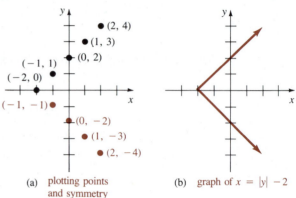

(a)   plotting points and symmetry

(b)   graph of $x = |y| - 2$

Figure 1.17

Intercepts   When you sketch the graph of an equation, it is always a good idea to ascertain whether the graph has any **intercepts**. The x-coordinate of a point where the graph crosses the x-axis is called an **x-intercept**. The y-coordinate of a point where the graph crosses the y-axis is called a **y-intercept**. In Figure 1.18 the x-intercepts of the graph are x_1, x_2, and x_3. The single y-intercept is y_1. Figure 1.19 shows a graph that has no x- or y-intercepts. In Example 4 the x-intercept of the graph is 2; the y-intercept is 1. In Example 6 the x-intercept of the graph is −2; the y-intercepts are −2 and 2.

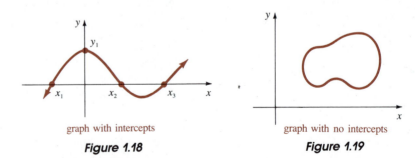

graph with intercepts

Figure 1.18

graph with no intercepts

Figure 1.19

Since $y = 0$ for any point on the x-axis and $x = 0$ for any point on the y-axis, we can determine the intercepts of the graph of an equation in the following manner:

x-intercepts: Set $y = 0$ in the equation and solve for x.

y-intercepts: Set $x = 0$ in the equation and solve for y.

(1.8)

_____ **Example 7** _____

Find the intercepts for the graph of (a) $y = 4x - 3$ and (b) $y = \dfrac{x^2 + 1}{x^2 + 5}$.

Solution

(a) Setting $y = 0$ yields $0 = 4x - 3$ or $x = \frac{3}{4}$. Setting $x = 0$ gives $y = -3$. The x- and y-intercepts are $\frac{3}{4}$ and -3, respectively.

(b) In the second equation, $y = 0$ if $x^2 + 1 = 0$ and $x^2 + 5 \neq 0$. Since $x^2 + 5 \neq 0$ for all real numbers, we have $x^2 + 1 = 0$ or $x^2 = -1$. But there are no real numbers that satisfy the last equation. Hence, the graph has no x-intercepts. Now, setting $x = 0$, the equation gives $y = \frac{1}{5}$. The y-intercept is $\frac{1}{5}$.

Circles   The distance formula (1.4) enables us to find an equation for a very familiar plane curve. A **circle** is the set of all points (x, y) in the Cartesian plane that are equidistant from a fixed point $C(h, k)$. If r is the fixed distance, then a point $P(x, y)$ is on the circle if and only if

$$d(C, P) = \sqrt{(x - h)^2 + (y - k)^2} = r$$

Equivalently, we have the **standard form** for the equation of a circle with **center** $C(h, k)$ and **radius r**:

$$(x - h)^2 + (y - k)^2 = r^2 \qquad (1.9)$$

See Figure 1.20. If $h = 0$ and $k = 0$, then the standard form for the equation of a circle with center at the origin is

$$x^2 + y^2 = r^2$$

See Figure 1.21.

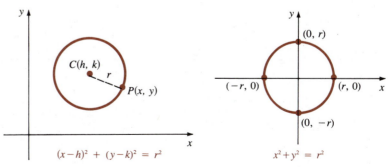

Figure 1.20   **Figure 1.21**

Example 8

Consider the line segment from $P_1(-2, 2)$ to $P_2(4, 5)$ in Example 2. Find an equation of the circle that passes through P_1 and P_2 with center at the midpoint M of P_1P_2.

Solution   Since the center of the circle is the midpoint $M(1, \frac{7}{2})$, we must have $h = 1$ and $k = \frac{7}{2}$. The radius r of the circle is either $d(M, P_1)$ or $d(M, P_2)$. Using

$$d(M, P_1) = \sqrt{(-2 - 1)^2 + \left(2 - \frac{7}{2}\right)^2} = \sqrt{\frac{45}{4}} = \frac{3\sqrt{5}}{2} = r$$

it follows from (1.9) that an equation for the circle is

$$(x - 1)^2 + \left(y - \frac{7}{2}\right)^2 = \frac{45}{4}$$

By squaring out the terms in (1.9), we see that every circle has an alternative equation of the form

$$Ax^2 + Ay^2 + Cx + Dy + E = 0, \qquad A \neq 0$$

However, the converse is not necessarily true; that is, not every equation of the form $Ax^2 + Ay^2 + Cx + Dy + E = 0$ is a circle. See Problems 43–48.

Example 9

Show that $x^2 + y^2 - 4x + 8y = 0$ is an equation of a circle. Find its center and radius.

Solution   By *completing the square* in both x and y we see that

$$(x^2 - 4x \quad) + (y^2 + 8y \quad) = 0$$

becomes

$$(x^2 - 4x + 4) + (y^2 + 8y + 16) = 20$$

or

$$(x - 2)^2 + (y + 4)^2 = 20$$

The latter is the standard form for the equation of a circle with center $(2, -4)$ and radius $2\sqrt{5}$.

Conic Sections   The circle is just one member of a class of curves known as **conic sections**. An equation of a conic section can always be expressed in the form $Ax^2 + By^2 + Cx + Dy + E = 0$, where A and B are not both zero.

_____ **Example 10** _____

Graph $9x^2 + 16y^2 = 25$.

Solution First, observe that replacing (x, y) in turn by $(-x, y)$, $(x, -y)$, and $(-x, -y)$ does not change the given equation. Hence, the graph is symmetric with respect to the y-axis, x-axis, and the origin. Furthermore, $9(1)^2 + 16(1)^2 = 25$ indicates that $(1, 1)$ is a point on the graph. Finally,

$$y = 0 \quad \text{implies} \quad 9x^2 = 25 \quad \text{or} \quad x = \pm\frac{5}{3}$$

$$x = 0 \quad \text{implies} \quad 16y^2 = 25 \quad \text{or} \quad y = \pm\frac{5}{4}$$

The x-intercepts are $-\frac{5}{3}$ and $\frac{5}{3}$; the y-intercepts are $-\frac{5}{4}$ and $\frac{5}{4}$. By connecting the points in Figure 1.22(a) with a smooth curve, we obtain the graph in Figure 1.22(b). Observe that the graph is *not* a circle.

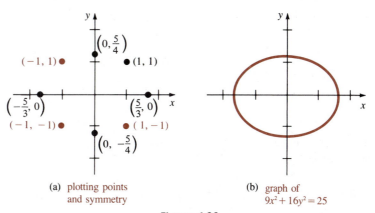

(a) plotting points
and symmetry

(b) graph of
$9x^2 + 16y^2 = 25$

Figure 1.22

The Ellipse When written in the equivalent form

$$\frac{x^2}{(5/3)^2} + \frac{y^2}{(5/4)^2} = 1$$

the equation in Example 10 is seen as a special case of the standard form

$$\frac{(x - h)^2}{a^2} + \frac{(y - k)^2}{b^2} = 1 \qquad (1.10)$$

For $a = b$ the graph of (1.10) is a circle of radius a. If $a \neq b$, the graph of (1.10) is called an **ellipse with center (h, k)**. Thus, the graph of $9x^2 + 16y^2 = 25$ is an ellipse with center at the origin. We shall study the ellipse, along with the other conic sections, in Chapter 12.

Exercises 1.2

Answers to odd-numbered problems begin on page A-16.

In Problems 1–6 the point (a, b) is in the first quadrant. Determine the quadrant of the given point.

1. $(a, -b)$ **2.** (b, a) **3.** $(-b, -a)$

4. $(-a, a)$ **5.** $(-a, b)$ **6.** $(-b, a)$

In Problems 7–10 find the distance between the given points.

7. $P_1(3, -1), P_2(7, -3)$ **8.** $P_1(0, 5), P_2(-8, -2)$

9. $P_1(\sqrt{3}, 0), P_2(0, -\sqrt{6})$ **10.** $P_1(\frac{5}{2}, 5), P_2(-\frac{3}{2}, 5)$

In Problems 11 and 12 determine whether the given points are the vertices of a right triangle.

11. $(16, 2), (-6, -2), (20, 10)$

12. $(-2, -8), (0, 3), (-6, -5)$

In Problems 13 and 14 use the distance formula to determine whether the given points are collinear.

13. $(1, 3), (-2, -3), (4, 9)$ **14.** $(0, 2), (1, 1), (5, -2)$

In Problems 15 and 16 solve for x.

15. $P_1(x, 2), P_2(1, 1), d(P_1, P_2) = \sqrt{10}$

16. $P_1(x, 0), P_2(-4, 3x), d(P_1, P_2) = 4$

17. Find an equation that relates x and y if it is known that the distance from (x, y) to $(0, 1)$ is the same as the distance from (x, y) to $(x, -1)$.

18. Show that the point $(-1, 5)$ is on the perpendicular bisector of the line segment from $P_1(1, 1)$ to $P_2(3, 7)$.

In Problems 19 and 20 find the midpoint of the line segment from P_1 to P_2.

19. $P_1(4, 7), P_2(8, -3)$ **20.** $P_1(-3, 5), P_2\left(\frac{1}{2}, \frac{3}{4}\right)$

21. If the coordinates of the midpoint of the line segment from $P_1(1, 3)$ to $P_2(x_2, y_2)$ are $(3, 4)$, what are the coordinates of P_2?

22. Figure 1.23 shows the midpoints of the sides of a triangle. Determine the coordinates of the vertices of the triangle.

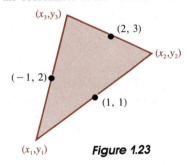

Figure 1.23

23. If M is the midpoint of the line segment from $P_1(2, 3)$ to $P_2(6, -9)$, find the midpoint of the line segment from P_1 to M and the midpoint of the line segment from M to P_2.

24. The point $P(x, y)$ with coordinates defined by $x = x_1 + r(x_2 - x_1), y = y_1 + r(y_2 - y_1)$ is on the line segment from $P_1(x_1, y_1)$ to $P_2(x_2, y_2)$. Use $P_1(2, 3), P_2(6, -9), r = \frac{1}{4}$, $r = \frac{1}{2}$, and $r = \frac{3}{4}$, and compare your results with those of Problem 23. Find a point $\frac{2}{3}$ of the way from P_1 to P_2.

In Problems 25–32 graph the given equation. Determine any symmetry.

25. $y = 2x + 1$ **26.** $y = x - 5$

27. $x = y^2$ **28.** $y = \sqrt{x}$

29. $y = |x|$ **30.** $4y - x^3 = 0$

31. $x^2 = y^2$ **32.** $y = x^2 + x$

In Problems 33–36 complete the given graph (Figures 1.24–1.27).

33. Graph is symmetric with respect to the x-axis.

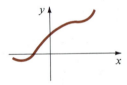

Figure 1.24

34. Graph is symmetric with respect to the y-axis.

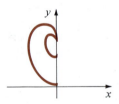

Figure 1.25

35. Graph is symmetric with respect to the origin.

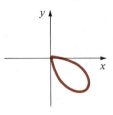

Figure 1.26

36. Graph is symmetric with respect to the x- and y-axes.

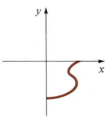

Figure 1.27

In Problems 37–42 find an equation in standard form of the circle that satisfies the given conditions.

37. Center $(4, -6)$, radius 8

38. Center $(-\frac{5}{2}, \frac{1}{2})$, radius $\sqrt{2}$

39. Center $(3, -4)$, passing through the origin

40. Center $(6, 2)$, tangent to the x-axis

41. Center $(1, 1)$, passing through $(5, 2)$

42. Center at the midpoint M of the line segment from $P_1(3, 8)$ to $P_2(5, 2)$, radius $\frac{1}{2}d(M, P_2)$

In Problems 43–48 determine whether the given equation is an equation of a circle. If so, give its center and radius.

43. $x^2 + y^2 + 8x - 6y = 0$

44. $2x^2 + 2y^2 - 16x - 40y = 37$

45. $3x^2 + 3y^2 - 18x + 6y = -2$

46. $x^2 + y^2 + 10y + 26 = 0$

47. $x^2 + y^2 - 12x + 8y + 52 = 0$

48. $x^2 + y^2 + \frac{1}{2}x + y = 0$

In Problems 49–52 the given equation is an equation of an ellipse. Graph.

49. $25x^2 + 4y^2 = 100$ **50.** $x^2 + 4y^2 = 36$

51. $\dfrac{x^2}{4^2} + \dfrac{y^2}{2^2} = 1$ **52.** $x^2 + \dfrac{y^2}{7} = 1$

In Problems 53 and 54 the given equation is an equation of an ellipse. Put the equation in standard form and give the center of the ellipse.

53. $x^2 + 4y^2 - 4x + 40y + 88 = 0$

54. $5x^2 + 2y^2 + 60x - 8y + 178 = 0$

Miscellaneous Problems

In Problems 55–58 graph the set of points (x, y) that satisfy the given equation or inequality.

55. $xy = 0$ **56.** $xy > 0$

57. $xy < 0$ **58.** $y < x$

59. Find an equation(s) for the circle(s) passing through $(1, 3)$ and $(-1, -3)$ that has radius 10.

60. Graph $y + |y| = x + |x|$.

1.3 Lines

Throughout the next several chapters we shall see how the notion of a line plays an important role in the study of differential calculus. Before discussing lines, however, it is convenient to introduce two special symbols.

Increments   If $P_1(x_1, y_1)$ and $P_2(x_2, y_2)$ are any two points in the plane, we define the **x-increment** to be the difference in x-coordinates

$$\Delta x = x_2 - x_1$$

and the **y-increment** to be the difference in y-coordinates

$$\Delta y = y_2 - y_1$$

We read the symbols "Δx" and "Δy" as "delta x" and "delta y," respectively.

_____ **Example 1** _____

Find the x- and y-increments for (a) $P_1(4, 6)$, $P_2(9, 7)$; (b) $P_1(10, -2)$, $P_2(3, 5)$; and (c) $P_1(8, 14)$, $P_2(8, -1)$.

Solution

(a) $\Delta x = 9 - 4 = 5,$ $\Delta y = 7 - 6 = 1$

(b) $\Delta x = 3 - 10 = -7,$ $\Delta y = 5 - (-2) = 7$

(c) $\Delta x = 8 - 8 = 0,$ $\Delta y = -1 - 14 = -15$

Example 1 shows that an increment can be positive, negative, or zero.

Slope   Suppose L denotes a nonvertical line in the Cartesian plane. Associated with such a line there is a number called the **slope** of the line. If $P_1(x_1, y_1)$ and $P_2(x_2, y_2)$ are distinct points on L, then the **slope** m of the line is defined to be the quotient

$$m = \frac{\Delta y}{\Delta x} = \frac{y_2 - y_1}{x_2 - x_1} \tag{1.11}$$

The increment $\Delta x = x_2 - x_1$ is said to be the *change in x* or **run** of the line; the corresponding *change in y*, or **rise** of the line, is $\Delta y = y_2 - y_1$. Thus,

$$m = \frac{\text{change in } y}{\text{change in } x} = \frac{\text{rise}}{\text{run}}$$

Assuming, for the sake of discussion, that $P_1(x_1, y_1)$ and $P_2(x_2, y_2)$ are chosen such that $x_1 < x_2$, we see that the slope of the line illustrated in Figure 1.28(a) is positive, whereas the slope of the line in Figure 1.28(b) is negative. If a line has positive slope, then the ordinates of points on the line increase as the abscissas increase; for lines with negative slope, ordinates decrease as abscissas increase. If a line is horizontal, the $\Delta y = 0$ and therefore its slope is zero. See Figure 1.28(c). The slope of a vertical line is undefined since (1.11) has no meaning when $\Delta x = 0$. See Figure 1.28(d).

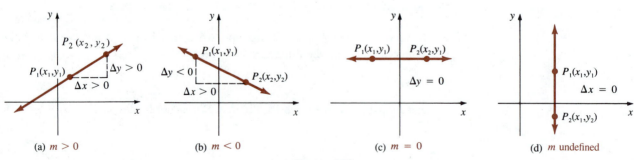

(a) $m > 0$ (b) $m < 0$ (c) $m = 0$ (d) m undefined

Figure 1.28

Example 2

Find the slope of the line that passes through the points (*a*) (6, 3), (−2, 5); (*b*) (7, −2), (4, −2); and (*c*) (1, 5), (1, −3).

Solution   Using (1.11) in the first two cases results in

(*a*) $m = \dfrac{5 - 3}{-2 - 6} = \dfrac{2}{-8} = -\dfrac{1}{4}$, and

(*b*) $m = \dfrac{-2 - (-2)}{4 - 7} = \dfrac{0}{-3} = 0$. The line through (7, −2) and (4, −2) is horizontal.

(*c*)   Since $\Delta x = 1 - 1 = 0$, the slope of the line through (1, 5) and (1, −3) is undefined. Hence, the line is vertical.

The slope of a line is *unique* in the sense that any pair of points on the line determines the same quotient *m*. This can be proved from the fact that ratios of corresponding sides of similar triangles are equal. If

$$\Delta x = x_2 - x_1,\ \Delta y = y_2 - y_1 \quad \text{and} \quad \Delta x' = x_2' - x_1',\ \Delta y' = y_2' - y_1'$$

then we see from Figure 1.29 that the right triangles $P_1 P_2 P_3$ and $P_1' P_2' P_3'$ are similar, and as a consequence

$$\frac{y_2 - y_1}{x_2 - x_1} = \frac{y_2' - y_1'}{x_2' - x_1'} \quad \text{or} \quad m = \frac{\Delta y}{\Delta x} = \frac{\Delta y'}{\Delta x'}$$

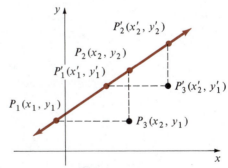

Figure 1.29

Parallel and Perpendicular Lines

In the next two theorems we shall assume that two lines L_1 and L_2 have slopes. This means that neither line is vertical since m_1 and m_2 are finite numbers. The first theorem relates the slopes of parallel lines.

THEOREM 1.1   Parallel Lines

Two lines L_1 and L_2 with slopes m_1 and m_2 are **parallel** if and only if $m_1 = m_2$.

Of course, two vertical lines, that is, lines parallel to the *y*-axis, are parallel but have undefined slopes.

THEOREM 1.2   **Perpendicular Lines**

Two lines L_1 and L_2 with slopes m_1 and m_2 are **perpendicular** if and only if $m_1 m_2 = -1$.

Proof   Assume L_1 and L_2 are perpendicular lines with slopes m_1 and m_2, respectively. From Figure 1.30 we see

$$m_1 = \frac{y_1 - y_0}{x_1 - x_0} \quad \text{and} \quad m_2 = \frac{y_2 - y_0}{x_2 - x_0}$$

Furthermore, the Pythagorean Theorem gives

$$c^2 = a^2 + b^2$$

The distance formula shows that the above expression is equivalent to

$$(x_2 - x_1)^2 + (y_2 - y_1)^2 = (x_2 - x_0)^2 + (y_2 - y_0)^2$$
$$+ (x_1 - x_0)^2 + (y_1 - y_0)^2$$

Simplifying and factoring the last result leads to

$$(y_1 - y_0)(y_2 - y_0) = -(x_1 - x_0)(x_2 - x_0)$$

$$\frac{y_1 - y_0}{x_1 - x_0} \cdot \frac{y_2 - y_0}{x_2 - x_0} = -1 \qquad (1.12)$$

or $\qquad\qquad m_1 m_2 = -1$

Conversely, if (1.12) holds, we can trace the argument backward to show that L_1 and L_2 are perpendicular. ∎

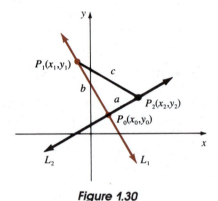

Figure 1.30

In other words, lines with slopes are perpendicular when one slope is the *negative reciprocal* of the other:

$$m_1 = -\frac{1}{m_2} \quad \text{and} \quad m_2 = -\frac{1}{m_1}$$

Equations of Lines   The concept of slope enables us to find **equations of lines**. Through a point $P_1(x_1, y_1)$ there passes only one line L with specified slope m. To find an equation of L, let us suppose that $P(x, y)$ denotes any point on the line for which $x \neq x_1$. By equating slopes,

$$\frac{y - y_1}{x - x_1} = m \quad \text{we obtain} \quad y - y_1 = m(x - x_1)$$

Since the coordinates of all points $P(x, y)$ on the line, *including $P_1(x_1, y_1)$*, satisfy the latter equation, we conclude that it is an equation of L. This particular equation is called the **point–slope form**. We summarize:

The Point–Slope Form for the Equation of a Line

$$y - y_1 = m(x - x_1) \qquad (1.13)$$

Example 3

Find an equation for the line through $(6, -2)$ with slope 4.

Solution From the point–slope form (1.13) we obtain

$$y - (-2) = 4(x - 6)$$

Equivalently,

$$y + 2 = 4x - 24 \quad \text{or} \quad y = 4x - 26$$

The point–slope form (1.13) yields two other important forms for equations of lines with slope. If the line L passes through the y-axis at $(0, b)$, then (1.13) gives

$$y - b = mx \quad \text{or} \quad y = mx + b$$

The number b is called the **y-intercept** of the line. Furthermore, if L is a horizontal line through $P_1(x_1, y_1)$, then setting $m = 0$ in (1.13) yields

$$y - y_1 = 0(x - x_1) \quad \text{or} \quad y = y_1$$

In summary:

The Slope–Intercept Form for the Equation of a Line

$$y = mx + b \qquad (1.14)$$

Equation of a Horizontal Line Through $P_1(x_1, y_1)$

$$y = y_1 \qquad (1.15)$$

Two distinct points $P_1(x_1, y_1)$ and $P_2(x_2, y_2)$ also determine a unique line. If the line has slope, then an equation can be obtained from (1.13) by first computing $m = (y_2 - y_1)/(x_2 - x_1)$ and using either the coordinates of P_1 or P_2. Finally, if the line through P_1 and P_2 is vertical, then any pair of points on the line has the same x-coordinate. Thus, if $P(x, y)$ is on the vertical line through $P_1(x_1, y_1)$ we must have

$$x = x_1$$

We summarize this last case:

Equation of a Vertical Line Through $P_1(x_1, y_1)$

$$x = x_1 \qquad (1.16)$$

Example 4

Find an equation for the line through $(2, -3)$ and $(-4, 1)$.

Solution By designating the first point as P_1, it follows from (1.11) that the slope of the line through the points is

$$\frac{1 - (-3)}{-4 - 2} = -\frac{2}{3}$$

Using the point–slope form (1.13), we get

$$y - (-3) = -\frac{2}{3}(x - 2) \quad \text{or} \quad y = -\frac{2}{3}x - \frac{5}{3}$$

Alternative Solution I   We can, of course, choose $(-4, 1)$ as P_1:

$$y - 1 = -\frac{2}{3}(x - (-4)) \quad \text{or} \quad y = -\frac{2}{3}x - \frac{5}{3}$$

Alternative Solution II   From the slope–intercept form (1.14) we can write $y = -\frac{2}{3}x + b$. Substituting $x = 2$ and $y = -3$ in the last equation gives $-3 = -\frac{4}{3} + b$ so $b = -\frac{5}{3}$. As before, $y = -\frac{2}{3}x - \frac{5}{3}$.

Example 5

(a)   An equation of a horizontal line through $(4, 9)$ is $y = 9$.

(b)   An equation of a vertical line through $(-1, -2)$ is $x = -1$.

Linear Equation   Any equation

$$ax + by + c = 0 \tag{1.17}$$

in which both x and y appear to the first power, and a, b, and c are constants, is a **linear equation**. As the name suggests, the graph of an equation of form (1.17) is a straight line. The following summarizes three special cases of (1.17):

(i) $a = 0$, $b \neq 0$,   horizontal line: $y = -\dfrac{c}{b}$

(ii) $a \neq 0$, $b = 0$,   vertical line: $x = -\dfrac{c}{a}$

(iii) $a \neq 0$, $b \neq 0$,   (1.16) can be written as

$$y = -\frac{a}{b}x - \frac{c}{b} \tag{1.18}$$

Comparing (1.18) with (1.14) we see that when $b \neq 0$ the line given by (1.18) has slope $-a/b$ and y-intercept $-c/b$.

Example 6

Find an equation for the line through $(-1, 5)$ perpendicular to the line with equation $2x + y + 4 = 0$.

Solution   Writing the given equation as $y = -2x - 4$, we see that the slope is -2. It follows from Theorem 1.2 that the line through $(-1, 5)$ perpendicular to $2x + y + 4 = 0$ has slope $\frac{1}{2}$. Hence, from (1.13) we obtain

$$y - 5 = \frac{1}{2}(x - (-1))$$

or equivalently,

$$y = \frac{1}{2}x + \frac{11}{2} \quad \text{or} \quad x - 2y + 11 = 0$$

Graphs   To graph an equation of a line, we need only determine two points whose coordinates satisfy the equation. When $a \neq 0$ and $b \neq 0$ in (1.17), the line must cross both coordinate axes. The **x-intercept** is the x-coordinate of the point where a line crosses the x-axis; the **y-intercept** is the y-coordinate of the point where a line crosses the y-axis. Since a point on the x-axis has coordinates $(x, 0)$, the x-intercept is found by setting $y = 0$ in the equation and solving for x. Similarly, we get the y-intercept by setting $x = 0$.

Example 7

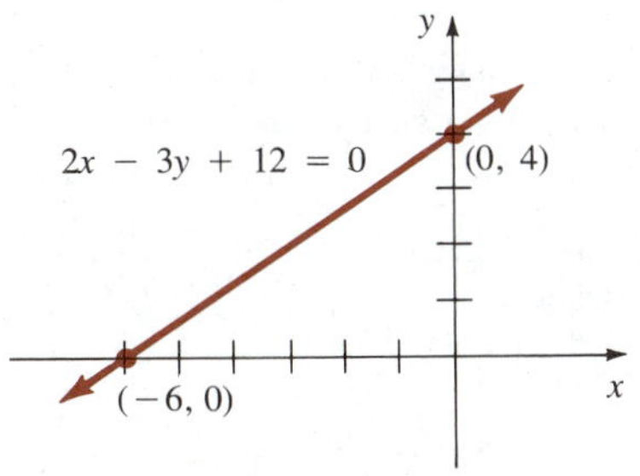

$2x - 3y + 12 = 0$

$(0, 4)$

$(-6, 0)$

Figure 1.31

Graph the line with equation $2x - 3y + 12 = 0$.

Solution   By setting $y = 0$ in the equation we obtain

$$2x + 12 = 0 \quad \text{or} \quad x = -6$$

The x-intercept is -6. Now when $x = 0$,

$$-3y + 12 = 0 \quad \text{implies} \quad y = 4$$

The y-intercept is 4. As shown in Figure 1.31 the line is drawn through the points $(-6, 0)$ and $(0, 4)$.

Exercises 1.3

Answers to odd-numbered problems begin on page A-16.

In Problems 1–4 find the slope of the line through the given points.

1. $(4, 1), (6, -2)$

2. $(3, 0), (6, 9)$

3. $\left(\frac{1}{2}, 4\right), \left(-\frac{3}{2}, 10\right)$

4. $(-3, 2), (11, 2)$

In Problems 5 and 6 find $P_2(x_2, y_2)$ using the given information.

5. $P_1(3, -2), \quad \Delta x = 4, \quad \Delta y = 5$

6. $P_1(0, 7), \quad \Delta x = 0, \quad \Delta y = -3$

In Problems 7–24 find an equation for the line that satisfies the given conditions.

7. Through $(5, -2)$, $(1, 2)$

8. Through $(0, 0)$, slope 8

9. Through $(1, 3)$, $\Delta x = 3$, $\Delta y = 9$

10. Through $(\frac{1}{4}, -\frac{1}{2})$, x-intercept $\frac{1}{8}$

11. y-intercept 8, slope 1

12. x-intercept -3, y-intercept $\frac{1}{2}$

13. Through $(10, -\frac{3}{2})$, slope 0

14. Through $(-\frac{7}{3}, \frac{3}{8})$, parallel to y-axis

15. Through $(1, 2)$, parallel to $4x + 2y = 1$

16. Through $(4, 4)$, parallel to $x - 3y = 0$

17. Through $(0, 0)$, perpendicular to $y = -\frac{1}{4}x + 7$

18. Through $(\frac{1}{2}, -1)$, perpendicular to $3x + 4y - 12 = 0$

19. Through $(2, 3)$ and the point common to $x + y = 1$ and $2x + y = 5$

20. Through the point common to $2x + 3 = 0$ and $y + 6 = 0$, slope -2

21. Through the midpoint of the line segment from $(-1, 3)$ to $(4, 8)$, perpendicular to the line segment

22. Through the midpoints of the line segments between the x- and y-intercepts of $3x + 4y = 12$ and $x + y = -6$

23. Through $(4, 2)$ parallel to the line through $(5, 1)$ and $(-1, 7)$

24. Through $(3, 9)$, slope undefined

In Problems 25–32 graph the line with the given equation. Give the slope and the x- and y-intercepts.

25. $y = 2x + 3$ **26.** $y = -3x$

27. $2y - 5 = 0$ **28.** $x = -4$

29. $5x + 3y = 15$ **30.** $x - y + 6 = 0$

31. $3x - 8y - 10 = 0$ **32.** $-\frac{1}{2}x + \frac{1}{4}y = 2$

In Problems 33 and 34 graph the line through $(3, 2)$ with the given slope.

33. $\frac{3}{4}$ **34.** $-\frac{1}{2}$

In Problems 35–40 find the value of k so that the graph of the given linear equation satisfies the indicated condition.

35. $kx + 3y = 1$, passes through $(5, 1)$

36. $-x + 7y = k$, x-intercept $\frac{3}{2}$

37. $x - ky + 3 = 0$, y-intercept -4

38. $2x + ky + 1 = 0$, perpendicular to $-5x + 10y = 3$

39. $kx + y = 0$, parallel to $3x - 7y = 12$

40. $kx + \sqrt{3}y = k$, slope $\sqrt{3}$

In Problems 41 and 42 determine whether the given points are collinear.

41. $(0, -4)$, $(1, -1)$, $(3, 5)$

42. $(-2, 3)$, $\left(1, \frac{3}{2}\right)$, $\left(-1, \frac{1}{2}\right)$

43. Find the coordinates of the point P in Figure 1.32.

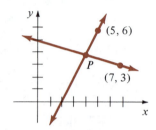

Figure 1.32

44. A line through $(2, 4)$ has slope 8. Without finding an equation of the line, determine whether the point $(1, -5)$ is on the line.

In Problems 45 and 46 use slopes to verify that P_1, P_2, and P_3 are vertices of a right triangle.

45. $P_1(8, 2)$, $P_2(1, -11)$, $P_3(-2, -1)$

46. $P_1(8, 2)$, $P_2(-3, 0)$, $P_3(5, 6)$

Miscellaneous Problems

47. Show that an equation of the line in Figure 1.33 is

$$\frac{x}{a} + \frac{y}{b} = 1$$

This is called the **intercept form** for the equation of the line.

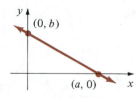

Figure 1.33

In Problems 48 and 49 find the intercept form for the equation of the line with the given intercepts.

48. x-intercept 3, y-intercept $-\frac{1}{2}$

49. x-intercept 5, y-intercept 2

50. Show that the **two-point form** for the equation of the line through $P_1(x_1, y_1)$ and $P_2(x_2, y_2)$, $x_1 \neq x_2$, is

$$y - y_1 = \left(\frac{y_2 - y_1}{x_2 - x_1}\right)(x - x_1)$$

The **distance** D **from a point** $P_1(x_1, y_1)$ **to a line** $ax + by + c = 0$ is

$$D = \frac{|ax_1 + by_1 + c|}{\sqrt{a^2 + b^2}}$$

Use this result in Problems 51–54.

51. Find the distance from $(2, 5)$ to the line $2x + 3y - 1 = 0$.

52. Find the distance from the origin to the line $y = \frac{3}{4}x - 10$.

53. Find the distance from the x-intercept of $-x + 3y + 4 = 0$ to the line through $(1, 1)$ and $(-1, 13)$.

54. Find the distance between the parallel lines $x + y = 1$ and $2x + 2y = 5$.

1.4 Functions

1.4.1 Definition and Graphs

Have you ever heard remarks such as "Success is a function of hard work" and "Demand is a function of price"? The word *function* is often used to suggest a relationship or a dependence of one quantity on another. In mathematics, the function concept has a similar, but slightly more specialized, interpretation. Before giving a precise definition, let us consider an example that uses the word *function* in a more restrictive sense.

_____ **Example 1** _____

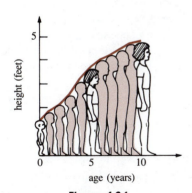

height (feet)

age (years)

Figure 1.34

(a) The area of a circle is a function of its radius.

(b) As shown in Figure 1.34, the height of a child, when measured at yearly intervals, is a function of the child's age.

(c) The first class postage for a letter is a function of its weight.

(d) The intensity of sound is a function of the distance from its source.

(e) The volume of a cubical box is a function of the length of one of its sides.

(f) The force between two particles of opposite charge is a function of the distance between them.

Rule or Correspondence   A function is a **rule**, or a **correspondence**, relating two sets in such a manner that each element in the first set corresponds to *one and only one* element in the second set. In other words, a functional relationship is a single-valued relationship. Thus, in Example 1, a circle of a given radius has only one area; at a specified instant in time a child can have only one height; and so on. As a further example, suppose four people are asked, first, to write their names and then their ages, and second, to write their names followed by the names of the cars that they own.

They respond:

Jackie—25	Jackie—Ford, Porsche
Bill —42	Bill —Plymouth
Irma —28	Irma —VW
Scott —36	Scott —Chevrolet, Oldsmobile, Buick

The first correspondence is a function since there is only one age associated with each name. The second correspondence is not a function because two elements in the first set of names (Jackie and Scott) are associated with more than one car name.

We summarize the preceding discussion with a formal definition.

> **DEFINITION 1.1   Function**
>
> A **function** f from a set X to a set Y is a rule that assigns to each element x in X a unique element y in Y. The set X is called the **domain** of f. The set of corresponding elements y in Y is called the **range** of f.

Unless stated to the contrary, we shall assume hereafter that the sets X and Y consist of real numbers.

Value of a Function

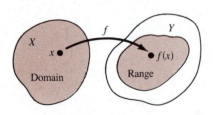

Figure 1.35

Let f be a function. The number y in the range that corresponds to a selected number x in the domain is said to be the *value* of the function at x, or **image** of x, and is written $f(x)$. The latter symbol is read "f of x" or "f at x," and we write $y = f(x)$. See Figure 1.35. Since the value of y depends on the choice for x, it is called the **dependent variable**; x is called the **independent variable**.

Functions are frequently defined by means of a formula or equation.

_____ **Example 2** _____

The rule for squaring a real number x is given by the equation

$$y = x^2 \quad \text{or} \quad f(x) = x^2$$

The values of f at, say, $x = -5$ and $x = \sqrt{7}$ are obtained by replacing x, in turn, by -5 and $\sqrt{7}$:

$$f(-5) = (-5)^2 = 25$$
$$f(\sqrt{7}) = (\sqrt{7})^2 = 7$$

_____ **Example 3** _____

In Example 2 since any real number can be squared, the domain of the function $f(x) = x^2$ is the set R of real numbers. Using interval notation the domain of f is also written as $(-\infty, \infty)$. Since $x^2 \geq 0$ for all x, it follows that the range of f is $[0, \infty)$.

For emphasis we could write the function in Example 2 as

$$f(\ \ ) = (\ \ )^2 \qquad\qquad\qquad\qquad (1.19)$$

This illustrates the fact that x is a *place holder* for any number in the domain of the function. Thus, if we wish to evaluate (1.19) at $3 + h$, where h is a real number, we put $3 + h$ into the parentheses:

$$f(3 + h) = (3 + h)^2 = 9 + 6h + h^2$$

Note that an inequality such as $y < x^2$ does not define a function. For any real number x there is no unique real number y that is less than x^2; for example, if $x = 4$, then $y = 3$, $y = 9$, and $y = 15.5$ are just some of the numbers that satisfy $y < 4^2$.

As a matter of course, the domain of a function defined by an equation is usually not specified. Unless stated to the contrary, it is understood that:

the domain of a function f is the largest set of real numbers for which the rule makes sense.*

For example, for $f(x) = 1/x$ we cannot compute $f(0)$ since $1/0$ is not defined. Thus, the domain of $f(x) = 1/x$ is the set of all real numbers except 0. By the same reasoning, we see that the domain of $f(x) = x/(x^2 - 4)$ is the set of all real numbers except -2 and 2. Similarly, $f(x) = \sqrt{x}$ requires that $x \geq 0$ and so the domain of the latter function is $[0, \infty)$.

A function is often compared to a computing machine. The "input" x is transformed by the "machine" f into the "output" $f(x)$:

Computers and calculators are programmed to recognize when a number is not in the set of allowable inputs of a function; for example, entering -4 in a calculator and pressing the $\sqrt{\ }$ key results in an error message.

_____

*This is sometimes referred to as the **natural domain** or **implicit domain** of the function.

Example 4

Determine the domain and range of the function $f(x) = 7 + \sqrt{3x - 6}$.

Solution   The radicand $3x - 6$ must be nonnegative. Solving $3x - 6 \geq 0$ gives $x \geq 2$ and so the domain of f is $[2, \infty)$. Now, by definition of the square root symbol $\sqrt{3x - 6} \geq 0$ for $x \geq 2$ and consequently, $y = 7 + \sqrt{3x - 6} \geq 7$. Since $3x - 6$ and $\sqrt{3x - 6}$ increase as x increases, we conclude that the range of f is $[7, \infty)$.

Note: Finding the range of a function by inspection is generally not an easy task.

Other Symbols   The use of f or $f(x)$ to represent a function is a natural notation. However, in different contexts such as mathematics, science, engineering, and business, functions are denoted by diverse symbols such as F, G, H, g, h, p, q, and so on. Different letters such as r, s, t, u, v, w, and z are often used for both the independent and dependent variables. Thus, a function could be written $w = G(z)$ or $v = h(t)$; for example, the area of a circle is $A = \pi r^2$, that is, $A = f(r)$ or $f(r) = \pi r^2$.

Example 5

(a)   The distance s that a freely falling body will travel is a function of time t. See Figure 1.36(a).

(b)   The minimum flying speed v of a bird is a function of its length L. See Figure 1.36(b).

(c)   In the analysis of walking, the period T of oscillation of a leg is a function of its length L. See Figure 1.36(c).

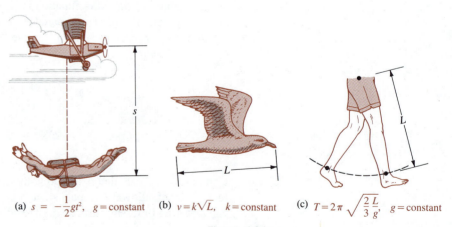

(a) $s = -\dfrac{1}{2}gt^2$, $g = $ constant   (b) $v = k\sqrt{L}$, $k = $ constant   (c) $T = 2\pi\sqrt{\dfrac{2}{3}\dfrac{L}{g}}$, $g = $ constant

Figure 1.36

_____ **Example 6** _____

Figure 1.37

Express the volume V of a cube as a function of its surface area S.

Solution   If we let x denote the length of one side of the cube, then its volume is given by $V = x^3$. Now as seen in Figure 1.37, a cube has six sides, and each side has area x^2. Thus, the surface area of a cube is $S = 6x^2$. The latter equation implies $x = \sqrt{S/6}$. Substituting this value of x into the formula for the volume then gives

$$V = \left(\sqrt{\frac{S}{6}}\right)^3 \quad \text{or} \quad V = \frac{\sqrt{6}}{36}S^{3/2}$$

Graphs   The **graph** of a function f is the set of points

$$\{(x, y) \mid y = f(x), \ x \text{ in the domain of } f\}$$

in the Cartesian plane. As a consequence of Definition 1.1, a function is characterized geometrically by the fact that *any* vertical line intersecting its graph does so *in exactly one point*. See Figure 1.38(a) and (b). Equivalently, if a vertical line intersects a graph in more than one point, it is *not* the graph of a function. See Figure 1.38(c) and (d).

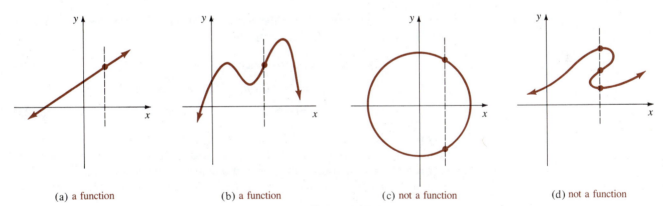

(a) a function   (b) a function   (c) not a function   (d) not a function

Figure 1.38

If a point (a, b) is on the graph of a function f, then the y-coordinate b is the value of the function at a; that is, $b = f(a)$. As we see in Figure 1.39, the value $f(a)$ is a *directed distance* from the x-axis to the point. Moreover, it is often possible to discern the domain and range of a function from its graph. Figure 1.40 shows that the domain of f is an interval on the x-axis, and the range of f is an interval on the y-axis.

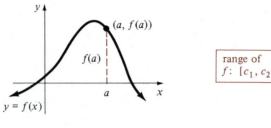

Figure 1.39

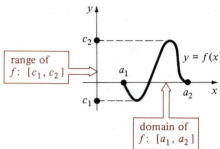

Figure 1.40

1.4.2 Types of Functions and More on Graphs

Throughout this text we shall encounter many different types of functions.

Polynomial Function   Recall from algebra that an expression such as $x^5 + 10x^2 - 2x + 1$ is called a **polynomial** of degree 5. In general, when $a_n \neq 0$

$$f(x) = a_n x^n + a_{n-1} x^{n-1} + \cdots + a_1 x + a_0 \qquad (1.20)$$

where n a nonnegative integer is said to be a **polynomial function** of degree n. The coefficients a_i, $i = 0, 1, \ldots, n$ are real numbers. The domain of any polynomial function (1.20) is the set R of real numbers. Polynomial functions of degree 0, 1, and 2 are, respectively,

$$f(x) = a_0, \quad \textbf{constant} \text{ function} \qquad (1.21)$$

$$f(x) = a_1 x + a_0, \qquad a_1 \neq 0, \quad \textbf{linear} \text{ function} \qquad (1.22)$$

$$f(x) = a_2 x^2 + a_1 x + a_0, \qquad a_2 \neq 0, \quad \textbf{quadratic} \text{ function} \qquad (1.23)$$

By comparing (1.22) with the slope–intercept form of a line, $y = mx + b$, and making the obvious identifications, we see that the graph of a linear function is a *straight line*. Of course, the graph of a constant function is a *horizontal line*. You may already know that the graph of a quadratic function is called a **parabola**.

Example 7

Graph the polynomial functions $f(x) = x^2$ and $f(x) = x^3$.

Solution   In Figure 1.41(a) and (b) we plotted the points that correspond to the values of x and $f(x)$ given in the accompanying tables. Since the domain of each function is the set of real numbers, we connect the points with a smooth and continuous curve.

x	$f(x)$
-2	4
-1	1
$-\frac{1}{2}$	$\frac{1}{4}$
0	0
$\frac{1}{2}$	$\frac{1}{4}$
1	1
2	4

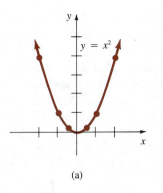

(a)

x	$f(x)$
-2	-8
-1	-1
$-\frac{1}{2}$	$-\frac{1}{8}$
0	0
$\frac{1}{2}$	$\frac{1}{8}$
1	1
2	8

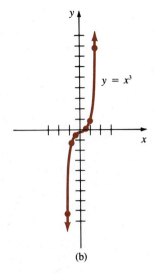

(b)

Figure 1.41

Figure 1.41(a) shows the *typical* shape of a parabola. However, Figure 1.41(b) shows one of *several possible graphs* of a third-degree, or **cubic**, polynomial function.

Rational Function   A function

$$f(x) = \frac{P(x)}{Q(x)}, \qquad \text{where } P \text{ and } Q \text{ are polynomial functions} \qquad (1.24)$$

is called a **rational function**. The domain of a rational function (1.24) consists of the set of all real numbers *except* those numbers for which $Q(x) = 0$.

___ **Example 8** _____

(a) $f(x) = \dfrac{x^3 + x + 5}{x^2 - 3x - 4}$ is a rational function. Since $x^2 - 3x - 4 = (x + 1)(x - 4)$ and $(x + 1)(x - 4) = 0$ for -1 and 4, the domain of f is the set of all real numbers except -1 and 4.

(b) $f(x) = x + x^{-1}$ is *not* a polynomial function because of the negative integer power -1. In fact, by using a common denominator, we see $f(x) = (x^2 + 1)/x$ is a rational function whose domain is the set of all real numbers except 0.

Power Function   A function

$$f(x) = kx^n, \qquad k \text{ a constant, } n \text{ a real number} \qquad (1.25)$$

is called a **power function**. Although we shall not prove it at this time, (1.25) defines a function for any real number exponent n. A power such as $x^{\sqrt{2}}$ does indeed make sense, and the rule $y = x^{\sqrt{2}}$ gives a single value of y for each nonnegative value of x. The domain of a power function depends on n; for example, when $k = 1$ and $n = \frac{1}{2}$, $y = x^{1/2} = \sqrt{x}$ is a function whose domain we have already seen to be $[0, \infty)$. When $k = 1$, the cases $n = 1, 2$, and 3 yield the simple linear, quadratic, and cubic polynomial functions $y = x$, $y = x^2$, and $y = x^3$, respectively. Figure 1.42 shows the graphs of the power functions that correspond to $k = 1$ and $n = -1$, $n = \frac{1}{2}$, $n = \frac{1}{3}$, and $n = \frac{2}{3}$, respectively.

Each of the functions given in Examples 5 and 6 is a power function.

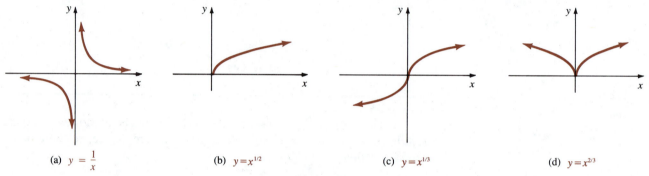

(a) $y = \dfrac{1}{x}$ (b) $y = x^{1/2}$ (c) $y = x^{1/3}$ (d) $y = x^{2/3}$

Figure 1.42

Example 9

(a)   The critical height h_{cr} of a solid column of radius r that supports its own weight is defined by the power function $h_{cr} = kr^{2/3}$. For heights greater than h_{cr}, the column will buckle.

(b)   The pulse rate p of an animal of mass m is given by the power function $p = km^{-1/4}$. Thus, a large animal such as an elephant will have a slow pulse rate.

Piecewise-defined Functions   A function need not be defined by a single formula. The following is an example of a **piecewise-defined function**.

Example 10

Graph the piecewise-defined function

$$f(x) = \begin{cases} -1, & x < 0 \\ 0 & x = 0 \\ x + 2, & x > 0 \end{cases}$$

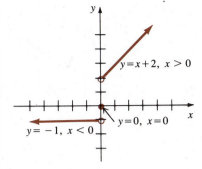

Figure 1.43

Solution   We note that f is *not* three functions but rather one function with domain the set of real numbers. However, the graph of f consists of three pieces obtained by drawing, in turn,

the graph of $y = -1$ for $x < 0$,

the point $(0, 0)$, and

the graph of $y = x + 2$ for $x > 0$.

See Figure 1.43.

Intercepts   If the graph of a function $y = f(x)$ crosses the y-axis, then its **y-intercept** is $f(0)$. The **x-intercepts** of the graph of f are the real solutions of the equation $f(x) = 0$. The values of x for which $f(x) = 0$ are also called the **zeros** of the function f.

Example 11

(a)   The graph of the polynomial function $f(x) = x^2 - x - 6$ has the y-intercept $f(0) = -6$. Also, $f(x) = 0$ when $x^2 - x - 6 = 0$ or $(x - 3)(x + 2) = 0$. Thus, 3 and -2 are x-intercepts.

(b)   The graph of the rational function $f(x) = (3x - 2)/x$ has no y-intercept since $f(0)$ is not defined (that is, 0 is not in the domain of f). Now, the only way a rational function $f(x) = P(x)/Q(x)$ can be zero is when $P(x) = 0$ and $Q(x) \neq 0$. Accordingly, for the given function, $3x - 2 = 0$ implies that the x-intercept is $\frac{2}{3}$.

Symmetry   Of the three symmetries of graphs discussed in Section 1.2, we note that the graph of a nonzero function cannot be symmetric with respect to the x-axis. This is because, in view of Definition 1.1, both points (x, y) and $(x, -y)$ cannot be on the graph of a function. A function whose graph is symmetric with respect to the y-axis is called an **even function**, whereas a function whose graph possesses symmetry with respect to the origin is called an **odd function**. The following two tests for symmetry are equivalent to tests (*i*) and (*iii*) of Section 1.2 (page 13).

The graph of $y = f(x)$ is **symmetric with respect to the y-axis**
if $f(-x) = f(x)$.   (1.26)

The graph of $y = f(x)$ is **symmetric with respect to the origin**
if $f(-x) = -f(x)$.   (1.27)

The graphs in Figure 1.44 illustrate each concept.

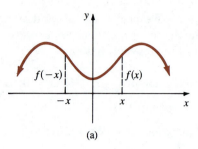

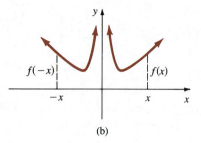

even functions

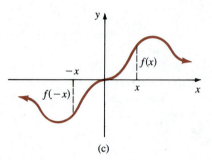

odd functions

Figure 1.44

Example 12

Determine whether the following functions are even or odd:

(a) $f(x) = \dfrac{3x}{x^2 + 1}$ (b) $f(x) = \dfrac{x^2}{x^3 + 1}$

Solution

(a) $f(-x) = \dfrac{3(-x)}{(-x)^2 + 1} = -\dfrac{3x}{x^2 + 1} = -f(x)$

The function is odd.

(b) $f(-x) = \dfrac{(-x)^2}{(-x)^3 + 1} = \dfrac{x^2}{-x^3 + 1} \neq \pm f(x)$

The function is neither even nor odd.

Example 13

Graph $f(x) = x^3 - 4x$.

Solution   The y-intercept is $f(0) = 0$. By writing $f(x) = x(x^2 - 4) = x(x + 2)(x - 2)$, we observe that the x-intercepts are -2, 0, and 2. Furthermore,

$$f(-x) = (-x)^3 - 4(-x) = -x^3 + 4x = -f(x)$$

shows that the graph of f is symmetric with respect to the origin; that is, f is an odd function. Now the three intercepts alone do not suggest the flow of the graph. But by plotting, say, $(1, f(1))$ or $(1, -3)$ we know from symmetry that $(-1, -f(1))$ or $(-1, 3)$ is also on the graph. After connecting the five points shown in Figure 1.45(a) by a smooth curve, it seems reasonable that the graph of f will be similar to that given in Figure 1.45(b).

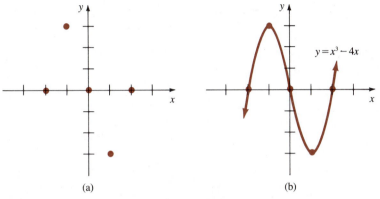

(a)   (b)

Figure 1.45

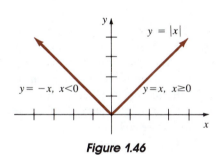

Figure 1.46

We note that *polynomial functions* that contain only *even powers* of x are necessarily *even functions*; polynomial functions that consist entirely of *odd powers* of x are *odd functions*. Hence, we can tell by inspection that the graphs of $f(x) = x^2$, $f(x) = x^4$, and $f(x) = x^6 - x^4 + x^2 + 1$ are symmetric with respect to the y-axis, whereas the graphs of $f(x) = x^3$, $f(x) = x^5$, and $f(x) = x^7 - x^3$ are symmetric with respect to the origin. The graph of a polynomial function such as $f(x) = x^3 + 5x^2 - x + 6$ that contains both even and odd powers has neither symmetry. You are cautioned not to generalize to other types of functions; for example, it could be argued that the **absolute value function** $f(x) = |x|$ contains only an odd power of x, but its graph (shown in Figure 1.46) is symmetric with respect to the y-axis. Of course, $f(x) = |x|$ is not a polynomial function.

_____ **Example 14** _____

Graph $f(x) = \dfrac{3 - x}{x + 2}$.

Solution   First, we see that the y-intercept is $f(0) = \frac{3}{2}$ and the x-intercept is 3. Next, since $f(-x)$ is equal to neither $f(x)$ nor $-f(x)$, we observe that the graph of f is not symmetric with respect to the y-axis or the origin. When you graph rational functions, it is important to take note of the domain; in this case it is evident that the domain of f is the set of all real numbers except -2. As shown in the accompanying table, when x is near -2, the values of the denominator $x + 2$ are close to zero. Hence, the corresponding functional values are large in absolute value. The graph of f is given in Figure 1.47.

x	$f(x)$
-5	$-\frac{8}{3}$
-3	-6
-2.1	-51
-1.9	49
-1	4
0	$\frac{3}{2}$
1	$\frac{2}{3}$
3	0

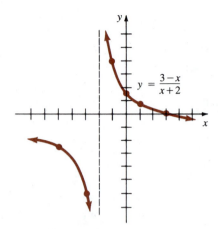

$$y = \frac{3-x}{x+2}$$

Figure 1.47

We shall examine the graphs of rational functions in greater detail in Section 2.3.

Remark

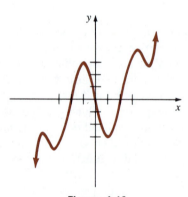

Figure 1.48

Throughout this text we shall use arrowheads on a graph to suggest that there are no surprises; that is, nothing new happens to the graph when the curve is extended in the indicated direction. But how do we know, without the drudgery of plotting more and more points, that a graph such as given in Figure 1.45 isn't really as shown in Figure 1.48? In Chapter 6 we shall see that one of the many powers of calculus enables us to sketch fairly accurate graphs of functions with a minimum amount of work and point-plotting. With calculus it is a simple matter to show that the graph of $f(x) = x^3 - 4x$ cannot be the graph given in Figure 1.48. Furthermore, Figure 1.45(b) is slightly inaccurate since the highest point of the graph on the interval $[-2, 0]$ can, with the aid of calculus, be shown to occur at $x = -2/\sqrt{3} \approx -1.15$ rather than at $x = -1$. Similarly, on the interval $[0, 2]$ the lowest point on the graph occurs not at $x = 1$ but at $x = 2/\sqrt{3} \approx 1.15$. This is admittedly difficult to see by just plotting points.

Exercises 1.4

Answers to odd-numbered problems begin on page A-17.

[1.4.1]

1. Given that $f(x) = x^2 - x$, find $f(-3)$, $f(1)$, $f(\sqrt{3})$, and $f(1 + a)$.

2. Given that $f(x) = \sqrt{x + 4}$, find $f(-3)$, $f(0)$, $f(1)$, and $f(5)$.

In Problems 3–6 find $\dfrac{f(a + h) - f(a)}{h}$ and simplify.

3. $f(x) = 6x - 9$

4. $f(x) = x^2 + 2x - 4$

5. $f(x) = x^3$

6. $f(x) = \dfrac{5}{x}$

In Problems 7–20 find the domain of the given function.

7. $f(x) = \sqrt{x + 1}$

8. $f(x) = x\sqrt{2x - 3}$

9. $f(x) = \dfrac{1 + x}{\sqrt{x}}$

10. $f(x) = \sqrt{x} + \sqrt{x - 2}$

11. $g(x) = \sqrt{25 - x^2}$

12. $g(x) = \sqrt{x^2 - 5x + 4}$

13. $F(x) = \dfrac{x^2 - 16}{x - 4}$

14. $G(x) = \dfrac{1}{x^2 + x - 6}$

15. $Q(x) = \dfrac{x}{2 - 1/x}$

16. $H(x) = 7x^3 - x^{-2} + 8$

17. $f(x) = x^{3/2}$

18. $f(x) = 1 + x^{2/3}$

19. $f(x) = \begin{cases} x, & x < 2 \\ x^2, & 2 \le x < 3 \end{cases}$

20. $g(x) = \begin{cases} x + 1, & x \le 0 \\ x - 6, & x \ge 1 \end{cases}$

In Problems 21–24 find the range of the given function.

21. $f(x) = 1 + x^2$

22. $g(x) = (2x + 1)^2$

23. $g(x) = 4 - \sqrt{x}$

24. $f(x) = 3 + \sqrt{4 - x^2}$

In Problems 25–28 determine whether the given graph (Figures 1.49–1.52) is the graph of a function.

25.

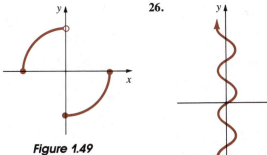

Figure 1.49

26.

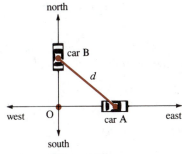

Figure 1.50

27.

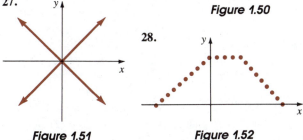

Figure 1.51

28.

Figure 1.52

29. Express the perimeter P of a square as a function of its area A.

30. Express the area A of a circle as a function of its circumference C.

31. Express the area A of an equilateral triangle as a function of the length s of one side.

32. In Problem 31 express A as a function of the height h of the triangle.

33. Express the volume V of a cube as a function of the area A of its base.

34. An open box is made from a cardboard square that is 40 cm on a side by cutting a square out from each corner and then turning up the remaining sides. Express the area A of the base of the box as a function of the length x of the side of the cutout square.

35. Car A passes point O heading east at a constant rate of 40 mph; car B passes the same point one hour later heading north at a constant rate of 60 mph. Express the distance d between the cars as a function of time t, where t is measured starting when car B passes O. See Figure 1.53.

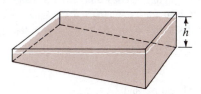

Figure 1.53

36. The swimming pool shown in Figure 1.54 is 3 ft deep at the shallow end, 8 ft deep at the deepest end, 40 ft long, 30 ft wide, and the bottom is an inclined plane. Express the volume V of water in the pool as a function of the height h of the water above the deep end. (*Hint*: V will be a piecewise-defined function.)

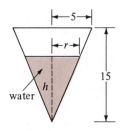

Figure 1.54

37. Water is being pumped into a conical tank whose height is 15 ft and whose radius is 5 ft. Express the volume V of the water at any time as a function of its depth h. Use the cross-sectional view given in Figure 1.55. (*Hint*: See page 191.)

Figure 1.55

38. Consider the circle of radius h with center (h, h) shown in Figure 1.56. Express the area A of the shaded region as a function of h.

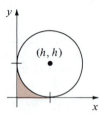

Figure 1.56

39. Express the area A of a rectangle whose lengths of sides satisfy the **golden ratio**

$$\frac{y}{x} = \frac{x}{x + y}$$

as a function of x. See Figure 1.57.

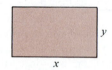

Figure 1.57

40. A wire of length L is cut x units from one end. One piece of the wire is bent into a square and the other piece is bent into a circle. Express the sum S of the areas of the square and circle as a function of x.

[1.4.2]

In Problems 41 and 42 find

(a) $f(-2)$ (b) $f(0)$

(c) $f(0.9)$ (d) $f(1 + h), h > 0$

for the given function.

41. $f(x) = \begin{cases} x + 3, & x \le 1 \\ x^2, & x > 1 \end{cases}$

42. $f(x) = \begin{cases} -x^2, & x < 0 \\ 3, & 0 \le x < 1 \\ 2x - 1, & x \ge 1 \end{cases}$

43. For what values of x is

$$f(x) = \begin{cases} x^3, & x < 0 \\ x^2, & x \ge 0 \end{cases}$$

equal to 25? to -64?

44. Determine whether the numbers -1 and 2 are in the range of the rational function $f(x) = (2x - 1)/(x + 4)$.

In Problems 45–52 determine whether the given function is even, odd, or neither even nor odd.

45. $f(x) = x^6 - x^2 + 5$

46. $g(x) = (x + 2)^2$

47. $g(x) = x^3 - 1$

48. $f(x) = 4x^5 + 8x^3$

49. $f(x) = \dfrac{1}{x(x^2 + 1)}$

50. $f(x) = x^{2/3} + 4$

51. $F(x) = \dfrac{|x|}{x}$

52. $G(x) = x^2 \sqrt{x^6 + 9}$

In Problems 53–70 graph the given function.

53. $f(x) = -x^2$

54. $f(x) = x^2 + 1$

55. $f(x) = (x + 1)^2$

56. $f(x) = 1 - x^2$

57. $f(x) = x^3 + 2$

58. $f(x) = -x^3$

59. $f(x) = (x + 1)(x - 3)$

60. $f(x) = x^3 - x$

61. $f(x) = x^4$

62. $f(x) = x^5$

63. $f(x) = \dfrac{1}{x^2}$

64. $f(x) = x^{-1/2}$

65. $f(x) = \dfrac{2}{1 + x^2}$

66. $f(x) = \dfrac{1}{x - 1}$

67. $f(x) = \begin{cases} 3, & x \le 1 \\ -3, & x > 1 \end{cases}$

68. $f(x) = \begin{cases} x^2, & x < 0 \\ x, & x \ge 0 \end{cases}$

69. $f(x) = \begin{cases} x^2, & x < -1 \\ 1, & -1 \le x \le 1 \\ 2 - x^2, & x > 1 \end{cases}$

70. $f(x) = \begin{cases} x + 2, & x < 0 \\ 2 - x, & 0 \le x < 2 \\ x - 2, & x \ge 2 \end{cases}$

In Problems 71–76 find the x- and y-intercepts of the graph of the given function. Do not graph.

71. $f(x) = x^2 - 2x - 8$

72. $f(x) = x^3 - 8x^2 + 15x$

73. $f(x) = (x - 1)(x - 6)(x^2 - 9)$

74. $f(x) = \sqrt{4x + 2}$

75. $f(x) = \dfrac{25x^2 - 4}{x + 3}$

76. $f(x) = \dfrac{x^4 - 16}{x}$

In Problems 77 and 78 determine the domain and range of the function whose graph is given (Figures 1.58 and 1.59).

77.

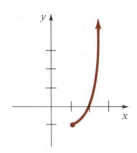

Figure 1.58

78.

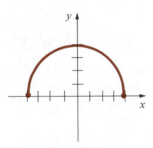

Figure 1.59

79. The International Whaling Commission has decreed that the weight W (in long tons) of a mature blue whale is given by the linear function $W = (3.51)L - 192$, where L is length and $L \geq 70$ ft.* Find the weight of a 90-foot blue whale.

80. The expected length (in cm) of a human fetus is given by the linear function $L = 1.53t - 6.7$, where $t \geq 12$ weeks. What is the expected length of a fetus at 36 weeks? What are the weekly increases in length? When is a fetus 1 ft (30.48 cm) in length?

81. The functional relationship between degrees Celsius T_c and degrees Fahrenheit T_f is linear. Express T_f as a function of T_c if (0° C, 32° F) and (60° C, 140° F) are on the graph of T_f. Show that 100° C is equivalent to the Fahrenheit boiling point 212° F. See Figure 1.60.

82. The functional relationship between degrees Celsius T_c and degrees Kelvin T_k is linear. Express T_k as a function of T_c given that (0° C, 273° K) and (27° C, 300° K) are on the graph of T_k. Express the boiling point 100° C in degrees Kelvin. Absolute zero is defined as 0° K. What is 0° K in degrees Celsius? Express T_k as a linear function of T_f. What is 0° K in degrees Fahrenheit? See Figure 1.60.

---

*For $0 < L < 70$ the whales are immature, and the formula is not used. A blue whale at birth is approximately 24 ft long.

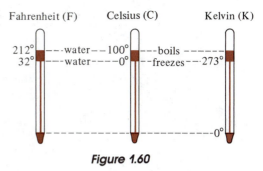

Figure 1.60

83. A ball is thrown upward from ground level with an initial velocity of 96 ft/sec. The height of the ball from the ground is given by the quadratic function $s(t) = -16t^2 + 96t$. At what times is the ball on the ground? Graph s over the time interval for which $s(t) \geq 0$.

84. In Problem 83 at what times is the ball 80 ft above the ground? How high does the ball go?

85. It is often assumed that the weight W of a body is a function of its length L. This assumption is given by the power function $W = kL^3$, where k is a constant. For a certain lizard $k = 400$ g/m^3. Determine the weight of a 0.5-meter-long lizard.

Calculator Problem

86. A relationship between weight W and length L of a sperm whale is given by the power function

$$W = 0.000137 \, L^{3.18}$$

where W is measured in tons and L in feet. Determine the weight of a 100-ft-long sperm whale.

Miscellaneous Problems

87. Any function f can be written as $f = f_e + f_o$, where $f_e = \frac{1}{2}[f(x) + f(-x)]$ and $f_o = \frac{1}{2}[f(x) - f(-x)]$. Show that f_e is an even function and f_o is an odd function.

If x changes from x_1 to x_2, the amount of change is $\Delta x = x_2 - x_1$ and the corresponding change in the functional value is $\Delta y = f(x_2) - f(x_1)$. In Problems 88 and 89 find the values of Δx and Δy.

88. $f(x) = x^2 + 4x$, $x_1 = 4$, $x_2 = 7$

89. $f(x) = \sqrt{x + 1}$, $x_1 = 3$, $x_2 = 8$

1.5 Combining Functions

A function f can be combined with another function g by means of arithmetic operations to form other functions. The **sum**, $f + g$; **difference**, $f - g$; **product**, fg; and **quotient**, f/g, are defined in the following manner.

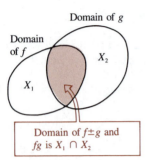

Domain of g

Domain of f

X_2

X_1

Domain of $f \pm g$ and fg is $X_1 \cap X_2$

Figure 1.61

> **DEFINITION 1.2**   Let f and g be functions.
>
> (*i*)   **Sum**: $(f + g)(x) = f(x) + g(x)$
>
> (*ii*) **Difference**: $(f - g)(x) = f(x) - g(x)$
>
> (*iii*) **Product**: $(fg)(x) = f(x)g(x)$
>
> (*iv*) **Quotient**: $(f/g)(x) = \dfrac{f(x)}{g(x)}$

The domain of $f + g$, $f - g$, and fg is the intersection of the domain of f with the domain of g. See Figure 1.61. The domain of the quotient f/g is the intersection of the domains of f and g *without* the numbers for which $g(x) = 0$.

Example 1

If $f(x) = 2x^2 - 5$ and $g(x) = 3x + 4$, find $f + g$, $f - g$, fg, and f/g.

Solution   From Definition 1.2,

$$(f + g)(x) = f(x) + g(x) = (2x^2 - 5) + (3x + 4) = 2x^2 + 3x - 1$$
$$(f - g)(x) = f(x) - g(x) = (2x^2 - 5) - (3x + 4) = 2x^2 - 3x - 9$$
$$(fg)(x) = f(x)g(x) = (2x^2 - 5)(3x + 4) = 6x^3 + 8x^2 - 15x - 20$$
$$(f/g)(x) = \frac{f(x)}{g(x)} = \frac{2x^2 - 5}{3x + 4}, \qquad x \neq -\frac{4}{3}$$

Example 2

If $f(x) = \sqrt{x - 1}$ and $g(x) = \sqrt{2 - x}$, find the domains of fg and f/g .

Solution   The domains of f and g are $[1, \infty)$ and $(-\infty, 2]$, respectively. Consequently, the domain of the product

$$(fg)(x) = \sqrt{x - 1}\,\sqrt{2 - x} = \sqrt{(x - 1)(2 - x)}$$

is the intersection of the domains: $[1, 2]$. However, the domain of the quotient

$$(f/g)(x) = \frac{\sqrt{x - 1}}{\sqrt{2 - x}} = \sqrt{\frac{x - 1}{2 - x}}$$

is $[1, 2)$ since $g(x) = 0$ at $x = 2$.

If f is a constant function, say $f = c$, then we see from Definition 1.2(iii) that the function cg is defined by

$$(cg)(x) = c \cdot g(x)$$

The domain of cg is the domain of g; for example, if $g(x) = 4x^3 - 3x$, then the function $5g$ is simply $(5g)(x) = 20x^3 - 15x$.

Composition　A different way of combining functions is by **composition**.

DEFINITION 1.3　　　　Let f and g be functions.

(i)　The **composition** of f and g, written $f \circ g$, is the function

$$(f \circ g)(x) = f(g(x))$$

(ii)　The **composition** of g and f, written $g \circ f$, is the function

$$(g \circ f)(x) = g(f(x))$$

A composite function such as $f \circ g$ is sometimes said to be "a function of a function." It is understood that the numbers represented by $g(x)$, in part (i) of Definition 1.3, must be in the domain of f. In other words, the domain of $f \circ g$ is that subset of the domain of g for which $g(x)$ is in the domain of f. See Figure 1.62. Of course, in part (ii) of Definition 1.3 $f(x)$ must be in the domain of g.

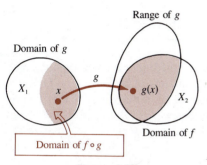

Range of g

Domain of g

g

X_1　x　$g(x)$　X_2

Domain of f

Domain of $f \circ g$

Figure 1.62

_____ **Example 3** _____

If $f(x) = x^2$ and $g(x) = x^2 + 1$, find $f \circ g$ and $g \circ f$.

Solution　From Definition 1.3 we find

$$(f \circ g)(x) = f(g(x)) = f(x^2 + 1)$$
$$= (x^2 + 1)^2 = x^4 + 2x^2 + 1$$

and　　　　$$(g \circ f)(x) = g(f(x)) = g(x^2)$$
$$= (x^2)^2 + 1 = x^4 + 1$$

Example 3 illustrates that, in general, $f \circ g \neq g \circ f$.

_____ **Example 4** _____

If $f(x) = 3x - \sqrt{x}$ and $g(x) = 2x + 1$, find the domain of $f \circ g$.

Solution　From Definition 1.3(i) it follows that

$$(f \circ g)(x) = f(g(x)) = f(2x + 1)$$
$$= 3(2x + 1) - \sqrt{2x + 1} = 6x + 3 - \sqrt{2x + 1}$$

For $f \circ g$ to be defined, we must demand that $g(x) \geq 0$ or $2x + 1 \geq 0$; that is, the domain of $f \circ g$ is given by $[-\frac{1}{2}, \infty)$.

Example 5

Express the function $F(x) = \sqrt{2x^2 + 5}$ as a composition $f \circ g$ of two functions f and g.

Solution   If we identify $f(x) = \sqrt{x}$ and $g(x) = 2x^2 + 5$, then

$$F(x) = (f \circ g)(x) = f(g(x))$$
$$= \sqrt{g(x)} = \sqrt{2x^2 + 5}$$

Alternative Solution   If we choose $f(x) = \sqrt{2x + 5}$ and $g(x) = x^2$, then the function F can also be written as

$$F(x) = (f \circ g)(x) = f(g(x))$$
$$= \sqrt{2g(x) + 5} = \sqrt{2x^2 + 5}$$

Shifted Graphs

Often we can sketch the graph of a function by **shifting** or **translating** the graph of a simpler function. If c is a constant, the graphs of the sum $y = f(x) + c$ and the difference $y = f(x) - c$ can be obtained from the graph of $y = f(x)$ by a **vertical shift**. The graphs of the compositions $y = f(x + c)$ and $y = f(x - c)$ correspond to **horizontal shifts** in the graph of $y = f(x)$. We summarize the results in the following table for $c > 0$.

Function	Graph
$y = f(x) + c$	Graph of $y = f(x)$ shifted **up** c units
$y = f(x) - c$	Graph of $y = f(x)$ shifted **down** c units
$y = f(x + c)$	Graph of $y = f(x)$ shifted to the **left** c units
$y = f(x - c)$	Graph of $y = f(x)$ shifted to the **right** c units

---

Example 6

The graphs of $y = x^2 + 1$, $y = x^2 - 1$, $y = (x + 1)^2$, and $y = (x - 1)^2$ given in Figure 1.63 are obtained from the graph of $f(x) = x^2$ by shifting the graph in Figure 1.41(a), in turn, 1 unit up, 1 unit down, 1 unit to the left, and 1 unit to the right.

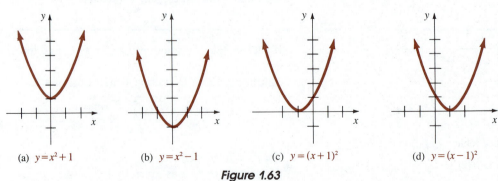

(a) $y = x^2 + 1$ (b) $y = x^2 - 1$ (c) $y = (x + 1)^2$ (d) $y = (x - 1)^2$

Figure 1.63

The graph of a function $y = f(x \pm c_1) \pm c_2$, $c_1 > 0$, $c_2 > 0$, combines both a horizontal shift (left or right) along with a vertical shift (up or down). For example, the graph of $y = f(x - c_1) + c_2$ is the graph of $y = f(x)$ shifted c_1 units to the right and then c_2 units up.

Example 7

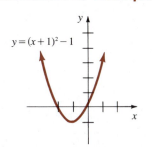

Figure 1.64

Graph $y = (x + 1)^2 - 1$.

Solution   From the foregoing discussion we identify the form $y = f(x + c_1) - c_2$, where $c_1 = 1$ and $c_2 = 1$. Thus, the graph of $y = (x + 1)^2 - 1$ is the graph of $f(x) = x^2$ shifted 1 unit to the left and then down 1 unit. The graph is given in Figure 1.64.

Reflections   For $c > 0$ the graph of the product $y = cf(x)$ retains, roughly, the same shape as the graph of $y = f(x)$. However, the graph of $y = -cf(x)$, $c > 0$, is a **reflection** of the graph of $y = cf(x)$ in the x-axis; that is, the graph is *turned upside down*.

Example 8

In Figure 1.65(a) we have compared the graphs of $y = x^2$ and $y = \frac{1}{4}x^2$. The graph of $y = -\frac{1}{4}x^2$ shown in Figure 1.65(b) is the graph of $y = \frac{1}{4}x^2$ reflected in the x-axis.

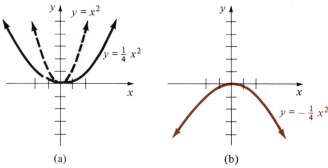

(a)   (b)

Figure 1.65

Exercises 1.5

Answers to odd-numbered problems begin on page A-17.

In Problems 1–10 find $f + g$, $f - g$, fg, and f/g.

1. $f(x) = 2x + 5$, $g(x) = -4x + 8$

2. $f(x) = 5x^2$, $g(x) = 7x - 9$

3. $f(x) = 3x^2$, $g(x) = 4x^3$

4. $f(x) = x^2 - 3x$, $g(x) = x + 1$

5. $f(x) = \dfrac{x}{x + 1}$, $g(x) = \dfrac{1}{x}$

6. $f(x) = \dfrac{2x - 1}{x + 3}$, $g(x) = \dfrac{x - 3}{4x + 2}$

7. $f(x) = x^2 + 2x - 3$, $g(x) = x^2 + 3x - 4$

8. $f(x) = x^2$, $g(x) = \sqrt{x}$

9. $f(x) = x^3 + x^2$, $g(x) = 4x^3 - 4x^2$

10. $f(x) = \sqrt{x + 3}$, $g(x) = \sqrt{x - 1}$

In Problems 11–20 find $f \circ g$ and $g \circ f$.

11. $f(x) = 3x - 2$, $g(x) = = x + 6$

12. $f(x) = 2x + 10$, $g(x) = \dfrac{1}{2}x - 5$

13. $f(x) = 4x + 1$, $g(x) = x^2$

14. $f(x) = x^2$, $g(x) = x^3 + x^2$

15. $f(x) = \dfrac{3}{x}$, $g(x) = \dfrac{x}{x + 1}$

16. $f(x) = 2x + 4$, $g(x) = \dfrac{1}{2x + 4}$

17. $f(x) = \dfrac{x - 1}{x + 2}$, $g(x) = \dfrac{x - 2}{x + 1}$

18. $f(x) = 6$, $g(x) = x^3 + 9$

19. $f(x) = x^3 - 5$, $g(x) = \sqrt[3]{x + 5}$

20. $f(x) = x^2 + \sqrt{x}$, $g(x) = x^2$

In Problems 21–24 find ff, $f \circ (2f)$, $f \circ (1/f)$, and $(1/f) \circ f$.

21. $f(x) = 2x^3$

22. $f(x) = x^2 + 1$

23. $f(x) = \dfrac{2}{x^2}$

24. $f(x) = \dfrac{1}{x - 1}$

In Problems 25–28 find $f(g(0))$, $f(g(\frac{1}{2}))$, $g(f(-2))$, and $g(f(g(1)))$.

25. $f(x) = 2x - 2$, $g(x) = 4x$

26. $f(x) = 3x + 1$, $g(x) = x^2 + 1$

27. $f(x) = x^2 + 1$, $g(x) = 2x^4 - 4x^2 + 3$

28. $f(x) = \dfrac{x}{x + 1}$, $g(x) = \dfrac{1}{2x + 1}$

In Problems 29–32 find the domains of fg, f/g, and $f \circ g$.

29. $f(x) = \sqrt{x}$, $g(x) = \sqrt{25 - x^2}$

30. $f(x) = \dfrac{1}{x^2 - x - 12}$, $g(x) = x^2$

31. $f(x) = \dfrac{x^2 - 4}{x}$, $g(x) = x^2 - 1$

32. $f(x) = \sqrt{x - 1}$, $g(x) = x^2 + 1$

In Problems 33–36 express the given function F as a composition $f \circ g$ of two functions f and g.

33. $F(x) = 2x^4 - x^2$

34. $F(x) = \dfrac{1}{x^2 + 9}$

35. $F(x) = (x - 1)^2 + 6\sqrt{x - 1}$

36. $F(x) = 5 + |4x - 1|$

The composition of three functions f, g, and h is the function

$$(f \circ g \circ h)(x) = f(g(h(x)))$$

In Problems 37 and 38 find $f \circ g \circ h$.

37. $f(x) = x^2 + 6$, $g(x) = 2x + 1$, $h(x) = 3x - 2$

38. $f(x) = \sqrt{x - 5}$, $g(x) = x^2 + 2$, $h(x) = \sqrt{2x + 1}$

In Problems 39 and 40 express the given function as a composition $f \circ g \circ h$ of three functions f, g, and h.

39. $F(x) = \sqrt{x^3 + 2}$

40. $F(x) = \left(\dfrac{x^2}{x^2 + 1}\right)^{2/3}$

In Problems 41 and 42 find a function g.

41. $f(x) = 2x - 5$, $(f \circ g)(x) = -4x + 13$

42. $f(x) = \sqrt{2x + 6}$, $(f \circ g)(x) = 4x^2$

In Problems 43–46 graph each function using the graph of $f(x) = \sqrt{x}$ given in Figure 1.66.

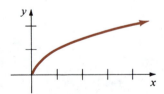

Figure 1.66

43. $y = \sqrt{x - 1}$

44. $y = \sqrt{x} + 3$

45. $y = -\sqrt{x}$

46. $y = 1 + \sqrt{x - 3}$

In Problems 47–50 graph each function using the graph of $f(x) = x^3$ given in Figure 1.67.

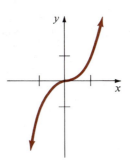

Figure 1.67

47. $y = (x - 2)^3$ **48.** $y = -(x + 1)^3$

49. $y = 1 - x^3$ **50.** $y = (x + 1)^3 - 1$

In Problems 51–54 each graph (Figures 1.68–1.71) is a shifted graph of the indicated function. Find an equation of the graph.

51. $y = |x|$ **52.** $y = \dfrac{2}{1 + x^2}$

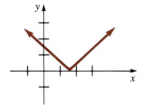

Figure 1.68

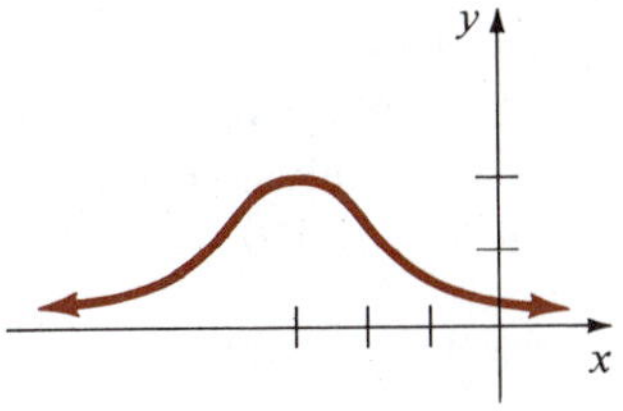

Figure 1.69

53. $y = -x^2$ **54.** $y = x^4$

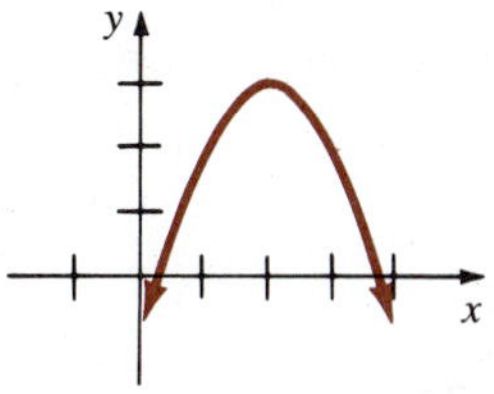

Figure 1.70

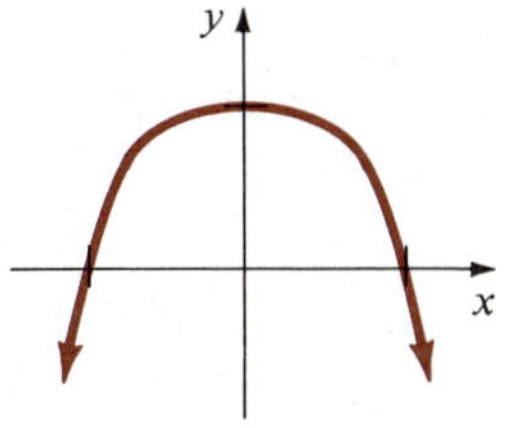

Figure 1.71

Complete the square in Problems 55 and 56, and express the given function in the form $y = f(x \pm c_1) \pm c_2$, where $f(x) = x^2$. Graph.

55. $y = x^2 + 4x$ **56.** $y = x^2 - 6x + 10$

Miscellaneous Problems

57. Determine whether $f \circ (g + h) = f \circ g + f \circ h$ is true or false.

58. Suppose $[-1, 1]$ is the domain of $f(x) = x^2$. What is the domain of $y = f(x - 2)$?

59. Given that f is odd and g is odd, determine whether the functions $f + g, f - g, fg$, and f/g are even, odd, or neither even nor odd.

60. Given that f is odd and g is even, determine whether the functions $f + g, f - g, fg$, and f/g are even, odd, or neither even nor odd.

Let U be the function defined by

$$U(x - a) = \begin{cases} 0, & 0 \le x < a \\ 1, & x \ge a \end{cases}$$

In Problems 61 and 62 graph the given function.

61. $y = U(x - 1)$ **62.** $y = U(x - 1) + U(x - 2)$

63. Find an equation in terms of $U(x - a)$ for the function illustrated in Figure 1.72.

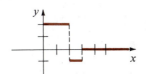

Figure 1.72

64. Given that $f(x) = x^2$, compare the graphs of $y = f(x - 3)$ and $y = f(x - 3)U(x - 3)$.

65. Suppose f and g are the piecewise-defined functions

$$f(x) = \begin{cases} x, & x < 0 \\ x + 2, & x \ge 0 \end{cases} \quad \text{and} \quad g(x) = \begin{cases} x^2, & x \le -1 \\ x - 5, & x > -1 \end{cases}$$

Find $f + g$ and fg.

1.6 Trigonometric Functions

Cosine and Sine   In trigonometry recall that the **cosine function** and **sine function** of an angle t, denoted by $\cos t$ and $\sin t$, respectively, can be interpreted in two ways:

(*i*)   as the x- and y-coordinates of a point on a unit circle, as shown in Figure 1.73(a), or as (*ii*)   the quotient of lengths of sides of a right triangle, as shown in Figure 1.73(b):

$$\cos t = \frac{\text{side adjacent}}{\text{hypotenuse}}, \quad \sin t = \frac{\text{side opposite}}{\text{hypotenuse}}$$

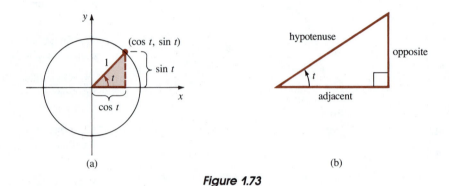

(a)   (b)

Figure 1.73

In the following discussion, we shall focus on the former interpretation.

Measure of Angles   An angle is measured in either **degrees** or **radians**. As shown in Figure 1.74(a), an angle of **one radian** subtends an arc of 1 unit on the circumference of a unit circle. The semicircumference is equivalent to an angle of π radians, where π denotes the irrational number 3.1415926 . . . . The equivalence

$$\pi \text{ radians} = 180 \text{ degrees} (180°)$$

can be used in the forms

$$1 \text{ radian} = \left(\frac{180}{\pi}\right)^° \quad \text{or} \quad 1° = \frac{\pi}{180} \text{ radians}$$

to convert from one measure to the other. By division we obtain the approximation 1 radian $\approx$ 57.296°.

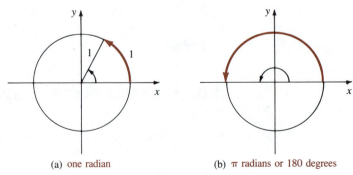

(a) one radian   (b) π radians or 180 degrees

Figure 1.74

_____ **Example 1** _____

(a) $\pi/9$ radians $= (\pi/9)(180/\pi) = 20°$

(b) $15° = 15(\pi/180) = \pi/12$ radians

Since radian measure is used almost exclusively throughout calculus, the reader is urged to become familiar with, and to be able to obtain, the radian equivalents of frequently occurring angles such as 30°, 45°, 90°, 120°, and so on. Some of these angles are given in the following table.

Degrees	0°	30°	45°	60°	90°	120°	135°	150°	180°	270°	360°
Radians	0	$\dfrac{\pi}{6}$	$\dfrac{\pi}{4}$	$\dfrac{\pi}{3}$	$\dfrac{\pi}{2}$	$\dfrac{2\pi}{3}$	$\dfrac{3\pi}{4}$	$\dfrac{5\pi}{6}$	π	$\dfrac{3\pi}{2}$	2π

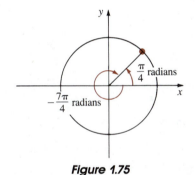

Figure 1.75

On a unit circle, a central angle t is said to be in **standard position** if its vertex is at the origin and its initial side coincides with the positive x-axis. Angles measured counterclockwise are given a **positive** measure, whereas angles measured clockwise are **negative**. Two angles in standard position are **coterminal** if they have the same terminal side. In Figure 1.75 the coterminal angles $\pi/4$ and $-7\pi/4$ determine the same point on the unit circle.

Additional Trigonometric Functions

Four additional trigonometric functions are defined in terms of $\sin t$ and $\cos t$:

$$\textbf{tangent: } \tan t = \frac{\sin t}{\cos t} \qquad \textbf{cotangent: } \cot t = \frac{\cos t}{\sin t} = \frac{1}{\tan t}$$

$$\textbf{secant: } \sec t = \frac{1}{\cos t} \qquad \textbf{cosecant: } \csc t = \frac{1}{\sin t}$$

Numerical Values

Using Figure 1.76 the following **numerical values** of $\sin t$ and $\cos t$ are evident.

$$\sin 0 = 0 \qquad\qquad \cos 0 = 1$$
$$\sin \frac{\pi}{2} = 1 \qquad\qquad \cos \frac{\pi}{2} = 0$$
$$\sin \pi = 0 \qquad\qquad \cos \pi = -1$$
$$\sin \frac{3\pi}{2} = -1 \qquad\qquad \cos \frac{3\pi}{2} = 0$$
$$\sin 2\pi = 0 \qquad\qquad \cos 2\pi = 1$$

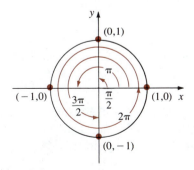

Figure 1.76

Some additional values of $\sin t$ and $\cos t$, given in the next table, will be used throughout this text. These values occur often enough to warrant their memorization.

$t(radians)$	0	$\dfrac{\pi}{6}$	$\dfrac{\pi}{4}$	$\dfrac{\pi}{3}$	$\dfrac{\pi}{2}$	$\dfrac{2\pi}{3}$	$\dfrac{3\pi}{4}$	$\dfrac{5\pi}{6}$	π	$\dfrac{3\pi}{2}$	2π
$\sin t$	0	$\dfrac{1}{2}$	$\dfrac{\sqrt{2}}{2}$	$\dfrac{\sqrt{3}}{2}$	1	$\dfrac{\sqrt{3}}{2}$	$\dfrac{\sqrt{2}}{2}$	$\dfrac{1}{2}$	0	-1	0
$\cos t$	1	$\dfrac{\sqrt{3}}{2}$	$\dfrac{\sqrt{2}}{2}$	$\dfrac{1}{2}$	0	$-\dfrac{1}{2}$	$-\dfrac{\sqrt{2}}{2}$	$-\dfrac{\sqrt{3}}{2}$	-1	0	1

The values of the other trigonometric functions can be obtained from the values of the sine and cosine.

Example 2

(a) $\tan \dfrac{\pi}{3} = \dfrac{\sin \dfrac{\pi}{3}}{\cos \dfrac{\pi}{3}} = \dfrac{\sqrt{3}/2}{1/2} = \sqrt{3}$

(b) $\sec \pi = \dfrac{1}{\cos \pi} = -1$

(c) $\cot 2\pi = \dfrac{\cos 2\pi}{\sin 2\pi}$ is not defined since $\sin 2\pi = 0$

Figure 1.77 shows the algebraic signs of $\sin t$ and $\cos t$ in the four quadrants of the Cartesian plane. With the aid of this figure you should be able to extend the preceding table to angles such as $7\pi/6$ radians, $5\pi/3$ radians, $7\pi/4$ radians, and so on.

Some Basic Identities   Since the angles t and $t + 2\pi$ are coterminal, the values of sine and cosine functions repeat every 2π radians:

$$\sin t = \sin(t + 2\pi), \qquad \cos t = \cos(t + 2\pi) \tag{1.28}$$

Because of the identities in (1.28), $\sin t$ and $\cos t$ are said to be **periodic** with **period** 2π. It can be proved that $\tan t$ is periodic with period π:

$$\tan t = \tan(t + \pi) \tag{1.29}$$

Moreover, the cosine and sine functions are related by the fundamental identity

$$\cos^2 t + \sin^2 t = 1 \tag{1.30}$$

where $\cos^2 t = (\cos t)^2$ and $\sin^2 t = (\sin t)^2$. This result follows from the fact that the coordinates of the point $(\cos t, \sin t)$ must satisfy an equation of a unit circle, $x^2 + y^2 = 1$. Dividing (1.30), in turn, by $\cos^2 t$ and $\sin^2 t$ yields

$$1 + \tan^2 t = \sec^2 t \tag{1.31}$$
$$1 + \cot^2 t = \csc^2 t \tag{1.32}$$

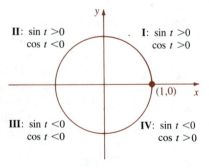

II: $\sin t > 0$
$\cos t < 0$

I: $\sin t > 0$
$\cos t > 0$

$(1,0)$ x

III: $\sin t < 0$
$\cos t < 0$

IV: $\sin t < 0$
$\cos t > 0$

Figure 1.77

There are many other identities that involve the trigonometric functions. For future reference some of the more important ones follow.

$$\sin(-t) = -\sin t \tag{1.33}$$

$$\cos(-t) = \cos t \tag{1.34}$$

Addition Formulas

$$\sin(t_1 \pm t_2) = \sin t_1 \cos t_2 \pm \cos t_1 \sin t_2 \tag{1.35}$$

$$\cos(t_1 \pm t_2) = \cos t_1 \cos t_2 \mp \sin t_1 \sin t_2 \tag{1.36}$$

Double-angle Formulas

$$\sin 2t = 2 \sin t \cos t \tag{1.37}$$

$$\cos 2t = \cos^2 t - \sin^2 t \tag{1.38}$$

Half-angle Formulas

$$\sin^2 \frac{t}{2} = \frac{1}{2}(1 - \cos t) \tag{1.39}$$

$$\cos^2 \frac{t}{2} = \frac{1}{2}(1 + \cos t) \tag{1.40}$$

Formulas (1.39) and (1.40) will be particularly useful later on in the equivalent forms

$$\sin^2 t = \frac{1}{2}(1 - \cos 2t) \quad \text{and} \quad \cos^2 t = \frac{1}{2}(1 + \cos 2t) \tag{1.41}$$

The sine and cosine functions are functions in the strictest sense of Definition 1.1; for each angle t there is only one value of $\sin t$ and one value of $\cos t$. Since arc length s on the circumference of a circle of radius r is related to its central angle t by $s = rt$, where t is measured in radians, it follows that, for a unit circle, $s = t$. In other words, for any real number s there is an angle whose measure is s radians. Thus, the sine and cosine functions have as their common domain the set R of real numbers; for example, when we write sin 2, it is understood that we mean sin(2 radians) and *not* sin 2°. By introducing the usual symbols x and y to denote the independent and dependent variables, we write

$$y = \sin x \quad \text{and} \quad y = \cos x$$

Pertinent information about the domain and range of each trigonometric function is summarized in the graphs given in Figure 1.78 (on page 52).

Examination of the graphs in Figure 1.78 clearly reveals the periodic nature of the trigonometric functions; for example, the portion of the graph of $y = \sin x$ on the interval $[0, 2\pi]$ repeats every 2π units. Also, the sine and cosine functions have **amplitude** 1 since the maximum distance a point on the graphs of $y = \sin x$ and $y = \cos x$ can be from the x-axis is 1 unit. In general, the functions

$$y = A \sin kx, \qquad k > 0$$

$$y = A \cos kx, \qquad k > 0$$

have amplitude $|A|$ and period $2\pi/k$. Note that the remaining four trigonometric functions do not have amplitudes.

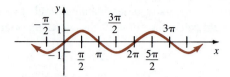

(a) $y = \sin x$
Domain: R
Range: $-1 \le y \le 1$

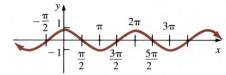

(b) $y = \cos x$
Domain: R
Range: $-1 \le y \le 1$

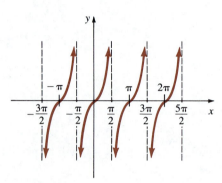

(c) $y = \tan x$
Domain: $x \ne (2n+1)\dfrac{\pi}{2}$, n an integer
Range: R

(d) $y = \cot x$
Domain: $x \ne n\pi$, n an integer
Range: R

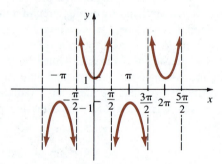

(e) $y = \sec x$
Domain: $x \ne (2n+1)\dfrac{\pi}{2}$, n an integer
Range: $y \ge 1, y \le -1$

(f) $y = \csc x$
Domain: $x \ne n\pi$, n an integer
Range: $y \ge 1, y \le -1$

Figure 1.78

Example 3

Compare the graphs of $y = 2 \sin x$ and $y = -2 \sin x$.

Solution   In each case the function has amplitude 2 and period 2π. For comparison, the dashed curve in Figure 1.79 is the graph of $y = \sin x$. Of course, the graph of $y = -2 \sin x$ is a reflection of the graph of $y = 2 \sin x$ in the x-axis.

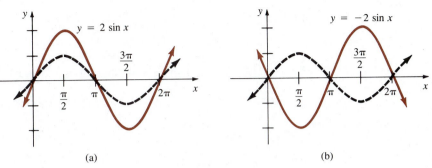

(a)

(b)

Figure 1.79

Example 4

Graph $y = \cos 4x$.

Solution The amplitude of the function is 1 and its period is $2\pi/4 = \pi/2$. For comparison, the dashed curve in Figure 1.80 is the graph of $y = \cos x$.

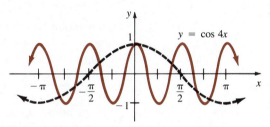

Figure 1.80

Example 5

Graph $y = \sin(x - \pi/2)$.

Solution From Section 1.5 we know that the graph of $y = \sin(x - \pi/2)$ is the graph of $y = \sin x$ shifted $\pi/2$ units to the right. Indeed, the graph of $y = \sin(x - \pi/2)$ given in Figure 1.81 is also the graph of $y = -\cos x$ since (1.35) implies $\sin(x - \pi/2) = -\cos x$.

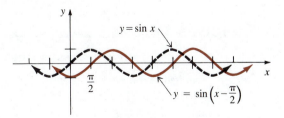

Figure 1.81

Example 6

If $f(x) = \sin x$ and $g(x) = x^2$, find fg, f/g, $f \circ g$, and $g \circ f$.

Solution From Definitions 1.2 and 1.3 we find

$$(fg)(x) = f(x)g(x) = x^2 \sin x$$

$$(f/g)(x) = \frac{f(x)}{g(x)} = \frac{\sin x}{x^2}$$

$$(f \circ g)(x) = f(g(x)) = \sin x^2$$

$$(g \circ f)(x) = g(f(x)) = (\sin x)^2 = \sin^2 x$$

Remarks

(*i*) In several instances we have used the notation $\sin^2 x$ for $(\sin x)^2$. As a rule, positive integer powers of trigonometric functions are written without parentheses; for example, $(\cos x)^3$ and $(\tan x)^5$ are the same as $\cos^3 x$ and $\tan^5 x$, respectively. Be careful not to confuse expressions such as $\sec^2 x$ and $\cot^6 x$ with $\sec x^2$ and $\cot x^6$.

(*ii*) Polynomial functions, rational functions, and power functions $y = kx^n$, where n is a rational number, belong to a class known as **algebraic functions**. An algebraic function involves a finite number of additions, subtractions, multiplications, divisions, and roots of polynomials; for example, $y = \sqrt{x} + \sqrt{x^2 + 5}$ is an algebraic function. The six trigonometric functions introduced in this section belong to a different class known as **transcendental functions**. A transcendental function is one that is not algebraic. We shall encounter other transcendental functions in Chapters 7 and 8.

(*iii*) We have seen that each of the trigonometric functions is periodic. In a more general context, a function f is **periodic** if there exists a positive number p such that $f(x + p) = f(x)$ for every number x in the domain of f. If p is the smallest positive number for which $f(x + p) = f(x)$, then p is the **period** of the function f. The function illustrated in Figure 1.82 has period 1.

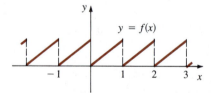

Figure 1.82

Exercises 1.6

Answers to odd-numbered problems begin on page A-18.

In Problems 1–6 convert from radian measure to degree measure.

1. $\pi/20$ **2.** $2\pi/9$ **3.** $11\pi/6$

4. $5\pi/18$ **5.** $-4\pi/3$ **6.** $-9\pi/4$

In Problems 7–12 convert from degree measure to radian measure.

7. $210°$ **8.** $225°$ **9.** $300°$

10. $315°$ **11.** $-150°$ **12.** $-420°$

In Problems 13–24 find the value of the given quantity.

13. $\sin(-\pi/6)$ **14.** $\cos(9\pi/4)$

15. $\sin(4\pi/3)$ **16.** $\tan(5\pi/4)$

17. $\sec(\pi/6)$ **18.** $\csc(-3\pi/4)$

19. $\cot(2\pi/3)$ **20.** $\sin(11\pi/6)$

21. $\tan(7\pi/6)$ **22.** $\csc(5\pi/6)$

23. $\cos(5\pi/2)$ **24.** $\cot(-\pi/3)$

25. Given that $\sin t = -2/\sqrt{5}$ and $\cos t = 1/\sqrt{5}$, find the values of the remaining four trigonometric functions.

26. Given that $\cos t = \frac{1}{10}$, find all possible values of $\sin t$.

27. Given that $\sin t = -\frac{1}{4}$ and the terminal side of the angle t is in the fourth quadrant, find the values of the remaining five trigonometric functions.

28. Given that $\tan t = -3$ and the terminal side of the angle t is in the second quadrant, find the values of the remaining five trigonometric functions.

29. Given that $\csc t = -\sqrt{2}$ and the terminal side of the angle t is in the third quadrant, find the values of the remaining five trigonometric functions.

30. Find all solutions of the equation $\sin t = -\sqrt{3}/2$ in the interval $[\pi, 3\pi/2]$.

31. Find all solutions of the equation $\sqrt{2}\cos^2 t - \cos t = 0$ in the interval $[0, 2\pi]$.

32. Find all solutions of the equation $\cos 2t + 3\sin t - 2 = 0$ in the interval $[0, 2\pi]$.

In Problems 33–42 use the identities in (1.35)–(1.40) to find the values of the given quantity or to prove the given identity.

33. $\sin(\pi/12)$ **34.** $\cos(5\pi/12)$

35. $\cos(5\pi/8)$ **36.** $\sin(3\pi/8)$

37. $\sin\left(t + \dfrac{\pi}{2}\right) = \cos t$ **38.** $\cos\left(t + \dfrac{3\pi}{2}\right) = \sin t$

39. $\cos(t + \pi) = -\cos t$ **40.** $\sin(\pi - t) = \sin t$

41. $\tan(t + \pi) = \tan t$ **42.** $\cot(t - \pi) = \cot t$

In Problems 43 and 44 explain why there is no angle t that satisfies the given equation.

43. $\sin t = \dfrac{4}{3}$ **44.** $\sec t = \dfrac{1}{2}$

In Problems 45–58 graph the given function.

45. $y = -\cos x$ **46.** $y = 2\cos x$

47. $y = 1 + \sin x$ **48.** $y = 1 + \cos x$

49. $y = 2 - \sin x$ **50.** $y = -1 + \sin x$

51. $y = -\tan x$ **52.** $y = 2 + \sec x$

53. $y = 3\cos 2x$ **54.** $y = \frac{5}{2}\sin 2\pi x$

55. $y = 2\sin(-\pi x)$ **56.** $y = \cos\frac{1}{2}x$

57. $y = \cos(x - \pi/2)$ **58.** $y = \sin(x + \pi)$

In Problems 59–62 the graph (Figures 1.83–1.86) is a shifted or reflected graph of the given function. Find an equation of the graph.

59. $y = \sin 4x$ **60.** $y = 3\sin x$

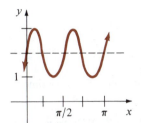

Figure 1.83

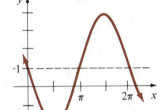

Figure 1.84

61. $y = \sin x$

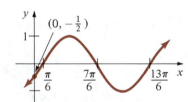

Figure 1.85

62. $y = \cos x$

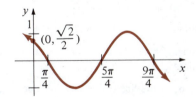

Figure 1.86

In Problems 63–68 find fg, f/g, $f \circ g$, and $g \circ f$.

63. $f(x) = 1 + 4x$
     $g(x) = \cos x$

64. $f(x) = x^2$
     $g(x) = \sin x^2$

65. $f(x) = \sin x$
 $g(x) = \cos x$

66. $f(x) = 1 + x^2$
 $g(x) = \tan x$

67. $f(x) = 1 + \cos 2x$
 $g(x) = \sqrt{x}$

68. $f(x) = \tan 2x$
 $g(x) = \cot 2x$

69. Trigonometric functions of the form $y = a + b \sin \omega(t - t_0)$, where a, b, ω, and t_0 are constants, are often used in ecological modeling to simulate periodic phenomena such as temperature variation, height of tides, and day length. Suppose

$$T(t) = 70 + 10 \sin \frac{\pi}{12}(t - 9)$$

represents temperature.

(a) Show that $T(t + 24) = T(t)$.

(b) At what time in $[0, 24]$ does $T(t) = 70$?

(c) What is the maximum temperature? At what time in $[0, 24]$ does the maximum occur?

(d) What is the minimum temperature? At what time in $[0, 24]$ does the minimum occur?

70. In the walk of a human the vertical component of the force of a foot on the ground can be approximated by a trigonometric function of the form $F(t) = K[\cos(\pi t/T) - q \cos(3\pi t/T)]$, where $K > 0$ and $0 < q < 1$. The constants K and q are dependent on the particular nature of the gait. The foot strikes the ground at time $t = -T/2$ and is lifted from the ground at time $t = T/2$.

(a) Show that $F(-T/2) = F(T/2) = 0$.

(b) Show that $F(t + 2T) = F(t)$.

(c) For $T = 1$, $K = 2$, and $q = \frac{1}{2}$, graph the function F on the interval $[-\frac{1}{2}, \frac{1}{2}]$.

In Problems 71–76 determine whether the given function is even, odd, or neither even nor odd.

71. $f(x) = \dfrac{\sin x}{x}$

72. $f(x) = x \cos x$

73. $f(x) = x + \tan x$

74. $f(x) = x^2 \sec x$

75. $f(x) = \cos(x + \pi)$

76. $f(x) = 2 + 5 \csc x$

Miscellaneous Problems

77. If $f(x) = A \sin kx$, $k > 0$, show that $f\left(x + \dfrac{2\pi}{k}\right) = f(x)$.

78. If $f(x) = A \tan kx$, $k > 0$, show that $f\left(x + \dfrac{\pi}{k}\right) = f(x)$.

In Problems 79–82 graph the given function.

79. $f(x) = x + \sin x$

80. $f(x) = \sin x + \cos x$

81. $f(x) = \sin \dfrac{1}{x}$

82. $f(x) = x \sin \dfrac{1}{x}$

——— Chapter 1 Review Exercises ———

Answers to odd-numbered problems begin on page A-19.

In Problems 1–16 answer true or false.

1. The interval $[a, a)$ contains no real numbers. _____

2. $22/7 = \pi$ _____

3. $\tan(\pi/6) = \cot(4\pi/3)$ _____

4. If $a < b$, then $a^2 < b^2$. _____

5. The solution of $|x + 1| > -2$ is $(-\infty, \infty)$. _____

6. The lines $2x + 3y = 5$ and $-2x + 3y = 1$ are perpendicular. _____

7. The slope of the line $y = -7$ is undefined. _____

8. If f is a function and $f(a) = f(b)$, then $a = b$. _____

9. The domain of the composition $f \circ g$ is the intersection of the domain of f and the domain of g. _____

10. The function $f(x) = x^5 - 4x^3 + 2$ is an odd function. _____

11. The function $f(x) = 5x^2 \cos x$ is an even function. _____

12. The graph of $y = f(x + 3)$ is the graph of $y = f(x)$ shifted 3 units to the right. _____

13. The function $y = -10 \sec x$ has amplitude 10. _____

14. If t_1 and t_2 are coterminal angles, then $\cos t_1 = \cos t_2$. _____

15. The range of the function $f(x) = 2 + \cos x$ is $[1, 3]$. _____

16. If $\cos t = \frac{1}{4}$, then $\cos(t + 3\pi) = -\frac{1}{4}$. _____

In Problems 17–34 fill in the blanks.

17. If $a < 0$, then $|-a| = $ _____ .

18. The solution of $|-x + 3| \geq 7$ is _____ .

19. If $(-2, 6)$ is the midpoint of the line segment from $P_1(x_1, 3)$ to $P_2(8, y_2)$, then $x_1 =$ _____ and $y_2 =$ _____ .

20. A line with x-intercept -4 and y-intercept 32 has slope _____ .

21. The lines $6x + 2y = 1$ and $kx - 9y = 5$ are parallel if $k =$ _____ .

22. The graph of the equation $5x = 2y^3 + y$ is symmetric with respect to _____ .

23. If (a, b) is a point in the third quadrant, then $(-a, b)$ is a point in the _____ quadrant.

24. The domain of the function $f(x) = \sqrt{x + 2}/x$ is _____ .

25. The range of the function $f(x) = 10/(x^2 + 1)$ is _____ .

26. The y-intercept of the graph of $f(x) = (2x - 4)/(5 - x)$ is _____ .

27. The x-intercepts of the graph of $f(x) = x^2 + 2x - 35$ are _____ .

28. If $f(x) = 4x^2 + 7$ and $g(x) = 2x + 3$, then $(f \circ g)(1) =$ _____ and $(g \circ f)(1) =$ _____ .

29. If $f(x) = 2x - 8$, then $(f \circ f)(x) =$ _____ and $(f \circ f \circ f)(x) =$ _____ .

30. The period of the function $f(x) = -5 \cos 6x$ is _____ .

31. In degree measure, $5\pi/18$ radians = _____ .

32. The set of points $P(x, y)$ on the circumference of the circle $(x + 5)^2 + (y - 2)^2 = 7$ are equidistant from the point _____ .

33. An equation of the line passing through $(-5, 11)$ and parallel to the line $y = -3$ is _____ .

34. A circle centered at the origin and passing through $(1, 1)$ has radius _____ .

35. Find an equation of the line that passes through $(3, -8)$ and is parallel to the line $2x + y = 10$.

36. Find an equation of the line that passes through the origin and through the point of intersection of the graphs of $x + y = 1$ and $2x - y = -7$.

37. Find an equation for the family of circles that passes through the origin and has centers on the line $y = x$.

38. Find an equation of the circle with center $(1, 2)$ that is tangent to the line $x = 7$.

39. Find the x- and y-intercepts and the center of the ellipse $x^2 + 9y^2 - 18y - 72 = 0$.

40. Find all numbers in the domain of $f(x) = 2x^2 - 7x$ that correspond to the number 4 in its range.

41. The width of a rectangular box is 3 times its length and its height is 2 times its length. Express the volume V of the box as a function of its length l. As function of its width w. As function of its height h.

42. A closed box, in the form of a cube, is to be constructed from two different materials. The material for the sides costs 1 cent per square centimeter and the material for the top and bottom costs 2.5 cents per square centimeter. Express the total cost C of construction as a function of the length x of a side.

43. In straight-line depreciation the annual amount of depreciation is defined as

$$\frac{\text{initial cost} - \text{salvage value}}{\text{estimated useful life}}$$

A machine that cost $50,000 when it was new has a salvage value of $10,000 in 10 years. What is the value of the machine in 4 years?

44. Suppose $f(x) = \sqrt{x + 4}$ and $g(x) = \sqrt{2 - x}$.

  (a) What is the domain of $f(-2x)$?

  (b) What is the domain of $f(x^2)$?

  (c) What is the domain of $g(x^2)$?

  (d) What is the domain of $(f + g)(x)$?

  (e) What is the domain of $(f/g)(x)$?

45. Suppose $\sin t = \frac{3}{5}$ and $\cos t = \frac{4}{5}$.

  (a) In which quadrant is the terminal side of the angle t?

  (b) Find the values of the remaining four trigonometric functions.

  (c) Find the values of $\sin 2t$ and $\cos 2t$.

46. If $f(x) = \cos x$, show that for any real number h,

$$\frac{f(x + h) - f(x)}{h} = \cos x\left(\frac{\cos h - 1}{h}\right) - \sin x\left(\frac{\sin h}{h}\right)$$

In Problems 47 and 48 solve the given equation on the indicated intervals.

47. $4 \cos^2 t - 1 = 0$; $[-\pi/2, \pi/2]$; $[0, \pi]$

48. $2 \sin^2 t + \sin t - 1 = 0$; $[0, \pi]$; $[\pi/2, 3\pi/2]$

In Problems 49 and 50 complete the graph (Figures 1.87 and 1.88).

49. f is an even function.

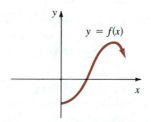

Figure 1.87

50. f is an odd function.

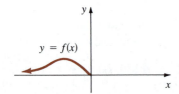

Figure 1.88

In Problems 51 and 52 connect the given points with the graph of a function (Figures 1.89 and 1.90).

51.

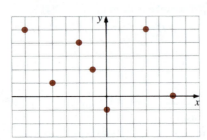

Figure 1.89

52.

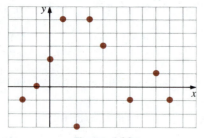

Figure 1.90

53. Fill in the blanks by referring to the graph of the function $y = f(x)$ given in Figure 1.91.

$f(-4) =$ _____ $f(-3) =$ _____

$f(-2) =$ _____ $f(-1) =$ _____

$f(0)\ \ =$ _____ $f(1)\ \ =$ _____

$f(1.5) =$ _____ $f(2)\ \ =$ _____

$f(3.5) =$ _____ $f(4)\ \ =$ _____

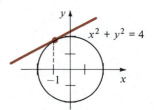

Figure 1.91

54. A tangent line to a circle is a line that touches its graph at one point on its circumference and is perpendicular to a radius. Find an equation of the tangent line in Figure 1.92.

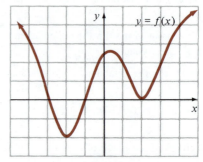

Figure 1.92

55. Consider the rectangle inscribed inside the ellipse $x^2/9 + y^2/16 = 1$ shown in Figure 1.93. Express the area A of the rectangle as a function of x.

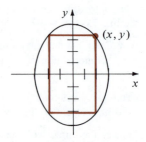

Figure 1.93

56. Using the graph of $y = f(x)$ given in Figure 1.94 sketch the graphs of

 (a) $y = f(x + 1)$ **(b)** $y = f(x) + 1$

 (c) $y = f(x) - 2$ **(d)** $y = f(x - 2)$

 (e) $y = -2f(x)$ **(f)** $y = f(x - \frac{1}{2}) + 1$

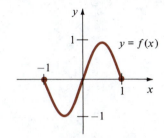

Figure 1.94

2

Limits of Functions

Two of the most fundamental concepts in the study of calculus are the notions of function and the *limit* of a function. In this chapter we shall be especially interested in determining whether the values $f(x)$ of a function f approach a fixed number L as x approaches a number a. Using the symbol $\to$ for the word "approach" we ask,

$$\text{Does } f(x) \to L \quad \text{as } x \to a?$$

2.1 Intuitive Notion of a Limit

Limit of a Function as x Approaches a Number

Consider the function

$$f(x) = \frac{16 - x^2}{4 + x}$$

whose domain is the set of all real numbers except -4. Although $f(-4)$ is not defined, nonetheless, $f(x)$ can be calculated for any value of x *near* -4. The table in Figure 2.1 shows that, as x approaches -4 from either the left or right, the functional values $f(x)$ are approaching 8; that is, when x is near -4, $f(x)$ is near 8. We say 8 is the **limit** of $f(x)$ as x approaches -4 and write,

$$f(x) \rightarrow 8 \quad \text{as} \quad x \rightarrow -4 \quad \text{or} \quad \lim_{x \to -4} \frac{16 - x^2}{4 + x} = 8$$

x	$f(x)$
-4.1	8.1
-4.01	8.01
-4.001	8.001
-3.9	7.9
-3.99	7.99
-3.999	7.999

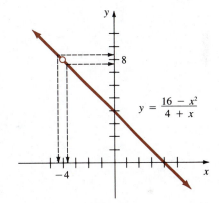

Figure 2.1

For $x \neq -4$, f can be simplified by cancellation:

$$f(x) = \frac{16 - x^2}{4 + x} = \frac{(4 + x)(4 - x)}{4 + x} = 4 - x$$

As seen in Figure 2.1, the graph of f is essentially the graph of $y = 4 - x$ with the exception that the graph of f has a hole at the point that corresponds to $x = -4$. As x gets closer and closer to -4, represented by the two arrowheads on the x-axis, the two arrowheads on the y-axis simultaneously get closer and closer to the number 8.

Intuitive Definition

In general, the notion of **$f(x)$ approaching a number L as x approaches a number a** is defined in the following manner:

If $f(x)$ can be made arbitrarily close to a finite number L by taking x sufficiently close to but different from a number a, from both the left and right side of a, then $\lim_{x \to a} f(x) = L$.

We shall use the notation $x \to a^-$ to denote that x approaches a from the *left* and $x \to a^+$ to mean x approaches a from the *right*. Thus, if the one-sided limits $\lim_{x \to a^-} f(x)$ and $\lim_{x \to a^+} f(x)$ have a common value L,

$$\lim_{x \to a^-} f(x) = \lim_{x \to a^+} f(x) = L$$

we then say $\lim_{x \to a} f(x)$ *exists* and write

$$\lim_{x \to a} f(x) = L$$

It is common practice to refer to the number L as the **limit of f at a.** However, we should observe:

> The existence of a limit of a function f at a does not depend on whether f is actually defined *at a* but only on whether f is defined for x *near a*.

Example 1

The graph of the function $f(x) = -x^2 + 2x + 2$ is shown in Figure 2.2. As seen from the graph and the accompanying tables, it seems plausible that

$$\lim_{x \to 4^-} f(x) = -6 \quad \text{and} \quad \lim_{x \to 4^+} f(x) = -6$$

and consequently

$$\lim_{x \to 4} f(x) = -6$$

$x \to 4^-$	$f(x)$
3.9	−5.41000
3.99	−5.94010
3.999	−5.99400

$x \to 4^+$	$f(x)$
4.1	−6.61000
4.01	−6.06010
4.001	−6.00600

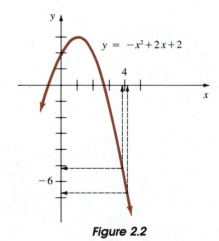

Figure 2.2

Note that in Example 1 the given function is certainly defined at $x = 4$, but at no time did we substitute $x = 4$ into the function to find the value of $\lim_{x \to 4} f(x)$. In the next example, we see that $\lim_{x \to 2} f(x)$ exists but $f(2)$ is not defined.

—— **Example 2** ——————————————————————————————

The graph of the piecewise-defined function

$$f(x) = \begin{cases} x^2, & x < 2 \\ -x + 6, & x > 2 \end{cases}$$

is given in Figure 2.3. From the graph and the accompanying table, we see that when x is close to 2, $f(x)$ is close to 4. That is, $\lim_{x \to 2} f(x) = 4$.

$x \to 2^-$	$f(x)$
1.9	3.61000
1.99	3.96010
1.999	3.99600

$x \to 2^+$	$f(x)$
2.1	3.90000
2.01	3.99000
2.001	3.99900

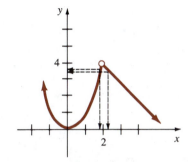

Figure 2.3

—— **Example 3** ——————————————————————————————

The graph of the piecewise-defined function

$$f(x) = \begin{cases} x + 2, & x \le 5 \\ -x + 10, & x > 5 \end{cases}$$

is given in Figure 2.4. From the graph and the accompanying table, we see that

$$\lim_{x \to 5^-} f(x) = 7 \quad \text{and} \quad \lim_{x \to 5^+} f(x) = 5$$

Since $\lim_{x \to 5^-} f(x) \ne \lim_{x \to 5^+} f(x)$, we conclude that $\lim_{x \to 5} f(x)$ does not exist.

$x \to 5^-$	$f(x)$
4.9	6.9
4.99	6.99
4.999	6.999

$x \to 5^+$	$f(x)$
5.1	4.9
5.01	4.99
5.001	4.999

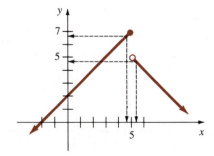

Figure 2.4

Example 4

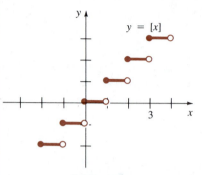

$y = [x]$

Figure 2.5

If x is any real number, the "output" $f(x)$ of the **greatest integer function** $f(x) = [x]$ is defined to be the greatest integer not exceeding x; for example,

$$f\left(\frac{1}{2}\right) = 0, \quad f\left(-\frac{1}{4}\right) = -1, \quad f(4.8) = 4, \quad \text{and} \quad f(3) = 3$$

The graph of f is given in Figure 2.5. From this graph we see that $f(n)$ is defined for every integer n; nonetheless, $\lim_{x \to n} f(x)$ does not exist. For example, as x approaches the number 3, the two one-sided limits exist but have different values:

$$\lim_{x \to 3^-} f(x) = 2 \quad \text{whereas} \quad \lim_{x \to 3^+} f(x) = 3$$

Example 5

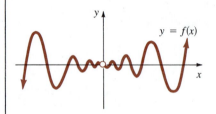

$y = f(x)$

Figure 2.6

In Figure 2.6 the graph of $y = f(x)$ shows that

$$\lim_{x \to 0^-} f(x) = 0 \quad \text{and} \quad \lim_{x \to 0^+} f(x) = 0$$

Hence,

$$\lim_{x \to 0} f(x) = 0$$

Example 6

In Figure 2.7 we see that there is a break in the graph of $y = f(x)$ at 2. If we let x approach 2 from, say, the left, the functional values $f(x)$ become larger and larger positive numbers. In other words, $f(x)$ is not approaching a finite number as $x \to 2^-$. Since $\lim_{x \to 2^-} f(x)$ does not exist, we conclude that $\lim_{x \to 2} f(x)$ does not exist.

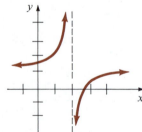

Figure 2.7

To determine whether $\lim_{x \to a} f(x)$ exists by graphing a function f is not always an easy task. In some instances numerical calculations can provide a convincing argument to the existence of a limit.

Example 7

Based solely on the numerical data in the accompanying tables, we naturally conclude that

$$\lim_{x \to 0} \frac{\sin x}{x} = 1$$

$x \to 0^-$	$\dfrac{\sin x}{x}$
-0.1	0.9983341
-0.01	0.9999833
-0.001	0.9999998

$x \to 0^+$	$\dfrac{\sin x}{x}$
0.1	0.9983341
0.01	0.9999833
0.001	0.9999998

Example 8

Careful inspection of the two tables below suggests that

$$\lim_{x \to 0} \frac{1 - \cos x}{x} = 0$$

$x \to 0^-$	$\dfrac{1 - \cos x}{x}$
-0.1	-0.0499583
-0.01	-0.0049999
-0.001	-0.0005001
-0.0001	-0.0000510

$x \to 0^+$	$\dfrac{1 - \cos x}{x}$
0.1	0.0499583
0.01	0.0049999
0.001	0.0005001
0.0001	0.0000510

BASIC Program   The tables in Examples 7 and 8 can easily be verified by means of a calculator or a computer. For those of you who have access to a computer, the following BASIC program may be useful in investigating the existence of $\lim_{x \to a} f(x)$.

```
10 REM LIMIT OF A FUNCTION
20 DEF FNY(X) = ...
30 INPUT "THE VALUE OF A IS ";A
40 INPUT "THE NUMBER OF TERMS IS ";K
50 PRINT "A - H ","FNY(A - H) ","A + H ",
   "FNY(A + H) "
60 FOR N = 1 TO K
70 LET H = 1/10^N
80 PRINT A - H,FNY(A - H),A + H,
   FNY(A + H)
90 NEXT N
100 END
```

To use this program you must supply A, define the function FNY(X), and specify K. In Example 7, A = 0, FNY(X) = (SIN (X))/X, and K = 3. A word of caution is in order here: For limits of the form $\lim_{x \to a} f(x)/g(x)$, where $f(x) \to 0$ as $x \to a$ and $g(x) \to 0$ as $x \to a$, the results from this program are heavily dependent on the computer used. In calculating numbers that are nearly 0/0, a computer may give a *division by zero* error message.*

Exercises 2.1

Answers to odd-numbered problems begin on page A-19.

In Problems 1–12 use a graph to find the given limit, if it exists.

1. $\lim_{x \to 2} (3x + 2)$

2. $\lim_{x \to -2} (x^2 - 1)$

3. $\lim_{x \to 0} \left(1 + \dfrac{1}{x}\right)$

4. $\lim_{x \to 5} \sqrt{x - 1}$

5. $\lim_{x \to 1} \dfrac{x^2 - 1}{x - 1}$

6. $\lim_{x \to 0} \dfrac{x^2 - 3x}{x}$

7. $\lim_{x \to 0} \dfrac{|x|}{x}$

8. $\lim_{x \to 2} |x - 2|$

9. $\lim_{x \to 0} f(x)$ where $f(x) = \begin{cases} x + 3, & x < 0 \\ -x + 3, & x \geq 0 \end{cases}$

10. $\lim_{x \to 2} f(x)$ where $f(x) = \begin{cases} x, & x < 2 \\ x + 1, & x \geq 2 \end{cases}$

11. $\lim_{x \to 2} f(x)$ where $f(x) = \begin{cases} x^2 - 2x, & x < 2 \\ 1, & x = 2 \\ x^2 - 6x + 8, & x > 2 \end{cases}$

12. $\lim_{x \to 0} f(x)$ where $f(x) = \begin{cases} x^2, & x < 0 \\ 2, & x = 0 \\ \sqrt{x} - 1, & x > 0 \end{cases}$

In Problems 13–16 use the given graph (Figures 2.8–2.11) to find each limit, if it exists.

(a) $\lim_{x \to 1^+} f(x)$ **(b)** $\lim_{x \to 1^-} f(x)$ **(c)** $\lim_{x \to 1} f(x)$

13.

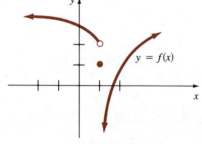

Figure 2.8

14.

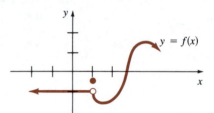

Figure 2.9

15.

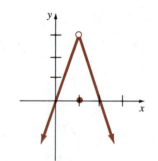

Figure 2.10

16.

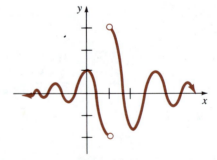

Figure 2.11

In Problems 17–20 use the given graph (Figures 2.12–2.15) to find each limit, if it exists.

(a) $\lim_{x \to -2} f(x)$ **(b)** $\lim_{x \to 0} f(x)$

(c) $\lim_{x \to 1} f(x)$ **(d)** $\lim_{x \to 2} f(x)$

*You might also try in line 70 of the program, H = 1/N^2 (1/n^2) or H = 1/2^N (1/2^n).

17.

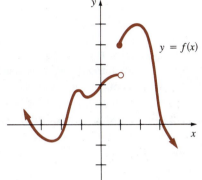

Figure 2.12

18.

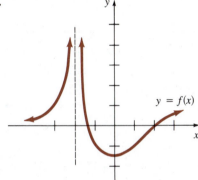

Figure 2.13

19.

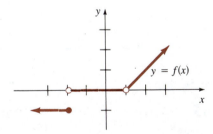

Figure 2.14

20.

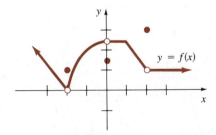

Figure 2.15

In Problems 21 and 22 use the graph of the given function (Figures 2.16 and 2.17) to determine whether $\lim_{x \to 0} f(x)$ exists.

21. $f(x) = \sin \dfrac{1}{x}$

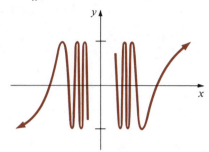

Figure 2.16

22. $f(x) = x \sin \dfrac{1}{x}$

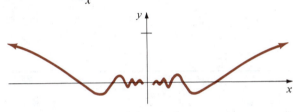

Figure 2.17

In Problems 23–26 use a graph to determine whether the given limit statement is correct. If incorrect, give a correct statement.

23. $\lim\limits_{x \to 0} \sqrt{x} = 0$

24. $\lim\limits_{x \to 1} \sqrt{1 - x} = 0$

25. $\lim\limits_{x \to 0} \sqrt[3]{x} = 0$

26. $\lim\limits_{x \to 2} \sqrt{4 - x^2} = 0$

Calculator/Computer Problems

In Problems 27–34 use a calculator or computer to investigate the given limit. Conjecture its value.

27. $\lim\limits_{x \to 0} \dfrac{2 - \sqrt{x + 4}}{x}$

28. $\lim\limits_{x \to 0} \dfrac{1 - \cos x}{x^2}$

29. $\lim\limits_{x \to 0} \dfrac{x}{\sin 3x}$

30. $\lim\limits_{x \to 0} \dfrac{\tan x}{x}$

31. $\lim\limits_{x \to 4} \dfrac{\sqrt{x} - 2}{x - 4}$

32. $\lim\limits_{x \to 3} \left[\dfrac{6}{x^2 - 9} - \dfrac{6\sqrt{x} - 2}{x^2 - 9} \right]$

33. $\lim\limits_{x \to 1} \dfrac{x^4 + x - 2}{x - 1}$

34. $\lim\limits_{x \to -2} \dfrac{x^3 + 8}{x + 2}$

2.2 Theorems on Limits

The intention of the informal discussion in Section 2.1 was to give you an intuitive grasp of when a limit does or does not exist. However, it is neither desirable nor practical, in every instance, to reach a conclusion about the existence of a limit based on a graph or table of functional values. We must be able to evaluate a limit, or discern its nonexistence, in a somewhat mechanical fashion. The theorems that we shall consider in this section establish such a means. The proofs of some of these results are given in Appendix II.

THEOREM 2.1   If c is a constant, then

$$\lim_{x \to a} c = c$$

Example 1

From Theorem 2.1,

(a) $\displaystyle\lim_{x \to 2} 10 = 10$ and   (b) $\displaystyle\lim_{x \to 0} \pi = \pi$

THEOREM 2.2

$$\lim_{x \to a} x = a$$

Example 2

From Theorem 2.2,

(a) $\displaystyle\lim_{x \to 2} x = 2$ and   (b) $\displaystyle\lim_{x \to 0} x = 0$

THEOREM 2.3   If c is a constant, then

$$\lim_{x \to a} c\, f(x) = c \lim_{x \to a} f(x)$$

Example 3

From Theorems 2.2 and 2.3,

(a) $\displaystyle\lim_{x \to 8} 5x = 5 \lim_{x \to 8} x = 5 \cdot 8 = 40$ and

(b) $\displaystyle\lim_{x \to -2} \left(-\tfrac{3}{2}x\right) = -\tfrac{3}{2} \lim_{x \to -2} x = \left(-\tfrac{3}{2}\right) \cdot (-2) = 3$

THEOREM 2.4   If $\lim_{x \to a} f(x)$ exists, then it is unique.

Limit of a Sum, Product, and Quotient

THEOREM 2.5   Let $\lim_{x \to a} f(x) = L_1$ and $\lim_{x \to a} g(x) = L_2$. Then

(*i*) $\lim_{x \to a} [f(x) + g(x)] = \lim_{x \to a} f(x) + \lim_{x \to a} g(x) = L_1 + L_2,$

(*ii*) $\lim_{x \to a} f(x) \cdot g(x) = \lim_{x \to a} f(x) \cdot \lim_{x \to a} g(x) = L_1 L_2,$ and

(*iii*) $\lim_{x \to a} \dfrac{f(x)}{g(x)} = \dfrac{\lim_{x \to a} f(x)}{\lim_{x \to a} g(x)} = \dfrac{L_1}{L_2},\qquad L_2 \neq 0.$

In other words, Theorem 2.5 can be stated as: When the limits exist,

(*i*)   *the limit of a sum is the sum of the limits,*

(*ii*)   *the limit of a product is the product of the limits,* and

(*iii*)   *the limit of a quotient is the quotient of the limits provided the limit of the denominator is not zero.*

Note: If the limits exist, then Theorem 2.5 is also applicable to one-sided limits. Moreover, Theorem 2.5 extends to differences as well as sums, products, and quotients that involve more than two functions.

___ **Example 4** _____

Evaluate $\lim_{x \to 5}(10x + 7).$

Solution   From Theorems 2.1, 2.2, and 2.3 we know $\lim_{x \to 5} 7$ and $\lim_{x \to 5} 10x$ exist. Hence, from Theorem 2.5(*i*)

$$\lim_{x \to 5}(10x + 7) = \lim_{x \to 5} 10x + \lim_{x \to 5} 7$$
$$= 10 \lim_{x \to 5} x + \lim_{x \to 5} 7$$
$$= 10 \cdot 5 + 7 = 57$$

Limit of a Power   Theorem 2.5(*ii*) can be used to calculate the limit of a positive integral power of a function; for example, if $\lim_{x \to a} f(x) = L$, then

$$\lim_{x \to a}[f(x)]^2 = \lim_{x \to a} f(x) \cdot f(x) = \lim_{x \to a} f(x) \cdot \lim_{x \to a} f(x) = L^2$$

The next theorem states the general result.

THEOREM 2.6   Let $\lim_{x \to a} f(x) = L$ and n be a positive integer, then

$$\lim_{x \to a}[f(x)]^n = L^n$$

For the special case $f(x) = x$, the result given in Theorem 2.6 yields

$$\lim_{x \to a} x^n = a^n \tag{2.1}$$

___

Example 5 _____

Evaluate $\lim_{x \to 10} x^3$.

Solution   From (2.1),

$$\lim_{x \to 10} x^3 = 10^3 = 1000$$

___

Example 6 _____

Evaluate $\lim_{x \to 3}(x^2 - 5x + 6)$.

Solution   Since all limits exist,

$$\lim_{x \to 3}(x^2 - 5x + 6) = \lim_{x \to 3} x^2 - \lim_{x \to 3} 5x + \lim_{x \to 3} 6$$
$$= 3^2 - 5 \cdot 3 + 6 = 0$$

___

Example 7 _____

Evaluate $\lim_{x \to 1}(4x - 1)^5$.

Solution   First, we see that

$$\lim_{x \to 1}(4x - 1) = \lim_{x \to 1} 4x - \lim_{x \to 1} 1 = 3$$

It then follows from Theorem 2.6 that

$$\lim_{x \to 1}(4x - 1)^5 = 3^5 = 243$$

___

Limit of a Polynomial
Function

We can use (2.1) and Theorem 2.5(i) to compute the limit of a general polynomial function. If

$$f(x) = c_n x^n + c_{n-1} x^{n-1} + \cdots + c_1 x + c_0$$

is a polynomial function, then

$$\lim_{x \to a} f(x) = \lim_{x \to a} (c_n x^n + c_{n-1} x^{n-1} + \cdots + c_1 x + c_0)$$
$$= \lim_{x \to a} c_n x^n + \lim_{x \to a} c_{n-1} x^{n-1} + \cdots + \lim_{x \to a} c_1 x + \lim_{x \to a} c_0$$
$$= c_n a^n + c_{n-1} a^{n-1} + \cdots + c_1 a + c_0$$

In other words, to compute a limit of a polynomial function f as x approaches a real number a, we need only evaluate the function *at a:*

$$\lim_{x \to a} f(x) = f(a) \qquad (2.2)$$

A reexamination of Example 6 shows that $\lim_{x \to 3} f(x)$, where $f(x) = x^2 - 5x + 6$, is given by $f(3) = 0$.

We can often use (2.2) in conjunction with Theorem 2.5(*iii*) to find a limit of a rational function.

Example 8

Evaluate $\lim\limits_{x \to -1} \dfrac{3x - 4}{6x + 2}$.

Solution   From (2.2), $\lim_{x \to -1}(3x - 4) = -7$ and $\lim_{x \to -1}(6x + 2) = -4$. Since the limit of the denominator is not zero, we see from Theorem 2.5(*iii*) that

$$\lim_{x \to -1} \frac{3x - 4}{6x + 2} = \frac{\lim\limits_{x \to -1} (3x - 4)}{\lim\limits_{x \to -1} (6x + 2)} = \frac{-7}{-4} = \frac{7}{4}$$

You should not get the impression that we can *always* find a limit of a function by substituting, or "plugging in," the real number a.

Example 9

Evaluate $\lim\limits_{x \to 1} \dfrac{x - 1}{x^2 + x - 2}$.

Solution   The limit of this quotient cannot be written immediately as the quotient of limits because $\lim_{x \to 1}(x^2 + x - 2) = 0$. However, by simplifying *first*, we can then apply Theorem 2.5(*iii*),

$$\lim_{x \to 1} \frac{x - 1}{x^2 + x - 2} = \lim_{x \to 1} \frac{x - 1}{(x - 1)(x + 2)}$$

$$= \lim_{x \to 1} \frac{1}{x + 2}, \qquad x \neq 1$$

$$= \frac{\lim\limits_{x \to 1} 1}{\lim\limits_{x \to 1}(x + 2)} = \frac{1}{3}$$

Limit of a Root   The limit of the nth root of a function is the nth root of the limit whenever the limit exists and has a real nth root.

> **THEOREM 2.7**   Let $\lim_{x \to a} f(x) = L$ and let n be a positive integer. If $L \geq 0$ when n is any positive integer or if $L < 0$ when n is an odd positive integer, then
> $$\lim_{x \to a} [f(x)]^{1/n} = L^{1/n}$$

Example 10

Evaluate $\lim_{x \to 9} \sqrt{x}$.

Solution   Since $\lim_{x \to 9} x = 9 > 0$, we know from Theorem 2.7 that

$$\lim_{x \to 9} \sqrt{x} = [\lim_{x \to 9} x]^{1/2} = 9^{1/2} = 3$$

Example 11

Evaluate $\lim_{x \to -8} \dfrac{x - \sqrt[3]{x}}{2x + 10}$.

Solution   Since $\lim_{x \to -8}(2x + 10) = -6 \neq 0$, we see from Theorems 2.5(*iii*) and 2.7 that

$$\lim_{x \to -8} \frac{x - \sqrt[3]{x}}{2x + 10} = \frac{\lim_{x \to -8} x - [\lim_{x \to -8} x]^{1/3}}{\lim_{x \to -8} (2x + 10)}$$

$$= \frac{-8 - (-8)^{1/3}}{-6}$$

$$= \frac{-6}{-6} = 1$$

Sometimes you can tell at a glance when *a limit does not exist*.

> **THEOREM 2.8**   If $\lim_{x \to a} f(x) = L_1 \neq 0$ and $\lim_{x \to a} g(x) = 0$, then $\lim_{x \to a} f(x)/g(x)$ does not exist.

Proof   We shall give an indirect proof of this result based on Theorem 2.5. Suppose $\lim_{x \to a} f(x) = L_1 \neq 0$, $\lim_{x \to a} g(x) = 0$, and that $\lim_{x \to a} f(x)/g(x)$ exists and equals L_2, then

$$L_1 = \lim_{x \to a} f(x)$$

$$= \lim_{x \to a} g(x) \cdot \frac{f(x)}{g(x)} \qquad [g(x) \neq 0]$$

$$= \lim_{x \to a} g(x) \cdot \lim_{x \to a} \frac{f(x)}{g(x)}$$

$$= 0 \cdot L_2 = 0$$

By contradicting the assumption that $L_1 \neq 0$, we have proved the theorem. ∎

Example 12

From Theorem 2.8, we can see that

$$\lim_{x \to 5} \frac{4x + 7}{x^2 - 25}$$

does not exist because $\lim_{x \to 5}(4x + 7) = 27 \neq 0$ and $\lim_{x \to 5}(x^2 - 25) = 0$.

The last theorem of this section has various names, **Squeeze Theorem, Pinching Theorem, Sandwiching Theorem, Squeeze Play Theorem,** and even the **Flyswatter Theorem.** As shown in Figure 2.18 if $f(x)$ is "squeezed" between $g(x)$ and $h(x)$ for all x close to a number a, and if we know that the functions g and h have a common limit L as $x \to a$, it stands to reason that f also approaches L as $x \to a$.

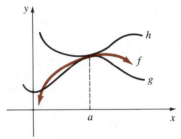

Figure 2.18

> **THEOREM 2.9   Squeeze Theorem**
>
> If f, g, and h are functions for which $g(x) \leq f(x) \leq h(x)$ for all x in an open interval that contains a number a, except possibly at a itself, and if $\lim_{x \to a} g(x) = \lim_{x \to a} h(x) = L$, then $\lim_{x \to a} f(x) = L$.

Example 13

Evaluate $\lim_{x \to 0} x^2 \sin \dfrac{1}{x}$.

Solution   For $x \neq 0$ we have $-1 \leq \sin(1/x) \leq 1$. Therefore,

$$-x^2 \leq x^2 \sin \frac{1}{x} \leq x^2$$

Now if we make the identifications $g(x) = -x^2$ and $h(x) = x^2$, it follows from (2.2) that $\lim_{x \to 0} g(x) = \lim_{x \to 0} h(x) = 0$. Thus, from the Squeeze Theorem we conclude that

$$\lim_{x \to 0} x^2 \sin \frac{1}{x} = 0$$

See Figure 2.19.

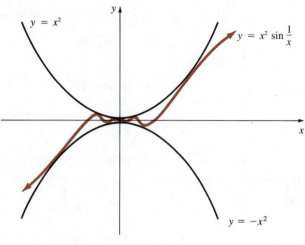

Figure 2.19

Remarks

(*i*) In mathematics a student should be aware of what a definition or theorem does *not* say. Thus, from time to time, we shall conclude the section discussion with a few words of caution on the possibility of giving a definition or theorem an unwarranted extension or limitation.

(*ii*) Theorem 2.5(*i*) does not say that a limit of a sum is *always* the sum of the limits. The graph of $f(x) = 1/x$ in Figure 1.42(a) shows $\lim_{x \to 0} 1/x$ does not exist. Therefore,

$$\lim_{x \to 0} \left[\frac{1}{x} - \frac{1}{x} \right] \neq \lim_{x \to 0} \frac{1}{x} - \lim_{x \to 0} \frac{1}{x}$$

Nonetheless, the limit of the difference exists:

$$\lim_{x \to 0} \left[\frac{1}{x} - \frac{1}{x} \right] = \lim_{x \to 0} 0 = 0$$

(*iii*) The limit of a product may exist and yet not be the product of the limits; for example,

$$\lim_{x \to 0} x \cdot \frac{1}{x} \neq \lim_{x \to 0} x \cdot \lim_{x \to 0} \frac{1}{x}$$

although

$$\lim_{x \to 0} x \cdot \frac{1}{x} = \lim_{x \to 0} 1 = 1$$

(*iv*) Theorem 2.8 does not say that the limit of a quotient fails to exist whenever the limit of the denominator is zero. Example 9 provides a counterexample to that interpretation. However, Theorem 2.8 states that a limit of a quotient does not exist whenever the limit of the denominator is zero *and* the limit of the numerator is not zero. Furthermore, to conclude that Theorem 2.8 provides the only way a limit can fail to exist would be a sweeping generalization.

―――――― **Exercises 2.2** ――――――――――――――――――――――

Answers to odd-numbered problems begin on page A-19.

In Problems 1–44 find the given limit, if it exists.

1. $\lim\limits_{x\to-5} 17$

2. $\lim\limits_{x\to7} \cos \pi$

3. $\lim\limits_{x\to3}(-4)x$

4. $\lim\limits_{x\to-2} x^2$

5. $\lim\limits_{x\to-1}(x^3 - 4x + 1)$

6. $\lim\limits_{x\to0} \dfrac{x + 5}{3x}$

7. $\lim\limits_{t\to1}(3t - 1)(5t^2 + 2)$

8. $\lim\limits_{t\to-2}(t + 4)^2$

9. $\lim\limits_{s\to7} \dfrac{s^2 - 21}{s + 2}$

10. $\lim\limits_{x\to6} \dfrac{x^2 - 6x}{x^2 - 7x + 6}$

11. $\lim\limits_{x\to6} \sqrt{2x - 5}$

12. $\lim\limits_{s\to8}(1 + \sqrt[3]{s})$

13. $\lim\limits_{t\to1} \dfrac{\sqrt{t}}{t^2 + t - 2}$

14. $\lim\limits_{x\to2} x^2\sqrt{x^2 + 5x + 2}$

15. $\lim\limits_{x\to1} \dfrac{x^3 - 1}{x - 1}$

16. $\lim\limits_{x\to-3} \dfrac{2x + 6}{4x^2 - 36}$

17. $\lim\limits_{x\to0} \left(x - \dfrac{1}{x - 1}\right)$

18. $\lim\limits_{t\to0} \left(\dfrac{4}{t} - 1\right)t$

19. $\lim\limits_{t\to-1} \dfrac{t^3 + 1}{t^2 - 1}$

20. $\lim\limits_{x\to-2} x\sqrt{x + 4}\,\sqrt[3]{x - 6}$

21. $\lim\limits_{x\to0^+} \dfrac{(x + 2)(x^5 - 1)^3}{(\sqrt{x} + 4)^2}$

22. $\lim\limits_{x\to0} x^3(x^4 + 2x^3)^{-1}$

23. $\lim\limits_{x\to0} \left[\dfrac{x^2 + 3x - 1}{x} + \dfrac{1}{x}\right]$

24. $\lim\limits_{x\to2} \left[\dfrac{1}{x - 2} - \dfrac{6}{x^2 + 2x - 8}\right]$

25. $\lim\limits_{x\to3^+} \dfrac{(x + 3)^2}{\sqrt{x - 3}}$

26. $\lim\limits_{x\to3}(x - 4)^{99}(x^2 - 7)^{10}$

27. $\lim\limits_{x\to10} \sqrt{\dfrac{10x}{2x + 5}}$

28. $\lim\limits_{r\to1} \dfrac{\sqrt{(r^2 + 3r - 2)^3}}{\sqrt[3]{(5r - 3)^2}}$

29. $\lim\limits_{h\to4} \sqrt{\dfrac{h}{h + 5}} \left(\dfrac{h^2 - 16}{h - 4}\right)^2$

30. $\lim\limits_{t\to2}(t + 2)^{3/2}(2t + 4)^{1/3}$

31. $\lim\limits_{t\to1^+} \dfrac{3t}{-1 + \sqrt{t}}$

32. $\lim\limits_{s\to8^-} \dfrac{16 - s^{4/3}}{4 - s^{2/3}}$

33. $\lim\limits_{x\to0^-} \sqrt[5]{\dfrac{x^3 - 64x}{x^2 + 2x}}$

34. $\lim\limits_{x\to-1^+} \left(8x + \dfrac{2}{x}\right)^5$

35. $\lim\limits_{t\to1}(at^2 - bt)^2$

36. $\lim\limits_{x\to-1} \sqrt{u^2x^2 + 2xu + 1}$

37. $\lim\limits_{h\to0} \dfrac{(x + h)^2 - x^2}{h}$

38. $\lim\limits_{h\to0} \dfrac{1}{h}[(x + h)^3 - x^3]$

39. $\lim\limits_{h\to0} \dfrac{1}{h}\left(\dfrac{1}{x + h} - \dfrac{1}{x}\right)$

40. $\lim\limits_{h\to0} \dfrac{\sqrt{x + h} - \sqrt{x}}{h} \quad (x > 0)$

41. $\lim\limits_{t\to1} \dfrac{\sqrt{t} - 1}{t - 1}$

42. $\lim\limits_{u\to5} \dfrac{\sqrt{u + 4} - 3}{u - 5}$

43. $\lim\limits_{v\to0} \dfrac{\sqrt{25 + v} - 5}{\sqrt{1 + v} - 1}$

44. $\lim\limits_{x\to1} \dfrac{4 - \sqrt{x + 15}}{x^2 - 1}$

In Problems 45 and 46 find $\lim\limits_{h\to0} \dfrac{f(2 + h) - f(2)}{h}$ for the given function.

45. $f(x) = 3x^2 + 1$

46. $f(x) = x^2 - 2x + 7$

In Problems 47 and 48 use the Squeeze Theorem to establish the given limit.

47. $\lim\limits_{x\to0} x^2\sin^2\dfrac{1}{x} = 0$

48. $\lim\limits_{x\to0} x \sin \dfrac{1}{x} = 0$

In Problems 49 and 50 use the Squeeze Theorem to evaluate the given limit.

49. $\lim\limits_{x \to 2} f(x)$ where $2x - 1 \le f(x) \le x^2 - 2x + 3, \quad x \ne 2$

50. $\lim\limits_{x \to 0} f(x)$ where $|f(x) - 1| \le x^2, \quad x \ne 0$

Miscellaneous Problem

51. If $|f(x)| \le B$ for all x, show that $\lim\limits_{x \to 0} x^2 f(x) = 0$.

2.3 Limits That Involve Infinity

Intuitive Definition The limit of a function f *will fail to exist* as x approaches a number a whenever the functional values increase or decrease without bound. In Figure 2.20(a) the fact that the functional values $f(x)$ increase without bound as x approaches a is denoted by

$$f(x) \to \infty \quad \text{as} \quad x \to a \quad \text{or} \quad \lim_{x \to a} f(x) = \infty$$

In Figure 2.20(b) the functional values decrease without bound as x approaches a, and so we write

$$f(x) \to -\infty \quad \text{as} \quad x \to a \quad \text{or} \quad \lim_{x \to \infty} f(x) = \infty$$

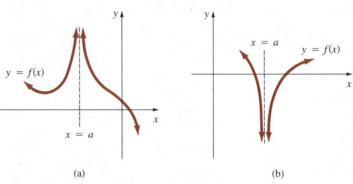

(a)	(b)

Figure 2.20

Similarly, Figure 2.21 shows the unbounded behavior of a function as x approaches a from one side.

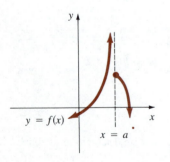

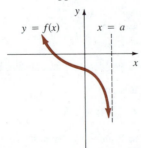

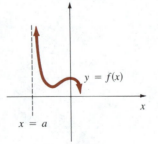

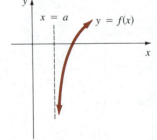

(a) $\lim\limits_{x \to a^-} f(x) = \infty$

(b) $\lim\limits_{x \to a^-} f(x) = -\infty$

(c) $\lim\limits_{x \to a^+} f(x) = \infty$

(d) $\lim\limits_{x \to a^+} f(x) = -\infty$

Figure 2.21

Vertical Asymptotes   In general, any limit of the type

$$\lim_{x \to a} f(x) = \infty, \quad \lim_{x \to a} f(x) = -\infty$$

$$\lim_{x \to a^-} f(x) = \infty, \quad \lim_{x \to a^-} f(x) = -\infty$$

and

$$\lim_{x \to a^+} f(x) = \infty, \quad \lim_{x \to a^+} f(x) = -\infty$$

is called an **infinite limit**. If any *one* of the foregoing conditions hold, then the line $x = a$ is a **vertical asymptote** for the graph of f.

_____ **Example 1** _____

From the accompanying tables and graph in Figure 2.22, we see that the values of the function $f(x) = 2/(x - 1)$ decrease without bound as $x \to 1^-$ and increase without bound as $x \to 1^+$; that is,

$$\lim_{x \to 1^-} f(x) = -\infty \quad \text{and} \quad \lim_{x \to 1^+} f(x) = \infty$$

The line $x = 1$ is a vertical asymptote.

$x \to 1^-$	$f(x)$
0.9	-20
0.99	-200
0.999	-2000

$x \to 1^+$	$f(x)$
1.1	20
1.01	200
1.001	2000

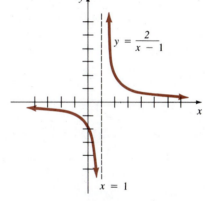

Figure 2.22

_____ **Example 2** _____

Graph the function $f(x) = \dfrac{x}{\sqrt{x + 2}}$.

Solution   Inspection of f reveals that its domain is $(-2, \infty)$, and the y-intercept is 0. From the accompanying table we conclude that f decreases without bound as x approaches -2 from the right:

$$\lim_{x \to -2^+} f(x) = -\infty$$

Hence, the line $x = -2$ is a vertical asymptote. The graph of f is given in Figure 2.23.

$x \to -2^+$	$f(x)$
-1.9	-6.01
-1.99	-19.90
-1.999	-63.21
-1.9999	-199.90

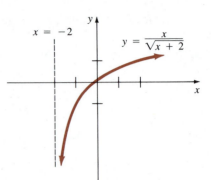

Figure 2.23

Example 3

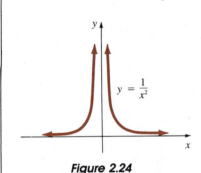

Figure 2.24

Since the domain of $f(x) = 1/x^2$ is the set of all real numbers except 0, the graph of f has no y-intercept. Also, $f(-x) = 1/(-x)^2 = 1/x^2 = f(x)$ implies that f is an even function, and so its graph is symmetric with respect to the y-axis. Finally, when $x \to 0^+$, the functional values $f(x)$ become larger and larger without bound. This fact, along with symmetry implies

$$\lim_{x \to 0} \frac{1}{x^2} = \infty$$

Therefore, the y-axis, that is, the line $x = 0$, is a vertical asymptote. The graph of f is given in Figure 2.24.

Example 4

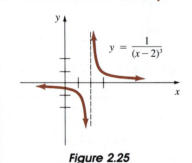

Figure 2.25

The graph of $f(x) = 1/(x - 2)^3$ given in Figure 2.25 shows that

$$\lim_{x \to 2^-} \frac{1}{(x - 2)^3} = -\infty$$

and

$$\lim_{x \to 2^+} \frac{1}{(x - 2)^3} = \infty$$

The line $x = 2$ is a vertical asymptote.

The preceding two examples illustrate the following theorem.

THEOREM 2.10

(*i*)   If n is an even positive integer, then

$$\lim_{x \to a} \frac{1}{(x - a)^n} = \infty$$

(*ii*)   If n is an odd positive integer, then

$$\lim_{x \to a^-} \frac{1}{(x-a)^n} = -\infty \quad \text{and} \quad \lim_{x \to a^+} \frac{1}{(x-a)^n} = \infty$$

For rational functions, vertical asymptotes can be found by inspection.

Suppose $f(x) = P(x)/Q(x)$ where P and Q are polynomial functions. If $Q(a) = 0$ and $P(a) \neq 0$, then $x = a$ is a vertical asymptote for the graph of f.

Example 5

Determine vertical asymptotes for $f(x) = \dfrac{x^2 + x + 1}{(x^2 + 1)(x^2 - 6x + 8)}$.

Solution f is a rational function with $P(x) = x^2 + x + 1$ and

$$Q(x) = (x^2 + 1)(x^2 - 6x + 8) = (x^2 + 1)(x - 2)(x - 4)$$

Since the denominator Q is zero only for 2 and 4 and since $P(2) \neq 0$ and $P(4) \neq 0$, it follows that the lines $x = 2$ and $x = 4$ are vertical asymptotes.

Example 6

The rational function $f(x) = (x^2 - 9)/(x - 3)$ does *not* have a vertical asymptote at $x = 3$. Why not?

Limits at Infinity   A function f might approach a constant value L as the independent variable x increases or decreases without bound. We write

$$\lim_{x \to \infty} f(x) = L \quad \text{or} \quad \lim_{x \to -\infty} f(x) = L$$

to denote a **limit at infinity**. Figure 2.26 shows four possibilities of the behavior of a function f as x becomes large in absolute value.

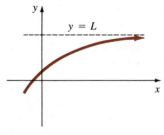

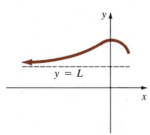

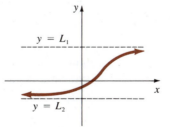

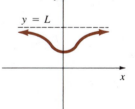

(a) $\displaystyle\lim_{x \to \infty} f(x) = L$

(b) $\displaystyle\lim_{x \to -\infty} f(x) = L$

(c) $\displaystyle\lim_{x \to \infty} f(x) = L_1$

$\displaystyle\lim_{x \to -\infty} f(x) = L_2$

(d) $\displaystyle\lim_{x \to \infty} f(x) = L$

$\displaystyle\lim_{x \to -\infty} f(x) = L$

Figure 2.26

Horizontal Asymptotes   If $f(x) \to L$ as either $x \to \infty$ or $x \to -\infty$, then we say that the line $y = L$ is a **horizontal asymptote** for the graph of f. In Figure 2.26(c) the function has two horizontal asymptotes, $y = L_1$ and $y = L_2$.

___

Example 7
___

Reinspection of Figures 2.22 and 2.24 shows that

$$\lim_{x \to \pm\infty} \frac{2}{x - 1} = 0 \quad \text{and} \quad \lim_{x \to \pm\infty} \frac{1}{x^2} = 0$$

Thus, $y = 0$ is a horizontal asymptote for the graphs of both $f(x) = 2/(x - 1)$ and $f(x) = 1/x^2$.

___

The next theorem is useful for evaluating limits at infinity.

THEOREM 2.11   Let t be a positive rational number. If x^t is defined, then

$$\lim_{x \to \infty} \frac{1}{x^t} = 0 \quad \text{and} \quad \lim_{x \to -\infty} \frac{1}{x^t} = 0$$

___

Example 8
___

(*a*)   From Theorem 2.11 we have

$$\lim_{x \to \infty} \frac{1}{\sqrt{x}} = \lim_{x \to \infty} \frac{1}{x^{1/2}} = 0$$

(*b*)   Since $\sqrt[3]{x} = x^{1/3}$ is defined for $x < 0$, it follows from Theorems 2.3 and 2.11 that

$$\lim_{x \to -\infty} \frac{6}{\sqrt[3]{x}} = 0$$

___

In general, if $F(x) = f(x)/g(x)$, then the following table summarizes the limit results for the forms $\lim_{x \to a} F(x)$, $\lim_{x \to \infty} F(x)$, and $\lim_{x \to -\infty} F(x)$. The symbol L denotes a finite number.

limit form: $x \to a$, ∞, *or* $-\infty$	$\dfrac{L}{\pm\infty}$	$\dfrac{\pm\infty}{L}$	$\dfrac{L}{0}$, $L \neq 0$
limit is	0	infinite	infinite

Furthermore, the limit results of Theorem 2.5 hold by replacing the symbol *a* by ∞ or $-\infty$ provided the limits exist; for example,

$$\lim_{x \to \infty} \frac{f(x)}{g(x)} = \frac{\lim\limits_{x \to \infty} f(x)}{\lim\limits_{x \to \infty} g(x)} \tag{2.3}$$

whenever $\lim_{x \to \infty} f(x)$ and $\lim_{x \to \infty} g(x)$ exist and $\lim_{x \to \infty} g(x) \neq 0$.

Example 9

Evaluate $\lim\limits_{x \to \infty} \dfrac{-6x^4 + x^2 + 1}{2x^4 - x}$.

Solution   We cannot apply (2.3) to the function as given since $\lim_{x \to \infty}(-6x^4 + x^2 + 1) = -\infty$ and $\lim_{x \to \infty}(2x^4 - x) = \infty$. However, by dividing the numerator and the denominator by x^4, we can write

$$\lim_{x \to \infty} \frac{-6x^4 + x^2 + 1}{2x^4 - x} = \lim_{x \to \infty} \frac{-6 + (1/x^2) + (1/x^4)}{2 - (1/x^3)}$$

$$= \frac{\lim\limits_{x \to \infty}[-6 + (1/x^2) + (1/x^4)]}{\lim\limits_{x \to \infty}[2 - (1/x^3)]}$$

$$= \frac{-6 + 0 + 0}{2 + 0} = -3$$

This means the line $y = -3$ is a horizontal asymptote for the graph of the function.

Example 9 illustrates a general procedure for determining the behavior of a rational function $f(x) = P(x)/Q(x)$ as $x \to \infty$ or $x \to -\infty$:

Divide the numerator and denominator by the highest power of x in the denominator.

Example 10

Evaluate $\lim\limits_{x \to \infty} \dfrac{1 - x^3}{3x + 2}$.

Solution   By dividing the numerator and the denominator by *x*,

$$\frac{1 - x^3}{3x + 2} = \frac{(1/x) - x^2}{3 + (2/x)}$$

we see that the function has the limit form $-\infty/3$ as $x \to \infty$. Thus,

$$\lim_{x \to \infty} \frac{1 - x^3}{x + 2} = -\infty$$

In other words, the limit does not exist.

Example 11

Graph the function $f(x) = \dfrac{x^2}{1 - x^2}$.

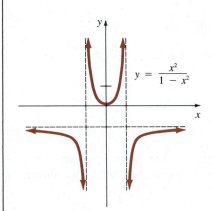

$y = \dfrac{x^2}{1 - x^2}$

Figure 2.27

Solution   Inspection of the function f reveals that its graph is symmetric with respect to the y-axis, the y-intercept is 0, and vertical asymptotes are $x = -1$ and $x = 1$. Now by dividing numerator and denominator by x^2 we get

$$f(x) = \frac{x^2}{1 - x^2} = \frac{1}{(1/x^2) - 1}$$

and consequently

$$\lim_{x \to \infty} f(x) = \frac{\lim\limits_{x \to \infty} 1}{\lim\limits_{x \to \infty}[(1/x^2) - 1]} = \frac{1}{0 - 1} = -1$$

Hence, the line $y = -1$ is a horizontal asymptote. The graph of f is given in Figure 2.27.

Example 12

Determine whether the graph of $f(x) = \dfrac{5x}{\sqrt{x^2 + 4}}$ has any horizontal asymptotes.

Solution   To investigate the limit of f as $x \to \infty$, we proceed as in Examples 9, 10, and 11. First, for $x \geq 0$, $\sqrt{x^2} = x$; therefore, the denominator of f can be written as

$$\sqrt{x^2 + 4} = \sqrt{x^2}\,\sqrt{1 + 4/x^2} = x\sqrt{1 + 4/x^2}$$

By dividing the numerator and denominator by x, we can then apply Theorems 2.5(iii), 2.7, and 2.11,

$$\lim_{x \to \infty} \frac{5x}{\sqrt{x^2 + 4}} = \lim_{x \to \infty} \frac{5}{\sqrt{1 + 4/x^2}}$$

$$= \frac{\lim\limits_{x \to \infty} 5}{\sqrt{\lim\limits_{x \to \infty} 1 + \lim\limits_{x \to \infty}(4/x^2)}} = \frac{5}{1} = 5$$

Now since $\sqrt{x^2} = |x|$, we have for $x < 0$, $\sqrt{x^2} = -x$ and so

$$\sqrt{x^2 + 4} = \sqrt{x^2}\,\sqrt{1 + 4/x^2} = -x\sqrt{1 + 4/x^2}$$

In this case, dividing the numerator and the denominator by $-x$ yields

$$\lim_{x \to -\infty} \frac{5x}{\sqrt{x^2 + 4}} = \lim_{x \to -\infty} \frac{-5}{\sqrt{1 + 4/x^2}} = \frac{-5}{1} = -5$$

Thus, the graph of f has the horizontal asymptotes $y = 5$ and $y = -5$. You should verify that the graph of f is that given in Figure 2.28.

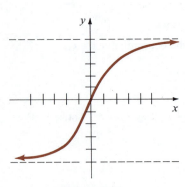

Figure 2.28

—————— **Example 13** ——————————————————————————

Alcohol is removed from the body by the lungs, the kidneys, and by chemical processes in the liver. At moderate concentration levels, the majority of work of removing the alcohol is done by the liver; less than 5% of the alcohol is eliminated by the lungs and kidneys. The rate r at which the liver processes alcohol from the bloodstream is related to the blood alcohol concentration x by a rational function of the form

$$r(x) = \frac{\alpha x}{x + \beta}$$

for some positive constants α and β. This is a special case of the so-called **Michaelis–Menten Law**. Note that $r(0) = 0$ and

$$\lim_{x \to \infty} \frac{\alpha x}{x + \beta} = \alpha$$

Since the values of r increase as x increases, we can interpret α as the maximum possible rate of removal. A typical value of α for humans is 0.22 gram/liter/minute.*

Remarks

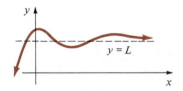

Figure 2.29

(*i*) You may have heard an asymptote described as "a line that the graph of a function approaches but never crosses." If a function possesses a vertical asymptote, then its graph can never cross it (why?). However, Figure 2.29 shows that a graph may cross a horizontal asymptote many times.

(*ii*) The graph of a function can have many vertical asymptotes. You should think about and supply an answer to the question: How many horizontal asymptotes can the graph of a function have?

(*iii*) We reemphasize that statements such as $\lim_{x \to a} f(x) = \infty$ and $\lim_{x \to \infty} f(x) = \infty$ mean that the limits *do not exist*. The symbol ∞ does not represent a number and should not be treated as a number.

—————— **Exercises 2.3** ——————————————————————————

Answers to odd-numbered problems begin on page A-19.

In Problems 1–16 sketch the graph of the given function. Identify any vertical and horizontal asymptotes.

1. $f(x) = \dfrac{1}{x + 3}$

2. $f(x) = \dfrac{3}{5 - x}$

3. $f(x) = \dfrac{x}{x - 2}$

4. $f(x) = \dfrac{-2x + 1}{x + 1}$

5. $f(x) = \dfrac{1}{(x - 2)^2}$

6. $f(x) = \dfrac{1}{x^3}$

7. $f(x) = \dfrac{1}{x^2 - 4}$

8. $f(x) = \dfrac{1}{x^2 - x - 6}$

9. $f(x) = \dfrac{1}{x^2 + 1}$

10. $f(x) = \dfrac{x}{x^2 + 1}$

—————————

*Based on F. Lundquist and H. Wolthers, "Kinetics of Alcohol Elimination in Man," *Acta Pharmacol. et Toxicol.*, (1958)14:265–289.

11. $f(x) = \dfrac{x^2}{x + 1}$

12. $f(x) = \dfrac{x^2 - x}{x^2 - 1}$

13. $f(x) = \dfrac{1}{x^2(x - 2)}$

14. $f(x) = \dfrac{4x^2}{x^2 + 4}$

15. $f(x) = \sqrt{\dfrac{x}{x - 1}}$

16. $f(x) = \dfrac{1 - \sqrt{x}}{\sqrt{x}}$

In Problems 17 and 18 use the given graph (Figures 2.30 and 2.31) to find the indicated quantity.

17.

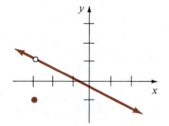

Figure 2.30

(a) $\lim\limits_{x \to 2^-} f(x)$

(b) $\lim\limits_{x \to 2^+} f(x)$

(c) $\lim\limits_{x \to 2} f(x)$

(d) $f(2)$

(e) $\lim\limits_{x \to -\infty} f(x)$

(f) $\lim\limits_{x \to \infty} f(x)$

18.

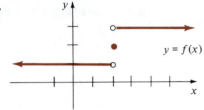

Figure 2.31

(a) $\lim\limits_{x \to -3^-} f(x)$

(b) $\lim\limits_{x \to -3^+} f(x)$

(c) $\lim\limits_{x \to -3} f(x)$

(d) $f(-3)$

(e) $\lim\limits_{x \to -\infty} f(x)$

(f) $\lim\limits_{x \to \infty} f(x)$

In Problems 19–22 use the given graph (Figures 2.32–2.35) to find

(a) $\lim\limits_{x \to 2^-} f(x)$

(b) $\lim\limits_{x \to 2^+} f(x)$

(c) $\lim\limits_{x \to -\infty} f(x)$

(d) $\lim\limits_{x \to \infty} f(x)$

19.

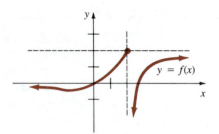

Figure 2.32

20.

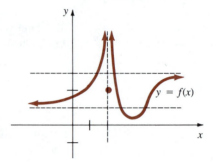

Figure 2.33

21.

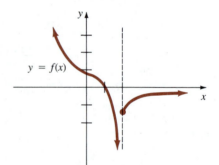

Figure 2.34

22.

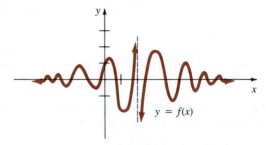

Figure 2.35

In Problems 23–50 find the given limit.

23. $\lim\limits_{x \to 5^-} \dfrac{1}{x - 5}$

24. $\lim\limits_{x \to 6} \dfrac{4}{(x - 6)^2}$

25. $\displaystyle\lim_{x\to-4^+}\frac{2}{(x+4)^3}$

26. $\displaystyle\lim_{x\to2^-}\frac{10}{x^2-4}$

27. $\displaystyle\lim_{x\to1}\frac{1}{(x-1)^4}$

28. $\displaystyle\lim_{x\to0^+}\frac{-1}{\sqrt{x}}$

29. $\displaystyle\lim_{x\to0^+}\frac{2+\sin x}{x}$

30. $\displaystyle\lim_{x\to\pi^+}\csc x$

31. $\displaystyle\lim_{x\to\infty}\frac{x^2-3x}{4x^2+5}$

32. $\displaystyle\lim_{x\to\infty}\frac{x^2}{1+x^{-2}}$

33. $\displaystyle\lim_{x\to\infty}\left(5-\frac{2}{x^4}\right)$

34. $\displaystyle\lim_{x\to-\infty}\left(\frac{6}{\sqrt[3]{x}}+\frac{1}{\sqrt[5]{x}}\right)$

35. $\displaystyle\lim_{x\to-\infty}\frac{2-1/x}{x^2+1}$

36. $\displaystyle\lim_{x\to-\infty}\frac{x-x^{-3}}{3x+x^{-2}}$

37. $\displaystyle\lim_{x\to\infty}\frac{8-\sqrt{x}}{1+4\sqrt{x}}$

38. $\displaystyle\lim_{x\to-\infty}\frac{1+7\sqrt[3]{x}}{2\sqrt[3]{x}}$

39. $\displaystyle\lim_{x\to\infty}\left(\frac{3x}{x+2}-\frac{x-1}{2x+6}\right)$

40. $\displaystyle\lim_{x\to\infty}\left(\frac{x}{3x+1}\right)\left(\frac{4x^2+1}{2x^2+x}\right)^3$

41. $\displaystyle\lim_{x\to\infty}\frac{4x+1}{\sqrt{x^2+1}}$

42. $\displaystyle\lim_{x\to\infty}\frac{\sqrt{9x^2+6}}{5x-1}$

43. $\displaystyle\lim_{x\to-\infty}\frac{2x+1}{\sqrt{3x^2+1}}$

44. $\displaystyle\lim_{x\to-\infty}\frac{-5x^2+6x+3}{\sqrt{x^4+x^2+1}}$

45. $\displaystyle\lim_{x\to\infty}\sqrt{\frac{3x+2}{6x-8}}$

46. $\displaystyle\lim_{x\to-\infty}\sqrt[3]{\frac{2x-1}{7-16x}}$

47. $\displaystyle\lim_{x\to\infty}(x-\sqrt{x^2+1})$

48. $\displaystyle\lim_{x\to\infty}(\sqrt{x^2+5x}-x)$

(*Hint:* Rationalize the numerator.)

49. $\displaystyle\lim_{x\to-\infty}\frac{|x-5|}{x-5}$

50. $\displaystyle\lim_{x\to\infty}\frac{|4x|+|x-1|}{x}$

In Problems 51–54 sketch a graph of a function f that satisfies the given conditions.

51. $\displaystyle\lim_{x\to1^+}f(x)=-\infty,\quad \lim_{x\to1^-}f(x)=-\infty,$
$f(2)=0,\ \displaystyle\lim_{x\to\infty}f(x)=0$

52. $f(0)=1,\quad \displaystyle\lim_{x\to-\infty}f(x)=3,\quad \lim_{x\to\infty}f(x)=-2$

53. $\displaystyle\lim_{x\to2}f(x)=\infty,\quad \lim_{x\to-\infty}f(x)=\infty,\quad \lim_{x\to\infty}f(x)=1$

54. $\displaystyle\lim_{x\to1^-}f(x)=2,\quad \lim_{x\to1^+}f(x)=-\infty,\quad f\!\left(\frac{3}{2}\right)=0,$
$f(3)=0,\quad \displaystyle\lim_{x\to-\infty}f(x)=0,\quad \lim_{x\to\infty}f(x)=0$

In Problems 55 and 56 find intercepts and any vertical and horizontal asymptotes. Sketch a graph of f.

55. $f(x)=\dfrac{x-2}{\sqrt{x^2+1}}$

56. $f(x)=\dfrac{x+3}{\sqrt{x^2-1}}$

57. According to Einstein's theory of relativity, the mass m of a body moving with velocity v is $m=m_0/\sqrt{1-v^2/c^2}$, where m_0 is the initial mass and c is the speed of light. What happens to m as $v\to c^-$?

58. An important problem in fishery science is to estimate the number of fish presently spawning in streams and use this information to predict the number of mature fish or "recruits" that will return to the rivers during the next reproductive period. If S is the number of spawners and R the number of recruits, the **Beverton–Holt spawner recruit function*** is $R(S)=S/(\alpha S+\beta)$, where α and β are positive constants. Show that this function predicts approximately constant recruitment when the number of spawners is sufficiently large.

Calculator/Computer Problems

In Problems 59–62 use a calculator or computer to investigate the given limit. Conjecture its value.

59. $\displaystyle\lim_{x\to\infty}x^2\sin\frac{2}{x^2}$

60. $\displaystyle\lim_{x\to\infty}x\sin\frac{3}{x}$

61. $\displaystyle\lim_{x\to\infty}\left(\cos\frac{1}{x}\right)^x$

62. $\displaystyle\lim_{x\to(\pi/2)^-}\frac{\tan x}{\tan 5x}$

63. Consider the function $f(x)=(1+x)^{1/x}$. Investigate the values $f(x)$ as (*a*) $x\to-1^+$, (*b*) $x\to0$, and (*c*) $x\to\infty$.

64. In Problem 63 sketch the graph of the function f for $x>-1$.

Miscellaneous Problems

In Problems 65–70 let $f(x)=P(x)/Q(x)$ be a rational function, where

$$P(x)=a_nx^n+a_{n-1}x^{n-1}+\cdots+a_0$$

and

$$Q(x)=b_mx^m+b_{m-1}x^{m-1}+\cdots+b_0$$

Find $\lim_{x\to\infty}f(x)$.

---

*Based on R. J. H. Beverton and S. J. Holt's, *On the Dynamics of Exploited Fish Populations* (London: H.M.S.O., 1957), pp. 44–50.

65. Degree of P < degree of Q

66. Degree of P = degree of Q

67. Degree of P > degree of Q

68. $P(x) = 1$,   degree of $Q \geq 1$

69. $Q(x) = 1$,   degree of $P \geq 1$

70. $Q(x) = xP(x)$

2.4 Continuity

In previous discussions about graphing, we used the phrase "connect the points with a smooth curve." This phrase invokes an image of a graph that is a nice *continuous* curve, that is, a curve with no gaps or breaks. Indeed, a **continuous function** is often described as one whose graph can be drawn without lifting pencil from paper.

Before stating the precise definition of continuity, we illustrate in Figure 2.36 some intuitive examples of graphs of functions that are not continuous, or **discontinuous**, at a number a.

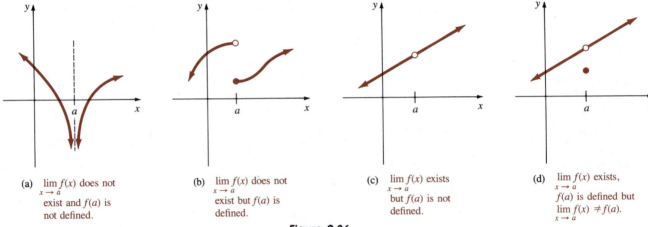

(a) $\lim\limits_{x \to a} f(x)$ does not exist and $f(a)$ is not defined.

(b) $\lim\limits_{x \to a} f(x)$ does not exist but $f(a)$ is defined.

(c) $\lim\limits_{x \to a} f(x)$ exists but $f(a)$ is not defined.

(d) $\lim\limits_{x \to a} f(x)$ exists, $f(a)$ is defined but $\lim\limits_{x \to a} f(x) \neq f(a)$.

Figure 2.36

Continuity at a Number a   Figure 2.36 suggests the following threefold condition of continuity of a function f at a number a.

DEFINITION 2.1   Continuity

A function f is said to be **continuous at a number a** if

(*i*) $f(a)$ is defined,

(*ii*) $\lim\limits_{x \to a} f(x)$ exists, and

(*iii*) $\lim\limits_{x \to a} f(x) = f(a)$

Example 1

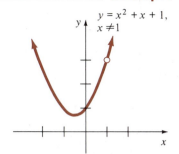

$y = x^2 + x + 1,$
$x \neq 1$

Figure 2.37

The rational function

$$f(x) = \frac{x^3 - 1}{x - 1}$$

$$= x^2 + x + 1, \qquad x \neq 1$$

is discontinuous at 1 since $f(1)$ is not defined. This is analogous to the case illustrated in Figure 2.36(c) since we have $\lim_{x \to 1} f(x) = 3$. As can be seen in Figure 2.37, f is continuous at any other number $x \neq 1$.

Example 2

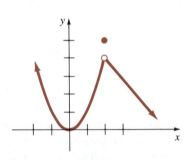

Figure 2.38

Figure 2.38 shows the graph of the piecewise-defined function

$$f(x) = \begin{cases} x^2, & x < 2 \\ 5, & x = 2 \\ -x + 6, & x > 2 \end{cases}$$

Now, $f(2)$ is defined and equals 5. Next

$$\left. \begin{array}{l} \lim_{x \to 2^-} f(x) = \lim_{x \to 2^-} x^2 = 4 \\[2mm] \lim_{x \to 2^+} f(x) = \lim_{x \to 2^+} (-x + 6) = 4 \end{array} \right\} \text{implies } \lim_{x \to 2} f(x) = 4$$

Since $\lim_{x \to 2} f(x) \neq f(2) = 5$, we see from (*iii*) of Definition 2.1 that f is discontinuous at 2.

Continuity on an Interval

A function f is said to be **continuous on an open interval** (a, b) if it is continuous at every number in the interval. A function f is **continuous on a closed interval** $[a, b]$ if it is continuous on (a, b) and, in addition,

$$\lim_{x \to a^+} f(x) = f(a) \quad \text{and} \quad \lim_{x \to b^-} f(x) = f(b)$$

Extensions of these concepts to intervals such as (a, ∞), $(-\infty, b)$, $(-\infty, \infty)$, $[a, b)$, $(a, b]$, $(-\infty, b]$, $[a, \infty)$ are made in the expected manner.

Example 3

(a)   As we see from Figure 2.39(a), $f(x) = 1/\sqrt{1 - x^2}$ is continuous on the open interval $(-1, 1)$ but is not continuous on the closed interval $[-1, 1]$ since neither $f(-1)$ nor $f(1)$ is defined.

(b) $f(x) = \sqrt{1 - x^2}$ is continuous on $[-1, 1]$. Observe from Figure 2.39(b) that

$$\lim_{x \to -1^+} f(x) = f(-1) = 0 \quad \text{and} \quad \lim_{x \to 1^-} f(x) = f(1) = 0$$

(c) $f(x) = \sqrt{x - 1}$ is continuous on $[1, \infty)$ since

$$\lim_{x \to a} f(x) = \lim_{x \to a} \sqrt{x - 1} = \sqrt{a - 1} = f(a), \qquad a > 1$$

and
$$\lim_{x \to 1^+} \sqrt{x - 1} = f(1) = 0$$

See Figure 2.39(c).

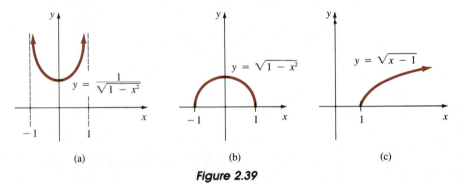

(a)   (b)   (c)

Figure 2.39

A review of the graphs in Figure 1.78 indicates that the sine and cosine functions are continuous on $(-\infty, \infty)$. The tangent and secant functions are discontinuous at $x = (2n + 1)\pi/2$, $n = 0, \pm 1, \pm 2, \ldots$. The cotangent and cosecant functions are discontinuous at $x = n\pi$, $n = 0, \pm 1, \pm 2, \ldots$.

Continuity of a Sum, Product, and Quotient

THEOREM 2.12   If f and g are functions continuous at a number a, then cf (c a constant), $f + g$, fg, and f/g ($g(a) \neq 0$) are also continuous at a.

Proof   Proof of the Continuity of *fg*

Since f and g are continuous at a number a, then

$$\lim_{x \to a} f(x) = f(a) \quad \text{and} \quad \lim_{x \to a} g(x) = g(a)$$

Hence, from Theorem 2.5(*ii*), we have

$$\lim_{x \to a} f(x)g(x) = (\lim_{x \to a} f(x)) \cdot (\lim_{x \to a} g(x))$$

$$= f(a)g(a) \qquad\qquad \blacksquare$$

The proofs of the remaining parts of Theorem 2.12 are obtained in a similar manner.

Since Definition 2.1 implies that $f(x) = x$ is continuous at any number x, we see from successive applications of Theorem 2.12 that the functions

$$x, x^2, x^3, \ldots, x^n$$

are also continuous on $(-\infty, \infty)$. Thus, another application of Theorem 2.12 shows that

a polynomial function is continuous on $(-\infty, \infty)$.*

Consequently,

a rational function $f(x) = P(x)/Q(x)$, where P and Q are polynomial functions, is continuous at all numbers x for which $Q(x) \neq 0$.

Limits of a Composite Function

The next theorem tells us that if a function f is continuous, then the limit of the function is the function of the limit.

THEOREM 2.13   If $\lim\limits_{x \to a} g(x) = L$ and f is continuous at L, then

$$\lim_{x \to a} f(g(x)) = f(\lim_{x \to a} g(x)) = f(L)$$

Theorem 2.13 is useful in proving other theorems. If the function g is continuous at a number a and f is continuous at $g(a)$, then we see that

$$\lim_{x \to a} f(g(x)) = f(\lim_{x \to a} g(x)) = f(g(a))$$

In other words,

the composite function $f \circ g$ of two continuous functions f and g is continuous.

We leave it as an exercise for you to prove Theorem 2.6 from Theorem 2.13.

Example 4

$f(x) = \sqrt{x}$ is continuous on $[0, \infty)$ and $g(x) = 2 + \sin x$ is continuous on $(-\infty, \infty)$. But, since $g(x) \geq 1 > 0$ for all x, the composite function

$$(f \circ g)(x) = f(g(x)) = \sqrt{2 + \sin x}$$

is continuous at any real number a.

---

*Functions such as polynomials that are continuous on $(-\infty, \infty)$ are sometimes said to be **continuous everywhere** or simply **continuous**.

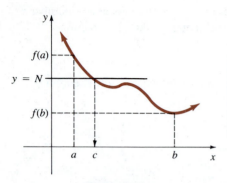

Figure 2.40

Figure 2.40 illustrates the plausibility of the next result about continuous functions.

THEOREM 2.14   **Intermediate Value Theorem**

If f denotes a function continuous on a closed interval $[a, b]$ for which $f(a) \neq f(b)$, and if N is any number between $f(a)$ and $f(b)$, then there exists at least one number c between a and b such that $f(c) = N$.

The Intermediate Value Theorem states that a function f continuous on a closed interval $[a, b]$ takes on all values between $f(a)$ and $f(b)$. Put another way, f does not "skip" any values.

Example 5

The polynomial function $f(x) = x^2 - x - 5$ is continuous on the interval $[-1, 4]$ and $f(-1) = -3$, $f(4) = 7$. For any number N for which $-3 \leq N \leq 7$, Theorem 2.14 guarantees that there is a solution to $c^2 - c - 5 = N$ in $[-1, 4]$. Specifically, if we choose $N = 1$, then $c^2 - c - 5 = 1$ is equivalent to

$$c^2 - c - 6 = 0 \quad \text{or} \quad (c - 3)(c + 2) = 0$$

Although the latter equation has two solutions, only the value $c = 3$ is between -1 and 4.

The foregoing example suggests a corollary to the Intermediate Value Theorem.

If f satisfies the hypotheses of Theorem 2.14 and $f(a)$ and $f(b)$ have opposite algebraic signs, then there exists some value of x between a and b for which $f(x) = 0$.

This fact is often used in locating real zeros of a continuous function f. Figure 2.41 shows several possibilities.

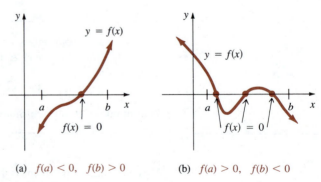

(a) $f(a) < 0$, $f(b) > 0$ (b) $f(a) > 0$, $f(b) < 0$

Figure 2.41

In Example 5, by knowing $f(-1) < 0$ and $f(4) > 0$, we are led to the fact that $x^2 - x - 5 = 0$ for at least one value of x between -1 and 4.

Remark

We often give a discontinuity of a function a special name.

- If $x = a$ is a vertical asymptote for the graph of $y = f(x)$, then f is said to have an **infinite discontinuity** at a.

- If $\lim_{x \to a^-} f(x) = L_1$ and $\lim_{x \to a^+} f(x) = L_2$ and $L_1 \neq L_2$, then f is said to have a **finite discontinuity** or a **jump discontinuity** at a. The function $y = f(x)$ given in Figure 2.42 has a jump discontinuity at 0 since $\lim_{x \to 0^-} f(x) = -1$ and $\lim_{x \to 0^+} f(x) = 1$. The greatest integer function* $f(x) = [x]$ has a jump discontinuity at every integer value of x.

- If $\lim_{x \to a} f(x)$ exists but f is either not defined at a or $f(a) \neq \lim_{x \to a} f(x)$, then f is said to have a **removable discontinuity** at a; for example, the function $f(x) = (x^2 - 1)/(x - 1)$ is not defined at 1 but $\lim_{x \to 1} f(x) = 2$. By *defining* $f(1) = 2$, the new function

$$f(x) = \begin{cases} \dfrac{x^2 - 1}{x - 1}, & x \neq 1 \\ 2, & x = 1 \end{cases}$$

is continuous at every number.

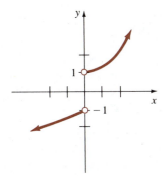

Figure 2.42

Exercises 2.4

Answers to odd-numbered problems begin on page A-20.

In Problems 1–10 determine the numbers, if any, at which the given function is discontinuous.

1. $f(x) = x^3 - 4x^2 + 7$

2. $f(x) = \dfrac{x}{x^2 + 4}$

3. $f(x) = (x^2 - 9x + 18)^{-1}$

4. $f(x) = \dfrac{x^2 - 1}{x^4 - 1}$

5. $f(x) = \dfrac{x - 1}{\sin 2x}$

6. $f(x) = \dfrac{\tan x}{x + 3}$

7. $f(x) = \begin{cases} x, & x < 0 \\ x^2, & 0 \leq x < 2 \\ x, & x > 2 \end{cases}$

8. $f(x) = \begin{cases} \dfrac{|x|}{x}, & x \neq 0 \\ 1, & x = 0 \end{cases}$

9. $f(x) = \begin{cases} \dfrac{\sin x}{x}, & x \neq 0 \\ \dfrac{1}{2}, & x = 0 \end{cases}$

10. $f(x) = \begin{cases} \dfrac{x^2 - 25}{x - 5}, & x \neq 5 \\ 10, & x = 5 \end{cases}$

---

*See Example 4 of Section 2.1.

In Problems 11–22 determine whether the given function is continuous on the indicated intervals. (For Problems 21 and 22, see Figures 2.43 and 2.44, respectively.)

11. $f(x) = x^2 + 1$

   (a) $[-1, 4]$ **(b)** $[5, \infty)$

12. $f(x) = \dfrac{1}{x}$

   (a) $(-\infty, \infty)$ **(b)** $(0, \infty)$

13. $f(x) = \dfrac{1}{\sqrt{x}}$

   (a) $(0, 4]$ **(b)** $[1, 9]$

14. $f(x) = \sqrt{x^2 - 9}$

   (a) $[-3, 3]$ **(b)** $[3, \infty)$

15. $f(x) = \tan x$

   (a) $[0, \pi]$ **(b)** $\left[-\dfrac{\pi}{2}, \dfrac{\pi}{2}\right]$

16. $f(x) = \csc x$

   (a) $(0, \pi)$ **(b)** $(2\pi, 3\pi)$

17. $f(x) = \dfrac{x}{x^3 + 8}$

   (a) $[-4, -3]$ **(b)** $(-\infty, \infty)$

18. $f(x) = \dfrac{1}{|x| - 4}$

   (a) $(-\infty, -1]$ **(b)** $[1, 6]$

19. $f(x) = \dfrac{x}{2 + \sec x}$

   (a) $(-\infty, \infty)$ **(b)** $\left[\dfrac{\pi}{2}, \dfrac{3\pi}{2}\right]$

20. $f(x) = \sin \dfrac{1}{x}$

   (a) $\left[\dfrac{1}{\pi}, \infty\right)$ **(b)** $\left[-\dfrac{2}{\pi}, \dfrac{2}{\pi}\right]$

21.

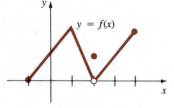

Figure 2.43

   (a) $[-1, 3]$ **(b)** $(2, 4]$

22.

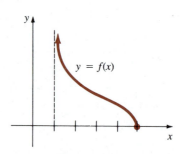

Figure 2.44

   (a) $[2, 4]$ **(b)** $[1, 5]$

In Problems 23–26 find values of m and n so that the given function is continuous.

23. $f(x) = \begin{cases} mx, & x < 4 \\ x^2, & x \geq 4 \end{cases}$

24. $f(x) = \begin{cases} \dfrac{x^2 - 4}{x - 2}, & x \neq 2 \\ m, & x = 2 \end{cases}$

25. $f(x) = \begin{cases} mx, & x < 3 \\ n, & x = 3 \\ -2x + 9, & x > 3 \end{cases}$

26. $f(x) = \begin{cases} mx - n, & x < 1 \\ 5, & x = 1 \\ 2mx + n, & x > 1 \end{cases}$

In Problems 27 and 28, $[x]$ denotes the greatest integer not exceeding x. Sketch a graph to determine the numbers at which the given function is discontinuous.

27. $f(x) = [2x - 1]$ **28.** $f(x) = [x] - x$

In Problems 29–36 use Theorem 2.13 to evaluate the given limit.

29. $\lim\limits_{x \to \pi/6} \sin(2x + \pi/3)$

30. $\lim\limits_{x \to \pi^2} \cos \sqrt{x}$

31. $\lim\limits_{x \to \pi/2} \sin(\cos x)$ **32.** $\lim\limits_{x \to \pi/2} (1 + \cos(\cos x))$

33. $\lim\limits_{t \to \pi} \cos\left(\dfrac{t^2 - \pi^2}{t - \pi}\right)$ **34.** $\lim\limits_{t \to 0} \tan\left(\dfrac{\pi t}{t^2 + 3t}\right)$

35. $\lim\limits_{t \to \pi} \sqrt{t - \pi + \cos^2 t}$

36. $\lim\limits_{t \to 1} (4t + \sin 2\pi t)^3$

In Problems 37 and 38 determine the interval(s) where $f \circ g$ is continuous.

37. $f(x) = \dfrac{1}{\sqrt{x-1}}, \quad g(x) = x + 4$

38. $f(x) = \dfrac{5x}{x-1}, \quad g(x) = (x-2)^2$

In Problems 39–42 verify the Intermediate Value Theorem for f on the given interval. Find a value of c in the interval for the indicated value of N.

39. $f(x) = x^2 - 2x, \ [1, 5]; \ N = 8$

40. $f(x) = x^2 + x + 1, \ [-2, 3]; \ N = 6$

41. $f(x) = x^3 - 2x + 1, \ [-2, 2]; \ N = 1$

42. $f(x) = \dfrac{10}{x^2 + 1}, \ [0, 1]; \ N = 8$

43. Given that f is continuous on $[a, b]$ and $f(a) = 5$ and $f(b) = 20$, prove that there is a number c in (a, b) such that $f(c) = 10$.

44. Given that $f(x) = x^5 + 2x - 7$, prove that there is a number c such that $f(c) = 50$.

45. Given that $f(x) = x^5 + x - 1$, prove that there is a number c such that $f(c) = 0$.

46. Given that f and g are continuous on $[a, b]$ such that $f(a) > g(a)$ and $f(b) < g(b)$; prove that there is a number c in (a, b) such that $f(c) = g(c)$.

47. Prove that the equation $2x^7 = 1 - x$ has a solution in $[0, 1]$.

48. Prove that the equation

$$\frac{x^2 + 1}{x + 3} + \frac{x^4 + 1}{x - 4} = 0$$

has a solution in the interval $(-3, 4)$.

Miscellaneous Problems

49. Let f be continuous and $f(x) > 0$ for all x in $(1, 2]$. Given that $f(1) = -3$, prove that f is discontinuous on $[1, 2]$.

50. Given that f and g are continuous at a number a, prove that $f + g$ is continuous at a.

51. Given that f and g are continuous at a number a, and $g(a) \neq 0$, prove that f/g is continuous at a.

52. Prove that the **Dirichlet function***

$$f(x) = \begin{cases} 1, & x \text{ rational} \\ 0, & x \text{ irrational} \end{cases}$$

is discontinuous at every real number a. What does the graph of f look like?

53. Prove Theorem 2.6.

54. Prove that $(\sin x)/x = \frac{1}{2}$ for some value of x between 0 and π.

55. Let $f(x) = [x]$ be the greatest integer function and $g(x) = \cos x$. Determine the numbers at which $f \circ g$ is discontinuous.

56. Can $f(x) = 1/(x - 1)$ be defined at 1 so that the resulting function is continuous at this number?

57. How should $f(x) = (x - 9)/(\sqrt{x} - 3)$ be defined at 9 so that the resulting function is continuous at this number?

58. Consider the functions

$$f(x) = |x| \quad \text{and} \quad g(x) = \begin{cases} x + 1, & x < 0 \\ x - 1, & x \geq 0 \end{cases}$$

Sketch the graphs of $f \circ g$ and $g \circ f$. Determine whether $f \circ g$ and $g \circ f$ are continuous at 0.

59. Suppose f is continuous and $f(x) \neq 0$ for all x in $[a, b]$. Prove that $f(x) < 0$ or $f(x) > 0$ for all x in the interval.

[O] 2.5 The Definition of a Limit

In this section we will consider an alternative approach to the notion of a limit that is based on analytical concepts rather than on intuitive concepts. While graphs and tables of functional values may be convincing for determining whether a limit does or does not exist, you are certainly aware that all calculators and

---

*The function is named after the German mathematician Peter Gustav Dirichlet (1805–1859).

computers work only with approximations and that graphs can be drawn inaccurately. A **proof** of the existence of a limit can never be based on one's ability to draw pictures. Although a good intuitive understanding of $\lim_{x \to a} f(x)$ and $\lim_{x \to \infty} f(x)$ is sufficient for proceeding with the study of the calculus in this text, an intuitive understanding is admittedly too vague to be of any use in proving theorems. To give a rigorous demonstration of the existence of a limit, or to prove the theorems of Section 2.2, we must first start with the precise definition of a limit.

2.5.1 The ε–δ Definition of $\lim\limits_{x \to a} f(x) = L$

Let us try to prove that

$$\lim_{x \to 2} (2x + 6) = 10$$

by elaborating on the following idea, "If $f(x) = 2x + 6$ can be made arbitrarily close to 10 by taking x sufficiently close to 2, from either side but different from 2, then $\lim_{x \to 2} f(x) = 10$." We need to make the concepts of "arbitrarily close" and "sufficiently close" precise. In order to set a standard of arbitrary closeness, let us demand that the distance between the numbers $f(x)$ and 10 be less than 0.1, that is,

$$|f(x) - 10| < 0.1 \quad \text{or} \quad 9.9 < f(x) < 10.1 \tag{2.4}$$

Then, how close must x be to 2 to accomplish (2.4)? To find out, we can solve the inequality

$$9.9 < 2x + 6 < 10.1$$

by ordinary algebra and discover that

$$1.95 < x < 2.05$$

Thus, for an "arbitrary closeness to 10" of 0.1, "sufficiently close to 2" means within 0.05 on either side of 2. In other words, if x is a number different from 2 in the open interval $(1.95, 2.05)$, then $f(x)$ is guaranteed to be in $(9.9, 10.1)$.

　　Using the same example, let us try to generalize. Suppose ε (epsilon) denotes *any small positive* number that is our measure of arbitrary closeness to the number 10. If we demand

$$|f(x) - 10| < \varepsilon \quad \text{or} \quad 10 - \varepsilon < f(x) < 10 + \varepsilon \tag{2.5}$$

then from $f(x) = 2x + 6$ and algebra, we find

$$2 - \frac{\varepsilon}{2} < x < 2 + \frac{\varepsilon}{2} \tag{2.6}$$

Using absolute values and a new symbol δ (delta), we can write (2.5) and (2.6) as,

$$|f(x) - 10| < \varepsilon \quad \text{whenever} \quad 0 < |x - 2| < \delta$$

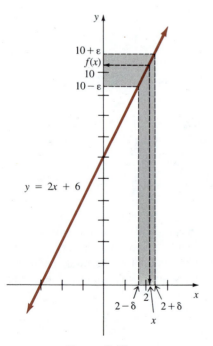

Figure 2.45

where $\delta = \varepsilon/2$. Thus, for a new value for ε, say $\varepsilon = 0.001$, $\delta = \varepsilon/2 = 0.0005$ tells us the corresponding closeness to 2. For any number x different from 2* in (1.9995, 2.0005), we can be sure $f(x)$ is in (9.999, 10.001). See Figure 2.45.

The Definition of a Limit   The foregoing discussion leads us to the following so-called ε–δ definition of a limit.

DEFINITION 2.2 ε–δ Definition of a Limit

Suppose a function f is defined everywhere on an open interval, except possibly at a number a in the interval. Then

$$\lim_{x \to a} f(x) = L$$

means for every $\varepsilon > 0$ there exists a $\delta > 0$ such that

$$|f(x) - L| < \varepsilon \quad \text{whenever} \quad 0 < |x - a| < \delta$$

Let $\lim_{x \to a} f(x) = L$ and suppose $\delta > 0$ is the number that "works" in the sense of Definition 2.2 for a given $\varepsilon > 0$. As shown in Figure 2.46(a), every x in $(a - \delta, a + \delta)$, with the possible exception of a itself, will then have an image $f(x)$ in $(L - \varepsilon, L + \varepsilon)$. Furthermore, as in Figure 2.46(b), a choice $\delta_1 < \delta$ for the same ε also "works" in that every x not equal to a in $(a - \delta_1, a + \delta_1)$ gives $f(x)$ in $(L - \varepsilon, L + \varepsilon)$. However, Figure 2.46(c) shows that choosing a smaller ε_1, $0 < \varepsilon_1 < \varepsilon$, will demand finding a new value of δ.

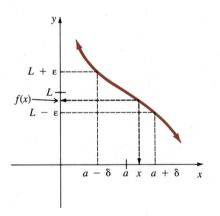

(a) A δ that works for a given ε.

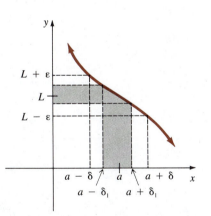

(b) A smaller δ_1 will also work for the same ε.

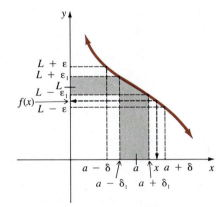

(c) A smaller ε_1 will require a $\delta_1 < \delta$. For x in $(a - \delta, a + \delta)$, $f(x)$ is not necessarily in $(L - \varepsilon_1, L + \varepsilon_1)$.

Figure 2.46

---

*This is why we use $0 < |x - 2| < \delta$ rather than $|x - 2| < \delta$. Keep in mind when considering $\lim_{x \to 2} f(x)$, we do not care about f at 2.

Example 1

Prove that $\lim_{x \to 3}(5x + 2) = 17$.

Solution   For any given $\varepsilon > 0$ regardless how small, we wish to find a δ so that

$$|(5x + 2) - 17| < \varepsilon \quad \text{whenever} \quad 0 < |x - 3| < \delta$$

To do this consider

$$|(5x + 2) - 17| = |5x - 15| = 5|x - 3|$$

Thus, to make $|(5x + 2) - 17| = 5|x - 3| < \varepsilon$, we need only make $0 < |x - 3| < \varepsilon/5$; that is, choose $\delta = \varepsilon/5$.

Verification   If $0 < |x - 3| < \varepsilon/5$, then

$$5|x - 3| < \varepsilon$$
$$|5x - 15| < \varepsilon$$
$$|(5x + 2) - 17| < \varepsilon$$

Example 2

Prove that $\lim_{x \to -4} \dfrac{16 - x^2}{4 + x} = 8$.

Solution

$$\left| \frac{16 - x^2}{4 + x} - 8 \right| = |4 - x - 8| = |-x - 4| = |x + 4| = |x - (-4)|$$

Thus,

$$\left| \frac{16 - x^2}{4 + x} - 8 \right| = |x - (-4)| < \varepsilon$$

whenever we have $0 < |x - (-4)| < \varepsilon$; that is, choose $\delta = \varepsilon$.

Example 3

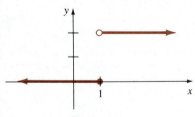

Figure 2.47

Consider the function

$$f(x) = \begin{cases} 0, & x \leq 1 \\ 2, & x > 1 \end{cases}$$

Intuitively, we can see from the graph of f in Figure 2.47 that $\lim_{x \to 1} f(x)$ does not exist. However, to *prove* this, we shall proceed indirectly.

Assume $\lim_{x \to 1} f(x) = L$. Then from Definition 2.2 we know that for the choice $\varepsilon = \frac{1}{2}$ there must exist a $\delta > 0$ so that

$$|f(x) - L| < \frac{1}{2} \quad \text{whenever} \quad 0 < |x - 1| < \delta$$

Now to the right of 1, let us choose $x = 1 + \delta/2$. Since

$$0 < \left| 1 + \frac{\delta}{2} - 1 \right| = \left| \frac{\delta}{2} \right| < \delta$$

we must have

$$\left| f\left(1 + \frac{\delta}{2} \right) - L \right| = |2 - L| < \frac{1}{2} \tag{2.7}$$

To the left of 1, choose $x = 1 - \delta/2$. But

$$0 < \left| 1 - \frac{\delta}{2} - 1 \right| = \left| -\frac{\delta}{2} \right| < \delta$$

implies

$$\left| f\left(1 - \frac{\delta}{2} \right) - L \right| = |0 - L| = |L| < \frac{1}{2} \tag{2.8}$$

From (2.7) and (2.8) we find, respectively

$$\frac{3}{2} < L < \frac{5}{2} \quad \text{and} \quad -\frac{1}{2} < L < \frac{1}{2}$$

Since no such L can satisfy both of these inequalities, we conclude that $\lim_{x \to 1} f(x)$ does not exist.

One-sided Limits   In conclusion, we state the definitions of the **one-sided limits**, $\lim_{x \to a^-} f(x)$ and $\lim_{x \to a^+} f(x)$.

DEFINITION 2.3   Suppose a function f is defined on an open interval (b, a). Then

$$\lim_{x \to a^-} f(x) = L$$

means for every $\varepsilon > 0$ there exists a $\delta > 0$ such that

$$|f(x) - L| < \varepsilon \quad \text{whenever} \quad a - \delta < x < a$$

DEFINITION 2.4   Suppose a function f is defined on an open interval (a, c). Then

$$\lim_{x \to a^+} f(x) = L$$

means for every $\varepsilon > 0$ there exists a $\delta > 0$ such that

$$|f(x) - L| < \varepsilon \quad \text{whenever} \quad a < x < a + \delta$$

Example 4

Prove $\lim\limits_{x \to 0^+} \sqrt{x} = 0$.

Solution

$$\left| \sqrt{x} - 0 \right| = \left| \sqrt{x} \right| = \sqrt{x}$$

Thus, $\left| \sqrt{x} - 0 \right| < \varepsilon$ whenever $0 < x < 0 + \varepsilon^2$; that is, choose $\delta = \varepsilon^2$.

Verification   If $0 < x < \varepsilon^2$, then

$$0 < \sqrt{x} < \varepsilon$$
$$\left| \sqrt{x} \right| < \varepsilon$$
$$\left| \sqrt{x} - 0 \right| < \varepsilon$$

2.5.2 The Definitions of $\lim\limits_{x \to a} f(x) = \infty$ and $\lim\limits_{x \to \infty} f(x) = L$

The two concepts

$$f(x) \to \infty \ (\text{or } -\infty) \text{ as } x \to a, \text{ and}$$
$$f(x) \to L \text{ as } x \to \infty \ (\text{or } -\infty)$$

are formalized in the next two definitions.

DEFINITION 2.5

(*i*) $\lim_{x \to a} f(x) = \infty$ means for each $M > 0$, there exists a $\delta > 0$ such that $f(x) > M$ whenever $0 < |x - a| < \delta$.

(*ii*) $\lim_{x \to a} f(x) = -\infty$ means for each $M < 0$, there exists a $\delta > 0$ such that $f(x) < M$ whenever $0 < |x - a| < \delta$.

Parts (*i*) and (*ii*) of Definition 2.5 are illustrated in Figures 2.48(a) and 2.48(b), respectively.

DEFINITION 2.6

(*i*) $\lim_{x \to \infty} f(x) = L$ if for each $\varepsilon > 0$ there exists an $N > 0$ such that $|f(x) - L| < \varepsilon$ whenever $x > N$.

(*ii*) $\lim_{x \to -\infty} f(x) = L$ if for each $\varepsilon > 0$ there exists an $N < 0$ such that $|f(x) - L| < \varepsilon$ whenever $x < N$.

Parts (*i*) and (*ii*) of Definition 2.6 are illustrated in Figures 2.49(a) and 2.49(b), respectively.

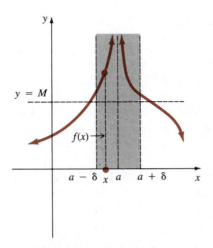

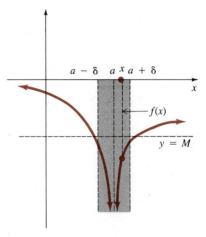

(a) For a given M, whenever
$a - \delta < x < a + \delta$, $x \neq a$,
then $f(x) > M$.

(b) For a given M, whenever
$a - \delta < x < a + \delta$, $x \neq a$,
then $f(x) < M$.

Figure 2.48

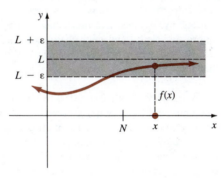

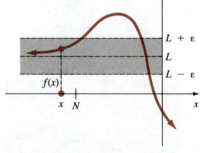

(a) For a given ε, $x > N$ implies
$L - \varepsilon < f(x) < L + \varepsilon$.

(b) For a given ε, $x < N$ implies
$L - \varepsilon < f(x) < L + \varepsilon$.

Figure 2.49

Example 5

Prove that $\lim\limits_{x \to \infty} \dfrac{3x}{x + 1} = 3$.

Solution By Definition 2.6(i) for any $\varepsilon > 0$, we must find an $N > 0$ such that

$$\left| \frac{3x}{x + 1} - 3 \right| < \varepsilon \quad \text{whenever } x > N$$

Now, by considering $x > 0$

$$\left| \frac{3x}{x + 1} - 3 \right| = \left| \frac{-3}{x + 1} \right| = \frac{3}{x + 1} < \frac{3}{x} < \varepsilon$$

whenever $x > 3/\varepsilon$. Hence, choose $N = 3/\varepsilon$; for example, if $\varepsilon = 0.01$, then
$N = 3/(0.01) = 300$ will guarantee that $|f(x) - 3| < 0.01$ whenever $x > 300$.

Remark

After this section you may agree with W. Whewell who wrote in 1858 that "A limit is a peculiar . . . conception." For many years after the invention of calculus in the seventeenth century, mathematicians argued and debated the nature of a limit. There was an awareness that intuition, graphs, and numerical examples of ratios of vanishing quantities provide at best a shaky foundation to such a fundamental concept. As you will see, beginning in the next chapter, the limit concept plays a central role in calculus. The study of calculus went through several periods of increased mathematical rigor beginning with the French mathematician Augustin-Louis Cauchy* and continuing later with the German mathematician Karl Wilhelm Weierstrass.

Augustin Cauchy

Karl Weierstrass

*Augustin-Louis Cauchy (1789–1857) Born during an era of upheaval in French history, Augustin-Louis Cauchy was destined to initiate a revolution in mathematics. For many original contributions, but especially for his efforts in clarifying mathematical obscurities, his incessant demand for satisfactory definitions and rigorous proofs of theorems, Cauchy is often called "the father of modern analysis." A prolific writer whose output has been surpassed by only a few, Cauchy produced nearly 800 papers in astronomy, physics, and mathematics. The same mind that was always open and inquiring in science and mathematics was narrow and unquestioning in many other areas. Outspoken and arrogant, Cauchy's passionate stands on political and religious issues often alienated him from his colleagues.

Karl Wilhelm Weierstrass (1815–1897) One of the foremost mathematical analysts of the nineteenth century never earned an academic degree! After majoring in law at the University of Bonn, but concentrating in fencing and beer drinking for four years, Karl Wilhelm Weierstrass "graduated" to real life with no degree. In need of a job, Weierstrass passed a state examination and received a teaching certificate in 1841. During a period of fifteen years as a secondary school teacher his dormant mathematical genius blossomed. Although the quantity of his research publications was modest, especially when compared to Cauchy, the quality of his works so impressed the German mathematical community that he was awarded a doctorate, *honoris causa*, from the University of Königsberg and eventually was appointed a professor at the University of Berlin. While there he achieved worldwide recognition both as a mathematician and as a teacher of mathematics. One of his students was Sonja Kowalewski, the greatest woman mathematician of the nineteenth century.

It was Karl Wilhelm Weierstrass who was responsible for putting the concept of a limit on a firm foundation with the $\varepsilon-\delta$ definition.

Exercises 2.5

Answers to odd-numbered problems begin on page A-20.

[2.5.1]

In Problems 1–20 use Definitions 2.2, 2.3, or 2.4 to prove the given limit result.

1. $\lim\limits_{x \to 5} 10 = 10$

2. $\lim\limits_{x \to -2} \pi = \pi$

3. $\lim\limits_{x \to 3} x = 3$

4. $\lim\limits_{x \to 4} 2x = 8$

5. $\lim\limits_{x \to -1} (x + 6) = 5$

6. $\lim\limits_{x \to 0}(x - 4) = -4$

7. $\lim\limits_{x \to 0}(3x + 7) = 7$

8. $\lim\limits_{x \to 1}(9 - 6x) = 3$

9. $\lim\limits_{x \to 2} \dfrac{2x - 3}{4} = \dfrac{1}{4}$

10. $\lim\limits_{x \to 1/2} 8(2x + 5) = 48$

11. $\lim\limits_{x\to -5} \dfrac{x^2 - 25}{x + 5} = -10$

12. $\lim\limits_{x\to 3} \dfrac{x^2 - 7x + 12}{2x - 6} = -\dfrac{1}{2}$

13. $\lim\limits_{x\to 0} \dfrac{8x^5 + 12x^4}{x^4} = 12$

14. $\lim\limits_{x\to 1} \dfrac{2x^3 + 5x^2 - 2x - 5}{x^2 - 1} = 7$

15. $\lim\limits_{x\to 0} x^2 = 0$

16. $\lim\limits_{x\to 0} 8x^3 = 0$

17. $\lim\limits_{x\to 0^+} \sqrt{5x} = 0$

18. $\lim\limits_{x\to (1/2)^+} \sqrt{2x - 1} = 0$

19. $\lim\limits_{x\to 0^-} f(x) = -1, f(x) = \begin{cases} 2x - 1, & x < 0 \\ 2x + 1, & x > 0 \end{cases}$

20. $\lim\limits_{x\to 1^+} f(x) = 3, \qquad f(x) = \begin{cases} 0, & x \le 1 \\ 3, & x > 1 \end{cases}$

In Problems 21–24 prove that $\lim_{x\to a} f(x)$ does not exist.

21. $f(x) = \begin{cases} 2, & x < 1 \\ 0, & x \ge 1 \end{cases}; \quad a = 1$

22. $f(x) = \begin{cases} 1, & x \le 3 \\ -1, & x > 3 \end{cases}; \quad a = 3$

23. $f(x) = \begin{cases} x, & x \le 0 \\ 2 - x, & x > 0 \end{cases}; \quad a = 0$

24. $f(x) = \dfrac{1}{x}; \quad a = 0$

Miscellaneous Problems

25. Prove that $\lim_{x\to 3} x^2 = 9$. (*Hint:* Consider only those numbers x for which $2 < x < 4$.)

26. Prove that $\lim_{x\to 2} 1/x = \frac{1}{2}$. (*Hint:* Consider only those numbers x for which $1 < x < 3$.)

27. Prove that $\lim_{x\to a} \sqrt{x} = \sqrt{a}, a > 0$. (*Hint:* Use the identity

$$|\sqrt{x} - \sqrt{a}| = |\sqrt{x} - \sqrt{a}| \cdot \frac{\sqrt{x} + \sqrt{a}}{\sqrt{x} + \sqrt{a}} = \frac{|x - a|}{\sqrt{x} + \sqrt{a}}$$

and the fact that $\sqrt{x} \ge 0$.)

28. Prove that $\lim_{x\to 0} f(x) = 0, f(x) = \begin{cases} x, & x \text{ rational} \\ 0, & x \text{ irrational} \end{cases}$

[2.5.2]

In Problems 29–32 use Definition 2.6 to prove the given limit result.

29. $\lim\limits_{x\to\infty} \dfrac{5x - 1}{2x + 1} = \dfrac{5}{2}$

30. $\lim\limits_{x\to\infty} \dfrac{2x}{3x + 8} = \dfrac{2}{3}$

31. $\lim\limits_{x\to -\infty} \dfrac{10x}{x - 3} = 10$

32. $\lim\limits_{x\to -\infty} \dfrac{x^2}{x^2 + 1} = 1$

——— Chapter 2 Review Exercises ———

Answers to odd-numbered problems begin on page A-20.

In Problems 1–18 answer true or false.

1. $\lim\limits_{x\to 2} \dfrac{x^3 - 8}{x - 2} = 12$ _____

2. $\lim\limits_{x\to 5} \sqrt{x - 5} = 0$ _____

3. $\lim\limits_{x\to 0} \dfrac{|x|}{x} = 1$ _____

4. $\lim\limits_{z\to 1} \dfrac{z^3 + 8z - 2}{z^2 + 9z - 10}$ does not exist. _____

5. If $\lim_{x\to a} f(x) = 3$ and $\lim_{x\to a} g(x) = 0$, then $\lim_{x\to a} f(x)/g(x)$ does not exist. _____

6. If $\lim_{x\to a} f(x)$ does not exist and $\lim_{x\to a} g(x)$ exists, then $\lim_{x\to a} f(x)g(x)$ does not exist. _____

7. An asymptote is a line that the graph of a function approaches but never crosses. _____

8. If $f(x) = P(x)/Q(x)$ is a rational function and $Q(a) = 0$, then the line $x = a$ is a vertical asymptote. _____

9. The graph of a rational function $f(x) = P(x)/Q(x)$ has a horizontal asymptote only if the degree of $P(x)$ equals the degree of $Q(x)$. _____

10. If $\lim_{x\to a} f(x) = \infty$ and $\lim_{x\to a} g(x) = \infty$, then $\lim_{x\to a} \dfrac{f(x)}{g(x)} = 1.$ _____

11. Any polynomial function is continuous on $(-\infty, \infty)$. _____

12. If $f(x) = x^5 + 3x - 1$, then there exists a number c in $(-2, 2)$ such that $f(c) = 0$. _____

13. If f and g are continuous at 2, then f/g is continuous at 2. _____

14. Suppose $f(x) = [x]$ is the greatest integer function. f is not continuous on the interval $[0, 1]$. _____

15. The graph of a function can have at most two horizontal asymptotes. _____

16. The function $f(x) = \begin{cases} \dfrac{x^2 - 6x + 5}{x - 5}, & x \neq 5 \\ 4, & x = 5 \end{cases}$

is discontinuous at 5. _____

17. The function $f(x) = \begin{cases} \dfrac{\sin x}{x}, & x \neq 0 \\ 1, & x = 0 \end{cases}$

is continuous at 0. _____

18. If a function f is discontinuous at 3, then $f(3)$ is not defined. _____

In Problems 19–32 fill in the blanks.

19. $\lim\limits_{x \to 2} (3x^2 - 4x) = $ _____

20. $\lim\limits_{x \to 3} (5x^2)^0 = $ _____

21. $\lim\limits_{t \to \infty} \dfrac{2t - 1}{3 - 10t} = $ _____

22. $\lim\limits_{x \to -\infty} \dfrac{\sqrt{x^2 + 1}}{2x + 1} = $ _____

23. $\lim\limits_{x \to \_\_\_} \dfrac{1}{x - 3} = -\infty$

24. $\lim\limits_{x \to \_\_\_} (5x + 2) = 22$

25. $\lim\limits_{x \to \_\_\_} x^3 = -\infty$

26. $\lim\limits_{x \to \_\_\_} \dfrac{1}{\sqrt{x}} = \infty$

27. If $f(x) = 2(x - 4)/|x - 4|$, $x \neq 4$, and $f(4) = 9$, then $\lim_{x \to 4^-} f(x) = $ _____.

28. Suppose $x^2 - x^4/3 \leq f(x) \leq x^2$ for all x. Then $\lim_{x \to 0} f(x)/x^2 = $ _____.

29. If f is continuous at a number a and $\lim_{x \to a} f(x) = 10$, then $f(a) = $ _____.

30. $f(x) = \dfrac{2x - 1}{4x^2 - 1}$ is discontinuous at $x = \frac{1}{2}$ because _____

_____.

31. $f(x) = \begin{cases} kx + 1, & x \leq 3 \\ 2 - kx, & x > 3 \end{cases}$ is continuous at 3 if

$k = $ _____.

32. If $\lim_{x \to -5} g(x) = -9$ and $f(x) = x^2$, then $\lim_{x \to -5} f(g(x))$ = _____.

In Problems 33–42 assume $\lim_{x \to a} f(x) = 4$ and $\lim_{x \to a} g(x) = 2$. Find the given limit, if it exists.

33. $\lim\limits_{x \to a} [5f(x) + 6]$

34. $\lim\limits_{x \to a} [f(x)]^3$

35. $\lim\limits_{x \to a} \dfrac{1}{g(x)}$

36. $\lim\limits_{x \to a} \sqrt{\dfrac{f(x)}{g(x)}}$

37. $\lim\limits_{x \to a} \dfrac{f(x)}{g(x) - 2}$

38. $\lim\limits_{x \to a} \dfrac{[f(x)]^2 - 4[g(x)]^2}{f(x) - 2g(x)}$

39. $\lim\limits_{x \to a^+} f(x)$

40. $\lim\limits_{x \to a^-} f(x)g(x)$

41. $\lim\limits_{x \to a} xg(x)$

42. $\lim\limits_{x \to a} \dfrac{6x + 3}{xf(x) + g(x)}$, $a \neq -\dfrac{1}{2}$

In Problems 43 and 44 sketch the graph of the given function. Determine the numbers, if any, at which f is discontinuous.

43. $f(x) = |x| + x$

44. $f(x) = \begin{cases} x + 1, & x < 2 \\ 3 & 2 < x < 4 \\ -x + 7, & x > 4 \end{cases}$

In Problems 45 and 46 determine intervals on which the given function is continuous.

45. $f(x) = \dfrac{\sqrt{4 - x^2}}{x^2 - 4x + 3}$

46. $f(x) = \dfrac{x + 1}{\sqrt{x} \sin x}$

47. In Example 13 of Section 2.2 we proved that

$$\lim\limits_{x \to 0} x^2 \sin \dfrac{1}{x}$$

exists. Does it follow from this that

$$\lim\limits_{x \to 0} x^3 \sin \dfrac{1}{x}$$

exists?

48. Sketch a graph of a function f that satisfies the following conditions:

$$f(0) = 1, \quad f(4) = 0, \quad f(6) = 0, \quad \lim\limits_{x \to 3^-} f(x) = 2,$$

$$\lim\limits_{x \to 3^+} f(x) = \infty, \quad \lim\limits_{x \to -\infty} f(x) = 0, \quad \lim\limits_{x \to \infty} f(x) = 2$$

In Problems 49 and 50 find the vertical and horizontal asymptotes for the graph of f.

49. $f(x) = \dfrac{2x^3 - 7}{(4x^2 - 25)(x + 9)}$

50. $f(x) = \sqrt{\dfrac{2x + 4}{x - 3}}$

Calculator/Computer Problem

51. Use a calculator or computer to investigate $\lim_{x \to 0^+} x^x$. Do you think $\lim_{x \to 0} x^x$ exists?

3

The Derivative

The foundations of calculus are rooted in the analysis of many geometric and physical problems. In Section 3.1 we will study the problems of finding a tangent line to a graph and the problem of finding the velocity of a moving body. These two apparently different problems are actually one and the same. Both solutions lead to the notion of the **instantaneous rate of change** of a function. This is what differential calculus is all about.

3.1 Rate of Change of a Function

3.1.1 Tangent to a Graph

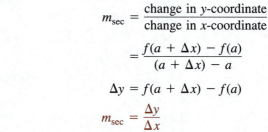

Figure 3.1

Suppose $y = f(x)$ is a continuous function. If, as illustrated in Figure 3.1, the graph of f possesses a tangent line L at a point P, then we would like to find its equation. To do so we need: (a) the coordinates of P and (b) the slope $m_{\tan}$ of L. The coordinates of P pose no difficulty since a point on a graph is obtained by specifying a value of x, say $x = a$, in the domain of f. The coordinates of the point of tangency are $(a, f(a))$.

As a means of *approximating* the slope $m_{\tan}$, we can readily find the slopes of *secant lines* that pass through the fixed point P and any other point Q on the graph. If P has coordinates $(a, f(a))$ and if we let Q have coordinates $(a + \Delta x, f(a + \Delta x))$, then, as shown in Figure 3.2(a), the slope of the secant line through P and Q is

$$m_{\text{sec}} = \frac{\text{change in } y\text{-coordinate}}{\text{change in } x\text{-coordinate}}$$

$$= \frac{f(a + \Delta x) - f(a)}{(a + \Delta x) - a}$$

If

$$\Delta y = f(a + \Delta x) - f(a)$$

then

$$m_{\text{sec}} = \frac{\Delta y}{\Delta x}$$

When the value of Δx is close to zero, either positive or negative, we get points Q and Q' on the graph of f on each side of, but close to, the point P. In turn, we expect that the slopes m_{PQ} and $m_{PQ'}$ are very close to the slope of the tangent line L. See Figure 3.2(b).

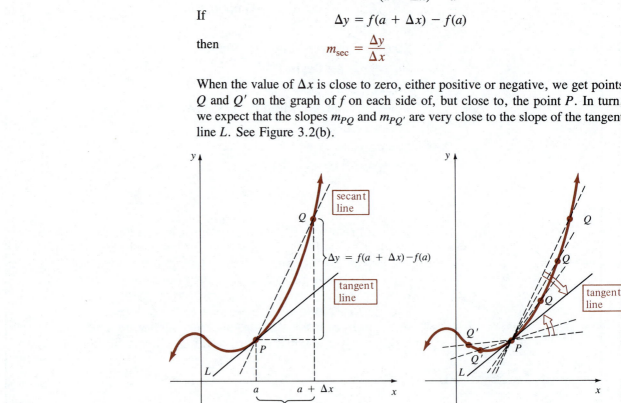

(a)

(b) Secant lines swing toward the tangent line as $Q \to P$ and as $Q' \to P$.

Figure 3.2

_____ **Example 1** _____

Find the slope of the tangent line to the graph of $f(x) = x^2$ at $(1, 1)$.

Solution As a start, let us choose $\Delta x = 0.1$ and find the slope of the secant line through $(1, 1)$ and $(1.1, (1.1)^2)$:

$$(i)\ \ f(1.1) = (1.1)^2$$
$$= 1.21$$

$$(ii)\ \ \Delta y = f(1.1) - f(1)$$
$$= 1.21 - 1$$
$$= 0.21$$

$$(iii)\ \ \frac{\Delta y}{\Delta x} = \frac{0.21}{0.1}$$
$$= 2.1$$

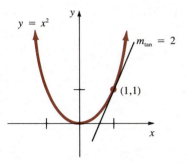

Figure 3.3

The right-most column of the following table should convince you that the slope of the tangent line shown in Figure 3.3 is $m_{\text{tan}} = 2$.

Δx	$1 + \Delta x$	$f(1)$	$f(1 + \Delta x)$	Δy	$\Delta y/\Delta x$
0.1	1.1	1	1.21	0.21	2.1
0.01	1.01	1	1.0201	0.0201	2.01
0.001	1.001	1	1.002001	0.002001	2.001
-0.1	0.9	1	0.81	-0.19	1.9
-0.01	0.99	1	0.9801	-0.0199	1.99
-0.001	0.999	1	0.998001	-0.001999	1.999

On the basis of Figure 3.2(b), Example 1, and our intuition, we are prompted to say that if a graph of a function $y = f(x)$ has a tangent line L at a point P, then L must be the line that is the *limit* of the secants through P and Q as $Q \to P$ and of the secants through P and Q' as $Q' \to P$. Moreover, the slope m_{tan} of L should be the *limiting value* of the values m_{sec} as $\Delta x \to 0$. This is summarized as follows.

DEFINITION 3.1   **Tangent Line**

Let $y = f(x)$ be a continuous function. At a point $(a, f(a))$, the **tangent line** to the graph is the line that passes through the point with slope

$$m_{\text{tan}} = \lim_{\Delta x \to 0} \frac{f(a + \Delta x) - f(a)}{\Delta x}$$

$$= \lim_{\Delta x \to 0} \frac{\Delta y}{\Delta x} \tag{3.1}$$

whenever the limit exists.

The slope of the tangent line at $(a, f(a))$ is also called the **slope of the curve** at the point. Definition 3.1 implies that a tangent at $(a, f(a))$ is *unique* since a point and a slope determine a single line.

We synthesize the application of Definition 3.1 to four steps.

(*i*)   Evaluate f at a and $a + \Delta x$: $f(a)$ and $f(a + \Delta x)$

(*ii*) Compute Δy: $\Delta y = f(a + \Delta x) - f(a)$

(*iii*) Divide Δy by $\Delta x \neq 0$: $\dfrac{\Delta y}{\Delta x} = \dfrac{f(a + \Delta x) - f(a)}{\Delta x}$

(*iv*) Compute the limit as $\Delta x \to 0$: $m_{\tan} = \lim\limits_{\Delta x \to 0} \dfrac{\Delta y}{\Delta x}$

_____ **Example 2** _____

Use Definition 3.1 to find the slope of the tangent line to the graph of $f(x) = x^2$ at $(1, f(1))$.

Solution

(*i*) $f(1) = 1^2 = 1$. For any $\Delta x \neq 0$,

$$f(1 + \Delta x) = (1 + \Delta x)^2$$
$$= 1 + 2\,\Delta x + (\Delta x)^2$$

(*ii*) $\Delta y = f(1 + \Delta x) - f(1)$
$$= [1 + 2\,\Delta x + (\Delta x)^2] - 1$$
$$= 2\,\Delta x + (\Delta x)^2$$
$$= \Delta x(2 + \Delta x)$$

(*iii*) $\dfrac{\Delta y}{\Delta x} = \dfrac{\Delta x(2 + \Delta x)}{\Delta x}$
$$= 2 + \Delta x$$

Thus, the slope of the tangent at $(1, f(1))$ is given by

(*iv*) $m_{\tan} = \lim\limits_{\Delta x \to 0} \dfrac{\Delta y}{\Delta x}$
$$= \lim_{\Delta x \to 0} (2 + \Delta x) = 2$$

_____ **Example 3** _____

Find the slope of the tangent line to the graph of $f(x) = 5x + 6$ at any point $(a, f(a))$.

Solution

(i) $f(a) = 5a + 6$. For any $\Delta x \neq 0$,

$$f(a + \Delta x) = 5(a + \Delta x) + 6 = 5a + 5\,\Delta x + 6$$

(ii) $\quad y = f(a + \Delta x) - f(a)$
$$= [5a + 5\,\Delta x + 6] - [5a + 6]$$
$$= 5\,\Delta x$$

(iii) $\quad \dfrac{\Delta y}{\Delta x} = \dfrac{5\,\Delta x}{\Delta x} = 5$

Thus, at any point on the graph of $f(x) = 5x + 6$, we have

(iv) $\quad m_{\tan} = \lim\limits_{\Delta x \to 0} \dfrac{\Delta y}{\Delta x}$
$$= \lim\limits_{\Delta x \to 0} 5 = 5$$

You should be able to explain why the answer in Example 3 is no surprise.

_____ **Example 4** _____

Find an equation of the tangent line to the graph of $f(x) = x^2$ at $(1, 1)$.

Solution   From Example 2 the slope of the tangent at $(1, 1)$ is $m_{\tan} = 2$. The point–slope form of a line then gives

$$y - 1 = 2(x - 1) \quad \text{or} \quad y = 2x - 1$$

_____ **Example 5** _____

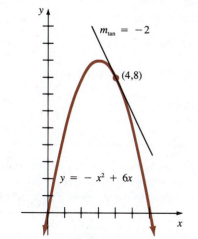

Figure 3.4

Find the slope of the tangent line to the graph of $f(x) = -x^2 + 6x$ at $(4, f(4))$.

Solution

(i) $f(4) = -4^2 + 6(4) = 8$. For any $\Delta x \neq 0$,

$$f(4 + \Delta x) = -(4 + \Delta x)^2 + 6(4 + \Delta x)$$
$$= -16 - 8\,\Delta x - (\Delta x)^2 + 24 + 6\,\Delta x$$
$$= 8 - 2\,\Delta x - (\Delta x)^2$$

(ii) $\Delta y = f(4 + \Delta x) - f(4)$
$$= [8 - 2\,\Delta x - (\Delta x)^2] - 8$$
$$= -2\,\Delta x - (\Delta x)^2$$
$$= \Delta x(-2 - \Delta x)$$

$$(iii) \quad \frac{\Delta y}{\Delta x} = \frac{\Delta x(-2 - \Delta x)}{\Delta x}$$

$$= -2 - \Delta x$$

$$(iv) \quad m_{\text{tan}} = \lim_{\Delta x \to 0} (-2 - \Delta x) = -2$$

Figure 3.4 shows the graph of f and the tangent line with slope -2 at $(4, 8)$.

Vertical Tangents The limit (3.1) can fail to exist for a function f at a number a and yet there may be a tangent at $(a, f(a))$. The tangent line to a graph at a point can be **vertical**, in which case its slope is undefined. We will consider the concept of a vertical tangent in Section 3.2.

Example 6

Although we shall not pursue the details at this time, it can be shown that the graph of $y = x^{1/3}$ possesses a vertical tangent line at the origin. In Figure 3.5 we see that the y-axis, that is, the line $x = 0$, is tangent to the graph at the point $(0, 0)$.

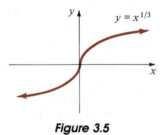

Figure 3.5

A Tangent May Not Exist The graph of a function f *will not* have a tangent line at a point whenever:

(*i*) f is discontinuous at $x = a$, or

(*ii*) the graph of f has a corner at $(a, f(a))$.

Moreover, the graph of f *may not* have a tangent line at a point where

(*iii*) the graph has a sharp peak.

Example 7

Figures 3.6(a) and 3.6(b) show the graphs of two functions that are discontinuous, although defined, at $x = a$. Neither graph has a tangent at $(a, f(a))$. In Figure 3.6(c) we would expect to be able to find a tangent to the graph of f everywhere except at $x = a$.

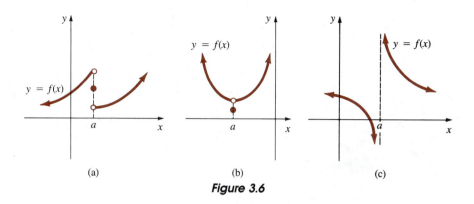

(a)   (b)   (c)

Figure 3.6

Figure 3.7 shows what may "go wrong" at a sharp peak or corner in the graph of a function f. The secant lines that pass through P and Q approach L_2 as $Q \to P$ but the secant lines through P and Q' approach a different line, L_1, as $Q' \to P$. Keep in mind, *when we say a tangent line exists at a point P on the graph of a function, we mean that there is only one tangent line.*

In the next example, you should try to visualize the behavior of the secant lines on either side of, but close to, the indicated corner or peak.

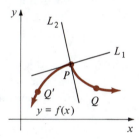

Figure 3.7

Example 8

Figure 3.8(a) shows a graph of a function at which the tangent line does not exist at two corners. In Figure 3.8(b) a tangent line does not exist at the peak. However, in Figure 3.8(c) the graph has a peak at which a vertical tangent exists.

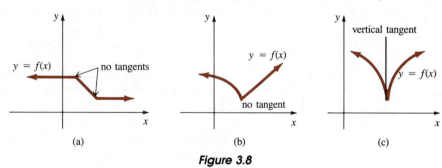

(a)   (b)   (c)

Figure 3.8

Example 9

y = |x|

Figure 3.9

Show that the graph of $f(x) = |x|$ does not have a tangent at $(0, 0)$.

Solution An inspection of the graph of f in Figure 3.9 reveals a corner at the origin. To prove that the tangent does not exist at $(0, 0)$, we must examine

$$\lim_{\Delta x \to 0} \frac{f(0 + \Delta x) - f(x)}{\Delta x} = \lim_{\Delta x \to 0} \frac{|0 + \Delta x| - |0|}{\Delta x}$$

$$= \lim_{\Delta x \to 0} \frac{|\Delta x|}{\Delta x} \qquad (3.2)$$

Now, for $\Delta x > 0$,

$$\frac{|\Delta x|}{\Delta x} = \frac{\Delta x}{\Delta x} = 1$$

whereas for $\Delta x < 0$,

$$\frac{|\Delta x|}{\Delta x} = \frac{-\Delta x}{\Delta x} = -1$$

Since $\lim_{\Delta x \to 0^+} |\Delta x|/\Delta x = 1$ and $\lim_{\Delta x \to 0^-} |\Delta x|/\Delta x = -1$, we conclude that the limit (3.2) does not exist. Thus, the graph possesses no tangent at $(0, 0)$.

Rate of Change

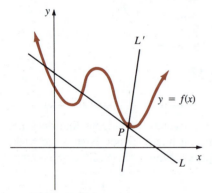

y = f(x)

Figure 3.10

The slope $\Delta y/\Delta x$ of a secant line through $(a, f(a))$ is also called the **average rate of change** of f at a. The slope $m_{\text{tan}} = \lim_{\Delta x \to 0} \Delta y/\Delta x$ is said to be the **instantaneous rate of change** of the function at a; for example, if $m_{\text{tan}} = \frac{1}{10}$ at a point $(a, f(a))$, we would not expect the values of f to change drastically for x-values near a.

Remark

You may recall from plane geometry that a tangent to a circle at a point P is a line that intersects, or touches, its graph only at the single point P. Figure 3.10 shows that this concept does not extend to the graph of a function. The line L is tangent at P but intersects the graph of f at three points. The line L' intersects the graph of f only at P but is not tangent to the graph.

3.1.2 Instantaneous Velocity

Almost everyone has an intuitive notion of speed or velocity as a rate at which a distance is covered in a certain length of time. When, say, a bus travels 60 miles in one hour, the *average velocity* of the bus must have been 60 mi/hr. Of course, it is difficult to maintain the rate of 60 mi/hr for the entire trip because the bus slows down for towns and speeds up when it passes cars. In other words, the velocity changes with time. If a bus company's schedule demands that the bus travel the 60 miles from one town to another in one hour, the driver knows instinctively that he must compensate for velocities or speeds below 60 mi/hr by traveling at speeds greater than this at other points in the journey. Knowing that the average velocity is 60 mi/hr does not, however, answer the question: What is the velocity of the bus at a particular instant?*

Average Velocity   In general, the **average velocity** or **average speed** of a moving object is the time rate of change of position defined by

$$v_{ave} = \frac{\text{distance traveled}}{\text{time of travel}} \tag{3.3}$$

Consider a runner who finishes a 10-km race in an elapsed time of 1 hr 15 min (1.25 hr). The runner's average velocity or average speed for the race was

$$v_{ave} = \frac{10}{1.25} = 8 \text{ km/hr}$$

But suppose we now wish to determine the runner's *exact* velocity v at the instant the runner is one-half hour into the race. If the distance run in the time interval from 0 hr to 0.5 hr is measured to be 5 km, then

$$v_{ave} = \frac{5}{0.5} = 10 \text{ km/hr}$$

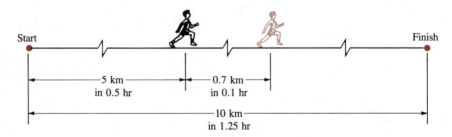

Figure 3.11

Again, this number is not a measure, or necessarily even a good indicator, of the instantaneous rate v at which the runner is moving 0.5 hr into the race. If we determine that at 0.6 hr the runner is 5.7 km from the starting line, then the average velocity from 0 hr to 0.6 hr is $v_{ave} = 5.7/0.6 = 9.5$ km/hr. However, during the time interval from 0.5 hr to 0.6 hr

$$v_{ave} = \frac{5.7 - 5}{0.6 - 0.5} = 7 \text{ km/hr}$$

The latter number is a more realistic measure of the rate v. See Figure 3.11. By "shrinking" the time interval between 0.5 hr and the time that corresponds to a measured position close to 5 km, we expect to obtain even better approximations to the runner's velocity at time 0.5 hr.

---

*The bus driver need only look at the speedometer and observe; for example, "I am now going 45 mi/hr." This may not be obvious to an observer who watches the bus move down the road.

Rectilinear Motion

To generalize the preceding concepts, let us suppose an object, or particle, at point P moves along either a vertical or horizontal coordinate line as shown in Figure 3.12. Furthermore, let the particle move in such a manner that its position, or coordinate, on the line is given by a function $s = f(t)$, where t represents time. The values of s are directed distances measured from O in units such as centimeters, meters, feet, miles, and so on. When P is to the right or above O, we take $s > 0$, whereas $s < 0$ when P is to the left or below O. Motion in a straight line is called **rectilinear motion**.

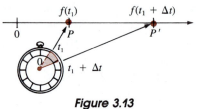

Figure 3.12

If a particle is at point P at time t_1 and at P' at time $t_1 + \Delta t$, then the coordinates of the points, shown in Figure 3.13, are $f(t_1)$ and $f(t_1 + \Delta t)$. By (3.3) the average velocity of the particle in the time interval $[t_1, t_1 + \Delta t]$ is

$$v_{ave} = \frac{\text{change in position}}{\text{change in time}}$$

$$= \frac{f(t_1 + \Delta t) - f(t_1)}{\Delta t}$$

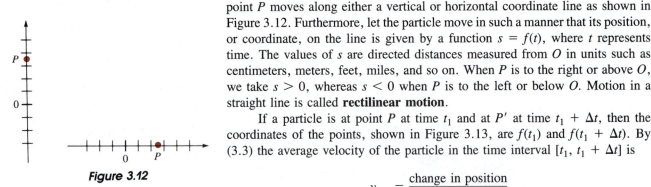

Figure 3.13

or $\qquad\qquad v_{ave} = \frac{\Delta s}{\Delta t} \qquad\qquad\qquad (3.4)$

This suggests that the limit of (3.4) as $\Delta t \to 0$ would give the **instantaneous rate of change** of $f(t)$ at t_1 or the **instantaneous velocity**.

DEFINITION 3.2   Instantaneous Velocity

Let $s = f(t)$ be a function that gives the position of an object moving in a straight line. The **instantaneous velocity** at time t_1 is

$$v(t_1) = \lim_{\Delta t \to 0} \frac{f(t_1 + \Delta t) - f(t_1)}{\Delta t}$$

$$= \lim_{\Delta t \to 0} \frac{\Delta s}{\Delta t} \qquad\qquad (3.5)$$

whenever the limit exists.

Note: Except for notation and interpretation, there is no mathematical difference between (3.1) and (3.5).

Example 10 _____

The height s above ground of a ball dropped from the top of the St. Louis Gateway Arch is given by $s = -4.9t^2 + 192$, where s is measured in meters and t in seconds (See Figure 3.14).* Find the instantaneous velocity of the falling ball at $t_1 = 3$ sec.

---

*This is not a made-up formula; the height of the Arch is 630 ft or 192 m. We shall see how the formula is derived in Section 6.8.

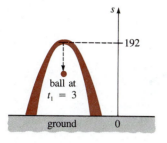

Figure 3.14

Solution We use the same four-step procedure as in the earlier examples.

(*i*) $f(3) = -4.9(9) + 192 = 147.9$. For any $\Delta t \neq 0$,

$$f(3 + \Delta t) = -4.9(3 + \Delta t)^2 + 192$$
$$= -4.9(\Delta t)^2 - 29.4\,\Delta t + 147.9$$

(*ii*) $\Delta s = f(3 + \Delta t) - f(3)$
$$= [-4.9(\Delta t)^2 - 29.4\,\Delta t + 147.9] - 147.9$$
$$= -4.9(\Delta t)^2 - 29.4\,\Delta t$$
$$= \Delta t(-4.9\,\Delta t - 29.4)$$

(*iii*) $\dfrac{\Delta s}{\Delta t} = \dfrac{\Delta t(-4.9\,\Delta t - 29.4)}{\Delta t}$
$$= -4.9\,\Delta t - 29.4$$

(*iv*) $v(3) = \lim\limits_{\Delta t \to 0} \dfrac{\Delta s}{\Delta t}$
$$= \lim\limits_{\Delta t \to 0} (-4.9\,\Delta t - 29.4) = -29.4 \text{ m/sec}$$

The minus sign is significant because the ball is moving opposite to the positive or upward direction. The number $f(3) = 147.9$ m is the height of the ball above the ground at 3 seconds.

We shall study rectilinear motion in greater detail in Section 4.1.

Remark

You have probably noticed that the goal in utilizing (3.1) and (3.5) is to cancel Δx and Δt in the quotients $\Delta y/\Delta x$ and $\Delta s/\Delta t$ *before* passing to the limit. If you are not able to do this in any of the following problems, you should double-check your algebra.

Exercises 3.1

Answers to odd-numbered problems begin on page A-20.

[3.1.1]

In Problems 1–6 sketch the graph of the function and the tangent line at the given point. Find the slope of the secant line through the points that correspond to the indicated values of x.

1. $f(x) = -x^2 + 9$, $(2, 5)$; $x = 2$, $x = 2.5$

2. $f(x) = x^2 + 4x$, $(0, 0)$; $x = -1/4$, $x = 0$

3. $f(x) = x^3$, $(-2, -8)$; $x = -2$, $x = -1$

4. $f(x) = 1/x$, $(1, 1)$; $x = 0.9$, $x = 1$

5. $f(x) = \sin x$, $(\pi/2, 1)$; $x = \pi/2$, $x = 2\pi/3$

6. $f(x) = \cos x$, $(-\pi/3, 1/2)$; $x = -\pi/2$, $x = -\pi/3$

In Problems 7–16 find the slope of the tangent line to the graph of the given function at the indicated point.

7. $f(x) = 2x - 1$; $(4, 7)$

8. $f(x) = -\dfrac{1}{2}x + 3$; $(a, f(a))$

9. $f(x) = x^2$; $(3, 9)$

10. $f(x) = x^2 + 4$; $(-1, 5)$

11. $f(x) = 2x^2 + 8x$; $(0, 0)$

12. $f(x) = x^2 - 5x + 4$; $(2, -2)$

13. $f(x) = x^3$; $(1, f(1))$

14. $f(x) = -x^3 + x^2$; $(2, f(2))$

15. $f(x) = 1/x$; $(1/3, f(1/3))$

16. $f(x) = 1/(x - 1)^2$; $(0, f(0))$

17. Find the slope of the tangent line to the graph of $f(x) = 1/x^2$ at the point $(x_0, f(x_0))$. What is the slope of the tangent line at the point where $x_0 = -2$? Where $x_0 = \frac{1}{3}$?

18. Find the slope of the tangent line to the graph of $f(x) = 5x^2 - x + 2$ at the point $(x, f(x))$. What is the slope of the tangent line at the point where $x = \frac{3}{5}$? Where $x = 4$?

In Problems 19–24 find an equation of the tangent line to the graph of the given function at the indicated value of x.

19. $f(x) = 1 - x^2$; $x = 5$

20. $f(x) = (x + 2)^2$; $x = -3$

21. $f(x) = (x - 1)^4 + 6x$; $x = 1$

22. $f(x) = 4x + 10$; $x = 7$

23. $f(x) = \dfrac{1}{1 + 2x}$; $x = 0$

24. $f(x) = 4 - \dfrac{8}{x}$; $x = -1$

In Problems 25 and 26 find the average rate of change of the given function on the indicated interval.

25. $f(x) = x^3 + 2x^2 - 4x$; $[-1, 2]$

26. $f(x) = \cos x$; $[-\pi, \pi]$

In Problems 27–30 (Figures 3.15–3.18) specify the values of x for which the tangent line to the graph is possibly horizontal, vertical, or does not exist.

27.

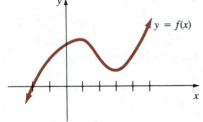

Figure 3.15

28.

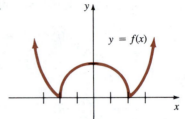

Figure 3.16

29.

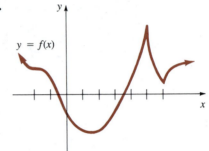

Figure 3.17

30.

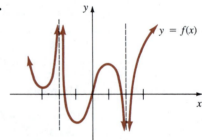

Figure 3.18

In Problems 31 and 32 (Figures 3.19 and 3.20) determine whether the line(s) through the given point(s) is (are) tangent to the graph of $f(x) = x^2$ at the point(s) indicated in color.

31.

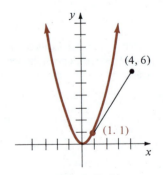

Figure 3.19

32.

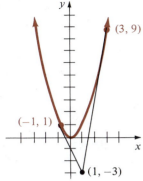

Figure 3.20

Miscellaneous Problems

33. Let $y = f(x)$ be an even function whose graph possesses a tangent line with slope m at (x_0, y_0). Show that the slope of the tangent at $(-x_0, y_0)$ is $-m$. [*Hint*: $f(-x_0 + \Delta x) = f(x_0 - \Delta x)$.]

34. Let $y = f(x)$ be an odd function whose graph possesses a tangent line with slope m at (x_0, y_0). Show that the slope of the tangent at $(-x_0, y_0)$ is m.

Calculator Problems

35. Consider the function $f(x) = \sqrt{3x + 1}$.

(a) What is $\Delta y / \Delta x$ at $x = 0$?

(b) Fill in the following table using values with four decimal places. Conjecture the slope of the tangent to the graph of f at $x = 0$.

Δx	Δy	$\Delta y / \Delta x$
0.2		
0.1		
0.01		
-0.2		
-0.1		
-0.01		

36. Consider the function $f(x) = \tan x$.

(a) Compute the quotient $(\tan 0.1)/0.1$.

(b) At what point on the graph of f is the number in part **(a)** an approximation to the slope of the tangent line?

[3.1.2]

37. A car travels the 290 mi between Los Angeles and Las Vegas in 5 hr. What is its average speed?

38. Two marks on a highway are $\frac{1}{2}$ mi apart. A Highway Patrol plane observes that a car traverses the distance between the marks in 40 sec. Will the car be stopped for speeding? (Assume the speed limit is 55 mph.)

39. A jet airplane averages 920 km/hr to fly the 3500 km between Hawaii and San Francisco. How many hours does the flight take?

40. A marathon race is run over a straight 26-mi course. The race begins at noon. At 1:30 P.M. a contestant passes the 10-mi mark and at 3:10 P.M. the contestant passes the 20-mi mark. What is the contestant's average running speed between 1:30 P.M. and 3:10 P.M.?

In Problems 41 and 42 the position of a particle on a horizontal coordinate line is given by the function. Find the instantaneous velocity of the particle at the indicated time.

41. $f(t) = -4t^2 + 10t + 6$; $t = 3$

42. $f(t) = t^2 + \dfrac{1}{5t + 1}$; $t = 0$

43. The height above ground of a ball dropped from an initial altitude of 122.5 m is given by $s(t) = 122.5 - 4.9t^2$, where s is measured in meters and t in seconds.

(a) What is the instantaneous velocity at $t = \frac{1}{2}$?

(b) At what time does the ball hit the ground?

(c) What is the impact velocity?

44. The height of a projectile shot from ground level is given by $s(t) = -16t^2 + 256t$, where s is measured in feet and t in seconds.

(a) Determine the height of the projectile at $t = 2$, $t = 6$, $t = 9$, and $t = 10$.

(b) What is the average velocity of the projectile between $t = 2$ and $t = 5$.

(c) Show that the average velocity between $t = 7$ and $t = 9$ is zero. Interpret physically.

(d) At what time does the projectile hit the ground?

(e) Determine the instantaneous velocity at time t_1.

(f) What is the impact velocity?

(g) What is the maximum height that the projectile attains?

3.2 The Derivative

In the last section we saw that if a graph of a function $y = f(x)$ possesses a tangent at a point $(a, f(a))$, then the slope of this tangent is

$$m_{\text{tan}} = \lim_{\Delta x \to 0} \frac{f(a + \Delta x) - f(a)}{\Delta x}$$

For a given function it is usually possible to obtain a general formula, or rule, that gives the value of the slope of a tangent line. This is accomplished by computing

$$\lim_{\Delta x \to 0} \frac{f(x + \Delta x) - f(x)}{\Delta x} \tag{3.6}$$

for *any* x (for which the limit exists). We then substitute a value of x *after* the limit has been found. The limit (3.6) is said to be the **derivative** of f and is denoted by f'.

DEFINITION 3.3   **Derivative of a Function**

The **derivative** of a function $y = f(x)$ with respect to x is

$$f'(x) = \lim_{\Delta x \to 0} \frac{f(x + \Delta x) - f(x)}{\Delta x}$$

$$= \lim_{\Delta x \to 0} \frac{\Delta y}{\Delta x} \tag{3.7}$$

whenever this limit exists.

The derivative $f'(x)$ is also said to be the **instantaneous rate of change** of the function $y = f(x)$ with respect to the variable x.

Let us now reconsider Examples 1 and 2 of Section 3.1.

Example 1

Find the derivative of $f(x) = x^2$.

Solution As before, the process consists of four steps:

$$(i)\ \ f(x + \Delta x) = (x + \Delta x)^2$$
$$= x^2 + 2x\,\Delta x + (\Delta x)^2$$

$$(ii)\ \ \Delta y = f(x + \Delta x) - f(x)$$
$$= [x^2 + 2x\,\Delta x + (\Delta x)^2] - x^2$$
$$= \Delta x[2x + \Delta x]$$

$$(iii)\ \ \frac{\Delta y}{\Delta x} = \frac{\Delta x[2x + \Delta x]}{\Delta x}$$
$$= 2x + \Delta x$$

$$(iv) \quad f'(x) = \lim_{\Delta x \to 0} \frac{\Delta y}{\Delta x}$$

$$= \lim_{\Delta x \to 0} [2x + \Delta x] = 2x$$

In the last example, $f'(x) = 2x$ is another function of x that is *derived* (whence the name derivative) from the original function. Also, observe that the result in Example 2 of Section 3.1 is obtained by evaluating $f'(x)$ *at $x = 1$.*

Example 2

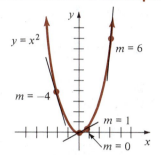

$y = x^2$

$m = 6$

$m = -4$

$m = 1$

$m = 0$

Figure 3.21

For $f(x) = x^2$, find $f'(-2)$, $f'(0)$, $f'(\frac{1}{2})$, and $f'(3)$. Interpret.

Solution   From Example 1, we know $f'(x) = 2x$. Hence,

$$f'(-2) = -4, \quad f'(0) = 0, \quad f'(\tfrac{1}{2}) = 1, \quad \text{and } f'(3) = 6$$

As shown in Figure 3.21, these values represent, in turn, the slopes of the tangent lines to the graph of $y = x^2$ at the points

$$(-2, 4), \quad (0, 0), \quad (\tfrac{1}{2}, \tfrac{1}{4}), \quad \text{and} \quad (3, 9)$$

Example 3

Find the derivative of $f(x) = -x^2 + 4x + 1$.

Solution   In this case you should be able to show that

$$\Delta y = f(x + \Delta x) - f(x) = \Delta x[-2x - \Delta x + 4]$$

Therefore, $$f'(x) = \lim_{\Delta x \to 0} \frac{\Delta y}{\Delta x}$$

$$= \lim_{\Delta x \to 0} \frac{\Delta x[-2x - \Delta x + 4]}{\Delta x}$$

$$= \lim_{\Delta x \to 0} [-2x - \Delta x + 4] = -2x + 4$$

Example 4

Find the derivative of $f(x) = x^3$.

Solution   To calculate $f(x + \Delta x)$, we use the Binomial Theorem.

$$(i) \quad f(x + \Delta x) = (x + \Delta x)^3$$

$$= x^3 + 3x^2\,\Delta x + 3x(\Delta x)^2 + (\Delta x)^3$$

$$(ii) \quad \Delta y = f(x + \Delta x) - f(x)$$

$$= [x^3 + 3x^2\,\Delta x + 3x(\Delta x)^2 + (\Delta x)^3] - x^3$$

$$= \Delta x[3x^2 + 3x\,\Delta x + (\Delta x)^2]$$

$$(iii) \quad \frac{\Delta y}{\Delta x} = \frac{\Delta x[3x^2 + 3x\,\Delta x + (\Delta x)^2]}{\Delta x}$$

$$= 3x^2 + 3x\,\Delta x + (\Delta x)^2$$

$$(iv) \quad f'(x) = \lim_{\Delta x \to 0} \frac{\Delta y}{\Delta x}$$

$$= \lim_{\Delta x \to 0} [3x^2 + 3x\,\Delta x + (\Delta x)^2] = 3x^2$$

Example 5

Find an equation of the tangent line to the graph of $f(x) = x^3$ at $x = \frac{1}{2}$.

Solution   Since we have just seen $f'(x) = 3x^2$, it follows that the slope of the tangent at $x = \frac{1}{2}$ is

$$f'\!\left(\frac{1}{2}\right) = 3\!\left(\frac{1}{2}\right)^2 = \frac{3}{4}$$

The y-coordinate of the point of tangency is $f(\frac{1}{2}) = (\frac{1}{2})^3 = \frac{1}{8}$. Thus, at $(\frac{1}{2}, \frac{1}{8})$ an equation of the tangent line is given by

$$y - \frac{1}{8} = \frac{3}{4}\!\left(x - \frac{1}{2}\right) \quad \text{or} \quad y = \frac{3}{4}x - \frac{1}{4}$$

The graph of the function and the tangent line are given in Figure 3.22.

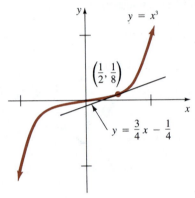

Figure 3.22

Symbols   The following is a list of some of the **symbols** used throughout mathematical literature to denote the derivative of a function:

$$f'(x), \quad \frac{dy}{dx}, \quad y', \quad Dy, \quad D_x y$$

The second symbol, obviously, has its origins in the Δy and Δx notation: $dy/dx = \lim_{\Delta x \to 0} \Delta y/\Delta x$. In this text we shall employ the first three symbols.* (The footnote is on page 119.) For a function such as $f(x) = x^2$, we write $f'(x) = 2x$; if the same function is written $y = x^2$, we then utilize $dy/dx = 2x$ or $y' = 2x$.

Value of a Derivative   At a specified number a, the **value of the derivative** is denoted by

$$f'(a), \quad \frac{dy}{dx}\bigg|_{x=a}, \quad \text{or} \quad y'(a)$$

Example 6

The value of the derivative of $y = x^2$ at 5 is

$$\frac{dy}{dx}\bigg|_{x=5} = 2x\,\bigg|_{x=5} = 10$$

Operation   The process of finding a derivative is called **differentiation**. If $f'(a)$ exists, the function f is said to be **differentiable** at a. The **operation** of differentiation with respect to the variable x is represented by the symbol d/dx; for example, we have seen

$$\frac{d}{dx}x^2 = 2x \quad \text{and} \quad \frac{d}{dx}(-x^2 + 4x + 1) = -2x + 4$$

Horizontal Tangents   If $y = f(x)$ is continuous at a number a and $f'(a) = 0$, then the tangent line at $(a, f(a))$ is said to be **horizontal**. In Example 2 we saw that for the continuous function $f(x) = x^2$, $f'(x) = 2x$, so that $f'(0) = 0$. Thus, the tangent line to the graph is horizontal at $(0, f(0))$ or $(0, 0)$. In Example 3 we saw that $f'(x) = -2x + 4$ for the continuous function $f(x) = -x^2 + 4x + 1$. Observe in this latter case that $f'(x) = 0$ when $-2x + 4 = 0$ or $x = 2$. There is a horizontal tangent at $(2, f(2))$ or $(2, 5)$.

Domain of f'   The domain of f', defined by (3.7), is the set of numbers x for which the limit exists. A derivative fails to exist at a number a for the same reasons a tangent to its graph fails to exist:

    (*i*)   the function is discontinuous at a, or

    (*ii*)   the graph could have a sharp peak or corner at $(a, f(a))$.

Isaac Newton

G. W. Leibniz

*Isaac Newton (1642–1727)   Newton used the notation $\dot{y}$ to represent a *fluxion*, or derivative of a function. This symbol never achieved overwhelming popularity among mathematicians and is used today primarily by physicists. For typographical reasons, the so-called fly-speck notation has been superseded by the prime notation.

    It is acknowledged that Newton, an English mathematician and physicist, was the first to set forth many of the basic principles of calculus in unpublished manuscripts on the *method of fluxions*, dated 1665. The word *fluxion* originated from the concept of quantities that "flow"—that is, quantities that change at a certain rate. Newton attained everlasting fame with the publication of his law of universal gravitation in his monumental treatise *Philosophiae Naturalis Principia Mathematica* in 1687. Newton was also the first to prove, using the calculus and his law of gravitation, Kepler's three empirical laws of planetary motion and was the first to prove that white light was composed of all colors. He was elected to Parliament, appointed Warden of the Mint, and was knighted in 1705. Sir Isaac Newton said about his many accomplishments: "If I have seen farther than others, it is by standing on shoulders of giants."

Gottfried Wilhelm Leibniz (1646–1716)   The dy/dx notation for a derivative of a function is due to Leibniz. In fact, it was Leibniz who introduced the word *function* into mathematical literature. A German mathematician, lawyer, and philosopher, Leibniz published a short version of his calculus in an article in a periodical journal in 1684. But, since it was well known that Newton's manuscripts on the *method of fluxions* dated from 1665, Leibniz was accused of appropriating his ideas from these unpublished works. Fueled by nationalistic prides, a controversy about who was the first to "invent" calculus raged for many years. Historians now agree that both Leibniz and Newton arrived at many of the major premises of calculus independently of one another. Leibniz and Newton are considered the "co-inventors" of the subject.

In addition, since the derivative gives slope, f' will fail to exist

(*iii*)   at a point $(a, f(a))$ at which the tangent line to the graph is vertical.

The domain of f' is necessarily a subset of the domain of f.

___ **Example 7** ___

(*a*)   The function $f(x) = 1/(2x - 1)$ is discontinuous at $\frac{1}{2}$. Therefore, f is not differentiable at this value.

(*b*)   The graph of $f(x) = |x|$ possesses a sharp corner at the origin. In Example 9 of Section 3.1, we saw that $\lim_{\Delta x \to 0} \Delta y / \Delta x$ failed to exist at 0 and so f is not differentiable there.

Vertical Tangents   Let $y = f(x)$ be continuous at a number a. If $\lim_{x \to a} |f'(x)| = \infty$, then the graph of f is said to have a **vertical tangent** at $(a, f(a))$. In Example 6 of Section 3.1 we mentioned the fact that the graph of $y = x^{1/3}$ possesses a vertical tangent line at $(0, 0)$. We prove this assertion in the next example.

___ **Example 8** ___

It is left as an exercise to use (3.7) to prove that the derivative of $f(x) = x^{1/3}$ is given by

$$f'(x) = \frac{1}{3x^{2/3}}$$

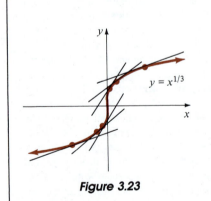

$y = x^{1/3}$

Figure 3.23

(See Problem 28 of Exercises 3.2.) Although f is continuous at 0, it is clear that f' is not defined at that value. In other words, f is not differentiable at 0. However, since

$$\lim_{x \to 0^+} f'(x) = \infty \quad \text{and} \quad \lim_{x \to 0^-} f'(x) = \infty$$

we have $|f'(x)| \to \infty$ as $x \to 0$. This is sufficient to say that there is a tangent line at $(0, f(0))$ or $(0, 0)$ and that it is vertical.

Figure 3.23 shows that the tangent lines to the graph on either side of the origin become steeper and steeper as $x \to 0$.

Differentiability on an Interval   A function f is said to be **differentiable**

(*i*)   **on an open interval** (a, b) when f' exists for every number in the interval,

(*ii*)   **on a closed interval** $[a, b]$ when f is differentiable on (a, b) and

$$f'_+(a) = \lim_{\Delta x \to 0^+} \frac{f(a + \Delta x) - f(a)}{\Delta x}, \qquad f'_-(b) = \lim_{\Delta x \to 0^-} \frac{f(b + \Delta x) - f(b)}{\Delta x} \qquad (3.8)$$

both exist.

The limits in (3.8) are called **right-hand** and **left-hand derivatives**, respectively. A function is differentiable on $[a, \infty)$ when it is differentiable on (a, ∞) and has a right-hand derivative at a. A similar definition in terms of a left-hand derivative holds for differentiability on $(-\infty, a]$. Moreover, it can be shown that

f is differentiable at a number c on an interval (a, b) if and only if $f'_+(c) = f'_-(c)$.

Example 9

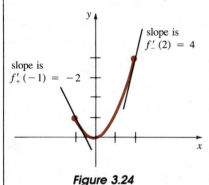

Figure 3.24

The function

$$f(x) = x^2, \qquad -1 \le x \le 2$$

is differentiable on $[-1, 2]$ since $f'(x) = 2x$ for every number x in $(-1, 2)$ and

$$f'_-(2) = 4 \quad \text{and} \quad f'_+(-1) = -2$$

See Figure 3.24.

Example 10

The function $f(x) = x^2$ is differentiable on $(-\infty, \infty)$.

Example 11

Since $f(x) = 1/x$ is discontinuous at $x = 0$, f is not differentiable on any interval containing 0.

Example 12

Show that the function

$$f(x) = \begin{cases} 2, & x < 1 \\ x + 1, & x \ge 1 \end{cases}$$

is not differentiable at $x = 1$.

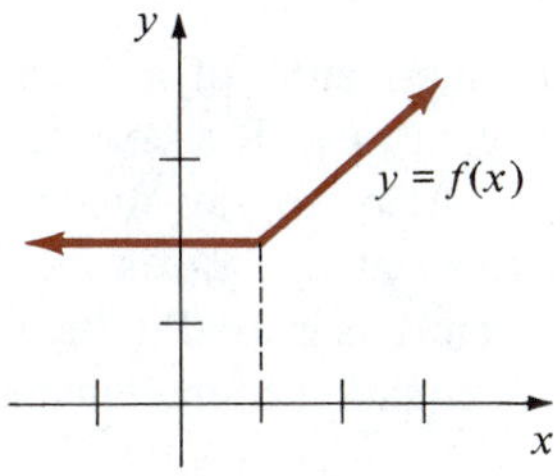

Solution   Figure 3.25 shows that the graph of f has a sharp corner at $(1, 2)$. Now,

$$f'_+(1) = \lim_{\Delta x \to 0^+} \frac{f(1 + \Delta x) - f(1)}{\Delta x} = \lim_{\Delta x \to 0^+} \frac{[(1 + \Delta x) + 1] - 2}{\Delta x} = 1$$

Figure 3.25

whereas

$$f'_-(1) = \lim_{\Delta x \to 0^-} \frac{f(1 + \Delta x) - f(1)}{\Delta x} = \lim_{\Delta x \to 0^-} \frac{2 - 2}{\Delta x} = 0$$

Since $f'_+(1) \neq f'_-(1)$, f is not differentiable at 1. Consequently, f is not differentiable on any interval that contains 1.

___

Example 13

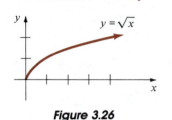

Figure 3.26

It is left as an exercise to show that the function $f(x) = \sqrt{x}$ is not differentiable on the interval $[0, \infty)$. (See Problem 27, Exercises 3.2.) The graph of f given in Figure 3.26 should provide you with a clue as to why $f'_+(0)$ does not exist. The function $f(x) = \sqrt{x}$ is, however, differentiable on $(0, \infty)$ and on any subinterval of $(0, \infty)$.

We have seen that $f(x) = |x|$ and $f(x) = x^{1/3}$ are continuous at 0 but not differentiable at that number. In Example 12 the function f is continuous at 1 but $f'(1)$ does not exist. In other words,

continuity does not imply differentiability

However, if the graph of a function f possesses a tangent line at a point $(a, f(a))$, then f must be continuous at the number a. That is,

differentiability implies continuity

We summarize this fact in the next theorem.

> **THEOREM 3.1**   If f is differentiable at a number a, then f is continuous at a.

This result is not difficult to prove and is left as an exercise. (See Problem 48, Exercises 3.2.)

Remarks

(*i*) In the preceding discussion, we saw that the derivative of a function is itself a function that gives the slope of a tangent line. The derivative is, however, *not* an equation of a tangent line. Also, to say that $y - y_0 = f'(x) \cdot (x - x_0)$ is an equation of the tangent at (x_0, y_0) is incorrect. Remember that $f'(x)$ must be evaluated at x_0 *before* it is used in the point–slope form. If f is differentiable at x_0, then an equation of the tangent line at (x_0, y_0) is $y - y_0 = f'(x_0)(x - x_0)$.

(*ii*)   Mathematicians from the seventeenth to the nineteenth centuries believed that a continuous function *usually* possessed a derivative. (We have noted exceptions in this section.) In 1872 the German mathematician Karl Weierstrass conclusively destroyed this tenet by publishing an example of a function that was continuous at every real number but nowhere differentiable.

—————— **Exercises 3.2** ——————————————————————————

Answers to odd-numbered problems begin on page A-21.

In Problems 1–20 use Definition 3.3 to find the derivative of the given function.

1. $f(x) = 10$

2. $f(x) = x - 1$

3. $f(x) = -3x + 5$

4. $f(x) = \pi x$

5. $f(x) = 3x^2$

6. $f(x) = -x^2 + 1$

7. $f(x) = 4x^2 - x + 6$

8. $f(x) = 3x^2 + 6x - 7$

9. $f(x) = (x + 1)^2$

10. $f(x) = (2x - 5)^2$

11. $f(x) = x^3 + x$

12. $f(x) = 2x^3 + x^2$

13. $y = -x^3 + 15x^2 - x$

14. $y = x^4$

15. $y = \dfrac{1}{x}$

16. $y = \dfrac{2}{x + 1}$

17. $y = \dfrac{x}{x - 1}$

18. $y = \dfrac{2x + 3}{x + 4}$

19. $f(x) = \dfrac{1}{x} + \dfrac{1}{x^2}$

20. $f(x) = \dfrac{4}{x^3}$

In Problems 21–24 find the derivative of the given function. Find an equation of the tangent line to the graph of the function at the indicated value of *x*.

21. $f(x) = 4x^2 + 7x;\ x = -1$

22. $f(x) = \dfrac{1}{3}x^3 + 2x - 4;\ x = 0$

23. $y = x - \dfrac{1}{x};\ x = 1$

24. $y = 2x + 1 + \dfrac{6}{x};\ x = 2$

In Problems 25 and 26 find the point(s) on the graph of the given function where the tangent line is horizontal.

25. $f(x) = x^2 + 8x + 10$

26. $f(x) = x^3 - x^2 + 1$

27. (a)   Find the derivative of $f(x) = \sqrt{x}$. [*Hint*: Multiply the numerator and denominator of $\Delta y/\Delta x$ by $(x + \Delta x)^{1/2} + x^{1/2}$.]

(b)   What is the domain of f'?

(c)   Show that $f'_+(0)$ does not exist.

28. (a)   Find the derivative of $f(x) = x^{1/3}$. [*Hint*: Multiply the numerator and denominator of $\Delta y/\Delta x$ by $(x + \Delta x)^{2/3} + (x + \Delta x)^{1/3}x^{1/3} + x^{2/3}$.]

(b)   What is the domain of f'?

In Problems 29–32 use the procedures outlined in Problems 27 and 28 to find the derivative of the given function. State the domain of f'.

29. $f(x) = \sqrt{2x + 1}$

30. $f(x) = 1/\sqrt{x}$

31. $f(x) = (x - 4)^{1/3}$

32. $f(x) = (4x)^{1/3} + 7$

33. Use the derivative obtained in Problem 31 to show that the graph of $f(x) = (x - 4)^{1/3}$ has a vertical tangent at $(4, 0)$.

34. Use the derivative obtained in Problem 32 to show that the graph of $f(x) = (4x)^{1/3} + 7$ has a vertical tangent at $(0, 7)$.

In Problems 35 and 36 show that the given function is not differentiable at the indicated value of *x*.

35. $f(x) = \begin{cases} -x + 2, & x \le 2 \\ 2x - 4, & x > 2 \end{cases};\quad x = 2$

36. $f(x) = \begin{cases} 3x, & x < 0 \\ -4x, & x \ge 0 \end{cases};\quad x = 0$

In Problems 37 and 38 determine whether the given function is differentiable at the indicated value of *x*.

37. $f(x) = x|x|;\ x = 0$

38. $f(x) = |x^2 - 1|;\ x = 1$

39. Show that if $f(x) = k$ is a constant function, then $f'(x) = 0$. Interpret geometrically.

40. Show that if $f(x) = ax + b$, $a \neq 0$, is a linear function, then $f'(x) = a$. Interpret geometrically.

In Problems 41–44 (Figures 3.27–3.30) sketch the graph of f' from the graph of f.

41.

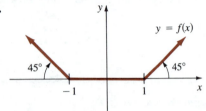

Figure 3.27

42.

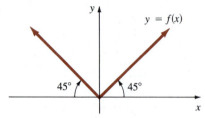

Figure 3.28

43.

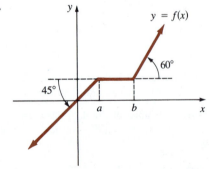

Figure 3.29

44.

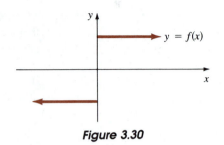

Figure 3.30

Miscellaneous Problems

45. An alternative formulation of the derivative of a function f, at a number a, is

$$f'(a) = \lim_{x \to a} \frac{f(x) - f(a)}{x - a} \qquad (3.9)$$

whenever the limit exists. Show that (3.9) is equivalent to (3.7) when $x = a$. (*Hint*: Let $\Delta x = x - a$.)

46. Use (3.9) to find the derivative of $f(x) = 6x^2$ at any number a.

47. Use (3.9) to find the derivative of $f(x) = x^{3/2}$ at any number a.

48. Use (3.9) to prove Theorem 3.1. [*Hint*: Take the limit of

$$f(x) - f(a) = \frac{f(x) - f(a)}{x - a} \cdot (x - a) \quad \text{as } x \to a$$

and use the fact that a limit of a difference is the difference of limits and the limit of a product is a product of the limits whenever all limits exist.]

3.3 Rules of Differentiation I: Power and Sum Rules

The definition of a derivative has the obvious drawback of being rather clumsy and tiresome to apply. We shall now see that the derivative of a function such as $f(x) = 6x^{100} + x^{35}$ can be obtained, so to speak, with a flick of a pencil.

Power Rule   The same procedure, outlined in Examples 1 and 4 of Section 3.2, to differentiate $f(x) = x^2$ and $f(x) = x^3$ can be employed to find the derivative of $f(x) = x^n$ for any positive integer n. The starting point is, again, the Binomial Theorem. Recall, for real numbers a and b, and n a positive integer

$$(a + b)^n = a^n + \frac{n}{1!}a^{n-1}b + \frac{n(n-1)}{2!}a^{n-2}b^2 + \cdots + \frac{n(n-1)\cdots(n-r+1)}{r!}a^{n-r}b^r + \cdots + b^n$$

where $r! = 1 \cdot 2 \cdot 3 \cdots (r-1)r$.

THEOREM 3.2   **Power Rule** (Positive Integer Exponents)

If n is a positive integer, then

$$\frac{d}{dx}x^n = nx^{n-1} \tag{3.10}$$

Proof   Let $f(x) = x^n$, n a positive integer. By the Binomial Theorem we can write

$$f(x + \Delta x) = (x + \Delta x)^n$$

$$= x^n + nx^{n-1}\,\Delta x + \frac{n(n-1)}{2}x^{n-2}(\Delta x)^2 + \cdots + (\Delta x)^n$$

Thus,

$$\Delta y = f(x + \Delta x) - f(x)$$

$$= \left[x^n + nx^{n-1}\,\Delta x + \frac{n(n-1)}{2}x^{n-2}\,(\Delta x)^2 + \cdots + (\Delta x)^n \right] - x^n$$

$$= \Delta x\left[nx^{n-1} + \frac{n(n-1)}{2}x^{n-2}\,\Delta x + \cdots + (\Delta x)^{n-1} \right]$$

and $$\frac{\Delta y}{\Delta x} = \frac{\Delta x\left[nx^{n-1} + \dfrac{n(n-1)}{2}x^{n-2}\,\Delta x + \cdots + (\Delta x)^{n-1} \right]}{\Delta x}$$

$$= nx^{n-1} + \frac{n(n-1)}{2}x^{n-2}\,\Delta x + \cdots + (\Delta x)^{n-1} \tag{3.11}$$

Since each term in (3.11) after the first contains a factor of Δx, it follows that

$$f'(x) = \lim_{\Delta x \to 0}\frac{\Delta y}{\Delta x} = nx^{n-1} \qquad\blacksquare$$

The Power Rule simply states that to differentiate x^n:

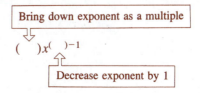

Observe that the derivatives of $y = x^2$ and $y = x^3$ follow from (3.10):

$$y = x^2, \qquad \frac{dy}{dx} = 2x^{2-1} = 2x$$

$$y = x^3, \qquad \frac{dy}{dx} = 3x^{3-1} = 3x^2$$

Example 1

Differentiate $y = x^6$.

Solution   From the Power Rule (3.10),

$$\frac{dy}{dx} = 6x^{6-1} = 6x^5$$

Example 2

Differentiate $y = x$.

Solution   Identifying $n = 1$, we have from (3.10),

$$\frac{dy}{dx} = 1x^{1-1} = 1x^0 = 1$$

The next result tells us that the derivative of a constant function is zero. The proof is immediate.

THEOREM 3.3   **Constant Function**

If $f(x) = k$ is a constant function, then $f'(x) = 0$.   (3.12)

Proof $\Delta y = f(x + \Delta x) - f(x) = k - k = 0$

Hence, $\displaystyle\lim_{\Delta x \to 0} \Delta y / \Delta x = 0$.   ∎

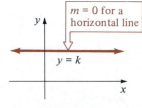

Figure 3.31

Theorem 3.3 has an obvious geometric interpretation. As shown in Figure 3.31 the slope m of horizontal line $y = k$ is, of course, zero.

Example 3

From (3.12) we have

$$(a)\ \frac{d}{dx}5 = 0, \quad \text{and} \quad (b)\ \frac{d}{dx}\pi^3 = 0.$$

Note the result in part (*b*) of Example 3. A *common mistake* is to apply the Power Rule and write $d/dx\,(\pi^3) = 3 \cdot \pi^2$. Bear in mind that (3.10) applies only to a *variable base x*. Furthermore, (3.12) implies that (3.10) holds for the case $n = 0$: If $x \neq 0$, $f(x) = x^0 = 1$ and $f'(x) = 0 = 0 \cdot x^{0-1}$.

THEOREM 3.4   **Constant Multiple of a Function**

If c is any constant and f is differentiable, then

$$\frac{d}{dx}[cf(x)] = cf'(x) \tag{3.13}$$

Proof   Let $G(x) = cf(x)$. Then

$$G'(x) = \lim_{\Delta x \to 0} \frac{G(x + \Delta x) - G(x)}{\Delta x} = \lim_{\Delta x \to 0} \frac{cf(x + \Delta x) - cf(x)}{\Delta x}$$

$$= \lim_{\Delta x \to 0} c\left[\frac{f(x + \Delta x) - f(x)}{\Delta x}\right]$$

$$= c \lim_{\Delta x \to 0} \frac{f(x + \Delta x) - f(x)}{\Delta x}$$

$$= cf'(x) \qquad \blacksquare$$

Example 4

Differentiate $y = 5x^3$.

Solution   From (3.10) and (3.13),

$$\frac{dy}{dx} = 5 \cdot \frac{d}{dx}x^3 = 5(3x^2) = 15x^2$$

THEOREM 3.5   **Sum Rule**

If f and g are differentiable functions, then

$$\frac{d}{dx}[f(x) + g(x)] = f'(x) + g'(x) \tag{3.14}$$

Proof   Let $G(x) = f(x) + g(x)$. Then

$$G'(x) = \lim_{\Delta x \to 0} \frac{G(x + \Delta x) - G(x)}{\Delta x}$$

$$= \lim_{\Delta x \to 0} \frac{[f(x + \Delta x) + g(x + \Delta x)] - [f(x) + g(x)]}{\Delta x}$$

$$= \lim_{\Delta x \to 0} \frac{f(x + \Delta x) - f(x) + g(x + \Delta x) - g(x)}{\Delta x}$$

$$= \lim_{\Delta x \to 0} \frac{f(x + \Delta x) - f(x)}{\Delta x} + \lim_{\Delta x \to 0} \frac{g(x + \Delta x) - g(x)}{\Delta x}$$

$$= f'(x) + g'(x) \qquad \blacksquare$$

Example 5

Differentiate $y = x^5 + x^2$.

Solution   From (3.10) and (3.14) we have

$$\frac{dy}{dx} = \frac{d}{dx}x^5 + \frac{d}{dx}x^2 = 5x^4 + 2x$$

The derivative of the function $f(x) = 6x^{100} + x^{35}$, mentioned in the introductory remark, is now readily seen to be $f'(x) = 600x^{99} + 35x^{34}$.

Since $f(x) - g(x) = f(x) + [-g(x)]$, the Sum Rule (3.14) also holds for the *difference* of two functions.

Example 6

Differentiate $y = 9x^7 - 3x^4$.

Solution   In view of (3.10), (3.13), and (3.14), we can write

$$\frac{dy}{dx} = 9\frac{d}{dx}x^7 - 3\frac{d}{dx}x^4 = 63x^6 - 12x^3$$

The Sum Rule (3.14) also extends to any finite sum of a differentiable function:

$$\frac{d}{dx}[f_1(x) + f_2(x) + \cdots + f_n(x)] = f_1'(x) + f_2'(x) + \cdots + f_n'(x)$$

From this latter result, (3.10), and (3.13), we see that any polynomial function is differentiable.

Example 7

Differentiate $y = 4x^5 - \frac{1}{2}x^4 + 9x^3 + 10x^2 - 13x + 6$.

Solution

$$\frac{dy}{dx} = 4\frac{d}{dx}x^5 - \frac{1}{2}\frac{d}{dx}x^4 + 9\frac{d}{dx}x^3 + 10\frac{d}{dx}x^2 - 13\frac{d}{dx}x + \frac{d}{dx}6$$

Since $\dfrac{d}{dx}6 = 0$, we obtain

$$\frac{dy}{dx} = 20x^4 - 2x^3 + 27x^2 + 20x - 13$$

Example 8

Find an equation of the tangent line to the graph of $f(x) = 3x^4 + 2x^3 - 7x$ at $x = -1$.

Solution   The y-coordinate of the point that corresponds to $x = -1$ is $f(-1) = 8$. Now,

$$f'(x) = 12x^3 + 6x^2 - 7 \quad \text{and so} \quad f'(-1) = -13$$

At $(-1, 8)$ the point–slope form gives an equation of the tangent line:

$$y - 8 = -13(x + 1) \quad \text{or} \quad y = -13x - 5$$

Normal Line   A **normal line** to a graph at a point P is one that is perpendicular to the tangent line at P.

Example 9

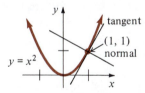

Figure 3.32

Find an equation of the normal line to the graph of $y = x^2$ at $x = 1$.

Solution   Since $dy/dx = 2x$, we know $m_{\text{tan}} = 2$ at $(1, 1)$. Thus, the slope of the normal line shown in Figure 3.32 is $m = -\frac{1}{2}$. Its equation is

$$y - 1 = -\frac{1}{2}(x - 1) \quad \text{or} \quad y = -\frac{1}{2}x + \frac{3}{2}$$

Remark

In the different contexts of science, engineering, and business, functions are often expressed in variables other than x and y. Correspondingly, we must adapt the derivative notation to new symbols; for example,

Function	*Derivative*
$v(t) = 32t$	$v'(t) = \dfrac{dv}{dt} = 32$
$A(r) = \pi r^2$	$A'(r) = \dfrac{dA}{dr} = 2\pi r$
$H(z) = \dfrac{1}{4}z^6$	$H'(z) = \dfrac{dH}{dz} = \dfrac{3}{2}z^5$
$D(p) = 800 - 120p + 5p^2$	$D'(p) = \dfrac{dD}{dp} = -120 + 10p$
$r(\theta) = 4\theta^2 - 3\theta$	$r'(\theta) = \dfrac{dr}{d\theta} = 8\theta - 3$

Exercises 3.3

Answers to odd-numbered problems begin on page A-21.

In Problems 1–20 find the derivative of the given function.

1. $y = x^9$

2. $y = 4x^{12}$

3. $y = \pi^6/12$

4. $y = -18$

5. $y = 7x^2 - 4x$

6. $y = 6x^3 + 3x^2 - 10$

7. $f(x) = \frac{1}{5}x^5 - 3x^4 + 9x^2 + 1$

8. $f(x) = -\frac{2}{3}x^6 + 4x^5 - 13x^2 + 8x + 2$

9. $f(x) = x^3(4x^2 - 5x - 6)$

10. $f(x) = \dfrac{2x^5 + 3x^4 - x^3}{x^2}$

11. $f(x) = (x + 1)^2$

12. $f(x) = (9 + x)(9 - x)$

13. $h(u) = (4u)^3$

14. $P(t) = (2t)^4 - (2t)^2$

15. $F(z) = 6z^3 + az^2 + a^3$, a a constant

16. $g(w) = \dfrac{w^n - 5^n}{n}$, n a positive integer

17. $G(\beta) = -3\beta^4 + 7\beta^3 - 5\beta^2 + 2$

18. $Q(u) = \dfrac{u^5 + 4u^2 - 3}{6}$

19. $f(x) = \dfrac{x^2}{(x^2 + 5)^{-2}}$

20. $f(x) = (x^3 + x^2)^3$

In Problems 21–24 find an equation of the tangent line to the graph of the given function at the indicated value of x.

21. $y = 2x^3 - 1$; $x = -1$

22. $y = \frac{1}{2}x^2 + 3$; $x = 2$

23. $y = 4x^2 - 4x - 20$; $x = 3$

24. $y = -x^3 + 6x^2$; $x = 1$

In Problems 25–28 find the point(s) on the graph of the given function at which the tangent is horizontal.

25. $y = x^2 - 8x + 5$

26. $y = \frac{1}{3}x^3 - \frac{1}{2}x^2$

27. $y = x^3 - 3x^2 - 9x + 2$

28. $y = x^4 - 4x^3$

In Problems 29–32 find an equation of the normal line to the graph of the given function at the indicated value of x.

29. $y = -x^2 + 1$; $x = 2$

30. $y = x^3$; $x = 1$

31. $y = \frac{1}{3}x^3 - 2x^2$; $x = 4$

32. $y = x^4 - x$; $x = -1$

33. Find the point on the graph of $f(x) = 2x^2 - 3x + 6$ at which the slope of the tangent line is 5.

34. Find the point on the graph of $f(x) = x^2 - x$ at which the tangent line is $3x - 9y - 4 = 0$.

35. Find the point on the graph of $f(x) = x^2 - x$ at which the slope of the normal line is 2.

36. Find the point(s) on the graph of $f(x) = x^2$ such that the tangent line at the point(s) has y-intercept -2.

37. Find the point(s) on the graph of $f(x) = x^2 - 5$ such that the tangent line at the point(s) has x-intercept -3.

38. Find the point on the graph of $f(x) = \frac{1}{4}x^2 - 2x$ at which the tangent line is parallel to the line $3x - 2y + 1 = 0$.

39. Find an equation of a tangent line to the graph of $f(x) = x^3$ that is perpendicular to the line $y = -3x$.

40. Find equations of the tangent lines that pass through $(0, -1)$ tangent to the graph of $f(x) = x^2 + 2x$.

41. Find a point on the graph of $f(x) = x^2 + x$ and a point on the graph of $g(x) = 2x^2 + 4x + 1$ at which the tangent lines are parallel.

42. Find values of a and b such that the slope of the tangent to the graph of $f(x) = ax^2 + bx$ at $(1, 4)$ is -5.

43. Find the values of b and c so that the graph of $f(x) = x^2 + bx$ possesses the tangent line $y = 2x + c$ at $x = -3$.

44. The height s above ground of a projectile at time t is given by

$$s(t) = -\frac{1}{2}gt^2 + v_0t + s_0$$

where g, v_0, and s_0 are constants. Find the instantaneous rate of change of s with respect to t at $t = 4$.

45. The volume V of a sphere of radius r is $V = (4\pi/3)r^3$. Find the surface area S of the sphere if S is the instantaneous rate of change of the volume with respect to the radius.

46. According to Poiseuille,* the velocity v of blood in an artery with a circular cross section radius R is $v(r) = (P/4vl)(R^2 - r^2)$, where P, v, and l are constants. What is the velocity of blood at the value of r for which $v'(r) = 0$? (See Figure 6.103 for an interpretation of the variable r.)

47. The potential energy of a spring-mass system when the spring is stretched a distance of x units is $U(x) = (k/2)x^2$, where k is the spring constant. The force exerted on the mass is $F = -dU/dx$. Find the force if the spring constant is 30 nt/m and the amount of stretch is $\frac{1}{2}$m.†

---

*Jean Louis Poiseuille (1799–1869), a French physician.
†See page 354 for a review of the units of force.

3.4 Rules of Differentiation II: Product and Quotient Rules

In the preceding section we saw that the Sum Rule (3.14) follows from the limit property that the limit of a sum is the sum of limits whenever the limits exist. We also know that when the limits exist, the limit of a product is the product of the limits. Therefore, a natural conjecture would be that the derivative of a product is the product of the derivatives. However, the rule for differentiating the product of two functions is *not* that simple.

THEOREM 3.6   **Product Rule**

If f and g are differentiable functions, then

$$\frac{d}{dx}[f(x)g(x)] = f(x)g'(x) + g(x)f'(x) \qquad (3.15)$$

Proof   Let $G(x) = f(x)g(x)$. Then

$$G'(x) = \lim_{\Delta x \to 0} \frac{G(x + \Delta x) - G(x)}{\Delta x}$$

$$= \lim_{\Delta x \to 0} \frac{f(x + \Delta x)g(x + \Delta x) - f(x)g(x)}{\Delta x}$$

$$= \lim_{\Delta x \to 0} \frac{f(x + \Delta x)g(x + \Delta x) \overbrace{- f(x + \Delta x)g(x) + f(x + \Delta x)g(x)}^{\text{zero}} - f(x)g(x)}{\Delta x}$$

$$= \lim_{\Delta x \to 0} \left[f(x + \Delta x)\frac{g(x + \Delta x) - g(x)}{\Delta x} + g(x)\frac{f(x + \Delta x) - f(x)}{\Delta x} \right]$$

$$= \lim_{\Delta x \to 0} f(x + \Delta x) \cdot \lim_{\Delta x \to 0} \frac{g(x + \Delta x) - g(x)}{\Delta x} + \lim_{\Delta x \to 0} g(x) \cdot \lim_{\Delta x \to 0} \frac{f(x + \Delta x) - f(x)}{\Delta x} \qquad (3.16)$$

Since f is differentiable, it is continuous and therefore

$$\lim_{\Delta x \to 0} f(x + \Delta x) = f(x)$$

Furthermore, $\lim_{\Delta x \to 0} g(x) = g(x)$. Hence, (3.16) becomes

$$G'(x) = f(x)g'(x) + g(x)f'(x).$$ ∎

The Product Rule is usually memorized in words:

The first function times the derivative of the second plus the second function times the derivative of the first.

_____ **Example 1** _____

Differentiate $y = (x^3 - 2x^2 + 4)(8x^2 + 5x)$.

Solution   From the Product Rule (3.15),

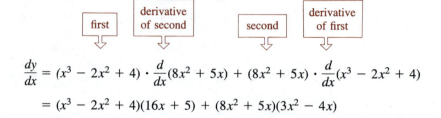

$$\frac{dy}{dx} = (x^3 - 2x^2 + 4) \cdot \frac{d}{dx}(8x^2 + 5x) + (8x^2 + 5x) \cdot \frac{d}{dx}(x^3 - 2x^2 + 4)$$

$$= (x^3 - 2x^2 + 4)(16x + 5) + (8x^2 + 5x)(3x^2 - 4x)$$

Although (3.15) is stated only for the product of two functions, it can also be applied to functions with a greater number of factors.

_____ **Example 2** _____

Differentiate $y = (4x + 1)(2x^2 - x)(x^3 - 8x)$.

Solution   We identify the first two factors as "the first function."

first	derivative of second	second	derivative of first

$$\frac{dy}{dx} = \overbrace{(4x + 1)(2x^2 - x)}\frac{d}{dx}(x^3 - 8x) + (x^3 - 8x)\frac{d}{dx}(4x + 1)(2x^2 - x)$$

To find the derivative of the first function, we must apply the Product Rule a second time. Thus,

> Product Rule again

$$\frac{dy}{dx} = (4x + 1)(2x^2 - x) \cdot (3x^2 - 8) + (x^3 - 8x) \cdot [(4x + 1)(4x - 1) + (2x^2 - x) \cdot 4]$$

$$= (4x + 1)(2x^2 - x)(3x^2 - 8) + (16x^2 - 1)(x^3 - 8x) - 4(2x^2 - x)(x^3 - 8x)$$

The derivative of the quotient of two functions is given by the following.

THEOREM 3.7   Quotient Rule

If f and g are differentiable functions and $g(x) \neq 0$, then

$$\frac{d}{dx}\left[\frac{f(x)}{g(x)}\right] = \frac{g(x)f'(x) - f(x)g'(x)}{[g(x)]^2} \qquad (3.17)$$

Proof Let $G(x) = f(x)/g(x)$. Then,

$$G'(x) = \lim_{\Delta x \to 0} \frac{G(x + \Delta x) - G(x)}{\Delta x}$$

$$= \lim_{\Delta x \to 0} \frac{\dfrac{f(x + \Delta x)}{g(x + \Delta x)} - \dfrac{f(x)}{g(x)}}{\Delta x}$$

$$= \lim_{\Delta x \to 0} \frac{g(x)f(x + \Delta x) - f(x)g(x + \Delta x)}{\Delta x\, g(x + \Delta x)g(x)}$$

$$= \lim_{\Delta x \to 0} \frac{g(x)f(x + \Delta x) \overbrace{- g(x)f(x) + g(x)f(x)}^{\text{zero}} - f(x)g(x + \Delta x)}{\Delta x\, g(x + \Delta x)g(x)}$$

$$= \lim_{\Delta x \to 0} \frac{g(x)\left[\dfrac{f(x + \Delta x) - f(x)}{\Delta x}\right] - f(x)\left[\dfrac{g(x + \Delta x) - g(x)}{\Delta x}\right]}{g(x + \Delta x)g(x)}$$

$$= \frac{g(x)f'(x) - f(x)g'(x)}{[g(x)]^2} \qquad\blacksquare$$

In words, the Quotient Rule is:

> The denominator times the derivative of the numerator minus the numerator times the derivative of the denominator all divided by the denominator squared.

Example 3 ─────────────────────────────

Differentiate $y = \dfrac{3x^2 - 1}{2x^3 + 5x^2 + 7}$.

Solution From the Quotient Rule (3.17),

denominator	derivative of numerator	numerator	derivative of denominator

$$\frac{dy}{dx} = \frac{(2x^3 + 5x^2 + 7) \cdot \dfrac{d}{dx}(3x^2 - 1) - (3x^2 - 1) \cdot \dfrac{d}{dx}(2x^3 + 5x^2 + 7)}{(2x^3 + 5x^2 + 7)^2}$$

denominator squared

$$= \frac{(2x^3 + 5x^2 + 7) \cdot 6x - (3x^2 - 1) \cdot (6x^2 + 10x)}{(2x^3 + 5x^2 + 7)^2}$$

$$= \frac{-6x^4 + 6x^2 + 52x}{(2x^3 + 5x^2 + 7)^2}$$

____ **Example 4** _____

Find an equation of the tangent line to the graph of $f(x) = \dfrac{6x^3}{x^3 + 1}$ at $x = 1$.

Solution   We use the Quotient Rule to find the derivative,

$$f'(x) = \frac{(x^3 + 1) \cdot \dfrac{d}{dx} 6x^3 - 6x^3 \cdot \dfrac{d}{dx} (x^3 + 1)}{(x^3 + 1)^2}$$

$$= \frac{(x^3 + 1) \cdot 18x^2 - 6x^3 \cdot (3x^2)}{(x^3 + 1)^2}$$

$$= \frac{18x^2}{(x^3 + 1)^2}$$

When $x = 1$, the slope of the tangent line is

$$f'(1) = \frac{18}{4} = \frac{9}{2}$$

The point of tangency is $(1, f(1))$ or $(1, 3)$. Hence, an equation of the tangent line is

$$y - 3 = \frac{9}{2}(x - 1) \quad \text{or} \quad y = \frac{9}{2}x - \frac{3}{2}$$

The next example shows that the derivative of a function may utilize a combination of rules.

____ **Example 5** _____

Differentiate $y = \dfrac{(x^2 + 1)(2x^2 + 1)}{(3x^2 + 1)}$.

Solution   We begin with the Quotient Rule, and then use the Product Rule when differentiating the numerator:

| Product Rule here |

$$\frac{dy}{dx} = \frac{(3x^2 + 1) \cdot \dfrac{d}{dx} [(x^2 + 1)(2x^2 + 1)] - (x^2 + 1)(2x^2 + 1) \cdot \dfrac{d}{dx} (3x^2 + 1)}{(3x^2 + 1)^2}$$

$$= \frac{(3x^2 + 1)[(x^2 + 1)4x + (2x^2 + 1)2x] - (x^2 + 1)(2x^2 + 1)6x}{(3x^2 + 1)^2}$$

$$= \frac{12x^5 + 8x^3}{(3x^2 + 1)^2}$$

So far the Power Rule, (3.10) of Section 3.3, is limited to the case where the exponent is a positive integer or zero. We shall now see that this rule is valid even when the exponent is a negative integer.

THEOREM 3.8   **Power Rule** (Negative Integer Exponents)

If n is a positive integer, then

$$\frac{d}{dx} x^{-n} = -nx^{-n-1} \qquad (3.18)$$

Proof If n denotes a positive integer, then $-n$ is a negative integer. Since $x^{-n} = 1/x^n$, it follows that we can obtain the derivative of x^{-n} by the Quotient Rule:

$$\frac{d}{dx} x^{-n} = \frac{d}{dx} \left[\frac{1}{x^n} \right]$$

$$= \frac{x^n \cdot \frac{d}{dx} 1 - 1 \cdot \frac{d}{dx} x^n}{(x^n)^2}$$

$$= \frac{0 - nx^{n-1}}{x^{2n}}$$

$$= -\frac{nx^{n-1}}{x^{2n}}$$

$$= -nx^{n-1-2n}$$

$$= -nx^{-n-1} \qquad \blacksquare$$

Example 6

Differentiate $y = x^{-2}$.

Solution As (3.18) shows, the procedure for differentiating a power with a negative integer exponent is the same as before: bring down the exponent as a multiple and decrease the exponent by 1. We have

$$\frac{dy}{dx} = -2x^{-2-1}$$

$$= -2x^{-3} = -\frac{2}{x^3}$$

Example 7

Differentiate $y = 5x^3 - 1/x^4$.

Solution   We first write the given function as $y = 5x^3 - x^{-4}$.

Thus,
$$\frac{dy}{dx} = 5 \cdot 3x^2 - (-4)x^{-5}$$

$$= 15x^2 + \frac{4}{x^5}$$

Remarks

(*i*) The Product and Quotient Rules will usually lead to expressions that demand simplification. If your answer to a problem does not look like the one in the text answer section, you may not have performed sufficient simplifications. Do not be content simply to carry through the mechanics of the various rules; it is always a good idea to practice your algebraic skills.

(*ii*) You should also note that the Quotient Rule is often used when it is not required; for example, although the Quotient Rule can be utilized in differentiating

$$y = \frac{x^5}{6} \quad \text{and} \quad y = \frac{10}{x^3}$$

it is simpler to write

$$y = \frac{1}{6}x^5 \quad \text{and} \quad y = 10x^{-3}$$

and then use the Power Rule:

$$\frac{dy}{dx} = \frac{5}{6}x^4 \quad \text{and} \quad \frac{dy}{dx} = -30x^{-4}$$

Exercises 3.4

Answers to odd-numbered problems begin on page A-21.

In Problems 1–30 find the derivative of the given function.

1. $y = 1/x$

2. $y = (2/x^3)^2$

3. $y = (5x)^{-2}$

4. $y = \pi^4 x^{-4}$

5. $y = 6x^2 + x^{-2}$

6. $y = 5x^4 - 1/2x^5$

7. $y = (x^2 - 7)(x^3 + 4x + 2)$

8. $y = (7x + 1)(x^4 - x^3 - 9x)$

9. $f(x) = \left(4 + \frac{1}{x}\right)\left(2x - \frac{1}{x^2}\right)$

10. $f(x) = \left(x^2 - \frac{1}{x^2}\right)\left(x^2 + \frac{1}{x^2}\right)$

11. $f(x) = \frac{10}{x^2 + 1}$

12. $f(x) = 5(4x - 3)^{-1}$

13. $G(x) = \dfrac{3x + 1}{2x - 5}$

14. $F(x) = \dfrac{2 - 3x}{7 - x}$

15. $y = (6x - 1)^2$

16. $y = (x^4 + 5x)^2$

17. $g(t) = \dfrac{t^2}{2t^2 + t + 1}$

18. $p(y) = \dfrac{y^2 - 10y + 2}{y(y^2 - 1)}$

19. $H(z) = (z + 1)(2z + 1)(3z + 1)$

20. $Q(r) = (r^2 + 1)(r^3 - r)(3r^4 + 2r - 1)$

21. $y = \dfrac{(2x + 1)(x - 5)}{3x + 2}$

22. $y = \dfrac{x^5}{(x^2 + 1)(x^3 + 4)}$

23. $y = \dfrac{2 - 1/x^3}{3 + 1/x^2}$

24. $y = \dfrac{x^{-2}}{x^{-3} + x^{-2} + 1}$

25. $f(u) = \dfrac{1}{u} + \dfrac{1}{u^2} + \dfrac{1}{u^3} + \dfrac{1}{u^4}$

26. $h(v) = \dfrac{1}{v + v^2 + v^3 + v^4}$

27. $y = \left(\dfrac{x + 1}{x + 3}\right)(x^2 - 2x - 1)$

28. $y = (x + 1)\left(x + 1 - \dfrac{1}{x + 2}\right)$

29. $f(x) = (3x + 1)^{-2}$

30. $g(x) = (x + 1)^3$

In Problems 31 and 32 find dy/dx without the aid of the Quotient Rule.

31. $y = \dfrac{6x^2 - 5x}{x}$

32. $y = \dfrac{x^4 + 2x^3 - 1}{x^2}$

In Problems 33–36 find an equation of the tangent line to the graph of the given function at the indicated value of x.

33. $y = 1/x^2$; $x = \frac{1}{2}$

34. $y = 4x - 1/x$; $x = -1$

35. $y = (2x^2 - 4)(x^3 + 5x + 3)$; $x = 0$

36. $y = \dfrac{5x}{x^2 + 1}$; $x = 2$

In Problems 37–40 find the point(s) on the graph of the given function at which the tangent is horizontal.

37. $y = (x^2 - 4)(x^2 - 6)$

38. $y = x(x - 1)^2$

39. $y = \dfrac{x^2}{x^4 + 1}$

40. $y = \dfrac{1}{x^2 - 6x}$

In Problems 41–44 find the point(s) on the graph of the given function at which the tangent has the indicated slope.

41. $y = 4x^{-1}$; $m_{\tan} = -64$

42. $y = x - 1/x^2$; $m_{\tan} = 55$

43. $y = \dfrac{x + 3}{x + 1}$; $m_{\tan} = -\dfrac{1}{8}$

44. $y = (x + 1)(2x + 5)$; $m_{\tan} = -3$

In Problems 45–48 use the information $f(1) = 2, f'(1) = -3$, and $g(1) = g'(1) = 6$ to evaluate the given derivative.

45. $\dfrac{d}{dx} f(x)g(x)\Big|_{x=1}$

46. $\dfrac{d}{dx} \dfrac{g(x)}{f(x)}\Big|_{x=1}$

47. $\dfrac{d}{dx} \dfrac{1 + 2f(x)}{x - g(x)}\Big|_{x=1}$

48. $\dfrac{d}{dx} \left(\dfrac{4}{x} + f(x)\right)g(x)\Big|_{x=1}$

In Problems 49–52 find the values of x for which $f'(x) > 0$.

49. $f(x) = 5/x^2$

50. $f(x) = \dfrac{x^2 + 3}{x + 1}$

51. $f(x) = (2x + 1)(x + 4)$

52. $f(x) = x + 4/x$

53. The Universal Law of Gravitation states that the force F between two bodies of masses m_1 and m_2 separated by a distance r is $F = km_1m_2/r^2$, where k is constant. What is the instantaneous rate of change of F with respect to r when $r = \frac{1}{2}$ km?

54. The potential energy U between two atoms in a diatomic molecule is given by $U(x) = q_1/x^{12} - q_2/x^6$, where q_1 and q_2 are positive constants and x is the distance between the atoms. The force between the atoms is defined as $F(x) = -U'(x)$. Show that $F(\sqrt[6]{2q_1/q_2}) = 0$.

Miscellaneous Problems

55. Verify that the function $y = 5x + 1 + 2/x$ satisfies the equation $xy' + y = 10x + 1$.

56. Let $y = f(x)g(x)h(x)$, where f, g, and h are differentiable functions. Use the Product Rule to show that

$$\frac{dy}{dx} = f(x)g(x)h'(x) + f(x)h(x)g'(x) + g(x)h(x)f'(x)$$

57. Let $y = f(x)$ be a differentiable function.

    (a) Find dy/dx for $y = [f(x)]^2$.

    (b) Find dy/dx for $y = [f(x)]^3$.

    (c) Conjecture a rule for finding the derivative of $y = [f(x)]^n$, where n is a positive integer.

58. Use the definition of the derivative to prove that when g is differentiable and $g(x) \neq 0$, then

$$\frac{d}{dx}\left[\frac{1}{g(x)}\right] = -\frac{g'(x)}{[g(x)]^2}$$

59. Use the result of Problem 58, the fact that $f(x)/g(x) = f(x) \cdot [1/g(x)]$, and the Product Rule to derive the Quotient Rule (3.17).

3.5 Derivatives of the Trigonometric Functions

3.5.1 Some Preliminary Limit Results

Recall from Sections 1.6 and 2.4 that the sine and cosine functions are continuous on $(-\infty, \infty)$. Hence, the following two limit results follow immediately from Definition 2.1:

$$\lim_{t \to 0} \sin t = 0 \tag{3.19}$$

$$\lim_{t \to 0} \cos t = 1 \tag{3.20}$$

Also, in Example 7 of Section 2.1, we illustrated the fact that

$$\lim_{t \to 0} \frac{\sin t}{t} = 1 \tag{3.21}$$

We now prove this result.

    Consider a circle centered at the origin with radius 1. As shown in Figure 3.33, let the shaded region OPR be a sector with central angle t such that $0 < t < \pi/2$. We see from the same figure that the areas of triangles OPR and OQR give lower and upper bounds, respectively, for the area of the sector OPR:

$$\text{area of } \triangle OPR < \text{area of sector } OPR < \text{area of } \triangle OQR \tag{3.22}$$

The height of $\triangle OPR$ is $\overline{OP} \sin t = 1 \cdot \sin t = \sin t$ and so

$$\text{area of } \triangle OPR = \frac{1}{2}\,\overline{OR} \cdot (\text{height})$$

$$= \frac{1}{2} \cdot 1 \cdot \sin t = \frac{1}{2}\sin t \tag{3.23}$$

In addition, $\overline{QR}/\overline{OR} = \tan t$ or $\overline{QR} = \tan t$ so that

$$\text{area of } \triangle OQR = \frac{1}{2}\,\overline{OR} \cdot \overline{QR}$$

$$= \frac{1}{2} \cdot 1 \cdot \tan t = \frac{1}{2}\tan t \tag{3.24}$$

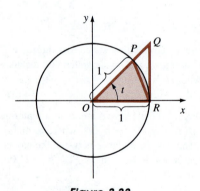

Figure 3.33

Finally, we know that the area of a sector of a circle is $\frac{1}{2}r^2\theta$, where r is the radius of the circle and θ is the central angle measured in radians. Thus,

$$\text{area of sector } OPR = \frac{1}{2} \cdot 1 \cdot t = \frac{1}{2}t \qquad (3.25)$$

Using (3.23), (3.24), and (3.25) in (3.22) gives

$$\frac{1}{2}\sin t < \frac{1}{2}t < \frac{1}{2}\tan t$$

or

$$1 < \frac{t}{\sin t} < \frac{t}{\cos t} \qquad (3.26)$$

From the properties of inequalities (3.26) is equivalent to

$$\cos t < \frac{\sin t}{t} < 1$$

We now let $t \to 0$ in the last result. Since $(\sin t)/t$ is "squeezed" between 1 and $\cos t$ (which is approaching 1), it follows from the Squeeze Theorem that $\lim_{t\to 0} (\sin t)/t = 1$. While we have assumed $0 < t < \pi/2$, the same result holds for $-\pi/2 < t < 0$. (See Problem 27, Exercises 3.5.)

_____ **Example 1** _____

Evaluate $\lim\limits_{t\to 0} \dfrac{\sin 4t}{t}$.

Solution   Let $u = 4t$ so that $t = u/4$. We see that as $t \to 0$, necessarily $u \to 0$, and hence,

$$\lim_{t\to 0} \frac{\sin 4t}{t} = \lim_{u\to 0} \frac{\sin u}{u/4}$$

$$= 4\lim_{u\to 0} \frac{\sin u}{u}$$

$$= 4 \cdot 1 = 4$$

Using an argument similar to that illustrated in Example 1 it follows that

$$\lim_{t\to 0} \frac{\sin kt}{t} = k \qquad (3.27)$$

Also,

$$\lim_{t\to 0} \frac{1}{(\sin t)/t} = \frac{\lim\limits_{t\to 0} 1}{\lim\limits_{t\to 0} (\sin t)/t} = 1$$

implies

$$\lim_{t\to 0} \frac{t}{\sin t} = 1 \qquad (3.28)$$

_____ **Example 2** _____

Evaluate $\lim\limits_{t \to 0} \dfrac{\sin^2 5t}{t^2}$.

Solution   We rewrite the function as

$$\frac{\sin 5t}{t} \cdot \frac{\sin 5t}{t}$$

and use (3.27):

$$\lim_{t \to 0} \frac{\sin^2 5t}{t^2} = \lim_{t \to 0} \frac{\sin 5t}{t} \cdot \frac{\sin 5t}{t}$$

$$= \lim_{t \to 0} \frac{\sin 5t}{t} \cdot \lim_{t \to 0} \frac{\sin 5t}{t}$$

$$= 5 \cdot 5 = 25$$

_____

Another limit result that we will use immediately is

$$\lim_{t \to 0} \frac{1 - \cos t}{t} = 0 \qquad\qquad (3.29)$$

To see this, we observe that

$$\frac{1 - \cos t}{t} = \frac{(1 - \cos t)(1 + \cos t)}{t(1 + \cos t)}$$

$$= \frac{1 - \cos^2 t}{t(1 + \cos t)}$$

$$= \frac{\sin^2 t}{t(1 + \cos t)}$$

$$= \frac{\sin t}{t} \cdot \frac{\sin t}{1 + \cos t}$$

and so

$$\lim_{t \to 0} \frac{1 - \cos t}{t} = \lim_{t \to 0} \frac{\sin t}{t} \cdot \lim_{t \to 0} \frac{\sin t}{1 + \cos t}$$

$$= 1 \cdot 0 = 0$$

3.5.2 Derivatives

sin x _and_ cos x   We find the derivative of $f(x) = \sin x$ by resorting to the definition $f'(x) = \lim_{\Delta x \to 0} \Delta y / \Delta x$. In this case

$$\Delta y = f(x + \Delta x) - f(x)$$

$$= \sin(x + \Delta x) - \sin x$$

$$= \sin x \cos \Delta x + \cos x \sin \Delta x - \sin x \qquad \boxed{\text{Addition Formula}}$$

$$= \sin x(\cos \Delta x - 1) + \cos x \sin \Delta x$$

and
$$\frac{\Delta y}{\Delta x} = \sin x \frac{\cos \Delta x - 1}{\Delta x} + \cos x \frac{\sin \Delta x}{\Delta x}$$

Employing Theorem 2.5(*ii*) and results (3.21) and (3.29) enables us to write

$$\begin{aligned} f'(x) &= \lim_{\Delta x \to 0} \frac{\Delta y}{\Delta x} = \sin x \cdot \lim_{\Delta x \to 0} \frac{\cos \Delta x - 1}{\Delta x} + \cos x \cdot \lim_{\Delta x \to 0} \frac{\sin \Delta x}{\Delta x} \\ &= \sin x \cdot 0 + \cos x \cdot 1 \end{aligned}$$

We conclude that
$$\frac{d}{dx} \sin x = \cos x \tag{3.30}$$

In a similar manner it can be shown that

$$\frac{d}{dx} \cos x = -\sin x \tag{3.31}$$

(See Problem 65, Exercises 3.5.)

Example 3

Find the slope of the tangent line to the graph of $f(x) = \sin x$ at $x = \pi/2$ and $x = 4\pi/3$.

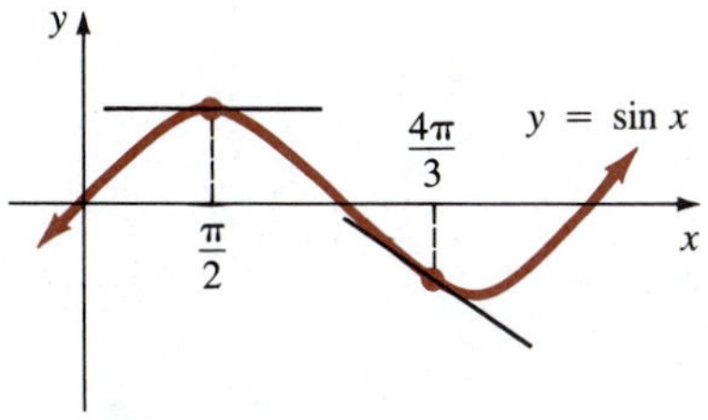

Figure 3.34

Solution We know from (3.30) that

$$f'(x) = \cos x$$

and so
$$f'(\pi/2) = \cos(\pi/2) = 0$$
$$f'(4\pi/3) = \cos(4\pi/3) = -\tfrac{1}{2}$$

In Figure 3.34 we see that the tangent line is horizontal at $(\pi/2, 1)$.

The Other Trigonometric Functions

The results in (3.30) and (3.31) can be used in conjunction with the rules of differentiation to find the derivatives of the tangent, cotangent, secant, and cosecant functions.

To differentiate $\tan x = (\sin x)/\cos x$, we can use the Quotient Rule

$$\begin{aligned} \frac{d}{dx} \frac{\sin x}{\cos x} &= \frac{\cos x \dfrac{d}{dx} \sin x - \sin x \dfrac{d}{dx} \cos x}{(\cos x)^2} \\ &= \frac{\cos x(\cos x) - \sin x(-\sin x)}{\cos^2 x} \\ &= \frac{\overbrace{\cos^2 x + \sin^2 x}^{1}}{\cos^2 x} \\ &= \frac{1}{\cos^2 x} \end{aligned}$$

Since $1/\cos^2 x = (1/\cos x)^2 = \sec^2 x$, we have the following result:

$$\frac{d}{dx} \tan x = \sec^2 x \tag{3.32}$$

The derivative formula for the cotangent is obtained in an analogous fashion and is left as an exercise. (See Problem 66 in Exercises 3.5.)

$$\frac{d}{dx} \cot x = -\csc^2 x \tag{3.33}$$

Now, $\sec x = 1/\cos x$. Therefore, we can use the Quotient Rule, again, to find the derivative of the secant function:

$$\frac{d}{dx} \sec x = \frac{d}{dx} \frac{1}{\cos x}$$

$$= \frac{\cos x \dfrac{d}{dx} 1 - 1 \cdot \dfrac{d}{dx} \cos x}{\cos^2 x}$$

$$= \frac{\sin x}{\cos^2 x} \tag{3.34}$$

By writing

$$\frac{\sin x}{\cos^2 x} = \frac{1}{\cos x} \cdot \frac{\sin x}{\cos x} = \sec x \tan x$$

we can express (3.34) as

$$\frac{d}{dx} \sec x = \sec x \tan x \tag{3.35}$$

The last result also follows immediately from the Quotient Rule:

$$\frac{d}{dx} \csc x = -\csc x \cot x \tag{3.36}$$

(See Problem 67, Exercises 3.5.)

Example 4 _____

Differentiate $y = x^2 \sin x$.

Solution   The Product Rule along with (3.30) yields

$$\frac{dy}{dx} = x^2 \frac{d}{dx} \sin x + \sin x \frac{d}{dx} x^2$$

$$= x^2 \cos x + 2x \sin x$$

Example 5

Differentiate $y = (\cos x)(x - \cot x)$.

Solution From the Product Rule, (3.31), and (3.33),

$$\frac{dy}{dx} = (\cos x)\frac{d}{dx}(x - \cot x) + (x - \cot x)\frac{d}{dx}\cos x$$

$$= (\cos x)(1 + \csc^2 x) + (x - \cot x)(-\sin x)$$

$$= 2\cos x - x\sin x + \cos x\csc^2 x$$

Example 6

Differentiate $y = \dfrac{\sin x}{2 + \sec x}$.

Solution From the Quotient Rule, (3.30), and (3.35),

$$\frac{dy}{dx} = \frac{(2 + \sec x)\dfrac{d}{dx}\sin x - \sin x\dfrac{d}{dx}(2 + \sec x)}{(2 + \sec x)^2}$$

$$= \frac{(2 + \sec x)\cos x - \sin x(\sec x\tan x)}{(2 + \sec x)^2}$$

$$= \frac{1 + 2\cos x - \tan^2 x}{(2 + \sec x)^2}$$

Exercises 3.5

Answers to odd-numbered problems begin on page A-21.

[3.5.1]

In Problems 1–26 find the value of each limit if it exists.

1. $\displaystyle\lim_{t\to 0}\frac{\sin 3t}{2t}$

2. $\displaystyle\lim_{t\to 0}\frac{\sin(-4t)}{t}$

3. $\displaystyle\lim_{x\to 0}\frac{\sin x}{4 + \cos x}$

4. $\displaystyle\lim_{x\to 0}\frac{1 + \sin x}{1 + \cos x}$

5. $\displaystyle\lim_{x\to 0}\frac{\cos 2x}{\cos 3x}$

6. $\displaystyle\lim_{x\to 0}\frac{\tan x}{3x}$

7. $\displaystyle\lim_{t\to 0}\frac{1}{t\sec t\csc 4t}$

8. $\displaystyle\lim_{t\to 0}5t\cot 2t$

9. $\displaystyle\lim_{t\to 0}\frac{2\sin^2 t}{t\cos^2 t}$

10. $\displaystyle\lim_{t\to 0}\frac{\sin^2(t/2)}{\sin t}$

11. $\displaystyle\lim_{t\to 0}\frac{\sin^2 6t}{t^2}$

12. $\displaystyle\lim_{t\to 0}\frac{t^3}{\sin^2 3t}$

13. $\displaystyle\lim_{x\to 1}\frac{\sin(x - 1)}{2x - 2}$

14. $\displaystyle\lim_{x\to 2\pi}\frac{x - 2\pi}{\sin x}$

15. $\displaystyle\lim_{x\to 0}\frac{\cos x}{x}$

16. $\displaystyle\lim_{t\to \pi/2}\frac{1 + \sin t}{\cos t}$

17. $\displaystyle\lim_{x\to 0}\frac{\cos(3x - \pi/2)}{x}$

18. $\displaystyle\lim_{x\to -2}\frac{\sin(5x + 10)}{4x + 8}$

19. $\displaystyle\lim_{t\to 0}\frac{\sin 3t}{\sin 7t}$

20. $\displaystyle\lim_{t\to 0}\sin 2t\csc 3t$

21. $\displaystyle\lim_{t\to 0^+}\frac{\sin t}{\sqrt{t}}$

22. $\displaystyle\lim_{t\to 0^+}\frac{1 - \cos\sqrt{t}}{\sqrt{t}}$

23. $\lim\limits_{x \to 0^+} \dfrac{(x + 2\sqrt{\sin x})^2}{x}$

24. $\lim\limits_{x \to 0} \dfrac{(1 - \cos x)^2}{x}$

25. $\lim\limits_{x \to 0} \dfrac{\cos x - 1}{\cos^2 x - 1}$.

26. $\lim\limits_{x \to 0} \dfrac{\sin x + \tan x}{x}$

Miscellaneous Problems

27. Prove (3.21) when $-\pi/2 < t < 0$.

28. Show that $\lim\limits_{x \to \infty} x \sin \dfrac{1}{x} = 1$.

Calculator Problems

29. (a) Let t be measured in degrees. Fill in the following table using values with four decimal places.

t	45	30	15	5	1	0.1
$\dfrac{\sin t}{t}$						

(b) Compute $\pi/180$ to four decimal places.

(c) Given that t is measured in degrees, make a conjecture about the value of $\lim\limits_{t \to 0} (\sin t)/t$.

30. Let t be measured in radians. Use a calculator to investigate whether $\lim\limits_{t \to 0} (1 - \cos t)/t^2$ exists.

[3.5.2]

In Problems 31–54 find the derivative of the given function.

31. $y = x^2 - \cos x$

32. $y = 4x^3 + x + \sin x$

33. $y = 1 + 7 \sin x - \tan x$

34. $y = 3 \cos x - 5 \cot x$

35. $y = x \sin x$

36. $y = (x^3 - 2)\tan x$

37. $y = \sin x \cos x$

38. $y = \cos x \cot x$

39. $y = (x^2 + \sin x)\sec x$

40. $y = \csc x \tan x$

41. $f(x) = (\csc x)^{-1}$

42. $f(x) = \dfrac{2}{\cos x \cot x}$

43. $f(x) = \dfrac{\cot x}{x + 1}$

44. $f(x) = \dfrac{x^2 - 6x}{1 + \cos x}$

45. $y = \dfrac{x^2}{1 + 2 \tan x}$

46. $y = \dfrac{2 + \sin x}{x}$

47. $F(\theta) = \dfrac{\sin \theta}{1 + \cos \theta}$

48. $g(z) = \dfrac{1 + \csc z}{1 + \sec z}$

49. $G(u) = \sin^2 u$

50. $H(v) = (1 + \cos v)(v - \sin v)$

51. $y = \cos^2 x + \sin^2 x$

52. $y = x^3 \cos x - x^3 \sin x$

53. $y = x^2 \sin x \tan x$

54. $y = \dfrac{1 + \sin x}{x \cos x}$

In Problems 55 and 56 consider the graph of the given function on the interval $[0, 2\pi]$. Find the point(s) at which the tangent is horizontal.

55. $f(x) = x + \cos x$

56. $f(x) = \sin x + \cos x$

In Problems 57–60 find an equation of the tangent line to the graph of the given function at the indicated value of x.

57. $f(x) = \cos x; x = \pi/3$

58. $f(x) = \tan x; x = \pi$

59. $f(x) = \sec x; x = \pi/6$

60. $f(x) = \csc x; x = \pi/2$

In Problems 61–64 find an equation of the normal line to the graph of the given function at the indicated value of x.

61. $f(x) = \sin x; x = 4\pi/3$

62. $f(x) = \tan^2 x; x = \pi/4$

63. $f(x) = x \cos x; x = \pi$

64. $f(x) = \dfrac{x}{1 + \sin x}; x = \pi/2$

Miscellaneous Problems

65. Prove (3.31).

66. Prove (3.33).

67. Prove (3.36).

68. If t is measured in degrees, then $\lim\limits_{t \to 0} (\sin t)/t = \pi/180$ and $\lim\limits_{t \to 0} (1 - \cos t)/t = 0$. Show that if x is measured in degrees, then

$$\frac{d}{dx} \sin x = \frac{\pi}{180} \cos x$$

and

$$\frac{d}{dx} \cos x = -\frac{\pi}{180} \sin x$$

In Problems 69 and 70 find the derivative of the given function by first employing a trigonometric identity.

69. $f(x) = \sin 2x$

70. $f(x) = \cos^2 \dfrac{x}{2}$

3.6 Rules of Differentiation III: The Chain Rule

Suppose we wish to differentiate

$$y = (x^5 + 1)^2 \tag{3.37}$$

By writing (3.37) as

$$y = (x^5 + 1) \cdot (x^5 + 1)$$

we can find the derivative using the Product Rule:

$$\begin{aligned}
\frac{dy}{dx} &= (x^5 + 1)\frac{d}{dx}(x^5 + 1) + (x^5 + 1)\frac{d}{dx}(x^5 + 1) \\
&= (x^5 + 1) \cdot 5x^4 + (x^5 + 1) \cdot 5x^4 \\
&= 2(x^5 + 1) \cdot 5x^4 \tag{3.38}
\end{aligned}$$

Similarly, to differentiate $y = (x^5 + 1)^3$, we can write $y = (x^5 + 1)^2 \cdot (x^5 + 1)$ and use both the Product Rule and the result given in (3.38). It is readily shown that

$$\frac{d}{dx}(x^5 + 1)^3 = 3(x^5 + 1)^2 \cdot 5x^4 \tag{3.39}$$

Power Rule for Functions Inspection of (3.38) and (3.39) reveals a pattern for differentiating a power of a function; for example, in (3.39) we see

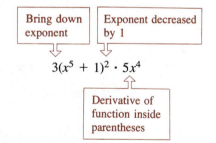

For emphasis, if we denote a differentiable function by [ ], it appears that

$$\frac{d}{dx}[\ \ ]^n = n[\ \ ]^{n-1}\frac{d}{dx}[\ \ ] \tag{3.40}$$

Later in this section, we will prove that (3.40) holds for *any* integer n. The general result is stated in the next theorem.

THEOREM 3.9   **Power Rule for Functions** (Integer Exponents)

If n is an integer and g is a differentiable function, then

$$\frac{d}{dx}[g(x)]^n = n[g(x)]^{n-1}\,g'(x) \tag{3.41}$$

Example 1

Differentiate $y = (4x)^{100}$.

Solution We first illustrate a *common mistake*:

$$\frac{dy}{dx} = 100(4x)^{99}$$

This result is incorrect because we failed to multiply the expression by the derivative of the function inside the parentheses. The correct procedure follows:

$$\frac{d}{dx} = \overbrace{100}^{n}\overbrace{(4x)^{99}}^{[g(x)]^{n-1}} \cdot \overbrace{\frac{d}{dx}4x}^{g'(x)}$$

$$= 100(4x)^{99} \cdot 4 = 400(4x)^{99}$$

Example 2

Differentiate $y = (2x^3 + 4x + 1)^4$.

Solution We identify $g(x) = 2x^3 + 4x + 1$ and $n = 4$. From (3.41) it then follows that

$$\frac{dy}{dx} = 4(2x^3 + 4x + 1)^3 \frac{d}{dx}(2x^3 + 4x + 1)$$

$$= 4(2x^3 + 4x + 1)^3(6x^2 + 4)$$

Example 3

Differentiate $y = \dfrac{(x^2 - 1)^3}{(5x + 1)^8}$.

Solution We first apply the Quotient Rule followed by the Power Rule for Functions:

$$\boxed{\text{Power Rule}}$$

$$\frac{dy}{dx} = \frac{(5x + 1)^8 \frac{d}{dx}(x^2 - 1)^3 - (x^2 - 1)^3 \frac{d}{dx}(5x + 1)^8}{(5x + 1)^{16}}$$

$$= \frac{(5x + 1)^8 \cdot 3(x^2 - 1)^2 \cdot 2x - (x^2 - 1)^3 \cdot 8(5x + 1)^7 \cdot 5}{(5x + 1)^{16}}$$

$$= \frac{6x(5x + 1)^8(x^2 - 1)^2 - 40(5x + 1)^7(x^2 - 1)^3}{(5x + 1)^{16}}$$

$$= \frac{(x^2 - 1)^2(-10x^2 + 6x + 40)}{(5x + 1)^9}$$

_____ **Example 4** _____

To differentiate $y = 1/(x^2 + 1)$, we could, of course, use the Quotient Rule. However, it is also possible to use the Power Rule for Functions with $n = -1$:

$$y = (x^2 + 1)^{-1}$$

$$\frac{dy}{dx} = (-1)(x^2 + 1)^{-2} \frac{d}{dx}(x^2 + 1)$$

$$= (-1)(x^2 + 1)^{-2} \, 2x$$

$$= \frac{-2x}{(x^2 + 1)^2}$$

_____ **Example 5** _____

Differentiate $y = \dfrac{1}{(7x^5 - x^4 + 2)^{10}}$.

Solution   Write the given function as

$$y = (7x^5 - x^4 + 2)^{-10}$$

Identify $n = -10$ and use the Power Rule (3.41):

$$\frac{dy}{dx} = -10(7x^5 - x^4 + 2)^{-11} \frac{d}{dx}(7x^5 - x^4 + 2)$$

$$= \frac{-10(35x^4 - 4x^3)}{(7x^5 - x^4 + 2)^{11}}$$

_____ **Example 6** _____

Differentiate $y = \tan^2 x$.

Solution   For emphasis, we first write $y = (\tan x)^2$ and then use (3.41):

$$\frac{dy}{dx} = 2(\tan x) \frac{d}{dx} \tan x$$

Employing (3.32) of Section 3.5 yields

$$\frac{dy}{dx} = 2 \tan x \, \sec^2 x$$

Chain Rule   A power of a function can be written as a composite function. If $f(x) = x^n$ and $u = g(x)$, then $f(u) = f(g(x)) = [g(x)]^n$. The Power Rule (3.41) is a special case of the **Chain Rule** for differentiating composite functions.

THEOREM 3.10   Chain Rule

If $y = f(u)$ is a differentiable function of u and $u = g(x)$ is a differentiable function of x, then

$$\frac{dy}{dx} = \frac{dy}{du} \cdot \frac{du}{dx}$$

$$= f'(g(x)) \cdot g'(x) \tag{3.42}$$

Proof for $\Delta u \neq 0$ For $\Delta x \neq 0$,

$$\Delta u = g(x + \Delta x) - g(x) \tag{3.43}$$

or $\qquad g(x + \Delta x) = g(x) + \Delta u = u + \Delta u$

In addition,

$$\Delta y = f(u + \Delta u) - f(u)$$

$$= f(g(x + \Delta x)) - f(g(x))$$

When x and $x + \Delta x$ are in some open interval for which $\Delta u \neq 0$, we can write

$$\frac{\Delta y}{\Delta x} = \frac{\Delta y}{\Delta u} \cdot \frac{\Delta u}{\Delta x}$$

Since g is assumed to be differentiable, it is continuous. Consequently, as $\Delta x \to 0$, $g(x + \Delta x) \to g(x)$ and so from (3.43) we see that $\Delta u \to 0$. Thus,

$$\lim_{\Delta x \to 0} \frac{\Delta y}{\Delta x} = \left(\lim_{\Delta x \to 0} \frac{\Delta y}{\Delta u} \right) \left(\lim_{\Delta x \to 0} \frac{\Delta u}{\Delta x} \right)$$

$$= \left(\lim_{\Delta u \to 0} \frac{\Delta y}{\Delta u} \right) \left(\lim_{\Delta x \to 0} \frac{\Delta u}{\Delta x} \right)$$

From the definition of the derivative, it follows that

$$\frac{dy}{dx} = \frac{dy}{du} \cdot \frac{du}{dx} \qquad \blacksquare$$

The assumption that $\Delta u \neq 0$ on some interval does not hold true for every differentiable function g. Although the result given in (3.42) remains valid when $\Delta u = 0$, the preceding proof does not.

Proof of the Power Rule for Functions   As noted previously, a power of a function can be written as $y = u^n$, where n is an integer and $u = g(x)$. Since $dy/du = nu^{n-1}$, $du/dx = g'(x)$, we see from the Chain Rule that

$$\frac{dy}{dx} = \frac{dy}{du} \cdot \frac{du}{dx} = nu^{n-1} \frac{du}{dx}$$

$$= n[g(x)]^{n-1} g'(x)$$

This is the Power Rule (3.41).

Trigonometric Functions   We obtain the derivatives of the trigonometric functions composed with a differentiable function g as another direct consequence of the Chain Rule. For example, if $y = \sin u$, where $u = g(x)$, then $dy/du = \cos u$. Hence, (3.42) implies

$$\frac{dy}{dx} = \frac{dy}{du} \cdot \frac{du}{dx} = \cos u \, \frac{du}{dx}$$

or equivalently

$$\frac{d}{dx} \sin [\;\;] = \cos [\;\;] \frac{d}{dx} [\;\;]$$

We summarize the six results:

I $\dfrac{d}{dx} \sin u = \cos u \, \dfrac{du}{dx}$ II $\dfrac{d}{dx} \cos u = -\sin u \, \dfrac{du}{dx}$

III $\dfrac{d}{dx} \tan u = \sec^2 u \, \dfrac{du}{dx}$ IV $\dfrac{d}{dx} \cot u = -\csc^2 u \, \dfrac{du}{dx}$

V $\dfrac{d}{dx} \sec u = \sec u \tan u \, \dfrac{du}{dx}$ VI $\dfrac{d}{dx} \csc u = -\csc u \cot u \, \dfrac{du}{dx}$

Example 7

Differentiate $y = \cos 4x$.

Solution   From II,

$$\frac{dy}{dx} = \overbrace{-\sin 4x}^{-\sin u} \overbrace{\frac{d}{dx}(4x)}^{\frac{du}{dx}} = -4 \sin 4x$$

Example 8

Differentiate $y = \tan(6x^2 + 1)$.

Solution   From III,

$$\frac{dy}{dx} = \overbrace{\sec^2(6x^2 + 1)}^{\sec^2 u} \cdot \overbrace{\frac{d}{dx}(6x^2 + 1)}^{\frac{du}{dx}}$$

$$= 12x \sec^2(6x^2 + 1)$$

_____ **Example 9** _____

Differentiate $y = (9x^3 + 1)^2 \sin 5x$.

Solution   We first use the Product Rule,

$$\frac{dy}{dx} = (9x^3 + 1)^2 \frac{d}{dx} \sin 5x + \sin 5x \frac{d}{dx} (9x^3 + 1)^2$$

followed by the Power Rule (3.41) and I,

$$\frac{dy}{dx} = (9x^3 + 1)^2 \cdot 5 \cos 5x + \sin 5x \cdot 2(9x^3 + 1) \cdot 27x^2$$

$$= (9x^3 + 1)(45x^3\cos 5x + 54x^2\sin 5x + 5 \cos 5x)$$

_____ **Example 10** _____

Differentiate $y = \cos^4(7x^3 + 6x - 1)$.

Solution   We use the Power Rule (3.41) and II:

$$\frac{dy}{dx} = 4 \cos^3(7x^3 + 6x - 1) \cdot \frac{d}{dx} \cos (7x^3 + 6x - 1)$$

$$= 4 \cos^3(7x^3 + 6x - 1)(-\sin(7x^3 + 6x - 1) \cdot \frac{d}{dx} (7x^3 + 6x - 1))$$

$$= -4(21x^2 + 6)\cos^3(7x^3 + 6x - 1)\sin(7x^3 + 6x - 1)$$

_____ **Exercises 3.6** _____

Answers to odd-numbered problems begin on page A-22.

In Problems 1–36 find the derivative of the given function.

1. $y = (-5x)^{30}$

2. $y = (3/x)^{14}$

3. $y = (2x^2 + x)^{200}$

4. $y = \left(x - \dfrac{1}{x^2}\right)^5$

5. $y = \dfrac{1}{(x^3 - 2x^2 + 7)^4}$

6. $y = \dfrac{1}{x^4 + x^2 + 1}$

7. $y = (3x - 1)^4(-2x + 9)^5$

8. $y = x^4(x^2 + 1)^6$

9. $y = \sin^3 x$

10. $y = \sec^2 x$

11. $f(x) = \left(\dfrac{x^2 - 1}{x^2 + 1}\right)^2$

12. $f(x) = \dfrac{3x - 4}{(5x + 2)^3}$

13. $f(x) = [x + (x^2 - 4)^3]^{10}$

14. $f(x) = \left[\dfrac{1}{(x^3 - x + 1)^2}\right]^4$

15. $g(t) = (t^{-1} + t^{-2} + t^{-3})^{-4}$

16. $F(\theta) = (2\theta + 1)^3\tan^2\theta$

17. $H(u) = (2 + u \sin u)^{-3}$

18. $q(t) = \dfrac{(1 + \cos t)^2}{(1 + \sin t)^3}$

19. $P(v) = \dfrac{v(2v - 5)^4}{(v + 1)^8}$

20. $R(s) = (s + 1)^2(s + 2)^3(s + 3)^4$

21. $y = \sin(\pi x + 1)$

22. $y = -2\cos(-3x + 7)$

23. $y = \sin^2 4x$

24. $y = \sin 4x^2$

25. $y = x^3\cos x^3$

26. $y = \sin 2x \cos 3x$

27. $y = \tan \dfrac{1}{x}$

28. $y = \cot^3 8x$

29. $F(\theta) = \dfrac{\sin 5\theta}{\cos 6\theta}$

30. $h(t) = \dfrac{t + \sin 4t}{10 + \cos 3t}$

31. $f(x) = (\sec 4x + \tan 2x)^5$

32. $f(x) = \csc^2 x - \csc x^2$

33. $f(x) = \sin(\sin x)$

34. $f(x) = \tan\left(\cos \dfrac{x}{2}\right)$

35. $f(x) = \sin^3(4x^2 - 1)$

36. $f(x) = [\cos(x^3 + x^2)]^{-4}$

In Problems 37–40 find the slope of the tangent line to the graph of the given function at the indicated value of x.

37. $y = (x^2 + 2)^3$; $x = -1$

38. $y = \dfrac{1}{(3x + 1)^2}$; $x = 0$

39. $y = \sin 3x + 4x \cos 5x$; $x = \pi$

40. $y = 50x - \tan^3 2x$; $x = \pi/6$

In Problems 41–44 find an equation of the tangent line to the graph of the given function at the indicated value of x.

41. $y = \left(\dfrac{x}{x + 1}\right)^2$; $x = -\dfrac{1}{2}$

42. $y = x^2(x - 1)^3$; $x = 2$

43. $y = \tan 3x$; $x = \pi/4$

44. $y = (-1 + \cos 4x)^3$; $x = \pi/8$

45. Find the point(s) on the graph of $f(x) = \dfrac{x}{(x^2 + 1)^2}$ where the tangent line is horizontal. Does the graph of f have any vertical tangents?

46. The function $H = (k/mg)\sin^2\theta$, k, m, and g constants, represents the height attained by a grasshopper whose takeoff angle is θ. Determine the values of θ at which $dH/d\theta = 0$.

47. The function $R = (v_0^2/g)\sin 2\theta$ gives the range of a projectile fired at an angle θ from the horizontal with an initial velocity v_0. If v_0 and g are constant, find those values of θ at which $dR/d\theta = 0$.

48. The volume of a spherical balloon of radius r is $V = (4\pi/3)r^3$. The radius is a function of time t and increases at a rate of 5 in/min. What is the instantaneous rate of change of V with respect to t?

49. Suppose a spherical balloon is being filled at a constant rate $dV/dt = 10$ in^3/min. At what rate is its radius increasing when $r = 2$ in?

50. Determine the values of t at which the instantaneous rate of change of $g(t) = \sin t + \frac{1}{2}\cos 2t$ is zero.

In Problems 51 and 52 let $y = f(u)$ and $u = g(x)$ be differentiable functions.

    (a) Compute dy/dx without forming $f(g(x))$.

    (b) Verify the answer in part **(a)** by finding $f(g(x))$ and $\dfrac{d}{dx}f(g(x))$.

51. $y = u^3 + 2u$, $u = x^9 + 4x^2$

52. $y = u^4 + 5u^3 - 7u^2 + 8u - 1$, $u = x^3$

53. Let F be a differentiable function. What is

$$\frac{d}{dx}F(3x)?$$

54. Let G be a differentiable function. What is

$$\frac{d}{dx}[G(-x^2)]^2?$$

55. Suppose $\dfrac{d}{du}f(u) = \dfrac{1}{u}$. What is $\dfrac{d}{dx}f(-10x + 7)$?

56. Suppose $\dfrac{d}{dx}f(x) = \dfrac{1}{1 + x^2}$. What is $\dfrac{d}{dx}f(x^3)$?

Miscellaneous Problems

57. Given that f is an odd differentiable function, use the Chain Rule to show that f' is an even function.

58. Given that f is an even differentiable function, use the Chain Rule to show that f' is an odd function.

59. Assuming differentiability of all functions, find dy/dx for $y = f(g(h(x)))$.

60. Suppose $g(t) = h(f(t))$, where $f(1) = 3$, $f'(1) = 6$, and $h'(3) = -2$. What is $g'(1)$?

3.7 Higher-order Derivatives

The Second Derivative The derivative $f'(x)$ is a function derived from a function $y = f(x)$. By differentiating the first derivative $f'(x)$, we obtain yet another function called the **second derivative**, which is denoted by $f''(x)$. In terms of the operation symbol d/dx, we define the second derivative with respect to x as the function obtained by differentiating $y = f(x)$ twice in succession:

$$\frac{d}{dx}\left(\frac{dy}{dx}\right)$$

The second derivative is commonly denoted by

$$f''(x), \qquad y'', \qquad \frac{d^2y}{dx^2}, \quad \text{or} \quad D_x^2 y$$

Normally, we shall use one of the first three symbols.

_____ **Example 1** _____

Find the second derivative of $y = x^3 - 2x^2$.

Solution   The first derivative is

$$\frac{dy}{dx} = 3x^2 - 4x$$

The second derivative follows from differentiating the first derivative:

$$\frac{d^2y}{dx^2} = \frac{d}{dx}(3x^2 - 4x) = 6x - 4$$

_____ **Example 2** _____

The first derivative of $f(x) = \sin x$ is

$$f'(x) = \cos x$$

Then, the second derivative is

$$f''(x) = -\sin x$$

_____ **Example 3** _____

Find the second derivative of $y = (x^3 + 1)^4$.

Solution   We obtain the first derivative from the Power Rule for Functions:

$$\frac{dy}{dx} = 4(x^3 + 1)^3 \frac{d}{dx} x^3$$

$$= 12x^2(x^3 + 1)^3$$

To find the second derivative, we will now use the Product and Power Rules:

$$\frac{d^2y}{dx^2} = 12x^2 \frac{d}{dx}(x^3 + 1)^3 + (x^3 + 1)^3 \frac{d}{dx} 12x^2$$

$$= 108x^4(x^3 + 1)^2 + 24x(x^3 + 1)^3$$

$$= (x^3 + 1)^2(132x^4 + 24x)$$

Assuming all derivatives exist, we can differentiate a function $y = f(x)$ as many times as we want. The **third derivative** is the derivative of the second derivative. The **fourth derivative** is the derivative of the third derivative, and so on. We denote the third and fourth derivatives by d^3y/dx^3 and d^4y/dx^4, respectively, and define them by

$$\frac{d^3y}{dx^3} = \frac{d}{dx}\left(\frac{d^2y}{dx^2}\right)$$

$$\frac{d^4y}{dx^4} = \frac{d}{dx}\left(\frac{d^3y}{dx^3}\right)$$

In general, if n is a positive integer, then the nth derivative is defined by

$$\frac{d^ny}{dx^n} = \frac{d}{dx}\left(\frac{d^{n-1}y}{dx^{n-1}}\right)$$

Other notations for the first n derivatives are

$$f'(x), \quad f''(x), \quad f'''(x), \quad f^{(4)}(x), \ldots, f^{(n)}(x)$$
$$y', \quad y'', \quad y''', \quad y^{(4)}, \ldots, y^{(n)}$$
$$D_xy, \quad D_x^2y, \quad D_x^3y, \quad D_x^4y, \ldots, D_x^ny$$

Example 4

Find the first five derivatives of

$$f(x) = 2x^4 - 6x^3 + 7x^2 + 5x - 10$$

Solution We have

$$f'(x) = 8x^3 - 18x^2 + 14x + 5$$
$$f''(x) = 24x^2 - 36x + 14$$
$$f'''(x) = 48x - 36$$
$$f^{(4)}(x) = 48$$
$$f^{(5)}(x) = 0$$

After reflecting a moment, you should be convinced that the $(n + 1)$st derivative of an nth-degree polynomial function is zero.

Example 5

Find the third derivative of $y = \dfrac{1}{x^3}$.

Solution   By writing $y = x^{-3}$, we have

$$\frac{dy}{dx} = -3x^{-4}$$

Hence

$$\frac{d^2y}{dx^2} = (-3)(-4)x^{-5} = 12x^{-5}$$

and

$$\frac{d^3y}{dx^3} = (12)(-5)x^{-6} = -\frac{60}{x^6}$$

Exercises 3.7

Answers to odd-numbered problems begin on page A-22.

In Problems 1–20 find the second derivative of the given function.

1. $y = -x^2 + 3x - 7$

2. $y = 15x^2 - \pi^2$

3. $y = (-4x + 9)^2$

4. $y = 2x^5 + 4x^3 - 6x^2$

5. $y = 10x^{-2}$

6. $y = \left(\dfrac{2}{x^2}\right)^3$

7. $y = x^3 + 8x^2 - \dfrac{2}{x^4}$

8. $y = \dfrac{x^6 - 7x^3 + 1}{x^2}$

9. $f(x) = x^2(3x - 4)^3$

10. $f(x) = (x^2 + 5x - 1)^4$

11. $g(t) = \dfrac{2t - 3}{t + 2}$

12. $h(z) = \dfrac{z^2}{z + 1}$

13. $f(x) = \cos 10x$

14. $f(x) = \tan \dfrac{x}{2}$

15. $f(x) = x \sin x$

16. $f(x) = \sin^2 5x$

17. $r(\theta) = \dfrac{1}{3 + 2\cos\theta}$

18. $H(\theta) = \dfrac{\cos\theta}{\theta}$

19. $f(x) = \sec x$

20. $f(x) = \cos x^2$

In Problems 21–24 find the indicated derivative.

21. $y = 4x^6 + x^5 - x^3; \dfrac{d^4y}{dx^4}$

22. $y = \dfrac{2}{x}; \dfrac{d^5y}{dx^5}$

23. $f(x) = \sin \pi x; f'''(x)$

24. $f(x) = \dfrac{1}{\sec(2x + 1)}; f^{(5)}(x)$

Given that n is a positive integer in Problems 25 and 26, find a formula for the given derivative.

25. $\dfrac{d^n}{dx^n} x^n$

26. $\dfrac{d^n}{dx^n}\left(\dfrac{1}{1 - 2x}\right)$

In Problems 27 and 28 find the point(s) on the graph of f at which $f''(x) = 0$.

27. $f(x) = x^3 + 12x^2 + 20x$

28. $f(x) = x^4 - 2x^3$

In Problems 29 and 30 determine intervals for which $f''(x) > 0$ and intervals for which $f''(x) < 0$.

29. $f(x) = (x - 1)^3$

30. $f(x) = x^3 + x^2$

31. Find an equation of the tangent line to the graph of $y = x^3 + 3x^2 - 4x + 1$ at the point where the value of the second derivative is zero.

32. Find an equation of the tangent line to the graph of $y = x^4$ at the point where the value of the third derivative is 12.

33. If $f(x) = \cos(x/3)$, what is the slope of the tangent line to the graph of f' at $x = 2\pi$?

34. Find the point(s) on the graph of $f(x) = \frac{1}{2}x^2 - 5x + 1$ at which **(a)** $f''(x) = f(x)$, and **(b)** $f''(x) = f'(x)$.

35. If $f(3) = -4$, $f'(3) = 2$, and $f''(3) = 5$, what is

$$\left.\frac{d^2}{dx^2} f^2(x)\right|_{x=3} ?$$

36. If $f'(0) = -1$ and $g'(0) = 6$, what is

$$\frac{d^2}{dx^2} [xf(x) + xg(x)] \Big|_{x=0} ?$$

Miscellaneous Problems

37. Show that for any constants C_1, C_2, and k,
$y = C_1 \cos kx + C_2 \sin kx$ satisfies the equation $y'' + k^2y = 0$.

38. Show that for any constants $C_1 \neq 0$ and C_2,
$y = (-1/C_1)(1 - C_1^2 x^2)^{1/2} + C_2$ satisfies the equation $xy'' = y' + (y')^3$.

39. Show that $y = \cos x + \sin x + x \cos x + x \sin x$ satisfies the equation $y^{(4)} + 2y'' + y = 0$.

40. Let $f(x) = x^3 + 2x$.

(a) Find $f'(x)$ and $f''(x)$.

(b) In general,

$$f''(x) = \lim_{\Delta x \to 0} \frac{f'(x + \Delta x) - f'(x)}{\Delta x}$$

provided this limit exists. Use $f'(x)$ obtained in part **(a)** and the foregoing definition to find $f''(x)$.

Suppose $y = f(x)$ is a twice differentiable function. The **curvature** of the graph of f at a point (x, y) is defined to be

$$\kappa = \frac{|f''(x)|}{[1 + (f'(x))^2]^{3/2}}$$

A small value of κ at a point indicates the graph is nearly straight near the point.

41. Show that the curvature of a graph of linear function is zero at every point.

42. Show that $\kappa = 1$ at each point of the semicircle defined by $y = \sqrt{1 - x^2}$.

43. Calculate the curvature of the graph of $f(x) = x^2$ at $x = 0$. As $x \to \infty$, what is the limiting value of κ? Interpret geometrically.

44. (a) Show that

$$\frac{d^2}{dx^2} (fg) = f''g + 2f'g' + fg''$$

$$\frac{d^3}{dx^3} (fg) = f'''g + 3f''g' + 3f'g'' + fg'''$$

(b) Discern the pattern of the derivatives in part **(a)** and then give

$$\frac{d^4}{dx^4} (fg)$$

3.8 Implicit Differentiation

Explicit and Implicit Functions

A function in which the dependent variable is expressed solely in terms of the independent variable x, namely $y = f(x)$, is said to be an **explicit** function; for example, $y = \frac{1}{2}x^3 - 1$ is an explicit function, whereas an equivalent equation $2y - x^3 + 2 = 0$ is said to define the function **implicitly** or y is an **implicit function** of x.

Now, as we know, the equation

$$x^2 + y^2 = 4 \tag{3.44}$$

describes a circle of radius 2 centered at the origin. Equation (3.44) is not a function since for any choice of x in the interval $-2 < x < 2$ there correspond two values of y. However, as shown in Figure 3.35, by considering either the top half, or the bottom half, of the circle, we obtain a function. We say that (3.44) defines *at least* two implicit functions of x on the interval $-2 \leq x \leq 2$. In this case, we observe that the top half of the circle is described by the function

$$f(x) = \sqrt{4 - x^2}, \qquad -2 \leq x \leq 2$$

whereas the bottom half is given by

$$g(x) = -\sqrt{4 - x^2}, \qquad -2 \leq x \leq 2$$

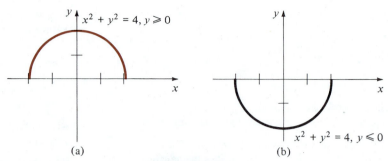

Figure 3.35

Note that both equations

$$x^2 + [f(x)]^2 = 4 \quad \text{and} \quad x^2 + [g(x)]^2 = 4$$

are identities on the interval $-2 \le x \le 2$.

In general, if an equation $F(x, y) = 0$ defines a function f implicitly on some interval, then $F(x, f(x)) = 0$ is an identity on the interval. The graph of f is a portion (or all) of the graph of the equation $F(x, y) = 0$.

A more complicated equation such as

$$x^4 + x^2y^3 - y^5 = 2x + 1$$

may determine several implicit functions on a suitably restricted interval of the x-axis and yet it may not be possible to solve for y in terms of x. However, in some cases we can determine the derivative dy/dx by a process known as **implicit differentiation**. This process consists of differentiating both sides of an equation with respect to x, using the rules of differentiation, and then solving for dy/dx. Since we think of y as being determined by the given equation as a differentiable function, the Chain Rule, in the form of the Power Rule for Functions, gives the useful result

$$\frac{d}{dx}y^n = ny^{n-1}\frac{dy}{dx} \qquad (3.45)$$

where n is an integer.

In the following examples we shall assume that the given equation determines at least one differentiable implicit function.

Example 1

Find dy/dx if $x^2 + y^2 = 4$.

Solution   We differentiate both sides of the equation and then utilize (3.45):

$$\boxed{\text{Power Rule}}$$

$$\frac{d}{dx}x^2 + \frac{d}{dx}y^2 = \frac{d}{dx}4$$

$$2x + 2y\frac{dy}{dx} = 0$$

Solving for the derivative yields

$$\frac{dy}{dx} = -\frac{x}{y} \qquad\qquad (3.46)$$

As illustrated in (3.46) of Example 1, implicit differentiation will usually yield a derivative that depends on both variables x and y. In our introductory discussion we saw that the equation $x^2 + y^2 = 4$ defines two differentiable implicit functions on the open interval $-2 < x < 2$. The symbolism $dy/dx = -x/y$ represents the derivative of either function on the interval. In general, implicit differentiation will yield the derivative of any differentiable implicit function defined by an equation $F(x, y) = 0$.

Example 2

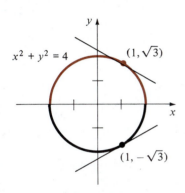

Figure 3.36

Find the slopes of the tangent lines to the graph of $x^2 + y^2 = 4$ at the points corresponding to $x = 1$.

Solution Substituting $x = 1$ into the given equation implies $y^2 = 3$ or $y = \pm\sqrt{3}$. Hence, there are tangent lines at $(1, \sqrt{3})$ and $(1, -\sqrt{3})$. Although $(1, \sqrt{3})$ and $(1, -\sqrt{3})$ are points on the graphs of two different implicit functions, indicated by the different colors in Figure 3.36, (3.46) of Example 1 gives the correct slope at each point. We have

$$\frac{dy}{dx}\bigg|_{(1,\ \sqrt{3})} = -\frac{1}{\sqrt{3}}$$

and

$$\frac{dy}{dx}\bigg|_{(1,\ -\sqrt{3})} = -\frac{1}{-\sqrt{3}} = \frac{1}{\sqrt{3}}$$

Example 3

Find dy/dx if $x^4 + x^2y^3 - y^5 = 2x + 1$.

Solution In this case, we use (3.45) and the Product Rule:

$$\frac{d}{dx}x^4 + \frac{d}{dx}x^2y^3 - \frac{d}{dx}y^5 = \frac{d}{dx}2x + \frac{d}{dx}1$$

Product Rule here

$$4x^3 + x^2 \cdot 3y^2\frac{dy}{dx} + 2xy^3 - 5y^4\frac{dy}{dx} = 2$$

$$(3x^2y^2 - 5y^4)\frac{dy}{dx} = 2 - 4x^3 - 2xy^3$$

$$\frac{dy}{dx} = \frac{2 - 4x^3 - 2xy^3}{3x^2y^2 - 5y^4}.$$

_____ **Example 4** _____

Find d^2y/dx^2 if $x^2 + y^2 = 4$.

Solution   From Example 1, we already know that the first derivative is

$$\frac{dy}{dx} = -\frac{x}{y}$$

Hence, by the Quotient Rule

$$\frac{d^2y}{dx^2} = -\frac{d}{dx}\left(\frac{x}{y}\right)$$

$$= -\frac{y \cdot 1 - x \cdot \dfrac{dy}{dx}}{y^2}$$

$$= -\frac{y - x\left(-\dfrac{x}{y}\right)}{y^2} \qquad \boxed{\text{Substituting for } dy/dx}$$

$$= -\frac{y^2 + x^2}{y^3}$$

Noting that $x^2 + y^2 = 4$ permits us to write the second derivative as

$$\frac{d^2y}{dx^2} = -\frac{4}{y^3}$$

_____ **Example 5** _____

Find dy/dx if $\sin y = y \cos 2x$.

Solution   From the Chain Rule and Product Rule we obtain

$$\frac{d}{dx}\sin y = \frac{d}{dx} y \cos 2x$$

$$\cos y \cdot \frac{dy}{dx} = y(-\sin 2x \cdot 2) + \cos 2x \cdot \frac{dy}{dx}$$

$$(\cos y - \cos 2x)\frac{dy}{dx} = -2y \sin 2x$$

$$\frac{dy}{dx} = -\frac{2y \sin 2x}{\cos y - \cos 2x}$$

Remark

To determine when an equation defines a function implicitly is not an easy matter. Thus, in the absence of any stated criteria, it is understood that finding dy/dx by implicit differentiation could, in some cases, be nothing more than formal symbol manipulation. For example, you should verify that $x^2 + y^2 = c$ will give $dy/dx = -x/y$ for any choice of the constant c. But for $c < 0$ the equation yields no real function and so dy/dx is meaningless for these values. (See Problems 35 and 36 in Exercises 3.8.)

Exercises 3.8

Answers to odd-numbered problems begin on page A-22.

In Problems 1–20 assume that the given equation defines at least one differentiable implicit function. Use implicit differentiation to find dy/dx.

1. $y^2 - 2y = x$

2. $4x^2 + y^2 = 8$

3. $xy^2 - x^2 + 4 = 0$

4. $(y - 1)^2 = 4(x + 2)$

5. $x + xy - y^2 - 20 = 0$

6. $y^3 - 2y + 3x^3 = 4x + 1$

7. $x^3y^2 = 2x^2 + y^2$

8. $x^5 - 6xy^3 + y^4 = 1$

9. $(x^2 + y^2)^6 = x^3 - y^3$

10. $y = (x - y)^2$

11. $y^{-3}x^6 + y^6x^{-3} = 2x + 1$

12. $y^4 - y^2 = 10x - 3$

13. $(x - 1)^2 + (y + 4)^2 = 25$

14. $\dfrac{x + y}{x - y} = x$

15. $y^2 = \dfrac{x - 1}{x + 2}$

16. $\dfrac{x}{y^2} + \dfrac{y^2}{x} = 5$

17. $xy = \sin(x + y)$

18. $x + y = \cos xy$

19. $x = \sec y$

20. $x \sin y - y \cos x = 1$

In Problems 21 and 22 find the indicated derivative.

21. $r^2 = \sin 2\theta$; $dr/d\theta$

22. $\pi r^2 h = 100$; dh/dr

In Problems 23 and 24 find dy/dx at the indicated point.

23. $xy^2 + 4y^3 + 3x = 0$; $(1, -1)$

24. $y = \sin xy$; $(\pi/2, 1)$

In Problems 25 and 26 find dy/dx at the points that correspond to the indicated value.

25. $2y^2 + 2xy - 1 = 0$; $x = 1/2$

26. $y^3 + 2x^2 = 11y$; $y = 1$

In Problems 27–30 find an equation of the tangent line at the indicated point or value.

27. $x^4 + y^3 = 24$; $(-2, 2)$

28. $\dfrac{1}{x} + \dfrac{1}{y} = 1$; $x = 3$

29. $\tan y = x$; $y = \pi/4$

30. $3y + \cos y = x^2$; $(1, 0)$

In Problems 31 and 32 determine the point(s) on the graph of the given equation where the tangent is horizontal.

31. $x^2 - xy + y^2 = 3$

32. $y^2 = x^2 - 4x + 7$

33. Find the point(s) on the graph of $x^2 + y^2 = 25$ at which the slope of the tangent is $\frac{1}{2}$.

34. Find the point of intersection of the tangents to the graph of $x^2 + y^2 = 25$ at $(-3, 4)$ and $(-3, -4)$.

In Problems 35 and 36 find dy/dx but show that the given equation does not define any real function.

35. $x^2 - 6x + y^2 + 8y + 27 = 0$ (*Hint*: Complete the square.)

36. $x^4 + 3x^2y^2 + 5 = 0$

In Problems 37–42 find d^2y/dx^2.

37. $4y^3 = 6x^2 + 1$

38. $xy^4 = 5$

39. $x + y = \sin y$

40. $y^2 - x^2 = \tan 2x$

41. $x^2 + 2xy - y^2 = 1$

42. $x^3 + y^3 = 27$

In Problems 43 and 44, first, use implicit differentiation to find dy/dx. Then, solve for y explicitly in terms of x and differentiate. Show that the two answers are equivalent.

43. $x^3y = x + 1$

44. $y \sin x = x - 2y$

In Problems 45–48 (Figures 3.37–3.40) determine an implicit function from the given equation such that its graph is the colored curve in the figure.

45. $(y - 1)^2 = x - 2$

46. $x^2 + xy + y^2 = 4$

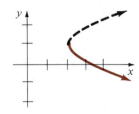

Figure 3.37

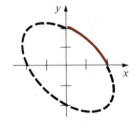

Figure 3.38

47. $x^2 + y^2 = 4$

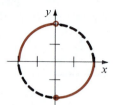

Figure 3.39

48. $y^2 = x^2(2 - x)$

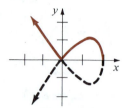

Figure 3.40

Determine the angle between the graphs of $x^2 + y^2 + 4y = 6$ and $x^2 + 2x + y^2 = 4$ at $(1, 1)$.

Calculator Problem

52. The graph of $(x^2 + y^2)^2 = 4(x^2 - y^2)$ shown in Figure 3.41 is called a **lemniscate**.

    (a) Find the points on the graph that correspond to $x = 1$. Use values with two decimal places.

    (b) Find an equation of the tangent line to the graph at each point found in part **(a)**.

    (c) Find the points on the graph at which the tangent is horizontal.

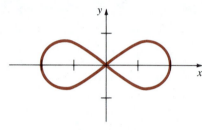

Figure 3.41

Miscellaneous Problems

In Problems 49 and 50 assume that both x and y are differentiable functions of a variable t. Find dy/dt in terms of x, y, and dx/dt.

49. $x^2 + y^2 = 25$ **50.** $x^2 + xy + y^2 - y = 9$

51. The angle θ $(0 < \theta < \pi)$ between two curves is defined to be the angle between their tangent lines at the point P of intersection. If m_1 and m_2 are the slopes of the tangent lines at P, it can be shown that $\tan \theta = (m_1 - m_2)/(1 + m_1 m_2)$.

3.9   Rules of Differentiation IV: Extended Power Rules

In Sections 3.3 and 3.6 we restricted the Power Rule and the Power Rule for Functions to *integer* exponents:

$$\frac{d}{dx} x^n = nx^{n-1} \tag{3.47}$$

$$\frac{d}{dx} u^n = nu^{n-1} \frac{du}{dx}, \qquad u = g(x) \tag{3.48}$$

Implicit differentiation provides a means of extending both (3.47) and (3.48) to *rational* exponents. If p and q are integers, $q \neq 0$, then for values of x for which $x^{p/q}$ is a real number,

$$y = x^{p/q} \tag{3.49}$$

is a function that yields

$$y^q = x^p \tag{3.50}$$

Now, assuming y' exists and $y \neq 0$, it follows from implicit differentiation that

$$\frac{d}{dx} y^q = \frac{d}{dx} x^p$$

$$q y^{q-1} \frac{dy}{dx} = p x^{p-1}$$

$$\frac{dy}{dx} = \frac{p}{q} \frac{x^{p-1}}{y^{q-1}}$$

$$= \frac{p}{q} \frac{x^{p-1}}{(x^{p/q})^{q-1}} \qquad \text{[from (3.49)]}$$

$$= \frac{p}{q} x^{(p/q)-1}$$

This last result leads to the following extension of (3.47).

THEOREM 3.11   Power Rule (Rational Exponents)

If p/q is a rational number, then

$$\frac{d}{dx} x^{p/q} = \frac{p}{q} x^{(p/q)-1} \qquad\qquad (3.51)$$

We note that (3.51) reduces to (3.47) when $p = n$ and $q = 1$. Furthermore, there is nothing new to memorize in (3.51); it is the same rule as before: *Bring down the exponent as a multiple and decrease the exponent by one.*

Example 1

Differentiate $y = \sqrt{x}$.

Solution   First, write the given function as $y = x^{1/2}$ and then use (3.51):

$$\frac{dy}{dx} = \frac{1}{2} x^{(1/2)-1}$$

$$= \frac{1}{2} x^{-1/2} = \frac{1}{2\sqrt{x}}$$

Example 2

Find an equation of the tangent line to the graph of $y = \sqrt{x}$ at $x = 4$.

Solution   When $x = 4$, $y = \sqrt{4} = 2$. It follows from Example 1 that the slope of the tangent line at $(4, 2)$ is

$$\frac{dy}{dx}\bigg|_{x=4} = \frac{1}{2\sqrt{4}} = \frac{1}{4}$$

Then an equation of the tangent line is

$$y - 2 = \frac{1}{4}(x - 4) \quad \text{or} \quad y = \frac{1}{4}x + 1$$

Figure 3.42 shows the graphs of the function and the tangent line.

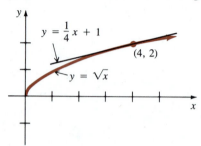

Figure 3.42

_____ **Example 3** _____

Differentiate $y = 1/\sqrt{x}$.

Solution   Since $y = x^{-1/2}$, it follows from (3.51) that

$$\frac{dy}{dx} = -\frac{1}{2}x^{(-1/2)-1} = -\frac{1}{2}x^{-3/2}$$

_____ **Example 4** _____

Differentiate $y = 9\sqrt[3]{x} + 4\sqrt{x^3}$.

Solution   Using rational exponents, the given function can be expressed as $y = 9x^{1/3} + 4x^{3/2}$. Thus,

$$\frac{dy}{dx} = 9(1/3)x^{(1/3)-1} + 4(3/2)x^{(3/2)-1}$$

$$= 3x^{-2/3} + 6x^{1/2}$$

We can show, in a manner similar to that leading to (3.51), that the Power Rule for Functions (3.48) is also true for rational exponents. This result is summarized in the next theorem.

> **THEOREM 3.12**   **Power Rule for Functions** (Rational Exponents)
>
> If p/q is a rational number and g is a differentiable function, then
>
> $$\frac{d}{dx}[g(x)]^{p/q} = \frac{p}{q}[g(x)]^{(p/q)-1} \cdot g'(x) \qquad (3.52)$$

Example 5

Differentiate $y = (4x + 1)^{1/3}$.

Solution From (3.52),

$$\frac{dy}{dx} = \frac{1}{3}(4x + 1)^{(1/3)-1} \frac{d}{dx}(4x + 1)$$

$$= \frac{4}{3}(4x + 1)^{-2/3}$$

Example 6

Differentiate $y = \sqrt{\dfrac{1 + x}{1 - x}}$.

Solution We use (3.52) followed by the Quotient Rule,

$$\frac{dy}{dx} = \frac{1}{2}\left[\frac{1 + x}{1 - x}\right]^{-1/2} \cdot \frac{d}{dx}\left(\frac{1 + x}{1 - x}\right)$$

$$= \frac{1}{2}\left[\frac{1 + x}{1 - x}\right]^{-1/2} \cdot \frac{(1 - x) \cdot 1 - (1 + x)(-1)}{(1 - x)^2}$$

$$= \frac{1}{(1 - x)^2}\left[\frac{1 + x}{1 - x}\right]^{-1/2}$$

$$= \frac{1}{(1 + x)^{1/2}(1 - x)^{3/2}}$$

Example 7

Differentiate $y = (3x^2 + \sqrt{x^2 + 1})^5$.

Solution The Power Rule for Functions (3.52) gives

$$\frac{dy}{dx} = 5(3x^2 + \sqrt{x^2 + 1})^4 \cdot \frac{d}{dx}(3x^2 + \sqrt{x^2 + 1})$$

$$= 5(3x^2 + \sqrt{x^2 + 1})^4 \cdot \left[6x + \frac{1}{2}(x^2 + 1)^{-1/2} \cdot 2x\right]$$

$$= 5(3x^2 + \sqrt{x^2 + 1})^4\left[6x + \frac{x}{\sqrt{x^2 + 1}}\right]$$

Recall from Section 3.2 that the graph of a function f continuous at a number a has a vertical tangent at $(a, f(a))$ if $\lim_{x \to a} |f'(x)| = \infty$. The graphs of many functions with rational exponents possess vertical tangents.

_____ **Example 8** _____

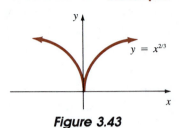

Figure 3.43

For $f(x) = x^{2/3}$, we have

$$f'(x) = \frac{2}{3}x^{-1/3} = \frac{2}{3x^{1/3}}$$

Notice that $\lim_{x \to 0^+} f'(x) = \infty$, whereas $\lim_{x \to 0^-} f'(x) = -\infty$. Since f is continuous at $x = 0$ and $|f'(x)| \to \infty$ as $x \to 0$, we conclude that the y-axis is a vertical tangent at $(0, 0)$. This fact is apparent from the graph in Figure 3.43.

Remark

In Chapter 8 the results in (3.47) and (3.48) will be extended even further to any real exponent (that is, rational *or* irrational.)

_____ **Exercises 3.9** _____

Answers to odd-numbered problems begin on page A-22.

In Problems 1–20 find the derivative of the given function.

1. $y = 10x^{3/2}$

2. $y = 8x^{-1.2}$

3. $y = \dfrac{1}{\sqrt[3]{x^4}}$

4. $y = \sqrt[5]{x^3}$

5. $y = \sqrt{x} + \dfrac{1}{\sqrt{x}}$

6. $y = \sqrt{4x^2 + 9}$

7. $y = (x^2 + 1)(x^2 - 4)^{2/3}$

8. $y = \dfrac{\sqrt{x}}{x^2 + 1}$

9. $y = \left(\dfrac{x - 9}{x + 2}\right)^{3/2}$

10. $y = \sqrt{2x + 1}\ \sqrt[3]{3x - 1}$

11. $y = \sin \sqrt{x}$

12. $y = x \cot \dfrac{1}{\sqrt{x}}$

13. $f(x) = x + \sqrt{x^2 + 1}$

14. $f(x) = \sqrt{x + \sqrt{x}}$

15. $g(t) = [(t^2 - 1)(t^3 + 4t)]^{1/3}$

16. $H(z) = \dfrac{\sqrt{z} + 1}{\sqrt{z} + 3}$

17. $q(\theta) = \dfrac{1}{\sqrt{\theta + \sin \theta}}$

18. $r(\theta) = \sqrt{\cos 4\theta}$

19. $F(s) = \sqrt[5]{(s^4 + 1)^2}$

20. $g(u) = u^{2/3}(u^{1/3} + 1)^3$

In Problems 21–26 find the second derivative of the given function.

21. $y = x + \sqrt{x}$

22. $y = 18x^{4/3}$

23. $y = \dfrac{x^{5/3} + 6x^{4/3} - 9x^{1/3}}{x}$

24. $y = (\sqrt{x} + 1)^4$

25. $F(\theta) = (\sin \theta)^{2.4}$

26. $f(t) = \left(\dfrac{t}{t + 6}\right)^{0.3}$

In Problems 27–30 find an equation of the tangent line to the graph of the given function at the indicated value of x.

27. $y = x^{1/3};\ x = 8$

28. $y = \sqrt{2x + 1};\ x = 4$

29. $y = \dfrac{x^2}{\sqrt{x^2 + 3}};\ x = 1$

30. $y = (\tan 2x)^{1/3};\ x = \pi/8$

In Problems 31–34 use implicit differentiation to find dy/dx.

31. $x + \sqrt{xy} + y = 1$

32. $xy^2 = \sqrt{x + 1}$

33. $2y^{3/2} = 3x(x^2 - 1)^{5/2}$

34. $(6x)^{1/2} + (8y)^{3/2} = 2$

35. The graph of $x^{2/3} + y^{2/3} = 1$, shown in Figure 3.44, is called a **hypocycloid**. Find equations of the tangent lines to the graph at the points corresponding to $x = \frac{1}{8}$.

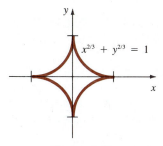

Figure 3.44

36. Find d^2y/dx^2 for the equation in Problem 35.

In Problems 37–42 determine whether the graph of the given function possesses any vertical tangents.

37. $y = (2x - 8)^{2/3}$

38. $y = 4x^2 + 6x^{1/3}$

39. $f(x) = x^{-1/3} + 1$

40. $f(x) = (x^2 + 9)^{1/3}$

41. $f(x) = \dfrac{1}{x^{1/3} + 1}$

42. $f(x) = (x + 1)^{1/3}(x - 5)^{2/3}$

43. The function $T = 2\pi\sqrt{L/g}$, g a constant, gives the period of a simple pendulum of length L. Given that $g = 32$ ft/sec^2, find dT/dL when $L = 2$ ft.

44. According to G.K. Zipf, the number N of cities in the United States that have a population over q million is estimated by $N = 24q^{-1.3}$. Find the instantaneous rate of change of N with respect to q.

45. The velocity v of a rocket y kilometers above the center of the earth is given by $v = \sqrt{2k/y - 2k/R + v_0^2}$, where k, R, and v_0 are constants. Find dv/dy.

46. According to the theory of relativity, the mass m of a body moving with velocity v is $m = m_0/\sqrt{1 - v^2/c^2}$, where m_0 is the initial mass and c is the speed of light. Find dm/dv.

47. In special circumstances the function

$$g(\gamma) = F_0/\sqrt{(\omega^2 - \gamma^2)^2 + 4\lambda^2\gamma^2}$$

where F_0, ω, and λ are constants, gives the amplitude of motion of a mass on a vibrating spring. Verify that $g'(\sqrt{\omega^2 - 2\lambda^2}) = 0$.

48. The amount of substance X present at time t during a third-order chemical reaction is given by

$$X(t) = \left(\frac{X_0^2}{2kX_0^2 t + 1}\right)^{1/2}$$

where X_0 and k are constants. Find $X'(1)$.

49. The surface area S of a human that has weight W is estimated by $S = 0.11W^{2/3}$. Find dS/dW.

Calculator Problem

50. An estimation of the percent saturation of hemoglobin in a human is given by

$$f(P) = \frac{0.013P^{2.7}}{1 + 0.00013P^{2.7}}$$

where P represents the partial pressure of oxygen in plasma. Find the instantaneous rate of change of f when $P = 40$.

3.10 Differentials

We started the discussion of the derivative with the problem of finding the slope of a tangent line to the graph of a function $y = f(x)$. As shown in Figure 3.45, the starting point for the solution of this problem was the consideration of

$$m_{\text{sec}} = \frac{f(x + \Delta x) - f(x)}{\Delta x} = \frac{\Delta y}{\Delta x}$$

For small values of Δx, $m_{\text{sec}} \approx m_{\text{tan}}$ or $\Delta y/\Delta x \approx m_{\text{tan}}$. But knowing that $m_{\text{tan}} = f'(x)$ enables us to write

$$\frac{\Delta y}{\Delta x} \approx f'(x)$$

or

$$\Delta y \approx f'(x)\,\Delta x \tag{3.53}$$

For convenience we rename the number Δx as follows.

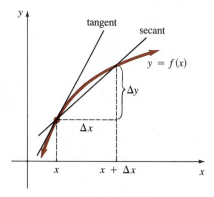

Figure 3.45

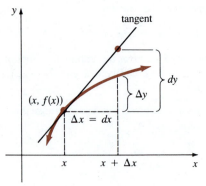

tangent

$(x, f(x))$

$\Delta x = dx$

Δy

dy

x $x + \Delta x$ x

Figure 3.46

> **DEFINITION 3.4**   The increment Δx is called the **differential of the independent variable** x and is denoted by dx; that is, $dx = \Delta x$.

We also rename the quantity $f'(x)\, \Delta x$ in (3.53).

> **DEFINITION 3.5**   The function $f'(x)\, \Delta x$ is called the **differential of the dependent variable** y and is denoted by dy; that is, $dy = f'(x)\, \Delta x = f'(x)\, dx$.

Since the slope of a tangent to a graph is

$$m_{\tan} = \frac{\text{rise}}{\text{run}} = f'(x) = \frac{f'(x)\, \Delta x}{\Delta x}, \qquad \Delta x \neq 0$$

it follows that the rise of the tangent line can be interpreted as dy.* From Figure 3.46 we see that when Δx is very small ($\Delta x \approx 0$),

$$\Delta y \approx dy \tag{3.54}$$

Example 1

(a)   Find Δy and dy for $y = 5x^2 + 4x + 1$.

(b)   Compare the values of Δy and dy for $x = 6$, $\Delta x = dx = 0.02$.

Solution

(a) $\Delta y = f(x + \Delta x) - f(x)$

$\qquad = [5(x + \Delta x)^2 + 4(x + \Delta x) + 1] - [5x^2 + 4x + 1]$

$\qquad = 10x\, \Delta x + 4\, \Delta x + 5(\Delta x)^2$

Now, by Definition 3.5,

$$dy = (10x + 4)\, dx$$

Since $dx = \Delta x$, observe that $\Delta y = (10x + 4)\, \Delta x + 5(\Delta x)^2$ and $dy = (10x + 4)\, \Delta x$ differ by the amount $5(\Delta x)^2$.

(b)   When $x = 6$, $\Delta x = 0.02$

$$\Delta y = 10(6)(0.02) + 4(0.02) + 5(0.02)^2$$

$$= 1.282$$

whereas $$dy = (10(6) + 4)(0.02)$$

$$= 1.28$$

The difference in answers is, of course, $5(0.02)^2 = 0.002$.

---

*For this reason, the derivative symbol dy/dx has the appearance of a quotient. To calculate dy, it looks as if both sides of the equality $dy/dx = f'(x)$ are multiplied by the denominator of the left member. While this is, strictly speaking, *not* the case, one can proceed *formally* in this manner.

Approximations

When $\Delta x \approx 0$, differentials give a means of "predicting" the value of $f(x + \Delta x)$ by knowing the value of the function and its derivative at x. As we see in Figure 3.47, if x changes by an amount Δx, then the corresponding change in the function is $\Delta y = f(x + \Delta x) - f(x)$ and so

$$f(x + \Delta x) = f(x) + \Delta y$$

But in light of (3.54), for a small change in x, we can then write

$$f(x + \Delta x) \approx f(x) + dy$$

That is,

$$f(x + \Delta x) \approx f(x) + f'(x)\, dx \qquad (3.55)$$

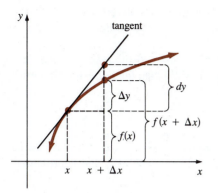

Figure 3.47

Example 2

Use (3.55) to find an approximation to $\sqrt{25.4}$.

Solution   First, identify the function $f(x) = \sqrt{x}$. We wish to calculate the approximate value of $f(x + \Delta x) = \sqrt{x + \Delta x}$ when $x = 25$ and $\Delta x = 0.4$. Now

$$dy = \frac{1}{2}x^{-1/2}\, dx = \frac{1}{2\sqrt{x}}\, \Delta x$$

so that (3.55) yields

$$\sqrt{x + \Delta x} \approx \sqrt{x} + \frac{1}{2\sqrt{x}}\, \Delta x \qquad (3.56)$$

When $x = 25$, $\Delta x = 0.4$, (3.56) becomes

$$\sqrt{25.4} \approx \sqrt{25} + \frac{1}{2\sqrt{25}}(0.4) = 5.04$$

Error

The **error** in a calculation is defined to be

$$\textbf{error} = \textbf{true value} - \textbf{approximate value} \qquad (3.57)$$

However, in practice the

$$\textbf{relative error} = \frac{\textbf{error}}{\textbf{true value}} \qquad (3.58)$$

is usually more important than the error. Moreover, (relative error) · 100 is called the **percentage error**. With the aid of a hand calculator, $\sqrt{25.4} = 5.03984$ is correct to five decimal places. Thus, in Example 2 the error is -0.00016, the relative error is -0.00003, and the percentage error is -0.003%.

Example 3

A side of a cube is measured to be 30 cm with a possible error of ± 0.02 cm. What is the approximate maximum possible error in the volume of the cube?

Solution The volume of a cube is $V = x^3$, where x is the length of one side. If Δx represents the error in the length of one side, then the corresponding error in the volume is

$$\Delta V = (x + \Delta x)^3 - x^3$$

To simplify matters, we utilize the differential $dV = 3x^2\,dx = 3x^2\,\Delta x$ as an approximation to ΔV. Thus, for $x = 30$ and $\Delta x = \pm 0.02$, the approximate maximum error is

$$dV = 3(30)^2(\pm 0.02) = \pm 54 \text{ cm}^3$$

In Example 3, an error of about 54 cm^3 in the volume for an error of 0.02 cm in the length of a side seems considerable. But, observe, if the relative error is $\Delta V/V$, then the approximate relative error is dV/V. When $x = 30$, $V = 27{,}000$, the approximate maximum relative error is $\pm 54/27{,}000 = \pm 1/500$, and the maximum percentage error is approximately $\pm 0.2\%$.

Rules for Differentials The rules for differentiation considered in this chapter can be rephrased in terms of differentials; for example, if $u = f(x)$ and $v = g(x)$ and $y = f(x) + g(x)$, then $dy/dx = f'(x) + g'(x)$. Hence, $dy = [f'(x) + g'(x)]\,dx = f'(x)\,dx + g'(x)\,dx = du + dv$. We summarize the equivalents of the sum, product, and quotient rules:

$$d(u + v) = du + dv \tag{3.59}$$

$$d(uv) = u\,dv + v\,du \tag{3.60}$$

$$d(u/v) = \frac{v\,du - u\,dv}{v^2} \tag{3.61}$$

As the next example shows, there is little need for memorizing (3.59), (3.60), and (3.61).

Example 4

Find dy for $y = x^2 \cos 3x$.

Solution To find the differential of a function, we can simply multiply its derivative by dx. Thus, by the Product Rule,

$$\frac{dy}{dx} = x^2(-\sin 3x \cdot 3) + \cos 3x\,(2x)$$

$$dy = \left(\frac{dy}{dx}\right) dx = (-3x^2 \sin 3x + 2x \cos 3x)\,dx \tag{3.62}$$

Alternative Solution   Applying (3.60) gives

$$dy = x^2 d(\cos 3x) + \cos 3x \, d(x^2)$$
$$= x^2(-\sin 3x \cdot 3 \, dx) + \cos 3x(2x \, dx) \qquad (3.63)$$

Factoring dx from (3.63) yields (3.62).

_____ **Exercises 3.10** _____

Answers to odd-numbered problems begin on page A-22.

In Problems 1–10 find the differential dy.

1. $y = 10$

2. $y = x^4 + 3x^2$

3. $y = \dfrac{1}{\sqrt{2x}}$

4. $y = \sqrt[4]{x^7}$

5. $y = 12(x^4 - 1)^{1/3}$

6. $y = x^2(1 - x)^5$

7. $y = \dfrac{x^2 - 1}{x^2 + 1}$

8. $y = \dfrac{x}{(3x - 1)^4}$

9. $y = x \cos x - \sin x$

10. $y = (2x + 1)\csc 2x$

In Problems 11–18 find Δy and dy.

11. $y = x^2 + 1$

12. $y = 3x^2 - 5x + 6$

13. $y = (x + 1)^2$

14. $y = x^3$

15. $y = \dfrac{3x + 1}{x}$

16. $y = \dfrac{1}{x^2}$

17. $y = \sin x$

18. $y = -4 \cos 2x$

In Problems 19 and 20 complete the following table for each function.

x	Δx	Δy	dy	$\Delta y - dy$
2	1			
2	0.5			
2	0.1			
2	0.01			

19. $y = 5x^2$

20. $y = \dfrac{1}{x}$

In Problems 21–30 use the concept of the differential to find an approximation to the given expression.

21. $\sqrt{37}$

22. $\dfrac{1}{\sqrt{96}}$

23. $(1.8)^5$

24. $9^{2/3}$

25. $\dfrac{(0.9)^4}{(0.9) + 1}$

26. $(1.1)^3 + 6(1.1)^2$

27. $\cos\left(\dfrac{\pi}{2} - 0.4\right)$

28. $\sin 1°$

29. $\sin 33°$

30. $\tan\left(\dfrac{\pi}{4} + 0.1\right)$

31. The area of a circle with radius r is $A = \pi r^2$.

   (a) Given that the radius of a circle changes from 4 cm to 5 cm, find the exact change in the area.

   (b) What is the approximate change in the area?

32. According to Poiseuille, the resistance R of a blood vessel of length l and radius r is $R = kl/r^4$, where k is a constant. Given that l is constant, find the approximate change in R when r changes from 0.2 mm to 0.3 mm.

33. Many golf balls consist of a spherical cover over a solid core. Find the exact volume of the cover if its thickness is t and the radius of the core is r. (*Hint*: The volume of a sphere is $V = \frac{4}{3}\pi r^3$. Consider concentric spheres having radii r and $r + \Delta r$.) Use differentials to find an approximation to the volume of the cover. See Figure 3.48. Find an approximation to the volume of the cover if $r = 0.8$ in. and $t = 0.04$ in.

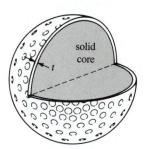

Figure 3.48

34. A hollow metal pipe is 1.5 m long. Find an approximation to the volume of the metal if the inner radius of the pipe is 2 cm and the thickness of the metal is 0.25 cm. See Figure 3.49.

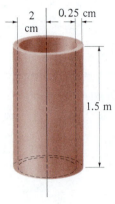

Figure 3.49

35. The side of a square is measured to be 10 cm with a possible error of ±0.3 cm. Use differentials to find an approximation to the maximum error in the area. Find the approximate relative error and the approximate percentage error.

36. An oil storage tank in the form of a circular cylinder has a height of 5 m. The radius is measured to be 8 m with a possible error of ±0.25 m. Use differentials to estimate the maximum error in the volume. Find the approximate relative error and the approximate percentage error.

37. In the study of some adiabatic processes, the pressure P of a gas is related to the volume V that it occupies by $P = c/V^\gamma$, where c and γ are constants. Show that the approximate relative error in P is proportional to the approximate relative error in V.

38. The range R of a projectile with an initial velocity v_0 and angle of elevation θ is given by $R = (v_0^2/g)\sin 2\theta$, where g is the acceleration of gravity. If v_0 and θ are held constant, then show that the percentage error in the range is proportional to the percentage error in g.

39. Use the formula in Problem 38 to determine the range of a projectile when the initial velocity is 256 ft/sec, the angle of elevation is 45°, and the acceleration of gravity is 32 ft/sec^2. What is the approximate change in the range of the projectile if the initial velocity is increased to 266 ft/sec?

40. The acceleration of gravity g is not constant but changes with altitude. For practical purposes, at the surface of the earth g is taken to be 32 ft/sec^2, 980 cm/sec^2, or 9.8 m/sec^2.

(a) From the Law of Universal Gravitation, the force F between a body of mass m_1 and the earth of mass m_2 is $F = km_1m_2/r^2$, where k is constant and r is the distance to the center of the earth. Alternatively, Newton's second law of motion implies $F = m_1g$. Show that $g = km_2/r^2$.

(b) Use part (a) to show $dg/g = -2dr/r$.

(c) Let $r = 6400$ km at the surface of the earth. Use part (b) to show that the approximate value of g at an altitude of 16 km is 9.75 m/sec^2.

Calculator Problems

41. In Problem 23 use a calculator to compute the error. Compute the relative error and the percentage error.

42. The period (in seconds) of a simple pendulum of length L is $T = 2\pi\sqrt{L/g}$, where g is the acceleration of gravity. Compute the exact change in the period if L is increased from 4 to 5 meters. Then, use differentials to find an approximation to the change in the period. Assume $g = 9.8$ m/sec^2.

43. In Problem 42, given that L is fixed at 4 meters, find an approximation to the change in the period if the pendulum is moved to an altitude where $g = 9.75$ m/sec^2.

3.11 Newton's Method

There are few straightforward methods for finding roots of an equation

$$f(x) = 0$$

For polynomial equations of degree 4 or less, we can always solve the equation by means of a formula that expresses the answers in terms of the coefficients of $f(x)$. We know, of course, that $ax^2 + bx + c = 0$, $a \neq 0$, can be solved by the quadratic formula. One of the major achievements in mathematics was the proof that polynomial equations of degree greater than 4 cannot be solved by means of a formula.* Thus, solving the algebraic equation

$$x^5 - 3x^2 + 4x - 6 = 0 \tag{3.64}$$

poses a quandary unless the polynomial factors. Furthermore, in scientific analyses, one is often asked to find roots of transcendental equations such as

$$2x = \tan x \tag{3.65}$$

In the case of problems such as (3.64) and (3.65), it is common practice to employ a technique that yields an *approximation* or *estimation* of the roots. One such procedure, known as **Newton's Method**,[†] employs the derivative of a function.

An Iterative Technique

Figure 3.50

Suppose f is differentiable and suppose c represents the unknown root of $f(x) = 0$; that is, $f(c) = 0$. Let x_0 denote a number that is chosen arbitrarily as a first guess to c, If $f(x_0) \neq 0$, compute $f'(x_0)$ and, as shown in Figure 3.50, construct a tangent to the graph of f at $(x_0, f(x_0))$. If we now let x_1 denote the x-intercept of this line, we must have

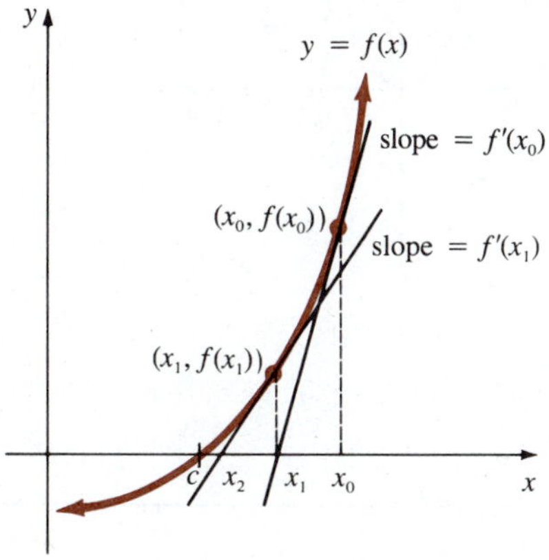

$$\text{slope of line} = f'(x_0) = \frac{f(x_0)}{x_0 - x_1}$$

Solving for x_1, then gives

$$x_1 = x_0 - \frac{f(x_0)}{f'(x_0)}$$

Repeat the procedure at $(x_1, f(x_1))$ and let x_2 be the x-intercept of the second tangent line. From

$$f'(x_1) = \frac{f(x_1)}{x_1 - x_2}$$

we find

$$x_2 = x_1 - \frac{f(x_1)}{f'(x_1)}$$

Continuing in this fashion, we determine x_{n+1} from

$$x_{n+1} = x_n - \frac{f(x_n)}{f'(x_n)} \tag{3.66}$$

The repetitive use, or **iteration**, of (3.66) yields a sequence $x_1, x_2, x_3, \ldots$ of approximations that we expect *converges* to the root c; that is, $x_n \to c$ as n increases.

---

*This was proved by the Norwegian mathematician, Niels Henrik Abel (1802–1829).
†This is also called the **Newton–Raphson Method**.

Graphical Analysis   Before applying (3.66) let's try to determine the existence and number of real roots of $f(x) = 0$ through graphical means; for example, the irrational number $\sqrt{3}$ can be interpreted as either

(*i*)   a root of the quadratic equation $x^2 - 3 = 0$, and, hence, as a zero of the continuous function $f(x) = x^2 - 3$, or

(*ii*)   as the x-coordinate of a point of intersection of the graphs of $y = x^2$ and $y = 3$.

Both interpretations are illustrated in Figure 3.51. Of course, another reason for a graph is to enable us to choose the initial guess x_0 so that it is close to the root c.

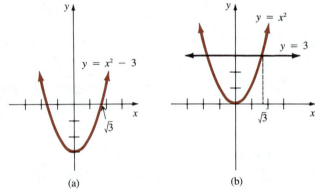

(a)   (b)

Figure 3.51

Example 1

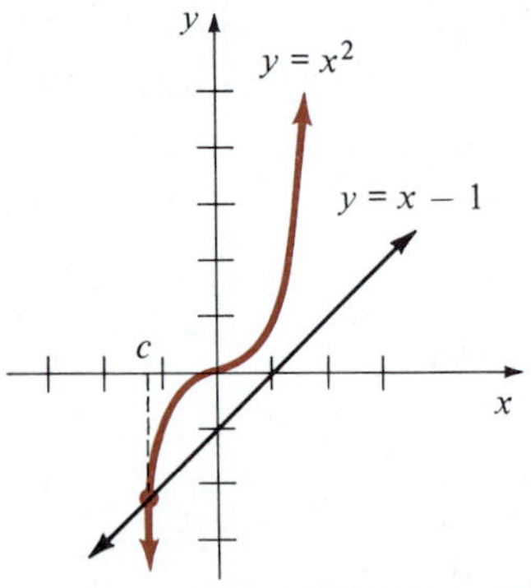

Determine the number of real roots of $x^3 - x + 1 = 0$.

Solution   From Figure 3.52 we see that the graphs of the functions

$$y = x^3 \quad \text{and} \quad y = x - 1$$

intersect at one point. Hence we conclude that the equation

$$x^3 = x - 1$$

or

$$x^3 - x + 1 = 0$$

possesses only one real root.*

Figure 3.52

---

*A cubic polynomial equation with real coefficients always has *at least* one real root, because complex roots of the form $a + bi$, $i^2 = -1$, $b \neq 0$, must always appear in pairs.

Example 2

Approximate $\sqrt{3}$ by Newton's Method.

Solution If we define $f(x) = x^2 - 3$, then $f'(x) = 2x$, and (3.66) becomes

$$x_{n+1} = x_n - \frac{x_n^2 - 3}{2x_n}$$

$$= \frac{1}{2}\left(x_n + \frac{3}{x_n}\right)$$

Since $1 < \sqrt{3} < 2$ it seems reasonable to choose $x_0 = 1$. Thus,

$$x_1 = \frac{1}{2}\left(x_0 + \frac{3}{x_0}\right) = \frac{1}{2}(1 + 3) = 2$$

$$x_2 = \frac{1}{2}\left(x_1 + \frac{3}{x_1}\right) = \frac{1}{2}\left(2 + \frac{3}{2}\right) = 1.75$$

$$x_3 = \frac{1}{2}\left(x_2 + \frac{3}{x_2}\right) = \frac{1}{2}\left(\frac{7}{4} + \frac{12}{7}\right) \approx 1.7321$$

$$x_4 = \frac{1}{2}\left(x_3 + \frac{3}{x_3}\right) \approx 1.7321$$

Since there is no significant difference in x_3 and x_4, it makes sense to stop the iteration. Indeed, we can prove that $\sqrt{3} = 1.73205$ is accurate to five decimal places.

Example 3

Use Newton's Method to find an approximation to the real root of the equation $x^3 - x + 1 = 0$.

Solution Let $f(x) = x^3 - x + 1$ so that $f'(x) = 3x^2 - 1$. Hence, (3.66) is

$$x_{n+1} = x_n - \frac{x_n^3 - x_n + 1}{3x_n^2 - 1}$$

$$= \frac{2x_n^3 - 1}{3x_n^2 - 1}$$

If we are interested in three and possibly four decimal place accuracy, we shall carry out the iteration until two successive iterants agree to four decimal places. Also, Figure 3.52 prompts us to make $x_0 = -1.5$ the initial guess. Consequently,

$$x_1 = \frac{2x_0^3 - 1}{3x_0^2 - 1} = \frac{2(-1.5)^3 - 1}{3(-1.5)^2 - 1} \approx -1.3478$$

$$x_2 = \frac{2x_1^3 - 1}{3x_1^2 - 1} \approx -1.3252$$

$$x_3 = \frac{2x_2^3 - 1}{3x_2^2 - 1} \approx -1.3247$$

$$x_4 = \frac{2x_3^3 - 1}{3x_3^2 - 1} \approx 1.3247$$

Hence, the root of the given equation is approximately -1.3247.*

Example 4

Find the smallest positive root of $2x = \tan x$.

Solution   Figure 3.53 shows that the equation has an infinite number of roots. With $f(x) = 2x - \tan x$ and $f'(x) = 2 - \sec^2 x$, (3.66) becomes

$$x_{n+1} = x_n - \frac{2x_n - \tan x_n}{2 - \sec^2 x_n}$$

Since calculators and computers do not possess a secant routine, we express the last equation in terms of $\sin x$ and $\cos x$:

$$x_{n+1} = x_n - \frac{2x_n \cos^2 x_n - \sin x_n \cos x_n}{2 \cos^2 x_n - 1} \tag{3.67}$$

It appears from Figure 3.53 that the first positive root is near $x_0 = 1$. Iteration of (3.67) then yields

$$x_1 \approx 1.3105$$
$$x_2 \approx 1.2239$$
$$x_3 \approx 1.1761$$
$$x_4 \approx 1.1659$$
$$x_5 \approx 1.1656$$
$$x_6 \approx 1.1656$$

We conclude that the first positive root is approximately 1.1656.

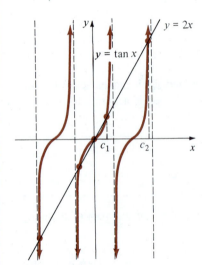

Figure 3.53

Example 4 illustrates the importance of the selection of the initial value x_0. You should verify that the choice $x_0 = \frac{1}{2}$ in (3.67) leads to a sequence of values $x_1, x_2, x_3, \ldots$ that converge to the one obvious root $c = 0$.

BASIC Program   A BASIC program for Newton's Method is listed below. In the program, FND(X) denotes the derivative of the function $y = f(x)$.

---

*If you are working through this example on a calculator, you may encounter difficulties when computing numbers such as $(-1.5)^3$ and $(-1.3478)^2$ using the $\boxed{y^x}$ key. The reason why this function usually demands $y > 0$ will be explained in Section 8.4. But, there is no real problem since $(-1.5)^3 = -(1.5)^3$ and $(-1.3478)^2 = (1.3478)^2$.

```
10 REM NEWTON'S METHOD FOR FINDING ROOTS
20 INPUT "THE INITIAL VALUE IS ";X0
30 DEF FNY(X) = ...
40 DEF FND(X) = ...
50 X1 = X0 - FNY(X0)/FND(X0)
60 PRINT X1
70 LET X0 = X1
80 GO TO 50
90 END
```

Those familiar with BASIC should provide an exit from the loop in the foregoing program.

Remark

There are problems with Newton's Method.

(*i*)   We must compute $f'(x)$. Needless to say, the form of $f'(x)$ could be formidable when the equation $f(x) = 0$ is complicated.

(*ii*) If the root c of $f(x) = 0$ is near a value for which $f'(x) = 0$, then the denominator in (3.66) is approaching zero. This necessitates a computation of $f(x_n)$ and $f'(x_n)$ to a high degree of accuracy. A calculation of this kind usually requires a computer with a double precision routine.

(*iii*) It is necessary to find an approximate location of a root of $f(x) = 0$ before x_0 is chosen. Attendant to this are the usual difficulties in graphing. But, worse, the iteration of (3.66) *may not converge* for an imprudent or perhaps blindly chosen x_0. In Figure 3.54(a) we see x_2 is undefined because $f'(x_1) = 0$. In Figure 3.54(b) we see what may happen to the tangent lines when x_0 is not close to c. In Figure 3.54(c) observe that when $f(x_0) = -f(x_1)$ and $f'(x_0) = f'(x_1)$, the tangent lines will "bounce" back and forth between two points $(x_0, f(x_0))$ and $(x_1, f(x_1))$. (See Problems 21 and 22, Exercises 3.11.)

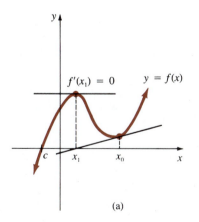

(a)

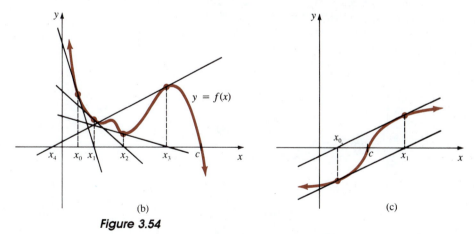

(b)   (c)

Figure 3.54

These three problems notwithstanding, the major advantage of Newton's Method is that when it converges to a root it usually does so rather rapidly. It can be shown that under certain conditions Newton's Method converges *quadratically*. Very roughly, this means that the number of places of accuracy can, but will not necessarily, double with each iteration.

_____ Exercises 3.11 _____

Answers to odd-numbered problems begin on page A-23.

Where appropriate carry out the iteration of (3.66) *until the successive iterants agree to four decimal places.*

In Problems 1–4 determine graphically whether the given equation possesses any real roots.

1. $x^3 = -2 + \sin x$ **2.** $x^3 - 3x = x^2 - 1$

3. $x^4 + x^2 - 2x + 3 = 0$ **4.** $\tan x = \cos x$

In Problems 5–8 use Newton's Method to find an approximation for the given number.

5. $\sqrt{10}$ **6.** $1 + \sqrt{5}$

7. $\sqrt[3]{4}$ **8.** $\sqrt[5]{2}$

In Problems 9–14 use Newton's Method, if necessary, to find approximations to all real roots of the given equations.

9. $x^3 = -x + 1$ **10.** $x^3 - x^2 + 1 = 0$

11. $x^4 + x^2 - 3 = 0$ **12.** $x^4 = 2x + 1$

13. $x^2 = \sin x$ **14.** $x + \cos x = 0$

15. Find the smallest positive x-intercept of the graph of $f(x) = 3 \cos x + 4 \sin x$.

16. Consider the function $f(x) = x^5 + x^2$. Use Newton's Method to approximate the smallest positive number for which $f(x) = 4$.

17. A cantilever beam 20 ft long with a load of 600 lb at its end is deflected by an amount $d = (60x^2 - x^3)/16,000$, where d is measured in inches and x in feet. See Figure 3.55. Use Newton's Method to approximate the value of x that corresponds to a deflection of 0.01 in.

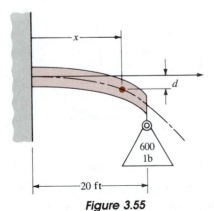

Figure 3.55

18. A vertical solid cylindrical column of fixed radius r that supports its own weight will eventually buckle when its height is increased. It can be proved that the maximum, or critical, height of such a column is $h_{cr} = kr^{2/3}$, where k is a constant and r is measured in meters. Use Newton's Method to approximate the diameter of a column for which $h_{cr} = 10$ m and $k = 35$.

19. A beam of light originating at point P in medium A, whose index of refraction is n_1, strikes the surface of medium B, whose index of refraction is n_2. We can prove from Snell's Law (see Exercises 4.7, Problem 53) that the beam is refracted tangent to the surface for the critical angle determined from $\sin \theta_c = n_2/n_1$, $0 < \theta_c < 90°$. For angles of incidence greater than the critical angle, all light is reflected internally to medium A. See Figure 3.56. If $n_2 = 1$ for air and $n_1 = 1.5$ for glass, use Newton's Method to approximate θ_c in radians.

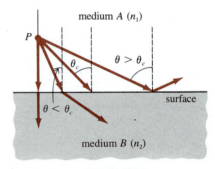

Figure 3.56

20. For a suspension bridge, the length s of a cable between two vertical supports whose span is l (horizontal distance) is related to the sag d of the cable by

$$s = l + \frac{8d^2}{3l} - \frac{32d^4}{5l^3}$$

See Figure 3.57. If $s = 404$ ft and $l = 400$ ft, use Newton's Method to approximate the sag. Round your answer to one decimal place.* (*Hint*: The root c satisfies $20 < c < 30$.)

_____

*The formula for s is itself only an approximation.

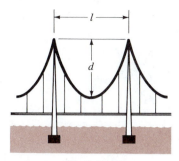

Figure 3.57

Miscellaneous Problems

21. Let f be a differentiable function. Show that if $f(x_0)$ $= -f(x_1)$ and $f'(x_0) = f'(x_1)$, then (3.66) implies $x_2 = x_0$.

22. Given

$$f(x) = \begin{cases} -\sqrt{4 - x}, & x < 4 \\ \sqrt{x - 4}, & x \geq 4 \end{cases}$$

observe $f(4) = 0$. Show that for any choice of x_0 Newton's Method will fail to converge to the root. (*Hint*: See Problem 21.)

23. Show that to find the rth root of a number N, (3.66) becomes

$$x_{n+1} = \frac{1}{r}\left[(r - 1)x_n + \frac{N}{x_n^{r-1}}\right]$$

‗‗‗‗ Chapter 3 Review Exercises ‗‗‗‗

Answers to odd-numbered problems begin on page A-23.

In Problems 1–16 answer true or false.

1. The instantaneous rate of change of $y = f(x)$ with respect to x at x_0 is the slope of the tangent line to the graph at $(x_0, f(x_0))$. ‗‗‗‗‗

2. If f is differentiable for every value of x, then f is continuous for every value of x. ‗‗‗‗‗

3. If f is not differentiable at $x = a$, there is no tangent to the graph at $(a, f(a))$. ‗‗‗‗‗

4. For $f(x) = -x^2 + 5x + 1$ an equation of the tangent line is $f'(x) = -2x + 5$. ‗‗‗‗‗

5. The function $f(x) = x/(x^2 + 9)$ is differentiable on the interval $[-3, 3]$. ‗‗‗‗‗

6. The function $f(x) = |x - 2|$ is not differentiable on $(-\infty, \infty)$. ‗‗‗‗‗

7. The derivative of a product is the product of the derivatives. ‗‗‗‗‗

8. A polynomial function has a tangent line at every point on its graph. ‗‗‗‗‗

9. If $f'(x) = g'(x)$, then $f(x) = g(x)$. ‗‗‗‗‗

10. The Power Rules considered in this chapter for differentiating $f(x) = x^n$ imply $\frac{d}{dx} x^{\sqrt{2}} = \sqrt{2}x^{\sqrt{2}-1}$. ‗‗‗‗‗

11. If $f(x) = \sqrt{2kx}$, k a constant, then $f'(x) = k/f(x)$. ‗‗‗‗‗

12. At $x = -1$ the tangent line to the graph of $f(x)$ $= x^3 - 3x^2 - 9x$ is parallel to the line $y = 2$. ‗‗‗‗‗

13. If f is continuous and $f(a)f(b) < 0$, there is a root of $f(x) = 0$ in the interval (a, b). ‗‗‗‗‗

14. The equation $x^4 = -x + 3$ possesses two real roots. ‗‗‗‗‗

15. Newton's Method always converges when the initial guess x_0 is chosen very close to the root c. ‗‗‗‗‗

16. Newton's Method will fail to converge to the real root of $x^3 - 2x^2 + x - 7 = 0$ for $x_0 = 1$. ‗‗‗‗‗

In Problems 17–30 fill in the blank.

17. If k is a constant and n a positive integer, then $\frac{d}{dx} k^n =$ ‗‗‗‗‗ .

18. If $y = f(x)$ is a continuous function, the average rate of change of f on the interval $[a, a + \Delta x]$ is ‗‗‗‗‗ .

19. For $f(x) = x^3$ the slope of the tangent at $(3, f(3))$ is

$$m_{\text{tan}} = \lim_{\Delta x \to 0} \frac{(\underline{} + \Delta x)^3 - (\underline{})}{\Delta x}$$

20. The function $f(x) = \cot x$ is not differentiable on $[0, \pi]$ because ‗‗‗‗‗‗‗‗‗‗‗‗‗‗‗‗‗ .

21. If $f'(x) = x^2$, then $\frac{d}{dx} f(x^3) =$ ‗‗‗‗‗ .

22. If $f(2) = 1, f'(2) = 5, g(2) = 2$, and $g'(2) = -3$, then

$$\frac{d}{dx} \frac{x^2 f(x)}{g(x)} \bigg|_{x=2} = \underline{}$$

23. If $y = \sin x$, then $\dfrac{d^4y}{dx^4} = $ _____ .

24. If $y = f(x)$ is a polynomial function of degree 3, then $\dfrac{d^4}{dx^4} f(x) = $ _____ .

25. If $f'(4) = 6$ and $g'(4) = 3$, then the slope of the tangent to the graph of $y = 2\,f(x) - 5\,g(x)$ at $x = 4$ is _____ .

26. $\displaystyle\lim_{x\to0} \dfrac{\sin 3x}{5x} = $ _____

27. $\displaystyle\lim_{t\to1} \dfrac{1 - \cos^2(t - 1)}{t - 1} = $ _____

28. The slope of the line perpendicular to the tangent line to the graph of $f(x) = \tan x$ at $x = \pi/3$ is _____ .

29. For $f(x) = 1/(1 - 3x)$ the instantaneous rate of change of f' at $x = 0$ is _____ .

30. The domain of f' for $f(x) = \sqrt{x} + \sqrt{x - 1}$ is _____ .

31. Given that $y = \cos x^2$, find all values of x for which $dy/dx = 0$.

32. Find an equation of the tangent line to the graph of $y = (x + 3)/(x - 2)$ at $x = 0$.

33. Find equations for the lines through $(0, -9)$ tangent to the graph of $y = x^2$.

34. Find Δy and dy for the function $y = x + 1/x$.

35. Use a differential to find an approximation for $1/\sqrt[3]{68}$.

36. Given that $x^2 - y^2 = x^3y$ defines a differentiable function, find dy/dx.

37. Given $x^{1/3} + y^{1/3} = 1$, find d^2y/dx^2.

38. Consider the equation $y^3 = 64x^2$. Use two different methods to find dy/dx.

In Problems 39–50 find the derivative of the given function.

39. $y = \dfrac{\cos 4x}{4x + 1}$

40. $y = 10 \cot 8x$

41. $f(x) = 2 + 2x + x^{-2} + x^2$

42. $f(x) = x^3 \sin^2 5x$

43. $F(t) = (t + \sqrt{t^2 + 1})^{10}$

44. $g(u) = \sqrt{\dfrac{6u - 1}{u + 7}}$

45. $G(x) = \dfrac{4x^{0.3}}{5x^{0.2}}$

46. $h(\theta) = \theta^{1.5}(\theta^2 + 1)^{0.5}$

47. $y = \sqrt[4]{x^4 + 16} \ \sqrt[3]{x^3 + 8}$

48. $y = \tan^2(\cos 2x)$

49. $y = \dfrac{\sqrt[3]{x^2}}{1 + \sqrt[3]{x}}$

50. $y = \dfrac{1}{x^3 + 4x^2 - 6x + 11}$

In Problems 51–54 find the indicated derivative.

51. $y = (3x)^{5/2}; \ \dfrac{d^3y}{dx^3}$

52. $y = \sin(x^3 - 2x); \ \dfrac{d^2y}{dx^2}$

53. $s = t^2 + \dfrac{1}{t^2}; \ \dfrac{d^2s}{dt^2}$

54. $W = \dfrac{v - 1}{v + 1}; \ \dfrac{d^3W}{dv^3}$

55. Find all points in the interval $[0, 2\pi]$ at which the tangent to the graph of $f(x) = 5 - 2\cos x$ is parallel to the line $y = \sqrt{3}x + 1$.

56. Find all points in the interval $[0, 2\pi]$ at which the tangent to the graph of $f(x) = 2\cos x + \cos 2x$ is horizontal.

57. If F is a differentiable function find

$$\dfrac{d^2}{dx^2} F(\sin 4x)$$

58. A mass m oscillates vertically when it is attached to a spring. The position of the mass at any time t is given by

$$x(t) = C_1 \cos \sqrt{\dfrac{k}{m}}\,t + C_2 \sin \sqrt{\dfrac{k}{m}}\,t$$

Verify that $x(t)$ satisfies the equation

$$\dfrac{d^2x}{dt^2} + \dfrac{k}{m}x = 0$$

In Problems 59 and 60 use Newton's Method to find the indicated root. Carry out the method until two successive iterants agree to four decimal places.

59. $x^3 - 4x + 2 = 0$, largest positive root

60. $\left(\dfrac{\sin x}{x}\right)^2 = \dfrac{1}{2}$, smallest positive root

61. A rectangular block of steel is hollowed out, making a tub with a uniform thickness t. The dimensions of the tub are shown in Figure 3.58(a). To float in water, as shown in Figure 3.58(b), the weight of the water displaced must equal the weight of the tub (Archimedes' principle). If the weight density of water is 62.4 lb/ft^3 and the weight density of the steel is 490 lb/ft^3, then

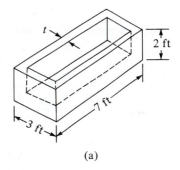

(a)

$$\begin{array}{l}\text{weight of water}\\ \text{displaced}\end{array} = 62.4 \times \begin{array}{l}\text{(volume of water}\\ \text{displaced)}\end{array}$$

$$\text{weight of tub} = 490 \times \text{(volume of steel in tub)}$$

(a) Show that t satisfies the equation

$$t^3 - 7t^2 + \frac{61}{4}t - \frac{1638}{1225} = 0$$

(b) Use Newton's Method to approximate the largest positive root of the equation in part **(a)**.

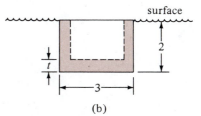

(b)

Figure 3.58

4

Applications of the Derivative

The derivative gives a rate of change. Geometrically, this rate of change is the slope of a tangent line to a graph. In Section 4.1 we elaborate on a concept discussed briefly in Section 3.1; namely, the rate of change with respect to time of a function that gives position of a moving object is the velocity of the object. However, in Section 4.2 we shall see that a time rate of change has other interpretations. The notion of rate along with the problem of finding the maximum and minimum values of a function are the central topics of study in this chapter.

4.1 Rectilinear Motion and the Derivative

In Section 3.1 motion of an object in a straight line, either horizontally or vertically, was said to be **rectilinear motion**. A function s that gives the coordinate of the object on a horizontal or vertical line is called a **position function**. The variable t represents time and $s(t)$ is a directed distance, which is measured in centimeters, meters, feet, miles, and so on, from a reference point $s = 0$. Recall that on a horizontal scale, we take the positive s-direction to be to the right of $s = 0$, and on a vertical scale we take the positive s-direction to be upward.

_____ **Example 1** _____

A particle moves on a horizontal line according to the position function $s(t) = -t^2 + 4t + 3$, where s is measured in meters and t in seconds. What is the position of the particle at 0, 2, and 6 seconds?

Solution   Substituting into the position function gives

$$s(0) = 3, \qquad s(2) = 7, \quad \text{and} \quad s(6) = -9$$

As shown in Figure 4.1, $s(6) = -9 < 0$ means that the position of the particle is to the left of the reference point $s = 0$.

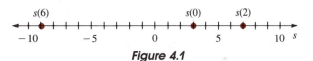

Figure 4.1

Velocity and Acceleration   If the **average velocity** of body in motion over a time interval of length Δt is

$$\frac{\text{change in position}}{\text{change in time}} = \frac{s(t + \Delta t) - s(t)}{\Delta t}$$

then the instantaneous rate of change is velocity given by

$$v(t) = \lim_{\Delta t \to 0} \frac{s(t + \Delta t) - s(t)}{\Delta t}$$

Thus, we have the following.

> If s is a position function of an object that moves rectilinearly, then its **velocity function** at time t is
>
> $$v(t) = \frac{ds}{dt}$$

The **speed** of the object at time t is $|v(t)|$.

Velocity is measured in centimeters per second (cm/sec), meters per second (m/sec), feet per second (ft/sec), kilometers per hour (km/hr), miles per hour (mi/hr), and so on.

In turn we can compute the rate of change of velocity.

> If $v(t)$ is the velocity of an object that moves rectilinearly, then its **acceleration function** at time t is

$$a(t) = \frac{dv}{dt} = \frac{d^2s}{dt^2}$$

Typical units for measuring acceleration are meters per second per second (m/sec^2), feet per second per second (ft/sec^2), miles per hour per hour (mi/hr^2), and so on. Often we read units of acceleration literally, as "meters per second squared."

Significance of Algebraic Signs

In Section 4.4 we shall see that whenever the derivative of a function f is *positive* on an interval I, then f is *increasing* on I. Geometrically, the graph of an increasing function rises as x increases. Similarly, if the derivative of a function f is *negative* on I, then f is *decreasing*, which means its graph goes down as x increases. Thus, when $v(t) = s'(t) > 0$, we can say $s(t)$ is increasing and the object is moving to the *right*. A negative velocity indicates motion to the *left*. Similarly, when $a(t) > 0$ the velocity is *increasing*, whereas if $a(t) < 0$, the velocity is *decreasing*. For example, an acceleration of -25 m/sec^2 means that the velocity is decreasing by 25 m/sec every second. Do not confuse "velocity decreasing" with the concept of "slowing down"; for example, consider a stone that is dropped from the top of a tall building. The acceleration of gravity is a negative constant, -9.8 m/sec^2. The negative sign means that the velocity of the stone decreases starting from zero. When the stone hits the ground its speed $|v(t)|$ is fairly large, but $v(t) < 0$. Note that an object that moves rectilinearly on a horizontal line will slow down when $v(t) > 0$ (motion to right) and $a(t) < 0$ (velocity decreasing) or when $v(t) < 0$ (motion to left) and $a(t) > 0$ (velocity increasing). In other words, an object is slowing down when its speed $|v(t)|$ is decreasing. In physics when a moving body is slowing down, the term *deceleration* is used.

Example 2

In Example 1 the velocity and acceleration functions for the particle are, respectively,

$$v(t) = \frac{ds}{dt} = -2t + 4 \quad \text{and} \quad a(t) = \frac{dv}{dt} = -2$$

At times 0, 2, and 6 seconds, the velocities are $v(0) = 4$ cm/sec, $v(2) = 0$ cm/sec, and $v(6) = -8$ cm/sec, respectively. Since the acceleration is always negative, the velocity is always decreasing. Notice that $v(t) = 2(-t + 2) > 0$ for $t < 2$ and $v(t) = 2(-t + 2) < 0$ for $t > 2$. If the time t is allowed to be negative as well as positive, then the particle is moving to the right for the time interval $(-\infty, 2)$ and moving to the left for the time interval $(2, \infty)$. The motion can be represented by the graph given in Figure 4.2(a). Since the motion actually takes place *on* the horizontal line, you should envision the movement of a point *P* that corresponds to the projection of a point on the graph onto the horizontal line. See Figure 4.2(b).

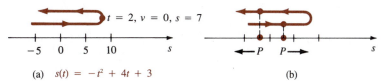

$t = 2, v = 0, s = 7$

(a) $s(t) = -t^2 + 4t + 3$

(b)

Figure 4.2

Example 3

A particle moves on a horizontal line according to the position function $s(t) = \frac{1}{3}t^3 - t$. Determine the time intervals on which the particle is slowing down.

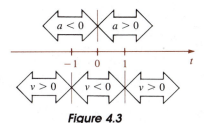

Figure 4.3

Solution   Inspection of the algebraic signs of

$$v(t) = t^2 - 1 = (t + 1)(t - 1)$$

and $$a(t) = 2t,$$

which are shown on the time scale in Figure 4.3, indicates that $v(t)$ and $a(t)$ have opposite signs; hence, the particle is slowing down on the time intervals $(-\infty, -1)$ and $(0, 1)$.

Example 4

An object moves on a horizontal line according to the position function $s(t) = t^4 - 18t^2 + 25$, where s is measured in centimeters and t in seconds. Use a graph to represent the motion during the time interval $[-4, 4]$.

Solution   The velocity function is

$$v(t) = \frac{ds}{dt} = 4t^3 - 36t$$

$$= 4t(t + 3)(t - 3)$$

and the acceleration function is

$$a(t) = \frac{d^2s}{dt^2} = 12t^2 - 36$$

$$= 12(t + \sqrt{3})(t - \sqrt{3})$$

Now, from the solutions of $v(t) = 0$, we can determine the time intervals for which $s(t)$ is increasing or decreasing. From the information given in the following tables, we construct the graph shown in Figure 4.4.

Time Interval	Sign of $v(t)$	Direction of Motion
$(-4, -3)$	$-$	left
$(-3, 0)$	$+$	right
$(0, 3)$	$-$	left
$(3, 4)$	$+$	right

Time	Position	Velocity	Acceleration
-4	-7	-112	156
-3	-56	0	72
0	25	0	-36
3	-56	0	72
4	-7	112	156

Time Interval	Sign of $a(t)$	Velocity
$(-4, -\sqrt{3})$	$+$	increasing
$(-\sqrt{3}, \sqrt{3})$	$-$	decreasing
$(\sqrt{3}, 4)$	$+$	increasing

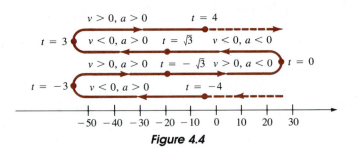

Figure 4.4

Inspection of Figure 4.4 shows that the particle slows down on the time intervals $(-4, -3)$, $(-\sqrt{3}, 0)$, and $(\sqrt{3}, 3)$.

Exercises 4.1

Answers to odd-numbered problems begin on page A-23.

In Problems 1–8 s is a position function of a particle that moves on a horizontal line. Find the position, velocity, speed, and acceleration of the particle at the indicated times.

1. $s(t) = 4t^2 - 6t + 1$; $t = \frac{1}{2}$, $t = 3$

2. $s(t) = (2t - 6)^2$; $t = 1$, $t = 4$

3. $s(t) = -t^3 + 3t^2 + t$; $t = -2$, $t = 2$

4. $s(t) = t^4 - t^3 + t$, $t = -1$, $t = 3$

5. $s(t) = t - \dfrac{1}{t}$; $t = \frac{1}{4}$, $t = 1$

6. $s(t) = \dfrac{t}{t + 2}$; $t = -1$, $t = 0$

7. $s(t) = t + \sin \pi t$; $t = 1$, $t = \frac{3}{2}$

8. $s(t) = t \cos \pi t$; $t = \frac{1}{2}$, $t = 1$

In Problems 9–12 s is a position function of a particle that moves on a horizontal line.

9. $s(t) = t^2 - 4t - 5$

   (a) What is the velocity of the particle when $s(t) = 0$?

   (b) What is the velocity of the particle when $s(t) = 7$?

10. $s(t) = t^3 - 3t^2 + 8$

   (a) What is the position of the particle when $v(t) = 0$?

   (b) What is the position of the particle when $a(t) = 0$?

11. $s(t) = t^3 - 4t$

   (a) What is the acceleration of the particle when $v(t) = 2$?

   (b) What is the position of the particle when $a(t) = 18$?

12. $s(t) = t^2 + 6t + 10$

   (a) What is the position of the particle when $s(t) = v(t)$?

   (b) What is the velocity of the particle when $v(t) = -a(t)$?

In Problems 13 and 14 s is a position function of a particle that moves on a horizontal line. Determine the time intervals for which the particle is slowing down.

13. $s(t) = t^3 - 27t$ **14.** $s(t) = t^4 - t^3$

In Problems 15–24 s is a position function of a particle that moves on a horizontal line. Find the velocity and acceleration functions. Represent the motion during the indicated time interval with a graph.

15. $s(t) = t^2$; $[-1, 3]$

16. $s(t) = t^3$; $[-2, 2]$

17. $s(t) = t^2 - 4t - 2$; $[-1, 5]$

18. $s(t) = (t + 3)(t - 1)$; $[-3, 1]$

19. $s(t) = 2t^3 - 6t^2$; $[-2, 3]$

20. $s(t) = (t - 1)^2(t - 2)$; $[-2, 3]$

21. $s(t) = 3t^4 - 8t^3$; $[-1, 3]$

22. $s(t) = t^4 - 4t^3 - 8t^2 + 60$; $[-2, 5]$

23. $s(t) = t - 4\sqrt{t}$; $[1, 9]$

24. $s(t) = 1 + \cos \pi t$; $[-\frac{1}{2}, \frac{5}{2}]$

25. The graph of a position function in the st-plane is given in Figure 4.5. Complete the accompanying table by stating whether $v(t)$ and $a(t)$ are positive, negative, or zero.

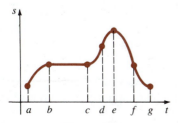

Figure 4.5

Interval	$v(t)$	$a(t)$
(a, b)		
(b, c)		
(c, d)		
(d, e)		
(e, f)		
(f, g)		

26. The graph of the velocity function v for a particle that moves on a horizontal line is given in Figure 4.6. Make a possible graph of the position function s.

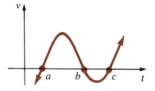

Figure 4.6

27. The height (in feet) of a projectile shot vertically upward from ground level is given by $s(t) = -16t^2 + 48t$.

   (a) Determine the time interval for which $v > 0$ and the time interval for which $v < 0$.

   (b) Find the maximum height attained by the projectile.

28. A particle moves on a horizontal line according to the position function $s(t) = -t^2 + 10t - 20$, where s is measured in centimeters and t in seconds. Determine the total distance traveled by the particle during the time interval $[-1, 6]$.

When friction is ignored, the distance s (in feet) that a body moves down an inclined plane of inclination θ is given by $s(t) = 16t^2 \sin \theta$, $[0, t_1]$, where $s(0) = 0$, $s(t_1) = L$, and t is measured in seconds. See Figure 4.7. Use this information in Problems 29 and 30.

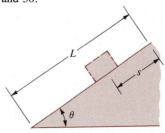

Figure 4.7

29. An object is sliding down a 256-ft-long hill whose inclination is 30°. What is the velocity and acceleration of the object at the bottom of the hill?

30. An entry in a soap box derby rolls down the hill shown in Figure 4.8. What is its velocity and acceleration at the bottom of the hill?

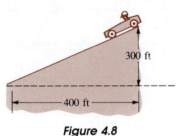

Figure 4.8

31. A bucket, attached to a circular windlass by a rope, is permitted to fall in a straight line under the influence of gravity.

If the distance the bucket falls is equal to the radian measure of the angle indicated in Figure 4.9, then $\theta = \frac{1}{2}gt^2$, where $g = 32$ ft/sec² is the acceleration of gravity. Find the rate at which the y-coordinate of a point P on the circumference of the windlass changes at $t = \sqrt{\pi}/4$ sec. Interpret the result.

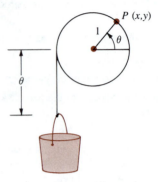

Figure 4.9

4.2 Related Rates

The derivative dy/dx of a function $y = f(x)$ is its instantaneous rate of change with respect to the variable x. When a function describes either position or distance, then its time rate of change is interpreted as velocity. In general, a time rate of change is the answer to the question "How *fast* is a quantity changing?" For example, if V stands for volume that is changing in time, then dV/dt is the rate, or how fast, the volume is changing with respect to time t. A rate of, say, $dV/dt = 10$ cm³/sec means that the volume is increasing 10 cubic centimeters each second. Similarly, if a person is walking *toward* the street lamp shown in Figure 4.10(a) at a constant rate of 3 ft/sec, then we know that $dx/dt = -3$ ft/sec. On the other hand, if the person is walking *away* from the street lamp then $dx/dt = 3$ ft/sec. The negative and positive rates mean, of course, that the distance x is decreasing and increasing, respectively.

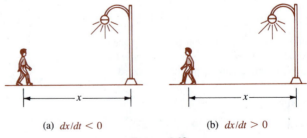

(a) $dx/dt < 0$ (b) $dx/dt > 0$

Figure 4.10

In this section we are concerned with **related rates**. Since the problems will be stated in words, you must interpret these words in terms of mathematical symbols. Usually it is a good idea to:

(*i*)   Draw a picture.

(*ii*)   Label all quantities that change in time with symbols.

(*iii*)   Analyze the words of the problem and discern which rates are *given* and what rate is *required*.

(*iv*)   Set up an equation that relates the variables.

(*v*)   Differentiate the equation found in step (*iv*) with respect to time *t*. This step requires the use of *implicit differentiation*. The resulting equation, after differentiation, is an equation that relates *the rates* at which the variables change.

Recall that if *y* denotes a function of *x*, then the Power Rule for Functions gives

$$\frac{d}{dx} y^n = ny^{n-1} \frac{dy}{dx} \tag{4.1}$$

where *n* is a rational number. Of course (4.1) is applicable to any function, say *r*, *z* or *x*, that depends on *t*:

$$\frac{d}{dt} r^n = nr^{n-1}\frac{dr}{dt}, \qquad \frac{d}{dt} x^n = nx^{n-1}\frac{dx}{dt}, \qquad \frac{d}{dt} z^n = nz^{n-1}\frac{dz}{dt} \tag{4.2}$$

and so on.

Example 1

A square is expanding with time. How is the rate at which the area increases related to the rate at which a side increases?

Solution   At any time the area *A* of a square is a function of the length of one side of *x*:

$$A = x^2 \tag{4.3}$$

Thus, the related rates are obtained from the time derivative of (4.3). With the help of the second result in (4.2), we see that

$$\frac{dA}{dt} = \frac{d}{dt} x^2$$

is the same as

$$\frac{dA}{dt} = 2x\frac{dx}{dt}$$

related rates

Example 2

Air is being pumped into a spherical balloon at a rate of 20 ft³/min. At what rate is the radius changing when the radius is 3 ft?

Solution   As shown in Figure 4.11, we denote the radius of the balloon by r and its volume by V. Now, the interpretation of "air is being pumped . . . at a rate of 20 ft³/min" means that we are

Given: $\dfrac{dV}{dt} = 20 \text{ ft}^3/\text{min}$

In addition, we

Want: $\dfrac{dr}{dt}\bigg|_{r=3}$

Finally, a relationship between V and r is given by the formula for the volume of a sphere.

Know: $V = \dfrac{4}{3}\pi r^3$ (4.4)

Differentiating (4.4) with respect to t and using the first result in (4.2) gives

$$\frac{dV}{dt} = \frac{4}{3}\pi \frac{d}{dt}r^3$$

$$= \frac{4}{3}\pi\left(3r^2 \frac{dr}{dt}\right)$$

$$= 4\pi r^2 \frac{dr}{dt}$$

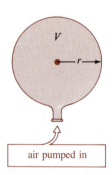

V

r

air pumped in

Figure 4.11

But $dV/dt = 20$; therefore, $20 = 4\pi r^2\,dr/dt$ yields

$$\frac{dr}{dt} = \frac{20}{4\pi r^2} = \frac{5}{\pi r^2}$$

Thus, $\dfrac{dr}{dt}\bigg|_{r=3} = \dfrac{5}{9\pi} \text{ ft/min} \approx 0.18 \text{ ft/min}$

Example 3

A woman jogging at a constant rate of 10 km/hr crosses a point P heading north. Ten minutes later a man jogging at a constant rate of 9 km/hr crosses the same point heading east. How fast is the distance between the joggers changing 20 minutes after the man crosses P?

Solution   Let time be measured in hours from the instant the man crosses point P. As shown in Figure 4.12, at $t > 0$ let the man M and woman W be located

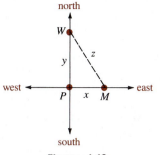

Figure 4.12

x and y kilometers, respectively, from point P. Let z be the corresponding distance between the two joggers. Now,

Given: $\quad\quad \dfrac{dx}{dt} = 9$ km/hr, $\quad \dfrac{dy}{dt} = 10$ km/hr

Want: $\quad\quad \dfrac{dz}{dt}\Big|_{t=1/3} \quad$ [20 min $= \frac{1}{3}$ hr]

Know: From the Pythagorean Theorem, the variables x, y, and z are related by

$$z^2 = x^2 + y^2 \tag{4.5}$$

Differentiating (4.5) with respect to t gives

$$2z\frac{dz}{dt} = 2x\frac{dx}{dt} + 2y\frac{dy}{dt} \tag{4.6}$$

Using the given rates in (4.6) then yields

$$z\frac{dz}{dt} = 9x + 10y$$

When $t = \frac{1}{3}$ hr we use distance = (rate) · (time) to obtain $x = 9 \cdot (\frac{1}{3}) = 3$ km. Since the woman has run $\frac{1}{6}$ hr (10 minutes) longer, we find $y = 10 \cdot (\frac{1}{3} + \frac{1}{6})$ $= 5$ km. At $t = \frac{1}{3}$ hr, it follows that $z = \sqrt{3^2 + 5^2} = \sqrt{34}$ km. Finally,

$$\sqrt{34}\,\frac{dz}{dt}\Big|_{t=1/3} = 9 \cdot 3 + 10 \cdot 5$$

or $\quad\quad \dfrac{dz}{dt}\Big|_{t=1/3} = \dfrac{77}{\sqrt{34}} \approx 13.21$ km/hr

---

Example 4

A lighthouse is located on a small island 2 mi off a straight shore. The beacon of the lighthouse revolves at a constant rate of 6 deg/sec. How fast is the light beam moving along the shore at a point 3 mi from a point on the shore closest to the lighthouse?

Solution We first introduce the variables θ and x as shown in Figure 4.13. In addition we change the information of θ to radian measure by recalling 1° is equivalent to $\pi/180$ radians.

Given: $\quad\quad \dfrac{d\theta}{dt} = 6 \cdot \dfrac{\pi}{180} = \dfrac{\pi}{30}$ rad/sec

Want: $\quad\quad \dfrac{dx}{dt}\Big|_{x=3}$

Know: $\quad\quad \dfrac{x}{2} = \tan\theta \quad$ or $\quad x = 2\tan\theta$

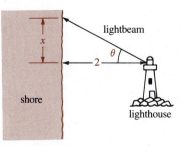

Figure 4.13

Differentiating the last equation with respect to t gives

$$\frac{dx}{dt} = 2 \sec^2\theta \frac{d\theta}{dt}$$

$$= \frac{\pi}{15} \sec^2\theta$$

At the instant $x = 3$, $\tan\theta = \frac{3}{2}$ so that from the trigonometric identity $1 + \tan^2\theta = \sec^2\theta$, we see $\sec^2\theta = \frac{13}{4}$. Hence,

$$\frac{dx}{dt}\bigg|_{x=3} = \frac{\pi}{15} \cdot \frac{13}{4} = \frac{13\pi}{60} \text{ mi/sec}$$

In physics the **momentum** of a body of mass m that moves in a straight line with velocity v is given by $p = mv$.

Example 5

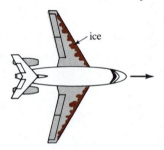

ice

Figure 4.14

An airplane of mass 10^5 kg flies in a straight line while ice builds up on the leading edges of its wings at a constant rate of 30 kg/hr. See Figure 4.14.

(a)   At what rate is the plane's momentum changing if it is flying at a constant rate of 800 km/hr?

(b)   At what rate is the momentum of the plane changing at $t = 1$ hr if at that instant its velocity is 750 km/hr and is increasing at a rate of 20 km/hr?

Solution

(a)   **Given:** $\dfrac{dm}{dt} = 30$ kg/hr

         Want: $\dfrac{dp}{dt}$

         Know: Momentum is $p = mv$.

Now, if v is constant, then

$$\frac{dp}{dt} = \frac{dm}{dt} v = 30 \cdot 800 = 2.4 \times 10^4 \text{ kg} \cdot \text{km/hr}^2$$

(b)   **Given:** $\dfrac{dm}{dt} = 30$ kg/hr, $v(1) = 750$,   and

$$\frac{dv}{dt}\bigg|_{t=1} = 20 \text{ km/hr}$$

         Want: $\dfrac{dp}{dt}\bigg|_{t=1}$

         Know: Momentum is $p = mv$.

When both m and v are changing, the Product Rule gives

$$\frac{dp}{dt} = m\frac{dv}{dt} + v\frac{dm}{dt}$$

At $t = 1$ hr the mass of the airplane has increased to $10^5 + 30$ kg. Therefore,

$$\left.\frac{dp}{dt}\right|_{t=1} = (10^5 + 30) \cdot 20 + 750 \cdot 30 = 2.0231 \times 10^6 \text{ kg km/hr}^2$$

Remark

These formulas may prove useful in the problems that follow:

Area of a circle: $A = \pi r^2$

Area of a trapezoid: $A = h(l_1 + l_2)/2$

Surface area of a sphere: $S = 4\pi r^2$

Volume of circular cylinder: $V = \pi r^2 h$

Volume of a cone: $V = \frac{1}{3}\pi r^2 h$

Volume of a frustum of a cone:* $V = \frac{\pi}{3}h(r_1^2 + r_1 r_2 + r_2^2)$

Volume of a sphere: $V = \frac{4}{3}\pi r^3$

———— Exercises 4.2 ————

Answers to odd-numbered problems begin on page A-23.

1. A cube is expanding with time. How is the rate at which the volume increases related to the rate at which a side increases?

2. The volume of a rectangular box is $V = xyz$. Given that each side expands at a constant rate of 10 cm/min, find the rate at which the volume is expanding when $x = 1$ cm, $y = 2$ cm, and $z = 3$ cm.

3. A plate in the shape of an equilateral triangle expands with time. A side increases at a constant rate of 2 cm/hr. At what rate is the area increasing when a side is 8 cm?

4. In Problem 3 at what rate is the area increasing at the instant when the area is $\sqrt{75}$ cm²?

5. A rectangle expands with time. The diagonal of the rectangle increases at a rate of 1 in/hr and the length increases at

a rate of $\frac{1}{4}$ in/hr. How fast is its width increasing when the width is 6 in and the length is 8 in?

6. The sides of a cube increase at a rate of 5 cm/hr. At what rate does the diagonal of the cube increase?

7. A boat is sailing toward the vertical cliff shown in Figure 4.15. How are the rates at which x, s, and θ change related?

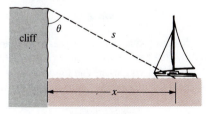

Figure 4.15

———

*See Figure 6.44

8. The total resistance R in a parallel circuit that contains two resistors of resistances R_1 and R_2 is given by $1/R = 1/R_1 + 1/R_2$. Each resistance changes with time. How are dR/dt, dR_1/dt, and dR_2/dt related?

9. A bug crawls along the graph of $y = x^2 + 4x + 1$, where x and y are measured in centimeters. If the abscissa x changes at a constant rate of 3 cm/min, how fast is the ordinate changing at the point $(2, 13)$?

10. In Problem 9, how fast is the ordinate changing when the bug is 6 cm above the x-axis?

11. A particle moves on the graph of $y^2 = x + 1$ so that $dx/dt = 4x + 4$. What is dy/dt when $x = 8$?

12. A particle in continuous motion moves on the graph of $4y = x^2 + x$. Find the point on the graph at which the rate of change of the abscissa and the rate of change of the ordinate are the same.

13. The x-coordinate of the point P shown in Figure 4.16 increases at a rate of $\frac{1}{3}$ cm/hr. How fast is the area of the right triangle OPA increasing when P has coordinates $(8, 2)$?

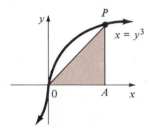

Figure 4.16

14. A suitcase is carried up the conveyor belt shown in Figure 4.17 at a rate of 2 ft/sec. How fast is the suitcase rising?

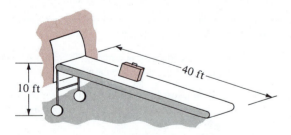

Figure 4.17

15. In the adiabatic expansion of air pressure, P and volume V are related by $PV^{1.4} = k$, where k is a constant. At a certain instant the pressure is 100 lb/in^2 and the volume is 32 in^3. At what rate is the pressure changing at that instant if the volume is decreasing at a rate of 2 in^3/sec?

16. A stone dropped into a still pond causes a circular wave. Assume the radius of the wave expands at a constant rate of 2 ft/sec.

(a) How fast does the diameter increase?

(b) How fast does the circumference increase?

(c) How fast does the area expand when the radius is 3 ft?

(d) How fast does the area expand when the area is 8π ft^2?

17. An oil tank in the shape of a circular cylinder of radius 8 m is being filled at a constant rate of 10 m^3/min. How fast is the level of the oil rising?

18. As a cylindrical water tank of diameter 40 ft is draining, the level of the water decreases at a constant rate of $\frac{3}{2}$ ft/min. How fast is the volume of the water decreasing?

19. Each vertical end of a 20-ft-long water trough is an equilateral triangle with vertex down. If water is being pumped in at a constant rate of 4 ft^3/min, how fast is the level of the water rising when the water is 1 ft deep?

20. A water trough with vertical ends in the form of isosceles trapezoids has dimensions as shown in Figure 4.18. If water is pumped in at a constant rate of $\frac{1}{2}$ m^3/sec, how fast is the level of the water rising when the water is $\frac{1}{4}$ m deep?

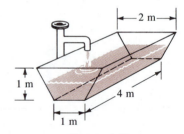

Figure 4.18

21. Water leaks out the bottom of the conical tank shown in Figure 4.19 at a constant rate of 1 ft^3/min. At what rate is the level of the water changing when the water is 6 ft deep? At what rate is the radius of the water changing at this instant?

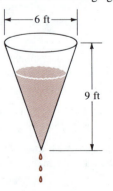

Figure 4.19

22. A 5-ft-tall person walks away from a 20-ft-tall street lamp at a constant rate of 3 ft/sec. See Figure 4.10(b).

    (a) At what rate is the length of the person's shadow increasing?

    (b) At what rate is the tip of the shadow moving away from the base of the street lamp?

23. A 15-ft ladder is leaning on a wall of a house. The bottom of the ladder is pulled away from the base of the wall at a constant rate of 2 ft/min. At what rate is the top of the ladder sliding down the wall when the bottom of the ladder is 5 ft from the wall?

24. A kite string is paid out at a constant rate of 3 ft/sec. If the wind carries the kite horizontally at an altitude of 200 ft, how fast is the kite moving when 400 ft of string have been paid out?

25. A plane flying parallel to level ground at a constant rate of 600 mi/hr approaches a radar station. If the altitude of the plane is 2 mi, how fast is the distance between the plane and the radar station decreasing when the horizontal distance between them is 1.5 mi? See Figure 4.20.

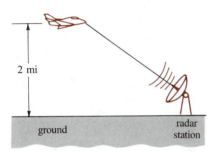

Figure 4.20

26. In Problem 25, at the point directly above the radar station, the plane goes into a 30° climb while retaining the same speed. How fast is the distance between the plane and radar station increasing one minute later? (*Hint:* Review the law of cosines.)

27. A plane at an altitude of 4 km passes directly over a tracking telescope on the ground. When the angle of elevation is 60°, it is observed that this angle is decreasing at a rate 30 deg/min. How fast is the plane traveling?

28. A rocket is traveling at a constant rate of 1000 mi/hr at an angle of 60° to the horizontal. See Figure 4.21.

    (a) At what rate is its altitude increasing?

    (b) What is the ground speed of the rocket?

29. A tracking telescope, located 1.25 km from the point of launching, follows a vertically ascending rocket. When the angle of elevation is 60°, the rate at which the angle is increasing is 3 deg/sec. At what rate is the rocket moving at that instant?

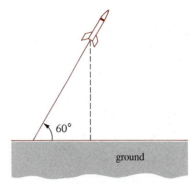

Figure 4.21

30. The volume V between two concentric spheres is expanding. The radius of the outer sphere increases at a constant rate of 2 m/hr, whereas the radius of the inner sphere decreases at a constant rate of $\frac{1}{2}$ m/hr. At what rate is V changing when the outer radius is 3 m and the inner radius is 1 m?

31. Many spherical objects such as raindrops, snowballs, and mothballs evaporate at a rate proportional to their surface areas. In this case show that the radius of the object decreases at a constant rate.

32. If the rate at which the volume of a sphere changes is constant, show that the rate at which its surface area changes is inversely proportional to the radius.

33. Two tankers depart from the same floating oil terminal. One tanker sails east at noon at a rate of 10 knots. (1 knot = 1 nautical mile/hr. A nautical mile is 6080 ft or 1.15 statute mile.) The other tanker sails north at 1:00 P.M. at a rate of 15 knots. At 2:00 P.M. at what rate is the distance between the two ships changing?

34. At 8:00 A.M. ship S_1 is 20 km due north of ship S_2. Ship S_1 sails south at a rate of 9 km/hr and ship S_2 sails west at a rate of 12 km/hr. At 9:20 A.M. at what rate is the distance between the two ships changing?

35. Sand flows from the top half of the conical hourglass shown in Figure 4.22 to the bottom half at a constant rate of

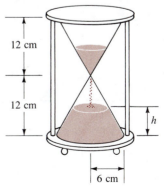

Figure 4.22

4 cm^3/sec. At any time, assume that the height of the sand is h and that the sand has the shape of a frustum of a cone. Express the rate at which h increases in terms of h.

36. The Ferris wheel shown in Figure 4.23 revolves clockwise once every two minutes. How fast is a passenger rising at the instant when she is 94 ft above the ground? How fast is she moving horizontally at the same instant?

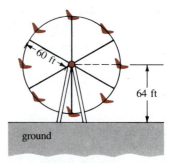

Figure 4.23

4.3 Extrema of Functions

Absolute Extrema   Suppose a function f is defined on an interval I. The **maximum** and **minimum** values of f on I (if there are any) are said to be **extrema** of the function. In the next two definitions, we distinguish two kinds of extrema.

> **DEFINITION 4.1**   **Absolute Extrema**
>
> (*i*)   A number $f(c_1)$ is an **absolute maximum** of a function f if $f(x) \le f(c_1)$ for every x in the domain of f.
>
> (*ii*)   A number $f(c_1)$ is an **absolute minimum** of a function f if $f(x) \ge f(c_1)$ for every x in the domain of f.

Absolute extrema are also called **global extrema**. Figure 4.24 shows several possibilities.

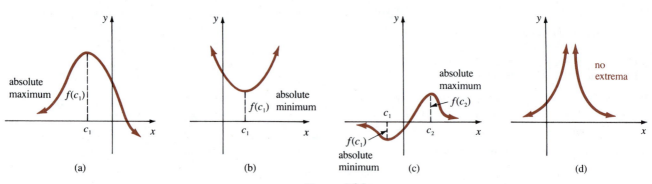

Figure 4.24

Example 1

(a) For $f(x) = \sin x$, $f(\pi/2) = 1$ is its absolute maximum and $f(3\pi/2) = -1$ is its absolute minimum.*

(b) The function $f(x) = x^2$ has the absolute minimum $f(0) = 0$ but has no absolute maximum.

(c) $f(x) = 1/x$ has neither an absolute maximum nor an absolute minimum.

The interval on which a function is defined is very important in the consideration of extrema.

Example 2

(a) $f(x) = x^2$, defined only on the *closed* interval [1, 2], has the absolute maximum $f(2) = 4$ and the absolute minimum $f(1) = 1$. See Figure 4.25(a).

(b) On the other hand, if $f(x) = x^2$ is defined on the *open* interval (1, 2), then f has no absolute extrema. In this case, $f(1)$ and $f(2)$ are not defined.

(c) $f(x) = x^2$, defined on [−1, 2], has the absolute maximum $f(2) = 4$ but now the absolute minimum is $f(0) = 0$. See Figure 4.25(b).

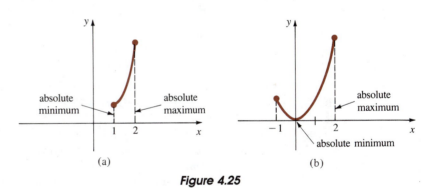

Figure 4.25

Parts (a) and (c) of Example 2 illustrate the following general result.

THEOREM 4.1   **Extreme Value Theorem**

A function f continuous on a closed interval [a, b] always has an absolute maximum and an absolute minimum on the interval.

---

*By periodicity, the maximum and minimum values also occur at $x = \pi/2 + 2n\pi$ and $x = 3\pi/2 + 2n\pi$, $n = \pm1, \pm2, \ldots$, respectively.

In other words, when f is continuous on $[a, b]$, there are numbers $f(c_1)$ and $f(c_2)$ such that $f(c_1) \le f(x) \le f(c_2)$ for all x in $[a, b]$. See Figure 4.26.

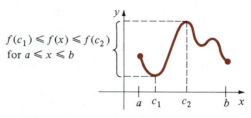

$f(c_1) \le f(x) \le f(c_2)$
for $a \le x \le b$

Figure 4.26

Endpoint Extrema   When an absolute extremum of a function occurs at an endpoint of an interval I, as in parts (*a*) and (*c*) of Example 2, we say it is an **endpoint extremum**. When I is not a closed interval such as $[a, b)$, $(-\infty, b]$, or $[a, \infty)$, then even when f is continuous there is no guarantee that an absolute extremum exists. See Figure 4.27.

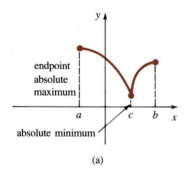

endpoint
absolute
maximum

absolute minimum

(a)

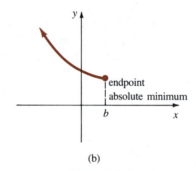

endpoint
absolute minimum

(b)

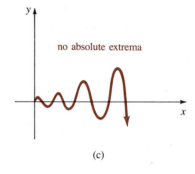

no absolute extrema

(c)

Figure 4.27

Relative Extrema   The function pictured in Figure 4.28(a) has no absolute extrema. However, suppose we focus our attention on values of x that are close to, or in a *neighborhood* of, the numbers c_1 and c_2. As shown in Figure 4.28(b), $f(c_1)$ is the maximum value of the function in the interval (a_1, b_1) and $f(c_2)$ is a minimum value in the interval (a_2, b_2). These **local** or **relative extrema** are defined as follows.

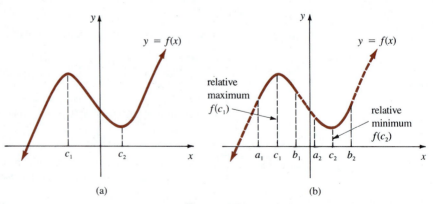

Figure 4.28

DEFINITION 4.2   Relative Extrema

(*i*)   A number $f(c_1)$ is a **relative maximum** of a function f if $f(x) \leq f(c_1)$ for every x in some open interval that contains c_1.

(*ii*)   A number $f(c_1)$ is a **relative minimum** of a function f if $f(x) \geq f(c_1)$ for every x in some open interval that contains c_1.

As a consequence of Definition 4.2, we can conclude that every absolute extremum, with the *exception* of an endpoint extremum, is also a relative extremum. An endpoint absolute extremum is precluded from being a relative extremum on the technicality that an open interval contained in the domain of the function cannot be found around an endpoint of the interval.

An examination of Figures 4.28 and 4.29 suggests that if c is a value at which a function f has a relative extremum, then either $f'(c) = 0$ or $f'(c)$ does not exist.

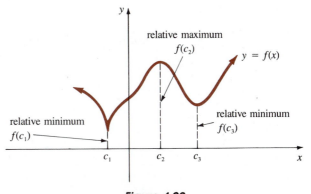

Figure 4.29

DEFINITION 4.3   Critical Values

A **critical value** of a function f is a number c in its domain for which $f'(c) = 0$ or $f'(c)$ does not exist.

Example 3

Find the critical values of $f(x) = x^3 - 15x + 6$.

Solution

$$f'(x) = 3x^2 - 15$$
$$= 3(x + \sqrt{5})(x - \sqrt{5})$$

The critical values are those numbers for which $f'(x) = 0$, namely, $-\sqrt{5}$ and $\sqrt{5}$.

_____ **Example 4** _____

Find the critical values of $f(x) = (x + 4)^{2/3}$.

Solution   By the Power Rule for Functions,

$$f'(x) = \frac{2}{3}(x + 4)^{-1/3}$$

$$= \frac{2}{3(x + 4)^{1/3}}$$

In this instance we see that $f'(x)$ does not exist when $x = -4$. Since -4 is in the domain of f, we conclude it is a critical value.

_____ **Example 5** _____

Find the critical values of $f(x) = \dfrac{x^2}{x - 1}$.

Solution   By the Quotient Rule, we find after simplifying

$$f'(x) = \frac{x(x - 2)}{(x - 1)^2}$$

Now $f'(x) = 0$ when $x = 0$ and $x = 2$, whereas $f'(x)$ does not exist when $x = 1$. However, inspection of f reveals $x = 1$ is not in its domain and so the only critical values are 0 and 2.

> **THEOREM 4.2**   If a function f has a relative extremum at a number c, then c is a critical value.

Proof   Assume $f(c)$ is a relative extremum.

(*i*)   If $f'(c)$ does not exist, then c is a critical value by Definition 4.3.

(*ii*)   If $f'(c)$ exists, there are three possibilities: $f'(c) > 0$, $f'(c) < 0$, or $f'(c) = 0$. For the sake of argument, let us further assume that $f(c)$ is a relative maximum. Hence, by Definition 4.2 there is some open interval that contains c in which

$$f(c + \Delta x) \le f(c) \tag{4.7}$$

where the number Δx is sufficiently small in absolute value. The inequality in (4.7) then implies

$$\frac{f(c + \Delta x) - f(c)}{\Delta x} \le 0 \quad \text{for } \Delta x > 0 \tag{4.8}$$

and $$\frac{f(c + \Delta x) - f(c)}{\Delta x} \ge 0 \quad \text{for } \Delta x < 0 \tag{4.9}$$

But since $\lim_{\Delta x \to 0}[f(c + \Delta x) - f(c)]/\Delta x$ exists and equals $f'(c)$, (4.8) and (4.9) show that $f'(c) \le 0$ *and* $f'(c) \ge 0$, respectively. The only way this can happen is to have $f'(c) = 0$. We leave the proof of the case when $f(c)$ is a relative minimum as an exercise.   ∎

We have seen that a function continuous on a *closed* interval has both an absolute maximum and an absolute minimum. The next theorem tells us where these extrema can occur.

> **THEOREM 4.3**   If f is continuous on a closed interval $[a, b]$, then an absolute extremum occurs either at an endpoint of the interval or at a critical value in the open interval (a, b).

Finding the Absolute Extrema

We summarize Theorem 4.3 in the following manner. To find an absolute extremum of a function f continuous on $[a, b]$:

(*i*)   Evaluate f at a and b.

(*ii*)   Find all critical values $c_1, c_2, \ldots, c_n$ in (a, b).

(*iii*)   Evaluate f at all critical values.

(*iv*)   The largest and smallest values in the list

$$f(a), f(b), f(c_1), \ldots, f(c_n)$$

are the absolute maximum and the absolute minimum, respectively, of f on the interval $[a, b]$.

Example 6 _____

Find the absolute extrema of $f(x) = x^3 - 3x^2 - 24x + 2$ on (*a*) $[-3, 1]$ and (*b*) $[-3, 8]$.

Solution   We need only evaluate f at the endpoints of each interval and at critical values within each open interval. From

$$f'(x) = 3x^2 - 6x - 24 = 3(x + 2)(x - 4)$$

we see that the critical values of the function are -2 and 4.

(*a*)   From the data in the accompanying table it is evident that the absolute maximum of f on $[-3, 1]$ is $f(-2) = 30$ and the absolute minimum is the endpoint extremum $f(1) = -24$.

on $[-3, 1]$		
x	-3	-2	1
$f(x)$	20	30	-24

(*b*)   On the interval $[-3, 8]$ we see that $f(4) = -78$ is an absolute minimum and $f(8) = 130$ is an endpoint absolute maximum.

on $[-3, 8]$				
x	-3	-2	4	8
$f(x)$	20	30	-78	130

Remarks

(*i*)   A function may, of course, assume its maximum and minimum values more than once on an interval. Figure 4.30 illustrates two possible cases.

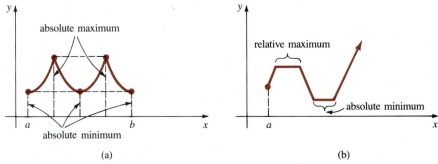

Figure 4.30

(*ii*)   The converse of Theorem 4.2 is not necessarily true; that is, a critical value of a function need not correspond to a relative extremum. Consider $f(x) = x^3$ and $g(x) = x^{1/3}$. The derivatives $f'(x) = 3x^2$ and $g(x) = (1/3)x^{-2/3}$ show that 0 is a critical value of both functions. But from the graphs of f and g in Figure 4.31, we see that neither function possesses any extrema.

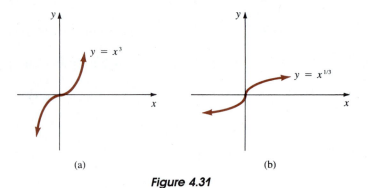

Figure 4.31

Exercises 4.3

Answers to odd-numbered problems begin on page A-23.

In Problems 1–14 find the critical values of the given function.

1. $f(x) = 2x^2 - 6x + 8$

2. $f(x) = x^3 + x - 2$

3. $f(x) = 2x^3 - 15x^2 - 36x$

4. $f(x) = x^4 - 4x^3 + 7$

5. $f(x) = (x - 2)^2(x - 1)$

6. $f(x) = x^2(x + 1)^3$

7. $f(x) = \dfrac{1 + x}{\sqrt{x}}$

8. $f(x) = \dfrac{x}{x^2 + 2}$

9. $f(x) = (4x - 3)^{1/3}$

10. $f(x) = x^{2/3} + x$

11. $f(x) = (x - 1)^2 \sqrt[3]{x + 2}$

12. $f(x) = \dfrac{x + 4}{\sqrt[3]{x + 1}}$

13. $f(x) = -x + \sin x$

14. $f(x) = \cos 4x$

In Problems 15–26 find the absolute extrema of the given function on the indicated interval.

15. $f(x) = -x^2 + 6x$; $[1, 4]$

16. $f(x) = (x - 1)^2$; $[2, 5]$

17. $f(x) = x^{2/3}$; $[-1, 8]$

18. $f(x) = x^{2/3}(x^2 - 1)$; $[-1, 1]$

19. $f(x) = x^3 - 6x^2 + 2$; $[-3, 2]$

20. $f(x) = -x^3 - x^2 + 5x$; $[-2, 2]$

21. $f(x) = x^3 - 3x^2 + 3x - 1$; $[-4, 3]$

22. $f(x) = x^4 + 4x^3 - 10$; $[0, 4]$

23. $f(x) = x^4(x - 1)^2$; $[-1, 2]$

24. $f(x) = \dfrac{\sqrt{x}}{x^2 + 1}$; $[\frac{1}{4}, \frac{1}{2}]$

25. $f(x) = 2 \cos 2x - \cos 4x$; $[0, 2\pi]$

26. $f(x) = 1 + 5 \sin 3x$; $[0, \dfrac{\pi}{2}]$

In Problems 27 and 28 find all critical values. Distinguish between absolute, endpoint absolute, and relative extrema.

27. $f(x) = x^2 - 2|x|$; $[-2, 3]$

28. $f(x) = \begin{cases} 4x + 12, & -5 \le x \le -2 \\ x^2, & -2 < x \le 1 \end{cases}$

29. Consider the continuous function f defined on $[a, b]$ shown in Figure 4.32. Given that c_1 through c_{10} are critical values:

   (a) List critical values at which $f'(x) = 0$.

   (b) List critical values at which $f'(x)$ is not defined.

   (c) Distinguish between the absolute and endpoint absolute extrema.

   (d) Distinguish between the relative maxima and the relative minima.

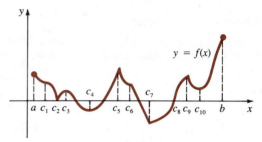

Figure 4.32

30. Consider the function $f(x) = x + 1/x$. Show that the relative minimum is greater than the relative maximum.

31. Draw a graph of a continuous function that possesses no absolute extrema but has a relative maximum and a relative minimum that are the same value.

32. Give an example of a continuous function, defined on a closed interval $[a, b]$, for which the absolute maximum is the same as the absolute minimum.

33. Let $f(x) = [x]$ be the greatest integer function. Show that every value of x is a critical value.

34. Show that $f(x) = (ax + b)/(cx + d)$ has no critical values when $ad - bc \ne 0$. What happens when $ad - bc = 0$?

35. The height of a projectile launched from ground level is given by $s(t) = -16t^2 + 320t$, where t is measured in seconds and s in feet.

   (a) $s(t)$ is defined only on the time interval $[0, 20]$. Why?

   (b) Use the results of Theorem 4.3 to determine the maximum height attained by the projectile.

36. The French physician Jean Louis Poiseuille discovered that the velocity v (in cm/sec) of blood flowing through an artery with circular cross section of radius R is given by $v(r) = (P/4\nu l)(R^2 - r^2)$, where P, ν, and l are positive constants. See Figure 4.33.

   (a) Determine a closed interval on which v is defined.

   (b) Determine the maximum and minimum velocities of the blood.

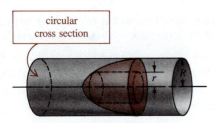

Figure 4.33

Miscellaneous Problems

37. Let $f(x) = x^n$, where n is a positive integer. Determine the values of n for which f has a relative extremum.

38. Prove that a polynomial function of degree n can have at most $n - 1$ critical values.

39. Suppose f is a continuous even function such that $f(a)$ is a relative minimum. What can be said about $f(-a)$?

40. Suppose f is a continuous odd function such that $f(a)$ is a relative maximum. What can be said about $f(-a)$?

41. Prove Theorem 4.2 in the case when $f(c)$ is a relative minimum.

42. The function f whose graph is given in Figure 4.34 is discontinuous on the interval $[a, b]$. Does f possess any absolute extrema?

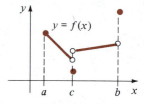

Figure 4.34

4.4 Rolle's Theorem and the Mean Value Theorem

When $y = f(x)$ is continuous and differentiable on an interval $[a, b]$, it seems plausible that if $f(a) = f(b) = 0$, then its graph must be as indicated in Figure 4.35. The graphs in turn suggest that there must be at least one point on the

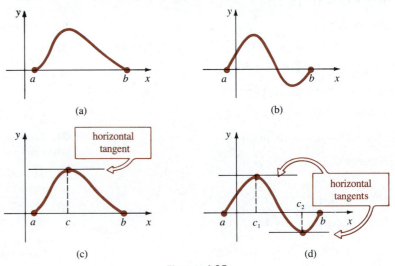

Figure 4.35

graph that corresponds to a number c in (a, b) at which the tangent is horizontal. See Figure 4.35(c) and (d). This result is formalized as Rolle's Theorem.*

THEOREM 4.4   **Rolle's Theorem**

Let f be continuous on $[a, b]$ and differentiable on (a, b) such that $f(a) = f(b) = 0$. Then there exists a number c in (a, b) such that $f'(c) = 0$.

Proof   Either f is a constant function on the interval $[a, b]$ or it is not.

(i)   If f is a constant function on $[a, b]$, then we must have $f'(c) = 0$ for every number c in (a, b).

(ii)   Now, if f is not a constant function on $[a, b]$ there must be some number x in (a, b) at which either $f(x) > 0$ or $f(x) < 0$. Suppose $f(x) > 0$. Since f is continuous on $[a, b]$, we know from the Extreme Value Theorem that f attains an absolute maximum at some number c in $[a, b]$. But from $f(a) = f(b) = 0$ and $f(x) > 0$ for some x in (a, b) we conclude that the number c cannot be an endpoint of $[a, b]$. Consequently, c is in (a, b). Since f is differentiable on (a, b), it is differentiable at c. Hence, from Theorem 4.2, we have $f'(c) = 0$. The proof of the case when $f(x) < 0$ follows in a similar manner.   ∎

_____ **Example 1** _____

Consider the function $f(x) = -x^3 + x$ defined on $[-1, 1]$. Since f is a polynomial function, it is continuous on $[-1, 1]$ and differentiable on $(-1, 1)$. Also, $f(-1) = f(1) = 0$. Thus, the hypotheses of Rolle's Theorem are satisfied. We conclude that there must be at least one number in $(-1, 1)$ for which $f'(x) = -3x^2 + 1$ is zero. To find this number, we solve $f'(c) = 0$ or $-3c^2 + 1 = 0$. The latter leads to *two* solutions in the interval, $c_1 = -\sqrt{3}/3$ and $c_2 = \sqrt{3}/3$.

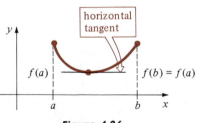

Figure 4.36

Note in the preceding example that the given function f satisfies the hypotheses of Rolle's Theorem on $[0, 1]$ as well as on $[-1, 1]$. In the case of the interval $[0, 1]$, $f'(c) = -3c^2 + 1 = 0$ yields the single solution $c = \sqrt{3}/3$.

The conclusion of Rolle's Theorem also holds when the condition $f(a) = f(b) = 0$ is replaced with $f(a) = f(b)$. The plausibility of this fact is illustrated in Figure 4.36.

Rolle's Theorem is helpful in proving the next important result.

THEOREM 4.5   **The Mean Value Theorem for Derivatives**

Let f be continuous on $[a, b]$ and differentiable on (a, b). Then there exists a number c in (a, b) such that

$$f'(c) = \frac{f(b) - f(a)}{b - a}$$

---

*Michel Rolle, a French mathematician (1652–1719).

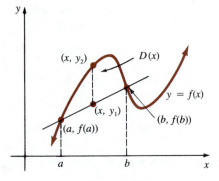

Figure 4.37

Proof As shown in Figure 4.37, let $D(x)$ denote the vertical distance between a point on the graph of $y = f(x)$ and the secant line through $(a, f(a))$ and $(b, f(b))$. Since the equation of the secant line is

$$y - f(b) = \frac{f(b) - f(a)}{b - a}(x - b)$$

we have, as shown in the figure,

$$D(x) = f(x) - \left[\frac{f(b) - f(a)}{b - a}(x - b) + f(b)\right] \qquad (4.10)$$

Since $D(a) = D(b) = 0$ and D is continuous on $[a, b]$ and differentiable on (a, b), Rolle's Theorem implies there is some number c in (a, b) for which $D'(c) = 0$. In view of (4.10),

$$D'(x) = f'(x) - \frac{f(b) - f(a)}{b - a}$$

and so $D'(c) = 0$ is the same as

$$f'(c) = \frac{f(b) - f(a)}{b - a} \qquad ■$$

Theorem 4.5 is also called the **Theorem of the Mean**.

Geometrically, the Mean Value Theorem asserts that the slope of the tangent line at $(c, f(c))$ is the same as the slope of the secant through $(a, f(a))$, $(b, f(b))$. See Figure 4.38(a). Also, as indicated in Figure 4.38(b), there may be more than one number c in (a, b) for which the tangent and secant lines are parallel.

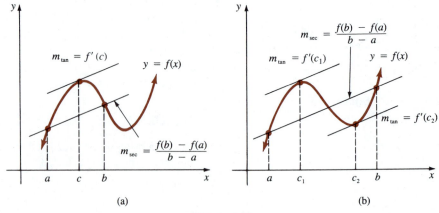

(a)   (b)

Figure 4.38

_____ **Example 2** _____

Given the function $f(x) = x^3 - 12x$ defined on $[-1, 3]$, does there exist a number c in $(-1, 3)$ that satisfies the conclusion of the Mean Value Theorem?

Solution Since f is a polynomial function, it is continuous on $[-1, 3]$ and differentiable on $(-1, 3)$. Now,

$$f(3) = -9 \quad \text{and} \quad f(-1) = 11$$
$$f'(x) = 3x^2 - 12 \quad \text{and} \quad f'(c) = 3c^2 - 12$$

and, hence, we must have

$$\frac{f(3) - f(-1)}{3 - (-1)} = \frac{-20}{4} = 3c^2 - 12$$

Thus, $3c^2 = 7$. Although the last equation has two solutions, the only solution in $(-1, 3)$ is $c = \sqrt{7/3} \approx 1.53$.

The Mean Value Theorem is very useful in proving other theorems. Recall from Section 3.3 that if $f(x) = k$ is a constant function, then $f'(x) = 0$. The converse of this result is given by the following.

THEOREM 4.6   If $f'(x) = 0$ for all x in an interval $[a, b]$, then $f(x)$ is a constant on the interval.

Proof Let x_1 and x_2 be any numbers in $[a, b]$ such that $x_1 < x_2$. By the Mean Value Theorem, there is a number c in (x_1, x_2) such that

$$\frac{f(x_2) - f(x_1)}{x_2 - x_1} = f'(c)$$

But, $f'(c) = 0$ by hypothesis. Hence, $f(x_2) - f(x_1) = 0$ or $f(x_1) = f(x_2)$. Since x_1 and x_2 are arbitrarily chosen, the function f has the same value at all points in the interval. Thus, f is constant. ∎

Increasing and Decreasing Functions

We shall use the Mean Value Theorem to relate the concepts of increasing and decreasing functions with the notion of a derivative.

DEFINITION 4.4   Let f be a function defined on an interval I and suppose x_1 and x_2 are numbers in I such that $x_1 < x_2$.

(i) f is said to be **increasing** on I if $f(x_1) < f(x_2)$.

(ii) f is said to be **decreasing** on I if $f(x_1) > f(x_2)$.

In other words, the graph of an increasing function rises as x increases, whereas the graph of a decreasing function falls as x increases. The graph in Figure 4.39 illustrates a function f that is increasing on $[b, c]$ and $[d, e]$ and decreasing on $[a, b]$, $[c, d]$, and $[e, h]$.

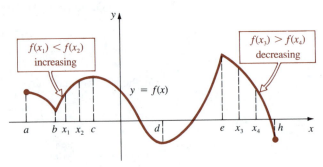

Figure 4.39

THEOREM 4.7   Let f be continuous on $[a, b]$ and differentiable on (a, b).

(*i*)   If $f'(x) > 0$ for all x in (a, b), then f is increasing on $[a, b]$.

(*ii*)   If $f'(x) < 0$ for all x in (a, b), then f is decreasing on $[a, b]$.

Proof of (*i*)   Let x_1 and x_2 be any numbers in $[a, b]$ such that $x_1 < x_2$. By the Mean Value Theorem, there is a number c in (x_1, x_2) such that $[f(x_2) - f(x_1)]/(x_2 - x_1) = f'(c)$. But, $f'(c) > 0$ by hypothesis. Hence, $f(x_2) - f(x_1) > 0$ or $f(x_1) < f(x_2)$. Since x_1 and x_2 are arbitrarily chosen, it follows from Definition 4.4 that f is increasing on $[a, b]$.   ∎

Example 3

Determine the intervals on which $f(x) = x^3 - 3x^2 - 24x$ is increasing and the intervals on which f is decreasing.

Solution   The derivative is

$$f'(x) = 3x^2 - 6x - 24 = 3(x + 2)(x - 4)$$

To determine when $f'(x) > 0$ and $f'(x) < 0$, we must solve

$$(x + 2)(x - 4) > 0 \quad \text{and} \quad (x + 2)(x - 4) < 0$$

respectively. One way of solving these inequalities is to examine the algebraic signs of the factors $(x + 2)$ and $(x - 4)$ in the intervals of the number line determined by the critical values -2 and 4: $(-\infty, -2]$, $[-2, 4]$, $[4, \infty)$. See Figure 4.40.

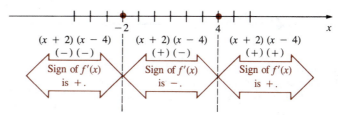

Figure 4.40

The information is summarized in the accompanying table.

Interval	Sign of $f'(x)$	$y = f(x)$
$(-\infty, -2)$	$+$	increasing on $(-\infty, -2]$
$(-2, 4)$	$-$	decreasing on $[-2, 4]$
$(4, \infty)$	$+$	increasing on $[4, \infty)$

Example 4

Determine the intervals on which $f(x) = x^{2/3}$ is increasing and the intervals on which f is decreasing.

Solution   Observe that

$$f'(x) = \frac{2}{3}x^{-1/3} = \frac{2}{3\sqrt[3]{x}}$$

is undefined at 0. Since 0 is in the domain of f, we conclude it is a critical value. Using the facts that $\sqrt[3]{x} < 0$ for $x < 0$ and $\sqrt[3]{x} > 0$ for $x > 0$, we are led to the information given in the accompanying table.

Interval	Sign of $f'(x)$	$y = f(x)$
$(-\infty, 0)$	$-$	decreasing on $(-\infty, 0]$
$(0, \infty)$	$+$	increasing on $[0, \infty)$

If a function f is discontinuous at one or both endpoints of $[a, b]$, then $f'(x) > 0$ (or $f'(x) < 0$) on (a, b) implies f is increasing (decreasing) on the open interval (a, b).

Remark

The converses of parts (*i*) and (*ii*) of Theorem 4.7 are not necessarily true. In other words, when f is an increasing (or decreasing) function on an interval it does not follow that $f'(x) > 0$ (or $f'(x) < 0$). A function could be, say, increasing but yet not differentiable.

Exercises 4.4

Answers to odd-numbered problems begin on page A-24.

In Problems 1–10 determine whether the given function satisfies the hypotheses of Rolle's Theorem on the indicated interval. If so, find all values of c that satisfy the conclusion of the theorem.

1. $f(x) = x^2 - 4$; $[-2, 2]$

2. $f(x) = x^2 - 6x + 5$; $[1, 5]$

3. $f(x) = x^3 + 27$; $[-3, -2]$

4. $f(x) = x^3 - 5x^2 + 4x$; $[0, 4]$

5. $f(x) = x^3 + x^2$; $[-1, 0]$

6. $f(x) = x(x - 1)^2$; $[0, 1]$

7. $f(x) = \sin x$; $[-\pi, 2\pi]$

8. $f(x) = \tan x$; $[0, \pi]$

9. $f(x) = x^{2/3} - 1$; $[-1, 1]$

10. $f(x) = x^{2/3} - 3x^{1/3} + 2$; $[1, 8]$

In Problems 11 and 12 (Figures 4.41 and 4.42) state why the function f whose graph is given fails the hypotheses of Rolle's Theorem on $[a, b]$.

11.
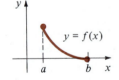

Figure 4.41

12.
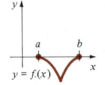

Figure 4.42

In Problems 13–22 determine whether the given function satisfies the hypotheses of the Mean Value Theorem on the indicated interval. If so, find all values of c that satisfy the conclusion of the theorem.

13. $f(x) = x^2$; $[-1, 7]$

14. $f(x) = -x^2 + 8x - 6$; $[2, 3]$

15. $f(x) = x^3 + x + 2$; $[2, 5]$

16. $f(x) = x^4 - 2x^2$; $[-3, 3]$

17. $f(x) = 1/x$; $[-10, 10]$

18. $f(x) = x + \dfrac{1}{x}$; $[1, 5]$

19. $f(x) = 1 + \sqrt{x}$; $[0, 9]$

20. $f(x) = \sqrt{4x + 1}$; $[2, 6]$

21. $f(x) = \dfrac{x + 1}{x - 1}$; $[-2, -1]$

22. $f(x) = x^{1/3} - x$; $[-8, 1]$

In Problems 23 and 24 (Figures 4.43 and 4.44) state why the function f whose graph is given fails the hypotheses of the Mean Value Theorem on $[a, b]$.

23.
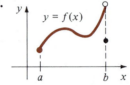

Figure 4.43

24.
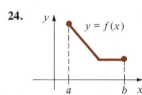

Figure 4.44

In Problems 25–44 determine the intervals on which the given function f is increasing and the intervals on which f is decreasing.

25. $f(x) = x^2 + 5$

26. $f(x) = x^3$

27. $f(x) = x^2 + 6x - 1$

28. $f(x) = -x^2 + 10x + 3$

29. $f(x) = x^3 - 3x^2$

30. $f(x) = \dfrac{1}{3}x^3 - x^2 - 8x + 1$

31. $f(x) = x^4 - 4x^3 + 9$

32. $f(x) = 4x^5 - 10x^4 + 2$

33. $f(x) = 1 - x^{1/3}$

34. $f(x) = x^{2/3} - 2x^{1/3}$

35. $f(x) = x + \dfrac{1}{x}$

36. $f(x) = \dfrac{1}{x} + \dfrac{1}{x^2}$

37. $f(x) = x\sqrt{8 - x^2}$

38. $f(x) = \dfrac{x + 1}{\sqrt{x^2 + 1}}$

39. $f(x) = \dfrac{5}{x^2 + 1}$

40. $f(x) = \dfrac{x^2}{x + 1}$

41. $f(x) = x(x - 3)^2$

42. $f(x) = (x^2 - 1)^3$

43. $f(x) = \sin x$

44. $f(x) = -x + \tan x$

45. A motorist enters a tollway and is given a stub stamped 1:15 P.M. Sixty miles down the road, when the motorist pays the toll at 2:15 P.M., he is also given a traffic ticket. Explain this by the Mean Value Theorem. Assume the speed limit is 55 mph.

46. In the analysis of the human cough, the trachea, or windpipe, is considered to be a cylindrical tube. The volume flow of air (in cm^3/sec) through the trachea during its contraction is given by

$$V(r) = kr^4(r_0 - r), \qquad r_0/2 \le r \le r_0$$

where k is a positive constant and r_0 is its radius when there is no pressure difference at the ends of the tracheal tube. Determine an interval for which V is increasing and an interval for which V is decreasing. What radius will give the maximum volume flow of air?

Miscellaneous Problems

47. Consider the function $f(x) = x^4 + x^3 - x - 1$. Use this function and Rolle's Theorem to show that the equation $4x^3 + 3x^2 - 1 = 0$ has at least one root in $[-1, 1]$.

48. Show that the equation $ax^3 + bx + c = 0$, $a > 0$, $b > 0$ cannot have two real roots. [*Hint:* Consider the function $f(x) = ax^3 + bx + c$. Suppose there are two numbers r_1 and r_2 such that $f(r_1) = f(r_2) = 0$.]

49. Show that the equation $ax^2 + bx + c = 0$ has at most two real roots. [*Hint:* Consider the function $f(x) = ax^2 + bx + c$. Suppose there are three distinct numbers r_1, r_2, and r_3 such that $f(r_1) = f(r_2) = f(r_3) = 0$.]

50. For a quadratic polynomial function $f(x) = ax^2 + bx + c$ show that the value of x_3 that satisfies the conclusion of the Mean Value Theorem on any interval $[x_1, x_2]$ is $x_3 = (x_1 + x_2)/2$.

Calculator Problems

In Problems 51 and 52 use a calculator to find a value of c that satisfies the conclusion of the Mean Value Theorem.*

51. $f(x) = \cos 2x$; $[0, \pi/4]$

52. $f(x) = 1 + \sin x$; $[\pi/4, \pi/2]$

4.5 Graphing and the First Derivative

Knowing that a function does, or does not, possess relative extrema is a great aid in drawing its graph.

Recall that when a function has a relative extremum it must occur at a critical value. By finding the critical values of a function, we have a *list of abscissas that possibly correspond to relative extrema*. We shall now combine the ideas of the two preceding sections to devise two tests for determining when a critical value actually is the *x*-coordinate of a relative extremum.

First Derivative Test   Suppose f is differentiable on (a, b) and that c is a critical value in the interval. If $f'(x) > 0$ for all x in (a, c) and $f'(x) < 0$ for all x in (c, b), then on the interval (a, b) the graph of f must be as indicated in Figure 4.45(a); that is, $f(c)$ is a relative maximum. On the other hand, when $f'(x) < 0$ for all x in (a, c) and $f'(x) > 0$ for all x in (c, b), then, as shown in Figure 4.45(b), $f(c)$ is a relative minimum. We have demonstrated a special case of the following theorem.

---

*On typical calculators this will require using the ⟨INV⟩ or ⟨f^{-1}⟩ keys. For example, if the value of $\sin x$ is specified, then the INV function will give a value of x in the interval $[-\pi/2, \pi/2]$. See Section 7.2.

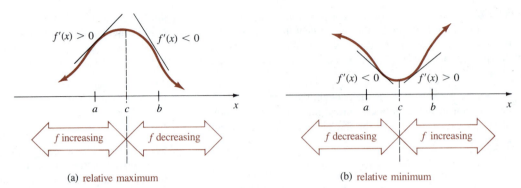

(a) relative maximum   (b) relative minimum

Figure 4.45

THEOREM 4.8   **First Derivative Test for Relative Extrema**

Let f be continuous on $[a, b]$ and differentiable on (a, b) except possibly at the critical value c.

 (*i*)   If $f'(x) > 0$ for $a < x < c$ and $f'(x) < 0$ for $c < x < b$, then $f(c)$ is a relative maximum.

 (*ii*)   If $f'(x) < 0$ for $a < x < c$ and $f'(x) > 0$ for $c < x < b$, then $f(c)$ is a relative minimum.

 (*iii*)   If $f'(x)$ has the same algebraic sign on $a < x < c$ and $c < x < b$, then $f(c)$ is not an extremum.

Example 1

Graph $f(x) = x^3 - 3x^2 - 9x + 2$.

Solution   The first derivative

$$f'(x) = 3x^2 - 6x - 9 = 3(x + 1)(x - 3) \tag{4.11}$$

yields the critical values -1 and 3. Now the first derivative test is essentially the procedure used in finding the intervals on which f is either increasing or decreasing. Using (4.11) we see in Figure 4.46(a) that $f'(x) > 0$ for $-\infty < x < -1$ and $f'(x) < 0$ for $-1 < x < 3$. It follows from part (*i*) of Theorem 4.8 that $f(-1) = 7$ is a relative maximum. Similarly, $f'(x) < 0$ for $-1 < x < 3$ and $f'(x) > 0$ for $3 < x < \infty$. Thus, from part (*ii*) of Theorem 4.8 $f(3) = -25$ is a relative minimum.

     Now, the graph of the function has the y-intercept $f(0) = 2$. Furthermore, inspection of the equation $x^3 - 3x^2 - 9x + 2 = 0$ indicates -2 is a real root, and, hence, an x-intercept. Division by the factor $x + 2$ then gives $(x + 2)$ $(x^2 - 5x + 1) = 0$. The quadratic formula reveals two additional x-intercepts:

$$\frac{5 - \sqrt{21}}{2} \approx 0.21 \quad \text{and} \quad \frac{5 + \sqrt{21}}{2} \approx 4.79$$

Putting all this information together leads to the graph given in Figure 4.46(b).

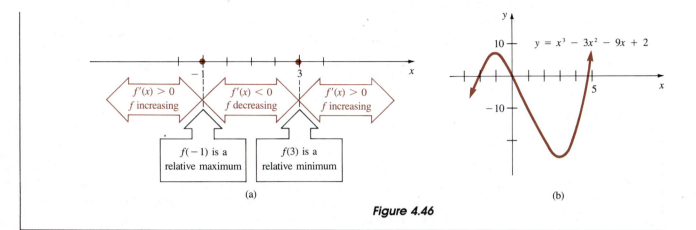

Figure 4.46

Example 2

Graph $f(x) = x + \dfrac{1}{x}$.

Solution   Inspection of the function in the form $f(x) = (x^2 + 1)/x$ reveals the graph of f has no intercepts, $x = 0$ is a vertical asymptote, and that the graph of f is symmetric with respect to the origin since $f(-x) = -f(x)$. Now,

$$f'(x) = 1 - \frac{1}{x^2} = \frac{x^2 - 1}{x^2} = \frac{(x + 1)(x - 1)}{x^2}$$

shows that -1 and 1 are critical values. In Figure 4.47(a) we see $f'(x) > 0$ for $-\infty < x < -1$ and $f'(x) < 0$ for $-1 < x < 0$. Thus, from the First Derivative Test, we conclude that $f(-1) = -2$ is a relative maximum. By symmetry it follows that $f(1) = 2$ is a relative minimum. The graph of f is given in Figure 4.47(b).

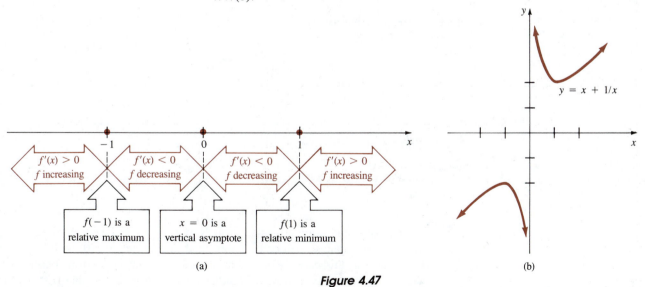

Figure 4.47

Example 3

Graph $f(x) = -x^{5/3} + 5x^{2/3}$.

Solution   The derivative is

$$f'(x) = -\frac{5}{3}x^{2/3} + \frac{10}{3}x^{-1/3}$$

$$= \frac{5}{3}x^{-1/3}(-x + 2)$$

Notice that f' does not exist at 0 but 0 is in the domain of the function since $f(0) = 0$. The critical values are 0 and 2. The First Derivative Test, illustrated in Figure 4.48(a), shows that $f(0) = 0$ is a relative minimum and that $f(2) = -(2)^{5/3} + 5(2)^{2/3} \approx 4.76$ is a relative maximum. Moreover, since $|f'(x)| \to \infty$ as $x \to 0$, there is a vertical tangent at $(0, 0)$. Finally, by writing $f(x) = x^{2/3}(-x + 5)$, we see that the x-intercepts are 0 and 5. The graph of f is given in Figure 4.48(b).

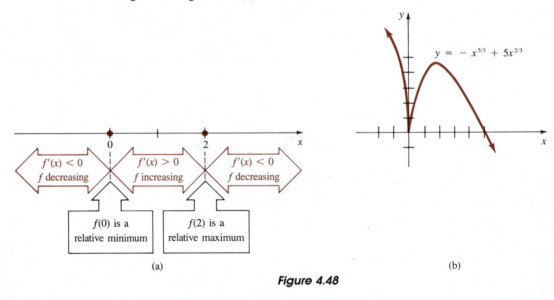

(a)   (b)

Figure 4.48

Example 4

Graph $f(x) = x^4 - 4x^3 + 10$.

Solution   The derivative

$$f'(x) = 4x^3 - 12x^2 = 4x^2(x - 3)$$

shows that 0 and 3 are critical values. Now, as seen in Figure 4.49(a), f' has the same algebraic sign in both $(-\infty, 0)$ and $(0, 3)$. Hence, $f(0) = 10$ is not an extremum. In this case $f'(0) = 0$ means there is only a horizontal tangent at $(0, 10)$. However, it is evident from the First Derivative Test that $f(3) = -17$

is a relative minimum. Indeed, the graph of f given in Figure 4.49(b) shows that $f(3)$ is also an absolute minimum. In conclusion, we see that the graph of f has two x-intercepts. Unfortunately, the real solutions of $x^4 - 4x^3 + 10 = 0$ are not obvious.

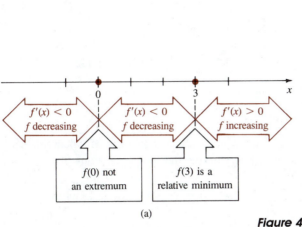

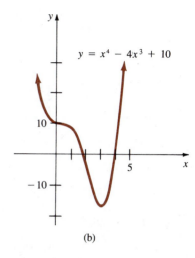

(a)   (b)

Figure 4.49

Example 5

Graph $f(x) = \dfrac{x^2 - 3}{x^2 + 1}$.

Solution

y-intercept: $f(0) = -3$

x-intercepts: $f(x) = 0$ when $x^2 - 3 = 0$. Thus, $-\sqrt{3}$ and $\sqrt{3}$ are x-intercepts.

Symmetry: y-axis since $f(-x) = f(x)$.

Vertical asymptotes: None since $x^2 + 1 \neq 0$ for all real numbers.

Horizontal asymptotes: $\lim_{x \to \infty} \dfrac{x^2 - 3}{x^2 + 1} = 1$. Symmetry implies

$$\lim_{x \to -\infty} \frac{x^2 - 3}{x^2 + 1} = 1.\ y = 1 \text{ is a horizontal asymptote.}$$

Derivative: $f'(x) = \dfrac{8x}{(x^2 + 1)^2}$

Critical values: $f'(x) = 0$ when $8x = 0$. Therefore, 0 is the only critical value.

First Derivative Test: See Figure 4.50(a). $f(0) = -3$ is a relative minimum.

Graph: See Figure 4.50(b).

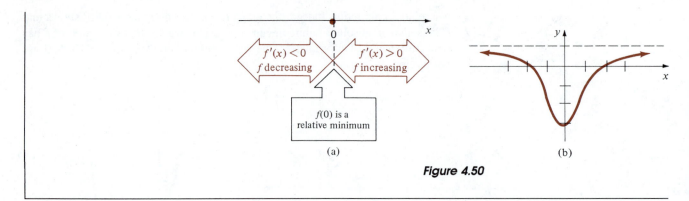

Figure 4.50

Remark

We conclude with a tabular summary of Theorem 4.8 you can use for easy reference while working the following problems.

First Derivative Test			
Critical value	Interval	Sign of $f'(x)$ on the interval	Conclusion
c	(a, c) (c, b)	+ −	 $f(c)$ is relative maximum
c	(a, c) (c, b)	− +	 $f(c)$ is relative minimum
c	(a, c) (c, b)	+ +	 no extremum
c	(a, c) (c, b)	− −	 no extremum

Exercises 4.5

Answers to odd-numbered problems begin on page A-24.

In Problems 1–26 use the First Derivative Test to find the relative extrema of the given function. Graph. Find intercepts when possible.

1. $f(x) = -x^2 + 2x + 1$

2. $f(x) = (x - 1)(x + 3)$

3. $f(x) = x^3 - 3x$

4. $f(x) = \frac{1}{3}x^3 - \frac{1}{2}x^2 + 1$

5. $f(x) = x(x - 2)^2$

6. $f(x) = -x^3 + 3x^2 + 9x - 1$

7. $f(x) = x^4 + 4x$

8. $f(x) = (x^2 - 1)^2$

9. $f(x) = \frac{1}{4}x^4 + \frac{4}{3}x^3 + 2x^2$

10. $f(x) = 2x^4 - 16x^2 + 3$

11. $f(x) = 4x^5 - 5x^4$

12. $f(x) = (x - 2)^2(x + 3)^3$

13. $f(x) = \dfrac{x^2 + 3}{x + 1}$

14. $f(x) = x + \dfrac{25}{x}$

15. $f(x) = \dfrac{1}{x} - \dfrac{1}{x^3}$

16. $f(x) = \dfrac{x^2}{x^2 - 4}$

17. $f(x) = \dfrac{10}{x^2 + 1}$

18. $f(x) = \dfrac{x^2}{x^4 + 1}$

19. $f(x) = (x - 4)^{2/3}$

20. $f(x) = (x^2 - 1)^{1/3}$

21. $f(x) = x\sqrt{1 - x^2}$

22. $f(x) = x(x^2 - 5)^{1/3}$

23. $f(x) = x - 12x^{1/3}$

24. $f(x) = x^{4/3} + 32x^{1/3}$

25. $f(x) = x^{2/3}(x^2 - 16)$

26. $f(x) = (x - 1)^{2/3}(x - 11)$

In Problems 27–30 (Figures 4.51–4.54) use the graph of f' to sketch a possible graph of f.

27.

Figure 4.51

28.

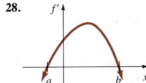

Figure 4.52

29.

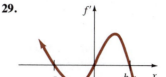

Figure 4.53

30.

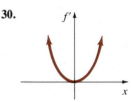

Figure 4.54

In Problems 31 and 32 (Figures 4.55 and 4.56) use the graph of f to sketch the graph of f'.

31.

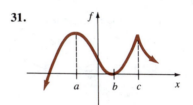

Figure 4.55

32.

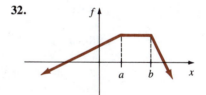

Figure 4.56

In Problems 33–36 sketch a graph of a function f that has the given properties.

33. $f(-1) = 0, f(0) = 1$
$f'(3)$ does not exist, $f'(5) = 0$
$f'(x) > 0, x < 3$ and $x > 5$
$f'(x) < 0, 3 < x < 5$

34. $f(0) = 0$
$f'(-1) = 0, f'(0) = 0, f'(1) = 0$
$f'(x) < 0, x < -1, -1 < x < 0$
$f'(x) > 0, 0 < x < 1, x > 1$

35. $f(-x) = f(x)$
$f(2) = 3$
$f'(x) < 0, 0 < x < 2$
$f'(x) > 0, x > 2$

36. $f(1) = -2, f(0) = -1$
$\lim_{x \to 3} f(x) = \infty$
$f'(4) = 0$
$f'(x) < 0, x < 1$
$f'(x) < 0, x > 4$

In Problems 37 and 38 determine where the slope of the tangent to the graph of the given function has a relative maximum or a relative minimum.

37. $f(x) = x^3 + 6x^2 - x$ **38.** $f(x) = x^4 - 6x^2$

39. (a) From the graph of $g(x) = \sin 2x$ determine the intervals for which $g(x) > 0$ and the intervals for which $g(x) < 0$.

(b) Find the critical values of $f(x) = \sin^2 x$. Use the First Derivative Test and the information in part **(a)** to find the relative extrema of f.

(c) Sketch the graph of the function f in part **(b)**.

40. (a) Find the critical values of $f(x) = x - \sin x$.

(b) Show that f has no relative extrema.

(c) Sketch the graph of f.

41. Find values of a, b, and c such that $f(x) = ax^2 + bx + c$ has a relative maximum 6 at $x = 2$ and the graph of f has y-intercept 4.

42. Find values of a, b, c, and d such that $f(x) = ax^3 + bx^2 + cx + d$ has a relative minimum -3 at $x = 0$ and a relative maximum 4 at $x = 1$.

Miscellaneous Problems

43. Suppose f is a differentiable function whose graph is symmetric about the y-axis. Prove that $f'(0) = 0$. Does f necessarily have a relative extremum at $x = 0$?

44. Let m and n denote positive integers. Show that $f(x) = x^m(x - 1)^n$ always has a relative minimum.

4.6 Graphing and the Second Derivative

In the discussion that follows, our goal is to relate the concept of the concavity of a graph with the second derivative of a function.

Concavity   You probably have an intuitive idea of what is meant by concavity. Figure 4.57(a) and (b) illustrates geometric shapes that are **concave upward** and **concave downward**, respectively.* Often a shape that is concave upward is said to "hold water," whereas a shape that is concave downward "spills water." The graph given in Figure 4.58 is concave upward on the interval (b, c) and concave downward on (a, b) and (c, d).

We state the definition of concavity in terms of the derivative.

"holds water"   "spills water"

(a) concave upward   (b) concave downward

Figure 4.57

DEFINITION 4.5   Let f be differentiable on (a, b).

(*i*)   If f' is an increasing function on (a, b), then the graph of f is **concave upward** on the interval.

(*ii*) If f' is a decreasing function on (a, b), then the graph of f is **concave downward** on the interval.

---

*The Gateway Arch in St. Louis is concave downward. The cables between the supports of the Golden Gate Bridge are concave upward.

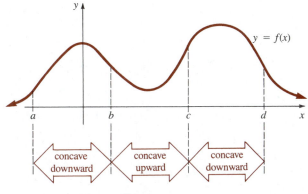

Figure 4.58

In other words, if the slope of the tangent line increases (decreases) on (a, b), then the graph of the function is concave upward (downward) on the interval. The plausibility of Definition 4.5 is illustrated in Figure 4.59. Equivalently, the graph of a function is concave upward on an interval if the graph at any point lies *above* the tangent at the point. A graph that is concave downward on an interval then lies *below* the tangent lines.

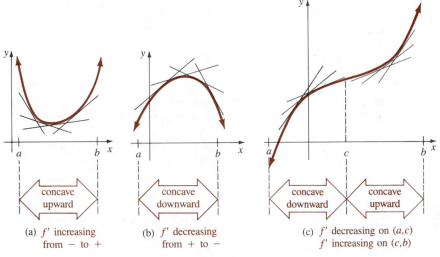

(a) f' increasing from $-$ to $+$

(b) f' decreasing from $+$ to $-$

(c) f' decreasing on (a,c)
f' increasing on (c,b)

Figure 4.59

Concavity and the Second Derivative

From Section 4.4 we remember that the algebraic sign of the derivative of a function indicates when the function is increasing or decreasing on an interval. Specifically, if the function referred to in the last sentence is the derivative f', then we can conclude that the algebraic sign of the derivative f'' indicates when f' is either increasing or decreasing on an interval. For example, if $f''(x) > 0$ on (a, b), then f' increases on (a, b). In view of Definition 4.5, if f' increases on (a, b), then the graph of f is concave upward on the interval. Therefore, we are led to the following test for concavity.

THEOREM 4.9   **Test for Concavity**

Let f be a function for which f'' exists on (a, b).

(i)   If $f''(x) > 0$ for all x in (a, b), then the graph of f is concave upward on (a, b).

(ii) If $f''(x) < 0$ for all x in (a, b), then the graph of f is concave downward on (a, b).

Example 1

Determine the intervals on which the graph of $f(x) = -x^3 + \frac{9}{2}x^2$ is concave upward and the intervals for which the graph is concave downward.

Solution From

$$f'(x) = -3x^2 + 9x$$

$$f''(x) = -6x + 9 = 6\left(-x + \frac{3}{2}\right)$$

we see that $f''(x) > 0$ when $6(-x + \frac{3}{2}) > 0$ or $x < \frac{3}{2}$ and that $f''(x) < 0$ when $6(-x + \frac{3}{2}) < 0$ or $x > \frac{3}{2}$. It follows from Theorem 4.9 that the graph of f is concave upward on $(-\infty, \frac{3}{2})$ and concave downward on $(\frac{3}{2}, \infty)$.

Point of Inflection The graph of the function in Example 1 changes concavity at the point that corresponds to $x = \frac{3}{2}$. As x increases through $\frac{3}{2}$, the graph of f changes from concave upward to concave downward *at the point* $(\frac{3}{2}, \frac{27}{4})$. A point on the graph of a function where the concavity changes from upward to downward, or vice versa, is called a **point of inflection**. More precisely, we have the following definition.

DEFINITION 4.6   Let f be continuous at c. A point $(c, f(c))$ is a **point of inflection** if there exists an open interval (a, b) that contains c such that the graph of f is either

(i)   concave upward on (a, c) and concave downward on (c, b), or

(ii) concave downward on (a, c) and concave upward on (c, b).

Figure 4.60 shows a graph of a function $y = f(x)$ that has three points of inflection: $(a, f(a))$, $(b, f(b))$, and $(c, f(c))$. Notice $(d, f(d))$ is not a point of inflection since the graph is concave upward on both intervals (c, d) and (d, e).

As a consequence of Definitions 4.5 and 4.6 we observe:

A point of inflection $(c, f(c))$ occurs at a number c for which $f''(c) = 0$ or $f''(c)$ does not exist.

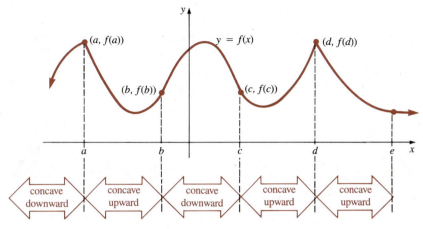

Figure 4.60

Example 2

Find any points of inflection of $f(x) = -x^3 + x^2$.

Solution The first and second derivatives of f are, respectively,

$$f'(x) = -3x^2 + 2x \quad \text{and} \quad f''(x) = -6x + 2$$

Since $f''(x) = 0$ at $\frac{1}{3}$, the point $(\frac{1}{3}, \frac{2}{27})$ is the only *possible* point of inflection. Now,

$$f''(x) = 6(-x + \tfrac{1}{3}) > 0 \quad \text{for } x < \tfrac{1}{3}$$
$$f''(x) = 6(-x + \tfrac{1}{3}) < 0 \quad \text{for } x > \tfrac{1}{3}$$

implies that the graph of f is concave upward on $(-\infty, \frac{1}{3})$ and concave downward on $(\frac{1}{3}, \infty)$. Thus, $(\frac{1}{3}, f(\frac{1}{3}))$ or $(\frac{1}{3}, \frac{2}{27})$ is a point of inflection.

Example 3

Find any points of inflection of $f(x) = 5x - (x - 4)^{1/3}$.

Solution From

$$f'(x) = 5 - \frac{1}{3}(x - 4)^{-2/3} \quad \text{and} \quad f''(x) = \frac{2}{9}(x - 4)^{-5/3}$$

we see that f'' does not exist at 4. Since $f''(x) < 0$ for $x < 4$ and $f''(x) > 0$ for $x > 4$, the graph of f is concave downward on $(-\infty, 4)$ and concave upward on $(4, \infty)$. Now, 4 is in the domain of f and so $(4, f(4))$ or $(4, 20)$ is a point of inflection.

_____ **Example 4** _____

An inspection of the graphs $y = \sin x$, $y = \cos x$, and $y = \tan x$ in Figure 4.61 would seem to indicate that the x-intercepts of the graph of each function are the abscissas of points of inflection. You are asked to prove that this is indeed the case. See Problems 19, 20, and 22 in Exercises 4.6.

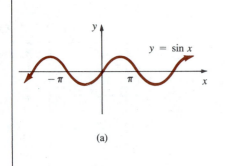

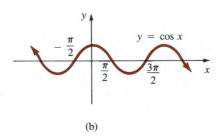

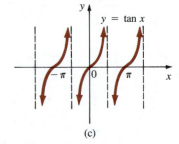

(a)   (b)   (c)

Figure 4.61

The concavity of a graph can be related to the notion of relative extrema.

Second Derivative Test   If c is a critical value of $y = f(x)$ and, say, $f''(c) > 0$, then the graph of f is concave upward on some interval (a, b) that contains c. Necessarily then, $f(c)$ is a relative minimum. Similarly, $f''(c) < 0$ at a critical value c implies $f(c)$ is a relative maximum. This so-called **Second Derivative Test** is illustrated in Figure 4.62.

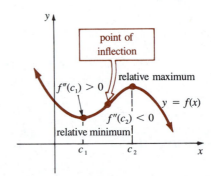

Figure 4.62

> **THEOREM 4.10**   **Second Derivative Test for Relative Extrema**
>
> Let f be a function for which f'' exists on an interval (a, b) that contains the critical number c.
>
> (*i*)   If $f''(c) > 0$, then $f(c)$ is a relative minimum.
>
> (*ii*)   If $f''(c) < 0$, then $f(c)$ is a relative maximum.

At this point one might ask, Why do we need another test for relative extrema when we already have the First Derivative Test? If the function under examination is a polynomial, it is very easy to compute the second derivative. In using Theorem 4.10 we need only determine the algebraic sign of f'' *at the* critical value. Contrast this with determining the sign of f' at numbers to the right and left of the critical value. If f' is not readily factored, the latter procedure may be somewhat difficult. On the other hand, it may be equally tedious to use Theorem 4.10 in the case of some functions that involve products, quotients, powers, and so on. In a word, Theorems 4.8 and 4.10 both have advantages and disadvantages.

Example 5

Graph $f(x) = x^4 - x^2$.

Solution
$$f'(x) = 4x^3 - 2x = 2x(2x^2 - 1)$$
$$f''(x) = 12x^2 - 2$$

Hence, the critical values of f are 0, $-\sqrt{2}/2$, and $\sqrt{2}/2$. The Second Derivative Test is summarized in the accompanying table.

x	Sign of $f''(x)$	$f(x)$	Conclusion
0	$-$	0	rel. max.
$\sqrt{2}/2$	$+$	$-1/4$	rel. min.
$-\sqrt{2}/2$	$+$	$-1/4$	rel. min.

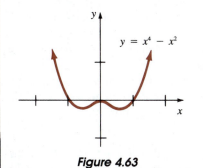

$y = x^4 - x^2$

Figure 4.63

Now from $f(x) = x^2(x^2 - 1) = x^2(x + 1)(x - 1)$, we see that the graph of f passes through $(0, 0)$, $(-1, 0)$, and $(1, 0)$. Furthermore, since f is a polynomial with only even powers, we conclude that its graph is symmetric with respect to the y-axis (even function). See Figure 4.63. You should also verify that the graph possesses two points of inflection: $(-\sqrt{6}/6, -\frac{5}{36})$ and $(\sqrt{6}/6, -\frac{5}{36})$.

Example 6

Graph $f(x) = 2 \cos x - \cos 2x$.

Solution
$$f'(x) = -2 \sin x + 2 \sin 2x$$
$$f''(x) = -2 \cos x + 4 \cos 2x$$

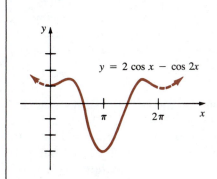

$y = 2 \cos x - \cos 2x$

Figure 4.64

Using the trigonometric identity $\sin 2x = 2 \sin x \cos x$, the equation $f'(x) = 0$ simplifies to $\sin x(1 - 2 \cos x) = 0$. The solutions of $\sin x = 0$ are 0, $\pm\pi$, $\pm 2\pi$, . . . and the solutions of $\cos x = \frac{1}{2}$ are $\pm\pi/3$, $\pm 5\pi/3$, ... But since f is 2π periodic (show this!), it suffices to consider only those critical values in $[0, 2\pi]$, namely, 0, $\pi/3$, π, $5\pi/3$, and 2π. The Second Derivative Test applied to these values is summarized in the accompanying table. The graph of f is given in Figure 4.64.

x	Sign of $f''(x)$	$f(x)$	Conclusion
0	+	1	rel. min.
$\pi/3$	−	$\frac{3}{2}$	rel. max.
π	+	−3	rel. min.
$5\pi/3$	−	$\frac{3}{2}$	rel. max.
2π	+	1	rel. min.

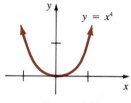

Figure 4.65

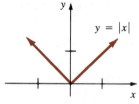

Figure 4.66

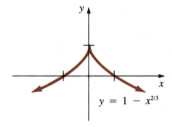

Figure 4.67

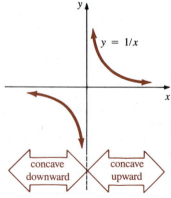

Figure 4.68

Remarks

(*i*)   One should not get the impression from Theorem 4.9 that when a graph is concave upward (or downward) on an interval (a, b) that $f''(x) > 0$ (or $f''(x) < 0$) for *all* x in the interval. The conditions stated in parts (*i*) and (*ii*) of Theorem 4.9 are sufficient but are not necessary. For example, it is clear from Figure 4.65 and Definition 4.5 that the twice-differentiable function $f(x) = x^4$ is concave upward on any interval that contains the origin. But from $f''(x) = 12x^2$ we see $f''(0) = 0$.

(*ii*)   The intuitive notions of "holding water" and "spilling water" are, of course, not synonymous with the concepts of concave upward and concave downward, respectively. We could argue from Figures 4.66 and 4.67 that the graph of $f(x) = |x|$ "holds water" on $(-1, 1)$ and the graph of $g(x) = 1 - x^{2/3}$ "spills water" on $(-1, 1)$. But we cannot assign any concavity to either graph on $(-1, 1)$ since both f and g are not differentiable on the interval. Furthermore, there are differentiable functions whose graphs possess no concavity. You are asked to supply an example of such a function in Problem 54 of the exercises. (This is not very difficult.)

(*iii*)   Recall, if $(c, f(c))$ is a point of inflection, then $f''(c) = 0$ or $f''(c)$ does not exist. The converse of this statement is not necessarily true. We cannot conclude, simply from the fact that $f''(c) = 0$ or $f''(c)$ does not exist, that $(c, f(c))$ is a point of inflection; for example, the second derivative of the function $f(x) = x^4$ is zero at $x = 0$. We see in Figure 4.65 that $(0, f(0))$ is not a point of inflection since the graph is concave upward on $(-\infty, 0)$ and on $(0, \infty)$. Also, for $f(x) = 1/x$, we see $f''(x) = 2/x^3$ is undefined at $x = 0$ and $f''(x) < 0$ for $x < 0$ and $f''(x) > 0$ for $x > 0$. However, 0 is not the x-coordinate of a point of inflection since f is not continuous at this value. See Figure 4.68.

(*iv*)   It is important to note that the Second Derivative Test does *not* state that if $f(c)$ is a relative extremum, then either $f''(c) > 0$ or $f''(c) < 0$. In Figure 4.67 we see that $f(x) = 1 - x^{2/3}$ has an absolute maximum at 0 but $f''(0)$ does not exist. Similarly, $f(x) = x^4$ has an absolute minimum at 0 but $f''(0) = 0$. Thus *the second derivative test can lead to no conclusion*. Whenever the second derivative test fails, the first derivative test should be used.

(*v*) In conclusion, we summarize Theorem 4.10 and the preceding remark.

Second Derivative Test		
Critical value	*f″(x) at critical value*	*Conclusion*
c	+	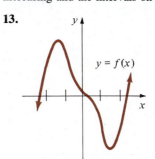 $f(c)$ is a relative minimum
c	−	$f(c)$ is a relative maximum
c	0	no conclusion

Exercises 4.6

Answers to odd-numbered problems begin on page A-25.

In Problems 1–12 use the second derivative to determine the intervals on which the given function is concave upward and the intervals on which it is concave downward.

1. $f(x) = -x^2 + 7x$

2. $f(x) = -(x + 2)^2 + 8$

3. $f(x) = -x^3 + 6x^2 + x - 1$

4. $f(x) = (x + 5)^3$

5. $f(x) = x(x - 4)^3$

6. $f(x) = 6x^4 + 2x^3 - 12x^2 + 3$

7. $f(x) = x^{1/3} + 2x$

8. $f(x) = x^{8/3} - 20x^{2/3}$

9. $f(x) = x + \dfrac{9}{x}$

10. $f(x) = \sqrt{x^2 + 10}$

11. $f(x) = \dfrac{1}{x^2 + 3}$

12. $f(x) = \dfrac{x - 1}{x + 2}$

In Problems 13 and 14 (Figures 4.69 and 4.70) estimate from the graph of the given function f the intervals on which f' is increasing and the intervals on which f' is decreasing.

13.

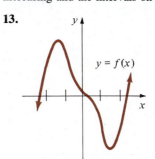

$y = f(x)$

Figure 4.69

14.

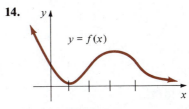

$y = f(x)$

Figure 4.70

15. Show that the graph of $f(x) = \sec x$ is concave upward on those intervals on which $\cos x > 0$, and concave downward on those intervals for which $\cos x < 0$.

16. Show that the graph of $f(x) = \csc x$ is concave upward on those intervals for which $\sin x > 0$, and concave downward on those intervals for which $\sin x < 0$.

In Problems 17–22 use the second derivative to locate all points of inflection.

17. $f(x) = x^4 - 12x^2 + x - 1$

18. $f(x) = x^{5/3} + 4x$

19. $f(x) = \sin x$

20. $f(x) = \cos x$

21. $f(x) = x - \sin x$

22. $f(x) = \tan x$

In Problems 23–38 use the Second Derivative Test, when applicable, to find the relative extrema of the given function. Graph. Find points of inflection and intercepts when possible.

23. $f(x) = -(2x - 5)^2$

24. $f(x) = \dfrac{1}{3}x^3 - 2x^2 - 12x$

25. $f(x) = x^3 + 3x^2 + 3x + 1$

26. $f(x) = \dfrac{1}{4}x^4 - 2x^2$

27. $f(x) = 6x^5 - 10x^3$

28. $f(x) = x^3(x + 1)^2$

29. $f(x) = \dfrac{x}{x^2 + 2}$

30. $f(x) = x^2 + \dfrac{1}{x^2}$

31. $f(x) = \sqrt{9 - x^2}$

32. $f(x) = x\sqrt{x - 6}$

33. $f(x) = x^{1/3}(x + 1)$

34. $f(x) = x^{1/2} - \dfrac{1}{4}x$

35. $f(x) = \cos 3x, \ [0, 2\pi]$

36. $f(x) = 2 + \sin 2x, \ [0, 2\pi]$

37. $f(x) = \cos x + \sin x, \ [0, 2\pi]$

38. $f(x) = 2 \sin x + \sin 2x, \ [0, 2\pi]$

In Problems 39–42 determine whether the given function has a relative extremum at the indicated critical value.

39. $f(x) = \sin x \cos x; \ \pi/4$

40. $f(x) = x \sin x; \ 0$

41. $f(x) = \tan^2 x; \ \pi$

42. $f(x) = (1 + \sin 4x)^3; \ \pi/8$

In Problems 43–46 sketch a graph of a function f that has the given properties.

43. $f(-2) = 0, f(4) = 0$
$f'(3) = 0, f''(1) = 0, f''(2) = 0$
$f''(x) < 0, x < 1, x > 2$
$f''(x) > 0, 1 < x < 2$

44. $f(0) = 5, f(2) = 0$
$f'(2) = 0, f''(3)$ does not exist
$f''(x) > 0, x < 3$
$f''(x) < 0, x > 3$

45. $f(0) = -1, f(\pi/2) > 0$
$f'(x) \geq 0$ for all x
$f''(x) > 0, (2n - 1)\dfrac{\pi}{2} < x < (2n + 1)\dfrac{\pi}{2}, \ n$ even
$f''(x) < 0, (2n - 1)\dfrac{\pi}{2} < x < (2n + 1)\dfrac{\pi}{2}, \ n$ odd

46. $f(-x) = -f(x)$
vertical asymptote $x = 2, \ \lim\limits_{x \to \infty} f(x) = 0$
$f''(x) < 0, 0 < x < 2$
$f''(x) > 0, x > 2$

47. Find values of a, b, and c such that the graph of $f(x) = ax^3 + bx^2 + cx$ passes through $(-1, 0)$ and has a point of inflection at $(1, 1)$.

48. Find values of a, b, and c such that the graph of $f(x) = ax^3 + bx^2 + cx$ has a horizontal tangent at the point of inflection $(1, 1)$.

Miscellaneous Problems

49. Use the Second Derivative Test as an aid in graphing $f(x) = \sin(1/x)$. Observe that f is discontinuous at $x = 0$.

50. Show that the graph of a general polynomial function

$$f(x) = a_n x^n + a_{n-1}x^{n-1} + \cdots + a_1 x + a_0, \qquad a_n \neq 0,$$

can have at most $n - 2$ points of inflection.

51. Let $f(x) = (x - x_0)^n$ where n is a positive integer.

(a) Show that $(x_0, 0)$ is a point of inflection of the graph of f if n is an odd integer.

(b) Show that $(x_0, 0)$ is not a point of inflection of the graph of f but corresponds to a relative minimum when n is an even integer.

52. Prove that the graph of a quadratic polynomial function $f(x) = ax^2 + bx + c$, $a \neq 0$ is concave upward on the x-axis when $a > 0$ and concave downward on the x-axis when $a < 0$.

53. Let f be a function for which f''' exists on an interval (a, b) that contains the number c. If $f''(c) = 0$ and $f'''(c) \neq 0$, what can be said about $(c, f(c))$?

54. Give an example of a differentiable function whose graph possesses no concavity.

55. Prove or disprove: A point of inflection for a function f must occur at a critical value of f'.

56. Prove or disprove: The function

$$f(x) = \begin{cases} 1/x, & x \neq 0 \\ 0, & x = 0 \end{cases}$$

has a point of inflection at $(0, 0)$.

4.7 Further Applications of Extrema

In science, engineering, and business one is often interested in the maximum and minimum values of functions; for example, a company is naturally interested in maximizing revenue while minimizing cost. The next time you go to a supermarket try this experiment: Take along a small ruler and measure the height and diameter of *all* cans that contain, say, 16 ounces of food (28.9 in^3). The fact that all cans of this specified volume have the same shape is no coincidence since there are specific dimensions that will minimize the amount of metal used, and, hence, minimize the cost of construction to a company. In the same vein, many of the so-called economy cars have appearances that are remarkably the same. This is not just a simple matter of one company copying the success of another company, but, rather, for a given volume engineers strive for a design that will minimize the amount of material used.

Helpful Hints   In the examples and problems that follow either we will be *given* a function or we will have to interpret the words to *set up* a function for which we seek a maximum or a minimum value. These are the kinds of word problems that show off the power of the calculus and provide one of many possible answers to the age-old question: "What's it good for?" Here are seven important steps in solving an applied max-min problem.

> (*i*) Develop a positive and analytical attitude. Read the problem slowly. Do not merely strive for an answer.
>
> (*ii*) Draw a picture when necessary.
>
> (*iii*) Introduce variables and note any relationship among the variables.
>
> (*iv*) Using all necessary variables, set up a function to be maximized or minimized. If more than one variable is used, then employ a relationship between the variables to reduce the function to one variable.
>
> (*v*) Make note of the interval on which the function is defined. Determine all critical values.

(vi) If the function to be maximized or minimized is continuous and defined on a closed interval $[a, b]$, then test for endpoint extrema. If the desired extremum does not occur at an endpoint, it must occur at a critical value within the open interval (a, b).

(vii) If the function to be maximized or minimized is defined on an interval that is not closed, then a derivative test should be used on each critical value.

In the first example we shall, this one time, label the corresponding steps.

Example 1

Find two nonnegative numbers whose sum is 15 such that the product of one with the square of the other is a maximum.

Solution

(ii) A picture is not possible.

(iii) Let x and y denote the two nonnegative numbers; that is, $x \geq 0$ and $y \geq 0$. Note, it is given that $x + y = 15$.

(iv) Let P denote the product:

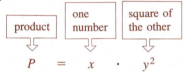

$$P = x \cdot y^2$$

From *(iii)* we can use $y = 15 - x$ to express P in terms of x alone:

$$P(x) = x(15 - x)^2$$

(v) The function $P(x)$ is defined only for $0 \leq x \leq 15$ since, if $x > 15$, then, contrary to the given conditions, $y = 15 - x$ would be negative. Now by the Product Rule,

$$P'(x) = x \cdot 2(15 - x)(-1) + (15 - x)^2$$
$$= (15 - x)(15 - 3x)$$

Thus, the only critical value in $(0, 15)$ is $x = 5$.

(vi) Testing the endpoints of the interval reveals $P(0) = P(15) = 0$ is the minimum value of the product. Hence, $P(5) = 5(10)^2 = 500$ must be the maximum value. The two nonnegative numbers are 5 and 10.

Example 2

In physics it is shown that when air resistance is ignored, the horizontal range R of a projectile is given by

$$R = \frac{v_0^2}{g} \sin 2\theta, \qquad 0 < \theta \leq \pi/2 \qquad (4.12)$$

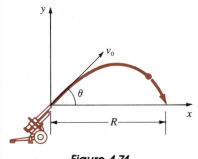

Figure 4.71

where v_0 is the constant initial velocity, g is the acceleration of gravity, and θ is the angle of elevation or departure. See Figure 4.71.

From
$$\frac{dR}{d\theta} = \frac{v_0^2}{g} 2 \cos 2\theta$$

we see that $dR/d\theta = 0$ when $\cos 2\theta = 0$ or $2\theta = \pi/2$. Hence, the only critical value in $(0, \pi/2)$ is $\theta = \pi/4$. As in Example 1, the endpoints give the minimum range: $R(0) = R(\pi/2) = 0$ and so $R(\pi/4) = v_0^2/g$ is the maximum range.* In other words, in order to achieve the maximum range, the projectile should be launched at an angle of $45°$ to the horizontal.

_____ **Example 3** _____

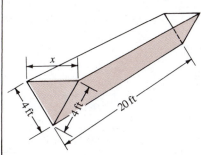

Figure 4.72

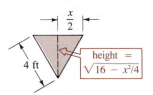

Figure 4.73

A 20-ft-long water trough has ends in the form of isosceles triangles whose equal sides are 4 ft long. Determine the dimension across the top of a triangular end so that the volume of the trough is a maximum.

Solution   The trough with the unknown dimension is illustrated in Figure 4.72. Its volume is

$$V = (\text{area of triangular end}) \cdot (\text{length})$$

From Figure 4.73 and the Pythagorean Theorem, the area of the triangular end is $\frac{1}{2}x\sqrt{16 - x^2/4}$. Consequently, the volume as a function of x is

$$V(x) = \left(\frac{1}{2}x\sqrt{16 - x^2/4}\right) \cdot 20$$
$$= 5x\sqrt{64 - x^2}$$

Furthermore, $V(x)$ only makes sense on the closed interval $[0, 8]$. (Why?) Taking the derivative and simplifying yields

$$V'(x) = -10\frac{x^2 - 32}{(64 - x^2)^{1/2}}$$

Although $V'(x) = 0$ for $x = \pm 4\sqrt{2}$, the only critical value of V in $(0, 8)$ is $4\sqrt{2}$. Since $V(0) = V(8) = 0$, we conclude that the maximum volume occurs when the width across the top of the trough is $4\sqrt{2}$ ft. The maximum volume is $V(4\sqrt{2}) = 160$ ft^3.

_____

*With air resistance, a projectile will fall short of this value. Remember to allow for this the next time you fire a cannon. Also, take a minute and compare the maximum distance a golf ball can be hit on earth when $v_0 = 160$ ft/sec and $g = 32$ ft/sec^2 with the moon, where $g = 5.4$ ft/sec^2.

Note: Often a problem can be solved in more than one way. In Example 1 we could just as well express the function *P*; that is, the product, in terms of the symbol *y*:

$$P = xy^2$$
$$= (15 - y)y^2$$
$$= 15y^2 - y^3$$

In this way P' can be found without the use of the Product Rule. Also, in hindsight, you should verify that the solution of Example 3 is slightly "cleaner" if the dimension across the top of the trough is labeled $2x$ instead of x. Indeed, as the next example shows, Example 3 can be solved utilizing an entirely different variable.

Example 4

Alternative Solution to Example 3 As shown in Figure 4.74, let θ denote the angle between the vertical and one of the sides. From trigonometry the height and base of the triangular end are $4 \cos \theta$ and $8 \sin \theta$, respectively. Expressed as a function of θ, V becomes

$$V(\theta) = \frac{1}{2}(4 \cos \theta)(8 \sin \theta) \cdot 20$$
$$= 320 \sin \theta \cos \theta$$
$$= 160(2 \sin \theta \cos \theta)$$
$$= 160 \sin 2\theta \qquad \boxed{\text{Double Angle Formula}}$$

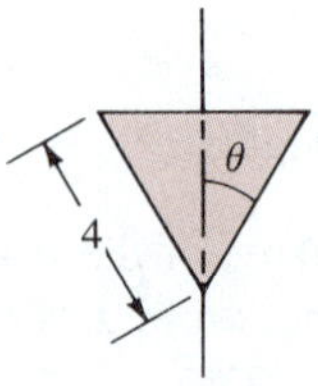

where $0 \leq \theta \leq \pi/2$. Proceeding as in Example 2, we find the maximum value $V = 160 \text{ ft}^3$ occurs at $\theta = \pi/4$. The dimension across the top of the trough, or the base of the isosceles triangle, is $8 \sin(\pi/4) = 4\sqrt{2}$ ft.

Figure 4.74

Example 5

Find the dimensions of a rectangle with greatest area that can be inscribed in the ellipse $x^2/4 + y^2/9 = 1$.

Solution As shown in Figure 4.75 let the corner of the rectangle in the first quadrant have coordinates (x, y). It is apparent from symmetry that the area of the inscribed rectangle is

$$A = 4xy$$

But solving $x^2/4 + y^2/9 = 1$ for y yields

$$y^2 = 9\left(1 - \frac{x^2}{4}\right) \quad \text{or} \quad y = \frac{3}{2}\sqrt{4 - x^2}$$

Hence, the area as a function of x is given by

$$A(x) = 6x\sqrt{4 - x^2}$$

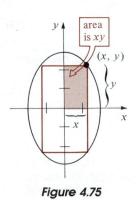

Figure 4.75

where $0 \leq x \leq 2$. Since this function is similar to $V(x)$ in Example 3, we leave it to the reader to show that $A(x)$ has a maximum at the critical value $\sqrt{2}$. Hence, the dimensions of the inscribed rectangle with greatest area are $2\sqrt{2} \times 3\sqrt{2}$. The corresponding area is

$$A(\sqrt{2}) = 12 \text{ square units}$$

Example 6

Find the point in the first quadrant on the circle $x^2 + y^2 = 1$ closest to $(2, 4)$.

Solution   Let (x, y), $x > 0$, $y > 0$, denote the point on the circle that is closest to the point $(2, 4)$. See Figure 4.76. The distance formula gives

$$d = \sqrt{(x - 2)^2 + (y - 4)^2} \quad \text{or} \quad d^2 = (x - 2)^2 + (y - 4)^2$$

Now the point that minimizes the square of the distance d^2 also minimizes the distance d. Let us write $D = d^2$. By expanding $(x - 2)^2$ and $(y - 4)^2$ and using $x^2 + y^2 = 1$ (or $y = \sqrt{1 - x^2}$) we find

$$D(x) = x^2 - 4x + 4 + \overbrace{(1 - x^2)}^{y^2} - 8\overbrace{\sqrt{1 - x^2}}^{y} + 16$$
$$= -4x - 8\sqrt{1 - x^2} + 21$$

where $0 \leq x \leq 1$. Differentiation then gives

$$D'(x) = -4 - 4(1 - x^2)^{-1/2}(-2x)$$
$$= \frac{-4\sqrt{1 - x^2} + 8x}{\sqrt{1 - x^2}}$$

After simplifying we find that $1/\sqrt{5}$ is a solution of the equation $D'(x) = 0$ in the interval $[0, 1]$. Since

$$D(0) = 13$$
$$D(1/\sqrt{5}) = 21 - 20/\sqrt{5} \approx 12.06$$
$$D(1) = 17$$

we conclude that D, and hence d, is a minimum when $x = 1/\sqrt{5}$. This means $(1/\sqrt{5}, 2/\sqrt{5})$ is the point on the circle in the first quadrant closest to $(2, 4)$.

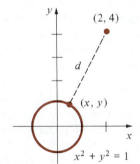

Figure 4.76

Example 7

A rectangular plot of land that contains 1500 m² will be fenced and divided into two equal portions by an additional fence parallel to two sides. Find the dimensions of the land that require the least amount of fencing.

Solution   As shown in Figure 4.77, let us introduce variables x and y so that $xy = 1500$. Then the function we wish to minimize is the sum of lengths of the five portions of fence:

$$L = 2x + 3y$$

But, $y = 1500/x$ enables us to write

$$L(x) = 2x + \frac{4500}{x}$$

where the only requirement on the variable x is that it be positive. Thus, unlike the prior examples, the function we are considering is *not defined on a closed interval*. Setting

$$L'(x) = 2 - \frac{4500}{x^2}$$

equal to zero we get $x^2 = 2250$ and conclude that the only critical value is $15\sqrt{10}$. Proceeding to the second derivative

$$L''(x) = \frac{13,500}{x^3}$$

we observe that $L''(15\sqrt{10}) > 0$. It follows from the Second Derivative Test that $L(15\sqrt{10}) = 2(15\sqrt{10}) + 4500/15\sqrt{10} = 60\sqrt{10}$ m is the required minimum amount of fencing. Returning to the relationship $xy = 1500$, we find the value of y and conclude that the dimensions of the land should be $15\sqrt{10}$ m × $10\sqrt{10}$ m.

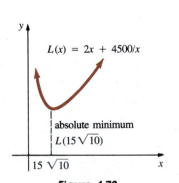

Figure 4.77

Remark

As an observant reader, you may question at least two aspects of Example 7. Where did the assumption that the land be divided into two equal portions enter into the solution? In point of fact, it did not. What is important is that the dividing fence be parallel to the two ends. Ask yourself what $L(x)$ would be if this were *not* the case. However, the actual positioning of the dividing fence between the ends is irrelevant as long as it is parallel to them.

In an applied problem we are naturally interested in only absolute extrema. Therefore, another question might be: Since the function L is not defined on a closed interval and since the Second Derivative Test does not guarantee absolute extrema, how can we be certain that $L(15\sqrt{10})$ is an absolute minimum? Sometimes we can argue to the existence of an absolute extremum from the physical context of the problem, but perhaps a better procedure is, when in doubt, draw a graph. Figure 4.78 answers the question for $L(x)$.

If it happens that a differentiable function f has only *one* critical value c in an open interval (a, b),* then you should be able to convince yourself of the validity of the following result.

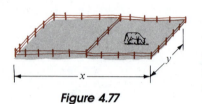

$L(x) = 2x + 4500/x$

absolute minimum

$L(15\sqrt{10})$

$15\sqrt{10}$

Figure 4.78

> *If a derivative test indicates that $f(c)$ is a relative maximum or a relative minimum, then $f(c)$ is an absolute maximum or an absolute minimum.*

*This includes (a, ∞), $(-\infty, b)$, and $(-\infty, \infty)$.

_____ **Exercises 4.7** _____

Answers to odd-numbered problems begin on page A-26.

1. Find two nonnegative numbers whose sum is 60 and whose product is a maximum.

2. Find two nonnegative numbers whose product is 50 and whose sum is a minimum.

3. Find a number that exceeds its square by the greatest amount.

4. Let m and n be positive integers. Find two nonnegative numbers whose sum is S such that the product of the mth power of one with the nth power of the other is a maximum.

5. Find two nonnegative numbers whose sum is 1 such that the sum of the square of one and twice the square of the other is a minimum.

6. Consider the graphs of $y = x^2 - 1$ and $y = 1 - x$ given in Figure 4.79. Find the maximum vertical distance between the graphs on the interval $-2 \leq x \leq 1$.

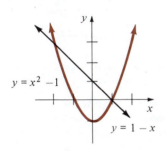

Figure 4.79

7. Find the point(s) on the graph of $y^2 = 6x$ closest to

    (a) $(5, 0)$ **(b)** $(3, 0)$

8. Find the point on the graph of $x + y = 1$ closest to $(2, 3)$.

9. Determine the point on the graph of $y = x^3 - 4x^2$ at which the tangent line has minimum slope.

10. Determine the point on the graph of $y = 8x^2 + 1/x$ at which the tangent line has a maximum slope.

11. A rancher has 3000 ft of fencing on hand. Determine the dimensions of a rectangular corral that encloses a maximum area.

12. A rectangular plot of land will be fenced into three equal portions by two dividing fences parallel to two sides. See Figure 4.80. If the area to be enclosed is 4000 m², find the dimensions of the land that require the least amount of fence.

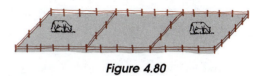

Figure 4.80

13. If the total fence to be used is 8000 m, find the dimensions of the enclosed land in Figure 4.80 that has the greatest area.

14. A rancher wishes to build a rectangular corral of 128,000 ft² with one side along a vertical cliff. The fencing along the cliff costs $1.50 per ft, whereas along the other three sides the fencing costs $2.50 per ft. Find the dimensions of the corral so that the cost of construction is a minimum.

In Problems 15–18 (Figures 4.81–4.84) find the dimensions of the shaded region such that its area is a maximum.

15.

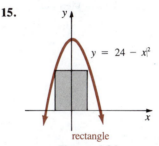

rectangle

Figure 4.81

16.

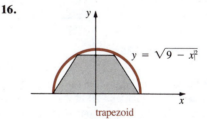

trapezoid

Figure 4.82

17.

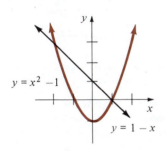

rectangle

Figure 4.83

18.

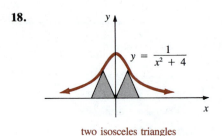

two isosceles triangles

Figure 4.84

19. An open rectangular box is to be constructed with a square base and a volume of 32,000 cm³. Find the dimensions of the box that require the least amount of material.

20. In Problem 19 find the dimensions of a closed box that require the least amount of material.

21. A box, open at the top, is to be made from a square piece of cardboard by cutting a square out of each corner and turning up the sides. Given that the cardboard measures 40 cm on a side, find the dimensions of the box that will give the maximum volume. What is the maximum volume? See Figure 4.85.

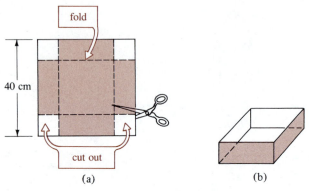

Figure 4.85

22. A box, open at the top, is to be made from a rectangular piece of cardboard that is 30 in long and 20 in wide. The box can hold itself together by cutting a square out of each corner, cutting on the interior solid lines, and then folding the cardboard on the dashed lines. See Figure 4.86(a) and (b). Express the

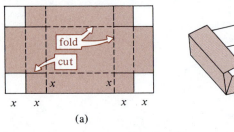

Figure 4.86

volume of the box as a function of the indicated variable x. Find the dimensions of the box that will give the maximum volume. What is the maximum volume?

23. A rectangular sheet of metal with perimeter 4 m will be rolled and formed into the lateral side of a cylindrical container. Find the dimensions of the container with largest volume.

24. A printed page will have 2-in. margins of white space on the sides and 1-in. margins of white space on the top and bottom. The area of the printed portion is 32 in.². Determine the dimensions of the page so that the least amount of paper is used.

25. Find the dimensions of the right circular cylinder with greatest volume that can be inscribed in a right circular cone of radius R and height H. See Figure 4.87.

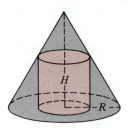

Figure 4.87

26. Find the dimensions of the right circular cone with greatest volume that can be inscribed in a sphere of radius R.

27. A gutter with a rectangular cross section is made by bending up equal amounts from the ends of a 30-cm-wide piece of tin. What are the dimensions of the cross section so that the volume is a maximum?*

28. A gutter will be made so that its cross section is an isosceles trapezoid with dimensions as indicated in Figure 4.88. Determine the value of θ so that the volume is a maximum.

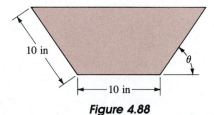

Figure 4.88

29. Find the dimensions of a cylindrical juice can that utilize the least amount of material when the volume of the can is 32 in³.

*Reread Example 3 and ponder why the length of the piece of metal does not have to be specified.

30. A plastic drinking cup in the shape of a right circular cone will have a volume of 24π cm^3. Find the dimensions that will minimize the amount of material used. (*Hint:* The lateral surface area of a cone is $A = \pi r L$, where r is its radius and L is its slant height.)

31. A conical cup is made from a circular piece of paper of radius R by cutting out a circular sector and then joining the dashed edges shown in Figure 4.89(a). Determine the value of r indicated in Figure 4.89(b) so that the volume of the cup is a maximum. What is the maximum volume of the cup?

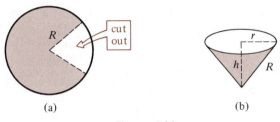

(a)   (b)

Figure 4.89

32. In Problem 31 find the central angle θ of the circular sector so that the volume of the conical cup is a maximum.

33. A Norman window consists of a rectangle surmounted by a semicircle. Find the dimensions of the window with largest area if its perimeter is 10 m. See Figure 4.90.

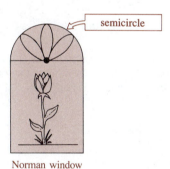

Norman window

Figure 4.90

34. Rework Problem 33 given that the rectangle is surmounted by an equilateral triangle.

35. A person would like to cut a 1-m-long piece of wire into two pieces. One piece will be bent into the shape of a circle and the other into the shape of a square. How should the wire be cut so that the sum of the areas is a maximum?

36. In Problem 35, suppose one piece of wire will be bent into the shape of a circle and the other into the shape of an equilateral triangle. How should the wire be cut so that the sum of the areas is a minimum? A maximum?

37. A cross section of a rectangular wooden beam cut from a circular log of diameter d has width x and depth y. See Figure 4.91. The strength of the beam varies directly as the product of the width and the square of the depth. Find the dimensions of the cross section of the beam of greatest strength.

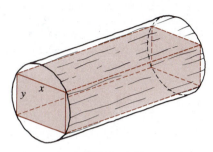

Figure 4.91

38. U.S. Postal Service regulations state that a rectangular box sent by fourth class mail must satisfy the requirement that the length plus the perimeter of one end must not exceed 100 in. Given that a box is to be constructed so that its height is one-half its width, find the dimensions of the box that has a maximum volume.

39. A metal container for transporting nuclear waste consists of a right circular cylinder with hemispherical ends. See Figure 4.92. The container is to have a volume of 30π ft^3. The cost of the metal per square foot for the ends is one and a half times the cost per square foot of the metal used in the cylindrical part. Find the dimensions of the container so that its cost of construction is a minimum.

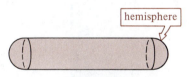

hemisphere

Figure 4.92

40. A 10-ft wall stands 5 ft away from a building, as shown in Figure 4.93. Find the length of the shortest ladder, supported by the wall, that reaches from the ground to the building.

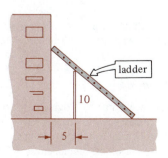

ladder

Figure 4.93

41. If the number of passengers on a bus tour of a city is exactly 30, the bus company charges $20 per person. For each additional passenger, the charge per person is reduced by $0.50. What is the number of passengers a bus should carry in order to maximize the company's revenue per bus?

42. If the U-Drive Truck Rental Company rents 50 trucks per day it makes a profit of $84 per truck. Because of increased maintenance and employee costs, for each additional truck rented per day the profit per truck is reduced by $1.00. Determine the number of trucks that should be rented so that the profit is a maximum. What is the maximum profit?

43. The potential energy between two atoms in a diatomic molecule is given by $U(x) = 2/x^{12} - 1/x^6$. Find the minimum potential energy between the two atoms.

44. The height of a projectile launched with a constant initial velocity v_0 at an angle of elevation θ_0 is given by

$$y = (\tan \theta_0)x - (g/2v_0^2 \cos^2\theta_0)x^2$$

where x is its horizontal displacement measured from the point of launch. Show that the maximum height attained by the projectile is $h = (v_0^2/2g)\sin^2\theta_0$.

45. When a hole is punched into the lateral side of a cylindrical tank full of water, the resulting stream hits the ground at a distance x ft from the base where $x = 2\sqrt{y(h - y)}$. See Figure 4.94. At what point should the hole be punched in the side so that the stream attains a maximum distance from the base? What is the maximum distance?

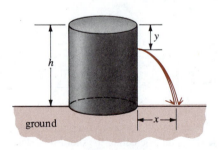

Figure 4.94

46. A company's total yearly cost for resupplying depleted inventories of a single item is sometimes given by $C(x) = (a/2)x + (b + cx)q/x$, where a, b, and c are positive constants and x represents the size of the reorder lot x. Determine the minimum yearly cost.

47. The illuminance E due to a light source of intensity I at a distance r from the source is given by $E = I/r^2$. The total illuminance from two light bulbs of intensities $I_1 = 125$ and $I_2 = 216$ is the sum of the illuminances. Find the point P

between the two light bulbs 10 m apart at which the total illuminance is a minimum. See Figure 4.95.

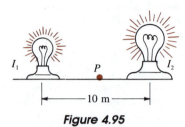

Figure 4.95

48. The illuminance E at any point P on the edge of a circular table due to a light placed directly above its center is given by $E = (I \cos \theta)/r^2$. See Figure 4.96. Given that the radius of the table is 1 m and $I = 100$, find the height at which the light should be placed so that E is a maximum.

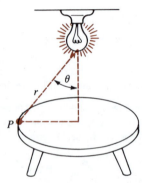

Figure 4.96

49. Determine the maximum length of a thin board that can be carried horizontally around the right-angle corner shown in Figure 4.97. (*Hint:* Use similar triangles.)

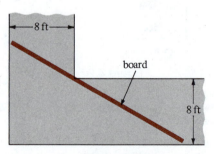

Figure 4.97

50. At midnight ship A is 50 km north of ship B. Ship A is sailing south at 20 km/hr and ship B is sailing west at 10 km/hr. At what time will the distance between the ships be a minimum? (*Hint:* Use distance = rate × time.)

51. A pipeline is to be constructed from a refinery across a swamp to storage tanks. See Figure 4.98. The cost of construction over the swamp is $25,000 per mile and $20,000 per mile over land. How should the pipeline be made so that the cost of construction is a minimum?

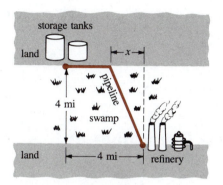

Figure 4.98

52. Rework Problem 51 given that the cost per mile across the swamp is twice the cost per mile over land.

53. Fermat's principle* in optics states that light travels from point A (in the xy-plane) in one medium to point B in another medium on a path that requires minimum time. Denote the speed of light in the medium that contains point A by c_1 and the speed of light in the medium that contains point B by c_2.

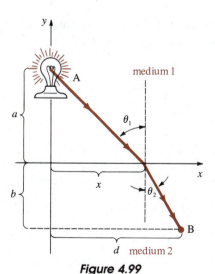

Figure 4.99

Show that the time of travel from A to B is a minimum when the angles θ_1 and θ_2, shown in Figure 4.99, satisfy Snell's law:†

$$\frac{\sin \theta_1}{c_1} = \frac{\sin \theta_2}{c_2}$$

54. Blood is carried throughout the body by the vascular system, which consists of capillaries, veins, arterioles, and arteries. One consideration of the problem of minimizing the energy expended in moving the blood through the various organs is to find an optimum angle θ for *vascular branching* such that the total resistance to the blood along a path from a larger blood vessel to a smaller blood vessel is a minimum. See Figure 4.100. Use Poiseuille's law, which is that the resistance R of a blood vessel of length l and radius r is $R = kl/r^4$ (see Problem 32, Exercises 3.10), where k is a constant, to show that the total resistance

$$R = k(x/r_1^4) + k(y/r_2^4)$$

along the path $P_1 P_2 P_3$ is a minimum when $\cos \theta = r_2^4/r_1^4$. (*Hint:* Express x and y in terms of θ and a.)

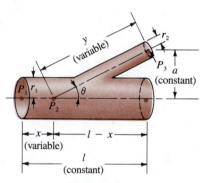

Figure 4.100

55. Several rules of thumb are used by physicians to calibrate a child's dose D_c of a particular drug in terms of an adult's dose D_a. Young's rule states that

$$D_c = \frac{t}{t + 12} D_a$$

where t is age in years, while Cowling's rule asserts that

$$D_c = \frac{t + 1}{24} D_a$$

At what age is the difference between the two rules a maximum? What is the maximum difference?

---

*Pierre de Fermat (1601–1665), a French mathematician, is famous for his many contributions to number theory.

†Willebord Snell (1591–1626), a Dutch astronomer and mathematician.

56. The rate P (in mg carbon/m^3/hr) at which photosynthesis takes place for a certain species of phytoplankton is related to the light intensity I (in 10^3 ft-candles) by the function

$$P = \frac{100I}{I^2 + I + 4}$$

At what light intensity is P the largest?

57. The long bones in mammals may be represented as hollow cylindrical tubes, filled with marrow, of outer radius R and inner radius r. Bones should be constructed to be lightweight yet capable of withstanding certain bending moments. In order to withstand a bending moment M, it can be shown that the mass m per unit length of the bone and marrow is given by

$$m = \pi\rho\left[\frac{M}{K(1 - x^4)}\right]^{2/3}\left(1 - \frac{1}{2}x^2\right)$$

where ρ is the density of the bone and K a positive constant. If $x = r/R$, show that m is a minimum when $r = 0.63R$ (approximately).

[O] 4.8 Applications of the Derivative in Economics

Revenue, Cost, and Profit When a company markets a product at p dollars per unit, the revenue realized in the production of x units is

$$R = px \tag{4.13}$$

Often the price itself depends on the number of units produced in a linear manner $p = ax + b$ so that (4.13) becomes

$$\begin{aligned} R(x) &= x(ax + b) \\ &= ax^2 + bx \end{aligned} \tag{4.14}$$

The function in (4.14) is an example of a **revenue function**. Furthermore, if $C(x)$ denotes the **cost** for producing x units, then the company's **profit** is defined to be

$$P(x) = R(x) - C(x)$$

In addition

$$Q(x) = \frac{C(x)}{x}$$

is said to be the **average cost** or cost per unit.

Naturally a company is interested in maximizing profit, $P(x)$, and minimizing the cost per unit, $Q(x)$. A typical cost function consists of

$$C = \text{variable costs} + \text{fixed costs}$$

Thus, in

$$C(x) = 200x + 600 \quad \text{and} \quad C(x) = x^2 + 640x + 950 \tag{4.15}$$

the constants 600 and 950 are the fixed costs and could represent rent, insurance premiums, and so on. Since it is assumed $x \geq 0$, note that the minimum value of each function in (4.15) is $C(0)$.

_____ **Example 1** _____

A company determines that in the production of x units of a commodity its revenue and cost functions are, respectively, $R(x) = -3x^2 + 970x$ and $C(x) = 2x^2 + 500$. Find the maximum profit and minimum average cost.

Solution   The profit for $x \geq 0$ is

$$P(x) = (-3x^2 + 970x) - (2x^2 + 500)$$
$$= -5x^2 + 970x - 500$$

From
$$P'(x) = -10x + 970$$
$$P''(x) = -10$$

we see that $x = 97$ is a critical value and that $P''(97) < 0$. Thus, the Second Derivative Test implies that $P(97) = 46{,}545$ units of dollars is a maximum.
    Now for $x > 0$ the average cost is

$$Q(x) = \frac{2x^2 + 500}{x} = 2x + \frac{500}{x}$$

so that
$$Q'(x) = 2 - \frac{500}{x^2}$$

$$Q''(x) = \frac{1000}{x^3}$$

Solving $Q'(x) = 0$ gives $x = \sqrt{250}$. Hence, $Q''(\sqrt{250}) > 0$ implies that $Q(\sqrt{250})$ is a minimum. Because the company can produce only an integral number of units, we use $\sqrt{250} \approx 16$ to find the *approximate* minimum average cost or $Q(16) = 43.25$ units of dollars.

    You should use the last example to verify that the maximum profit does not necessarily occur at the same production level that corresponds to a maximum revenue.

Marginal Functions   In economics the term **marginal function** usually refers to the derivative of that function.

DEFINITION 4.7

Marginal Revenue: $MR = R'(x)$

Marginal Cost: $MC = C'(x)$

Marginal Profit: $MP = P'(x)$

    To get a feeling for the use of these functions, consider the case of revenue function R. When $\Delta x = 1$, the quotient

$$\frac{\Delta R}{\Delta x} = \frac{R(x + \Delta x) - R(x)}{\Delta x} = R(x + 1) - R(x)$$

gives the slope of the secant line through the points $(x, R(x))$ and $(x + 1, R(x + 1))$ on the graph of R. See Figure 4.101. Since the slope of this secant is an approximation to the slope of the tangent at $(x, R(x))$, the derivative $R'(x)$ gives the *approximate value of the change in revenue for a unit increase in production*.

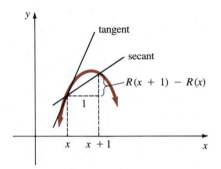

Figure 4.101

_____ **Example 2** _____

For the revenue function given in Example 1, find the revenue gained from the production of the forty-first unit. Approximate this value by means of the marginal revenue.

Solution   From $R(x) = -3x^2 + 970x$, we see that the revenue from producing 40 units is $R(40) = 34,000$ and the revenue from producing 41 units is $R(41) = 34,727$. Hence, the revenue from the production of the forty-first, or one more unit, is

$$R(41) - R(40) = 727 \text{ units of dollars}$$

By way of comparison,

$$MR = R'(x) = -6x + 970$$

and $$MR(40) = 730 \text{ units of dollars}$$

You should realize that all we are doing in the preceding example is approximating the change in a function by means of a differential; namely, $\Delta R \approx dR$ when $\Delta x = 1$.

_____ **Example 3** _____

The cost for producing x units of a product is given by $C(x) = x^2 + 560x + 1000$. Find the approximate cost for producing the fiftieth unit.

Solution   Rather than computing the exact cost $C(50) - C(49)$, we utilize the concept of the marginal cost. The derivative

$$MC = C'(x) = 2x + 560$$

evaluated at $x = 49$ gives an approximation to the cost of producing one more unit (the fiftieth),

$$MC(49) = 658 \text{ units of dollars}$$

Demand   The number D of units of a product **demanded** by consumers is a function of the price p of each unit. Intuitively, we would expect $D(p)$ to be small when p is large.

For a change Δp in price, the quotient

$$\frac{\dfrac{D(p + \Delta p) - D(p)}{D(p)}}{\dfrac{\Delta p}{p}} \tag{4.16}$$

is the proportionate change in demand divided by the proportionate change in price. Simplifying (4.16) and taking the limit as $\Delta p \to 0$ gives

$$\lim_{\Delta p \to 0} \frac{p}{D(p)} \cdot \frac{D(p + \Delta p) - D(p)}{\Delta p} = \frac{p}{D(p)} \cdot D'(p)$$

Since $D(p)$ is a decreasing function, we expect $D'(p) < 0$. For this reason it is common practice to introduce a minus sign and to refer to the resulting quantity

$$\eta = -\frac{p}{D(p)} \cdot D'(p)$$

as the **elasticity of demand**.

Cases   When $\eta < 1$ economists say that the demand is **inelastic**; in this case the percentage change in demand is less than the percentage change in price. If $\eta > 1$ the demand is said to be **elastic** and the percentage change in demand is greater than the percentage change in price. When $\eta = 1$ the percentage change in demand equals the percentage change in price.

Example 4

If $D(p) = -p^2 + 400$, $0 \le p \le 20$, determine whether the demand is elastic or inelastic at $p = 6$.

Solution $D'(p) = -2p$, $D'(6) = -12$, and $D(6) = 364$. Thus,

$$\eta = -\frac{6}{D(6)} \cdot D'(6) = \frac{72}{364} \approx 0.2 < 1$$

implies the demand is inelastic.

The result of the last example can be interpreted to mean that at $p = 6$ there is approximately a 0.2% decrease in demand for a 1% increase in price. If we suppose the price increases, say 10%, then the demand would decrease by approximately $10(0.2) = 2\%$.

In general, when a demand is either elastic or inelastic at a price level p, then a percentage increase in price brings with it a decrease in demand. This decrease is greater for an elastic demand than it is for an inelastic demand. In either case the revenue must decrease. However, when $\eta = 1$ an increase in price results in no change in revenue.

____ Exercises 4.8 ____

Answers to odd-numbered problems begin on page A-27.

In Problems 1 and 2 find the maximum revenue, maximum profit, and minimum average cost.

1. $R(x) = -x^2 + 400x$, $C(x) = x^2 + 40x + 100$

2. $R(x) = x(-2x + 60)$, $C(x) = 2x^2 + 12x + 18$

3. Given the revenue function $R(x) = -x^2 + 80x$,

   (a) Find the marginal revenue at $x = 10$.

   (b) Compare the result of part **(a)** with $R(11) - R(10)$.

4. A company finds that its cost for producing x units of a commodity is $C(x) = 3x^2 + 5x + 10$. Find the approximate cost for making the twenty-first unit.

5. Let $C(x) = 3x^2 + 100$ denote the cost for making x units of a product. Compare the exact cost for producing the thirty-first unit with the marginal cost at $x = 30$ and $x = 31$.

6. Suppose $R(x) = -x^2 + 1000x$ and $C(x) = 20x + 600$ are revenue and cost functions, respectively, for producing x units of a commodity. What profit is realized from the sale of 50 items? What is the approximate amount the profit changes from the sale of one more item?

7. Given that $R(x) = x(-x + 300)$ is a revenue function, show that the marginal revenue is always decreasing.

8. Given that $C(x) = 2x^3 - 21x^2 + 36x + 1000$ is a cost function, determine the interval(s) for which the cost is increasing. Determine whether there are any intervals on which the marginal cost increases.

9. Show that the maximum profit occurs when the marginal revenue equals the marginal cost.

10. Show that the minimum average cost occurs when the average cost equals the marginal cost.

In Problems 11–16 compute the elasticity of demand for the given demand function at the indicated price. State whether the demand is elastic or inelastic.

11. $D(p) = -4p + 500$, $0 \le p \le 125$; $p = 50$

12. $D(p) = -10p + 850$, $0 \le p \le 85$; $p = 40$

13. $D(p) = -2p^2 + 200$, $0 \le p \le 10$; $p = 6$

14. $D(p) = (20 - p)^2$, $0 \le p \le 20$; $p = 10$

15. $D(p) = 800\sqrt{30 - p}$, $0 \le p \le 30$; $p = 14$

16. $D(p) = 1000 + \dfrac{200}{\sqrt{p + 4}}$, $p \ge 0$; $p = 21$

17. Compute the elasticity of demand for $D(p) = -2p^2 + 800$, $0 \le p \le 20$ at $p = 15$. If the price increases by 6%, determine the approximate change in demand.

18. Compute the elasticity of demand for $D(p) = \sqrt{25 - p^2}$, $0 \le p \le 5$ at $p = 2$. If the price decreases by 21%, determine the approximate change in demand.

19. For the demand function in Problem 18 find the price level for which $\eta = 4$.

20. For the demand function $D(p) = -4p + 1000$, $0 \le p \le 250$, determine the prices for which the demand is elastic.

____ Chapter 4 Review Exercises ____

Answers to odd-numbered problems begin on page A-27.

In Problems 1–10 answer true or false.

1. If f is increasing on an interval, then $f'(x) > 0$ on the interval. _____

2. A function f has an extremum at a number c when $f'(c) = 0$. _____

3. A particle moving rectilinearly slows down when the velocity $v(t)$ decreases. _____

4. For a particle moving rectilinearly, acceleration is the first derivative of the velocity. _____

5. If $f''(x) < 0$ for all x in interval (a, b), then the graph of f is concave downward on the interval. _____

6. If $f''(c) = 0$, then $(c, f(c))$ is a point of inflection. _____

7. If f is continuous on $[a, b]$ and $f(a) = f(b) = 0$, then there exists some c in (a, b) such that $f'(c) = 0$. _____

8. The graph of a cubic polynomial can have at most one point of inflection. _____

9. A function continuous on a closed interval $[a, b]$ has both an absolute maximum and an absolute minimum. _____

10. Every absolute extremum is also a relative extremum. _____

In Problems 11 and 12 supply the reason(s) why the given statement is false.

11. If $f(c)$ is a relative maximum, then $f'(c) = 0$ and $f'(x) > 0$ for $x < c$ and $f'(x) < 0$ for $x > c$.

12. If $f(c)$ is a relative minimum, then $f''(c) > 0$.

In Problems 13–16 find the absolute extrema of the given function on the indicated interval.

13. $f(x) = x^3 - 75x + 150$; $[-3, 4]$

14. $f(x) = 4x^2 - \dfrac{1}{x}$; $[\frac{1}{4}, 1]$

15. $f(x) = \dfrac{x^2}{x + 4}$; $[-1, 3]$

16. $f(x) = (x^2 - 3x + 5)^{1/2}$; $[1, 3]$

17. Sketch a graph of a continuous function that has the properties:

$$f(0) = 1, \qquad f(2) = 3$$
$$f'(0) = 0, \qquad f'(2) \text{ does not exist}$$
$$f'(x) > 0, \qquad x < 0$$
$$f'(x) > 0, \qquad 0 < x < 2$$
$$f'(x) < 0, \qquad x > 2$$

18. Use the first and second derivatives as an aid in comparing the graphs of

$$y = x + \sin x \quad \text{and} \quad y = x + \sin 2x$$

19. The position of a particle moving on a horizontal line is given by $s(t) = -t^3 + 6t^2$. Graph the motion on the time

interval $[-1, 5]$. At what point is the velocity function a maximum? Does this point correspond to the maximum speed?

20. The height above ground of a projectile fired vertically is $s(t) = -4.9t^2 + 14.7t + 49$, where s is measured in meters and t in seconds. What is the maximum height attained by the projectile? At what speed does the projectile strike the ground?

21. Consider the function $f(x) = x \sin x$. Use f and Rolle's Theorem to show that the equation $\cot x = -1/x$ has a solution on the interval $(0, \pi)$.

22. Show that the function $f(x) = x^{1/3}$ does not satisfy the hypothesis of the Mean Value Theorem on the interval $[-1, 8]$ but yet a number c can be found in $(-1, 8)$ such that $f'(c) = [f(b) - f(a)]/(b - a)$. Explain.

In Problems 23–26 find the relative extrema of the given function. Graph.

23. $f(x) = 2x^3 + 3x^2 - 36x$

24. $f(x) = x^5 - \dfrac{5}{3}x^3 + 2$

25. $f(x) = 4x - 6x^{2/3} + 2$

26. $f(x) = \dfrac{x^2 - 2x + 2}{x - 1}$

In Problems 27–30 find the relative extrema and the points of inflection of the given function. Do not graph.

27. $f(x) = x^4 + 8x^3 + 18x^2$

28. $f(x) = x^6 - 3x^4 + 5$

29. $f(x) = 10 - (x - 3)^{1/3}$

30. $f(x) = x(x - 1)^{5/2}$

31. Let a, b, and c be real numbers. Find the x-coordinate of the point of inflection for the graph of

$$f(x) = (x - a)(x - b)(x - c)$$

32. A triangle is expanding with time. The area of the triangle is increasing at a rate of 15 in^2/min, whereas its base is decreasing at a rate of $\frac{1}{2}$ in/min. At what rate is the altitude of the triangle changing when the altitude is 8 in and the base is 6 in?

33. A pulley is secured to the edge of a dock that is 15 ft above the surface of the water. A small boat is being pulled toward the dock by means of a rope on the pulley. The rope is attached to the bow of the boat 3 ft above the water line. If the rope is pulled in at a constant rate of 1 ft/sec, how fast does the boat approach the dock when it is 16 ft from the dock?

34. Water drips into a hemispherical tank of radius 10 m at a rate $\frac{1}{10}$ m^3/min and drips out a hole in the bottom of the tank at a rate of $\frac{1}{5}$ m³/min. It can be shown that the volume of the water in the tank at anytime is

$$.V = 10\pi h^2 - \frac{\pi}{3}h^3*$$

See Figure 4.102. Is the depth of the water increasing or decreasing? At what rate is the depth of the water changing when the depth is 5 m?

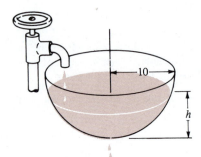

Figure 4.102

35. Consider the ladder whose bottom is sliding away from the base of the vertical wall shown in Figure 4.103. Show that the rate at which θ_1 is increasing is the same as the rate at which θ_2 is decreasing.

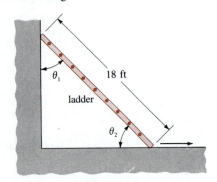

Figure 4.103

36. Find two nonnegative numbers whose sum is 8 such that the sum of their squares is a maximum.

37. Find the minimum value of the sum of a nonnegative number and its reciprocal.

38. A battery with constant emf E and constant internal resistance r is wired in series with a resistor that has resistance R. The current in the circuit is then $I = E/(r + R)$. Find the value of R for which the power $P = RI^2$ dissipated in the external load is a maximum. This is called *impedance matching*.

---

*You will be asked to derive this formula. See Problem 36 in Exercises 6.3.

39. Two coils that carry the same current produce a magnetic field at point Q on the x-axis of strength

$$B = \frac{1}{2}\mu_0 r_0^2 I\left\{\left[r_0^2 + \left(x + \frac{r_0}{2}\right)^2\right]^{-3/2}\right.$$
$$\left. + \left[r_0^2 + \left(x - \frac{r_0}{2}\right)^2\right]^{-3/2}\right\}$$

where μ_0, r_0, and I are constants. See Figure 4.104. Show that the maximum value of B occurs at $x = 0$.

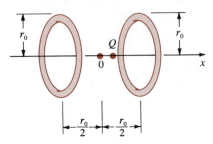

Figure 4.104

40. The velocity of air through the trachea (in cm/sec) of radius r is

$$v = V(r)/\pi r^2, \qquad r_0/2 \le r \le r_0$$

where

$$V(r) = kr^4(r_0 - r), \qquad k > 0, \quad r_0 > 0$$

is the volume flow of air. What radius will give the maximum velocity of air?

41. Some birds fly more slowly over water than over land. A bird flies at constant rates of 6 km/hr over water and 10 km/hr over land. Use the information in Figure 4.105 to find the path the bird should take to minimize the total flying time between the shore of one island and its nest on the shore of another island.

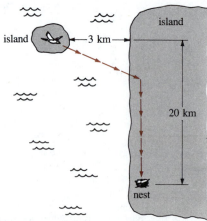

Figure 4.105

42. The area of a circular sector of radius r and arc length s is $A = \frac{1}{2}rs$. See Figure 4.106. Find the maximum area of a sector enclosed by a perimeter of 60 cm.

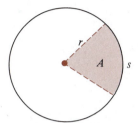

Figure 4.106

43. A rectangular yard is to be enclosed with a fence by attaching it to a house whose length is 40 ft. See Figure 4.107. The amount of fence to be used is 160 ft. Describe how the fence should be used so that the greatest area is enclosed.

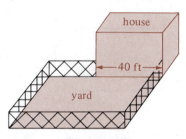

Figure 4.107

44. Solve Problem 43 if the amount of fence to be used is 80 ft.

45. A pigpen, attached to a barn, is enclosed using fence on two sides, as shown in Figure 4.108. The amount of fence to be used is 585 ft. Find the values of x and y indicated in the figure so that the greatest area is enclosed. What is the greatest area?

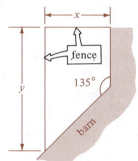

Figure 4.108

46. A rancher wants to use 100 m of fence to construct a diagonal fence connecting two existing walls that meet at a right angle. How should this be done so that the area enclosed by the walls and the fence is a maximum?

47. Two television masts on a roof are secured by wires that are attached at a single point between the masts. See Figure 4.109. Where should the point be located to minimize the amount of wire used?

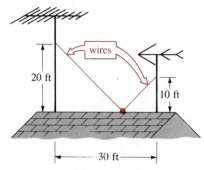

Figure 4.109

48. A statue is placed on a pedestal as shown in Figure 4.110. How far should a person stand from the pedestal to maximize the viewing angle θ? [*Hint:* Review the trigonometric identity for $\tan(\theta_2 - \theta_1)$. Also, it suffices to maximize $\tan \theta$ rather than θ. Why?]

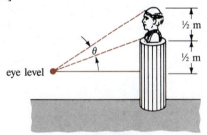

Figure 4.110

49. According to Fermat's principle, a ray of light originating at point A and reflected from a plane surface to point B travels on a path requiring the least time. See Figure 4.111. Assume that the speed of light c as well as h_1, h_2, and d are constants. Show that the time is a minimum when $\tan \theta_1 = \tan \theta_2$. Since $0 < \theta_1 < \pi/2$ and $0 < \theta_2 < \pi/2$, it follows that $\theta_1 = \theta_2$. In other words, the angle of incidence equals the angle of reflection. (*Note:* Figure 4.111 is inaccurate on purpose.)

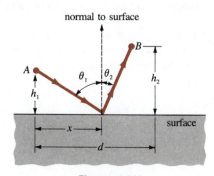

Figure 4.111

50. Determine the dimensions of a right circular cone having minimum volume V that circumscribes a sphere of radius r. See Figure 4.112. (*Hint:* Use similar triangles.)

Figure 4.112

51. A container in the form of a right circular cylinder has a volume of 100 in³. The top of the container costs 3 times as much per unit area as the bottom and the sides. Show that the dimension that gives the least cost of construction is a height that is 4 times the radius.

52. A piece of paper is 8 in wide. One corner is folded over to the other edge of the paper as shown in Figure 4.113. Find the width x of the fold so that the length L of the crease is a minimum.

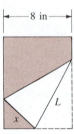

Figure 4.113

53. A box with a cover is to be made from a rectangular piece of cardboard 30 in long and 15 in wide by cutting a square out of each corner at one end of the cardboard and cutting a rectangle out of each corner at the other end. The cardboard is then folded on the dashed lines, as shown in Figure 4.114. Find the dimensions of the box that will give the maximum volume. What is the maximum volume?

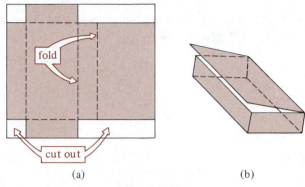

(a)   (b)

Figure 4.114

54. The running track shown in Figure 4.115 is to consist of two parallel straight parts and two semicircular parts. The length of the track is to be 2 km. Find the design of the track so that the rectangular plot of land enclosed by the track is a maximum.

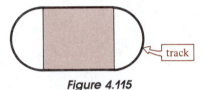

Figure 4.115

55. A company determines that in the production and sale of x units of a commodity that its revenue and cost functions are $R(x) = x(-3x + 660)$ and $C(x) = x^2 + 196x + 400$, respectively. Find

(a) The maximum revenue

(b) The maximum profit

(c) The minimum average cost

(d) The approximate revenue from the sale of the eleventh item

(e) The approximate profit from the sale of the thirty-first item

(f) How much does the company lose if no units are sold?

56. Suppose the function $D(p) = -2p + 50$, $0 \leq p \leq 25$, represents the demand for a product. Approximately how much does the demand change when the price of each unit of the product changes from \$10.00 to \$11.80?

5

The Integral

In the last several chapters, we have been concerned with the definition, properties, and applications of the derivative. We turn now from differential to integral calculus. Leibniz originally called this second of the two major divisions of calculus, *calculus summatorius*. In 1696, at the persuasion of the Swiss mathematician Jakob Bernoulli, Leibniz changed its name to *calculus integralis*. As the original Latin name suggests, the notion of a sum will be important in the full development of the integral.

5.1 Antiderivatives

In Chapters 3 and 4 we were concerned only with the basic problem:

Given a function f find its derivative f'.

In this chapter and in subsequent chapters of this text, we shall see that an equally important problem is:

Given a function f, find a function whose derivative is the same as f.

That is, for a given function f, we wish to find another function F for which $F'(x) = f(x)$ for all x on some interval.

DEFINITION 5.1   Antiderivative of a Function

A function F is said to be an **antiderivative** of a function f on some interval if $F'(x) = f(x)$.

Example 1

An antiderivative of $f(x) = 2x$ is $F(x) = x^2$ since $F'(x) = 2x$.

There is always more than one antiderivative of a function. For instance, in the foregoing example, $F_1(x) = x^2 - 1$ and $F_2(x) = x^2 + 10$ are also antiderivatives of $f(x) = 2x$ since $F_1'(x) = F_2'(x) = f(x)$. Indeed, if F is an antiderivative of a function f, then so is $G(x) = F(x) + C$, for any constant C. This is a consequence of the fact that

$$G'(x) = \frac{d}{dx}(F(x) + C) = F'(x) + 0 = F'(x) = f(x)$$

Thus, $F(x) + C$ stands for a *set of functions* each member of which has a derivative equal to $f(x)$. We shall now prove that any antiderivative of f must be of the form $G(x) = F(x) + C$; that is, *two antiderivatives of the same function can differ by at most a constant.* Hence, $F(x) + C$ is *the most general antiderivative* of $f(x)$.

THEOREM 5.1   If $G'(x) = F'(x)$ for all x in some interval $[a, b]$, then

$$G(x) = F(x) + C$$

for all x in the interval.

Proof   Suppose we define $g(x) = G(x) - F(x)$. Then, since $G'(x) = F'(x)$, it follows that $g'(x) = G'(x) - F'(x) = 0$ for all x in $[a, b]$. If x_1 and x_2 are any two numbers that satisfy $a \le x_1 < x_2 \le b$, it follows from the

Mean Value Theorem (Theorem 4.5) that a number k exists in the open interval (x_1, x_2) for which

$$g'(k) = \frac{g(x_2) - g(x_1)}{x_2 - x_1}$$

or

$$g(x_2) - g(x_1) = g'(k)(x_2 - x_1)$$

But $g'(x) = 0$ for all x in $[a, b]$; in particular $g'(k) = 0$. Hence, $g(x_2) - g(x_1) = 0$ or $g(x_2) = g(x_1)$. Now, by assumption, x_1 and x_2 are any two, but different, numbers in the interval. Since the function values $g(x_1)$ and $g(x_2)$ are the same, we must conclude that the function $g(x)$ is a constant C. Thus, $g(x) = C$ implies

$$G(x) - F(x) = C \quad \text{or} \quad G(x) = F(x) + C \qquad \blacksquare$$

_____ **Example 2** _____

(a)   The most general antiderivative of $f(x) = 2x$ is $G(x) = x^2 + C$.

(b)   The most general antiderivative of $f(x) = 2x + 5$ is $G(x) = x^2 + 5x + C$ since $G'(x) = 2x + 5$.

Indefinite Integral Notation   For convenience, let us introduce a notation for an antiderivative of a function. If $F'(x) = f(x)$, we shall represent the most general antiderivative of f by

$$\int f(x)\, dx = F(x) + C$$

The symbol $\int$ was introduced by Leibniz and is called an **integral sign**. The notation $\int f(x)\, dx$ is called the **indefinite integral** of $f(x)$ with respect to x. The function $f(x)$ is called the **integrand**. The process of finding an antiderivative is called **antidifferentiation** or **integration**. The number C is called a **constant of integration**. Just as $d/dx\,(\;\;)$ denotes differentiation *with respect to x*, the symbolism $\int (\;\;)\, dx$ denotes integration *with respect to x*.

The Indefinite Integral of a   When differentiating the power x^n, we multiply by the exponent n and decrease
Power   the exponent by 1. To find an antiderivative of x^n, the reverse of the differentiation rule would be: *Increase the exponent by 1 and divide by the new exponent* $n + 1$. The indefinite integral analogue of the Power Rule of differentiation is given by the following.

If n is a rational number, then for $n \neq -1$

$$\int x^n\, dx = \frac{x^{n+1}}{n + 1} + C \qquad (5.1)$$

Proof   From Theorem 3.11,

$$\frac{d}{dx}\left(\frac{x^{n+1}}{n+1} + C\right) = (n + 1)\frac{x^{(n+1)-1}}{n+1} + 0$$

$$= x^n \qquad \blacksquare$$

Note: The result given in (5.1) does not include the case when $n = -1$. We have not as yet encountered any function whose derivative is $x^{-1} = 1/x$. The evaluation of

$$\int \frac{1}{x}\, dx$$

will be considered in Chapter 8.

Example 3

Evaluate (a) $\displaystyle\int x^6\, dx$ and (b) $\displaystyle\int \frac{1}{x^5}\, dx$.

Solution

(a)   With $n = 6$ it follows from (5.1) that

$$\int x^6\, dx = \frac{x^7}{7} + C$$

(b)   By writing $1/x^5$ as x^{-5} and identifying $n = -5$, we have from (5.1)

$$\int x^{-5}\, dx = \frac{x^{-4}}{-4} + C$$

$$= -\frac{1}{4x^4} + C$$

Example 4

Evaluate $\displaystyle\int \sqrt{x}\, dx$.

Solution   We first write

$$\int \sqrt{x}\, dx = \int x^{1/2}\, dx$$

and with $n = \frac{1}{2}$ we obtain from (5.1),

$$\int x^{1/2}\, dx = \frac{x^{3/2}}{3/2} + C$$

$$= \frac{2}{3}x^{3/2} + C$$

It should always be kept in mind that the *results of integration can always be checked by differentiation*; for example,

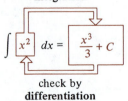

integration

check by
differentiation

_____ **Example 5** _____

Evaluate $\int dx$.

Solution   Since $\int dx = \int 1 \cdot dx$, and since $d/dx \, (x + C) = 1 + 0 = 1$, it follows from the definition of an antiderivative that

$$\int dx = x + C$$

The result in Example 5 can also be obtained from (5.1) with $n = 0$.

The following property of indefinite integrals is an immediate consequence of the fact that the derivative of a sum is the sum of the derivatives.

THEOREM 5.2   If $F'(x) = f(x)$ and $G'(x) = g(x)$, then

$$\int [f(x) \pm g(x)] \, dx = \int f(x) \, dx \pm \int g(x) \, dx$$
$$= F(x) \pm G(x) + C$$

Observe that there is no reason to use two constants of integration since

$$\int [f(x) \pm g(x)] \, dx = (F(x) + C_1) \pm (G(x) + C_2)$$
$$= F(x) \pm G(x) + (C_1 \pm C_2)$$
$$= F(x) \pm G(x) + C$$

where we have replaced $C_1 \pm C_2$ by the single constant C.

_____ **Example 6** _____

Evaluate $\int (x^{-1/2} + x^4) \, dx$.

Solution   From Theorem 5.2 and (5.1) we can write

$$\int (x^{-1/2} + x^4) \, dx = \int x^{-1/2} \, dx + \int x^4 \, dx$$
$$= \frac{x^{1/2}}{1/2} + \frac{x^5}{5} + C$$
$$= 2x^{1/2} + \frac{x^5}{5} + C$$

THEOREM 5.3   If $F'(x) = f(x)$, then

$$\int k \, f(x) \, dx = k \int f(x) \, dx$$

for any constant k.

The antiderivative, or indefinite integral, of any finite sum can be obtained by integrating each term.

Example 7

Evaluate $\displaystyle\int \left(4x - 2x^{-1/3} + \frac{5}{x^2} \right) dx.$

Solution From the preceding discussion, it follows that

$$\int \left(4x - 2x^{-1/3} + \frac{5}{x^2} \right) dx = 4 \int x \, dx - 2 \int x^{-1/3} + 5 \int x^{-2} \, dx$$

$$= 4 \cdot \frac{x^2}{2} - 2 \cdot \frac{x^{2/3}}{2/3} + 5 \cdot \frac{x^{-1}}{-1} + C$$

$$= 2x^2 - 3x^{2/3} - 5x^{-1} + C$$

Example 8

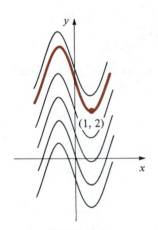

Figure 5.1

Find a function $y = f(x)$ whose graph passes through the point $(1, 2)$ that also satisfies $dy/dx = 3x^2 - 3$.

Solution From the definition of an antiderivative, if

$$\frac{dy}{dx} = 3x^2 - 3 \quad \text{then} \quad y = \int (3x^2 - 3) \, dx$$

That is, $y = x^3 - 3x + C$. Now when $x = 1$, $y = 2$ so that $2 = 1 - 3 + C$ or $C = 4$. Hence, $y = x^3 - 3x + 4$. Thus, out of the family of antiderivatives of $3x^2 - 3$ we see in Figure 5.1 that there is only one whose graph passes through $(1, 2)$.

When solving an equation such as $dy/dx = 3x^2 - 3$ in Example 8, the side condition $f(1) = 2$ is called an **initial condition**.

Example 9

Verify that $\displaystyle\int \sin x \, dx = -\cos x + C.$

Solution We know from Section 3.5 that

$$\frac{d}{dx} (-\cos x + C) = -(-\sin x) + 0 = \sin x$$

Hence, by Definition 5.1 it follows that

$$\int \sin x \, dx = -\cos x + C$$

Exercises 5.1

Answers to odd-numbered problems begin on page A-27.

In Problems 1–20 evaluate the given indefinite integral.

1. $\displaystyle\int 3\,dx$

2. $\displaystyle\int (4x - 1)\,dx$

3. $\displaystyle\int x^5\,dx$

4. $\displaystyle\int 5x^{1/4}\,dx$

5. $\displaystyle\int \frac{dx}{\sqrt[3]{x}}$

6. $\displaystyle\int \sqrt[3]{x^2}\,dx$

7. $\displaystyle\int (1 - t^{-0.52})\,dt$

8. $\displaystyle\int 10w\sqrt{w}\,dw$

9. $\displaystyle\int (3x^2 + 2x - 1)\,dx$

10. $\displaystyle\int \left(2\sqrt{t} - t - \frac{9}{t^2}\right)dt$

11. $\displaystyle\int (4x + 1)^2\,dx$

12. $\displaystyle\int (\sqrt{x} - 1)^2\,dx$

13. $\displaystyle\int (x + 2)(x - 2)\,dx$

14. $\displaystyle\int \frac{x^3 + 8}{x + 2}\,dx$

15. $\displaystyle\int \frac{r - 10}{r^3}\,dr$

16. $\displaystyle\int \frac{(x + 1)^2}{\sqrt{x}}\,dx$

17. $\displaystyle\int \frac{x^{-1} - x^{-2} + x^{-3}}{x^2}\,dx$

18. $\displaystyle\int \left(\frac{5}{\sqrt[3]{s^2}} + \frac{2}{\sqrt{s^3}}\right)ds$

19. $\displaystyle\int (4w - 1)^3\,dw$

20. $\displaystyle\int (5u - 1)(3u^2 + 2)\,du$

21. Find a function f whose graph passes through the point $(2, 3)$ that also satisfies $f'(x) = 2x - 1$.

22. Find a function f so that $f'(x) = 1/\sqrt{x}$ and $f(9) = 1$.

23. If $f''(x) = 2x$, find $f'(x)$ and $f(x)$.

24. Find a function f such that $f''(x) = 1, f'(-1) = 2$, and $f(-1) = 0$.

25. Find a function f such that $f''(x) = 12x^2 + 2$ for which the slope of the tangent line to its graph at $(1, 1)$ is 3.

26. If $f^{(n)}(x) = 0$, what is f?

In Problems 27–32 verify the given result by differentiation.

27. $\displaystyle\int \frac{1}{\sqrt{2x + 1}}\,dx = \sqrt{2x + 1} + C$

28. $\displaystyle\int (2x^2 - 4x)^9(x - 1)\,dx = \frac{1}{40}(2x^2 - 4x)^{10} + C$

29. $\displaystyle\int \cos x\,dx = \sin x + C$

30. $\displaystyle\int \sec^2 x\,dx = \tan x + C$

31. $\displaystyle\int x \sin x^2\,dx = -\frac{1}{2}\cos x^2 + C$

32. $\displaystyle\int \frac{\cos x}{\sin^3 x}\,dx = -\frac{1}{2\sin^2 x} + C$

In Problems 33 and 34 perform the indicated operations.

33. $\displaystyle\frac{d}{dx}\int (x^2 - 4x + 5)\,dx$ **34.** $\displaystyle\int \frac{d}{dx}(x^2 - 4x + 5)\,dx$

35. A bucket that contains liquid is rotating about a vertical axis at a constant angular velocity ω. The shape of the cross section of the rotating liquid in the xy-plane is determined from

$$\frac{dy}{dx} = \frac{\omega^2}{g}x$$

With coordinate axes as shown in Figure 5.2, find $y = f(x)$.

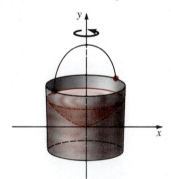

Figure 5.2

36. The ends of a beam of length L rest on two supports as shown in Figure 5.3. With a uniform load on the beam its shape (or elastic curve) is determined from

$$EIy'' = \frac{qL}{2}x - \frac{q}{2}x^2$$

where E, I, and q are constants. Find $y = f(x)$ if $f(0) = 0$ and $f'(L/2) = 0$.

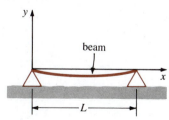

Figure 5.3

5.2 Indefinite Integrals and the *u*-Substitution

So far we have considered only antiderivatives of rational powers of x:

$$\int x^n \, dx = \frac{x^{n+1}}{n+1} + C, \qquad n \neq -1 \qquad (5.2)$$

In the present exposition, we shall examine the indefinite analogues of the Power Rule for Functions (Theorem 3.12) and the indefinite integral analogues of the derivative formulas for the trigonometric functions (I–VI of Section 3.6).

The Indefinite Integral of a Power of a Function

If we wish to find a function F such that

$$\int (5x + 1)^{1/2} \, dx = F(x) + C$$

we must have

$$F'(x) = (5x + 1)^{1/2}$$

By reasoning "backward," we could argue that to obtain $(5x + 1)^{1/2}$ we must have differentiated $(5x + 1)^{3/2}$. It would then seem that we could proceed as in (5.2):

$$\int (5x + 1)^{1/2} \, dx = \frac{(5x + 1)^{3/2}}{3/2} + C \; = \frac{2}{3}(5x + 1)^{3/2} + C \qquad (5.3)$$

Regrettably the "answer" in (5.3) does not check since the Power Rule for Functions gives

$$\frac{d}{dx}\left[\frac{2}{3}(5x + 1)^{3/2} + C\right] = \frac{2}{3}\frac{3}{2}(5x + 1)^{1/2} \cdot 5$$

$$= 5(5x + 1)^{1/2} \neq (5x + 1)^{1/2}$$

To account for the missing factor of 5 in (5.3) we use Theorem 5.3 and a little bit of cleverness:

$$\int (5x + 1)^{1/2} \, dx = \int (5x + 1)^{1/2} \boxed{\frac{1}{5} \cdot 5} \, dx \qquad \boxed{\text{equal to 1}}$$

$$= \frac{1}{5} \int \boxed{(5x + 1)^{1/2}5} \, dx \qquad \boxed{\begin{array}{c}\text{derivative of} \\ (5x + 1)^{3/2} \\ \hline 3/2 \end{array}}$$

$$= \frac{1}{5}\frac{2}{3}(5x + 1)^{3/2} + C$$

$$= \frac{2}{15}(5x + 1)^{3/2} + C$$

You should now verify by differentiation that the last function is indeed an antiderivative of $(5x + 1)^{1/2}$.

The key to evaluating indefinite integrals such as

$$\int \frac{x}{(4x^2 + 3)^6} \, dx \quad \text{and} \quad \int \sin 10x \, dx$$

lies in the next result, which is the indefinite integration form of the Chain Rule.

THEOREM 5.4   If F is an antiderivative of f, then

$$\int f(g(x))g'(x)\, dx = F(g(x)) + C \tag{5.4}$$

Proof   By the Chain Rule,

$$\frac{d}{dx} F(g(x)) = F'(g(x))g'(x)$$

$$= f(g(x))g'(x)$$

Hence, from the definition of an antiderivative,

$$\int f(g(x))g'(x)\, dx = F(g(x)) + C \qquad\blacksquare$$

To apply (5.4) we must be certain that we have the exact form

Derivative of function inside f

$$\int f(\boxed{g(x)})\boxed{g'(x)}\, dx$$

In particular, if $F(x) = x^{n+1}/(n + 1)$, n a rational number, $n \ne -1$, and if $u = g(x)$ is a differentiable function, then

$$F(g(x)) = \frac{[g(x)]^{n+1}}{n + 1} \quad \text{and} \quad \frac{d}{dx} F(g(x)) = [g(x)]^n g'(x)$$

Hence, Theorem 5.4 immediately implies

$$\int [g(x)]^n g'(x)\, dx = \frac{[g(x)]^{n+1}}{n + 1} + C \tag{5.5}$$

On a practical level it is often helpful to *change the variable* in an integration problem by employing the **substitutions**

$$u = g(x), \qquad du = g'(x)\, dx$$

in (5.4). Thus, (5.5) can be summarized in the following manner:

If n is a rational number and $u = g(x)$ is a differentiable function, then for $n \ne -1$

$$\int u^n\, du = \frac{u^{n+1}}{n + 1} + C \tag{5.6}$$

Example 1

Evaluate $\displaystyle\int \frac{x}{(4x^2 + 3)^6}\, dx$.

Solution   Let us write the integral as

$$\int (4x^2 + 3)^{-6} x\, dx$$

and make the identifications

$$u = 4x^2 + 3 \quad \text{and} \quad du = 8x\, dx$$

Now, to get the precise form $\int u^{-6}\, du$ we must adjust the integrand by multiplying and dividing by 8. By (5.6) we obtain

$$\int (4x^2 + 3)^{-6} x\, dx = \frac{1}{8} \int \overbrace{(4x^2 + 3)^{-6}}^{u^{-6}} \overbrace{(8x\, dx)}^{du}$$

$$= \frac{1}{8} \int u^{-6}\, du$$

$$= \frac{1}{8} \cdot \frac{u^{-5}}{-5} + C$$

$$= -\frac{1}{40}(4x^2 + 3)^{-5} + C$$

Check By the Power Rule for Functions,

$$\frac{d}{dx}\left[-\frac{1}{40}(4x^2 + 3)^{-5} + C \right] = \left(-\frac{1}{40}\right)(-5)(4x^2 + 3)^{-6}(8x)$$

$$= \frac{x}{(4x^2 + 3)^6}$$

Example 2

Evaluate $\int (x^2 + 2)^3 x\, dx$.

Solution If

$$u = x^2 + 2 \quad \text{then} \quad du = 2x\, dx$$

Thus from (5.6),

$$\int (x^2 + 2)^3 x\, dx = \frac{1}{2} \int \overbrace{(x^2 + 2)^3}^{u^3} \overbrace{(2x\, dx)}^{du}$$

$$= \frac{1}{2} \int u^3\, du$$

$$= \frac{1}{2} \cdot \frac{u^4}{4} + C_1$$

$$= \frac{1}{8}(x^2 + 2)^4 + C_1 \tag{5.7}$$

Alternative Solution   If we use the Binomial Theorem before integrating, we have

$$\int (x^2 + 2)^3 \, x \, dx = \int (x^6 + 6x^4 + 12x^2 + 8)x \, dx$$

$$= \int (x^7 + 6x^5 + 12x^3 + 8x) \, dx$$

$$= \frac{x^8}{8} + 6 \cdot \frac{x^6}{6} + 12 \cdot \frac{x^4}{4} + 8 \cdot \frac{x^2}{2} + C_2$$

$$= \frac{1}{8}x^8 + x^6 + 3x^4 + 4x^2 + C_2 \tag{5.8}$$

Note that (5.7) can be written as

$$\frac{1}{8}(x^2 + 2)^4 + C_1 = \frac{1}{8}x^8 + x^6 + 3x^4 + 4x^2 + 2 + C_1$$

Although (5.7) and (5.8) are not exactly the same, the two results differ only by a constant.

Example 3

Evaluate $\int \sqrt[3]{(7 - 2x^3)^4} \, x^2 \, dx$.

Solution   We first write the integral as

$$\int (7 - 2x^3)^{4/3}x^2 \, dx$$

and then make the identifications

$$u = 7 - 2x^3, \qquad du = -6x^2 \, dx$$

Hence,

$$\int (7 - 2x^3)^{4/3}x^2 \, dx = -\frac{1}{6}\int (7 - 2x^3)^{4/3}(-6x^2 \, dx)$$

$$= -\frac{1}{6}\int u^{4/3} \, du$$

$$= -\frac{1}{6}\frac{u^{7/3}}{7/3} + C$$

$$= -\frac{3}{6 \cdot 7}u^{7/3} + C$$

$$= -\frac{1}{14}(7 - 2x^3)^{7/3} + C$$

Indefinite Integrals of Trigonometric Functions

If $u = g(x)$ is a differentiable function, then the differentiation formulas

$$\frac{d}{dx} \sin u = \cos u \frac{du}{dx} \quad \text{and} \quad \frac{d}{dx}(-\cos u) = \sin u \frac{du}{dx}$$

yield, in turn, the integration formulas

$$\int \cos u \frac{du}{dx}\, dx = \sin u + C \qquad (5.9)$$

$$\int \sin u \frac{du}{dx}\, dx = -\cos u + C \qquad (5.10)$$

Since $du = g'(x)\, dx = \dfrac{du}{dx}\, dx$, (5.9) and (5.10) are equivalent to

$$\int \cos u\, du = \sin u + C$$

and

$$\int \sin u\, du = -\cos u + C$$

In general, I–VI of Section 3.6 give the following integration results.

I′ $\displaystyle\int \cos u\, du = \sin u + C$ 　　　II′ $\displaystyle\int \sin u\, du = -\cos u + C$

III′ $\displaystyle\int \sec^2 u\, du = \tan u + C$ 　　IV′ $\displaystyle\int \csc^2 u\, du = -\cot u + C$

V′ $\displaystyle\int \sec u \tan u\, du = \sec u + C$ 　VI′ $\displaystyle\int \csc u \cot u\, du = -\csc u + C$

Example 4

Evaluate $\displaystyle\int 3 \cos 3x\, dx$.

Solution If

$$u = 3x \quad \text{then} \quad du = 3dx$$

Hence, we recognize the given problem as being exactly of form I′:

$$\int \cos \overbrace{3x}^{u}\, \overbrace{(3\, dx)}^{du} = \int \cos u\, du$$

$$= \sin u + C$$

$$= \sin 3x + C$$

As in the discussion of (5.6), the differential du is probably the most important part in each of I′–VI′. Before applying one of these results, we may need to "fix up" or adjust an integrand by multiplying and dividing by a constant in order to obtain the appropriate du.

Example 5

Evaluate $\int \sin 10x \, dx$.

Solution   If

$$u = 10x \quad \text{then we need} \quad du = 10dx$$

Accordingly, we write

$$\int \sin 10x \, dx = \frac{1}{10} \int \sin \overbrace{10x}^{u} \overbrace{(10 \, dx)}^{du}$$

$$= \frac{1}{10} \int \sin u \, du$$

$$= \frac{1}{10}(-\cos u) + C \qquad \text{(from II}')$$

$$= -\frac{1}{10} \cos 10x + C$$

After a while, try to perform an integration without using the *u*-substitution.

Example 6

Evaluate $\int \sec^2(1 - 4x) \, dx$.

Solution

$$\int \sec^2(1 - 4x) \, dx = -\frac{1}{4} \int \sec^2(1 - 4x)(-4 \, dx)$$

$$= -\frac{1}{4}\tan(1 - 4x) + C \qquad \text{(from III}')$$

Check

$$\frac{d}{dx}\left[-\frac{1}{4}\tan(1 - 4x) + C\right] = -\frac{1}{4}\sec^2(1 - 4x)\frac{d}{dx}(1 - 4x)$$

$$= -\frac{1}{4}\sec^2(1 - 4x)(-4) = \sec^2(1 - 4x)$$

The next example shows that not every indefinite integral of a trigonometric function is one of the types I'–VI'.

_____ **Example 7** _____

Evaluate $\displaystyle\int \cos^4 x \sin x \, dx$.

Solution   For emphasis we rewrite the problem as

$$\int (\cos x)^4 \sin x \, dx$$

With the identifications

$$u = \cos x \quad \text{and} \quad du = -\sin x \, dx$$

we recognize

$$\int (\cos x)^4 \sin x \, dx = -\int (\cos x)^4 (-\sin x \, dx)$$

$$= -\int u^4 \, du$$

Hence, from (5.6) we obtain

$$\int (\cos x)^4 \sin x \, dx = -\frac{u^5}{5} + C$$

$$= -\frac{1}{5} \cos^5 x + C$$

Sometimes it may be necessary to use a trigonometric identity to solve a problem. The half-angle formulas

$$\cos^2 x = \frac{1 + \cos 2x}{2} \quad \text{and} \quad \sin^2 x = \frac{1 - \cos 2x}{2}$$

are particularly useful in problems that require the antiderivatives of $\cos^2 x$ and $\sin^2 x$.

_____ **Example 8** _____

Evaluate $\displaystyle\int \cos^2 x \, dx$.

Solution   It should be verified that the integral is *not* of the form $\int u^2 \, du$. Now using the half-angle formula $\cos^2 x = (1 + \cos 2x)/2$ we obtain

$$\int \cos^2 x \, dx = \int \frac{1 + \cos 2x}{2} \, dx$$

$$= \frac{1}{2} \int [1 + \cos 2x] \, dx$$

$$= \frac{1}{2}\left[\int dx + \frac{1}{2}\int \cos 2x(2 \, dx) \right] \qquad \text{[from Theorem 5.2]}$$

$$= \frac{1}{2}\left[x + \frac{1}{2}\sin 2x\right] + C \qquad \text{[from (5.2) and I']}$$

$$= \frac{1}{2}x + \frac{1}{4}\sin 2x + C$$

We shall consider substitutions and integrals of powers of trigonometric functions in greater detail in Chapter 9.

Remark

The following example illustrates a common, but *totally incorrect*, procedure for evaluating an indefinite integral:

$$\int (4 + x^2)^{1/2} \, dx = \frac{1}{2x} \int (4 + x^2)^{1/2} 2x \, dx \qquad \left(\frac{1}{2x} \cdot 2x = 1\right)$$

$$= \frac{1}{2x} \int u^{1/2} \, du$$

$$= \frac{1}{2x} \cdot \frac{3}{2}(4 + x^2)^{3/2} + C$$

You should verify that differentiation of the latter function does *not* yield $(4 + x^2)^{1/2}$. The mistake is in the first line of the "solution." *Variables*, in this case $2x$, *cannot be brought outside an integral symbol*. If $u = x^2 + 4$, then the integrand lacks the function $du = 2x \, dx$; in fact, there is no way of adjusting the problem to fit the form given in (5.6). At this juncture the integral $\int (4 + x^2)^{1/2} \, dx$ simply cannot be evaluated.

Exercises 5.2

Answers to odd-numbered problems begin on page A-27.

In Problems 1–44 evaluate the given indefinite integral.

1. $\displaystyle\int \sqrt{1 - 4x} \, dx$

2. $\displaystyle\int (8x + 2)^{1/3} \, dx$

3. $\displaystyle\int \frac{dx}{(5x + 1)^3}$

4. $\displaystyle\int (7 - x)^{49} \, dx$

5. $\displaystyle\int \sqrt[5]{(3 - 4x)^3} \, dx$

6. $\displaystyle\int \frac{dx}{\sqrt[3]{(2x + 1)^2}}$

7. $\displaystyle\int (x^2 - 2x + 1)^3 \, dx$

8. $\displaystyle\int (4y^2 + 4y + 1)^{2/3} \, dy$

9. $\displaystyle\int 2x\sqrt{x^2 + 4} \, dx$

10. $\displaystyle\int x\sqrt[3]{7x^2 + 1} \, dx$

11. $\displaystyle\int \frac{z}{\sqrt[3]{z^2 + 9}} \, dz$

12. $\displaystyle\int \frac{x}{\sqrt[3]{(1 - x^2)^2}} \, dx$

13. $\displaystyle\int (4x^2 - 16x + 7)^4(x - 2) \, dx$

14. $\displaystyle\int (x^2 + 2x - 10)^{2/3}(5x + 5) \, dx$

15. $\displaystyle\int \frac{x^2 + 1}{\sqrt[3]{x^3 + 3x - 16}} \, dx$

16. $\displaystyle\int \frac{s(s^3 - 4)}{\sqrt{s^5 - 10s^2 + 6}} \, ds$

17. $\displaystyle\int \sqrt{3 - \frac{2}{v}} \frac{dv}{v^2}$

18. $\displaystyle\int \sqrt{\frac{x^3 + 1}{x^3}} \frac{dx}{x^4}$

19. $\displaystyle\int \sqrt[3]{\frac{1 - \sqrt[3]{x}}{x^2}} \, dx$

20. $\displaystyle\int \sqrt{\frac{2 + 3\sqrt{x}}{x}} \, dx$

21. $\displaystyle\int \sin 4x \, dx$

22. $\displaystyle\int 5 \cos \frac{x}{2} \, dx$

23. $\displaystyle\int x \cos x^2 \, dx$　　**24.** $\displaystyle\int x^2 \sec^2 x^3 \, dx$

25. $\displaystyle\int \frac{1}{\sec(5x + 1)} \, dx$.

26. $\displaystyle\int \sec x(\sec x + \tan x) \, dx$

27. $\displaystyle\int \frac{\sin 2\theta}{\cos \theta} \, d\theta$　　**28.** $\displaystyle\int (1 + \cot^2 x) \, dx$

29. $\displaystyle\int \frac{\csc \sqrt{x} \cot \sqrt{x}}{\sqrt{x}} \, dx$　　**30.** $\displaystyle\int \frac{\sin(1/x)}{x^2} \, dx$

31. $\displaystyle\int \sin^5 3x \cos 3x \, dx$　　**32.** $\displaystyle\int \frac{\sin t}{\sqrt{4 + \cos t}} \, dt$

33. $\displaystyle\int \tan^2 2x \sec^2 2x \, dx$　　**34.** $\displaystyle\int \tan x \sec^2 x \, dx$

35. $\displaystyle\int \tan^2 7x \, dx$　　**36.** $\displaystyle\int \tan 5v \sec 5v \, dv$

37. $\displaystyle\int \frac{2 + \cos x}{\sin^2 x} \, dx$　　**38.** $\displaystyle\int \frac{(1 + \sin x)^4}{\sec x + \tan x} \, dx$

39. $\displaystyle\int \sin^2 x \, dx$　　**40.** $\displaystyle\int \cos^2 \pi x \, dx$

41. $\displaystyle\int (z + 1)\csc^2(z^2 + 2z) \, dz$

42. $\displaystyle\int \frac{\cos \sqrt[3]{1 - 3x}}{\sqrt[3]{(1 - 3x)^2}} \, dx$

43. $\displaystyle\int (1 + \cos 2x)^2 \, dx$　　**44.** $\displaystyle\int \cos 2x \sin^2 x \, dx$

45. Find a function f whose graph passes through the point $(\pi, -1)$ that also satisfies $f'(x) = 1 - \sin x$.

46. Find a function f such that $f''(x) = (1 + 2x)^5$, $f(0) = 0$, and $f'(0) = 0$.

Miscellaneous Problems

47. Show that

(a) $\displaystyle\int \sin x \cos x \, dx = \frac{1}{2} \sin^2 x + C_1$

(b) $\displaystyle\int \sin x \cos x \, dx = -\frac{1}{2} \cos^2 x + C_2$

(c) $\displaystyle\int \sin x \cos x \, dx = -\frac{1}{4} \cos 2x + C_3$

48. In Problem 47:

(a) Verify that the derivative of each answer in parts (a), (b), and (c) is $\sin x \cos x$.

(b) By a trigonometric identity, show how the result in (b) can be obtained from the answer in part (a).

(c) By adding the results in (a) and (b), obtain the result in part (c).

In Problems 49–52 evaluate the given indefinite integral.

49. $\displaystyle\int \cos^3 x \, dx$　　(*Hint:* $\cos^3 x = \cos^2 x \cdot \cos x$)

50. $\displaystyle\int \sin^3 2x \, dx$

51. $\displaystyle\int \frac{t}{\sqrt{t + 2}} \, dt$　　(*Hint:* $t = t + 2 - 2$)

52. $\displaystyle\int \frac{4z + 3}{(4z + 5)^3} \, dz$

5.3　Sigma Notation

In Section 5.1 we defined the *indefinite integral*. Later on we shall define a related but different concept, the *definite integral*. We shall see that the definite integral is defined as the limit of a certain kind of *sum*. Therefore, it is helpful to introduce a special notation that enables us to write an indicated sum of constants such as

$$1 + 2 + 3 + \cdots + n$$
$$2^2 + 4^2 + 6^2 + \cdots + (2n)^2$$

and

$$\frac{1}{3} + \frac{1}{5} + \frac{1}{7} + \cdots + \frac{1}{2n + 1}$$

in a concise manner.

Let a_k be a real number depending on an integer k. We denote the sum $a_1 + a_2 + a_3 + \cdots + a_n$ by the symbol $\sum_{k=1}^{n} a_k$. That is,

$$\sum_{k=1}^{n} a_k = a_1 + a_2 + a_3 + \cdots + a_n \qquad (5.11)$$

Since Σ is the capital Greek letter sigma, (5.11) is called **sigma notation** or **summation notation**. The variable k is called the **index of summation**. Thus, $\sum_{k=1}^{n} a_k$ is the sum of all numbers of the form a_k as k takes on the successive values $k = 1$, $k = 2$, . . . , and concludes with $k = n$.

_____ **Example 1** _____

(a) $\displaystyle\sum_{k=1}^{5} (3k - 1) = [3(1) - 1] + [3(2) - 1] + [3(3) - 1] + [3(4) - 1]$
$$+ [3(5) - 1]$$
$$= 2 + 5 + 8 + 11 + 14$$

(b) $\displaystyle\sum_{k=1}^{4} \frac{1}{(k + 1)^2} = \frac{1}{2^2} + \frac{1}{3^2} + \frac{1}{4^2} + \frac{1}{5^2}$

(c) $\displaystyle\sum_{k=1}^{100} k^3 = 1^3 + 2^3 + 3^3 + \cdots + 98^3 + 99^3 + 100^3$

_____ **Example 2** _____

(a) The sum of the first ten positive odd integers

$$1 + 3 + 5 + 7 + \cdots + 19$$

can be written succinctly as $\displaystyle\sum_{k=1}^{10} (2k - 1)$.

(b) It is also easily verified that the sum of the first ten positive even integers

$$2 + 4 + 6 + 8 + \cdots + 20$$

is $\displaystyle\sum_{k=1}^{10} 2k$.

The index of summation need not start at the value $k = 1$; for example,

$$\sum_{k=3}^{5} 2^k = 2^3 + 2^4 + 2^5 \quad \text{and} \quad \sum_{k=0}^{5} 2^k = 2^0 + 2^1 + 2^2 + 2^3 + 2^4 + 2^5$$

Note that the sum in part (a) of Example 2 can also be written as

$$\sum_{k=0}^{9} (2k + 1)$$

However, in a general discussion we shall always assume that the summation index starts at $k = 1$. This assumption is for convenience rather than necessity.

The index of summation is often called a **dummy variable** since the symbol itself is not important; it is the successive integer values of the index and the corresponding sum that are important. In general,

$$\sum_{k=1}^{n} a_k = \sum_{i=1}^{n} a_i = \sum_{j=1}^{n} a_j = \sum_{m=1}^{n} a_m$$

and so on.

Example 3

$$\sum_{k=1}^{10} 4^k = \sum_{i=1}^{10} 4^i = \sum_{j=1}^{10} 4^j = 4^1 + 4^2 + 4^3 + \cdots + 4^{10}$$

The following is a list of some of the important properties of the sigma notation.

THEOREM 5.5   For positive integers m and n,

(i) $\displaystyle\sum_{k=1}^{n} c\, a_k = c \sum_{k=1}^{n} a_k,$ where c is any constant

(ii) $\displaystyle\sum_{k=1}^{n} (a_k \pm b_k) = \sum_{k=1}^{n} a_k \pm \sum_{k=1}^{n} b_k$

(iii) $\displaystyle\sum_{k=1}^{n} a_k = \sum_{k=1}^{m} a_k + \sum_{k=m+1}^{n} a_k,$ $m < n$

The proof of formula (i) is an immediate consequence of the distributive law. The proofs of (ii) and (iii) are left as exercises.

Example 4

(a)   From Theorems 5.5(i) and 5.5(ii), we find

$$\sum_{k=1}^{20} (3k^2 + 4k) = 3 \sum_{k=1}^{20} k^2 + 4 \sum_{k=1}^{20} k$$

(b)   From Theorem 5.5(iii), we can write

$$\sum_{k=1}^{50} k^2 = \sum_{k=1}^{3} k^2 + \sum_{k=4}^{50} k^2$$

$$= (1^2 + 2^2 + 3^2) + (4^2 + 5^2 + 6^2 + \cdots + 50^2)$$

If c is a constant—that is, independent of a summation index k—then $\sum_{k=1}^{n} c$ means

$$c + c + c + \cdots + c$$

Since there are n c's in this sum, we have

$$\sum_{k=1}^{n} c = n \cdot c \qquad (5.12)$$

Example 5

From (5.12),

$$\sum_{k=1}^{75} 6 = 75 \cdot 6 = 450$$

The sum of the first n positive integers can be written $\sum_{k=1}^{n} k$. If this sum is denoted by S, then

$$S = 1 + 2 + 3 + \cdots + (n - 1) + n \qquad (5.13)$$

can also be written as

$$S = n + (n - 1) + (n - 2) + \cdots + 1 \qquad (5.14)$$

If we add (5.13) and (5.14), then

$$2S = \underbrace{(n + 1) + (n + 1) + (n + 1) + \cdots + (n + 1)}_{n \text{ terms of } n + 1}$$

$$= n(n + 1)$$

Solving for S gives

$$S = \frac{n(n + 1)}{2}$$

or

$$\sum_{k=1}^{n} k = \frac{n(n + 1)}{2} \qquad (5.15)$$

Example 6

Find the sum of the first 100 consecutive positive integers.

Solution   The required sum is

$$1 + 2 + 3 + \cdots + 99 + 100 = \sum_{k=1}^{100} k$$

With $n = 100$, it follows from (5.15) that

$$\sum_{k=1}^{100} k = \frac{n(n + 1)}{2} = \frac{100(101)}{2}$$

$$= 50(101) = 5050$$

Summation Formulas   Formulas (5.12) and (5.15) are two of several **summation formulas** that will be of use in the succeeding sections. For completeness, we include them in the following list. The number n is a positive integer.

$$\text{I} \quad \sum_{k=1}^{n} c = nc$$

$$\text{II} \quad \sum_{k=1}^{n} k = \frac{n(n + 1)}{2}$$

$$\text{III} \quad \sum_{k=1}^{n} k^2 = \frac{n(n + 1)(2n + 1)}{6}$$

$$\text{IV} \quad \sum_{k=1}^{n} k^3 = \frac{n^2(n + 1)^2}{4}$$

$$\text{V} \quad \sum_{k=1}^{n} k^4 = \frac{n(n + 1)(6n^3 + 9n^2 + n - 1)}{30}$$

As we have already seen, I and II can be derived readily; derivations of the remaining formulas are not quite so simple. You should be able to derive III with the aid of hints supplied in Problems 43 and 44.

_____ **Example 7** _____

Evaluate $\sum_{k=1}^{10} k^2$.

Solution   With $n = 10$, from formula III we have

$$\sum_{k=1}^{10} k^2 = \frac{10(11)(21)}{6} = 385$$

_____ **Example 8** _____

Evaluate $\sum_{k=1}^{10} (k + 2)^3$.

Solution   By the Binomial Theorem and Theorem 5.5(ii), we can write

$$\sum_{k=1}^{10} (k + 2)^3 = \sum_{k=1}^{10} (k^3 + 6k^2 + 12k + 8)$$

$$= \sum_{k=1}^{10} k^3 + 6 \sum_{k=1}^{10} k^2 + 12 \sum_{k=1}^{10} k + \sum_{k=1}^{10} 8$$

With $n = 10$, it follows from summation formulas IV, III, II, and I, respectively, that

$$\sum_{k=1}^{10} (k + 2)^3 = \frac{10^2 11^2}{4} + 6\frac{10(11)(21)}{6} + 12\frac{10(11)}{2} + 10 \cdot 8$$

$$= 3025 + 2310 + 660 + 80 = 6075$$

Exercises 5.3

Answers to odd-numbered problems begin on page A-28.

In Problems 1–10 expand the indicated sum.

1. $\sum_{k=1}^{5} 3k$

2. $\sum_{k=1}^{5} (2k - 3)$

3. $\sum_{k=1}^{4} \frac{2^k}{k}$

4. $\sum_{k=1}^{4} \frac{3^k}{k}$

5. $\sum_{k=1}^{10} \frac{(-1)^k}{2k + 5}$

6. $\sum_{k=1}^{10} \frac{(-1)^{k-1}}{k^2}$

7. $\sum_{j=2}^{5} (j^2 - 2j)$

8. $\sum_{m=0}^{4} (m + 1)^2$

9. $\sum_{k=1}^{5} \cos k\pi$

10. $\sum_{k=1}^{5} \frac{\sin(k\pi/2)}{k}$

In Problems 11–20 write the given sum using sigma notation.

11. $2^2 + 4^2 + 6^2 + 8^2 + 10^2 + 12^2$

12. $1 + 2^2 + 3^2 + 4^2 + 5^2$

13. $3 + 5 + 7 + 9 + 11 + 13 + 15$

14. $2 + 4 + 8 + 16 + 32 + 64$

15. $1 + 4 + 7 + 10 + \cdots + 37$

16. $2 + 6 + 10 + 14 + \cdots + 38$

17. $1 - \frac{1}{2} + \frac{1}{3} - \frac{1}{4} + \frac{1}{5}$

18. $-\frac{1}{2} + \frac{2}{3} - \frac{3}{4} + \frac{4}{5} - \frac{5}{6}$

19. $6 + 6 + 6 + 6 + 6 + 6 + 6 + 6$

20. $1 + \sqrt{2} + \sqrt{3} + 2 + \sqrt{5} + \cdots + 3$

In Problems 21–30 find the value of the given sum.

21. $\sum_{k=1}^{20} 2k$

22. $\sum_{k=0}^{50} (-3k)$

23. $\sum_{k=1}^{10} (k + 1)$

24. $\sum_{k=1}^{1000} (2k - 1)$

25. $\sum_{k=1}^{6} (k^2 + 3)$

26. $\sum_{k=1}^{5} (6k^2 - k)$

27. $\sum_{p=0}^{10} (p^3 + 4)$

28. $\sum_{i=-1}^{10} (2i^3 - 5i + 3)$

29. $\sum_{j=1}^{10} (4j^4 - 7)$

30. $\sum_{k=1}^{5} (k^2 + 1)^2$

31. Find the value of $\sum_{k=21}^{60} k^2$. (*Hint*: Examine
$\sum_{k=1}^{60} k^2 - \sum_{k=1}^{20} k^2$.)

32. Find the value of $\sum_{k=-30}^{30} k^2$.

33. Find the value of $\sum_{k=1}^{400} (\sqrt{k} - \sqrt{k - 1})$.

34. Find the value of $\sum_{k=1}^{100} \frac{1}{k(k + 1)}$. (*Hint*:
$\frac{1}{k(k + 1)} = \frac{1}{k} - \frac{1}{k + 1}$)

35. (a) Evaluate $\sum_{k=1}^{n} [f(k) - f(k - 1)]$.

   (b) In Problems 33 and 34 identify a function f so that the given sums are of the form given in part (a). A sum of the form given in part (a) is said to **telescope**.

36. Show that $\displaystyle\sum_{k=3}^{10} (2k - 5)$ and $\displaystyle\sum_{j=0}^{7} (2j + 1)$ are equal.

37. Determine whether $\displaystyle\sum_{k=1}^{n} (k + 1)(k + 2)$ and

$\displaystyle\sum_{j=4}^{n+3} (j - 1)(j - 2)$ are equal.

38. Derive a formula for the sum of the first n positive even integers.

39. In his experiments on gravity, Galileo* found that the distance a mass moves down an inclined plane in consecutive time intervals is proportional to a positive odd integer. Hence, the total distance s a mass moves in n seconds, n a positive integer, is proportional to $1 + 3 + 5 + \cdots + 2n - 1$. Show that the total distance a mass moves down an inclined plane is proportional to the square of the elapsed time n.

Miscellaneous Problems

40. Prove Theorem 5.5(ii).

41. Prove Theorem 5.5(iii).

42. Consider the ratio

$$\frac{1^2 + 2^2 + 3^2 + \cdots + n^2}{1 + 2 + 3 + \cdots + n}$$

for $n = 1, 2, 3, 4,$ and 5. Discern the value of the ratio for any positive integer value of n. Use this result to obtain summation formula III.

43. (a) Use Problem 35, part **(a)** to show that

$$\sum_{k=1}^{n} [(k + 1)^2 - k^2] = -1 + (n + 1)^2 = n^2 + 2n$$

(b) Use the fact that $(k + 1)^2 - k^2 = 2k + 1$ to show that

$$\sum_{k=1}^{n} [(k + 1)^2 - k^2] = n + 2 \sum_{k=1}^{n} k$$

(c) Compare the results in parts **(a)** and **(b)** to derive summation formula II.

44. Apply the procedure outlined in Problem 43 to the sum $\sum_{k=1}^{n} [(k + 1)^3 - k^3]$ to derive summation formula III.

5.4 Area under a Graph

As the derivative is motivated by the geometric problem of constructing a tangent to a curve, the historical problem leading to the definition of a definite integral is the problem of finding area. Specifically, we are interested in finding the area A of a region bounded by the x-axis, the graph of a *nonnegative* function[†] $y = f(x)$ defined on some interval $[a, b]$, and

(i) the vertical lines $x = a$ and $x = b$, as shown in Figure 5.4(a), or

(ii) the x-intercepts of the graph shown in Figure 5.4(b).

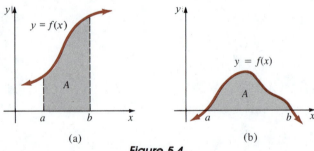

(a)   (b)

Figure 5.4

We shall call this area the **area under the graph** of f on the interval $[a, b]$.

---

*Galileo Galilei (1564–1642), an Italian astronomer and physicist.

[†]The requirement that f be nonnegative on $[a, b]$ means that no portion of its graph on the interval is below the x-axis.

Assume for the moment that we do not know a formula for calculating the area A of the right triangle given in Figure 5.5(a). By superimposing a Cartesian coordinate system on the triangle, as shown in Figure 5.5(b), we see that the problem is the same as finding the area in the first quadrant bounded by the straight lines $y = (h/b)x$, $y = 0$ (the x-axis), and $x = b$. In other words, we wish to find the area under the graph of $y = (h/b)x$ on the interval $[0, b]$.

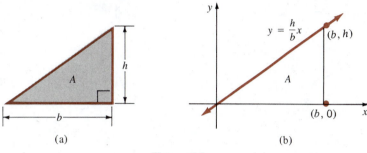

(a)   (b)

Figure 5.5

Using rectangles, Figure 5.6 indicates three different ways of *approximating* the area A.

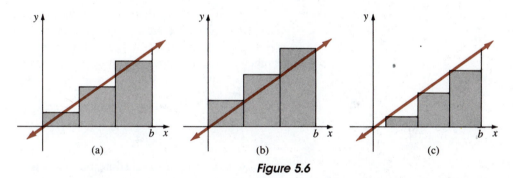

(a)   (b)   (c)

Figure 5.6

For convenience, let us pursue the procedure hinted at in Figure 5.6(b) in greater detail. We begin by dividing the interval $[0, b]$ into n subintervals of equal length $\Delta x = b/n$. If the right endpoint of each of these intervals is denoted by x_k^*, then

$$x_1^* = \Delta x = \frac{b}{n}$$

$$x_2^* = 2\,\Delta x = 2\left(\frac{b}{n}\right)$$

$$\vdots$$

$$x_k^* = k\,\Delta x = k\left(\frac{b}{n}\right)$$

$$\vdots$$

$$x_n^* = n\,\Delta x = n\left(\frac{b}{n}\right) = b$$

As shown in Figure 5.7(a) we now construct a rectangle of length $f(x_k^*)$ and width Δx on each of the n subintervals. Since the area of a rectangle is *length × width*, the area of each rectangle is $f(x_k^*)\,\Delta x$. See Figure 5.7(b).

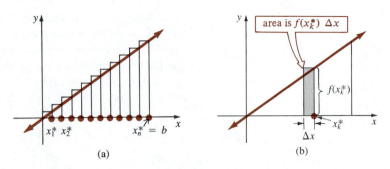

Figure 5.7

Adding the area of the n rectangles gives an approximation to the number A. We write

$$A \approx f(x_1^*) \, \Delta x + f(x_2^*) \, \Delta x + \cdots + f(x_n^*) \, \Delta x$$

or using sigma notation,

$$A \approx \sum_{k=1}^{n} f(x_k^*) \, \Delta x \tag{5.16}$$

It seems plausible that we can reduce the error introduced by this method of approximation (the area of each rectangle is larger than the area under the graph on a subinterval $[x_{k-1}, x_k]$) by **partitioning** $[0, b]$ into finer subdivisions. In other words, we expect that a better approximation to A can be obtained by using more and more rectangles ($n \to \infty$) of decreasing widths ($\Delta x \to 0$).

Now,

$$f(x) = \frac{h}{b} x, \qquad x_k^* = k\left(\frac{b}{n}\right), \qquad f(x_k^*) = \frac{h}{n} \cdot k, \quad \text{and} \quad \Delta x = \frac{b}{n}$$

so that with the aid of summation formula II of Section 5.3, (5.16) becomes

$$A \approx \sum_{k=1}^{n} \left(\frac{h}{n} \cdot k\right) \frac{b}{n}$$

$$= \frac{bh}{n^2} \sum_{k=1}^{n} k$$

$$= \frac{bh}{n^2} \cdot \frac{n(n + 1)}{2} = \frac{bh}{2}\left(1 + \frac{1}{n}\right)$$

Finally, as $n \to \infty$, we obtain the familiar formula

$$A = \frac{1}{2}bh \lim_{n \to \infty} \left(1 + \frac{1}{n}\right) = \frac{1}{2}bh$$

The General Problem   Now, let us turn from the preceding specific example to the general problem of finding the area A under the graph of a continuous function $y = f(x)$ on an interval $[a, b]$. As shown in Figure 5.8(a), we shall also assume that $f(x) \geq 0$ for all x in the interval. As suggested in Figure 5.8(b), the area A can be approximated by adding the areas of n rectangles that are constructed on the interval.

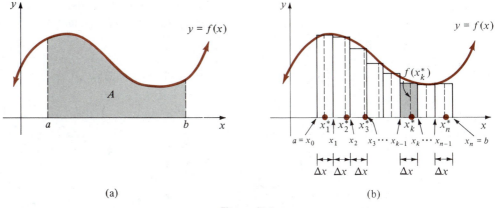

(a)   (b)

Figure 5.8

One possible procedure for determining A is summarized as follows.

1. Partition the interval $[a, b]$ into n subintervals $[x_{k-1}, x_k]$ of equal length, $k = 1, 2, \ldots , n$, where $x_0 = a$, $x_n = b$, and

    $$a = x_0 < x_1 < x_2 < \cdots < x_{n-1} < x_n = b$$

2. Denote the length of each subinterval by Δx, where $\Delta x = x_k - x_{k-1}$.

3. Choose any number x_k^* in each subinterval $[x_{k-1}, x_k]$ and form the product $f(x_k^*)\,\Delta x$. This represents the area of a rectangle on the kth subinterval.

4. Form the sum $\sum_{k=1}^{n} f(x_k^*)\,\Delta x$. This is the sum of the areas of the n rectangles and represents an approximation to the value of A.

With these preliminaries, we are now in a position to define the concept of area under a graph.

DEFINITION 5.2   **Area**

Let f be continuous on $[a, b]$ and $f(x) \geq 0$ for all x in the interval. We define the area A under the graph on the interval to be

$$A = \lim_{n \to \infty} \sum_{k=1}^{n} f(x_k^*)\,\Delta x \tag{5.17}$$

It is usually proved in advanced calculus that when f is continuous, the limit in (5.17) always exists regardless of the manner used to partition $[a, b]$; that is, the subintervals may or may not be taken of equal length, and the points x_k^* can be chosen quite arbitrarily in the subintervals $[x_{k-1}, x_k]$. However, if the subintervals are not of equal length, then a different kind of limiting process is necessary in (5.17). We must replace $n \to \infty$ with the requirement that the length of the longest subinterval approach zero.

To use (5.17), suppose we choose x_k^* as we did in the discussion of Figure 5.6; namely, let x_k^* be the right endpoint of each subinterval. Since the length of each of the n equal subintervals is $\Delta x = (b - a)/n$ we have

$$x_1^* = x_0 + \Delta x = a + \frac{b - a}{n}$$
$$x_2^* = x_0 + 2\,\Delta x = a + 2\left(\frac{b - a}{n}\right)$$
$$\vdots$$
$$x_k^* = x_0 + k\,\Delta x = a + k\left(\frac{b - a}{n}\right)$$
$$\vdots$$
$$x_n^* = x_0 + n\,\Delta x = a + n\left(\frac{b - a}{n}\right) = b$$

It follows, by substituting $a + k\left(\dfrac{b - a}{n}\right)$ for x_k^* and $(b - a)/n$ for Δx in (5.17), that the area A is also given by

$$A = \lim_{n \to \infty} \sum_{k=1}^{n} f\left(a + k\frac{b - a}{n}\right) \cdot \frac{b - a}{n} \qquad (5.18)$$

We note that since $\Delta x = (b - a)/n$, $n \to \infty$ is the same as $\Delta x \to 0$.

Example 1

Find the area A under the graph of $f(x) = x + 2$ on the interval $[0, 4]$.

Solution   The area is bounded by the trapezoid indicated in Figure 5.9(a). By identifying $a = 0$ and $b = 4$, we find

$$\Delta x = \frac{4 - 0}{n} = \frac{4}{n}$$

Thus, (5.18) becomes

$$A = \lim_{n \to \infty} \sum_{k=1}^{n} f\left(0 + k\frac{4}{n}\right)\frac{4}{n}$$
$$= \lim_{n \to \infty} \frac{4}{n} \sum_{k=1}^{n} f\left(\frac{4k}{n}\right)$$
$$= \lim_{n \to \infty} \frac{4}{n} \sum_{k=1}^{n} \left(\frac{4k}{n} + 2\right)$$
$$= \lim_{n \to \infty} \frac{4}{n}\left[\frac{4}{n} \sum_{k=1}^{n} k + 2 \sum_{k=1}^{n} 1\right]$$

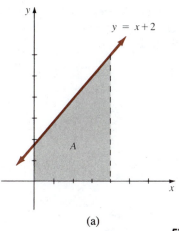
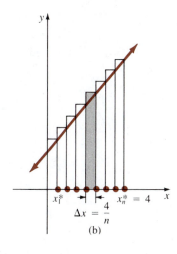

(a)

(b)

Figure 5.9

Now, by summation formulas I and II of Section 5.3, we can write

$$A = \lim_{n \to \infty} \frac{4}{n} \left[\frac{4}{n} \cdot \frac{n(n+1)}{2} + 2n \right]$$

$$= \lim_{n \to \infty} \left[\frac{16n(n+1)}{2 \quad n^2} + 8 \right]$$

$$= \lim_{n \to \infty} \left[8\left(1 + \frac{1}{n}\right) + 8 \right]$$

$$= 8 \lim_{n \to \infty} \left(1 + \frac{1}{n}\right) + 8 \lim_{n \to \infty} 1$$

$$= 8 + 8 = 16 \text{ square units}$$

Example 2

Find the area A under the graph of $f(x) = 4 - x^2$ on the interval $[-1, 2]$.

Solution   The area is indicated in Figure 5.10(a). Since $a = -1$ and $b = 2$, it follows that

$$\Delta x = \frac{2 - (-1)}{n} = \frac{3}{n}$$

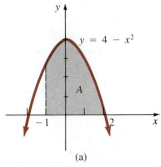
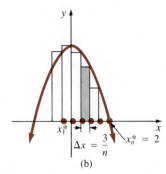

(a)

(b)

Figure 5.10

Let us review the steps leading up to formula (5.18). The width of each rectangle is given by $\Delta x = 3/n$. Now, starting at $x = -1$, the right endpoint of each subinterval is

$$x_1^* = -1 + \frac{3}{n}$$

$$x_2^* = -1 + 2\left(\frac{3}{n}\right)$$

$$\vdots$$

$$x_k^* = -1 + k\left(\frac{3}{n}\right)$$

$$\vdots$$

$$x_n^* = -1 + n\left(\frac{3}{n}\right) = 2$$

The length of each rectangle is then

$$f(x_1^*) = f\left(-1 + \frac{3}{n}\right) = 4 - \left[-1 + \frac{3}{n}\right]^2$$

$$f(x_2^*) = f\left(-1 + 2\left(\frac{3}{n}\right)\right) = 4 - \left[-1 + 2\left(\frac{3}{n}\right)\right]^2$$

$$\vdots$$

$$f(x_k^*) = f\left(-1 + k\left(\frac{3}{n}\right)\right) = 4 - \left[-1 + k\left(\frac{3}{n}\right)\right]^2$$

$$\vdots$$

$$f(x_n^*) = f\left(-1 + n\left(\frac{3}{n}\right)\right) = f(2) = 4 - (2)^2 = 0$$

The area of the kth rectangle is length $\times$ width:

$$f(x_k^*)\frac{3}{n} = \left(4 - \left[-1 + k\frac{3}{n}\right]^2\right)\frac{3}{n}$$

$$= \left(3 + 6\frac{k}{n} - 9\frac{k^2}{n^2}\right)\frac{3}{n}$$

Adding the area of the n rectangles gives an approximation to the area under the graph on the interval: $A \approx \sum_{k=1}^{n} f(x_k^*)(3/n)$. As the number n of rectangles increases without bound, we obtain

$$A = \lim_{n\to\infty} \sum_{k=1}^{n} \left(3 + 6\frac{k}{n} - 9\frac{k^2}{n^2}\right)\frac{3}{n}$$

$$= \lim_{n\to\infty} \frac{3}{n} \sum_{k=1}^{n} \left(3 + 6\frac{k}{n} - 9\frac{k^2}{n^2}\right)$$

$$= \lim_{n\to\infty} \frac{3}{n}\left[3 \sum_{k=1}^{n} 1 + \frac{6}{n} \sum_{k=1}^{n} k - \frac{9}{n^2} \sum_{k=1}^{n} k^2\right]$$

Using the summation formulas I, II, and III of the last section, it follows that

$$A = \lim_{n \to \infty} \frac{3}{n}\left[3n + \frac{6}{n} \cdot \frac{n(n+1)}{2} - \frac{9}{n^2} \cdot \frac{n(n+1)(2n+1)}{6}\right]$$

$$= \lim_{n \to \infty} \left[9 + 9\left(1 + \frac{1}{n}\right) - \frac{9}{2}\left(1 + \frac{1}{n}\right)\left(2 + \frac{1}{n}\right)\right]$$

$$= 9 + 9 - 9 = 9 \text{ square units}$$

Remark

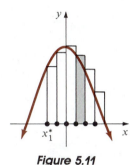

Figure 5.11

There is nothing special about choosing x_k^* to be the right endpoint of each subinterval. We reemphasize that x_k^* can be taken to be any convenient number in $[x_{k-1}, x_k]$. Had we chosen x_k^* to be the left endpoint of each subinterval in Example 2, then the corresponding rectangles would be as shown in Figure 5.11. In this case, we would have $x_k^* = -1 + (k-1)(3/n)$, $k = 1, 2, 3, \ldots, n$. In Problems 24 and 25 of Exercises 5.4 you are asked to solve the area problem in Example 2 by first choosing x_k^* to be the left endpoint and then the midpoint of each subinterval $[x_{k-1}, x_k]$.

Exercises 5.4

Answers to odd-numbered problems begin on page A-28.

In Problems 1–20 find the area under the graph of the given function on the indicated interval.

1. $f(x) = 4$, $[2, 5]$

2. $f(x) = 7$, $[-4, 8]$

3. $f(x) = x$, $[0, 6]$

4. $f(x) = 2x$, $[1, 3]$

5. $f(x) = 2x + 1$, $[1, 5]$

6. $f(x) = 3x - 6$, $[2, 4]$

7. $f(x) = x^2$, $[0, 2]$

8. $f(x) = x^2$, $[-2, 1]$

9. $f(x) = 1 - x^2$, $[-1, 1]$

10. $f(x) = 2x^2 + 3$, $[-3, -1]$

11. $f(x) = x^2 + 2x$, $[1, 2]$

12. $f(x) = (x - 1)^2$, $[0, 2]$

13. $f(x) = x^3$, $[0, 1]$

14. $f(x) = -x^3$, $[-3, 0]$

15. $f(x) = x^3 - 3x^2 + 4$, $[0, 2]$

16. $f(x) = (x + 1)^3$, $[-1, 1]$

17. $f(x) = x^4$, $[0, 2]$

18. $f(x) = 16 - x^4$, $[-1, 0]$

19. $f(x) = |x|$, $[-1, 3]$

20. $f(x) = \begin{cases} 2, & 0 \le x < 1 \\ x + 1, & 1 \le x \le 4 \end{cases}$

21. Sketch the graph of $y = 1/x$ on $\frac{1}{2} \le x \le \frac{5}{2}$. By dividing the interval into four subintervals of equal lengths, construct rectangles that approximate the area A under the graph. First use the right endpoint of each subinterval, and then use the left endpoint.

22. Repeat Problem 21 for $y = \cos x$ on $-\pi/2 \le x \le \pi/2$.

23. Derive a formula analogous to (5.18) in which x_k^* is chosen as **(a)** the left endpoint of each subinterval, **(b)** the midpoint of each subinterval.

24. Rework Example 2 by choosing x_k^* to be the left endpoint of each subinterval.

25. Rework Example 2 by choosing x_k^* to be the midpoint of each subinterval.

26. Find the area of the region in the first quadrant bounded by the graphs of $x = 0$, $y = 9$, and $y = x^2$.

27. Find the area of the region in the first quadrant bounded by the graphs of $y = x$ and $y = x^2$.

28. The area of the trapezoid given in Figure 5.12 is

$$A = \left(\frac{h_1 + h_2}{2}\right)b$$

Derive this formula.

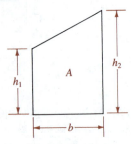

Figure 5.12

In Problems 29 and 30 determine a region whose area A is given by the formula. Do not try to evaluate.

29. $A = \lim\limits_{n \to \infty} \sum\limits_{k=1}^{n} \sqrt{4 - \dfrac{4k^2}{n^2}} \dfrac{2}{n}$

30. $A = \lim\limits_{n \to \infty} \sum\limits_{k=1}^{n} \left(\sin \dfrac{k\pi}{n}\right)\dfrac{\pi}{n}$

31. Find the area under the graph of $y = \sqrt{x}$ on $0 \le x \le 1$ by considering the area under the graph of $y = x^2$ on $0 \le x \le 1$.

32. Find the area under the graph of $y = \sqrt[3]{x}$ on $0 \le x \le 8$ by considering the area under the graph of $y = x^3$ on $0 \le x \le 2$.

5.5 The Definite Integral

5.5.1 The Definition of a Definite Integral

We are now in a position to define the concept of a **definite integral**. To do this, consider the following five steps.

$$y = f(x)$$

1. Let f be defined on a closed interval $[a, b]$.
2. Partition the interval $[a, b]$ into n subintervals $[x_{k-1}, x_k]$ of length $\Delta x_k = x_k - x_{k-1}$. Let P denote the partition

$$a = x_0 < x_1 < x_2 < \cdots < x_{n-1} < x_n = b$$

3. Let $\|P\|$ be the length of the longest subinterval. The number $\|P\|$ is called the **norm** of the partition P.
4. Choose a number x_k^* in each subinterval.
5. Form the sum

$$\sum_{k=1}^{n} f(x_k^*)\Delta x_k \qquad\qquad (5.19)$$

Sums such as (5.19) for the various partitions of $[a, b]$ are known as **Riemann sums** and are named for the famous German mathematician, Georg Friedrich Bernhard Riemann .*

Although the procedure looks very similar to the five steps leading up to the definition of area under a graph, there are some important differences. Observe that a Riemann sum does not require that f be either continuous or nonnegative on the interval $[a, b]$. Thus, (5.19) does not necessarily represent an approximation to the area under a graph. Keep in mind that "area under a graph" refers to *the area bounded between the graph of a nonnegative function and the x-axis*. As shown in Figure 5.13, if $f(x) < 0$ for some x in $[a, b]$, a Riemann sum could contain terms $f(x_k^*)\Delta x_k$, where $f(x_k^*) < 0$. In this case the products $f(x_k^*)\Delta x_k$ are numbers that are the negatives of the areas of rectangles drawn below the x-axis.

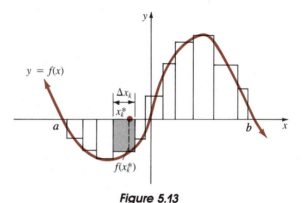

Figure 5.13

Bernhard Riemann

*__Georg Friedrich Bernhard Riemann (1826–1866)__   Bernhard Riemann, born in Hanover, Germany, in 1826, was the son of a Lutheran minister. Although a devote Christian, Riemann was disinclined to follow his father's vocation and abandoned the study of theology at the University of Göttingen in favor of a course of studies in which his genius was obvious: mathematics. It is likely that the concept of Riemann sums grew out of a course on the definite integral that he had at the university; this concept reflects his attempt to give a precise mathematical meaning to the definite integral of Newton and Leibniz. After submitting his doctoral dissertation on the foundations of functions of complex variables to the examining committee at the University of Göttingen, Carl Friedrich Gauss, the "prince of mathematicians," paid Riemann a very rare compliment: "The dissertation offers convincing evidence . . . of a creative, active, truly mathematical mind . . . of glorious fertile originality." Riemann, like so many other promising scholars of that time, possessed a fragile constitution. He died at age 39 of pleurisy. His original contributions to differential geometry, topology, non-Euclidean geometry, and his bold investigations into the nature of space, electricity, and magnetism, foreshadowed the work of Einstein in the next century.

Example 1

Compute the Riemann sum for $f(x) = x^2 - 4$ on $[-2, 3]$ with five subintervals determined by

$$x_0 = -2, \qquad x_1 = -\frac{1}{2}, \qquad x_2 = 0, \qquad x_3 = 1, \qquad x_4 = \frac{7}{4}, \qquad x_5 = 3$$

and $\quad x_1^* = -1, \qquad x_2^* = -\frac{1}{4}, \qquad x_3^* = \frac{1}{2}, \qquad x_4^* = \frac{3}{2}, \qquad x_5^* = \frac{5}{2}$

Solution   Figure 5.14 shows the various points x_k and x_k^* on the interval.

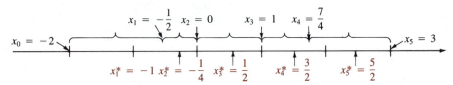

Figure 5.14

Now,
$$\Delta x_1 = x_1 - x_0 = -\frac{1}{2} - (-2) = \frac{3}{2}$$

$$\Delta x_2 = x_2 - x_1 = 0 - \left(-\frac{1}{2}\right) = \frac{1}{2}$$

$$\Delta x_3 = x_3 - x_2 = 1 - 0 = 1$$

$$\Delta x_4 = x_4 - x_3 = \frac{7}{4} - 1 = \frac{3}{4}$$

$$\Delta x_5 = x_5 - x_4 = 3 - \frac{7}{4} = \frac{5}{4}$$

and
$$f(x_1^*) = f(-1) = -3$$

$$f(x_2^*) = f\left(-\frac{1}{4}\right) = -\frac{63}{16}$$

$$f(x_3^*) = f\left(\frac{1}{2}\right) = -\frac{15}{4}$$

$$f(x_4^*) = f\left(\frac{3}{2}\right) = -\frac{7}{4}$$

$$f(x_5^*) = f\left(\frac{5}{2}\right) = \frac{9}{4}$$

Therefore, the Riemann sum is

$$f(x_1^*)\,\Delta x_1 + f(x_2^*)\,\Delta x_2 + f(x_3^*)\,\Delta x_3 + f(x_4^*)\,\Delta x_4 + f(x_5^*)\,\Delta x_5$$

$$= (-3)\left(\frac{3}{2}\right) + \left(-\frac{63}{16}\right)\left(\frac{1}{2}\right) + \left(-\frac{15}{4}\right)(1) + \left(-\frac{7}{4}\right)\left(\frac{3}{4}\right) + \left(\frac{9}{4}\right)\left(\frac{5}{4}\right)$$

$$= -\frac{279}{32} \approx -8.72$$

For a function f defined on an interval $[a, b]$, there are an infinite number of possible Riemann sums for a given partition P of the interval since the numbers x_k^* can be chosen arbitrarily in each subinterval $[x_{k-1}, x_k]$.

_____ **Example 2** _____

Compute the Riemann sum for the function and partition of $[-2, 3]$ in Example 1 if $x_1^* = -\frac{3}{2}$, $x_2^* = -\frac{1}{8}$, $x_3^* = \frac{3}{4}$, $x_4^* = \frac{3}{2}$, and $x_5^* = 2.1$.

Solution

$$f(x_1^*) = f\left(-\frac{3}{2}\right) = -\frac{7}{4}$$

$$f(x_2^*) = f\left(-\frac{1}{8}\right) = -\frac{255}{64}$$

$$f(x_3^*) = f\left(\frac{3}{4}\right) = -\frac{55}{16}$$

$$f(x_4^*) = f\left(\frac{3}{2}\right) = -\frac{7}{4}$$

$$f(x_5^*) = f(2.1) = 0.41$$

Since the numbers Δx_k are the same as before, we have

$$f(x_1^*) \, \Delta x_1 + f(x_2^*) \, \Delta x_2 + f(x_3^*) \, \Delta x_3 + f(x_4^*) \, \Delta x_4 + f(x_5^*) \, \Delta x_5$$

$$= \left(-\frac{7}{4}\right)\left(\frac{3}{2}\right) + \left(-\frac{255}{64}\right)\left(\frac{1}{2}\right) + \left(-\frac{55}{16}\right)(1) + \left(-\frac{7}{4}\right)\left(\frac{3}{4}\right) + (0.41)\left(\frac{5}{4}\right)$$

$$\approx -8.85$$

We are interested in a special kind of limit of (5.19). If the Riemann sums $\sum_{k=1}^{n} f(x_k^*) \, \Delta x_k$ are close to a number L for *every* partition P of $[a, b]$ for which the norm $\|P\|$ is close to zero, we then write

$$\lim_{\|P\|\to 0} \sum_{k=1}^{n} f(x_k^*) \, \Delta x_k = L \qquad (5.20)$$

and say that L is the **definite integral** of f on the interval $[a, b]$. If the limit in (5.20) exists, the function f is said to be **integrable** on the interval. In the following definition we introduce a new symbol for the number L.

DEFINITION 5.3   The Definite Integral

Let f be a function defined on closed interval $[a, b]$. Then the **definite integral of f from a to b**, denoted by $\int_a^b f(x) \, dx$, is defined to be

$$\int_a^b f(x) \, dx = \lim_{\|P\|\to 0} \sum_{k=1}^{n} f(x_k^*) \, \Delta x_k \qquad (5.21)$$

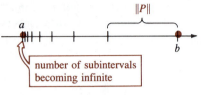

number of subintervals
becoming infinite

Figure 5.15

The numbers a and b in the preceding definition are called the **lower** and **upper limits of integration**, respectively. The integral symbol $\int$, as used by Leibniz, is an elongated S for the word "sum." Also, note that $\|P\| \to 0$ always implies that the number of subintervals n becomes infinite in number ($n \to \infty$). However, as shown in Figure 5.15, the fact that $n \to \infty$ does not necessarily imply $\|P\| \to 0$.

The following important result is stated without proof.

THEOREM 5.6   If f is continuous on $[a, b]$, then $\int_a^b f(x)\, dx$ exists; that is, f is integrable on the interval.

Regular Partition   When a definite integral exists,

the limit in (5.21) exists for every possible way of partitioning $[a, b]$ and for every way of choosing x_k^* in the subintervals $[x_{k-1}, x_k]$.

In particular, by choosing the subintervals of equal length

$$\Delta x = \frac{b - a}{n} \quad \text{and} \quad x_k^* = a + k\frac{b - a}{n}, \qquad k = 1, 2, \ldots, n$$

we can write

$$\int_a^b f(x)\, dx = \lim_{n \to \infty} \sum_{k=1}^{n} f\left(a + k\frac{b - a}{n}\right)\frac{b - a}{n} \tag{5.22}$$

A partition P of $[a, b]$ in which the subintervals are of the same length is called a **regular partition**.

You might conclude that the formulations of $\int_a^b f(x)\, dx$ given in (5.21) and (5.22) are exactly the same as (5.17) and (5.18) of Section 5.4 for the general case of finding the area under the curve $y = f(x)$ on $[a, b]$. In a way this is correct; however, Definition 5.3 is a more general concept since, as noted before, we are not requiring that f be continuous on $[a, b]$ or that $f(x) \geq 0$ on the interval. Thus, a *definite integral need not be area.* What then is a definite integral? For now, accept the fact that a definite integral is simply a real number. Contrast this with the indefinite integral, which is a function (or a set of functions). Is the area under the graph of a continuous nonnegative function a definite integral? The answer is yes.

THEOREM 5.7   If f is continuous on $[a, b]$ and $f(x) \geq 0$ for all x in the interval, then the area A under the graph of f on $[a, b]$ is

$$A = \int_a^b f(x)\, dx$$

We shall return to the question of finding areas after we have studied some properties of the definite integral in the next section.

_____ **Example 3** _____

Evaluate $\int_{-2}^{1} x^3 \, dx$.

Solution Since $f(x) = x^3$ is continuous on $[-2, 1]$, we know from Theorem 5.6 that the definite integral exists. We use a regular partition and the result given in (5.22). Choosing

$$\Delta x = \frac{1 - (-2)}{n} = \frac{3}{n} \quad \text{and} \quad x_k^* = -2 + k \cdot \frac{3}{n}$$

we have

$$f\left(-2 + \frac{3k}{n}\right) = \left(-2 + \frac{3k}{n}\right)^3$$

$$= -8 + 36\left(\frac{k}{n}\right) - 54\left(\frac{k^2}{n^2}\right) + 27\left(\frac{k^3}{n^3}\right)$$

It then follows from (5.22) and summation formulas I–IV that

$$\int_{-2}^{1} x^3 \, dx = \lim_{n \to \infty} \sum_{k=1}^{n} f\left(-2 + \frac{3k}{n}\right)\frac{3}{n}$$

$$= \lim_{n \to \infty} \frac{3}{n} \sum_{k=1}^{n} \left[-8 + 36\left(\frac{k}{n}\right) - 54\left(\frac{k^2}{n^2}\right) + 27\left(\frac{k^3}{n^3}\right)\right]$$

$$= \lim_{n \to \infty} \frac{3}{n}\left[-8n + \frac{36}{n} \cdot \frac{n(n + 1)}{2} - \frac{54}{n^2} \cdot \frac{n(n + 1)(2n + 1)}{6}\right.$$

$$\left. + \frac{27}{n^3} \cdot \frac{n^2(n + 1)^2}{4}\right]$$

$$= \lim_{n \to \infty} \left[-24 + 54\left(1 + \frac{1}{n}\right) - 27\left(1 + \frac{1}{n}\right)\left(2 + \frac{1}{n}\right)\right.$$

$$\left. + \frac{81}{4}\left(1 + \frac{1}{n}\right)\left(1 + \frac{1}{n}\right)\right]$$

$$= -24 + 54 - 27(2) + \frac{81}{4} = -\frac{15}{4}$$

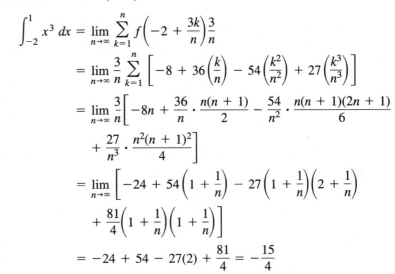

y = x^3

Figure 5.16

Figure 5.16 shows that we are not considering area.

_____ **Example 4** _____

The values of the Riemann sums in Examples 1 and 2 are approximations to the value of the definite integral $\int_{-2}^{3} (x^2 - 4) \, dx$. It is left as an exercise to show

$$\int_{-2}^{3} (x^2 - 4) \, dx = -\frac{25}{3} \approx -8.33$$

See Problem 18 in Exercises 5.5.

BASIC Program   The following listing is a BASIC program for the approximation

$$\int_a^b f(x)\, dx \approx \sum_{k=1}^{n} f\left(a + k\frac{b-a}{n}\right)\frac{b-a}{n}$$

```
10  REM EVALUATION OF DEFINITE INTEGRALS
    VIA RIEMANN SUMS
20  DEF FNY(X) = ...
30  INPUT "WHAT IS THE INTERVAL? ";A,B
40  INPUT "HOW MANY SUBDIVISIONS? ";N
50  LET H = (B - A)/N
60  FOR X = A + H TO B STEP H
70  LET R = R + FNY(X)
80  NEXT X
90  LET R = R * H
100 PRINT "USING ";N; " SUBDIVISIONS, THE
    RIEMANN SUM YIELDS ";R
110 END
```

[O] 5.5.2 An ε–δ Definition

Let f be a function defined on $[a, b]$ and let L denote a real number. The intuitive concept that Riemann sums are close to L whenever the norm $\|P\|$ of a partition P is close to zero can be expressed in a precise manner using the ε–δ symbols introduced in Section 2.5. To say f is integrable on $[a, b]$, we mean that for every real number $\varepsilon > 0$ there exists a real number $\delta > 0$ such that

$$\left| \sum_{k=1}^{n} f(x_k^*)\, \Delta x_k - L \right| < \varepsilon \tag{5.23}$$

whenever P is a partition of $[a, b]$ for which $\|P\| < \delta$ and the x_k^* are numbers in $[x_{k-1}, x_k]$. In other words,

$$\lim_{\|P\| \to 0} \sum_{k=1}^{n} f(x_k^*)\, \Delta x_k$$

exists and is equal to the number L.

Remarks

(*i*)   There are functions that are defined for every value of x in $[a, b]$ for which the limit in (5.21) does not exist. Also, if the function f is not defined for all values of x in the interval, the definite integral *may* not exist; for example, later on we shall see why an integral such as $\int_{-3}^{2} (1/x)\, dx$ does not exist. Notice that $y = 1/x$ is discontinuous at $x = 0$ and is unbounded on the interval. However, one should not conclude from this one example that when a function f has a discontinuity in $[a, b]$ that $\int_a^b f(x)\, dx$ necessarily does not exist. (See Problem 34 in Exercises 5.5.) Continuity of a function f on $[a, b]$ is sufficient but not necessary to guarantee the existence of

$\int_a^b f(x)\,dx$. The set of functions continuous on $[a, b]$ is a subset of the set of functions that are integrable on the interval.

It is important that you be aware of further sufficient conditions of integrability.

If a function f is bounded on $[a, b]$, that is, if there exists a positive constant B such that $-B \le f(x) \le B$ for all x in the interval, and has a finite number of discontinuities in $[a, b]$, then f is integrable on $[a, b]$.

For example, the function

$$f(x) = \begin{cases} 4, & 0 \le x < 2 \\ 1, & 2 \le x \le 3 \end{cases}$$

is discontinuous at $x = 2$ but is bounded on $[0, 3]$. Hence, $\int_0^3 f(x)\,dx$ exists. See Problem 53 of Exercises 5.7.

(*ii*) The procedure outlined in (5.22) has limited utility as a practical means of computing a definite integral. We shall see in the next several sections that sometimes there is an easier way of finding the number $\int_a^b f(x)\,dx$.

_____ **Exercises 5.5** _____

Answers to odd-numbered problems begin on page A-28.

[5.5.1]

In Problems 1–8 compute the Riemann sum for the given function on the indicated interval. Specify $\|P\|$.

1. $f(x) = x$, $[0, 4]$, two subintervals;

$x_0 = 0$, $x_1 = \dfrac{5}{2}$, $x_2 = 4$;

$x_1^* = 2$, $x_2^* = 3$

2. $f(x) = -2x$, $[-1, 2]$, three subintervals;

$x_0 = -1$, $x_1 = -\dfrac{1}{2}$, $x_2 = 1$, $x_3 = 2$;

$x_1^* = -\dfrac{1}{2}$, $x_2^* = 0$, $x_3^* = \dfrac{3}{2}$

3. $f(x) = 3x + 1$, $[0, 3]$, four subintervals;

$x_0 = 0$, $x_1 = 1$, $x_2 = \dfrac{5}{3}$, $x_3 = \dfrac{7}{3}$, $x_4 = 3$;

$x_1^* = \dfrac{1}{2}$, $x_2^* = \dfrac{4}{3}$, $x_3^* = 2$, $x_4^* = \dfrac{8}{3}$

4. $f(x) = x - 4$, $[-2, 5]$, five subintervals;

$x_0 = -2$, $x_1 = -1$, $x_2 = -\dfrac{1}{2}$, $x_3 = \dfrac{1}{2}$, $x_4 = 3$, $x_5 = 5$;

$x_1^* = -\dfrac{3}{2}$, $x_2^* = -\dfrac{1}{2}$, $x_3^* = 0$, $x_4^* = 2$, $x_5^* = 4$

5. $f(x) = x^2$, $[-1, 1]$, four subintervals;

$x_0 = -1$, $x_1 = -\dfrac{1}{4}$, $x_2 = \dfrac{1}{4}$, $x_3 = \dfrac{3}{4}$, $x_4 = 1$;

$x_1^* = -\dfrac{3}{4}$, $x_2^* = 0$, $x_3^* = \dfrac{1}{2}$, $x_4^* = \dfrac{7}{8}$

6. $f(x) = x^2 + 1$, $[1, 3]$, three subintervals;

$x_0 = 1$, $x_1 = \dfrac{3}{2}$, $x_2 = \dfrac{5}{2}$, $x_3 = 3$;

$x_1^* = \dfrac{5}{4}$, $x_2^* = \dfrac{7}{4}$, $x_3^* = 3$

7. $f(x) = \sin x$, $[0, 2\pi]$, three subintervals;

$x_0 = 0$, $x_1 = \pi$, $x_2 = \dfrac{3\pi}{2}$, $x_3 = 2\pi$;

$x_1^* = \dfrac{\pi}{2}$, $x_2^* = \dfrac{7\pi}{6}$, $x_3^* = \dfrac{7\pi}{4}$

8. $f(x) = \cos x$, $\left[-\dfrac{\pi}{2}, \dfrac{\pi}{2}\right]$, four subintervals;

$x_0 = -\dfrac{\pi}{2}$, $x_1 = -\dfrac{\pi}{4}$, $x_2 = 0$, $x_3 = \dfrac{\pi}{3}$, $x_4 = \dfrac{\pi}{2}$;

$x_1^* = -\dfrac{\pi}{3}$, $x_2^* = -\dfrac{\pi}{6}$, $x_3^* = \dfrac{\pi}{4}$, $x_4^* = \dfrac{\pi}{3}$

9. Given $f(x) = x - 2$ on $[0, 5]$, compute the Riemann sum using a partition with five subintervals of equal length. Let x_k^*, $k = 1, 2, \ldots, 5$ be the right endpoint of each subinterval.

10. Given $f(x) = x^2 - x + 1$ on $[0, 1]$, compute the Riemann sum using a partition with three subintervals of equal length. Let x_k^*, $k = 1, 2, 3$ be the left endpoint of each subinterval.

In Problems 11–20 use (5.22) to evaluate the given definite integral.

11. $\displaystyle\int_1^9 3\ dx$

12. $\displaystyle\int_{-1}^4 (-2)\ dx$

13. $\displaystyle\int_{-3}^1 x\ dx$

14. $\displaystyle\int_0^3 x\ dx$

15. $\displaystyle\int_0^2 (x^2 - 1)\ dx$

16. $\displaystyle\int_0^3 (x^2 - 2x)\ dx$

17. $\displaystyle\int_1^2 (x^2 - x)\ dx$

18. $\displaystyle\int_{-2}^3 (x^2 - 4)\ dx$

19. $\displaystyle\int_0^1 (x^3 - 1)\ dx$

20. $\displaystyle\int_0^2 (3 - x^3)\ dx$

21. Let $f(x) = k$ be a constant function on $[a, b]$. Use (5.22) to show that $\int_a^b k\ dx = k(b - a)$.

22. Let P be any partition of $[a, b]$. Use Problem 21 to show that

$$\lim_{\|P\| \to 0} \sum_{k=1}^n \Delta x_k = b - a$$

Interpret this result geometrically.

In Problems 23 and 24 use (5.22) to obtain the given result.

23. $\displaystyle\int_a^b x\ dx = \frac{1}{2}(b^2 - a^2)$

24. $\displaystyle\int_a^b x^2\ dx = \frac{1}{3}(b^3 - a^3)$

25. Evaluate the definite integral $\int_0^1 \sqrt{x}\ dx$ by using a partition of $[0, 1]$ in which the subintervals $[x_{k-1}, x_k]$ are defined by $[(k - 1)^2/n^2, k^2/n^2]$ and choosing x_k^* to be the right endpoint of each subinterval.

26. Evaluate the definite integral $\int_0^{\pi/2} \cos x\ dx$ by using a regular partition of $[0, \frac{\pi}{2}]$ and choosing x_k^* to be the midpoint of each subinterval $[x_{k-1}, x_k]$. Use the results

(*i*) $\cos \theta + \cos 3\theta + \cdots + \cos(2n - 1)\theta$
$$= \frac{\sin 2n\theta}{2 \sin \theta}$$

(*ii*) $\displaystyle\lim_{n \to \infty} \frac{1}{n \sin(\pi/4n)} = \frac{4}{\pi}$

In Problems 27 and 28 let P be a partition of the indicated interval and x_k^* a number in the kth subinterval. Write the given sum as a definite integral.

27. $\displaystyle\lim_{\|P\| \to 0} \sum_{k=1}^n \sqrt{9 + (x_k^*)^2}\ \Delta x_k;\ [-2, 4]$

28. $\displaystyle\lim_{\|P\| \to 0} \sum_{k=1}^n (\tan x_k^*)\ \Delta x_k;\ \left[0, \frac{\pi}{4}\right]$

In Problems 29 and 30 let P be a regular partition of the indicated interval and x_k^* the right endpoint of each subinterval. Write the given sum as a definite integral.

29. $\displaystyle\lim_{n \to \infty} \sum_{k=1}^n \frac{4k + 2n}{n^2};\ [0, 2]$

30. $\displaystyle\lim_{n \to \infty} \sum_{k=1}^n \frac{3}{n + 3k};\ [1, 4]$

Calculator Problem

31. Find an approximate value for $\int_0^1 (x^3 + 1)^{1/2}\ dx$ by using a Riemann sum, a regular partition of $[0, 1]$ into five subintervals, and x_k^* the right endpoint of each subinterval.

Computer Problem

32. Use the BASIC program of this section to obtain approximate values for $\int_0^1 (x^3 + 1)^{1/2}\ dx$ by choosing $n = 500$ and $n = 1000$.

Miscellaneous Problems

33. Consider the function defined for all x in the interval $[-1, 1]$:

$$f(x) = \begin{cases} 0, & x \text{ rational} \\ 1, & x \text{ irrational} \end{cases}$$

Show that $\int_{-1}^1 f(x)\ dx$ does not exist. (*Hint*: The result in Problem 22 may be useful.)

[5.5.2]

34. Consider the discontinuous function defined for all x in the interval $[0, 2]$:

$$f(x) = \begin{cases} 0, & x \ne 1 \\ 1, & x = 1 \end{cases}$$

Show that f is integrable on the interval and $\int_0^2 f(x)\ dx = 0$. [*Hint*: Use (5.23).]

5.6 Properties of the Definite Integral

The following two definitions are useful when working with definite integrals.

DEFINITION 5.4   If $f(a)$ exists, then

$$\int_a^a f(x)\ dx = 0$$

DEFINITION 5.5   If f is integrable on $[a, b]$, then

$$\int_b^a f(x)\ dx = -\int_a^b f(x)\ dx$$

In the definition of $\int_a^b f(x)\ dx$ it was assumed that $a < b$, and so the usual "direction" of definite integration is left to right. Definition 5.5 states that reversing this direction of integration results in the negative of the integral.

Example 1

By Definition 5.4,

$$\int_1^1 (x^3 + 3x)\ dx = 0$$

Example 2

In Example 3 of Section 5.5 we saw that $\int_{-2}^1 x^3\ dx = -\frac{15}{4}$. It follows from Definition 5.5 that

$$\int_1^{-2} x^3\ dx = -\int_{-2}^1 x^3\ dx$$

$$= -\left(-\frac{15}{4}\right) = \frac{15}{4}$$

The next theorem gives some of the basic properties of the definite integral. These properties are analogous to the properties of the sigma notation given in Theorem 5.5 as well as the properties of the indefinite integral, or antiderivative, discussed in Section 5.1.

THEOREM 5.8   Let f and g be integrable functions on $[a, b]$. Then,

(*i*) $\int_a^b k\, f(x)\ dx = k \int_a^b f(x)\ dx$, where k is any constant

(*ii*) $\int_a^b [f(x) \pm g(x)]\ dx = \int_a^b f(x)\ dx \pm \int_a^b g(x)\ dx$

We note that Theorem 5.8(*ii*) extends to any finite sum of integrable functions on the interval.

The independent variable x in a definite integral is called a **dummy variable** of integration. The value of the integral does not depend on the symbol used. In other words,

$$\int_a^b f(x)\ dx = \int_a^b f(r)\ dr = \int_a^b f(s)\ ds = \int_a^b f(t)\ dt$$

and so on.

Example 3

$$\int_{-2}^1 x^3\ dx = \int_{-2}^1 r^3\ dr = \int_{-2}^1 t^3\ dt = -\frac{15}{4}$$

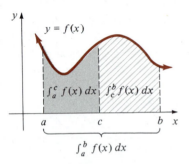

Figure 5.17

THEOREM 5.9   Let f be an integrable function on $[a, b]$. If c is any number in $[a, b]$, then

$$\int_a^b f(x)\ dx = \int_a^c f(x)\ dx + \int_c^b f(x)\ dx$$

It is easy to interpret the additive property given in Theorem 5.9 in the special case when f is continuous on $[a, b]$ and $f(x) \geq 0$ for all x in the interval. As seen in Figure 5.17 the area under the graph of f on $[a, c]$ plus the area under the graph on $[c, b]$ is the same as the area under the graph on the entire interval $[a, b]$.

Note: The conclusion of Theorem 5.9 holds when a, b, and c are *any* three numbers in some closed interval. In other words, it is not necessary to have $a < c < b$.

For a given partition P of an interval $[a, b]$, intuitively, it seems that the sum

$$\lim_{\|P\| \to 0} \sum_{k=1}^n \Delta x_k$$

is simply the length of the interval, $b - a$. Thus, we have the following. (See Problem 22 in Exercises 5.5.)

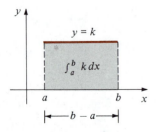

Figure 5.18

THEOREM 5.10   For any constant k,

$$\int_a^b k\ dx = k \int_a^b dx$$
$$= k(b - a)$$

If $k > 0$, then Theorem 5.10 implies that $\int_a^b k\ dx$ is simply the area of a rectangle of width $b - a$ and height k. See Figure 5.18.

_____ **Example 4** _____

From Theorem 5.10,

$$\int_2^8 5\, dx = 5 \int_2^8 dx$$

$$= 5(8 - 2) = 30$$

_____ **Example 5** _____

Evaluate $\int_{-2}^1 (x^3 + 4)\, dx$.

Solution   From Theorem 5.8(ii) we can write

$$\int_{-2}^1 (x^3 + 4)\, dx = \int_{-2}^1 x^3\, dx + \int_{-2}^1 4\, dx$$

Now from Theorem 5.10 and the known result

$$\int_{-2}^1 x^3\, dx = -\frac{15}{4}$$

it follows that

$$\int_{-2}^1 (x^3 + 4)\, dx = \left(-\frac{15}{4}\right) + 4[1 - (-2)]$$

$$= 12 - \frac{15}{4} = \frac{33}{4}$$

Finally, when $f(x) \geq 0$ on $[a, b]$ the following result is not surprising.

THEOREM 5.11   Let f be integrable on $[a, b]$ and $f(x) \geq 0$ for all x in $[a, b]$, then

$$\int_a^b f(x)\, dx \geq 0$$

_____ **Exercises 5.6** _____

Answers to odd-numbered problems begin on page A-28.

In Problems 1–40 use the definitions and theorems of this section to evaluate the given definite integral. Where appropriate use the known results

$$\int_{-1}^3 x^3\, dx = 20, \qquad \int_{-1}^3 x^2\, dx = \frac{28}{3}, \qquad \int_{-1}^3 x\, dx = 4,$$

$$\int_0^{\pi/3} \sin x\, dx = \frac{1}{2}, \qquad \text{and} \qquad \int_0^{\pi/2} \cos x\, dx = 1$$

1. $\int_3^6 4\, dx$

2. $\int_{-2}^5 (-2)\, dx$

3. $\int_{-1}^1 5\, dx$

4. $\int_{-3}^{-3} 6\, dx$

5. $\int_1^{-2} \left(\frac{1}{2}\right) dx$

6. $\int_4^{-4} (-7)\, dx$

7. $\int_{3}^{-1} x^3 \, dx$

8. $-\int_{3}^{-1} 10x \, dx$

9. $\int_{-1}^{3} 6x^2 \, dx$

10. $\int_{-1}^{3} \frac{1}{5}x^3 \, dx$

11. $\int_{-1}^{3} (-x)^3 \, dx$

12. $\int_{-1}^{3} (2x)^3 \, dx$

13. $\int_{-1}^{3} (x^3 + 1) \, dx$

14. $\int_{-1}^{3} (x^3 - 2) \, dx$

15. $\int_{-1}^{3} (2x^3 - 10) \, dx$

16. $\int_{-1}^{3} \left(5 - \frac{1}{4}x^3\right) \, dx$

17. $\int_{-1}^{3} t^3 \, dt$

18. $\int_{-1}^{3} (u^3 + u) \, du$

19. $\int_{-1}^{3} (-x^3 + 2x - 11) \, dx$

20. $\int_{-1}^{3} x(x - 1)(x + 2) \, dx$

21. $\int_{-1}^{3} (x + 1)^3 \, dx$

22. $\int_{-1}^{3} (t + 2)(t^2 - 2t + 4) \, dt$

23. $\int_{-1}^{0} x^3 \, dx + \int_{0}^{3} x^3 \, dx$

24. $\int_{-1}^{2} x^3 \, dx + \int_{2}^{3} x^3 \, dx$

25. $\int_{-1}^{1/2} 2t \, dt - \int_{3}^{1/2} 2t \, dt$

26. $\int_{-1}^{1} (5u^3 + 1) \, du - \int_{3}^{1} (5u^3 + 1) \, du$

27. $\int_{0}^{3} x^3 \, dx + \int_{3}^{0} x^3 \, dx$

28. $\int_{-1}^{-1} x^3 \, dx + \int_{-1}^{3} (x^3 - 4) \, dx$

29. $\int_{0}^{\pi/3} 6 \sin \theta \, d\theta$

30. $\int_{0}^{\pi/3} (9 - 2 \sin x) \, dx$

31. $\int_{0}^{\pi/4} \sin x \, dx + \int_{\pi/4}^{\pi/3} \sin x \, dx$

32. $\int_{1}^{5} \cos^2 x \, dx + \int_{1}^{5} \sin^2 x \, dx$

33. $\int_{0}^{\pi/2} \cos^2 \frac{x}{2} \, dx$

34. $\int_{0}^{\pi/3} \frac{\sin 2x}{\cos x} \, dx$

35. $\int_{\pi/4}^{\pi/4} \frac{dt}{\sin t}$

36. $\int_{0}^{\pi/3} (1 + \cos x) \, dx + \int_{\pi/3}^{\pi/2} \cos x \, dx$

37. $\int_{0}^{4} x \, dx + \int_{0}^{4} (9 - x) \, dx$

38. $\int_{-1}^{0} t^2 \, dt + \int_{0}^{2} x^2 \, dx + \int_{2}^{3} u^2 \, du$

39. $\int_{\pi/2}^{0} 10 \cos(t + \pi) \, dt$

40. $\int_{0}^{\pi/3} (1 + \cos x \tan x) \, dx$

Miscellaneous Problems

41. If f is integrable on $[a, b]$, then so is f^2. Show that $\int_{a}^{b} [f(x)]^2 \, dx \geq 0$.

42. If f and g are integrable on $[a, b]$ and $f(x) \geq g(x)$ for all x in $[a, b]$, show that $\int_{a}^{b} f(x) \, dx \geq \int_{a}^{b} g(x) \, dx$. (*Hint*: Consider $f(x) - g(x)$ and Theorem 5.11.)

In Problems 43 and 44, use Problem 42 to establish the given result.

43. $\int_{0}^{1} x^3 \, dx \leq \int_{0}^{1} x^2 \, dx$

44. $\int_{0}^{\pi/4} (\cos x - \sin x) \, dx \geq 0$

45. By using areas, Figure 5.19 illustrates the plausibility of the following **comparison property**.

If f is continuous on $[a, b]$ and $m \leq f(x) \leq M$ for all x in the interval, then

$$m(b - a) \leq \int_{a}^{b} f(x) \, dx \leq M(b - a)$$

Use this property to show

$$1 \leq \int_{0}^{1} (x^3 + 1)^{1/2} \, dx \leq 1.42$$

(See Problem 31, Exercises 5.5.)

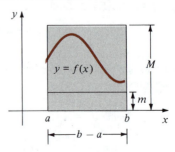

Figure 5.19

5.7 The Fundamental Theorem of Calculus

At the end of Section 5.5, we indicated that there is an easier way of evaluating a definite integral than by computing a limit of a sum. This "easier way" is by means of the so-called *Fundamental Theorem of Calculus*. In this theorem we shall see that the concept of an antiderivative of a continuous function provides the bridge between the differential calculus and the integral calculus.

Fundamental Theorem of Calculus—First Form

Suppose f is continuous and $f(t) \geq 0$ for all t on some interval $[a, b]$. Thus, the integral

$$\int_a^b f(t)\ dt \tag{5.24}$$

exists and represents the area under the graph of f on the interval. Now, if x is any number in $[a, b]$, then the function

$$A(x) = \int_a^x f(t)\ dt \tag{5.25}$$

gives the area under the graph on the interval $[a, x]$. See Figure 5.20. If $\Delta x > 0$, then

$$A(x + \Delta x) = \int_a^{x+\Delta x} f(t)\ dt \tag{5.26}$$

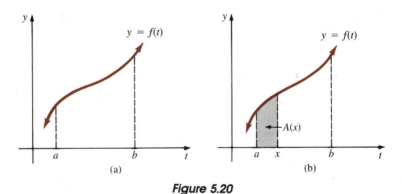

Figure 5.20

is the area indicated in Figure 5.21(a), whereas the difference

$$A(x + \Delta x) - A(x) \tag{5.27}$$

is the area shown in Figure 5.21(b).

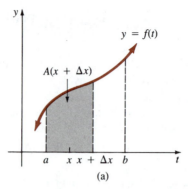
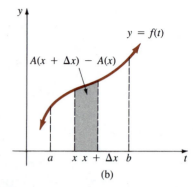

(a)   (b)

Figure 5.21

Since f is continuous on the closed interval $[x, x + \Delta x]$, we know from the Extreme Value Theorem, Theorem 4.1, that it attains a minimum value m and a maximum value M on the interval. As suggested in Figure 5.22, the difference (5.27) is bracketed between the numbers $m\Delta x$ and $M\Delta x$:

$$m\Delta x \le A(x + \Delta x) - A(x) \le M\Delta x*$$

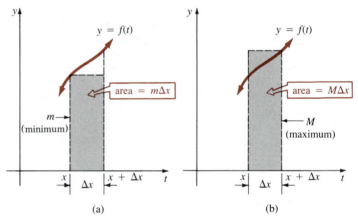

(a)   (b)

Figure 5.22

Dividing by Δx then gives

$$m \le \frac{A(x + \Delta x) - A(x)}{\Delta x} \le M$$

Using the fact that $\lim_{\Delta x \to 0^+} m = \lim_{\Delta x \to 0^+} M = f(x)$, it follows that

$$\lim_{\Delta x \to 0^+} \frac{A(x + \Delta x) - A(x)}{\Delta x} = f(x) \tag{5.28}$$

---

*Although, for the sake of illustration, we have shown f to be an increasing function in Figure 5.22, the inequalities hold for any function continuous on $[x, x + \Delta x]$.

From the definition of a derivative, we can write (5.28) as

$$A'(x) = f(x) \tag{5.29}$$

A similar argument holds for $\Delta x < 0$.

The foregoing discussion suggests the following theorem known as the first form of the **Fundamental Theorem of Calculus**.

THEOREM 5.12   **Fundamental Theorem of Calculus— Derivative Form**

Let f be continuous on $[a, b]$ and let x be any number in the interval. If $G(x)$ is a function defined by

$$G(x) = \int_a^x f(t)\, dt \quad \text{then} \quad G'(x) = f(x)$$

Here we have chosen to denote the integral $\int_a^x f(t)\, dt$ by $G(x)$, since in the context of Theorem 5.12, the function f need not be nonnegative on $[a, b]$ and so the integral need not represent a measure of area as did $A(x)$ in (5.25).

Example 1

(a) $\dfrac{d}{dx} \displaystyle\int_{-2}^x t^3\, dt = x^3$ (b) $\dfrac{d}{dx} \displaystyle\int_1^x \sqrt{t^2 + 1}\, dt = \sqrt{x^2 + 1}$

Fundamental Theorem of Calculus—Second Form

For a continuous function f, the statement $G'(x) = f(x)$ for $G(x) = \int_a^x f(t)\, dt$, means that $G(x)$ is an antiderivative of the integrand. If F is any antiderivative of f, we know from Theorem 5.1 that $G(x) - F(x) = C$ or $G(x) = F(x) + C$, where C is any arbitrary constant. It follows for any x in $[a, b]$ that

$$F(x) + C = \int_a^x f(t)\, dt \tag{5.30}$$

If we substitute $x = a$ in (5.30), then

$$F(a) + C = \int_a^a f(t)\, dt$$

implies $C = -F(a)$ since $\int_a^a f(t)\, dt = 0$. Thus, (5.30) becomes

$$F(x) - F(a) = \int_a^x f(t)\, dt$$

Since the latter equation is valid at $x = b$, we find

$$F(b) - F(a) = \int_a^b f(t)\, dt \tag{5.31}$$

We have proved an exceedingly useful second form of the **Fundamental Theorem of Calculus**.

> **THEOREM 5.13   Fundamental Theorem of Calculus—**
> **Antiderivative Form**
>
> Let f be continuous on $[a, b]$ and let F be any function for which $F'(x) = f(x)$. Then,
>
> $$\int_a^b f(x)\ dx = F(b) - F(a) \qquad (5.32)$$

The difference in (5.32) is usually written

$$F(x)\Big]_a^b$$

That is, $$\underbrace{\int_a^b f(x)\ dx}_{\substack{\text{definite}\\ \text{integral}}} = \underbrace{\int f(x)\ dx\Big]_a^b}_{\substack{\text{indefinite}\\ \text{integral or}\\ \text{antiderivative}}} = F(x)\Big]_a^b$$

Since Theorem 5.13 indicates that we may use *any* antiderivative, we may always choose the constant of integration C to be zero. Observe if $C \neq 0$, then

$$(F(x) + C)\Big]_a^b = (F(b) + C) - (F(a) + C) = F(b) - F(a) = F(x)\Big]_a^b$$

Alternative Proof of Theorem 5.13   It is worthwhile to examine yet another proof of Theorem 5.13 using the basic premise that a definite integral is a limit of a sum. If F is an antiderivative of f, then $F'(x) = f(x)$. Since F is differentiable on (a, b), the Mean Value Theorem for derivatives guarantees that there exists an x_k^* in each subinterval (x_{k-1}, x_k) of the partition P: $a = x_0 < x_1 < x_2 < \cdots < x_{n-1} < x_n = b$

such that $$F(x_k) - F(x_{k-1}) = F'(x_k^*)(x_k - x_{k-1})$$
$$= F'(x_k^*)\Delta x_k$$
$$= f(x_k^*)\ \Delta x_k$$

Now,

$$F(b) - F(a) = F(x_n) - F(x_0)$$
$$= F(x_n) + \underbrace{F(x_1) - F(x_1) + F(x_2) - F(x_2) + \cdots + F(x_{n-1}) - F(x_{n-1})}_{\text{zero}} - F(x_0)$$

$$= [F(x_1) - F(x_0)] + [F(x_2) - F(x_1)] + \cdots + [F(x_n) - F(x_{n-1})]$$
$$= \sum_{k=1}^n [F(x_k) - F(x_{k-1})]$$
$$= \sum_{k=1}^n F'(x_k^*)\ \Delta x_k$$
$$= \sum_{k=1}^n f(x_k^*)\ \Delta x_k \qquad (5.33)$$

But, $\lim_{\|P\|\to 0} [F(b) - F(a)] = F(b) - F(a)$ and so the limit of (5.33) as $\|P\| \to 0$ is

$$F(b) - F(a) = \lim_{\|P\|\to 0} \sum_{k=1}^{n} f(x_k^*) \, \Delta x_k \qquad (5.34)$$

From Definition 5.3, the right-hand member of (5.34) is $\int_a^b f(x) \, dx$.

Example 2

In Example 3 of Section 5.5 we resorted to the rather lengthy definition of the definite integral to show

$$\int_{-2}^{1} x^3 \, dx = -\frac{15}{4}$$

Since $F(x) = x^4/4$ is an antiderivative of $f(x) = x^3$, we now obtain immediately

$$\int_{-2}^{1} x^3 \, dx = \frac{x^4}{4}\Bigg]_{-2}^{1}$$

$$= \frac{1}{4} - \frac{(-2)^4}{4}$$

$$= \frac{1}{4} - \frac{16}{4} = -\frac{15}{4}$$

Example 3

Evaluate $\int_1^3 x \, dx$.

Solution An antiderivative of $f(x) = x$ is $F(x) = x^2/2$. Consequently, Theorem 5.13 gives

$$\int_1^3 x \, dx = \frac{x^2}{2}\Bigg]_1^3$$

$$= \frac{9}{2} - \frac{1}{2} = 4$$

Example 4

Evaluate $\int_{-2}^{2} (3x^2 - x + 1) \, dx$.

Solution   We apply (5.1) of Section 5.1 to each term of the integrand and then use the Fundamental Theorem:

$$\int_{-2}^{2} (3x^2 - x + 1)\, dx = \left(x^3 - \frac{x^2}{2} + x \right) \Big]_{-2}^{2}$$

$$= (8 - 2 + 2) - (-8 - 2 - 2) = 20$$

Example 5

Evaluate $\displaystyle\int_{\pi/6}^{\pi} \cos x\, dx$.

Solution   An antiderivative of $f(x) = \cos x$ is $F(x) = \sin x$. Therefore,

$$\int_{\pi/6}^{\pi} \cos x\, dx = \sin x \Big]_{\pi/6}^{\pi}$$

$$= \sin \pi - \sin \frac{\pi}{6}$$

$$= 0 - \frac{1}{2} = -\frac{1}{2}$$

Example 6

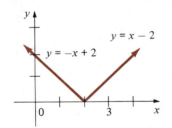

Figure 5.23

Evaluate $\displaystyle\int_{0}^{3} |x - 2|\, dx$.

Solution   From the definition of absolute value we know

$$|x - 2| = \begin{cases} x - 2 & \text{if } x - 2 \geq 0 \\ -(x - 2) & \text{if } x - 2 < 0 \end{cases} \quad \text{or} \quad |x - 2| = \begin{cases} x - 2 & \text{if } x \geq 2 \\ -x + 2 & \text{if } x < 2 \end{cases}$$

The graph of $f(x) = |x - 2|$ is given in Figure 5.23.
Now in view of Theorem 5.9 we can write

$$\int_{0}^{3} |x - 2|\, dx = \int_{0}^{2} |x - 2|\, dx + \int_{2}^{3} |x - 2|\, dx$$

$$= \int_{0}^{2} (-x + 2)\, dx + \int_{2}^{3} (x - 2)\, dx$$

$$= \left(-\frac{x^2}{2} + 2x \right) \Big]_{0}^{2} + \left(\frac{x^2}{2} - 2x \right) \Big]_{2}^{3}$$

$$= (-2 + 4) + \left(\frac{9}{2} - 6 \right) - (2 - 4) = \frac{5}{2}$$

Substitution in a Definite Integral

Recall from Section 5.2 that we sometimes used a substitution as an aid in evaluating an indefinite integral of the form $\int f(g(x))g'(x)\, dx$. Care should be exercised when using a substitution in a definite integral $\int_a^b f(g(x))g'(x)\, dx$ since we can proceed in *two ways*:

(*i*)   Evaluate the definite integral by means of the substitution $u = g(x)$. Resubstitute $u = g(x)$ in the antiderivative, and then apply the Fundamental Theorem of Calculus by using the original limits of integration $x = a$ and $x = b$.

(*ii*)   Alternatively, the resubstitution can be avoided by changing the limits of integration to correspond to the value of u at $x = a$ and u at $x = b$.

The latter method, which is usually quicker, is summarized in the next theorem.

THEOREM 5.14   Let $u = g(x)$ be a function that has a continuous derivative on the interval $[a, b]$, and let f be a function that is continuous on the range of g. If $F'(u) = f(u)$ and $c = g(a)$, $d = g(b)$, then

$$\int_a^b f(g(x))g'(x)\, dx = F(d) - F(c)$$

Proof
$$\int_a^b f(g(x))g'(x)\, dx = \int_c^d f(u)\frac{du}{dx}\, dx$$
$$= \int_c^d f(u)\, du$$
$$= F(u)\Big]_c^d = F(d) - F(c) \quad\blacksquare$$

Example 7

Evaluate $\int_0^2 \sqrt{2x^2 + 1}\, x\, dx$.

Solution   We shall illustrate the procedure outlined in (*i*) above. To evaluate the indefinite integral $\int \sqrt{2x^2 + 1}\, x\, dx$, we use

$$u = 2x^2 + 1 \quad\text{and}\quad du = 4x\, dx$$

Thus, $\int \sqrt{2x^2 + 1}\, x\, dx = \frac{1}{4}\int \sqrt{2x^2 + 1}\,(4x\, dx)$ [Substitution]

$$= \frac{1}{4}\int u^{1/2}\, du$$
$$= \frac{1}{4}\frac{u^{3/2}}{3/2} + C \quad\text{[Resubstitution]}$$
$$= \frac{1}{6}(2x^2 + 1)^{3/2} + C$$

Therefore, by Theorem 5.13,

$$\int_0^2 \sqrt{2x^2 + 1} \; x \; dx = \frac{1}{6}(2x^2 + 1)^{3/2}\Big]_0^2$$

$$= \frac{1}{6}[9^{3/2} - 1^{3/2}]$$

$$= \frac{1}{6}[27 - 1] = \frac{13}{3}$$

The second procedure is illustrated in the following example.

Example 8

In Example 7, if $u = 2x^2 + 1$, notice that $x = 0$ implies $u = 1$ and that $x = 2$ gives $u = 9$. Thus, by Theorem 5.14,

$$\int_0^2 \sqrt{2x^2 + 1} \; x \; dx = \frac{1}{4} \int_1^9 u^{1/2} \; du \qquad \begin{array}{l} u \text{ limits:} \\ \text{integration with} \\ \text{respect to } u \end{array}$$

$$= \frac{1}{4}\frac{u^{3/2}}{3/2}\Big]_1^9$$

$$= \frac{1}{6}[9^{3/2} - 1^{3/2}] = \frac{13}{3}$$

When the graph of a function $y = f(x)$ is symmetric with respect to either the y-axis (even function) or the origin (odd function) on a symmetric interval $[-a, a]$, then the definite integral $\int_{-a}^a f(x) \; dx$ can be evaluated by means of a "shortcut."

THEOREM 5.15   Even Function Rule

Let f be an even integrable function on $[-a, a]$. Then

$$\int_{-a}^a f(x) \; dx = 2 \int_0^a f(x) \; dx$$

THEOREM 5.16   Odd Function Rule

Let f be an odd integrable function on $[-a, a]$. Then

$$\int_{-a}^a f(x) \; dx = 0$$

While we shall leave the proofs of these two theorems as exercises, the geometric motivations for the results are given in Figure 5.24. The point in Theorem 5.16 is this: When we integrate an odd integrable function f on a symmetric interval $[-a, a]$, there is no need to find an antiderivative of f in order to utilize Theorem 5.13; the value of the integral is always zero.

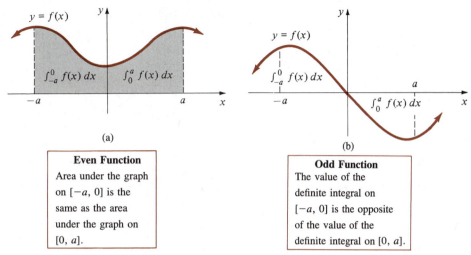

Figure 5.24

Even Function
Area under the graph on $[-a, 0]$ is the same as the area under the graph on $[0, a]$.

Odd Function
The value of the definite integral on $[-a, 0]$ is the opposite of the value of the definite integral on $[0, a]$.

Example 9

Evaluate $\displaystyle\int_{-1}^{1} (x^4 + x^2) \, dx$.

Solution   Since $f(-x) = (-x)^4 + (-x)^2 = x^4 + x^2 = f(x)$, the integrand is an even function on the symmetric interval $[-1, 1]$. It follows from Theorem 5.15 that

$$\int_{-1}^{1} (x^4 + x^2) \, dx = 2 \int_{0}^{1} (x^4 + x^2) \, dx$$

$$= 2 \left(\frac{x^5}{5} + \frac{x^3}{3} \right) \Bigg]_{0}^{1}$$

$$= 2 \left(\frac{1}{5} + \frac{1}{3} \right) = \frac{16}{15}$$

Example 10

Evaluate $\displaystyle\int_{-\pi/2}^{\pi/2} \sin x \, dx$.

Solution   In this case $f(x) = \sin x$ is an odd function on the symmetric interval $[-\pi/2, \pi/2]$. Thus, by Theorem 5.16 we have immediately

$$\int_{-\pi/2}^{\pi/2} \sin x \, dx = 0$$

Piecewise Continuous Functions

A function f is said to be **piecewise continuous** on an interval $[a, b]$ if there are at most a finite number of points c_k, $k = 1, 2, \ldots, n$, $(c_{k-1} < c_k)$ at which f has a finite, or jump, discontinuity and f is continuous on each open interval (c_{k-1}, c_k). See Figure 5.25.

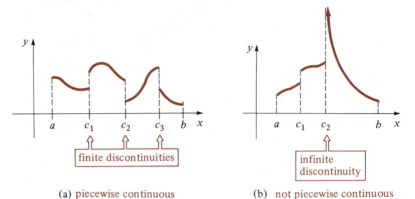

(a) piecewise continuous   (b) not piecewise continuous

Figure 5.25

By the criterion on page 281, a piecewise continuous function is integrable. A definite integral of a piecewise continuous function on $[a, b]$ can be evaluated with the help of Theorem 5.9:

$$\int_a^b f(x)\,dx = \int_a^{c_1} f(x)\,dx + \int_{c_1}^{c_2} f(x)\,dx + \cdots + \int_{c_n}^b f(x)\,dx$$

and by simply treating the integrands of the definite integrals on the right side of the above equation as if they were continuous on the closed intervals $[a, c_1]$, $[c_1, c_2], \ldots, [c_n, b]$.

Example 11

Evaluate $\displaystyle\int_{-1}^4 f(x)\,dx$, where $f(x) = \begin{cases} x + 1, & -1 \le x < 0 \\ x, & 0 \le x < 2 \\ 3, & 2 \le x \le 4 \end{cases}$

Solution   The graph of the piecewise continuous function f is given in Figure 5.26. Now, from the preceding discussion,

$$\int_{-1}^4 f(x)\,dx = \int_{-1}^0 f(x)\,dx + \int_0^2 f(x)\,dx + \int_2^4 f(x)\,dx$$

$$= \int_{-1}^0 (x + 1)\,dx + \int_0^2 x\,dx + \int_2^4 3\,dx$$

$$= \left(\frac{x^2}{2} + x\right)\Bigg]_{-1}^0 + \frac{x^2}{2}\Bigg]_0^2 + 3x\Bigg]_2^4$$

$$= \frac{17}{2}$$

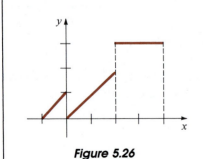

Figure 5.26

Remark

The antiderivative form of the Fundamental Theorem of Calculus is an extremely important and powerful tool for evaluating definite integrals. Why should we bother with a clumsy limit of a sum when the value of $\int_a^b f(x)\,dx$ can be found by computing $\int f(x)\,dx$ at the two numbers a and b? This is true up to a point—however, it is time to learn another fact of mathematical life: There are functions for which the antiderivative $\int f(x)\,dx$ *cannot* be expressed in terms of *elementary functions*.* The simple continuous function $f(x) = \sqrt{x^3 + 1}$ possesses no anti-derivative that is an elementary function. Although, in view of Theorem 5.6, we can say that the definite integral $\int_0^1 \sqrt{x^3 + 1}\,dx$ exists, Theorem 5.13 provides no help in finding its value. We will consider problems of this sort in the next section. (See Problem 70 in Exercises 5.7.)

Exercises 5.7

Answers to odd-numbered problems begin on page A-28.

In Problems 1–44 evaluate the given expression.

1. $\dfrac{d}{dx}\displaystyle\int_0^x (3t^2 - 2t)\,dt$

2. $\dfrac{d}{dx}\displaystyle\int_x^9 \sqrt{u + 2}\,du$

3. $\displaystyle\int_3^7 dx$

4. $\displaystyle\int_2^{10} (-4)\,dx$

5. $\displaystyle\int_{-1}^2 (2x + 3)\,dx$

6. $\displaystyle\int_{-5}^4 t^2\,dt$

7. $\displaystyle\int_1^3 (6x^2 - 4x + 5)\,dx$

8. $\displaystyle\int_{-2}^1 (12x^5 - 36)\,dx$

9. $\displaystyle\int_0^{\pi/2} \sin x\,dx$

10. $\displaystyle\int_{-\pi/3}^{\pi/4} \cos\theta\,d\theta$

11. $\displaystyle\int_{1/2}^{3/4} \dfrac{1}{u^2}\,du$

12. $\displaystyle\int_{1/4}^1 \dfrac{2}{\sqrt{x}}\,dx$

13. $\displaystyle\int_{-1}^1 \sqrt[3]{x}\,dx$

14. $\displaystyle\int_0^8 \sqrt[3]{t^2}\,dt$

15. $\displaystyle\int_0^2 x(1 - x)\,dx$

16. $\displaystyle\int_3^2 x(x - 2)(x + 2)\,dx$

17. $\displaystyle\int_{-1}^1 (7x^3 - 2x^2 + 5x - 4)\,dx$

18. $\displaystyle\int_{-3}^{-1} (x^2 - 4x + 8)\,dx$

19. $\displaystyle\int_1^4 \dfrac{x - 1}{\sqrt{x}}\,dx$

20. $\displaystyle\int_2^4 \dfrac{x^2 + 8}{x^2}\,dx$

21. $\displaystyle\int_0^5 (2 - \sqrt{x})^2\,dx + \int_5^9 (2 - \sqrt{x})^2\,dx$

22. $\displaystyle\int_{-2}^0 (x + 1)^2\,dx + \int_0^3 (x + 1)^2\,dx$

23. $\displaystyle\int_{-4}^2 \left(\dfrac{x}{2} + 1\right)^5\,dx$

24. $\displaystyle\int_1^3 \dfrac{1}{x^2 + 10x + 25}\,dx$

25. $\displaystyle\int_{-4}^{12} \sqrt{z + 4}\,dz$

26. $\displaystyle\int_0^{7/2} (2x + 1)^{-1/3}\,dx$

27. $\displaystyle\int_0^3 \dfrac{x}{\sqrt{x^2 + 16}}\,dx$

28. $\displaystyle\int_{-2}^1 \dfrac{t}{(t^2 + 1)^2}\,dt$

29. $\displaystyle\int_{1/2}^1 \left(1 + \dfrac{1}{x}\right)^3 \dfrac{1}{x^2}\,dx$

30. $\displaystyle\int_1^4 \dfrac{\sqrt[3]{1 + 4\sqrt{x}}}{\sqrt{x}}\,dx$

31. $\displaystyle\int_0^1 \dfrac{x + 1}{\sqrt{x^2 + 2x + 3}}\,dx$

32. $\displaystyle\int_{-1}^1 \dfrac{u^3 + u}{(u^4 + 2u^2 + 1)^5}\,du$

33. $\displaystyle\int_0^{\pi/8} \sec^2 2x\,dx$

34. $\displaystyle\int_{\sqrt{\pi/2}}^{\sqrt{\pi/2}} x\csc x^2 \cot x^2\,dx$

35. $\displaystyle\int_{-1/2}^{3/2} (x - \cos\pi x)\,dx$

36. $\displaystyle\int_1^4 \dfrac{\cos\sqrt{x}}{2\sqrt{x}}\,dx$

---

*The *elementary functions* that we have seen so far are sums, products, quotients, and powers of polynomial and trigonometric functions. Later on we shall add to this list the inverse trigonometric functions and the logarithmic and exponential functions.

37. $\displaystyle\int_0^{\pi/2} \sqrt{\cos x}\, \sin x\, dx$ **38.** $\displaystyle\int_{\pi/6}^{\pi/3} \sin x \cos x\, dx$

39. $\displaystyle\int_{\pi/6}^{\pi/2} \frac{1 + \cos\theta}{(\theta + \sin\theta)^2}\, d\theta$

40. $\displaystyle\int_{-\pi/4}^{\pi/4} (\sec x + \tan x)^2\, dx$

41. $\displaystyle\int_{-2}^{2} \sin 3x\, dx$ **42.** $\displaystyle\int_{-\pi/2}^{\pi/2} \cos^2 x\, dx$

43. $\displaystyle\int_{-5}^{5} (x^3 + x)^9\, dx$ **44.** $\displaystyle\int_{-1}^{1} \tan x\, dx$

In Problems 45–50 evaluate the given definite integral.

45. $\displaystyle\int_{-3}^{1} |x|\, dx$ **46.** $\displaystyle\int_0^4 |2x - 6|\, dx$

47. $\displaystyle\int_{-8}^{3} \sqrt{|x| + 1}\, dx$ **48.** $\displaystyle\int_0^2 |x^2 - 1|\, dx$

49. $\displaystyle\int_{-\pi}^{\pi} |\sin x|\, dx$ **50.** $\displaystyle\int_0^{\pi} |\cos x|\, dx$

In Problems 51 and 52 evaluate $\int_{-1}^{2} f(x)\, dx$ for the given function.

51. $f(x) = \begin{cases} -x, & x < 0 \\ x^2, & x \geq 0 \end{cases}$

52. $f(x) = \begin{cases} 2x + 3, & x \leq 0 \\ 3, & x > 0 \end{cases}$

In Problems 53–56 evaluate the definite integral of the given piecewise continuous function.

53. $\displaystyle\int_0^3 f(x)\, dx$, where $f(x) = \begin{cases} 4, & 0 \leq x < 2 \\ 1, & 2 \leq x \leq 3 \end{cases}$

54. $\displaystyle\int_0^{\pi} f(x)\, dx$, where $f(x) = \begin{cases} \sin x, & 0 \leq x < \pi/2 \\ \cos x, & \pi/2 < x \leq \pi \end{cases}$

55. $\displaystyle\int_{-2}^{2} f(x)\, dx$, where $f(x) = \begin{cases} x^2, & -2 \leq x < -1 \\ 4, & -1 \leq x < 1 \\ x^2, & 1 \leq x \leq 2 \end{cases}$

56. $\displaystyle\int_0^4 f(x)\, dx$, where $f(x) = [x]$ is the greatest integer function.

57. Use the substitution $u = t + 1$ to evaluate

$$\int_0^3 t\sqrt{t + 1}\, dt$$

58. Use the substitution $u = 2x + 1$ to evaluate

$$\int_0^4 \frac{x^2}{\sqrt{2x + 1}}\, dx$$

In Problems 59 and 60 evaluate the given definite integral.

59. $\displaystyle\int_{-1}^{2} \left\{ \int_1^x 12t^2\, dt \right\} dx$ **60.** $\displaystyle\int_0^{\pi/2} \left\{ \int_0^t \sin x\, dx \right\} dt$

In Problems 61 and 62 let $G(x) = \int_a^x f(t)\, dt$ and $G'(x) = f(x)$.

61. (a) What is $G(x^2)$?   (b) What is $[G(x^2)]'$?

62. (a) What is $G(x^3 + 2x)$?
    (b) What is $[G(x^3 + 2x)]'$?

In Problems 63 and 64 determine $F'(x)$.

63. $F(x) = \displaystyle\int_{6x-1}^{0} \sqrt{4t + 9}\, dt$

64. $F(x) = \displaystyle\int_{3x}^{x^2} \frac{1}{t^2 + 1}\, dt$ (*Hint:* Use two integrals.)

Miscellaneous Problems

In Problems 65 and 66 let P be a partition of the indicated interval and x_k^* a number in the kth subinterval. Determine the value of the given limit.

65. $\displaystyle\lim_{\|P\| \to 0} \sum_{k=1}^{n} (2x_k^* + 5)\, \Delta x_k;\ [-1, 3]$

66. $\displaystyle\lim_{\|P\| \to 0} \sum_{k=1}^{n} \cos\frac{x_k^*}{4}\, \Delta x_k;\ [0, 2\pi]$

In Problems 67 and 68 let P be a regular partition of the indicated interval and x_k^* a number in the kth subinterval. Establish the given result.

67. $\displaystyle\lim_{n \to \infty} \frac{\pi}{n} \sum_{k=1}^{n} \sin x_k^* = 2;\ [0, \pi]$

68. $\displaystyle\lim_{n \to \infty} \frac{2}{n} \sum_{k=1}^{n} x_k^* = 0;\ [-1, 1]$

69. Reread Theorem 5.13 and give a reason why the following procedure is *incorrect*:

$$\int_{-1}^{1} x^{-2}\, dx = -x^{-1}\Big]_{-1}^{1} = -1 + (-1) = -2$$

70. Accept (on faith) the following approximation

$$(1 + x^3)^{1/2} \approx 1 + \frac{1}{2}x^3 - \frac{1}{8}x^6 + \frac{1}{16}x^9 \qquad (5.35)$$

Use (5.35) to obtain an approximation for $\int_0^1 (x^3 + 1)^{1/2}\, dx$.

71. Prove the Even Function Rule, Theorem 5.15. (*Hint:* $\int_{-a}^{a} f(x)\, dx = \int_{-a}^{0} f(x)\, dx + \int_0^a f(x)\, dx$. Let $t = -x$ in $\int_{-a}^{0} f(x)\, dx$ and use $f(-x) = f(x)$.)

72. Prove the Odd Function Rule, Theorem 5.16. (*Hint:* Proceed as in Problem 71.)

5.8 Approximate Integration

Life in mathematics would be extremely pleasant if the antiderivative of every continuous function could be expressed in terms of elementary functions such as polynomial, rational, or trigonometric functions. This is not the case. Hence, Theorem 5.13 cannot be used to evaluate every definite integral. Sometimes the best we can do is to *approximate* the value of $\int_a^b f(x)\,dx$. In this concluding section, we shall consider two such numerical procedures.

In the following discussion it is again useful to consider the definite integral $\int_a^b f(x)\,dx$ as area under the graph of f on $[a, b]$. Although continuity of f is essential, there is no actual requirement that $f(x) \geq 0$ on the interval.

One way of approximating a definite integral is to proceed in the same manner as we did in the initial discussion about finding area under a graph—namely, construct rectangular elements and add their areas. See Figure 5.27(a). This is basically the idea of a Riemann sum. However, it seems plausible from Figure 5.27(b) that a better estimate of $\int_a^b f(x)\,dx$ can be obtained by adding the areas of trapezoids instead of the areas of rectangles.

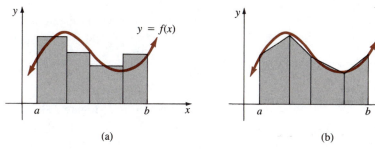

(a)   (b)

Figure 5.27

Trapezoidal Rule   The area of a trapezoid shown in Figure 5.28(a) is

$$h\frac{l_1 + l_2}{2}$$

Thus, for a trapezoidal element shown in Figure 5.28(b) its area A_k is

$$A_k = \Delta x_k \frac{f(x_{k-1}) + f(x_k)}{2}$$

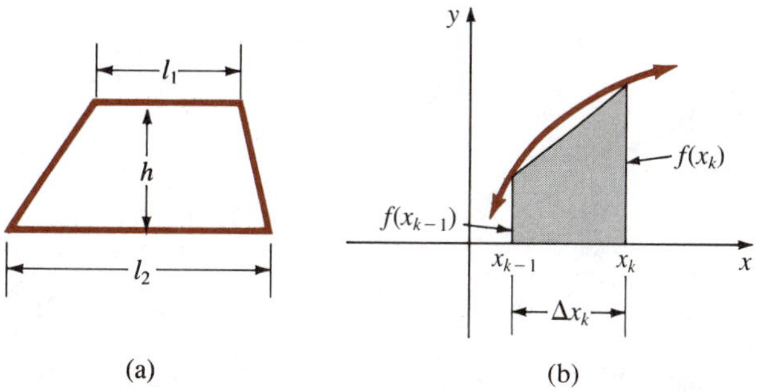

Figure 5.28

In general, for a regular partition of an interval $[a, b]$ on which a function f is continuous, the so-called **Trapezoidal Rule** is given by

$$\int_a^b f(x)\, dx \approx \Delta x \frac{f(x_0) + f(x_1)}{2} + \Delta x \frac{f(x_1) + f(x_2)}{2} + \cdots + \Delta x \frac{f(x_{n-1}) + f(x_n)}{2} \quad (5.36)$$

Since $\Delta x = (b - a)/n$, (5.36) simplifies to

$$\int_a^b f(x)\, dx \approx \frac{b - a}{2n}[f(x_0) + 2f(x_1) + 2f(x_2) + \cdots + 2f(x_{n-1}) + f(x_n)] \quad (5.37)$$

where $x_0 = a$, $x_n = b$, and $x_k = a + k\,\Delta x$, $k = 0, 1, \ldots, n$.

For instance, since the function $f(x) = 1/x$ is continuous on any interval $[a, b]$ not including the origin, we know that $\int_a^b dx/x$ exists. However, we have not seen, as yet, any function F such that $F'(x) = 1/x$; that is, we do not know an antiderivative of f.

Example 1

Approximate $\int_1^2 dx/x$ by the Trapezoidal Rule for $n = 1$, $n = 2$, and $n = 6$.

Solution   As shown in Figure 5.29, the case $n = 1$ is one trapezoid in which $\Delta x = 1$, $f(1) = 1$, and $f(2) = \frac{1}{2}$. Therefore, from (5.37),

$$\int_1^2 \frac{dx}{x} \approx \frac{1}{2}[1 + \frac{1}{2}] = 0.7500$$

When $n = 2$, Figure 5.30 shows $\Delta x = \frac{1}{2}$, $x_0 = 1$, $x_1 = 1 + \Delta x = \frac{3}{2}$, $x_2 = 1 + 2\Delta x = 2$, and $f(x_0) = 1$, $f(x_1) = \frac{2}{3}$, $f(x_2) = \frac{1}{2}$ Hence, (5.37) gives

$$\int_1^2 \frac{dx}{x} \approx \frac{1}{4}\left[1 + 2\left(\frac{2}{3}\right) + \frac{1}{2}\right] \approx 0.7083$$

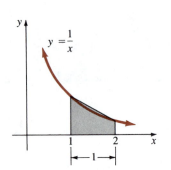

$y = \dfrac{1}{x}$

Figure 5.29

Finally, when $n = 6$,

$$\Delta x = \frac{1}{6}$$

$$x_0 = 1$$

$$x_1 = 1 + \Delta x = \frac{7}{6}$$

$$x_2 = 1 + 2\Delta x = \frac{4}{3}$$

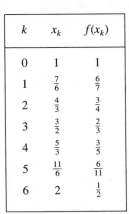

k	x_k	$f(x_k)$
0	1	1
1	$\frac{7}{6}$	$\frac{6}{7}$
2	$\frac{4}{3}$	$\frac{3}{4}$
3	$\frac{3}{2}$	$\frac{2}{3}$
4	$\frac{5}{3}$	$\frac{3}{5}$
5	$\frac{11}{6}$	$\frac{6}{11}$
6	2	$\frac{1}{2}$

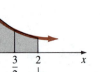

$y = \dfrac{1}{x}$

Figure 5.30

and so on. Using the information in the accompanying table, (5.37) gives

$$\int_1^2 \frac{dx}{x} \approx \frac{1}{12}\left[1 + 2\left(\frac{6}{7}\right) + 2\left(\frac{3}{4}\right) + 2\left(\frac{2}{3}\right) + 2\left(\frac{3}{5}\right) + 2\left(\frac{6}{11}\right) + \frac{1}{2}\right]$$

$$\approx 0.6949$$

Error for Trapezoidal Rule   Suppose $I = \int_a^b f(x)\,dx$ and T_n is an approximation to I using n trapezoids. We define the **error** in the method to be

$$E_n = |I - T_n|$$

An upper bound for the error can be obtained by means of the next result. The proof is omitted.

> **THEOREM 5.17**   If there exists a number $M > 0$ such that $|f''(x)| \le M$ for all x in $[a, b]$, then
>
> $$E_n \le \frac{M(b - a)^3}{12n^2} \qquad\qquad (5.38)$$

Observe that this upper bound for the error E_n is inversely proportional to n^2. Thus, if the number of trapezoids is doubled, the error E_{2n} is less than $\frac{1}{4}$ the error bound for E_n. The next example shows how (5.38) can be utilized in determining the number of trapezoids that will yield a specified accuracy.

___

Example 2

Determine a value of n so that (5.37) will give an approximation to $\int_1^2 dx/x$ that is accurate to two decimal places.

Solution   The Trapezoidal Rule will be accurate to two decimal places for those values of n for which the upper bound $M(b - a)^3/12n^2$ for the error is strictly less than 0.005.*

For $f(x) = 1/x$ we have $f''(x) = 2/x^3$. Since f'' decreases on $[1, 2]$, it follows that $f''(x) \le f''(1) = 2$ for all x in the interval. Thus, with $M = 2$, $b - a = 1$ we want

$$\frac{2(1)^3}{12n^2} < 0.005 \quad \text{or} \quad n^2 > \frac{100}{3} \approx 33$$

By taking $n \ge 6$ we obtain the desired accuracy.

___

Example 2 indicates that the third approximation $\int_1^2 dx/x \approx 0.6949$ obtained in Example 1 is accurate to two decimal places. By way of comparison, it is known that the estimate $\int_1^2 dx/x \approx 0.6931$ is correct to four decimal places.

___

Example 3

Approximate $\int_{1/2}^1 \cos \sqrt{x}\,dx$ by the Trapezoidal Rule so that the error is less than 0.001.

___

*If we want accuracy to three decimal places, we use 0.0005, and so on.

Solution   The second derivative of $f(x) = \cos \sqrt{x}$ is

$$f''(x) = \frac{1}{4x}\left(\frac{\sin \sqrt{x}}{\sqrt{x}} - \cos \sqrt{x}\right)$$

For x in the interval $[\frac{1}{2}, 1]$ we have $0 < (\sin \sqrt{x})/\sqrt{x} \le 1$ and $0 < \cos \sqrt{x} \le 1$ and consequently $|f''(x)| \le 1/4x$. Therefore, on the interval, $|f''(x)| \le \frac{1}{2}$. Thus, with $M = \frac{1}{2}$ and $b - a = \frac{1}{2}$ it follows from (5.38) that we want

$$\frac{\frac{1}{2}\left(\frac{1}{2}\right)^3}{12n^2} < 0.001 \quad \text{or} \quad n^2 > \frac{125}{24} \approx 5.21$$

Hence, to obtain the desired accuracy it suffices to choose $n = 3$ and $\Delta x = \frac{1}{6}$. With the aid of a calculator to obtain the information in the accompanying table, we find from (5.37) that

$$\int_{1/2}^{1} \cos \sqrt{x}\, dx \approx \frac{1}{12}\left[\cos \sqrt{\frac{1}{2}} + 2 \cos \sqrt{\frac{2}{3}} + 2 \cos \sqrt{\frac{5}{6}} + \cos 1\right]$$

$$\approx 0.3244$$

k	x_k	$f(x_k)$
0	$\frac{1}{2}$	0.7602
1	$\frac{2}{3}$	0.6848
2	$\frac{5}{6}$	0.6115
3	1	0.5403

Although not obvious from a figure, an improved method of approximating a definite integral $\int_a^b f(x)\, dx$ can be obtained by considering a series of parabolic arcs instead of a series of chords used in the Trapezoidal Rule. It can be proved that a parabolic arc that passes through *three* specified points will "fit" the graph of f better than a single straight line. See Figure 5.31. By adding the areas under the parabolic arcs, we obtain an approximation to the integral.

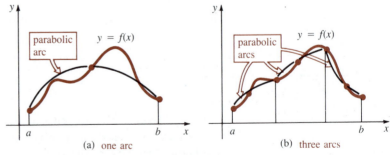

(a) one arc   (b) three arcs

Figure 5.31

To begin, let us find the area under an arc of a parabola that passes through three points $P_0(x_0, y_0)$, $P_1(x_1, y_1)$, and $P_2(x_2, y_2)$, where $x_0 < x_1 < x_2$ and $x_1 - x_0 = x_2 - x_1 = h$. As shown in Figure 5.32, this can be done by finding the area under the graph of $y = Ax^2 + Bx + C$ on the interval $[-h, h]$ so that P_0, P_1, and P_2 have coordinates $(-h, y_0)$, $(0, y_1)$, and (h, y_2), respectively. The

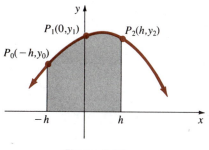

Figure 5.32

interval $[-h, h]$ is chosen for simplicity; the area in question does not depend on the location of the y-axis. Clearly,

$$\text{area} = \int_{-h}^{h} (Ax^2 + Bx + C) \, dx$$

$$= \left(A\frac{x^3}{3} + B\frac{x^2}{2} + Cx \right)\Bigg]_{-h}^{h}$$

$$= \frac{h}{3}(2Ah^2 + 6C) \tag{5.39}$$

But, since the graph is to pass through $(-h, y_0)$, $(0, y_1)$, and (h, y_2), we must have

$$y_0 = Ah^2 - Bh + C \tag{5.40}$$

$$y_1 = C \tag{5.41}$$

$$y_2 = Ah^2 + Bh + C \tag{5.42}$$

By adding (5.40) and (5.42) and using (5.41), we find $2Ah^2 = y_0 + y_2 - 2y_1$. Thus, (5.39) can be expressed as

$$\text{area} = \frac{h}{3}(y_0 + 4y_1 + y_2) \tag{5.43}$$

Simpson's Rule Now let us suppose $y = f(x)$ is continuous on $[a, b]$ and the interval is partitioned into n subintervals of equal width $\Delta x = (b - a)/n$, where *n is an even integer*. As shown in Figure 5.33, on each subinterval $[x_{k-2}, x_k]$ of width $2\,\Delta x$ we approximate the graph of f by an arc of a parabola through points P_{k-2}, P_{k-1}, and P_k on the graph that corresponds to the endpoints and midpoint of the subinterval. If A_k denotes the area under the parabola on $[x_{k-2}, x_k]$, it follows from (5.43) that

$$A_k = \frac{\Delta x}{3}[f(x_{k-2}) + 4f(x_{k-1}) + f(x_k)]$$

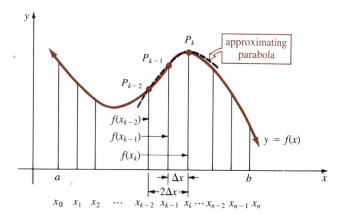

Figure 5.33

Thus, **Simpson's Rule*** consists of summing all the A_k:

$$\int_a^b f(x)\, dx \approx \frac{\Delta x}{3}[f(x_0) + 4f(x_1) + f(x_2)] + \frac{\Delta x}{3}[f(x_2) + 4f(x_3) + f(x_4)]$$

$$+ \cdots + \frac{\Delta x}{3}[f(x_{n-2}) + 4f(x_{n-1}) + f(x_n)] \qquad (5.44)$$

Using $\Delta x = (b - a)/n$, (5.44) becomes

$$\int_a^b f(x)\, dx \approx \frac{b - a}{3n}[f(x_0) + 4f(x_1) + 2f(x_2) + 4f(x_3) + \cdots + 2f(x_{n-2}) + 4f(x_{n-1}) + f(x_n)] \quad (5.45)$$

where $x_0 = a$, $x_n = b$, and $x_k = a + k\,\Delta x$, $k = 0, 1, \ldots, n$. We note again that the integer n in (5.45) must be even since each A_k represents area under a parabolic arc on a subinterval of width $2\,\Delta x$.

Error for Simpson's Rule   If $I = \int_a^b f(x)\, dx$, then, as before, the error in the method is

$$E_n = |I - S_n|$$

where S_n denotes the right-hand member of (5.45). The following theorem establishes an upper bound for E_n using an upper bound on the fourth derivative.

THEOREM 5.18   If there exists a number $M > 0$ such that $|f^{(4)}(x)| \le M$ for all x in $[a, b]$, then

$$E_n \le \frac{M(b - a)^5}{180n^4} \qquad (5.46)$$

Example 4

Determine a value of n so that (5.45) will give an approximation to $\int_1^2 dx/x$ that is accurate to two decimal places.

Solution   For $f(x) = 1/x$, $f^{(4)}(x) = 24/x^5$ and on $[1, 2]$, $f^{(4)}(x) \le f^{(4)}(1) = 24$. Thus, with $M = 24$ it follows from (5.46) that

$$\frac{24(1)^6}{180n^4} < 0.005 \quad \text{or} \quad n^4 > \frac{80}{3} \approx 26.67$$

and so $n > 2.27$. Since n must be an even integer, it suffices to take $n \ge 4$.

---

*Named after the English mathematician Thomas Simpson (1710–1761).

_____ **Example 5** _____

k	x_k	$f(x_k)$
0	1	1
1	$\frac{5}{4}$	$\frac{4}{5}$
2	$\frac{3}{2}$	$\frac{2}{3}$
3	$\frac{7}{4}$	$\frac{4}{7}$
4	2	$\frac{1}{2}$

Approximate $\int_1^2 dx/x$ by Simpson's Rule for $n = 4$.

Solution   When $n = 4$ we have $\Delta x = \frac{1}{4}$. From (5.45) and the accompanying table we obtain

$$\int_1^2 \frac{dx}{x} \approx \frac{1}{12}\left[1 + 4\left(\frac{4}{5}\right) + 2\left(\frac{2}{3}\right) + 4\left(\frac{4}{7}\right) + \frac{1}{2}\right]$$

$$\approx 0.6933$$

In Example 5, keep in mind that even though we are using $n = 4$ the function $f(x) = 1/x$ is being approximated by only two parabolic arcs. Also, recall that the Trapezoidal Rule gave $\int_1^2 dx/x \approx 0.6949$ with $n = 6$ and that 0.6931 is an estimation of the integral correct to four decimal places.

Remarks

(*i*)   Under some circumstances the Trapezoidal Rule will give the exact value of an integral $\int_a^b f(x)\,dx$. It stands to reason that the right-hand member of (5.37) will yield the precise value whenever f is a linear function. See Problem 31.

(*ii*)   In general, Simpson's Rule will yield greater accuracy than the Trapezoidal Rule. So why should we even bother with this latter rule? In some instances the slightly simpler Trapezoidal Rule gives accuracy that is sufficient for the purpose on hand. Furthermore, the requirement that n must be an even integer in Simpson's Rule may prevent its application to a given problem. See Problem 16. Also, to find an error bound for Simpson's Rule we must compute and then find an upper bound for the fourth derivative. The expression for $f^{(4)}(x)$ can, of course, be very complicated. The error bound for the Trapezoidal Rule depends on the second derivative.

(*iii*)   Since the discussion of this section is based upon fitting linear and quadratic functions to the graph of a given function f, you might question whether the next step is to try to fit the graph of f with arcs of cubic or even quartic functions. Indeed, you can. A cubic approximation would utilize four points over three intervals of equal width Δx. A quartic arc would use five points, and so on. In this manner, you generate a sequence of approximation formulas known as **Newton–Cotes formulas**. Because of increasing complexity, and other inherent problems, Newton–Cotes formulas of order higher than 2 are seldom used.

(*iv*)   Finally, it is interesting to note that Simpson's Rule will give the exact value of $\int_a^b f(x)\,dx$ whenever f is a linear, quadratic, or even a cubic polynomial function. See Problem 33.

x	$Q(x)$
2	334.5
2.4	195.2
2.8	146.7
3.2	108.6

(v) In some applications it may only be possible to obtain numerical values of a quantity $Q(x)$—say, by measurements in an experiment—at specific points in some interval $[a, b]$ and yet it may be necessary to have some idea of the value of the definite integral $\int_a^b Q(x)\, dx$. Even though $Q(x)$ is not defined by means of an explicit formula we may still be able to apply the Trapezoidal Rule or Simpson's Rule to approximate the integral. For example, from the data in the accompanying table and the Trapezoidal Rule we get

$$\int_2^{3.2} Q(x)\, dx \approx \frac{(3.2 - 2)}{6}[Q(2) + 2Q(2.4) + 2Q(2.8) + Q(3.2)] \approx 225.4$$

You are encouraged to work Problems 15, 16, 29, and 30.

BASIC Programs   The following listings are BASIC programs for the Trapezoidal Rule and Simpson's Rule:

```
10 REM EVALUATION OF DEFINITE INTEGRALS
   VIA THE TRAPEZOIDAL RULE
20 DEF FNY(X) = ...
30 INPUT "WHAT IS THE INTERVAL? ";A,B
40 INPUT "HOW MANY SUBDIVISIONS? ";N
50 LET H = (B - A)/N
60 LET T = FNY(A) + FNY(B)
70 FOR X = A + H TO B - H/2 STEP H
80 LET T = T + 2 * FNY(X)
90 NEXT X
100 LET T = T * H/2
110 PRINT "USING ";N; " SUBDIVISIONS, THE
    TRAPEZOID RULE YIELDS ";T
120 END
```

```
10 REM EVALUATION OF DEFINITE INTEGRALS
   USING SIMPSON'S RULE
20 DEF FNY(X) = ...
30 INPUT "WHAT IS THE INTERVAL? ";A,B
40 INPUT "HOW MANY SUBDIVISIONS? ";N
50 LET H = (B - A)/N
60 LET S = FNY(A) + FNY(B)
70 FOR X = A + H TO B - H/2 STEP 2 * H
80 LET S = S + 4 * FNY(X)
90 NEXT X
100 FOR X = A + 2 * H TO B - 3 * H/2 STEP
    2 * H
110 LET S = S + 2 * FNY(X)
120 NEXT X
130 LET S = H * S/3
140 PRINT "USING ";N; " SUBDIVISIONS,
    SIMPSON'S RULE YIELDS ";S
150 END
```

Exercises 5.8

Answers to odd-numbered problems begin on page A-28.

In Problems 1 and 2 compare the exact value of the integral with the approximation obtained from the Trapezoidal Rule for the indicated value of n.

1. $\int_1^3 (x^3 + 1)\, dx;\; n = 4$ **2.** $\int_0^2 \sqrt{x + 1}\, dx;\; n = 6$

In Problems 3–10 use the Trapezoidal Rule to obtain an approximation to the given integral for the indicated value of n.

3. $\int_1^6 \dfrac{dx}{x};\; n = 5$ **4.** $\int_0^2 \dfrac{dx}{3x + 1};\; n = 4$

5. $\int_0^1 \sqrt{x^2 + 1}\, dx;\; n = 10$ **6.** $\int_1^2 \dfrac{dx}{\sqrt{x^3 + 1}};\; n = 5$

7. $\int_0^\pi \dfrac{\sin x}{x + \pi}\, dx;\; n = 6$ **8.** $\int_0^{\pi/4} \tan x\, dx;\; n = 3$

9. $\int_0^2 \cos x^2\, dx;\; n = 6$

10. $\int_0^1 \dfrac{\sin x}{x}\, dx;\; n = 5$ [*Hint*: Define $f(0) = 1$.]

11. Determine the number of trapezoids needed so that an approximation to $\int_{-1}^2 dx/(x + 3)$ is accurate to two decimal places.

12. Determine the number of trapezoids needed so that the error in an approximation to $\int_0^{1.5} \sin^2 x\, dx$ is less than 0.0001.

13. Use the Trapezoidal Rule so that an approximation to the area under the graph of $f(x) = 1/(1 + x^2)$ on $[0, 2]$ is accurate to two decimal places. (*Hint*: Examine $f'''(x)$.)

14. The domain of $f(x) = 10^x$ is the set of real numbers and $f(x) > 0$ for all x. Use the Trapezoidal Rule to approximate the area under the graph of f on $[-2, 2]$ with $n = 4$.

15. Use the data given in Figure 5.34 and the Trapezoidal Rule to find an approximation to the area under the graph of the continuous function f on the interval $[1, 4]$.

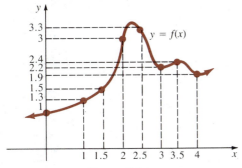

Figure 5.34

16. The so-called moment of inertia I of a three-bladed ship's propeller whose dimensions are shown in Figure 5.35(a) is given by

$$I = \frac{3\rho\pi}{2g} + \frac{3\rho}{g} \int_1^{4.5} r^2 A\, dr$$

where ρ is the density of the metal, g is the acceleration of gravity, and A is the area of a cross section of the propeller at a distance r ft from the center of the hub. If $\rho = 570$ lb/ft^3 for bronze, use the data in Figure 5.35(b) and the Trapezoidal Rule to find an approximation to I.

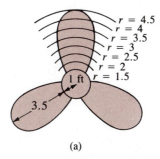

(a)

r (ft)	1	1.5	2	2.5	3	3.5	4	4.5
A (ft)	0.3	0.50	0.62	0.70	0.60	0.50	0.27	0

(b)

Figure 5.35

In Problems 17 and 18 compare the exact value of the integral with the approximation obtained from Simpson's Rule for the indicated value of n.

17. $\int_0^4 \sqrt{2x + 1}\, dx;\; n = 4$

18. $\int_0^{\pi/2} \sin^2 x\, dx;\; n = 2$

In Problems 19–26 use Simpson's Rule to obtain an approximation to the given integral for the indicated value of n.

19. $\int_{1/2}^{5/2} \dfrac{dx}{x};\; n = 4$ **20.** $\int_0^5 \dfrac{dx}{x + 2};\; n = 6$

21. $\int_0^1 \dfrac{dx}{1 + x^2};\; n = 4$

22. $\int_{-1}^1 \sqrt{x^2 + 1}\, dx;\; n = 2$

23. $\int_0^\pi \dfrac{\sin x}{x + \pi}\, dx;\ n = 6$ **24.** $\int_0^1 \cos \sqrt{x}\, dx;\ n = 4$

25. $\int_2^4 \sqrt{x^3 + x}\, dx;\ n = 4$

26. $\int_{\pi/4}^{\pi/2} \dfrac{dx}{2 + \sin x};\ n = 2$

27. Using Simpson's Rule, determine n so that the error in approximating $\int_1^3 dx/x$ is less than 10^{-5}. Compare with the n needed in the Trapezoidal Rule to give the same accuracy.

28. Find an upper bound for the error in approximating $\int_0^3 dx/(2x + 1)$ by Simpson's Rule with $n = 6$.

29. Use Simpson's Rule to find an approximation to the area of the archway shown in Figure 5.36. (*Hint*: The shape of the archway above the 9-ft level is not parabolic.)

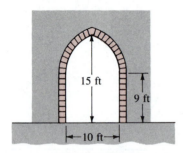

15 ft

9 ft

←10 ft→

Figure 5.36

30. Use Simpson's Rule to find an approximation to the volume of the rocket whose cross-sectional areas, shown in Figure 5.37, are spaced at intervals of width 4 m.

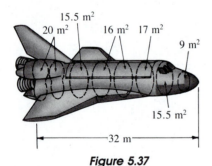

15.5 m²

20 m²

16 m²   17 m²

9 m²

15.5 m²

32 m

Figure 5.37

Miscellaneous Problems

31. Prove that the Trapezoidal Rule will give the exact value of

$$\int_a^b f(x)\, dx$$

when $f(x) = c_1 x + c_0$, c_0 and c_1 constants. Geometrically, why does this make sense?

32. Verify the result of Problem 31 by comparing the exact value of

$$\int_0^4 (2x + 5)\, dx$$

with the values obtained from the Trapezoidal Rule with $n = 2$ and $n = 4$.

33. Prove that Simpson's Rule will give the exact value of

$$\int_a^b f(x)\, dx$$

where $f(x) = c_3 x^3 + c_2 x^2 + c_1 x + c_0$, c_0, c_1, c_2, and c_3 constants.

34. Verify the result of Problem 33 by comparing the exact value of

$$\int_{-1}^3 (x^3 - x^2)\, dx$$

with the values obtained from Simpson's Rule with $n = 2$ and $n = 4$.

$\underline{\qquad}$ Chapter 5 Review Exercises $\underline{\qquad}$

Answers to odd-numbered problems begin on page A-28.

In Problems 1–20 answer true or false.

1. If $f'(x) = 3x^2 + 2x$, then, necessarily, $f(x) = x^3 + x^2$. $\underline{\qquad}$

2. $\sum_{k=2}^{6} (2k - 3) = \sum_{j=0}^{4} (2j + 1)$ $\underline{\qquad}$

3. $\sum_{k=1}^{40} 5 = \sum_{k=1}^{20} 10$ $\underline{\qquad}$

4. $\int_{1}^{3} \sqrt{t^2 + 7} \, dt = -\int_{3}^{1} \sqrt{t^2 + 7} \, dt$ $\underline{\qquad}$

5. $\int_{0}^{1} t^2 \, dt + \int_{1}^{0} x^2 \, dx = 0$ $\underline{\qquad}$

6. If f is integrable, then f is continuous. $\underline{\qquad}$

7. A definite integral is always area under a graph. $\underline{\qquad}$

8. $\int \sin x \, dx = \cos x + C$ $\underline{\qquad}$

9. If P is a partition of $[a, b]$ into n subintervals, then $n \to \infty$ implies $\|P\| \to 0$. $\underline{\qquad}$

10. If $\int_{a}^{b} f(x) \, dx > 0$, then $f(x) > 0$ for all x in $[a, b]$. $\underline{\qquad}$

11. If $F'(x) = 0$ for all x, then $F(x) = C$ for all x. $\underline{\qquad}$

12. If f is an odd integrable function on $[-\pi, \pi]$, then $\int_{-\pi}^{\pi} f(x) \, dx = 0$. $\underline{\qquad}$

13. $\int_{-1}^{1} |x| \, dx = 2 \int_{0}^{1} x \, dx$ $\underline{\qquad}$

14. $\int x \cos x \, dx = x \sin x + \cos x + C$ $\underline{\qquad}$

15. Simpson's Rule generally gives a better approximation to a definite integral than the Trapezoidal Rule. $\underline{\qquad}$

16. The Trapezoidal Rule can be used only for integrals $\int_{a}^{b} f(x) \, dx$ for which $f(x) \geq 0$ on $[a, b]$. $\underline{\qquad}$

17. To use Simpson's Rule the interval $[a, b]$ must be partitioned into an even number of subintervals. $\underline{\qquad}$

18. Simpson's Rule will give the exact value of

$$\int_{0}^{5} (x^3 - 4x^2 + 8x + 10) \, dx$$

with $n = 100$. $\underline{\qquad}$

19. If $|f^{(4)}(x)| \leq M$ for all x in $[a, b]$, then the error E_{2n} in Simpson's Rule is less than or equal to $\frac{1}{16}$ the error bound for E_n. $\underline{\qquad}$

20. To approximate

$$\int_{-1}^{2} f(x) \, dx$$

using a regular partition for which $\Delta x = \frac{3}{10}$, Simpson's Rule will utilize five parabolic arcs. $\underline{\qquad}$

In Problems 21–30 fill in the blanks.

21. If G is an antiderivative of f, then $G'(x) = \underline{\qquad}$.

22. $\int \frac{d}{dx} x^2 \, dx = \underline{\qquad}$

23. The value of $\frac{d}{dx} \int_{3}^{x} \sqrt{t^2 + 5} \, dt$ at $x = 1$ is $\underline{\qquad}$.

24. Using sigma notation, the sum $\frac{1}{3} + \frac{2}{5} + \frac{3}{7} + \frac{4}{9} + \frac{5}{11}$ can be expressed as $\underline{\qquad}$.

25. The value of $\sum_{k=1}^{15} 3k^2$ is $\underline{\qquad}$.

26. If $u = t^2 + 1$, then the definite integral

$$\int_{2}^{4} t(t^2 + 1)^{1/3} \, dt \quad \text{becomes} \quad \frac{1}{2} \int_{-}^{-} u^{1/3} \, du$$

27. The area under the graph of $f(x) = 5x$ on the interval $[0, 10]$ is $\underline{\qquad}$.

28. If the interval $[1, 6]$ is partitioned into four subintervals determined by $x_0 = 1$, $x_1 = 2$, $x_2 = \frac{5}{2}$, $x_3 = 5$, and $x_4 = 6$, the norm of the partition is $\underline{\qquad}$.

29. A partition of an interval in which the subintervals are $\underline{\qquad}$ is said to be a regular partition.

30. If P is a partition of $[0, 4]$ and x_k^* is a number in the kth subinterval, then

$$\lim_{\|P\| \to 0} \sum_{k=1}^{n} \sqrt{x_k^*} \, \Delta x_k$$

is the definition of the definite integral $\underline{\qquad}$. By the Fundamental Theorem of Calculus, the value of this definite integral is $\underline{\qquad}$.

In Problems 31–40 evaluate the given integral.

31. $\int_{-1}^{1} (4x^3 - 6x^2 + 2x - 1)\, dx$

32. $\int_{1}^{9} \dfrac{6}{\sqrt{x}}\, dx$

33. $\int (5t + 1)^{100}\, dt$

34. $\int \dfrac{w}{\sqrt{3w^2 - 1}}\, dw$

35. $\int_{0}^{\pi/4} (\sin 2x - 5 \cos 4x)\, dx$

36. $\int_{\pi^2/9}^{\pi^2} \dfrac{\sin \sqrt{z}}{\sqrt{z}}\, dz$

37. $\int_{4}^{4} (-2x^2 + x^{1/2})\, dx$

38. $\int_{-\pi/4}^{\pi/4} dx + \int_{-\pi/4}^{\pi/4} \tan^2 x\, dx$

39. $\int \cot^6 8x \, \csc^2 8x\, dx$

40. $\int \csc 3x \cot 3x\, dx$

41. If $\int_{1}^{4} f(x)\, dx = 2$ and $\int_{4}^{9} f(x)\, dx = -8$, evaluate $\int_{1}^{9} f(x)\, dx$.

42. If $\int_{0}^{5} f(x)\, dx = -3$ and $\int_{0}^{7} f(x)\, dx = 2$, evaluate $\int_{5}^{7} f(x)\, dx$.

43. Evaluate $\int_{0}^{3} (1 + |x - 1|)\, dx$.

44. If

$$f(x) = \begin{cases} x^3, & x \le 0 \\ x^2, & 0 < x \le 1 \\ x, & x > 1 \end{cases}$$

evaluate $\int_{-2}^{2} f(x)\, dx$.

45. A bucket with dimensions (in ft) shown in Figure 5.38 is filled at a constant rate of $dV/dt = \frac{1}{4}$ ft³/min. At $t = 0$ the scale reads 31.2 lb. If water weighs 62.4 lb/ft³, what does the scale read at the end of 8 minutes? When is the bucket full? (*Hint*: See page 191 for the formula for the volume of a frustum of a cone. Also, ignore the weight of the bucket.)

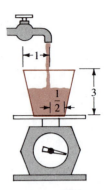

Figure 5.38

46. Evaluate:

(a) $\int_{0}^{1} \dfrac{d}{dt}\left[\dfrac{10t^4}{(2t^3 + 6t + 1)^2} \right] dt$

(b) $\int_{\pi/2}^{\pi/2} \dfrac{\sin^{10} t}{16t^7 + 1}\, dt$

47. Suppose the Trapezoidal Rule is used to approximate the integral

$$\int_{0}^{2} x^2\, dx$$

with $n = 4$. What is the error in the method?

48. Suppose it is desired to approximate

$$\int_{1/2}^{3} \dfrac{dx}{x}$$

to an accuracy of three decimal places. Compare the value of n needed for the Trapezoidal Rule with that needed for Simpson's Rule.

In Problems 49 and 50 use **(a)** the Trapezoidal Rule, and **(b)** Simpson's Rule to obtain an approximation to the given integral for the indicated value of n.

49. $\int_{0}^{1} \sin x^2\, dx, \quad n = 4$

50. $\int_{1}^{7} x^2 \sqrt{2x - 1}\, dx, \quad n = 6$

6

Applications of the Integral

Although we return to the problem of finding areas through the use of the definite integral in Section 6.1, you will see in the subsequent sections of this chapter that integrals, definite as well as indefinite, have many interpretations besides area.

6.1 Area and Area Bounded by Two Graphs

Area   If f is a nonnegative continuous function on $[a, b]$, then, as we have already seen, the area under the graph of f on the interval is

$$A = \int_a^b f(x)\, dx$$

Suppose now $f(x) \leq 0$ for all x in $[a, b]$ as shown in Figure 6.1(a). Since $-f(x) \geq 0$, we define the area bounded by the graph of $y = f(x)$ and the x-axis from $x = a$ to $x = b$ to be the area A under the graph of $y = -f(x)$ on $[a, b]$. The area A is shown in Figure 6.1(b). Written as a definite integral, we have

$$A = \int_a^b -f(x)\, dx$$

$$= -\int_a^b f(x)\, dx$$

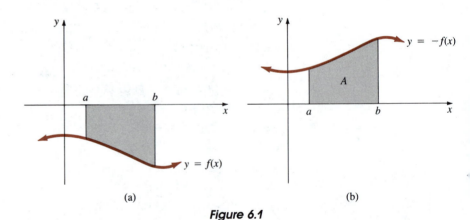

(a)   (b)

Figure 6.1

Since $-f(x) = |f(x)|$ on $[a, b]$, we are prompted to give the following definition.

DEFINITION 6.1   Area

If $y = f(x)$ is continuous on $[a, b]$, then the area A bounded by its graph on the interval and the x-axis is given by

$$A = \int_a^b |f(x)|\, dx \qquad\qquad (6.1)$$

_____ **Example 1** _____

Find the area bounded by the graph of $y = x^3$ and the x-axis on $[-2, 1]$.

Solution   From (6.1) we have

$$A = \int_{-2}^{1} |x^3|\, dx$$

In Figure 6.2 we have compared the graph of $y = x^3$ with the graph of $y = |x^3|$. Since $x^3 < 0$ for $x < 0$, we have

$$|f(x)| = \begin{cases} x^3, & x \geq 0 \\ -x^3, & x < 0 \end{cases}$$

Thus, by Definition 6.1 and Theorem 5.9

$$
\begin{aligned}
A &= \int_{-2}^{1} |x^3|\, dx \\
&= \int_{-2}^{0} |x^3|\, dx + \int_{0}^{1} |x^3|\, dx \\
&= \int_{-2}^{0} -x^3\, dx + \int_{0}^{1} x^3\, dx \\
&= -\frac{x^4}{4}\bigg]_{-2}^{0} + \frac{x^4}{4}\bigg]_{0}^{1} \\
&= 0 - \left(-\frac{16}{4}\right) + \frac{1}{4} - 0 = \frac{17}{4} \text{ square units}
\end{aligned}
$$

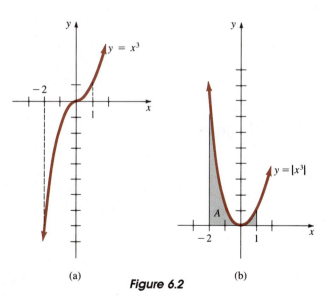

(a)   (b)

Figure 6.2

Example 2

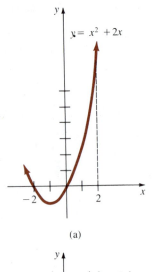

(a)

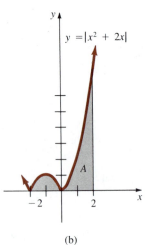

(b)

Figure 6.3

Find the area bounded by the graph of $y = x^2 + 2x$ and the x-axis on $[-2, 2]$.

Solution   The graphs of $y = f(x)$ and $y = |f(x)|$ on the interval $[-2, 2]$ are given in Figure 6.3. Now, from Figure 6.3(a) we see that

$$|f(x)| = \begin{cases} x^2 + 2x, & x \geq 0 \\ -(x^2 + 2x), & -2 \leq x < 0 \end{cases}$$

Therefore,

$$A = \int_{-2}^{2} |x^2 + 2x| \, dx$$

$$= \int_{-2}^{0} |x^2 + 2x| \, dx + \int_{0}^{2} |x^2 + 2x| \, dx$$

$$= \int_{-2}^{0} -(x^2 + 2x) \, dx + \int_{0}^{2} (x^2 + 2x) \, dx$$

$$= \left(-\frac{x^3}{3} - x^2 \right) \Big]_{-2}^{0} + \left(\frac{x^3}{3} + x^2 \right) \Big]_{0}^{2}$$

$$= 0 - \left(\frac{8}{3} - 4 \right) + \left(\frac{8}{3} + 4 \right) - 0$$

$$= 8 \text{ square units}$$

Example 3

Find the area bounded by the graph of $y = \sin x$ and the x-axis on $[0, 2\pi]$.

Solution   From (6.1) we see that

$$A = \int_{0}^{2\pi} |\sin x| \, dx$$

As indicated in Figure 6.4(a), $\sin x < 0$ on the interval $(\pi, 2\pi)$ and so

$$|f(x)| = \begin{cases} \sin x, & 0 \leq x \leq \pi \\ -\sin x, & \pi < x \leq 2\pi \end{cases}$$

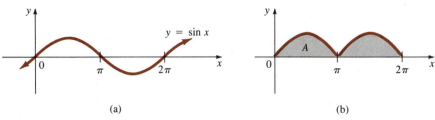

(a)

(b)

Figure 6.4

Therefore,

$$A = \int_0^\pi |\sin x| \, dx + \int_\pi^{2\pi} |\sin x| \, dx$$

$$= \int_0^\pi \sin x \, dx + \int_\pi^{2\pi} (-\sin x) \, dx$$

$$= -\cos x \Big]_0^\pi + \cos x \Big]_\pi^{2\pi}$$

$$= -\cos \pi + \cos 0 + (\cos 2\pi - \cos \pi)$$

$$= -(-1) + 1 + (1 - (-1)) = 4 \text{ square units}$$

Area Bounded by Two Graphs

The foregoing discussion is a special case of the more general problem of finding the area of a region bounded by two graphs. The area *under* the graph of a continuous nonnegative function $y = f(x)$ on $[a, b]$ is the area of the region bounded by its graph and the graph of the function $y = 0$ (the x-axis) from $x = a$ to $x = b$.

Suppose $y = f(x)$ and $y = g(x)$ are continuous on $[a, b]$ and $f(x) \geq g(x)$ for all x in the interval. See Figure 6.5. Let P be a partition of the interval

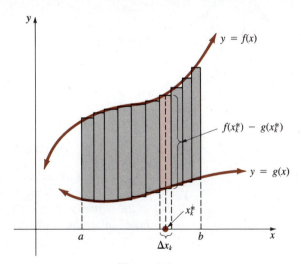

Figure 6.5

$[a, b]$ into n subintervals $[x_{k-1}, x_k]$. If we choose an x_k^* in each subinterval, we can then construct n corresponding rectangles that have area

$$\Delta A_k = [f(x_k^*) - g(x_k^*)]\, \Delta x_k$$

The area of the region bounded by the two graphs is approximately

$$\sum_{k=1}^{n} \Delta A_k = \sum_{k=1}^{n} [f(x_k^*) - g(x_k^*)]\, \Delta x_k$$

and this in turn suggests that the exact area is

$$A = \lim_{\|P\|\to 0} \sum_{k=1}^{n} [f(x_k^*) - g(x_k^*)]\, \Delta x_k$$

Since f and g are continuous, so is $f - g$. Hence, the above limit exists and is, by definition, the definite integral $\int_a^b [f(x) - g(x)]\, dx$.

In general, we have the following definition.

DEFINITION 6.2   Area Bounded by Two Graphs

If f and g are continuous functions on an interval $[a, b]$ and $f(x) \geq g(x)$ for all x in the interval, then the area A of the region bounded by their graphs on the interval is given by

$$A = \int_a^b [f(x) - g(x)]\, dx \tag{6.2}$$

Note that (6.2) reduces to (6.1) when $g(x) = 0$ for all x in $[a, b]$. Also, (6.2) applies to regions for which one or both of the functions f and g have negative values. See Figure 6.6.

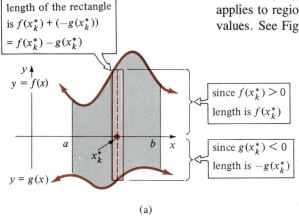

length of the rectangle is $f(x_k^*) + (-g(x_k^*))$ $= f(x_k^*) - g(x_k^*)$

since $f(x_k^*) > 0$ length is $f(x_k^*)$

since $g(x_k^*) < 0$ length is $-g(x_k^*)$

(a)

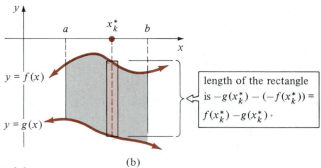

length of the rectangle is $-g(x_k^*) - (-f(x_k^*)) = f(x_k^*) - g(x_k^*)$.

(b)

Figure 6.6

You are urged *not to memorize* a formula such as (6.2), but to sketch the necessary graphs. If the curves intersect on the interval, then the relative position of the curves may change. In any event, on any subinterval of $[a, b]$ the appropriate integrand is always the

(upper graph ordinate) − *(lower graph ordinate)*

— **Example 4** —————————————————————————————

Find the area of the region bounded by the graphs of $y = \sqrt{x}$ and $y = x^2$.

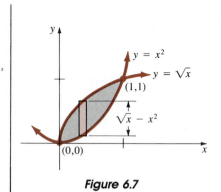

Figure 6.7

Solution   As shown in Figure 6.7, the area in question is located in the first quadrant, and the graphs intersect at the points $(0, 0)$ and $(1, 1)$; (that is, 0 and 1 are the solutions of $x^2 = \sqrt{x}$). Since $y = \sqrt{x}$ is the upper graph on the interval $(0, 1)$, it follows from (6.2) that

$$A = \int_0^1 (\sqrt{x} - x^2) \, dx$$

$$= \left(\frac{2}{3} x^{3/2} - \frac{x^3}{3} \right) \Big]_0^1$$

$$= \frac{2}{3} - \frac{1}{3} - 0 = \frac{1}{3} \text{ square units}$$

_____ **Example 5** _____

Find the area of the region bounded by the graphs of $y = x^2 + 2x$ and $y = -x + 4$ on $[-4, 2]$.

Solution   Let us denote the given functions by

$$y_1 = x^2 + 2x \quad \text{and} \quad y_2 = -x + 4$$

It is easily verified that the graphs intersect at the points $(-4, 8)$ and $(1, 3)$. In addition, inspection of Figure 6.8 shows that on the interval $(-4, 1)$, $y_2 = -x + 4$ is the upper graph, whereas on the interval $(1, 2)$, $y_1 = x^2 + 2x$ is the upper graph. Hence, the total area is the sum of

$$A_1 = \int_{-4}^1 (y_2 - y_1) \, dx \quad \text{and} \quad A_2 = \int_1^2 (y_1 - y_2) \, dx$$

Thus,

$$A = A_1 + A_2$$

$$= \int_{-4}^1 (y_2 - y_1) \, dx + \int_1^2 (y_1 - y_2) \, dx$$

$$= \int_{-4}^1 [(-x + 4) - (x^2 + 2x)] \, dx + \int_1^2 [(x^2 + 2x) - (-x + 4)] \, dx$$

$$= \int_{-4}^1 (-x^2 - 3x + 4) \, dx + \int_1^2 (x^2 + 3x - 4) \, dx$$

$$= \left(-\frac{x^3}{3} - \frac{3}{2} x^2 + 4x \right) \Big]_{-4}^1 + \left(\frac{x^3}{3} + \frac{3}{2} x^2 - 4x \right) \Big]_1^2$$

$$= \left(-\frac{1}{3} - \frac{3}{2} + 4 \right) - \left(\frac{64}{3} - 24 - 16 \right) + \left(\frac{8}{3} + 6 - 8 \right) - \left(\frac{1}{3} + \frac{3}{2} - 4 \right)$$

$$= \frac{71}{3} \text{ square units}$$

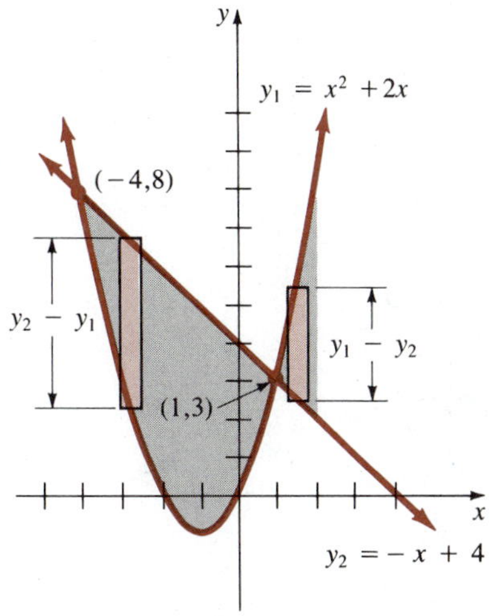

Figure 6.8

Example 6

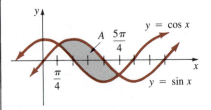

Figure 6.9

Find the area of the region bounded by the graphs of $y = \sin x$ and $y = \cos x$ on $[\pi/4, 5\pi/4]$.

Solution   From Figure 6.9 we see that $y = \sin x$ is the upper graph on the entire interval. Therefore,

$$A = \int_{\pi/4}^{5\pi/4} (\sin x - \cos x)\, dx$$

$$= (-\cos x - \sin x)\Big]_{\pi/4}^{5\pi/4}$$

$$= \left(\frac{\sqrt{2}}{2} + \frac{\sqrt{2}}{2}\right) - \left(-\frac{\sqrt{2}}{2} - \frac{\sqrt{2}}{2}\right)$$

$$= 2\sqrt{2} \approx 2.83 \text{ square units}$$

In finding the area bounded by two graphs, it is sometimes inconvenient to integrate with respect to the variable x.

Example 7

Find the area of the region bounded by the graphs of $y^2 = 1 - x$ and $2y = x + 2$.

Solution   We note that the equation $y^2 = 1 - x$ is equivalent to two functions, $y_2 = \sqrt{1 - x}$ and $y_1 = -\sqrt{1 - x}$ for $x \le 1$. If we define $y_3 = \frac{1}{2}x + 1$, we see from Figure 6.10 that the height of an element of area on the interval $(-8, 0)$ is $y_3 - y_1$, whereas the height of an element on the interval $(0, 1)$ is $y_2 - y_1$. Thus, if we integrate with respect to x, the area is the sum of

$$A_1 = \int_{-8}^{0} (y_3 - y_1)\, dx \quad \text{and} \quad A_2 = \int_{0}^{1} (y_2 - y_1)\, dx$$

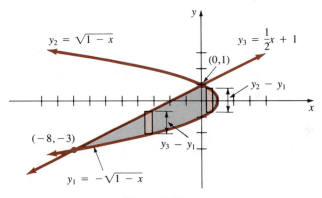

Figure 6.10

That is,

$$A = \int_{-8}^{0} \left[\left(\frac{1}{2}x + 1 \right) - (-\sqrt{1-x}) \right] dx + \int_{0}^{1} [\sqrt{1-x} - (-\sqrt{1-x})]\, dx$$

$$= \int_{-8}^{0} \left(\frac{1}{2}x + 1 + \sqrt{1-x} \right) dx + 2 \int_{0}^{1} \sqrt{1-x}\, dx$$

$$= \left(\frac{1}{4}x^2 + x - \frac{2}{3}(1-x)^{3/2} \right)\Big]_{-8}^{0} - \frac{4}{3}(1-x)^{3/2}\Big]_{0}^{1}$$

$$= -\frac{2}{3} \cdot 1^{3/2} - \left(16 - 8 - \frac{2}{3} \cdot 9^{3/2} \right) - \frac{4}{3} \cdot 0 + \frac{4}{3} \cdot 1^{3/2}$$

$$= \frac{32}{3} \text{ square units}$$

Example 8

Alternative Solution to Example 7   The necessity of using two integrals in Example 7 to find the area is avoided by constructing horizontal rectangles and using y as the independent variable. If we define $x_2 = 1 - y^2$ and $x_1 = 2y - 2$, then, as shown in Figure 6.11, the area of a horizontal element is

$$\Delta A_k = [(\text{right graph abscissa}) - (\text{left graph abscissa})] \cdot \text{width}$$

That is, $$\Delta A_k = [x_2^* - x_1^*]\, \Delta y_k$$

where

$$x_2^* = 1 - (y_k^*)^2, \qquad x_1^* = 2y_k^* - 2, \quad \text{and} \quad \Delta y_k = y_k - y_{k-1}$$

Summing the rectangles in the positive y-direction leads to

$$A = \lim_{\|P\| \to 0} \sum_{k=1}^{n} [x_2^*(y) - x_1^*(y)]\, \Delta y_k$$

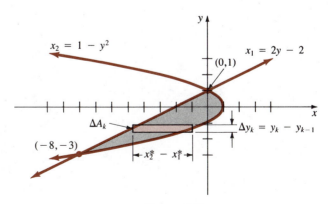

Figure 6.11

where $\|P\|$ is the norm of a partition P of the interval $-3 \leq y \leq 1$. In other words,

$$A = \int_{-3}^{1} (x_2 - x_1)\, dy$$

where the lower limit -3 and the upper limit 1 are the y-coordinates of the points of intersection $(-8, -3)$ and $(0, 1)$, respectively. Substituting the indicated values of x_2 and x_1 then gives

$$A = \int_{-3}^{1} [(1 - y^2) - (2y - 2)]\, dy$$

$$= \int_{-3}^{1} (-y^2 - 2y + 3)\, dy$$

$$= \left(-\frac{y^3}{3} - y^2 + 3y \right) \Big]_{-3}^{1}$$

$$= \left(-\frac{1}{3} - 1 + 3 \right) - \left(\frac{27}{3} - 9 - 9 \right) = \frac{32}{3} \text{ square units}$$

—————— **Exercises 6.1** ——————

Answers to odd-numbered problems begin on page A-29

In Problems 1–20 find the area bounded by the graph of the given function and the x-axis on the indicated interval.

1. $y = x^2 - 1$; $[-1, 1]$

2. $y = x^2 - 1$; $[0, 2]$

3. $y = x^3$; $[-3, 0]$

4. $y = 1 - x^3$; $[0, 2]$

5. $y = x^2 - 3x$; $[0, 3]$

6. $y = -(x + 1)^2$; $[-1, 0]$

7. $y = x^3 - 6x$; $[-1, 1]$

8. $y = x^3 - 3x^2 + 2$; $[0, 2]$

9. $y = (x - 1)(x - 2)(x - 3)$; $[0, 3]$

10. $y = x(x + 1)(x - 1)$; $[-1, 1]$

11. $y = \dfrac{x^2 - 1}{x^2}$; $[\frac{1}{2}, 3]$

12. $y = \dfrac{x^2 - 1}{x^2}$; $[1, 2]$

13. $y = \sqrt{x} - 1$; $[0, 4]$

14. $y = 2 - \sqrt{x}$; $[0, 9]$

15. $y = \sqrt[3]{x}$; $[-2, 3]$

16. $y = 2 - \sqrt[3]{x}$; $[-1, 8]$

17. $y = \sin x$; $[-\pi, \pi]$

18. $y = 1 + \cos x$; $[0, 3\pi]$

19. $y = -1 + \sin x$; $[-3\pi/2, \pi/2]$

20. $y = \sec^2 x$; $[0, \pi/3]$

In Problems 21–48 find the area of the region bounded by the graphs of the given equations.

21. $y = x$, $y = -2x$, $x = 3$

22. $y = x$, $y = 4x$, $x = 2$

23. $y = x^2$, $y = 4$

24. $y = x^2$, $y = x$

25. $y = x^3$, $y = 8$, $x = -1$

26. $y = x^3$, $y = \sqrt[3]{x}$, first quadrant

27. $y = 4(1 - x^2)$, $y = 1 - x^2$

28. $y = 2(1 - x^2)$, $y = x^2 - 1$

29. $y = x$, $y = 1/x^2$, $x = 3$

30. $y = x^2$, $y = 1/x^2$, $y = 9$, first quadrant

31. $y = -x^2 + 6$, $y = x^2 + 4x$

32. $y = x^2$, $y = -x^2 + 3x$

33. $y = x^{2/3}$, $y = 4$

34. $y = 1 - x^{2/3}$, $y = x^{2/3} - 1$

35. $y = x^2 - 2x - 3$, $y = 2x + 2$, on $[-1, 6]$

36. $y = -x^2 + 4x$, $y = \dfrac{3}{2}x$

37. $x = 3y^2$, $x = 6$

38. $x = y^2$, $x = 0$, $y = 1$

39. $x = -y$, $x = 2 - y^2$

40. $x = y^2$, $x = 6 - y^2$

41. $x = y^2 + 2y + 2$, $x = -y^2 - 2y + 2$

42. $x = y^2 - 6y + 1$, $x = -y^2 + 2y + 1$

43. $y = x^3 - x$, $y = x + 4$, $x = -1$, $x = 1$

44. $x = y^3 - y$, $x = 0$

45. $y = \cos x$, $y = \sin x$, $x = 0$, $x = \pi/2$

46. $y = 2 \sin x$, $y = -x$, $x = \pi/2$

47. $y = 4 \sin x$, $y = 2$, on $[\pi/6, 5\pi/6]$

48. $y = 2 \cos x$, $y = -\cos x$, on $[-\pi/2, \pi/2]$

In Problems 49 and 50 (Figures 6.12 and 6.13) find the area of the shaded region.

49.

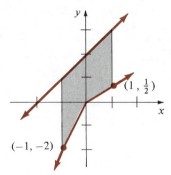

Figure 6.12

50.

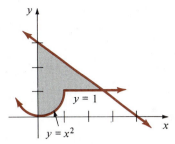

Figure 6.13

51. Find the area of the triangle with vertices at $(1, 1)$, $(2, 4)$, and $(3, 2)$.

52. Set up a definite integral that represents the area of an ellipse $x^2/a^2 + y^2/b^2 = 1$. Do not try to evaluate.

In Problems 53 and 54 use the fact that the area of a circle of radius a is πa^2 to evaluate the given definite integral.

53. $\displaystyle\int_0^a \sqrt{a^2 - x^2}\, dx$

54. $\displaystyle\int_{-a}^a \sqrt{a^2 - x^2}\, dx$

55. Consider the region bounded by the graphs of $y^2 = -x - 2$, $y = 2$, $y = -2$, and $y = 2(x - 1)$. Compute the area of the region by integrating with respect to x.

56. Compute the area of the region given in Problem 55 by integrating with respect to y.

Miscellaneous Problems

57. A trapezoid is bounded by the graphs of $f(x) = Ax + B$, $x = a$, $x = b$, and $x = 0$. Show that the area of the trapezoid is $\dfrac{f(a) + f(b)}{2}(b - a)$.

58. Given that a function f is continuous on an interval $[a, b]$, it can be proved that there exists a number c in the open interval (a, b) for which the number $f(c)(b - a)$ is the same as $\int_a^b f(x)\, dx$. Interpret this result geometrically in terms of area under the graph of f on the interval.

6.2 Volumes by Slicing

In this section and the following two, we shall show how the definite integral can be used to compute the volumes of certain solids.

Volume of a Solid: Slicing Method

Suppose V is the volume of the solid shown in Figure 6.14 bounded by planes that are perpendicular to the x-axis at $x = a$ and $x = b$. Furthermore, suppose we know a continuous function $A(x)$ that gives the area of a plane cross-sectional region, or **slice**, which is formed by cutting the solid by a plane perpendicular to the *x*-axis. For example, for $a < x_1 < x_2 < b$ the areas of the cross sections shown in Figure 6.14 are $A(x_1)$ and $A(x_2)$.

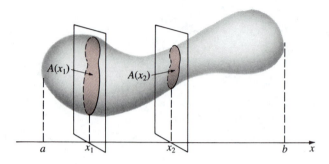

Figure 6.14

Now, let P be the partition

$$a = x_0 < x_1 < x_2 < \cdots < x_n = b$$

and x_k^* any number in $[x_{k-1}, x_k]$. Then an approximation to the volume of the solid on this subinterval is the volume of the right cylinder shown in the enlargement in Figure 6.15. Since the volume of a right cylinder is

$$\Delta V_k = (\text{area of base}) \cdot (\text{height}) = A(x_k^*)(x_k - x_{k-1}) = A(x_k^*)\, \Delta x_k$$

it follows that an approximation to the volume of the solid on $[a, b]$ is

$$\sum_{k=1}^{n} \Delta V_k = \sum_{k=1}^{n} A(x_k^*)\, \Delta x_k$$

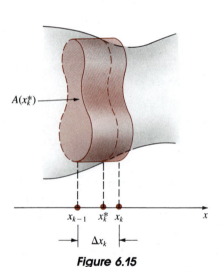

Figure 6.15

Thus, we conclude that the exact volume is given by the definite integral

$$V = \lim_{\|P\| \to 0} \sum_{k=1}^{n} A(x_k^*)\, \Delta x_k = \int_a^b A(x)\, dx$$

We state the formal result as a definition.

DEFINITION 6.3   Let V be the volume of a solid bounded by planes that are perpendicular to the *x*-axis at $x = a$ and $x = b$. If $A(x)$ is a continuous function that gives the area of a cross section of the solid formed by a plane perpendicular to the *x*-axis at any point in $[a, b]$, then the volume of the solid is

$$V = \int_a^b A(x)\, dx \tag{6.3}$$

Example 1

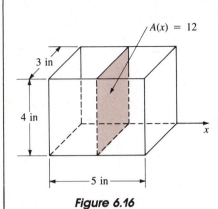

$A(x) = 12$

3 in

4 in

5 in

Figure 6.16

Consider the rather simple example of finding the volume of the rectangular box shown in Figure 6.16. The cross-sectional area is the constant function

$$A(x) = 12 \text{ in}^2$$

Although we know the volume is $V = 3 \times 4 \times 5 = 60 \text{ in}^3$, it also follows from Definition 6.3 that

$$V = \int_0^5 12 \, dx$$

$$= 12x \Big]_0^5$$

$$= 12[5 - 0] = 60 \text{ in}^3$$

Example 2

Find the volume V of the right circular cone shown in Figure 6.17(a).

Solution   The area of a cross section, obtained by a horizontal slice x units from the bottom of the cone, is πR^2. Inspection of Figure 6.17(b) shows that R is related to x by similar triangles. We have

$$\frac{h}{r} = \frac{h - x}{R} \qquad \text{so that} \qquad R = \frac{r}{h}(h - x)$$

Therefore,

$$A(x) = \pi R^2$$

$$= \pi \frac{r^2}{h^2}(h - x)^2$$

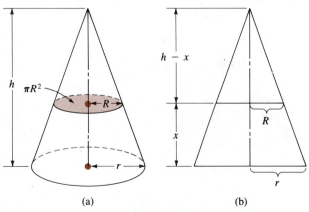

(a)   (b)

Figure 6.17

By Definition 6.3, the volume of the cone is given by

$$V = \int_0^h \pi \frac{r^2}{h^2}(h - x)^2 \, dx$$

$$= -\pi \frac{r^2}{h^2} \int_0^h (h - x)^2(-dx)$$

$$= -\pi \frac{r^2}{h^2} \frac{(h - x)^3}{3} \Big]_0^h$$

$$= -\pi \frac{r^2}{h^2} \left[0 - \frac{h^3}{3} \right]$$

$$= \frac{\pi}{3} r^2 h$$

Exercises 6.2

Answers to odd-numbered problems begin on page A-29

1. A 75-foot high power-line pole has a cross section in the form of an equilateral triangle. Given that the length of a side is $(75 - x)/10$, where x is the distance in feet from the ground, find the volume of the pole.

2. The cross section of a pyramid is a square x feet on a side, x feet from its top. Given that the pyramid is 100 feet high, find its volume.

3. The cross sections perpendicular to a diameter of a circular base are squares. Given that the radius of the base is 4 ft, find the volume of the solid. See Figure 6.18.

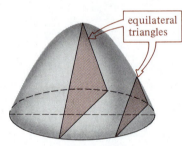

Figure 6.19

5. The base of a solid is bounded by the curves $x = y^2$, $x = 4$ in the xy-plane. The cross sections perpendicular to the x-axis are rectangles for which the height is four times the base. Find the volume of the solid.

6. The base of a solid is bounded by the curve $y = 4 - x^2$ and the x-axis. The cross sections perpendicular to the x-axis are equilateral triangles. Find the volume of the solid.

7. The base of a solid is an isosceles triangle whose base is 4 feet and height is 5 feet. The cross sections perpendicular to the altitude are semicircles. Find the volume of the solid.

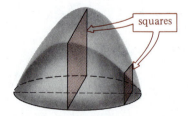

Figure 6.18

4. The cross sections perpendicular to a diameter of a circular base are equilateral triangles. Given that the radius of the base is 4 ft, find the volume of the solid. See Figure 6.19.

8. The axes of two right circular cylinders, each having radius $r = 3$ feet, intersect at right angles. Find the value of the resulting volume.

9. The base of a solid is a right isosceles triangle that is formed by the coordinate axes and the line $x + y = 3$. The cross sections perpendicular to the y-axis are squares. Find the volume of the solid.

10. A hole of radius 1 foot is drilled through the middle of the solid sphere of radius $r = 2$ feet. Find the volume of the remaining solid.

11. Consider the right circular cylinder of radius a shown in Figure 6.20. A plane inclined at an angle of 45° to the base of the cylinder passes through a diameter of the base. Find the volume of the resulting wedge cut from the cylinder.

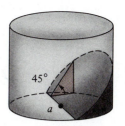

Figure 6.20

12. Rework Problem 11 if the plane passing through the diameter of the base is inclined at an angle of 60°.

6.3 Solids of Revolution: Disk and Washer Method

If a region R in the xy-plane is revolved about an axis L, it will generate a solid called a **solid of revolution**. See Figure 6.21. As shown in Section 6.2, we can find the volume V of a solid by means of a definite integral provided that we know the function $A(x)$ that gives the cross-sectional area of a slice formed by passing a plane through the solid perpendicular to an axis. In the case of finding the volume of a solid of revolution, it is always possible to find $A(x)$; the axis in question is the axis of revolution. The two methods considered in this section are just special cases of Definition 6.3.

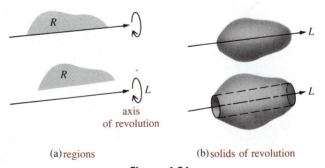

(a) regions   (b) solids of revolution

Figure 6.21

The Disk Method   Let R be the region bounded by the graph of a nonnegative continuous function $y = f(x)$, the x-axis, and the vertical lines $x = a$ and $x = b$, as shown in Figure 6.22. If this region is revolved about the x-axis, let us find the volume V of the resulting solid of revolution.

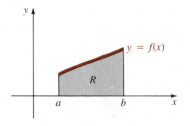

Figure 6.22

Let P be a partition of $[a, b]$, and let x_k^* be any number in the kth subinterval $[x_{k-1}, x_k]$. As the rectangular element of width $\Delta x_k = x_k - x_{k-1}$ and height $f(x_k^*)$ is revolved about the x-axis, it generates a **disk** as shown in Figure 6.23(b). Now the volume of a right circular cylinder, or disk, of radius r and height h is

$$\text{(area of base)} \cdot \text{(height)} = \pi r^2 h$$

Thus, if we make the identification $r = f(x_k^*)$, $h = \Delta x_k$, the volume of a representative disk is

$$\Delta V_k = \pi [f(x_k^*)]^2 \, \Delta x_k$$

As seen in Figure 6.23(c) a partition with n subintervals yields n such disks. Hence, the sum

$$\sum_{k=1}^{n} \Delta V_k = \sum_{k=1}^{n} \pi [f(x_k^*)]^2 \, \Delta x_k$$

represents an approximation to the volume of the solid shown in Figure 6.23(d). This suggests that the exact volume is

$$V = \lim_{\|P\| \to 0} \sum_{k=1}^{n} \pi [f(x_k^*)]^2 \, \Delta x_k$$

or

$$V = \pi \int_a^b [f(x)]^2 \, dx \qquad (6.4)$$

Note that (6.4) is a special case of Definition 6.3 with $A(x) = \pi [f(x)]^2$.

If a region R is revolved about some other axis, then (6.4) may simply not be applicable to the problem of finding the volume of the resulting solid. Rather than apply a formula blindly, you should set up an appropriate integral by carefully analyzing the geometry of each problem.

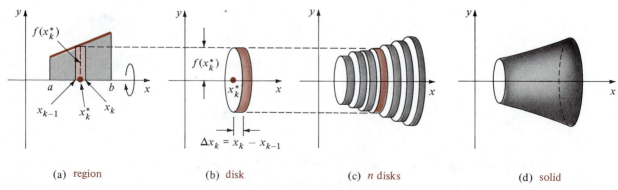

(a) region   (b) disk   (c) *n* disks   (d) solid

Figure 6.23

_____ **Example 1** _____

Find the volume V of the solid formed by revolving the region bounded by the graphs of $y = \sqrt{x}$, $y = 0$, and $x = 4$ about the x-axis.

Solution Figure 6.24(a) shows the region in question. Now, the volume of the disk shown in Figure 6.24(b) is

$$\Delta V_k = \text{(area of base)} \cdot \text{(height)}$$
$$= \pi [f(x_k^*)]^2 \, \Delta x_k$$
$$= \pi [(x_k^*)^{1/2}]^2 \, \Delta x_k$$
$$= \pi x_k^* \, \Delta x_k$$

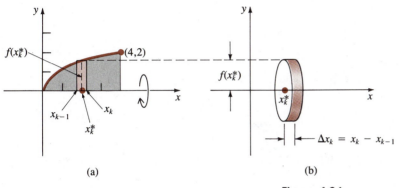

Figure 6.24

Then, the approximate volume is

$$\sum_{k=1}^{n} \Delta V_k = \sum_{k=1}^{n} \pi x_k^* \, \Delta x_k$$

Taking the limit of the last sum as $\|P\| \to 0$ yields

$$V = \pi \int_0^4 x \, dx$$
$$= \pi \frac{x^2}{2} \Big]_0^4 = 8\pi \text{ cubic units}$$

The Washer Method Let the region R bounded by the graphs of the continuous functions $y = f(x)$, $y = g(x)$, and the lines $x = a$ and $x = b$, as shown in Figure 6.25(a), be revolved about the x-axis. Then the rectangular element between the two graphs on $[x_{k-1}, x_k]$ will generate a circular ring or **washer** as shown in Figure 6.25(b). The volume of the washer is

$$\Delta V_k = \text{(volume of disk)} - \text{(volume of hole)}$$
$$= \pi [f(x_k^*)]^2 \, \Delta x_k - \pi [g(x_k^*)]^2 \, \Delta x_k$$
$$= \pi ([f(x_k^*)]^2 - [g(x_k^*)]^2) \, \Delta x_k$$

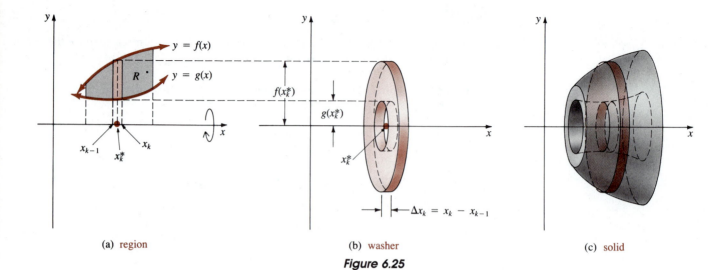

(a) region   (b) washer   (c) solid

Figure 6.25

An approximation to the volume is

$$\sum_{k=1}^{n} \Delta V_k = \sum_{k=1}^{n} \pi([f(x_k^*)]^2 - [g(x_k^*)]^2) \, \Delta x_k$$

This suggests that as $\|P\| \to 0$ the exact volume of the solid is given by

$$V = \pi \int_a^b ([f(x)]^2 - [g(x)]^2) \, dx \tag{6.5}$$

Formula (6.5) is a special case of Definition 6.3 with $A(x) = \pi([f(x)]^2 - [g(x)]^2)$. In addition, (6.5) reduces to (6.4) when $g(x) = 0$.

Example 2

Find the volume V of the solid formed by revolving the region bounded by the graphs of $y = \sqrt{x}$, $y = 0$, and $x = 4$ about the y-axis.

Solution   The region is shown in Figure 6.26(a). As the horizontal rectangular element of height $\Delta y_k = y_k - y_{k-1}$ is revolved about the y-axis, we obtain the washer illustrated in Figure 6.26(b). As in the general discussion above, the volume of the washer is

$$\Delta V_k = (\text{volume of disk}) - (\text{volume of hole})$$

$$= \pi 4^2 \, \Delta y_k - \pi(x_k^*)^2 \, \Delta y_k$$

$$= \pi[16 - (x_k^*)^2] \, \Delta y_k$$

In this case the given function implies $x_k^* = (y_k^*)^2$. Thus, the approximate volume of the solid is

$$\sum_{k=1}^{n} \Delta V_k = \sum_{k=1}^{n} \pi[16 - (y_k^*)^4] \, \Delta y_k$$

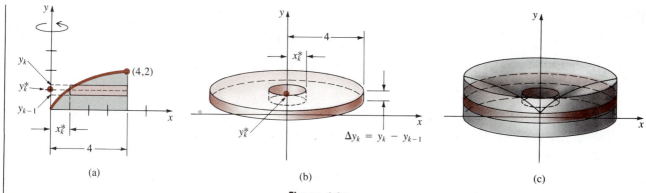

Figure 6.26

The use of a horizontal rectangle of height Δy_k corresponds to a partition of the interval $[0, 2]$ on the y-axis. Hence, we conclude that the exact volume is given by the definite integral

$$V = \pi \int_0^2 (16 - y^4) \, dy$$

$$= \pi \left(16y - \frac{y^5}{5} \right) \Big]_0^2 = \frac{128\pi}{5} \text{ cubic units}$$

Example 3

Find the volume V of the solid formed by revolving the region bounded by the graphs of $y = x + 2$, $y = x$, $x = 0$, and $x = 3$ about the x-axis.

Solution As seen in Figure 6.27, a vertical rectangular element of width Δx_k, when revolved about the x-axis, yields a washer having volume

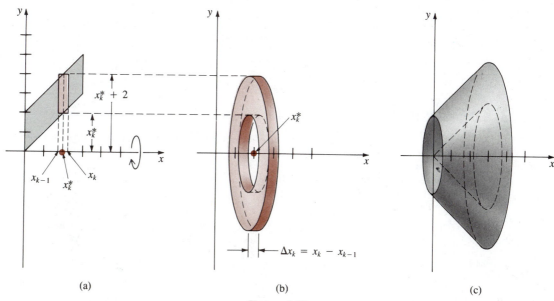

Figure 6.27

$$\Delta V_k = \pi(x_k^* + 2)^2 \, \Delta x_k - \pi(x_k^*)^2 \, \Delta x_k$$
$$= \pi(4x_k^* + 4) \, \Delta x_k$$

The usual summing and limiting process yields

$$V = \pi \int_0^3 (4x + 4) \, dx$$

$$= \pi(2x^2 + 4x) \Big]_0^3 = 30\pi \text{ cubic units}$$

Revolution about a Line

The next example shows how to find the volume of a solid of revolution when a region is revolved about an axis that is not a coordinate axis.

Example 4

Find the volume V of the solid that is formed by revolving the region given in Example 2 about the line $x = 4$.

Solution   From inspection of Figure 6.28, we see that a horizontal rectangular element of width Δy_k generates a solid disk of volume

$$\Delta V_k = \pi(4 - x_k^*)^2 \, \Delta y_k$$

where, as before, $x_k^* = (y_k^*)^2$. This leads to the integral

$$V = \pi \int_0^2 (4 - y^2)^2 \, dy$$

$$= \pi \int_0^2 (16 - 8y^2 + y^4) \, dy$$

$$= \pi \left(16y - \frac{8}{3}y^3 + \frac{y^5}{5} \right) \Big]_0^2 = \frac{256\pi}{15} \text{ cubic units}$$

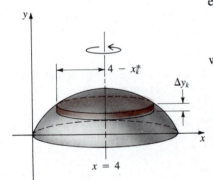

Figure 6.28

Exercises 6.3

Answers to odd-numbered problems begin on page A-29

In Problems 1–6 refer to Figure 6.29. Use the disk or washer method to find the volume of the solid of revolution that is formed by revolving the given region about the indicated line.

1. R_1 about OC

2. R_1 about OA

3. R_2 about OA

4. R_2 about OC

5. R_1 about AB

6. R_2 about AB

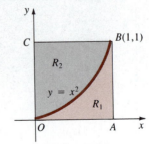

Figure 6.29

In Problems 7–32 find the volume of the solid of revolution that is formed by revolving the region bounded by the graphs of the given equations about the indicated line or axis.

7. $y = 9 - x^2$, $y = 0$; x-axis

8. $y = x^2 + 1$, $x = 0$, $y = 5$; y-axis

9. $y = \dfrac{1}{x}$, $x = 1$, $y = \dfrac{1}{2}$; y-axis

10. $y = \dfrac{1}{x}$, $x = \dfrac{1}{2}$, $x = 3$, $y = 0$; x-axis

11. $y = x^2$, $y = 2x$; y-axis

12. $y = x^2$, $y = \sqrt{x}$; x-axis

13. $y = (x - 2)^2$, $x = 0$, $y = 0$; x-axis

14. $y = (x + 1)^2$, $x = 0$, $y = 0$; y-axis

15. $y = 4 - x^2$, $y = 1 - \dfrac{1}{4}x^2$; x-axis

16. $y = 1 - x^2$, $y = x^2 - 1$; y-axis

17. $x + y = 3$, $y = 2x$, $y = 0$; y-axis

18. $x + y = 2$, $x = 0$, $y = 0$, $y = 1$; x-axis

19. $y = \sqrt{x - 1}$, $x = 5$, $y = 0$; $x = 5$

20. $x = y^2$, $x = 1$; $x = 1$

21. $y = x^{1/3}$, $x = 0$, $y = 1$; $y = 2$

22. $x = -y^2 + 2y$, $x = 0$; $x = 2$

23. $x^2 - y^2 = 16$, $x = 5$; y-axis

24. $y = x^2 - 6x + 9$, $y = 9 - \dfrac{1}{2}x^2$; x-axis

25. $x = y^2$, $y = x - 6$; y-axis

26. $y = x^3 + 1$, $x = 0$, $y = 9$; y-axis

27. $y = x^3 - x$, $y = 0$; x-axis

28. $y = \sin x$, $y = 0$, $0 \le x \le \pi$; x-axis

29. $y = |\cos x|$, $y = 0$, $0 \le x \le 2\pi$; x-axis

30. $y = \sec x$, $x = -\dfrac{\pi}{4}$, $x = \dfrac{\pi}{4}$, $y = 0$; x-axis

31. $y = \tan x$, $y = 0$, $x = \dfrac{\pi}{4}$; x-axis

32. $y = \sin x$, $y = \cos x$, $x = 0$, x-axis

33. Find the volume of the solid of revolution that is formed by revolving the region bounded by the graphs of $y = \sin^2 x$, $y = 0$, $0 \le x \le \pi$, about the x-axis. (*Hint:* $\sin^4 x = \sin^2 x \cdot \sin^2 x$)

34. Prove that the volume of a right circular cone of radius r and height h is $V = (\pi/3)r^2 h$. (*Hint:* Consider $y = (r/h)x$, $0 \le x \le h$.)

35. Prove that the volume of a sphere of radius r is $V = (\frac{4}{3})\pi r^3$. (*Hint:* Consider $x^2 + y^2 = r^2$.)

36. A hemispherical water trough of radius r is filled with water to a depth h. Show that the volume of the water is $V = \pi r h^2 - (\pi/3)h^3$.

37. Find the volume of the solid of revolution that is formed by revolving the region bounded by the graph of the ellipse $x^2/a^2 + y^2/b^2 = 1$, $a > b > 0$, in the first and second quadrants, and $y = 0$ about the x-axis. The resulting solid is known as a **prolate spheroid**.

38. Derive the formula for the volume of a frustum of a cone given on page 191. (*Hint:* Consider the line through $(0, r_1)$ and (h, r_2).)

39. The region shown in Figure 6.30 is revolved about the y-axis forming a bowl. Determine the amount of holding capacity of the bowl.

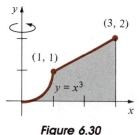

Figure 6.30

40. The portion of the sphere of radius r shown in Figure 6.31 is called a **spherical sector**. Find the volume of this sector.

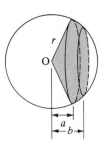

Figure 6.31

Calculator Problem

41. The function $y = f(x)$ is positive and continuous on the interval [0.6, 1]. It is known that

x	0.6	0.7	0.8	0.9	1.0
$f(x)$	1.8	2.0	2.2	2.5	2.7

The region that is bounded by the graph of f on the interval is revolved about the x-axis. Use an appropriate numerical technique to approximate, to one decimal place, the volume of the solid of revolution.

6.4 Solids of Revolution: The Shell Method

In Section 6.3, we saw that a rectangular element of area that is perpendicular to an axis of revolution will generate a disk or circular ring (washer). Note, however, that if we were to revolve the rectangular element shown in Figure 6.32(a) about a line parallel to the element, in this case the y-axis, we generate a hollow **shell** as shown in Figure 6.32(b).

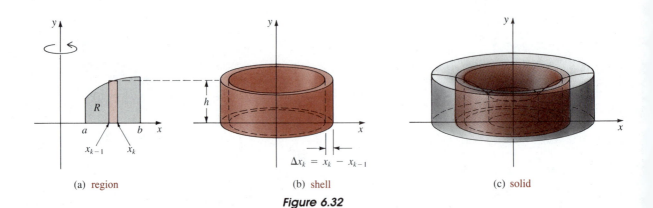

(a) region   (b) shell   (c) solid

Figure 6.32

To find the volume of the solid of revolution shown in Figure 6.32(c), we form a partition P of $[a, b]$ into n subintervals $[x_{k-1}, x_k]$. Since the volume of a shell is the difference

$$\Delta V_k = (\text{volume of outer cylinder}) - (\text{volume of inner cylinder})$$

it follows from Figure 6.32(b) that

$$
\begin{aligned}
\Delta V_k &= \pi x_k^2 h - \pi x_{k-1}^2 h \\
&= \pi [x_k^2 - x_{k-1}^2]h \\
&= \pi [x_k + x_{k-1}][x_k - x_{k-1}]h
\end{aligned}
\tag{6.6}
$$

If we define $x_k^* = \frac{1}{2}(x_k + x_{k-1})$, then x_k^* is the midpoint of the subinterval of length $\Delta x_k = x_k - x_{k-1}$. In addition, $x_k + x_{k-1} = 2x_k^*$. Therefore, by making the identification $h = f(x_k^*)$, the volume (6.6) of the shell can be written as

$$\Delta V_k = 2\pi x_k^* f(x_k^*) \, \Delta x_k$$

An approximation to the volume is

$$\sum_{k=1}^{n} \Delta V_k = \sum_{k=1}^{n} 2\pi x_k^* f(x_k^*) \, \Delta x_k \tag{6.7}$$

As the norm $\|P\|$ of the partition approaches zero, we expect that the limit of (6.7) is the definite integral

$$V = 2\pi \int_a^b xf(x) \, dx \tag{6.8}$$

Since it is impossible to analyze every possible case, we urge you, again, not to memorize a particular formula such as (6.8). On the other hand, it is important to be able to set up the integral for a given problem without the necessity of going through a lengthy analysis. To accomplish this, imagine that a shell is cut down its side and flattened out to form a thin rectangular solid as in Figure 6.33(b). The volume of the shell is then

volume = (length) · (width) · (thickness)

= (circumference of the cylinder) · (height) · (thickness)

$$= 2\pi r \, h \, t \tag{6.9}$$

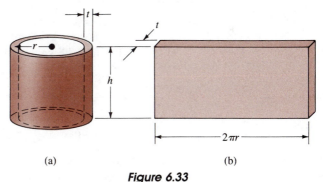

(a)   (b)

Figure 6.33

Example 1

Use the shell method to find the volume V of the solid of revolution that is given in Example 2 of Section 6.3.

Solution   From Figure 6.34 we can make the identification $r = x_k^*$, $h = y_k^*$, $t = \Delta x_k$. Thus, from (6.9),

$$\Delta V_k = 2\pi x_k^* y_k^* \, \Delta x_k$$

$$= 2\pi x_k^* \sqrt{x_k^*} \, \Delta x_k$$

and so
$$V = 2\pi \int_0^4 x \cdot x^{1/2} \, dx$$

$$= 2\pi \int_0^4 x^{3/2} \, dx$$

$$= 2\pi \left. \frac{2}{5} x^{5/2} \right]_0^4$$

$$= \frac{4\pi}{5} 4^{5/2} = \frac{128\pi}{5} \text{ cubic units}$$

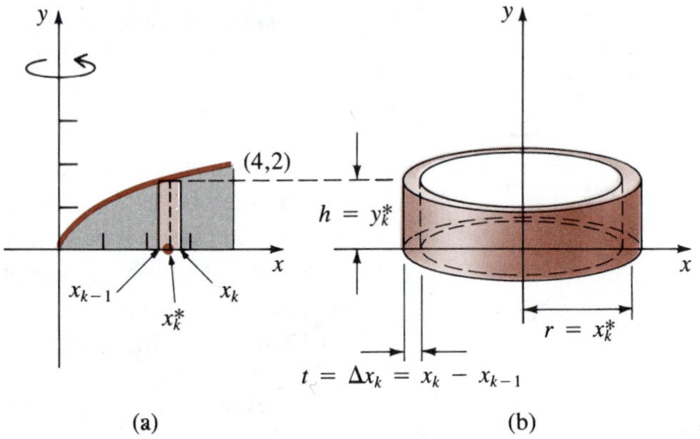

Figure 6.34

It is not always convenient or even possible to use the disk or washer method to find the volume of a solid of revolution.

Example 2

Find the volume V of the solid that is formed by revolving the region bounded by the graphs of $x = y^2 - 2y$ and $x = 3$ about the line $y = 1$.

Solution In this case a rectangular element of area that is perpendicular to a horizontal line and revolved about the line $y = 1$ would generate a disk. Since the radius of the disk is not measured from the x-axis but from the line $y = 1$, it would be necessary to solve $x = y^2 - 2y$ for y in terms of x. We can avoid this inconvenience by using horizontal elements of area, which then generate the shell indicated in Figure 6.35(b). Note that when $x = 3$, the equation $3 = y^2 - 2y$, or equivalently $(y + 1)(y - 3) = 0$, has solutions -1 and 3. Thus, we need only partition the interval $[1, 3]$ on the y-axis. After making the identifications $r = y_k^* - 1$, $h = 3 - x_k^*$, and $t = \Delta y_k$, it follows from (6.9) that the volume of a shell is

$$\Delta V_k = 2\pi(y_k^* - 1)(3 - x_k^*) \, \Delta y_k$$
$$= 2\pi(y_k^* - 1)(3 - [(y_k^*)^2 - 2y_k^*]) \, \Delta y_k$$

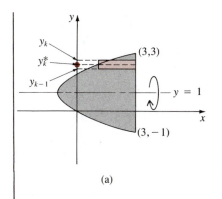

(a)

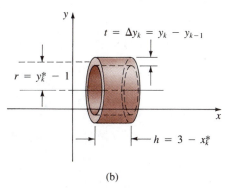

(b)

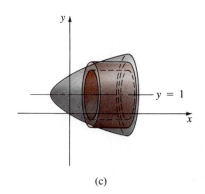

(c)

Figure 6.35

The volume is

$$V = 2\pi \int_1^3 (y - 1)(3 - y^2 + 2y)\, dy$$

$$= 2\pi \int_1^3 (-y^3 + 3y^2 + y - 3)\, dy$$

$$= 2\pi \left(-\frac{y^4}{4} + y^3 + \frac{y^2}{2} - 3y \right) \Big]_1^3$$

$$= 2\pi \left[\left(-\frac{81}{4} + 27 + \frac{9}{2} - 9 \right) - \left(-\frac{1}{4} + 1 + \frac{1}{2} - 3 \right) \right]$$

$$= 8\pi \text{ cubic units}$$

Example 3

Find the volume V of the solid that is formed by revolving the region in the first quadrant bounded by the graphs of $y = \sqrt{4 - x^2}$, $y = x(x - 1)(x - 2)$, and $x = 0$ about the y-axis.

Solution As seen in Figure 6.36, a vertical element of area of width Δx_k generates a shell of volume

$$\Delta V_k = 2\pi x_k^*[\sqrt{4 - (x_k^*)^2} - x_k^*(x_k^* - 1)(x_k^* - 2)]\, \Delta x_k$$

Hence,

$$V = 2\pi \int_0^2 [x\sqrt{4 - x^2} - x^2(x - 1)(x - 2)]\, dx$$

$$= 2\pi \int_0^2 \left[-\frac{1}{2}(4 - x^2)^{1/2}(-2x) - x^4 + 3x^3 - 2x^2 \right] dx$$

$$= 2\pi \left[-\frac{1}{3}(4 - x^2)^{3/2} - \frac{x^5}{5} + \frac{3}{4}x^4 - \frac{2}{3}x^3 \right]_0^2$$

$$= \frac{88\pi}{15} \text{ cubic units}$$

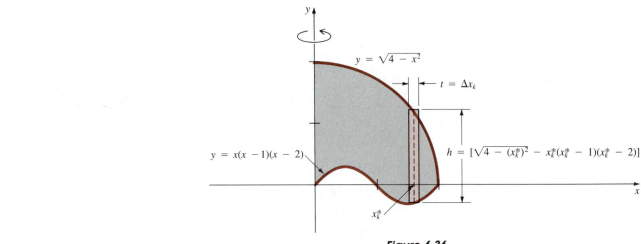

Figure 6.36

Exercises 6.4

Answers to odd-numbered problems begin on page A-29

In Problems 1–6 refer to Figure 6.37. Use the shell method to find the volume of the solid of revolution that is formed by revolving the given region about the indicated line.

1. R_1 about OC

2. R_1 about OA

3. R_2 about BC

4. R_2 about OA

5. R_1 about AB

6. R_2 about AB

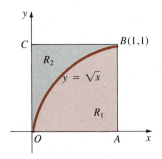

Figure 6.37

In Problems 7–30 use the shell method to find the volume of the solid of revolution that is formed by revolving the region bounded by the graphs of the given equations about the indicated line or axis.

7. $y = x$, $x = 0$, $y = 5$; x-axis

8. $y = 1 - x$, $x = 0$, $y = 0$; $y = -2$

9. $y = x^2$, $x = 0$, $y = 3$ (first quadrant); x-axis

10. $y = x^2$, $x = 2$, $y = 0$; y-axis

11. $y = x^2$, $x = 1$, $y = 0$; $x = 3$

12. $y = x^2$, $y = 9$; x-axis

13. $y = x^2 + 4$, $x = 0$, $x = 2$, $y = 2$; y-axis

14. $y = x^2 - 5x + 4$, $y = 0$; y-axis

15. $y = (x - 1)^2$, $y = 1$; x-axis

16. $y = (x - 2)^2$, $y = 4$; $x = 4$

17. $y = x^{1/3}$, $x = 1$, $y = 0$; $y = -1$

18. $y = x^{1/3} + 1$, $y = -x + 1$, $x = 1$; $x = 1$

19. $y = x^2$, $y = x$; y-axis

20. $y = x^2$, $y = x$; $x = 2$

21. $y = x^3$, $y = x + 6$, $x = 0$; y-axis

22. $y = x^3 - x$, $y = 0$ (second quadrant); y-axis

23. $y = x^2 - 2$, $y = -x^2 + 2$, $x = 0$ (second and third quadrants); y-axis

24. $y = x^2 - 4x$, $y = -x^2 + 4x$; $x = -1$

25. $x = y^2 - 5y$, $x = 0$; x-axis

26. $x = y^2 + 2$, $y = x - 4$, $y = 1$; x-axis

27. $y = \sqrt{x - 1}$, $x = 0$, $y = 0$, $y = 3$; $y = 3$

28. $y = \sqrt{x}$, $y = \sqrt{1-x}$, $y = 0$; x-axis

29. $y = \sin x^2$, $x = 0$, $y = 1$; y-axis

30. $x^2 - y^2 = 1$, $x = \sqrt{5}$, $y = 0$ (first quadrant); y-axis

31. Use the shell method to derive the formula for the volume of a right circular cone. (See Problem 34 in Exercises 6.3.)

32. Use the shell method to derive the formula for the volume of a sphere.

33. The region bounded by the graph of the ellipse $x^2/a^2 + y^2/b^2 = 1$, $a > b > 0$, in the first and fourth quadrants, and $x = 0$ is revolved about the y-axis. Use the shell method to find the volume of the solid of revolution. The resulting solid is known as an **oblate spheroid**.

34. A cylindrical bucket of radius 2 ft that contains liquid is rotating about the y-axis with a constant angular velocity ω. The surface of the liquid has a parabolic cross section given by $y = \omega^2 x^2/2g$, $-2 \le x \le 2$. Use the shell method to find the volume of the liquid in the rotating bucket given that the height of the bucket is 3 ft. See Figure 6.38.

35. The shaded regions in Figure 6.39 are revolved about the y-axis. Use the shell method to find the volume of the solid of revolution.

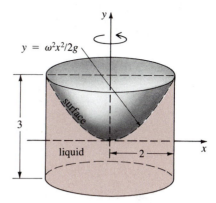

Figure 6.38

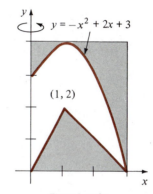

Figure 6.39

6.5 Arc Length

Smooth Functions   If $y = f(x)$ has a continuous first derivative on an interval $[a, b]$, then its graph is said to be **smooth** and f is called a **smooth function**. As the name implies, a smooth graph has no sharp points. In this section we will find the length of a smooth graph.

Let f have a smooth graph on $[a, b]$, and let P denote the arbitrary partition

$$a = x_0 < x_1 < x_2 < \cdots < x_{n-1} < x_n = b$$

As usual, let the length of each subinterval be given by $\Delta x_k = x_k - x_{k-1}$ and let $\|P\|$ be the length of the longest subinterval. As shown in Figure 6.40, the length of the chord between $(x_{k-1}, f(x_{k-1}))$ and $(x_k, f(x_k))$ is an approximation to the length of the graph between these points. The length of the chord is

$$\Delta s_k = \sqrt{(x_k - x_{k-1})^2 + (f(x_k) - f(x_{k-1}))^2} \tag{6.10}$$

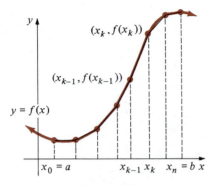

Figure 6.40

By the Mean Value Theorem, we know there exists an x_k^* in each open subinterval (x_{k-1}, x_k) such that

$$\frac{f(x_k) - f(x_{k-1})}{x_k - x_{k-1}} = f'(x_k^*)$$

or

$$f(x_k) - f(x_{k-1}) = f'(x_k^*)(x_k - x_{k-1})$$

Substituting from this last equation into (6.10) gives

$$\begin{aligned} \Delta s_k &= \sqrt{(x_k - x_{k-1})^2 + [f'(x_k^*)]^2(x_k - x_{k-1})^2} \\ &= \sqrt{1 + [f'(x_k^*)]^2}(x_k - x_{k-1}) \\ &= \sqrt{1 + [f'(x_k^*)]^2}\,\Delta x_k \end{aligned}$$

The sum

$$\sum_{k=1}^{n} \Delta s_k = \sum_{k=1}^{n} \sqrt{1 + [f'(x_k^*)]^2}\,\Delta x_k$$

gives an approximation to the total length of the graph on $[a, b]$. As $\|P\| \to 0$, we obtain

$$\lim_{\|P\| \to 0} \sum_{k=1}^{n} \sqrt{1 + [f'(x_k^*)]^2}\,\Delta x_k = \int_a^b \sqrt{1 + [f'(x)]^2}\,dx \qquad (6.11)$$

The foregoing discussion prompts us to use equation (6.11) as the definition of the length of the graph on the interval.

DEFINITION 6.4   Arc Length

Let f be a function for which f' is continuous on an interval $[a, b]$. The length s of the graph on the interval is given by

$$s = \int_a^b \sqrt{1 + [f'(x)]^2}\,dx \qquad (6.12)$$

A graph that has arc length is said to be **rectifiable**.

Example 1

Find the length of the graph of $y = 4x^{3/2}$ from the origin $(0, 0)$ to the point $(1, 4)$.

Solution   The graph of the function on $[0, 1]$ is given in Figure 6.41. Now,

$$\frac{dy}{dx} = f'(x) = 6x^{1/2}$$

is continuous on the interval; therefore, it follows from (6.12) that

$$s = \int_0^1 \sqrt{1 + [6x^{1/2}]^2}\; dx$$

$$= \int_0^1 \sqrt{1 + 36x}\; dx$$

$$= \int_0^1 (1 + 36x)^{1/2}\; dx$$

$$= \frac{1}{36} \int_0^1 (1 + 36x)^{1/2}(36\; dx)$$

$$= \frac{1}{54}(1 + 36x)^{3/2}\Big]_0^1$$

$$= \frac{1}{54}[37^{3/2} - 1] \approx 4.1493$$

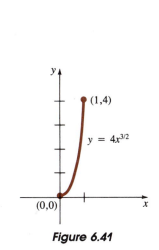

(1,4)

$y = 4x^{3/2}$

(0,0)

Figure 6.41

Differential of Arc Length

If $s = \int_a^x \sqrt{1 + [f'(t)]^2}\; dt$, then by Theorem 5.12

$$\frac{ds}{dx} = \sqrt{1 + [f'(x)]^2}$$

and, consequently,

$$ds = \sqrt{1 + [f'(x)]^2}\; dx \qquad (6.13)$$

The latter function is called the **differential of the arc length** and can be used to approximate lengths of curves. Using $dy = f'(x)\; dx$, (6.13) is often written as

$$ds = \sqrt{(dx)^2 + (dy)^2} \quad \text{or} \quad (ds)^2 = (dx)^2 + (dy)^2 \qquad (6.14)$$

Figure 6.42 shows that the differential ds can be interpreted as the hypotenuse of a right triangle with sides dx and dy.

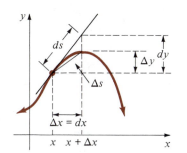

ds

Δy dy

Δs

$\Delta x = dx$

$x \quad x + \Delta x$

Figure 6.42

Remark

The integral in (6.12) often leads to problems in which special techniques of integration are necessary. See Chapter 9. But even with these subsequent procedures, it is not *always* possible to evaluate the indefinite integral $\int \sqrt{1 + [f'(x)]^2}\; dx$ in terms of the familiar elementary functions even for some of the simplest functions. See Problem 21 in Exercises 6.5.

Exercises 6.5

Answers to odd-numbered problems begin on page A-29

In Problems 1–10 find the length of the graph of the given equation on the indicated interval.

1. $y = x$; $[-1, 1]$

2. $y = 2x + 1$; $[0, 3]$

3. $y = x^{3/2} + 4$; $(0, 4)$ to $(1, 5)$

4. $y = 3x^{2/3}$; $[1, 8]$

5. $y = \frac{2}{3}(x^2 + 1)^{3/2}$; $[1, 4]$

6. $(y + 1)^2 = 4(x + 1)^3$; $(-1, -1)$ to $(0, 1)$

7. $y = \frac{1}{3}x^{3/2} - x^{1/2}$; $[1, 4]$

8. $y = \frac{1}{6}x^3 + \frac{1}{2x}$; $[2, 4]$

9. $y = \frac{1}{4}x^4 + \frac{1}{8x^2}$; $[2, 3]$

10. $y = \frac{1}{5}x^5 + \frac{1}{12x^3}$; $[1, 2]$

In Problems 11–14 set up, but do not evaluate, an integral for the length of the graph of the given function on the indicated interval.

11. $y = x^2$; $[-1, 3]$

12. $y = 2\sqrt{x + 1}$; $[-1, 3]$

13. $y = \sin x$; $[0, \pi]$

14. $y = \tan x$; $[-\pi/4, \pi/4]$

Let $x = g(y)$, where g' is continuous on an interval $[c, d]$ on the y-axis. The length of the graph on the interval is then

$$s = \int_c^d \sqrt{1 + [g'(y)]^2}\, dy \qquad (6.15)$$

In Problems 15 and 16 use formula (6.15) to find the length of the graph of the given equation on the indicated interval.

15. $x = 4 - y^{2/3}$; $[0, 8]$

16. $5x = y^{5/2} + 5y^{-1/2}$; $[4, 9]$

Miscellaneous Problems

17. Consider the length of the graph of $x^{2/3} + y^{2/3} = 1$ in the first quadrant.

  (a) Show that use of equation (6.12) leads to a discontinuous integrand.

  (b) By assuming that the Fundamental Theorem of Calculus can be used to evaluate the integral obtained in part (a), find the total length of the graph.

18. Set up, but make no attempt to evaluate, an integral that gives the total length of the ellipse $x^2/a^2 + y^2/b^2 = 1$, $a > b > 0$.

19. Given that the circumference of a circle of radius r is $2\pi r$, find the value of the integral

$$\int_0^1 \frac{dx}{\sqrt{1 - x^2}}$$

20. Use (6.15) to approximate the length of the graph of $y = x^4/4$ from $(2, 4)$ to $(2.1, 4.862025)$.

Calculator Problems

21. Use the Trapezoidal Rule with $n = 10$ to find an approximation to the length of the graph of $y = x^2$ on the interval $[0, 1]$.

22. Use Simpson's Rule with $n = 4$ to find an approximation to the length of the graph of $y = \frac{1}{3}x^3 + 1$ from $(0, 1)$ to $(2, \frac{11}{3})$.

6.6 Surfaces of Revolution

Surface of Revolution—
x-axis

As we have seen in Sections 6.3 and 6.4, when the graph of a continuous function $y = f(x)$ on an interval $[a, b]$ is revolved about the x-axis, it generates a solid of revolution. In this section, we are interested in finding the area of the corresponding surface—that is, a **surface of revolution** as shown in Figure 6.43(b).

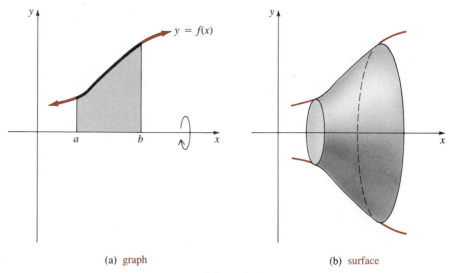

(a) graph   (b) surface

Figure 6.43

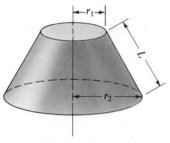

Figure 6.44

To derive the formula for the area of a surface of revolution, we first note that the lateral area (top and bottom excluded) of a *frustum* of a right circular cone shown in Figure 6.44 is given by

$$\pi[r_1 + r_2]L \tag{6.16}$$

where r_1 and r_2 are the radii of the top and bottom and L is the slant height. See Problem 16 in Exercises 6.6.

Now, suppose $y = f(x)$ is a smooth function and $f(x) \geq 0$ on the interval $[a, b]$. Let P be a partition of the interval:

$$a = x_0 < x_1 < x_2 < \cdots < x_{n-1} < x_n = b$$

Now, if we connect the points $(x_{k-1}, f(x_{k-1}))$ and $(x_k, f(x_k))$ shown in Figure 6.45(a) by a chord, we form a trapezoid. When revolved about the x-axis, this trapezoid generates a frustum of a cone with radii $f(x_{k-1})$ and $f(x_k)$. See Figure 6.45(b). As shown in cross section in Figure 6.46, the slant height can be obtained from the Pythagorean Theorem:

$$\sqrt{(x_k - x_{k-1})^2 + (f(x_k) - f(x_{k-1}))^2}$$

Thus, from (6.16) the surface area of this element is

$$\Delta S_k = \pi[f(x_k) + f(x_{k-1})] \sqrt{(x_k - x_{k-1})^2 + (f(x_k) - f(x_{k-1}))^2}$$

$$= \pi[f(x_k) + f(x_{k-1})] \sqrt{1 + \left(\frac{f(x_k) - f(x_{k-1})}{x_k - x_{k-1}}\right)^2} (x_k - x_{k-1})$$

$$= \pi[f(x_k) + f(x_{k-1})] \sqrt{1 + \left(\frac{f(x_k) - f(x_{k-1})}{x_k - x_{k-1}}\right)^2} \Delta x_k$$

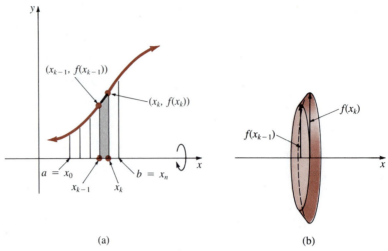

(a)   (b)

Figure 6.45

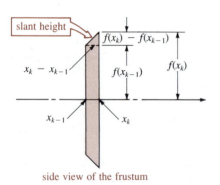

side view of the frustum

Figure 6.46

where $\Delta x_k = x_k - x_{k-1}$. This last quantity is an approximation to the actual area of the surface of the revolution on the subinterval $[x_{k-1}, x_k]$.

Now, as in the discussion of arc length, we invoke the Mean Value Theorem for derivatives to assert that there exists an x_k^* in (x_{k-1}, x_k) such that

$$f'(x_k^*) = \frac{f(x_k) - f(x_{k-1})}{x_k - x_{k-1}}$$

This suggests that the surface area is given by

$$\lim_{\|P\| \to 0} \pi \sum_{k=1}^{n} [f(x_k) + f(x_{k-1})] \sqrt{1 + [f'(x_k^*)]^2}\, \Delta x_k$$

Since we also expect $f(x_{k-1})$ and $f(x_k)$ to approach the common limit $f(x)$ as $\|P\| \to 0$, the last equation becomes

$$2\pi \int_a^b f(x) \sqrt{1 + [f'(x)]^2}\, dx$$

The foregoing is a plausibility argument for the following definition.

DEFINITION 6.5   Area of a Surface of Revolution

Let f be a function for which f' is continuous and $f(x) \geq 0$ for all x in the interval $[a, b]$. The area S of the surface that is obtained by revolving the graph of f on the interval about the x-axis is given by

$$S = 2\pi \int_a^b f(x) \sqrt{1 + [f'(x)]^2}\, dx \qquad (6.17)$$

Example 1

Find the area S of the surface that is formed by revolving the graph of $y = \sqrt{x}$ on the interval $[1, 4]$ about the x-axis.

Solution   We have $f(x) = x^{1/2}$, $f'(x) = \frac{1}{2}x^{-1/2} = 1/(2\sqrt{x})$ and from (6.17) of Definition 6.5

$$S = 2\pi \int_1^4 \sqrt{x}\,\sqrt{1 + \left(\frac{1}{2\sqrt{x}}\right)^2}\,dx$$

$$= 2\pi \int_1^4 \sqrt{x}\,\sqrt{1 + \frac{1}{4x}}\,dx$$

$$= 2\pi \int_1^4 \sqrt{x}\,\sqrt{\frac{4x + 1}{4x}}\,dx$$

$$= \pi \int_1^4 \sqrt{4x + 1}\,dx$$

$$= \frac{\pi}{4} \int_1^4 (4x + 1)^{1/2}(4\,dx)$$

$$= \frac{\pi}{6}(4x + 1)^{3/2}\Big]_1^4$$

$$= \frac{\pi}{6}\,[17^{3/2} - 5^{3/2}] \approx 30.85 \text{ square units}$$

See Figure 6.47.

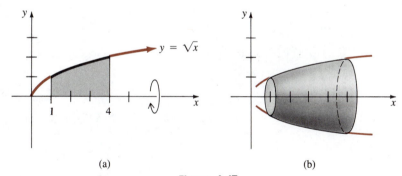

(a)   (b)

Figure 6.47

Surface of Revolution— y-axis   It can be shown that if the graph of a continuous function $y = f(x)$ on $[a, b]$, $0 \leq a < b$, is revolved about the y-axis, then the area S of the resulting surface of revolution is given by

$$S = 2\pi \int_a^b x\sqrt{1 + [f'(x)]^2}\,dx \tag{6.18}$$

As in (6.17) we assume in (6.18) that $f'(x)$ is also continuous on the interval $[a, b]$.

Example 2

Find the area S of the surface formed by revolving the graph of $y = 3x^{1/3}$ on the interval $[1, 8]$ about the y-axis.

Solution We have $f'(x) = x^{-2/3}$ so that from (6.18) it follows that

$$S = 2\pi \int_1^8 x\sqrt{1 + x^{-4/3}}\, dx$$

$$= 2\pi \int_1^8 x^{1/3}\sqrt{x^{4/3} + 1}\, dx$$

$$= 2\pi\left(\frac{3}{4}\right)\int_1^8 (x^{4/3} + 1)^{1/2}\left(\frac{4}{3}x^{1/3}\, dx\right)$$

$$= \pi(x^{4/3} + 1)^{3/2}\Big]_1^8$$

$$= \pi[17^{3/2} - 2^{3/2}] \approx 211.32 \text{ square units}$$

See Figure 6.48.

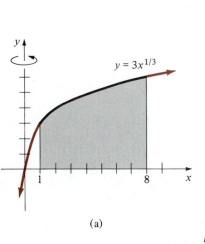

(a)

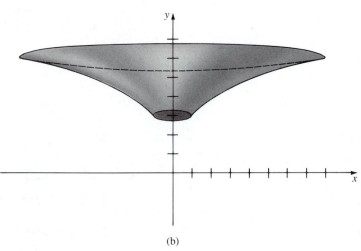

(b)

Figure 6.48

Exercises 6.6

Answers to odd-numbered problems begin on page A-29

In Problems 1–10 find the area of the surface that is formed by revolving each graph on the given interval about the indicated axis.

1. $y = 2\sqrt{x}$, $[0, 8]$; x-axis

2. $y = \sqrt{x + 1}$, $[1, 5]$; x-axis

3. $y = x^3$, $[0, 1]$; x-axis

4. $y = x^{1/3}$, $[1, 8]$; y-axis

5. $y = x^2 + 1$, $[0, 3]$; y-axis

6. $y = 4 - x^2$, $[0, 2]$; y-axis

7. $y = 2x + 1$, $[2, 7]$; x-axis

8. $y = \sqrt{16 - x^2}$, $[0, \sqrt{7}]$; y-axis

9. $y = \frac{1}{4}x^4 + \frac{1}{8x^2}$, $[1, 2]$; y-axis

10. $y = \frac{1}{3}x^3 + \frac{1}{4x}$, $[1, 2]$; x-axis

11. The surface formed by two parallel planes cutting a sphere of radius r is called a **spherical zone**. Find the area of the spherical zone shown in Figure 6.49.

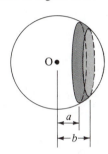

Figure 6.49

12. The graph of $y = |x + 2|$ on $[-4, 2]$, shown in Figure 6.50, is revolved about the x-axis. Find the area S of the surface of revolution.

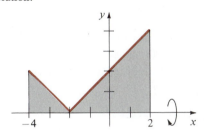

Figure 6.50

Miscellaneous Problems

13. Find the area of the surface that is formed by revolving $x^{2/3} + y^{2/3} = a^{2/3}$, $[-a, a]$, about the x-axis.

14. Show that the lateral surface area of a right circular cone of radius r and slant height L is $\pi r L$. (*Hint:* A cone cut down its side and flattened forms a circular sector with area $\frac{1}{2}L^2\theta$.)

15. Use Problem 14 to show that the lateral surface area of a right circular cone of radius r and height h is given by $\pi r \sqrt{r^2 + h^2}$. Derive the same result using (6.16) or (6.17).

16. Use the result of Problem 14 to derive formula (6.16). (*Hint:* Consider a complete cone of radius r_2 and slant height L_2. Cut the conical top off. Similar triangles might help.)

17. Show that the surface area of a frustum of a right circular cone of radii r_1 and r_2 and height h is given by $\pi(r_1 + r_2)\sqrt{h^2 + (r_2 - r_1)^2}$.

18. Let $y = f(x)$ be a continuous nonnegative function $[a, b]$ that has a continuous first derivative on the interval. Prove that if the graph of f is revolved around a horizontal line $y = L$, then the area S of the resulting surface of revolution is given by

$$S = 2\pi \int_a^b |f(x) - L| \sqrt{1 + [f'(x)]^2} \, dx$$

19. Use the result of Problem 18 to find a definite integral that gives the area of the surface that is formed by revolving $y = x^{2/3}$, $[1, 8]$, about the line $y = 4$. Do not evaluate.

Calculator Problem

20. Use the Trapezoidal Rule with $n = 5$ to find an approximation to the area of the surface that is formed by revolving the graph of $y = \frac{1}{2}x^2$ on the interval $[0, 2]$ about the x-axis.

6.7 Average Value of a Function and the Mean Value Theorem

Averages   Every student is aware of averages. If a student takes four examinations in a semester and scores 80%, 75%, 85%, and 92% on them, then the student's average score is

$$\frac{80 + 75 + 85 + 92}{4}$$

or 83%. In general, given n numbers $a_1, a_2, \ldots, a_n$, we say that their **arithmetic mean**, or **average**, is

$$\frac{a_1 + a_2 + \cdots + a_n}{n} = \frac{1}{n}\sum_{k=1}^{n} a_k$$

Suppose now that we have a continuous function f defined on an interval $[a, b]$. For the arbitrarily chosen numbers x_i, $i = 1, 2, \ldots, n$ such that $a < x_1 < x_2 < \cdots < x_n < b$, the average of the set of corresponding functional values is

$$\frac{f(x_1) + f(x_2) + \cdots + f(x_n)}{n} = \frac{1}{n} \sum_{k=1}^{n} f(x_k) \tag{6.19}$$

Example 1

Consider the function $f(x) = x^2$ on $[0, 3]$. If $x_1 = \frac{1}{2}$, $x_2 = 1$, $x_3 = \frac{3}{2}$, $x_4 = 2$, then

$$\frac{f\left(\frac{1}{2}\right) + f(1) + f\left(\frac{3}{2}\right) + f(2)}{4} = \frac{\frac{1}{4} + 1 + \frac{9}{4} + 4}{4} = \frac{15}{8}$$

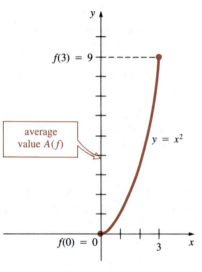

Figure 6.51

If we now consider the set of values $f(x)$ that correspond to *all* points x in an interval, it should be clear that we cannot use a discrete sum as in (6.19). For example, for $f(x) = x^2$ on $[0, 3]$ the values of the function range from a minimum of $f(0) = 0$ to a maximum of $f(3) = 9$. As indicated in Figure 6.51, we intuitively expect that there exists an average value $A(f)$ such that $f(0) \leq A(f) \leq f(3)$.

Returning to the general case of a continuous function defined on a closed interval $[a, b]$, let P be a partition of the interval into n subintervals of equal length $\Delta x = (b - a)/n$. If x_k^* is a number chosen in each subinterval, then the average

$$\frac{f(x_1^*) + f(x_2^*) + \cdots + f(x_n^*)}{n}$$

can be written as

$$\frac{f(x_1^*) + f(x_2^*) + \cdots + f(x_n^*)}{\dfrac{b - a}{\Delta x}} \tag{6.20}$$

since $n = (b - a)/\Delta x$. Rewriting (6.20) as

$$\frac{1}{b - a} \sum_{k=1}^{n} f(x_k^*) \, \Delta x$$

and taking the limit of this last expression as $\|P\| = \Delta x \to 0$, we obtain the definite integral

$$\frac{1}{b - a} \int_a^b f(x) \, dx \tag{6.21}$$

Since we have assumed that f is continuous on $[a, b]$, let us denote its minimum and maximum on the interval by m and M, respectively. If we multiply

$$m \leq f(x_k^*) \leq M$$

by $\Delta x > 0$ and sum, we obtain

$$\sum_{k=1}^{n} m \, \Delta x \leq \sum_{k=1}^{n} f(x_k^*) \, \Delta x \leq \sum_{k=1}^{n} M \, \Delta x$$

Since $\sum_{k=1}^{n} \Delta x = b - a$, this last expression can be written as

$$(b - a)m \leq \sum_{k=1}^{n} f(x_k^*) \, \Delta x_k \leq (b - a)M$$

And so as $\Delta x \to 0$, it follows that

$$(b - a)m \leq \int_a^b f(x) \, dx \leq (b - a)M$$

We conclude that the number (6.21) satisfies

$$m \leq \frac{1}{b - a} \int_a^b f(x) \, dx \leq M$$

By the Intermediate Value Theorem, f takes on all values between m and M. Hence, the number given by (6.21) actually corresponds to a value of the function on the interval. This prompts us to state the following definition.

DEFINITION 6.6   **Average Value of a Function**

Let f be continuous on $[a, b]$. The average value of f on the interval is the number

$$A(f) = \frac{1}{b - a} \int_a^b f(x) \, dx \tag{6.22}$$

Although we are interested primarily in continuous functions, Definition 6.6 is valid for any integrable function on the interval.

Example 2

Find the average value of $f(x) = x^2$ on $[0, 3]$.

Solution   From (6.22) of Definition 6.6, we obtain

$$A(f) = \frac{1}{3 - 0} \int_0^3 x^2 \, dx$$

$$= \frac{1}{3} \left(\frac{x^3}{3} \right) \Big]_0^3$$

$$= \frac{1}{3} \left(\frac{27}{3} \right) = 3$$

It is sometimes possible to determine the value of x in the interval that corresponds to the average value of a function.

_____ **Example 3** _____

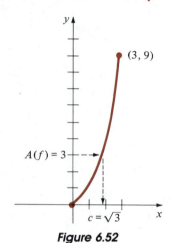

Figure 6.52

Determine the value of x in the interval $[0, 3]$ that corresponds to the average value of the function $f(x) = x^2$.

Solution   Since the function $f(x) = x^2$ is continuous on the closed interval $[0, 3]$, we know from Theorem 2.14 that there exists a number c between 0 and 3 so that

$$f(c) = c^2 = A(f)$$

But, from Example 2, we know $A(f) = 3$. Thus, the equation

$$c^2 = 3 \quad \text{has solutions} \quad c = \pm\sqrt{3}$$

As shown in Figure 6.52 the only solution of this equation in $[0, 3]$ is $c = \sqrt{3}$.

Mean Value Theorem for Definite Integrals

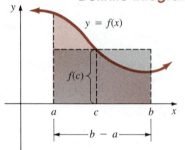

Figure 6.53

The following is an immediate consequence of the foregoing discussion.

THEOREM 6.1   The Mean Value Theorem for Definite Integrals

Let f be continuous on $[a, b]$. Then there exists a number c in the open interval (a, b) such that

$$f(c)(b - a) = \int_a^b f(x)\, dx$$

If $f(x) \geq 0$ for all x in $[a, b]$, then Theorem 6.1 simply states that there is some value c in (a, b) for which the area of a rectangle of height $f(c)$ and width $b - a$ is the same as the area under the graph indicated in Figure 6.53.

_____ **Example 4** _____

Find the height $f(c)$ of a rectangle so that the area A under the graph of $y = x^2 + 1$, on $[-2, 2]$ is the same as $f(c)[2 - (-2)] = 4f(c)$.

Solution   This is basically the same type of problem as illustrated in Example 3. Now, the area under the graph is

$$A = \int_{-2}^{2} (x^2 + 1)\, dx$$

$$= \left(\frac{x^3}{3} + x\right)\Bigg]_{-2}^{2} = \frac{28}{3}$$

Also, $4f(c) = 4(c^2 + 1)$ so that

$$4(c^2 + 1) = \frac{28}{3}$$

implies $c^2 = \frac{4}{3}$. The solutions $c = 2/\sqrt{3}$ and $c = -2/\sqrt{3}$ are both in the interval $(-2, 2)$. The height of the rectangle is $f(c) = (\pm 2/\sqrt{3})^2 + 1 = \frac{7}{3}$. See Figure 6.54.

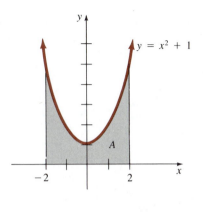

(a)

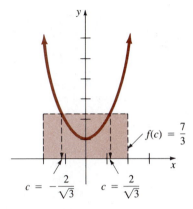

(b)

Figure 6.54

Exercises 6.7

Answers to odd-numbered problems begin on page A-29

In Problems 1–20 find the average value $A(f)$ of the given function on the indicated interval.

1. $f(x) = 4x$; $[-3, 1]$

2. $f(x) = 2x + 3$; $[-2, 5]$

3. $f(x) = x^2 + 10$; $[0, 2]$

4. $f(x) = 2x^3 - 3x^2 + 4x - 1$; $[-1, 1]$

5. $f(x) = 3x^2 - 4x$; $[-1, 3]$

6. $f(x) = (x + 1)^2$; $[0, 2]$

7. $f(x) = x^3$; $[-2, 2]$

8. $f(x) = x(3x - 1)^2$; $[0, 1]$

9. $f(x) = \sqrt{x}$; $[0, 9]$

10. $f(x) = \sqrt{5x + 1}$; $[0, 3]$

11. $f(x) = x\sqrt{x^2 + 16}$; $[0, 3]$

12. $f(x) = \left(1 + \dfrac{1}{x}\right)^{1/3} \dfrac{1}{x^2}$; $\left[\dfrac{1}{2}, 1\right]$

13. $f(x) = \dfrac{1}{x^3}$; $\left[\dfrac{1}{4}, \dfrac{1}{2}\right]$

14. $f(x) = x^{2/3} - x^{-2/3}$; $[1, 4]$

15. $f(x) = \dfrac{2}{(x + 1)^2}$; $[3, 5]$

16. $f(x) = \dfrac{(\sqrt{x} - 1)^3}{\sqrt{x}}$; $[4, 9]$

17. $f(x) = \sin x$; $[-\pi, \pi]$

18. $f(x) = \cos 2x$; $\left[0, \dfrac{\pi}{4}\right]$

19. $f(x) = \csc^2 x$; $\left[\dfrac{\pi}{6}, \dfrac{\pi}{2}\right]$

20. $f(x) = \dfrac{\sin \pi x}{\cos^2 \pi x}$; $\left[-\dfrac{1}{3}, \dfrac{1}{3}\right]$

21. For $f(x) = x^2 + 2x$, find a value of c in the interval $[-1, 1]$ for which $f(c) = A(f)$.

22. For $f(x) = \sqrt{x + 3}$, find a value of c in the interval $[1, 6]$ for which $f(c) = A(f)$.

23. The average value of a continuous nonnegative function $y = f(x)$ on the interval $[1, 5]$ is $A(f) = 3$. What is the area under the graph on the interval?

24. The area under the graph of a continuous nonnegative function $y = f(x)$ on the interval $[-3, 4]$ is 21 square units. What is the average value of the function on the interval?

25. The function $T(t) = 100 + 3t - \frac{1}{2}t^2$ approximates the temperature at t hours past noon on a typical August day in Las Vegas. Find the average temperature between noon and 6 P.M.

26. A company determines that the revenue obtained after the sale of x units of a product is given by $R(x) = 50 + 4x + 3x^2$. Find the average revenue for sales $x = 1$ to $x = 5$. Compare the result with the average $\frac{1}{5}\Sigma_{k=1}^{5} R(k)$.

27. Let $s(t)$ denote the position of a particle on a horizontal axis as a function of time t. The average velocity v_{ave} during the time interval $[t_1, t_2]$ is $v_{\text{ave}} = [s(t_2) - s(t_1)]/(t_2 - t_1)$. Show that $v_{\text{ave}} = A(v)$.

28. In the absence of damping, the position of a mass m on a freely vibrating spring is given by the function $x(t) = A\cos(\omega t + \phi)$, where A, ω, and ϕ are constants. The period of oscillation is $2\pi/\omega$. The potential energy of the system is $U(x) = \frac{1}{2}kx^2$, where k is the so-called spring constant. The kinetic energy of the system is $K = \frac{1}{2}mv^2$, where $v = dx/dt$. If $\omega^2 = k/m$, show that the average potential energy and average kinetic energy over one period are the same and that each equals $\frac{1}{4}kA^2$.

29. The *impulse-momentum theorem* states that the change in momentum of a body in a time interval $[t_0, t_1]$ is $mv_1 - mv_0 = (t_1 - t_0)\overline{F}$, where mv_0 is the initial momentum, mv_1 is the final momentum, and $\overline{F}$ is the average force acting on the body during the interval. Find the change in momentum of a pile driver dropped on a piling between times $t = 0$ and $t = t_1$ if

$$F(t) = k\left[1 - \left(\frac{2t}{t_1} - 1\right)^2\right]$$

where k is a constant.

30. In a small artery the velocity of blood, in cm/sec, is given by $v(r) = (P/4\nu l)(R^2 - r^2)$, $0 \le r \le R$, where P is blood pressure, ν the viscosity of the blood, l the length of the artery, and R the radius of the artery. Find the average of $v(r)$ on the interval $[0, R]$.

Miscellaneous Problems

31. If $y = f(x)$ is a differentiable function, find the average value of f' on the interval $[x, x + h]$, where $h > 0$.

32. For a linear function $f(x) = ax + b$, $a > 0$, $b > 0$, show that the average value of the function on $[x_1, x_2]$ is $A(f) = aX + b$, where X is the x-coordinate of the midpoint of the interval.

33. Given that n is a positive integer and $a > 1$, show that the average value of $f(x) = (n + 1)x^n$ on the interval $[1, a]$ is $A(f) = a^n + a^{n-1} + \cdots + a + 1$.

34. The **mean square** of a continuous function on an interval $[a, b]$ is defined as

$$\frac{1}{b - a}\int_a^b [f(x)]^2 \, dx$$

Compute the average value and the mean square of $f(x) = (x + 1)^{-1/3}$ on $[0, 7]$.

35. Figure 6.55 shows a car's velocity in miles per hour as a function of miles traveled. Interpret the graph. Find the average velocity of the car with respect to distance. What is the average velocity of the car with respect to time?

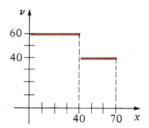

Figure 6.55

36. Show that the average value of the greatest integer function $f(x) = [x]$ on the interval $[0, n]$, where n is a positive integer, is $\frac{1}{2}(n - 1)$.

6.8 Rectilinear Motion and the Integral

If $s = f(t)$ is the position function of an object moving rectilinearly—that is, in a straight line, then we know

$$\text{velocity} = v(t) = \frac{ds}{dt}$$

$$\text{acceleration} = a(t) = \frac{dv}{dt}$$

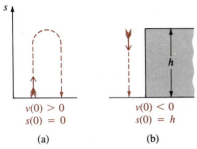

$v(0) > 0$
$s(0) = 0$

(a)

$v(0) < 0$
$s(0) = h$

(b)

Figure 6.56

As an immediate consequence of the definition of an antiderivative, the quantities s and v can be written as indefinite integrals

$$s(t) = \int v(t)\, dt \tag{6.23}$$

$$v(t) = \int a(t)\, dt \tag{6.24}$$

By knowing the **initial position** $s(0)$ and the **initial velocity** $v(0)$, we can find specific values of the constants of integration used in (6.23) and (6.24).

Recall that when a body moves horizontally on a line, the positive direction is to the right. For motion in a vertical line, the positive direction is upward. As shown in Figure 6.56, if an arrow is shot upward from ground level, then $s(0) = 0$, $v(0) > 0$, whereas if the arrow is shot downward from some initial height, say h meters off the ground, then $s(0) = h$, $v(0) < 0$. A body that moves in a vertical line close to the surface of the earth, such as the arrow shot upward, is acted upon by the force of gravity. This force causes a body to accelerate. Near the surface of the earth, the acceleration of gravity $a(t)$ is assumed to be a constant given by

$$-32 \text{ ft/sec}^2, \qquad -9.8 \text{m/sec}^2, \quad \text{or} \quad -980 \text{ cm/sec}^2$$

___ **Example 1** _____

A projectile is shot vertically upward from ground level with an initial velocity of 49 m/sec. What is its velocity at $t = 2$ sec? What is the maximum height attained by the projectile? How long is the projectile in the air? What is its impact velocity?

Solution   We use $a(t) = -9.8$ and

$$v(t) = \int (-9.8)\, dt$$
$$= -9.8t + C_1 \tag{6.25}$$

From the given initial condition $v(0) = 49$, we see that (6.25) implies $C_1 = 49$. Hence,

$$v(t) = -9.8t + 49$$

and so $v(2) = -9.8(2) + 49 = 29.4$ m/sec. Notice that $v(2) > 0$ implies the projectile is traveling upward.

Now, the height of the projectile, measured from ground level, is

$$s(t) = \int v(t)\, dt$$

$$= \int (-9.8t + 49)\, dt$$

$$= -4.9t^2 + 49t + C_2 \tag{6.26}$$

Since the projectile starts from ground level $s(0) = 0$ and (6.26) gives $C_2 = 0$. Hence,

$$s(t) = -4.9t^2 + 49t \qquad (6.27)$$

When the projectile attains its maximum height, $v(t) = 0$. Solving $-9.8t + 49 = 0$ then gives $t = 5$. From (6.27) we find the corresponding height to be $s(5) = 122.5$ m.

Finally, to find the time that the projectile hits the ground, we solve $s(t) = 0$ or $-4.9t^2 + 49t = 0$. Writing the latter equation as $-4.9t(t - 10) = 0$, we see the projectile is in the air for 10 sec. The impact velocity is $v(10) = -49$ m/sec.*

Example 2

Figure 6.57

A tennis ball is thrown vertically downward from a height of 54 ft with an initial velocity of 8 ft/sec. What is its impact velocity if it hits a 6-ft-tall person on the head? See Figure 6.57.

Solution   In this case $a(t) = -32$, $s(0) = 54$ and, since the ball is thrown downward, $v(0) = -8$. Now

$$v(t) = \int (-32)\, dt + C_1$$
$$= -32t + C_1$$

Using the initial velocity $v(0) = -8$, we find $C_1 = -8$. Therefore,

$$v(t) = -32t - 8$$

Continuing, we find

$$s(t) = \int (-32t - 8)\, dt$$
$$= -16t^2 - 8t + C_2$$

When $t = 0$ we know $s = 54$ and so the last equation implies $C_2 = 54$. Hence,

$$s(t) = -16t^2 - 8t + 54$$

To determine the time that corresponds to $s = 6$ we solve

$$-16t^2 - 8t + 54 = 6$$

Simplifying gives $-8(2t - 3)(t + 2) = 0$ and $t = \frac{3}{2}$ sec. The velocity of the ball, when it hits the person, is then $v(\frac{3}{2}) = -56$ ft/sec.

---

*When air resistance is ignored, the magnitude of the impact velocity (speed) is the same as the initial upward velocity from ground level. This is not true when air resistance is taken into consideration. See Problem 30 in Exercises 6.8.

Distance   The total **distance** an object travels in a straight line in a time interval $[t_1, t_2]$ is given by the definite integral

$$\int_{t_1}^{t_2} |v(t)| \, dt \qquad (6.28)$$

The absolute value is necessary in (6.28) since the object may be moving to the left, and hence has negative velocity for some part of the time.

Example 3

The position function of an object that moves on a coordinate line is $s(t) = t^2 - 6t$, where s is measured in centimeters and t in seconds. Find the distance traveled in the time interval $[0, 9]$.

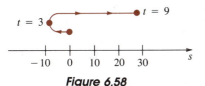

Figure 6.58

Solution   The velocity function $v(t) = ds/dt = 2t - 6 = 2(t - 3)$ shows that the motion is as indicated in Figure 6.58; namely, $v < 0$ for $0 \le t < 3$ and $v \ge 0$ for $3 \le t \le 9$. Hence, from (6.28) the distance traveled is

$$\int_0^9 |2t - 6| \, dt = \int_0^3 |2t - 6| \, dt + \int_3^9 |2t - 6| \, dt$$

$$= \int_0^3 -(2t - 6) \, dt + \int_3^9 (2t - 6) \, dt$$

$$= (-t^2 + 6t) \Big]_0^3 + (t^2 - 6t) \Big]_3^9$$

$$= (-9 + 18) + (81 - 54) - (9 - 18) = 45 \text{ cm}$$

Of course, the last result must check with the number obtained by simply counting units in Figure 6.58 between $s(0)$ and $s(3)$, and between $s(3)$ and $s(9)$.

Exercises 6.8

Answers to odd-numbered problems begin on page A-29

In Problems 1–6 a body moves in a straight line with velocity $v(t)$. Find the position function $s(t)$.

1. $v(t) = 6$; $s = 5$ when $t = 2$

2. $v(t) = 2t + 1$; $s = 0$ when $t = 1$

3. $v(t) = t^2 - 4t$; $s = 6$ when $t = 3$

4. $v(t) = \sqrt{4t + 5}$; $s = 2$ when $t = 1$

5. $v(t) = -10 \cos(4t + \pi/6)$; $s = \frac{5}{4}$ when $t = 0$

6. $v(t) = 2 \sin 3t$; $s = 0$ when $t = \pi$

In Problems 7–12 a body moves in a straight line with acceleration $a(t)$. Find $v(t)$ and $s(t)$.

7. $a(t) = -5$; $v = 4$ and $s = 2$ when $t = 1$

8. $a(t) = 6t$; $v = 0$ and $s = -5$ when $t = 2$

9. $a(t) = 3t^2 - 4t + 5$; $v = -3$ and $s = 10$ when $t = 0$

10. $a(t) = (t - 1)^2$; $v = 4$ and $s = 6$ when $t = 1$

11. $a(t) = 7t^{1/3} - 1$; $v = 50$ and $s = 0$ when $t = 8$

12. $a(t) = 100 \cos 5t$; $v = -20$ and $s = 15$ when $t = \pi/2$

13. A driver of a car that is traveling at a constant 88 km/hr takes his eyes off the road for 2 sec. How far does the car move in this time?

14. A ball is dropped (released from rest) from a height of 144 ft. How long does it take for the ball to hit the ground? At what speed does it hit the ground?

15. An egg is dropped from the top of a building and impacts 4 sec from release. How tall is the building?

16. A stone is dropped into a well and the splash is heard 2 sec later. If the speed of sound in air is 1080 ft/sec, find the depth of the well.

17. An arrow is projected vertically upward from ground level with an initial velocity of 24.5 m/sec. How high does it rise?

18. How high would the arrow in Problem 17 rise on the planet Mars where the acceleration of gravity is 3.6 m/sec^2?

19. A golf ball is thrown vertically upward from the roof of a 384-ft-high building with an initial velocity of 32 ft/sec. What is the maximum height attained by the ball? At what time does the ball hit the ground?

20. In Problem 19, what is the velocity of the golf ball as it passes an observer in a window that is 256 ft off the ground?

21. A person throws a marshmallow vertically downward with an initial velocity of 16 ft/sec from a window that is 102 ft off the ground. If the marshmallow hits a 6-ft-tall person on the head, what is the impact velocity?

22. The person hit on the head in Problem 21 climbs to the top of a 22-ft-high ladder and throws a stone vertically upward with an initial velocity of 96 ft/sec. If the stone hits the culprit at the 102-ft level, what is the impact velocity?

In Problems 23–28 an object moves in a straight line according to the given position function. If s is measured in centimeters, find the distance traveled by the object in the indicated time interval.

23. $s(t) = t^2 - 2t$; [0, 5]

24. $s(t) = -t^2 + 4t + 7$; [0, 6]

25. $s(t) = t^3 - 3t^2 - 9t$; [0, 4]

26. $s(t) = t^4 - 32t^2$; [1, 5]

27. $s(t) = 6 \sin \pi t$; [1, 3]

28. $s(t) = (t - 3)^2$; [2, 7]

Miscellaneous Problems

29. If a body is moving rectilinearly with a constant acceleration a and $v = v_0$ when $s = 0$, show that

$$v^2 = v_0^2 + 2as \qquad \left(Hint: \frac{dv}{dt} = \frac{dv}{ds}\frac{ds}{dt} = \frac{dv}{ds}v \right)$$

30. Prove that, when air resistance is ignored, a projectile shot vertically upward from ground level hits the ground again with a speed equal to the initial velocity v_0.

31. Suppose the acceleration of gravity on a planet is one-half that of the earth. Prove that a ball tossed vertically upward from the surface of the planet would attain a maximum height twice that on the earth when the same initial velocity is used.

32. In Problem 31, suppose the initial velocity of the ball on the planet is v_0 and the initial velocity of the ball on the earth is $2v_0$. Compare the maximum heights attained. Determine the initial velocity of the ball on the earth (in terms of v_0) so that the maximum height attained is the same as on the planet.

6.9 Work

In physics, when a *constant* force F moves an object a distance d in the same direction of the force, the **work** done is defined to be the product

$$W = Fd \qquad (6.29)$$

For example, if a 10-lb force moves an object 7 ft in the same direction as the force, then the work done is 70 ft-lb.

Units   Commonly used **units** are listed in the following table.

Quantity	Engineering System	mks	cgs
Force	pound (lb)	newton (nt)	dyne
Distance	foot (ft)	meter (m)	centimeter (cm)
Work	foot-pound (ft-lb)	newton-meter (joule)	dyne-centimeter (erg)

Thus, if a force of 300 nt moves an object 15 m, the work done is 4500 nt-m or 4500 joules. For comparison, we note that

$$1 \text{ nt} = 10^5 \text{ dynes} = 0.2247 \text{ lb}$$
$$1 \text{ ft-lb} = 1.356 \text{ joules} = 1.356 \times 10^7 \text{ ergs}$$

Hence, 70 ft-lb is equivalent to 94.92 joules and 4500 joules is equivalent to 3318.58 ft-lb.

Now, if $F(x)$ is a continuous *variable* force acting across an interval $[a, b]$, the work is not simply a product as in (6.29). Suppose P is the partition

$$a = x_0 < x_1 < x_2 < \cdots < x_n = b$$

and Δx_k is the length of the kth subinterval $[x_{k-1}, x_k]$. Let x_k^* be chosen arbitrarily in each subinterval. If the numbers Δx_k are small, we can consider the force that acts over each subinterval as constant. Hence, the work done from x_{k-1} to x_k is given by the approximation

$$\Delta W_k = F(x_k^*) \, \Delta x_k$$

Thus, an approximation to the work done from a to b is

$$F(x_1^*) \, \Delta x_1 + F(x_2^*) \, \Delta x_2 + \cdots + F(x_n^*) \, \Delta x_n = \sum_{k=1}^{n} F(x_k^*) \, \Delta x_k = \sum_{k=1}^{n} \Delta W_k$$

It is natural to assume that the exact work done by F over the interval is

$$W = \lim_{\|P\| \to 0} \sum_{k=1}^{n} F(x_k^*) \, \Delta x_k$$

We summarize the foregoing discussion with the following definition.

DEFINITION 6.7   Work

Let F be continuous on the interval $[a, b]$ and let $F(x)$ represent force at a value x in the interval. Then the work W done by the force in moving an object from a to b is

$$W = \int_a^b F(x) \, dx \tag{6.30}$$

Note: If F is constant, $F(x) = k$ for all x in the interval, then (6.30) becomes $W = \int_a^b k \, dx = kx]_a^b = k(b - a)$, which is consistent with (6.29).

Spring Problems   Hooke's Law* states that, when a spring is stretched (or compressed) beyond its natural length, the restoring force exerted by the spring is directly proportional to the amount of elongation (or compression) x. Thus, in order to stretch a spring x units beyond its natural length we need to apply the force

$$F(x) = kx \tag{6.31}$$

---

*Published by the English physicist Robert Hooke (1635–1703) in 1678.

where k is a constant of proportionality called the **spring constant**. See Figure 6.59.

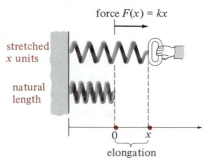

Figure 6.59

Example 1

It takes a force of 130 nt to stretch a spring 50 cm. Find the work done in stretching the spring 20 cm beyond its natural (unstretched) length.

Solution   When a force is measured in newtons, distances are commonly expressed in meters. Since $x = 50$ cm $= \frac{1}{2}$ m when $F = 130$ nt, (6.31) becomes

$$130 = k\left(\frac{1}{2}\right)$$

which implies the spring constant is $k = 260$ nt/m. Thus, $F = 260x$. Now, 20 cm $= \frac{1}{5}$ m, so that the work done in stretching the spring by this amount is

$$W = \int_0^{1/5} 260x \, dx$$

$$= 130x^2 \Big]_0^{1/5} = \frac{26}{5} = 5.2 \text{ joules}$$

Note: Suppose the natural length of the spring in Example 1 is 40 cm. An equivalent way of stating the problem is: Find the work done in stretching the spring to a length of 60 cm. Since the elongation is $60 - 40 = 20$ cm $= \frac{1}{5}$ m, we still integrate $F = 260x$ on the interval $[0, \frac{1}{5}]$. However, if the problem were to find the work done in stretching the same spring from 50 cm to 60 cm, we then integrate on the interval $[\frac{1}{10}, \frac{1}{5}]$. In this situation we are starting from a position where the spring is already stretched 10 cm $(\frac{1}{10}$ m).

Work Done against Gravity   From the Universal Law of Gravitation, the force between a planet (or moon) of mass m_1 and a body of mass m_2 is given by

$$F = k \frac{m_1 m_2}{r^2} \tag{6.32}$$

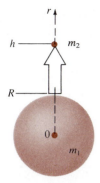

Figure 6.60

where k is a constant and r is the distance from the center of the planet to mass m_2. See Figure 6.60. In lifting the mass m_2 off the surface of a planet of radius R to a height h, the work done can be obtained by using (6.32) in (6.30):

$$W = \int_R^{R+h} k\, \frac{m_1 m_2}{r^2}\, dr$$

$$= km_1 m_2 \left(-\frac{1}{r}\right) \Big]_R^{R+h}$$

$$= km_1 m_2 \left(\frac{1}{R} - \frac{1}{R+h}\right) \qquad (6.33)$$

In the metric system $k = 6.67 \times 10^{-11}$ nt $\cdot$ m^2/kg^2. Some masses and values of R are given in the accompanying table.

	m_1 (in kg)	R (in m)
Venus	4.9×10^{24}	6.2×10^6
Earth	6.0×10^{24}	6.4×10^6
Moon	7.3×10^{22}	1.7×10^6
Mars	6.4×10^{23}	3.3×10^6

Example 2

The work done in lifting a 5000-kg payload from the surface of the earth to a height of 30,000 m (0.03×10^6 m) follows from (6.33) and the preceding table:

$$W = (6.67 \times 10^{-11})(6.0 \times 10^{24})(5000)\left(\frac{1}{6.4 \times 10^6} - \frac{1}{6.43 \times 10^6}\right)$$

$$\approx 1.46 \times 10^9 \text{ joules}$$

Pump Problems

When a liquid that weighs ρ lb per cubic foot is pumped from a tank, the work done in moving a fixed volume or layer of liquid d ft in a vertical direction is

$$W = \text{(force)} \cdot \text{(distance)}$$

$$= \text{(weight per unit volume)} \cdot \text{(volume)} \cdot \text{(distance)}$$

$$= \rho \cdot \text{(volume)} \cdot d \qquad (6.34)$$

In physics the quantity ρ is called the **weight density** of the fluid. For water $\rho = 62.4$ lb/ft^3 or 9800 nt/m^3.

Example 3

A hemispherical tank of radius 20 ft is filled with water to a 15-ft depth. Find the work done in pumping all the water to the top of the tank.

Solution   As shown in Figure 6.61, let the positive x-axis be directed *downward* and let the origin be at the center-top of the tank. Since the cross section of the tank is a semicircle, x and y are related by $x^2 + y^2 = (20)^2$, $0 \le x \le 20$. Now, suppose the subinterval $[5, 20]$ is partitioned into n subintervals $[x_{k-1}, x_k]$ of length Δx_k. Let x_k^* be any point in the kth subinterval and let ΔW_k denote an

Figure 6.61

approximation to the work done by the pump in lifting a layer of water of thickness Δx_k to the top of the tank. It follows from (6.34) that

$$\Delta W_k = \underbrace{(62.4\pi(y_k^*)^2\,\Delta x_k)}_{\text{force}} \cdot \underbrace{x_k^*}_{\text{distance}}$$

where $(y_k^*)^2 = 400 - (x_k^*)^2$. Hence, the work done by the pump is approximately

$$\sum_{k=1}^{n} \Delta W_k = \sum_{k=1}^{n} 62.4\pi[400 - (x_k^*)^2]x_k^*\,\Delta x_k$$

The work done in pumping all the water to the top of the tank is the limit of this last expression as $\|P\| \to 0$; that is,

$$W = \int_5^{20} 62.4\pi(400 - x^2)x\,dx$$

$$= 62.4\pi\left(200x^2 - \frac{x^4}{4}\right)\Bigg]_5^{20}$$

$$= (62.4\pi)(35{,}156.25) \approx 6{,}891{,}869 \text{ ft-lb}$$

It is worth pursuing the analysis of Example 3 for the case where the positive x-axis is taken in the *upward* direction and the origin is at the center-bottom of the tank.

_____ **Example 4** _____

Alternative Solution to Example 3 With the axes as shown in Figure 6.62, we see that a layer of water must be lifted a distance of $20 - x_k^*$ ft. But, since the center of the semicircle is at $(20, 0)$, x and y are now related by $(x - 20)^2 + y^2 = 400$. Hence,

$$\Delta W_k = \underbrace{(62.4\pi(y_k^*)^2 \, \Delta x_k)}_{\text{force}} \cdot \underbrace{(20 - x_k^*)}_{\text{distance}}$$

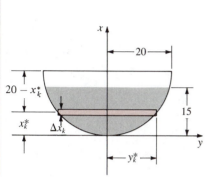

Figure 6.62

$$= 62.4\pi[400 - (x - 20)^2](20 - x_k^*) \, \Delta x_k$$

and

$$W = 62.4\pi \int_0^{15} [400 - (x - 20)^2](20 - x) \, dx$$

$$= 62.4\pi \int_0^{15} (x^3 - 60x^2 + 800x) \, dx$$

Note the new limits of integration. You should verify that the value of W in this case is the same as in Example 3.

_____ **Example 5** _____

In Example 3, find the work done in pumping all the water to a point 10 ft above the hemispherical tank.

Solution   As in Figure 6.61 we position the positive x-axis downward. Now, from Figure 6.63 we see

$$\Delta W_k = (62.4\pi(y_k^*)^2 \, \Delta x_k) \cdot (10 + x_k^*)$$

$$= 62.4\pi[400 - (x_k^*)^2](10 + x_k^*) \, \Delta x_k$$

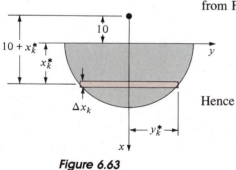

Figure 6.63

Hence

$$W = 62.4\pi \int_5^{20} (400 - x^2)(10 + x) \, dx$$

$$= 62.4\pi \int_5^{20} (-x^3 - 10x^2 + 400x + 4000) \, dx$$

$$= 62.4\pi \left(-\frac{x^4}{4} - \frac{10}{3}x^3 + 200x^2 + 4000x \right) \Big]_5^{20}$$

$$\approx 13{,}508{,}063 \text{ ft-lb}$$

Cable Problems   The next example illustrates the fact that when calculating the work done in lifting an object by means of a cable (heavy rope or chain) the weight of the cable must be taken into consideration.

_____ **Example 6** _____

A cable weighing 6 lb/ft is connected to a construction elevator weighing 1500 lb. Find the work done in lifting the elevator to a height of 500 ft.

Solution   The work done in lifting the elevator a distance of 500 ft is simply

$$W_e = (1500) \cdot (500) = 750{,}000 \text{ ft-lb}$$

Now let W_c denote the work done in lifting the cable. As shown in Figure 6.64 suppose the positive x-axis is directed upward and the interval $[0, 500]$ is partitioned into n subintervals with lengths Δx_k. At a height of x_k^* ft off the ground a segment of cable corresponding to the subinterval $[x_{k-1}, x_k]$ weighs $6\Delta x_k$ and must be pulled up an additional $500 - x_k^*$ ft. Hence, we can write

$$(\Delta W_c)_k = \underbrace{(6\Delta x_k)}_{\text{force}} \cdot \underbrace{(500 - x_k^*)}_{\text{distance}}$$

$$= (3000 - 6x_k^*)\,\Delta x_k$$

and so

$$W_c = \int_0^{500} (3000 - 6x)\,dx$$

$$= (3000x - 3x^2)\Big]_0^{500}$$

$$= 750{,}000 \text{ ft-lb}$$

Thus, the total work done in lifting the elevator is

$$W = W_e + W_c$$

$$= 1{,}500{,}000 \text{ ft-lb}$$

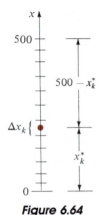

Figure 6.64

_____ **Example 7** _____

Alternative Solution to Example 6   This is a slightly faster analysis of Example 6. As shown in Figure 6.65, when the elevator is at a height x ft, it must be pulled up an additional $500 - x$ ft. The lifting force needed at that height is

$$\underbrace{1500}_{\substack{\text{weight} \\ \text{of} \\ \text{elevator}}} + \underbrace{6(500 - x)}_{\substack{\text{weight} \\ \text{of} \\ \text{cable}}} = 4500 - 6x$$

Thus, by (6.30) the work done is

$$W = \int_0^{500} (4500 - 6x)\,dx$$

$$= 1{,}500{,}000 \text{ ft-lb}$$

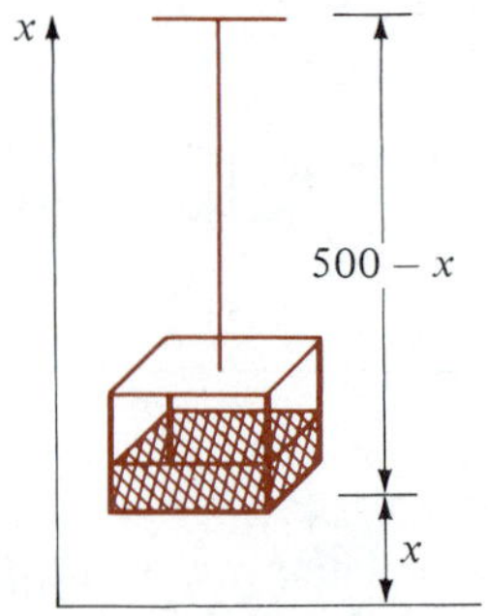

Figure 6.65

Exercises 6.9

Answers to odd-numbered problems begin on page A-29

1. Find the work done when a 55-lb force moves an object 20 yd in the same direction of the force.

2. A force of 100 nt is applied to an object at an angle of 30° measured from the horizontal. If the object moves 8 m horizontally, find the work done by the force.

3. In 1977 George Willig, who weighed 165 lb, scaled the outside of the World Trade Center building in New York City to a height of 1350 ft, in 3.5 hours at a rate of 6.4 ft/min. How much work did George do?

4. A person pushes against an immovable wall with a horizontal force of 75 lb. How much work is done?

5. A mass weighing 10 lb stretches a spring $\frac{1}{2}$ ft. How much will a mass weighing 8 lb stretch the same spring?

6. A spring has a natural length of 0.5 m. A force of 50 nt stretches the spring to a length of 0.6 m.

(a) What force is needed to stretch the spring x meters?

(b) What force is required to stretch the spring to a length of 1 m?

(c) How long is the spring when stretched by a force of 200 nt?

7. In Problem 6,

(a) Find the work done in stretching the spring 0.2 m.

(b) Find the work done in stretching the spring from a length of 1 m to a length of 1.1 m.

8. A force of $F = \frac{3}{2}x$ lb is needed to stretch a 10-in spring an additional x in.

(a) Find the work done in stretching the spring to a length of 16 in.

(b) Find the work done in stretching the spring 16 in.

9. A mass weighing 10 lb is suspended from a 2-ft spring. The spring is stretched 8 in and then the mass is removed.

(a) Find the work done in stretching the spring to a length of 3 ft.

(b) Find the work done in stretching the spring from a length of 4 ft to a length of 5 ft.

10. A 50-lb force compresses a 15-in-long spring by 3 in. Find the work done in compressing the spring to a final length of 5 in.

11. Find the work done in lifting a mass of 10,000 kg from the surface of the earth to a height of 500 km.

12. Find the work done in lifting a mass of 50,000 kg from the surface of the moon to a height of 200 km.

13. A right-cylindrical tank of height 12 ft and radius 3 ft is filled with water. Find the work done in pumping all the water to the top of the tank.

14. A tank in the form of a right circular cone, vertex down, is filled with water to a depth of one-half its height. If the height of the tank is 20 ft, and its diameter is 8 ft, find the work done in pumping all the water to the top of the tank. (*Hint:* Assume that the origin is at the vertex of the cone.)

15. For the tank in Problem 14, find the work done in pumping all the water to a point 5 ft above the top of the tank.

16. A horizontal trough with semicircular cross sections contains oil whose weight density is 80 lb/ft³. The dimensions of the tank (in feet) are shown in Figure 6.66. If the depth of the oil is 3 ft, find the work done in pumping all the oil to the top of the tank.

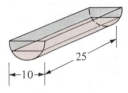

25

10

Figure 6.66

17. A tank has cross sections in the form of isosceles triangles, vertex down. The top of the tank is 6 ft wide, its height is 4 ft, and its length is 10 ft. Find the work done in filling the tank with water, through a hole in its bottom, by a pump located 5 ft below its vertex.

18. A 100-ft anchor chain, weighing 20 lb/ft, is hanging vertically over the side of a boat. How much work is performed by pulling in 40 ft of the chain?

19. A ship is anchored in 200 ft of water. In water the ship's anchor weighs 3000 lb and its anchor chain weighs 40 lb/ft. If the anchor chain hangs vertically, how much work is done in pulling in 100 ft of the chain?

20. A bucket of sand weighing 80 lb is lifted vertically by means of a rope and pulley to a height of 65 ft. Find the work done if **(a)** the weight of the rope is negligible; **(b)** the rope weighs $\frac{1}{2}$ lb/ft.

21. A bucket, initially containing 20 ft^3 of water, is lifted vertically from ground level. If the water leaks out at a rate of $\frac{1}{2}$ ft^3 per vertical foot, find the work done in lifting the bucket to a height at which it is empty.

22. The force of attraction between an electron and the nucleus of an atom is inversely proportional to the square of the distance separating them. If the initial distance between nucleus and electron is 1 unit, find the work done by an external force that moves the electron out to a distance 4 times the initial distance.

23. A rocket weighing 2,500,000 lb when fueled carries a 200,000-lb shuttle orbiter. Assume, in the early stages of the launch, that the rocket burns fuel at a rate of 100 lb per ft.

(a) Express the total weight of the system in terms of its altitude above the surface of the earth. See Figure 6.67.

(b) Find the work done in lifting the system to an altitude of 1000 ft.

Figure 6.67

24. In thermodynamics, if a gas enclosed in a cylinder expands against a piston so that the volume of the gas changes from v_1 to v_2, then the work done on the piston is given by $W = \int_{v_1}^{v_2} p\, dv$, where p is pressure (force per unit area). See Figure 6.68. In an adiabatic expansion* of an ideal gas, pressure and volume are related by $pv^\gamma = k$, where γ and k are constants. Show that if $\gamma \neq 1$ then

$$W = \frac{p_2 v_2 - p_1 v_1}{1 - \gamma}$$

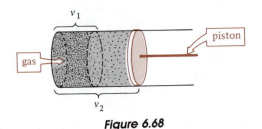

Figure 6.68

*This means that no heat is entering or leaving the system.

Miscellaneous Problems

25. As shown in Figure 6.69, a body of mass m is moved by a horizontal force F on a frictionless surface from a position at x_1 to a position at x_2. At these respective points, the body is moving at velocities v_1 and v_2, where $v_2 > v_1$. Show that the work done by the force is the increase in kinetic energy

$$W = \frac{1}{2}mv_2^2 - \frac{1}{2}mv_1^2$$

(*Hint:* Use Newton's second law $F = ma$, and express the acceleration a in terms of velocity v.)

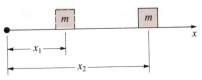

Figure 6.69

26. Prove that when a body of weight mg is lifted vertically from a point y_1 to a point y_2, $y_2 > y_1$, the work done is the change in potential energy $W = mg\, y_2 - mg\, y_1$.

Calculator Problems

27. **(a)** Use (6.33) to show that the work required to lift a mass m_2 to a point an "infinite distance" from the surface of the planet is $W = km_1 m_2 / R$.

(b) If the mass m_2 is imparted a velocity v_0 at the surface of the planet to enable it to attain an "infinite distance" from its surface, then we must have

$$\frac{1}{2}m_2 v_0^2 = k\frac{m_1 m_2}{R}$$

Use this relation to find the "escape velocity" v_0 of any planet. Use the data in the table on page 357 to find v_0 for Earth and for the planet Mars.

28. The graph of a variable force F is given in Figure 6.70. Using rectangular elements of area, find an approximation to the work done in moving a particle from $x = 1$ to $x = 5$. Then use the Trapezoidal Rule.

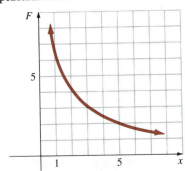

Figure 6.70

29. A continuous variable force $F(x)$ acts over the interval $[0, 1]$, where F is measured in newtons and x in meters. It is determined that

x(m)	0	0.2	0.4	0.6	0.8	1
$F(x)$(nt)	0	50	90	150	210	260

Use an appropriate numerical technique to approximate the work done over the interval.

6.10 Liquid Pressure and Force

When a *horizontal* flat plate is submerged below the surface of a liquid, the **force** exerted on the plate by the liquid above it is given by

$$F = \text{(pressure of liquid)} \cdot \text{(area of surface)}$$
$$= \text{(force per unit area)} \cdot \text{(area of surface)} \qquad (6.35)$$

If ρ is the weight density of the liquid (weight per unit volume) and A is the area of the horizontal plate submerged to a depth h, shown in Figure 6.71(a), then (6.35) is the same as

$$F = \underbrace{\text{(weight per unit volume)} \cdot \text{(depth)}}_{\text{pressure}} \cdot \text{(area of surface)}$$

$$= \rho h A \qquad (6.36)$$

However, when a *vertical* flat plate is submerged, the pressure of the liquid varies with the depth. See Figure 6.71(b). For example, the liquid pressure on a vertical dam is less at the top than at its base.

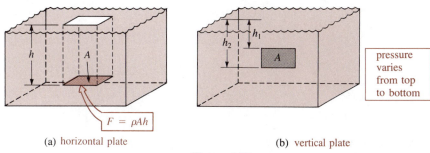

(a) horizontal plate   (b) vertical plate

Figure 6.71

Example 1

A flat rectangular plate with dimensions 5 ft × 6 ft is submerged horizontally in water at a depth of 10 ft. Determine the pressure and the force exerted on the plate by the water above it.

Solution   Recall that the weight density for water is 62.4 lb/ft^3. Hence,

$$\text{pressure} = \rho h = (62.4) \cdot 10 = 624 \text{ lb/ft}^2$$

Since the surface area of the plate is 30 ft^2, it follows from (6.36) that the force on the plate is

$$F = 624 \cdot 30 = 18{,}720 \text{ lb}$$

To determine the total force F exerted by a liquid on one side of a vertically submerged flat surface, we employ one form of Pascal's principle:*

Pressure exerted by a liquid at a depth h is the same in all directions.

Thus, if a large container with a flat bottom and vertical sidewalls is filled with water to a depth of 10 ft, the pressure of 624 lb/ft^2 at its bottom applies equally to the sidewalls.

Now, suppose a vertical flat plate, bounded by the horizontal lines $x = a$ and $x = b$ and the graphs of $y = f(x)$ and $y = g(x)$, is submerged in a liquid as shown in Figure 6.72. Let $f(x) - g(x)$ denote the width of the plate at any point x in $[a, b]$ and let P be any partition of the interval. If x_k^* is any point in the kth subinterval $[x_{k-1}, x_k]$, we conclude from (6.36) that the force exerted by the liquid on the corresponding rectangular element is approximated by

$$\Delta F_k = \rho \cdot x_k^* \cdot [f(x_k^*) - g(x_k^*)] \, \Delta x_k$$

where, as before, ρ denotes the weight density of the liquid. Thus, an approximation to the force on one side of the plate is

$$\sum_{k=1}^{n} \Delta F_k = \sum_{k=1}^{n} \rho x_k^* [f(x_k^*) - g(x_k^*)] \, \Delta x_k$$

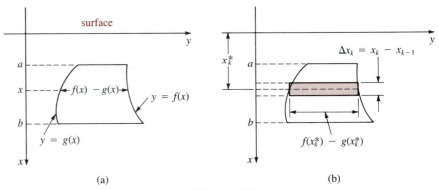

(a)   (b)

Figure 6.72

---

*Blaise Pascal (1623–1662), a French mathematician, scientist, and philosopher.

This suggests that the total force is

$$F = \lim_{\|P\| \to 0} \sum_{k=1}^{n} \rho x_k^* [f(x_k^*) - g(x_k^*)] \, \Delta x_k$$

DEFINITION 6.8   Force Exerted by a Liquid

Let ρ be the weight density of a liquid and let f and g be functions continuous on $[a, b]$. The **force** F exerted by the liquid on one side of the submerged plate shown in Figure 6.72 is

$$F = \int_a^b \rho x [f(x) - g(x)] \, dx$$

Example 2

A plate in the shape of an isosceles triangle 3 ft high and 4 ft wide is submerged vertically, base downward, with the base 5 ft below the surface of the water. Find the force exerted by the water on one side of the plate.

Solution   For convenience, we place the positive x-axis along the axis of symmetry of the triangular plate with the origin at the surface of the water. As indicated in Figure 6.73, we partition the interval $[2, 5]$ into n subintervals $[x_{k-1}, x_k]$ and choose a point x_k^* in each subinterval. Since the equation of the straight line that contains points $(2, 0)$ and $(5, 2)$ is

$$y = \frac{2}{3}x - \frac{4}{3}$$

we conclude by symmetry that the width of the rectangular element shown in Figure 6.73 is

$$2y_k^* = 2\left(\frac{2}{3}x_k^* - \frac{4}{3}\right)$$

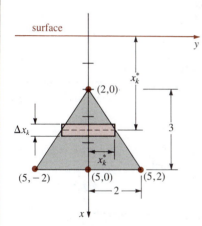

Figure 6.73

Now $\rho = 62.4 \text{ lb/ft}^3$ so that the force on that portion of the plate that corresponds to the kth subinterval is approximated by

$$\Delta F_k = (62.4) \cdot x_k^* \cdot 2\left(\frac{2}{3}x_k^* - \frac{4}{3}\right) \Delta x_k$$

Forming the sum $\sum_{k=1}^{n} \Delta F_k$ and passing to the limit as $\|P\| \to 0$ gives

$$F = \int_2^5 (62.4)2x\left(\frac{2}{3}x - \frac{4}{3}\right) dx$$

$$= 124.8 \int_2^5 \left(\frac{2}{3}x^2 - \frac{4}{3}x\right) dx$$

$$= 124.8\left(\frac{2}{9}x^3 - \frac{2}{3}x^2\right)\Big]_2^5$$

$$= 124.8\left(\frac{108}{9}\right) \approx 1497.6 \text{ lb}$$

Exercises 6.10

Answers to odd-numbered problems begin on page A-29

1. Consider the tanks with flat circular bottoms shown in Figure 6.74. Each tank is full of water whose weight density is 9800 nt/m³. Find the pressure and force exerted by the water on the bottom of each tank.

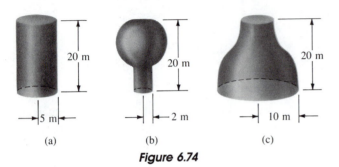

(a) (b) (c)

Figure 6.74

2. A rectangular swimming pool in the form of a rectangular parallelepiped has dimensions of 30 ft × 15 ft × 9 ft. Find the pressure and force exerted on the flat bottom if the pool is filled with water to a depth of 8 ft. See Figure 6.75.

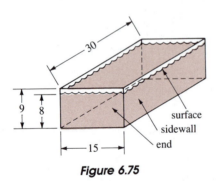

Figure 6.75

3. For the swimming pool in Problem 2, find the force exerted by the water on a vertical sidewall and on a vertical end. See Figure 6.75.

4. A plate in the shape of an equilateral triangle $\sqrt{3}$ ft on a side is submerged vertically, base downward, with vertex 1 ft below the surface of the water. Find the force exerted by the water on one side of the plate.

5. Find the force on one side of the plate in Problem 4 if it is suspended with base upward 1 ft below the surface of the water.

6. A triangular plate is submerged vertically in water as shown in Figure 6.76. Find the force exerted by the water on one side of the plate.

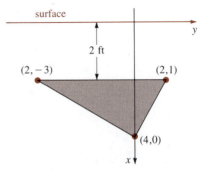

Figure 6.76

7. Assuming the positive x-axis is downward, a plate bounded by the parabola $x = y^2$ and the line $x = 4$ is submerged vertically in oil that has weight density 50 lb/ft³. If the vertex of the parabola is at the surface, find the force exerted by the oil on one side of the plate.

8. Assuming the positive x-axis is downward, a plate bounded by the parabola $x = y^2$ and the line $y = -x + 2$ is submerged vertically in water. If the vertex of the parabola is at the surface, find the force exerted by the water on one side of the plate.

9. A full water trough has vertical ends in the form of trapezoids as shown in Figure 6.77. Find the force exerted by the water on one end of the trough.

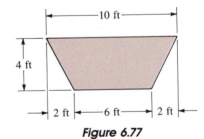

Figure 6.77

10. A full water trough has vertical ends in the form shown in Figure 6.78. Find the force exerted by the water on one end of the trough.

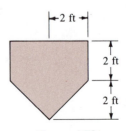

Figure 6.78

11. A vertical end of a full swimming pool has the shape given in Figure 6.79. Find the force exerted by the water on this end of the pool.

Figure 6.79

12. A tank in the shape of a right circular cylinder of diameter 10 ft is lying on its side. The tank is half full of oil that has weight density 60 lb/ft³. Find the force exerted by the oil on one end of the tank.

13. A circular plate of radius 4 ft is submerged vertically so that the center of the plate is 10 ft below the surface of the water. Find the force exerted by the water on one side of the plate. (*Hint:* For simplicity, take the origin to be the center of the plate, positive x-axis downward. See Problem 54, Exercises 6.1.)

14. A tank whose ends are in the form of an ellipse $x^2/4 + y^2/9 = 1$ is submerged in a liquid that has weight density ρ so that the end plates are vertical. Find the force exerted by the liquid on one end if its center is 10 ft below the surface of the liquid. (*Hint:* Proceed as in Problem 13 and use the fact that the area of an ellipse $x^2/a^2 + y^2/b^2 = 1$ is πab.)

15. A solid block in the shape of a cube 2 ft on a side is submerged in a large tank of water. The top of the block is horizontal and is 3 ft below the surface of the water. Find the total force on the block (six sides) that is caused by liquid pressure. See Figure 6.80.

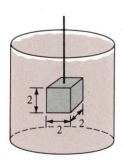

Figure 6.80

16. In Problem 15, what is the difference between the force on the bottom of the block and the force on the top of the block? This difference is the buoyant force of the water and, by Archimedes' principle, is equal to the weight of the water displaced. What is the weight of the water displaced by the block?

Miscellaneous Problems

17. Consider the rectangular swimming pool shown in Figure 6.81(a) whose ends are trapezoids. The pool is full of water. By taking the positive x-axis, as shown in Figure 6.81(b), find the force exerted by the water on the bottom of the pool. (*Hint:* Express the depth d in terms of x.)

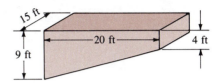

(a)

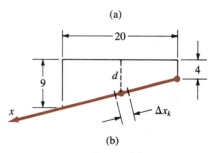

(b)

Figure 6.81

18. An earth dam is constructed with dimensions as shown in Figure 6.82(a). By taking the positive x-axis, as shown in Figure 6.82(b), find the force exerted by the water on the slanted wall of the dam.

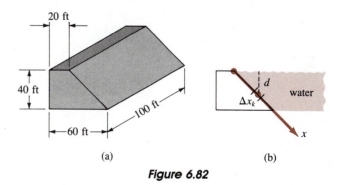

(a)   (b)

Figure 6.82

19. Analyze problem 18 with the positive x-axis shown in Figure 6.83.

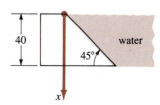

Figure 6.83

6.11 Center of Mass of a Rod

Moment and Mass   If x denotes the directed distance from the origin O to a mass m, as shown in Figure 6.84, we say that the product mx is the **moment of the mass** about the origin. Some units are summarized in the following table.

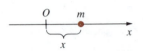

$$O \quad m$$
$$x$$

Figure 6.84

Quantity	Engineering System	mks	cgs
Mass	slug	kilogram (kg)	gram (g)
Moment of mass	slug-feet	kilogram-meter	gram-centimeter

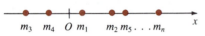

$$m_3 \quad m_4 \quad O \ m_1 \quad m_2 \ m_5 \ldots m_n \qquad x$$

Figure 6.85

Now, for n masses $m_1, m_2, \ldots, m_n$ at directed distances $x_1, x_2, \ldots, x_n$, respectively from O, as in Figure 6.85, we say that

$$m = m_1 + m_2 + \cdots + m_n = \sum_{k=1}^{n} m_k$$

is the **total mass of the system**, and

$$M_O = m_1 x_1 + m_2 x_2 + \cdots + m_n x_n = \sum_{k=1}^{n} m_k x_k$$

is the **moment of the system about the origin**. If

$$\sum_{k=1}^{n} m_k x_k = 0$$

the system is said to be in **equilibrium**. See Figure 6.86. If the system of masses in Figure 6.85 is not in equilibrium, there is a point P with coordinate $\bar{x}$ such that

$$\sum_{k=1}^{n} m_k(x_k - \bar{x}) = 0 \quad \text{or} \quad \sum_{k=1}^{n} m_k x_k - \bar{x} \sum_{k=1}^{n} m_k = 0$$

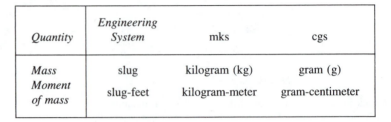

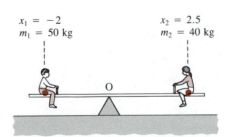

$x_1 = -2$
$m_1 = 50$ kg

$x_2 = 2.5$
$m_2 = 40$ kg

(a) seesaw in equilibrium
since $m_1 x_1 + m_2 x_2 = 0$

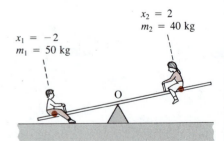

$x_1 = -2$
$m_1 = 50$ kg

$x_2 = 2$
$m_2 = 40$ kg

(b) seesaw not in equilibrium
since $m_1 x_1 + m_2 x_2 \neq 0$

Figure 6.86

Solving for $\bar{x}$ gives

$$\bar{x} = \frac{M_O}{m} = \frac{\displaystyle\sum_{k=1}^{n} m_k x_k}{\displaystyle\sum_{k=1}^{n} m_k} \tag{6.37}$$

The point with coordinate $\bar{x}$ is called the **center of mass** or the **center of gravity** of the system.*

Since (6.37) implies

$$\bar{x}\left(\sum_{k=1}^{n} m_k\right) = \sum_{k=1}^{n} m_k x_k$$

it follows that $\bar{x}$ is the directed distance from the origin to a point at which the total mass of the system can be considered to be concentrated.

Example 1

Three bodies of mass 4 kg, 6 kg, and 10 kg are located at $x_1 = -2$, $x_2 = 4$, and $x_3 = 9$, respectively. Distances are measured in meters. Find the center of mass.

Solution From (6.37)

$$\bar{x} = \frac{4 \cdot (-2) + 6 \cdot 4 + 10 \cdot 9}{4 + 6 + 10}$$

$$= \frac{106}{20} = 5.3$$

Figure 6.87

Figure 6.87 shows that the center of mass is 5.3 m to the right of the origin.

Rod with Variable Density

Now, let us consider the problem of finding the center of mass of a rod of length L that has a **variable linear density** ρ. We assume that the rod coincides with the x-axis on the interval $[0, L]$, as shown in Figure 6.88, and that the density is a continuous function $\rho(x)$ measured in slugs/ft, kg/m, or g/cm. After forming a partition P of the interval, we choose a point x_k^* in $[x_{k-1}, x_k]$. The number

Figure 6.88

$$\Delta m_k = \rho(x_k^*) \, \Delta x_k$$

is an approximation to mass of that portion of the rod on the subinterval. Also, the moment of this element of mass about the origin is approximated by

$$(\Delta M_O)_k = x_k^* \rho(x_k^*) \, \Delta x_k$$

---

*In a system in which the acceleration of gravity varies from mass to mass, the center of gravity is not the same as the center of mass.

Thus, we conclude that

$$m = \lim_{\|P\| \to 0} \sum_{k=1}^{n} \rho(x_k^*) \, \Delta x_k = \int_0^L \rho(x) \, dx$$

and

$$M_O = \lim_{\|P\| \to 0} \sum_{k=1}^{n} x_k^* \rho(x_k^*) \, \Delta x_k = \int_0^L x\rho(x) \, dx$$

are the **mass of the rod** and its **moment about the origin**, respectively.

It then follows from $\bar{x} = M_O/m$ that the center of mass of the rod is given by

$$\bar{x} = \frac{\displaystyle\int_0^L x\rho(x) \, dx}{\displaystyle\int_0^L \rho(x) \, dx} \qquad (6.38)$$

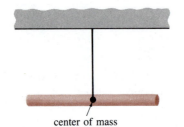

center of mass

Figure 6.89

As shown in Figure 6.89, a rod suspended by a string attached to its center of mass would hang in perfect balance.

Example 2

Show that if a rod has a constant linear density, then the center of mass is at its geometric center.

Solution Let L denote the length of the rod and let $\rho = k$ kg/m be the density. Its mass in kilograms is

$$m = \int_0^L k \, dx$$
$$= kx \Big]_0^L = kL$$

and its moment about the origin in kg-m is

$$M_O = \int_0^L kx \, dx$$
$$= \frac{k}{2}x^2 \Big]_0^L = \frac{k}{2}L^2$$

Thus, from (6.38) we have

$$\bar{x} = \frac{\dfrac{k}{2}L^2}{kL} = \frac{L}{2}$$

___ **Example 3** _____

A 16-cm-long rod has a linear density, measured in g/cm, given by $\rho(x) = \sqrt{x}, \; 0 \le x \le 16$. Find its center of mass.

Solution   The mass of the rod in grams is

$$m = \int_0^{16} x^{1/2} \, dx$$

$$= \frac{2}{3} x^{3/2} \Big]_0^{16} = \frac{128}{3}$$

The moment about the origin in g-cm is

$$M_O = \int_0^{16} x \cdot x^{1/2} \, dx$$

$$= \int_0^{16} x^{3/2} \, dx$$

$$= \frac{2}{5} x^{5/2} \Big]_0^{16} = \frac{2048}{5}$$

From (6.38) we find

$$\bar{x} = \frac{2048/5}{128/3} = 9.6$$

That is, the center of mass of the rod is 9.6 cm from the left end of the rod that coincides with the origin.

_____ **Exercises 6.11** _____

Answers to odd-numbered problems begin on page A-29

In Problems 1–4 find the center of mass of the given system of masses. The mass m_k is located on the x-axis at a point whose directed distance from the origin is x_k. Assume that mass is measured in grams and that distance is measured in centimeters.

1. $m_1 = 2, \; m_2 = 5; \; x_1 = 4, \; x_2 = -2$

2. $m_1 = 6, \; m_2 = 1, \; m_3 = 3; \; x_1 = -\dfrac{1}{2}, \; x_2 = -3, \; x_3 = 8$

3. $m_1 = 10, \; m_2 = 5, \; m_3 = 8, \; m_4 = 7; \; x_1 = -5, \; x_2 = 2,$
$x_3 = 6, \; x_4 = -3$

4. $m_1 = 2, \; m_2 = \dfrac{3}{2}, \; m_3 = \dfrac{7}{2}, \; m_4 = \dfrac{1}{2}; \; x_1 = 9, \; x_2 = -4,$
$x_3 = -6, \; x_4 = -10$

5. Two masses are placed at the ends of a uniform board of negligible mass, as shown in Figure 6.90. Where should a fulcrum be placed so that the system is in balance? (*Hint:* Although the origin can be placed anywhere, let us agree to choose it halfway between the masses.)

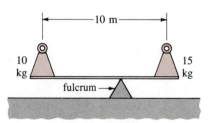

Figure 6.90

6. A uniform metal bar of mass 4 kg and length 2 m supports two masses, as shown in Figure 6.91. Where should the wire be attached to the bar so that the system hangs in balance?

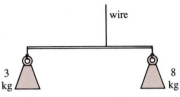

Figure 6.91

In Problems 7–14 a rod of linear density $\rho(x)$ kg/m coincides with the x-axis on the interval indicated. Find its center of mass.

7. $\rho(x) = 2x + 1$; [0, 5]

8. $\rho(x) = -x^2 + 2x$; [0, 2]

9. $\rho(x) = x^{1/3}$; [0, 1]

10. $\rho(x) = -x^2 + 1$; [0, 1]

11. $\rho(x) = |x - 3|$; [0, 4]

12. $\rho(x) = 1 + |x - 1|$; [0, 3]

13. $\rho(x) = \begin{cases} x^2, & 0 \le x \le 1 \\ 2 - x, & 1 \le x \le 2 \end{cases}$; [0, 2]

14. $\rho(x) = \begin{cases} x, & 0 \le x \le 2 \\ 2, & 2 \le x \le 3 \end{cases}$; [0, 3]

15. The density of a 10-ft rod varies as the square of the distance from the left end. Find its center of mass if the density at its center is 12.5 slug/ft.

16. The linear density of a 3-m-long rod varies as the distance from the right end. Find the linear density at the center of the rod if its total mass is 6 kg.

17. A rod of linear density $\rho(x)$ kg/m coincides with the x-axis on the interval [0, 6]. If $\rho(x) = x(6 - x) + 1$, where would one intuitively expect the center of mass to be? (*Hint:* Graph $\rho(x)$.) Prove your assertion.

18. The linear density of a rod, in kg/m, is given by $\rho(x) = \sqrt{2x + 1}$. Find the average density on the interval [0, 4].

Miscellaneous Problems

19. Find the center of mass of the three masses m_1, m_2, and m_3 located at the vertices of the equilateral triangle shown in Figure 6.92. (*Hint:* First find the center of mass of m_1 and m_2.)

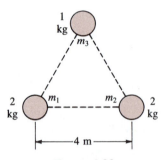

Figure 6.92

20. A rod of linear density $\rho(x) = \sqrt{4 - x^2}$ coincides with the x-axis on [0, 2]. Find its center of mass. (*Hint:* To evaluate $\int_0^2 \sqrt{4 - x^2}\, dx$, consider the geometric interpretation of an integral as area. See Problem 53 in Exercises 6.1.)

6.12 Centroid of a Plane Region

For *n* masses located in the xy-plane, as indicated in Figure 6.93, the **center of mass of the system** is defined to be the point $(\bar{x}, \bar{y})$, where

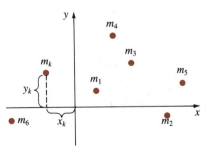

Figure 6.93

$$\bar{x} = \frac{M_y}{m} = \frac{\displaystyle\sum_{k=1}^{n} m_k x_k}{\displaystyle\sum_{k=1}^{n} m_k} = \frac{\text{moment of system about y-axis}}{\text{total mass}}$$

$$\bar{y} = \frac{M_x}{m} = \frac{\displaystyle\sum_{k=1}^{n} m_k y_k}{\displaystyle\sum_{k=1}^{n} m_k} = \frac{\text{moment of system about x-axis}}{\text{total mass}}$$

Lamina   Now let's turn to the problem of finding the center of mass, or balancing point, of a thin two-dimensional smear of matter, or **lamina**, that has a constant density ρ (mass per unit area). See Figure 6.94. When ρ is constant, the lamina is said to be **homogeneous**. As shown in Figure 6.95(a) let us suppose that the lamina coincides with a region R in the xy-plane bounded by the graph of a continuous nonnegative function $y = f(x)$, the x-axis, and the vertical lines $x = a$ and $x = b$. If P is a partition of the interval $[a, b]$, then the mass of the rectangular element shown in Figure 6.95(b) is

$$\Delta m_k = \rho \, \Delta A_k$$
$$= \rho f(x_k^*) \, \Delta x_k$$

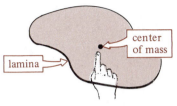

center of mass

lamina

Figure 6.94

where, in this case, we take x_k^* to be the midpoint of the subinterval $[x_{k-1}, x_k]$ and ρ is the constant density. The moment of this element about the y-axis is

$$(\Delta M_y)_k = x_k^* \, \Delta m_k$$
$$= x_k^*(\rho \, \Delta A_k)$$
$$= \rho x_k^* f(x_k^*) \, \Delta x_k$$

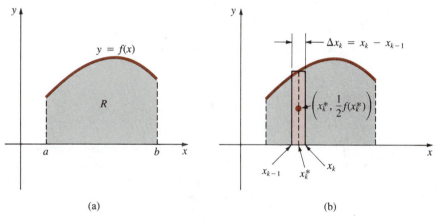

(a)

(b)

Figure 6.95

Since the density is constant, the center of mass of the element is necessarily at its geometric center $(x_k^*, \frac{1}{2}f(x_k^*))$. Hence, the moment of the element about the x-axis is

$$(\Delta M_x)_k = \frac{1}{2}f(x_k^*)[\rho \, \Delta A_k]$$

$$= \frac{1}{2}\rho[f(x_k^*)]^2 \, \Delta x_k$$

We conclude that

$$m = \lim_{\|P\| \to 0} \sum_{k=1}^{n} \rho f(x_k^*) \, \Delta x_k = \int_a^b \rho f(x) \, dx$$

$$M_y = \lim_{\|P\| \to 0} \sum_{k=1}^{n} \rho x_k^* f(x_k^*) \, \Delta x_k = \int_a^b \rho x f(x) \, dx$$

and $\qquad M_x = \lim_{\|P\| \to 0} \dfrac{1}{2} \sum_{k=1}^{n} \rho [f(x_k^*)]^2 \, \Delta x_k = \dfrac{1}{2} \int_a^b \rho [f(x)]^2 \, dx$

Thus, the coordinates of the center of mass of the lamina are defined to be

$$\bar{x} = \frac{M_y}{m} = \frac{\displaystyle\int_a^b \rho x f(x) \, dx}{\displaystyle\int_a^b \rho f(x) \, dx}$$

$$\bar{y} = \frac{M_x}{m} = \frac{\dfrac{1}{2} \displaystyle\int_a^b \rho [f(x)]^2 \, dx}{\displaystyle\int_a^b \rho f(x) \, dx}$$

(6.39)

Centroid We note that the constant density ρ will cancel in the above equations for $\bar{x}$ and $\bar{y}$, and that the denominator $\int_a^b f(x) \, dx$ is then the area A of the region R. In other words, the center of mass depends only on the shape of R:

$$\bar{x} = \frac{M_y}{A} = \frac{\displaystyle\int_a^b x f(x) \, dx}{\displaystyle\int_a^b f(x) \, dx}$$

$$\bar{y} = \frac{M_x}{A} = \frac{\dfrac{1}{2} \displaystyle\int_a^b [f(x)]^2 \, dx}{\displaystyle\int_a^b f(x) \, dx}$$

(6.40)

To emphasize the distinction, albeit minor, between the physical object, which is the homogeneous lamina, and the geometric object, which is the plane region R, we say the equations in (6.40) define the coordinates of the **centroid** of the region.

Example 1

Find the centroid of the region in the first quadrant bounded by the graph of $y = 9 - x^2$, the x-axis, and the y-axis.

Solution The region is shown in Figure 6.96. Now, if $f(x) = 9 - x^2$, then

$$\Delta A_k = f(x_k^*) \, \Delta x_k$$

$$(\Delta M_y)_k = x f(x_k^*) \, \Delta x_k$$

$$(\Delta M_x)_k = \frac{1}{2} f(x_k^*) [f(x_k^*) \, \Delta x_k]$$

$$= \frac{1}{2} [f(x_k^*)]^2 \, \Delta x_k$$

Hence,

$$A = \int_0^3 (9 - x^2)\, dx$$

$$= \left(9x - \frac{x^3}{3}\right)\Big]_0^3 = 18$$

$$M_y = \int_0^3 x(9 - x^2)\, dx$$

$$= \left(\frac{9}{2}x^2 - \frac{x^4}{4}\right)\Big]_0^3 = \frac{81}{4}$$

$$M_x = \frac{1}{2}\int_0^3 (9 - x^2)^2\, dx$$

$$= \frac{1}{2}\int_0^3 (81 - 18x^2 + x^4)\, dx$$

$$= \frac{1}{2}\left(81x - 6x^3 + \frac{x^5}{5}\right)\Big]_0^3 = \frac{324}{5}$$

It follows from (6.40) that the coordinates of the centroid are

$$\bar{x} = \frac{M_y}{A} = \frac{81/4}{18} = \frac{9}{8}$$

$$\bar{y} = \frac{M_x}{A} = \frac{324/5}{18} = \frac{54}{15}$$

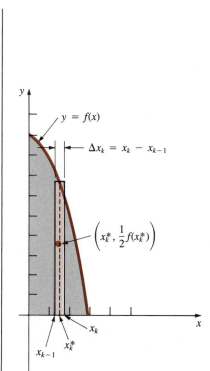

$y = f(x)$

$\Delta x_k = x_k - x_{k-1}$

$\left(x_k^*, \frac{1}{2}f(x_k^*)\right)$

x_k

x_{k-1} x_k^*

Figure 6.96

Example 2

Find the centroid of the region bounded by the graphs of $x = y^2 + 1$, $x = 0$, $y = 2$, and $y = -2$.

Solution   The region is shown in Figure 6.97. Inspection of the figure suggests that we use horizontal rectangular elements. If $f(y) = y^2 + 1$, then

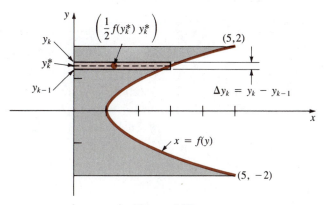

y_k

$\left(\frac{1}{2}f(y_k^*)\, y_k^*\right)$

$(5,2)$

y_k^*

y_{k-1}

$\Delta y_k = y_k - y_{k-1}$

$x = f(y)$

$(5, -2)$

Figure 6.97

$$\Delta A_k = f(y_k^*)\, \Delta y_k$$

$$(\Delta M_x)_k = y_k^* f(y_k^*)\, \Delta y_k$$

$$(\Delta M_y)_k = \frac{1}{2} f(y_k^*)[f(y_k^*)\, \Delta y_k]$$

$$= \frac{1}{2}[f(y_k^*)]^2\, \Delta y_k$$

and so, $$A = \int_{-2}^{2} (y^2 + 1)\, dy$$

$$= \left(\frac{y^3}{3} + y \right) \Bigg]_{-2}^{2} = \frac{28}{3}$$

$$M_x = \int_{-2}^{2} y(y^2 + 1)\, dy$$

$$= \left(\frac{y^2}{4} + \frac{y^2}{2} \right) \Bigg]_{-2}^{2} = 0$$

$$M_y = \frac{1}{2} \int_{-2}^{2} (y^2 + 1)^2\, dy$$

$$= \frac{1}{2} \int_{-2}^{2} (y^4 + 2y^2 + 1)\, dy$$

$$= \frac{1}{2} \left(\frac{y^5}{5} + \frac{2}{3}y^3 + y \right) \Bigg]_{-2}^{2} = \frac{206}{15}$$

Thus we have $$\bar{x} = \frac{M_y}{A} = \frac{206/15}{28/3} = \frac{103}{60}$$

$$\bar{y} = \frac{M_x}{A} = \frac{0}{28/3} = 0$$

As we would expect, since the lamina is symmetric with respect to the x-axis, the centroid is on the axis of symmetry. We also note that the centroid is outside the region.

Example 3

Find the centroid of the region bounded by the graphs of $y = -x^2 + 3$ and $y = x^2 - 2x - 1$.

Solution   Figure 6.98 shows the region in question. We note that the points of intersection of the graphs are $(-1, 2)$ and $(2, -1)$. Now, if $f(x) = -x^2 + 3$ and $g(x) = x^2 - 2x - 1$, then the area of the region is

$$A = \int_{-1}^{2} [f(x) - g(x)]\, dx$$

$$= \int_{-1}^{2} (-2x^2 + 2x + 4)\, dx$$

$$= \left(-\frac{2}{3}x^3 + x^2 + 4x \right) \Bigg]_{-1}^{2} = 9$$

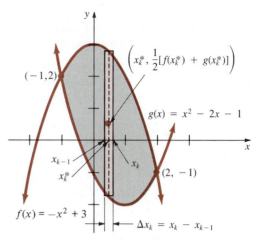

Figure 6.98

Since the coordinates of the midpoint of the indicated element are $(x_k^*, \frac{1}{2}[f(x_k^*) + g(x_k^*)])$, it follows that

$$M_y = \int_{-1}^{2} x[f(x) - g(x)]\, dx$$

$$= \int_{-1}^{2} (-2x^3 + 2x^2 + 4x)\, dx$$

$$= \left(-\frac{1}{2}x^4 + \frac{2}{3}x^3 + 2x^2 \right) \Bigg]_{-1}^{2} = \frac{9}{2}$$

$$M_x = \frac{1}{2}\int_{-1}^{2} [f(x) + g(x)][f(x) - g(x)]\, dx$$

$$= \frac{1}{2}\int_{-1}^{2} ([f(x)]^2 - [g(x)]^2)\, dx$$

$$= \frac{1}{2}\int_{-1}^{2} [(-x^2 + 3)^2 - (x^2 - 2x - 1)^2]\, dx$$

$$= \frac{1}{2}\int_{-1}^{2} (4x^3 - 8x^2 - 4x + 8)\, dx$$

$$= \frac{1}{2}\left(x^4 - \frac{8}{3}x^3 - 2x^2 + 8x \right) \Bigg]_{-1}^{2} = \frac{9}{2}$$

Thus, the coordinates of the centroid are

$$\bar{x} = \frac{M_y}{A} = \frac{9/2}{9} = \frac{1}{2}$$

$$\bar{y} = \frac{M_x}{A} = \frac{9/2}{9} = \frac{1}{2}$$

Exercises 6.12

Answers to odd-numbered problems begin on page A-30

In Problems 1–4 find the center of mass of the given system of masses. The mass m_k is located at the point P_k. Assume that mass is measured in grams and that distance is measured in centimeters.

1. $m_1 = 3$, $m_2 = 4$; $P_1 = (-2, 3)$, $P_2 = (1, 2)$

2. $m_1 = 1$, $m_2 = 3$, $m_3 = 2$; $P_1 = (-4, 1)$, $P_2 = (2, 2)$, $P_3 = (5, -2)$

3. $m_1 = 4$, $m_2 = 8$, $m_3 = 10$; $P_1 = (1, 1)$, $P_2 = (-5, 2)$, $P_3 = (7, -6)$

4. $m_1 = 1$, $m_2 = \frac{1}{2}$, $m_3 = 4$, $m_4 = \frac{5}{2}$; $P_1 = (9, 3)$, $P_2 = (-4, -6)$, $P_3 = (\frac{3}{2}, -1)$, $P_4 = (-2, 10)$.

In Problems 5–24 find the centroid of the region bounded by the graphs of the given equations.

5. $y = 2x + 4$, $y = 0$, $x = 0$, $x = 2$

6. $y = x + 1$, $y = 0$, $x = 3$

7. $y = x^2$, $y = 0$, $x = 1$

8. $y = x^2 + 2$, $y = 0$, $x = -1$, $x = 2$

9. $y = x^3$, $y = 0$, $x = 3$

10. $y = x^3$, $y = 8$, $x = 0$

11. $y = \sqrt{x}$, $y = 0$, $x = 1$, $x = 4$

12. $x = y^2$, $x = 1$

13. $y = x^2$, $y - x = 2$

14. $y = x^2$, $y = \sqrt{x}$

15. $y = x^3$, $y = x^{1/3}$, first quadrant

16. $y = 4 - x^2$, $y = 0$, $x = 0$, second quadrant

17. $x^3 y = 1$, $y = 0$, $x = 1$, $x = 3$

18. $y = x^2 - 2x + 1$, $y = -4x + 9$

19. $x = y^2 - 1$, $y = -1$, $y = 2$, $x = -2$

20. $y = x^2 - 4x + 6$, $y = 0$, $x = 0$, $x = 4$

21. $y = 4 - 4x^2$, $y = 1 - x^2$

22. $y^2 + x = 1$, $y + x = -1$

23. $y = x$, $y = -2x + 6$, $y = 0$

24. $y = x$, $y = -2x + 6$, $x = 0$

In Problems 25 and 26 use symmetry to locate $\bar{x}$ and integration to find $\bar{y}$ of the region bounded by the graphs of the given functions.

25. $y = 1 + \cos x$, $y = 1$, $-\pi/2 \le x \le \pi/2$

26. $y = 4 \sin x$, $y = -\sin x$, $0 \le x \le \pi$

Miscellaneous Problems

27. Find the centroid of the region bounded by the graphs of $y = \sqrt{25 - x^2}$ and $y = 0$. (*Hint:* The area of a circle is πr^2.)

28. Find the centroid of the region in the first quadrant bounded by the graphs of $y = 3\sqrt{1 - x^2/4}$, $y = 0$, and $x = 0$. (*Hint:* The area of an ellipse $x^2/a^2 + y^2/b^2 = 1$ is πab.)

29. Let L be an axis in a plane and R a region in the same plane that does not intersect L. The **First Theorem of Pappus*** states that when R is revolved about L, the volume V of the resulting solid of revolution is equal to the area A of R times the length of path transversed by the centroid of R. As shown in Figure 6.99, let the region R be bounded by the graphs of $y = f_1(x)$ and $y = f_2(x)$. Show that if R is revolved about the x-axis, then $V = (2\pi\bar{y})A$, where A is the area of the region.

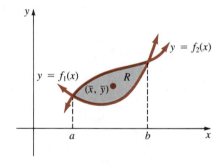

Figure 6.99

30. Verify the First Theorem of Pappus by revolving the region bounded by $y = x^2 + 1$, $y = 1$, $x = 2$ about the x-axis.

*Pappus of Alexandria (c. A.D. 350).

In Problems 31 and 32 (Figures 6.100 and 6.101) find the centroid of the plane region shown in the figure. (*Hint:* Find the centroid of each part of the region.)

31.

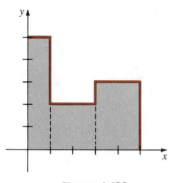

Figure 6.100

32.

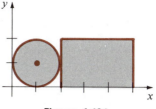

Figure 6.101

[O] 6.13 **Further Applications**

Cardiac Output   In physiology, the cardiac output R of the heart is defined as the volume of blood that the heart pumps per unit time. An abnormal cardiac output is an indicator of disease.

One way of measuring this output is called the **dye dilution method**. As shown in Figure 6.102, an amount D, measured in milligrams, of dye is injected into the pulmonary artery near the heart. The dye flows through the lungs and the pulmonary veins into the left atrium of the heart and eventually through the aorta. A probe inserted into the aorta monitors the amount of dye leaving the heart at equally spaced values of time over an interval $[0, T]$ (say, every second up to 30 seconds).

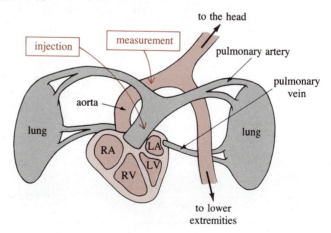

Right Atrium (RA)   Left Atrium (LA)
Right Ventricle (RV)   Left Ventricle (LV)

Figure 6.102

In determining R, let us assume that the concentration of the dye at any time is given by a continuous function $c(t)$. The interval $[0, T]$ is then subdivided into n subintervals of equal length $\Delta t = t_k - t_{k-1} = T/n$. The amount of dye that flows past the measuring point over a time interval $[t_{k-1}, t_k]$ is given by the approximation

$$\Delta D_k = \text{(concentration)} \cdot \text{(volume)} = \text{(concentration)} \cdot \text{(rate} \cdot \text{time)}$$
$$= \text{(concentration)} \cdot \text{(volume/time)} \cdot \text{(time)}$$
$$= c(t_k^*) \cdot R \cdot \Delta t$$

where t_k^* is some value in $[t_{k-1}, t_k]$. The approximate amount of dye that flows past the monitoring point in the aorta over the time interval $[0, T]$ is

$$\sum_{k=1}^{n} \Delta D_k = \sum_{k=1}^{n} c(t_k^*)R \, \Delta t = R \sum_{k=1}^{n} c(t_k^*) \, \Delta t \qquad (6.41)$$

Letting $n \to \infty$ in (6.35) suggests that the total amount of dye that flows past the monitoring point is

$$D = R \int_0^T c(t) \, dt \qquad (6.42)$$

Solving (6.42) for the flow rate R yields

$$R = \frac{D}{\displaystyle\int_0^T c(t) \, dt} \qquad * \qquad (6.43)$$

___

Example 1

Five milligrams of dye are injected into the pulmonary artery. Find the cardiac output of the heart over a period of 30 seconds if the concentration of the dye is $c(t) = -(1/100) \, t(t - 30)$ milligrams/liter.

Solution   From (6.43), we have

$$R = \frac{5}{\displaystyle\int_0^{30} -\frac{1}{100} t(t - 30) \, dt}$$

$$= \frac{5}{\left(-\dfrac{t^3}{300} + \dfrac{3}{20}t^2 \right)\Big]_0^{30}}$$

$$= \frac{1}{9} \text{ liters/second} = \frac{1}{9}(60) \approx 6.67 \text{ liters/minute}$$

___

*The value of R given in (6.43) is actually an approximation to the cardiac output. In practice, the graph of $c(t)$ is one that is fitted to approximate the set of data points $(0, c(0))$, $(1, c(1))$, ...

Flow of Blood in an Artery   The velocity of blood (cm/sec) in an arteriole (small artery) that has a circular cross section of radius R was given by Poiseuille to be

$$v(r) = \frac{P}{4\nu l}(R^2 - r^2), \qquad 0 \le r \le R \tag{6.44}$$

where P is the blood pressure, ν the viscosity of the blood, and l the length of the arteriole. Flow characterized by (6.44) is called *laminar flow* since, as shown in Figure 6.103, blood is thought to flow in cylindrical shells or *laminae* parallel to the walls of the arteriole at a distance r from the center.

Figure 6.103

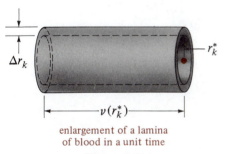

enlargement of a lamina
of blood in a unit time

Figure 6.104

We can find the volume of blood F, or total flow, inside an artery in a unit time by definite integration. To see this, subdivide $[0, R]$ into n subintervals $[r_{k-1}, r_k]$. From Figure 6.104, we see that the volume of blood per unit time inside a lamina of radius r_k^* and thickness Δr_k is

$$\Delta F_k = 2\pi r_k^*\, v(r_k^*)\, \Delta r_k$$

Taking the limit as $n \to \infty$ of the sum

$$\sum_{k=1}^{n} \Delta F_k = \sum_{k=1}^{n} 2\pi r_k^* v(r_k^*)\, \Delta r_k$$

suggests

$$F = \int_0^R 2\pi r v(r)\, dr$$

$$= \frac{\pi P}{2\nu l} \int_0^R r(R^2 - r^2)\, dr$$

$$= \frac{\pi P}{2\nu l}\left(R^2 \frac{r^2}{2} - \frac{r^4}{4} \right)\Bigg]_0^R$$

or

$$F = \frac{\pi P R^4}{8\nu l} \tag{6.45}$$

In other words, Poiseuille's law (6.45) indicates that the volume of blood in an arteriole per unit time is proportional to the fourth power of the radius. Formula (6.45) is used for laminar flow; this usually means that R ranges from $R = 2.5$ cm for the aorta down to $R = 1/1000$ cm for a capillary.

Total Revenue   If a landlord receives income at a rate of $18,000 a year from an apartment house, the revenue that is collected in, say, 15 years is

$$\text{rate} \times \text{time} = \$18{,}000 \times 15 = \$270{,}000$$

Observe that the total revenue, in this case, can also be written as the definite integral

$$\int_0^{15} 18{,}000 \, dt = 18{,}000t \Big]_0^{15} = \$270{,}000$$

Suppose now that revenue flows continuously from some source at a rate $f(t)$ dollars per year over a period of T years. What is the **total revenue** obtained over the interval $[0, T]$? As suggested in the preceding simple example, the answer can be expressed as a definite integral. The analysis is very similar to that of the cardiac output model.

Partition the interval $[0, T]$ into n subintervals of equal length $\Delta t = t_k - t_{k-1} = T/n$. The approximate revenue obtained over a period $[t_{k-1}, t_k]$ is

$$\Delta r_k = f(t_k^*) \, \Delta t$$

where t_k^* is some value in $[t_{k-1}, t_k]$. The sum

$$\sum_{k=1}^n \Delta r_k = \sum_{k=1}^n f(t_k^*) \, \Delta t$$

is approximately the revenue r obtained in T years. The total revenue r is obtained from this last expression by letting $n \to \infty$:

$$r = \int_0^T f(t) \, dt$$

Consumer and Producer   Economists have used the definite integral to define the concepts of **consumer**
Surpluses   and **producer surpluses**.

The demand for a commodity by consumers, as well as the amount supplied to the market by the manufacturers, can often be expressed as a function of the per unit price. Let $D(x)$ and $S(x)$ be the number of units demanded and the number of units supplied, respectively, when the commodity sells at a price x per unit. If the demand equals the supply,

$$D(x) = S(x)$$

the market is said to be in **equilibrium** and the corresponding price of the commodity is called the **equilibrium price**. If p is the equilibrium price and b is the price at which the demand for the commodity is zero ($D(b) = 0$), the integral

$$CS = \int_p^b D(x) \, dx \tag{6.46}$$

is called the **consumer surplus**. When the market is in equilibrium, the fact that there is a demand for the commodity at even higher prices means that those consumers who would have been willing to pay the higher price will benefit.

The *CS* represents, in units of money, the combined savings realized by these consumers.

The integral

$$PS = \int_c^p S(x) \, dx \qquad (6.47)$$

where $S(c) = 0$, is called the **producer surplus**. With the price of the commodity set at the equilibrium price p, the *PS* is the combined savings realized by those producers who would have been willing to supply the commodity at even lower prices.

___ **Example 2** ___

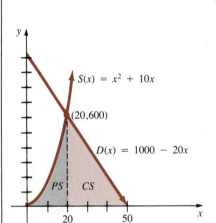

Figure 6.105

Suppose the demand and supply of a commodity selling for x dollars a unit are $D(x) = 1000 - 20x$ and $S(x) = x^2 + 10x$, respectively. Find the consumer and producer surplus.

Solution It is apparent that $D(x) = 0$ when $b = 50$, $S(x) = 0$ when $c = 0$, and $D(x) = S(x)$ for $p = 20$. As shown in Figure 6.105, *CS* represents the area under the graph of $D(x)$ on the interval [20, 50] and *PS* is the area under the graph of $S(x)$ on [0, 20].

From (6.46), we obtain

$$CS = \int_{20}^{50} (1000 - 20x) \, dx$$

$$= (1000x - 10x^2) \Big]_{20}^{50} = \$9000$$

and from (6.47),

$$PS = \int_0^{20} (x^2 + 10x) \, dx$$

$$= \left(\frac{x^3}{3} + 5x^2 \right) \Big]_0^{20} \approx \$4666.67$$

___ **Exercises 6.13** ___

Answers to odd-numbered problems begin on page A-30

1. Find the cardiac output of a heart over 24 seconds if 5 milligrams of dye are used and

$$c(t) = \begin{cases} 0, & 0 \le t < 2 \\ -\dfrac{1}{40}(t^2 - 26t + 48), & 2 \le t \le 24 \end{cases}$$

where $c(t)$ is measured in milligrams/liter.

2. Find the cardiac output of a heart over 30 seconds if 5 milligrams of dye are used and

$$c(t) = \begin{cases} \dfrac{4}{21}t, & 0 \le t < 15, \\ -\dfrac{4}{21}t + \dfrac{40}{7}, & 15 \le t \le 30 \end{cases}$$

where $c(t)$ is measured in milligrams/liter.

3. Find the average concentration in Problem 1 on [0, 24].

4. Find the average concentration in Problem 2 on [0, 30].

5. Assuming blood pressure is constant, determine the percentage decrease in the volume of blood that flows through an arteriole per unit time when the radius of the arteriole is constricted from 1 cm to 0.5 cm.

6. The average velocity of blood through an arteriole at a circular cross section of radius R and area A is defined to be

$$\bar{v} = \frac{1}{A} \int_0^R 2\pi r v(r) \, dr$$

where $v(r)$ is given by (6.44). **(a)** Show that $\bar{v} = PR^2/8vl$. **(b)** Show that $\bar{v}$ is one-half the maximum velocity of the blood.

7. Find the total revenue obtained in 4 years if the rate of income in dollars/year is $f(t) = 1200(t - 5)^2$.

8. Find the total revenue obtained in 8 years if the rate of income in dollars/year is $f(t) = 600\sqrt{1 + 3t}$.

9. Find the average rate of flow of revenue over 10 years if the rate of income in dollars/year is given by $f(t) = 3t^2 + 4t + 200$.

10. Find the average rate of flow of revenue over 8 years if the rate of income is as given in Problem 8.

In Problems 11–16 find the consumer and producer surpluses.

11. $S(x) = 2x$, $D(x) = 100 - 2x$

12. $S(x) = \frac{3}{2}x$, $D(x) = 3000 - 6x$

13. $S(x) = x^2 - 4$, $D(x) = -x + 8$

14. $S(x) = 4x^2 + 4x$, $D(x) = -12x + 48$

15. $S(x) = 2x^2 + 3x$, $D(x) = 36 - x^2$

16. $S(x) = x^2 + 16x$, $D(x) = x^2 - 16x + 64$, $0 \le x \le 8$

Miscellaneous Problems

17. Show that if the marginal revenue is $MR = f(x)$ and the marginal cost is $MC = g(x)$, then the change in the profit P between a production level of $x = a$ units to a production level of $x = b$ units is $\int_a^b [f(x) - g(x)] \, dx$.

18. Use Problem 17 to find the change in the profit when $f(x) = -6x + 60$, $g(x) = 15$, and the production level changes from 3 to 7 units.

19. Pareto's law of income states that the number of people with incomes between $x = a$ dollars and $x = b$ dollars is $N = \int_a^b Ax^{-k} \, dx$, where A is a constant and k is an empirical rational constant greater than 1. The average income of all these people is defined to be

$$\bar{x} = \frac{1}{N} \int_a^b Ax^{1-k} \, dx$$

Evaluate $\bar{x}$.

_____ Chapter 6 Review Exercises _____

Answers to odd-numbered problems begin on page A-30.

In Problems 1–12 answer true or false.

1. When $\int_a^b f(x) \, dx > 0$, the integral gives the area under the graph of $y = f(x)$ on the interval $[a, b]$. _____

2. $\int_0^3 (x - 1) \, dx$ is the area under the graph of $y = x - 1$ on [0, 3]. _____

3. The integral $\int_a^b [f(x) - g(x)] \, dx$ gives the area between the graphs of the continuous functions f and g whenever $f(x) \ge g(x)$ for every x in $[a, b]$. _____

4. The disk and washer methods for finding volumes of solids of revolution are special cases of the slicing method. _____

5. The average value $A(f)$ of a continuous function on an interval $[a, b]$ is necessarily a number that satisfies $m \le A(f) \le M$, where m and M are the maximum and minimum values of f on the interval, respectively. _____

6. If f and g are continuous on $[a, b]$, then the average value of $f + g$ is $A(f + g) = A(f) + A(g)$. _____

7. The center of mass of a pencil with a constant linear density ρ is at its geometric center. _____

8. The center of mass of a lamina that coincides with a plane region R is a point in R at which the lamina would hang in balance. _____

9. The pressure on the flat bottom of a swimming pool is the same as the horizontal pressure on the vertical sidewalls at the same depth. _____

10. Consider a circular tin can with radius 6 in and a circular reservoir with radius 50 ft. If each has a flat bottom and is filled with water to a depth of 1 ft, then the liquid pressure on the bottom of the reservoir is greater than the pressure on the bottom of the tin can. _____

11. If $s(t)$ is the position function of a body that moves in a straight line, then $\int_{t_1}^{t_2} v(t)\, dt$ is the distance the body moves in the interval $[t_1, t_2]$. _____

12. In the absence of air resistance, when dropped simultaneously from the same height, a cannonball will hit the ground before a marshmallow. _____

In Problems 13–18 fill in the blanks.

13. The unit of work in the mks system of units is _____.

14. To warm up, a 200-lb jogger pushes against a tree for 5 min with a constant force of 60 lb and then runs 2 mi in 10 min. The total work done is _____.

15. The work done by a 100-lb constant force applied at an angle of 60° to the horizontal over a distance of 50 ft is _____.

16. If 80 nt of force stretches a spring that is initially 1 m long into a spring that is 1.5 m long, then the spring will measure _____ m long when 100 nt of force is applied.

17. The coordinates of the centroid of a region R are $(2, 5)$ and the moment of the region about the x-axis is 30. Hence, the area of R is _____ square units.

18. The graph of a function with a continuous first derivative is said to be _____.

In Problems 19–28 (Figures 6.106–6.115) set up, but do not evaluate, the integral(s) that give the area of the shaded region in each figure.

19.

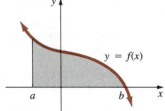

Figure 6.106

20.

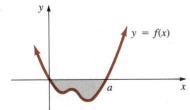

Figure 6.107

21.

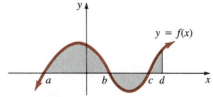

Figure 6.108

22.

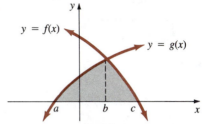

Figure 6.109

23.

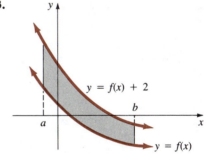

Figure 6.110

24.

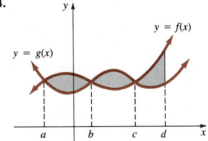

Figure 6.111

25.

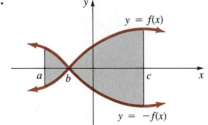

Figure 6.112

26.

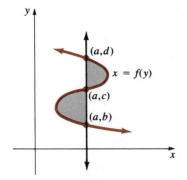

Figure 6.113

27.

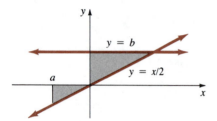

Figure 6.114

28.

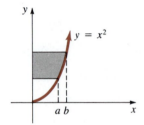

Figure 6.115

In Problems 29–33 consider the shaded region R in Figure 6.116. Set up, but do not evaluate, the integral(s) that give the indicated quantity.

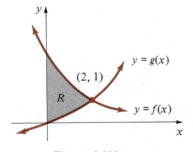

Figure 6.116

29. The centroid of the region

30. The volume of the solid of revolution that is formed by revolving R about the x-axis

31. The volume of the solid of revolution that is formed by revolving R about the y-axis

32. The volume of the solid of revolution that is formed by revolving R about the line $x = 2$

33. The volume of the solid with R as its base such that the cross sections of the solid parallel to the y-axis are squares

34. A solid has as its base the region bounded by the graph of $y = \sin x$ and the x-axis on the interval $[0, \pi]$. As shown in Figure 6.117 cross sections of the solid perpendicular to the x-axis are right isosceles triangles. Find the volume of the solid.

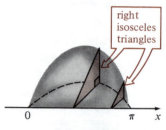

Figure 6.117

35. Find the average value of $f(x) = x^{3/2} + x^{1/2}$ on $[1, 4]$.

36. Find a value of x in the interval $[0, 3]$ that corresponds to the average value of the function $f(x) = 2x - 1$.

37. A spring whose unstretched length is $\frac{1}{2}$ m is stretched to a length of 1 m by a force of 50 nt. Find the work done in stretching the spring from a length of 1 m to a length of 1.5 m.

38. The work done in stretching a spring 6 inches beyond its natural length is 10 ft-lb. Find the spring constant.

39. A water tank, in the form of a cube that is 10 ft on a side, is filled with water. Find the work done in pumping all the water to a point 5 ft above the tank.

40. A bucket weighing 2 lb contains 30 lb of liquid. As the bucket is raised vertically at a rate of 1 ft/sec, the liquid leaks out at a rate of $\frac{1}{4}$ lb/sec. Find the work done in lifting the bucket a distance of 5 ft.

41. In Problem 40, find the work done in lifting the bucket to a point where it is empty.

42. In Problem 40, find the work done in lifting the leaking bucket a distance of 5 ft if the rope attached to the bucket weighs $\frac{1}{8}$ lb/ft.

43. A tank on top of a tower 15 ft high consists of a frustum of a cone surmounted by a right circular cylinder. The dimensions (in feet) are given in Figure 6.118. Find the work done in filling the tank with water from ground level.

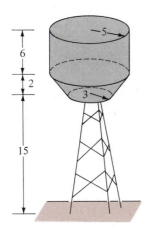

Figure 6.118

44. A rock is thrown vertically upward from the surface of the moon with an initial velocity of 44 ft/sec.

(a) If the acceleration of gravity on the moon is 5.5 ft/sec^2, find the maximum height attained. Compare with the earth.

(b) On the way down, the rock hits a 6-ft-tall astronaut on the head. What is its impact velocity?

45. Find the length of the graph of $y = (x - 1)^{3/2}$ from (1, 0) to (5, 8).

46. The linear density of a 6-m-long rod is a linear function of the distance from its left end. The density in the middle of the rod is 11 kg/m and at the right end the density is 17 kg/m. Find the center of mass of the rod.

47. A flat plate, in the form of a quarter circle, is submerged vertically in oil as shown in Figure 6.119. If the weight density of the oil is 800 kg/m^3, find the force exerted by the oil on one side of the plate.

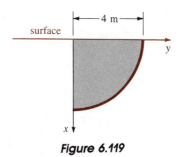

Figure 6.119

48. Three masses are suspended from uniform rigid bars of negligible mass as shown in Figure 6.120. Determine where the indicated wires should be attached so that the entire system hangs in balance.

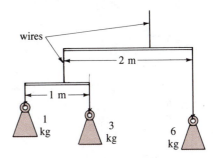

Figure 6.120

7

Inverse Trigonometric Functions

Our goal in this chapter is to study the general concept of an inverse of a function along with the derivatives and integrals that involve six new functions; namely, the inverse trigonometric functions.

7.1 Inverse Functions

One-to-One Functions   Recall that a function is a rule of correspondence that assigns to each value x, in its domain X, a single, or unique, value y in its range. This rule does not preclude having the same number y associated with *two different* values of x. For example, for $f(x) = x^2$, the value $y = 9$ occurs at either $x = -3$ or $x = 3$; that is, $f(-3) = f(3) = 9$. On the other hand, for the function $f(x) = x^3$ the value $y = 64$ occurs *only* at $x = 4$. Functions of the latter kind are given the special name **one-to-one**.

> **DEFINITION 7.1**   A function f is said to be **one-to-one** if every element in its range corresponds to exactly one element in its domain X.

Horizontal Line Test   Interpreted geometrically, this means that a horizontal line ($y = $ constant) can intersect the graph of a one-to-one function in at most one point. Furthermore, if every horizontal line that intersects the graph of a function does so in at most one point, then the function is necessarily one-to-one. A function is *not* one-to-one if *some* horizontal line intersects its graph more than once.

___ **Example 1** ___

The graphs of $g(x) = -x^2 + 2x + 4$ and $f(x) = -4x + 3$, shown in Figure 7.1, indicate that there are two elements x_1 and x_2 in the domain of g for which $g(x_1) = g(x_2) = c$ but only one element x_1 in the domain of f for which $f(x_1) = c$. Hence, g is not one-to-one but f is one-to-one.

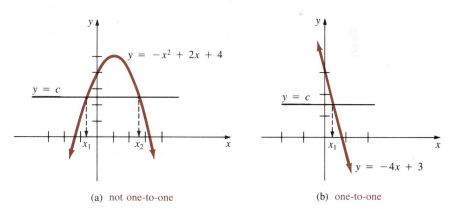

(a) not one-to-one   (b) one-to-one

Figure 7.1

Inverse of a One-to-One Function   Suppose f is a one-to-one function that has domain X and range Y. Since every element y of Y corresponds with precisely one element x of X, the function f must actually determine a "reverse" function g whose domain is Y and range is X. As shown in Figure 7.2, f and g must satisfy

$$f(x) = y \quad \text{and} \quad g(y) = x$$
or
$$f(g(y)) = y \quad \text{and} \quad g(f(x)) = x \tag{7.1}$$

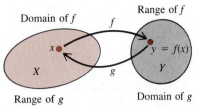

Domain of f　　　f　　Range of f

X　　g　　Y　　$y = f(x)$

Range of g　　　Domain of g

Figure 7.2

The function g is given the formal name **inverse of** f. Denoting each independent variable as x, we summarize the results in (7.1).

> **DEFINITION 7.2**　　　Let f be a one-to-one function with domain X and range Y. The inverse of f is a function g with domain Y and range X for which
>
> $$f(g(x)) = x \quad \text{for every } x \text{ in } Y$$
> $$\text{and} \quad g(f(x)) = x \quad \text{for every } x \text{ in } X \tag{7.2}$$

Symbolically, the inverse of a function f is usually written f^{-1} and is read "f inverse." This latter notation, although standard, is somewhat unfortunate. We hasten to point out that $f^{-1}(x)$ is *not* the same as $[f(x)]^{-1} = 1/f(x)$. In terms of this new notation, (7.2) becomes

$$f(f^{-1}(x)) = x \quad \text{and} \quad f^{-1}(f(x)) = x \tag{7.3}$$

Finding f^{-1}　　The first equation in (7.3) can be used explicitly to find the inverse of a one-to-one function.

Example 2

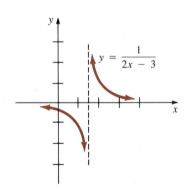

$y = \dfrac{1}{2x - 3}$

Figure 7.3

Find the inverse of $f(x) = \dfrac{1}{2x - 3}$, $x \neq \dfrac{3}{2}$.

Solution　Inspection of Figure 7.3 shows that f is one-to-one. Hence, from (7.3) we have

$$f(f^{-1}(x)) = \frac{1}{2f^{-1}(x) - 3} = x$$

Solve this last equation for $f^{-1}(x)$:

$$2f^{-1}(x) - 3 = \frac{1}{x}$$

$$2f^{-1}(x) = \frac{3x + 1}{x}$$

$$f^{-1}(x) = \frac{3x + 1}{2x}$$

Check　$f(f^{-1}(x)) = \dfrac{1}{2\left(\dfrac{3x + 1}{2x}\right) - 3} = \dfrac{x}{3x + 1 - 3x} = x$

$$f^{-1}(f(x)) = \frac{3\left(\dfrac{1}{2x - 3}\right) + 1}{2\left(\dfrac{1}{2x - 3}\right)} = \frac{\dfrac{2x}{2x - 3}}{\dfrac{2}{2x - 3}} = x$$

Observe in Example 2 that, as discussed, the domain of f (all real numbers except $\frac{3}{2}$), and the range of f (all real numbers except 0), are the range and domain of f^{-1}, respectively.

An Alternative Manner of Finding f⁻¹   The inverse of a function f can be found in a different manner. If g is the inverse of f, then $x = g(y)$. Thus, we need only do the following:

(*i*)   Solve $y = f(x)$ for the symbol x, and following convention,

(*ii*)   Relabel the dependent variable x as y and the independent variable y as x.

The next example illustrates this procedure.

___

Example 3

Find the inverse of $f(x) = -4x + 3$.

Solution   From Example 1 we know f is one-to-one and hence has an inverse. Now, write f as $y = -4x + 3$ and solve for x in terms of y:

$$x = \frac{-y + 3}{4}$$

Relabeling variables then yields

$$y = \frac{-x + 3}{4} \quad \text{or} \quad f^{-1}(x) = -\frac{1}{4}x + \frac{3}{4}$$

___

Graphs of f and f⁻¹   Let (a, b) be any point on the graph of a one-to-one function f. Then, $f(a) = b$ and

$$f^{-1}(b) = f^{-1}(f(a)) = a$$

implies that (b, a) is on the graph of f^{-1}. Since the points (a, b) and (b, a) are symmetric with respect to the line $y = x$, we see in Figure 7.4(b) that the graph of f^{-1} is a reflection of the graph of f in the line $y = x$.

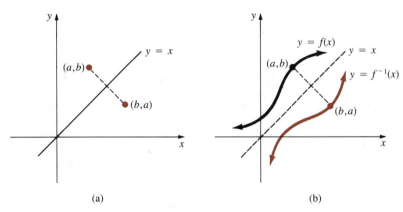

(a)   (b)

Figure 7.4

Example 4

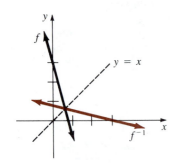

Figure 7.5

Compare the graphs of f and f^{-1} from Example 3.

Solution   The graphs of $f(x) = -4x + 3$ and $f^{-1}(x) = -\frac{1}{4}x + \frac{3}{4}$ are straight lines. The x-intercept and y-intercept of the graph of f are $\frac{3}{4}$ and 3, respectively. On the other hand, the x- and y-intercepts of the graph of f^{-1} are, in turn, 3 and $\frac{3}{4}$. Drawing the lines through the corresponding points yields the graphs shown in Figure 7.5.

Example 5

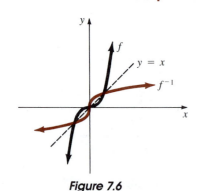

Figure 7.6

Find the inverse of $f(x) = x^3$ and compare the graphs of f and f^{-1}.

Solution   By solving the equation $y = x^3$ for x we obtain $x = y^{1/3}$. With the usual relabeling of variables we see that $f^{-1}(x) = x^{1/3}$. The fact that the graphs of f and f^{-1} are reflections of one another in the line $y = x$ should be apparent in Figure 7.6.

Continuity of f^{-1}   We state the next two theorems without proof.

> **THEOREM 7.1**   Let f be a continuous one-to-one function on an interval $[a, b]$. Then, f^{-1} is continuous on the interval $[f(a), f(b)]$.

> **THEOREM 7.2**   Let f be continuous and increasing on an interval $[a, b]$. Then, f^{-1} exists and is continuous and increasing on $[f(a), f(b)]$.

Example 6

Prove that $f(x) = 5x^3 + 8x - 9$ has an inverse.

Solution   Since f is a polynomial function, it is continuous. Also,

$$f'(x) = 15x^2 + 8 > 0 \qquad \text{for all } x$$

implies that f is increasing on $(-\infty, \infty)$. It follows from Theorem 7.2 that f^{-1} exists.

Note: Theorem 7.2 also holds when the word *increasing* is replaced with the word *decreasing*.

Derivative of f⁻¹

In the next result it is convenient to denote, as we originally did, the inverse of a function $y = f(x)$ by $x = g(y)$.

THEOREM 7.3   Let f be a differentiable function that has an inverse g. Then, g is differentiable when $dy/dx = f'(x) \neq 0$ and

$$\frac{dx}{dy} = \frac{1}{dy/dx} \qquad (7.4)$$

Proof Since f is differentiable, it is continuous so that

$$\Delta y = f(x + \Delta x) - f(x) \to 0 \quad \text{when} \quad \Delta x \to 0$$

By Theorem 7.1, g is continuous; therefore, $\Delta x \to 0$ whenever $\Delta y \to 0$. Thus,

$$\frac{dx}{dy} = \lim_{\Delta y \to 0} \frac{\Delta x}{\Delta y}$$

$$= \lim_{\Delta x \to 0} \frac{1}{\Delta y / \Delta x}$$

$$= \frac{1}{\displaystyle\lim_{\Delta x \to 0} \frac{\Delta y}{\Delta x}}$$

$$= \frac{1}{dy/dx} \qquad \blacksquare$$

Although the notation in (7.4) facilitated its proof, and has obvious mnemonic significance, it is desirable to write the result as

$$g'(y) = \frac{1}{f'(x)}$$

$$= \frac{1}{f'(g(y))} \qquad (7.5)$$

With relabeling of variables, (7.5) becomes

$$g'(x) = \frac{1}{f'(g(x))} \qquad (7.6)$$

Now if f^{-1} denotes the inverse instead of g, then (7.6) finally becomes

$$(f^{-1})'(x) = \frac{1}{f'(f^{-1}(x))} \qquad (7.7)$$

Equation (7.7) clearly shows that to compute the derivative of f^{-1} we must know $f^{-1}(x)$ explicitly. However, if (a, b) is a known point on the graph of f, (7.7) does enable us to evaluate the derivative of f^{-1} at (b, a) without an equation that defines $f^{-1}(x)$.

Example 7

For the function f in Example 6, find the slope of the tangent to the graph of f^{-1} at $(f(1), 1)$.

Solution   Since $f(x) = 5x^3 + 8x - 9$, we have $f(1) = 4$,

$$f'(x) = 15x^2 + 8 \quad \text{and} \quad f'(1) = 23$$

But $f(1) = 4$ implies that $f^{-1}(4) = 1$. Thus, (7.7) gives

$$(f^{-1})'(4) = \frac{1}{f'(f^{-1}(4))}$$

$$= \frac{1}{f'(1)} = \frac{1}{23}$$

In other words, the slope of the tangent to the graph of f at $(1, 4)$ is 23; the slope of the tangent to the graph of f^{-1} at $(4, 1)$ is the reciprocal $\frac{1}{23}$.

In subsequent sections, we shall see that it is often possible to find the derivative of an inverse function by resorting to implicit differentiation.

If a function f is one-to-one, then it has an inverse. Conversely, if f has an inverse, then it is one-to-one. Thus, if a function f is *not* one-to-one, it does *not* possess an inverse. Nonetheless, as the next example shows, it may be possible to restrict the domain of a function that it is not one-to-one in such a manner that the newly defined function does, in fact, have an inverse.

Example 8

In our initial discussion, we saw that $f(x) = x^2$ is not one-to-one. However, by simply requiring x to be nonnegative, we see from Figure 7.7(a) and Theorem 7.2 that the new function

$$F(x) = x^2, x \geq 0 \quad \text{has the inverse} \quad F^{-1}(x) = \sqrt{x}, x \geq 0$$

which is shown in Figure 7.7(b).

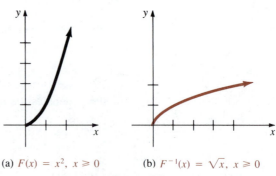

(a) $F(x) = x^2,\ x \geqslant 0$ (b) $F^{-1}(x) = \sqrt{x},\ x \geqslant 0$

Figure 7.7

Remark

A one-to-one function f can have only one inverse. In other words, f^{-1} is unique.

Exercises 7.1

Answers to odd-numbered problems begin on page A-30

In Problems 1–12 determine whether the given function is one-to-one by examining its graph. If the function is one-to-one, find its inverse.

1. $f(x) = 5$

2. $f(x) = 6x - 9$

3. $f(x) = \frac{1}{3}x + 3$

4. $f(x) = |x + 1|$

5. $f(x) = x(x - 5)$

6. $f(x) = (x + 1)^2$

7. $f(x) = x^3 - 8$

8. $f(x) = x^3 - 3x$

9. $f(x) = \frac{4}{x}$

10. $f(x) = \frac{1}{3x + 6}$

11. $f(x) = x^4 + 2$

12. $f(x) = x^5$

In Problems 13–22 each function is one-to-one. Find its inverse.

13. $f(x) = x + 5$

14. $f(x) = -2x + 8$

15. $f(x) = 3x^3 + 7$

16. $f(x) = (-x + 9)^3$

17. $f(x) = \sqrt[3]{2x - 4}$

18. $f(x) = 6 - (10x - 2)^{1/2}$

19. $f(x) = \frac{1}{1 + 8x^3}$

20. $f(x) = 5 - \frac{2}{x}$

21. $f(x) = \frac{2 - x}{1 - x}$

22. $f(x) = 10 + x^{3/5}$

In Problems 23–26 determine, without graphing, whether the given function has an inverse.

23. $f(x) = 10x^3 + 8x + 12$

24. $f(x) = -7x^5 - 6x^3 - 2x + 17$

25. $f(x) = 2x^3 + 3x^2 - 12x$

26. $f(x) = x^4 - 2x^2$

In Problems 27–32, without finding the inverse, state the domain and range of f^{-1}.

27. $f(x) = \sqrt{x + 2}$

28. $f(x) = 3 + \sqrt{2x - 1}$

29. $f(x) = \frac{1}{x + 3}$

30. $f(x) = \frac{x - 1}{x - 4}$

31. $f(x) = (x - 5)^2, \, x \geq 5$

32. $f(x) = x^2 + 2x + 6, \, x \geq -1$

In Problems 33 and 34 (Figures 7.8 and 7.9) sketch the graph of f^{-1} from the graph of f.

33.

34.

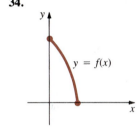

Figure 7.8   Figure 7.9

In Problems 35 and 36 (Figures 7.10 and 7.11) sketch the graph of f from the graph of f^{-1}.

35.

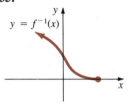

36.

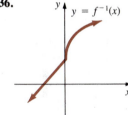

Figure 7.10   Figure 7.11

In Problems 37 and 38 use (7.7) to find the derivative of the inverse function at the indicated point.

37. $f(x) = 2x^3 + 8; \, (f(\frac{1}{2}), \frac{1}{2})$

38. $f(x) = -x^3 - 3x + 7; \, (f(-1), -1)$

In Problems 39–42, without finding the inverse, find, at the indicated value of x, the corresponding point on the graph of f^{-1} and an equation of the tangent line at this point.

39. $f(x) = \frac{1}{3}x^3 + x - 7; \, x = 3$

40. $f(x) = \frac{2x + 1}{4x - 1}; \, x = 0$

41. $f(x) = (x^5 + 1)^3; \, x = 1$

42. $f(x) = 8 - 6\sqrt[3]{x + 2}; \, x = -3$

In Problems 43 and 44 show, without integrating or graphing, that the given function is one-to-one on the indicated interval. Without finding f^{-1}, find $(f^{-1})'(0)$ in each case.

43. $f(x) = \displaystyle\int_1^x \frac{dt}{t+1}, \; x > -1$

44. $f(x) = \displaystyle\int_1^x \sqrt{t^2 + 4} \; dt, \; (-\infty, \infty)$

In Problems 45 and 46 find f^{-1}. Find $(f^{-1})'$ using (7.7). Verify the result by direct differentiation of f^{-1}.

45. $f(x) = \dfrac{2x + 1}{x}$ **46.** $f(x) = (5x + 7)^3$

In Problems 47–50 define a new function F, by restricting the domain of f, that is one-to-one and has the same range as f. Find F^{-1}.

47. $f(x) = (5 - 2x)^2$ **48.** $f(x) = 3x^2 + 9$

49. $f(x) = x^2 + 2x + 4$ **50.** $f(x) = -x^2 + 8x$

Miscellaneous Problems

53. An equivalent definition of a one-to-one function is: A function f is one-to-one if and only if $x_1 \neq x_2$ implies that $f(x_1) \neq f(x_2)$ whenever x_1 and x_2 are in the domain of x. Use this definition to prove that $f(x) = 3x + 7$ is one-to-one.

54. Use the definition of a one-to-one function given in Problem 53 to prove that a function that is either increasing or decreasing is one-to-one. (*Hint*: $x_1 \neq x_2$ implies $x_1 < x_2$ or $x_1 > x_2$.)

51. Consider the one-to-one function $f(x) = x^3 + x$ on the interval $[1, 2]$. See Figure 7.12. Without finding f^{-1}, determine the value of $\int_{f(1)}^{f(2)} f^{-1}(x) \, dx$.

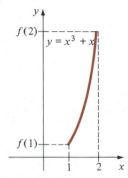

Figure 7.12

52. Suppose a function f is one-to-one on $(-\infty, \infty)$ and has the properties $f(0) = 1; f(-2) = 0; f''(x) > 0, x < 0, f''(x) < 0, x > 0; \lim_{x \to -\infty} f(x) = -1;$ and $\lim_{x \to \infty} f(x) = 2$. Sketch a graph of f^{-1}.

55. Suppose f is a periodic function with period $p, f(x + p) = f(x)$. Can f have an inverse function?

56. Let f^{-1} be the inverse of a function f. If f and f^{-1} are differentiable, use implicit differentiation and $f(f^{-1}(x)) = x$ to derive the result given in (7.7).

57. If f and $(f^{-1})'$ are differentiable, use (7.7) to show that $(f^{-1})''(x) = -f''(f^{-1}(x))/[f'(f^{-1}(x))]^3$.

7.2 Inverse Trigonometric Functions

Inspection of their graphs and observations such as

$$\sin(x + 2\pi) = \sin x$$
$$\cos(x + 2\pi) = \cos x$$
$$\tan(x + \pi) = \tan x$$

$$\sin \frac{\pi}{6} = \sin \frac{5\pi}{6} = \frac{1}{2}$$

$$\cos \frac{\pi}{4} = \cos \frac{7\pi}{4} = \frac{\sqrt{2}}{2}$$

and $$\cot \frac{\pi}{2} = \cot \frac{3\pi}{2} = 0$$

should convince you that none of the trigonometric functions are one-to-one. Nevertheless, by proceeding as in Example 8 of Section 7.1, we can find the inverse of a trigonometric function that is defined *on a restricted domain*.

Inverse Sine By considering $y = \sin x$ only on the closed interval $-\pi/2 \le x \le \pi/2$, it is apparent from Figure 7.13(a) that for each value of y in $[-1, 1]$ there is only one corresponding x in $[-\pi/2, \pi/2]$. In addition, we notice that this new function fulfills the criteria of Theorem 7.2 and thus has an inverse. Recall, the domain and range of the original function become, respectively, the range and domain of the inverse. Accordingly,

the **inverse sine**, written $\sin^{-1}x$, is defined by

$$y = \sin^{-1}x \quad \text{if and only if} \quad x = \sin y \tag{7.8}$$

where $-1 \le x \le 1$ and $-\pi/2 \le y \le \pi/2$.

In other words,

the inverse sine of a number is the angle (in radians) whose sine is x,

provided that this angle satisfies $-\pi/2 \le \sin^{-1}x \le \pi/2$. The graph of $y = \sin^{-1}x$ is given in Figure 7.13(b).

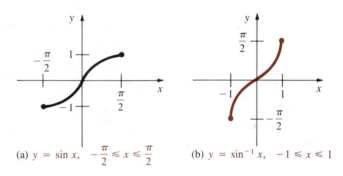

(a) $y = \sin x, \quad -\dfrac{\pi}{2} \le x \le \dfrac{\pi}{2}$ (b) $y = \sin^{-1}x, \quad -1 \le x \le 1$

Figure 7.13

In view of (7.3) of Section 7.1 we have

$$\sin(\sin^{-1}x) = x \quad \text{when} \quad -1 \le x \le 1$$

$$\sin^{-1}(\sin x) = x \quad \text{when} \quad -\frac{\pi}{2} \le x \le \frac{\pi}{2} \tag{7.9}$$

_____ **Example 1** _____

Evaluate $\sin^{-1}\left(\dfrac{\sqrt{2}}{2}\right)$.

Solution Let $y = \sin^{-1}(\sqrt{2}/2)$. One angle y for which $\sin y = \sqrt{2}/2$ is $y = \pi/4$. Since this is the only number in $[-\pi/2, \pi/2]$ for which this is true, we can write

$$\sin^{-1}\left(\frac{\sqrt{2}}{2}\right) = \frac{\pi}{4}$$

Notation   In order not to confuse $\sin^{-1}x$ with the reciprocal $(\sin x)^{-1} = 1/\sin x$, an alternative notation is often used:

$$\arcsin x = \sin^{-1}x$$

and is read as it appears "arc sine of x."

Example 2

Evaluate $\arcsin\left(\dfrac{1}{2}\right)$.

Solution   If $y = \arcsin \frac{1}{2}$, then $\frac{1}{2} = \sin y$. Then only number y in $[-\pi/2, \pi/2]$ for which this is true is $\pi/6$, that is,

$$\arcsin\left(\frac{1}{2}\right) = \frac{\pi}{6}$$

Note: Although some calculators give

$$\arcsin\left(\frac{1}{2}\right) = 30°$$

this is, strictly speaking, *incorrect*. The output of an inverse trigonometric function is an angle in *radian measure*.

Inverse Cosine   The function $y = \cos x$ is continuous and decreasing on the interval $[0, \pi]$. Hence,

the **inverse cosine** is defined by

$$y = \cos^{-1}x \quad \text{if and only if} \quad x = \cos y \qquad (7.10)$$

where $-1 \leq x \leq 1$ and $0 \leq y \leq \pi$.

The graphs of $y = \cos x$ and $y = \cos^{-1}x$ are illustrated in Figure 7.14.

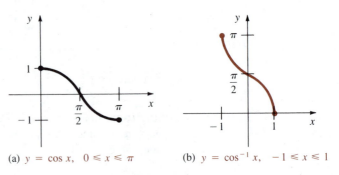

(a) $y = \cos x, \quad 0 \leq x \leq \pi$ (b) $y = \cos^{-1} x, \quad -1 \leq x \leq 1$

Figure 7.14

Example 3

Evaluate $\cos^{-1}(-1)$.

Solution   Let $y = \cos^{-1}(-1)$. From (7.10) we see that the only number y in the closed interval $[0, \pi]$ for which $\cos y = -1$ is $y = \pi$. Thus,

$$\cos^{-1}(-1) = \pi$$

Analogous to (7.9) we have

$$\cos(\cos^{-1}x) = x \quad \text{when} \quad -1 \le x \le 1$$
$$\cos^{-1}(\cos x) = x \quad \text{when} \quad 0 \le x \le \pi$$

Inverse Tangent   On the open interval $(-\pi/2, \pi/2)$, $y = \tan x$ is continuous and increasing. Thus,

the **inverse tangent** is defined by

$$y = \tan^{-1}x \quad \text{if and only if} \quad x = \tan y \qquad (7.11)$$

where $-\infty < x < \infty$ and $-\pi/2 < y < \pi/2$.

See Figure 7.15.

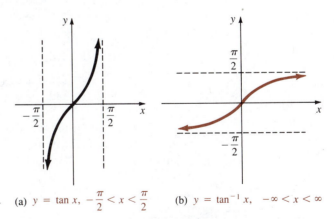

(a) $y = \tan x,\ -\dfrac{\pi}{2} < x < \dfrac{\pi}{2}$ (b) $y = \tan^{-1}x,\ -\infty < x < \infty$

Figure 7.15

With their domains suitably restricted, the remaining trigonometric functions also have inverses. Definition 7.3 summarizes (7.8), (7.10), and (7.11) along with the definitions of the inverses of the cotangent, secant, and cosecant. We urge you to compare the graphs of $y = \cot^{-1}x$, $y = \sec^{-1}x$, and $y = \csc^{-1}x$, given in Figure 7.16, with the pertinent portions of the graphs of $y = \cot x$, $y = \sec x$, and $y = \csc x$ (see page 52).

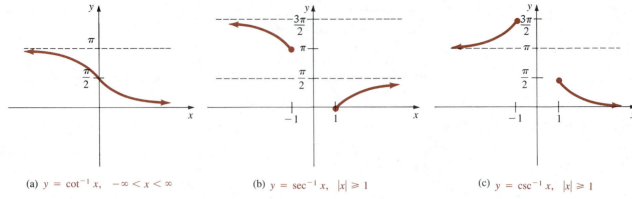

(a) $y = \cot^{-1} x, \quad -\infty < x < \infty$

(b) $y = \sec^{-1} x, \quad |x| \ge 1$

(c) $y = \csc^{-1} x, \quad |x| \ge 1$

Figure 7.16

DEFINITION 7.3   Inverse Trigonometric Functions

(*i*) $y = \sin^{-1}x$ if and only if $x = \sin y$,
   $|x| \le 1$ and $-\pi/2 \le y \le \pi/2$

(*ii*) $y = \cos^{-1}x$ if and only if $x = \cos y$,
   $|x| \le 1$ and $0 \le y \le \pi$

(*iii*) $y = \tan^{-1}x$ if and only if $x = \tan y$,
   $-\infty < x < \infty$ and $-\pi/2 < y < \pi/2$

(*iv*) $y = \cot^{-1}x$ if and only if $x = \cot y$,
   $-\infty < x < \infty$ and $0 < y < \pi$

(*v*) $y = \sec^{-1}x$ if and only if $x = \sec y$,
   $|x| \ge 1$ and y is a number in $[0, \pi/2)$ or $[\pi, 3\pi/2)$

(*vi*) $y = \csc^{-1}x$ if and only if $x = \csc y$,
   $|x| \ge 1$ and y is a number in $(0, \pi/2]$ or $(\pi, 3\pi/2]$

As it did for the sine function, the prefix "arc" also denotes the inverse of any of the other five trigonometric functions. We shall use both notations interchangeably.

____ **Example 4** _____

Evaluate $\operatorname{arcsec}(-\sqrt{2})$.

Solution   Let $y = \operatorname{arcsec}(-\sqrt{2})$ so that $-\sqrt{2} = \sec y$ or $\cos y = -1/\sqrt{2}$. The only number y in $[0, \pi/2)$ or $[\pi, 3\pi/2)$ that satisfies these conditions is $y = 5\pi/4$. Since $\sec 5\pi/4 = -\sqrt{2}$, we have

$$\operatorname{arcsec}(-\sqrt{2}) = \frac{5\pi}{4}$$

___ **Example 5** _____

Evaluate $\cos\left(\tan^{-1}\dfrac{2}{3}\right)$.

Solution Let $y = \tan^{-1}\frac{2}{3}$ so that $\frac{2}{3} = \tan y$. Using the triangle in Figure 7.17 as an aid, we see

$$\cos\left(\tan^{-1}\frac{2}{3}\right) = \cos\ y$$

$$= \frac{3}{\sqrt{13}}$$

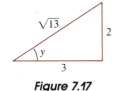

Figure 7.17

___ **Example 6** _____

Evaluate $\cos(\sin^{-1}1)$.

Solution Since $\sin^{-1}1 = \pi/2$, we have

$$\cos(\sin^{-1}1) = \cos\frac{\pi}{2} = 0$$

The inverse tangent is especially important in applications.

___ **Example 7** _____

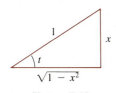

Figure 7.18

On a microcomputer the arctangent function is intrinsic, or built into, the BASIC programming language, whereas the arcsine and arccosine functions are not. In order to write a program that involves, say, the arcsine, the writer must utilize what is known as a "user-defined function" that expresses the arcsine in terms of the arctangent.

If $t = \sin^{-1}x$, we can interpret x as the side of a right triangle that is opposite t and 1 as the hypotenuse of the triangle. In Figure 7.18, we see that the side adjacent to the angle t is $\sqrt{1 - x^2}$. Hence,

$$\tan t = \frac{x}{\sqrt{1 - x^2}}$$

$$t = \tan^{-1}\frac{x}{\sqrt{1 - x^2}}$$

or $$\sin^{-1}x = \tan^{-1}\frac{x}{\sqrt{1 - x^2}} \qquad (7.12)$$

A similar result can be derived for the inverse cosine. See Problem 44, Exercises 7.2.

Although we derived formula (7.12) from a triangle in which $0 < t < \pi/2$, the result is valid as well for $-\pi/2 < t < \pi/2$.

Example 8

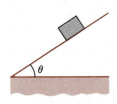

Figure 7.19

The so-called *angle of repose* of a block at the verge of slipping down an inclined plane is defined by

$$\theta = \tan^{-1}f \tag{7.13}$$

where f is the coefficient of friction. (For wood on wood $0.3 \leq f \leq 0.6$.) See Figure 7.19. For an angle greater than $\tan^{-1}f$ the block will slide down the inclined plane.

Remark

The range of an inverse trigonometric function is arbitrary up to a point since a trigonometric function has an inverse on *any* interval of the *x*-axis for which the function is one-to-one. For example, we could just as well limit the domain of $y = \sin x$ to the interval $[\pi/2, 3\pi/2]$, thereby yielding $y = \sin^{-1}x$, $|x| \leq 1$ whose range is given by $\pi/2 \leq y \leq 3\pi/2$. The ranges specified in *(i)*–*(iv)* of Definition 7.3 are universally agreed upon and grew out of the most logical and most convenient limitation of the original function. Thus, when one sees $\arccos x$ or $\tan^{-1}x$ in any context, we know

$$0 \leq \arccos x \leq \pi \quad \text{and} \quad -\pi/2 < \tan^{-1}x < \pi/2$$

The conventions in *(i)*–*(iii)* are the same that are used in hand calculators when the $\boxed{\sin^{-1}x}$, $\boxed{\cos^{-1}x}$, and $\boxed{\tan^{-1}x}$ keys are employed. You may be interested to know that there has been no agreement on the ranges of either $y = \sec^{-1}x$ or $y = \csc^{-1}x$. For example, in contrast to *(v)* and *(vi)* of Definition 7.3, some calculus texts will define

$$y = \sec^{-1}x \quad \text{if and only if} \quad x = \sec y$$
$$|x| \geq 1 \quad \text{and} \quad y \text{ is a number in } [0, \pi/2) \quad \text{or} \quad (\pi/2, \pi] \tag{7.14}$$

and

$$y = \csc^{-1}x \quad \text{if and only if} \quad x = \csc y$$
$$|x| \geq 1 \quad \text{and} \quad y \text{ is a number in } [-\pi/2, 0) \quad \text{or} \quad (0, \pi/2]$$

Exercises 7.2

Answers to odd-numbered problems begin on page A-31

In Problems 1–30 obtain the exact value of the given expression. Do not use a calculator or tables.

1. $\sin^{-1}(0)$

2. $\cos^{-1}\left(\dfrac{1}{2}\right)$

3. $\arccos\left(-\dfrac{\sqrt{2}}{2}\right)$

4. $\sin^{-1}(-1)$

5. $\arctan(1)$

6. $\tan^{-1}(\sqrt{3})$

7. $\cot^{-1}(-1)$

8. $\sec^{-1}(-1)$

9. $\arcsin\left(-\dfrac{\sqrt{3}}{2}\right)$

10. $\text{arccot}(-\sqrt{3})$

11. $\text{arcsec}\left(-\dfrac{2}{\sqrt{3}}\right)$

12. $\csc^{-1}(-\sqrt{2})$

13. $\cos\left(\sin^{-1}\dfrac{1}{2}\right)$

14. $\sin(\cos^{-1}0)$

15. $\tan^{-1}(\cos \pi)$

16. $\cos^{-1}\left(\sin \dfrac{5\pi}{4}\right)$

17. $\sin\left(\arctan \dfrac{4}{3}\right)$

18. $\cos\left(\sin^{-1}\dfrac{2}{5}\right)$

19. $\tan\left(\cot^{-1}\dfrac{1}{2}\right)$

20. $\csc\left(\tan^{-1}\dfrac{2}{3}\right)$

21. $\sec(\tan^{-1}1)$

22. $\arcsin\left(\cot\left(-\dfrac{\pi}{4}\right)\right)$

23. $\sin^{-1}\left(\sin \dfrac{2\pi}{3}\right)$

24. $\cos^{-1}\left(\cos \dfrac{4\pi}{3}\right)$

25. $\cos^{-1}(\cos 3\pi)$

26. $\sin^{-1}(\sin \pi)$

27. $\tan\left(\sin^{-1}\left(-\dfrac{1}{2}\right)\right)$

28. $\sec\left(\tan^{-1}\left(-\dfrac{1}{\sqrt{3}}\right)\right)$

29. $\cos(2 \sin^{-1}(-1))$

30. $\sin\left(2 \cos^{-1}\dfrac{1}{2}\right)$

In Problems 31–36 evaluate the given expression by means of an appropriate trigonometric identity.

31. $\sin\left(2 \sin^{-1}\dfrac{1}{3}\right)$

32. $\cos\left(2 \cos^{-1}\dfrac{3}{4}\right)$

33. $\sin\left(\arcsin \dfrac{1}{\sqrt{3}} + \arccos \dfrac{2}{3}\right)$

34. $\sin(\arctan 2 - \arctan 1)$

35. $\cos(\tan^{-1}4 - \tan^{-1}3)$

36. $\cos\left(\sin^{-1}\dfrac{1}{2} + \cos^{-1}\dfrac{1}{2}\right)$

In Problems 37–42 write the given expression as an algebraic quantity in x.

37. $\sin(\arccos x)$

38. $\cos\left(\tan^{-1}\dfrac{x}{2}\right)$

39. $\tan(\sec^{-1}x)$

40. $\sec(\cot^{-1}x)$

41. $\sin(2 \tan^{-1}x)$

42. $\tan(\arccos x)$

In Problems 43 and 44 verify the identities.

43. $\sin^{-1}x + \cos^{-1}x = \dfrac{\pi}{2}$

44. $\cos^{-1}x = \dfrac{\pi}{2} - \tan^{-1}\dfrac{x}{\sqrt{1 - x^2}}$

In Problems 45 and 46 solve for x.

45. $2 \arcsin(2x - 5) = \pi$

46. $3x + \dfrac{\pi}{3} = \cos^{-1}\left(-\dfrac{\sqrt{3}}{2}\right)$

47. If $t = \sin^{-1}(-2/\sqrt{5})$, find $\cos t$, $\tan t$, $\cot t$, $\sec t$, and $\csc t$.

48. If $\theta = \arctan(0.6)$, find $\sin \theta$, $\cos \theta$, $\cot \theta$, $\sec \theta$, and $\csc \theta$.

Calculator Problems

49. Use a calculator to verify that

    (a) $\tan(\tan^{-1}1.3) = 1.3$ and $\tan^{-1}(\tan 1.3) = 1.3$

    (b) $\tan(\tan^{-1}5) = 5$ and $\tan^{-1}(\tan 5) = -1.2832$

Explain why $\tan^{-1}(\tan 5) \neq 5$.

50. Let $x = 1.7$ radians. Compare, if possible, the values of $\sin^{-1}(\sin x)$ and $\sin(\sin^{-1}x)$. Explain any differences.

51. **(a)** For a car moving at velocity v, the banking angle of a curve at which there is no side thrust on the wheels is given by

$$\phi = \tan^{-1}\dfrac{v^2}{Rg} \qquad (7.15)$$

where g is the acceleration of gravity and R is the radius of the curve. See Figure 7.20. At what angle should a road be banked for a car moving at 55 mph around a curve of radius 700 ft? (*Hint:* Use consistent units.)

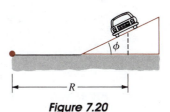

Figure 7.20

    (b) If f is the coefficient of friction between the car and the road and $\theta = \tan^{-1}f$ is defined by (7.13), then the maximum velocity that a car can travel around the curve without slipping is $v_m^2 = gR \tan(\theta + \phi)$, where ϕ is a given banking angle. If $f = 0.25$, find v_m for the road banked at the angle determined in part **(a)**.

52. Consider a ladder of length L leaning against a house with a load at point P shown in Figure 7.21. The angle β, at which the ladder is at the verge of slipping, is defined by

$$\dfrac{x}{L} = \dfrac{f}{1 + f^2}(f + \tan \beta)$$

where f is the coefficient of friction between the ladder and the ground.

    (a) Find β when $f = 1$ and the load is at the top of the ladder.

    (b) Find β when $f = 0.5$ and the load is $\frac{3}{4}$ of the way up the ladder.

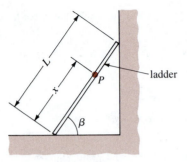

Figure 7.21

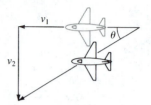

Find the course of a plane flying west at 300 km/hr if a wind from the north blows at 60 km/hr.

Figure 7.22

53. An airplane flies west at a constant speed v_1 and a wind blows from the north at a constant speed v_2. The plane's course south of west is given by $\theta = \tan^{-1}(v_2/v_1)$. See Figure 7.22.

Miscellaneous Problem

54. Evaluate $\sin^{-1}(\sin(-2))$ without the aid of a calculator.

7.3   Derivatives and Integrals That Involve Inverse Trigonometric Functions

Derivatives The derivative of an inverse trigonometric function can be obtained, with minimum effort, through the use of implicit differentiation. A review of Figures 7.15(b), 7.16(a), and Theorem 7.3 reveals that the inverse tangent and inverse cotangent are differentiable for all x. However, the remaining four inverse trigonometric functions are not differentiable at either $x = -1$ or $x = 1$. We shall confine our attention to the derivations of the derivative formulas for the inverse sine, inverse tangent, and inverse secant and leave the others as exercises.

Inverse Sine For $-1 < x < 1$ and $-\pi/2 < y < \pi/2$,

$$y = \sin^{-1}x \quad \text{if and only if} \quad x = \sin y$$

Therefore, implicit differentiation yields

$$\frac{d}{dx}x = \frac{d}{dx}\sin y$$

$$1 = \cos y \frac{dy}{dx}$$

$$\frac{dy}{dx} = \frac{1}{\cos y} \tag{7.16}$$

For the given restriction on the variable y, $\cos y > 0$ and $\cos y = \sqrt{1 - \sin^2 y} = \sqrt{1 - x^2}$. By substituting this quantity in (7.16), we have shown

$$\frac{d}{dx}\sin^{-1}x = \frac{1}{\sqrt{1 - x^2}} \tag{7.17}$$

As predicted, note that (7.17) is not defined at $x = -1$ and $x = 1$.

Inverse Tangent   For $-\infty < x < \infty$ and $-\pi/2 < y < \pi/2$,

$$y = \tan^{-1}x \quad \text{if and only if} \quad x = \tan y$$

and so

$$\frac{d}{dx}x = \frac{d}{dx}\tan y$$

$$1 = \sec^2 y \frac{dy}{dx}$$

$$\frac{dy}{dx} = \frac{1}{\sec^2 y} \tag{7.18}$$

Since $\sec^2 y = 1 + \tan^2 y = 1 + x^2$, (7.18) becomes

$$\frac{d}{dx}\tan^{-1}x = \frac{1}{1 + x^2} \tag{7.19}$$

Inverse Secant   For $|x| > 1$ and $0 < y < \pi/2$ or $\pi < y < 3\pi/2$

$$y = \sec^{-1}x \quad \text{if and only if} \quad x = \sec y$$

Differentiating the last equation implicitly gives

$$\frac{dy}{dx} = \frac{1}{\sec y \tan y} \tag{7.20}$$

In view of the restrictions on y, we have

$$\tan y = \sqrt{\sec^2 y - 1} = \sqrt{x^2 - 1}, \qquad |x| > 1$$

Thus (7.20) becomes

$$\frac{d}{dx}\sec^{-1}x = \frac{1}{x\sqrt{x^2 - 1}} \tag{7.21}$$

Of course, (7.21) is consistent with Figure 7.16(b); the slope of the tangent line to the graph is negative for $x < -1$ and positive for $x > 1$.

The derivative of the composition of an inverse trigonometric function with a differentiable function $u = g(x)$ is obtained from the Chain Rule.

I $\quad\dfrac{d}{dx}\sin^{-1}u = \dfrac{1}{\sqrt{1 - u^2}}\dfrac{du}{dx}$ II $\quad\dfrac{d}{dx}\cos^{-1}u = \dfrac{-1}{\sqrt{1 - u^2}}\dfrac{du}{dx}$

III $\quad\dfrac{d}{dx}\tan^{-1}u = \dfrac{1}{1 + u^2}\dfrac{du}{dx}$ IV $\quad\dfrac{d}{dx}\cot^{-1}u = \dfrac{-1}{1 + u^2}\dfrac{du}{dx}$

V $\quad\dfrac{d}{dx}\sec^{-1}u = \dfrac{1}{u\sqrt{u^2 - 1}}\dfrac{du}{dx}$ VI $\quad\dfrac{d}{dx}\csc^{-1}u = \dfrac{-1}{u\sqrt{u^2 - 1}}\dfrac{du}{dx}$

Example 1

Differentiate $y = \sin^{-1}5x$.

Solution　From I,

$$\frac{dy}{dx} = \frac{1}{\sqrt{1 - (5x)^2}}\frac{d}{dx}5x$$

$$= \frac{5}{\sqrt{1 - 25x^2}}$$

Example 2

Differentiate $y = \tan^{-1}\sqrt{2x + 1}$.

Solution　From III,

$$\frac{dy}{dx} = \frac{1}{1 + (\sqrt{2x + 1})^2}\frac{d}{dx}(2x + 1)^{1/2}$$

$$= \frac{1}{1 + (2x + 1)} \cdot \frac{1}{2}(2x + 1)^{-1/2} \cdot 2$$

$$= \frac{1}{(2x + 2)\sqrt{2x + 1}}$$

Example 3

Differentiate $y = \sec^{-1}x^2$.

Solution　For $x^2 > 1$, we have from V,

$$\frac{dy}{dx} = \frac{1}{x^2\sqrt{(x^2)^2 - 1}}\frac{d}{dx}x^2$$

$$= \frac{2x}{x^2\sqrt{x^4 - 1}}$$

$$= \frac{2}{x\sqrt{x^4 - 1}}$$

Integrals　From the differentiation formulas I–VI, we obtain equivalent integration formulas. For the sake of brevity, we shall, however, consider only the analogues of I, III, and V:

$$\int \frac{du}{\sqrt{1 - u^2}} = \sin^{-1}u + C \qquad (7.22)$$

$$\int \frac{du}{1 + u^2} = \tan^{-1}u + C \qquad (7.23)$$

$$\int \frac{du}{u\sqrt{u^2 - 1}} = \sec^{-1}u + C \qquad (7.24)$$

It is understood that the variable u is appropriately restricted in (7.22) and (7.24). Bear in mind that in (7.22) we have $|u| < 1$ and in (7.24), $|u| > 1$.

Example 4

Evaluate $\displaystyle\int \frac{dx}{\sqrt{100 - x^2}}$.

Solution   By factoring 100 from the radical and identifying

$$u = \frac{x}{10}, \qquad du = \frac{dx}{10}$$

the result is obtained from (7.22):

$$\int \frac{dx}{\sqrt{100 - x^2}} = \frac{1}{10} \int \frac{dx}{\sqrt{1 - \left(\dfrac{x}{10}\right)^2}}$$

$$= \int \frac{\dfrac{dx}{10}}{\sqrt{1 - \left(\dfrac{x}{10}\right)^2}}$$

$$= \int \frac{du}{\sqrt{1 - u^2}}$$

$$= \sin^{-1}u + C$$

$$= \sin^{-1}\frac{x}{10} + C$$

Example 5

Find the area under the graph of $f(x) = 1/(1 + x^2)$ on the interval $[-1, 1]$.

Solution   From Figure 7.23 we see f is a nonnegative function and so

$$A = \int_{-1}^{1} \frac{dx}{1 + x^2}$$

$$= \tan^{-1}x \Big]_{-1}^{1}$$

$$= \tan^{-1}1 - \tan^{-1}(-1)$$

$$= \frac{\pi}{4} - \left(-\frac{\pi}{4}\right) = \frac{\pi}{2} \text{ square units}$$

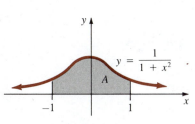

Figure 7.23

Example 6

Evaluate $\displaystyle\int \frac{(\tan^{-1}x)^2}{1 + x^2}\, dx$.

Solution   It should be apparent that the integral does not fall into any of the three forms (7.22)–(7.24). But if we identify

$$u = \tan^{-1}x \quad\text{and}\quad du = \frac{dx}{1 + x^2}$$

then

$$\int \frac{(\tan^{-1}x)^2}{1 + x^2}\, dx = \int u^2\, du$$

$$= \frac{u^3}{3} + C$$

$$= \frac{(\tan^{-1}x)^3}{3} + C$$

Long Division   When faced with the problem of evaluating $\int f(x)\, dx$, where $f(x) = P(x)/Q(x)$ is a rational function, a working rule is:

If the degree of $P(x)$ is greater than or equal to the degree of $Q(x)$, use long division before integration; that is, write

$$\frac{P(x)}{Q(x)} = a\ polynomial + \frac{R(x)}{Q(x)}$$

where the degree of $R(x)$ is less than the degree of $Q(x)$.

Example 7

Evaluate $\displaystyle\int \frac{x^2}{1 + x^2}\, dx$.

Solution   The integrand calls for long division:

$$\frac{x^2}{1 + x^2} = 1 - \frac{1}{1 + x^2}$$

In view of this result, we have immediately

$$\int \frac{x^2}{1 + x^2}\, dx = x - \tan^{-1}x + C$$

For convenience, we will extend (7.22), (7.23), and (7.24) in the following manner. For $a > 0$,

$$\int \frac{du}{\sqrt{a^2 - u^2}} = \sin^{-1}\frac{u}{a} + C \tag{7.25}$$

$$\int \frac{du}{a^2 + u^2} = \frac{1}{a} \tan^{-1}\frac{u}{a} + C \tag{7.26}$$

$$\int \frac{du}{u\sqrt{u^2 - a^2}} = \frac{1}{a} \sec^{-1}\frac{u}{a} + C \tag{7.27}$$

Notice that the result of Example 4 is obtained by identifying $a = 10$ and $u = x$ in (7.25).

Example 8

Evaluate $\displaystyle\int \frac{dx}{x\sqrt{x^4 - 16}}$.

Solution On first inspection, the integral does not appear to be any one of the forms (7.25–7.27). But, by multiplying the numerator and the denominator of the integrand by x, we recognize

$$u = x^2, \qquad du = 2x\,dx, \quad \text{and} \quad a = 4$$

Hence, from (7.27),

$$\int \frac{dx}{x\sqrt{x^4 - 16}} = \frac{1}{2} \int \frac{2x\,dx}{x^2\sqrt{x^4 - 16}}$$

$$= \frac{1}{2} \int \frac{du}{u\sqrt{u^2 - 4^2}}$$

$$= \frac{1}{2} \cdot \frac{1}{4} \sec^{-1}\frac{u}{4} + C$$

$$= \frac{1}{8} \sec^{-1}\frac{x^2}{4} + C$$

Completing the Square To utilize the integration formulas (7.25–7.27), it may be necessary to **complete the square**. Recall from algebra that to complete the square for a quadratic expression $x^2 + bx$, we add and subtract the square of one-half the coefficient of x:

$$x^2 + bx = x^2 + bx + \underbrace{\left(\frac{b}{2}\right)^2 - \left(\frac{b}{2}\right)^2}_{\text{zero}}$$

$$= \left(x^2 + bx + \frac{b^2}{4}\right) - \frac{b^2}{4}$$

$$= \left(x + \frac{b}{2}\right)^2 - \frac{b^2}{4}$$

_____ **Example 9** _____

Evaluate $\displaystyle\int \frac{dx}{x^2 + 12x + 37}$.

Solution   First complete the square:

$$
\begin{aligned}
x^2 + 12x + 37 &= (x^2 + 12x) + 37 \\
&= (x^2 + 12x + 36 - 36) + 37 \\
&= (x^2 + 12x + 36) + 1 \\
&= (x + 6)^2 + 1
\end{aligned}
$$

With $u = x + 6$ and $du = dx$

it follows from (7.23) that

$$
\begin{aligned}
\int \frac{dx}{x^2 + 12x + 37} &= \int \frac{dx}{(x + 6)^2 + 1} \\
&= \int \frac{du}{u^2 + 1} \\
&= \tan^{-1}u + C \\
&= \tan^{-1}(x + 6) + C
\end{aligned}
$$

To complete the square for a quadratic expression $ax^2 + bx$, $a \neq 1$, we first write

$$
ax^2 + bx = a\left(x^2 + \frac{b}{a}x\right)
$$

and then add and subtract $(b/2a)^2$ *inside* the parentheses.

_____ **Example 10** _____

Evaluate $\displaystyle\int \frac{dx}{\sqrt{4 - 6x - 9x^2}}$.

Solution   By completing the square

$$
\begin{aligned}
4 - 6x - 9x^2 &= 4 - (9x^2 + 6x) \\
&= 4 - 9\left(x^2 + \frac{2}{3}x + \frac{1}{9} - \frac{1}{9}\right) \\
&= 5 - 9\left(x + \frac{1}{3}\right)^2 \\
&= 5 - (3x + 1)^2
\end{aligned}
$$

and identifying

$$
u = 3x + 1, \qquad du = 3\ dx, \quad \text{and} \quad a = \sqrt{5}
$$

it follows from (7.25) that

$$\int \frac{dx}{\sqrt{4 - 6x - 9x^2}} = \frac{1}{3} \int \frac{3\, dx}{\sqrt{5 - (3x + 1)^2}}$$

$$= \frac{1}{3} \int \frac{du}{\sqrt{(\sqrt{5})^2 - u^2}}$$

$$= \frac{1}{3} \sin^{-1} \frac{u}{\sqrt{5}} + C$$

$$= \frac{1}{3} \sin^{-1} \frac{3x + 1}{\sqrt{5}} + C$$

Exercises 7.3

Answers to odd-numbered problems begin on page A-31

In Problems 1–20 find the derivative of the given function.

1. $y = \sin^{-1}(5x - 1)$

2. $y = \cos^{-1} \frac{x + 1}{3}$

3. $y = 4 \cot^{-1} \frac{x}{2}$

4. $y = 2x - 10 \sec^{-1} 5x$

5. $y = 2\sqrt{x} \tan^{-1} \sqrt{x}$

6. $y = (\tan^{-1} x)(\cot^{-1} x)$

7. $y = \dfrac{\sin^{-1} 2x}{\cos^{-1} 2x}$

8. $y = \dfrac{\sin^{-1} x}{\sin x}$

9. $y = \dfrac{1}{\tan^{-1} x^2}$

10. $y = \dfrac{\sec^{-1} x}{x}$

11. $y = 2 \sin^{-1} x + x \cos^{-1} x$

12. $y = \cot^{-1} x - \tan^{-1} \dfrac{x}{\sqrt{1 - x^2}}$

13. $y = \left(x^2 - 9 \tan^{-1} \dfrac{x}{3} \right)^3$

14. $y = \sqrt{x} - \cos^{-1}(x + 1)$

15. $F(t) = \tan^{-1} \dfrac{t - 1}{t + 1}$

16. $g(t) = \arccos \sqrt{3t + 1}$

17. $f(x) = \arcsin(\cos 4x)$

18. $f(x) = \tan^{-1} \left(\dfrac{\sin x}{2} \right)$

19. $f(x) = \tan(\sin^{-1} x^2)$

20. $f(x) = \cos(x \sin^{-1} x)$

In Problems 21 and 22 use implicit differentiation to find dy/dx.

21. $\tan^{-1} y = x^2 + y^2$

22. $\sin^{-1} y - \cos^{-1} x = 1$

In Problems 23–44 evaluate the given integral.

23. $\displaystyle\int \frac{dx}{1 + 25x^2}$

24. $\displaystyle\int \frac{dx}{\sqrt{9 - x^2}}$

25. $\displaystyle\int_0^{1/4} (1 - 4x^2)^{-1/2}\, dx$

26. $\displaystyle\int \frac{dx}{x\sqrt{x^6 - 4}}$

27. $\displaystyle\int \frac{2x - 3}{\sqrt{1 - x^2}}\, dx$

28. $\displaystyle\int_{-1}^1 \frac{dt}{4 + (t - 1)^2}$

29. $\displaystyle\int \frac{1 - x^2}{1 + x^2}\, dx$

30. $\displaystyle\int \frac{x^4 - x^3 + 3x^2 - 3x + 4}{x^2 + 3}\, dx$

31. $\displaystyle\int \frac{4t}{\sqrt{2 - 3t^2}}\, dt$

32. $\displaystyle\int \frac{\theta}{\sqrt{1 - \theta^4}}\, d\theta$

33. $\displaystyle\int \frac{dx}{5 + 2x^2}$

34. $\displaystyle\int \frac{(3x^2 - 7)^{-1/2}}{x}\, dx$

35. $\displaystyle\int_0^{\sqrt{2}/2} \sqrt{\frac{\arcsin x}{1 - x^2}}\, dx$

36. $\displaystyle\int_0^{\sqrt{3}} \frac{\tan^{-1} x}{1 + x^2}\, dx$

37. $\displaystyle\int \frac{dx}{(x + 1)\sqrt{x^2 + 2x}}$

38. $\displaystyle\int \frac{dx}{16x^2 - 8x + 26}$

39. $\displaystyle\int \frac{x^3 + 4x^2 + 8x + 1}{x^2 + 4x + 8}\, dx$

40. $\displaystyle\int \frac{dx}{\sqrt{1 - 6x - x^2}}$

41. $\displaystyle\int \frac{t}{4t^4 + 4t^2 + 2}\, dt$

42. $\displaystyle\int \frac{2t + 1}{\sqrt{9 - (2t + 1)^2}}\, dt$

43. $\displaystyle\int_{\pi/12}^{\pi/6} \frac{\cos 2x}{1 + 4 \sin^2 2x}\, dx$

44. $\displaystyle\int \frac{\sin x}{\sqrt{49 - \cos^2 x}}\, dx$

In Problems 45 and 46 find the slope of the tangent line to the graph of the given function at the indicated value of x.

45. $y = \sin^{-1}\dfrac{x}{2}$; $x = 1$

46. $y = (\cos^{-1}x)^2$; $x = -1/\sqrt{2}$

In Problems 47 and 48 find an equation of the tangent line to the graph of the given function at the indicated value of x.

47. $f(x) = x \tan^{-1}x$; $x = 1$

48. $f(x) = \sin^{-1}(x - 1)$; $x = \frac{1}{2}$

In Problems 49 and 50 find the relative extrema of the given function.

49. $f(x) = -2x + 3 \tan^{-1}x$

50. $f(x) = 2x - \sin^{-1}x$

51. Compare the graphs of $f(x) = \tan^{-1}x^2$ and $f(x) = (\tan^{-1}x)^2$.

52. Determine whether the graph of $y = x \arcsin x$ possesses any points of inflection.

53. Find the area under the graph of

$$y = (24 + 2x - x^2)^{-1/2}$$

on the interval $[2, 4]$.

54. The region bounded by the graphs of $y = (4 + x^2)^{-1/2}$, $x = 1$, $x = 3$, and $y = 0$ is revolved about the x-axis. Find the volume of the solid of revolution.

55. The region bounded by the graphs of

$$y = \frac{1}{x^2\sqrt{x^4 - 16}}, \qquad x = \frac{5}{2}, \qquad x = 4, \quad \text{and} \quad y = 0$$

is revolved around the y-axis. Find the volume of the solid of revolution.

56. Find the length of the graph of $y = \sqrt{2 - x^2}$ on $[0, 1]$.

57. A boat is being pulled toward a dock by means of a winch. The winch is located at the end of the dock and is 10 ft above the level at which the tow rope is attached to the bow of the boat. The rope is pulled in at a constant rate of 1 ft/sec. Use an inverse trigonometric function to determine the rate at which the angle of elevation between the bow of the boat and the end of the dock is changing when 30 ft of tow rope are out.

58. A searchlight on a patrol boat that is situated $\frac{1}{2}$ km offshore follows a dune buggy that moves parallel to the water along a straight beach. The dune buggy is traveling at a constant rate of 15 km/hr. Use an inverse trigonometric function to determine the rate at which the searchlight is rotating when the dune buggy is $\frac{1}{2}$ km from the point on the shore nearest the boat.

59. A stained glass window measuring 20 ft high is 10 ft above the eye level of an observer. Use inverse trigonometric functions to determine how far the observer should stand from the wall so that the viewing angle at eye level between the bottom and the top of the window is a maximum.

60. Use differentials to approximate $\cos^{-1}(0.47)$.

Calculator Problems

61. Use Newton's Method to find an approximation to the positive root of $\tan^{-1}x = x/4$.

62. Use Newton's Method to approximate the x-coordinate of the point of intersection of the graphs of $y = \tan^{-1}x$ and $y = \cos^{-1}x$.*

63. Use Simpson's Rule with $n = 4$ to find an approximation to $\int_0^1 \tan^{-1}x\, dx$.

64. Consider the region bounded by the graphs of

$$y = \frac{18}{9 + x^2}, \qquad x = 0, \qquad x = 3, \quad \text{and} \quad y = 0$$

Approximate, to one decimal place, the coordinates of the centroid of the region. (*Hint:* At this time, we do not know the value of either

$$\int_0^3 \frac{dx}{(9 + x^2)^2} \quad \text{or} \quad \int_0^3 \frac{x\, dx}{9 + x^2}$$

Use Simpson's Rule with $n = 4$.)

Miscellaneous Problems

65. For $|x| < 1$ and $0 < \cos^{-1}x < \pi$, prove that

$$\frac{d}{dx}\cos^{-1}x = \frac{-1}{\sqrt{1 - x^2}}$$

66. For $-\infty < x < \infty$ and $0 < \cot^{-1}x < \pi$, prove that

$$\frac{d}{dx}\cot^{-1}x = \frac{-1}{1 + x^2}$$

67. For $|x| > 1$ and $0 < \csc^{-1}x < \pi/2$ or $\pi < \csc^{-1}x < 3\pi/2$, prove that

$$\frac{d}{dx}\csc^{-1}x = \frac{-1}{x\sqrt{x^2 - 1}}$$

68. Show that if $f(x) = \sin^{-1}x + \cos^{-1}x$, then $f'(x) = 0$. Interpret the result.

69. Repeat Problem 68 for $f(x) = \tan^{-1}x + \tan^{-1}(1/x)$.

---

*If you are working on a computer, you might reexamine Example 7 and Problem 44 of Section 7.2.

—— Chapter 7 Review Exercises ——

Answers to odd-numbered problems begin on page A-31

In Problems 1–20 answer true or false.

1. If f is not one-to-one, then f^{-1} does not exist. ——

2. If $(3, 1)$ is on the graph of a one-to-one function f, then $(1, 3)$ is on the graph of f^{-1}. ——

3. The graph of f is a reflection of the graph of f^{-1} in the line $y = x$. ——

4. Every linear function $f(x) = ax + b$, $a \neq 0$, is one-to-one. ——

5. If f is continuous but not one-to-one, then some horizontal line must intersect its graph in at least two points. ——

6. The function $y = \sin x$, $0 \leq x \leq \pi$, does not have an inverse. ——

7. The function $f(x) = x^5 + x^3 + x$ does not have an inverse. ——

8. If f is continuous and decreasing on $(-\infty, \infty)$, then f^{-1} exists. ——

9. The slope of the tangent line to the graph of $f(x) = \tan^{-1}x$ at $x = 1$ is $\frac{1}{2}$. ——

10. If f is one-to-one, then the domain of f is the range of f^{-1}. ——

11. $\dfrac{d}{dx} \tan(\tan^{-1}x) = 1$ for all x ——

12. If f is one-to-one, then $x_1 \neq x_2$ implies $f(x_1) \neq f(x_2)$. ——

13. The functions $f(x) = \dfrac{x + 1}{x - 1}$ and $g(x) = \dfrac{x + 1}{x - 1}$ are inverses of each other. ——

14. Let P be a point on the graph of a differentiable one-to-one function f and let P' be the corresponding point on the graph of f^{-1}. If $m \neq 0$ is the slope of the tangent at P, then $1/m$ is the slope of the tangent at P'. ——

15. $f(x) = \sin^{-1}x$ is not differentiable at $x = 1$. ——

16. If $t = \tan^{-1}(-\frac{1}{2})$, then $\cos t = -2/\sqrt{5}$. ——

17. $\arcsin (1) = 90°$ ——

18. $\tan^{-1}x = 1/\tan x$ ——

19. $\sin^{-1}\left(\sin \dfrac{5\pi}{6}\right) = \dfrac{5\pi}{6}$ ——

20. $\sec(\arctan(-1)) = -\sqrt{2}$ ——

21. Given that $f(x) = 8/(1 - x^3)$ is a one-to-one function, find f^{-1} and $(f^{-1})'$.

22. Given that $f(x) = 10 - \sqrt{x + 3}$ is a one-to-one function, without finding an explicit equation for f^{-1}, find

    (a) Its domain and range, and

    (b) An equation of the tangent line to its graph at $(f(6), 6)$.

In Problems 23–32 find dy/dx.

23. $y = \sin^{-1}\dfrac{3}{x}$

24. $y = \cos x \cos^{-1}x$

25. $y = (\cot^{-1}x)^{-1}$

26. $y = x \operatorname{arcsec}(2x - 1)$

27. $y = 2 \cos^{-1}x + 2x\sqrt{1 - x^2}$

28. $y = x^2\tan^{-1}\sqrt{x^2 - 1}$

29. $y = (\tan^{-1}(\sin x^2))^2$

30. $y = \displaystyle\int_0^x \dfrac{\sin^{-1}t}{t + 1}\, dt$, $x > 0$

31. $y - x = \sin^{-1}y$

32. $(1 + 9x^2)y^2 = \cot^{-1}3x$

In Problems 33–42 evaluate the given integral.

33. $\displaystyle\int \dfrac{dx}{\sqrt{9 - 4x^2}}$

34. $\displaystyle\int \dfrac{x^{-1/2}}{\sqrt{1 - x}}\, dx$

35. $\displaystyle\int_1^3 \dfrac{dx}{\sqrt{x}\,(1 + x)}$

36. $\displaystyle\int_0^2 \dfrac{x^4 - 15}{x^2 + 4}\, dx$

37. $\displaystyle\int \dfrac{dx}{\sqrt{7 - 6x - x^2}}$

38. $\displaystyle\int \dfrac{dx}{(x + 2)\sqrt{x(x + 4)}}$

39. $\displaystyle\int_0^{1/2} \dfrac{dx}{(\cos^{-1}x)^2\sqrt{1 - x^2}}$

40. $\displaystyle\int \dfrac{\cos x}{\sqrt{25 - \sin^2x}}\, dx$

41. $\displaystyle\int_0^{1/\sqrt[3]{2}} \dfrac{x^2}{\sqrt{1 - x^6}}\, dx$

42. $\displaystyle\int \dfrac{x}{\sqrt{1 - x^2}}\, dx$

In Problems 43 and 44 find the exact value of the given expression.

43. $\cos\left(\sin^{-1}\left(-\dfrac{12}{13}\right)\right)$

44. $\sin\left(2\cos^{-1}\left(-\dfrac{4}{5}\right)\right)$

45. From (7.9) of Section 7.2 we know that $\sin^{-1}(\sin x) = x$ whenever x is a number in the interval $[-\pi/2, \pi/2]$. Graph the function $f(x) = \sin^{-1}(\sin x)$, where x is any real number.

46. Graph the function $f(x) = \cos^{-1}(\cos x)$.

47. Find the area of the region in the first quadrant bounded between the graphs of $y = 1/(1 + x^2)$, $y = 1/\sqrt{1 - x^2}$, and $x = \frac{1}{2}$.

48. The region bounded by the graphs of $y = 32/(16 + x^4)$, $y = 0$, $x = 0$, and $x = 2$ is revolved around the y-axis. Find the volume of the solid of revolution.

49. Show that the area of the shaded region in Figure 7.24 is $2 \sin a - a^2$.

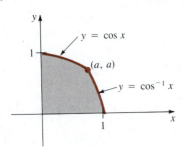

Figure 7.24

50. Consider the plane pendulum, shown in Figure 7.25, that swings between points A and C. If B is midway between A and C, it can be shown that the time the pendulum takes to travel between points B and P is

$$t = \int_0^s \sqrt{\frac{L}{g(s_C^2 - x^2)}}\, dx$$

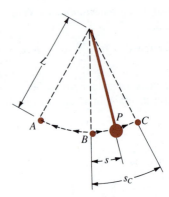

Figure 7.25

(a) Show that $t = \sqrt{\dfrac{L}{g}}\,\sin^{-1}\dfrac{s}{s_C}$.

(b) Use the result in part **(a)** to determine the time of travel from B to C.

(c) Use **(b)** to determine the period T of the pendulum—that is, the time of an oscillation from A to C and back to A.

8

Logarithmic and Exponential Functions

There are still a few gaps in our knowledge of the derivative. Up to now, we have only been able to differentiate $y = x^n$ when the exponent n is a rational number. Moreover, we have not encountered any function whose derivative is $1/x = x^{-1}$.

This means we can evaluate $\int x^n \, dx$ only when n is rational, with the *exception* of the rational number $n = -1$. In this chapter, we shall study two more transcendental functions; these enable us to answer the following questions: Can meaning be given to expressions such as x^π or $x^{\sqrt{2}}$? Is $y = x^{\sqrt{2}}$ a function? If so, what is its derivative? What is $\int x^{-1} dx$? What is $\int x^n \, dx$ when n is irrational?

8.1 The Natural Logarithmic Function

The function $y = 1/t$, which is discontinuous at $t = 0$, is shown in Figure 8.1(a). As indicated in Figure 8.1(b), $1/t > 0$ for $t > 0$ so that the area under the graph on the interval $[1, x]$ is given by $\int_1^x dt/t$. This seemingly innocent-looking integral defines a function of x of such great importance that it is given a special name.

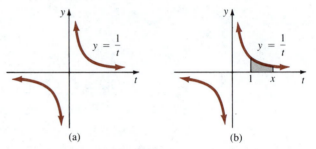

(a)   (b)

Figure 8.1

DEFINITION 8.1   **Natural Logarithmic Function**

The **natural logarithmic function**, denoted by $\ln x$, is defined by

$$\ln x = \int_1^x \frac{dt}{t} \tag{8.1}$$

for all $x > 0$.

Usually the symbol $\ln x$ is pronounced phonetically as "ell-en of x."

_____ **Example 1** _____

Estimate the value of $\ln 2$.

Solution   In Section 5.8 we used the Trapezoidal Rule to estimate $\int_1^2 dt/t$. Example 1 of that section reveals

$$\ln 2 = \int_1^2 \frac{dt}{t} \approx 0.6949$$

Consideration of the areas shown in Figure 8.2 establishes an inequality for the value of $\ln 2$:

$$\frac{1}{2} < \ln 2 < 1 \tag{8.2}$$

The result in (8.2) will prove to be useful in graphing the natural logarithmic function.

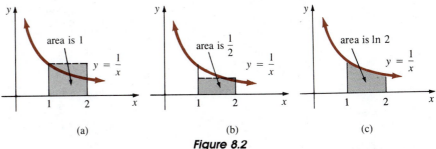

Figure 8.2

The word *logarithm* undoubtedly brings back memories from precalculus mathematics of formidable computations with **common**, or **base 10**, logarithms. It is important to realize that (8.1) is *not* the base 10 logarithm. Although it may seem like strange advice, it is best to forget about the notion of the base of a logarithm for the discussion of this section. Our immediate goal is to show that ln x "behaves like," or more precisely, has the same properties as, the common logarithm. However, you will not be told at this point why (8.1) is called "a natural" function. By the end of this chapter the answer should be obvious.

Derivative of ln *x*   The derivative of the natural logarithmic function can be found readily from the derivative form of the Fundamental Theorem of Calculus. Recall, $\frac{d}{dx} \int_a^x f(t)\, dt = f(x)$. Hence, $\frac{d}{dx} \int_1^x (1/t)\, dt = 1/x$. In other words,

$$\frac{d}{dx} \ln x = \frac{1}{x}, \qquad x > 0 \tag{8.3}$$

Note: Inherent in (8.1) and (8.3) is the answer to one of the questions posed in the prologue to this chapter—namely, the natural logarithmic function is an antiderivative of x^{-1}.

The result given in (8.3) generalizes by the Chain Rule to the composition of the natural logarithmic function and a positive differentiable function $u = g(x)$:

$$\frac{d}{dx} \ln u = \frac{1}{u} \frac{du}{dx}, \qquad u > 0 \tag{8.4}$$

Laws of the Natural Logarithm   We put the derivative result in (8.4) to immediate use in proving the **laws of the natural logarithm**, which are summarized in the next theorem.

THEOREM 8.1   Let a and b be positive real numbers and let t be a rational number.

Then,

$$(i) \quad \ln ab = \ln a + \ln b$$

$$(ii) \quad \ln \frac{a}{b} = \ln a - \ln b$$

$$(iii) \quad \ln a^t = t \ln a$$

Proof of (i)   Define a function $F(x) = \ln ax$ and let $f(x) = \ln x$. From (8.3) and (8.4),

$$F'(x) = \frac{1}{ax} \cdot \frac{d}{dx} ax$$

$$= \frac{a}{ax}$$

$$= \frac{1}{x} = f'(x)$$

Thus, from the definition of an antiderivative

$$F(x) = f(x) + C$$

But, since $\ln 1 = \int_1^1 dt/t = 0$, we have

$$F(1) = f(1) + C$$
$$\ln a = \ln 1 + C$$

and so $C = \ln a$. Hence, $F(x) = \ln x + \ln a$. Substituting $x = b$ gives

$$\ln ab = \ln b + \ln a$$
$$= \ln a + \ln b$$

Proof of (iii)   Define two functions $F(x) = \ln x^t$ and $G(x) = t \ln x$, where t is a rational number. Now from (8.4)

$$F'(x) = \frac{1}{x^t} \cdot tx^{t-1} = \frac{t}{x}$$

and from (8.3) $\qquad G'(x) = \dfrac{t}{x}$

As before, we can write

$$F(x) = G(x) + C$$
$$F(1) = G(1) + C$$
$$\ln 1 = t \ln 1 + C$$

In this case $C = 0$. Substituting $x = a$ in $\ln x^t = t \ln x$ gives

$$\ln a^t = t \ln a \qquad \blacksquare$$

The proof of property (ii) can be obtained from (i) and (iii) and is left as an exercise.

Graph of ln x First, we observe that

$$\ln 1 = 0 \qquad (8.5)$$

means the graph of $y = \ln x$ crosses the x-axis at $(1, 0)$. Next,

$$\frac{d}{dx} \ln x = \frac{1}{x} > 0 \qquad \text{for } x > 0 \qquad (8.6)$$

implies $y = \ln x$ is an increasing function on the interval $(0, \infty)$. Furthermore,

$$\frac{d^2}{dx^2} \ln x = \frac{d}{dx} \frac{1}{x} = -\frac{1}{x^2} < 0 \qquad \text{for } x > 0 \qquad (8.7)$$

shows that the graph of $y = \ln x$ must be concave downward on $(0, \infty)$. Also, when n is an integer, (*iii*) of the laws of logarithms and the inequality in (8.2) give

$$\ln 2^n = n \ln 2 > \frac{n}{2} \qquad (8.8)$$

From (8.8) we conclude that $\ln x$ can be made arbitrarily large by choosing x to be large; that is

$$\lim_{x \to \infty} \ln x = \infty \qquad (8.9)$$

Finally, since $-\ln(1/x) = \ln x$ (why?) and $\lim_{x \to 0^+} 1/x = \infty$, we have

$$\lim_{x \to 0^+} \ln x = -\infty \qquad (8.10)$$

Putting the information in (8.5), (8.6), (8.7), (8.9), and (8.10) together yields the graph of $y = \ln x$ shown in Figure 8.3(a). Of course, one can obtain (approximate) points on the graph using the information in Example 1 and the property $\ln 2^n = n \ln 2$. Some numerical values are given in the table accompanying the graph.

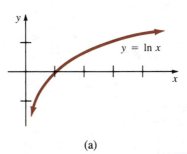

x	$\ln x$
$\frac{1}{4}$	$-2 \ln 2 \approx -1.39$
$\frac{1}{2}$	$-\ln 2 \approx -0.69$
4	$2 \ln 2 \approx 1.39$
8	$3 \ln 2 \approx 2.08$
64	$6 \ln 2 \approx 4.17$

(a)　　　　　　　　　　(b)

Figure 8.3

y = ln x has an Inverse A closer inspection of its graph should convince you that the natural logarithmic function is one-to-one. For any real number r, there is only one value of x_0 for which $r = \ln x_0$. See Figure 8.4. Also, in view of the fact that it is an increasing function for $x > 0$, it follows from Theorem 7.2 that the natural logarithmic function has an inverse. This inverse function will be studied in the next section.

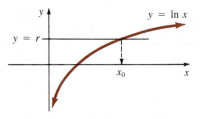

Figure 8.4

Although the domain of the natural logarithmic function $y = \ln x$ is the set of positive real numbers, the domain of $y = \ln|x|$ extends to the set of real numbers except $x = 0$. Furthermore,

$$\text{for } x > 0, \qquad \frac{d}{dx}\ln x = \frac{1}{x}$$

$$\text{for } x < 0, \qquad \frac{d}{dx}\ln(-x) = \frac{1}{(-x)}\frac{d}{dx}(-x)$$

$$= \frac{-1}{-x} = \frac{1}{x}$$

We have shown

$$\frac{d}{dx}\ln|x| = \frac{1}{x}, \qquad x \neq 0 \tag{8.11}$$

_____ **Example 2** _____

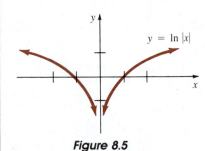

Figure 8.5

Find the slope of the tangent line to the graph of $y = \ln|x|$ at $x = 2$ and at $x = -2$.

Solution   Since (8.11) gives $dy/dx = 1/x$, we have

$$\left.\frac{dy}{dx}\right|_{x=2} = \frac{1}{2} \quad \text{and} \quad \left.\frac{dy}{dx}\right|_{x=-2} = -\frac{1}{2}$$

Observe in Figure 8.5 that the graph of $y = \ln|x|$ is symmetric with respect to the y-axis.

When $u = g(x)$ is a differentiable function, the Chain Rule gives, additionally,

$$\frac{d}{dx}\ln|u| = \frac{1}{u}\frac{du}{dx}, \qquad u \neq 0 \tag{8.12}$$

_____ **Example 3** _____

Differentiate (*a*) $y = \ln(2x - 3)$, and (*b*) $y = \ln|2x - 3|$.

Solution

(*a*) For $2x - 3 > 0$, we have from (8.4),

$$\frac{dy}{dx} = \frac{1}{2x - 3}\frac{d}{dx}(2x - 3)$$

$$= \frac{2}{2x - 3} \tag{8.13}$$

(*b*) For $2x - 3 \neq 0$, we have from (8.12),

$$\frac{dy}{dx} = \frac{1}{2x - 3} \frac{d}{dx}(2x - 3)$$

$$= \frac{2}{2x - 3} \tag{8.14}$$

Although (8.13) and (8.14) *appear* to be equal, they are definitely not the same function. The difference is simply that the domain of (8.13) is $(\frac{3}{2}, \infty)$, whereas the domain of (8.14) is the set of real numbers except $x = \frac{3}{2}$.

Example 4

Differentiate $y = \ln|\sin x|$.

Solution   For $\sin x \neq 0$ it follows from (8.12) that

$$\frac{dy}{dx} = \frac{1}{\sin x} \frac{d}{dx} \sin x$$

$$= \frac{\cos x}{\sin x} = \cot x$$

Example 5

Differentiate $y = \ln x^3$.

Solution   Because x^3 must be positive, it is understood that $x > 0$. Hence, from (8.4),

$$\frac{dy}{dx} = \frac{1}{x^3} \frac{d}{dx} x^3$$

$$= \frac{3x^2}{x^3} = \frac{3}{x}$$

Alternative Solution   From property (*iii*) we can first write,

$$y = 3 \ln x$$

and then differentiate to obtain the same result as above.

Example 6

The functions $f(x) = \ln x^4$ and $g(x) = 4 \ln x$ are not the same. Since $x^4 > 0$ for all $x \neq 0$, the domain of f is the set of real numbers except $x = 0$. The domain of g is $(0, \infty)$. Thus,

$$f'(x) = \frac{4}{x}, \quad x \neq 0 \quad \text{whereas} \quad g'(x) = \frac{4}{x}, \quad x > 0$$

___ **Example 7** ___

Differentiate $y = \ln\dfrac{x^{1/2}(2x + 7)^4}{(3x^2 + 1)^2}$.

Solution   Using the properties of logarithms, we can write for $x > 0$,

$$
\begin{aligned}
y &= \ln x^{1/2}(2x + 7)^4 - \ln(3x^2 + 1)^2 \\
&= \ln x^{1/2} + \ln(2x + 7)^4 - \ln(3x^2 + 1)^2 \\
&= \frac{1}{2}\ln x + 4\ln(2x + 7) - 2\ln(3x^2 + 1)
\end{aligned}
$$

so that

$$
\begin{aligned}
\frac{dy}{dx} &= \frac{1}{2}\cdot\frac{1}{x} + 4\cdot\frac{1}{2x + 7}\cdot 2 - 2\cdot\frac{1}{3x^2 + 1}\cdot 6x \\
&= \frac{1}{2x} + \frac{8}{2x + 7} - \frac{12x}{3x^2 + 1}
\end{aligned}
$$

___ **Example 8** ___

Differentiate $y = \ln(\ln x)$.

Solution   From (8.4),

$$
\begin{aligned}
\frac{dy}{dx} &= \frac{1}{\ln x}\frac{d}{dx}(\ln x) \\
&= \frac{1}{\ln x}\cdot\frac{1}{x} = \frac{1}{x\ln x}
\end{aligned}
$$

Remark

You should exercise care when working with logarithms. Note that

$\ln x^2$	*is not the same as*	$(\ln x)^2$
$\ln(x^2 + 4)$	*is not the same as*	$\ln x^2 + \ln 4$
$\dfrac{\ln(x + 1)}{\ln(3x + 2)}$	*is not the same as*	$\ln(x + 1) - \ln(3x + 2)$

and $\ln\dfrac{x}{x + 1}$ *is not the same as* $\dfrac{\ln x}{\ln(x + 1)}$

Exercises 8.1

Answers to odd-numbered problems begin on page A-31

In Problems 1–4 state the domain of the given function.

1. $f(x) = \ln(x + 1)$

2. $f(x) = \ln(3 - x)$

3. $f(x) = \ln|x^2 - 1|$

4. $f(x) = \ln(x^2 - 1)$

In Problems 5–10 use Theorem 8.1 to determine whether f and g are the same functions.

5. $f(x) = \ln x^6$
   $g(x) = 6 \ln x$

6. $f(x) = \ln x^{1/3}$
   $g(x) = \frac{1}{3} \ln x$

7. $f(x) = \ln x(x^4 + 3)$
   $g(x) = \ln x + \ln(x^4 + 3)$

8. $f(x) = \ln(2x + 7)$
   $g(x) = \ln 2x + \ln 7$

9. $f(x) = \dfrac{\ln(x^2 + 9)}{\ln(x^2 + 1)}$
   $g(x) = \ln(x^2 + 9) - \ln(x^2 + 1)$

10. $f(x) = \ln(1/x)$
   $g(x) = -\ln x$

In Problems 11–34 find the derivative of the given function.

11. $y = 10 \ln x$

12. $y = \ln 10 x$

13. $y = \ln x^{1/2}$

14. $y = (\ln x)^{1/2}$

15. $y = \ln(x^4 + 3x^2 + 1)$

16. $y = \ln(x^2 + 1)^{20}$

17. $y = x^2 \ln x^3$

18. $y = x - \ln|5x + 1|$

19. $y = \dfrac{\ln x}{x}$

20. $y = x(\ln x)^2$

21. $y = \ln \dfrac{x}{x + 1}$

22. $y = \dfrac{\ln 4x}{\ln 2x}$

23. $y = -\ln|\cos x|$

24. $y = \ln(x + \sqrt{x^2 - 1})$

25. $y = \dfrac{1}{\ln x}$

26. $y = \ln \dfrac{1}{x}$

27. $f(x) = \ln(x \ln x)$

28. $g(x) = \sqrt{\ln \sqrt{x}}$

29. $f(x) = \ln(\ln(\ln x))$

30. $w(\theta) = \theta \sin(\ln 5\theta)$

31. $H(t) = \ln t^2(3t^2 + 6)$

32. $G(t) = \ln\sqrt{5t + 1}(t^3 + 4)^6$

33. $f(x) = \ln \dfrac{(x + 1)(x + 2)}{x + 3}$

34. $f(x) = \ln \sqrt{\dfrac{(3x + 2)^5}{x^4 + 7}}$

In Problems 35–40 use implicit differentiation to find dy/dx.

35. $y^2 = \ln xy$

36. $y = \ln(x + y)$

37. $x + y^2 = \ln \dfrac{x}{y}$

38. $y = \ln xy^2$

39. $xy = \ln(x^2 + y^2)$

40. $x^2 + y^2 = \ln(x + y)^2$

41. Find an equation of the tangent line to the graph of $y = \ln x$ at $x = 1$.

42. Find the slope of the tangent to the graph of $y = (\ln|x|)^2$ at $x = 1$.

43. Find an equation of the tangent line to the graph of $y = \ln(x^2 - 3)$ at $x = 2$.

44. Find the slope of the tangent to the graph of y' at the point where the slope of the tangent to the graph of $y = \ln x^2$ is 4.

45. Determine the point on the graph of $y = \ln 2x$ at which the tangent line is perpendicular to $x + 4y = 1$.

46. If $y = \ln x$, find $d^n y/dx^n$.

In Problems 47–50 sketch the graph of the given function.

47. $y = -\ln x$

48. $y = 2 + \ln x$

49. $y = \ln(x - 2)$

50. $y = \ln|x + 1|$

51. Answer the following questions about the graph of $f(x) = \ln(x^2 + 1)$: Intercepts? Symmetry? Asymptotes? Relative extrema? Concavity? Sketch the graph of f.

52. Compare the graphs of $y = \ln x^2$ and $y = 2 \ln x$.

53. For $x > 0$ verify that both $y = x^{-1/2}$ and $y = x^{-1/2} \ln x$ satisfy the equation $4x^2 \, d^2y/dx^2 + 8x \, dy/dx + y = 0$.

54. For $x > 0$ verify that $y = C_1 x^{-1} \cos(\sqrt{2} \ln x) + C_2 x^{-1} \sin(\sqrt{2} \ln x)$, where C_1 and C_2 are constants, satisfies the equation $x^2 y'' + 3xy' + 3y = 0$.

Calculator Problems

In Problems 55 and 56 show graphically that the given equation possesses only one real root. Use Newton's Method to approximate the root to three decimal places.

55. $\ln x = 2$

56. $x + \ln x - 3 = 0$

Miscellaneous Problems

57. Use (*i*) and (*iii*) of Theorem 8.1 to prove (*ii*) of Theorem 8.1. (*Hint:* $\ln a/b = \ln(a \cdot 1/b)$.)

8.2 The Exponential Function

In Section 8.1 we observed that $y = \ln x$ is a one-to-one function and, consequently, possesses an inverse. The inverse of the natural logarithmic function is denoted by $y = \exp x$ and is called the **exponential function**, or sometimes, the **natural exponential function**.

DEFINITION 8.2	$y = \exp x$ if and only if $x = \ln y$

Range and Domain

Since the domain of the natural logarithmic function is the set of positive real numbers, it follows that the range of the exponential function is the same set. In other words,

$$\exp x > 0$$

for all x. Similarly, since the value of $y = \ln x$ can be any real number, the domain of $y = \exp x$ is the set of real numbers.

From (7.2) of Section 7.1, we have

$$\ln(\exp x) = x \quad \text{for all } x \quad \text{and} \quad \exp(\ln x) = x, \quad x > 0 \qquad (8.15)$$

The Number e

The number $\exp 1$ is important in mathematics and is denoted by the special symbol e in honor of the great Swiss mathematician Leonhard Euler.* That is,

$$e = \exp 1$$

It can be shown that to twelve decimal places

$$e = 2.718281828459. . . \qquad (8.16)$$

The number e, like the number π, is an irrational number. You are asked to prove in Problem 70 in Exercises 8.2 that $2.7 < e < 2.8$.

Leonhard Euler

*Leonhard Euler (1707–1783) A man with a prodigious memory and phenomenal powers of concentration, Euler's interests, like those of Descartes, were almost universal: He was a theologian, physicist, astronomer, linguist, physiologist, classical scholar, and a foremost mathematician. Considered to be a true genius in mathematics, he made lasting contributions to algebra, trigonometry, analytic geometry, calculus, calculus of variations, differential equations, complex variables, number theory, and topology. The volume of his mathematical output did not seem to be affected by the distractions of thirteen children or the fact that he was totally blind for the last 17 years of his life. Euler wrote over 700 papers and 32 books on mathematics and was responsible for introducing many of the symbols (such as e, π, $i = \sqrt{-1}$) and notations that are still used (such as $f(x)$, Σ, $\sin x$, $\cos x$). Euler was born in Basle, Switzerland, on April 15, 1707, and died of a stroke in St. Petersburg on September 18, 1783, while serving in the court of the Russian empress Catherine the Great.

Graph of y = eˣ   Since $y = e^x$ is the inverse of the natural logarithmic function, its graph can be obtained by reflection of the graph of $y = \ln x$ through the line $y = x$. Specifically we note that

$$\lim_{x \to 0^+} \ln x = -\infty \quad \text{implies} \quad \lim_{x \to -\infty} e^x = 0$$

and

$$\lim_{x \to \infty} \ln x = \infty \quad \text{implies} \quad \lim_{x \to \infty} e^x = \infty$$

The graphs of $y = \ln x$ and $y = e^x$ are compared in Figure 8.6.

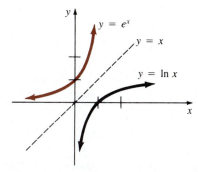

Figure 8.6

Law of Exponents   The following laws for e^x are the familiar laws of exponents.

THEOREM 8.2   Let r and s be any real number and t be a rational number. Then,

$$(i)\ e^0 = 1 \qquad\qquad (ii)\ e^1 = e$$

$$(iii)\ e^r e^s = e^{r+s} \qquad (iv)\ \frac{e^r}{e^s} = e^{r-s}$$

$$(v)\ (e^r)^t = e^{rt} \qquad (vi)\ e^{-r} = \frac{1}{e^r}$$

Proof of (i) $e^0 = 1$ since $\ln 1 = 0$

Proof of (iv)   Let $M = e^r$ and $N = e^s$ so that $r = \ln M$ and $s = \ln N$, respectively. From (ii) of the laws of the natural logarithm,

$$\ln\frac{M}{N} = \ln M - \ln N$$

$$= r - s$$

Thus, from (8.20)

$$\frac{M}{N} = e^{r-s} \quad \text{or} \quad \frac{e^r}{e^s} = e^{r-s}$$

Proof of (vi) Let $M = e^{-r}$. Hence, (8.20) gives

$$\ln M = -r$$

$$-\ln M = r$$

$$\ln M^{-1} = r$$

$$\ln \frac{1}{M} = r$$

$$\frac{1}{M} = e^r$$

$$M = \frac{1}{e^r} \quad \text{or} \quad e^{-r} = \frac{1}{e^r} \qquad \blacksquare$$

The proofs of (*iii*) and (*v*) are left as exercises.

Now, from (8.15) we know $\ln(\exp x) = x$ and therefore $\ln(\exp 1) = 1$. The latter result is equivalent to

$$\ln e = 1 \tag{8.17}$$

If t is a rational number, we have from (8.17) and (*iii*) of the laws of the natural logarithm that

$$\ln e^t = t \ln e$$

$$= t \cdot 1 = t \tag{8.18}$$

But, in view of Definition 8.2, (8.18) is the same as

$$e^t = \exp t \tag{8.19}$$

The result given in (8.19) suggests that e^x be defined *for any real number x* in the following manner.

> **DEFINITION 8.3**   For any real number x,
>
> $$e^x = \exp x$$

Subsequently, we shall drop the notation $\exp x$ and use e^x exclusively. From now on, the function $y = e^x$ will be called the **exponential function with base** e. The number x is also called the **exponent** of the base e.

At this point, let us summarize the preceding discussion using the symbol e.

> - $y = e^x$ if and only if $x = \ln y$ (8.20)
> - $y = e^x$ is the inverse of $y = \ln x$.   (8.21)
> - The domain of $y = e^x$ is $(-\infty, \infty)$.   (8.22)
> - The range of $y = e^x$ is $(0, \infty)$. This means $e^x > 0$ for all x. (8.23)
> - $\ln e^x = x$ for all x (8.24)
> - $e^{\ln x} = x,$ $x > 0$ (8.25)

Numerical values of e^x have been extensively tabulated (see Table II) and most scientific hand calculators have a key labeled $\boxed{e^x}$.

Example 1

From a calculator we find

$$e^2 \approx 7.3891$$

Example 2

If $y > 0$ is a number such that $\ln y = -1$, then (8.20) yields $y = e^{-1}$. From (*vi*) of Theorem 8.2

$$y = \frac{1}{e}$$

and from (8.16) it follows that $e^{-1} \approx 0.3679$.

Derivative of e^x We know that if

$$y = e^x \quad \text{then} \quad \ln y = x$$

Differentiating the last equation implicitly with respect to x gives

$$\frac{1}{y}\frac{dy}{dx} = 1 \quad \text{or} \quad \frac{dy}{dx} = y$$

Since $y = e^x$, we obtain the remarkable result*

$$\frac{d}{dx}e^x = e^x \tag{8.26}$$

Using the Chain Rule, (8.26) immediately generalizes to

$$\frac{d}{dx}e^u = e^u\frac{du}{dx} \tag{8.27}$$

where $u = g(x)$ is a differentiable function.

Example 3

Differentiate $y = e^{4x}$.

Solution From (8.27),

$$\frac{dy}{dx} = e^{4x} \cdot \frac{d}{dx}(4x)$$

$$= e^{4x}(4) = 4e^{4x}$$

---

*It can be shown that $y = e^x$ is the *only* function whose graph passes through $(0, 1)$ for which the derivative is the function itself. Any function in the family $y = Ce^x$, C any constant, also satisfies $y' = y$.

Example 4

Differentiate $y = e^{1/x^3}$.

Solution   From (8.27),

$$\frac{dy}{dx} = e^{1/x^3} \cdot \frac{d}{dx}(x^{-3})$$

$$= e^{1/x^3}(-3x^{-4})$$

$$= -\frac{3e^{1/x^3}}{x^4}$$

Example 5

Differentiate $y = \ln(e^{4x} + e^{-4x})$.

Solution   From (8.4) of Section 8.1 and (8.27), we obtain

$$\frac{dy}{dx} = \frac{1}{e^{4x} + e^{-4x}} \cdot \frac{d}{dx}(e^{4x} + e^{-4x})$$

$$= \frac{1}{e^{4x} + e^{-4x}} \cdot (4e^{4x} - 4e^{-4x})$$

$$= \frac{4(e^{4x} - e^{-4x})}{e^{4x} + e^{-4x}}$$

Example 6

Find the slope of the tangent line to the graph of $y = e^{2\sqrt{x}} \ln 3x$ at $x = 1$.

Solution   By the Product Rule, we have

$$\frac{dy}{dx} = e^{2\sqrt{x}} \cdot \frac{1}{3x} \cdot 3 + \ln 3x \cdot e^{2\sqrt{x}} \cdot 2 \cdot \frac{1}{2}x^{-1/2}$$

Thus, $\dfrac{dy}{dx}\Big|_{x=1} = e^2 + e^2\ln 3 = e^2(1 + \ln 3)$

Example 7

The derivative of $f(x) = e^{kx}$ is $f'(x) = ke^{kx}$. Since $e^{kx} > 0$, we see

$$f'(x) < 0 \quad \text{for } k < 0 \quad \text{and} \quad f'(x) > 0 \quad \text{for } k > 0$$

This shows that f is an increasing function on $(-\infty, \infty)$ when $k > 0$ and a decreasing function on the interval when $k < 0$. The graph of f for the case $k = -1$ is given in Figure 8.7.

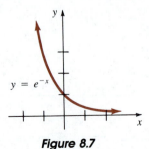

$y = e^{-x}$

Figure 8.7

Example 8

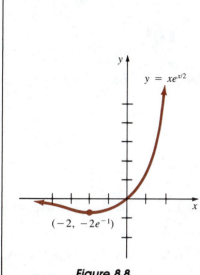

Figure 8.8

Graph $y = xe^{x/2}$.

Solution   We note that since $f(0) = 0$, the graph passes through the origin. There are no other x-intercepts. Also, when $x < 0$, $y < 0$, and for $x > 0$, $y > 0$. Now, the derivative is

$$\frac{dy}{dx} = xe^{x/2} \cdot \frac{1}{2} + e^{x/2}$$

$$= \frac{1}{2}e^{x/2}(x + 2)$$

Setting this last result equal to zero yields the critical value -2. Since

$$\frac{dy}{dx} < 0 \quad \text{for } x < -2 \quad \text{and} \quad \frac{dy}{dx} > 0 \quad \text{for } x > -2$$

we conclude from the First Derivative Test that $f(-2) = -2e^{-1}$ is a relative minimum. Inspection of Figure 8.8 also indicates that this value is an absolute minimum of the function.

e as a Limit   The number e can be expressed as a limit. In Chapter 10 we shall prove with the aid of (8.4) that

$$\lim_{h \to 0} (1 + h)^{1/h} = e \tag{8.28}$$

The following table (obtained by means of a calculator) merely suggests the foregoing result:

h	$(1 + h)^{1/h}$
0.1	2.5937425
0.01	2.7048138
0.001	2.7169238
0.0001	2.7181459
0.00001	2.7182546
-0.0001	2.7184177
-0.00001	2.7182818

If we let $n = 1/h$, $h > 0$, then as $h \to 0$ we have $n \to \infty$; therefore, (8.28) can be written in the alternative form

$$\lim_{n \to \infty} \left(1 + \frac{1}{n}\right)^n = e \tag{8.29}$$

These limit results are important and should be remembered. Indeed, (8.28) is often taken as the *definition* of the number e.

Remark

The numbers e and π are **transcendental** as well as irrational numbers. A transcendental number is one that is *not* a root of a polynomial equation with integer coefficients. For example, $\sqrt{2}$ is irrational but is not transcendental since it is a root of the polynomial equation $x^2 - 2 = 0$. The number e was proved to be transcendental by the French mathematician Charles Hermite (1822–1901) in 1873, whereas π was proved to be transcendental nine years later by the German mathematician Ferdinand Lindemann. The latter proof showed conclusively that "squaring a circle" with a rule and a compass was impossible.

_____ Exercises 8.2 _____

Answers to odd-numbered problems begin on page A-32

In Problems 1–30 find the derivative of the given function.

1. $y = e^{-x}$

2. $y = e^{2x+3}$

3. $y = e^{\sqrt{x}}$

4. $y = e^{\cos 10x}$

5. $y = \dfrac{e^{-2x}}{x}$

6. $y = x^3 e^{4x}$

7. $y = \sqrt{1 + e^{-5x}}$

8. $y = \dfrac{1}{(e^{2x} - e^{-2x})^2}$

9. $y = \ln(x^4 + e^{x^2})$

10. $y = \ln\sqrt{e^x + e^{-x}}$

11. $y = \dfrac{e^x + e^{-x}}{e^x - e^{-x}}$

12. $y = \dfrac{e^{7x}}{e^{-x}}$

13. $y = e^{\ln x}$

14. $y = \ln e^x$

15. $y = e^{3x}\ln(x^2 + 1)$

16. $y = \dfrac{\ln x}{e^x}$

17. $y = e^{\frac{x+2}{x-2}}$

18. $y = \ln\left|\dfrac{1 + e^{2x}}{1 - e^{2x}}\right|$

19. $y = e^{e^{x^2}}$

20. $y = (e^3)^{x-1}$

21. $y = e^{2x}e^{3x}e^{4x}$

22. $y = e^x + e^{x+e^x}$

23. $F(t) = e^{t^{1/3}} + (e^t)^{1/3}$

24. $g(t) = e^{-t}\tan e^t$

25. $f(x) = (2x + 1)^3 e^{-(1-x)^4}$

26. $f(x) = e^{x\sqrt{x^2+1}}$

27. $f(x) = \dfrac{xe^x}{x + e^x}$

28. $f(x) = xe^{2x}\ln x$

29. $f(x) = \tan^{-1}e^{2x}$

30. $f(x) = x \sec e^{-x}$

In Problems 31–36 use implicit differentiation to find dy/dx.

31. $y = e^{x+y}$

32. $\ln y = x + e^y$

33. $y = \cos e^{xy}$

34. $y = e^{(x+y)^2}$

35. $x + y^2 = e^{x/y}$

36. $e^x + e^y = y$

In Problems 37–40 find the indicated derivative.

37. $y = e^{-4x}$; $\dfrac{d^3y}{dx^3}$

38. $y = \sin e^{2x}$; $\dfrac{d^2y}{dx^2}$

39. $y = \ln(e^x + 1)$; $\dfrac{d^2y}{dx^2}$

40. $y = xe^x$; $\dfrac{d^4y}{dx^4}$

41. Find an equation of the tangent line to the graph of $y = e^x$ at $x = 1$.

42. Find an equation of the tangent line to the graph of $y = \ln(e^x + 1)$ at $x = 0$.

43. Find the slope of the normal line to the graph of $y = (x - 1)e^{-x}$ at $x = 0$.

44. Find the point on the graph of $y = e^x$ at which the tangent line is parallel to $3x - y = 7$.

In Problems 45–48 sketch the graph of the given function.

45. $y = -e^x$

46. $y = 1 + e^{-x}$

47. $y = 2 - e^{-x}$

48. $y = e^x + e^{-x}$

In Problems 49–54 find the relative extrema of each function. Sketch the graph.

49. $y = xe^{-x}$

50. $y = \dfrac{e^x}{x}$

51. $y = e^{-x^2}$

52. $y = e^{(x-2)^2}$

53. $y = x \ln x$

54. $y = \dfrac{\ln x}{x}$

55. Consider the function $f(x) = e^{2/x}$, $x > 0$.

   (a) Prove that f has an inverse.

   (b) Sketch a graph of f.

   (c) What is the range of f?

   (d) Find f^{-1}.

   (e) What is the domain of f^{-1}?

56. Consider the function $f(x) = \sqrt{1 + \ln x}$.

   (a) What is the domain and range of f?

   (b) Prove that f has an inverse.

   (c) Sketch a graph of f.

   (d) Find f^{-1}.

57. Sketch the graph of $y = e^{|x|}$. Find dy/dx. Is the function differentiable at 0?

58. Show that the function $f(x) = e^{\sin x}$ is periodic with period 2π. Find the relative extrema and points of inflection of f. Sketch the graph of f.

59. Find a function that satisfies $dy/dx = y$ whose graph passes through $(0, 5)$.

60. If $y = xe^x$, find $d^n y/dx^n$.

61. Verify that $y = C_1 e^{-3x} + C_2 e^{2x}$, where C_1 and C_2 are constants, satisfies the equation $y'' + y' - 6y = 0$.

62. Verify that $y = e^{-2x}$ and $y = xe^{-2x}$ both satisfy the equation $y'' + 4y' + 4y = 0$.

63. The current $i(t)$ in a series circuit that contains an inductor and a resistor is given by

$$i(t) = \frac{E_0}{R} + Ce^{-(R/L)t}$$

where E_0, R, L, and C are constants. Show that i satisfies the equation $L\, di/dt + Ri = E_0$.

64. A graph of $P(t) = ac/(bc + e^{-at})$, where a, b, and c are constants, is called a **logistic curve** and occurs in one mathematical model of an expanding but limited population. Show that P satisfies the **logistic equation** $dP/dt = P(a - bP)$.

65. Show that the logistic curve $P(t) = 2/(1 + e^{-2t})$ does not possess any relative extrema. Find any points of inflection. Graph $P(t)$.

66. The graph of $P(t) = e^{a/b} e^{-ce^{-bt}}$, where a, b, and c are constants, is called a **Gompertz curve**.* Show that P satisfies $dP/dt = P(a - b \ln P)$.

Calculator Problems

67. The Jenss model (1937) represents the most accurate empirically devised formula for predicting the height h (in centimeters) in terms of age t (in years) for preschool-aged children (3 months to 6 years):

$$h = 79.04 + 6.39t - e^{3.26 - 0.99t}$$

   (a) What height does this model predict for a two-year-old?

   (b) How fast is a two-year-old increasing in height?

   (c) At what age is the rate of growth most rapid? ·(*Hint:* $h'(t)$ is a continuous function defined on the closed interval $[\frac{1}{4}, 6]$.)

68. The weight of many animals can be modeled by a Von Bertalanffy function $W(t) = a[1 - be^{-ct}]^3$ for some positive constants a, b, and c. For a population of female elephants, the weight (in kilograms) at age t (in years) is given by

$$W(t) = 2600[1 - 0.51e^{-0.075t}]^3$$

   (a) Show that $W(t)$ is increasing for $t > 0$.

   (b) Compute and interpret $\lim_{t \to \infty} W(t)$.

   (c) How fast is a newborn female elephant increasing in weight?

   (d) An adult elephant weighs 1600 kg. Determine her age.

   (e) At what age is the rate of growth of an elephant a maximum?

69. Consider the function $f(n) = (1 + 1/n)^n$. Use a calculator to fill in the following table.

n	$f(n)$
100	
1000	
10,000	
100,000	
1,000,000	

---

*Named after Benjamin Gompertz (1779–1865), an English mathematician.

70. (a) Use the Trapezoidal Rule with $n = 6$ to establish the inequality

$$\int_1^{2.7} \frac{dt}{t} < 1 < \int_1^{2.8} \frac{dt}{t}$$

(b) Use part **(a)** to show

$$\ln 2.7 < \ln e < \ln 2.8$$

(c) Show that $2.7 < e < 2.8$.

71. Find all intercepts of the graph of the function $f(x) = e^{2x} - e^x - 12$.

72. Show graphically that the equation $e^{-x} = 3x$ possesses only one real root. Use Newton's Method to approximate the root to three decimal places.

Miscellaneous Problems

73. Prove (*iii*) of Theorem 8.2.

74. Prove (*v*) of Theorem 8.2.

75. Prove that the x-intercept of the tangent line to the graph of $y = e^{-x}$ at $x = x_0$ is one unit to the right of x_0.

76. Given $f(x) = \ln x$ and $f^{-1}(x) = e^x$, use (7.7) of Section 7.1 to show that $(d/dx)e^x = e^x$.

77. Prove that for $x > 0$, $e^x > x + 1$. [*Hint:* Consider $f(x) = e^x - x - 1$.]

78. Use (8.28) of this section to show that for any real number r

$$e^r = \lim_{n \to \infty} \left(1 + \frac{r}{n} \right)^n$$

79. If P dollars are invested at an annual rate of interest r compounded t times a year, the return S after m years is

$$S = P \left(1 + \frac{r}{t} \right)^{tm}$$

When interest is **compounded continuously**, the return is defined to be

$$S = P \lim_{t \to \infty} \left(1 + \frac{r}{t} \right)^{tm}$$

Use the results of Problem 78 to show $S = Pe^{rm}$.

Calculator Problem

80. Use the result of Problem 79 to compare the return when $5000 is invested at 10% annual interest compounded quarterly for three years with the return that results when continuous compounding of interest is used.

8.3 Integrals Involving Logarithmic and Exponential Functions

As a consequence of the derivative formulas (8.11) and (8.12) of Section 8.1, we obtain the antiderivative or indefinite integral formulas that yield the natural logarithm

$$\int \frac{dx}{x} = \ln|x| + C \qquad\qquad (8.30)$$

$$\int \frac{du}{u} = \ln|u| + C \qquad\qquad (8.31)$$

Similarly, (8.26) and (8.27) of the preceding section translate into

$$\int e^x \, dx = e^x + C \qquad\qquad (8.32)$$

$$\int e^u \, du = e^u + C \qquad\qquad (8.33)$$

Example 1

Find the area A bounded between the graph of $y = 1/x$ and the x-axis on the interval $[-2, -\frac{1}{2}]$.

Solution Since $f(x) = 1/x < 0$ on the interval, it follows from (8.30) and Definition 6.1 that

$$A = \int_{-2}^{-1/2} \left| \frac{1}{x} \right| \, dx$$

$$= -\int_{-2}^{-1/2} \frac{1}{x} \, dx$$

$$= -\ln|x| \Big]_{-2}^{-1/2}$$

$$= -\ln\left| -\frac{1}{2} \right| + \ln|-2|$$

$$= \ln 2 - (\ln 1 - \ln 2)$$

$$= 2 \ln 2 \approx 1.3863 \text{ square units}$$

The area is indicated in Figure 8.9.

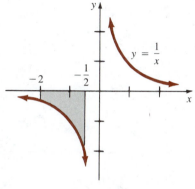

Figure 8.9

Example 2

Evaluate $\int \dfrac{dx}{x - 7}$.

Solution If $\qquad u = x - 7 \quad$ then $\quad du = dx$

It follows from (8.31) that

$$\int \frac{dx}{x - 7} = \int \frac{du}{u}$$

$$= \ln|u| + C$$

$$= \ln|x - 7| + C$$

Example 3

Evaluate $\int \dfrac{x^2}{x^3 + 5} \, dx$.

Solution If $\qquad u = x^3 + 5 \quad$ then $\quad du = 3x^2 \, dx$

Hence, from (8.31) we have

$$\int \frac{x^2}{x^3 + 5} \, dx = \frac{1}{3} \int \frac{3x^2 \, dx}{x^3 + 5}$$

$$= \frac{1}{3} \int \frac{du}{u}$$

$$= \frac{1}{3} \ln|u| + C$$

$$= \frac{1}{3} \ln|x^3 + 5| + C$$

Example 4

Evaluate $\displaystyle\int \frac{dx}{1 + e^{-2x}}$.

Solution The given integral is not of form (8.31); however, if we multiply the numerator and denominator by e^{2x}, we have

$$\int \frac{dx}{1 + e^{-2x}} = \int \frac{e^{2x}}{e^{2x} + 1} \, dx$$

Now if $\qquad u = e^{2x} + 1 \quad$ then $\quad du = 2e^{2x} \, dx$

and, therefore,

$$\int \frac{dx}{1 + e^{-2x}} = \int \frac{e^{2x} \, dx}{e^{2x} + 1}$$

$$= \frac{1}{2} \int \frac{2e^{2x} \, dx}{e^{2x} + 1}$$

$$= \frac{1}{2} \int \frac{du}{u}$$

$$= \frac{1}{2} \ln|u| + C$$

$$= \frac{1}{2} \ln|e^{2x} + 1| + C$$

$$= \frac{1}{2} \ln(e^{2x} + 1) + C$$

Note that the absolute value can be dropped because $e^{2x} + 1 > 0$.

Example 5

Evaluate $\displaystyle\int \frac{e^x}{1 + e^{2x}}\, dx$.

Solution   Although similar in appearance to the integral in the preceding example, observe that $e^{2x} = (e^x)^2$ and if $u = e^x$, $du = e^x\, dx$, then

$$\int \frac{e^x}{1 + e^{2x}}\, dx = \int \frac{du}{1 + u^2}$$

$$= \tan^{-1}u + C$$

$$= \tan^{-1}e^x + C$$

Example 6

Evaluate $\displaystyle\int e^{5x}\, dx$.

Solution   Let $u = 5x$ so that $du = 5\, dx$

Then from (8.33), $\displaystyle\int e^{5x}\, dx = \frac{1}{5}\int e^{5x}(5\, dx)$

$$= \frac{1}{5}e^u + C$$

$$= \frac{1}{5}e^{5x} + C$$

Example 7

Evaluate $\displaystyle\int \frac{e^{4/x}}{x^2}\, dx$.

Solution   Using $u = \dfrac{4}{x}$ then $du = -\dfrac{4}{x^2}\, dx$

From (8.33) we have

$$\int \frac{e^{4/x}}{x^2}\, dx = -\frac{1}{4}\int e^{4/x}\left(-\frac{4}{x^2}\, dx\right)$$

$$= -\frac{1}{4}\int e^u\, du$$

$$= -\frac{1}{4}e^u + C$$

$$= -\frac{1}{4}e^{4/x} + C$$

Example 8

Evaluate $\int \dfrac{e^x}{\sqrt{1 + e^x}}\, dx$.

Solution   Write the integral as

$$\int (1 + e^x)^{-1/2} e^x\, dx$$

and let $\qquad\qquad u = 1 + e^x \qquad du = e^x\, dx$

We recognize

$$\int (1 + e^x)^{-1/2} e^x\, dx = \int u^{-1/2}\, du$$
$$= 2u^{1/2} + C$$
$$= 2(1 + e^x)^{1/2} + C$$

Example 9

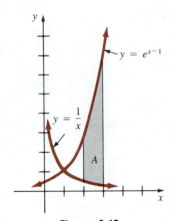

Figure 8.10

Find the area A of the region bounded between the graphs of $y = e^{x-1}$ and $y = 1/x$ on the interval $[2, 3]$.

Solution   If we denote the given functions by $f(x) = e^{x-1}$ and $g(x) = 1/x$, then inspection of Figure 8.10 shows that $f(x) - g(x) \geq 0$ on the interval. It follows from Definition 6.2 that

$$A = \int_2^3 \left(e^{x-1} - \frac{1}{x}\right) dx$$

$$= (e^{x-1} - \ln x)\Big]_2^3$$

$$= e^2 - e - \ln 3 + \ln 2 \approx 4.2653 \text{ square units}$$

The following integration formulas, which relate some trigonometric functions with the natural logarithmic functions, occur often enough in practice to merit special attention:

$$\int \tan x\, dx = -\ln|\cos x| + C \tag{8.34}$$

$$\int \cot x\, dx = \ln|\sin x| + C \tag{8.35}$$

$$\int \sec x\, dx = \ln|\sec x + \tan x| + C \tag{8.36}$$

$$\int \csc x\, dx = \ln|\csc x - \cot x| + C \tag{8.37}$$

To obtain (8.34) we write

$$\int \tan x \, dx = \int \frac{\sin x}{\cos x} \, dx$$

$$= -\int \frac{(-\sin x) \, dx}{\cos x}$$

$$= -\int \frac{du}{u}$$

$$= -\ln|u| + C$$

$$= -\ln|\cos x| + C$$

We shall verify (8.36) by differentiation:

$$\frac{d}{dx} \ln|\sec x + \tan x| = \frac{1}{\sec x + \tan x} \frac{d}{dx}(\sec x + \tan x)$$

$$= \frac{\sec x \tan x + \sec^2 x}{\sec x + \tan x}$$

$$= \frac{\sec x(\tan x + \sec x)}{\sec x + \tan x}$$

$$= \sec x$$

By definition, $\ln|\sec x + \tan x|$ is an antiderivative of $\sec x$. Also, each of the preceding formulas can be written in a general form such as

$$\int \tan u \, du = -\ln|\cos u| + C$$

where $u = g(x)$ and $du = g'(x) \, dx$.

_____ **Example 10** _____

Evaluate $\int x \sec x^2 \, dx$.

Solution Let $\qquad\qquad u = x^2 \quad$ and $\quad du = 2x \, dx$

Then from (8.36),

$$\int x \sec x^2 \, dx = \frac{1}{2} \int \sec x^2 (2x \, dx)$$

$$= \frac{1}{2} \int \sec u \, du$$

$$= \frac{1}{2} \ln|\sec u + \tan u| + C$$

$$= \frac{1}{2} \ln|\sec x^2 + \tan x^2| + C$$

Exercises 8.3

Answers to odd-numbered problems begin on page A-33

In Problems 1–50 evaluate the given integral.

1. $\displaystyle\int \frac{dx}{3x}$

2. $\displaystyle\int \frac{dx}{x+4}$

3. $\displaystyle\int \frac{dx}{2x-1}$

4. $\displaystyle\int (5x+6)^{-1}\,dx$

5. $\displaystyle\int \frac{x}{x^2+1}\,dx$

6. $\displaystyle\int \frac{x^2}{5x^3+8}\,dx$

7. $\displaystyle\int \frac{x+2}{x^2+4x-3}\,dx$

8. $\displaystyle\int \frac{2x^2+1}{2x^3+3x-1}\,dx$

9. $\displaystyle\int \frac{x}{x+1}\,dx$

10. $\displaystyle\int \frac{x^2-2x+5}{x}\,dx$

11. $\displaystyle\int \frac{(x+3)^2}{x+2}\,dx$

12. $\displaystyle\int \frac{dx}{x^{1/3}(x^{2/3}+1)}$

13. $\displaystyle\int \frac{dx}{x\ln x}$

14. $\displaystyle\int \frac{dx}{x\ln\sqrt{x}}$

15. $\displaystyle\int \frac{\ln x}{x}\,dx$

16. $\displaystyle\int \frac{dx}{x(\ln x)^2}$

17. $\displaystyle\int x^{-1}\sqrt[3]{1+\ln x}\,dx$

18. $\displaystyle\int \frac{\sin(\ln x)}{x}\,dx$

19. $\displaystyle\int \frac{\cos t}{1+\sin t}\,dt$

20. $\displaystyle\int \frac{1-\sin\theta}{\theta+\cos\theta}\,d\theta$

21. $\displaystyle\int e^{10x}\,dx$

22. $\displaystyle\int \frac{dx}{e^{4x}}$

23. $\displaystyle\int x^2 e^{-2x^3}\,dx$

24. $\displaystyle\int \frac{e^{1/x^3}}{x^4}\,dx$

25. $\displaystyle\int \frac{e^{\sqrt{x}}}{\sqrt{x}}\,dx$

26. $\displaystyle\int e^x \tan e^x\,dx$

27. $\displaystyle\int e^{-2x^2+4}x\,dx$

28. $\displaystyle\int (2-e^{3x})^2\,dx$

29. $\displaystyle\int (1+e^x)^3\,dx$

30. $\displaystyle\int \frac{e^{2x}}{(1+e^{2x})^3}\,dx$

31. $\displaystyle\int \frac{1+e^t}{e^t}\,dt$

32. $\displaystyle\int \frac{e^\theta}{1+e^\theta}\,d\theta$

33. $\displaystyle\int \frac{(e^{2x}-e^{-2x})^2}{e^x}\,dx$

34. $\displaystyle\int \frac{e^x[\ln(1+e^x)]^2}{1+e^x}\,dx$

35. $\displaystyle\int \frac{e^x-e^{-x}}{e^x+e^{-x}}\,dx$

36. $\displaystyle\int \frac{dx}{e^x+1}$

37. $\displaystyle\int \frac{e^{2x}}{e^x+1}\,dx$

38. $\displaystyle\int \frac{e^{3x}+e^{2x}}{e^x-1}\,dx$

39. $\displaystyle\int \sqrt{e^x}\,dx$

40. $\displaystyle\int e^{\cos 2x}\sin 2x\,dx$

41. $\displaystyle\int \tan 5x\,dx$

42. $\displaystyle\int x(1-\cot x^2)\,dx$

43. $\displaystyle\int (1+\sec\theta)^2\,d\theta$

44. $\displaystyle\int \frac{dx}{\sin 2x}$

45. $\displaystyle\int_0^4 \frac{dx}{2x+1}$

46. $\displaystyle\int_{-2}^{-1} \frac{dt}{t-3}$

47. $\displaystyle\int_{-3}^3 xe^{-x^2}\,dx$

48. $\displaystyle\int_{-1}^0 \frac{4+e^{x+1}}{e^x}\,dx$

49. $\displaystyle\int_1^{e^2} \frac{\ln(2/x)}{x}\,dx$

50. $\displaystyle\int_{\ln 2}^{\ln 3} e^{-x}\,dx$

51. Find the area under the graph of $y = 2x^{-1}$ on the interval $[\frac{1}{2}, 4]$.

52. Find the area of the region bounded by the graphs of $y = x$, $y = 1/x$, and $x = 3$.

53. Find the area of the region bounded by the graphs of $y = 1/x$, and $3x + 3y + 10 = 0$.

54. Find the area bounded by the graph of $y = (x-1)^{-1}$ and the x-axis on the interval $[-4, -2]$.

55. Find the area under the graph of $y = e^{-2x}$ on the interval $[0, 2]$.

56. Find the area of the region bounded by the graphs of $y = e^x$, $y = -e^{-x}$, $x = 0$, and $x = 1$.

57. Evaluate $\int_{-1}^1 (e^x - 1)\,dx$ and determine whether the integral represents area. If not, find the area bounded by the graph of the integrand and the x-axis on the given interval.

58. The region in the first quadrant bounded by the graphs of $y = 1/(1+x^2)$, $x = 0$, $x = \sqrt{3}$, and $y = 0$ is revolved about the y-axis. Find the volume of the solid of revolution.

59. The region in the first quadrant bounded by the graphs $y = e^x$, $x = 0$, $x = 2$, and $y = 0$ is revolved about the x-axis. Find the volume of the solid of revolution.

60. The region in the first quadrant bounded by the graphs of $y = 1/\sqrt{x}$, $x = 1$, $x = 4$, and $y = 0$ is revolved about the x-axis. Find the volume of the solid of revolution.

61. The region in the first quadrant bounded by the graphs of $y = e^{-x^2}$, $x = 0$, $x = 1$, and $y = 0$ is revolved about the y-axis. Find the volume of the solid of revolution.

62. Find the length of the graph of $y = \ln(\cos x)$ on the interval $[0, \pi/4]$.

63. Find the length of the graph of $y = \frac{1}{2}(e^x + e^{-x})$ on the interval $[0, \ln 2]$.

64. Find the length of the graph of $y = \frac{1}{2}x^2 - \frac{1}{4}\ln x$ on the interval $[1, 4]$.

65. Find the area of the surface that is formed by revolving the graph of the function in Problem 63 on the indicated interval about the x-axis.

66. When gas enclosed in a cylinder expands against a piston so that the volume of the gas changes from v_1 to v_2, the work done on the piston is $W = \int_{v_1}^{v_2} p \, dv$, where p is pressure. If pressure p and volume v are related by $pv = k$, show that the work done is $k \ln(v_2/v_1)$.

67. Evaluate

(a) $e^{4\int dx/x}$

(b) $e^{\int \frac{x-1}{x+1} dx}$

68. If $f(x) = \ln x$, show that the average of f' on $[a, b]$, $0 < a < b$, is $\dfrac{1}{b-a} \ln \dfrac{b}{a}$.

Calculator Problems

69. (a) Use the Trapezoidal Rule to obtain an approximation of $\int_0^1 e^{-x^2} \, dx$ that has an error less than 0.01.

(b) How many trapezoids are required to yield an approximation accurate to two decimal places? Four decimal places?

70. Repeat Problem 69 using Simpson's Rule.

Miscellaneous Problems

In Problems 71–74 verify the given result.

71. $\displaystyle\int \ln x \, dx = x \ln x - x + C$

72. $\displaystyle\int xe^x \, dx = xe^x - e^x + C$

73. $\displaystyle\int \frac{dx}{a^2 - x^2} = \frac{1}{2a} \ln \left| \frac{a+x}{a-x} \right| + C$

74. $\displaystyle\int \frac{\sqrt{x^2 + a^2}}{x} \, dx$
$$= \sqrt{x^2 + a^2} - a \ln \left| \frac{a + \sqrt{x^2 + a^2}}{x} \right| + C$$

75. Consider the area under the graph of $y = \ln x$ on the interval $[1, e]$.

(a) Use the result given in Problem 71 to find the area.

(b) Determine an alternative integral method for finding this same area.

76. Verify the formula given in (8.35).

77. Verify the formula given in (8.37).

78. Use differentiation to show that
$$\int \csc x \, dx = -\ln|\csc x + \cot x| + C$$
Reconcile this result with (8.37).

79. If $\dfrac{d}{dx}\left[\dfrac{x^2}{2} \ln x - \dfrac{x^2}{4} \right] = x \ln x$, what is $\displaystyle\int x \ln x \, dx$?

80. If $\dfrac{d}{dx}\left[x^2 e^x - 2xe^x + 2e^x \right] = x^2 e^x$, what is $\displaystyle\int x^2 e^x \, dx$?

8.4 Exponential and Logarithmic Functions to Other Bases

Inspired by the property
$$a^t = e^{\ln a^t} = e^{t \ln a}$$
where t is a rational number and $a > 0$, we make the following definition.

DEFINITION 8.4	If r is any real number and $a > 0$, then
	$$a^r = e^{r \ln a} \qquad\qquad (8.38)$$

The explicit use of (8.38) is to give meaning to a^r when r is an irrational number.

_____ **Example 1** _____

(a) $5^\pi = e^{\pi \ln 5}$ (b) $10^{\sqrt{3}} = e^{\sqrt{3} \ln 10}$

_____

You may have found the foregoing example interesting but still have the legitimate question: What does one do with a quantity such as $10^{\sqrt{3}} = e^{\sqrt{3} \ln 10}$? Curious individuals may skip ahead and read the remark that concludes this section.

Laws of Exponents

The following laws of exponents are immediate results of Definition 8.4. We shall prove the third property and leave some of the others as exercises.

> **THEOREM 8.3**   Let r and s be any real numbers. For $a > 0$
>
> (i) $a^0 = 1$ (ii) $a^1 = a$
>
> (iii) $a^r a^s = a^{r+s}$ (iv) $(ab)^r = a^r b^r$
>
> (v) $\dfrac{a^r}{a^s} = a^{r-s}$ (vi) $\left(\dfrac{a}{b}\right)^r = \dfrac{a^r}{b^r}$
>
> (vii) $(a^r)^s = a^{rs}$ (viii) $a^{-r} = \dfrac{1}{a^r}$

Proof of (iii)   From (8.38) we can write

$$a^r = e^{r \ln a} \quad \text{and} \quad a^s = e^{s \ln a}$$

Therefore, from property (iii) of the laws of exponents given in Section 8.2,

$$
\begin{aligned}
a^r a^s &= e^{r \ln a} e^{s \ln a} \\
&= e^{r \ln a + s \ln a} \\
&= e^{(r+s) \ln a} \\
&= a^{r+s}
\end{aligned}
$$ ∎

As a further consequence of Definition 8.4, we can now extend property (iii) of the laws of the natural logarithm to include the case when the exponent is an irrational number.

> **THEOREM 8.4**   Let r be any real number. For $a > 0$,
>
> $$\ln a^r = r \ln a$$

Proof   Since $y = e^x$ is equivalent to $\ln y = x$, we see that $a^r = e^{r \ln a}$ yields $\ln a^r = r \ln a$. ∎

Exponential Function with Base a

Recall, $y = e^x$ is a differentiable function whose domain is the set of real numbers. Likewise, Definition 8.4 implies that

$$f(x) = a^x, \qquad a > 0 \tag{8.39}$$

is also a differentiable function that has the same domain. We say that f is an **exponential function with base a**. Furthermore, since $y = e^{kx}$ increases for $k > 0$, it follows from (8.38) that whenever $a > 1$, $\ln a > 0$, that $f(x) = a^x$ is an increasing function on $(-\infty, \infty)$. For $0 < a < 1$, $\ln a < 0$ and so $f(x) = a^x$ is a decreasing function. Typical graphs of exponential functions, which illustrate these two cases, are similar to the graphs of e^x and e^{-x}, respectively. The graphs of $y = 3^x$ and $y = (\frac{1}{3})^x = 3^{-x}$ are shown in Figure 8.11.

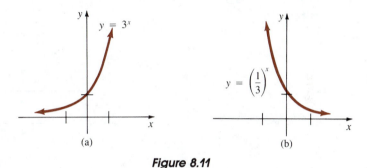

(a)   (b)

Figure 8.11

Derivative of a^x

The derivative of a^x is obtained from the derivative of e^u:

$$\frac{d}{dx}a^x = \frac{d}{dx}e^{x\ln a}$$

$$= e^{x\ln a}\frac{d}{dx}x \ln a$$

$$= e^{x\ln a}\ln a \tag{8.40}$$

Using (8.38) in (8.40) then yields

$$\frac{d}{dx}a^x = a^x \ln a \tag{8.41}$$

In (8.41) we can see, again, the importance of the number e; the derivative of an exponential function with a base other than e carries with it the extra factor of $\ln a$.

The Chain Rule gives the general result. If $u = g(x)$ is a differentiable function, then

$$\frac{d}{dx}a^u = a^u\frac{du}{dx}\ln a \tag{8.42}$$

Note that when $a = e$, $\ln e = 1$ and (8.41) and (8.42) reduce, in turn, to (8.26) and (8.27) of Section 8.2.

Example 2

Differentiate $y = 10^x$.

Solution   From (8.41),

$$\frac{dy}{dx} = 10^x \ln 10$$

Example 3

Differentiate $y = 3^{\cos 5x}$.

Solution   From (8.42),

$$\frac{dy}{dx} = 3^{\cos 5x}\left(\frac{d}{dx} \cos 5x\right) \ln 3$$
$$= 3^{\cos 5x}(-5 \sin 5x) \ln 3$$

Example 4

Differentiate $y = 5^{x^3} e^{-x^2}$.

Solution   From the Product Rule,

$$\frac{dy}{dx} = 5^{x^3}\frac{d}{dx}e^{-x^2} + e^{-x^2}\frac{d}{dx}5^{x^3}$$
$$= 5^{x^3}(-2xe^{-x^2}) + e^{-x^2}(5^{x^3}3x^2 \ln 5)$$
$$= 5^{x^3}e^{-x^2}(-2x + 3x^2 \ln 5)$$

You should verify by differentiation that

$$\int a^x \, dx = \frac{a^x}{\ln a} + C \qquad\qquad (8.43)$$

$$\int a^u \, du = \frac{a^u}{\ln a} + C \qquad\qquad (8.44)$$

are the integral forms of (8.41) and (8.42), respectively.

Example 5

Evaluate $\int 8^x \, dx$.

Solution   From (8.43),

$$\int 8^x \, dx = \frac{8^x}{\ln 8} + C$$

Example 6 _____

Evaluate $\int \dfrac{2^{\sqrt{x}}}{\sqrt{x}}\,dx$

Solution Identifying

$$a = 2, \quad u = x^{1/2}, \quad \text{and} \quad du = \frac{1}{2}x^{-1/2}\,dx$$

we see from (8.44) that

$$\int \frac{2^{\sqrt{x}}}{\sqrt{x}}\,dx = 2 \int 2^{\sqrt{x}}\left(\frac{1}{2}x^{-1/2}\,dx\right)$$

$$= 2 \int 2^{u}\,du$$

$$= 2\left(\frac{2^{u}}{\ln 2}\right) + C$$

$$= 2\left(\frac{2^{\sqrt{x}}}{\ln 2}\right) + C$$

$$= \frac{2^{1+\sqrt{x}}}{\ln 2} + C$$

Power Rule Revisited   At last we are able to state and prove the Power Rule of differentiation for any real exponent, *rational or irrational*.

THEOREM 8.5   **Power Rule** (Real Exponents)

If r is any real number, then

$$\frac{d}{dx}x^{r} = rx^{r-1} \tag{8.45}$$

Proof   In view of Definition 8.4, we can write

$$x^{r} = e^{r\ln x}$$

Therefore, $$\frac{d}{dx}x^{r} = e^{r\ln x}\frac{d}{dx}(r\ln x)$$

$$= \frac{r}{x}e^{r\ln x}$$

$$= \frac{r}{x}x^{r}$$

$$= rx^{r-1} \qquad\blacksquare$$

Alternatively, (8.45) can be derived by taking the logarithm of both sides of $y = x^r$ and then using implicit differentiation.

Example 7

Differentiate (*a*) $y = \sqrt{3}^x$ and (*b*) $y = x^{\sqrt{3}}$.

Solution

(*a*) From (8.41),

$$\frac{dy}{dx} = \sqrt{3}^x \ln \sqrt{3}$$

(*b*) From (8.45),

$$\frac{dy}{dx} = \sqrt{3}x^{\sqrt{3}-1}$$

The result given in (8.45) also extends to a power of a function.

Example 8

Differentiate $y = (x^4 + e^{-3x})^e$.

Solution

$$\frac{dy}{dx} = e(x^4 + e^{-3x})^{e-1} \frac{d}{dx}(x^4 + e^{-3x})$$

$$= e(x^4 + e^{-3x})^{e-1}(4x^3 - 3e^{-3x})$$

The result given in (8.45) along with the fundamental definition of the natural logarithms are the "missing links" that we need to evaluate the indefinite integral $\int x^r \, dx$ *for any real exponent r:*

$$\int x^r \, dx = \begin{cases} \dfrac{x^{r+1}}{r+1} + C, & r \neq -1 \\[2mm] \ln|x| + C, & r = -1 \end{cases} \tag{8.46}$$

Example 9

Evaluate $\int x^\pi \, dx$.

Solution   From (8.46),

$$\int x^\pi \, dx = \frac{x^{\pi+1}}{\pi + 1} + C$$

Check From (8.45),

$$\frac{d}{dx}\left(\frac{x^{\pi+1}}{\pi+1}+C\right) = (\pi+1)\frac{x^{\pi}}{\pi+1} = x^{\pi}$$

Logarithmic Function with Base a

When the base a is positive and $a \neq 1$, the exponential function $y = a^x$ is one-to-one and, hence, possesses an inverse function denoted by $y = \log_a x$. The latter function is called the **logarithmic function with base a**. Analogous to (8.20) of Section 8.2,

$$y = \log_a x \quad \text{if and only if} \quad x = a^y \tag{8.47}$$

This is the interpretation of the logarithm that one normally uses in a course in precalculus mathematics. Since the range of the exponential function is the domain of its inverse, the domain of the logarithmic function is the set of positive real numbers.

Because of (8.47) and (8.20) of Section 8.2, we are finally able to state that the base of the natural logarithm is the number e:

$$\ln x = \log_e x$$

Laws of Logarithms

The laws of logarithms stated for $\ln x$ also hold for $\log_a x$. Of course, as in Theorem 8.4, for $b > 0$, $a > 0$, $a \neq 1$ the property

$$\log_a b^r = r \log_a b$$

is valid for any real number r.

Derivative of $\log_a x$

To find the derivative of the logarithmic function $y = \log_a x$, we take the natural logarithm of both sides of $x = a^y$:

$$\ln x = y \ln a$$

and use implicit differentiation,

$$\frac{1}{x} = \frac{dy}{dx} \ln a$$

$$\frac{dy}{dx} = \frac{1}{x \ln a}$$

In other words,

$$\frac{d}{dx} \log_a x = \frac{1}{x \ln a} \tag{8.48}$$

As usual, the general case can be obtained from the Chain Rule. If $u = g(x)$ is a differentiable function, then

$$\frac{d}{dx} \log_a |u| = \frac{1}{u \ln a} \frac{du}{dx} \tag{8.49}$$

When $a = 10$ in (8.47) we say that $\log_{10} x$ is the **common logarithm** of the number x.

____ **Example 10** _____

Find the slope of the tangent to the graph of $y = \log_{10} x$ at $x = \frac{1}{2}$.

Solution   From (8.48),

$$\frac{dy}{dx} = \frac{1}{x \ln 10}$$

With the aid of a calculator, we find

$$\left.\frac{dy}{dx}\right|_{x=1/2} = \frac{2}{\ln 10} \approx 0.8686$$

____ **Example 11** _____

Differentiate $y = \log_5 |x^3 - x|$.

Solution   From (8.49),

$$\frac{dy}{dx} = \frac{1}{(x^3 - x)\ln 5} \cdot \frac{d}{dx}(x^3 - x)$$

$$= \frac{3x^2 - 1}{(x^3 - x)\ln 5}$$

____ **Example 12** _____

Differentiate $y = \log_2 [x^4(x^2 + 1)^\pi]$.

Solution   By the laws of logarithms, we first write

$$y = 4 \log_2 |x| + \pi \log_2(x^2 + 1)$$

Observe that $x^4 = |x|^4$ and $x^2 + 1 > 0$. Thus, from (8.49) we have

$$\frac{dy}{dx} = \frac{4}{x \ln 2} + \frac{2\pi x}{(x^2 + 1)\ln 2}$$

Earthquakes   The American geologist/seismologist Charles F. Richter (1900–1985) devised a scale for comparing energies of various earthquakes. The so-called **Richter scale**, proposed in 1935, is defined in terms of a common logarithm:

$$R = \log_{10} \frac{A}{A_0} \tag{8.50}$$

where R represents the magnitude of the earthquake, A is the amplitude of the largest seismic wave that occurs, and A_0 is a reference amplitude corresponding to the magnitude $R = 0$.

Example 13

The Richter magnitude of the 1906 San Francisco earthquake is estimated to have been 8.3. In 1985, the destructive Mexico City earthquake had a magnitude of 7.8 on the Richter scale. How much greater was the intensity of the San Francisco earthquake than that of the Mexico City earthquake?

Solution   From (8.50) we have

$$8.3 = \log_{10}\left(\frac{A}{A_0}\right)_{1906} \quad \text{and} \quad 7.8 = \log_{10}\left(\frac{A}{A_0}\right)_{1985}$$

In view of (8.47) this means

$$\left(\frac{A}{A_0}\right)_{1906} = 10^{8.3} \quad \text{and} \quad \left(\frac{A}{A_0}\right)_{1985} = 10^{7.8}$$

Using the laws of exponents and a calculator we find that

$$\left(\frac{A}{A_0}\right)_{1906} = 10^{8.3}$$

$$= 10^{0.5}\, 10^{7.8}$$

$$= 10^{0.5}\left(\frac{A}{A_0}\right)_{1985}$$

$$\approx 3.16\left(\frac{A}{A_0}\right)_{1985}$$

Hence, the 1906 San Francisco earthquake was approximately 3 times as strong as the 1985 Mexico City earthquake.

Remark

Definition 8.4 provides a means for estimating the numerical value of a number such as $10^{\sqrt{3}}$ either through tabulated or calculator values of e^x and $\ln x$. Employing a calculator,

$$10^{\sqrt{3}} = e^{(1.7321\ldots)(2.3026)}$$

$$\approx e^{3.9882} \approx 53.9574$$

Many scientific calculators also possess the $\boxed{y^x}$ key. By entering $y = 10$ and calculating $\sqrt{3}$, the y^x function gives what appears to be an alternative and direct way of obtaining the value of $10^{\sqrt{3}}$. In fact, a calculator uses the routine $e^{x\ln y}$ to compute y^x. This is shown by the simple experiment: We know $(-5)^2 = 25$ but when $\boxed{y^x}$ is utilized with $y = -5$ and $x = 2$, many calculators display an error message. A careful rereading of Definition 8.4 will indicate why.*

---

*Some of the better calculators have a built-in check to determine whether the exponent is an integer. If this is the case, the calculator will then use a different routine to compute y^x.

Exercises 8.4

Answers to odd-numbered problems begin on page A-33

In Problems 1–6 write each number as a power of e. Obtain an approximation with at least two significant figures.

1. 2^π

2. $5^{\sqrt{2}}$

3. 10^e

4. $6^{-2.3}$

5. $7^{-\sqrt{5}}$

6. $\left(\dfrac{3}{2}\right)^{2\pi}$

In Problems 7–36 find the derivative of the given function.

7. $y = 4^x$

8. $y = 2 - 3^x$

9. $y = 10^{-2x}$

10. $y = 9^{3x+1}$

11. $y = 20^{x^2}$

12. $y = 5^{\sqrt{x}}$

13. $y = \pi^{\sqrt{2}}$

14. $y = 2^{\sin 5x}$

15. $y = x^2 2^x$

16. $y = e^x x^e$

17. $y = x^{\sqrt{5}}$

18. $y = \left(\dfrac{1}{3x}\right)^{\sqrt{3}}$

19. $y = x^{\sin 2}$

20. $y = (\cos x)^{\pi/2}$

21. $f(t) = (t^4 + 3t^2)^{\ln 4}$

22. $g(t) = (t^2)^e$

23. $f(x) = \dfrac{4^x}{1 + e^x}$

24. $f(x) = \dfrac{(x + 1)^\pi}{\pi^{x+1}}$

25. $f(x) = \sqrt{5^x}$

26. $f(x) = \dfrac{(25)^x}{5^{-x}}$

27. $H(x) = \dfrac{1}{(\ln x)^{1/e}}$

28. $P(v) = \tan(2^{-v^2})$

29. $y = \log_4 x$

30. $y = (\ln 5)^x$

31. $y = (\ln x)(\log_{10} x)$

32. $y = \ln(\log_{10} x)$

33. $y = x^\pi \log_3 |6x - 4|$

34. $y = \log_{10}\left(\dfrac{x^2 + 1}{x^4 + 9}\right)$

35. $f(x) = 3^{x^3}\log_3 x^3$

36. $f(x) = (\log_5 3x)^7$

In Problems 37–40 find the indicated derivative.

37. $y = 9^x; \dfrac{d^2 y}{dx^2}$

38. $y = x3^x; \dfrac{d^2 y}{dx^2}$

39. $y = (2x)^e; \dfrac{d^3 y}{dx^3}$

40. $y = (\log_{10} x)^2; \dfrac{d^2 y}{dx^2}$

In Problems 41–44 use implicit differentiation to find dy/dx.

41. $2^y = xy$

42. $x^2 + y^2 = 4^x - 4^{-x}$

43. $y + e^y = \log_{10} |x|$

44. $\log_2 xy = \sqrt{2x + 1}$

In Problems 45–60 evaluate the given integral.

45. $\displaystyle\int 7^x \, dx$

46. $\displaystyle\int \dfrac{dx}{7^{2x}}$

47. $\displaystyle\int (10^x)^{1/3} \, dx$

48. $\displaystyle\int 2^x 4^x \, dx$

49. $\displaystyle\int_2^3 10^{-x} \, dx$

50. $\displaystyle\int_0^1 t^2 3^{-t^3} \, dt$

51. $\displaystyle\int (e^x + x^e + e^e) \, dx$

52. $\displaystyle\int (x + 1)^e \, dx$

53. $\displaystyle\int 2^t (1 + 2^t)^{20} \, dt$

54. $\displaystyle\int (1 + 2^t)^2 \, dt$

55. $\displaystyle\int_{-1}^1 e^t 2^{e^t} \, dt$

56. $\displaystyle\int_0^\pi \dfrac{\sin 2x}{2^{\cos 2x}} \, dx$

57. $\displaystyle\int (1 + 2^{\tan\theta})^2 \sec^2\theta \, d\theta$

58. $\displaystyle\int_1^{10} \dfrac{(\log_{10} x)^3}{x} \, dx$

59. $\displaystyle\int \dfrac{5^x}{1 + 5^x} \, dx$

60. $\displaystyle\int \dfrac{5^x}{1 + 25^x} \, dx$

61. Find the slope of the tangent line to the graph of $y = 4^x$ at $x = 3$.

62. Find an equation of the tangent line to the graph of $y = x + x^\pi$ at $x = 1$.

63. Find the point(s) on the graph of $f(x) = x4^x$ where the tangent line is horizontal.

64. Determine the intervals on which $f(x) = x \log_2 |x - 1|$ is concave upward and the intervals on which it is concave downward.

65. Find the area of the region in the first quadrant bounded by the graphs of $y = 2^x$, $y = 2^{-x}$, and $x = 4$.

66. Find the volume of the solid of revolution that is formed by revolving the region bounded by the graphs of $y = 2^{x^2}$, $y = 0$, $x = 0$, and $x = 1$ about the y-axis.

67. In 1979 an earthquake occurred in San Francisco of magnitude 5.9 on the Richter scale. How much greater was the intensity of that city's 1906 earthquake of magnitude 8.3? How does the intensity of the 1979 earthquake compare with the 1971 San Fernando Valley earthquake of magnitude 6.6?

68. An earthquake has a magnitude of 4.7 on the Richter scale. What is the magnitude of an earthquake on the Richter scale if its intensity is 50 times greater?

69. The **pH**, or hydrogen potential, of a solution is defined by $\text{pH} = -\log_{10}[\text{H}^+]$, where $[\text{H}^+]$ is the concentration of hydrogen ions in moles/liter. If $0 < \text{pH} < 7$, the solution is said to be *acid* and if pH > 7, the solution is *base* or *alkaline*. Find the pH of a solution for which $[H^+] = 3.9 \times 10^{-8}$ moles/liter. Is the solution acid or base?

70. Two liquids have pH values of 4.6 and 6.1. Use Problem 69 to determine how much more acidic the first solution is than the second.

71. The intensity level b of a sound is defined by $b = 10 \log_{10}(I/I_0)$, where I is the intensity of the sound (in watts/cm^2) and I_0 is a reference intensity corresponding to $b = 0$. An intensity level is measured in decibels (dB). Now the intensity I of a sound is inversely proportional to the square of its distance d from its source. Use this fact to show that if a sound has an intensity level b_1 at a distance d_1 from its source, then at a distance d_2 the intensity level b_2 is given by $b_2 = b_1 + 20 \log_{10}(d_1/d_2)$.

72. The intensity level of a plane at an altitude of 1500 ft and passing a point 5 mi out from an airport is 80 dB. Use Problem 71 to determine the intensity level of another plane passing the same point but at an altitude of 2700 ft.

Miscellaneous Problems

73. Prove (*i*) of Theorem 8.3.

74. Prove (*v*) of Theorem 8.3.

75. Prove (*vii*) of Theorem 8.3.

76. Use (8.47) to prove the **change of base formula**

$$\log_a x = \frac{\log_b x}{\log_b a}$$

77. Use the result of Problem 76 to show that $\log_{10} e = 1/\ln 10$.

78. The **Naperian logarithm*** of a number x is defined as

$$10^7 \log_{1/e}\left(\frac{x}{10^7}\right)$$

Use the change of base formula in Problem 76 to express the Naperian logarithm in terms of the natural logarithm.

[O] 8.5 An Alternative Approach to the Natural Logarithmic Function

You may have noticed that the considerations of the logarithmic and exponential functions over the last four sections have been subtle and bear little resemblance to those concepts studied in a course in precalculus mathematics. In an elementary course, first one normally encounters an exponential function

$$y = a^x, \qquad a > 0 \qquad a \neq 1, \qquad x \text{ any real number}$$

and shows that this function possesses an inverse called the logarithmic function. Recall, *logarithm* is a word meaning the exponent of the base a that gives the number x; that is,

$$y = \log_a x \quad \text{is equivalent to} \quad x = a^y$$

John Napier

***John Napier (1550–1617)**   The first table of logarithms was published in 1614 by the Scotsman John Napier. He used his invention as a means of converting unwieldy calculations involving products and quotients into simpler calculations involving sums and differences. Napier is also credited with popularizing the use of decimal fractions through the notational device known as a decimal point. For the wealthy Napier, mathematics was simply a hobby or diversion in his life devoted to the polemics of politics and religion. His friend and collaborator, the English mathematician Henry Briggs (1561–1639), modified the base of the Naperian logarithm (a number which is close to $1/e$) and published the first tables of the common, or base 10, logarithms. These logarithms are sometimes called Briggsian logarithms.

From the laws of exponents, one then proceeds to deduce the properties and the laws of logarithms. Just the reverse of this method is presented in Sections 8.1 and 8.2. Why the difference? The answer is simple. In precalculus mathematics, it is *assumed and never proved* that a^x makes sense, or is well defined, for $a > 0$ and for every real number x. For a rational exponent p/q, where p and q are integers, $a^{p/q} = (a^{1/q})^p$ when $a^{1/q}$ is a real number. But how does one interpret a number with an irrational exponent such as 2^π?

Which interpretation should you use in calculus? To start with the integral definition of the logarithmic function *or* to start with an intuitive notion of a^x and its inverse, raises a certain amount of controversy among teachers of mathematics.

You should not get the impression from the foregoing remarks that a^x cannot be defined first in a rigorous manner. Indeed, for any real number x, a^x is defined in the following manner.

> **DEFINITION 8.5**   Suppose a is any positive real number, x a fixed real number, and t is a rational number. Then,
>
> $$a^x = \lim_{t \to x} a^t$$

_____ **Example 1** _____

It can be proved that the limit of the sequence of numbers with rational exponents

$$2^3, \; 2^{3.1}, \; 2^{3.14}, \; 2^{3.141}, \; 2^{3.1416}, \; \ldots$$

is the number 2^π.

_____

With Definition 8.5 as a foundation, the many properties of the exponential function can now be proved. However, the proofs—for example of the laws of exponents—are beyond the level of a first course in calculus.

Derivative of $\log_a x$ Although we have studied two new functions in Sections 8.1 and 8.2, we obtained the derivatives of these functions without recourse to first principles, namely, $f'(x) = \lim_{\Delta x \to 0} \Delta y / \Delta x$. Independent of prior considerations, let us assume for the remainder of the discussion that $y = a^x$, $a > 0$, $a \neq 1$ is well defined and is a continuous one-to-one function whose inverse is the continuous function $y = \log_a x$, $x > 0$. The following equations show how to find the derivative of the logarithm from the definition of the derivative:

$$\frac{dy}{dx} = \lim_{\Delta x \to 0} \frac{\log_a(x + \Delta x) - \log_a x}{\Delta x}$$

$$= \lim_{\Delta x \to 0} \frac{1}{\Delta x} \log_a \frac{x + \Delta x}{x} \qquad \text{(algebra and laws of logarithms)}$$

$$= \lim_{\Delta x \to 0} \frac{1}{\Delta x} \log_a \left(1 + \frac{\Delta x}{x}\right) \qquad \text{(algebra)}$$

$$= \lim_{\Delta x \to 0} \frac{1}{x} \frac{x}{\Delta x} \log_a\left(1 + \frac{\Delta x}{x}\right) \qquad \text{(algebra, } x/x = 1\text{)}$$

$$= \lim_{\Delta x \to 0} \frac{1}{x} \log_a\left(1 + \frac{\Delta x}{x}\right)^{x/\Delta x} \qquad \text{(laws of logarithms)}$$

$$= \frac{1}{x} \log_a\left[\lim_{\Delta x \to 0}\left(1 + \frac{\Delta x}{x}\right)^{x/\Delta x}\right] \tag{8.51}$$

The last step is justified by invoking continuity of the function and assuming that the limit inside the brackets exists. Let us make the change of variable $h = \Delta x/x$ in (8.51). Since x is fixed, $\Delta x \to 0$ implies $h \to 0$. Consequently,

$$\lim_{\Delta x \to 0}\left(1 + \frac{\Delta x}{x}\right)^{x/\Delta x} = \lim_{h \to 0}(1 + h)^{1/h}$$

Now using the concepts of Section 11.2 it can be proved that

$$\lim_{h \to 0}(1 + h)^{1/h} = e$$

where $e = 2.71828 \ldots$. Hence, (8.51) becomes

$$\frac{d}{dx} \log_a x = \frac{1}{x} \log_a e \tag{8.52}$$

When the "natural" choice of $a = e$ is made, (8.52) simplifies to

$$\frac{d}{dx} \log_e x = \frac{1}{x}$$

since $\log_e e = 1.$* Accordingly, we say $\log_e x$ is the "natural logarithm" and abbreviate this function by

$$\log_e x = \ln x$$

Exercises 8.5

Answers to odd-numbered problems begin on page A-33

1. Conjecture the value of

$$\lim_{\Delta x \to 0} \frac{e^{\Delta x} - 1}{\Delta x}$$

by filling in the table on the right.

2. Use the definition of the derivative along with the result obtained in Problem 1 to find dy/dx for $y = e^x$.

3. Use a carefully drawn graph to approximate the value of $2^{\sqrt{3}}$.

Δx	$(e^{\Delta x} - 1)/\Delta x$
0.1	
0.01	
0.001	
0.0001	
−0.0001	

*Those with sharp eyes and long memories will have noticed that (8.52) is not the same as (8.48) of Section 8.4. The results are equivalent since $\log_a e = 1/\log_e a$. See Problem 76, Exercises 8.4.

8.6 Logarithmic Differentiation

The elementary property, if $a = b$, $a > 0$, $b > 0$, then $\ln a = \ln b$, is surprisingly useful in finding the derivatives of complicated expressions.

_____ **Example 1** _____

Find the derivative of

$$y = \frac{\sqrt[3]{x^4 + 6x^2}(8x + 3)^5}{(2x^2 + 7)^{2/3}}$$

Solution While we could find dy/dx through application of the Quotient, Product, and Power Rules, this entire procedure can be avoided by first taking the logarithm of the absolute value of both sides of the given equation, simplifying, and *then* differentiating implicitly. We have

$$\ln|y| = \ln\sqrt[3]{x^4 + 6x^2} + \ln|8x + 3|^5 - \ln(2x^2 + 7)^{2/3}$$

$$= \frac{1}{3}\ln(x^4 + 6x^2) + 5\ln|8x + 3| - \frac{2}{3}\ln(2x^2 + 7)$$

Taking the derivative with respect to x gives

$$\frac{1}{y}\frac{dy}{dx} = \frac{1}{3}\cdot\frac{1}{x^4 + 6x^2}\cdot(4x^3 + 12x) + 5\cdot\frac{1}{8x + 3}\cdot 8 - \frac{2}{3}\cdot\frac{1}{2x^2 + 7}\cdot 4x$$

$$\frac{dy}{dx} = y\left[\frac{4x^3 + 12x}{3(x^4 + 6x^2)} + \frac{40}{8x + 3} - \frac{8x}{3(2x^2 + 7)}\right]$$

$$= \frac{\sqrt[3]{x^4 + 6x^2}(8x + 3)^5}{(2x^2 + 7)^{2/3}}\left[\frac{4x^3 + 12x}{3(x^4 + 6x^2)} + \frac{40}{8x + 3} - \frac{8x}{3(2x^2 + 7)}\right]$$

The process illustrated in the preceding example is called **logarithmic differentiation**.*

In Section 8.4 we learned that a function that has the form

$$y = (\text{variable})^{\text{constant}}$$

where the constant is any real number, can be differentiated by the Power Rule. Logarithmic differentiation provides a means of finding the derivative of an expression of the form

$$y = (\text{variable})^{\text{variable}}$$

_____

*This method of differentiation is attributed to the Swiss mathematician, Johann Bernoulli (1667–1758).

_____ **Example 2** _____

Differentiate $y = x^{\sqrt{x}}$, $x > 0$.

Solution   Taking the logarithm of both sides of the given equation yields

$$\ln y = \ln x^{\sqrt{x}}$$
$$= \sqrt{x} \ln x$$

Implicit differentiation then yields

$$\frac{1}{y}\frac{dy}{dx} = \sqrt{x} \cdot \frac{1}{x} + \frac{1}{2}x^{-1/2} \cdot \ln x$$

$$\frac{dy}{dx} = y\left[\frac{1}{\sqrt{x}} + \frac{\ln x}{2\sqrt{x}}\right]$$

$$= x^{\sqrt{x}}\left[\frac{1}{\sqrt{x}} + \frac{\ln x}{2\sqrt{x}}\right]$$

$$= x^{\sqrt{x}-1/2}\left(1 + \frac{1}{2}\ln x\right)$$

It may be necessary to take the logarithm of the members of a given equation more than once in order to compute the derivative.

_____ **Example 3** _____

Differentiate $y = (x^2 + 1)^{(x+1)^x}$, $x > -1$.

Solution   Since the equation

$$\ln y = (x + 1)^x \ln(x^2 + 1)$$

still involves a variable base with a variable exponent, we take the logarithm a second time:

$$\ln(\ln y) = \ln((x + 1)^x \ln(x^2 + 1))$$
$$= x \ln(x + 1) + \ln(\ln(x^2 + 1))$$

Finally, differentiation with respect to x gives

$$\frac{1}{\ln y} \cdot \frac{1}{y} \cdot \frac{dy}{dx} = x \cdot \frac{1}{x + 1} + \ln(x + 1) + \frac{1}{\ln(x^2 + 1)} \cdot \frac{1}{x^2 + 1} \cdot 2x$$

$$\frac{dy}{dx} = y \ln y\left[\frac{x}{x + 1} + \ln(x + 1) + \frac{2x}{(x^2 + 1)\ln(x^2 + 1)}\right]$$

$$= (x^2 + 1)^{(x+1)^x}(x + 1)^x \ln(x^2 + 1)\left[\frac{x}{x + 1} + \ln(x + 1) + \frac{2x}{(x^2 + 1)\ln(x^2 + 1)}\right]$$

Remark

Note carefully that the Power Rule is not applicable to $y = x^x$. In other words, the derivative is *not* xx^{x-1}. It is left as an exercise to show that the derivative of $y = x^x$ is $dy/dx = x^x(1 + \ln x)$. See Problem 8 in Exercises 8.6.

Exercises 8.6

Answers to odd-numbered problems begin on page A-33

In Problems 1–20 use logarithmic differentiation to find dy/dx.

1. $y = \sqrt{\dfrac{(2x + 1)(3x + 2)}{4x + 3}}$

2. $y = \dfrac{x^{10}\sqrt{x^2 + 5}}{\sqrt[3]{8x^2 + 2}}$

3. $y = \dfrac{(x^3 - 1)^5(x^4 + 3x^3)^4}{(7x + 5)^9}$

4. $y = x\sqrt{x + 1}\;\sqrt[3]{x^2 + 2}$

5. $y = \dfrac{x^{2/3}}{\sqrt{(x^2 + 2x + 3)(2x^4 + x^2 + 1)}}$

6. $y = (x + 2)^3(x^2 - 1)^2(4 - 5x)^6(2x^3 - x)^7$

7. $y = (x^2 + 4)^{2x}$

8. $y = x^x$

9. $y = x^x 2^x$

10. $y = x^{2^x}$

11. $y = x(x - 1)^x$

12. $y = \dfrac{(x^2 + 1)^x}{x^2}$

13. $y = (x^6 + x^4)^{\ln x}$

14. $y = x^{(x+1)\ln x}$

15. $y = (\ln x)^x$

16. $y = (x^2 + 4x + 7)^{x^2 + \ln x}$

17. $y = x^{\cos x}$

18. $y = x(\sin x)^x$

19. $y = x^{e^x}$

20. $y = x^{x^x}$

21. Find an equation of the tangent line to the graph of $y = x^{x+2}$ at $x = 1$.

22. Find an equation of the tangent line to the graph of $y = x(\ln x)^x$ at $x = e$.

23. For $x > 0$, find the relative extrema of $f(x) = x^{3x}$.

24. For $x > 0$, find the relative extrema of $f(x) = x^{1/x}$.

In Problems 25 and 26 find dy/dx.

25. $y = \tan x^x$

26. $y = x^x e^{x^x}$

In Problems 27 and 28 find d^2y/dx^2.

27. $y = x^{2x}$

28. $y = (\sin x)^{e^x}$

Miscellaneous Problem

29. Derive a general formula for the derivative of $y = u^v$, where $u = f(x)$ and $v = g(x)$ are differentiable functions.

8.7 Separable Differential Equations and Their Applications

Before considering some applications of the functions e^x and $\ln x$, we need to examine the concept of a differential equation. Roughly, a **differential equation** is an equation that involves derivatives or differentials of an unknown function. For example,

$$\frac{d^2y}{dx^2} + 4\frac{dy}{dx} + 8y = 0$$

is a differential equation.

In this section we are going to confine our attention to a special kind of differential equation that involves only the first derivative dy/dx.

> **DEFINITION 8.6**   A **separable first-order differential equation** is an equation $F(x, y, y') = 0$ that can be put into the form
>
> $$h(y)\frac{dy}{dx} = g(x) \qquad (8.53)$$

An equation of form (8.53) is also said to have *separable variables*; for example,

$$\frac{dy}{dx} = -\frac{x}{y} \quad \text{and} \quad y' = y$$

are separable since they can be written respectively as

$$y\frac{dy}{dx} = -x \quad \text{and} \quad \frac{1}{y}y' = 1 \qquad (8.54)$$

You should verify that the first-order differential equation

$$\frac{dy}{dx} + 2y = x$$

is *not* separable.

It is usual practice to write a separable equation in terms of differentials. The equations in (8.54) can be written alternatively as

$$y\,dy = -x\,dx \quad \text{and} \quad \frac{dy}{y} = dx$$

In general, (8.53) is written $h(y)\,dy = g(x)\,dx$.

Solutions   We are interested in *solving* an equation of form (8.53). A **solution** is any differentiable function, defined explicitly or implicitly, which, when substituted into the differential equation, reduces it to an identity.

If $y = f(x)$ denotes a solution of (8.53), we must have

$$h(f(x))f'(x) = g(x)$$

and therefore, by integration,

$$\int h(f(x))f'(x)\,dx = \int g(x)\,dx \qquad (8.55)$$

But, $dy = f'(x)\,dx$ so (8.55) is the same as

$$\int h(y)\,dy = \int g(x)\,dx \qquad (8.56)$$

Method of Solution   Equation (8.56) indicates the procedure for solving separable differential equations: Integrate both sides of

$$h(y)\,dy = g(x)\,dx$$

Example 1

Solve $\dfrac{dy}{dx} = -\dfrac{x}{y}$.

Solution Write the given equation in differential form

$$y \, dy = -x \, dx$$

and integrate both sides

$$\int y \, dy = -\int x \, dx$$

$$\frac{y^2}{2} = -\frac{x^2}{2} + C$$

Thus, a family of solutions is defined by $x^2 + y^2 = C_1$, where we have replaced $2C$ by C_1.

Example 2

Solve $\dfrac{dy}{dt} = ky$, where k is a constant.

Solution Write the equation as

$$\frac{dy}{y} = k \, dt$$

and integrate

$$\int \frac{dy}{y} = k \int dt$$

By assuming $y > 0$, we have

$$\ln y = kt + C_1$$

or

$$y = e^{kt + C_1} = e^{C_1} e^{kt}$$

Relabeling e^{C_1} as C, then yields

$$y = Ce^{kt} \tag{8.57}$$

Growth and Decay   The simple differential equation in Example 2,

$$\frac{dy}{dt} = ky \tag{8.58}$$

has many applications involving either growth or decay. For example, in biology it is often observed that the rate dN/dt, at which certain bacteria grow, is proportional to the number N of bacteria present at any time:

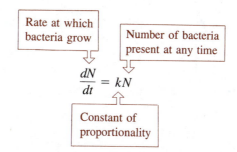

Also, over short periods of time, the population $P(t)$ of small animals such as rodents can be predicted fairly accurately by the solution of the differential equation $dP/dt = kP$. In physics, the solution $A(t)$ of the differential equation $dA/dt = kA$ provides a means for approximating the amount remaining of a substance that is decaying or disintegrating through radioactivity. Thus, one differential equation (8.58) can serve as the mathematical model for many diverse phenomena. In chemistry, the amount of a substance remaining during a first-order reaction is described by (8.58).

Example 3

Initially, the number of bacteria present in a culture is N_0. At $t = 1$ hour the number of bacteria is measured to be $(\frac{3}{2})N_0$. If the rate of growth is assumed to be proportional to the number of bacteria present, determine the time necessary for the number of bacteria to triple.

Solution   We first solve the differential equation

$$\frac{dN}{dt} = kN$$

subject to $N(0) = N_0$.

From (8.57) of Example 2, we can write

$$N(t) = Ce^{kt} \tag{8.59}$$

At $t = 0$ it follows from (8.59) that $N_0 = Ce^0 = C$ and so $N(t) = N_0 e^{kt}$. Now at $t = 1$ we have

$$N(1) = \frac{3}{2}N_0 = N_0 e^k \quad \text{or} \quad e^k = \frac{3}{2}$$

To four decimal places we have

$$k = \ln \frac{3}{2} = 0.4055$$

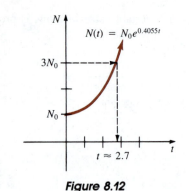

Figure 8.12

Thus, for any time $t \geq 0$

$$N(t) = N_0 e^{0.4055t}$$

To find the time at which the bacteria have tripled we solve

$$3N_0 = N_0 e^{0.4055t}$$

for t:

$$0.4055t = \ln 3$$

$$t = \frac{\ln 3}{0.4055} \approx 2.7 \text{ hr}$$

See Figure 8.12.

Note: We can write the function $N(t)$ obtained in the preceding example in an alternative form. From the laws of exponents,

$$N(t) = N_0 e^{kt}$$
$$= N_0 (e^k)^t$$
$$= N_0 \left(\frac{3}{2}\right)^t$$

since $e^k = \frac{3}{2}$. This latter solution provides a convenient method for computing $N(t)$ for small positive integral values of t; it also clearly shows the influence of the subsequent experimental observation at $t = 1$ on the solution for all time. We notice too that the actual number of bacteria present initially, that is, at time $t = 0$, is quite irrelevant in finding the time required to triple the number in the culture. The necessary time to triple, say, 100 or 10,000 bacteria is still approximately 2.7 hours.

Cooling   Newton's law of cooling states that the rate at which the temperature $T(t)$ changes in a cooling body is proportional to the difference between the temperature in the body and the constant temperature T_0 of the surrounding medium; that is,

$$\frac{dT}{dt} = k(T - T_0) \tag{8.60}$$

where k is a constant of proportionality.

Example 4

When a cake is removed from a baking oven, its temperature is measured at 300°F. Three minutes later its temperature is 200°F. Determine the temperature of the cake at any time after leaving the oven if the room temperature is 70°F.

Solution   We must solve the problem

$$\frac{dT}{dt} = k(T - 70), \qquad T(0) = 300$$

and determine the value of k so that $T(3) = 200$.

Assuming $T > 70$, it follows by separation of variables that

$$\frac{dT}{T - 70} = k \, dt$$

$$\ln(T - 70) = kt + C_1$$

$$T - 70 = C_2 e^{kt} \qquad (C_2 = e^{C_1})$$

$$T = 70 + C_2 e^{kt}$$

When $t = 0$, $T = 300$ so that $300 = 70 + C_2$ gives $C_2 = 230$ and therefore $T = 70 + 230 e^{kt}$.

From $T(3) = 200$ we find

$$e^{3k} = \frac{13}{23}$$

and so, to four decimal places, a calculator gives

$$k = \frac{1}{3} \ln \frac{13}{23} = -0.1902$$

Thus $\qquad\qquad T(t) = 70 + 230 e^{-0.1902t}$

The graph of T along with some calculated values are given in Figure 8.13.

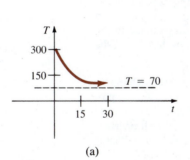

$T(t)$	t (minutes)
75°	20.1
74°	21.3
73°	22.8
72°	24.9
71°	28.6
70.5°	32.3

(a)   (b)

Figure 8.13

Exercises 8.7

Answers to odd-numbered problems begin on page A-34

In Problems 1–16 solve the given differential equation by separation of variables.

1. $\dfrac{dy}{dx} = \cos 2x$

2. $\dfrac{dy}{dx} = (x + 1)^2$

3. $dx - x^2 \, dy = 0$

4. $dx + e^{3x} \, dy = 0$

5. $\dfrac{dy}{dx} = \dfrac{y^3}{x^2}$

6. $xy' = 4y$

7. $xy' - y = 1$

8. $\dfrac{dy}{dx} = \dfrac{x^2 y^2}{1 + x}$

9. $\dfrac{dy}{dx} = e^{3x+2y}$

10. $\dfrac{dy}{dx} + y = ye^{x+2}$

11. $\dfrac{dy}{dx} = \left(\dfrac{2y + 3}{4x + 5}\right)^2$

12. $\dfrac{dx}{dy} = \dfrac{1 + 2y^2}{y \sin x}$

13. $\sec^2 x \, dy + \csc y \, dx = 0$

14. $\sin 3x \, dx + 2y \cos^3 3x \, dy = 0$

15. $(1 + x^2 + y^2 + x^2 y^2) \, dy = y \, dx$

16. $\dfrac{dy}{dx} = \dfrac{xy + 3x - y - 3}{xy - 2x + 4y - 8}$

17. The population of a certain community is known to increase at a rate proportional to the number of people present at any time. Find the population $P(t)$ at any time. If the population has doubled in 5 years, how long will it take to triple? To quadruple?

18. Suppose we know that the population of the community in Problem 17 is 10,000 after 3 years. What was the initial population? What will the population be in 10 years?

19. Initially there were 100 milligrams of a radioactive substance present. After 6 hours the mass decreased by 3%. If the rate of decay is proportional to the amount of the substance present at any time, find the amount $A(t)$ remaining at any time. What amount remains after 24 hours?

20. The **half-life** of a radioactive substance is the time it takes for one-half of the atoms in an initial amount to disintegrate. Determine the half-life of the substance in Problem 19.

21. A breeder reactor converts the relatively stable uranium 238 into the isotope plutonium 239. After 15 years, it is determined that 0.043% of the initial amount A_0 of the plutonium has disintegrated. If the rate of decay is proportional to the amount of the substance present, determine the half-life of the isotope. (See Problem 20.)

22. When a vertical beam of light passes through a transparent substance, the rate at which its intensity I decreases is proportional to $I(t)$, where t represents the thickness of the medium in feet. In clear seawater the intensity 3 feet below the surface is 25% of the initial intensity I_0 of the incident beam. What is the intensity of the beam 15 feet below the surface?

23. When interest is compounded *continuously*, the amount of money S increases at a rate proportional to the amount present at any time: $dS/dt = rS$, where r is the annual rate of interest.

    (a) Find the amount of money accrued at the end of 5 years when $5000 is deposited in a savings account drawing $5\frac{3}{4}\%$ annual interest compounded continuously.

    (b) In how many years will the initial deposit be doubled?

    (c) Use a hand calculator to compare the number obtained in part **(a)** with

$$S = 5000\left(1 + \frac{0.0575}{4}\right)^{5(4)}$$

This value represents the amount that would be accrued when interest is compounded quarterly.

24. A thermometer is removed from a room where the air temperature is 70°F, and placed outside where the temperature is 10°F. After $\frac{1}{2}$ minute the thermometer reads 50°F. What is the reading at $t = 1$ minute? How long will it take for the thermometer to reach 15°F?

25. A thermometer is taken from an inside room and placed outside where the air temperature is 5°F. After 1 minute the thermometer reads 55°F and after 5 minutes the reading is 30°F. What is the initial temperature of the room?

26. The differential equation in (8.60) also holds for an object that absorbs heat from the surrounding medium. If a small metal bar, whose initial temperature is 20°C, is dropped into a container of boiling water, how long will it take for the bar to reach 90°C if it is known that its temperature increased 2° in 1 second? How long will it take the bar to reach 98°C?

27. In a series circuit that contains only a resistor and an inductor, Kirchhoff's second law states the sum of the voltage drop across the inductor $(L(di/dt))$ and the voltage drop across the resistor (iR) is the same as the impressed voltage (E) on the circuit. See Figure 8.14. Thus, we obtain the differential equation for the current $i(t)$

$$L\frac{di}{dt} + Ri = E$$

where L and R are constants known as the inductance and the resistance, respectively. Determine the current i if E is 12 volts, the inductance is $\frac{1}{2}$ henry, the resistance is 10 ohms, and $i(0) = 0$.

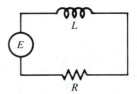

Figure 8.14

28. A 30-volt battery is connected to a series circuit in which the inductance is 0.1 henry and the resistance is 50 ohms. Find the current $i(t)$ if $i(0) = 0$. Determine the behavior of the current for large values of time. (See Problem 27.)

29. Under some circumstances, a falling body B of mass m (such as a person hanging from a parachute) encounters air resistance proportional to its instantaneous velocity $v(t)$. See Figure 8.15. Equating the sum of the forces acting on the body, $mg - kv$, with Newton's second law results in the differential equation

$$m\frac{dv}{dt} = mg - kv, \qquad k > 0$$

Solve the equation subject to $v(0) = v_0$ and determine the limiting velocity of the weight. If distance s is related to velocity $ds/dt = v$, find an explicit expression for s if it is further known that $s(0) = s_0$.

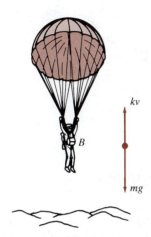

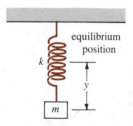

Figure 8.15

30. The rate at which a drug disseminates into the bloodstream is governed by the differential equation

$$\frac{dX}{dt} = A - BX$$

where A and B are positive constants. The function $X(t)$ describes the concentration of the drug in the bloodstream at any time t. Find the limiting value of X as $t \to \infty$. At what time is the concentration one-half this limiting value? Assume that $X(0) = 0$.

31. The height h of water that is flowing through an orifice at the bottom of a cylindrical tank is given by

$$\frac{dh}{dt} = -\frac{A_2}{A_1}\sqrt{2gh}, \qquad g = 32 \text{ ft/sec}^2$$

where A_1 and A_2 are the cross-sectional areas of the tank and orifice, respectively. See Figure 8.16. Solve the differential equation if the height of the water at $t = 0$ is 20 ft and $A_1 = 50$ ft^2 and $A_2 = \frac{1}{4}$ ft^2. At what time is the tank empty?

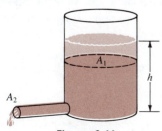

Figure 8.16

32. The energy of the spring-mass system shown in Figure 8.17 is the sum of potential and kinetic energies:

$$E = \frac{1}{2}ky^2 + \frac{1}{2}m\left(\frac{dy}{dt}\right)^2$$

where k is the spring constant, m is mass, and y is the displacement of the mass from an equilibrium position. If E is constant, use separation of variables to show that

$$y = \sqrt{\frac{2E}{k}} \sin\left(\pm\sqrt{\frac{k}{m}}t + C\sqrt{\frac{k}{2E}}\right)$$

where C is an arbitrary constant.

Figure 8.17

Calculator Problem

33. A rocket is shot vertically upward from the ground with an initial velocity v_0. See Figure 8.18. If the positive direction is taken to be upward, then in the absence of air resistance, the differential equation for the velocity v after fuel burnout is

$$v\frac{dv}{dy} = -\frac{k}{y^2}$$

where y is a positive constant.

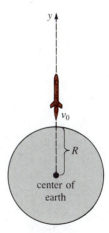

Figure 8.18

(a)   Solve the differential equation.

(b)   If $k = gR^2$, $R = 4000$ miles, use a calculator to show that the "escape velocity" of a rocket is approximately $v_0 = 25,000$ mi/hr.

Miscellaneous Problems

34. If P_0 is the initial population of a community, show that if P is governed by $dP/dt = kP$, then

$$\left(\frac{P_1}{P_0}\right)^{t_2} = \left(\frac{P_2}{P_0}\right)^{t_1}$$

where $P_1 = P(t_1)$ and $P_2 = P(t_2)$, $t_1 < t_2$.

35. Assume that the rate of decay of a radioactive substance is proportional to the amount $A(t)$ present at any time. If $A_1 = A(t_1)$ and $A_2 = A(t_2)$, $t_1 < t_2$, show that the half-life of the substance is given by

$$t = \frac{(t_2 - t_1)\ln 2}{\ln(A_1/A_2)}$$

8.8 The Hyperbolic Functions

Photograph by Frank Siteman/
Stock, Boston, Inc.   Gateway Arch

If you have ever toured the 640-foot-high Gateway Arch in St. Louis, Missouri, you may have asked the question: What is the shape of the arch? and received the rather cryptic answer: The shape of an inverted catenary. The word *catenary* stems from the Latin word *catena* and literally means a hanging chain (the Romans used the catena as a dog leash). It can be demonstrated that the shape assumed by a long flexible wire, cable, or rope hanging under its own weight between two points is that shape of the graph of the function $f(x) = (k/2) \cdot (e^{cx} + e^{-cx})$ for appropriate choices of the constants c and k. Combinations such as this involving e^x and e^{-x} occur so often in the study of applied mathematics that they warrant special definitions.

DEFINITION 8.7   For any real number x, the **hyperbolic sine** of x is

$$\sinh x = \frac{e^x - e^{-x}}{2}$$

and the **hyperbolic cosine** of x is

$$\cosh x = \frac{e^x + e^{-x}}{2}$$

Analogous to the trigonometric functions tan x, cot x, sec x, and csc x, which are defined in terms of sin x and cos x, we define four additional hyperbolic functions in terms of sinh x and cosh x.

DEFINITION 8.8   For a real number x, the

(*i*)   **hyperbolic tangent** of x is

$$\tanh x = \frac{\sinh x}{\cosh x} = \frac{e^x - e^{-x}}{e^x + e^{-x}}$$

(*ii*) **hyperbolic cotangent** of x is

$$\coth x = \frac{\cosh x}{\sinh x} = \frac{e^x + e^{-x}}{e^x - e^{-x}}, \qquad x \neq 0$$

(*iii*) **hyperbolic secant** of x is

$$\text{sech } x = \frac{1}{\cosh x} = \frac{2}{e^x + e^{-x}}$$

(*iv*) **hyperbolic cosecant** of x is

$$\text{csch } x = \frac{1}{\sinh x} = \frac{2}{e^x - e^{-x}}, \qquad x \neq 0$$

Graphs   The graphs of the hyperbolic sine and hyperbolic cosine can be obtained by adding ordinates. As illustrated in Figure 8.19(a) the graph of $y = \cosh x$ is gotten by first graphing $\frac{1}{2}e^x$ and $\frac{1}{2}e^{-x}$ and then adding y-coordinates at each point. Similarly, the graph of $y = \sinh x$ shown in Figure 8.19(b) is found by adding the y-coordinates of points on the graphs of $\frac{1}{2}e^x$ and $-\frac{1}{2}e^{-x}$.

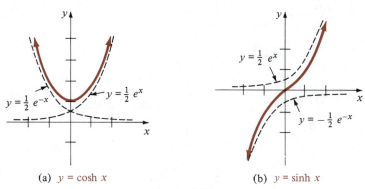

(a) $y = \cosh x$ (b) $y = \sinh x$

Figure 8.19

The graphs of $\tanh x$, $\coth x$, $\text{sech } x$, and $\text{csch } x$ are given in Figure 8.20.

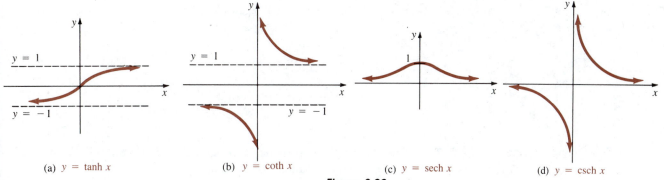

(a) $y = \tanh x$ (b) $y = \coth x$ (c) $y = \text{sech } x$ (d) $y = \text{csch } x$

Figure 8.20

Identities   Hyperbolic functions possess many identities that are similar to those of the trigonometric functions. Notice that the graphs in Figure 8.19(a) and (b) are symmetric with respect to the y-axis and the origin, respectively. In other words, $y = \cosh x$ is an even function and $y = \sinh x$ is an odd function:

$$\cosh(-x) = \cosh x \tag{8.61}$$

$$\sinh(-x) = -\sinh x \tag{8.62}$$

In trigonometry a fundamental identity is

$$\cos^2 x + \sin^2 x = 1$$

For hyperbolic functions the analogue of this identity is

$$\cosh^2 x - \sinh^2 x = 1 \tag{8.63}$$

To prove this we resort to Definition 8.7:

$$\cosh^2 x - \sinh^2 x = \left(\frac{e^x + e^{-x}}{2}\right)^2 - \left(\frac{e^x - e^{-x}}{2}\right)^2$$

$$= \frac{e^{2x} + 2 + e^{-2x}}{4} - \frac{e^{2x} - 2 + e^{-2x}}{4}$$

$$= \frac{e^{2x} - e^{2x} + 2 + 2 + e^{-2x} - e^{-2x}}{4}$$

$$= \frac{4}{4} = 1$$

The proofs of (8.61), (8.62), and the following list of identities are left as exercises.

$$1 - \tanh^2 x = \operatorname{sech}^2 x \tag{8.64}$$

$$\coth^2 x - 1 = \operatorname{csch}^2 x \tag{8.65}$$

$$\sinh(x + y) = \sinh x \cosh y + \cosh x \sinh y \tag{8.66}$$

$$\sinh(x - y) = \sinh x \cosh y - \cosh x \sinh y \tag{8.67}$$

$$\cosh(x + y) = \cosh x \cosh y + \sinh x \sinh y \tag{8.68}$$

$$\cosh(x - y) = \cosh x \cosh y - \sinh x \sinh y \tag{8.69}$$

$$\sinh 2x = 2 \sinh x \cosh x \tag{8.70}$$

$$\cosh 2x = \cosh^2 x + \sinh^2 x \tag{8.71}$$

$$\cosh^2 x = \frac{1}{2}(1 + \cosh 2x) \tag{8.72}$$

$$\sinh^2 x = \frac{1}{2}(-1 + \cosh 2x) \tag{8.73}$$

Derivatives of Hyperbolic Functions   The derivatives of the hyperbolic functions follow from (8.27) of Section 8.2 and the rules of differentiation; for example,

$$\frac{d}{dx} \sinh x = \frac{d}{dx} \frac{e^x - e^{-x}}{2}$$

$$= \frac{1}{2} \left[\frac{d}{dx} e^x - \frac{d}{dx} e^{-x} \right]$$

$$= \frac{e^x + e^{-x}}{2}$$

$$= \cosh x$$

Similarly, it should be apparent from the definition of the hyperbolic cosine that

$$\frac{d}{dx} \cosh x = \sinh x$$

To differentiate, say, the hyperbolic tangent, we use the Quotient Rule and the identity given in (8.63):

$$\frac{d}{dx} \tanh x = \frac{d}{dx} \frac{\sinh x}{\cosh x}$$

$$= \frac{\cosh x \cdot \frac{d}{dx} \sinh x - \sinh x \cdot \frac{d}{dx} \cosh x}{\cosh^2 x}$$

$$= \frac{\cosh^2 x - \sinh^2 x}{\cosh^2 x}$$

$$= \frac{1}{\cosh^2 x}$$

$$= \operatorname{sech}^2 x$$

The derivatives of the six hyperbolic functions in the most general case follow from the Chain Rule. We assume $u = g(x)$ is a differentiable function.

$$\text{I} \quad \frac{d}{dx} \sinh u = \cosh u \frac{du}{dx}$$

$$\text{II} \quad \frac{d}{dx} \cosh u = \sinh u \frac{du}{dx}$$

$$\text{III} \quad \frac{d}{dx} \tanh u = \operatorname{sech}^2 u \frac{du}{dx}$$

$$\text{IV} \quad \frac{d}{dx} \coth u = -\operatorname{csch}^2 u \frac{du}{dx}$$

$$\text{V} \quad \frac{d}{dx} \operatorname{sech} u = -\operatorname{sech} u \tanh u \frac{du}{dx}$$

$$\text{VI} \quad \frac{d}{dx} \operatorname{csch} u = -\operatorname{csch} u \coth u \frac{du}{dx}$$

You should take careful note of the slight difference in the results given by II and V and the analogous formulas for the trigonometric functions.

Example 1

Differentiate $y = \sinh\sqrt{2x + 1}$.

Solution From I,

$$\frac{dy}{dx} = \cosh\sqrt{2x + 1}\left(\frac{1}{2}(2x + 1)^{-1/2} \cdot 2\right)$$

$$= \frac{\cosh\sqrt{2x + 1}}{\sqrt{2x + 1}}$$

Example 2

Differentiate $y = \coth x^3$.

Solution From IV,

$$\frac{dy}{dx} = -\operatorname{csch}^2 x^3 \cdot 3x^2$$

Example 3

Evaluate the derivative of $y = \dfrac{3x}{4 + \cosh 2x}$ at $x = 0$.

Solution From the Quotient Rule,

$$\frac{dy}{dx} = \frac{(4 + \cosh 2x) \cdot 3 - 3x(\sinh 2x \cdot 2)}{(4 + \cosh 2x)^2}$$

When $x = 0$, it is seen from Definition 8.7 (see also Figure 8.19) that $\sinh 0 = 0$ and $\cosh 0 = 1$. Thus,

$$\frac{dy}{dx}\bigg|_{x=0} = \frac{15}{25} = \frac{3}{5}$$

Integrals of Hyperbolic Functions The integral forms of the preceding differentiation formulas are summarized as follows:

$$\text{I}' \quad \int \cosh u \, du = \sinh u + C$$

$$\text{II}' \quad \int \sinh u \, du = \cosh u + C$$

$$\text{III}' \quad \int \text{sech}^2 u \; du = \tanh u + C$$

$$\text{IV}' \quad \int \text{csch}^2 u \; du = -\coth u + C$$

$$\text{V}' \quad \int \text{sech } u \tanh u \; du = -\text{sech } u + C$$

$$\text{VI}' \quad \int \text{csch } u \coth u \; du = -\text{csch } u + C$$

Example 4

Evaluate $\int \cosh 5x \; dx$.

Solution   If $\qquad u = 5x \quad$ then $\quad du = 5 \; dx$

From I$'$,

$$\int \cosh 5x \; dx = \frac{1}{5} \int \cosh 5x(5 \; dx)$$

$$= \frac{1}{5} \int \cosh u \; du$$

$$= \frac{1}{5} \sinh u + C$$

$$= \frac{1}{5} \sinh 5x + C$$

Example 5

Evaluate $\int x \; \text{csch } x^2 \coth x^2 \; dx$.

Solution   If $\qquad u = x^2 \quad$ then $\quad du = 2x \; dx$

From VI$'$,

$$\int x \; \text{csch } x^2 \coth x^2 \; dx = \frac{1}{2} \int \text{csch } x^2 \coth x^2(2x \; dx)$$

$$= \frac{1}{2} \int \text{csch } u \coth u \; du$$

$$= -\frac{1}{2} \text{csch } x^2 + C$$

_____ **Example 6** _____

Evaluate $\int \tanh x \, dx$.

Solution   First, notice that the given integral does not possess any of the six forms given in I′–VI′. However, if we write

$$\int \tanh x \, dx = \int \frac{\sinh x}{\cosh x} \, dx$$

we can then make the identifications

$$u = \cosh x \quad \text{and} \quad du = \sinh x \, dx$$

Therefore,
$$\int \tanh x \, dx = \int \frac{\sinh x}{\cosh x} \, dx$$
$$= \int \frac{du}{u}$$
$$= \ln|u| + C$$
$$= \ln(\cosh x) + C$$

since $\cosh x > 0$ for all real numbers x.

Remarks

(*i*)   As mentioned in the introduction to this section, the graph of any function of the form $f(x) = (k/2)(e^{cx} + e^{-cx}) = k \cosh cx$, k and c constants, is called a **catenary**. The shape assumed by a wire or heavy rope strung between two posts has the basic shape of a graph of a hyperbolic cosine. Furthermore, if two circular rings are held vertically and are not too far apart, then a soap film stretched between the rings will assume a surface having minimum area. The surface is a portion of a **catenoid**, which is the surface obtained by revolving a catenary about the x-axis. See Figure 8.21 and Problem 70.

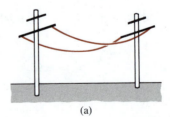
(a)

soap film

(b)

Figure 8.21

(*ii*)   The similarity between trigonometric and hyperbolic functions extends beyond the derivative formulas and basic identities.

If t is an angle measured in radians whose terminal side is OP, then the coordinates of P on a unit circle $x^2 + y^2 = 1$ are $(\cos t, \sin t)$. Now,

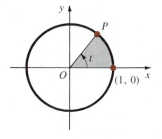

Figure 8.22

the area of the shaded circular sector shown in Figure 8.22 is $A = \frac{1}{2}t$ and so $t = 2A$.* In this manner, the *circular functions* cos t and sin t can be considered functions of the area A.

You might already know that the graph of the equation $x^2 - y^2 = 1$ is called a *hyperbola*. Because cosh $t \geq 1$ and $\cosh^2 t - \sinh^2 t = 1$, it follows that the coordinates of a point P on the right-hand branch of the hyperbola are (cosh t, sinh t). Furthermore, it is not very difficult to show that the area of the hyperbolic sector shown in Figure 8.23 is related to the number t by $t = 2A$. See Problems 77 and 78 in Exercises 8.8. Whence, we see the origin of the name "hyperbolic function."

(*iii*)   Unlike the trigonometric functions, the hyperbolic functions are *not* periodic.

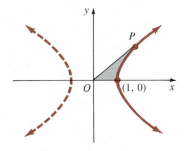

Figure 8.23

Exercises 8.8

Answers to odd-numbered problems begin on page A-34

1. If sinh $x = -\frac{1}{2}$, find the values of the remaining hyperbolic functions.

2. If cosh $x = 3$, find the values of the remaining hyperbolic functions.

3. Prove (8.61).   **4.** Prove (8.62).

5. Prove (8.64).   **6.** Prove (8.65).

7. Prove (8.66).   **8.** Prove (8.67).

9. Prove (8.68).   **10.** Prove (8.69).

11. Prove (8.70).   **12.** Prove (8.71).

13. Prove (8.72).   **14.** Prove (8.73).

In Problems 15–18 prove the given identity.

15. cosh x + sinh $x = e^x$

16. cosh x - sinh $x = e^{-x}$

17. $\tanh(x + y) = \dfrac{\tanh x + \tanh y}{1 + \tanh x \tanh y}$

18. $\tanh 2x = \dfrac{2 \tanh x}{1 + \tanh^2 x}$

In Problems 19–42 find the derivative of the given function.

19. $y = \cosh 10x$ **20.** $y = \operatorname{sech} 8x$

21. $y = \tanh \sqrt{x}$ **22.** $y = \operatorname{csch} \dfrac{1}{x}$

23. $y = \operatorname{sech}(3x - 1)^2$ **24.** $y = \sinh e^{x^2}$

25. $y = \coth(\cosh 3x)$ **26.** $y = \tanh(\sinh x^3)$

27. $y = \sinh 2x \cosh 3x$ **28.** $y = \operatorname{sech} x \coth 4x$

29. $y = x \cosh x^2$ **30.** $y = \dfrac{\sinh x}{x}$

31. $y = \sinh^3 x$ **32.** $y = \cosh^4 \sqrt{x}$

------

*Recall the area of a circular sector is $A = \frac{1}{2}r^2\theta$, where θ is measured in radians. Set $r = 1$ and $\theta = t$.

33. $f(x) = (x - \cosh x)^{2/3}$

34. $f(x) = \sqrt{4 + \tanh 6x}$

35. $f(x) = \ln(\cosh 4x)$

36. $f(x) = (\ln (\operatorname{sech} x))^2$

37. $f(x) = \dfrac{e^x}{1 + \cosh x}$

38. $f(x) = \dfrac{\ln x}{x^2 + \sinh x}$

39. $F(t) = e^{\sinh t}$

40. $H(t) = e^t e^{\operatorname{csch} t^2}$

41. $g(t) = \dfrac{\sin^{-1} t}{1 + \sinh 2t}$

42. $w(t) = \dfrac{\tanh t}{(1 + \cosh t)^2}$

43. Find an equation of the tangent line to the graph of $y = \sinh 3x$ at $x = 0$.

44. Find an equation of the tangent line to the graph of $y = \cosh x$ at $x = 1$.

In Problems 45–62 evaluate the given integral.

45. $\displaystyle\int \sinh 8x \, dx$

46. $\displaystyle\int \left(x^2 + \cosh\frac{x}{6}\right) dx$

47. $\displaystyle\int \cosh(5x - 4) \, dx$

48. $\displaystyle\int x \sinh(1 - x^2) \, dx$

49. $\displaystyle\int x^2 \operatorname{sech}^2 x^3 \, dx$

50. $\displaystyle\int \dfrac{\operatorname{csch}^2 \sqrt{x}}{\sqrt{x}} \, dx$

51. $\displaystyle\int \dfrac{\operatorname{csch} \sqrt[3]{x} \coth \sqrt[3]{x}}{(\sqrt[3]{x})^2} \, dx$

52. $\displaystyle\int \operatorname{sech} 2x \tanh 2x \, dx$

53. $\displaystyle\int \cosh^2 x \sinh x \, dx$

54. $\displaystyle\int \sqrt{1 + \sinh 2x} \cosh 2x \, dx$

55. $\displaystyle\int \dfrac{\sinh 5x}{7 + \cosh 5x} \, dx$

56. $\displaystyle\int x \coth x^2 \, dx$

57. $\displaystyle\int e^{-\cosh 3x} \sinh 3x \, dx$

58. $\displaystyle\int \dfrac{e^{\tanh x}}{\cosh^2 x} \, dx$

59. $\displaystyle\int (\cosh^2 x - 1)^3 \cosh x \, dx$

60. $\displaystyle\int \tanh x \operatorname{sech}^2 x \, dx$

61. $\displaystyle\int e^x \cosh e^x \, dx$

62. $\displaystyle\int e^x \cosh x \, dx$

63. Find the area under the graph of $y = \cosh x$ on the interval $[-1, 1]$.

64. Find the area of the region that is bounded by the graph of $y = \sinh x$ and the x-axis on $[-1, 1]$.

65. Find the area of the region that is bounded by the graphs of $y = \cosh x$, $y = x$, $x = -1$, and $x = 3$.

66. Find the volume of the solid of revolution that is formed by revolving the region bounded by the graphs of $y = \operatorname{sech} x$, $y = 0$, $x = 0$, and $x = 1$ about the x-axis.

67. Find the volume of the solid of revolution that is formed by revolving the region bounded by the graphs of $y = \sinh x^2$, $y = 0$, and $x = \sqrt{3}$ about the y-axis.

68. Find the length of the graph of $y = \cosh x$ on the interval $[0, 2]$.

69. Find the volume of the solid of revolution that is formed by revolving the region bounded by the graphs of $y = \cosh x$, $y = 0$, $x = a$, and $x = b$ about the x-axis. (*Hint:* See (8.72).)

70. The surface generated by revolving the catenary in Problem 69 about the x-axis is called a **catenoid**. Of all the surfaces generated on $a \leq x \leq b$ by revolving curves in this manner about the x-axis, the catenoid has the least surface area. Find the surface area of the catenoid.

In Problems 71 and 72 evaluate the given function.

71. $\cosh(\ln x)$

72. $\sinh(\ln x)$

In Problems 73 and 74 find d^2y/dx^2 for the given function.

73. $y = \tanh x$

74. $y = \operatorname{sech} x$

75. Verify that $y = C_1 \cosh kx + C_2 \sinh kx$ satisfies the equation $y'' - k^2 y = 0$ for any constants C_1 and C_2.

Miscellaneous Problems

76. Prove IV in the form

$$\frac{d}{dx} \coth x = -\operatorname{csch}^2 x$$

77. Prove V in the form

$$\frac{d}{dx} \operatorname{sech} x = -\operatorname{sech} x \tanh x$$

78. Prove VI in the form

$$\frac{d}{dx} \operatorname{csch} x = -\operatorname{csch} x \coth x$$

79. Show that the area of the shaded region in Figure 8.24 is given by

$$A(t) = \frac{1}{2} \cosh t \sinh t - \int_1^{\cosh t} \sqrt{x^2 - 1} \, dx$$

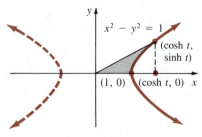

Figure 8.24

80. (a)   Use the result given in Problem 79 to show that $A(0) = 0$ and $A'(t) = \frac{1}{2}$.

(b)   Use part **(a)** to show $A(t) = \frac{1}{2}t$.

81.   Show that for any positive integer n,

$$(\cosh x + \sinh x)^n = \cosh nx + \sinh nx$$

8.9   The Inverse Hyperbolic Functions

For any real number y in the range of the hyperbolic sine, there corresponds only one real number x in its domain. In other words, $y = \sinh x$ is a one-to-one function and, hence, has an inverse function that is written $y = \sinh^{-1}x$. As in our earlier discussion of the inverse trigonometric functions in Section 7.2, this latter notation is equivalent to $x = \sinh y$.

The graphs of $y = \sinh x$ and $y = \sinh^{-1}x$ are compared in Figure 8.25.

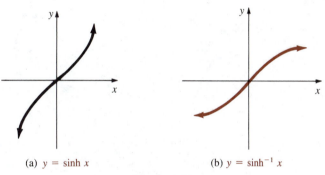

(a) $y = \sinh x$ (b) $y = \sinh^{-1} x$

Figure 8.25

The hyperbolic cosine is not a one-to-one function and therefore does not possess an inverse function unless its domain is restricted. Inspection of Figure 8.26 shows that when the domain of $y = \cosh x$ is restricted to the values $x \geq 0$ ($y \geq 1$), its inverse function $y = \cosh^{-1}x$ is defined for $x \geq 1$ ($y \geq 0$).

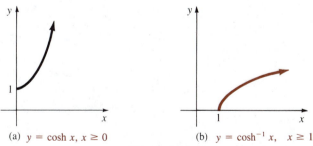

(a) $y = \cosh x,\ x \geq 0$ (b) $y = \cosh^{-1} x,\ \ x \geq 1$

Figure 8.26

Definitions of the six inverse hyperbolic functions, along with appropriate restrictions on the variable x, are summarized as follows.

DEFINITION 8.9   Inverse Hyperbolic Functions

(*i*) $y = \sinh^{-1} x$ if and only if $x = \sinh y$

(*ii*) $y = \cosh^{-1} x$ if and only if $x = \cosh y,$ $x \geq 1$

(*iii*) $y = \tanh^{-1} x$ if and only if $x = \tanh y,$ $|x| < 1$

(*iv*) $y = \coth^{-1} x$ if and only if $x = \coth y,$ $|x| > 1$

(*v*) $y = \operatorname{sech}^{-1} x$ if and only if $x = \operatorname{sech} y,$ $0 < x \leq 1$

(*vi*) $y = \operatorname{csch}^{-1} x$ if and only if $x = \operatorname{csch} y,$ $x \neq 0$

The graphs of the last four functions in Definition 8.9 and given in Figure 8.27.

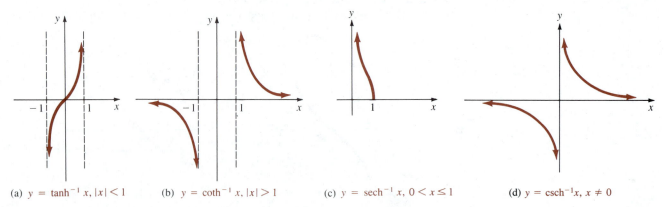

(a) $y = \tanh^{-1} x, |x| < 1$ (b) $y = \coth^{-1} x, |x| > 1$ (c) $y = \operatorname{sech}^{-1} x, 0 < x \leq 1$ (d) $y = \operatorname{csch}^{-1} x, x \neq 0$

Figure 8.27

Inverse Hyperbolic Functions as Logarithms   Since all the hyperbolic functions are defined in terms of combinations of e^x, it should not come as any surprise to find that the inverse hyperbolic functions can be expressed in terms of the natural logarithm. For example, $y = \sinh^{-1} x$ is equivalent to

$$x = \sinh y$$

so that

$$x = \frac{e^y - e^{-y}}{2}$$

$$2x = \frac{e^{2y} - 1}{e^y}$$

or

$$e^{2y} - 2xe^y - 1 = 0$$

Since this last equation is quadratic in e^y, the quadratic formula gives

$$e^y = \frac{2x \pm \sqrt{4x^2 + 4}}{2}$$

$$= x \pm \sqrt{x^2 + 1}$$

Now the solution corresponding to the minus sign must be rejected since $e^y > 0$ but $x - \sqrt{x^2 + 1} < 0$. Thus, we have

$$e^y = x + \sqrt{x^2 + 1}$$

or

$$y = \sinh^{-1}x = \ln(x + \sqrt{x^2 + 1})$$

Similarly, for $y = \tanh^{-1}x$, $|x| < 1$, then

$$x = \tanh y$$

$$= \frac{e^y - e^{-y}}{e^y + e^{-y}}$$

$$e^y(1 - x) = (1 + x)e^{-y}$$

$$e^{2y} = \frac{1 + x}{1 - x}$$

$$2y = \ln\left(\frac{1 + x}{1 - x}\right)$$

or

$$y = \tanh^{-1}x = \frac{1}{2}\ln\left(\frac{1 + x}{1 - x}\right)$$

We have proved two of the results in the next theorem.

THEOREM 8.6

(i) $\qquad$ $\sinh^{-1}x = \ln(x + \sqrt{x^2 + 1})$

(ii) $\qquad$ $\cosh^{-1}x = \ln(x + \sqrt{x^2 - 1})$, $\qquad x \geq 1$

(iii) $\qquad$ $\tanh^{-1}x = \frac{1}{2}\ln\left(\frac{1 + x}{1 - x}\right)$, $\qquad |x| < 1$

(iv) $\qquad$ $\coth^{-1}x = \frac{1}{2}\ln\left(\frac{x + 1}{x - 1}\right)$, $\qquad |x| > 1$

(v) $\qquad$ $\operatorname{sech}^{-1}x = \ln\left(\frac{1}{x} + \frac{\sqrt{1 - x^2}}{x}\right)$, $\qquad 0 < x \leq 1$

(vi) $\qquad$ $\operatorname{csch}^{-1}x = \ln\left(\frac{1}{x} + \frac{\sqrt{1 + x^2}}{|x|}\right)$, $\qquad x \neq 0$

Derivatives $\quad$ To find the derivative of an inverse hyperbolic function, we can proceed in two different ways. For example, if

$$y = \sinh^{-1}x$$

then by definition

$$x = \sinh y$$

Using implicit differentiation, we can write

$$\frac{d}{dx}(x) = \frac{d}{dx}(\sinh\ y)$$

$$1 = \cosh\ y\frac{dy}{dx}$$

$$\frac{dy}{dx} = \frac{1}{\cosh\ y}$$

$$= \frac{1}{\sqrt{\sinh^2 y + 1}}$$

$$= \frac{1}{\sqrt{x^2 + 1}}$$

On the other hand, we know from Theorem 8.6(*i*) that

$$y = \ln(x + \sqrt{x^2 + 1})$$

Therefore, from the derivative of the logarithm, we obtain

$$\frac{dy}{dx} = \frac{1}{x + \sqrt{x^2 + 1}}\left(1 + \frac{1}{2}(x^2 + 1)^{-1/2} \cdot 2x\right)$$

$$= \frac{1}{x + \sqrt{x^2 + 1}}\frac{\sqrt{x^2 + 1} + x}{\sqrt{x^2 + 1}}$$

$$= \frac{1}{\sqrt{x^2 + 1}}$$

We have proved a special case of part I in the next result. The proofs of the remaining parts are left as exercises. The function $u = g(x)$ is differentiable.

$$\text{I}\quad \frac{d}{dx}\sinh^{-1}u = \frac{1}{\sqrt{u^2 + 1}}\frac{du}{dx}$$

$$\text{II}\quad \frac{d}{dx}\cosh^{-1}u = \frac{1}{\sqrt{u^2 - 1}}\frac{du}{dx},\qquad u > 1$$

$$\text{III}\quad \frac{d}{dx}\tanh^{-1}u = \frac{1}{1 - u^2}\frac{du}{dx},\qquad |u| < 1$$

$$\text{IV}\quad \frac{d}{dx}\coth^{-1}u = \frac{1}{1 - u^2}\frac{du}{dx},\qquad |u| > 1$$

$$\text{V}\quad \frac{d}{dx}\text{sech}^{-1}u = \frac{-1}{u\sqrt{1 - u^2}}\frac{du}{dx},\qquad 0 < u < 1$$

$$\text{VI}\quad \frac{d}{dx}\text{csch}^{-1}u = \frac{-1}{|u|\sqrt{1 + u^2}}\frac{du}{dx},\qquad u \neq 0$$

Example 1

Differentiate $y = \cosh^{-1}(x^2 + 5)$.

Solution From II,

$$\frac{dy}{dx} = \frac{1}{\sqrt{(x^2 + 5)^2 - 1}} \cdot 2x$$

$$= \frac{2x}{\sqrt{x^4 + 10x^2 + 24}}$$

Example 2

Differentiate $y = \tanh^{-1}4x$.

Solution From III,

$$\frac{dy}{dx} = \frac{1}{1 - (4x)^2} \cdot 4$$

$$= \frac{4}{1 - 16x^2}$$

Example 3

Differentiate $y = e^{x^2} \operatorname{sech}^{-1}x$.

Solution By the Product Rule and V, we have

$$\frac{dy}{dx} = e^{x^2}\left(\frac{-1}{x\sqrt{1 - x^2}}\right) + 2xe^{x^2}\operatorname{sech}^{-1}x$$

$$= -\frac{e^{x^2}}{x\sqrt{1 - x^2}} + 2xe^{x^2}\operatorname{sech}^{-1}x$$

Integrals   As in Section 7.3, we limit ourselves to three integral formulas. From I–IV it is readily shown that when a is a positive constant

$$\text{I}' \quad \int \frac{du}{\sqrt{u^2 + a^2}} = \sinh^{-1}\frac{u}{a} + C$$

$$\text{II}' \quad \int \frac{du}{\sqrt{u^2 - a^2}} = \cosh^{-1}\frac{u}{a} + C, \qquad u > a > 0$$

$$\text{III}' \quad \int \frac{du}{a^2 - u^2} = \begin{cases} \dfrac{1}{a}\tanh^{-1}\dfrac{u}{a} + C, & |u| < a \\[2mm] \dfrac{1}{a}\coth^{-1}\dfrac{u}{a} + C, & |u| > a \end{cases}$$

Example 4

Evaluate $\displaystyle\int \frac{dx}{\sqrt{3x^2 + 1}}$.

Solution If we identify

$$a = 1, \quad u = \sqrt{3}x, \quad \text{and} \quad du = \sqrt{3}\, dx$$

it then follows from I′ that

$$\int \frac{dx}{\sqrt{3x^2 + 1}} = \frac{1}{\sqrt{3}} \int \frac{\sqrt{3}\, dx}{\sqrt{(\sqrt{3}x)^2 + 1}}$$

$$= \frac{1}{\sqrt{3}} \int \frac{du}{\sqrt{u^2 + 1}}$$

$$= \frac{1}{\sqrt{3}} \sinh^{-1} u + C$$

$$= \frac{1}{\sqrt{3}} \sinh^{-1} \sqrt{3}x + C$$

Example 5

Evaluate $\displaystyle\int \frac{dx}{9 - x^2}$.

Solution With $a = 3$ and $u = x$, we obtain from III′

$$\int \frac{dx}{9 - x^2} = \begin{cases} \dfrac{1}{3} \tanh^{-1} \dfrac{x}{3} + C, & |x| < 3 \\[2mm] \dfrac{1}{3} \coth^{-1} \dfrac{x}{3} + C, & |x| > 3 \end{cases}$$

Sometimes we would like to write the integral formulas I′–III′ in terms of logarithms rather than inverse hyperbolic functions. With the aid of parts (*i*), (*ii*), (*iii*), and (*iv*) of Theorem 8.6, it follows that when a is a positive constant

$$\int \frac{du}{\sqrt{u^2 + a^2}} = \ln(u + \sqrt{u^2 + a^2}) + C \tag{8.74}$$

$$\int \frac{du}{\sqrt{u^2 - a^2}} = \ln(u + \sqrt{u^2 - a^2}) + C, \qquad u > a > 0 \tag{8.75}$$

$$\int \frac{du}{a^2 - u^2} = \begin{cases} \dfrac{1}{2a} \ln\left(\dfrac{a + u}{a - u}\right) + C, & |u| < a \\[3mm] \dfrac{1}{2a} \ln\left(\dfrac{u + a}{u - a}\right) + C, & |u| > a \end{cases} \tag{8.76}$$

Note that the last integral can be written compactly as

$$\int \frac{du}{a^2 - u^2} = \frac{1}{2a} \ln\left|\frac{a + u}{a - u}\right| + C, \qquad |u| \neq a \qquad (8.77)$$

Example 6

In view of (8.77) the result in Example 5 can be written as

$$\int \frac{dx}{9 - x^2} = \frac{1}{6} \ln\left|\frac{3 + x}{3 - x}\right| + C$$

Example 7

Find the area A under the graph of $y = 1/\sqrt{x^2 + 1}$ on the interval $[0, 2]$.

Solution   Figure 8.28 shows the area in question. Since the function is non-negative, it follows immediately from (8.74) that

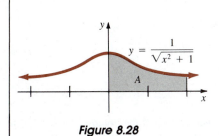

$$y = \frac{1}{\sqrt{x^2 + 1}}$$

Figure 8.28

$$A = \int_0^2 \frac{dx}{\sqrt{x^2 + 1}}$$

$$= \ln(x + \sqrt{x^2 + 1})\Big]_0^2$$

$$= \ln(2 + \sqrt{5}) \approx 1.4436 \text{ square units}$$

Remark

In Section 1.6, we pointed out that functions can be categorized as either algebraic or transcendental. The functions studied in Chapters 7 and 8:

- trigonometric, inverse trigonometric,
- logarithmic, exponential, hyperbolic, inverse hyperbolic, and the
- power function $f(x) = x^r$, r irrational,

are transcendental since they cannot be constructed out of real numbers and a variable x by a finite number of additions, subtractions, multiplications, divisions, and extraction of roots. Note that $y = x^{1/2}$ is an algebraic function but $y = x^{\pi}$ is transcendental.

Exercises 8.9

In Problems 1–20 find the derivative of the given function.

1. $y = \sinh^{-1}3x$

2. $y = \cosh^{-1}\dfrac{x}{2}$

3. $y = \tanh^{-1}(1 - x^2)$

4. $y = \coth^{-1}\dfrac{1}{x}$

5. $y = \coth^{-1}(\csc x)$

6. $y = \sinh^{-1}(\sin x)$

7. $y = x \sinh^{-1}x^3$

8. $y = x^2\operatorname{csch}^{-1}x$

9. $y = \dfrac{\operatorname{sech}^{-1}x}{x}$

10. $y = \dfrac{\coth^{-1}e^{2x}}{e^{2x}}$

11. $y = \ln(\text{sech}^{-1}x)$

12. $y = \ln|\text{csch}^{-1}x|$

13. $y = (\cosh^{-1}6x)^{1/2}$

14. $y = \dfrac{1}{(\tanh^{-1}2x)^3}$

15. $y = \sqrt{x^2 - 1} + \cosh^{-1}x$

16. $y = \dfrac{1}{1 - x^2} + \coth^{-1}4x$

17. $y = \sinh^{-1}(\cosh x)$

18. $y = \sinh(\tanh^{-1}x)$

19. $y = \ln\sqrt{1 - x^2} + x\coth^{-1}x$

20. $y = -\dfrac{1}{3}\sqrt{9x^2 - 1} + x\cosh^{-1}3x$

In Problems 21–40 evaluate the given integral.

21. $\displaystyle\int \frac{dx}{\sqrt{4x^2 + 1}}$

22. $\displaystyle\int \frac{dx}{\sqrt{x^2 + 4}}$

23. $\displaystyle\int \frac{dx}{1 - 4x^2}$

24. $\displaystyle\int \frac{dx}{4 - x^2}$

25. $\displaystyle\int \frac{dx}{\sqrt{9x^2 - 16}}$

26. $\displaystyle\int \frac{dx}{\sqrt{25x^2 + 9}}$

27. $\displaystyle\int \frac{x\,dx}{\sqrt{x^4 + 1}}$

28. $\displaystyle\int \frac{x\,dx}{1 - x^4}$

29. $\displaystyle\int \frac{dx}{x^2 + 2x}$

30. $\displaystyle\int \frac{dx}{\sqrt{x^2 + 2x}}$

31. $\displaystyle\int \frac{dx}{\sqrt{x^2 + 4x + 5}}$

32. $\displaystyle\int \frac{dx}{-x^2 + 6x - 8}$

33. $\displaystyle\int \frac{e^x}{\sqrt{e^{2x} - 1}}\,dx$

34. $\displaystyle\int \frac{dx}{e^{-x} - e^x}$

35. $\displaystyle\int \frac{\cos\theta}{\sqrt{1 + \sin^2\theta}}\,d\theta$

36. $\displaystyle\int \frac{\sin t}{\sqrt{2 - \sin^2 t}}\,dt$

37. $\displaystyle\int \frac{t}{\sqrt{t^2 + 16}}\,dt$

38. $\displaystyle\int \frac{r}{1 - r^2}\,dr$

39. $\displaystyle\int \frac{x^2 + 1}{x^2 - 1}\,dx$

40. $\displaystyle\int \frac{2 + x}{\sqrt{5x^2 + 1}}\,dx$

In Problems 41 and 42 express the antiderivative of the integral as a natural logarithm. Find the value of the integral.

41. $\displaystyle\int_{-1}^{3} \frac{dx}{\sqrt{x^2 + 1}}$

42. $\displaystyle\int_{0}^{4} \frac{dt}{\sqrt{t^2 + 9}}$

43. Find the area under the graph of $y = 1/(1 - x^2)$ on the interval $[-\frac{3}{4}, 0]$.

44. Find the area bounded by the graph of $y = 1/(1 - x^2)$ and the x-axis on $[2, 5]$.

45. Consider the graph of the function $y = 1/\sqrt{1 - x^2}$ shown in Figure 8.29.

(a) Find the area under the graph on the interval $[\frac{1}{2}, \sqrt{3}/2]$.

(b) The region bounded by the graphs of $y = 1/\sqrt{1 - x^2}$, $y = 0$, $x = \frac{1}{2}$, and $x = \sqrt{3}/2$ is revolved about the y-axis. Find the volume of the solid of revolution.

(c) The region in part (b) is revolved about the x-axis. Find the volume of the solid of revolution.

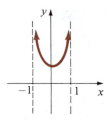

Figure 8.29

46. Use the shell method to find the volume of the solid of revolution that is formed by revolving the region bounded by the graphs of $y = 1/(1 - x^4)$, $x = 0$, $x = \frac{1}{2}$, and $y = 0$ about the y-axis.

47. The differential equation that describes the shape of a wire of constant density w hanging under its own weight is

$$\frac{d^2y}{dx^2} = \frac{w}{T}\sqrt{1 + \left(\frac{dy}{dx}\right)^2}$$

where T is a constant. Use separation of variables and the substitution $p = dy/dx$ to solve this equation subject to $y(0) = 1$, $y'(0) = 0$.

48. The velocity $v(t)$ of a body of mass m, which is falling through a viscous medium that imparts a resisting force proportional to the square of the velocity, satisfies the differential equation

$$m\frac{dv}{dt} = mg - kv^2$$

where g is the acceleration of gravity and k is a positive constant. Use separation of variables to show that a solution of this equation that satisfies $v(0) = 0$ is

$$v(t) = \sqrt{\frac{mg}{k}}\,\tanh\left(\sqrt{\frac{gk}{m}}\,t\right)$$

What is $\lim_{t\to\infty} v(t)$? (*Hint:* See Figure 8.20(a).)

Miscellaneous Problems

49. Prove II in the form

$$\frac{d}{dx}\cosh^{-1}x = \frac{1}{\sqrt{x^2-1}}$$

50. Prove III in the form

$$\frac{d}{dx}\tanh^{-1}x = \frac{1}{1-x^2}$$

51. Prove IV in the form

$$\frac{d}{dx}\coth^{-1}x = \frac{1}{1-x^2}$$

52. Prove V in the form

$$\frac{d}{dx}\operatorname{sech}^{-1}x = \frac{-1}{x\sqrt{1-x^2}}$$

53. Prove VI in the form

$$\frac{d}{dx}\operatorname{csch}^{-1}x = \frac{-1}{|x|\sqrt{1+x^2}}$$

54. Use I′ and Theorem 8.6(i) to prove (8.74).

55. Use II′ and Theorem 8.6(ii) to prove (8.75).

56. Use III′ and Theorems 8.6(iii) and 8.6(iv) to prove (8.76).

57. Use the identity $\dfrac{1}{1-x^2} = \dfrac{\frac{1}{2}}{1+x} + \dfrac{\frac{1}{2}}{1-x}$ to evaluate $\int dx/(1-x^2)$.

58. Use the form $\int du/\sqrt{u^2+1}$ with $u = \tan\theta$ to evaluate $\int \sec\theta\,d\theta$ in terms of a natural logarithm.

Chapter 8 Review Exercises

Answers to odd-numbered problems begin on page A-35

In Problems 1–20 answer true or false.

1. The natural logarithmic function is an antiderivative of x^{-1}. _____

2. $\dfrac{d}{dx}\log_{10}x = \dfrac{1}{x}$ _____

3. $\dfrac{d}{dx}2^x = x2^{x-1}$ _____

4. $\displaystyle\int_1^3 \frac{dx}{x} = \ln 3$ _____

5. $\displaystyle\int 10^\pi\,dx = \frac{10^{\pi+1}}{\pi+1} + C$ _____

6. The inverse of $y = e^x$ is $y = \ln x$. _____

7. If $\ln x = 1$, then $x = e$. _____

8. If $0 < a < b$, then $\ln a < \ln b$. _____

9. For $a > 0$ and $b > 0$, $\ln(a+b) = \ln a + \ln b$. _____

10. If x is an irrational number and $a > 0$, then $\ln a^x = x\ln a$. _____

11. $\ln x^2 = 2\ln x$ for all x. _____

12. $e^{\sqrt{x}} = \sqrt{e^x}$ for all x. _____

13. $\exp(x+y) = (\exp x)\cdot(\exp y)$ _____

14. If $f(x) = \ln|x+1|$ and $g(x) = \ln(x+1)$, then $f'(x) = g'(x) = 1/(x+1)$ for all $x \neq -1$. _____

15. $\dfrac{d}{dx}\cosh x = -\sinh x$ _____

16. The hyperbolic cosine is an even function. _____

17. The hyperbolic cosine is never negative. _____

18. $\cosh^2 x + \sinh^2 x = 1$ _____

19. The hyperbolic sine function is one-to-one. _____

20. Every inverse hyperbolic function can be expressed as a logarithm. _____

In Problems 21–30 fill in the blank.

21. $\ln e^3 =$ _____

22. $\ln(\ln e) =$ _____

23. $\ln\dfrac{e^a}{e^b} =$ _____

24. $e^{-2\ln 3} =$ _____

25. The slope of the tangent line to the graph of $y = \ln x$ at $x = \frac{1}{2}$ is _____ .

26. For $f(x) = \ln|2x-4|$, the domain of f' is _____ .

27. To differentiate $f(x) = x^x$, we use a process called _____ .

28. If $e^{-x} = 5$, then $x = $ _____.

29. The graph of the hyperbolic cosine is called a _____.

30. The value of $\cosh^{-1}1$ is _____.

In Problems 31–50 find dy/dx.

31. $y = \ln(x\sqrt{4x - 1})$

32. $y = (\ln \cos^2 x)^2$

33. $y = \sin(e^x \ln x)$

34. $y = x^3 \log_3 \sqrt{x}$

35. $y = e^x + e^{2x}$

36. $y = (e + e^2)^x$

37. $y = x^7 + 7^x + 7^\pi$

38. $y = (e^x + 1)^{-e}$

39. $y = (x^2 + e^{2x})^{\ln x}$

40. $y = 5^{x^2} x^{\sin x}$

41. $xy^2 = e^x - e^y$

42. $y = \ln xy$

43. $y = \sinh e^{x^3}$

44. $y = (\tanh 5x)^{-1}$

45. $y = \cosh(\cosh x)$

46. $y = (\text{sech } \sqrt{2}x)^{\sqrt{2}}$

47. $y = \sinh^{-1}(\sin^{-1}x)$

48. $y = (\tan^{-1}x)(\tanh^{-1}x)$

49. $y = xe^{x\cosh^{-1}x}$

50. $y = \sinh^{-1}\sqrt{x^2 - 1}$

In Problems 51–70 evaluate the given integral.

51. $\displaystyle\int \frac{x}{1 - 5x^2}\, dx$

52. $\displaystyle\int_1^2 \left[\frac{1}{x} - \frac{1}{x + 1}\right] dx$

53. $\displaystyle\int \frac{x + 1}{x(x + 2)}\, dx$

54. $\displaystyle\int \frac{3x^2 - 4x + 2}{x}\, dx$

55. $\displaystyle\int_0^1 (x^7 + 7^x + 7^\pi)\, dx$

56. $\displaystyle\int 3^{\cos x}\sin x\, dx$

57. $\displaystyle\int \cot 4x\, dx$

58. $\displaystyle\int x^2 \sec x^3\, dx$

59. $\displaystyle\int_2^5 \frac{dx}{e^{3x}}$

60. $\displaystyle\int \frac{dx}{e^{-x}(4 + e^x)}$

61. $\displaystyle\int 2^t 3^t 4^t\, dt$

62. $\displaystyle\int 4^{e^\theta} e^\theta\, d\theta$

63. $\displaystyle\int \frac{\sinh(1/x)}{x^2}\, dx$

64. $\displaystyle\int \frac{\cosh(\ln 2x)}{x}\, dx$

65. $\displaystyle\int \frac{\cosh 3x}{\sinh^4 3x}\, dx$

66. $\displaystyle\int \frac{\text{sech}^2 x}{1 + \tanh x}\, dx$

67. $\displaystyle\int \frac{dx}{-x^2 + 14x - 45}$

68. $\displaystyle\int \frac{6t^3}{2t + 1}\, dt$

69. $\displaystyle\int_0^1 \frac{dx}{\sqrt{9x^2 + 1}}$

70. $\displaystyle\int_{2e}^{3e} \frac{dx}{x\sqrt{(\ln x)^2 - 1}}$

71. Sketch the graph of $y = \ln(x^2 - 1)$.

72. Find the area under the graph of $y = 1/\sqrt{x^2 - 1}$ on the interval $[2, 5]$.

73. The region in the first quadrant bounded by the graphs of $y = e^x$, $y = e^{-x}$, and $x = \ln 6$ is revolved about the x-axis. Find the volume of the solid of revolution.

74. Find the point on the graph of $y = \ln 2x$ such that the tangent line passes through the origin.

In Problems 75–78 solve the given differential equation by separation of variables.

75. $\cos x \dfrac{dy}{dx} + (\sin x)y = 0$

76. $x\, dx - 3y^2(x + 2)\, dy = 0$

77. $\dfrac{dy}{dx} = 4 - y^2$

78. $(e^y + 1)^2 e^{-y}\, dx + (e^x + 1)^3 e^{-x}\, dy = 0$

79. Carbon 14 decays at a rate proportional to the amount present at any time. The half-life of C-14 is known to be 5600 years. Determine the age of a fossilized bone that contains $\frac{1}{1000}$ of its original amount of C-14.

80. A metal bar is taken out of a furnace whose temperature is 150°C and put into a tank of water whose temperature is maintained at a constant 30°C. After $\frac{1}{4}$ hour in the tank, the temperature of the bar is 90°C. What is the temperature of the bar in $\frac{1}{2}$ hour? In 1 hour?

9

Techniques of Integration

Often one encounters an integral that cannot be categorized as a general form such as $\int u^n \, du$ or $\int e^u \, du$. For example, it is not possible to evaluate $\int x^2 \sqrt{x+1} \, dx$ by an immediate application of any one of the formulas listed on the following page. However, by applying a *technique of integration*, an integral such as this can sometimes be reduced to one or more of these familiar forms.

A Review of Integration Formulas

$$\int u^n \, du = \frac{u^{n+1}}{n+1} + C, \qquad n \neq -1$$

$$\int \frac{du}{u} = \ln|u| + C$$

$$\int e^u \, du = e^u + C$$

$$\int a^u \, du = \frac{a^u}{\ln a} + C$$

$$\int \cos u \, du = \sin u + C$$

$$\int \sin u \, du = -\cos u + C$$

$$\int \sec^2 u \, du = \tan u + C$$

$$\int \csc^2 u \, du = -\cot u + C$$

$$\int \sec u \tan u \, du = \sec u + C$$

$$\int \csc u \cot u \, du = -\csc u + C$$

$$\int \cosh u \, du = \sinh u + C$$

$$\int \sinh u \, du = \cosh u + C$$

$$\int \tan u \, du = -\ln|\cos u| + C$$

$$\int \cot u \, du = \ln|\sin u| + C$$

$$\int \sec u \, du = \ln|\sec u + \tan u| + C$$

$$\int \csc u \, du = \ln|\csc u - \cot u| + C$$

$$\int \frac{du}{\sqrt{a^2 - u^2}} = \sin^{-1}\frac{u}{a} + C$$

$$\int \frac{du}{a^2 + u^2} = \frac{1}{a}\tan^{-1}\frac{u}{a} + C$$

$$\int \frac{du}{u\sqrt{u^2 - a^2}} = \frac{1}{a}\sec^{-1}\frac{u}{a} + C$$

$$\int \frac{du}{\sqrt{u^2 + a^2}} = \sinh^{-1}\frac{u}{a} + C$$

$$\int \frac{du}{\sqrt{u^2 - a^2}} = \cosh^{-1}\frac{u}{a} + C$$

$$\int \frac{du}{u^2 - a^2} = \begin{cases} \dfrac{1}{a}\tanh^{-1}\dfrac{u}{a} + C, & |u| < a \\[2ex] \dfrac{1}{a}\coth^{-1}\dfrac{u}{a} + C, & |u| > a \end{cases}$$

9.1 Algebraic Substitutions

Throughout Chapters 5, 6, 7, and 8, we used a substitution $u = g(x)$ on many occasions to evaluate an integral. For example, $\int e^{x^2} x \, dx$ is recognized as $(\frac{1}{2}) \int e^u \, du$ when $u = x^2$. We now extend the idea of the u-substitution to integrals that are not of the precise form $\int f(g(x))g'(x) \, dx$.

Example 1

Evaluate $\displaystyle\int x^2 \sqrt{x+1} \, dx$.

Solution If we let

$$u = x + 1, \quad \text{then} \quad x = u - 1, \qquad dx = du$$

and

$$x^2 = (u-1)^2 = u^2 - 2u + 1$$

$$\sqrt{x+1} = u^{1/2}$$

Thus,

$$\int x^2\sqrt{x+1}\,dx = \int (u^2 - 2u + 1)u^{1/2}\,du$$

$$= \int (u^{5/2} - 2u^{3/2} + u^{1/2})\,du$$

$$= \frac{2}{7}u^{7/2} - \frac{4}{5}u^{5/2} + \frac{2}{3}u^{3/2} + C$$

$$= \frac{2}{7}(x+1)^{7/2} - \frac{4}{5}(x+1)^{5/2} + \frac{2}{3}(x+1)^{3/2} + C$$

You should verify that the derivative of the last line actually is $x^2\sqrt{x+1}$.

The choice of which, if any, substitution to use is not always obvious. Generally, if the integrand contains a power of a function, then it is a good idea to try to let u be that function *or the power of the function itself.* In Example 1, the alternative substitution $u = \sqrt{x+1}$ or $u^2 = x + 1$ leads to the different integral $2\int (1 - u^2)^2 u^2\,du$. The latter can be evaluated by expanding the integrand and integrating each term.

Example 2

Evaluate $\displaystyle\int \frac{dx}{x + \sqrt{x}}$.

Solution Let

$$u = \sqrt{x} \quad \text{so that} \quad x = u^2 \quad \text{and} \quad dx = 2u\,du$$

Hence,

$$\int \frac{dx}{x + \sqrt{x}} = \int \frac{2u\,du}{u^2 + u}$$

$$= \int \frac{2\,du}{u + 1}$$

$$= 2\ln|u + 1| + C$$

$$= 2\ln(\sqrt{x} + 1) + C$$

Integrands Containing a Quadratic Expression

As we saw in Chapters 7 and 8, if an integrand contains a quadratic expression, $ax^2 + bx + c$, *completion of the square* may lead to an integral that can be expressed as an inverse trigonometric function or an inverse hyperbolic function. Of course, more complicated integrals can yield other functions as well.

Example 3

Evaluate $\displaystyle\int \frac{x+4}{x^2 + 6x + 18}\,dx$.

Solution   After completing the square, the given integral can be written as

$$\int \frac{x+4}{x^2 + 6x + 18}\,dx = \int \frac{x+4}{(x+3)^2 + 9}\,dx$$

Now, if $u = x + 3$, then $x = u - 3$ and $dx = du$. Therefore,

$$\int \frac{x + 4}{x^2 + 6x + 18}\, dx = \int \frac{u + 1}{u^2 + 9}\, du$$

$$= \int \frac{u}{u^2 + 9}\, du + \int \frac{du}{u^2 + 9}$$

$$= \frac{1}{2} \int \frac{2u}{u^2 + 9}\, du + \int \frac{du}{u^2 + 9}$$

$$= \frac{1}{2} \ln(u^2 + 9) + \frac{1}{3} \tan^{-1}\frac{u}{3} + C$$

$$= \frac{1}{2} \ln[(x + 3)^2 + 9] + \frac{1}{3} \tan^{-1}\frac{x + 3}{3} + C$$

$$= \frac{1}{2} \ln(x^2 + 6x + 18) + \frac{1}{3} \tan^{-1}\frac{x + 3}{3} + C$$

Example 4

Evaluate $\displaystyle\int_0^2 \frac{6x + 1}{\sqrt[3]{3x + 2}}\, dx$.

Solution   If

$$u = 3x + 2 \quad \text{then} \quad x = \frac{1}{3}(u - 2) \quad \text{and} \quad dx = \frac{1}{3}\, du$$

$$6x + 1 = 2(u - 2) + 1 = 2u - 3$$

$$\sqrt[3]{3x + 2} = u^{1/3}$$

Now, observe that when $x = 0$, $u = 2$, and when $x = 2$, $u = 8$. Therefore, in terms of integrating on the variable u, we obtain

$$\int_0^2 \frac{6x + 1}{\sqrt[3]{3x + 2}}\, dx = \int_2^8 \frac{2u - 3}{u^{1/3}} \frac{1}{3}\, du$$

$$= \int_2^8 \left(\frac{2}{3}u^{2/3} - u^{-1/3}\right) du$$

$$= \left(\frac{2}{5}u^{5/3} - \frac{3}{2}u^{2/3}\right)\Bigg]_2^8$$

$$= \left(\frac{2}{5} \cdot 2^5 - \frac{3}{2} \cdot 2^2\right) - \left(\frac{2}{5} \cdot 2^{5/3} - \frac{3}{2} \cdot 2^{2/3}\right)$$

$$= \frac{34}{5} - \frac{2}{5} \cdot 2^{5/3} + \frac{3}{2} \cdot 2^{2/3} \approx 7.9112$$

Remarks

(*i*) When working the exercises throughout this chapter, do not be overly disturbed if you do not always obtain the same answer as given in the text. Different techniques applied to the same problem can lead to answers that look different. Remember, two antiderivatives of the same function can differ at most by a constant. Try to resolve any conflicts.

(*ii*) It might also prove helpful at this point to recall that integration of the quotient of two polynomial functions, $P(x)/Q(x)$, usually begins with long division if the degree of P is greater than or equal to the degree of Q.

(*iii*) Look for problems that can be solved by previous methods.

────── **Exercises 9.1** ──────

Answers to odd-numbered problems begin on page A-35

In Problems 1–40 evaluate the given integral.

1. $\displaystyle\int x(x + 1)^3 \, dx$

2. $\displaystyle\int \frac{x^2 - 3}{(x + 1)^3} \, dx$

3. $\displaystyle\int (2x + 1)\sqrt{x - 5} \, dx$

4. $\displaystyle\int (x^2 - 1)\sqrt{2x + 1} \, dx$

5. $\displaystyle\int \frac{x}{\sqrt{x - 1}} \, dx$

6. $\displaystyle\int \frac{x^2}{\sqrt{x + 2}} \, dx$

7. $\displaystyle\int \frac{x + 3}{(3x - 4)^{3/2}} \, dx$

8. $\displaystyle\int (x^2 + x)\sqrt[3]{x + 7} \, dx$

9. $\displaystyle\int \frac{\sqrt{x}}{x + 1} \, dx$

10. $\displaystyle\int \frac{t}{\sqrt{t} + 1} \, dt$

11. $\displaystyle\int \frac{\sqrt{t} - 3}{\sqrt{t} + 1} \, dt$

12. $\displaystyle\int \frac{\sqrt{r} + 3}{r + 3} \, dr$

13. $\displaystyle\int \frac{x^3}{\sqrt[3]{x^2 + 1}} \, dx$

14. $\displaystyle\int \frac{x^5}{\sqrt[5]{x^2 + 4}} \, dx$

15. $\displaystyle\int \frac{x^2}{(x - 1)^4} \, dx$

16. $\displaystyle\int \frac{2x + 1}{(x + 7)^2} \, dx$

17. $\displaystyle\int \frac{dx}{\sqrt{x} - \sqrt[3]{x}}$ (*Hint:* Let $u = x^{1/6}$.)

18. $\displaystyle\int \frac{\sqrt[6]{t}}{\sqrt[3]{t} + 1} \, dt$

19. $\displaystyle\int \sqrt{e^x - 1} \, dx$

20. $\displaystyle\int \frac{dx}{\sqrt{e^x - 1}}$

21. $\displaystyle\int \sqrt{1 - \sqrt{v}} \, dv$

22. $\displaystyle\int \frac{\sqrt{w}}{\sqrt{1 - \sqrt{w}}} \, dw$

23. $\displaystyle\int \frac{\sqrt{1 + \sqrt{t}}}{\sqrt{t}} \, dt$

24. $\displaystyle\int \sqrt{t} \sqrt{1 + t\sqrt{t}} \, dt$

25. $\displaystyle\int \frac{2x + 7}{x^2 + 2x + 5} \, dx$

26. $\displaystyle\int \frac{6x - 1}{4x^2 + 4x + 10} \, dx$

27. $\displaystyle\int \frac{2x + 5}{\sqrt{16 - 6x - x^2}} \, dx$

28. $\displaystyle\int \frac{4x - 3}{\sqrt{11 + 10x - x^2}} \, dx$

29. $\displaystyle\int_0^1 x\sqrt{5x + 4} \, dx$

30. $\displaystyle\int_{-1}^0 x\sqrt[3]{x + 1} \, dx$

31. $\displaystyle\int_1^{16} \frac{dx}{10 + \sqrt{x}}$

32. $\displaystyle\int_4^9 \frac{\sqrt{x} - 1}{\sqrt{x} + 1} \, dx$

33. $\displaystyle\int_2^9 \frac{5x - 6}{\sqrt[3]{x - 1}} \, dx$

34. $\displaystyle\int_{-\sqrt{3}}^0 \frac{2x^3}{\sqrt{x^2 + 1}} \, dx$

35. $\displaystyle\int_0^1 (1 - \sqrt{x})^{50} \, dx$

36. $\displaystyle\int_0^4 \frac{dx}{(1 + \sqrt{x})^3}$

37. $\displaystyle\int_1^8 \frac{dx}{x^{1/3} + x^{2/3}}$

38. $\displaystyle\int_1^{64} \frac{x^{1/3}}{x^{2/3} + 2} \, dx$

39. $\displaystyle\int_0^1 x^2(1 - x)^5 \, dx$

40. $\displaystyle\int_0^6 \frac{2x + 5}{\sqrt{2x + 4}} \, dx$

41. Find the area under the graph of $y = 1/(x^{1/3} + 1)$ on the interval $[0, 1]$.

42. Find the area bounded by the graph of $y = x^3\sqrt{x + 1}$ and the x-axis on the interval $[-1, 1]$.

43. Find the volume of the solid of revolution that is formed by revolving the region bounded by the graphs of $y = 1/(\sqrt{x} + 1)$, $x = 0$, $x = 4$, and $y = 0$ about the y-axis.

44. Find the volume of the solid of revolution that is formed by revolving the region in Problem 43 about the x-axis.

45. Find the length of the graph of $y = \dfrac{4}{5}x^{5/4}$ on the interval $[0, 9]$.

9.2 Integration by Parts

Suppose $u = f(x)$ and $v = g(x)$ are differentiable functions. Then by the Product Rule, we have

$$\frac{d}{dx}[f(x)g(x)] = f(x)g'(x) + g(x)f'(x) \tag{9.1}$$

In turn, integration of (9.1)

$$f(x)g(x) = \int f(x)g'(x)\,dx + \int g(x)f'(x)\,dx$$

produces a formula

$$\int f(x)g'(x)\,dx = f(x)g(x) - \int g(x)f'(x)\,dx \tag{9.2}$$

which is extremely useful in integrating certain products. The procedure is known as **integration by parts**. The basic idea behind (9.2) is to evaluate the integral $\int f(x)g'(x)\,dx$ by means of evaluating another, and it is hoped simpler, integral $\int g(x)f'(x)\,dx$.

The formula in (9.2) is usually written in terms of the differentials $du = f'(x)\,dx$ and $dv = g'(x)\,dx$:

$$\int u\,dv = uv - \int v\,du \tag{9.3}$$

To apply this result, we start with an integration followed by a differentiation:

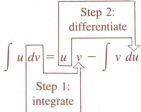

The last step is, of course, the evaluation of $\int v\,du$.

—————— **Example 1** —————————————————————————————————

Evaluate $\displaystyle\int \frac{x}{\sqrt{x+1}}\,dx$.

Solution   First, we write the integral as

$$\int x(x+1)^{-1/2}\,dx$$

In this latter form, there are several possible choices for the function dv. We could have $dv = (x+1)^{-1/2}\,dx$, $dv = x\,dx$, or simply $dv = dx$. As a practical matter, the choice of dv is dictated by what happens in the second integral in (9.3). Specifically, if we choose

$$dv = (x+1)^{-1/2}\,dx \qquad\qquad u = x$$

| integrate | | differentiate |

then $\qquad\qquad\qquad v = 2(x+1)^{1/2} \qquad\qquad du = dx$

Substituting these functions into (9.3) yields

$$\int \overbrace{x}^{u}\overbrace{(x+1)^{-1/2}\,dx}^{dv} = \overbrace{x}^{u}\overbrace{[2(x+1)^{1/2}]}^{v} - \int \overbrace{2(x+1)^{1/2}}^{v}\overbrace{dx}^{du}$$

$$= 2x(x+1)^{1/2} - 2\cdot\frac{2}{3}(x+1)^{3/2} + C$$

$$= 2x(x+1)^{1/2} - \frac{4}{3}(x+1)^{3/2} + C$$

Note that a constant is not needed in the integration of dv. The constant affixed at the finish of the problem is a "collective" constant. Furthermore, knowledge that the "right" choice for dv has been made is based on pragmatic hindsight: Did the procedure work? To see what happens when the "wrong" choice is made, consider Example 1 if this time we select

$$dv = x\,dx \qquad u = (x+1)^{-1/2}$$

$$v = \frac{1}{2}x^2 \qquad du = -\frac{1}{2}(x+1)^{-3/2}\,dx$$

Applying (9.3) in this instance gives

$$\int x(x+1)^{-1/2}\,dx = \frac{1}{2}x^2(x+1)^{-1/2} + \frac{1}{4}\int x^2(x+1)^{-3/2}\,dx$$

The problem here is apparent; the second integral $\int v\,du$ is more complicated than the original $\int u\,dv$. The alternative selection $dv = dx$ also leads to an impasse.

Example 2

Evaluate $\int x \tan^{-1}x \; dx$.

Solution   By choosing

$$dv = x \; dx \qquad u = \tan^{-1}x$$

$$v = \frac{x^2}{2} \qquad du = \frac{dx}{1 + x^2}$$

we see that (9.3) gives

$$\int \overbrace{(\tan^{-1}x)}^{u}\overbrace{(x \; dx)}^{dv} = \overbrace{(\tan^{-1}x)}^{u}\overbrace{\frac{x^2}{2}}^{v} - \int \overbrace{\frac{x^2}{2}}^{v} \cdot \overbrace{\frac{dx}{1 + x^2}}^{du}$$

To evaluate $\int x^2 \; dx/(1 + x^2)$, we use long division (see Example 7 of Section 7.3). Hence,

$$\int x \tan^{-1}x \; dx = \frac{x^2}{2}\tan^{-1}x - \frac{1}{2}\int \left(1 - \frac{1}{1 + x^2}\right) dx$$

$$= \frac{x^2}{2}\tan^{-1}x - \frac{1}{2}x + \frac{1}{2}\tan^{-1}x + C$$

Example 3

Evaluate $\int \sec^3 x \; dx$.

Solution   Inspection of the integral reveals no obvious choice for dv. However, by writing $\sec^3 x = \sec x \cdot \sec^2 x$, we can identify

$$dv = \sec^2 x \; dx \qquad u = \sec x$$

$$v = \tan x \qquad du = \sec x \tan x \; dx$$

It follows from (9.3) and a trigonometric identity that

$$\int \sec^3 x \; dx = \sec x \tan x - \int \tan^2 x \sec x \; dx$$

$$= \sec x \tan x - \int (\sec^2 x - 1)\sec x \; dx$$

$$= \sec x \tan x + \int \sec x \; dx - \int \sec^3 x \; dx$$

$$= \sec x \tan x + \ln|\sec x + \tan x| - \int \sec^3 x \; dx$$

At this point, it may appear that we are going around in circles. Actually, the problem is complete; we solve the last equation for $\int \sec^3 x \, dx$ and add a constant of integration:

$$2 \int \sec^3 x \, dx = \sec x \tan x + \ln|\sec x + \tan x|$$

$$\int \sec^3 x \, dx = \frac{1}{2}\sec x \tan x + \frac{1}{2}\ln|\sec x + \tan x| + C$$

Example 4

Evaluate $\int x^3 \ln x \, dx$.

Solution Let

$$dv = x^3 \, dx \qquad u = \ln x$$

$$v = \frac{x^4}{4} \qquad du = \frac{1}{x} \, dx$$

Integrating by parts then gives

$$\int x^3 \ln x \, dx = \frac{x^4}{4} \ln x - \frac{1}{4} \int x^4 \cdot \frac{1}{x} \, dx$$

$$= \frac{x^4}{4} \ln x - \frac{1}{4} \int x^3 \, dx$$

$$= \frac{x^4}{4} \ln x - \frac{x^4}{16} + C$$

A problem may require integration by parts several times in succession.

Example 5

Evaluate $\int x^2 e^{-x} \, dx$.

Solution Let

$$dv = e^{-x} \, dx \qquad u = x^2$$

$$v = -e^{-x} \qquad du = 2x \, dx$$

so that

$$\int x^2 e^{-x} \, dx = -x^2 e^{-x} + 2 \int x e^{-x} dx$$

In $\int x e^{-x} \, dx$, we use integration by parts a second time with

$$dv = e^{-x} \, dx \qquad u = x$$

$$v = -e^{-x} \qquad du = dx$$

Thus,
$$\int x^2 e^{-x}\, dx = -x^2 e^{-x} + 2\left[-xe^{-x} + \int e^{-x}\, dx\right]$$
$$= -x^2 e^{-x} - 2xe^{-x} - 2e^{-x} + C$$

As a rule, integrals of the type $\int x^k (\ln x)^n\, dx$, $\int x^n e^{kx}\, dx$, and $\int x^n \sin kx\, dx$, where n is a positive integer and k a constant, will require integration by parts n times.

___ **Example 6** _____

Evaluate $\displaystyle\int x \sin 3x\, dx$.

Solution   The selection

$$dv = \sin 3x\, dx \qquad u = x$$
$$v = -\frac{1}{3}\cos 3x \qquad du = dx$$

leads to
$$\int x \sin 3x\, dx = -\frac{1}{3} x \cos 3x + \frac{1}{3}\int \cos 3x\, dx$$
$$= -\frac{1}{3} x \cos 3x + \frac{1}{9}\sin 3x + C$$

___ **Example 7** _____

Evaluate $\displaystyle\int e^{2x} \cos 3x\, dx$.

Solution   Let

$$dv = e^{2x}\, dx \qquad u = \cos 3x$$
$$v = \frac{1}{2}e^{2x} \qquad du = -3 \sin 3x\, dx$$

Hence,
$$\int e^{2x} \cos 3x\, dx = \frac{1}{2}e^{2x} \cos 3x + \frac{3}{2}\int e^{2x} \sin 3x\, dx \qquad (9.4)$$

We apply integration by parts, again, in $\int e^{2x} \sin 3x\, dx$ by selecting

$$dv = e^{2x}\, dx \qquad u = \sin 3x$$
$$v = \frac{1}{2}e^{2x} \qquad du = 3 \cos 3x\, dx$$

Thus, (9.4) becomes

$$\int e^{2x} \cos 3x \, dx = \frac{1}{2}e^{2x} \cos 3x + \frac{3}{2}\left[\frac{1}{2}e^{2x} \sin 3x - \frac{3}{2}\int e^{2x} \cos 3x \, dx\right]$$

$$= \frac{1}{2}e^{2x} \cos 3x + \frac{3}{4}e^{2x} \sin 3x - \frac{9}{4}\int e^{2x} \cos 3x \, dx$$

Proceeding as in Example 3, we solve for the original integral $\int e^{2x} \cos 3x \, dx$

$$\frac{13}{4}\int e^{2x} \cos 3x \, dx = \frac{1}{2}e^{2x} \cos 3x + \frac{3}{4}e^{2x} \sin 3x$$

$$\int e^{2x} \cos 3x \, dx = \frac{2}{13}e^{2x} \cos 3x + \frac{3}{13}e^{2x} \sin 3x + C$$

You are urged to rework Example 7 using

(*i*) $dv = \cos 3x \, dx$ in the original integral and
$dv = \sin 3x \, dx$ in the second; and

(*ii*) $dv = e^{2x} \, dx$ in the original integral and
$dv = \sin 3x \, dx$ in the second.

Definite Integrals   A definite integral can be evaluated using integration by parts in the following manner:

$$\int_a^b f(x)g'(x) \, dx = f(x)g(x)\Big]_a^b - \int_a^b g(x)f'(x) \, dx$$

_____ Example 8 _____

Find the area under the graph of $y = \ln x$ on the interval $[1, e]$.

Solution   From Figure 9.1 we see that the area A is given by

$$A = \int_1^e \ln x \, dx$$

Choosing
$$dv = dx \qquad u = \ln x$$
$$v = x \qquad du = \frac{1}{x} \, dx$$

we have
$$A = x \ln x\Big]_1^e - \int_1^e x \cdot \frac{1}{x} \, dx$$

$$= x \ln x\Big]_1^e - \int_1^e dx$$

$$= x \ln x\Big]_1^e - x\Big]_1^e$$

$$= e \ln e - \ln 1 - e + 1 = 1 \text{ square unit}$$

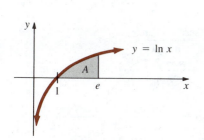

Figure 9.1

since $\ln e = 1$ and $\ln 1 = 0$.

Exercises 9.2

Answers to odd-numbered problems begin on page A-35

In Problems 1–44 evaluate the given integral using integration by parts.

1. $\displaystyle\int x\sqrt{x+3}\,dx$

2. $\displaystyle\int \frac{x}{\sqrt{2x-5}}\,dx$

3. $\displaystyle\int \ln 4x\,dx$

4. $\displaystyle\int \ln(x+1)\,dx$

5. $\displaystyle\int x\ln 2x\,dx$

6. $\displaystyle\int x^{1/2}\ln x\,dx$

7. $\displaystyle\int \frac{\ln x}{x^2}\,dx$

8. $\displaystyle\int \frac{\ln x}{\sqrt{x^3}}\,dx$

9. $\displaystyle\int (\ln t)^2\,dt$

10. $\displaystyle\int (t\ln t)^2\,dt$

11. $\displaystyle\int_0^2 x\ln(x+1)\,dx$

12. $\displaystyle\int_1^e x^2\ln x^2\,dx$

13. $\displaystyle\int_0^1 \tan^{-1}x\,dx$

14. $\displaystyle\int_1^4 \frac{\tan^{-1}\sqrt{x}}{\sqrt{x}}\,dx$

15. $\displaystyle\int \sin^{-1}x\,dx$

16. $\displaystyle\int x^2\tan^{-1}x\,dx$

17. $\displaystyle\int xe^{3x}\,dx$

18. $\displaystyle\int x^2e^{5x}\,dx$

19. $\displaystyle\int x^3e^{x^2}\,dx$

20. $\displaystyle\int x^5e^{2x^3}\,dx$

21. $\displaystyle\int_2^4 xe^{-x/2}\,dx$

22. $\displaystyle\int_0^1 (x^2-x)e^{-x}\,dx$

23. $\displaystyle\int t\cos 8t\,dt$

24. $\displaystyle\int x\sinh x\,dx$

25. $\displaystyle\int x^2\sin x\,dx$

26. $\displaystyle\int_0^\pi x^2\cos x\,dx$

27. $\displaystyle\int e^x\sin 4x\,dx$

28. $\displaystyle\int_{-\pi}^\pi e^x\cos x\,dx$

29. $\displaystyle\int \theta\sec\theta\tan\theta\,d\theta$

30. $\displaystyle\int e^{2t}\cos e^t\,dt$

31. $\displaystyle\int \sin x\cos 2x\,dx$

32. $\displaystyle\int \cosh x\cosh 2x\,dx$

33. $\displaystyle\int x^3\sqrt{x^2+4}\,dx$

34. $\displaystyle\int \frac{t^5}{(t^3+1)^2}\,dt$

35. $\displaystyle\int \sin(\ln x)\,dx$

36. $\displaystyle\int \cos x\ln(\sin x)\,dx$

37. $\displaystyle\int \csc^3 x\,dx$

38. $\displaystyle\int x\sec^2 x\,dx$

39. $\displaystyle\int x\sin^2 x\,dx$

40. $\displaystyle\int x\tan^2 x\,dx$

41. $\displaystyle\int (\sin^{-1}x)^2\,dx$

42. $\displaystyle\int \ln(x+\sqrt{x^2+1})\,dx$

43. $\displaystyle\int \frac{x^2e^x}{(x+2)^2}\,dx$

44. $\displaystyle\int e^{2x}\tan^{-1}e^x\,dx$

In Problems 45–48 use a substitution first, followed by integration by parts.

45. $\displaystyle\int_0^4 \tan^{-1}\sqrt{x}\,dx$

46. $\displaystyle\int xe^{\sqrt{x}}\,dx$

47. $\displaystyle\int \sin\sqrt{x+2}\,dx$

48. $\displaystyle\int_0^{\pi^2} \cos\sqrt{t}\,dt$

49. Evaluate $\displaystyle\int \frac{\sin^{-1}x}{\sqrt{1-x^2}}\,dx$ using two different methods.

50. Evaluate $\int \sin mx\cos nx\,dx$ using integration by parts.

51. Find the area under the graph of $y = 1 + \ln x$ on the interval $[e^{-1}, 3]$.

52. Find the area bounded by the graph of $y = \tan^{-1}x$ and the x-axis on the interval $[-1, 1]$.

53. The region in the first quadrant bounded by the graphs of $y = \ln x$, $x = 5$, and $y = 0$ is revolved about the x-axis. Find the volume of the solid of revolution.

54. The region in the first quadrant bounded by the graphs of $y = e^x$, $x = 0$, and $y = 3$ is revolved about the y-axis. Find the volume of the solid of revolution.

55. The region in the first quadrant bounded by the graphs of $y = \sin x$, $y = 0$, $0 \le x \le \pi$, is revolved about the y-axis. Find the volume of the solid of revolution.

56. (a) Show that integration by parts applied to $\int dx/(x\ln x)$ with $dv = dx/x$, $u = 1/\ln x$, $v = \ln x$, and $du = -dx/x(\ln x)^2$ leads to $0 = 1$. Explain.

(b) Evaluate the integral in part **(a)** by an alternative method.

Miscellaneous Problems

In Problems 57–60 establish the given formula.

57. $\displaystyle\int (\ln x)^n\,dx = x(\ln x)^n - n\int (\ln x)^{n-1}\,dx$

58. $\displaystyle\int \sin^n x\,dx = -\frac{\sin^{n-1}x\cos x}{n} + \frac{n-1}{n}\int \sin^{n-2}x\,dx$

59. $\displaystyle\int \sec^n x\,dx = \frac{\sec^{n-2}x\tan x}{n-1}$
$+ \frac{n-2}{n-1}\int \sec^{n-2}x\,dx, \qquad n \ne 1$

60. $\displaystyle\int e^{ax}\sin bx\,dx = \frac{e^{ax}(a\sin bx - b\cos bx)}{a^2+b^2} + C$

61. Evaluate $\displaystyle\int \sin^3 x\,dx$.

62. Evaluate $\displaystyle\int \sec^4 x\,dx$.

9.3 Integration of Powers of Trigonometric Functions

Integrals of the Form

$$\int \sin^m x \cos^n x \, dx$$

With the aid of trigonometric identities, it is possible to evaluate integrals of the type

$$\int \sin^m x \cos^n x \, dx \tag{9.5}$$

We distinguish two cases.

Case I *m* or *n* is an odd positive integer.

Let us first assume that *m* is an odd positive integer. By writing

$$\sin^m x = \sin^{m-1} x \sin x$$

where $m - 1$ is even, and using $\sin^2 x = 1 - \cos^2 x$ the integrand in (9.5) can be expressed as a *sum* of powers of cos *x* times sin *x*. The original integral can then be expressed as a sum of integrals, each having the recognizable form

$$\int \cos^k x \sin x \, dx = -\int \overbrace{\cos^k x}^{u^k} \overbrace{(-\sin x) \, dx}^{du} = -\int u^k \, du$$

We note that the exponent *k* need not be an integer.

Example 1

Evaluate $\displaystyle\int \sin^5 x \cos^2 x \, dx$.

Solution

$$\int \sin^5 x \cos^2 x \, dx = \int \cos^2 x \sin^4 x \sin x \, dx$$

$$= \int \cos^2 x (\sin^2 x)^2 \sin x \, dx$$

$$= \int \cos^2 x (1 - \cos^2 x)^2 \sin x \, dx$$

$$= \int \cos^2 x (1 - 2\cos^2 x + \cos^4 x) \sin x \, dx$$

$$= -\int \overbrace{\cos^2 x}^{u^2} \overbrace{(-\sin x) \, dx}^{du} + 2\int \overbrace{\cos^4 x}^{u^4} \overbrace{(-\sin x) \, dx}^{du}$$

$$\quad - \int \overbrace{\cos^6 x}^{u^6} \overbrace{(-\sin x) \, dx}^{du}$$

$$= -\frac{1}{3}\cos^3 x + \frac{2}{5}\cos^5 x - \frac{1}{7}\cos^7 x + C$$

Example 2

Evaluate $\int \sin^3 x\, dx$.

Solution

$$\int \sin^3 x\, dx = \int \sin^2 x \sin x\, dx$$

$$= \int (1 - \cos^2 x)\sin x\, dx$$

$$= \int \sin x\, dx + \int \cos^2 x(-\sin x)\, dx$$

$$= -\cos x + \frac{1}{3}\cos^3 x + C$$

If n is an odd positive integer, the procedure for evaluation is the same except that we seek an integrand that is a sum of powers of $\sin x$ times $\cos x$.

Example 3

Evaluate $\int \sin^4 x \cos^3 x\, dx$.

Solution

$$\int \sin^4 x \cos^3 x\, dx = \int \sin^4 x \cos^2 x \cos x\, dx$$

$$= \int \sin^4 x(1 - \sin^2 x)\cos x\, dx$$

$$= \int \overbrace{\sin^4 x}^{u^4}\overbrace{(\cos x)\, dx}^{du} - \int \overbrace{\sin^6 x}^{u^6}\overbrace{(\cos x)\, dx}^{du}$$

$$= \frac{1}{5}\sin^5 x - \frac{1}{7}\sin^7 x + C$$

Case II m and n are both even nonnegative integers.

When both m and n are even nonnegative integers, the evaluation of (9.5) relies heavily on the identities

$$\sin x \cos x = \frac{1}{2}\sin 2x, \qquad \sin^2 x = \frac{1 - \cos 2x}{2}, \qquad \cos^2 x = \frac{1 + \cos 2x}{2}$$

We have already seen the special cases

$$\int \sin^2 x\, dx \quad \text{and} \quad \int \cos^2 x\, dx$$

several times.

___ **Example 4** _____

Evaluate $\int \sin^2 x \cos^2 x \, dx$.

Solution

$$\int \sin^2 x \cos^2 x \, dx = \int \frac{1 - \cos 2x}{2} \cdot \frac{1 + \cos 2x}{2} \, dx$$

$$= \frac{1}{4} \int (1 - \cos^2 2x) \, dx$$

$$= \frac{1}{4} \int \left(1 - \frac{1 + \cos 4x}{2} \right) dx$$

$$= \frac{1}{4} \int \left(\frac{1}{2} - \frac{1}{2} \cos 4x \right) dx$$

$$= \frac{1}{8} x - \frac{1}{32} \sin 4x + C$$

Alternative Solution

$$\int \sin^2 x \cos^2 x \, dx = \int (\sin x \cos x)^2 \, dx$$

$$= \int \left(\frac{\sin 2x}{2} \right)^2 dx$$

$$= \frac{1}{4} \int \frac{1 - \cos 4x}{2} \, dx$$

The remainder of the solution is the same as before.

___ **Example 5** _____

Evaluate $\int \cos^4 x \, dx$.

Solution

$$\int \cos^4 x \, dx = \int (\cos^2 x)^2 \, dx$$

$$= \int \left(\frac{1 + \cos 2x}{2} \right)^2 dx$$

$$= \frac{1}{4} \int (1 + 2 \cos 2x + \cos^2 2x) \, dx$$

$$= \frac{1}{4} \int \left(1 + 2 \cos 2x + \frac{1 + \cos 4x}{2} \right) dx$$

$$= \frac{1}{4} \int \left(\frac{3}{2} + 2 \cos 2x + \frac{1}{2} \cos 4x \right) dx$$

$$= \frac{3}{8} x + \frac{1}{4} \sin 2x + \frac{1}{32} \sin 4x + C$$

The foregoing procedures are summarized in the following table.

<table>
<tr><td colspan="3" align="center">Evaluation of $\displaystyle\int \sin^m x \cos^n x \, dx$</td></tr>
<tr><td align="center">*Case*</td><td align="center">*Procedure*</td><td align="center">*Identities Used*</td></tr>
<tr>
<td>I *m* odd
    n odd</td>
<td>$u = \cos x$
$u = \sin x$</td>
<td>$\sin^2 x = 1 - \cos^2 x$
$\cos^2 x = 1 - \sin^2 x$</td>
</tr>
<tr>
<td>II *m* and *n* even</td>
<td>Reduce the powers of sin *x*
and cos *x* in the integrand
by using the identities</td>
<td>$\left\{\begin{array}{l} \sin x \cos x = \dfrac{1}{2}\sin 2x \\[2mm] \sin^2 x = \dfrac{1 - \cos 2x}{2} \\[2mm] \cos^2 x = \dfrac{1 + \cos 2x}{2} \end{array}\right.$</td>
</tr>
</table>

Integrals of the Form

$$\int \tan^m x \, \sec^n x \, dx$$

To evaluate an integral of the type

$$\int \tan^m x \, \sec^n x \, dx \tag{9.6}$$

we shall consider three cases.

Case I   *m* is an odd positive integer.

When *m* is an odd positive integer, then $m - 1$ is even. Using

$$\tan^m x \, \sec^n x = \tan^{m-1} x \, \sec^{n-1} x \, \sec x \, \tan x$$

and $\qquad\qquad\qquad \tan^2 x = \sec^2 x - 1$

the given integral can be written as a sum of integrals each having the form

$$\int \overbrace{\sec^k x}^{u^k} \, \overbrace{\sec x \, \tan x \, dx}^{du} = \int u^k \, du$$

Example 6

Evaluate $\displaystyle\int \tan^3 x \, \sec^7 x \, dx$.

Solution

$$\int \tan^3 x \, \sec^7 x \, dx = \int \tan^2 x \, \sec^6 x \, \sec x \, \tan x \, dx$$

$$= \int (\sec^2 x - 1)\sec^6 x \, \sec x \, \tan x \, dx$$

$$= \int \overbrace{\sec^8 x}^{u^8}(\underbrace{\sec x \tan x) \, dx}_{du} - \int \overbrace{\sec^6 x}^{u^6}(\underbrace{\sec x \tan x) \, dx}_{du}$$

$$= \frac{1}{9} \sec^9 x - \frac{1}{7} \sec^7 x + C$$

Case II   *n* is an even positive integer.

If n is an even positive integer, the evaluation procedure is similar to case I for integrals of the type given in (9.5). Employing

$$\sec^n x = \sec^{n-2} x \, \sec^2 x$$

and the identity $1 + \tan^2 x = \sec^2 x$, we can write the given integral as a sum of integrals of the form

$$\int \overbrace{\tan^k x}^{u^k} \underbrace{\sec^2 x \, dx}_{du} = \int u^k \, du$$

___ **Example 7** ___

Evaluate $\int \sqrt{\tan x} \, \sec^4 x \, dx$.

Solution

$$\int \sqrt{\tan x} \, \sec^4 x \, dx = \int (\tan x)^{1/2} \sec^2 x \, \sec^2 x \, dx$$

$$= \int (\tan x)^{1/2} (1 + \tan^2 x) \sec^2 x \, dx$$

$$= \int \overbrace{(\tan x)^{1/2}}^{u^{1/2}} \underbrace{\sec^2 x \, dx}_{du} + \int \overbrace{(\tan x)^{5/2}}^{u^{5/2}} \underbrace{\sec^2 x \, dx}_{du}$$

$$= \frac{2}{3} (\tan x)^{3/2} + \frac{2}{7} (\tan x)^{7/2} + C$$

Case III   *m* is even and *n* is odd.

Finally, if m is an even positive integer and n is an odd positive integer, we write the integrand in terms of sec x and use integration by parts.

Example 8

Evaluate $\int \tan^2 x \sec x \, dx$.

Solution By writing

$$\int \tan^2 x \sec x \, dx = \int (\sec^2 x - 1) \sec x \, dx$$

$$= \int \sec^3 x \, dx - \int \sec x \, dx$$

we have two integrals previously evaluated. Integration by parts gives (see Example 3 in Section 9.2)

$$\int \sec^3 x \, dx = \frac{1}{2} \sec x \tan x + \frac{1}{2} \ln|\sec x + \tan x| + C_1 \qquad (9.7)$$

Also, $$\int \sec x \, dx = \ln|\sec x + \tan x| + C_2 \qquad (9.8)$$

Subtracting the results in (9.7) and (9.8) finally yields

$$\int \tan^2 x \sec x \, dx = \frac{1}{2} \sec x \tan x - \frac{1}{2} \ln|\sec x + \tan x| + C$$

Although the next example does not, strictly speaking, fall into any of the three cases listed under (9.6) the procedure is similar to that of case I.

Example 9

Evaluate $\int \tan^4 x \, dx$.

Solution

$$\int \tan^4 x \, dx = \int \tan^2 x \tan^2 x \, dx$$

$$= \int \tan^2 x (\sec^2 x - 1) \, dx$$

$$= \int (\tan x)^2 \sec^2 x \, dx - \int \tan^2 x \, dx$$

$$= \int (\tan x)^2 \sec^2 x \, dx - \int (\sec^2 x - 1) \, dx$$

$$= \int (\tan x)^2 \sec^2 x \, dx - \int \sec^2 x \, dx + \int dx$$

$$= \frac{1}{3} \tan^3 x - \tan x + x + C$$

The three cases considered in evaluating the integral $\int \tan^m x \sec^n x \, dx$ are summarized in the following table.

	Case	Procedure	Identities Used
	Evaluation of $\int \tan^m x \sec^n x \, dx$		
I	m odd	$u = \sec x$	$\tan^2 x = \sec^2 x - 1$
II	n even	$u = \tan x$	$\sec^2 x = 1 + \tan^2 x$
III	m even and n odd	Change integrand to powers of $\sec x$ alone. Integration by parts may be required.	$\tan^2 x = \sec^2 x - 1$

Remark

Integrals of the type $\int \cot^m x \csc^n x \, dx$ are handled in a manner analogous to (9.6). In this case the identity $\csc^2 x = 1 + \cot^2 x$ will be useful.

—————— **Exercises 9.3** ——————

Answers to odd-numbered problems begin on page A-35

In Problems 1–36 evaluate the given integral.

1. $\int (\sin x)^{1/2} \cos x \, dx$

2. $\int \cos^4 5x \sin 5x \, dx$

3. $\int \cos^3 x \, dx$

4. $\int \sin^5 t \, dt$

5. $\int \sin^3 x \cos^3 x \, dx$

6. $\int \sin^5 2x \cos^2 2x \, dx$

7. $\int_{\pi/3}^{\pi/2} \sin^3 \theta \sqrt{\cos \theta} \, d\theta$

8. $\int \frac{\cos^3 x}{\sin^2 x} \, dx$

9. $\int_0^{\pi/2} \sin^5 x \cos^5 x \, dx$

10. $\int_0^{\pi} \sin^3 2t \, dt$

11. $\int \sin^4 t \, dt$

12. $\int \cos^6 \theta \, d\theta$

13. $\int \sin^2 x \cos^4 x \, dx$

14. $\int_{-\pi}^{\pi} \sin^4 x \cos^2 x \, dx$

15. $\int \sin^4 x \cos^4 x \, dx$

16. $\int \sin^2 3x \cos^2 3x \, dx$

17. $\int_0^{\pi/3} \tan^2 x \, dx$

18. $\int (\tan x + \cot x)^2 \, dx$

19. $\int \tan^3 2t \sec^4 2t \, dt$

20. $\int (2 - \sqrt{\tan x})^2 \sec^2 x \, dx$

21. $\int \frac{dx}{\cos^4 x}$

22. $\int_{-\pi/4}^{\pi/4} \tan y \sec^4 y \, dy$

23. $\int \cot^{10} x \csc^4 x \, dx$

24. $\int (1 + \csc^2 t)^2 \, dt$

25. $\int \tan^5 x \, dx$

26. $\int \cot^6 x \, dx$

27. $\int \tan^3 x (\sec x)^{-1/2} \, dx$

28. $\int \left(\tan \frac{x}{2} \sec \frac{x}{2} \right)^3 dx$

29. $\int \tan^2 x \sec^3 x \, dx$

30. $\int (1 + \tan x)^2 \sec x \, dx$

31. $\int \cos^2 x \cot x \, dx$

32. $\int \sin x \sec^7 x \, dx$

33. $\int \frac{\sec^4(1 - t)}{\tan^8(1 - t)} \, dt$

34. $\int \frac{\sin^3 \sqrt{t} \cos^2 \sqrt{t}}{\sqrt{t}} \, dt$

35. $\int \cot^3 t \, dt$

36. $\int \csc^5 t \, dt$

In Problems 37 and 38 find the volume of the solid of revolution that is formed by revolving the region bounded by the graphs of the given equations about the x-axis.

37. $y = \cos 2x, y = 0, 0 \le x \le \pi/6$

38. $y = \tan^2 x, y = 0, 0 \le x \le \pi/4$

Miscellaneous Problems

In Problems 39–44 use the trigonometric identities.

$$\sin mx \cos nx = \frac{1}{2}[\sin(m + n)x + \sin(m - n)x]$$

$$\sin mx \sin nx = \frac{1}{2}[\cos(m - n)x - \cos(m + n)x]$$

$$\cos mx \cos nx = \frac{1}{2}[\cos(m - n)x + \cos(m + n)x]$$

to evaluate the given integrals.

39. $\displaystyle\int \sin x \cos 2x \, dx$

40. $\displaystyle\int \cos 3x \cos 5x \, dx$

41. $\displaystyle\int \sin 2x \sin 4x \, dx$

42. $\displaystyle\int \frac{5 - 3 \sin 2x}{\sec 6x} \, dx$

43. $\displaystyle\int_0^{\pi/6} \cos 2x \cos x \, dx$

44. $\displaystyle\int_0^{\pi/2} \sin \frac{3}{2}x \sin \frac{1}{2}x \, dx$

45. Show that

$$\int_{-\pi}^{\pi} \sin mx \sin nx \, dx = \begin{cases} 0, & m \neq n \\ \pi, & m = n \end{cases}$$

46. Evaluate $\displaystyle\int_{-\pi}^{\pi} \sin mx \cos nx \, dx$.

9.4 Trigonometric Substitutions

When an integrand contains integer powers of x and integer powers of

$$\sqrt{a^2 - x^2}, \qquad \sqrt{a^2 + x^2}, \quad \text{or} \quad \sqrt{x^2 - a^2}, \qquad a > 0$$

we may be able to evaluate the integrals by means of a trigonometric substitution. The three cases that we now consider depend, in turn, on the fundamental identities:

$$1 - \sin^2\theta = \cos^2\theta$$
$$1 + \tan^2\theta = \sec^2\theta$$
$$\sec^2\theta - 1 = \tan^2\theta$$

Case I Integrands Containing $\sqrt{a^2 - x^2}, a > 0$

If we let $x = a \sin \theta, -\pi/2 \leq \theta \leq \pi/2$, then

$$\sqrt{a^2 - x^2} = \sqrt{a^2 - a^2 \sin^2\theta}$$
$$= \sqrt{a^2(1 - \sin^2\theta)}$$
$$= \sqrt{a^2 \cos^2\theta}$$
$$= a \cos \theta$$

When $\sqrt{a^2 - x^2}$ appears in the denominator of an integrand, there is the further restriction $-\pi/2 < \theta < \pi/2$.

Example 1

Evaluate $\displaystyle \int \frac{x^2}{\sqrt{9 - x^2}}\, dx$.

Solution   Identifying $a = 3$ leads to the substitutions

$$x = 3 \sin \theta \qquad dx = 3 \cos \theta\, d\theta$$

where $-\pi/2 < \theta < \pi/2$. The integral becomes

$$\int \frac{x^2}{\sqrt{9 - x^2}}\, dx = \int \frac{9 \sin^2\theta}{\sqrt{9 - 9 \sin^2\theta}}\, (3 \cos \theta\, d\theta)$$

$$= 9 \int \sin^2\theta\, d\theta$$

Recall, to evaluate this last trigonometric integral, we make use of the identity $\sin^2\theta = (1 - \cos 2\theta)/2$:

$$\int \frac{x^2}{\sqrt{9 - x^2}}\, dx = \frac{9}{2} \int (1 - \cos 2\theta)\, d\theta$$

$$= \frac{9}{2}\theta - \frac{9}{4} \sin 2\theta + C$$

In order to express this result back in terms of the variable x, we note that $\sin \theta = x/3$, $\cos \theta = \sqrt{1 - \sin^2\theta} = \sqrt{9 - x^2}/3$, and $\theta = \sin^{-1}(x/3)$. Since $\sin 2\theta = 2 \sin \theta \cos \theta$, it follows that

$$\int \frac{x^2}{\sqrt{9 - x^2}}\, dx = \frac{9}{2} \sin^{-1}\frac{x}{3} - \frac{1}{2}x\sqrt{9 - x^2} + C$$

Example 2

Evaluate $\displaystyle \int \frac{\sqrt{1 - x^2}}{x}\, dx$.

Solution   Let

$$x = \sin \theta \qquad dx = \cos \theta\, d\theta$$

then

$$\int \frac{\sqrt{1 - x^2}}{x}\, dx = \int \frac{\sqrt{1 - \sin^2\theta}}{\sin \theta}(\cos \theta\, d\theta)$$

$$= \int \frac{\cos^2\theta}{\sin \theta}\, d\theta$$

$$= \int \frac{1 - \sin^2\theta}{\sin \theta}\, d\theta$$

$$= \int (\csc \theta - \sin \theta)\, d\theta$$

$$= \ln|\csc \theta - \cot \theta| + \cos \theta + C \qquad\qquad (9.9)$$

Since $\cos\theta = \sqrt{1 - \sin^2\theta} = \sqrt{1 - x^2}$, $\csc\theta = 1/\sin\theta = 1/x$, and $\cot\theta = \cos\theta/\sin\theta = \sqrt{1 - x^2}/x$, (9.9) can be written as

$$\int \frac{\sqrt{1 - x^2}}{x}\,dx = \ln\left|\frac{1 - \sqrt{1 - x^2}}{x}\right| + \sqrt{1 - x^2} + C$$

In Examples 1 and 2 the return to the variable x can be accomplished in an alternative manner. If we construct a right triangle, as shown in Figure 9.2, such that $\sin\theta = x/a$, then the other trigonometric functions can be readily expressed in terms of x. For instance, in Example 2 $\sin\theta = x/1$ and so we see from Figure 9.3 that $\cos\theta = \sqrt{1 - x^2}$ and $\cot\theta = \cos\theta/\sin\theta = \sqrt{1 - x^2}/x$.

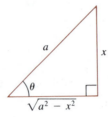

Figure 9.2

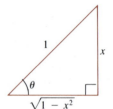

Figure 9.3

Case II   Integrands Containing $\sqrt{a^2 + x^2}$, $a > 0$

Suppose $x = a\tan\theta$, where $-\pi/2 < \theta < \pi/2$. Then,

$$\begin{aligned}\sqrt{a^2 + x^2} &= \sqrt{a^2 + a^2\tan^2\theta}\\ &= \sqrt{a^2(1 + \tan^2\theta)}\\ &= \sqrt{a^2\sec^2\theta}\\ &= a\sec\theta\end{aligned}$$

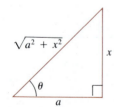

Figure 9.4

As in the preceding discussion, an integral that involves an algebraic term $\sqrt{a^2 + x^2}$ is transformed into a trigonometric integral. After integration we can eliminate the variable θ by employing a right triangle where $\tan\theta = x/a$. See Figure 9.4.

Example 3

Evaluate $\displaystyle\int \frac{dx}{(4 + x^2)^{3/2}}$.

Solution   Observe that the integrand is an integer power of $\sqrt{4 + x^2}$ since $(4 + x^2)^{3/2} = (\sqrt{4 + x^2})^3$. Now, when

$$x = 2\tan\theta \qquad dx = 2\sec^2\theta\,d\theta$$

$\sqrt{4 + x^2} = 2 \sec \theta$, and $(4 + x^2)^{3/2} = 8 \sec^3 \theta$. Thus,

$$\int \frac{dx}{(4 + x^2)^{3/2}} = \int \frac{2 \sec^2 \theta \, d\theta}{8 \sec^3 \theta}$$

$$= \frac{1}{4} \int \cos \theta \, d\theta$$

$$= \frac{1}{4} \sin \theta + C$$

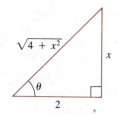

Figure 9.5

From the triangle in Figure 9.5, we see that $\sin \theta = x/\sqrt{4 + x^2}$. Hence,

$$\int \frac{dx}{(4 + x^2)^{3/2}} = \frac{1}{4} \frac{x}{\sqrt{4 + x^2}} + C$$

Case III   Integrands Containing $\sqrt{x^2 - a^2}$, $a > 0$

In this last case, if we use the substitution $x = a \sec \theta$, where $0 \le \theta < \pi/2$ or $\pi \le \theta < 3\pi/2$, then

$$\sqrt{x^2 - a^2} = \sqrt{a^2 \sec^2 \theta - a^2}$$

$$= \sqrt{a^2 (\sec^2 \theta - 1)}$$

$$= \sqrt{a^2 \tan^2 \theta}$$

$$= a \tan \theta$$

_____ **Example 4** _____

Evaluate $\displaystyle\int \frac{\sqrt{x^2 - 16}}{x^4} \, dx$.

Solution   Setting

$$x = 4 \sec \theta \qquad dx = 4 \sec \theta \tan \theta \, d\theta$$

yields

$$\int \frac{\sqrt{x^2 - 16}}{x^4} \, dx = \int \frac{\sqrt{16 \sec^2 \theta - 16}}{256 \sec^4 \theta} (4 \sec \theta \tan \theta \, d\theta)$$

$$= \frac{1}{16} \int \frac{\tan^2 \theta}{\sec^3 \theta} \, d\theta$$

$$= \frac{1}{16} \int \frac{\sin^2 \theta}{\cos^2 \theta} \cos^3 \theta \, d\theta$$

$$= \frac{1}{16} \int \sin^2 \theta (\cos \theta \, d\theta)$$

$$= \frac{1}{48} \sin^3 \theta + C$$

Referring to the triangle in Figure 9.6, we see that if $\sec \theta = x/4$, then $\cos \theta = 4/x$ and $\sin \theta = \sqrt{x^2 - 16}/x$. It follows that

$$\int \frac{\sqrt{x^2 - 16}}{x^4}\, dx = \frac{1}{48} \frac{(x^2 - 16)^{3/2}}{x^3} + C$$

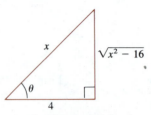

Figure 9.6

Example 5

Find the length of the graph of $y = \frac{1}{2}x^2 + 3$ on the interval $[0, 1]$.

Solution Recall the formula for arc length is $s = \int_a^b \sqrt{1 + [f'(x)]^2}\, dx$. Since $dy/dx = x$, we have

$$s = \int_0^1 \sqrt{1 + x^2}\, dx$$

Now, if we substitute

$$x = \tan \theta \qquad dx = \sec^2\theta\, d\theta$$

the limits of integration in the resulting definite integral are $\theta = \tan^{-1}0 = 0$ and $\theta = \tan^{-1}1 = \pi/4$. Therefore,

$$s = \int_0^{\pi/4} \sqrt{1 + \tan^2\theta}\, \sec^2\theta\, d\theta$$
$$= \int_0^{\pi/4} \sec^3\theta\, d\theta$$

The antiderivative of $\sec^3\theta$ was found in Example 3 of Section 9.2 using integration by parts:

$$s = \left(\frac{1}{2}\sec\theta \tan\theta + \frac{1}{2}\ln|\sec\theta + \tan\theta|\right)\Bigg]_0^{\pi/4}$$
$$= \frac{1}{2}\sec\frac{\pi}{4}\tan\frac{\pi}{4} + \frac{1}{2}\ln\left|\sec\frac{\pi}{4} + \tan\frac{\pi}{4}\right|$$
$$= \frac{\sqrt{2}}{2} + \frac{1}{2}\ln(\sqrt{2} + 1) \approx 1.1478$$

Integrands Containing a Quadratic Expression

By completion of the square, it is possible to express an integrand that contains a quadratic expression in one of the following forms:

$$a^2 - u^2, \quad a^2 + u^2, \quad \text{or} \quad u^2 - a^2$$

The appropriate substitutions are summarized in the accompanying table.

Form	Substitution
$\sqrt{a^2 - u^2}$	$u = a \sin \theta$
$\sqrt{a^2 + u^2}$	$u = a \tan \theta$
$\sqrt{u^2 - a^2}$	$u = a \sec \theta$

Example 6

Evaluate $\displaystyle\int \frac{dx}{(x^2 + 8x + 25)^{3/2}}$.

Solution   Since

$$\int \frac{dx}{(x^2 + 8x + 25)^{3/2}} = \int \frac{dx}{[9 + (x + 4)^2]^{3/2}}$$

we identify $a^2 + u^2$ with $a = 3$ and $u = x + 4$. Using

$$x + 4 = 3 \tan \theta \qquad dx = 3 \sec^2 \theta \, d\theta$$

we find

$$\int \frac{dx}{(x^2 + 8x + 25)^{3/2}} = \int \frac{3 \sec^2 \theta \, d\theta}{[9 + 9 \tan^2 \theta]^{3/2}}$$

$$= \frac{1}{9} \int \frac{\sec^2 \theta}{\sec^3 \theta} \, d\theta$$

$$= \frac{1}{9} \int \cos \theta \, d\theta$$

$$= \frac{1}{9} \sin \theta + C$$

Inspection of the triangle in Figure 9.7 indicates how to express $\sin \theta$ in terms of x. It follows that

$$\int \frac{dx}{(x^2 + 8x + 25)^{3/2}} = \frac{1}{9} \frac{x + 4}{\sqrt{(x + 4)^2 + 9}} + C$$

$$= \frac{x + 4}{9 \sqrt{x^2 + 8x + 25}} + C$$

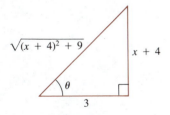

Figure 9.7

Remark

In the three cases just considered, other substitutions are possible although not necessarily desirable. For example, $x = a \cos \theta$, $0 \le \theta \le \pi$, can be used to eliminate the radical $\sqrt{a^2 - x^2}$, $a > 0$:

$$\sqrt{a^2 - x^2} = \sqrt{a^2(1 - \cos^2 \theta)}$$

$$= \sqrt{a^2 \sin^2 \theta}$$

$$= a \sin \theta$$

Likewise, we can use the *hyperbolic substitution* $x = a \sinh t$ for $\sqrt{a^2 + x^2}$, $a > 0$:

$$\sqrt{a^2 + x^2} = \sqrt{a^2(1 + \sinh^2 t)}$$
$$= \sqrt{a^2 \cosh^2 t}$$
$$= a \cosh t$$

See Problems 57 and 58 in Exercises 9.4.

Exercises 9.4

Answers to odd-numbered problems begin on page A-36.

In Problems 1–44 evaluate the given integral by a trigonometric substitution when appropriate.*

1. $\displaystyle \int \frac{\sqrt{1 - x^2}}{x^2} \, dx$

2. $\displaystyle \int \frac{x^2}{\sqrt{x^2 - 4}} \, dx$

3. $\displaystyle \int \frac{dx}{\sqrt{x^2 - 36}}$

4. $\displaystyle \int \sqrt{3 - x^2} \, dx$

5. $\displaystyle \int \sqrt{4 + x^2} \, dx$

6. $\displaystyle \int (1 - x^2)^{3/2} \, dx$

7. $\displaystyle \int x^3 \sqrt{1 - x^2} \, dx$

8. $\displaystyle \int x^3 \sqrt{x^2 - 1} \, dx$

9. $\displaystyle \int \frac{dx}{(x^2 - 4)^{3/2}}$

10. $\displaystyle \int (9 - x^2)^{-3/2} \, dx$

11. $\displaystyle \int x \sqrt{x^2 + 7} \, dx$

12. $\displaystyle \int \frac{x}{25 + x^2} \, dx$

13. $\displaystyle \int \frac{dx}{\sqrt{25 - x^2}}$

14. $\displaystyle \int \frac{dx}{x \sqrt{x^2 - 25}}$

15. $\displaystyle \int \frac{dx}{x \sqrt{16 - x^2}}$

16. $\displaystyle \int \frac{dx}{x^2 \sqrt{16 - x^2}}$

17. $\displaystyle \int \frac{dx}{x \sqrt{1 + x^2}}$

18. $\displaystyle \int \frac{dx}{x^2 \sqrt{1 + x^2}}$

19. $\displaystyle \int \frac{\sqrt{1 - x^2}}{x^4} \, dx$

20. $\displaystyle \int \frac{\sqrt{x^2 - 1}}{x^4} \, dx$

21. $\displaystyle \int \frac{x^2}{(9 - x^2)^{3/2}} \, dx$

22. $\displaystyle \int \frac{x^2}{(4 + x^2)^{3/2}} \, dx$

23. $\displaystyle \int \frac{dx}{(1 + x^2)^2}$

24. $\displaystyle \int \frac{x^2}{(x^2 - 1)^2} \, dx$

25. $\displaystyle \int \frac{dx}{(4 + x^2)^{5/2}}$

26. $\displaystyle \int \frac{x^3}{(1 - x^2)^{5/2}} \, dx$

27. $\displaystyle \int \frac{dx}{\sqrt{x^2 + 2x + 10}}$

28. $\displaystyle \int \frac{x}{\sqrt{4x - x^2}} \, dx$

29. $\displaystyle \int \frac{dx}{(x^2 + 6x + 13)^2}$

30. $\displaystyle \int \frac{dx}{(11 - 10x - x^2)^2}$

31. $\displaystyle \int \frac{x - 3}{(5 - 4x - x^2)^{3/2}} \, dx$

32. $\displaystyle \int \frac{dx}{(x^2 + 2x)^{3/2}}$

33. $\displaystyle \int \frac{2x + 4}{x^2 + 4x + 13} \, dx$

34. $\displaystyle \int \frac{dx}{4 + (x - 3)^2}$

35. $\displaystyle \int \frac{x^2}{x^2 + 16} \, dx$

36. $\displaystyle \int \frac{\sqrt{4 - 9x^2}}{x} \, dx$

37. $\displaystyle \int_{-1}^{1} \sqrt{4 - x^2} \, dx$

38. $\displaystyle \int_{-1}^{\sqrt{3}} \frac{x^2}{\sqrt{4 - x^2}} \, dx$

39. $\displaystyle \int_{0}^{5} \frac{dx}{(x^2 + 25)^{3/2}}$

40. $\displaystyle \int_{\sqrt{2}}^{2} \frac{dx}{x^3 \sqrt{x^2 - 1}}$

41. $\displaystyle \int_{1}^{6/5} \frac{16 \, dx}{x^4 \sqrt{4 - x^2}}$

42. $\displaystyle \int_{0}^{1/2} x^3 (1 + x^2)^{-1/2} \, dx$

43. $\displaystyle \int \frac{dx}{\sqrt{e^{2x} - 1}}$

44. $\displaystyle \int \sqrt{e^{2x} - 1} \, dx$

45. Find the area under the graph of $y = 1/(x\sqrt{3 + x^2})$ on the interval $[1, \sqrt{3}]$.

46. Find the area under the graph of $y = x^5 \sqrt{1 - x^2}$ on the interval $[0, 1]$.

47. Show that the area of a circle given by $x^2 + y^2 = a^2$ is πa^2.

48. Show that the area of an ellipse given by $a^2 x^2 + b^2 y^2 = a^2 b^2$ is πab.

49. The region described in Problem 45 is revolved about the x-axis. Find the volume of the solid of revolution.

50. The region in the first quadrant bounded by the graphs of $y = 4/(4 + x^2)$, $x = 2$, and $y = 0$ is revolved about the x-axis. Find the volume of the solid of revolution.

*Look before you leap.

51. The region in the first quadrant bounded by the graphs of $y = x\sqrt{4 + x^2}$, $x = 2$, and $y = 0$ is revolved about the y-axis. Find the volume of the solid of revolution.

52. The region in the first quadrant bounded by the graphs of $y = x/\sqrt{4 - x^2}$, $x = 1$, and $y = 0$ is revolved about the y-axis. Find the volume of the solid of revolution.

53. Find the length of the graph of $y = \ln x$ on the interval $[1, \sqrt{3}]$.

54. Find the length of the graph of $y = -\frac{1}{2}x^2 + 2x$ on the interval $[1, 2]$.

In Problems 55 and 56 use integration by parts followed by a trigonometric substitution.

55. $\displaystyle\int x^2 \sin^{-1}x \, dx$ **56.** $\displaystyle\int x \cos^{-1}x \, dx$

Miscellaneous Problems

In Problems 57 and 58 use the indicated substitution to evaluate the given integral.

57. $\displaystyle\int \frac{dx}{x^2\sqrt{9 + x^2}}; \; x = 3 \sinh t$

58. $\displaystyle\int \frac{(1 + x)^2}{\sqrt{1 - x^2}} \, dx; \; x = \cos \theta$

59. Establish the formula

$$\int \sqrt{u^2 \pm a^2} \, du = \frac{1}{2}u\sqrt{u^2 \pm a^2}$$

$$\pm \frac{a^2}{2}\ln\left|u + \sqrt{u^2 \pm a^2}\right| + C, \qquad a > 0$$

60. The region bounded by the graph of $(x - a)^2 + y^2 = r^2$, $r < a$, is revolved about the y-axis. Find the volume of the solid of revolution or **torus**. See Figure 9.8.

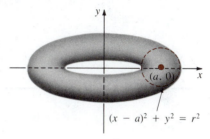

Figure 9.8

9.5   Partial Fractions

9.5.1 Denominators Containing Linear Factors

When the terms in the sum

$$\frac{2}{x + 5} + \frac{1}{x + 1} \tag{9.10}$$

are combined by means of a common denominator, we obtain the single rational expression

$$\frac{3x + 7}{(x + 5)(x + 1)} \tag{9.11}$$

Now suppose that we are faced with the problem of evaluating the integral $\int (3x + 7) \, dx/[(x + 5)(x + 1)]$. Of course, the solution is obvious: We use the equality of (9.10) and (9.11) to write

$$\int \frac{3x + 7}{(x + 5)(x + 1)} \, dx = \int \left[\frac{2}{x + 5} + \frac{1}{x + 1}\right] dx$$

$$= 2 \ln|x + 5| + \ln|x + 1| + C$$

This example illustrates a procedure for integrating certain rational functions $P(x)/Q(x)$, where the degree of $P(x)$ is less than the degree of $Q(x)$. This method, known as **partial fractions**, consists of decomposing such a rational function into simpler component fractions, and then evaluating the integral term-by-term. In this section, we shall study four cases of partial fraction decomposition.

Case I   Nonrepeated Linear Factors

We state the following fact from algebra without proof. If

$$\frac{P(x)}{Q(x)} = \frac{P(x)}{(a_1x + b_1)(a_2x + b_2) \cdots (a_nx + b_n)}$$

where all the factors $a_ix + b_i$, $i = 1, 2, \ldots, n$ are distinct and the degree of $P(x)$ is less than n, then unique real constants $C_1, C_2, \ldots, C_n$ exist such that

$$\frac{P(x)}{Q(x)} = \frac{C_1}{a_1x + b_1} + \frac{C_2}{a_2x + b_2} + \cdots + \frac{C_n}{a_nx + b_n}$$

Example 1

Evaluate $\displaystyle\int \frac{2x + 1}{(x - 1)(x + 3)}\, dx$.

Solution   We make the assumption that the integrand can be written as

$$\frac{2x + 1}{(x - 1)(x + 3)} = \frac{A}{x - 1} + \frac{B}{x + 3}$$

Combining the terms of the right-hand member of the equation over a common denominator gives

$$\frac{2x + 1}{(x - 1)(x + 3)} = \frac{A(x + 3) + B(x - 1)}{(x - 1)(x + 3)}$$

Since the numerators are identical,

$$\begin{aligned} 2x + 1 &= A(x + 3) + B(x - 1) \\ &= (A + B)x + (3A - B) \end{aligned} \qquad (9.12)$$

the coefficients of the powers of x are the same:

$$\begin{aligned} 2 &= A + B \\ 1 &= 3A - B \end{aligned}$$

These simultaneous equations can now be solved for A and B. The results are $A = \frac{3}{4}$ and $B = \frac{5}{4}$. Therefore,

$$\int \frac{2x + 1}{(x - 1)(x + 3)}\, dx = \int \left[\frac{3/4}{x - 1} + \frac{5/4}{x + 3} \right] dx$$

$$= \frac{3}{4} \ln|x - 1| + \frac{5}{4} \ln|x + 3| + C$$

Note: In the preceding example, the numbers A and B can be determined in an alternative manner. Since (9.12) is an identity, that is, the equality is true for every value of x, it holds for $x = 1$ and $x = -3$ (the zeros of the denominator). Setting $x = 1$ in (9.12) gives $3 = 4A$, from which it follows that $A = \frac{3}{4}$. Similarly, by setting $x = -3$ in (9.12), we obtain $-5 = (-4)B$ or $B = \frac{5}{4}$.

Example 2

Evaluate $\int \dfrac{x^3 - 2x}{x^2 + 3x + 2}\, dx$.

Solution We first observe that the degree of the numerator is greater than the degree of the denominator. Hence, long division is called for:

$$\int \frac{x^3 - 2x}{x^2 + 3x + 2}\, dx = \int \left[x - 3 + \frac{5x + 6}{x^2 + 3x + 2} \right] dx$$

Since $x^2 + 3x + 2 = (x + 1)(x + 2)$, we write

$$\frac{5x + 6}{(x + 1)(x + 2)} = \frac{A}{x + 1} + \frac{B}{x + 2}$$

and
$$5x + 6 = A(x + 2) + B(x + 1) \qquad (9.13)$$

If we set $x = -2$ and $x = -1$ in (9.13), we see immediately that $B = 4$ and $A = 1$, respectively. Thus,

$$\int \frac{x^3 - 2x}{x^2 + 3x + 2}\, dx = \int \left[x - 3 + \frac{1}{x + 1} + \frac{4}{x + 2} \right] dx$$

$$= \frac{x^2}{2} - 3x + \ln|x + 1| + 4 \ln|x + 2| + C$$

Example 3

Find the area A under the graph of $y = 1/x(x + 1)$ on the interval $[\frac{1}{2}, 2]$.

Solution The area in question is shown in Figure 9.9. We have

$$A = \int_{1/2}^{2} \frac{dx}{x(x + 1)}$$

Using partial fractions

$$\frac{1}{x(x + 1)} = \frac{A}{x} + \frac{B}{x + 1}$$

$$= \frac{A(x + 1) + Bx}{x(x + 1)}$$

it follows that
$$1 = A(x + 1) + Bx$$
$$= (A + B)x + A$$

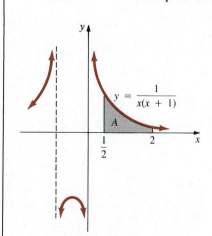

$$y = \frac{1}{x(x + 1)}$$

Figure 9.9

The solution of the system

$$0 = A + B$$
$$1 = A$$

is immediate: $A = 1$, $B = -1$. Therefore,

$$A = \int_{1/2}^{2} \left[\frac{1}{x} - \frac{1}{x + 1} \right] dx$$

$$= (\ln|x| - \ln|x + 1|) \Big]_{1/2}^{2}$$

$$= \ln \left| \frac{x}{x + 1} \right| \Big]_{1/2}^{2} = \ln 2 \approx 0.6931 \text{ square units}$$

Case II   Repeated Linear Factors

If

$$\frac{P(x)}{Q(x)} = \frac{P(x)}{(ax + b)^n}$$

where $n > 1$ and the degree of $P(x)$ is less than n, then unique real constants $C_1, C_2, \ldots, C_n$ can be found such that

$$\frac{P(x)}{(ax + b)^n} = \frac{C_1}{ax + b} + \frac{C_2}{(ax + b)^2} + \cdots + \frac{C_n}{(ax + b)^n}$$

_____ **Example 4** _____

Evaluate $\displaystyle\int \frac{x^2 + 2x + 4}{(x + 1)^3} \, dx$.

Solution   The decomposition of the integrand is

$$\frac{x^2 + 2x + 4}{(x + 1)^3} = \frac{A}{x + 1} + \frac{B}{(x + 1)^2} + \frac{C}{(x + 1)^3}$$

By equating numerators,

$$x^2 + 2x + 4 = A(x + 1)^2 + B(x + 1) + C$$
$$= Ax^2 + (2A + B)x + (A + B + C)$$

we obtain the system of equations

$$1 = A$$
$$2 = 2A + B$$
$$4 = A + B + C$$

Solving the equations yields $A = 1$, $B = 0$, and $C = 3$. Therefore,

$$\int \frac{x^2 + 2x + 4}{(x + 1)^3} \, dx = \int \left[\frac{1}{x + 1} + \frac{3}{(x + 1)^3} \right] dx$$

$$= \int \left[\frac{1}{x + 1} + 3(x + 1)^{-3} \right] dx$$

$$= \ln|x + 1| - \frac{3}{2}(x + 1)^{-2} + D$$

Combining the Cases   When the denominator $Q(x)$ contains distinct as well as repeated linear factors, we combine the two cases.

_____ **Example 5** _____

Evaluate $\int \dfrac{6x - 1}{x^3(2x - 1)} \, dx$.

Solution   We write

$$\frac{6x - 1}{x^3(2x - 1)} = \frac{A}{x} + \frac{B}{x^2} + \frac{C}{x^3} + \frac{D}{2x - 1}$$

from which it follows that

$$6x - 1 = Ax^2(2x - 1) + Bx(2x - 1) + C(2x - 1) + Dx^3 \qquad (9.14)$$

$$= (2A + D)x^3 + (-A + 2B)x^2 + (-B + 2C)x - C \qquad (9.15)$$

If we set $x = 0$ and $x = \frac{1}{2}$ in (9.14), we find $C = 1$ and $D = 16$, respectively. Now, by equating the coefficients of x^3 and x^2 in (9.15), we get

$$0 = 2A + D$$
$$0 = -A + 2B$$

Since we know the value of D, the first equation yields $A = -D/2 = -8$. The second then gives $B = A/2 = -4$. Therefore,

$$\int \frac{6x - 1}{x^3(2x - 1)} \, dx = \int \left[-\frac{8}{x} - \frac{4}{x^2} + \frac{1}{x^3} + \frac{16}{2x - 1} \right] dx$$

$$= -8 \ln|x| + 4x^{-1} - \frac{1}{2}x^{-2} + 8 \ln|2x - 1| + E$$

$$= 8 \ln \left| \frac{2x - 1}{x} \right| + 4x^{-1} - \frac{1}{2}x^{-2} + E$$

9.5.2 Denominators Containing Irreducible* Quadratic Factors

Case III   Nonrepeated Quadratic Factors

Suppose the denominator of the rational function $P(x)/Q(x)$ can be expressed as a product of distinct irreducible quadratic factors $a_i x^2 + b_i x + c_i$, $i = 1, 2, \ldots, n$. If the degree of $P(x)$ is less than $2n$, we can find unique real constants $A_1, A_2, \ldots, A_n, B_1, B_2, \ldots, B_n$ such that

$$\frac{P(x)}{(a_1 x^2 + b_1 x + c_1)(a_2 x^2 + b_2 x + c_2) \cdots (a_n x^2 + b_n x + c_n)}$$

$$= \frac{A_1 x + B_1}{a_1 x^2 + b_1 x + c_1} + \frac{A_2 x + B_2}{a_2 x^2 + b_2 x + c_2} + \cdots + \frac{A_n x + B_n}{a_n x^2 + b_n x + c_n}$$

___Example 6___

Evaluate $\displaystyle\int \frac{4x}{(x^2 + 1)(x^2 + 2x + 3)}\, dx$.

Solution   We write

$$\frac{4x}{(x^2 + 1)(x^2 + 2x + 3)} = \frac{Ax + B}{x^2 + 1} + \frac{Cx + D}{x^2 + 2x + 3}$$

from which we find

$$4x = (Ax + B)(x^2 + 2x + 3) + (Cx + D)(x^2 + 1)$$
$$= (A + C)x^3 + (2A + B + D)x^2 + (3A + 2B + C)x + (3B + D)$$

Since the denominator of the integrand has no real roots, we compare coefficients of powers of x:

$$0 = A + C$$
$$0 = 2A + B + D$$
$$4 = 3A + 2B + C$$
$$0 = 3B + D$$

Solving the equations yields $A = 1$, $B = 1$, $C = -1$, and $D = -3$. Therefore,

$$\int \frac{4x}{(x^2 + 1)(x^2 + 2x + 3)}\, dx = \int \left[\frac{x + 1}{x^2 + 1} - \frac{x + 3}{x^2 + 2x + 3} \right] dx$$

Now, the integral of each term still presents a slight problem. First, we write

$$\frac{x + 1}{x^2 + 1} = \frac{1}{2}\frac{2x}{x^2 + 1} + \frac{1}{x^2 + 1} \tag{9.16}$$

---

*The word *irreducible* means that the quadratic expression $ax^2 + bx + c$ does not factor over the set of real numbers. This situation occurs when $b^2 - 4ac < 0$.

and then, after completing the square,

$$\frac{x + 3}{x^2 + 2x + 3} = \frac{x + 1 + 2}{(x + 1)^2 + 2} = \frac{1}{2} \frac{2(x + 1)}{(x + 1)^2 + 2} + \frac{2}{(x + 1)^2 + 2} \quad (9.17)$$

In the right-hand member of (9.16) and (9.17), we recognize that the integrals of the first and second terms are, respectively, of the forms $\int du/u$ and $\int du/(u^2 + a^2)$. Finally, we obtain

$$\int \frac{4x}{(x^2 + 1)(x^2 + 2x + 3)} \, dx = \int \left[\frac{1}{2} \frac{2x}{x^2 + 1} + \frac{1}{x^2 + 1} - \frac{1}{2} \frac{2(x + 1)}{(x + 1)^2 + 2} - \frac{2}{(x + 1)^2 + (\sqrt{2})^2} \right] dx$$

$$= \frac{1}{2} \ln (x^2 + 1) + \tan^{-1}x - \frac{1}{2} \ln [(x + 1)^2 + 2] - \sqrt{2} \tan^{-1}\frac{x + 1}{\sqrt{2}} + E$$

$$= \frac{1}{2} \ln \left(\frac{x^2 + 1}{x^2 + 2x + 3} \right) + \tan^{-1}x - \sqrt{2} \tan^{-1}\frac{x + 1}{\sqrt{2}} + E$$

Case IV   Repeated Quadratic Factors

We now consider the case when the integrand is $P(x)/(ax^2 + bx + c)^n$, where $ax^2 + bx + c$ is irreducible and $n > 1$. If the degree of $P(x)$ is less than $2n$, we can find the unique real constants $A_1, A_2, \ldots, A_n, B_1, B_2, \ldots, B_n$ such that

$$\frac{P(x)}{(ax^2 + bx + c)^n} = \frac{A_1x + B_1}{ax^2 + bx + c} + \frac{A_2x + B_2}{(ax^2 + bx + c)^2}$$

$$+ \cdots + \frac{A_nx + B_n}{(ax^2 + bx + c)^n}$$

Example 7

Evaluate $\int \frac{x^2}{(x^2 + 4)^2} \, dx$.

Solution   The partial fraction decomposition of the integrand

$$\frac{x^2}{(x^2 + 4)^2} = \frac{Ax + B}{x^2 + 4} + \frac{Cx + D}{(x^2 + 4)^2}$$

leads to
$$x^2 = (Ax + B)(x^2 + 4) + Cx + D$$
$$= Ax^3 + Bx^2 + (4A + C)x + (4B + D)$$

and
$$0 = A$$
$$1 = B$$
$$0 = 4A + C$$
$$0 = 4B + D$$

We find that $A = 0$, $B = 1$, $C = 0$, and $D = -4$. Consequently,

$$\int \frac{x^2}{(x^2 + 4)^2} \, dx = \int \left[\frac{1}{x^2 + 4} - \frac{4}{(x^2 + 4)^2} \right] dx$$

The integral of the first term is an inverse tangent. However, to evaluate the integral of the second term, we employ the trigonometric substitution $x = 2 \tan \theta$:

$$\int \frac{dx}{(x^2 + 4)^2} = \int \frac{2 \sec^2\theta \, d\theta}{(4 \tan^2\theta + 4)^2}$$

$$= \frac{1}{8} \int \frac{\sec^2\theta}{\sec^4\theta} \, d\theta$$

$$= \frac{1}{8} \int \cos^2\theta \, d\theta$$

$$= \frac{1}{16} \int (1 + \cos 2\theta) \, d\theta$$

$$= \frac{1}{16} \left(\theta + \frac{1}{2} \sin 2\theta \right)$$

$$= \frac{1}{16} (\theta + \sin \theta \cos \theta)$$

$$= \frac{1}{16} \left[\tan^{-1}\frac{x}{2} + \frac{x}{\sqrt{x^2 + 4}} \cdot \frac{2}{\sqrt{x^2 + 4}} \right]$$

$$= \frac{1}{16} \left[\tan^{-1}\frac{x}{2} + \frac{2x}{x^2 + 4} \right]$$

Therefore, the original integral is

$$\int \frac{x^2}{(x^2 + 4)^2} \, dx = \frac{1}{2} \tan^{-1}\frac{x}{2} - 4 \left[\frac{1}{16} \tan^{-1}\frac{x}{2} + \frac{1}{8} \frac{x}{x^2 + 4} \right] + E$$

$$= \frac{1}{4} \tan^{-1}\frac{x}{2} - \frac{x}{2(x^2 + 4)} + E$$

The next example combines the four preceding cases.

Example 8

Determine the form of the partial fraction decomposition for the integrand of

$$\int \frac{dx}{(3x + 5)(x - 2)^2(x^2 + 6)(x^2 + x + 1)^2}.$$

Solution   Since $3x + 5$ and $x - 2$ are linear factors, whereas both $x^2 + 6$ and $x^2 + x + 1$ are irreducible quadratic factors, we can write

$$\frac{1}{(3x + 5)(x - 2)^2(x^2 + 6)(x^2 + x + 1)^2}$$

$$= \frac{A}{3x + 5} + \frac{B}{x - 2} + \frac{C}{(x - 2)^2} + \frac{Dx + E}{x^2 + 6} + \frac{Fx + G}{x^2 + x + 1} + \frac{Hx + K}{(x^2 + x + 1)^2}$$

Example 9

Evaluate $\displaystyle\int \frac{x + 3}{x^4 + 9x^2}\,dx$.

Solution From $x^4 + 9x^2 = x^2(x^2 + 9)$, we see that the problem combines the quadratic factor $x^2 + 9$ with the repeated linear factor x. Accordingly, the partial fraction decomposition is

$$\frac{x + 3}{x^2(x^2 + 9)} = \frac{A}{x} + \frac{B}{x^2} + \frac{Cx + D}{x^2 + 9}$$

Proceeding as usual, we find

$$x + 3 = (A + C)x^3 + (B + D)x^2 + 9Ax + 9B$$

and

$$0 = A + C$$
$$0 = B + D$$
$$1 = 9A$$
$$3 = 9B$$

Hence, $A = \frac{1}{9}$, $B = \frac{1}{3}$, $C = -\frac{1}{9}$, and $D = -\frac{1}{3}$. This gives us

$$\int \frac{x + 3}{x^2(x^2 + 9)}\,dx = \int \left[\frac{1/9}{x} + \frac{1/3}{x^2} - \frac{x/9 + 1/3}{x^2 + 9} \right] dx$$

$$= \int \left[\frac{1/9}{x} + \frac{1/3}{x^2} - \frac{1}{18}\frac{2x}{x^2 + 9} - \frac{1}{3}\frac{1}{x^2 + 9} \right] dx$$

$$= \frac{1}{9}\ln|x| - \frac{1}{3}x^{-1} - \frac{1}{18}\ln(x^2 + 9) - \frac{1}{9}\tan^{-1}\frac{x}{3} + E$$

$$= \frac{1}{18}\ln\left(\frac{x^2}{x^2 + 9} \right) - \frac{1}{3}x^{-1} - \frac{1}{9}\tan^{-1}\frac{x}{3} + E$$

Remark

Integrals such as $\int dx/(x + 2)^4$ and $\int (2x + 1)\,dx/(x^2 + 1)^2$ *appear* to be candidates for partial fractions. However, this is not the case. Why? You should be able to evaluate these integrals by alternative means.

_____ **Exercises 9.5** _____

Answers to odd-numbered problems begin on page A-36

[9.5.1]

In Problems 1–40 use partial fractions when appropriate to evaluate the given integral.

1. $\displaystyle\int \frac{dx}{x(x - 2)}$

2. $\displaystyle\int \frac{dx}{x(2x + 3)}$

3. $\displaystyle\int \frac{x + 2}{2x^2 - x}\,dx$

4. $\displaystyle\int \frac{3x + 10}{x^2 + 2x}\,dx$

5. $\displaystyle\int \frac{dx}{x^2 - 9}$

6. $\displaystyle\int \frac{dx}{4x^2 - 25}$

7. $\int \dfrac{x+1}{x^2-16}\,dx$

8. $\int \dfrac{x+5}{(x+4)(x^2-1)}\,dx$

9. $\int \dfrac{dx}{x^2+4x+3}$

10. $\int \dfrac{dx}{x^2+x-2}$

11. $\int \dfrac{x}{2x^2+5x+2}\,dx$

12. $\int \dfrac{x+7}{x^2-3x-10}\,dx$

13. $\int \dfrac{x^2+2x-6}{x^3-x}\,dx$

14. $\int \dfrac{5x^2-x+1}{x^3-4x}\,dx$

15. $\int \dfrac{dx}{(x+1)(x+2)(x+3)}$

16. $\int \dfrac{dx}{(4x^2-1)(x+7)}$

17. $\int \dfrac{4t^2+3t-1}{t^3-t^2}\,dt$

18. $\int \dfrac{2x-11}{x^3+2x^2}\,dx$

19. $\int \dfrac{dx}{x^3+2x^2+x}$

20. $\int \dfrac{t-1}{t^4+6t^3+9t^2}\,dt$

21. $\int \dfrac{dx}{(x-3)^4}$

22. $\int \dfrac{4x^2-5x+7}{x^3}\,dx$

23. $\int \dfrac{2x-1}{(x+1)^3}\,dx$

24. $\int \dfrac{x^2+2x-6}{(x-1)^3}\,dx$

25. $\int \dfrac{x}{(x^2-1)^2}\,dx$

26. $\int \dfrac{dx}{x^2(x^2-4)^2}$

27. $\int \dfrac{dx}{(x^2+6x+5)^2}$

28. $\int \dfrac{dx}{(x^2-x-6)(x^2-2x-8)}$

29. $\int \dfrac{x^4+2x^2-x+9}{x^5+2x^4}\,dx$

30. $\int \dfrac{5x-1}{x(x-3)^2(x+2)^2}\,dx$

31. $\int \dfrac{x^4+3x^2+4}{(x+1)^2}\,dx$

32. $\int \dfrac{x^5-10x^3}{x^4-10x^2+9}\,dx$

33. $\displaystyle\int_2^4 \dfrac{dx}{x^2-6x+5}$

34. $\displaystyle\int_0^1 \dfrac{dx}{x^2-4}$

35. $\displaystyle\int_0^2 \dfrac{2x-1}{(x+3)^2}\,dx$

36. $\displaystyle\int_1^5 \dfrac{2x+6}{x(x+1)^2}\,dx$

37. $\int \dfrac{\cos x}{\sin^2 x+3\sin x+2}\,dx$

38. $\int \dfrac{\sin x}{\cos^2 x-\cos^3 x}\,dx$

39. $\int \dfrac{e^t}{(e^t+1)^2(e^t-2)}\,dt$

40. $\int \dfrac{e^{2t}}{(e^t+1)^3}\,dt$

In Problems 41 and 42 use the indicated substitution to evaluate the given integral.

41. $\int \dfrac{\sqrt{1-x^2}}{x^3}\,dx;\ u^2=1-x^2$

42. $\int \sqrt{\dfrac{x-1}{x+1}}\,dx;\ u^2=\dfrac{x-1}{x+1}$

43. Find the area under the graph of $y=1/(x^2+2x-3)$ on the interval $[2,4]$.

44. Find the area bounded by the graph of $y=x/(x+2)(x+3)$ and the x-axis on the interval $[-1,1]$.

45. The region in the first quadrant bounded by the graphs of $y=2/x(x+1)$, $x=1$, $x=3$, and $y=0$ is revolved about the x-axis. Find the volume of the solid of revolution.

46. The region in the first quadrant bounded by the graphs of $y=1/\sqrt{(x+1)(x+4)}$, $x=0$, $x=2$, and $y=0$ is revolved about the x-axis. Find the volume of the solid of revolution.

47. The region in the first quadrant bounded by the graphs of $y=4/(x+1)^2$, $x=0$, $x=1$, and $y=0$ is revolved about the y-axis. Find the volume of the solid of revolution.

48. Find the length of the graph of $y=e^x$ on the interval $[0,\ln 2]$. (*Hint:* Let $u^2=1+e^{2x}$.)

[9.5.2]

In Problems 49–78 use partial fractions when appropriate to evaluate the given integral.

49. $\int \dfrac{dx}{x^4+5x^2+4}$

50. $\int \dfrac{dx}{x^4+13x^2+36}$

51. $\int \dfrac{x-15}{(x^2+2x+5)(x^2+6x+10)}\,dx$

52. $\int \dfrac{x^2}{(x^2+8x+20)(x^2+4x+6)}\,dx$

53. $\int \dfrac{x-1}{x(x^2+1)}\,dx$

54. $\int \dfrac{dx}{(x-1)(x^2+3)}$

55. $\int \dfrac{2x-3}{x^3-3x^2+9x-27}\,dx$

56. $\int \dfrac{x+4}{x^4+9x^2}\,dx$

57. $\int \dfrac{x}{(x+1)^2(x^2+1)}\,dx$

58. $\displaystyle\int \frac{x^2}{(x-1)^3(x^2+4)}\,dx$

59. $\displaystyle\int \frac{dt}{t^4-1}$

60. $\displaystyle\int \frac{t^3}{t^4-16}\,dt$

61. $\displaystyle\int \frac{2x+1}{(x^2+4)^2}\,dx$

62. $\displaystyle\int \frac{dx}{(x^4-16)^2}$

63. $\displaystyle\int \frac{3x^2-x+1}{(x+1)(x^2+2x+2)}\,dx$

64. $\displaystyle\int \frac{4x-5}{(x-2)(x^2+4x+8)}\,dx$

65. $\displaystyle\int \frac{dx}{x^3-1}$

66. $\displaystyle\int \frac{dx}{x^4+27x}$

67. $\displaystyle\int \frac{dx}{(x^3+x)^2}$

68. $\displaystyle\int \frac{dx}{x^3(x^2+1)^2}$

69. $\displaystyle\int \frac{x^3-2x^2+x-3}{x^4+8x^2+16}\,dx$

70. $\displaystyle\int \frac{3x^2+2x-4}{x^4+6x^2+9}\,dx$

71. $\displaystyle\int \frac{2x}{(4x^2+5)^2}\,dx$

72. $\displaystyle\int \frac{x^2}{(x^2+1)^3}\,dx$

73. $\displaystyle\int \frac{x^2-2x+3}{x(x^2+2x+2)^2}\,dx$

74. $\displaystyle\int \frac{x}{(x-1)(x^2+4x+5)^2}\,dx$

75. $\displaystyle\int_0^1 \frac{dx}{x^3+x^2+2x+2}$

76. $\displaystyle\int_0^1 \frac{x^2}{x^4+8x^2+16}\,dx$

77. $\displaystyle\int_{-1}^1 \frac{2x^3+5x}{x^4+5x^2+6}\,dx$

78. $\displaystyle\int_1^2 \frac{1}{x^5+4x^4+5x^3}\,dx$

In Problems 79 and 80 use the indicated substitution to evaluate the given integral.

79. $\displaystyle\int \frac{\sqrt[3]{x+1}}{x}\,dx; \; u^3 = x+1$

80. $\displaystyle\int \frac{dx}{\sqrt{x}(1+\sqrt[3]{x})^2}; \; u^6 = x$

81. Find the area under the graph of $y = \dfrac{x^3}{(x^2+1)(x^2+2)}$ on the interval $[0, 4]$.

82. Find the area bounded by the graph of $y = 3x^2/(x^3-1)$ and the x-axis on the interval $[-1, \frac{1}{2}]$.

83. The region in the first quadrant bounded by the graphs of $y = 2x/(x^2+1)$, $x = 1$, and $y = 0$ is revolved about the x-axis. Find the volume of the solid of revolution.

84. The region in the first quadrant bounded by the graphs of

$$y = \frac{8}{(x^2+1)(x^2+4)}$$

$x = 0$, $x = 1$, and $y = 0$ is revolved about the y-axis. Find the volume of the solid of revolution.

9.6 Integration of Rational Functions of Sine and Cosine

Integrals of rational expressions that involve $\sin x$ and $\cos x$ can be reduced to integrals of quotients of polynomials by means of the substitution

$$u = \tan \frac{x}{2}, \qquad -\pi < x < \pi \tag{9.18}$$

Since $\tan^{-1} u = x/2$, we see from Figure 9.10 that

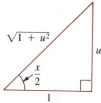

Figure 9.10

$$\sin \frac{x}{2} = \frac{u}{\sqrt{1+u^2}} \quad \text{and} \quad \cos \frac{x}{2} = \frac{1}{\sqrt{1+u^2}}$$

From the trigonometric identities for double angles, it follows that

$$\sin x = 2 \sin \frac{x}{2} \cos \frac{x}{2} = \frac{2u}{1 + u^2}$$

$$\cos x = \cos^2 \frac{x}{2} - \sin^2 \frac{x}{2} = \frac{1 - u^2}{1 + u^2}$$

Furthermore, from $x = 2 \tan^{-1}u$, we immediately obtain

$$dx = \frac{2}{1 + u^2} \, du$$

In summary, the substitution $u = \tan(x/2)$ leads to

$$\sin x = \frac{2u}{1 + u^2}, \qquad \cos x = \frac{1 - u^2}{1 + u^2}, \qquad dx = \frac{2}{1 + u^2} \, du \qquad (9.19)$$

___ **Example 1** _____

Evaluate $\displaystyle \int \frac{dx}{2 + 2 \sin x + \cos x}$.

Solution   Using (9.19) and simplifying, the given integral becomes

$$\int \frac{dx}{2 + 2 \sin x + \cos x} = \int \frac{2 \, du}{u^2 + 4u + 3}$$

Since $u^2 + 4u + 3 = (u + 1)(u + 3)$, we use partial fractions:

$$\int \frac{dx}{2 + 2 \sin x + \cos x} = \int \left[\frac{1}{u + 1} - \frac{1}{u + 3} \right] du$$

$$= \ln|u + 1| - \ln|u + 3| + C$$

$$= \ln\left| \frac{u + 1}{u + 3} \right| + C$$

$$= \ln\left| \frac{1 + \tan(x/2)}{3 + \tan(x/2)} \right| + C$$

_____ **Exercises 9.6** _____

Answers to odd-numbered problems begin on page A-36

In Problems 1–16 evaluate the given integral by substitution.

1. $\displaystyle \int \frac{dx}{1 + \sin x + \cos x}$

2. $\displaystyle \int \frac{dx}{2 + \sin x + \cos x}$

7. $\displaystyle \int \frac{dx}{\tan x + \sin x}$

8. $\displaystyle \int \frac{dx}{\cot x + \cos x}$

3. $\displaystyle \int \frac{dx}{2 + \cos x}$

4. $\displaystyle \int \frac{dx}{4 - 5 \sin x}$

9. $\displaystyle \int \frac{\sec x}{1 + \cos x} \, dx$

10. $\displaystyle \int \frac{\csc x}{1 + \sin x} \, dx$

5. $\displaystyle \int \frac{dx}{1 + \sec x}$

6. $\displaystyle \int \frac{\sec x}{\sec x + \tan x - 1} \, dx$

11. $\displaystyle \int \frac{dx}{1 - 2 \sin x}$

12. $\displaystyle \int \frac{dx}{2 \sec x - 1}$

13. $\displaystyle\int_0^{\pi/3} \frac{dx}{1 + \sin x}$

14. $\displaystyle\int_0^{2\pi/3} \frac{\cos x}{\cos x + \sin x}\, dx$

15. $\displaystyle\int_0^{\pi/2} \frac{dx}{3 + 2\cos x + 3\sin x}$

16. $\displaystyle\int_0^{\pi/2} \frac{1 + \sin x}{1 + \cos x}\, dx$

17. Evaluate the integral in Problem 13 by an alternative method. (*Hint*: Multiply numerator and denominator by $1 - \sin x$.)

18. Find the area under the graph of $y = 1/(1 + \cos x)$ on the interval $[0, \pi/2]$.

19. **(a)** Use the method of this section to show that

$$\int \sec x\, dx = \ln\left|\frac{1 + \tan(x/2)}{1 - \tan(x/2)}\right| + C$$

(b) Show that the result in part **(a)** is equivalent to (8.36) of Section 8.3.

20. **(a)** Use the method of this section to show that

$$\int \csc x\, dx = \frac{1}{2}\ln\left(\frac{1 - \cos x}{1 + \cos x}\right) + C$$

(b) Show that the result in part **(a)** is equivalent to (8.37) of Section 8.3.

[O] 9.7 A Review of Applications

The purpose of this brief section is twofold: (1) to review some applications of the integral that we have seen in prior chapters while (2) utilizing the techniques of the present chapter. It will be up to you to decide which method of integration is appropriate in a problem.

Exercises 9.7

Answers to odd-numbered problems begin on page A-37

Surface area

1. Find the area of the surface that is formed by revolving $y = x^2/2$ on the interval $[0, 1]$ about the x-axis.

2. Find the area of the surface that is formed by revolving $y = e^x$ on the interval $[0, \ln\sqrt{2}]$ about the x-axis. (*Hint*: First let $u = e^x$.)

Average value

In Problems 3–6 find the average value of the given function on the indicated interval.

3. $f(x) = \tan^{-1}\dfrac{x}{2}$; $[0, 2]$

4. $f(x) = x\sin x$; $[0, \pi]$

5. $f(x) = \dfrac{1}{x^2 - 3x - 4}$; $[1, 2]$

6. $f(x) = (4 - x^2)^{-3/2}$; $[-1, 1]$

Rectilinear motion

7. A body moves in a straight line with velocity $v(t) = e^{-t}\sin t$, where v is measured in cm/sec. Find the position function $s(t)$ if it is known that $s = 0$ when $t = 0$.

8. A body moves in a straight line with acceleration $a(t) = te^{-t}$, where a is measured in cm/sec^2. Find the velocity function $v(t)$ and position function $s(t)$ if $v(0) = 1$ and $s(0) = -1$.

Pump problems

9. A water tank is formed by revolving the region bounded by the graphs of $y = \sin \pi x$, $y = 0$, $0 \le x \le 1$ about the x-axis, which is taken in the downward direction. The tank is filled to a depth of $\frac{1}{2}$ ft. Determine the work done in pumping all the water to the top of the tank.

10. A water tank is formed by revolving the region bounded by the graphs of $y = \ln x$, $y = 0$, $y = 2$, and $x = 0$ about the y-axis, which is taken in the upward direction. Dimensions are in feet. Given that the tank is full, find the work done in pumping all the water to its top.

Liquid pressure

In Problems 11 and 12 (Figures 9.11 and 9.12) find the force caused by liquid pressure on one side of the given vertical plate. Assume that the plate is submerged in water and that dimensions are in feet.

11.

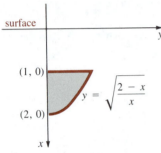

Figure 9.11

12.

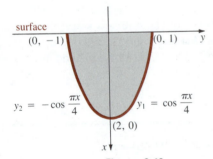

Figure 9.12

Center of mass of a rod

In Problems 13 and 14 a rod of linear density $\rho(x)$ kg/m coincides with the x-axis on the interval indicated. Find its center of mass.

13. $\rho(x) = (9 - x^2)^{-3/2}$; $[0, \sqrt{5}]$

14. $\rho(x) = \sqrt{16 + x^2}$; $[0, 3]$

Centroid of a plane region

In Problems 15–18 find the centroid of the region bounded by the graphs of the given equations.

15. $y = \sin x$, $y = 0$, $x = \pi/2$

16. $y = e^x$, $y = 0$, $x = 0$, $x = \ln 2$

17. $y = \dfrac{1}{\sqrt{1 + x^2}}$, $y = 0$, $x = 0$, $x = \sqrt{3}$

18. $y = \ln x$, $y = 0$, $x = e$

Separation of variables

19. Solve the **logistic equation*** $dP/dt = P(a - bP)$, $a > 0$, $b > 0$, by separation of variables subject to $P(0) = P_0$.

20. The rate at which a chemical is formed during a second-order chemical reaction is given by $dX/dt = k(a - X)(b - X)$, where k, a, and b are constants. Use separation of variables to solve the differential equation in the case $a \ne b$.

21. A woman, W, starting at the origin, moves in the direction of the positive x-axis pulling a weight along the curve C, called a **tractrix**, indicated in Figure 9.13. The weight, initially located on the y-axis at $(0, s)$, is pulled by a rope of constant length s that is kept taut throughout the motion. Solve the differential equation of the tractrix

$$\frac{dy}{dx} = -\frac{y}{\sqrt{s^2 - y^2}}$$

by separation of variables. Assume that the initial point on the y-axis is $(0, 10)$ and the length of the rope is $s = 10$ ft.

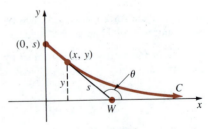

Figure 9.13

---

*Around 1840, the Belgian mathematician-biologist P. F. Verhulst used the logistic equation as a mathematical model for predicting the human population of various countries.

[O] 9.8 A Word on the Use of Integral Tables

There is a time and place to use a table of integrals. Some functions defy the conventional methods of integration considered in this chapter and require advanced methods such as the use of complex variables. On the other hand, some functions merely present difficulties and given time, energy, and a modicum of ingenuity an elementary antiderivative can be found.

Listed on the inside of the front and back covers, you will find an extensive list of integration formulas. Some of these formulas are familiar but most are new. A table of integrals is not a cure for all problems. A person can often waste an inordinate amount of time hunting for the answers for integrals such as

$$\int \frac{(4 - e^{-x})^{5/3}}{e^x} \, dx, \qquad \int \frac{x^3 + 4x}{(x - 1)(x + 5)} \, dx, \qquad \int e^{\sin\theta} \sin 2\theta \, d\theta$$

when actually a few minutes of analysis will "crack" all three problems. In a word, a table of integrals should be a last rather than a first resort.

Example 1

Evaluate $\displaystyle\int \frac{x^3}{\sqrt{3 + 2x}} \, dx$ from tables.

Solution   With $u = x$, $a = 3$, $b = 2$, and $n = 3$ we see from formula 61 of the Table of Integrals that

$$\int \frac{x^3}{\sqrt{3 + 2x}} \, dx = \frac{2x^3\sqrt{3 + 2x}}{2 \cdot 7} - \frac{2 \cdot 3 \cdot 3}{2 \cdot 7} \int \frac{x^2}{\sqrt{3 + 2x}} \, dx$$

Continuing, we apply formula 56 to the second integral:

$$\int \frac{x^3}{\sqrt{3 + 2x}} \, dx = \frac{1}{7}x^3\sqrt{3 + 2x} - \frac{9}{7}\left[\frac{2}{15 \cdot 8}(8 \cdot 9 + 3 \cdot 4x^2 - 4 \cdot 6x)\sqrt{3 + 2x}\right] + C$$

$$= \frac{1}{7}x^3\sqrt{3 + 2x} - \frac{54}{35}\sqrt{3 + 2x} - \frac{9}{35}x^2\sqrt{3 + 2x} + \frac{18}{35}x\sqrt{3 + 2x} + C$$

Alternative Solution I   Let $u = 3 + 2x$ and proceed as in Section 9.1.

Alternative Solution II   Let $dv = (3 + 2x)^{-1/2} \, dx$, $u = x^3$ and use integration by parts.

Example 2

Evaluate $\displaystyle\int \sqrt{4x - x^2} \, dx$ from tables.

Solution   From formula 120 with $u = x$ and $a = 2$, we get

$$\int \sqrt{4x - x^2} \, dx = \frac{x - 2}{2}\sqrt{4x - x^2} + 2 \cos^{-1}\left(\frac{2 - x}{2}\right) + C$$

Alternative Solution   Write $4x - x^2 = 4 - (2 - x)^2$ and use a trigonometric substitution.

Example 3

Evaluate $\displaystyle\int \frac{dx}{1 + e^x}$ from tables.

Solution   From formula 109 with $u = x$, $a = 1$, and $b = 1$,

$$\int \frac{dx}{1 + e^x} = x - \ln|1 + e^x| + C$$

Alternative Solution

Write $\displaystyle\frac{1}{1 + e^x} = \frac{1 + e^x - e^x}{1 + e^x} = 1 - \frac{e^x}{1 + e^x}$

and integrate term-by-term.

Exercises 9.8

Answers to odd-numbered problems begin on page A-37

In Problems 1–22 evaluate the given integral using the Table of Integrals given on the inside front and back covers.

1. $\displaystyle\int \frac{dx}{x^2\sqrt{9 + x^2}}$

2. $\displaystyle\int x^2\sqrt{25 - x^2}\, dx$

3. $\displaystyle\int \frac{\sqrt{x^2 - 5}}{x}\, dx$

4. $\displaystyle\int (4 - x^2)^{-3/2}\, dx$

5. $\displaystyle\int \frac{dx}{x(4 + 5x)^2}$

6. $\displaystyle\int \frac{x^4}{\sqrt{1 + x}}\, dx$

7. $\displaystyle\int t^2\sqrt{1 + 2t}\, dt$

8. $\displaystyle\int \frac{\sqrt{1 + u}}{u}\, du$

9. $\displaystyle\int \frac{x^2}{(3 - x)^2}\, dx$

10. $\displaystyle\int \frac{dx}{x\sqrt{5 - x}}$

11. $\displaystyle\int \tan^5\theta\, d\theta$

12. $\displaystyle\int \cos 6y \cos 2y\, dy$

13. $\displaystyle\int \ln(x^2 + 16)\, dx$

14. $\displaystyle\int e^x\ln|e^{2x} - 1|\, dx$

15. $\displaystyle\int \frac{dx}{1 + \sin 2x}$

16. $\displaystyle\int \frac{x}{1 - \sin 4x}\, dx$

17. $\displaystyle\int \frac{x}{\sqrt{2x - x^2}}\, dx$

18. $\displaystyle\int \frac{\sqrt{6x - x^2}}{x^2}\, dx$

19. $\displaystyle\int_0^{\pi/2} \sin^{10}x\, dx$

20. $\displaystyle\int_1^e x^9 \ln x\, dx$

21. $\displaystyle\int_0^\pi e^{2t} \sin 3t\, dt$

22. $\displaystyle\int_0^1 \frac{dt}{2 + e^{4t}}$

Miscellaneous Problems

In Problems 23 and 24 for the given integral, derive the general formula found in the Table of Integrals.

23. $\displaystyle\int \frac{u}{a + bu}\, du$

24. $\displaystyle\int \frac{u}{(a + bu)^2}\, du$

Chapter 9 Review Exercises

Answers to odd-numbered problems begin on page A-37

In Problems 1–10 answer true or false.

1. Under the change of variable $u = 2x + 3$, the integral $\int_1^5 \dfrac{4x}{\sqrt{2x + 3}}\,dx$ becomes $\int_5^{13} (u^{1/2} - 3u^{-1/2})\,du$. _____

2. The trigonometric substitution $u = a \sec \theta$ is appropriate for integrals that contain $\sqrt{a^2 + u^2}$. _____

3. The method of integration by parts is derived from the Product Rule for differentiation. _____

4. $\displaystyle\int_1^e 2x \ln x^2\,dx = e^2 + 1$ _____

5. Partial fractions is not applicable to $\displaystyle\int \dfrac{dx}{(x - 1)^3}$. _____

6. A partial fraction decomposition of $x^2/(x + 1)^2$ can be found having the form $A/(x + 1) + B/(x + 1)^2$, where A and B are constants. _____

7. To evaluate $\displaystyle\int \dfrac{dx}{(x^2 - 1)^2}$, we assume constants A, B, C, and D can be found such that

$$\dfrac{1}{(x^2 - 1)^2} = \dfrac{Ax + B}{x^2 - 1} + \dfrac{Cx + D}{(x^2 - 1)^2}$$ _____

8. To evaluate $\int x^n e^x\,dx$, n a positive integer, integration by parts is used $n - 1$ times. _____

9. To evaluate $\displaystyle\int \dfrac{x}{\sqrt{9 - x^2}}\,dx$, it is necessary to use $x = 3 \sin \theta$. _____

10. When evaluated, the integral $\int \sin^3 x \cos^2 x\,dx$ can be expressed as a sum of powers of $\cos x$. _____

In Problems 11–90 use the methods of this chapter, or previous chapters, to evaluate the given integral.

11. $\displaystyle\int \dfrac{dx}{\sqrt{x} + 9}$

12. $\displaystyle\int e^{\sqrt{x+1}}\,dx$

13. $\displaystyle\int \dfrac{x}{\sqrt{x^2 + 4}}\,dx$

14. $\displaystyle\int \dfrac{dx}{\sqrt{x^2 + 4}}$

15. $\displaystyle\int \dfrac{dx}{(x^2 + 4)^3}$

16. $\displaystyle\int \dfrac{x^2}{x^2 + 4}\,dx$

17. $\displaystyle\int \dfrac{x^2 + 4}{x^2}\,dx$

18. $\displaystyle\int \dfrac{3x - 1}{x(x^2 - 4)}\,dx$

19. $\displaystyle\int \dfrac{x - 5}{x^2 + 4}\,dx$

20. $\displaystyle\int \dfrac{\sqrt[3]{x + 27}}{x}\,dx$

21. $\displaystyle\int \dfrac{(\ln x)^9}{x}\,dx$

22. $\displaystyle\int (\ln 3x)^2\,dx$

23. $\displaystyle\int t \sin^{-1} t\,dt$

24. $\displaystyle\int \dfrac{\ln x}{(x - 1)^2}\,dx$

25. $\displaystyle\int (x + 1)^3(x - 2)\,dx$

26. $\displaystyle\int \dfrac{dx}{(x + 1)^3(x - 2)}$

27. $\displaystyle\int \ln(x^2 + 4)\,dx$

28. $\displaystyle\int 8t\, e^{2t^2}\,dt$

29. $\displaystyle\int \dfrac{dx}{x^4 + 10x^3 + 25x^2}$

30. $\displaystyle\int \dfrac{dx}{x^2 + 8x + 25}$

31. $\displaystyle\int \dfrac{x}{x^3 + 3x^2 - 9x - 27}\,dx$

32. $\displaystyle\int \dfrac{x + 1}{(x^2 - x)(x^2 + 3)}\,dx$

33. $\displaystyle\int \dfrac{\sin^2 t}{\cos^2 t}\,dt$

34. $\displaystyle\int \dfrac{\sin^3 \theta}{(\cos \theta)^{3/2}}\,d\theta$

35. $\displaystyle\int \tan^{10} x \sec^4 x\,dx$

36. $\displaystyle\int \dfrac{x \tan x}{\cos x}\,dx$

37. $\displaystyle\int y \cos y\,dy$

38. $\displaystyle\int x^2 \sin x^3\,dx$

39. $\displaystyle\int (1 + \sin^2 t)\cos^3 t\,dt$

40. $\displaystyle\int \dfrac{\sec^3 \theta}{\tan \theta}\,d\theta$

41. $\displaystyle\int e^w(1 + e^w)^5\,dw$

42. $\displaystyle\int (x - 1)e^{-x}\,dx$

43. $\displaystyle\int \cot^3 4x\,dx$

44. $\displaystyle\int (3 - \sec x)^2\,dx$

45. $\displaystyle\int_0^{\pi/4} \cos^2 x \tan x\,dx$

46. $\displaystyle\int_0^{\pi/3} \sin^4 x \tan x\,dx$

47. $\displaystyle\int \dfrac{\sin x}{1 + \sin x}\,dx$

48. $\displaystyle\int \dfrac{\cos x}{1 + \sin x}\,dx$

49. $\displaystyle\int_0^1 \dfrac{dx}{(x + 1)(x + 2)(x + 3)}$

50. $\displaystyle\int_{\ln 3}^{\ln 2} \sqrt{e^x + 1}\,dx$

51. $\displaystyle\int e^x \cos 3x\,dx$

52. $\displaystyle\int x(x - 5)^9\,dx$

53. $\displaystyle\int \cos(\ln t)\,dt$

54. $\displaystyle\int \sec^2 x \ln(\tan x)\,dx$

55. $\int \cos \sqrt{x} \, dx$

56. $\int \frac{\cos \sqrt{x}}{\sqrt{x}} \, dx$

57. $\int \cos x \sin 2x \, dx$

58. $\int (\cos^2 x - \sin^2 x) \, dx$

59. $\int \sqrt{x^2 + 2x + 5} \, dx$

60. $\int \frac{dx}{(8 - 2x - x^2)^{3/2}}$

61. $\int \tan^5 x \sec^3 x \, dx$

62. $\int \cos^4 \frac{x}{2} \, dx$

63. $\int \frac{t^5}{1 + t^2} \, dt$

64. $\int \frac{dx}{\sqrt{1 - x^2}}$

65. $\int \frac{5x^3 + x^2 + 6x + 1}{(x^2 + 1)^2} \, dx$

66. $\int \frac{\sqrt{x^2 + 9}}{x^2} \, dx$

67. $\int x \sin^2 x \, dx$

68. $\int (t + 1)^2 e^{3t} \, dt$

69. $\int \sin x \cos x \, e^{\sin x} \, dx$

70. $\int e^x \tan^2 e^x \, dx$

71. $\int_0^{\pi/6} \frac{\cos x}{\sqrt{1 + \sin x}} \, dx$

72. $\int_0^{\pi/2} \frac{dx}{\sin x + \cos x}$

73. $\int \sinh^{-1} t \, dt$

74. $\int x \cot x^2 \, dx$

75. $\int_3^8 \frac{dx}{x \sqrt{x + 1}}$

76. $\int \frac{t + 3}{t^2 + 2t + 1} \, dt$

77. $\int \frac{\sec^4 3u}{\cot^{12} 3u} \, du$

78. $\int_0^2 x^5 \sqrt{x^2 + 4} \, dx$

79. $\int \frac{3 + \sin x}{\cos^2 x} \, dx$

80. $\int \frac{\sin 2x}{5 + \cos^2 x} \, dx$

81. $\int x(1 + \ln x)^2 \, dx$

82. $\int x \cos^2 x \, dx$

83. $\int e^x e^{e^x} \, dx$

84. $\int \frac{dx}{\sqrt{x + 1} - \sqrt{x}}$

85. $\int \frac{2 \, dx}{\sin x - \tan x}$

86. $\int \cos x \cos 2x \, dx$

87. $\int \frac{dx}{\sqrt{1 - (5x + 2)^2}}$

88. $\int (\ln 2x) \ln x \, dx$

89. $\int \cos x \ln |\sin x| \, dx$

90. $\int \ln \left(\frac{x + 1}{x - 1} \right) \, dx$

10

Indeterminate Forms and Improper Integrals

The material that follows in Section 10.2 and Chapter 11 demands that we know more about computing limits. Therefore, in Section 10.1 we consider a fairly simple but very useful rule for computing certain limits by taking derivatives.

   As a preparation for this chapter, you are encouraged to review Sections 2.1–2.4 and 5.5.

10.1 L'Hôpital's Rule

10.1.1 The Indeterminate Forms 0/0 and ∞/∞

In Chapter 1 we considered limits of quotients such as

$$\lim_{x \to 1} \frac{x^2 + 3x - 4}{x - 1} \quad \text{and} \quad \lim_{x \to \infty} \frac{2x^2 - x}{3x^2 + 1} \tag{10.1}$$

where in the first limit both the numerator and denominator are approaching zero as $x \to 1$, and in the second both the numerator and denominator are approaching ∞ as $x \to \infty$.

Terminology   In general, we say that the limit

$$\lim_{x \to a} \frac{f(x)}{g(x)}$$

has the **indeterminate form 0/0 at $x = a$** if

$$f(x) \to 0 \quad \text{and} \quad g(x) \to 0 \quad \text{as} \quad x \to a$$

and has the **indeterminate form ∞/∞ at $x = a$** if

$$|f(x)| \to \infty \quad \text{and} \quad |g(x)| \to \infty \quad \text{as} \quad x \to a*$$

A limit can also have an indeterminate form as

$$x \to a^-, \quad x \to a^+, \quad x \to -\infty, \quad \text{or} \quad x \to \infty$$

Example 1

(a) $\displaystyle \lim_{x \to 0} \frac{\sin x}{x}$ has the indeterminate form 0/0 at $x = 0$ since

$$\sin x \to 0 \quad \text{and} \quad x \to 0 \quad \text{as} \quad x \to 0$$

(b) $\displaystyle \lim_{x \to 3^+} \frac{1/(3 - x)}{1/(3 - x)^2}$ has the indeterminate form ∞/∞ at $x = 3$ since

$$\frac{1}{3 - x} \to -\infty \quad \text{and} \quad \frac{1}{(3 - x)^2} \to \infty \quad \text{as} \quad x \to 3^+$$

(c) $\displaystyle \lim_{x \to \infty} \frac{\ln x}{e^x}$ has the indeterminate form ∞/∞ since

$$\ln x \to \infty \quad \text{and} \quad e^x \to \infty \quad \text{as} \quad x \to \infty$$

---

*The absolute value signs mean that as x approaches a we could have, say, $f(x) \to \infty$, $g(x) \to \infty$; or $f(x) \to -\infty$, $g(x) \to \infty$; or $f(x) \to -\infty$, $g(x) \to -\infty$; and so on.

Note: Limits of the form

$$0/k, \qquad k/0, \qquad \infty/k, \quad \text{and} \quad k/\infty$$

where k is a nonzero constant, are *not* indeterminate forms. The value of a limit whose form is $0/k$ or k/∞ is zero, whereas a limit whose form is either $k/0$ or ∞/k does not exist.

In establishing whether limits of quotients such as those given in (10.1) exist, we resorted to the algebraic manipulations of factoring, canceling, and division. However, recall that the proof of $\lim_{x \to 0}(\sin x)/x = 1$ used an elaborate geometric argument. But, algebra and geometric intuition fail miserably when confronted with a problem of the type

$$\lim_{x \to 0} \frac{\sin x}{e^x - e^{-x}}$$

which has the indeterminate form $0/0$. The next theorem will aid us in proving a rule that is extremely helpful in evaluating many limits which have an indeterminate form.

THEOREM 10.1   The Extended Mean Value Theorem

Let f and g be continuous on $[a, b]$ and differentiable on (a, b) and $g'(x) \neq 0$ for all x in (a, b). Then there exists a number c in (a, b) such that

$$\frac{f(b) - f(a)}{g(b) - g(a)} = \frac{f'(c)}{g'(c)}$$

Observe that Theorem 10.1 reduces to the Mean Value Theorem when $g(x) = x$. A proof of this theorem, which is reminiscent of the proof of Theorem 4.5, is outlined in Problems 43 and 44 in Exercises 10.1.

The following rule is named after the French mathematician G. F. A. L'Hôpital. We assume f and g are differentiable on the open intervals (r, a) and (a, s) and that $g'(x) \neq 0$ for $x \neq a$.

L'Hôpital's Rule*

Suppose $\lim\limits_{x \to a} \dfrac{f(x)}{g(x)}$ has an indeterminate form at $x = a$ and $\lim\limits_{x \to a} \dfrac{f'(x)}{g'(x)} = L$ or $\pm\infty$. Then

$$\lim_{x \to a} \frac{f(x)}{g(x)} = \lim_{x \to a} \frac{f'(x)}{g'(x)} \tag{10.2}$$

---

*It is questionable whether the Marquis Guillaume François Antoine de L'Hôpital (1661–1704) discovered the rule bearing his name. The result is probably due to Johann Bernoulli. However, L'Hôpital was the first to publish the rule in his text *Analyse des Infiniment Petits*. The book was published in 1696 and is considered to be the first textbook on calculus.

Proof of the case 0/0   Since

$$\lim_{x \to a} f(x) = 0 \quad \text{and} \quad \lim_{x \to a} g(x) = 0$$

it can be assumed that $f(a) = 0$ and $g(a) = 0$. It follows that f and g are continuous at a. Moreover, since f and g are differentiable, these functions are continuous on both (r, a) and (a, s). Consequently, f and g are continuous on the interval (r, s). Now, for any $x \neq a$ in the interval, Theorem 10.1 is applicable to either $[x, a]$ or $[a, x]$. In either case, there exists a number c between x and a such that

$$\frac{f(x) - f(a)}{g(x) - g(a)} = \frac{f(x)}{g(x)} = \frac{f'(c)}{g'(c)}$$

Letting $x \to a$ implies $c \to a$ and so

$$\lim_{x \to a} \frac{f(x)}{g(x)} = \lim_{x \to a} \frac{f'(c)}{g'(c)} = \lim_{c \to a} \frac{f'(c)}{g'(c)} = L \qquad \blacksquare$$

Example 2

Evaluate $\lim\limits_{x \to 0} \dfrac{\sin x}{x}$ by L'Hôpital's Rule.

Solution   In Example 1 we saw that the given limit has the indeterminate form $0/0$ at $x = 0$. Thus, from (10.2) we can write

$$\lim_{x \to 0} \frac{\sin x}{x} = \lim_{x \to 0} \frac{\cos x}{1} = \frac{1}{1} = 1$$

derivative

derivative

Example 3

Evaluate $\lim\limits_{x \to 0} \dfrac{\sin x}{e^x - e^{-x}}$.

Solution   Since the given limit has the indeterminate form $0/0$ at $x = 0$, we apply (10.2),

$$\lim_{x \to 0} \frac{\sin x}{e^x - e^{-x}} = \lim_{x \to 0} \frac{\dfrac{d}{dx} \sin x}{\dfrac{d}{dx}(e^x - e^{-x})}$$

$$= \lim_{x \to 0} \frac{\cos x}{e^x + e^{-x}} = \frac{1}{1 + 1} = \frac{1}{2}$$

The result given in (10.2) remains valid when $x \to a$ is replaced by one-sided limits or by $x \to \infty$, $x \to -\infty$. The proof of the case $x \to \infty$ can be obtained by using the substitution $x = 1/t$ in $\lim_{x \to \infty} f(x)/g(x)$ and noting that $x \to \infty$ is equivalent to $t \to 0^+$.

Example 4

Evaluate $\lim\limits_{x \to \infty} \dfrac{\ln x}{e^x}$.

Solution   The limit has the indeterminate form ∞/∞. Thus, from L'Hôpital's Rule we have

$$\lim_{x \to \infty} \frac{\ln x}{e^x} = \lim_{x \to \infty} \frac{1/x}{e^x} = \lim_{x \to \infty} \frac{1}{xe^x}$$

(derivative / derivative)

In this latter limit, $xe^x \to \infty$ as $x \to \infty$, whereas 1 remains constant. Consequently,

$$\lim_{x \to \infty} \frac{\ln x}{e^x} = \lim_{x \to \infty} \frac{1}{xe^x} = 0$$

It may be necessary to apply L'Hôpital's Rule several times in the course of solving a problem.

Example 5

Evaluate $\lim\limits_{x \to \infty} \dfrac{6x^2 + 5x + 7}{4x^2 + 2x}$.

Solution   The indeterminate form is clearly ∞/∞, and so by (10.2),

$$\lim_{x \to \infty} \frac{6x^2 + 5x + 7}{4x^2 + 2x} = \lim_{x \to \infty} \frac{12x + 5}{8x + 2}$$

Since the new limit still has the indeterminate form ∞/∞, we apply (10.2) a second time:

$$\lim_{x \to \infty} \frac{12x + 5}{8x + 2} = \lim_{x \to \infty} \frac{12}{8} = \frac{3}{2}$$

Thus,
$$\lim_{x \to \infty} \frac{6x^2 + 5x + 7}{4x^2 + 2x} = \frac{3}{2}$$

Example 6

Evaluate $\lim\limits_{x\to\infty} \dfrac{e^{3x}}{x^2}$.

Solution   The given limit, and the limit obtained after one application of L'Hôpital's Rule, have the indeterminate form ∞/∞,

$$\lim_{x\to\infty} \frac{e^{3x}}{x^2} = \lim_{x\to\infty} \frac{3e^{3x}}{2x} = \lim_{x\to\infty} \frac{9e^{3x}}{2}$$

After the second application of (10.2) we notice $e^{3x} \to \infty$ while the denominator remains constant. From this we conclude that

$$\lim_{x\to\infty} \frac{e^{3x}}{x^2} = \infty$$

In other words, the given limit does not exist.

Example 7

Evaluate $\lim\limits_{x\to\infty} \dfrac{x^4}{e^{2x}}$.

Solution   We apply (10.2) four times:

$$\lim_{x\to\infty} \frac{x^4}{e^{2x}} = \lim_{x\to\infty} \frac{4x^3}{2e^{2x}} \qquad (\infty/\infty)$$

$$= \lim_{x\to\infty} \frac{12x^2}{4e^{2x}} \qquad (\infty/\infty)$$

$$= \lim_{x\to\infty} \frac{6x}{2e^{2x}} \qquad (\infty/\infty)$$

$$= \lim_{x\to\infty} \frac{6}{4e^{2x}} = 0$$

In successive applications of L'Hôpital's Rule, it is sometimes possible to change a limit form from, say, ∞/∞ to $0/0$.

Example 8

Evaluate $\lim\limits_{t\to\pi/2^+} \dfrac{\tan t}{\tan 3t}$.

Solution   We observe $\tan t \to -\infty$ and $\tan 3t \to -\infty$ as $t \to \pi/2^+$. Hence, from (10.2),

$$\lim_{t\to\pi/2^+} \frac{\tan t}{\tan 3t} = \lim_{t\to\pi/2^+} \frac{\sec^2 t}{3\sec^2 3t} \qquad (\infty/\infty)$$

$$= \lim_{t\to\pi/2^+} \frac{\cos^2 3t}{3\cos^2 t}$$

The latter form, which results from using $\sec t = 1/\cos t$, is now $0/0$. Proceeding we find,

$$\lim_{t \to \pi/2^+} \frac{\tan t}{\tan 3t} = \lim_{t \to \pi/2^+} \frac{2 \cos 3t(-3 \sin 3t)}{6 \cos t(-\sin t)}$$

$$= \lim_{t \to \pi/2^+} \frac{2 \sin 3t \cos 3t}{2 \sin t \cos t}$$

$$= \lim_{t \to \pi/2^+} \frac{\sin 6t}{\sin 2t} \qquad \boxed{\text{Double Angle Formula}}$$

$$= \lim_{t \to \pi/2^+} \frac{6 \cos 6t}{2 \cos 2t} = \frac{-6}{-2} = 3$$

___

Example 9

___

Evaluate $\displaystyle\lim_{x \to 1^+} \frac{\ln x}{\sqrt{x - 1}}$.

Solution The given limit has the indeterminate form $0/0$ at $x = 1$. Hence, by L'Hôpital's Rule

$$\lim_{x \to 1^+} \frac{\ln x}{\sqrt{x - 1}} = \lim_{x \to 1^+} \frac{1/x}{(1/2)(x - 1)^{-1/2}} = \lim_{x \to 1^+} \frac{2\sqrt{x - 1}}{x} = \frac{0}{1} = 0$$

___

Remarks

(*i*) In the application of L'Hôpital's Rule, many students will misinterpret

$$\lim_{x \to a} \frac{f'(x)}{g'(x)} \quad \text{as} \quad \lim_{x \to a} \frac{d}{dx} \frac{f(x)}{g(x)}$$

The rule utilizes the *quotient of derivatives* and *not* the *derivative of the quotient*.

(*ii*) Inspect a problem before you leap to its solution. The limit $\lim_{x \to 0}(\cos x)/x$ is the form $1/0$ and, as a consequence, does not exist. Lack of mathematical forethought in writing

$$\lim_{x \to 0} \frac{\cos x}{x} = \lim_{x \to 0} \frac{\sin x}{1} = 0$$

is an incorrect application of L'Hôpital's Rule. Of course, the "answer" has no significance.

(*iii*) L'Hôpital's rule is not a cure-all for every indeterminate form. For example, $\lim_{x \to \infty} e^x/e^{x^2}$ is certainly of the form ∞/∞, but

$$\lim_{x \to \infty} \frac{e^x}{e^{x^2}} = \lim_{x \to \infty} \frac{e^x}{2xe^{x^2}}$$

is of no immediate help. Nor does L'Hôpital's Rule provide any help in a problem such as

$$\lim_{x \to 0} \frac{e^{1/x}}{x}$$

Why not?

10.1.2 The Indeterminate Forms $\infty - \infty$, $0 \cdot \infty$, 0^0, ∞^0, and 1^∞

There are five additional indeterminate forms:

$$\infty - \infty, \qquad 0 \cdot \infty, \qquad 0^0, \qquad \infty^0, \quad \text{and} \quad 1^\infty$$

By a combination of algebra and a little cleverness we can often convert one of these new limit forms to either $0/0$ or ∞/∞.

The Form $\infty - \infty$ The first example illustrates a limit that has the indeterminate form $\infty - \infty$. This example should destroy any unwarranted convictions that $\infty - \infty = 0$.

_____ **Example 10** _____

Evaluate $\displaystyle\lim_{x \to 0^+} \left[\frac{1 + 3x}{\sin x} - \frac{1}{x} \right]$.

Solution   We note that $(1 + 3x)/\sin x \to \infty$ and $1/x \to \infty$ as $x \to 0^+$. However, after writing the difference as a single fraction, we recognize the form $0/0$:

$$\lim_{x \to 0^+} \left[\frac{1 + 3x}{\sin x} - \frac{1}{x} \right] = \lim_{x \to 0^+} \frac{3x^2 + x - \sin x}{x \sin x}$$

$$= \lim_{x \to 0^+} \frac{6x + 1 - \cos x}{x \cos x + \sin x} \qquad \text{[applying (10.2)]}$$

$$= \lim_{x \to 0^+} \frac{6 + \sin x}{-x \sin x + 2 \cos x} \qquad \text{[(10.2) again]}$$

$$= \frac{6 + 0}{0 + 2} = 3$$

The Form $0 \cdot \infty$ If

$$f(x) \to 0 \quad \text{and} \quad |g(x)| \to \infty \quad \text{as} \quad x \to a$$

then $\lim_{x \to a} f(x)g(x)$ has the indeterminate form $0 \cdot \infty$. We can change a limit that has this form to one with the form $0/0$ or ∞/∞ by writing, in turn,

$$f(x)g(x) = \frac{f(x)}{1/g(x)} \quad \text{or} \quad f(x)g(x) = \frac{g(x)}{1/f(x)}$$

___ **Example 11** _____

Evaluate $\displaystyle\lim_{x\to\infty} x \sin \frac{1}{x}$.

Solution   Since $\sin(1/x) \to 0$ as $x \to \infty$, the limit has the indeterminate form $0 \cdot \infty$. By writing

$$\lim_{x\to\infty} \frac{\sin(1/x)}{1/x}$$

we now have the form $0/0$. Hence,

$$\lim_{x\to\infty} \frac{\sin(1/x)}{1/x} = \lim_{x\to\infty} \frac{(-x^{-2})\cos(1/x)}{(-x^{-2})}$$

$$= \lim_{x\to\infty} \cos \frac{1}{x} = 1$$

The Forms 0^0, ∞^0, 1^∞
Suppose $y = f(x)^{g(x)}$ tends toward 0^0, ∞^0, or 1^∞ as $x \to a$. By taking the natural logarithm of y:

$$\ln y = \ln f(x)^{g(x)}$$
$$= g(x) \ln f(x)$$

we see
$$\lim_{x\to a} \ln y = \lim_{x\to a} g(x) \ln f(x)$$

has the form $0 \cdot \infty$. If it is assumed that $\lim_{x\to a} \ln y = \ln(\lim_{x\to a} y) = L$, then $\lim_{x\to a} y = e^L$ or

$$\lim_{x\to a} f(x)^{g(x)} = e^L$$

Of course, the procedure just outlined is applicable to limits involving $x \to a^-$, $x \to a^+$, $x \to \infty$, or $x \to -\infty$.

___ **Example 12** _____

Evaluate $\displaystyle\lim_{x\to 0^+} x^{1/\ln x}$.

Solution   The limit has the indeterminate form 0^0. Now, if we set $y = x^{1/\ln x}$, then

$$\ln y = \frac{1}{\ln x} \ln x = 1$$

Notice that we do not need L'Hôpital's Rule in this case since

$$\lim_{x\to 0^+} \ln y = 1 \quad \text{or} \quad \ln(\lim_{x\to 0^+} y) = 1$$

Hence, $\displaystyle\lim_{x\to 0^+} y = e^1$ or equivalently $\displaystyle\lim_{x\to 0^+} x^{1/\ln x} = e$

In the next example we consider an important limit result whose proof was postponed in Section 8.2.

Example 13

Evaluate $\lim\limits_{x \to 0}(1 + x)^{1/x}$.

Solution   The limit has the indeterminate form 1^{∞}. If $y = (1 + x)^{1/x}$, then

$$\ln y = \frac{1}{x} \ln(1 + x)$$

Now, $\lim_{x \to 0} \ln(1 + x)/x$ has the form $0/0$ and so by (10.2)

$$\lim_{x \to 0} \frac{\ln(1 + x)}{x} = \lim_{x \to 0} \frac{1/(1 + x)}{1}$$

$$= \lim_{x \to 0} \frac{1}{1 + x} = 1$$

Thus, $\lim\limits_{x \to 0}(1 + x)^{1/x} = e$

Example 14

Evaluate $\lim\limits_{x \to \infty}\left(1 - \dfrac{3}{x}\right)^{2x}$.

Solution   As in the preceding example, the indeterminate form is 1^{∞}. If

$$y = \left(1 - \frac{3}{x}\right)^{2x} \quad \text{then} \quad \ln y = 2x \ln\left(1 - \frac{3}{x}\right)$$

Observe that the form of $\lim_{x \to \infty} 2x \ln(1 - 3/x)$ is $\infty \cdot 0$, whereas the form of $\lim_{x \to \infty} 2 \ln(1 - 3/x)/(1/x)$ is $0/0$. Applying (10.2) to the latter limit gives

$$\lim_{x \to \infty} 2\frac{\ln(1 - 3/x)}{1/x} = \lim_{x \to \infty} 2\frac{\dfrac{3/x^2}{(1 - 3/x)}}{-1/x^2}$$

$$= \lim_{x \to \infty} \frac{-6}{(1 - 3/x)} = -6$$

Finally, we conclude that

$$\lim_{x \to \infty}\left(1 - \frac{3}{x}\right)^{2x} = e^{-6}$$

Exercises 10.1

Answers to odd-numbered problems begin on page A-37.

[10.1.1]

In Problems 1–40 use L'Hôpital's Rule where appropriate to find the limit if it exists.

1. $\lim\limits_{x \to 0} \dfrac{\cos x - 1}{x}$

2. $\lim\limits_{t \to 3} \dfrac{t^3 - 27}{t - 3}$

3. $\lim\limits_{x \to 1} \dfrac{2x - 2}{\ln x}$

4. $\lim\limits_{x \to 0^+} \dfrac{\ln 2x}{\ln 3x}$

5. $\lim\limits_{x \to 0} \dfrac{e^x - 1}{3x + x^2}$

6. $\lim\limits_{x \to 0} \dfrac{\tan x}{2x}$

7. $\lim\limits_{t \to \pi} \dfrac{5 \sin^2 t}{1 + \cos t}$

8. $\lim\limits_{\theta \to 1} \dfrac{\theta^2 - 1}{e^{\theta^2} - e}$

9. $\lim\limits_{x \to 0} \dfrac{6 + 6x + 3x^2 - 6e^x}{x - \sin x}$

10. $\lim\limits_{x \to \infty} \dfrac{3x^2 - 4x^3}{5x + 7x^3}$

11. $\lim\limits_{x \to 0^+} \dfrac{\cot 2x}{\cot x}$

12. $\lim\limits_{x \to 0} \dfrac{\arcsin(x/6)}{\arctan(x/2)}$

13. $\lim\limits_{t \to 2} \dfrac{t^2 + 3t - 10}{t^3 - 2t^2 + t - 2}$

14. $\lim\limits_{r \to -1} \dfrac{r^3 - r^2 - 5r - 3}{(r + 1)^2}$

15. $\lim\limits_{x \to 0} \dfrac{x - \sin x}{x^3}$

16. $\lim\limits_{x \to 1} \dfrac{x^2 + 4}{x^2 + 1}$

17. $\lim\limits_{x \to 0} \dfrac{\cos 2x}{x^2}$

18. $\lim\limits_{x \to \infty} \dfrac{2e^{4x} + x}{e^{4x} + 3x}$

19. $\lim\limits_{x \to 1^+} \dfrac{\ln\sqrt{x}}{x - 1}$

20. $\lim\limits_{x \to \infty} \dfrac{\ln(3x^2 + 5)}{\ln(5x^2 + 1)}$

21. $\lim\limits_{x \to 2} \dfrac{e^{x^2} - e^{2x}}{x}$

22. $\lim\limits_{x \to 0} \dfrac{4^x - 3^x}{x}$

23. $\lim\limits_{x \to \infty} \dfrac{x \ln x}{x^2 + 1}$

24. $\lim\limits_{t \to 0} \dfrac{1 - \cosh t}{t^2}$

25. $\lim\limits_{x \to 0} \dfrac{\sin 5x}{x}$

26. $\lim\limits_{x \to 0} \dfrac{(\sin 2x)^2}{x^2}$

27. $\lim\limits_{x \to \infty} \dfrac{e^x}{x^4}$

28. $\lim\limits_{x \to \infty} \dfrac{e^{1/x}}{\sin(1/x)}$

29. $\lim\limits_{x \to 0} \dfrac{x - \tan^{-1}x}{x - \sin^{-1}x}$

30. $\lim\limits_{t \to 1} \dfrac{t^{1/3} - t^{1/2}}{t - 1}$

31. $\lim\limits_{u \to \pi/2} \dfrac{\ln(\sin u)}{(2u - \pi)^2}$

32. $\lim\limits_{\theta \to \pi/2} \dfrac{\tan \theta}{\ln(\cos \theta)}$

33. $\lim\limits_{x \to -\infty} \dfrac{1 + e^{-2x}}{1 - e^{-2x}}$

34. $\lim\limits_{x \to 0} \dfrac{e^x - x - 1}{2x^2}$

35. $\lim\limits_{r \to 0} \dfrac{r - \cos r}{r - \sin r}$

36. $\lim\limits_{t \to \pi} \dfrac{\csc 7t}{\csc 2t}$

37. $\lim\limits_{x \to 0^+} \dfrac{x^2}{\ln^2(1 + 3x)}$

38. $\lim\limits_{x \to 3} \left(\dfrac{\ln x - \ln 3}{x - 3}\right)^2$

39. $\lim\limits_{x \to 0} \dfrac{3x^2 + e^x - e^{-x} - 2 \sin x}{x \sin x}$

40. $\lim\limits_{x \to 8} \dfrac{\sqrt{x + 1} - 3}{x^2 - 64}$

Miscellaneous Problems

41. Prove that for any integer n, $\lim\limits_{x \to \infty} \dfrac{e^x}{x^n} = \infty$.

42. Prove that for any positive constant k, $\lim\limits_{x \to \infty} \dfrac{\ln x}{x^k} = 0$.

43. Review Rolle's Theorem and show that, under the hypotheses of Theorem 10.1, $g(b) - g(a) \neq 0$. [*Hint:* Suppose $g(b) - g(a) = 0$ or $g(b) = g(a)$. What does Rolle's Theorem say?]

44. Define the function

$$\phi(x) = f(x) - f(a) - \dfrac{f(b) - f(a)}{g(b) - g(a)} \, [g(x) - g(a)]$$

(a) Show that $\phi(a) = \phi(b) = 0$.

(b) Apply Rolle's Theorem to ϕ to obtain the result in Theorem 10.1.

[10.1.2]

In Problems 45–80 use L'Hôpital's Rule where appropriate to find the limit if it exists.

45. $\lim\limits_{x \to 0} \left(\dfrac{1}{e^x - 1} - \dfrac{1}{x}\right)$

46. $\lim\limits_{x \to 0^+} (\cot x - \csc x)$

47. $\lim\limits_{x \to \infty} x(e^{1/x} - 1)$

48. $\lim\limits_{x \to 0^+} x \ln x$

49. $\lim\limits_{x \to 0^+} x^x$

50. $\lim\limits_{x \to 1^-} x^{1/(1-x)}$

51. $\lim\limits_{x\to 0}\left[\dfrac{1}{x} - \dfrac{1}{\sin x}\right]$

52. $\lim\limits_{x\to 0}\left[\dfrac{1}{x} - \sin^{-1}x\right]$

53. $\lim\limits_{x\to 0^+} x\left[\dfrac{1}{\tan x} - \dfrac{5}{x}\right]$

54. $\lim\limits_{x\to 0^+}\left[\dfrac{1}{x} - \dfrac{1}{\ln(x+1)}\right]$

55. $\lim\limits_{\theta\to 0}\theta\csc 4\theta$

56. $\lim\limits_{x\to \pi/2^+}(\sin^2 x)^{\tan x}$

57. $\lim\limits_{x\to\infty}(2 + e^x)^{e^{-x}}$

58. $\lim\limits_{x\to 0^-}(1 - e^x)^{x^2}$

59. $\lim\limits_{t\to\infty}\left(1 + \dfrac{3}{t}\right)^t$

60. $\lim\limits_{h\to 0}(1 + 2h)^{4/h}$

61. $\lim\limits_{x\to 0} x^{(1-\cos x)}$

62. $\lim\limits_{\theta\to 0}(\cos 2\theta)^{1/\theta^2}$

63. $\lim\limits_{x\to\infty}\dfrac{1}{x^2\sin^2(2/x)}$

64. $\lim\limits_{x\to 1}(x^2 - 1)^{x^2}$

65. $\lim\limits_{x\to 0}\left[\dfrac{x}{1+x} - \sin x\right]$

66. $\lim\limits_{t\to\pi}(\cos 2t)^{t-\pi}$

67. $\lim\limits_{x\to 1}\left[\dfrac{1}{x-1} - \dfrac{5}{x^2+3x-4}\right]$

68. $\lim\limits_{x\to 0}\left[\dfrac{1}{x^2} - \dfrac{1}{x}\right]$

69. $\lim\limits_{x\to\infty} x^2 e^{-x}$

70. $\lim\limits_{x\to\infty}(x + e^x)^{2/x}$

71. $\lim\limits_{x\to\infty} x\left(\dfrac{\pi}{2} - \arctan x\right)$

72. $\lim\limits_{t\to\pi/4}(t - \pi/4)\tan 2t$

73. $\lim\limits_{t\to 3}\left[\dfrac{\sqrt{t+1}}{t^2-9} - \dfrac{2}{t^2-9}\right]$

74. $\lim\limits_{x\to 0^+} x\ln(\sin x)$

75. $\lim\limits_{x\to -\infty}\left[\dfrac{1}{e^x} - x^2\right]$

76. $\lim\limits_{x\to 0}(1 + 5\sin x)^{\cot x}$

77. $\lim\limits_{x\to\infty}\left(\dfrac{3x}{3x+1}\right)^x$

78. $\lim\limits_{\theta\to\pi/2^-}(\sec^3\theta - \tan^3\theta)$

79. $\lim\limits_{x\to 0}(\sinh x)^{\tan x}$

80. $\lim\limits_{x\to 0^+} x^{(\ln x)^2}$

Calculator Problems

In Problems 81 and 82 fill in the given table.

81.

x	$(1+x)^{1/x^2}$
0.1	
0.01	

82.

x	$(1+x)^{1/x^2}$
-0.1	
-0.01	

Miscellaneous Problems

83. Use L'Hôpital's Rule to evaluate

(a) $\lim\limits_{x\to 0^+}(1+x)^{1/x^2}$ (b) $\lim\limits_{x\to 0^-}(1+x)^{1/x^2}$

Reexamine Problems 81 and 82.

84. Evaluate $\lim\limits_{x\to\infty}[\sqrt{x^2+x} - x]$ without the aid of L'Hôpital's Rule.

In Problems 85 and 86 find all positive integers n for which the given limit exists.

85. $\lim\limits_{x\to 0^+}\dfrac{\ln(x+1)}{x^n}$

86. $\lim\limits_{x\to 1^+}\dfrac{e^{x-1} - x}{(x-1)^n}$

10.2 Improper Integrals

Up to now in our study of the definite integral $\int_a^b f(x)\,dx$, it was understood that

(*i*)   the limits of integration were finite numbers, and

(*ii*)   the function f was either continuous on $[a, b]$ or, if discontinuous, was bounded on the interval.

When either of these two conditions is dropped, the resulting integral is said to be an **improper integral**. In the first subsection that follows, we shall consider integrals of functions that are defined on unbounded intervals. In the second subsection, we shall examine integrals on bounded intervals of functions that are unbounded. In the latter type of improper integral, an integrand f has an *infinite discontinuity* at some number in the interval of integration.

10.2.1 Infinite Limits of Integration

Improper Integral—
First Kind

If f is defined on an unbounded interval, then there are three possible **improper** integrals with infinite limits of integration. Their definitions are summarized as follows:

(*i*) If f is continuous on $[a, \infty)$, then

$$\int_a^\infty f(x)\ dx = \lim_{t \to \infty} \int_a^t f(x)\ dx \qquad (10.3)$$

(*ii*) If f is continuous on $(-\infty, a]$, then

$$\int_{-\infty}^a f(x)\ dx = \lim_{s \to -\infty} \int_s^a f(x)\ dx \qquad (10.4)$$

(*iii*) If f is continuous for all x, and a is any real number, then

$$\int_{-\infty}^\infty f(x)\ dx = \int_{-\infty}^a f(x)\ dx + \int_a^\infty f(x)\ dx \qquad (10.5)$$

When the limits in (10.3) and (10.4) exist, the integrals are said to **converge**. If the limit fails to exist, the integral is said to **diverge**. In (10.5) the integral $\int_{-\infty}^\infty f(x)\ dx$ converges if and only if *both* $\int_{-\infty}^a f(x)\ dx$ and $\int_a^\infty f(x)\ dx$ converge. In other words, if, say, $\int_{-\infty}^a f(x)\ dx$ diverges, then $\int_{-\infty}^\infty f(x)\ dx$ diverges regardless of whether the other integral $\int_a^\infty f(x)\ dx$ converges.

Example 1

Evaluate $\displaystyle\int_1^\infty x^2\ dx$ if possible.

Solution   By (10.3),

$$\int_1^\infty x^2\ dx = \lim_{t \to \infty} \int_1^t x^2\ dx = \lim_{t \to \infty} \frac{x^3}{3}\Big]_1^t = \lim_{t \to \infty} \left(\frac{t^3}{3} - \frac{1}{3}\right)$$

Since $\displaystyle\lim_{t \to \infty} \left(\frac{t^3}{3} - \frac{1}{3}\right) = \infty$, we conclude that the integral diverges.

_____ **Example 2** _____

Evaluate $\displaystyle\int_{-\infty}^{\infty} x^2\, dx$ if possible.

Solution   Since a can be chosen arbitrarily, we pick $a = 1$ and write

$$\int_{-\infty}^{\infty} x^2\, dx = \int_{-\infty}^{1} x^2\, dx + \int_{1}^{\infty} x^2\, dx$$

But, in Example 1 we saw that $\int_{1}^{\infty} x^2\, dx$ diverges. This is sufficient to conclude that $\int_{-\infty}^{\infty} x^2\, dx$ also diverges.

_____ **Example 3** _____

Evaluate $\displaystyle\int_{2}^{\infty} \frac{dx}{x^3}$ if possible.

Solution   By (10.3),

$$\int_{2}^{\infty} \frac{dx}{x^3} = \lim_{t\to\infty} \int_{2}^{t} x^{-3}\, dx = \lim_{t\to\infty} \frac{x^{-2}}{-2}\Big]_{2}^{t} = \lim_{t\to\infty} \left[\frac{1}{8} - \frac{1}{2t^2} \right]$$

Since $\displaystyle\lim_{t\to\infty} \left[\frac{1}{8} - \frac{1}{2t^2} \right] = \frac{1}{8}$, the integral converges and

$$\int_{2}^{\infty} \frac{dx}{x^3} = \frac{1}{8}$$

_____ **Example 4** _____

Evaluate $\displaystyle\int_{-1}^{\infty} e^{-x}\, dx$ if possible. Interpret geometrically.

Solution   By (10.3),

$$\int_{-1}^{\infty} e^{-x}\, dx = \lim_{t\to\infty} \int_{-1}^{t} e^{-x}\, dx = \lim_{t\to\infty}(-e^{-x})\Big]_{-1}^{t} = \lim_{t\to\infty}[e - e^{-t}]$$

Since $\lim_{t\to\infty} e^{-t} = 0$, $\lim_{t\to\infty}[e - e^{-t}] = e$; and so the given integral converges to e. In Figure 10.1(a) we see that the area under the graph of the nonnegative function $f(x) = e^{-x}$ on $[-1, t]$ is $e - e^{-t}$. But, by taking $t \to \infty$, $e^{-t} \to 0$, and hence, as shown in Figure 10.1(b), we can interpret $\int_{-1}^{\infty} e^{-x}\, dx = e$ as a measure of the area under the graph of f on $[-1, \infty)$.

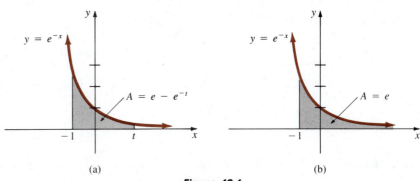

(a)   (b)

Figure 10.1

Example 5

Evaluate $\int_{-\infty}^{0} \cos x \, dx$ if possible.

Solution   By (10.4),

$$\int_{-\infty}^{0} \cos x \, dx = \lim_{s \to -\infty} \int_{s}^{0} \cos x \, dx = \lim_{s \to -\infty} \sin x \Big]_{s}^{0} = \lim_{s \to -\infty} (-\sin s)$$

Since $\sin s$ oscillates between -1 and 1, we conclude that $\lim_{s \to -\infty}(-\sin s)$ does not exist. Hence, $\int_{-\infty}^{0} \cos x \, dx$ diverges.

Example 6

Evaluate $\int_{-\infty}^{\infty} \dfrac{e^x}{e^x + 1} \, dx$ if possible.

Solution   Choosing $a = 0$ we can write

$$\int_{-\infty}^{\infty} \frac{e^x}{e^x + 1} \, dx = \int_{-\infty}^{0} \frac{e^x}{e^x + 1} \, dx + \int_{0}^{\infty} \frac{e^x}{e^x + 1} \, dx$$

$$= I_1 + I_2$$

First, let us examine I_1.

$$I_1 = \lim_{s \to -\infty} \int_{s}^{0} \frac{e^x}{e^x + 1} \, dx = \lim_{s \to -\infty} \ln(e^x + 1) \Big]_{s}^{0} = \lim_{s \to -\infty} [\ln 2 - \ln(e^s + 1)]$$

Since $e^s + 1 \to 1$ as $s \to -\infty$, $\ln(e^s + 1) \to 0$. Hence, $I_1 = \ln 2$.
    Second, we have

$$I_2 = \lim_{t \to \infty} \int_{0}^{t} \frac{e^x}{e^x + 1} \, dx = \lim_{t \to \infty} \ln(e^x + 1) \Big]_{0}^{t} = \lim_{t \to \infty} [\ln(e^t + 1) - \ln 2]$$

However, $e^t + 1 \to \infty$ as $t \to \infty$, so $\ln(e^t + 1) \to \infty$. Hence, I_2 diverges. It follows that $\int_{-\infty}^{\infty} e^x \, dx/(e^x + 1)$ is divergent.

_____ **Example 7** _____

It is left as an exercise to show that $\int_{-\infty}^{\infty} \dfrac{dx}{1 + x^2}$ converges and

$$\int_{-\infty}^{\infty} \frac{dx}{1 + x^2} = \int_{-\infty}^{0} \frac{dx}{1 + x^2} + \int_{0}^{\infty} \frac{dx}{1 + x^2}$$

$$= -\left(-\frac{\pi}{2}\right) + \frac{\pi}{2} = \pi$$

_____ **Example 8** _____

In (6.33) of Section 6.9, we saw that the work done in lifting a mass m_2 off the surface of a planet of mass m_1 to a height h is given by

$$W = \int_{R}^{R+h} \frac{km_1m_2}{r^2} \, dr$$

where R is the radius of the planet. Hence, the amount of work done in lifting m_2 to an unlimited or "infinite distance" from the surface of the planet is

$$W = \int_{R}^{\infty} \frac{km_1m_2}{r^2} \, dr$$

$$= \lim_{t \to \infty} \int_{R}^{t} \frac{km_1m_2}{r^2} \, dr = \frac{km_1m_2}{R}$$

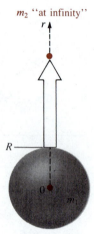

m_2 "at infinity"

Figure 10.2

See Figure 10.2. From the data in Example 2 of Section 6.9, it follows that work done in lifting a payload of 5000 kg to an "infinite distance" from the surface of the earth is

$$W = \frac{(6.67 \times 10^{-11})(6.0 \times 10^{24})(5000)}{6.4 \times 10^6} \approx 3.13 \times 10^{11} \text{ joules}$$

Remarks

(*i*)   You should verify that $\int_{-\infty}^{\infty} x \, dx = \int_{-\infty}^{0} x \, dx + \int_{0}^{\infty} x \, dx$ diverges since both $\int_{-\infty}^{0} x \, dx$ and $\int_{0}^{\infty} x \, dx$ diverge. A *common mistake* when working with integrals with doubly infinite limits is to use *one* limit:

$$\int_{-\infty}^{\infty} x \, dx = \lim_{t \to \infty} \int_{-t}^{t} x \, dx = \lim_{t \to \infty} \frac{x^2}{2}\bigg]_{-t}^{t} = \lim_{t \to \infty} \left[\frac{t^2}{2} - \frac{t^2}{2}\right] = 0$$

Of course, this "answer" is incorrect. Integrals of the type $\int_{-\infty}^{\infty} f(x) \, dx$ require the evaluation of *two independent* limits.

(*ii*)   In our previous work we often wrote without thinking,

$$\int_{a}^{b} [f(x) + g(x)] \, dx = \int_{a}^{b} f(x) \, dx + \int_{a}^{b} g(x) \, dx \qquad (10.6)$$

For improper integrals one should proceed with a bit more caution. For example, the integral $\int_1^\infty [1/x - 1/(x + 1)]\, dx$ converges (see Problem 25 in Exercises 10.2), but

$$\int_1^\infty \left[\frac{1}{x} - \frac{1}{x + 1}\right] dx \neq \int_1^\infty \frac{dx}{x} - \int_1^\infty \frac{dx}{x + 1}$$

The property in (10.6) remains valid for improper integrals whenever the integrals on the right side of the equality converge.

10.2.2 Integrals with an Integrand That Becomes Infinite

Recall, if f is continuous on $[a, b]$, then the definite integral $\int_a^b f(x)\, dx$ exists. Moreover, if $F'(x) = f(x)$, then

$$\int_a^b f(x)\, dx = F(b) - F(a) \tag{10.7}$$

However, we cannot evaluate an integral such as

$$\int_{-2}^1 \frac{1}{x^2}\, dx \tag{10.8}$$

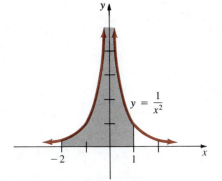

$y = \dfrac{1}{x^2}$

Figure 10.3

by the simple and traditional method given in (10.7) since $f(x) = 1/x^2$ possesses an infinite discontinuity in $[-2, 1]$. See Figure 10.3. In other words, for the integral in (10.8) the "procedure"

$$-x^{-1}\Big]_{-2}^1 = (-1) - \left(\frac{1}{2}\right) = -\frac{3}{2}$$

is just meaningless scratchings on paper. Thus, we have another type of integral that demands special handling.

Improper Integral— Second Kind

An integral $\int_a^b f(x)\, dx$ is also said to be **improper** if f has an infinite discontinuity at some number in the interval of integration. We again distinguish three cases.

(*i*)   If f is continuous on $[a, b)$ and $|f(x)| \to \infty$ as $x \to b^-$, then

$$\int_a^b f(x)\, dx = \lim_{t \to b^-} \int_a^t f(x)\, dx \tag{10.9}$$

(*ii*)   If f is continuous on $(a, b]$ and $|f(x)| \to \infty$ as $x \to a^+$, then

$$\int_a^b f(x)\, dx = \lim_{s \to a^+} \int_s^b f(x)\, dx \tag{10.10}$$

(*iii*)   If $|f(x)| \to \infty$ as $x \to c$ for some c in (a, b) and f is continuous at all other numbers in $[a, b]$, then

$$\int_a^b f(x)\, dx = \int_a^c f(x)\, dx + \int_c^b f(x)\, dx \tag{10.11}$$

As before, we say an improper integral **converges** or **diverges** depending on whether the defining limit exists or does not exist. In (10.11) the integral $\int_a^b f(x)\,dx$ converges if and only if $\int_a^c f(x)\,dx$ and $\int_c^b f(x)\,dx$ converge in the sense of (10.9) and (10.10), respectively.

_____ **Example 9** _____

Evaluate $\displaystyle\int_0^4 \frac{dx}{\sqrt{x}}$ if possible.

Solution   Observe that $f(x) = 1/\sqrt{x} \to \infty$ as $s \to 0^+$. Thus, by (10.10),

$$\int_0^4 \frac{dx}{\sqrt{x}} = \lim_{s \to 0^+} \int_s^4 x^{-1/2}\,dx = \lim_{s \to 0^+} 2x^{1/2}\Big]_s^4 = \lim_{s \to 0^+} [4 - 2s^{1/2}]$$

Since $\lim_{s \to 0^+} [4 - 2s^{1/2}] = 4$, the integral converges and

$$\int_0^4 \frac{dx}{\sqrt{x}} = 4$$

As seen in Figure 10.4, the number 4 can be regarded as a measure of the area under the graph of f on the interval $[0, 4]$.

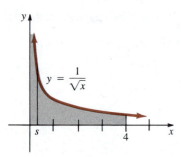

$y = \dfrac{1}{\sqrt{x}}$

Figure 10.4

_____ **Example 10** _____

Evaluate $\displaystyle\int_0^e \ln x\,dx$ if possible.

Solution   In this case we know $f(x) = \ln x \to -\infty$ as $x \to 0^+$. Using (10.10) and integration by parts gives

$$\int_0^e \ln x\,dx = \lim_{s \to 0^+} \int_s^e \ln x\,dx = \lim_{s \to 0^+} (x \ln x - x)\Big]_s^e = \lim_{s \to 0^+} s(1 - \ln s)$$

Now, when the limit is written

$$\lim_{s \to 0^+} \frac{1 - \ln s}{1/s}$$

we recognize the indeterminate form ∞/∞. Thus, by L'Hôpital's Rule

$$\lim_{s \to 0^+} \frac{1 - \ln s}{1/s} = \lim_{s \to 0^+} \frac{-1/s}{-1/s^2} = \lim_{s \to 0^+} s = 0$$

The integral converges and

$$\int_0^e \ln x\,dx = 0$$

--- **Example 11** ---

Evaluate $\int_1^5 \dfrac{dx}{(x-2)^{1/3}}$ if possible.

Solution   In the interval $[1, 5]$ the integrand has an infinite discontinuity at 2. Consequently, from (10.11) we write

$$\int_1^5 \frac{dx}{(x-2)^{1/3}} = \int_1^2 (x-2)^{-1/3}\,dx + \int_2^5 (x-2)^{-1/3}\,dx = I_1 + I_2$$

Now,

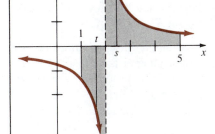

$$I_1 = \lim_{t \to 2^-} \int_1^t (x-2)^{-1/3}\,dx = \lim_{t \to 2^-} \frac{3}{2}(x-2)^{2/3}\Big]_1^t$$

$$= \frac{3}{2} \lim_{t \to 2^-}[(t-2)^{2/3} - 1] = -\frac{3}{2}$$

$$I_2 = \lim_{s \to 2^+} \int_s^5 (x-2)^{-1/3}\,dx = \lim_{s \to 2^+} \frac{3}{2}(x-2)^{2/3}\Big]_s^5$$

$$= \frac{3}{2} \lim_{s \to 2^+}[3^{2/3} - (s-2)^{2/3}] = \frac{3^{5/3}}{2}$$

Since both I_1 and I_2 converge, the given integral converges and

$$\int_1^5 \frac{dx}{(x-2)^{1/3}} = -\frac{3}{2} + \frac{3^{5/3}}{2} \approx 1.62$$

Figure 10.5

Note from Figure 10.5 that this last number is not a measure of area. Why?

--- **Example 12** ---

Evaluate $\int_{-2}^1 \dfrac{dx}{x^2}$ if possible.

Solution   This is the integral discussed in (10.8). Since, in the interval $[-2, 1]$, the integrand has an infinite discontinuity at 0 we write

$$\int_{-2}^1 \frac{1}{x^2}\,dx = \int_{-2}^0 \frac{dx}{x^2} + \int_0^1 \frac{dx}{x^2} = I_1 + I_2$$

Now the result

$$I_1 = \int_{-2}^0 \frac{dx}{x^2} = \lim_{t \to 0^-} \int_{-2}^t x^{-2}\,dx = \lim_{t \to 0^-} -x^{-1}\Big]_{-2}^t = \lim_{t \to 0^-}\left[-\frac{1}{t} - \frac{1}{2}\right] = \infty$$

indicates there is no need to evaluate $I_2 = \int_0^1 dx/x^2$. The integral $\int_{-2}^1 dx/x^2$ diverges.

Remarks

(*i*)   It is possible for an integral to have infinite limits of integration *and* an integrand with an infinite discontinuity. To determine whether an integral such as

$$\int_1^\infty \frac{dx}{x\sqrt{x^2 - 1}}$$

converges, we break up the integration at some convenient point of continuity of the integrand, say, $x = 2$:

$$\int_1^\infty \frac{dx}{x\sqrt{x^2 - 1}} = \int_1^2 \frac{dx}{x\sqrt{x^2 - 1}} + \int_2^\infty \frac{dx}{x\sqrt{x^2 - 1}}$$

$$= I_1 + I_2$$

I_1 and I_2 are improper integrals; I_1 is of the type given in (10.10), and I_2 is of the type given in (10.3). If both I_1 *and* I_2 converge, then the original integral is convergent. See Problems 87 and 88 in Exercises 10.2.

(*ii*)   The integrand of $\int_a^b f(x)\, dx$ can also have infinite discontinuities at both $x = a$ and $x = b$. In this case the improper integral is defined by means of (10.11). Lastly, if an integrand f has an infinite discontinuity at several numbers in (a, b), then the improper integral is defined by a natural extension of (10.11). See Problems 89 and 90 in Exercises 10.2.

_____ Exercises 10.2 _____

Answers to odd-numbered problems begin on page A-37.

[10.2.1]

In Problems 1–30 evaluate the given improper integral or show that it diverges.

1. $\displaystyle\int_3^\infty \frac{dx}{x^4}$

2. $\displaystyle\int_{-\infty}^{-1} \frac{dx}{\sqrt[3]{x}}$

3. $\displaystyle\int_1^\infty \frac{dx}{x^{0.99}}$

4. $\displaystyle\int_1^\infty \frac{dx}{x^{1.01}}$

5. $\displaystyle\int_{-\infty}^3 e^{2x}\, dx$

6. $\displaystyle\int_{-\infty}^\infty e^{-x}\, dx$

7. $\displaystyle\int_1^\infty \frac{\ln x}{x}\, dx$

8. $\displaystyle\int_1^\infty \frac{\ln t}{t^2}\, dt$

9. $\displaystyle\int_e^\infty \frac{dx}{x(\ln x)^3}$

10. $\displaystyle\int_e^\infty \ln x\, dx$

11. $\displaystyle\int_{-\infty}^\infty t e^{-t^2}\, dt$

12. $\displaystyle\int_{-\infty}^\infty \frac{dx}{1 + x^2}$

13. $\displaystyle\int_{-\infty}^0 \frac{x}{(x^2 + 9)^2}\, dx$

14. $\displaystyle\int_5^\infty \frac{dx}{\sqrt[4]{3x + 1}}$

15. $\displaystyle\int_2^\infty u e^{-u}\, du$

16. $\displaystyle\int_{-\infty}^3 \frac{x^3}{x^4 + 1}\, dx$

17. $\displaystyle\int_{2/\pi}^\infty \frac{\sin(1/x)}{x^2}\, dx$

18. $\displaystyle\int_{-\infty}^\infty \cos 3\theta\, d\theta$

19. $\displaystyle\int_{-1}^\infty \frac{dx}{x^2 + 2x + 2}$

20. $\displaystyle\int_{-\infty}^0 \frac{dx}{x^2 + 2x + 3}$

21. $\displaystyle\int_0^\infty e^{-x} \sin x\, dx$

22. $\displaystyle\int_{-\infty}^0 e^x \cos 2x\, dx$

23. $\displaystyle\int_{1/2}^\infty \frac{x + 1}{x^3}\, dx$

24. $\displaystyle\int_0^\infty (e^{-x} - e^{-2x})^2\, dx$

25. $\displaystyle\int_1^\infty \left[\frac{1}{x} - \frac{1}{x + 1} \right] dx$

26. $\displaystyle\int_3^\infty \left[\frac{1}{x} + \frac{1}{x^2 + 9} \right] dx$

27. $\displaystyle\int_{2}^{\infty} \frac{dx}{x^2 + 6x + 5}$

28. $\displaystyle\int_{-\infty}^{-3} \frac{dx}{x^2 - 4}$

29. $\displaystyle\int_{-\infty}^{-2} \frac{x^2}{(x^3 + 1)^2}\, dx$

30. $\displaystyle\int_{0}^{\infty} \frac{dx}{e^x + e^{-x}}$

In Problems 31 and 32 find the area under the graph of the given function on the indicated interval.

31. $f(x) = \dfrac{1}{(2x + 1)^2};\ [1, \infty)$

32. $f(x) = \dfrac{10}{x^2 + 25};\ (-\infty, 5]$

33. Consider the region that is bounded by the graphs of $y = 1/x$ and $y = 0$ on the interval $[1, \infty)$.

   (a) Show that the region does not have finite area.

   (b) Show that the solid of revolution that is formed by revolving the region around the x-axis has finite volume. See Figure 10.6.

Figure 10.6

34. Find the volume of the solid of revolution that is formed by revolving the region bounded by the graphs of $y = xe^{-x}$ and $y = 0$ on $[0, \infty)$ around the x-axis.

35. Find the work done against gravity in lifting a 10,000-kg payload to an infinite distance above the surface of the moon. (*Hint:* Review page 357 of Section 6.9.)

36. The work done by an external force in moving a test charge q_0 radially from point A to point B in the electric field of a charge q is defined to be:

$$W = -\frac{qq_0}{4\pi e_0} \int_{r_A}^{r_B} \frac{dr}{r^2}$$

See Figure 10.7.

   (a) Show that $W = \dfrac{qq_0}{4\pi e_0}\left(\dfrac{1}{r_B} - \dfrac{1}{r_A}\right)$.

   (b) Find the work done in bringing the test charge in from an infinite distance to point B.

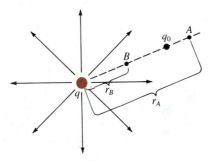

Figure 10.7

37. A **probability density function** is any nonnegative function f defined on an interval $[a, b]$ for which $\int_a^b f(x)\, dx = 1$. Verify that

$$f(x) = \begin{cases} 0, & x < 0 \\ ke^{-kx}, & x \geq 0, \end{cases} \quad k > 0$$

is a probability density function on $(-\infty, \infty)$.

38. The **capital value** of a perpetual stream of income is defined to be $V = \int_0^\infty f(t)e^{-rt}\, dt$, where f is the rate at which income flows per year and r is the annual rate of interest compounded continuously. Find the capital value of a stream of income whose rate is $f(t) = 500t$ if the interest rate is 10%.

The **Laplace transform*** of a function $y = f(x)$ defined by

$$\mathscr{L}\{f(x)\} = \int_0^\infty e^{-st}f(t)\, dt$$

***Pierre Simon Marquis de Laplace (1749–1827)** A noted mathematician, physicist, and astronomer, Laplace was called by some of his more enthusiastic contemporaries the "Newton of France." Although Laplace made use of this transform in his work in probability theory, it is likely that the integral was first discovered by Euler. Laplace's noted treatises were *Mécanique Céleste* and *Théorie Analytique des Probabilités*. Born into a poor farming family, Laplace became a friend of Napoleon but was elevated to the nobility by Louis XVIII after the restoration.

P. S. Laplace

is very useful in some areas of applied mathematics. In Problems 39–44 find the Laplace transform of the given function and state a restriction on s for which the integral converges.

39. $f(x) = 1$ **40.** $f(x) = x$

41. $f(x) = e^x$ **42.** $f(x) = e^{-5x}$

43. $f(x) = \sin x$ **44.** $f(x) = \cos 2x$

Miscellaneous Problems

In Problems 45–48 determine all values of k such that the given integral is convergent.

45. $\displaystyle\int_1^\infty \frac{dx}{x^k}$ **46.** $\displaystyle\int_{-\infty}^1 x^{2k}\, dx$

47. $\displaystyle\int_0^\infty e^{kx}\, dx$ **48.** $\displaystyle\int_1^\infty \frac{(\ln x)^k}{x}\, dx$

The following is a **comparison test** for convergence. Suppose f and g are continuous and $0 \le f(x) \le g(x)$ for $x \ge a$. If $\int_a^\infty g(x)\, dx$ converges, then $\int_a^\infty f(x)\, dx$ also converges. In Problems 49–52 use this result to show that the given integral converges.

49. $\displaystyle\int_1^\infty \frac{\sin^2 x}{x^2}\, dx$ **50.** $\displaystyle\int_2^\infty \frac{dx}{x^3 + 4}$

51. $\displaystyle\int_0^\infty \frac{dx}{x + e^x}$ **52.** $\displaystyle\int_0^\infty e^{-x^2}\, dx$

53. Prove that the surface area of the solid of revolution described in Problem 33(b) is not finite. (*Hint:* Review Section 6.6 and consider the inequality $\sqrt{x^4 + 1}/x^3 > 1/x$.)

54. Another integral in applied mathematics is the so-called **gamma function**:

$$\Gamma(\alpha) = \int_0^\infty t^{\alpha-1} e^{-t}\, dt, \qquad \alpha > 0$$

Show that

(a) $\Gamma(\alpha + 1) = \alpha\Gamma(\alpha)$, and that

(b) $\Gamma(n + 1) = 1 \cdot 2 \cdot 3 \cdots (n - 1) \cdot n = n!.$*

55. The **gamma distribution** plays an important role in the modeling of arrival times of customers at a check-out counter. In addition, this distribution is used in traffic control studies. In either case a probability density function (see Problem 37) takes the form $f(x) = cx^n e^{-\beta x}$, for $x > 0$, $\beta > 0$, and n a positive integer. Find c so that $\int_0^\infty f(x)\, dx = 1$. (*Hint:* Let $t = \beta x$ and use the result of Problem 54.)

---

*A rigorous demonstration of this fact requires mathematical induction. The symbol $n!$ is read "n factorial." See page 556.

Calculator Problems

56. **(a)** Show that the convergent integral $\int_1^\infty e^{1/x}\, dx/x^{5/2}$ can be written as $\int_0^1 t^{1/2} e^t\, dt$.

(b) Use the result of part **(a)** and Simpson's Rule with $n = 4$ to find an approximation to the original integral.

[10.2.2]
In Problems 57–78 evaluate the given improper integral or show that it diverges.

57. $\displaystyle\int_0^5 \frac{dx}{x}$ **58.** $\displaystyle\int_0^8 \frac{dx}{x^{2/3}}$

59. $\displaystyle\int_0^1 \frac{dx}{x^{0.99}}$ **60.** $\displaystyle\int_0^1 \frac{dx}{x^{1.01}}$

61. $\displaystyle\int_0^2 \frac{dx}{\sqrt{2 - x}}$ **62.** $\displaystyle\int_1^3 \frac{dx}{(x - 1)^2}$

63. $\displaystyle\int_{-1}^1 \frac{dx}{x^{5/3}}$ **64.** $\displaystyle\int_0^2 \frac{dx}{\sqrt[3]{x - 1}}$

65. $\displaystyle\int_0^2 (x - 1)^{-2/3}\, dx$ **66.** $\displaystyle\int_0^{27} \frac{e^{x^{1/3}}}{x^{2/3}}\, dx$

67. $\displaystyle\int_0^1 x \ln x\, dx$ **68.** $\displaystyle\int_1^e \frac{dx}{x \ln x}$

69. $\displaystyle\int_0^{\pi/2} \tan t\, dt$ **70.** $\displaystyle\int_0^{\pi/4} \frac{\sec^2\theta}{\sqrt{\tan\theta}}\, d\theta$

71. $\displaystyle\int_0^\pi \frac{\sin x}{1 + \cos x}\, dx$ **72.** $\displaystyle\int_0^\pi \frac{\cos x}{\sqrt{1 - \sin x}}\, dx$

73. $\displaystyle\int_{-1}^0 \frac{x}{\sqrt{1 + x}}\, dx$ **74.** $\displaystyle\int_0^3 \frac{dx}{x^2 - 1}$

75. $\displaystyle\int_0^1 \frac{x^2}{\sqrt{1 - x^2}}\, dx$ **76.** $\displaystyle\int_0^2 \frac{e^w}{\sqrt{e^w - 1}}\, dw$

77. $\displaystyle\int_1^3 \frac{dx}{\sqrt{3 + 2x - x^2}}$

78. $\displaystyle\int_0^1 \left[\frac{1}{\sqrt{x}} + \frac{1}{\sqrt{1 - x}}\right] dx$

In Problems 79 and 80 determine whether the area under the graph of the given function on the indicated interval is finite.

79. $f(x) = \dfrac{x}{\sqrt{16 - x^2}}$, $[0, 4]$

80. $f(x) = \sec x$, $[0, \pi/2]$

81. Consider the region that is bounded by the graphs of $y = 1/\sqrt{x + 2}$ and $y = 0$ on the interval $[-2, 1]$.

(a) Show that the region has finite area.

(b) Show that the solid of revolution that is formed by revolving the region around the x-axis has infinite volume.

82. The region that is bounded by the graphs of $y = 1/x^{3/2}$ and $y = 0$ on the interval $[0, 4]$ is revolved around the y-axis. Determine whether the volume of the solid of revolution is finite.

83. Determine whether the area of the region that is bounded by the graphs of

$$y = \frac{1}{x} \quad \text{and} \quad y = \frac{1}{x(x^2 + 1)}$$

on the interval $[0, 1]$ is finite.

84. Find the area of the region that is bounded by the graphs of $y = 1/\sqrt{x-1}$ and $y = -1/\sqrt{x-1}$ on the interval $[1, 5]$.

In Problems 85 and 86 determine all values of k such that the given integral is convergent.

85. $\displaystyle\int_0^1 \frac{dx}{x^k}$ **86.** $\displaystyle\int_0^1 x^k \ln x \, dx$

In Problems 87–90 determine whether the given integral converges or diverges.

87. $\displaystyle\int_1^\infty \frac{dx}{x\sqrt{x^2-1}}$ **88.** $\displaystyle\int_{-\infty}^4 \frac{dx}{(x-1)^{2/3}}$

89. $\displaystyle\int_{-1}^1 \frac{dx}{\sqrt{1-x^2}}$ **90.** $\displaystyle\int_0^2 \frac{2x-1}{\sqrt[3]{x^2-x}}\, dx$

91. Discuss whether $\displaystyle\int_0^{\pi/2} \frac{\sin x}{x}\, dx$ is an improper integral.

Chapter 10 Review Exercises

Answers to odd-numbered problems begin on page A-38.

In Problems 1–20 answer true or false.

1. A limit of the form $\infty - \infty$ always has the value 0. _____

2. A limit of the form ∞/∞ is indeterminate. _____

3. A limit of the form $0/\infty$ is indeterminate. _____

4. If $\displaystyle\lim_{x\to a}\frac{f(x)}{g(x)}$ and $\displaystyle\lim_{x\to a}\frac{f'(x)}{g'(x)}$ are both of the form ∞/∞, then the first limit does not exist. _____

5. A limit of the form 1^∞ is always 1. _____

6. For an indeterminate form, L'Hôpital's Rule states that the limit of a quotient is the same as the derivative of the quotient. _____

7. $\displaystyle\lim_{x\to\infty}\frac{x^n}{e^x} = 0$ for every integer n. _____

8. $\displaystyle\lim_{x\to 0}\frac{1-\cos^2 x}{x^2} = \lim_{x\to 0}\frac{1-\cos x}{x}\cdot\lim_{x\to 0}\frac{1+\cos x}{x}$ _____

9. If $\int_a^\infty f(x)\, dx$ and $\int_a^\infty g(x)\, dx$ converge, then $\int_a^\infty [f(x) + g(x)]\, dx$ converges. _____

10. If $\int_a^\infty [f(x) + g(x)]\, dx$ converges, then $\int_a^\infty f(x)\, dx$ converges. _____

11. If f is continuous for all x and $\int_{-\infty}^a f(x)\, dx$ diverges, then $\int_{-\infty}^\infty f(x)\, dx$ diverges. _____

12. The integral $\int_{-\infty}^\infty f(x)\, dx$ is defined by $\displaystyle\lim_{t\to\infty}\int_{-t}^t f(x)\, dx$. _____

13. $\displaystyle\int_0^4 x^{-0.999}\, dx$ converges. _____

14. $\displaystyle\int_1^\infty x^{-0.999}\, dx$ diverges. _____

15. $\displaystyle\int_2^\infty \left[\frac{e^x}{e^x+1} - \frac{e^x}{e^x-1}\right] dx$ diverges since $\displaystyle\int_2^\infty \frac{e^x}{e^x+1}\, dx$ diverges. _____

16. If $f(x) \to 0$ as $x \to \infty$, then $\int_a^\infty f(x)\, dx$ converges. _____

17. If a positive function f has an infinite discontinuity at a number in $[a, b]$, then the area under the graph on the interval is also infinite. _____

18. A region with infinite area, when revolved about an axis, always generates a solid of revolution of infinite volume. _____

19. $\displaystyle\int_0^\pi \sec^2 x\, dx = \tan x \Big]_0^\pi = 0$ _____

20. $\displaystyle\lim_{x\to\pi/2}\frac{\cos x}{(x-\pi/2)^2} = \lim_{x\to\pi/2}\frac{\sin x}{2(x-\pi/2)} = \lim_{x\to\pi/2}\frac{\cos x}{2}$

$= 0$ _____

In Problems 21–36 evaluate the given limit if it exists.

21. $\displaystyle\lim_{x\to-2}\frac{x^2+4x+4}{x^2+x-2}$

22. $\displaystyle\lim_{x\to1}\frac{x^2-3x+5}{x^2-1}$

23. $\displaystyle\lim_{x\to\sqrt{3}}\frac{\sqrt{3}-\tan(\pi/x^2)}{x-\sqrt{3}}$

24. $\displaystyle\lim_{\theta\to-\pi}\frac{\sin\theta}{\theta+\pi}$

25. $\displaystyle\lim_{x\to2}\frac{3x^2-4x-2}{x-2}$

26. $\displaystyle\lim_{x\to-\infty}\frac{-x^2+3x+1}{4x^2-2x}$

27. $\displaystyle\lim_{\theta\to0}\frac{10\theta-5\sin2\theta}{10\theta-2\sin5\theta}$

28. $\displaystyle\lim_{x\to\infty}\frac{x}{\ln x}$

29. $\displaystyle\lim_{x\to\infty}x\left(\cos\frac{1}{x}-e^{2/x}\right)$

30. $\displaystyle\lim_{y\to0}\left[\frac{1}{y}-\frac{1}{\ln(y+1)}\right]$

31. $\displaystyle\lim_{t\to0}\frac{(\sin t)^2}{\sin t^2}$

32. $\displaystyle\lim_{x\to0}\frac{\tan5x}{e^{3x/2}-e^{-x/2}}$

33. $\displaystyle\lim_{x\to0^+}(3x)^{-1/\ln x}$

34. $\displaystyle\lim_{x\to0}(2x+e^{3x})^{4/x}$

35. $\displaystyle\lim_{x\to\infty}\ln\left(\frac{x+e^{2x}}{1+e^{4x}}\right)$

36. $\displaystyle\lim_{x\to0^+}x(\ln x)^2$

In Problems 37–48 evaluate the given integral or show that it diverges.

37. $\displaystyle\int_0^3 x(x^2-9)^{-2/3}\,dx$

38. $\displaystyle\int_0^5 x(x^2-9)^{-2/3}\,dx$

39. $\displaystyle\int_{-\infty}^0 (x+1)e^x\,dx$

40. $\displaystyle\int_0^\infty \frac{e^{2x}}{e^{4x}+1}\,dx$

41. $\displaystyle\int_3^\infty \frac{dx}{1+5x}$

42. $\displaystyle\int_0^\infty \frac{x}{(x^2+4)^2}\,dx$

43. $\displaystyle\int_0^e \ln\sqrt{x}\,dx$

44. $\displaystyle\int_0^{\pi/2} \frac{\sec^2 t}{\tan^3 t}\,dt$

45. $\displaystyle\int_0^{\pi/2} \frac{dx}{1-\cos x}$

46. $\displaystyle\int_0^\infty \frac{x}{x+1}\,dx$

47. $\displaystyle\int_0^1 \frac{dx}{\sqrt{x}\,e^{\sqrt{x}}}$

48. $\displaystyle\int_0^\infty \frac{dx}{\sqrt{x}\,e^{\sqrt{x}}}$

49. Find the area of the region that is bounded by the graphs of $y=e^{-x}$ and $y=e^{-3x}$ on $[0,\infty)$.

50. Consider the region that is bounded by the graphs of $y=1/(1-x)^{1/3}$ and $y=0$ on the interval $[0,1]$.

(a)   Find the area of the region.

(b)   Find the volume of the solid of revolution that is formed by revolving the region about the x-axis.

(c)   Find the volume of the solid of revolution that is formed by revolving the region about the line $x=1$.

51. Consider the graph of $f(x)=(x^2-1)/(x^2+1)$ given in Figure 10.8.

(a)   Determine whether the region R_1, which is bounded between the graph of f and its horizontal asymptote, is finite.

(b)   Determine whether the regions R_2 and R_3 have finite areas.

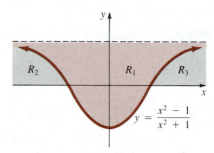

Figure 10.8

52. Prove that $\displaystyle\int_1^\infty \frac{\sqrt{x}}{(1+x)^2}\,dx=\frac{\pi}{4}+\frac{1}{2}$.

53. Prove that $\displaystyle\int_0^\infty \frac{dx}{\sqrt{x}(x+1)}=\pi$.

54. Prove that if $a>0$ and $ax^2+bx+c>0$ for all x, then

$$\int_{-\infty}^\infty \frac{dx}{ax^2+bx+c}=\frac{2\pi}{\sqrt{4ac-b^2}}.$$

In Problems 55 and 56 use the fact that $\lim_{x\to\infty}\int_0^x e^{t^2}\,dt=\infty$ to evaluate the given limit.

55. $\displaystyle\lim_{x\to\infty}\frac{x\int_0^x e^{t^2}\,dt}{e^{x^2}}$

56. $\displaystyle\lim_{x\to\infty}\frac{\int_0^x e^{t^2}\,dt}{xe^{x^2}}$

57. In population biology, the collection of young that are the results of an annual reproductive period is called a *year class*. In many animal population models, it is assumed that the number $N(t)$ of a year class that are still alive after t years is given by $N(t)=N_0 e^{-kt}$, for some $k>0$. It can be further shown that the **average life length** of an animal is given by $E=(-1/N_0)\int_0^\infty tN'(t)\,dt$. Determine the average life length for Pacific halibut if $k=0.2$.

11

Sequences and Series

Everyday experience gives one a feeling for the notion of a **sequence**. For example, the words *sequence of events* or *sequence of numbers* intuitively suggests an arrangement whereby the events E or numbers n are set down in some order: $E_1, E_2, E_3, \ldots$ or $n_1, n_2, n_3, \ldots$.

    Every student of mathematics is also familiar with the fact that any real number can be written as a decimal. For example, the rational number $\frac{1}{3}$ has the familiar representation

$$\frac{1}{3} = 0.3333\ldots$$

where the mysterious three dots ... (an ellipsis) signify that the 3s go on *forever*. This means 0.3333 ... is an infinite sum or the **infinite series**

$$\frac{3}{10} + \frac{3}{100} + \frac{3}{1000} + \frac{3}{10,000} + \cdots$$

    In this chapter we shall see that the concepts of sequence and infinite series are related.

11.1 Sequences

If the domain of a function f is the set of positive integers, then the elements $f(n)$ in its range can be arranged in order of increasing n:

$$f(1), f(2), f(3), f(4), \ldots, f(n), \ldots$$

Example 1

If n is a positive integer, then the first several elements in the range of $f(n) = (1 + 1/n)^n$ are

$$f(1) = 2, \qquad f(2) = \frac{9}{4}, \qquad f(3) = \frac{64}{27}, \ldots \qquad (11.1)$$

Functions whose domains are the entire set of positive integers are given a special name.

DEFINITION 11.1   Sequence

A **sequence** is a function whose domain is the set of positive integers.*

Terms Rather than using the customary function notation $f(n)$, a sequence is usually denoted by the symbol $\{a_n\}$. The **terms** of the sequence are formed by letting n take on the values 1, 2, 3, . . . in the **general term** a_n. Thus, $\{a_n\}$ is equivalent to

$$a_1, a_2, a_3, \ldots, a_n, \ldots \quad \longleftarrow \boxed{\text{elements in range}}$$
$$\;\uparrow\;\; \uparrow\;\; \uparrow \qquad\qquad\qquad$$
$$\;1\;\;\; 2\;\;\; 3 \;\; \ldots \;\; n \;\; \ldots \quad \longleftarrow \boxed{\text{elements in domain}}$$

Example 2

Write out the first four terms of the sequences

(a) $\left\{\dfrac{1}{\sqrt{n}}\right\}$,   (b) $\{n^2 + n\}$,   (c) $\{(-1)^n\}$

Solution By substituting $n = 1, 2, 3, 4$ in the respective general terms, we obtain

(a) $1, \dfrac{1}{\sqrt{2}}, \dfrac{1}{\sqrt{3}}, \dfrac{1}{2}, \cdots$ (b) $2, 6, 12, 20, \ldots$ (c) $-1, 1, -1, 1, \ldots$

---

*Some texts use the words *infinite sequence*. When the domain of a function is a finite subset of the set of positive integers, we get a *finite sequence*. All the sequences considered in this discussion will be infinite.

Convergent Sequences

For the sequence in (*a*) of Example 2, we see that as *n* becomes progressively larger, the values $a_n = 1/\sqrt{n}$ do not increase without bound. Indeed, as $n \to \infty$, $1/\sqrt{n} \to 0$. *We say that the terms $1/\sqrt{n}$ approach the* **limit** 0 and that the sequence $\{1/\sqrt{n}\}$ **converges** to 0. In contrast, the terms of the sequences in (*b*) and (*c*) do not approach a limit as $n \to \infty$.

In general we have the following definition.

DEFINITION 11.2   Convergent Sequences

A sequence $\{a_n\}$ is said to **converge** to a number L if for every $\varepsilon > 0$ there exists a positive number N such that

$$|a_n - L| < \varepsilon \quad \text{whenever} \quad n > N \tag{11.2}$$

Divergent Sequences

If $\{a_n\}$ is a convergent sequence, (11.2) means that the terms a_n can be made arbitrarily close to L for n sufficiently large. We indicate that a sequence converges to a number L by writing

$$\lim_{n \to \infty} a_n = L$$

When $\lim_{n \to \infty} a_n$ does not exist, we say that the sequence **diverges**. Figure 11.1 illustrates several ways in which a sequence $\{a_n\}$ can converge to a number L. As indicated in Figure 11.1(a), if $\{a_n\}$ converges, then *all but a finite* number of the terms a_n are in the interval $(L - \varepsilon, L + \varepsilon)$.

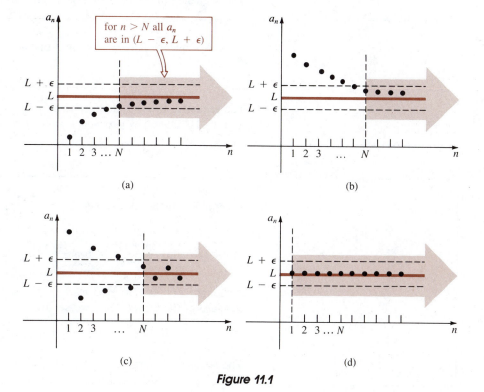

(a)   (b)

(c)   (d)

Figure 11.1

Example 3

Using Definition 11.2, prove that $\{1/\sqrt{n}\}$ converges to 0.

Solution   Let $\varepsilon > 0$ be given. Since the terms of the sequence are positive, the inequality $|a_n - 0| < \varepsilon$ is the same as

$$\frac{1}{\sqrt{n}} < \varepsilon$$

This is equivalent to $\sqrt{n} > 1/\varepsilon$ or $n > 1/\varepsilon^2$. Hence, we need only choose N to be the first positive integer greater than or equal to $1/\varepsilon^2$.

For instance, had we chosen $\varepsilon = 0.01$ in Example 3, then $|1/\sqrt{n} - 0| = 1/\sqrt{n} < 0.01$ whenever $n > 10{,}000$. That is, we choose $N = 10{,}000$.

In practice, to determine whether a sequence $\{a_n\}$ converges or diverges, we work directly with $\lim_{n\to\infty} a_n$ and proceed as we would in the examination of $\lim_{x\to\infty} f(x)$. If a_n either increases or decreases without bound as $n \to \infty$, then $\{a_n\}$ is necessarily divergent and we write, respectively,

$$\lim_{n\to\infty} a_n = \infty \quad \text{or} \quad \lim_{n\to\infty} a_n = -\infty \qquad (11.3)$$

Example 4

The sequence $\{n^2 + n\}$ is divergent since $\lim_{n\to\infty} (n^2 + n) = \infty$.

A sequence may diverge in a manner other than that given in (11.3).

Example 5

The sequence $\{(-1)^n\}$ diverges since $\lim_{n\to\infty} (-1)^n$ does not exist. The general term $(-1)^n$ alternates between 1 and -1 as $n \to \infty$.

Example 6

Determine whether the sequence $\left\{\dfrac{3n(-1)^n}{n+1}\right\}$ converges or diverges.

Solution $\displaystyle\lim_{n\to\infty} \frac{3n(-1)^n}{n+1} = \lim_{n\to\infty} \frac{3(-1)^n}{1 + 1/n}$

Although $3/(1 + 1/n) \to 3$ as $n \to \infty$, the foregoing limit still does not exist. Because of the term $(-1)^n$, we see that as $n \to \infty$

$$a_n \to 3, \quad n \text{ even} \quad \text{and} \quad a_n \to -3, \quad n \text{ odd}$$

The sequence diverges.

Constant Sequence A sequence of constants $\qquad c, c, c, \ldots$

is written $\{c\}$ and converges to c. See Figure 11.1(d).

Example 7

The sequence $\{\pi\}$ converges to π.

Example 8

Show that the sequence $\{(n + 1)^{1/n}\}$ converges to 1.

Solution If $y = (x + 1)^{1/x}$, we then recognize the indeterminate form ∞^0 as $x \to \infty$. Hence, by L'Hôpital's Rule,

$$\lim_{x \to \infty} \ln y = \lim_{x \to \infty} \frac{\ln(x + 1)}{x} = \lim_{x \to \infty} \frac{\dfrac{1}{x + 1}}{1} = 0$$

Consequently,

$$\lim_{n \to \infty} (n + 1)^{1/n} = e^0 = 1$$

Example 9

Determine whether the sequence $\left\{ \sqrt{\dfrac{n}{9n + 1}} \right\}$ converges or diverges.

Solution Since L'Hôpital's Rule shows that $n/(9n + 1) \to \frac{1}{9}$ as $n \to \infty$, we may write

$$\lim_{n \to \infty} \sqrt{\frac{n}{9n + 1}} = \sqrt{\lim_{n \to \infty} \frac{n}{9n + 1}} = \sqrt{\frac{1}{9}} = \frac{1}{3}$$

The sequence converges to $\frac{1}{3}$.

Properties The following **properties** of sequences are analogous to those given in Theorems 2.3 and 2.5.

THEOREM 11.1 Let $\{a_n\}$ and $\{b_n\}$ be convergent sequences. If $\lim_{n \to \infty} a_n = L_1$ and $\lim_{n \to \infty} b_n = L_2$, then

(*i*) $\displaystyle\lim_{n \to \infty} ka_n = k \lim_{n \to \infty} a_n = kL_1,$ k a constant

(*ii*) $\displaystyle\lim_{n \to \infty} (a_n + b_n) = \lim_{n \to \infty} a_n + \lim_{n \to \infty} b_n = L_1 + L_2$

(iii) $\displaystyle\lim_{n\to\infty} a_n b_n = \lim_{n\to\infty} a_n \cdot \lim_{n\to\infty} b_n = L_1 \cdot L_2$

(iv) $\displaystyle\lim_{n\to\infty} \frac{a_n}{b_n} = \frac{\displaystyle\lim_{n\to\infty} a_n}{\displaystyle\lim_{n\to\infty} b_n} = \frac{L_1}{L_2},\qquad L_2 \neq 0$

Example 10

In view of Example 3 and Theorem 11.1(i), we see that the sequence $\{5/\sqrt{n}\}$ converges to $5 \cdot 0 = 0$.

Example 11

Use Theorem 11.1 to show that $\{1/n\}$ converges to 0.

Solution   Using Example 3 and Theorem 11.1(iii), we can write

$$\lim_{n\to\infty} \frac{1}{n} = \lim_{n\to\infty} \frac{1}{\sqrt{n}} \cdot \frac{1}{\sqrt{n}} = \lim_{n\to\infty} \frac{1}{\sqrt{n}} \cdot \lim_{n\to\infty} \frac{1}{\sqrt{n}} = 0 \cdot 0 = 0$$

The next two theorems should seem plausible.

THEOREM 11.2

(i) The sequence $\{r^n\}$ converges to 0 if $|r| < 1$.

(ii) The sequence $\{r^n\}$ diverges if $|r| > 1$.

THEOREM 11.3   For any positive rational number r, the sequence $\left\{\dfrac{1}{n^r}\right\}$ converges to 0.

The proof of Theorem 11.2(i) follows from Definition 11.2 and is left as an exercise.

Example 12

(a) The sequence $\left\{\left(\dfrac{3}{2}\right)^n\right\}$, or $\dfrac{3}{2}, \dfrac{9}{4}, \dfrac{27}{8}, \ldots$ diverges by Theorem 11.2(ii) since $r = \dfrac{3}{2} > 1$.

(b) The sequence $\{e^{-n}\}$ converges to 0 by Theorem 11.2(i) since $\{e^{-n}\} = \left\{\left(\dfrac{1}{e}\right)^n\right\}$ and $\dfrac{1}{e} < 1$.

Example 13

Determine whether the sequence $\left\{\dfrac{e^n}{n + 4e^n}\right\}$ converges or diverges.

Solution Observe that $e^n \to \infty$ and $n + 4e^n \to \infty$ as $n \to \infty$. If $f(x) = e^x/(x + 4e^x)$, then $\lim_{x \to \infty} f(x)$ has the indeterminate form ∞/∞. By L'Hôpital's Rule,

$$\lim_{x \to \infty} \frac{e^x}{x + 4e^x} = \lim_{x \to \infty} \frac{e^x}{1 + 4e^x} = \lim_{x \to \infty} \frac{e^x}{4e^x} = \frac{1}{4}$$

Thus,
$$\lim_{n \to \infty} \frac{e^n}{n + 4e^n} = \frac{1}{4}$$

The sequence converges to $\frac{1}{4}$.

Example 14

Determine whether the sequence $\left\{\dfrac{2 - 3e^{-n}}{6 + 4e^{-n}}\right\}$ converges or diverges.

Solution Observe that $2 - 3e^{-n} \to 2$ and $6 + 4e^{-n} \to 6$ as $n \to \infty$. According to Theorem 11.1(iv), we have

$$\lim_{n \to \infty} \frac{2 - 3e^{-n}}{6 + 4e^{-n}} = \frac{\lim_{n \to \infty} (2 - 3e^{-n})}{\lim_{n \to \infty} (6 + 4e^{-n})} = \frac{2}{6} = \frac{1}{3}$$

The sequence converges to $\frac{1}{3}$.

Example 15

From Theorems 11.1(ii) and 11.3 we see that $\{10 + 4/n^{3/2}\}$ converges to 10.

Remarks

(i) In 1772 the German astronomer Johann Elert Bode studied the sequence

$$0, 3, 6, 12, 24, 48, 96, \ldots$$

By adding 4 to each term and dividing the result by 10, he obtained

$$0.4, 0.7, 1.0, 1.6, 2.8, 5.2, 10.0, \ldots$$

If the number 1 represents an astronomical unit,* then 0.4, 0.7, 1.0, 1.6, 5.2, and 10 predict fairly accurately the distances of the planets Mercury,

---

*Ninety-three million miles or the mean distance from the earth to the sun.

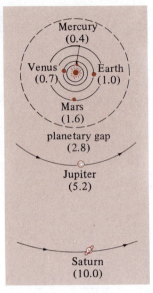

Figure 11.2

Figure 11.3

Venus, Earth, Mars, Jupiter, and Saturn from the sun. However, at that time no planet was known to exist at 2.8 astronomical units from the sun. The discovery of the planet Uranus in 1781, at a distance from the sun consistent with the next term of Bode's sequence (see Problem 62 in Exercises 11.1), brought about a flurry of observational activity by astronomers seeking the missing planet. In 1801 the asteroid Ceres was the first of thousands of asteroids discovered that filled the so-called planetary gap between Mars and Jupiter. Early speculation centered on the belief that the asteroids were remnants of an exploded planet. See Figure 11.2.

Much earlier, Leonardo da Vinci (1452–1519) was able to discern the velocity of a falling body by examining a sequence. Leonardo permitted water drops to fall, at equally spaced intervals of time, between two boards covered with blotting paper. When a spring mechanism was disengaged, the boards were clapped together. See Figure 11.3. By inspecting the sequence of water blots, Leonardo discovered that the distances between consecutive drops increased in "a continuous arithmetic proportion." In this manner he discovered the formula $v = gt$.

(ii) The symbol $n!$, read "n factorial," occurs frequently enough in this chapter to warrant a brief review of its definition. If n is a positive integer, then $n!$ is the product of the first n positive integers:

$$n! = 1 \cdot 2 \cdot 3 \cdots (n - 1) \cdot n$$

For example, $5! = 1 \cdot 2 \cdot 3 \cdot 4 \cdot 5 = 120$. An important property of the factorial is given by

$$n! = (n - 1)!n$$

To see this, consider the case when $n = 6$:

$$6! = 1 \cdot 2 \cdot 3 \cdot 4 \cdot 5 \cdot 6 = (1 \cdot 2 \cdot 3 \cdot 4 \cdot 5)6 = 5!6$$

Stated in a slightly different manner, the property $n! = (n - 1)!n$ is equivalent to

$$(n + 1)! = n!(n + 1)$$

One last point, for purposes of convenience and to ensure that the formula $n! = (n - 1)!n$ is valid when $n = 1$ we define $0!$ as follows:

$$0! = 1$$

Exercises 11.1

Answers to odd-numbered problems begin on page A-38.

In Problems 1–12 list the first four terms of the sequence whose general term is a_n.

1. $a_n = \dfrac{1}{2n + 1}$

2. $a_n = \dfrac{3}{4n - 2}$

3. $a_n = \dfrac{(-1)^n}{n}$

4. $a_n = \dfrac{(-1)^n n^2}{n + 1}$

5. $a_n = 3(-1)^{n-1}$

6. $a_n = 10(-1)^n$

7. $a_n = 10^n$

8. $a_n = 10^{-n}$

9. $a_n = \displaystyle\sum_{k=1}^{n} \frac{1}{k}$

10. $a_n = \displaystyle\sum_{k=1}^{n} 2^{-k}$

11. $a_n = 2n!$

12. $a_n = (2n)!$

In Problems 13–18 use Definition 11.2 to show that each sequence converges to the given number L.

13. $\left\{\dfrac{1}{n}\right\}; L = 0$

14. $\left\{\dfrac{1}{n^2}\right\}; L = 0$

15. $\left\{\dfrac{n}{n+1}\right\}; L = 1$

16. $\left\{\dfrac{4n}{2n-1}\right\}; L = 2$

17. $\{10^{-n}\}; L = 0$

18. $\left\{\dfrac{e^n + 1}{e^n}\right\}; L = 1$

In Problems 19–52 find $\lim_{n\to\infty} a_n$, if it exists, of each sequence having the given general term a_n.

19. $a_n = \dfrac{10}{\sqrt{n} + 1}$

20. $a_n = \dfrac{1}{n^{3/2}}$

21. $a_n = \dfrac{1}{5n + 6}$

22. $a_n = \dfrac{4}{2n + 7}$

23. $a_n = \dfrac{3n - 2}{6n + 1}$

24. $a_n = \dfrac{n}{1 - 2n}$

25. $a_n = 20(-1)^{n+1}$

26. $a_n = \left(-\dfrac{1}{3}\right)^n$

27. $a_n = \dfrac{n^2 - 1}{2n}$

28. $a_n = \dfrac{7n}{n^2 + 1}$

29. $a_n = \dfrac{n^2 - 3}{4n^2 + n}$

30. $a_n = \dfrac{n^2}{1 + 2n^2}$

31. $a_n = ne^{-n}$

32. $a_n = n^3 e^{-n}$

33. $a_n = \dfrac{\sqrt{n} + 1}{n}$

34. $a_n = \dfrac{n}{\sqrt{n} + 1}$

35. $a_n = \cos n\pi$

36. $a_n = \sin n\pi$

37. $a_n = \dfrac{\ln n}{n}$

38. $a_n = \dfrac{e^n}{\ln(n + 1)}$

39. $a_n = \dfrac{5 - 2^{-n}}{7 + 4^{-n}}$

40. $a_n = \dfrac{2^n}{3^n + 1}$

41. $a_n = \dfrac{2^n + 1}{2^n}$

42. $a_n = 4 + \dfrac{3^n}{2^n}$

43. $a_n = n \sin \dfrac{6}{n}$

44. $a_n = \left(1 - \dfrac{5}{n}\right)^n$

45. $a_n = \dfrac{e^n - e^{-n}}{e^n + e^{-n}}$

46. $a_n = \dfrac{\pi}{4} - \arctan(n)$

47. $a_n = n^{2/(n+1)}$

48. $a_n = 10^{(n+1)/n}$

49. $a_n = \ln\left(\dfrac{4n + 1}{3n - 1}\right)$

50. $a_n = \dfrac{\ln n}{\ln 3n}$

51. $a_n = \sqrt{n + 1} - \sqrt{n}$

52. $a_n = \sqrt{n}(\sqrt{n + 1} - \sqrt{n})$

In Problems 53–58 write the given sequence in the form $\{a_n\}$.

53. $\dfrac{2}{1}, \dfrac{4}{3}, \dfrac{6}{5}, \dfrac{8}{7}, \ldots$

54. $1 + \dfrac{1}{2}, \dfrac{1}{2} + \dfrac{1}{3}, \dfrac{1}{3} + \dfrac{1}{4}, \dfrac{1}{4} + \dfrac{1}{5}, \ldots$

55. $3, -5, 7, -9, \ldots$

56. $-2, 2, -2, 2, \ldots$

57. $2, \dfrac{2}{3}, \dfrac{2}{9}, \dfrac{2}{27}, \ldots$

58. $\dfrac{1}{1 \cdot 4}, \dfrac{1}{2 \cdot 8}, \dfrac{1}{3 \cdot 16}, \dfrac{1}{4 \cdot 32}, \ldots$

59. A ball is dropped from an initial height of 15 ft onto a concrete slab. Each time it bounces, it reaches a height of $\frac{2}{3}$ its preceding height. See Figure 11.4. What height does it reach on its third bounce? On its nth bounce?

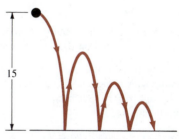

15

Figure 11.4

60. A ball, falling from a great height, travels 16 ft during the first second, 48 ft during the second, 80 ft during the third, and so on. How far does the ball travel during the sixth second?

61. A patient takes 15 milligrams of a drug each day. Suppose 80% of the drug accumulated is excreted each day by bodily functions. Write out the first six terms of the sequence $\{A_n\}$, where A_n is the amount of the drug present in the patient's body immediately after the nth dose.

62. The mean distances from the sun to the planets Uranus, Neptune, and Pluto are 19.19, 30.07, and 39.46 astronomical units, respectively. Determine how well these numbers agree with the next three terms of Bode's sequence given on page 555.

63. One dollar is deposited in a savings account that pays an annual rate r of interest. If no money is withdrawn, what is the amount accrued in the account after the first, second, and third years?

64. Each person has two parents. Determine how many great-great-great grandparents each person has.

Miscellaneous Problems

If $\{a_n\}$, $\{b_n\}$, and $\{c_n\}$ are sequences for which $a_n \le b_n \le c_n$ for all n and $\lim_{n\to\infty} a_n = \lim_{n\to\infty} c_n = L$, then $\lim_{n\to\infty} b_n = L$. In Problems 65–68 use this **Squeeze Theorem** to establish convergence of the given sequence.

65. $\left\{ \dfrac{\sin^2 n}{4^n} \right\}$ **66.** $\left\{ \dfrac{|\cos n|}{n^2} \right\}$

67. $\left\{ \sqrt{16 + \dfrac{1}{n^2}} \right\}$ **68.** $\left\{ \dfrac{\ln n}{n(n+2)} \right\}$

Sequences can be defined **recursively**. In Problems 69–72, for each sequence, write out the next four terms after the indicated initial term(s).

69. $a_{n+1} = \dfrac{1}{2} a_n;\ a_1 = -1$

70. $a_{n+1} = 2a_n - 1;\ a_1 = 2$

71. $a_{n+1} = \dfrac{a_n}{a_{n-1}};\ a_1 = 1,\ a_2 = 3$

72. $a_{n+1} = 2a_n - 3a_{n-1};\ a_1 = 2,\ a_2 = 4$

73. In his work *Liber Abacci*, published in 1202, Leonardo Fibonacci of Pisa speculated on the reproduction of rabbits:

> How many pairs of rabbits will be produced in a year beginning with a single pair, if in every month each pair bears a new pair which become productive from the second month on?

Discern the pattern in the following table and complete it.

	Start	After each month												
		1	2	3	4	5	6	7	8	9	10	11	12
Adult pairs	1	1	2	3	5	8	13	21						
Baby pairs	0	1	1	2	3	5	8	13						
Total pairs	1	2	3	5	8	13	21	34						

74. Write out the five terms, after the initial two, of the sequence defined recursively by $F_{n+1} = F_n + F_{n-1}$, $F_1 = 1$, $F_2 = 1$. Reexamine Problem 73.

75. Use Definition 11.2 to prove Theorem 11.2(i) under the assumption $0 < \varepsilon < 1$.

76. Show that for any real number x, the sequence $\{(1 + x/n)^n\}$ converges to e^x.

Calculator Problems

77. Consider the sequence whose first three terms are

$$1 + \frac{1}{2}, \quad 1 + \cfrac{1}{2 + \cfrac{1}{2}}, \quad 1 + \cfrac{1}{2 + \cfrac{1}{2 + \cfrac{1}{2}}}, \quad \dots$$

  (a) What are the fourth and fifth terms?

  (b) Calculate the numerical values of the first five terms of the sequence.

  (c) Make a conjecture about the convergence or divergence of the sequence.

78. The sequence

$$\left\{ 1 + \frac{1}{2} + \frac{1}{3} + \cdots + \frac{1}{n} - \ln n \right\}$$

is known to converge to the so-called **Euler–Mascheroni constant** γ. Calculate the first ten terms of the sequence.

79. Conjecture the limit of the convergent sequence $\sqrt{3}$, $\sqrt{3\sqrt{3}}$, $\sqrt{3\sqrt{3\sqrt{3}}}$, $\dots$

80. A regular n-sided polygon is circumscribed by a circle of radius r. See Figure 11.5.

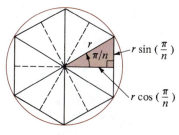

Figure 11.5

(a)   Using $2n$ congruent right triangles, show that the area of the circle is approximated by

$$A_n = \frac{n}{2}r^2 \sin\left(\frac{2\pi}{n}\right)$$

(b)   Compute A_{100} and A_{1000}.

(c)   Show that the sequence $\{A_n\}$ converges to πr^2.

11.2 Monotonic Sequences

In the preceding section we showed that a sequence $\{a_n\}$ converged by finding $\lim_{n\to\infty} a_n$. However, it is not always easy nor even possible to determine whether a sequence $\{a_n\}$ converges by seeking the exact value of $\lim_{n\to\infty} a_n$. For example, does the sequence

$$\left\{1 + \frac{1}{2} + \frac{1}{3} + \cdots + \frac{1}{n} - \ln n\right\}$$

converge? It turns out that this sequence does indeed converge; it is a special type of sequence that can be proved convergent without finding the value of

$$\lim_{n\to\infty}\left(1 + \frac{1}{2} + \frac{1}{3} + \cdots + \frac{1}{n} - \ln n\right)$$

See Problem 28 of Exercises 11.2.
We begin with a definition.

DEFINITION 11.3   **Monotonic Sequence**

A sequence $\{a_n\}$ is said to be **monotonic** if it is

  (*i*)   **increasing**: $a_1 < a_2 < a_3 < \cdots < a_n < a_{n+1} < \cdots$, or

 (*ii*)   **decreasing**: $a_1 > a_2 > a_3 > \cdots > a_n > a_{n+1} > \cdots$, or

(*iii*)   **nondecreasing**: $a_1 \leq a_2 \leq a_3 \leq \cdots \leq a_n \leq a_{n+1} \leq \cdots$, or

(*iv*)   **nonincreasing**: $a_1 \geq a_2 \geq a_3 \geq \cdots \geq a_n \geq a_{n+1} \geq \cdots$.

Example 1

(a)   The sequences

$$4, 6, 8, 10 \ldots$$

$$1, \frac{1}{2}, \frac{1}{4}, \frac{1}{8}, \ldots$$

$$5, 5, 4, 4, 4, 3, 3, 3, 3, \ldots$$

are monotonic. They are, respectively, increasing, decreasing, and nonincreasing.

(b)   The sequence $-1, \frac{1}{2}, -\frac{1}{3}, \frac{1}{4}, -\frac{1}{5}, \ldots$ is not monotonic.

It is not always evident whether a sequence is increasing, decreasing, and so on. The following table illustrates three ways of demonstrating monotonicity.

Form	*If*	*Then*
A **function** $f(x)$ such that $f(n) = a_n$	$f'(x) > 0$ for all x $f'(x) < 0$ for all x	$\{a_n\}$ is increasing $\{a_n\}$ is decreasing
The **ratio** a_{n+1}/a_n where $a_n > 0$ for all n	$a_{n+1}/a_n > 1$ for all n $a_{n+1}/a_n < 1$ for all n	$\{a_n\}$ is increasing $\{a_n\}$ is decreasing
The **difference** $a_{n+1} - a_n$	$a_{n+1} - a_n > 0$ for all n $a_{n+1} - a_n < 0$ for all n	$\{a_n\}$ is increasing $\{a_n\}$ is decreasing

Example 2

Show that $\{n/e^n\}$ is a decreasing sequence.

Solution   If we define $f(x) = x/e^x$, then $f(n) = a_n$. Now,

$$f'(x) = \frac{1 - x}{e^x} < 0 \qquad \text{for } x > 1$$

implies that f is decreasing on $[1, \infty)$. Thus, the given sequence is decreasing.

Alternative Solution

$$\frac{a_{n+1}}{a_n} = \frac{n + 1}{e^{n+1}} \cdot \frac{e^n}{n} = \frac{n + 1}{ne} = \frac{1}{e} + \frac{1}{ne} < 1 \qquad \text{for } n \geq 1$$

implies that $a_{n+1} < a_n$ for $n \geq 1$.

Example 3

The sequence $\left\{\dfrac{2n + 1}{n + 1}\right\}$ or $\dfrac{3}{2}, \dfrac{5}{3}, \dfrac{7}{4}, \dfrac{9}{5}, \ldots$ appears to be increasing. Since

$$a_{n+1} - a_n = \frac{2n + 3}{n + 2} - \frac{2n + 1}{n + 1} = \frac{1}{(n + 2)(n + 1)} > 0 \qquad \text{for all } n$$

we have $a_{n+1} > a_n$ for all n.

DEFINITION 11.4   Bounded Sequence

A sequence $\{a_n\}$ is said to be **bounded** if there exists a positive number B such that $|a_n| \le B$ for all n.

Example 4

The sequence $\left\{\dfrac{2n + 1}{n + 1}\right\}$ is bounded above by 2 since

$$\frac{2n + 1}{n + 1} \le \frac{2n + 2}{n + 1} = \frac{2(n + 1)}{n + 1} = 2$$

Furthermore, $(2n + 1)/(n + 1) \ge 0$ shows that the sequence is bounded below by 0.* Thus, $0 \le (2n + 1)/(n + 1) \le 2$ for all n implies that the sequence is bounded. Of course, the terms of the sequence are also bounded below by -2. This enables us to write $|(2n + 1)/(n + 1)| \le 2$ for all n.

The next result will be useful in subsequent sections of this chapter.

THEOREM 11.4   A bounded monotonic sequence converges.

Proof   We prove the theorem in the case of a nondecreasing sequence. By assumption $\{a_n\}$ is bounded and so $|a_n| \le B$, for all n. In turn, this means the infinite set of terms $S = \{a_1, a_2, a_3, \ldots, a_n, \ldots\}$ is bounded above and therefore has a least upper bound $L^\dagger$. The sequence actually converges to L. For $\varepsilon > 0$ we know that $L - \varepsilon < L$, and consequently $L - \varepsilon$ is not an upper bound for S (there are no upper bounds smaller than the least upper bound). Hence, there exists a positive integer N such that $a_N > L - \varepsilon$. But, since $\{a_n\}$ is nondecreasing

$$L - \varepsilon \le a_N \le a_{N+1} \le a_{N+2} \le a_{N+3} \le \cdots \le L + \varepsilon$$

---

*Indeed, from Example 3 we see that the terms are bounded below by the first term of the sequence.

†This is one of the basic axioms in mathematics. It is known as the **completeness property** of the real number system.

It follows that for $n > N$, $L - \varepsilon \le a_n \le L + \varepsilon$ or $|a_n - L| < \varepsilon$. From Definition 11.2 we conclude that $\lim_{n \to \infty} a_n = L$.   ∎

Example 5

The sequence $\left\{ \dfrac{2n + 1}{n + 1} \right\}$ was shown to be monotonic (Example 3) and bounded (Example 4). Hence, by Theorem 11.4 the sequence is convergent.

Example 6

Show that the sequence $\left\{ \dfrac{1 \cdot 3 \cdot 5 \cdots (2n - 1)}{2 \cdot 4 \cdot 6 \cdots (2n)} \right\}$ converges.

Solution   First, the ratio

$$\frac{a_{n+1}}{a_n} = \frac{1 \cdot 3 \cdot 5 \cdots (2n - 1)(2n + 1)}{2 \cdot 4 \cdot 6 \cdots (2n)(2n + 2)} \cdot \frac{2 \cdot 4 \cdot 6 \cdots (2n)}{1 \cdot 3 \cdot 5 \cdots (2n - 1)}$$

$$= \frac{2n + 1}{2n + 2} < 1$$

shows that $a_{n+1} < a_n$ for all n. The sequence is monotonic since it is decreasing. Next, from

$$0 < \frac{1 \cdot 3 \cdot 5 \cdot 7 \cdots (2n - 1)}{2 \cdot 4 \cdot 6 \cdot 8 \cdots (2n)} = \frac{1}{2} \cdot \frac{3}{4} \cdot \frac{5}{6} \cdot \frac{7}{8} \cdots \frac{2n - 1}{2n} < 1 \qquad \text{(Why?)}$$

we see that the sequence is bounded. It follows from Theorem 11.4 that the sequence is convergent.

Remark

Every convergent sequence $\{a_n\}$ is necessarily bounded. But it does not follow that every bounded sequence is convergent. You are asked to supply an example that illustrates this last statement in Problem 25 of Exercises 11.2. On the other hand, if a sequence $\{a_n\}$ is unbounded, then it is necessarily divergent.

Exercises 11.2

Answers to odd-numbered problems begin on page A-38.

In Problems 1–12 determine whether the given sequence is monotonic. If so, state whether it is increasing, decreasing, nondecreasing, or nonincreasing.

1. $\left\{ \dfrac{n}{3n + 1} \right\}$

2. $\left\{ \dfrac{10 + n}{n^2} \right\}$

3. $\{(-1)^n \sqrt{n}\}$

4. $\{(n - 1)(n - 2)\}$

5. $\left\{ \dfrac{e^n}{n} \right\}$

6. $\left\{ \dfrac{e^n}{n^5} \right\}$

7. $\left\{ \dfrac{2^n}{n!} \right\}$

8. $\left\{ \dfrac{2^{2n}(n!)^2}{(2n)!} \right\}$

9. $\left\{n + \dfrac{1}{n}\right\}$

10. $\{n^2 + (-1)^n n\}$

11. $\{(\sin 1)(\sin 2) \cdots (\sin n)\}$

12. $\left\{\ln\left(\dfrac{n+2}{n+1}\right)\right\}$

In Problems 13–24 use Theorem 11.4 to show that the given sequence converges.*

13. $\left\{\dfrac{4n-1}{5n+2}\right\}$

14. $\left\{\dfrac{6-4n^2}{1+n^2}\right\}$

15. $\left\{\dfrac{3^n}{1+3^n}\right\}$

16. $\{n5^{-n}\}$

17. $\{e^{1/n}\}$

18. $\left\{\dfrac{n!}{n^n}\right\}$

19. $\left\{\dfrac{n!}{1 \cdot 3 \cdot 5 \cdots (2n-1)}\right\}$

20. $\left\{\dfrac{2 \cdot 4 \cdot 6 \cdots (2n)}{3 \cdot 5 \cdot 7 \cdots (2n+1)}\right\}$

21. $\{\tan^{-1}n\}$

22. $\left\{\dfrac{\ln(n+3)}{n+3}\right\}$

23. $(0.8), (0.8)^2, (0.8)^3, \ldots$

24. $\sqrt{3}, \sqrt{\sqrt{3}}, \sqrt{\sqrt{\sqrt{3}}}, \ldots$

25. Give an example of a bounded sequence that is not convergent.

26. Consider the sequence $\{a_n\}$ defined by

$$a_1 = 2, \quad a_2 = 1, \quad a_{n+1} = \left(1 - \dfrac{1}{n^2}\right)a_n, \qquad n \ge 2$$

Show that $\{a_n\}$ converges.

27. Show that $\{\int_1^n e^{-t^2}\, dt\}$ converges. (*Hint:* For $x \ge 1$, $e^{-x^2} \le e^{-x}$.)

28. Prove that the sequence

$$\left\{1 + \dfrac{1}{2} + \dfrac{1}{3} + \cdots + \dfrac{1}{n} - \ln n\right\}$$

is bounded and monotonic and hence convergent.† (*Hint:* First prove the inequality

$$\dfrac{1}{2} + \dfrac{1}{3} + \cdots + \dfrac{1}{n-1} + \dfrac{1}{n} < \ln n < 1 + \dfrac{1}{2} + \dfrac{1}{3} + \cdots + \dfrac{1}{n-1}$$

by considering the area under the graph of $y = 1/x$ on $[1, n]$.)

11.3 Infinite Series

The concept of a *series* is closely related to the concept of a *sequence*. If $\{a_n\}$ is the sequence $a_1, a_2, a_3, \ldots, a_n, \ldots$, then the indicated sum

$$a_1 + a_2 + a_3 + \cdots + a_n + \cdots \tag{11.4}$$

is called an **infinite series** or simply a **series**. The a_k, $k = 1, 2, 3, \ldots$ are called the **terms** of the series; a_n is called the **general term**. We write (11.4) compactly as $\sum_{k=1}^{\infty} a_k$, or for convenience $\sum a_k$.

_____ **Example 1** _____

In the opening remarks to this chapter we noted that the decimal representation for the rational number $\frac{1}{3}$ is, in fact, an infinite series

$$\dfrac{3}{10} + \dfrac{3}{10^2} + \dfrac{3}{10^3} + \cdots = \sum_{k=1}^{\infty} \dfrac{3}{10^k}$$

_____

*Of course, in some of these problems the actual limit can be obtained by other methods.
†This limit is denoted by γ. From Problem 78 of Exercises 11.1, $\gamma \approx 0.5772. \ldots$

The question we seek to answer in this and the next several sections is:

When does an infinite series "add up" to a number?

Intuitively, we expect that $\frac{1}{3}$ is the sum of the series $\sum_{k=1}^{\infty} 3/10^k$. But, just as intuitively, we expect that an infinite series such as

$$100 + 1000 + 10{,}000 + 100{,}000 + \cdots$$

where the terms are becoming larger and larger, would have no sum. In other words, we do not expect the latter series to "add up" or *converge* to any number. The concept of convergence of an infinite series is defined in terms of the convergence of a special kind of sequence.

Sequence of Partial Sums   Associated with every infinite series $\sum a_k$, there is a **sequence of partial sums** $\{S_n\}$ whose terms are defined by

$$S_1 = a_1$$
$$S_2 = a_1 + a_2$$
$$S_3 = a_1 + a_2 + a_3$$
$$\vdots$$
$$S_n = a_1 + a_2 + a_3 + \cdots + a_n$$
$$\vdots$$

The general term $S_n = a_1 + a_2 + \cdots + a_n$ of this sequence is called the **nth partial sum** of the series.

Example 2

The sequence of partial sums for $\displaystyle\sum_{k=1}^{\infty} \frac{3}{10^k}$ is

$$S_1 = \frac{3}{10}$$

$$S_2 = \frac{3}{10} + \frac{3}{10^2}$$

$$S_3 = \frac{3}{10} + \frac{3}{10^2} + \frac{3}{10^3}$$

$$\vdots$$

$$S_n = \frac{3}{10} + \frac{3}{10^2} + \frac{3}{10^3} + \cdots + \frac{3}{10^n}$$

$$\vdots$$

In Example 2, when n is very large, S_n will give a good approximation to $\frac{1}{3}$, and so it seems reasonable to write

$$\frac{1}{3} = \lim_{n \to \infty} S_n = \lim_{n \to \infty} \sum_{k=1}^{n} \frac{3}{10^k} = \sum_{k=1}^{\infty} \frac{3}{10^k}$$

This leads to the following definition.

DEFINITION 11.5   Convergent Series

An infinite series $\sum_{k=1}^{\infty} a_k$ is said to be **convergent** if the sequence of partial sums $\{S_n\}$ converges; that is,

$$\lim_{n \to \infty} S_n = \lim_{n \to \infty} \sum_{k=1}^{n} a_k = S$$

The number S is the **sum** of the series. If $\lim_{n \to \infty} S_n$ does not exist, the series is said to be **divergent**.

Example 3

Show that the series $\displaystyle\sum_{k=1}^{\infty} \frac{1}{(k+4)(k+5)}$ is convergent.

Solution   By partial fractions the general term of the series can be written as

$$a_n = \frac{1}{n+4} - \frac{1}{n+5}$$

Thus, the nth partial sum of the series is

$$S_n = \left[\frac{1}{5} - \frac{1}{6}\right] + \left[\frac{1}{6} - \frac{1}{7}\right] + \left[\frac{1}{7} - \frac{1}{8}\right] + \cdots + \left[\frac{1}{n+4} - \frac{1}{n+5}\right]$$

$$= \frac{1}{5} - \frac{1}{6} + \frac{1}{6} - \frac{1}{7} + \frac{1}{7} - \frac{1}{8} + \frac{1}{8} - \cdots - \frac{1}{n+4} + \frac{1}{n+4} - \frac{1}{n+5}$$

$$= \frac{1}{5} - \frac{1}{n+5}$$

Since $\lim_{n \to \infty} 1/(n+5) = 0$, we see that $\lim_{n \to \infty} S_n = \frac{1}{5}$. Hence, the series converges and we write

$$\sum_{k=1}^{\infty} \frac{1}{(k+4)(k+5)} = \frac{1}{5}$$

Telescoping Series   Because of the manner in which the general term of the sequence of partial sums "collapses" to two terms, the series in Example 3 is said to be a **telescoping** series. See Problem 55 in Exercises 11.3.

Geometric Series A series of the form

$$a + ar + ar^2 + \cdots + ar^{n-1} + \cdots = \sum_{k=1}^{\infty} ar^{k-1} \qquad (11.5)$$

is called a **geometric series**.

> **THEOREM 11.5**   A geometric series $\displaystyle\sum_{k=1}^{\infty} ar^{k-1}$, $a \neq 0$, converges to
>
> $\dfrac{a}{1-r}$ if $|r| < 1$ and diverges if $|r| \geq 1$.

Proof Consider the general term of the sequence of partial sums of (11.5):

$$S_n = a + ar + ar^2 + \cdots + ar^{n-1} \qquad (11.6)$$

Multiplying both sides of (11.6) by r gives

$$rS_n = ar + ar^2 + ar^3 + \cdots + ar^n \qquad (11.7)$$

We subtract (11.7) from (11.6) and solve for S_n:

$$S_n - rS_n = a - ar^n$$
$$(1 - r)S_n = a(1 - r^n)$$
$$S_n = \frac{a(1 - r^n)}{1 - r}, \qquad r \neq 1 \qquad (11.8)$$

Now, from Theorem 11.2 we know that $\lim_{n \to \infty} r^n = 0$ for $|r| < 1$. Consequently,

$$\lim_{n \to \infty} S_n = \lim_{n \to \infty} \frac{a(1 - r^n)}{1 - r} = \frac{a}{1 - r}, \qquad |r| < 1$$

If $|r| > 1$, $\lim_{n \to \infty} r^n$ does not exist and so the limit of (11.8) fails to exist. The proof that a geometric series diverges when $r = \pm 1$ is left as an exercise. See Problem 57 in Exercises 11.3. ∎

Example 4

In the geometric series

$$\sum_{k=1}^{\infty} \left(-\frac{1}{3}\right)^{k-1} = 1 - \frac{1}{3} + \frac{1}{9} - \frac{1}{27} + \cdots$$

we identify $a = 1$ and $r = -\frac{1}{3}$. Since $\left|-\frac{1}{3}\right| < 1$, the series converges. From Theorem 11.5 the sum of the series is

$$\sum_{k=1}^{\infty} \left(-\frac{1}{3}\right)^{n-1} = \frac{1}{1 - (-1/3)} = \frac{3}{4}$$

Example 5

The geometric series

$$\sum_{k=1}^{\infty} 5\left(\frac{3}{2}\right)^{k-1} = 5 + \frac{15}{2} + \frac{45}{4} + \frac{135}{8} + \cdots$$

diverges because $|r| = \frac{3}{2} > 1$.

Example 6

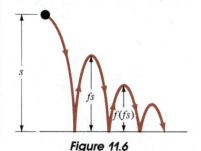

Figure 11.6

When a ball is dropped from a height of s ft off the ground, it takes $t = \sqrt{2s/g}$ sec to reach the ground. (See Section 6.8.) Suppose the ball always rebounds to a certain fixed fraction f $(0 < f < 1)$ of its prior height. Find a formula for the time T it takes for the ball to come to rest. See Figure 11.6.

Solution

The time to fall from a height of s ft to the ground is: $\sqrt{2s/g}$

The time to rise fs ft and then fall fs ft to the ground is: $2\sqrt{2fs/g}$

The time to rise $f(fs)$ ft and then fall $f(fs)$ ft to the ground is: $2\sqrt{2f^2s/g}$

and so on. Thus, the total time T is given by the infinite series

$$T = \sqrt{2s/g} + 2\sqrt{2fs/g} + 2\sqrt{2f^2s/g} + \cdots + 2\sqrt{2f^ns/g} + \cdots$$

$$= \sqrt{2s/g} \, [1 + 2\sum_{k=1}^{\infty} (\sqrt{f})^k]$$

Because $0 < f < 1$, the series $\sum_{k=1}^{\infty} (\sqrt{f})^k$ is a convergent geometric series with $a = \sqrt{f}$ and $r = \sqrt{f}$. Consequently

$$T = \sqrt{2s/g}\left[1 + 2\frac{\sqrt{f}}{1 - \sqrt{f}}\right] \quad \text{or} \quad T = \sqrt{2s/g}\left[\frac{1 + \sqrt{f}}{1 - \sqrt{f}}\right]$$

Harmonic Series   Another example of a divergent series is the so-called **harmonic series**:

$$1 + \frac{1}{2} + \frac{1}{3} + \cdots + \frac{1}{n} + \cdots = \sum_{k=1}^{\infty} \frac{1}{k} \qquad (11.9)$$

The general term of the sequence of partial sums for (11.9) is given by

$$S_n = 1 + \frac{1}{2} + \frac{1}{3} + \cdots + \frac{1}{n}$$

Thus,

$$S_{2n} = 1 + \frac{1}{2} + \frac{1}{3} + \cdots + \frac{1}{n} + \frac{1}{n+1} + \frac{1}{n+2} + \cdots + \frac{1}{2n}$$

$$= S_n + \frac{1}{n+1} + \frac{1}{n+2} + \cdots + \frac{1}{2n}$$

$$\geq S_n + \underbrace{\frac{1}{2n} + \frac{1}{2n} + \cdots + \frac{1}{2n}}_{n \text{ terms}} = S_n + n \cdot \frac{1}{2n} = S_n + \frac{1}{2}$$

The inequality $S_{2n} \geq S_n + \frac{1}{2}$ implies that the sequence of partial sums is unbounded. To see this, we observe that

$$S_2 \geq S_1 + \frac{1}{2} = 1 + \frac{1}{2} = \frac{3}{2}$$

$$S_4 \geq S_2 + \frac{1}{2} \geq \frac{3}{2} + \frac{1}{2} = 2$$

$$S_8 \geq S_4 + \frac{1}{2} \geq 2 + \frac{1}{2} = \frac{5}{2}$$

$$S_{16} \geq S_8 + \frac{1}{2} \geq \frac{5}{2} + \frac{1}{2} = 3$$

and so on. Hence, we conclude that the harmonic series is divergent.

If a_n and S_n are the general terms of a series and the corresponding sequence of partial sums, respectively, then $a_n = S_n - S_{n-1}$. Now, if the series converges to a number S, we have $\lim_{n \to \infty} S_n = S$ and $\lim_{n \to \infty} S_{n-1} = S$. This implies that $\lim_{n \to \infty} a_n = \lim_{n \to \infty} (S_n - S_{n-1}) = S - S = 0$. We have established the next theorem.

THEOREM 11.6   If $\sum_{k=1}^{\infty} a_k$ converges, then $\lim_{n \to \infty} a_n = 0$.

Test for a Divergent Series   Theorem 11.6 simply states that if an infinite series converges, it is necessary that the nth, or general, term of the series approach zero. Equivalently, we conclude that:

> If the nth term of an infinite series does *not* approach zero as $n \to \infty$, then the series does *not* converge.

We formalize this result as a test for divergence.

THEOREM 11.7   **The nth Term Test for Divergence**

If $\lim_{n \to \infty} a_n \neq 0$, then $\sum_{k=1}^{\infty} a_k$ diverges.

_____ **Example 7** _____

Consider the series $\sum_{k=1}^{\infty} \dfrac{4k - 1}{5k + 3}$. Since

$$\lim_{n \to \infty} \frac{4n - 1}{5n + 3} = \lim_{n \to \infty} \frac{4 - 1/n}{5 + 3/n} = \frac{4}{5} \neq 0$$

it follows from Theorem 11.7 that the series diverges.

We state the following theorems without proof.

> **THEOREM 11.8**   If c is any constant, then $\sum_{k=1}^{\infty} a_k$ and $\sum_{k=1}^{\infty} ca_k$ both converge or both diverge.

> **THEOREM 11.9**   If $\sum_{k=1}^{\infty} a_k$ and $\sum_{k=1}^{\infty} b_k$ converge to S_1 and S_2, respectively, then $\sum_{k=1}^{\infty} (a_k + b_k)$ converges to $S_1 + S_2$.

> **THEOREM 11.10**   If $\sum_{k=1}^{\infty} a_k$ is convergent and $\sum_{k=1}^{\infty} b_k$ is divergent, then $\sum_{k=1}^{\infty} (a_k + b_k)$ is divergent.

_____ **Example 8** _____

With the aid of Theorem 11.5, we see that the geometric series $\sum_{k=1}^{\infty} (\frac{1}{2})^{k-1}$ and $\sum_{k=1}^{\infty} (\frac{1}{3})^{k-1}$ converge to 2 and $\frac{3}{2}$, respectively. Hence, from Theorem 11.9, $\sum_{k=1}^{\infty} [(\frac{1}{2})^{k-1} + (\frac{1}{3})^{k-1}]$ converges to $2 + \frac{3}{2} = \frac{7}{2}$.

_____ **Example 9** _____

From Example 3 we know that $\sum_{k=1}^{\infty} 1/(k + 4)(k + 5)$ converges. Since $\sum_{k=1}^{\infty} 1/k$ is the divergent harmonic series, it follows from Theorem 11.10 that

$$\sum_{k=1}^{\infty} \left[\frac{1}{(k + 4)(k + 5)} + \frac{1}{k} \right]$$

diverges.

Remarks

(*i*) When written in terms of summation notation, a geometric series may not be immediately recognizable, or if it is, the values of a and r may not be apparent. For example, to see that $\sum_{n=3}^{\infty} 4(\frac{1}{2})^{n+2}$ is a geometric series, it is best to write out two or three terms:

$$\sum_{n=3}^{\infty} 4\left(\frac{1}{2}\right)^{n+2} = \overbrace{4\left(\frac{1}{2}\right)^5}^{a} + \overbrace{4\left(\frac{1}{2}\right)^6}^{ar} + \overbrace{4\left(\frac{1}{2}\right)^7}^{ar^2} + \cdots$$

From the right side of the last equality, we can make the identifications $a = 4(\frac{1}{2})^5$ and $r = \frac{1}{2} < 1$. Consequently, the series converges to $\frac{4(\frac{1}{2})^5}{1 - \frac{1}{2}} = \frac{1}{4}$. If desired, although there is no real need to do this, we can express $\sum_{n=3}^{\infty} 4(\frac{1}{2})^{n+2}$ in the more familiar form $\sum_{k=1}^{\infty} ar^{k-1}$ by letting $k = n - 2$. The result is

$$\sum_{n=3}^{\infty} 4\left(\frac{1}{2}\right)^{n+2} = \sum_{k=1}^{\infty} 4\left(\frac{1}{2}\right)^{k+4}$$

$$= \sum_{k=1}^{\infty} 4\overbrace{\left(\frac{1}{2}\right)^5}^{a} \overbrace{\left(\frac{1}{2}\right)^{k-1}}^{r^{k-1}}$$

(*ii*) Every rational number is either a terminating decimal or a repeating decimal.* Every repeating decimal is a geometric series. Thus, $\sum_{k=1}^{\infty} 3/10^k$ converges since $r = \frac{1}{10} < 1$. With $a = \frac{3}{10}$ we find that

$$\sum_{k=1}^{\infty} \frac{3}{10^k} = \frac{3/10}{1 - 1/10} = \frac{3/10}{9/10} = \frac{1}{3}$$

(*iii*) Note carefully how Theorems 11.6 and 11.7 are stated. Specifically, Theorem 11.6 does *not* say if $\lim_{n \to \infty} a_n = 0$, then $\sum a_k$ converges. In fact, if $\lim_{n \to \infty} a_n = 0$, the series may either converge or diverge. For example, in the harmonic series $\sum_{k=1}^{\infty} 1/k$, $a_n = 1/n$ and $\lim_{n \to \infty} 1/n = 0$, but the series diverges.

(*iv*) When determining convergence, it is possible, and sometimes convenient, to **delete** or ignore the first several terms of a series. In other words, infinite series $\sum_{k=1}^{\infty} a_k$ and $\sum_{k=N}^{\infty} a_k$, $N > 1$, that differ by at most a finite number of terms, are either both convergent or both divergent. Of course, deleting the first $N - 1$ terms of a convergent series usually does affect the sum of the series.

Exercises 11.3

Answers to odd-numbered problems begin on page A-38.

In Problems 1–10 write out the first four terms in each series.

1. $\displaystyle\sum_{k=1}^{\infty} \frac{2k + 1}{k}$

2. $\displaystyle\sum_{k=1}^{\infty} \frac{2^k}{k}$

3. $\displaystyle\sum_{k=1}^{\infty} \frac{(-1)^{k-1}}{k(k + 1)}$

4. $\displaystyle\sum_{k=1}^{\infty} \frac{(-1)^{k+1}}{k3^k}$

5. $\displaystyle\sum_{n=0}^{\infty} \frac{n + 1}{n!}$

6. $\displaystyle\sum_{n=1}^{\infty} \frac{(2n)!}{n^2 + 1}$

7. $\displaystyle\sum_{m=1}^{\infty} \frac{2 \cdot 4 \cdot 6 \cdots (2m)}{1 \cdot 3 \cdot 5 \cdots (2m - 1)}$

8. $\displaystyle\sum_{m=1}^{\infty} \frac{1 \cdot 3 \cdot 5 \cdots (2m - 1)}{m!}$

9. $\displaystyle\sum_{j=3}^{\infty} \frac{\cos j\pi}{2j + 1}$

10. $\displaystyle\sum_{i=5}^{\infty} i \sin \frac{i\pi}{2}$

---

*Even a terminating decimal such as 0.5 is a repeating decimal in the sense that $0.5 = 0.5000 \ldots$.

In Problems 11–20 determine whether the given geometric series converges or diverges. If convergent, find the sum of the series.

11. $\displaystyle\sum_{k=1}^{\infty} 3\left(\frac{1}{5}\right)^{k-1}$

12. $\displaystyle\sum_{k=1}^{\infty} 10\left(\frac{3}{4}\right)^{k-1}$

13. $\displaystyle\sum_{k=1}^{\infty} \frac{(-1)^{k-1}}{2^{k-1}}$

14. $\displaystyle\sum_{k=1}^{\infty} \left(\frac{\pi}{3}\right)^{k-1}$

15. $\displaystyle\sum_{r=1}^{\infty} 5^r 4^{-r}$

16. $\displaystyle\sum_{s=1}^{\infty} (-3)^s 7^{-s}$

17. $\displaystyle\sum_{n=1}^{\infty} 1000(0.9)^n$

18. $\displaystyle\sum_{n=1}^{\infty} \frac{(1.1)^n}{1000}$

19. $\displaystyle\sum_{k=0}^{\infty} \frac{1}{(\sqrt{3} - \sqrt{2})^k}$

20. $\displaystyle\sum_{k=0}^{\infty} \left(\frac{\sqrt{5}}{1 + \sqrt{5}}\right)^k$

In Problems 21–26 write each repeating decimal number as a quotient of integers.

21. $0.222 \ldots$

22. $0.555 \ldots$

23. $0.616161 \ldots$

24. $0.393939 \ldots$

25. $1.314314 \ldots$

26. $0.5262626 \ldots$

In Problems 27–32 find the sum of each convergent series.

27. $\displaystyle\sum_{k=1}^{\infty} \frac{1}{k(k + 1)}$

28. $\displaystyle\sum_{k=1}^{\infty} \frac{1}{(k + 1)(k + 2)}$

29. $\displaystyle\sum_{k=1}^{\infty} \frac{1}{4k^2 - 1}$

30. $\displaystyle\sum_{k=1}^{\infty} \frac{1}{k^2 + 7k + 12}$

31. $\displaystyle\sum_{k=1}^{\infty} \left[\left(\frac{1}{3}\right)^{k-1} + \left(\frac{1}{4}\right)^{k-1}\right]$

32. $\displaystyle\sum_{k=1}^{\infty} \frac{2^k - 1}{4^k}$

In Problems 33–42 show that each series is divergent.

33. $\displaystyle\sum_{k=1}^{\infty} 10$

34. $\displaystyle\sum_{k=1}^{\infty} (5k + 1)$

35. $\displaystyle\sum_{k=1}^{\infty} \frac{k}{2k + 1}$

36. $\displaystyle\sum_{k=1}^{\infty} \frac{k^2 + 1}{k^2 + 2k + 3}$

37. $\displaystyle\sum_{k=1}^{\infty} (-1)^k$

38. $\displaystyle\sum_{k=1}^{\infty} \ln\left(\frac{k}{3k + 1}\right)$

39. $\displaystyle\sum_{k=1}^{\infty} \frac{10}{k}$

40. $\displaystyle\sum_{k=1}^{\infty} \frac{1}{6k}$

41. $\displaystyle\sum_{k=1}^{\infty} \left[\frac{1}{2^{k-1}} + \frac{1}{k}\right]$

42. $\displaystyle\sum_{k=1}^{\infty} k \sin\frac{1}{k}$

In Problems 43–46 determine the values of x for which the given series converges.

43. $\displaystyle\sum_{k=1}^{\infty} \left(\frac{x}{2}\right)^{k-1}$

44. $\displaystyle\sum_{k=1}^{\infty} \left(\frac{1}{x}\right)^{k-1}$

45. $\displaystyle\sum_{k=1}^{\infty} (x + 1)^k$

46. $\displaystyle\sum_{k=0}^{\infty} 2^k x^{2k}$

47. A ball is dropped from an initial height of 15 ft onto a concrete slab. Each time the ball bounces, it reaches a height of $\frac{2}{3}$ its preceding height. Use geometric series to determine the distance the ball travels before it comes to rest.

48. In Problem 47, determine the time it takes for the ball to come to rest.

49. To eradicate agricultural pests (such as the Medfly), sterilized male flies are released into the general population at regular time intervals. Let N_0 be the number of flies released each day and let s be the proportion that survive a given day. Of the original N_0 sterilized males, $N_0 s^n$ will survive n successive weeks. Hence, the total number of such males that survives n weeks after the program has begun is $N_0 + N_0 s + N_0 s^2 + \cdots + N_0 s^n$. What does this sum approach as $n \to \infty$? Suppose $s = 0.9$ and 10,000 sterilized males are needed to control the population in a certain area. Determine the number that should be released each day.

50. In some circumstances the amount of a drug that will accumulate in a patient's body after a long period of time is $A_0 + A_0 e^{-k} + A_0 e^{-2k} + \cdots$, where $k > 0$ is a constant and A_0 is the daily dose of the drug. Find the sum of the series.

51. A patient takes 15 milligrams of a drug each day. If 80% of the drug accumulated is excreted each day by bodily functions, how much of the drug will accumulate after a long period of time (that is, as $n \to \infty$)? (Assume that the measurement of the accumulation is made immediately after each dose. See Problem 61 in Exercises 11.1.)

52. A force is applied to a particle, which moves in a straight line, in such a fashion that after each second the particle moves only one-half the distance it moved in the preceding second. If the particle moves 20 cm in the first second, how far will it move?

Miscellaneous Problems

53. Prove that $\sum_{k=1}^{\infty} 1/\sqrt{k}$ is divergent by showing $S_n \geq \sqrt{n}$.

54. Use the fact that $k! \geq 2^{k-1}$, $k = 1, 2, 3, \ldots$ to prove that $\sum_{k=1}^{\infty} 1/k!$ converges.

55. Prove that if $\lim_{n \to \infty} f(n + 1) = L$, where L is a number, then $\sum_{k=1}^{\infty} [f(k + 1) - f(k)] = L - f(1)$.

56. Find the sum of the series

$$\sum_{k=1}^{\infty} \left(\int_{k}^{k+1} xe^{-x}\, dx \right)$$

57. Prove that a geometric series $\sum_{k=1}^{\infty} ar^{k-1}$ diverges if $r = \pm 1$.

58. Determine whether the sum of two divergent series is necessarily divergent.

59. Suppose the sequence $\{a_n\}$ converges to a number $L \neq 0$. Show that $\sum_{k=1}^{\infty} a_k$ diverges.

60. Find all values of x in $(-\pi/2, \pi/2)$ for which

$$\lim_{n \to \infty} \left(\frac{1}{1 - \tan x} - \sum_{k=0}^{n} \tan^k x \right) = 0$$

61. Determine whether the following argument is valid: If $S = 1 + 2 + 4 + 8 + \cdots$, then $2S = 2 + 4 + 8 + \cdots = S - 1$. Solving $2S = S - 1$ gives $S = -1$.

62. Determine whether $\sum_{n=1}^{\infty} \left(\sum_{k=1}^{n} \frac{1}{k} \right)$ converges or diverges.

11.4 Series of Positive Terms

Unless $\sum_{k=1}^{\infty} a_k$ is a telescoping series or a geometric series, it is a difficult, if not futile, task to prove convergence or divergence directly from the sequence of partial sums. However, it is usually possible to determine whether a series converges or diverges by means of a *test* that utilizes only the terms of the series. In this section we examine five such tests that are applicable to infinite series of *positive terms*.

11.4.1 Integral and Comparison Tests

Integral Test The first test that we shall consider relates the concepts of convergence and divergence of an improper integral to convergence and divergence of an infinite series.

THEOREM 11.11   **Integral Test**

Suppose f is a continuous function that is nonnegative and decreasing for $x \geq 1$ such that $f(k) = a_k$ for $k \geq 1$.

(i) If $\int_{1}^{\infty} f(x)\, dx$ converges, then $\sum_{k=1}^{\infty} a_k$ converges.

(ii) If $\int_{1}^{\infty} f(x)\, dx$ diverges, then $\sum_{k=1}^{\infty} a_k$ diverges.

Proof If the graph of f is given as in Figure 11.7, then by considering the areas of the rectangles shown in the figure, we see

$$0 \leq a_2 + a_3 + a_4 + \cdots + a_n \leq \int_{1}^{n} f(x)\, dx \leq a_1 + a_2 + a_3 + \cdots + a_{n-1}$$

or

$$S_n - a_1 \leq \int_{1}^{n} f(x)\, dx \leq S_{n-1}$$

From the inequality $S_n - a_1 \leq \int_1^n f(x)\, dx$, it is apparent that $\lim_{n \to \infty} S_n$ exists whenever $\lim_{n \to \infty} \int_1^n f(x)\, dx$ exists. On the other hand, from $S_{n-1} \geq \int_1^n f(x)\, dx$, we conclude that $\lim_{n \to \infty} S_{n-1}$ fails to exist whenever $\int_1^{\infty} f(x)\, dx$ diverges.

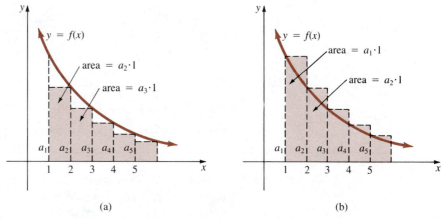

Figure 11.7

Note: In the Integral Test, if the positive-term series is of the form $\sum_{k=N}^{\infty} a_k$, we then use

$$\int_{N}^{\infty} f(x)\, dx \quad \text{where} \quad f(k) = a_k$$

Example 1

Test for convergence $\displaystyle\sum_{k=3}^{\infty} \frac{\ln k}{k}$.

Solution   The function $f(x) = (\ln x)/x$ is continuous, nonnegative, and decreasing* on $[3, \infty)$ and $f(k) = a_k = (\ln k)/k$. Now,

$$\int_{3}^{\infty} \frac{\ln x}{x}\, dx = \lim_{t \to \infty} \int_{3}^{t} \frac{\ln x}{x}\, dx$$

$$= \lim_{t \to \infty} \frac{1}{2} (\ln x)^2 \Big]_{3}^{t}$$

$$= \lim_{t \to \infty} \frac{1}{2} [(\ln t)^2 - (\ln 3)^2] = \infty$$

shows that the series diverges.

p-series   The integral test is particularly useful on any **p-series**:

$$1 + \frac{1}{2^p} + \frac{1}{3^p} + \cdots = \sum_{k=1}^{\infty} \frac{1}{k^p} \qquad (11.10)$$

The divergent harmonic series $\sum_{k=1}^{\infty} 1/k$ is a p-series with $p = 1$. The next result follows immediately from the Integral Test and is left as an exercise.

*Show this by examining $f'(x)$.

> **THEOREM 11.12**   The p-series $\sum_{k=1}^{\infty} 1/k^p$ converges if $p > 1$ and diverges if $p \leq 1$.

Example 2

(a) The series $\sum_{k=1}^{\infty} 1/k^{1/2}$ diverges since $p = \frac{1}{2} < 1$.

(b) The series $\sum_{k=1}^{\infty} 1/k^2$ converges since $p = 2 > 1$.

Comparison Tests   It is often possible to determine convergence or divergence of a series $\sum a_k$ by *comparing* its terms with the terms of a *test series* $\sum b_k$ that is known to be either convergent or divergent.

> **THEOREM 11.13**   **Comparison Test**
>
> Suppose $\sum_{k=1}^{\infty} a_k$ and $\sum_{k=1}^{\infty} b_k$ are series of positive terms.
>
> (i)   If $\sum_{k=1}^{\infty} b_k$ converges and $a_k \leq b_k$ for every positive integer k, then $\sum_{k=1}^{\infty} a_k$ converges.
>
> (ii) If $\sum_{k=1}^{\infty} b_k$ diverges and $a_k \geq b_k$ for every positive integer k, then $\sum_{k=1}^{\infty} a_k$ diverges.

Proof   Let $S_n = a_1 + a_2 + \cdots + a_n$ and $T_n = b_1 + b_2 + \cdots + b_n$ be the general terms of the sequences of partial sums for $\sum a_k$ and $\sum b_k$, respectively.

(i)   If $\sum b_k$ is a convergent series for which $a_k \leq b_k$, then $S_n \leq T_n$. Since $\lim_{n \to \infty} T_n$ exists, $\{S_n\}$ is a bounded increasing sequence and, hence, convergent by Theorem 11.4. Therefore, $\sum a_k$ is convergent.

(ii) If $\sum b_k$ diverges and $a_k \geq b_k$, then $S_n \geq T_n$. Since T_n increases without bound, so does S_n. Hence, $\sum a_k$ is divergent. ∎

In general, if $\sum c_k$ and $\sum d_k$ are two series for which $c_k \leq d_k$ for all k, we say that the series $\sum c_k$ is **dominated** by the series $\sum d_k$. Thus, for positive-term series, Theorem 11.13(i) indicates that a series $\sum a_k$ is convergent if it is dominated by a convergent series. In Theorem 11.13(ii) we see that a series $\sum a_k$ diverges if it dominates a divergent series.

Example 3

Test for convergence $\displaystyle \sum_{k=1}^{\infty} \frac{k}{k^3 + 4}$.

Solution   We observe that

$$\frac{k}{k^3 + 4} \leq \frac{k}{k^3} = \frac{1}{k^2}$$

Because the given series is dominated by the convergent *p*-series $\sum_{k=1}^{\infty} 1/k^2$ (see Example 2), it follows from Theorem 11.13(*i*) that the given series is also convergent.

_____

Example 4

Test for convergence $\displaystyle\sum_{k=1}^{\infty} \frac{\ln(k + 2)}{k}$.

Solution   Since $\ln(k + 2) > 1$ for $k \geq 1$, we have

$$\frac{\ln(k + 2)}{k} > \frac{1}{k}$$

In this case the given series has been shown to dominate the divergent harmonic series $\sum_{k=1}^{\infty} 1/k$. Hence, by Theorem 11.13(*ii*) the given series diverges.

_____

Another kind of comparison test involves taking the limit of the ratio of the general term of a series to the general term of a test series that is known to be convergent or divergent.

> **THEOREM 11.14**   **Limit Comparison Test**
>
> Suppose $\sum_{k=1}^{\infty} a_k$ and $\sum_{k=1}^{\infty} b_k$ are series of positive terms and that
>
> $$\lim_{n \to \infty} \frac{a_n}{b_n} = L$$
>
> (*i*)   If L is a positive constant, then the two series are either both convergent or both divergent.
>
> (*ii*)   If $L = 0$ and $\sum_{k=1}^{\infty} b_k$ converges, then $\sum_{k=1}^{\infty} a_k$ converges.
>
> (*iii*)   If $\lim_{n \to \infty} a_n/b_n = \infty$ and $\sum_{k=1}^{\infty} b_k$ diverges, then $\sum_{k=1}^{\infty} a_k$ diverges.

Proof of (*i*)   Since $\lim_{n \to \infty} a_n/b_n = L > 0$, we can choose *n* so large, say $n \geq N$ for some positive integer *N*, that

$$\frac{1}{2}L \leq \frac{a_n}{b_n} \leq \frac{3}{2}L$$

This inequality implies that $a_n \leq \frac{3}{2}Lb_n$ for $n \geq N$. If $\sum_{k=1}^{\infty} b_k$ converges, it follows from the Comparison Test that $\sum_{k=N}^{\infty} a_k$ and, therefore, $\sum_{k=1}^{\infty} a_k$ is convergent. Furthermore, since $\frac{1}{2}Lb_n \leq a_n$ for $n \geq N$, we see that if $\sum_{k=1}^{\infty} b_k$ diverges, then $\sum_{k=N}^{\infty} a_k$ and $\sum_{k=1}^{\infty} a_k$ diverge. ∎

The Limit Comparison Test is often applicable to series for which the Comparison Test is inconvenient.

Example 5

You should convince yourself that it is difficult to apply the Comparison Test to the series $\sum_{k=1}^{\infty} 1/(k^3 - 5k^2 + 1)$. However, we know $\sum_{k=1}^{\infty} 1/k^3$ is a convergent *p*-series ($p = 3 > 1$). Hence, with

$$a_n = \frac{1}{n^3 - 5n^2 + 1} \quad \text{and} \quad b_n = \frac{1}{n^3}$$

we have

$$\lim_{n \to \infty} \frac{a_n}{b_n} = \lim_{n \to \infty} \frac{n^3}{n^3 - 5n^2 + 1} = 1$$

From part (*i*) of Theorem 11.14, it follows that the given series converges.

If the general term a_n of a series $\sum a_k$ is a quotient of either rational powers of *n* or roots of polynomials in *n*, it is possible to discern the general term b_n of a test series $\sum b_k$ by examining the "degree behavior" of a_n for large values of *n*. In other words, to find a candidate for b_n we need only examine the quotient of the *highest powers of n* in the numerator and denominator of a_n.

Example 6

Test for convergence $\displaystyle\sum_{k=1}^{\infty} \frac{k}{\sqrt[3]{8k^5 + 7}}$.

Solution   For large values of *n*, $a_n = n/\sqrt[3]{8n^5 + 7}$ "behaves" like a constant multiple of

$$\frac{n}{\sqrt[3]{n^5}} = \frac{n}{n^{5/3}} = \frac{1}{n^{2/3}}$$

Thus, we try the divergent *p*-series

$$\sum_{k=1}^{\infty} \frac{1}{k^{2/3}} \quad \left(p = \frac{2}{3} < 1\right)$$

as a test series

$$\lim_{n \to \infty} \frac{a_n}{b_n} = \lim_{n \to \infty} \frac{n/\sqrt[3]{8n^5 + 7}}{1/n^{2/3}}$$

$$= \lim_{n \to \infty} \left(\frac{n^5}{8n^5 + 7}\right)^{1/3} = \frac{1}{2}$$

Thus, from part (*i*) of Theorem 11.14 the given series diverges.

Remarks

(*i*) When applying the Integral Test, you should be aware that the value of the convergent improper integral $\int_1^{\infty} f(x) \, dx$ is not equal to the sum of the series.

(*ii*) The results of the Integral Test for $\Sigma_{k=n}^{\infty} a_k$ hold even if the continuous nonnegative function f does not begin to decrease until $x \geq N \geq n$. For the series $\Sigma_{k=1}^{\infty} (\ln k)/k$, the function $f(x) = (\ln x)/x$ decreases on the interval $[3, \infty)$. Nonetheless, in the Integral Test we may use $\int_1^{\infty} (\ln x) \, dx/x$.

(*iii*) The hypotheses in the Comparison Test can also be weakened, giving a stronger theorem. It is only required that $a_k \leq b_k$ or $a_k \geq b_k$ for k sufficiently large and not for all positive integers.

(*iv*) In the application of the basic Comparison Test, it is often easy to reach a point where the given series is dominated by a divergent series. For example, $1/(5^k + \sqrt{k}) \leq 1/\sqrt{k}$ is certainly true and $\Sigma_{k=1}^{\infty} 1/\sqrt{k}$ diverges. This kind of reasoning proves nothing about the series $\Sigma_{k=1}^{\infty} 1/(5^k + \sqrt{k})$. But this series does converge. Why? Similarly, no conclusion can be reached by showing that a given series dominates a convergent series.

The following table summarizes the Comparison Test. Let $\Sigma \, a_k$ be a series of positive terms.

Comparison of Terms	Test series $\Sigma \, b_k$	Conclusion about $\Sigma \, a_k$
$a_k \leq b_k$	converges	converges
$a_k \leq b_k$	diverges	none
$a_k \geq b_k$	diverges	diverges
$a_k \geq b_k$	converges	none

11.4.2 Ratio and Root Tests

Ratio Test   The next test that we shall consider employs a limit of the **ratio** of the $(n + 1)$st term to the nth term of the series. This test is especially useful when a_k involves factorials, kth powers of a constant, and sometimes, kth powers of k.

THEOREM 11.15   Ratio Test

Suppose $\Sigma_{k=1}^{\infty} a_k$ is a series of positive terms such that

$$\lim_{n \to \infty} \frac{a_{n+1}}{a_n} = L$$

(*i*)   If $L < 1$, the series is convergent.

(*ii*)   If $L > 1$, or if $\lim_{n \to \infty} a_{n+1}/a_n = \infty$, the series is divergent.

(*iii*)   If $L = 1$, the test is inconclusive.

Proof of (*i*)   Let r be a positive number such that $0 \leq L \leq r < 1$. For n sufficiently large, say $n \geq N$ for some positive integer N, $a_{n+1}/a_n < r$, that is,

$$a_{n+1} < r \, a_n \qquad n \geq N$$

This inequality implies

$$a_{N+1} < r \, a_N$$
$$a_{N+2} < r \, a_{N+1} < a_N r^2$$
$$a_{N+3} < r \, a_{N+2} < a_N r^3 \; .$$

and so on. Thus, the series $\sum_{k=N+1}^{\infty} a_k$ converges by comparison with the convergent geometric series $\sum_{k=1}^{\infty} a_N r^k$. Since $\sum_{k=1}^{\infty} a_k$ differs from $\sum_{k=N+1}^{\infty} a_k$ by at most a finite number of terms, we conclude the former series also converges.

Proof of (*ii*)   Let r be a finite number such that $1 < r < L$. Then for n sufficiently large, say $n \geq N$ for some positive integer N, $a_{n+1}/a_n > r$ or $a_{n+1} > r \, a_n$. For $r > 1$ this last inequality implies $a_{n+1} > a_n$ and so $\lim_{n \to \infty} a_n \neq 0$. From Theorem 11.7 we see that $\sum_{k=1}^{\infty} a_k$ diverges.

In the case when $L = 1$, we must apply another test to the series to determine its convergence or divergence.

Example 7

Test for convergence $\displaystyle\sum_{k=1}^{\infty} \frac{5^k}{k!}$.

Solution

$$\lim_{n \to \infty} \frac{a_{n+1}}{a_n} = \lim_{n \to \infty} \frac{5^{n+1}}{(n+1)!} \cdot \frac{n!}{5^n}$$

$$= \lim_{n \to \infty} 5 \frac{n!}{(n+1)!}$$

$$= \lim_{n \to \infty} 5 \frac{n!}{n!(n+1)}$$

$$= \lim_{n \to \infty} \frac{5}{n+1} = 0$$

Since $L = 0 < 1$, it follows from Theorem 11.15(*i*) that the series is convergent.

Example 8

Test for convergence $\displaystyle\sum_{k=1}^{\infty} \frac{k^k}{k!}$.

Solution

$$\lim_{n \to \infty} \frac{a_{n+1}}{a_n} = \lim_{n \to \infty} \frac{(n + 1)^{n+1}}{(n + 1)!} \cdot \frac{n!}{n^n}$$

$$= \lim_{n \to \infty} \frac{(n + 1)^{n+1}}{n + 1} \cdot \frac{1}{n^n}$$

$$= \lim_{n \to \infty} \left(\frac{n + 1}{n}\right)^n$$

$$= \lim_{n \to \infty} \left(1 + \frac{1}{n}\right)^n = e$$

The last limit result follows from (8.29) of Section 8.2. Since $L = e > 1$, it follows from Theorem 11.15(ii) that the series is divergent.

Root Test   If the terms of a series $\Sigma \, a_k$ consist only of kth powers, then the following test, which involves taking the nth **root** of the nth term, may be applicable.

THEOREM 11.16   **Root Test**

Suppose $\Sigma_{k=1}^{\infty} \, a_k$ is a series of positive terms such that

$$\lim_{n \to \infty} \sqrt[n]{a_n} = L$$

(i)   If $L < 1$, the series is convergent.

(ii)   If $L > 1$, or if $\lim_{n \to \infty} \sqrt[n]{a_n} = \infty$, the series is divergent.

(iii)   If $L = 1$, the test is inconclusive.

The proof of the Root Test is very similar to the proof of the Ratio Test and will not be given.

Example 9

Test for convergence $\displaystyle\sum_{k=1}^{\infty} \left(\frac{5}{k}\right)^k$.

Solution   By the Root Test we see that

$$\lim_{n \to \infty} \left[\left(\frac{5}{n}\right)^n\right]^{1/n} = \lim_{n \to \infty} \frac{5}{n} = 0$$

Since $L = 0 < 1$, we conclude from (i) of Theorem 11.16 that the series converges.

Remarks

(i)   The Ratio Test will always lead to the inconclusive case when applied to a p-series. Try it on $\Sigma_{k=1}^{\infty} \, 1/k^2$ and see what happens.

(ii)   The tests examined in this section tell us when a series has a sum but none of these tests gives so much as a clue as to what the actual sum is. Knowing that a series converges, we can now add up five, a hundred, or a thousand terms on a computer to obtain an approximation of its sum.

Exercises 11.4

Answers to odd-numbered problems begin on page A-38.

[11.4.1]

In Problems 1–40 use the Integral Test, Comparison Test, or Limit Comparison Test to determine whether the given series converges or diverges.

1. $\displaystyle\sum_{k=1}^{\infty} \frac{1}{k^{1.1}}$

2. $\displaystyle\sum_{k=1}^{\infty} \frac{1}{k^{0.99}}$

3. $\displaystyle\sum_{k=1}^{\infty} \frac{1}{2k + 7}$

4. $\displaystyle\sum_{k=1}^{\infty} \frac{1}{10 + \sqrt{k}}$

5. $\displaystyle\sum_{k=1}^{\infty} \frac{k}{3k + 1}$

6. $\displaystyle\sum_{k=1}^{\infty} \frac{1}{k^2 + 5}$

7. $\displaystyle\sum_{k=1}^{\infty} ke^{-k^2}$

8. $\displaystyle\sum_{k=1}^{\infty} \frac{\arctan k}{1 + k^2}$

9. $\displaystyle\sum_{k=1}^{\infty} \frac{1}{(k + 1)(k + 2)}$

10. $\displaystyle\sum_{k=1}^{\infty} \frac{1}{k + \sqrt{k}}$

11. $\displaystyle\sum_{k=2}^{\infty} \frac{1}{k \ln k}$

12. $\displaystyle\sum_{k=3}^{\infty} \frac{\ln k}{k^5}$

13. $\displaystyle\sum_{k=2}^{\infty} \frac{(\ln k)^{-2}}{k}$

14. $\displaystyle\sum_{k=2}^{\infty} \frac{1}{k\sqrt{\ln k}}$

15. $\displaystyle\sum_{n=2}^{\infty} \frac{1}{n\sqrt{n^2 - 1}}$

16. $\displaystyle\sum_{n=1}^{\infty} \frac{1}{\sqrt{(n + 1)(n + 2)}}$

17. $\displaystyle\sum_{n=1}^{\infty} \frac{n^2 - n + 2}{3n^5 + n^2}$

18. $\displaystyle\sum_{n=2}^{\infty} \frac{n}{(4n + 1)^{3/2}}$

19. $\displaystyle\sum_{k=1}^{\infty} \frac{2 + \sin k}{\sqrt[3]{k^4 + 1}}$

20. $\displaystyle\sum_{k=1}^{\infty} \frac{3}{2 + \sin k}$

21. $\displaystyle\sum_{k=1}^{\infty} \frac{1 + 8^k}{3 + 10^k}$

22. $\displaystyle\sum_{k=1}^{\infty} \frac{(1.1)^k}{4k}$

23. $\displaystyle\sum_{k=1}^{\infty} \frac{1}{3^k + k}$

24. $\displaystyle\sum_{k=1}^{\infty} \frac{1 + 3^k}{2^k}$

25. $\displaystyle\sum_{k=2}^{\infty} \frac{1}{\ln k}$

26. $\displaystyle\sum_{k=2}^{\infty} \frac{2k + 1}{k \ln k}$

27. $\displaystyle\sum_{j=1}^{\infty} \frac{j + e^{-j}}{5^j(j + 9)}$

28. $\displaystyle\sum_{n=1}^{\infty} \frac{5n^2 + 2n}{3n(n^2 + 1)}$

29. $\displaystyle\sum_{n=1}^{\infty} \frac{1 + 1/n}{10^n}$

30. $\displaystyle\sum_{i=1}^{\infty} \frac{ie^{-i}}{i + 1}$

31. $\displaystyle\sum_{j=1}^{\infty} \ln\left(5 + \frac{1}{5^j}\right)$

32. $\displaystyle\sum_{j=1}^{\infty} \ln\left(1 + \frac{1}{3^j}\right)$

33. $\displaystyle\sum_{k=2}^{\infty} \frac{k + \ln k}{k^3 + 2k - 1}$

34. $\displaystyle\sum_{k=1}^{\infty} \frac{\sin(1/k)}{k}$

35. $\displaystyle\sum_{k=1}^{\infty} \frac{\sqrt{k + 1}}{\sqrt[3]{64k^9 + 40}}$

36. $\displaystyle\sum_{k=1}^{\infty} \frac{5k^2 + k - k^{-1}}{2k^3 + 2k^2 + 8}$

37. $\displaystyle\sum_{j=1}^{\infty} \frac{e^{1/j}}{j^2}$

38. $\displaystyle\sum_{k=1}^{\infty} \tan \frac{1}{k}$

39. $\displaystyle\frac{1}{2 \cdot 3} + \frac{2}{3 \cdot 4} + \frac{3}{4 \cdot 5} + \frac{4}{5 \cdot 6} + \cdots$

40. $\displaystyle\frac{1}{1 \cdot 3} + \frac{1}{2 \cdot 9} + \frac{1}{3 \cdot 27} + \frac{1}{4 \cdot 81} + \cdots$

Miscellaneous Problems

41. Prove Theorem 11.12.

42. Suppose $a_k > 0$ for $k = 1, 2, 3, \ldots$. Prove that if $\sum_{k=1}^{\infty} a_k$ converges, then $\sum_{k=1}^{\infty} a_k^2$ converges. Is the converse true?

[11.4.2]

In Problems 43–62 use the Ratio Test or Root Test to determine whether the given series converges or diverges.

43. $\displaystyle\sum_{k=1}^{\infty} \frac{1}{k!}$

44. $\displaystyle\sum_{k=1}^{\infty} \frac{2^k}{k!}$

45. $\displaystyle\sum_{k=1}^{\infty} \frac{k!}{1000^k}$

46. $\displaystyle\sum_{k=1}^{\infty} k\left(\frac{2}{3}\right)^k$

47. $\displaystyle\sum_{j=1}^{\infty} \frac{j^{10}}{(1.1)^j}$

48. $\displaystyle\sum_{j=1}^{\infty} \frac{1}{j^5(0.99)^j}$

49. $\displaystyle\sum_{k=1}^{\infty} \frac{1}{k^k}$

50. $\displaystyle\sum_{k=1}^{\infty} \left(\frac{ke}{k + 1}\right)^k$

51. $\displaystyle\sum_{n=1}^{\infty} \frac{4^{n-1}}{n3^{n+2}}$

52. $\displaystyle\sum_{n=1}^{\infty} \frac{n^3 2^{n+3}}{7^{n-1}}$

53. $\displaystyle\sum_{k=1}^{\infty} \frac{k!}{(2k)!}$

54. $\displaystyle\sum_{k=1}^{\infty} \frac{(2k)!}{k!(2k)^k}$

55. $\displaystyle\sum_{k=1}^{\infty} \frac{5^{2k+1}}{k^k}$

56. $\displaystyle\sum_{k=2}^{\infty} \frac{1}{(\ln k)^k}$

57. $\displaystyle\sum_{k=1}^{\infty} \left(\frac{k}{k+1}\right)^{k^2}$

58. $\displaystyle\sum_{k=1}^{\infty} \left(1 - \frac{2}{k}\right)^{k^2}$

59. $\displaystyle\sum_{k=1}^{\infty} \frac{1 \cdot 3 \cdot 5 \cdots (2k-1)}{k!}$

60. $\displaystyle\sum_{k=1}^{\infty} \frac{k!}{2 \cdot 4 \cdot 6 \cdots (2k)}$

61. $\displaystyle\sum_{k=1}^{\infty} \frac{k^k}{e^k}$

62. $\displaystyle\sum_{k=1}^{\infty} \frac{k! 3^k}{k^k}$

In Problems 63 and 64 determine the nonnegative values of p for which the given series converges.

63. $\displaystyle\sum_{k=1}^{\infty} kp^k$

64. $\displaystyle\sum_{k=1}^{\infty} k^2 \left(\frac{2}{p}\right)^k$

In Problems 65 and 66 determine all real values of p for which the given series converges.

65. $\displaystyle\sum_{k=1}^{\infty} \frac{k^p}{k!}$

66. $\displaystyle\sum_{k=2}^{\infty} \frac{\ln k}{k^p}$

Miscellaneous Problems

67. The Fibonacci sequence* $\{F_n\}$,

$$1, 1, 2, 3, 5, 8, \ldots$$

is defined by the recursion formula $F_{n+1} = F_n + F_{n-1}$, where $F_1 = 1$, $F_2 = 1$. Verify that the general term of the sequence is

$$F_n = \frac{1}{\sqrt{5}}\left(\frac{1 + \sqrt{5}}{2}\right)^n - \frac{1}{\sqrt{5}}\left(\frac{1 - \sqrt{5}}{2}\right)^n$$

by showing that this result satisfies the recursion formula.

68. Let F_n be the general term of the Fibonacci sequence given in Problem 67. Prove that

$$\lim_{n \to \infty} \frac{F_{n+1}}{F_n} = \frac{1 + \sqrt{5}}{2}$$

69. Let $\{F_n\}$ be the Fibonacci sequence given in Problem 67. Prove that the series

$$\frac{1}{1} + \frac{1}{1} + \frac{1}{2} + \frac{1}{3} + \frac{1}{5} + \frac{1}{8} + \cdots = \sum_{n=1}^{\infty} \frac{1}{F_n}$$

converges.

---

*See Problems 73 and 74 of Exercises 11.1.

11.5 Alternating Series and Absolute Convergence

A series having either form

$$a_1 - a_2 + a_3 - a_4 + \cdots + (-1)^{n+1}a_n + \cdots = \sum_{k=1}^{\infty} (-1)^{k+1}a_k$$

or

$$-a_1 + a_2 - a_3 + a_4 - \cdots + (-1)^n a_n + \cdots = \sum_{k=1}^{\infty} (-1)^k a_k$$

where $a_k > 0$ for $k = 1, 2, 3, \ldots$, is said to be an **alternating series**. Since $\sum_{k=1}^{\infty} (-1)^k a_k$ is just a multiple of $\sum_{k=1}^{\infty} (-1)^{k+1}a_k$, we shall limit our discussion to the latter series.

_____ **Example 1** _____

$$1 - \frac{1}{2} + \frac{1}{3} - \frac{1}{4} + \cdots = \sum_{k=1}^{\infty} \frac{(-1)^{k+1}}{k}$$

and $$\frac{\ln 2}{4} - \frac{\ln 3}{8} + \frac{\ln 4}{16} - \frac{\ln 5}{32} + \cdots = \sum_{k=2}^{\infty} (-1)^k \frac{\ln k}{2^k}$$

are examples of alternating series.

Alternating Series Test

The first series of Example 1, $1 - \frac{1}{2} + \frac{1}{3} - \frac{1}{4} + \cdots$, is called the **alternating harmonic series**. Although the harmonic series $\sum_{k=1}^{\infty} 1/k$ diverges, the introduction of positive and negative terms in the sequence of partial sums for the alternating harmonic series is sufficient to produce a convergent series. We will prove that $\sum_{k=1}^{\infty} (-1)^{k+1}/k$ converges by means of the next test.

THEOREM 11.17   Alternating Series Test

If $\lim_{n \to \infty} a_n = 0$ and $0 < a_{k+1} \le a_k$ for every positive integer k, then $\sum_{k=1}^{\infty} (-1)^{k+1} a_k$ converges.

Proof   Consider the partial sums that contain $2n$ terms:

$$S_{2n} = a_1 - a_2 + a_3 - a_4 + \cdots + a_{2n-1} - a_{2n}$$

$$= (a_1 - a_2) + (a_3 - a_4) + \cdots + (a_{2n-1} - a_{2n}) \quad (11.11)$$

Since $a_k - a_{k+1} \ge 0$ for $k = 1, 2, 3, \ldots$, we have

$$S_2 \le S_4 \le S_6 \le \cdots \le S_{2n} \le \cdots$$

Thus, the sequence $\{S_{2n}\}$, whose general term S_{2n} contains an even number of terms of the series, is a monotonic sequence. Rewriting (11.11) as

$$S_{2n} = a_1 - (a_2 - a_3) - \cdots - a_{2n}$$

shows that $S_{2n} < a_1$ for every positive integer n. Hence, $\{S_{2n}\}$ is bounded. By Theorem 11.4 it follows that $\{S_{2n}\}$ converges to a limit S. Now,

$$S_{2n+1} = S_{2n} + a_{2n+1}$$

implies that $$\lim_{n \to \infty} S_{2n+1} = \lim_{n \to \infty} S_{2n} + \lim_{n \to \infty} a_{2n+1}$$

$$= S + 0 = S$$

This shows that the sequence of partial sums $\{S_{2n+1}\}$, whose general term S_{2n+1} contains an odd number of terms, also converges to S. Since both $\{S_{2n}\}$ and $\{S_{2n+1}\}$ both converge to S we conclude that $\{S_n\}$ converges to S.   ∎

_____ **Example 2** _____

Show that the alternating harmonic series $\displaystyle\sum_{k=1}^{\infty} \frac{(-1)^{k+1}}{k}$ converges.

Solution   With the identification $a_n = 1/n$, we have immediately

$$\lim_{n \to \infty} a_n = \lim_{n \to \infty} \frac{1}{n} = 0 \quad \text{and} \quad a_{k+1} \le a_k$$

since $1/(k + 1) \le 1/k$ for $k \ge 1$. It follows from Theorem 11.17 that the alternating series converges.

_____ **Example 3** _____

The alternating series $\displaystyle\sum_{k=1}^{\infty} (-1)^{k+1} \frac{2k + 1}{3k - 1}$ diverges since

$$\lim_{n \to \infty} a_n = \lim_{n \to \infty} \frac{2n + 1}{3n - 1} = \frac{2}{3}$$

Recall from Theorem 11.6 that it is necessary that $\lim_{n \to \infty} a_n = 0$ for the convergence of *any* infinite series.

Although showing that $a_{k+1} \le a_k$ may seem a straightforward task, this is often not the case.

_____ **Example 4** _____

Test for convergence $\displaystyle\sum_{k=1}^{\infty} (-1)^{k+1} \frac{\sqrt{k}}{k + 1}$.

Solution   In order to show that the terms of the series satisfy the condition $a_{k+1} \le a_k$, let us consider the function $f(x) = \sqrt{x}/(x + 1)$ for which $f(k) = a_k$. From

$$f'(x) = -\frac{x - 1}{2\sqrt{x}(x + 1)^2} < 0 \qquad \text{for } x > 1$$

we see that the function f decreases for $x > 1$. Hence, $a_{k+1} \le a_k$ is true for $k \ge 1$. Now, L'Hôpital's Rule shows that

$$\lim_{x \to \infty} f(x) = 0 \quad \text{and so} \quad \lim_{n \to \infty} f(n) = \lim_{n \to \infty} a_n = 0$$

Hence, the alternating series converges by Theorem 11.17.

Error in Approximating the
Sum of an Alternating Series

Suppose the alternating series $\sum_{k=1}^{\infty} (-1)^{k+1}a_k$ converges to a number S. The partial sums

$$S_1 = a_1, \quad S_2 = a_1 - a_2, \quad S_3 = a_1 - a_2 + a_3, \quad S_4 = a_1 - a_2 + a_3 - a_4, \ldots$$

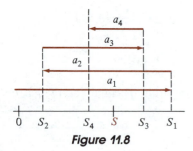

Figure 11.8

can be represented on a number line as shown in Figure 11.8. The sequence $\{S_n\}$ converges in the manner illustrated in Figure 11.1(c); that is, the terms S_n get closer and closer to S as $n \to \infty$ although they oscillate on either side of S. As indicated in Figure 11.6, the even-numbered partial sums are less than S and the odd-numbered partial sums are greater than S. Roughly, the even-numbered partial sums increase to the number S, and, in turn, the odd-numbered partial sums decrease to S. Because of this, *the sum S of the series must lie between consecutive partial sums S_n and S_{n+1}:*

$$S_n \leq S \leq S_{n+1}, \qquad \text{if } n \text{ is even} \tag{11.12}$$

or

$$S_{n+1} \leq S \leq S_n, \qquad \text{if } n \text{ is odd} \tag{11.13}$$

Now (11.12) yields $0 \leq S - S_n \leq S_{n+1} - S_n$ for n even, and (11.13) implies $0 \leq S_n - S \leq S_n - S_{n+1}$ for n odd. Thus, in either case $|S_n - S| \leq |S_{n+1} - S_n|$. But $S_{n+1} - S_n = a_{n+1}$ for n even and $S_{n+1} - S_n = -a_{n+1}$ for n odd. Thus $|S_n - S| \leq a_{n+1}$ for all n. We state this result as our next theorem.

THEOREM 11.18   Suppose the alternating series $\sum_{k=1}^{\infty} (-1)^{k+1}a_k$, $a_k > 0$, converges to a number S. If S_n is the nth partial sum of the series and $a_{k+1} \leq a_k$ for all k, then $|S_n - S| \leq a_{n+1}$ for all n.

Theorem 11.18 is useful in approximating the sum of a convergent alternating series. It states that the **error** $|S_n - S|$ between the nth partial sum and the sum of the series is less than the absolute value of the $(n + 1)$st term of the series.

___ **Example 5** ___

Approximate the sum of the convergent series $\sum_{k=1}^{\infty} \dfrac{(-1)^{k+1}}{(2k)!}$ to four decimal places.

Solution   Theorem 11.18 indicates that we must have

$$a_{n+1} = \frac{1}{(2n + 2)!} < 0.00005$$

From $n = 1$, $a_2 = \dfrac{1}{4!} \approx 0.041667$

$n = 2$, $a_3 = \dfrac{1}{6!} \approx 0.001389$

$n = 3$, $a_4 = \dfrac{1}{8!} \approx 0.000025 < 0.00005$

we see that $S_3 = \dfrac{1}{2!} - \dfrac{1}{4!} + \dfrac{1}{6!} \approx 0.4597$

has the desired accuracy.

> **DEFINITION 11.6   Absolute Convergence**
>
> A series $\sum_{k=1}^{\infty} a_k$ is said to be **absolutely convergent** if $\sum_{k=1}^{\infty} |a_k|$ converges.

___ **Example 6** _____

The alternating series $\displaystyle\sum_{k=1}^{\infty} \frac{(-1)^{k+1}}{k^2}$ is absolutely convergent since

$$\sum_{k=1}^{\infty} \left| \frac{(-1)^{k+1}}{k^2} \right| = \sum_{k=1}^{\infty} \frac{1}{k^2}$$

is a convergent p-series.

> **DEFINITION 11.7   Conditional Convergence**
>
> A series $\sum_{k=1}^{\infty} a_k$ is said to be **conditionally convergent** if $\sum_{k=1}^{\infty} |a_k|$ diverges and $\sum_{k=1}^{\infty} a_k$ converges.

___ **Example 7** _____

In Example 2 we saw that the alternating harmonic series $\sum_{k=1}^{\infty} (-1)^{k+1}/k$ is convergent. But taking the absolute value of each term gives the divergent harmonic series $\sum_{k=1}^{\infty} 1/k$. Thus, $\sum_{k=1}^{\infty} (-1)^{k+1}/k$ is conditionally convergent.

The next result shows that every absolutely convergent series is also convergent.

> **THEOREM 11.19**   If $\sum_{k=1}^{\infty} |a_k|$ converges, then $\sum_{k=1}^{\infty} a_k$ converges.

Proof If $c_k = a_k + |a_k|$, then $c_k \leq 2|a_k|$. Since $\Sigma |a_k|$ converges, it follows from the Comparison Test that Σc_k converges. Furthermore,

$$\sum_{k=1}^{\infty} (c_k - |a_k|)$$

converges since both Σc_k and $\Sigma |a_k|$ converge. But

$$\sum_{k=1}^{\infty} a_k = \sum_{k=1}^{\infty} (c_k - |a_k|)$$

Therefore, Σa_k converges.   ■

Note that since $\Sigma |a_k|$ is a series of positive terms, the tests of the preceding section can be utilized to determine whether a series converges absolutely.

_____ **Example 8** _____

Test for convergence $\displaystyle\sum_{k=1}^{\infty} \frac{(-1)^{k+1}}{1 + k^2}$.

Solution   It follows from the Integral Test that $\sum_{k=1}^{\infty} 1/(1 + k^2)$ is convergent. (Show this.) Hence, by Definition 11.6, the alternating series is absolutely convergent. From Theorem 11.19 we conclude that the given series is convergent.

The following modified forms of the **Ratio Test** and the **Root Test** can be applied directly to an alternating series.

THEOREM 11.20   Ratio Test

Suppose $\sum_{k=1}^{\infty} a_k$ is a series of nonzero terms such that

$$\lim_{n \to \infty} \left| \frac{a_{n+1}}{a_n} \right| = L$$

(*i*) If $L < 1$, the series is absolutely convergent.

(*ii*) If $L > 1$, or if $\lim_{n\to\infty} |a_{n+1}/a_n| = \infty$, the series is divergent.

(*iii*) If $L = 1$, the test is inconclusive.

_____ **Example 9** _____

Test for convergence $\displaystyle\sum_{k=1}^{\infty} \frac{(-1)^{k+1}2^{2k-1}}{k3^k}$.

Solution

$$\lim_{n \to \infty} \left| \frac{a_{n+1}}{a_n} \right| = \lim_{n \to \infty} \left| \frac{(-1)^{n+2}2^{2n+1}}{(n + 1)3^{n+1}} \cdot \frac{n3^n}{(-1)^{n+1}2^{2n-1}} \right|$$

$$= \lim_{n \to \infty} \frac{4n}{3(n + 1)} = \frac{4}{3}$$

Since $L = \frac{4}{3} > 1$, we see from Theorem 11.20(*ii*) that the alternating series diverges.

THEOREM 11.21   Root Test

Suppose $\sum_{k=1}^{\infty} a_k$ is an infinite series such that

$$\lim_{n \to \infty} \sqrt[n]{|a_n|} = L$$

(*i*) If $L < 1$, the series is absolutely convergent.

(*ii*) If $L > 1$, or if $\lim_{n\to\infty} \sqrt[n]{|a_n|} = \infty$, the series is divergent.

(*iii*) If $L = 1$, the test is inconclusive.

Remarks

(*i*)   The conclusion of Theorem 11.17 remains true when the hypothesis "$a_{k+1} \leq a_k$ for every positive integer k" is replaced with the statement "$a_{k+1} \leq a_k$ for k sufficiently large." For the alternating series $\sum_{k=1}^{\infty} (-1)^{k+1}(\ln k)/k^{1/3}$, it is readily shown by the procedure used in Example 4 that $a_{k+1} \leq a_k$ for $k \geq 21$. Moreover, $\lim_{n \to \infty} a_n = 0$. Hence, the series converges by the Alternating Series Test.

(*ii*) If the series of absolute values $\sum |a_k|$ is found to be divergent, then no conclusion can be drawn concerning the convergence or divergence of the series $\sum a_k$.

(*iii*) If $\sum a_k$ is absolutely convergent, then the terms of the series can be rearranged or regrouped in any manner and the resulting series will converge to the same number as the original series. In contrast, if the terms of a conditionally convergent series are written in a different order, the new series may diverge or converge to an entirely different number. It is left as an exercise to show that if S is the sum of the convergent alternating harmonic series,

$$S = 1 - \frac{1}{2} + \frac{1}{3} - \frac{1}{4} + \frac{1}{5} - \frac{1}{6} + \cdots$$

then the rearranged series

$$1 + \frac{1}{3} - \frac{1}{2} + \frac{1}{5} + \frac{1}{7} - \frac{1}{4} + \cdots$$

converges to $\frac{3}{2}S$. See Problem 52 in Exercises 11.5. You are also encouraged to reflect on the following "reasoning":

$$2S = 2\left[1 - \frac{1}{2} + \frac{1}{3} - \frac{1}{4} + \frac{1}{5} - \frac{1}{6} + \frac{1}{7} - \frac{1}{8} + \frac{1}{9} - \cdots \right]$$

$$= 2 - 1 + \frac{2}{3} - \frac{1}{2} + \frac{2}{5} - \frac{1}{3} + \frac{2}{7} - \frac{1}{4} + \frac{2}{9} - \cdots$$

$$= (2 - 1) - \frac{1}{2} + \left(\frac{2}{3} - \frac{1}{3} \right) - \frac{1}{4} + \left(\frac{2}{5} - \frac{1}{5} \right) - \frac{1}{6} + \cdots$$

$$= S$$

By dividing by S, we obtain the interesting result that $2 = 1$.

Exercises 11.5

Answers to odd-numbered problems begin on page A-39.

In Problems 1–14 use the Alternating Series Test to determine whether the given series is convergent.

1. $\displaystyle\sum_{k=1}^{\infty} \frac{(-1)^{k+1}}{k + 2}$

2. $\displaystyle\sum_{k=1}^{\infty} \frac{(-1)^{k-1}}{\sqrt{k}}$

3. $\displaystyle\sum_{k=1}^{\infty} (-1)^{k-1} \frac{k}{k + 1}$

4. $\displaystyle\sum_{k=1}^{\infty} (-1)^{k} \frac{k}{k^2 + 1}$

5. $\displaystyle\sum_{k=1}^{\infty} (-1)^{k+1} \frac{k^2 + 2}{k^3}$

6. $\displaystyle\sum_{k=1}^{\infty} (-1)^{k+1} \frac{3k - 1}{k + 5}$

7. $\displaystyle\sum_{k=1}^{\infty} (-1)^{k-1}\left(\frac{1}{k} + \frac{1}{3^k}\right)$

8. $\displaystyle\sum_{k=1}^{\infty} (-1)^{k+1}\frac{k+1}{4^k}$

9. $\displaystyle\sum_{n=1}^{\infty} (-1)^{n+1}\frac{4\sqrt{n}}{2n+1}$

10. $\displaystyle\sum_{n=1}^{\infty} (-1)^{n-1}\frac{\sqrt[3]{n}}{n+1}$

11. $\displaystyle\sum_{n=2}^{\infty} \left(\cos n\pi\right)\frac{\sqrt{n+1}}{n+2}$

12. $\displaystyle\sum_{k=2}^{\infty} (-1)^k\frac{\sqrt{k^2+1}}{k^3}$

13. $\displaystyle\sum_{k=2}^{\infty} (-1)^k\frac{k}{\ln k}$

14. $\displaystyle\sum_{k=3}^{\infty} (-1)^{k-1}\frac{\ln k^{10}}{k}$

In Problems 15–34 determine whether the given series is absolutely convergent, conditionally convergent, or divergent.

15. $\displaystyle\sum_{k=1}^{\infty} \frac{(-1)^{k+1}}{2k+1}$

16. $\displaystyle\sum_{k=1}^{\infty} \frac{(-1)^{k-1}}{\sqrt{k}+5}$

17. $\displaystyle\sum_{k=1}^{\infty} (-1)^{k+1}\left(\frac{2}{3}\right)^k$

18. $\displaystyle\sum_{k=1}^{\infty} (-1)^{k+1}\frac{2^{2k}}{3^k}$

19. $\displaystyle\sum_{k=1}^{\infty} (-1)^k\frac{k}{5^k}$

20. $\displaystyle\sum_{k=1}^{\infty} (-1)^k(k2^{-k})^2$

21. $\displaystyle\sum_{k=1}^{\infty} \frac{(-1)^k}{k!}$

22. $\displaystyle\sum_{k=1}^{\infty} (-1)^k\frac{(k!)^2}{(2k)!}$

23. $\displaystyle\sum_{k=1}^{\infty} (-1)^{k+1}\frac{k!}{100^k}$

24. $\displaystyle\sum_{k=1}^{\infty} (-1)^{k-1}\frac{5^{2k-3}}{10^{k+2}}$

25. $\displaystyle\sum_{k=1}^{\infty} (-1)^{k-1}\frac{k}{1+k^2}$

26. $\displaystyle\sum_{k=1}^{\infty} (-1)^{k+1}\frac{k}{1+k^4}$

27. $\displaystyle\sum_{k=1}^{\infty} \cos k\pi$

28. $\displaystyle\sum_{k=1}^{\infty} \frac{\sin\left(\frac{2k+1}{2}\pi\right)}{\sqrt{k}+1}$

29. $\displaystyle\sum_{k=1}^{\infty} (-1)^{k-1}\sin\left(\frac{1}{k}\right)$

30. $\displaystyle\sum_{k=1}^{\infty} (-1)^{k-1}\frac{\sin\left(\frac{1}{k}\right)}{k^2}$

31. $\displaystyle\sum_{k=1}^{\infty} (-1)^k\left[\frac{1}{k+1} - \frac{1}{k}\right]$

32. $\displaystyle\sum_{k=1}^{\infty} (-1)^k[\sqrt{k+1} - \sqrt{k}]$

33. $\displaystyle\sum_{k=1}^{\infty} (-1)^k\left(\frac{2k}{k+50}\right)^k$

34. $\displaystyle\sum_{k=1}^{\infty} (-1)^{k+1}\frac{6^{3k}}{k^k}$

In Problems 35 and 36 approximate the sum of the convergent series to the indicated number of decimal places.

35. $\displaystyle\sum_{k=1}^{\infty} \frac{(-1)^{k+1}}{(2k-1)!}$; five

36. $\displaystyle\sum_{k=1}^{\infty} \frac{(-1)^{k+1}}{k!}$; three

In Problems 37 and 38 find the smallest positive integer n so that S_n approximates the sum of the convergent series to the indicated number of decimal places.

37. $\displaystyle\sum_{k=1}^{\infty} \frac{(-1)^{k+1}}{k^3}$; two

38. $\displaystyle\sum_{k=1}^{\infty} \frac{(-1)^{k+1}}{\sqrt{k}}$; three

In Problems 39 and 40 approximate the sum of the convergent series so that the error is less than the indicated amount.

39. $1 - \dfrac{1}{4^2} + \dfrac{1}{4^3} - \dfrac{1}{4^4} + \cdots$; 10^{-3}

40. $1 - \dfrac{2}{5^2} + \dfrac{3}{5^3} - \dfrac{4}{5^4} + \cdots$; 10^{-4}

In Problems 41 and 42 estimate the error in using the indicated partial sum as an approximation to the sum of the convergent series.

41. $\displaystyle\sum_{k=1}^{\infty} \frac{(-1)^{k+1}}{k}$; S_{100}

42. $\displaystyle\sum_{k=1}^{\infty} \frac{(-1)^{k+1}}{k\,2^k}$; S_6

Miscellaneous Problems

In Problems 43–48 state why the Alternating Series Test is not applicable to the given series. Determine whether the given series converges or diverges.

43. $\displaystyle\sum_{k=1}^{\infty} \frac{\sin(k\pi/6)}{\sqrt{k^4+1}}$

44. $\displaystyle\sum_{k=1}^{\infty} \frac{100+(-1)^k2^k}{3^k}$

45. $1 - \dfrac{1}{2} - \dfrac{1}{4} + \dfrac{1}{8} + \dfrac{1}{16} - - + + \cdots$

46. $\dfrac{1}{1} - \dfrac{1}{4} - \dfrac{1}{9} + \dfrac{1}{16} + \dfrac{1}{25} + \dfrac{1}{36} - - - + + + + \cdots$

47. $\dfrac{2}{1} - \dfrac{1}{1} + \dfrac{2}{2} - \dfrac{1}{2} + \dfrac{2}{3} - \dfrac{1}{3} + \dfrac{2}{4} - \dfrac{1}{4} + \cdots$ (*Hint:* Consider the partial sums S_{2n} for $n = 1, 2, 3, \ldots$.)

48. $\dfrac{1}{2} + \dfrac{1}{2} - \dfrac{1}{3} - \dfrac{1}{3} - \dfrac{1}{3} + \dfrac{1}{4} + \dfrac{1}{4} + \dfrac{1}{4} + \dfrac{1}{4} - - - - \cdots$

49. Determine whether each of the following series converges or diverges.

  (a) $1 - 1 + 1 - 1 + \cdots$

  (b) $(1 - 1) + (1 - 1) + (1 - 1) + \cdots$

  (c) $1 + (-1 + 1) + (-1 + 1) + \cdots$

  (d) $1 + (-1 + 1) + (-1 + 1 - 1) + \cdots$

50. The series $1 - \dfrac{1}{3} + \dfrac{1}{9} - \dfrac{1}{27} + \cdots$ is an absolutely convergent geometric series. Show that its rearrangement

$$-\dfrac{1}{3} + \dfrac{1}{1} - \dfrac{1}{27} + \dfrac{1}{9} - \cdots$$

is convergent. Try the Ratio Test and Root Test. (*Hint*: Examine $3^{k+(-1)^k}$, $k = 0, 1, 2, \ldots$.)

In Problems 51 and 52 use

$$S = 1 - \frac{1}{2} + \frac{1}{3} - \frac{1}{4} + \frac{1}{5} - \frac{1}{6} + \cdots$$

to prove the given result.

51. $\dfrac{1}{2}S = 0 + \dfrac{1}{2} + 0 - \dfrac{1}{4} + 0 + \dfrac{1}{6} - \cdots$

52. $\dfrac{3}{2}S = 1 + \dfrac{1}{3} - \dfrac{1}{2} + \dfrac{1}{5} + \dfrac{1}{7} - \dfrac{1}{4} + \cdots$

53. If $\Sigma\, a_k$ is absolutely convergent, prove that $\Sigma\, a_k^2$ converges. (*Hint*: For n sufficiently large, $|a_n| < 1$. Why?)

11.6 Power Series

A series containing nonnegative integral powers of a variable x

$$c_0 + c_1 x + c_2 x^2 + \cdots + c_n x^n + \cdots = \sum_{k=0}^{\infty} c_k x^k \qquad (11.14)$$

where the c_k are constants depending on k, is called a **power series in x**. The series in (11.14) is just a particular case of the more general form

$$c_0 + c_1(x - a) + c_2(x - a)^2 + \cdots + c_n(x - a)^n + \cdots = \sum_{k=0}^{\infty} c_k(x - a)^k \qquad (11.15)$$

which is called a **power series in $x - a$**. The problem we face in this section is:

Find the values of x for which a power series converges.

Observe that (11.14) and (11.15) converge to c_0 when $x = 0$ and $x = a$, respectively.*

Example 1

The power series

$$1 + x + x^2 + \cdots + x^n + \cdots = \sum_{k=0}^{\infty} x^k$$

is recognized as a geometric series with $r = x$. Hence, the series converges for those values of x that satisfy $|x| < 1$ or $-1 < x < 1$.

Interval of Convergence The set of all real numbers x for which a power series converges is said to be its **interval of convergence**. Considering the general case, a power series in $x - a$ may converge

---

*It is convenient to define $x^0 = 1$ and $(x - a)^0 = 1$ even when $x = 0$ and $x = a$, respectively.

(*i*)   on a *finite interval* centered at *a*: $(a - r, a + r)$, $[a - r, a + r)$, $(a - r, a + r]$, or $[a - r, a + r]$; or

(*ii*)   on the *infinite interval* $(-\infty, \infty)$; or

(*iii*)   at the *single point* $x = a$.

In the respective cases, we say that the **radius of convergence** is r, ∞, or 0. Figure 11.9 illustrates the case $(a - r, a + r)$.

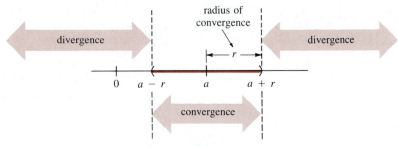

Figure 11.9

The Ratio Test, as stated in Theorem 11.20, is especially useful in finding an interval of convergence.

Example 2

Find the interval of convergence for $\displaystyle\sum_{k=0}^{\infty} \frac{x^k}{2^k(k + 1)^2}$.

Solution

$$\lim_{n \to \infty} \left| \frac{a_{n+1}}{a_n} \right| = \lim_{n \to \infty} \left| \frac{x^{n+1}}{2^{n+1}(n + 2)^2} \cdot \frac{2^n(n + 1)^2}{x^n} \right|$$

$$= \lim_{n \to \infty} \left(\frac{n + 1}{n + 2} \right)^2 \frac{|x|}{2} = \frac{|x|}{2}$$

From part (*i*) of Theorem 11.20, we have absolute convergence whenever this limit is strictly less than 1. Thus, the series is absolutely convergent for those values of x that satisfy $|x|/2 < 1$ or $|x| < 2$; that is, the series will converge for any number x in the open interval $(-2, 2)$. However, if $|x|/2 = 1$, or when $x = 2$ and $x = -2$, the Ratio Test gives no information. We must perform separate checks of the series for convergence at these endpoints. Substituting 2 for x gives $\sum_{k=1}^{\infty} 1/(k + 1)^2$, which is convergent by comparison with the convergent *p*-series $\sum_{k=1}^{\infty} 1/k^2$. Similarly, substituting -2 for x yields the alternating series $\sum_{k=1}^{\infty} (-1)^k/(k + 1)^2$, which is evidently convergent. (Why?) We conclude that the interval of convergence is the closed interval $[-2, 2]$. The radius of convergence is 2. The series diverges if $|x| > 2$.

Example 3

Find the interval of convergence for $\displaystyle\sum_{k=0}^{\infty} \frac{x^k}{k!}$.

Solution   By Theorem 11.20 we have

$$\lim_{n\to\infty} \left| \frac{a_{n+1}}{a_n} \right| = \lim_{n\to\infty} \left| \frac{x^{n+1}}{(n+1)!} \cdot \frac{n!}{x^n} \right|$$

$$= \lim_{n\to\infty} \frac{n!}{(n+1)!} |x|$$

$$= \lim_{n\to\infty} \frac{|x|}{n+1}$$

Since $\lim_{n\to\infty} |x|/(n+1) = 0$ for any choice of x, the series converges absolutely for every real number. Thus, the interval of convergence is $(-\infty, \infty)$ and the radius of convergence is ∞.

Example 4

Find the interval of convergence for $\displaystyle\sum_{k=1}^{\infty} \frac{(x-5)^k}{k3^k}$.

Solution $\displaystyle\lim_{n\to\infty} \left| \frac{a_{n+1}}{a_n} \right| = \lim_{n\to\infty} \left| \frac{(x-5)^{n+1}}{(n+1)3^{n+1}} \cdot \frac{n3^n}{(x-5)^n} \right|$

$$= \lim_{n\to\infty} \left(\frac{n}{n+1} \right) \frac{|x-5|}{3} = \frac{|x-5|}{3}$$

Divergent   Convergent   Divergent

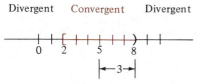

Figure 11.10

The series converges absolutely if $|x-5|/3 < 1$ or $|x-5| < 3$. This last inequality yields the open interval $(2, 8)$. At $x = 2$ and $x = 8$, we obtain, in turn, $\sum_{k=1}^{\infty} (-1)^k/k$ and $\sum_{k=1}^{\infty} 1/k$. The former series converges by the Alternating Series Test; the latter series is the divergent harmonic series. Consequently, the interval of convergence is $[2, 8)$. The radius of convergence is 3. The series diverges if $x < 2$ or $x \geq 8$. See Figure 11.10.

Example 5

Find the interval of convergence for $\displaystyle\sum_{k=1}^{\infty} k!(x+10)^k$.

Solution $\displaystyle\lim_{n\to\infty} \left| \frac{a_{n+1}}{a_n} \right| = \lim_{n\to\infty} \left| \frac{(n+1)!(x+10)^{n+1}}{n!(x+10)^n} \right|$

$$= \lim_{n\to\infty} (n+1)|x+10|$$

$$= \begin{cases} \infty, & x \neq -10 \\ 0, & x = -10 \end{cases}$$

The series diverges for every real number x, *except* for $x = -10$. At $x = -10$, we obtain a convergent series consisting of all zeros. The radius of convergence is 0.

Exercises 11.6

Answers to odd-numbered problems begin on page A-39.

In Problems 1–24 use the Ratio Test to find the interval of convergence of the given power series.

1. $\displaystyle\sum_{k=1}^{\infty} \frac{(-1)^k}{k} x^k$

2. $\displaystyle\sum_{k=1}^{\infty} \frac{x^k}{k^2}$

3. $\displaystyle\sum_{k=1}^{\infty} \frac{2^k}{k} x^k$

4. $\displaystyle\sum_{k=0}^{\infty} \frac{5^k}{k!} x^k$

5. $\displaystyle\sum_{k=1}^{\infty} \frac{(x-3)^k}{k^3}$

6. $\displaystyle\sum_{k=1}^{\infty} \frac{(x+7)^k}{\sqrt{k}}$

7. $\displaystyle\sum_{k=1}^{\infty} \frac{(-1)^k}{10^k} (x-5)^k$

8. $\displaystyle\sum_{k=1}^{\infty} \frac{k}{(k+2)^2} (x-4)^k$

9. $\displaystyle\sum_{k=0}^{\infty} k! 2^k x^k$

10. $\displaystyle\sum_{k=0}^{\infty} \frac{k-1}{k^{2k}} x^k$

11. $\displaystyle\sum_{k=1}^{\infty} \frac{(3x-1)^k}{k^2+k}$

12. $\displaystyle\sum_{k=0}^{\infty} \frac{(4x-5)^k}{3^k}$

13. $\displaystyle\sum_{k=2}^{\infty} \frac{x^k}{\ln k}$

14. $\displaystyle\sum_{k=2}^{\infty} \frac{(-1)^k x^k}{k \ln k}$

15. $\displaystyle\sum_{k=1}^{\infty} \frac{k^2}{3^{2k}} (x+7)^k$

16. $\displaystyle\sum_{k=1}^{\infty} k^3 2^{4k} (x-1)^k$

17. $\displaystyle\sum_{k=1}^{\infty} \frac{2^{5k}}{5^{2k}} \left(\frac{x}{3}\right)^k$

18. $\displaystyle\sum_{k=1}^{\infty} \frac{1000^k}{k^k} x^k$

19. $\displaystyle\sum_{k=0}^{\infty} \frac{(-3)^k}{(k+1)(k+2)} (x-1)^k$

20. $\displaystyle\sum_{k=1}^{\infty} \frac{3^k}{(-2)^k k(k+1)} (x+5)^k$

21. $\displaystyle\sum_{k=1}^{\infty} \frac{(-1)^{k+1}}{(k!)^2} \left(\frac{x-2}{3}\right)^k$

22. $\displaystyle\sum_{k=0}^{\infty} \frac{(6-x)^{k+1}}{\sqrt{2k+1}}$

23. $\displaystyle\sum_{k=0}^{\infty} \frac{(-1)^k}{9^k} x^{2k+1}$

24. $\displaystyle\sum_{k=1}^{\infty} \frac{5^k}{(2k)!} x^{2k}$

In Problems 25–28 use the Root Test, Theorem 11.21, to find the interval of convergence of the given power series.

25. $\displaystyle\sum_{k=2}^{\infty} \frac{x^k}{(\ln k)^k}$

26. $\displaystyle\sum_{k=1}^{\infty} (k+1)^k (x+1)^k$

27. $\displaystyle\sum_{k=1}^{\infty} \left(\frac{4}{3}\right)^k (x+3)^k$

28. $\displaystyle\sum_{k=1}^{\infty} \left(\frac{k}{k+1}\right)^{k^2} (x-e)^k$

In Problems 29 and 30 find the radius of convergence of the given power series.

29. $\displaystyle\sum_{k=1}^{\infty} \frac{k!}{1 \cdot 3 \cdot 5 \cdots (2k-1)} \left(\frac{x}{2}\right)^k$

30. $\displaystyle\sum_{k=2}^{\infty} \frac{1 \cdot 3 \cdot 5 \cdots (2k-3)}{3^k k!} (x-1)^k$

Miscellaneous Problems

In Problems 31–38 the given series is not a power series. Nonetheless, find all values of x for which the given series converges.

31. $\displaystyle\sum_{k=1}^{\infty} \frac{1}{x^k}$

32. $\displaystyle\sum_{k=1}^{\infty} \frac{7^k}{x^{2k}}$

33. $\displaystyle\sum_{k=1}^{\infty} \left(\frac{x+1}{x}\right)^k$

34. $\displaystyle\sum_{k=1}^{\infty} \frac{1}{2^k} \left(\frac{x}{x+2}\right)^k$

35. $\displaystyle\sum_{k=0}^{\infty} \left(\frac{x^2+2}{6}\right)^{k^2}$

36. $\displaystyle\sum_{k=1}^{\infty} \frac{k!}{(kx)^k}$

37. $\displaystyle\sum_{k=0}^{\infty} e^{kx}$

38. $\displaystyle\sum_{k=0}^{\infty} k! e^{-kx^2}$

39. Find all values of x in $[0, 2\pi]$ for which $\sum_{k=0}^{\infty} (2/\sqrt{3})^k \sin^k x$ converges.

40. Show that $\sum_{k=1}^{\infty} (\sin kx)/k^2$ converges for all real values of x.

11.7 Differentiation and Integration of Power Series

A Power Series Represents a Function

For each x in its interval of convergence, a power series* $\Sigma\, c_k x^k$ converges to one number. Thus, a power series defines or *represents* a function f with domain the interval of convergence. For each x in the interval of convergence, we define the functional value $f(x)$ by the sum of the series:

$$f(x) = c_0 + c_1 x + c_2 x^2 + \cdots + c_n x^n + \cdots = \sum_{k=0}^{\infty} c_k x^k \quad (11.16)$$

The next three theorems, which we state without proof, answer some fundamental questions about a function represented by a power series.

> **THEOREM 11.22**   If $f(x) = \sum_{k=0}^{\infty} c_k x^k$ converges on an interval $(-r, r)$ for which the radius of convergence is either positive or ∞, then f is continuous at each x in $(-r, r)$.

Differentiation of a Power Series

> **THEOREM 11.23**   If $f(x) = \sum_{k=0}^{\infty} c_k x^k$ converges on an interval $(-r, r)$ for which the radius of convergence is either positive or ∞, then f is differentiable at each x in $(-r, r)$ and
>
> $$f'(x) = \sum_{k=1}^{\infty} k c_k x^{k-1} \quad (11.17)$$

The result in (11.17) simply states that a power series can be differentiated term-by-term as we would a polynomial:

$$f'(x) = \frac{d}{dx} c_0 + \frac{d}{dx} c_1 x + \frac{d}{dx} c_2 x^2 + \frac{d}{dx} c_3 x^3 + \cdots + \frac{d}{dx} c_n x^n + \cdots$$

$$= c_1 + 2 c_2 x + 3 c_3 x^2 + \cdots + n c_n x^{n-1} + \cdots = \sum_{k=1}^{\infty} k c_k x^{k-1}$$

The radius of convergence of (11.17) is the same as that of $\Sigma\, c_k x^k$. Thus, by applying Theorem 11.23 to f' defined in (11.17), we can say f' is differentiable at each x in $(-r, r)$; that is,

$$f''(x) = 2 c_2 + 3 \cdot 2 c_3 x + \cdots + n(n-1) c_n x^{n-2} + \cdots$$

$$= \sum_{k=2}^{\infty} k(k-1) c_k x^{k-2}$$

---

*For convenience, we limit our discussion to power series in x. The results of this section apply equally to power series in $x - a$.

Continuing in this manner it follows that:

A function f represented by a power series on $(-r, r)$, $r > 0$, possesses derivatives of all orders in the interval.

Integration of a Power Series

As in (11.17), the process of integration of a power series may also be carried out term-by-term:

$$\int f(x)\, dx = \int c_0\, dx + \int c_1 x\, dx + \int c_2 x^2\, dx + \cdots + \int c_n x^n\, dx + \cdots$$

$$= \left(c_0 x + \frac{c_1}{2} x^2 + \frac{c_2}{3} x^3 + \cdots + \frac{c_n}{n + 1} x^{n+1} + \cdots \right) + C$$

$$= \sum_{k=0}^{\infty} \frac{c_k}{k + 1} x^{k+1} + C$$

This is summarized in the next theorem.

THEOREM 11.24   If $f(x) = \sum_{k=0}^{\infty} c_k x^k$ converges on an interval $(-r, r)$ for which the radius of convergence is either positive or ∞, then

$$\int f(x)\, dx = \sum_{k=0}^{\infty} \frac{c_k}{k + 1} x^{k+1} + C \qquad (11.18)$$

The radius of convergence of (11.18) is again the same as that of $\sum c_k x^k$.

For definite integrals, (11.18) implies that

$$\int_a^b f(x)\, dx = \sum_{k=0}^{\infty} c_k \left(\int_a^b x^k\, dx \right) = \sum_{k=0}^{\infty} c_k \frac{b^{k+1} - a^{k+1}}{k + 1}$$

for any numbers a and b in $(-r, r)$.

Example 1

Find a power series representation for the function $f(x) = 1/(1 + x)$.

Solution   Recall that a geometric series converges to $a/(1 - r)$ if $|r| < 1$. Identifying $a = 1$, $r = -x$, we see that

$$\frac{1}{1 + x} = 1 - x + x^2 - x^3 + \cdots + (-1)^n x^n + \cdots = \sum_{k=0}^{\infty} (-1)^k x^k \qquad (11.19)$$

for any x in $(-1, 1)$.

Example 2

Term-by-term differentiation of (11.19) yields a power series representation for $1/(1 + x)^2$ on $(-1, 1)$:

$$\frac{-1}{(1 + x)^2} = -1 + 2x - 3x^2 + \cdots + (-1)^n nx^{n-1} + \cdots$$

or

$$\frac{1}{(1 + x)^2} = 1 - 2x + 3x^2 + \cdots + (-1)^{n+1} nx^{n-1} + \cdots$$

$$= \sum_{k=1}^{\infty} (-1)^{k+1} k x^{k-1}$$

Example 3

Find a power series representation for $\ln(1 + x)$ on $(-1, 1)$.

Solution Substituting $x = t$ in (11.19) gives

$$\frac{1}{1 + t} = 1 - t + t^2 - t^3 + \cdots + (-1)^n t^n + \cdots$$

Thus, for any x in $(-1, 1)$,

$$\int_0^x \frac{dt}{1 + t} = \int_0^x dt - \int_0^x t \, dt + \int_0^x t^2 \, dt - \cdots + (-1)^n \int_0^x t^n \, dt + \cdots$$

$$= x - \frac{x^2}{2} + \frac{x^3}{3} - \cdots + (-1)^n \frac{x^{n+1}}{n + 1} + \cdots$$

But, $\int_0^x dt/(1 + t) = \ln(1 + t)]_0^x = \ln(1 + x)$, and so

$$\ln(1 + x) = x - \frac{x^2}{2} + \frac{x^3}{3} - \cdots + (-1)^n \frac{x^{n+1}}{n + 1} + \cdots$$

$$= \sum_{k=0}^{\infty} \frac{(-1)^k}{k + 1} x^{k+1} \qquad (11.20)$$

Example 4

Approximate $\ln(1.2)$ to four decimal places.

Solution Substituting $x = 0.2$ in (11.20) gives

$$\ln(1.2) = 0.2 - \frac{(0.2)^2}{2} + \frac{(0.20)^3}{3} - \frac{(0.2)^4}{4} + \frac{(0.2)^5}{5} - \frac{(0.2)^6}{6} + \cdots \qquad (11.21)$$

$$= 0.2 - 0.02 + 0.00267 - 0.0004 + 0.000064 - 0.00001067 + \cdots$$

$$\approx 0.1823 \qquad (11.22)$$

If the sum of the series (11.21) in Example 4 is denoted by S, then we know from Theorem 11.18 that $|S_n - S| \leq a_{n+1}$. The number given in (11.22) is accurate to four decimal places, since, for the fifth partial sum, $|S_5 - S| \leq 0.00001067 < 0.00005$.

Remark

It is interesting to note that if the interval of convergence of a power series representation of a function f is the open interval $(-r, r)$, then the power series representation for $\int_0^x f(t)\, dt$ may converge at one or both endpoints of the interval. You should check that the series in (11.20) diverges at $x = -1$ but converges at $x = 1$. At the latter value we discover that the sum of the alternating harmonic series is given by:

$$\ln 2 = 1 - \frac{1}{2} + \frac{1}{3} - \frac{1}{4} + \cdots$$

_____ Exercises 11.7 _____

Answers to odd-numbered problems begin on page A-39.

In Problems 1–14 find a power series representation of the given function. Give the interval of convergence.

1. $f(x) = \dfrac{1}{1 - x}$

2. $f(x) = \dfrac{1}{(1 - x)^2}$

3. $f(x) = \dfrac{1}{5 + 3x}$

4. $f(x) = \dfrac{6}{2 - x}$

5. $f(x) = \dfrac{1}{(1 - x)^3}$

6. $f(x) = \dfrac{x^2}{(1 + x)^3}$

7. $f(x) = \ln(4 + x)$

8. $f(x) = \ln(1 + 2x)$

9. $f(x) = \dfrac{1}{1 + x^2}$

10. $f(x) = \dfrac{x}{1 + x^2}$

11. $f(x) = \tan^{-1} x$

12. $f(x) = \ln(1 + x^2)$

13. $f(x) = \displaystyle\int_0^x \ln(1 + t^2)\, dt$

14. $f(x) = \displaystyle\int_0^x \tan^{-1} t\, dt$

In Problems 15 and 16 find the domain of the given function.

15. $f(x) = \dfrac{x}{3} - \dfrac{x^2}{2 \cdot 3^2} + \dfrac{x^3}{3 \cdot 3^3} - \dfrac{x^4}{4 \cdot 3^4} + \cdots$

16. $f(x) = 1 + 2x + \dfrac{4x^2}{1 \cdot 2} + \dfrac{8x^3}{1 \cdot 2 \cdot 3} + \cdots$

In Problems 17–22 use infinite series to approximate the given quantity to four decimal places.

17. $\ln(1.1)$

18. $\tan^{-1}(0.2)$

19. $\displaystyle\int_0^{1/2} \frac{dx}{1 + x^3}$

20. $\displaystyle\int_0^{1/3} \frac{x}{1 + x^4}\, dx$

21. $\displaystyle\int_0^{0.3} x \tan^{-1} x\, dx$

22. $\displaystyle\int_0^{1/2} \tan^{-1} x^2\, dx$

Miscellaneous Problems

23. If $f(x) = \displaystyle\sum_{k=0}^{\infty} \frac{(-1)^k}{(2k + 1)!} x^{2k+1}$

for $-\infty < x < \infty$, show that $f''(x) + f(x) = 0$.

24. (a) Show that if $f(x) = \displaystyle\sum_{k=0}^{\infty} \frac{x^k}{k!}$,

for $-\infty < x < \infty$, then $f'(x) = f(x)$.

(b) Show that $\qquad e^x = \displaystyle\sum_{k=0}^{\infty} \frac{x^k}{k!}$

by solving the differential equation in part **(a)** by separation of variables. [*Hint:* $f(0) = 1$.]

In Problems 25 and 26 use the result of Problem 24(**b**) to find a power series representation of the given function.

25. $f(x) = e^{-x}$

26. $f(x) = e^{x/5}$

In Problems 27–30 use the result of Problem 24(**b**) to approximate the given quantity to four decimal places.

27. $\dfrac{1}{e}$

28. $e^{-1/2}$

29. $\displaystyle\int_0^{0.2} e^{-x^2}\, dx$

30. $\displaystyle\int_0^{1} e^{-x^2/2}\, dx$

31. Use Problem 11 to show

$$\frac{\pi}{4} = 1 - \frac{1}{3} + \frac{1}{5} - \frac{1}{7} + \cdots$$

32. The series in Problem 31 is known to converge very slowly. Show this by finding the smallest positive integer n so that S_n approximates $\pi/4$ to four decimal places.

11.8 Taylor Series

For a power series representing a function f on an interval $(a - r, a + r)$, $r > 0$,

$$f(x) = c_0 + c_1(x - a) + c_2(x - a)^2 + c_3(x - a)^3 + \cdots + c_n(x - a)^n + \cdots$$

$$= \sum_{k=0}^{\infty} c_k(x - a)^k \qquad (11.23)$$

there is a relationship between the coefficients c_k and the derivatives of f. By Theorem 11.23 we can write

$$f'(x) = c_1 + 2c_2(x - a) + 3c_3(x - a)^2 + \cdots \qquad (11.24)$$

$$f''(x) = 2c_2 + 3 \cdot 2c_3(x - a) + \cdots \qquad (11.25)$$

$$f'''(x) = 3 \cdot 2 \cdot 1c_3 + \cdots \qquad (11.26)$$

and so on. Evaluating (11.23), (11.24), (11.25), and (11.26) at $x = a$ gives

$$f(a) = c_0, \qquad f'(a) = 1!c_1, \qquad f''(a) = 2!c_2, \quad \text{and} \quad f'''(a) = 3!c_3$$

respectively. In general, $f^{(n)}(a) = n!c_n$, or

$$c_n = \frac{f^{(n)}(a)}{n!}, \qquad n \geq 0 \qquad (11.27)$$

When $n = 0$ we interpret the zeroth derivative as $f(a)$ and $0! = 1$. Substituting (11.27) in (11.23) yields

$$f(x) = \sum_{k=0}^{\infty} \frac{f^{(k)}(a)}{k!}(x - a)^k \qquad (11.28)$$

which is valid for all values for x in $(a - r, a + r)$, $r > 0$. This series is called the **Taylor series for f at a.*** The special case of a Taylor series, when $a = 0$,

$$f(x) = \sum_{k=0}^{\infty} \frac{f^{(k)}(0)}{k!}x^k \qquad (11.29)$$

is called the **Maclaurin series for f.**†

On the other hand, if we are given a differentiable function f, the natural question arises:

Can we expand f into a Taylor series (11.28)?

Formally, the answer is yes—by simply calculating the coefficients as dictated by (11.27).

---

*Named in honor of the English mathematician Brook Taylor (1685–1731), who published this result in 1715.

†Named after the Scottish mathematician, and former student of Isaac Newton, Colin Maclaurin (1698–1746). It is not clear why Maclaurin's name is associated with this series.

Example 1

Find the Taylor series for $f(x) = \ln x$ at $a = 1$.

Solution   We have

$$f(x) = \ln x, \qquad\qquad f(1) = 0$$

$$f'(x) = \frac{1}{x}, \qquad\qquad f'(1) = 1$$

$$f''(x) = -\frac{1}{x^2}, \qquad\qquad f''(1) = -1$$

$$f'''(x) = \frac{1 \cdot 2}{x^3}, \qquad\qquad f'''(1) = 2!$$

$$\cdot$$
$$\cdot$$
$$\cdot$$

$$f^{(n)}(x) = (-1)^{n-1}\frac{(n-1)!}{x^n}, \qquad f^{(n)}(1) = (-1)^{n-1}(n-1)!$$

Thus, (11.27) and (11.28) yield

$$(x - 1) - \frac{1}{2}(x - 1)^2 + \frac{1}{3}(x - 1)^3 - \cdots = \sum_{k=1}^{\infty} \frac{(-1)^{k-1}}{k}(x - 1)^k \quad (11.30)$$

With the aid of the Ratio Test, it is found that this series converges for all values of x in the interval $(0, 2]$.

Taylor's Theorem   It is apparent from (11.28) that a function f can have a Taylor series at a only if it possesses finite derivatives of all orders at this value. Thus, $f(x) = \ln x$ does not possess a Maclaurin series. (Why?) Moreover, it should be noted that even if f possesses derivatives of all orders and generates a Taylor series convergent on some interval, we do not know whether the series converges to $f(x)$ at every value of x in the interval. (See Problem 43 in Exercises 11.8.) If it does, we say that the series *represents* the given function on the interval. At this point it has not been shown that the series given in (11.30) represents $\ln x$ on $(0, 2]$. This problem can be resolved by considering **Taylor's Theorem**.

THEOREM 11.25   Taylor's Theorem

Let f be a function such that $f^{(n+1)}(x)$ exists for every x in the interval $(a - r, a + r)$. Then for all x in the interval

$$f(x) = P_n(x) + R_n(x)$$

where

$$P_n(x) = f(a) + \frac{f'(a)}{1!}(x - a) + \cdots + \frac{f^{(n)}(a)}{n!}(x - a)^n \quad (11.31)$$

is called the nth degree **Taylor polynomial of f at a**, and

$$R_n(x) = \frac{f^{(n+1)}(c)}{(n+1)!}(x-a)^{n+1} \tag{11.32}$$

is called the **Lagrange form of the remainder**.* The number c is between a and x.

Since the proof of this theorem would deflect us from the main thrust of our discussion, it is given in Appendix III. The importance of Theorem 11.25 lies in the fact that the $P_n(x)$ are the partial sums of the Taylor series and that

$$P_n(x) = f(x) - R_n(x)$$

Hence, from

$$\lim_{n\to\infty} P_n(x) = f(x) - \lim_{n\to\infty} R_n(x)$$

we see that if $R_n(x) \to 0$ as $n \to \infty$, then the sequence of partial sums converges to $f(x)$. We summarize the result.

THEOREM 11.26   If f has derivatives of all orders at every x in the interval $(a - r, a + r)$ and if $\lim_{n\to\infty} R_n(x) = 0$ for every x in the interval, then

$$f(x) = \sum_{k=0}^{\infty} \frac{f^{(k)}(a)}{k!}(x-a)^k$$

In practice, the proof that the remainder R_n approaches zero as $n \to \infty$ often depends on the fact that

$$\lim_{n\to\infty} \frac{|x|^n}{n!} = 0 \tag{11.33}$$

This latter result follows from applying Theorem 11.6 to the series $\sum_{k=0}^{\infty} x^k/k!$, which is known to be absolutely convergent for all real numbers. (See Example 3 in Section 11.6.)

Example 2

Represent $f(x) = \cos x$ by a Maclaurin series.

Solution   We have,
$$\begin{aligned}
f(x) &= \cos x, & f(0) &= 1 \\
f'(x) &= -\sin x, & f'(0) &= 0 \\
f''(x) &= -\cos x, & f''(0) &= -1 \\
f'''(x) &= \sin x, & f'''(0) &= 0
\end{aligned}$$

and so on. From (11.29) we obtain

$$1 - \frac{x^2}{2!} + \frac{x^4}{4!} - \frac{x^6}{6!} + \cdots = \sum_{k=0}^{\infty} \frac{(-1)^k}{(2k)!} x^{2k} \tag{11.34}$$

---

*There are several forms of the remainder. This form is due to the French mathematician, Joseph Louis Lagrange (1736–1813).

The Ratio Test shows that (11.34) converges absolutely for all real values of x. To prove that $\cos x$ is indeed represented by the series (11.34), we must show that $\lim_{n \to \infty} R_n(x) = 0$.

To this end, we note that the derivatives of f satisfy

$$|f^{(n+1)}(x)| = \begin{cases} |\sin x|, & n \text{ even} \\ |\cos x|, & n \text{ odd} \end{cases}$$

In either case, $|f^{(n+1)}(c)| \le 1$ for any real number c, and so

$$|R_n(x)| = \frac{|f^{(n+1)}(c)|}{(n+1)!}|x|^{n+1} \le \frac{|x|^{n+1}}{(n+1)!}$$

For any fixed, but arbitrary choice of x, $\lim_{n \to \infty} |x|^{n+1}/(n+1)! = 0$ by (11.33). But, $\lim_{n \to \infty} |R_n(x)| = 0$ implies that $\lim_{n \to \infty} R_n(x) = 0$. Therefore,

$$\cos x = 1 - \frac{x^2}{2!} + \frac{x^4}{4!} - \frac{x^6}{6!} + \cdots + (-1)^n \frac{x^{2n}}{(2n)!} + \cdots$$

is a valid representation of $\cos x$ for every real number x.

Example 3

Represent $f(x) = \sin x$ in a Taylor series at $a = \pi/3$.

Solution   We have

$$f(x) = \sin x, \qquad f\left(\frac{\pi}{3}\right) = \frac{\sqrt{3}}{2}$$

$$f'(x) = \cos x, \qquad f'\left(\frac{\pi}{3}\right) = \frac{1}{2}$$

$$f''(x) = -\sin x, \qquad f''\left(\frac{\pi}{3}\right) = -\frac{\sqrt{3}}{2}$$

$$f'''(x) = -\cos x, \qquad f'''\left(\frac{\pi}{3}\right) = -\frac{1}{2}$$

and so on. Hence, the Taylor series at $\pi/3$ that corresponds to $\sin x$ is

$$\frac{\sqrt{3}}{2} + \frac{1}{2 \cdot 1!}\left(x - \frac{\pi}{3}\right) - \frac{\sqrt{3}}{2 \cdot 2!}\left(x - \frac{\pi}{3}\right)^2 - \frac{1}{2 \cdot 3!}\left(x - \frac{\pi}{3}\right)^3 + \cdots \quad (11.35)$$

Again, from the Ratio Test it follows that (11.35) converges absolutely for all real values of x. To show that

$$\sin x = \frac{\sqrt{3}}{2} + \frac{1}{2 \cdot 1!}\left(x - \frac{\pi}{3}\right) - \frac{\sqrt{3}}{2 \cdot 2!}\left(x - \frac{\pi}{3}\right)^2 - \frac{1}{2 \cdot 3!}\left(x - \frac{\pi}{3}\right)^3 + \cdots$$

for every real x, we note that, as in the preceding example, $|f^{(n+1)}(c)| \le 1$. This implies that $|R_n(x)| \le |x - \pi/3|^{n+1}/(n+1)!$ from which we see, with the help of (11.33), that $\lim_{n \to \infty} R_n(x) = 0$.

Example 4

Prove that the series (11.30) represents $f(x) = \ln x$ on the interval $(0, 2]$.

Solution   For $f(x) = \ln x$, the nth derivative is given by

$$f^{(n)}(x) = \frac{(-1)^{n-1}(n-1)!}{x^n}$$

Therefore

$$f^{(n+1)}(c) = \frac{(-1)^n n!}{c^{n+1}}$$

and

$$|R_n(x)| = \left| \frac{(-1)^n n!}{c^{n+1}} \cdot \frac{(x-1)^{n+1}}{(n+1)!} \right| = \frac{1}{n+1} \left| \frac{x-1}{c} \right|^{n+1}$$

where c is some number in $(0, 2]$ between 1 and x.

If $1 \le x \le 2$, then $0 < x - 1 \le 1$. Since $1 < c < x$, we must have $0 < x - 1 \le 1 < c$ and, consequently, $(x-1)/c < 1$. Hence

$$|R_n(x)| \le \frac{1}{n+1} \quad \text{and} \quad \lim_{n \to \infty} R_n(x) = 0$$

In the case where $0 < x < 1$, we can also show that $\lim_{n \to \infty} R_n(x) = 0$. We omit the proof.* Hence,

$$\ln x = (x-1) - \frac{1}{2}(x-1)^2 + \frac{1}{3}(x-1)^3 - \cdots$$

for all values of x in the interval $(0, 2]$.

We summarize some important Maclaurin series.

$$e^x = 1 + x + \frac{x^2}{2!} + \frac{x^3}{3!} + \cdots = \sum_{k=0}^{\infty} \frac{x^k}{k!}, \qquad -\infty < x < \infty \qquad (11.36)$$

$$\cos x = 1 - \frac{x^2}{2!} + \frac{x^4}{4!} - \frac{x^6}{6!} + \cdots = \sum_{k=0}^{\infty} \frac{(-1)^k}{(2k)!} x^{2k}, \qquad -\infty < x < \infty \qquad (11.37)$$

$$\sin x = x - \frac{x^3}{3!} + \frac{x^5}{5!} - \frac{x^7}{7!} + \cdots = \sum_{k=0}^{\infty} \frac{(-1)^k}{(2k+1)!} x^{2k+1}, \qquad -\infty < x < \infty \qquad (11.38)$$

$$\cosh x = 1 + \frac{x^2}{2!} + \frac{x^4}{4!} + \frac{x^6}{6!} + \cdots = \sum_{k=0}^{\infty} \frac{x^{2k}}{(2k)!}, \qquad -\infty < x < \infty \qquad (11.39)$$

$$\sinh x = x + \frac{x^3}{3!} + \frac{x^5}{5!} + \frac{x^7}{7!} + \cdots = \sum_{k=0}^{\infty} \frac{x^{2k+1}}{(2k+1)!}, \qquad -\infty < x < \infty \qquad (11.40)$$

$$\ln(1+x) = x - \frac{x^2}{2} + \frac{x^3}{3} - \frac{x^4}{4} + \cdots = \sum_{k=0}^{\infty} \frac{(-1)^k}{(k+1)} x^{k+1}, \qquad -1 < x \le 1 \qquad (11.41)$$

---

*This part of the proof is usually based on an integral form of the remainder $R_n(x)$ that we shall not consider.

You are asked to demonstrate the validity of the representations (11.36), (11.38), (11.39), and (11.40) as exercises.

Some Graphs of Taylor Polynomials

In Example 2 we saw that the Taylor series of $f(x) = \cos x$ at $a = 0$ represents the function for all x since $\lim_{n \to \infty} R_n(x) = 0$. It is always of interest to see graphically how partial sums of the Taylor series, which are the Taylor polynomials defined in (11.31), converge to the function. In Figure 11.11 the graphs of the Taylor polynomials $P_0(x)$, $P_2(x)$, $P_4(x)$, and $P_{10}(x)$ at $a = 0$ are compared with the graph $f(x) = \cos x$. A comparison of numerical values is given in Figure 11.11(e).

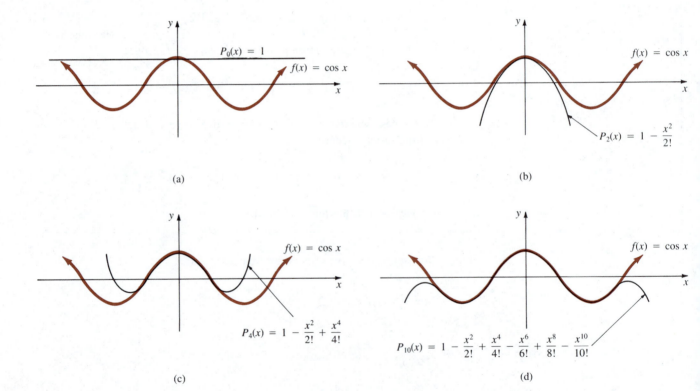

x	$P_2(x)$	$P_4(x)$	$P_{10}(x)$	$\cos x$
$\pi/6$	0.86292	0.86605	0.86603	0.86603
$\pi/4$	0.69157	0.70743	0.70711	0.70711
$\pi/3$	0.45169	0.50180	0.50000	0.5
$\pi/2$	−0.23370	0.01997	0.00000	0

(e)

Figure 11.11

Approximations with Taylor Polynomials   When the value of x is close to the number a ($x \approx a$), the Taylor polynomial $P_n(x)$ of a function f at a can be used to approximate the functional value $f(x)$. The error in this approximation is given by

$$|f(x) - P_n(x)| = |R_n(x)|$$

Example 5

Approximate $e^{-0.2}$ by $P_3(x)$. Determine the accuracy of the approximation.

Solution   Because the value $x = -0.2$ is close to zero, we use the Taylor polynomial $P_3(x)$ of $f(x) = e^x$ at $a = 0$. It follows from $f(x) = f'(x) = f''(x) = f'''(x) = e^x$ and (11.31) that

$$P_3(x) = f(0) + \frac{f'(0)}{1!}x + \frac{f''(0)}{2!}x^2 + \frac{f'''(0)}{3!}x^3$$

$$= 1 + x + \frac{1}{2}x^2 + \frac{1}{6}x^3$$

and $P_3(-0.2) = 1 + (-0.2) + \frac{1}{2}(-0.2)^2 + \frac{1}{6}(-0.2)^3 \approx 0.81867$.

Consequently, $\qquad\qquad e^{-0.2} \approx 0.81867 \qquad\qquad$ (11.42)

Now, from (11.30) we can write

$$|R_3(x)| = \frac{e^c}{4!}|x|^4 < \frac{|x|^4}{4!}$$

since $-0.2 < c < 0$ and $e^c < 1$. Thus,

$$|R_3(-0.2)| < \frac{|-0.2|^4}{24} < 0.0001$$

implies that the result in (11.42) is accurate to three decimal places.

Remarks

(*i*)   The Taylor series method of finding a power series for a function, and then proving that the series represents the function, has one big and obvious drawback. Obtaining a general expression for the nth derivative for most functions is nearly impossible. Thus, we are often limited to finding just the first few coefficients c_n. See Problems 25 and 26 in Exercises 11.8.

(*ii*)   Taylor's Theorem is also called the **Generalized Mean Value Theorem**.* The case $n = 0$ reduces to the usual Mean Value Theorem given on page 203. Those who read Appendix III will see that the proof of Theorem 11.25, like the proof of Theorem 3.5, relies on the construction of a special function and Rolle's Theorem.

---

*Do not confuse this with Theorem 10.1.

(*iii*) It is easy to pass over the significance of the results in (11.28) and (11.29). Suppose we wish to find the Maclaurin series for $f(x) = 1/(2 - x)$. We can, of course, use (11.29)—and you are asked to do so in Problem 1. On the other hand, you should also recognize, from the discussion of the preceding section, that a power series representation of f can be obtained utilizing geometric series. The point is: *The representation is unique. Thus, on its interval of convergence, a power series representing a function, regardless of how it is obtained, is the Taylor or Maclaurin series of that function.*

Exercises 11.8

Answers to odd-numbered problems begin on page A-39.

In Problems 1–10 use (11.29) to find the Maclaurin series for the given function.

1. $f(x) = \dfrac{1}{2 - x}$

2. $f(x) = \dfrac{1}{1 + 5x}$

3. $f(x) = \ln(1 + x)$

4. $f(x) = \ln(1 + 2x)$

5. $f(x) = \sin x$

6. $f(x) = \cos 2x$

7. $f(x) = e^x$

8. $f(x) = e^{-x}$

9. $f(x) = \sinh x$

10. $f(x) = \cosh x$

11. Prove that the series obtained in Problem 5 represents $\sin x$ for every real value of x.

12. Prove that the series obtained in Problem 7 represents e^x for every real value of x.

13. Prove that the series obtained in Problem 9 represents $\sinh x$ for every real value of x.

14. Prove that the series obtained in Problem 10 represents $\cosh x$ for every real value of x.

In Problems 15–24 use (11.28) to find the Taylor series for the given function at the indicated value of a.

15. $f(x) = \dfrac{1}{1 + x}$, $a = 4$

16. $f(x) = \sqrt{x}$, $a = 1$

17. $f(x) = \sin x$, $a = \pi/4$

18. $f(x) = \sin x$, $a = \pi/2$

19. $f(x) = \cos x$, $a = \pi/3$

20. $f(x) = \cos x$, $a = \pi/6$

21. $f(x) = \ln x$, $a = 2$

22. $f(x) = \ln(x + 1)$, $a = 2$

23. $f(x) = e^x$, $a = 1$

24. $f(x) = e^{-2x}$, $a = 1/2$

In Problems 25 and 26 use (11.29) to find the first four nonzero terms of the Maclaurin series for the given function.

25. $f(x) = \tan x$

26. $f(x) = \sin^{-1}x$

In Problems 27–34 use previous results or problems to find the Maclaurin series for the given function.

27. $f(x) = e^{-x^2}$

28. $f(x) = x^2 e^{-3x}$

29. $f(x) = x \cos x$

30. $f(x) = \sin x^3$

31. $f(x) = \ln(1 - x)$

32. $f(x) = \ln\left(\dfrac{1 + x}{1 - x}\right)$

33. $f(x) = \sec^2 x$

34. $f(x) = \ln(\cos x)$

In Problems 35–38 approximate the given quantity using the Taylor polynomial $P_n(x)$ for the indicated values of n and a. Determine the accuracy of the approximation.

35. $\sin 46°$, $n = 2$, $a = \pi/4$ (*Hint*: Convert 46° to radian measure.)

36. $\cos 29°$, $n = 2$, $a = \pi/6$

37. $e^{0.3}$, $n = 4$, $a = 0$ **38.** $\sinh(0.1)$, $n = 3$, $a = 0$

In Problems 39 and 40 use infinite series to approximate the given quantity to four decimal places.

39. $\displaystyle\int_0^1 \sin x^2 \, dx$

40. $\displaystyle\int_0^1 \dfrac{\sin x}{x} \, dx$

Miscellaneous Problems

41. Use $i^2 = -1$ and (11.36) to derive **Euler's formula**

$$e^{i\theta} = \cos\theta + i \sin\theta$$

42. Use Euler's formula given in Problem 41 to show that

(a) $\sin\theta = \dfrac{e^{i\theta} - e^{-i\theta}}{2i}$ (b) $\cos\theta = \dfrac{e^{i\theta} + e^{-i\theta}}{2}$

43. Use (11.29) to find the Maclaurin series for

$$f(x) = \begin{cases} e^{-1/x^2}, & x \neq 0 \\ 0, & x = 0 \end{cases}$$

[*Hint*: $f'(0) = \displaystyle\lim_{\Delta x \to 0} \dfrac{f(0 + \Delta x) - f(0)}{\Delta x}$. Let $t = \Delta x$.]

44. Use the Maclaurin series for $\cos x$ and a trigonometric identity to find the Maclaurin series for $\sin^2 x$.

45. In leveling a long roadway of length L, an allowance must be made for the curvature of the earth.

(a) Use (11.29) to show that the first three nonzero terms of the Maclaurin series for $f(x) = \sec x$ are

$$1 + \frac{1}{2}x^2 + \frac{5}{24}x^4 + \cdots$$

(b) For small values of x, use the approximation

$$\sec x = 1 + \frac{1}{2}x^2$$

and Figure 11.12 to show that the leveling correction is $y = L^2/2R$, where R is the radius of the earth.

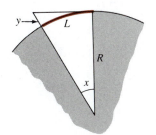

Figure 11.12

(c) Find the number of inches of leveling correction needed for a length of 1 mile of roadway. Use $R = 4000$ miles.

46. A wave of length L is traveling left to right across water of depth d (in ft), as illustrated in Figure 11.13. We can show that the speed v of the wave is related to L and d by the function $v = \sqrt{(gL/2\pi)}\tanh(2\pi d/L)$.

(a) Show that in deep water $v \approx \sqrt{gL/2\pi}$.

(b) Use (11.29) to find the first three nonzero terms of the Maclaurin series for $f(x) = \tanh x$. Show that when d/L is small, then $v \approx \sqrt{gd}$. In other words, in shallow water the speed of a wave is independent of wave length.

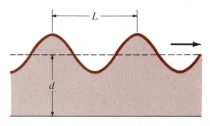

Figure 11.13

11.9 Binomial Series

Binomial Theorem   From basic mathematics we know that

$$(1 + x)^2 = 1 + 2x + x^2$$
$$(1 + x)^3 = 1 + 3x + 3x^2 + x^3$$

and, in general, if m is a nonnegative integer,

$$(1 + x)^m = 1 + mx + \frac{m(m - 1)}{2!}x^2 + \cdots + \frac{m(m - 1)\cdots(m - n + 1)}{n!}x^n + \cdots + x^m \quad (11.43)$$

The expansion of $(1 + x)^m$ in (11.43) is called the **Binomial Theorem**. The result in (11.43) inspires the following definition.

> **DEFINITION 11.8**   **Binomial Series**
>
> For any real number r, the series
>
> $$1 + rx + \frac{r(r - 1)}{2!}x^2 + \cdots + \frac{r(r - 1)\cdots(r - n + 1)}{n!}x^n + \cdots \quad (11.44)$$
>
> is called the **binomial series**.

Note that (11.44) only terminates when r is a nonnegative integer. In this case (11.44) reduces to (11.43).

The Ratio Test shows that the binomial series converges if $|x| < 1$ and diverges if $|x| > 1$. Thus, on the interval $(-1, 1)$ the binomial series defines an infinitely differentiable function f. It should come as no big surprise to learn that the function represented by (11.44) is $f(x) = (1 + x)^r$. Since the proof of this leads to a separable differentiable equation, you can practice your skills by following the guided steps given in Problem 18 of Exercises 11.9.

THEOREM 11.27   If $|x| < 1$, then for any real number r

$$(1 + x)^r = 1 + rx + \frac{r(r - 1)}{2!}x^2 + \cdots + \frac{r(r - 1) \cdots (r - n + 1)}{n!}x^n + \cdots \qquad (11.45)$$

Example 1

Find a power series representation for $\sqrt{1 + x}$.

Solution   With $r = \frac{1}{2}$ it follows from (11.45) that for $|x| < 1$

$$\sqrt{1 + x} = 1 + \frac{1}{2}x + \frac{\frac{1}{2}\left(\frac{1}{2} - 1\right)}{2!}x^2 + \frac{\frac{1}{2}\left(\frac{1}{2} - 1\right)\left(\frac{1}{2} - 2\right)}{3!}x^3 + \cdots$$

$$+ \frac{\frac{1}{2}\left(\frac{1}{2} - 1\right) \cdots \left(\frac{1}{2} - n + 1\right)}{n!}x^n + \cdots$$

$$= 1 + \frac{1}{2}x - \frac{1}{2^2 2!}x^2 + \frac{1 \cdot 3}{2^3 3!}x^3 + \cdots$$

$$+ (-1)^{n+1}\frac{1 \cdot 3 \cdot 5 \cdots (2n - 3)}{2^n n!}x^n + \cdots$$

In science a binomial series is often used to find approximations.

Example 2

In Einstein's theory of relativity, the mass of a particle moving at a velocity v relative to an observer is

$$m = \frac{m_0}{\sqrt{1 - v^2/c^2}} \qquad (11.46)$$

where m_0 is the rest mass and c is the speed of light.

Many of the results from classical physics do not hold for particles, such as electrons, which may move close to the speed of light. Kinetic energy is no longer $K = \frac{1}{2}m_0 v^2$, but is

$$K = mc^2 - m_0 c^2 \qquad (11.47)$$

If we identify $r = -\frac{1}{2}$ and $x = -v^2/c^2$ in (11.46), we have $|x| < 1$ since no particle can surpass the speed of light. Hence, (11.47) can be written:

$$
\begin{aligned}
K &= \frac{m_0 c^2}{\sqrt{1 + x}} - m_0 c^2 \\
&= m_0 c^2 [(1 + x)^{-1/2} - 1] \\
&= m_0 c^2 \left[\left(1 - \frac{1}{2}x + \frac{3}{8}x^2 - \frac{5}{16}x^3 + \cdots \right) - 1 \right] \\
&= m_0 c^2 \left[\frac{1}{2}\left(\frac{v^2}{c^2}\right) + \frac{3}{8}\left(\frac{v^4}{c^4}\right) + \frac{5}{16}\left(\frac{v^6}{c^6}\right) + \cdots \right]
\end{aligned}
\tag{11.48}
$$

In the case where v is very much smaller than c, terms beyond the first in (11.48) are negligible. This leads to the well-known result

$$
K \approx m_0 c^2 \left[\frac{1}{2}\left(\frac{v^2}{c^2}\right) \right] = \frac{1}{2} m_0 v^2
$$

Exercises 11.9

Answers to odd-numbered problems begin on page A-39.

In Problems 1–10 use (11.45) to find the first four terms of a power series representation of the given function. Give the radius of convergence.

1. $f(x) = \sqrt[3]{1 + x}$

2. $f(x) = \sqrt{1 - x}$

3. $f(x) = \sqrt{9 - x}$

4. $f(x) = \dfrac{1}{\sqrt{1 + 5x}}$

5. $f(x) = \dfrac{1}{\sqrt{1 + x^2}}$

6. $f(x) = \dfrac{x}{\sqrt[3]{1 - x^2}}$

7. $f(x) = (4 + x)^{3/2}$

8. $f(x) = \dfrac{x}{\sqrt{(1 + x)^5}}$

9. $f(x) = \dfrac{x}{(2 + x)^2}$

10. $f(x) = x^2(1 - x^2)^{-3}$

In Problems 11 and 12 explain why the error in the given approximation is less than the indicated amount. (*Hint*: Review Theorem 11.18.)

11. $(1 + x)^{1/3} \approx 1 + \dfrac{x}{3}; \quad \dfrac{1}{9}x^2, \quad x > 0$

12. $(1 + x^2)^{-1/2} \approx 1 - \dfrac{x^2}{2} + \dfrac{3}{8}x^4; \quad \dfrac{5}{16}x^6$

13. Find a power series representation for $\sin^{-1}x$ using

$$
\sin^{-1}x = \int_0^x \frac{dt}{\sqrt{1 - t^2}}
$$

14. (a) Show that the length of one-quarter of the ellipse $x^2/a^2 + y^2/b^2 = 1$ is given by

$$
a \int_0^{\pi/2} \sqrt{1 - k^2 \sin^2\theta} \, d\theta
$$

where $k^2 = (a^2 - b^2)/a^2 < 1$. This integral is called the **complete elliptic integral of the second kind**.

(b) Show that

$$
a \int_0^{\pi/2} \sqrt{1 - k^2 \sin^2\theta} \, d\theta = a\frac{\pi}{2} - \frac{a}{2}\frac{\pi}{4}k^2 - \frac{a}{8}\frac{3\pi}{16}k^4 - \cdots
$$

15. In Figure 11.14 a hanging cable is supported at points A and B and carries a uniformly distributed load (such as the floor of a bridge). If $y = (4d/l^2)x^2$ is the equation of the cable, show that its length is given by

$$
s = l + \frac{8d^2}{3l} - \frac{32d^4}{5l^3} + \cdots
$$

See Problem 20 in Exercises 3.11.

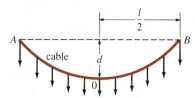

uniform load distributed horizontally
Figure 11.14

16. Approximate the following integrals to three decimal places.

(a) $\displaystyle\int_0^{0.2} \sqrt{1 + x^3}\, dx$ **(b)** $\displaystyle\int_0^{1/2} \sqrt[3]{1 + x^4}\, dx$

Miscellaneous Problems

17. By the law of cosines the potential at point A in Figure 11.15 due to a unit charge at point B is $1/R = (1 - 2xr + r^2)^{-1/2}$, where $x = \cos\theta$. The expression $(1 - 2xr + r^2)^{-1/2}$ is said to be the **generating function** for the so-called **Legendre polynomials** $P_k(x)$ since

$$(1 - 2xr + r^2)^{-1/2} = \sum_{k=0}^{\infty} P_k(x) r^k$$

Use (11.45) to find $P_0(x)$, $P_1(x)$, and $P_2(x)$.

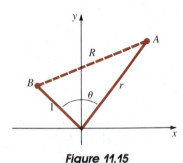

Figure 11.15

18. (a) Suppose

$$f(x) = 1 + rx + \frac{r(r - 1)}{2!}x^2 + \cdots$$
$$+ \frac{r(r - 1) \cdots (r - n + 1)}{n!}x^n + \cdots$$

for $|x| < 1$. Find $f'(x)$ and $x\,f'(x)$.

(b) Show that

$$(n + 1)\frac{r(r - 1) \cdots (r - n)}{(n + 1)!}$$
$$+ n\frac{r(r - 1) \cdots (r - n + 1)}{n!}$$
$$= r\frac{r(r - 1) \cdots (r - n + 1)}{n!}$$

(c) Show that $f'(x) + x\,f'(x) = r\,f(x)$.

(d) Solve the differential equation $(1 + x)f'(x) = r\,f(x)$ by separation of variables subject to $f(0) = 1$.

In Problems 19 and 20 use (11.45) to find a power series representation in $x - 1$ of the given function. (*Hint*: $1 + x = 2 + (x - 1)$.)

19. $f(x) = \sqrt{1 + x}$ **20.** $f(x) = (1 + x)^{-2}$

Chapter 11 Review Exercises

Answers to odd-numbered problems begin on page A-40.

In Problems 1–26 answer true or false.

1. Every bounded sequence converges. _____

2. If a sequence is not monotonic, it is not convergent. _____

3. The sequence $\left\{\dfrac{10^n}{2^{n^2}}\right\}$ is not monotonic. _____

4. If $a_n \leq B$ for all n and $a_{n+1}/a_n \geq 1$ for all n, then $\{a_n\}$ converges. _____

5. $\displaystyle\lim_{n\to\infty} \frac{|x|^n}{n!} = 0$ for every value of x. _____

6. If $\{a_n\}$ is a convergent sequence, then $\displaystyle\sum_{k=1}^{\infty} a_k$ always converges. _____

7. If $a_n \to 0$ as $n \to \infty$, then $\Sigma\, a_k$ converges. _____

8. If $\Sigma\, a_k^2$ converges, then $\Sigma\, a_k$ converges. _____

9. $\displaystyle\sum_{k=1}^{\infty} \frac{1}{k^p}$ converges for $p = 1.0001$. _____

10. The series $2 + \frac{2}{2} + \frac{2}{3} + \frac{2}{4} + \cdots$ diverges. _____

11. If $\Sigma\, |a_k|$ diverges, then $\Sigma\, a_k$ diverges. _____

12. If $\displaystyle\sum_{k=1}^{\infty} a_k$, $a_k > 0$, converges, then $\displaystyle\sum_{k=1}^{\infty} (-1)^{k+1} a_k$ converges. _____

13. If $\displaystyle\sum_{k=1}^{\infty} (-1)^{k+1} a_k$ converges absolutely, then $\displaystyle\sum_{k=1}^{\infty} (-1)^{k+1} \frac{a_k}{k}$ converges. _____

14. If $\Sigma\, b_k$ converges and $a_k \geq b_k$ for every positive integer k, then $\Sigma\, a_k$ converges. _____

15. If $\lim\limits_{n\to\infty}\left|\dfrac{a_{n+1}}{a_n}\right| = 1$, then $\Sigma\, a_k$ converges absolutely. _____

16. Every power series has a nonzero radius of convergence. _____

17. A power series converges absolutely at every value of x in its interval of convergence. _____

18. A power series $\Sigma_{k=0}^{\infty}\, c_k x^k$ represents an infinitely differentiable function on an interval $(-r,\ r)$ of convergence. _____

19. A power series $\Sigma_{k=0}^{\infty}\, c_k x^k$ converges for $-1 < x < 1$ and is convergent at $x = 1$. The series must also converge at $x = -1$. _____

20. If $\Sigma\, a_k x^k,\ a_k > 0$, has the interval of convergence $[-r,\ r)$, series converges conditionally, but not absolutely, at $-r$. _____

21. Since $\int_0^{\infty} e^{-x}\, dx = 1$, the series $\Sigma_{k=0}^{\infty}\, e^{-k}$ also converges to 1. _____

22. The sequence $\left\{\dfrac{(-1)^n n}{n+1}\right\}$ converges. _____

23. The series

$$1 - \frac{1}{2^2} + \frac{1}{3^2} + \frac{1}{4^2} - \frac{1}{5^2} - \frac{1}{6^2} + + + - - - \cdots$$

converges. _____

24. $f(x) = \ln x$ cannot be represented by a Maclaurin series. _____

25. $f(x) = \displaystyle\sum_{k=0}^{\infty} \frac{f^{(k)}(a)}{k!}(x-a)^k$ on an interval if $\lim\limits_{n\to\infty} R_n(x) = 0$. _____

26. The interval of convergence of the power series $x - \dfrac{x^2}{2} + \dfrac{x^3}{3} - \dfrac{x^4}{4} + \cdots$ is $(-1,\ 1)$. _____

In Problems 27–34 fill in the blanks.

27. If $\{a_n\}$ converges to 4 and $\{b_n\}$ converges to 5, then $\{a_n b_n\}$ converges to _____, $\{a_n + b_n\}$ converges to _____, $\{a_n/b_n\}$ converges to _____, and $\{a_n^2\}$ converges to _____.

28. To approximate the sum of the series $\displaystyle\sum_{k=1}^{\infty} \frac{(-1)^{k+1}}{10^k}$ to four decimal places, we need only use the _____th partial sum.

29. The sum of the series $\displaystyle\sum_{k=0}^{\infty} 4\left(\frac{2}{3}\right)^k$ is _____.

30. If $\Sigma\, a_k$ converges and $\Sigma\, b_k$ diverges, then $\Sigma\, (a_k + b_k)$ _____.

31. The series $\displaystyle\sum_{k=1}^{\infty} [\tan^{-1}k - \tan^{-1}(k+1)]$ converges to _____.

32. The power series $\displaystyle\sum_{k=0}^{\infty} \frac{x^k}{k!}$ represents the function $f(x) =$ _____ for all x.

33. The binomial series representation of $f(x) = (4+x)^{1/2}$ has the radius of convergence _____.

34. The geometric series $\displaystyle\sum_{k=1}^{\infty} \left(\frac{5}{x}\right)^{k-1}$ converges for the following values of x: _____.

In Problems 35–44 determine whether the given series converges or diverges.

35. $\displaystyle\sum_{k=1}^{\infty} \frac{k}{(k^2+1)^2}$

36. $\displaystyle\sum_{k=1}^{\infty} \frac{1}{1+e^{-k}}$

37. $\displaystyle\sum_{k=1}^{\infty} \frac{k!}{e^{k^2}}$

38. $\displaystyle\sum_{k=1}^{\infty} \frac{1}{3k^2+4k+6}$

39. $\displaystyle\sum_{k=1}^{\infty} \frac{\sqrt{k}\,\ln k}{k^4+4}$

40. $\displaystyle\sum_{k=1}^{\infty} \frac{\sin k}{k^{3/2}}$

41. $\displaystyle\sum_{k=2}^{\infty} \frac{k}{\sqrt[3]{k^6-4k}}$

42. $\displaystyle\sum_{k=2}^{\infty} \frac{1}{k\sqrt{\ln k}}$

43. $\displaystyle\sum_{k=1}^{\infty} \frac{1+(-1)^k}{\sqrt{k}}$

44. $\displaystyle\sum_{k=1}^{\infty} \frac{(k^2)!}{(k!)^2}$

In Problems 45 and 46 find the sum of the given convergent series.

45. $\displaystyle\sum_{k=1}^{\infty} \frac{(-1)^{k-1}+3}{(1.01)^{k-1}}$

46. $\displaystyle\sum_{k=1}^{\infty} \frac{1}{k^2+11k+30}$

In Problems 47–50 find the interval of convergence of the given power series.

47. $\displaystyle\sum_{k=1}^{\infty} \frac{3^k}{k^3}x^k$

48. $\displaystyle\sum_{k=1}^{\infty} \frac{k}{4^k}(2x-1)^k$

49. $\displaystyle\sum_{k=1}^{\infty} k!(x+5)^k$

50. $\displaystyle\sum_{k=2}^{\infty} \frac{(2x)^k}{\ln k}$

51. Find the radius of convergence for

$$\sum_{k=1}^{\infty} \frac{2\cdot 5\cdot 8\cdots(3k-1)}{3\cdot 7\cdot 11\cdots(4k-1)}x^k$$

52. Find all values of x for which $\displaystyle\sum_{k=1}^{\infty} (\cos x)^k$ converges.

In Problems 53–56 find, by any method, the first three nonzero terms of the Maclaurin series for the given function.

53. $f(x) = \dfrac{1}{\sqrt[3]{1 + x^5}}$

54. $f(x) = \dfrac{x}{2 - x}$

55. $f(x) = \sin x \cos x$

56. $f(x) = \displaystyle\int_0^x e^{t^2}\, dt$

57. Find the Taylor series for $f(x) = \cos x$ at $a = \pi/2$.

58. Prove that the series in Problem 57 represents the function by showing $R_n(x) \to 0$ as $n \to \infty$.

12

Analytic Geometry in the Plane

To the ancient Greek geometers such as Euclid (c. 300 B.C.) and Archimedes (c. 287–212 B.C.), a *conic section* was a curve of intersection of a plane and a double-napped right circular cone. However, in this chapter we shall see how the *parabola*, *ellipse*, and *hyperbola* are defined by means of distance. Using a Cartesian coordinate system and the distance formula, we obtain equations for the conics. In Figure 12.1 we have illustrated the four conic sections along with the so-called *degenerate conics*.

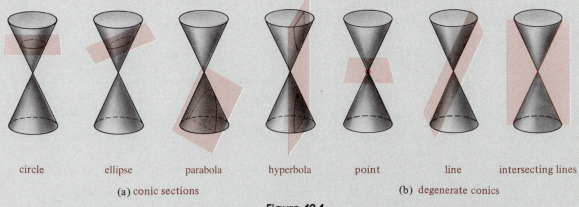

circle   ellipse   parabola   hyperbola   point   line   intersecting lines

(a) conic sections   (b) degenerate conics

Figure 12.1

12.1 The Parabola

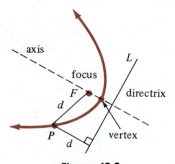

Figure 12.2

The graph of a quadratic function $f(x) = ax^2 + bx + c$, $a \neq 0$, is a parabola. However, not every parabola is the graph of a function of x. In general, a parabola is defined in the following manner.

> **DEFINITION 12.1**   A **parabola** is the set of all points P in the plane that are equidistant from a fixed line L, called the **directrix**, and a fixed point F, called the **focus**.

As shown in Figure 12.2, a parabola has an **axis** of symmetry that passes through the focus F and is perpendicular to the directrix. The point on the axis midway between F and L is called the **vertex** of the parabola.

Equation of a Parabola   To describe a parabola analytically, let us assume for the sake of discussion that the directrix is the horizontal line $y = -p$ and that the focus is $F(0, p)$. Using Definition 12.1 and Figure 12.3(a), we see that $d(F, P) = d(P, Q)$ is the same as

$$\sqrt{x^2 + (y - p)^2} = y + p$$

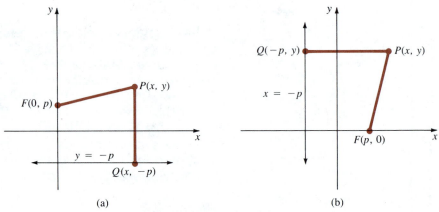

Figure 12.3

Squaring both sides and simplifying leads to

$$x^2 = 4py \tag{12.1}$$

We say that (12.1) is the **standard form** for the equation of a parabola with focus $F(0, p)$ and directrix $y = -p$. In like manner, if the directrix and focus are $x = -p$ and $F(p, 0)$, respectively, we find that the standard form for the equation of the parabola is

$$y^2 = 4px \tag{12.2}$$

See Figure 12.3(b).

Although we assume $p > 0$ in Figure 12.3, this need not, of course, be the case. The following table and Figure 12.4 summarize information about equations (12.1) and (12.2).

Equation	Vertex	Axis	Focus	Directrix	Graph Opens
$x^2 = 4py$	$(0, 0)$	$x = 0$	$(0, p)$	$y = -p$	up if $p > 0$; down if $p < 0$
$y^2 = 4px$	$(0, 0)$	$y = 0$	$(p, 0)$	$x = -p$	right if $p > 0$; left if $p < 0$

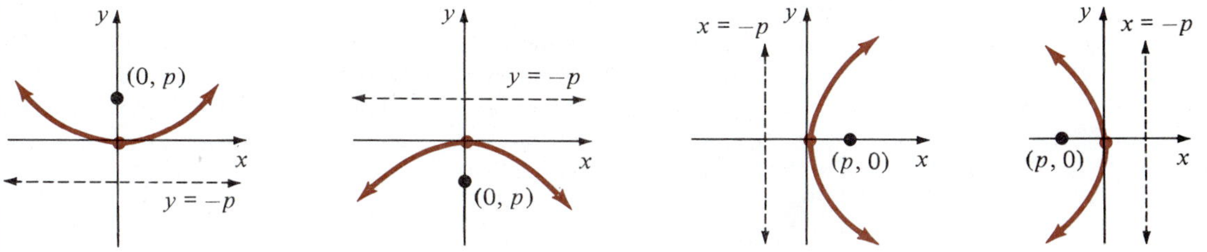

$x^2 = 4py, p > 0$ $x^2 = 4py, p < 0$ $y^2 = 4px, p > 0$ $y^2 = 4px, p < 0$

(a)   (b)   (c)   (d)

Figure 12.4

Example 1

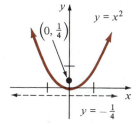

Figure 12.5

Find the focus and directrix of the parabola whose equation is $y = x^2$.

Solution   Comparison of the equation $y = x^2$ with (12.1) enables us to identify $4p = 1$ and so $p = \frac{1}{4}$. Hence, the focus of the parabola is $(0, \frac{1}{4})$ and its directrix is the horizontal line $y = -\frac{1}{4}$. The familiar graph, along with the focus and directrix, are given in Figure 12.5.

Example 2

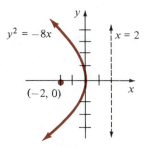

Figure 12.6

Find the focus and directrix of the parabola whose equation is $x = -\frac{1}{8}y^2$. Graph.

Solution   By writing the given equation as $y^2 = -8x$, we see from (12.2) that $4p = -8$ and $p = -2$. Hence, the focus and directrix are $(-2, 0)$ and $x = 2$, respectively. Since $p < 0$, the parabola opens to the left and must have the same basic shape as the graph given in Figure 12.4(d). See Figure 12.6.

Vertex at (h, k)   In general, the standard form for the equation of a parabola with vertex (h, k) is given by either

$$(x - h)^2 = 4p(y - k) \qquad\qquad (12.3)$$

or

$$(y - k)^2 = 4p(x - h) \qquad\qquad (12.4)$$

Figure 12.7 illustrates two possible graphs of (12.3) and (12.4). As in (12.1) and (12.2), to locate the focus and directrix we need only measure $|p|$ units along the axis of the parabola. We measure either above and below the vertex or to each side of the vertex. The next table summarizes the information about (12.3) and (12.4).

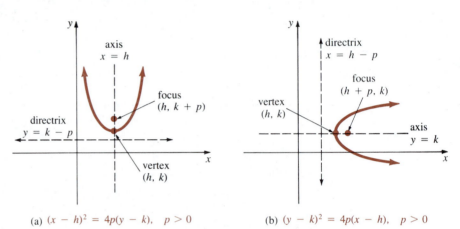

(a) $(x - h)^2 = 4p(y - k), \quad p > 0$ (b) $(y - k)^2 = 4p(x - h), \quad p > 0$

Figure 12.7

Equation	Vertex	Axis	Focus	Directrix	Graph Opens
$(x - h)^2 = 4p(y - k)$	(h, k)	$x = h$	$(h, k + p)$	$y = k - p$	up if $p > 0$; down if $p < 0$
$(y - k)^2 = 4p(x - h)$	(h, k)	$y = k$	$(h + p, k)$	$x = h - p$	right if $p > 0$; left if $p < 0$

Example 3

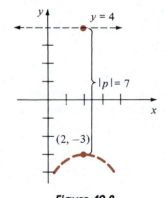

Figure 12.8

Find an equation of the parabola with vertex $(2, -3)$ and directrix $y = 4$. Determine the focus.

Solution   With the vertex at $(2, -3)$ it follows that $h = 2$ and $k = -3$. Since the directrix of the parabola is $y =$ constant, the equation we seek must be of form (12.3). In addition, in Figure 12.8 we see that $|p| = 7$. Furthermore, $p = -7$ because the graph must open down. (We can also find the value of p by simply solving $k - p = 4$.) Thus, the standard form for the equation of the parabola is

$$(x - 2)^2 = 4(-7)(y - (-3)) \quad \text{or} \quad (x - 2)^2 = -28(y + 3)$$

The focus is $(2, -3 + (-7))$ or $(2, -10)$.

Example 4

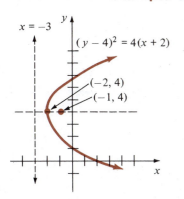

$x = -3$

$(y - 4)^2 = 4(x + 2)$

$(-2, 4)$

$(-1, 4)$

Figure 12.9

Find the vertex, axis, focus, directrix, and graph of the parabola whose equation is $y^2 - 8y = 4x - 8$.

Solution   Completing the square in y yields

$$y^2 - 8y + 16 = 4x + 8 \quad \text{or} \quad (y - 4)^2 = 4(x + 2)$$

Identifying the last equation with (12.4) shows that the vertex is $(-2, 4)$ and that the axis is $y = 4$. Now $4p = 4$ implies that $p = 1 > 0$ and so the parabola opens to the right and its focus and directrix are $(-1, 4)$ and $x = -3$, respectively. You should be able to verify that the graph given in Figure 12.9 has y-intercepts $4 - 2\sqrt{2}$ and $4 + 2\sqrt{2}$, and x-intercept 2.

Remark

The designs of objects such as searchlights, automobile headlights, reflecting telescopes, and microwave antennas are based on a pleasant property of the parabola. As shown in Figure 12.10(a), if parallel beams of light, say from a distant star, enter a reflecting telescope with a parabolic mirror, then it can be proved that all beams will be reflected to the focus. See Problem 42 in Exercises 12.1. On the other hand, if a light source is placed at the focus of a parabolic reflecting surface, then a beam of light will be reflected from the surface parallel to the axis of the parabola. Other examples of the parabolic form are the cables of a suspension bridge and the trajectory of an obliquely launched projectile. See Figure 12.10(b) and 12.10(c). Under some circumstances, the pursuit curve of a shark seeking its prey is also parabolic.

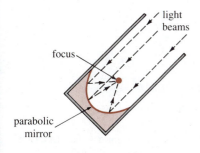

focus

parabolic mirror

(a) reflecting telescope

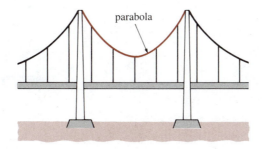

parabola

(b) suspension bridge

parabola

(c) trajectory

Figure 12.10

_____ **Exercises 12.1** _____

Answers to odd-numbered problems begin on page A-40.

In Problems 1–20 find the vertex, focus, directrix, and graph of the parabola whose equation is given.

1. $-2x^2 = y$

2. $x^2 = 8y$

3. $y^2 = 12x$

4. $\frac{1}{4}x^2 = -y$

5. $y^2 = -10x$

6. $y^2 = 5x$

7. $(x - 1)^2 = 4y$

8. $(y + 2)^2 = x$

9. $(y - 3)^2 + 8 = 2x$

10. $-(x + 2)^2 = 4y - 16$

11. $(x - 4)^2 = 4(y + 3)$

12. $(2y + 3)^2 = -16(2x + 1)$

13. $y = x^2 + 4x + 6$

14. $3y = -x^2 + 3x + 7$

15. $y^2 - 6y = 4x + 3$

16. $x = y^2 + 10y + 27$

17. $6x^2 + 24x - 8y + 19 = 0$

18. $-x^2 + 6x - 4y - 9 = 0$

19. $2(x + 3) = -y(y + 5)$

20. $2y = x(2 - x)$

In Problems 21–30 find an equation of the parabola that satisfies the given conditions.

21. Vertex $(0, 0)$, axis $x = 0$, through $(-1, 4)$

22. Vertex $(1, 2)$, axis $y = 2$, through $(0, 0)$

23. Vertex $(1, 1)$, directrix $x = 4$

24. Vertex $(3, -2)$, directrix $y = 2$

25. Focus $(-2, 4)$, vertex $(1, 4)$

26. Focus $(0, 0)$, vertex $(0, \frac{3}{2})$

27. Focus $(0, 4)$, directrix $x = -5$

28. Focus $(2, -\frac{5}{2})$, directrix $2y - 1 = 0$

29. Axis $x = 0$, through $(1, 3)$ and $(2, 6)$

30. Axis $y = 0$, through $(2, 1)$ and $(11, -2)$

In Problems 31 and 32 (Figures 12.11 and 12.12) find an equation of the parabola in part (a) of the figure. Use part (a) to find an equation of the shifted parabola in part (b).

31.

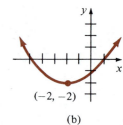
(a)

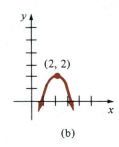
(b)

Figure 12.11

32.

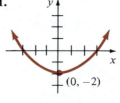
(a)

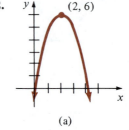
(b)

Figure 12.12

33. Find an equation of the tangent line to the parabola whose equation is $x = y^2 + 2y + 1$ at $y = 3$.

34. In Problem 33, find the area of the region that is bounded by the parabola and the graph of $x = 4$.

35. Show that the vertex is the point on a parabola closest to the focus. (*Hint:* Consider $x^2 = 4py$.)

36. Oil is gushing from an end of a horizontal pipe 30 m above ground level. Assume the oil stream is an arc of a parabola with vertex at the end of the pipe. Ten meters down from the end of the pipe the horizontal distance to the oil stream is 15 m. See Figure 12.13. Find, relative to the end of the pipe, where the oil hits the ground.

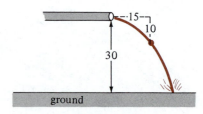

Figure 12.13

37. The distance between two vertical supports of a suspension bridge is 100 m and the sag of the cable is 15 m.

(a) Assume that the shape of the cable is a parabola. Find its equation if the vertex is in the center at the lowest point on the cable.

(b) What is the height of the cable 30 m from the center?

38. The reflecting telescope at Mount Palomar uses a circular mirror 200 inches in diameter. A cross section of the mirror through a diameter is a parabola whose focal length is 55.5 ft. (Focal length is the distance from the focus to the vertex.)

(a) Find an equation of the parabolic cross section.

(b) What is the maximum depth of the mirror?

Miscellaneous Problems

39. The **focal chord*** is the line segment through the focus, perpendicular to the axis, with endpoints on the parabola. See Figure 12.14. Show that the length of the focal chord for both $x^2 = 4py$ and $y^2 = 4px$ is $4|p|$.

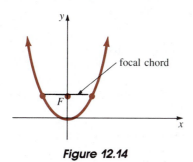

Figure 12.14

40. Use Problem 39 to find the length of the focal chord for $y = \frac{1}{3}x^2$.

41. Show that the tangent lines at the endpoints of the focal chord for $x^2 = 4py$ are perpendicular to each other.

42. As shown in Figure 12.15, let L be the tangent line to the graph of $y^2 = 4px$ at $P(x_0, y_0)$.

(a) Find an equation of the tangent line at P.

(b) Show that the x-intercept of L is $(-x_0, 0)$.

(c) Show that the line segments QF and PF have the same length.

(d) Show that $\alpha = \beta = \gamma$.

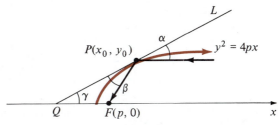

Figure 12.15

43. Find the vertex and focus of a parabola whose equation is $y = ax^2 + bx + c$, $a \neq 0$.

44. Using $y = ax^2 + bx + c$, find an equation of the parabola passing through the points $(0, 5)$, $(1, 4)$, and $(-2, 13)$.

45. The distance d from a point (x_0, y_0) to a line $ax + by + c = 0$ is given by $d = |ax_0 + by_0 + c|/\sqrt{a^2 + b^2}$. Use this fact and Definition 12.1 to find an equation of the parabola that has $(1, 2)$ as its focus and the line $x - y - 3 = 0$ as its directrix.

12.2 The Ellipse

The ellipse is defined as follows.

> **DEFINITION 12.2**   An **ellipse** is the set of points P in the plane such that the sum of the distances between P and two fixed points F_1 and F_2, called **foci**, is a constant.

As shown in Figure 12.16(a), if P is a point on the ellipse and $d_1 = d(F_1, P)$, $d_2 = d(F_2, P)$, and $k > 0$ is a constant, then Definition 12.2 asserts that

$$d_1 + d_2 = k \tag{12.5}$$

*The focal chord is also known as the *latus rectum*.

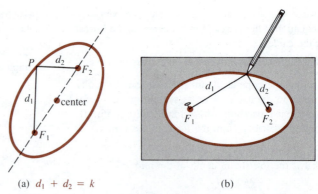

(a) $d_1 + d_2 = k$ (b)

Figure 12.16

The midpoint of the line segment F_1F_2 is called the **center** of the ellipse. On a practical level, (12.5) can be used to sketch an ellipse. Figure 12.16(b) shows that if a string of length k is attached to a paper by two tacks, then an ellipse can be traced by inserting a pencil against the string and moving it in such a manner that the string remains taut.

Equation of an Ellipse

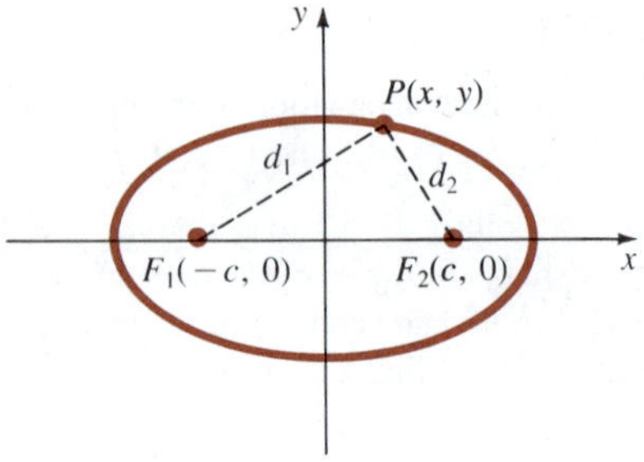

For convenience, let us choose $k = 2a$ and put the foci on the x-axis with coordinates $F_1(-c, 0)$ and $F_2(c, 0)$. See Figure 12.17. It follows from (12.5) that

$$\sqrt{(x + c)^2 + y^2} + \sqrt{(x - c)^2 + y^2} = 2a$$

or $$\sqrt{(x + c)^2 + y^2} = 2a - \sqrt{(x - c)^2 + y^2} \qquad (12.6)$$

We square both sides of (12.6) and simplify:

$$(x + c)^2 + y^2 = 4a^2 - 4a\sqrt{(x - c)^2 + y^2} + (x - c)^2 + y^2$$
$$a\sqrt{(x - c)^2 + y^2} = a^2 - cx$$

Squaring again gives

$$a^2[(x - c)^2 + y^2] = a^4 - 2a^2cx + c^2x^2$$
$$(a^2 - c^2)x^2 + a^2y^2 = a^2(a^2 - c^2)$$

or $$\frac{x^2}{a^2} + \frac{y^2}{a^2 - c^2} = 1 \qquad (12.7)$$

Figure 12.17

Referring to Figure 12.17, we see that the points F_1, F_2, and P form a triangle. Since the so-called triangle inequality states that the sum of the lengths of any two sides of a triangle is greater than the length of the remaining side, we must have $2a > 2c$ or $a > c$. Hence, $a^2 - c^2 > 0$. By writing

$$b^2 = a^2 - c^2 \qquad (12.8)$$

the **standard form** for the equation of an ellipse with foci $F_1(-c, 0)$ and $F_2(c, 0)$ becomes

$$\frac{x^2}{a^2} + \frac{y^2}{b^2} = 1 \qquad (12.9)$$

The **major axis** of an ellipse is the line segment through its center, containing the foci, and with endpoints on the ellipse. The line segment through the center, perpendicular to the major axis, and with endpoints on the ellipse is called the **minor axis**. Now, setting $y = 0$ in equation (12.9) reveals that $-a$ and a are the x-intercepts, whereas $x = 0$ leads to the y-intercepts $-b$ and b. The corresponding points $(\pm a, 0)$ and $(0, \pm b)$ are called **vertices**. Since $a > b$, the major axis of an ellipse is necessarily longer than its minor axis.

If the foci are on the y-axis at $F_1(0, -c)$ and $F_2(0, c)$, then a repetition of this analysis will show that an equation of the ellipse is

$$\frac{x^2}{b^2} + \frac{y^2}{a^2} = 1 \qquad (12.10)$$

where, as before, $b^2 = a^2 - c^2$.

Example 1

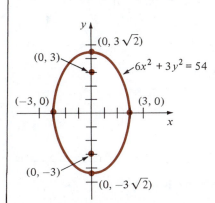

(0, 3√2)

(0, 3)

$6x^2 + 3y^2 = 54$

(−3, 0)   (3, 0)

(0, −3)

(0, −3√2)

Figure 12.18

Find the vertices and foci of the ellipse whose equation is $6x^2 + 3y^2 = 54$. Graph.

Solution By dividing both sides by 54,

$$\frac{x^2}{9} + \frac{y^2}{18} = 1$$

we see that $18 > 9$ and so we identify the equation with (12.10). From $a^2 = 18$, $b^2 = 9$ we conclude that the vertices are $(0, \pm 3\sqrt{2})$ and $(\pm 3, 0)$. The major axis is vertical with endpoints at $(0, -3\sqrt{2})$ and $(0, 3\sqrt{2})$. The minor axis is horizontal with endpoints $(-3, 0)$ and $(3, 0)$. Now, $c^2 = a^2 - b^2 = 9$ implies that $c = 3$. Hence, the foci are on the y-axis at $(0, -3)$ and $(0, 3)$. The graph is given in Figure 12.18.

Example 2

Find an equation of an ellipse that has $(2, 0)$ as a focus and x-intercept 5.

Solution Since the given focus is on the x-axis, we can find an equation of form (12.9). Consequently, $c = 2$, $a = 5$, and from (12.8), $b^2 = 25 - 4 = 21$. The desired equation is then

$$\frac{x^2}{25} + \frac{y^2}{21} = 1$$

Center at (h, k) When the center is at (h, k), the standard form for the equation of an ellipse is either

$$\frac{(x - h)^2}{a^2} + \frac{(y - k)^2}{b^2} = 1 \qquad (12.11)$$

or

$$\frac{(x - h)^2}{b^2} + \frac{(y - k)^2}{a^2} = 1 \qquad (12.12)$$

In each case we still have $a > c$ and $b^2 = a^2 - c^2$. But, for (12.11) the foci are located on the horizontal line $y = k$ at $(h \pm c, k)$. For (12.12) the foci are on the vertical line $x = h$ at $(h, k \pm c)$. To find the vertices of an ellipse with center (h, k), we set, in turn, $x = h$ and $y = k$ in its equation and solve for y and x. The coordinates of the vertices are given in Figure 12.19.

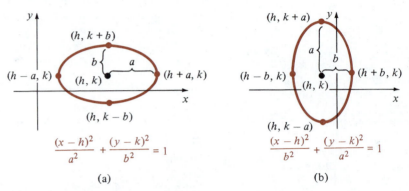

Figure 12.19

Example 3

Find the vertices and foci of the ellipse whose equation is

$$9x^2 - 36x + 25y^2 + 150y + 36 = 0$$

Graph.

Solution   Completing the square in *both* x and y and simplifying leads to

$$\frac{(x - 2)^2}{25} + \frac{(y + 3)^2}{9} = 1$$

The center of the ellipse is $(2, -3)$ and, because $25 > 9$, we know from (12.11) that its major and minor axes lie along the lines $y = -3$ and $x = 2$, respectively. Now, solving

$$\frac{(x - 2)^2}{25} = 1 \quad \text{and} \quad \frac{(y + 3)^2}{9} = 1$$

gives $x = 2 \pm 5$ and $y = -3 \pm 3$. Hence, the vertices are $(7, -3)$, $(-3, -3)$, $(2, 0)$, and $(2, -6)$. Alternatively, the vertices can be found by moving $a = 5$ units to the right and left of the center along $y = -3$ and $b = 3$ units above and below the center along $x = 2$. Finally, $c^2 = a^2 - b^2 = 25 - 9 = 16$ shows that $c = 4$ and that the foci are $(-2, -3)$ and $(6, -3)$. The graph of the equation is given in Figure 12.20.

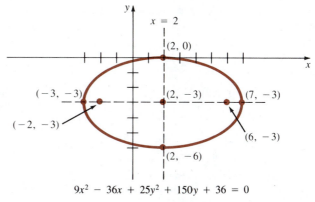

$$9x^2 - 36x + 25y^2 + 150y + 36 = 0$$

Figure 12.20

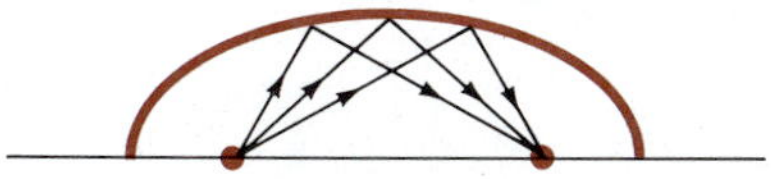

Remark

The ellipse has a reflecting property similar to that of the parabola. If a light or sound source is placed at one focus of an ellipse, say F_1 in Figure 12.21, then all rays or waves will be reflected to the other focus F_2. For example, if a pool table is constructed in the form of an ellipse with a pocket at one focus, then any shot originating at the other focus will never miss the pocket. Similarly, if a ceiling is elliptical with two foci on the floor, but considerably distant from each other, then anyone whispering at one focus will be heard at the other. Two famous "whispering galleries" are located in Statuary Hall at the Capitol in Washington, D.C., and in the Mormon Tabernacle in Salt Lake City, Utah. See Problem 50 in Exercises 12.2.

Figure 12.21

Exercises 12.2

Answers to odd-numbered problems begin on page A-40.

In Problems 1–20 find the vertices, foci, and graph of the ellipse whose equation is given.

1. $\dfrac{x^2}{16} + \dfrac{y^2}{25} = 1$

2. $\dfrac{x^2}{3^2} + \dfrac{y^2}{2^2} = 1$

3. $x^2 + \dfrac{y^2}{10} = 1$

4. $\dfrac{x^2}{6} + y^2 = 1$

5. $\dfrac{x^2}{8} + \dfrac{y^2}{4} = 2$

6. $9x^2 + y^2 = 9$

7. $4x^2 + 7y^2 = 28$

8. $2x^2 + 4y^2 = 1$

9. $\dfrac{(x-1)^2}{25} + \dfrac{(y-3)^2}{36} = 1$

10. $\dfrac{(x+1)^2}{9} + \dfrac{(y-4)^2}{4} = 1$

11. $\dfrac{(x-2)^2}{64} + \dfrac{(y-2)^2}{36} = 1$

12. $\dfrac{(2x+1)^2}{4} + \dfrac{(3y-2)^2}{16} = 1$

13. $2\left(x + \dfrac{1}{2}\right)^2 + (y-1)^2 = 4$

14. $x^2 + 9(y-2)^2 = 81$

15. $3x^2 + y^2 - 6y = 0$

16. $y^2 = 2x(1 - x)$

17. $9x^2 + 25y^2 + 18x + 50y = 191$

18. $x^2 + 2y^2 + x + y = 0$

19. $5x^2 + y^2 - 40x - 4y + 83 = 0$

20. $4x^2 + 12y^2 - 4x - 24y + 1 = 0$

In Problems 21–32 find an equation of the ellipse that satisfies the given conditions.

21. Center $(0, 0)$, vertices $(5, 0)$ and $(0, -2)$

22. Vertices $(\pm 4, 0)$ and $(0, \pm 7)$

23. Vertices $(0, \pm 4)$, foci $(0, \pm 2)$

24. Vertices $(-1, 1)$ and $(5, 1)$, foci $(4, 1)$ and $(0, 1)$

25. Vertices $(-5, -2)$ and $(3, -2)$, foci $(-1, -5)$ and $(-1, 1)$

26. Vertices $(-3, 3)$, $(9, 3)$, $(3, -7)$, and $(3, 13)$

27. Foci $(\pm\sqrt{7}, 0)$, length of minor axis $\sqrt{3}$

28. Foci $(1 \pm \sqrt{3}, 1)$, length of major axis 6

29. Vertices $(\pm 2\sqrt{2}, 0)$, through $(2, -1)$

30. Vertices $(\pm 6, 2)$, through $(4, 4)$

31. Center $(4, -2)$, focus $(4, 4)$, through $(4, 8)$

32. Set of all points in the plane the sum of whose distances from $(0, 1)$ and $(0, -1)$ is 4

33. Find the slope(s) of the tangent line(s) to the ellipse whose equation is $9x^2 + 5y^2 + 18x - 10y - 27 = 0$ at $x = 1$.

34. Find an equation of the normal line to the ellipse whose equation is $x^2 + 4(y - 1)^2 = 16$ at $(2\sqrt{3}, 2)$.

35. Show that the point on the ellipse whose equation is $x^2/a^2 + y^2/b^2 = 1$, which is closest to the focus $(c, 0)$, is the vertex $(a, 0)$.

36. The planet Mercury has an elliptical orbit with the sun located at one focus. The length of the major axis is 7.2×10^7 miles and the length of the minor axis is 7.04×10^7 miles. What is Mercury's perihelion distance (the least distance between the planet and the sun)? What is Mercury's aphelion distance (the greatest distance between the planet and the sun)?

37. The ellipse whose equation is $x^2/4 + (y - 1)^2/9 = 1$ is shifted 4 units to the right. What are the center, vertices, and foci of the shifted ellipse?

38. Without solving, determine how many solutions there are to the system of equations

$$25x^2 + y^2 = 25$$
$$9(x - 1)^2 + 4y^2 = 36$$

39. The arch of a masonry bridge is a semi-ellipse whose span is 90 ft and whose height in the center is 30 ft. What is the height of the arch 15 ft from the center?

40. A satellite is put into an elliptical orbit around the earth. The radius of the earth is 6000 km (approximately) and its center is located at one focus of the orbit.

(a) Use the information given in Figure 12.22 to find an equation of the orbit.

(b) What is the height of the satellite above the surface of the earth at point P?

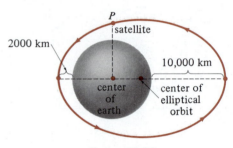

Figure 12.22

Miscellaneous Problems

41. The line segment perpendicular to the major axis through a focus with endpoints on the ellipse is called a **focal chord**. See Figure 12.23. Show that the length of a focal chord of an ellipse with center at the origin is $2b^2/a$.

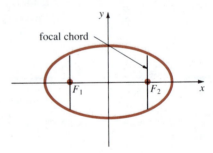

Figure 12.23

42. Rework part (b) of Problem 40 using the information in Problem 41.

The **eccentricity** of an ellipse is defined to be c/a. In Problems 43 and 44 find the eccentricity of the ellipse whose equation is given.

43. $\dfrac{(x - 3)^2}{25} + \dfrac{(y + 4)^2}{16} = 1$

44. $x^2 + 4y^2 - 2x - 24y + 21 = 0$

45. Find an equation of form (12.9) of an ellipse whose major axis has length 8 and whose eccentricity is $\frac{1}{3}$.

46. Find an equation of form (12.12) of an ellipse whose foci are $(3, -1)$, $(3, 11)$, and whose eccentricity is $\frac{3}{4}$.

47. The orbit of Halley's comet is an ellipse whose major axis is 3.34×10^9 miles long and whose minor axis is 8.5×10^8 miles long. What is the eccentricity of the comet's orbit?

48. What is the eccentricity of the satellite's orbit in Problem 40?

49. Show that the circumference of an ellipse whose equation is $x^2/a^2 + y^2/b^2 = 1$ is

$$4a \int_0^{\pi/2} \sqrt{1 - e^2 \sin^2\theta} \, d\theta$$

where $e = c/a$.

50. As shown in Figure 12.24, let L be the tangent line to the graph of $x^2/a^2 + y^2/b^2 = 1$ at $P(x_0, y_0)$. Show that $\alpha = \beta$.

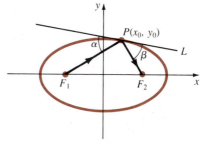

Figure 12.24

Calculator Problem

51. Use the result of Problem 49 and Simpson's Rule* with $n = 4$ to find an approximation to the circumference of the ellipse whose equation is $x^2/25 + y^2/9 = 1$.

12.3 The Hyperbola

The definition of a hyperbola is the same as that of the ellipse with one exception: The word *sum* is replaced by *difference*.

> **DEFINITION 12.3**   A **hyperbola** is the set of points P in the plane such that the difference of the distances between P and two fixed points F_1 and F_2, called **foci**, is a constant.

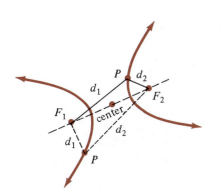

Figure 12.25

As shown in Figure 12.25, a hyperbola consists of two **branches**. The midpoint of the line segment F_1F_2 is the **center** of the hyperbola. If P is a point on the hyperbola, then

$$|d_1 - d_2| = k \tag{12.13}$$

where $d_1 = d(F_1, P)$ and $d_2 = d(F_2, P)$.

Equation of a Hyperbola   Proceeding as we did for the ellipse, the foci are placed on the x-axis at $F_1(-c, 0)$ and $F_2(c, 0)$ and the constant k is chosen to be $2a$ for algebraic convenience. Writing (12.13) as

$$d_1 - d_2 = \pm 2a \tag{12.14}$$

or $\sqrt{(x + c)^2 + y^2} - \sqrt{(x - c)^2 + y^2} = \pm 2a$

$$\sqrt{(x + c)^2 + y^2} = \pm 2a + \sqrt{(x - c)^2 + y^2}$$

---

*The antiderivative $\int \sqrt{1 - e^2 \sin^2\theta} \, d\theta$ cannot be expressed as an elementary function.

we square, simplify, and square again:

$$(x + c)^2 + y^2 = 4a^2 \pm 4a\sqrt{(x - c)^2 + y^2} + (x - c)^2 + y^2$$

$$\pm a\sqrt{(x - c)^2 + y^2} = cx - a^2$$

$$a^2[(x - c)^2 + y^2] = c^2x^2 - 2a^2cx + a^4$$

$$x^2(c^2 - a^2) - a^2y^2 = a^2(c^2 - a^2)$$

$$\frac{x^2}{a^2} - \frac{y^2}{c^2 - a^2} = 1 \qquad (12.15)$$

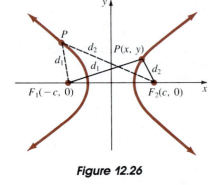

Figure 12.26

From Figure 12.26, we see that the triangle inequality gives

$$d_1 < d_2 + 2c \quad \text{and} \quad d_2 < d_1 + 2c$$

or equivalently,

$$d_1 - d_2 < 2c \quad \text{and} \quad d_2 - d_1 < 2c$$

Using (12.14) the foregoing inequalities imply $2a < 2c$ or $a < c$. Thus, if b^2 denotes the positive constant

$$b^2 = c^2 - a^2 \qquad (12.16)$$

in (12.15), we obtain the **standard form** for the equation of a hyperbola with foci $F_1(-c, 0)$ and $F_2(c, 0)$:

$$\frac{x^2}{a^2} - \frac{y^2}{b^2} = 1 \qquad (12.17)$$

Notice that the graph of (12.17) has x-intercepts $\pm a$, but has no y-intercepts since $-y^2/b^2 = 1$ has no real solution. The points $(-a, 0)$ and $(a, 0)$ are **vertices** of the hyperbola. The line segment through the center with the vertices as endpoints is called the **transverse axis**. Also, it is common practice to refer to the line segment with $(0, -b)$ and $(0, b)$ as endpoints and that passes through the center perpendicular to the transverse axis as the **conjugate axis**. See Figure 12.27(a).

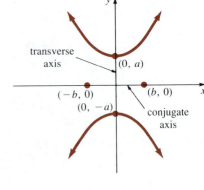

(a) $\dfrac{x^2}{a^2} - \dfrac{y^2}{b^2} = 1$ (b) $\dfrac{y^2}{a^2} - \dfrac{x^2}{b^2} = 1$

Figure 12.27

When the foci are placed on the y-axis at $F_1(0, -c)$ and $F_2(0, c)$, then the standard form for the equation of the hyperbola is

$$\frac{y^2}{a^2} - \frac{x^2}{b^2} = 1 \tag{12.18}$$

where, again, $b^2 = c^2 - a^2$. As indicated in Figure 12.27(b), the transverse axis has endpoints $(0, -a)$ and $(0, a)$ and the conjugate axis has endpoints $(-b, 0)$ and $(b, 0)$.

Asymptotes   Solving (12.17) for y in terms of x gives

$$y = \pm\frac{b}{a}x\sqrt{1 - \frac{a^2}{x^2}}$$

Observe that when the value of $|x|$ becomes very large, the value of the term a^2/x^2 becomes very small. Geometrically, this means that, for large values of $|x|$, the hyperbola is close to the lines

$$y = \frac{b}{a}x \quad \text{and} \quad y = -\frac{b}{a}x$$

These lines, which pass through the center of the hyperbola, are called **asymptotes**. In the case of (12.18), the asymptotes are $y = (a/b)x$ and $y = -(a/b)x$. Rather than just memorizing formulas, one can find equations for the asymptotes by the following mnemonic: Replace 1 with 0 in the standard equations (12.17) and (12.18) and factor the difference of two squares,

$$\frac{x^2}{a^2} - \frac{y^2}{b^2} = 0 \quad \text{or} \quad \frac{y^2}{a^2} - \frac{x^2}{b^2} = 0$$

Then set each factor equal to zero and solve for y.

As shown in Figure 12.28, by drawing the asymptotes first, we have a guideline for drawing a fairly accurate hyperbola.

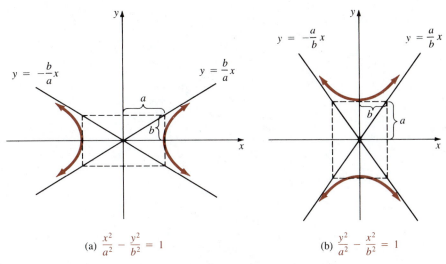

(a) $\dfrac{x^2}{a^2} - \dfrac{y^2}{b^2} = 1$ (b) $\dfrac{y^2}{a^2} - \dfrac{x^2}{b^2} = 1$

Figure 12.28

_____ **Example 1** _____

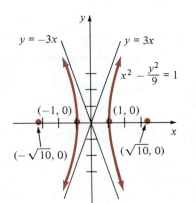

Figure 12.29

Find the vertices, foci, and asymptotes of the hyperbola whose equation is $x^2 - y^2/9 = 1$. Graph.

Solution   By writing $\dfrac{x^2}{1^2} - \dfrac{y^2}{3^2} = 1$

we identify the equation with (12.17) and conclude that the transverse axis is horizontal. Setting $y = 0$ gives $x^2 = 1$ and so the vertices are $(-1, 0)$ and $(1, 0)$. From $a^2 = 1$, $b^2 = 9$, (12.16) implies that $c^2 = a^2 + b^2 = 10$ and $c = \sqrt{10}$. The coordinates of the foci are $(-\sqrt{10}, 0)$ and $(\sqrt{10}, 0)$. Finally, from

$$x^2 - \frac{y^2}{3^2} = 0 \quad \text{or} \quad \left(x + \frac{y}{3}\right)\left(x - \frac{y}{3}\right) = 0$$

we see that the asymptotes are $y = -3x$ and $y = 3x$. Graphing the asymptotes and vertices leads to the hyperbola in Figure 12.29.

Center at (h, k)   The analogues of (12.17) and (12.18) are

$$\frac{(x - h)^2}{a^2} - \frac{(y - k)^2}{b^2} = 1 \tag{12.19}$$

and

$$\frac{(y - k)^2}{a^2} - \frac{(x - h)^2}{b^2} = 1 \tag{12.20}$$

when the center of the hyperbola is at (h, k). For each equation, $b^2 = c^2 - a^2$. The asymptotes, which pass through (h, k), can be found from the factors of

$$\frac{(x - h)^2}{a^2} - \frac{(y - k)^2}{b^2} = 0 \quad \text{or} \quad \frac{(y - k)^2}{a^2} - \frac{(x - h)^2}{b^2} = 0$$

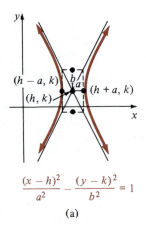

$$\frac{(x - h)^2}{a^2} - \frac{(y - k)^2}{b^2} = 1$$

(a)

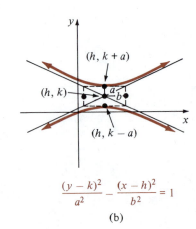

$$\frac{(y - k)^2}{a^2} - \frac{(x - h)^2}{b^2} = 1$$

(b)

Figure 12.30

To find the two vertices, we set $y = k$ in (12.19) and solve for x, or $x = h$ in (12.20) and solve for y. On a practical level, to find the vertices, we need only measure a units, either horizontally or vertically, from the center. Similarly, the foci are c units from the center: $(h \pm c, k)$ for (12.19) and $(h, k \pm c)$ for (12.20). The coordinates of the vertices are given in Figure 12.30.

Example 2

Find the vertices, foci, and asymptotes of the hyperbola whose equation is $(y - 2)^2/4 - (x + 2)^2/4 = 1$. Graph.

Solution   Identifying the equation with the standard form (12.20), we see immediately that the center is $(-2, 2)$, and $a^2 = 4$, $b^2 = 4$, $c^2 = 8$. In addition, we conclude the following:

Transverse axis: vertical

Vertices: $a = 2$ units above and below the center along the line $x = -2$; that is, $(-2, 4)$ and $(-2, 0)$. (Check this by solving $(y - 2)^2/4 = 1$.)

Foci: $c = 2\sqrt{2}$ units above and below the center along the line $x = -2$; that is, $(-2, 2 + 2\sqrt{2})$ and $(-2, 2 - 2\sqrt{2})$.

Asymptotes: $\dfrac{(y - 2)^2}{4} - \dfrac{(x + 2)^2}{4} = 0$ implies $y - 2 = -(x + 2)$ and $y - 2 = x + 2$, or $y = -x$ and $y = x + 4$.

By using the asymptotes and vertices, we find the graph shown in Figure 12.31.

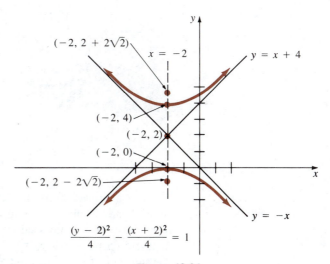

Figure 12.31

Example 3

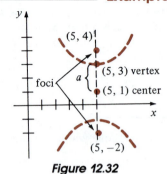

Figure 12.32

Find an equation of the hyperbola with foci $(5, -2)$, $(5, 4)$ and one vertex $(5, 3)$.

Solution Since the center must be at the midpoint of the line segment between $(5, -2)$ and $(5, 4)$, we have $h = 5$ and $k = (-2 + 4)/2 = 1$. The center is $(5, 1)$. Furthermore, from Figure 12.32 we see that $a = 2$ and $a^2 = 4$. With $c = 3$ and $b^2 = c^2 - a^2 = 9 - 4 = 5$, it follows from (12.20) that an equation is

$$\frac{(y - 1)^2}{4} - \frac{(x - 5)^2}{5} = 1$$

Remark

As shown in Figure 12.33(a), a plane flying at a supersonic speed parallel to level ground leaves a hyperbolic sonic "footprint" on the ground. The intersection of a wall and the cone of light emanating from a lamp is a hyperbola. See Figure 12.33(b). The Cassegranian reflecting telescope shown in Figure 12.33(c) utilizes a convex hyperbolic secondary mirror to reflect a ray of light back through a hole to an eyepiece behind the parabolic primary reflector. This telescope construction utilizes the fact that a beam of light directed along a line through one focus of a hyperbolic mirror will be reflected on a line through the other focus. See Figure 12.37 and Problem 42 in Exercises 12.3.

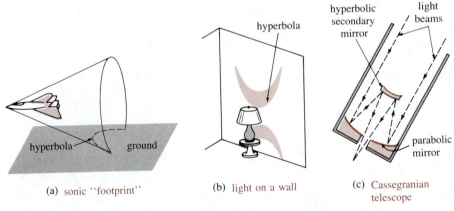

(a) sonic "footprint"   (b) light on a wall   (c) Cassegranian telescope

Figure 12.33

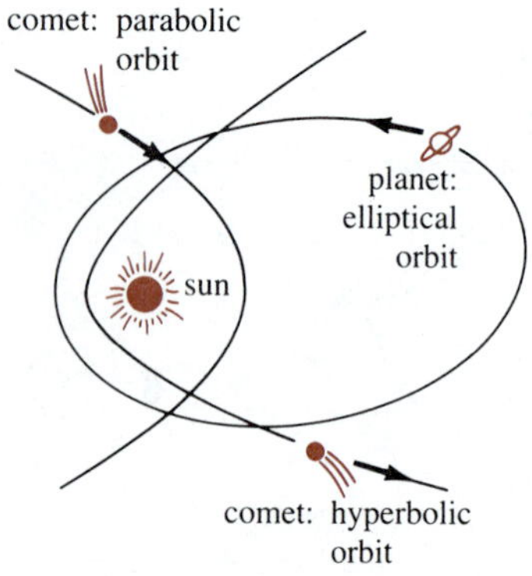

In intermediate mechanics, it is usually proved that the orbit of a body or atomic particle that moves in a central force field whose magnitude varies as $1/r^2$ must be a conic section. **Kepler's first law of planetary motion*** (see footnote on page 629) states that the orbit of a planet must be an ellipse with the sun at one focus. The orbit of a "short-period" comet, such as Halley's comet, is also elliptical. The "period" of Halley's comet, or the time it takes the comet to travel once around its orbit, is approximately 76 years. As indicated in Figure 12.34, the orbit of a "nonperiodic" comet, or a comet that never returns to the

Figure 12.34

solar system, can be either parabolic or hyperbolic. A modern hypothesis is that the orbits of all comets are elliptical since they originate in a gigantic spherical cloud of comet nuclei that surrounds the solar system. This cloud, not yet observed, is called the "Oort cloud" after its proposer, the Dutch astronomer Jan Hendrik Oort. The Oort cloud supposedly exists far beyond the orbit of Pluto, extending outward from about 1000 astronomical units to 100,000 astronomical units (1 A.U. is 93 million miles or the distance from the earth to the sun; the mean distance to the planet Pluto is only 39.5 A.U.).

When a star passes close enough to another star to have its orbit disturbed, the orbit will be hyperbolic.

Exercises 12.3

Answers to odd-numbered problems begin on page A-40.

In Problems 1–20 find the vertices, foci, asymptotes, and graph of the hyperbola whose equation is given.

1. $\dfrac{x^2}{4^2} - \dfrac{y^2}{3^2} = 1$

2. $\dfrac{y^2}{16} - \dfrac{x^2}{9} = 1$

3. $y^2 - x^2 = 9$

4. $2x^2 - y^2 = 4$

5. $4x^2 - 25y^2 = 100$

6. $y^2 - x^2 + 10 = 0$

7. $16x^2 - 9y^2 + 144 = 0$

8. $4y^2 - 9x^2 = 1$

9. $(x - 1)^2 - (y - 2)^2 = 1$

10. $\dfrac{(y - 3)^2}{9} - x^2 = 1$

11. $\dfrac{(y + 1)^2}{36} - \dfrac{(x - 4)^2}{4} = 1$

12. $\dfrac{(x + 3/2)^2}{25} - \dfrac{(y - 5/2)^2}{9} = 1$

13. $\dfrac{(2x - 1)^2}{16} - \dfrac{(3y + 4)^2}{36} = 1$

14. $\dfrac{4}{9}(x + 2)^2 - \dfrac{16}{9}(y + 3)^2 + 1 = 0$

15. $25(y - 5)^2 - 4(x + 2)^2 = 100$

16. $x^2 + 5x - y^2 + 3y = 1$

17. $x(x - 4) = y(y - 2)$

18. $(3x - y - 10)(3x + y - 8) = 9$

19. $x^2 - y^2 + 8x - 2y - 10 = 0$

20. $9y^2 - 64x^2 + 90y - 128x = 415$

Johannes Kepler

*Johannes Kepler (1571–1630)** This law is one of three laws of planetary motion advanced by the German astronomer, and successor to the great astronomical observer Tycho Brahe, Johannes Kepler. The other two laws are:

• As a planet revolves, a line joining it to the sun sweeps out equal areas in equal time intervals.

• The square of the period of a planet—that is, the time it takes for a planet to revolve around the sun—is proportional to the cube of the planet's mean distance from the sun.

These laws, based on observations, were subsequently proved by Isaac Newton using his newly formulated law of gravitation. In his writings Kepler hinted at the law of gravitation by arguing that the earth and the moon were held in their orbits by some "vital force." He was also the first to advance the idea that tides were caused by lunar attraction on the oceans. Kepler was also a mathematician, astrologer, and a mystic. His mother was accused of being a witch.

In Problems 21–32 find an equation of the hyperbola that satisfies the given conditions.

21. Center $(0, 0)$, vertices $(\pm 3, 0)$, one focus $(5, 0)$

22. Foci $(0, \pm 3)$, one vertex $(0, 1)$

23. Foci $(2 \pm 5\sqrt{2}, -7)$, length of transverse axis 10

24. Center $(0, 0)$, one focus $(5, 0)$, length of conjugate axis 8

25. Vertices $(1, -1)$ and $(1, 5)$, one focus $(1, -2)$

26. Center $(-2, -5)$, length of horizontal transverse axis 14, length of conjugate axis 12

27. Vertices $(0, \pm 4)$, passing through $(\frac{1}{2}, -3\sqrt{2})$

28. Vertices $(\pm\frac{1}{2}, 0)$, asymptotes $y = \pm\frac{3}{2}x$

29. Foci $(\pm 3, 0)$, asymptotes $y = \pm\sqrt{3}x$

30. Center $(2, 4)$, one vertex $(2, 5)$, one asymptote $2y - x - 6 = 0$

31. Center $(-1, 2)$, focus $(5, 2)$, through $(3, 2)$

32. Set of all points in the plane the difference of whose distances from $(1, 3)$ and $(1, -1)$ is 2

33. Determine the point(s) on the hyperbola whose equation is $2x^2 - y^2 = 14$ at which the slope of the tangent is 4.

34. Find an equation of the tangent line to the hyperbola whose equation is $(x - 1)^2 - (y + 2)^2 = 1$ at $x = 2$. At $x = 3$.

35. Tangent lines to the graph of the hyperbola $x^2 - y^2 = 40$ intersect at $(0, 5)$. Find the points of tangency.

36. For the equation $x^2/a^2 - y^2/b^2 = 1$ show that the area A of the shaded region in Figure 12.35 is

$$A = \frac{1}{2}ab \ln\left(\frac{x_0}{a} + \frac{y_0}{b}\right)$$

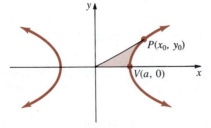

Figure 12.35

37. The region in the first quadrant that is bounded by the graphs of $x^2/4 - y^2 = 1$, an asymptote, $y = 3$, and $y = 0$ is revolved about the y-axis. Find the volume of the solid of revolution.

38. A cannon is heard at three points that are given the coordinates $P_1(0, 3)$, $P_2(12, 6)$, and $P_3(0, -3)$, where distance is measured in kilometers. By sound equipment it is determined that the cannon is 2 km closer to P_1 than to P_3.

(a)   Show that the position of the cannon must be on a branch of a hyperbola.

(b)   Suppose it is determined that the cannon lies on the line through P_2 and P_3. Find its position.*

Miscellaneous Problems

39. The line segment that is perpendicular to a line containing the transverse axis through a focus with endpoints on the hyperbola is called a **focal chord**. See Figure 12.36. Show that the length of a focal chord of a hyperbola with center at the origin is $2b^2/a$.

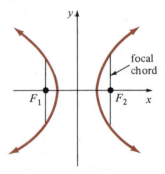

Figure 12.36

40. **Conjugate hyperbolas** are two hyperbolas such that the transverse axis of each is the conjugate axis of the other. Using the same coordinate axes, graph $x^2/36 - y^2/4 = 1$ and its conjugate.

41. A **rectangular hyperbola** is one for which the asymptotes are perpendicular. Which of the hyperbolas whose equations are given in Problems 1–20 are rectangular?

---

*During wartime this method was employed to locate enemy artillery. The same idea is used in LORAN (long-range navigation), where the position of a ship can be determined by calculating the time difference between the radio signals from two known stations.

42. As shown in Figure 12.37, let L be the tangent line to the graph of $x^2/a^2 - y^2/b^2 = 1$ at $P(x_0, y_0)$. Show that $\alpha = \beta$.

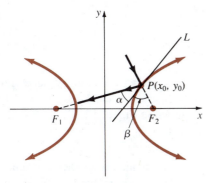

Figure 12.37

The **eccentricity** of a hyperbola is defined to be c/a. In Problems 43 and 44 find the eccentricity of the hyperbola whose equation is given.

43. $y^2 - \dfrac{x^2}{8} = 1$

44. $20(x + 1)^2 - 16(y - 1)^2 = 320$

45. Find an equation of form (12.18) of a hyperbola whose foci are $(0, -10)$ and $(0, 10)$ and whose eccentricity is $\frac{5}{3}$.

46. Find an equation of a hyperbola such that the endpoints of its conjugate axis are $(-5, 4)$ and $(-5, 10)$ and whose eccentricity is $\sqrt{10}$.

[O] 12.4 Translation and Rotation of Axes

Consider the ellipse whose equation is

$$\frac{(x - 3)^2}{6^2} + \frac{(y - 4)^2}{3^2} = 1$$

If we let $X = x - 3$ and $Y = y - 4$, the equation becomes

$$\frac{X^2}{6^2} + \frac{Y^2}{3^2} = 1$$

In terms of a new XY-coordinate system, the graph of the equation is as shown in Figure 12.38(a). The center of the ellipse corresponds to the simultaneous condition $X = 0$, $Y = 0$, which, in turn, is the point $(3, 4)$ in the xy-plane. By superimposing the xy-plane over the XY-plane, we obtain the points indicated in Figure 12.38(b).

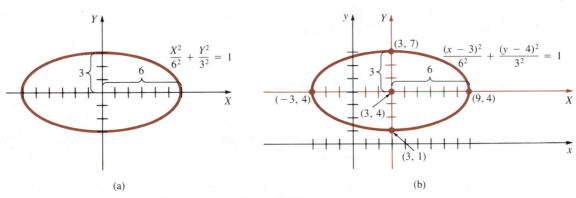

(a)   (b)

Figure 12.38

Translation of Axes

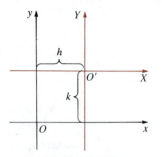

Figure 12.39

Equations such as $(x - h)^2 = 4p(y - k)$ and $(x - h)^2/a^2 + (y - k)^2/b^2 = 1$ can be simplified by means of the substitution $X = x - h$, $Y = y - k$ and readily graphed in terms of X and Y. In general, the equations

$$\begin{array}{ccc} X = x - h & & x = X + h \\ & \text{or} & \\ Y = y - k & & y = Y + k \end{array}$$

define a **translation** of axes or a translation between the xy-plane and the XY-plane. Relative to the xy-plane, the equation of the Y-axis is $x = h$ and the equation of the X-axis is $y = k$. As shown in Figure 12.39, the origin O' in the XY-plane is the point (h, k) in the xy-plane.

Rotation of Axes

In contrast to the translation of axes, we obtain yet another coordinate system by a **rotation** of axes. If the positive x-axis is rotated by an amount θ, $0 < \theta < 90°$, about the origin, we obtain the X- and Y- axes shown in Figure 12.40. A straightforward exercise in trigonometry (see Problem 25 of Exercises 12.4) shows that if (X, Y) are the coordinates of a point in the new coordinate system, then its xy-coordinates are

$$\begin{aligned} x &= X \cos \theta - Y \sin \theta \\ y &= X \sin \theta + Y \cos \theta \end{aligned} \tag{12.21}$$

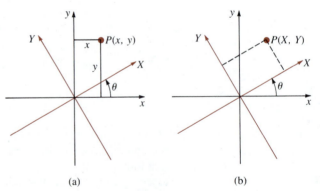

Figure 12.40

Solving (12.21) for X and Y gives a way of converting xy-coordinates into the XY-coordinates:

$$\begin{aligned} X &= x \cos \theta + y \sin \theta \\ Y &= -x \sin \theta + y \cos \theta \end{aligned} \tag{12.22}$$

Example 1

The positive x-axis is rotated by 30°. Find the XY-coordinates of the point whose xy-coordinates are $(4, 6)$.

Solution   With $\theta = 30°$, $x = 4$, and $y = 6$, we have, from (12.22),

$$X = 4 \cos 30° + 6 \sin 30°$$
$$Y = -4 \sin 30° + 6 \cos 30°$$

or
$$X = 2\sqrt{3} + 3 \approx 6.5$$
$$Y = -2 + 3\sqrt{3} \approx 3.2$$

The point and the rotated axes are shown in Figure 12.41.

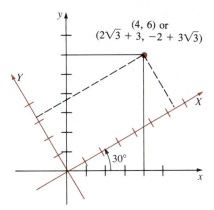

Figure 12.41

Elimination of the xy-Term   An appropriate rotation of axes enables us to eliminate the xy-term in

$$ax^2 + bxy + cy^2 + dx + ey + f = 0 \qquad (12.23)$$

and obtain an equation of the form

$$AX^2 + CY^2 + DX + EY + F = 0 \qquad (12.24)$$

If (12.24) defines a real locus of points other than a point, a line, or a pair of intersecting lines, its graph will then be a conic section. The cases: a point, a line, and a pair of intersecting lines are said to be **degenerate conics**. See Figure 12.1(b).

Substituting the equations in (12.21) into (12.23) and simplifying reveals that the coefficient of the XY-term is zero provided that

$$2(c - a)\sin \theta \cos \theta + b(\cos^2\theta - \sin^2\theta) = 0$$

or equivalently
$$(a - c)\sin 2\theta = b \cos 2\theta \qquad (12.25)$$

Notice that if $a = c$ in (12.23), then $a - c = 0$ in (12.25). The resulting equation, $\cos 2\theta = 0$, implies that a rotation by an amount $\theta = 45°$ will eliminate the xy-term in (12.23). However, if $a \neq c$, then this elimination can be accomplished by choosing θ to be an angle for which

$$\tan 2\theta = \frac{b}{a - c} \qquad (12.26)$$

Example 2

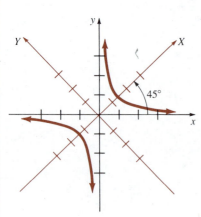

Figure 12.42

The simple equation $xy = 1$ can be rewritten in terms of X and Y without the product xy. Since $a = c = 0$, we use $\theta = 45°$ in (12.21):

$$x = \frac{X}{\sqrt{2}} - \frac{Y}{\sqrt{2}}$$

$$y = \frac{X}{\sqrt{2}} + \frac{Y}{\sqrt{2}}$$

Thus, $xy = 1$ is the same as

$$\left(\frac{X}{\sqrt{2}} - \frac{Y}{\sqrt{2}}\right)\left(\frac{X}{\sqrt{2}} + \frac{Y}{\sqrt{2}}\right) = 1 \quad \text{or} \quad \frac{X^2}{2} - \frac{Y^2}{2} = 1$$

As illustrated in Figure 12.42, the equation is a standard form for a hyperbola in the XY-coordinate system with vertices $(\pm\sqrt{2}, 0)$.

In the slightly more complicated case of $a - c \neq 0$, the identities

$$\cos 2\theta = \frac{\pm 1}{\sqrt{1 + \tan^2 2\theta}}, \quad \sin \theta = \sqrt{\frac{1 - \cos 2\theta}{2}}, \quad \cos \theta = \sqrt{\frac{1 + \cos 2\theta}{2}} \quad (12.27)$$

prove to be very useful.*

Example 3

$Y = \sqrt{13}X^2$

Figure 12.43

By rotation of axes, eliminate the xy-term in the equation

$$9x^2 + 12xy + 4y^2 + 2x - 3y = 0$$

Identify and graph.

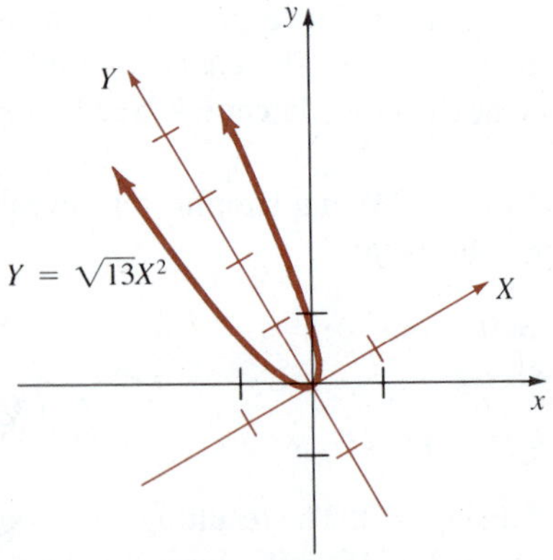

Solution   Identifying $a = 9$, $b = 12$, and $c = 4$, we have from (12.26), $\tan 2\theta = \frac{12}{5}$. Since 2θ is an angle in the first quadrant, (12.27) yields

$$\cos 2\theta = \frac{5}{13}, \quad \sin \theta = \frac{2}{\sqrt{13}}, \quad \cos \theta = \frac{3}{\sqrt{13}}$$

Hence, the equations in (12.21) become

$$x = \frac{3}{\sqrt{13}}X - \frac{2}{\sqrt{13}}Y$$

$$y = \frac{2}{\sqrt{13}}X + \frac{3}{\sqrt{13}}Y$$

---

*The first identity comes from $1 + \tan^2 2\theta = \sec^2 2\theta$. The second two are the half-angle formulas given in (1.41).

Substituting these equations into $9x^2 + 12xy + 4y^2 + 2x - 3y = 0$ and simplifying gives

$$Y = \sqrt{13}X^2$$

Thus, the graph of the given equation is a parabola. From $\theta = \frac{1}{2} \tan^{-1}(\frac{12}{5})$, we see that the positive x-axis is rotated by $\theta \approx 33.7°$. The graph of the equation is given in Figure 12.43.

_____ Exercises 12.4 _____

Answers to odd-numbered problems begin on page A-41.

In Problems 1–4 the two points are given in xy-coordinates. Find the XY-coordinates of the first point if the translated origin O' is the second point.

1. $(3, 2); (1, 3)$

2. $(-1, 7); (-4, 3)$

3. $(\frac{1}{2}, -\frac{3}{2}); (2, -5)$

4. $(-6, -8); (2, -8)$

In Problems 5–10 use translation of axes to discuss the difference in the graphs of the given pairs of equations.

5. $y = |x|$
  $y = |x - 1| + 4$

6. $y = x^{2/3}$
  $y = (x + 2)^{2/3} - 1$

7. $y = \sin x$
  $y = \sin(x + \pi/2)$

8. $y = x^2$
  $y = x^2 - 6x + 7$

9. $x^2 - y^2 = 4$
  $(x + 1)^2 - (y - 1)^2 = 4$

10. $4x^2 + y^2 = 16$
  $4(x - 5)^2 + y^2 = 16$

In Problems 11–14 the positive x-axis is rotated by the indicated amount. Find the XY-coordinates of the point with the given xy-coordinates.

11. $(6, 2); 45°$

12. $(-2, 8); 30°$

13. $(-1, -1); 60°$

14. $(5, 3); 15°$

In Problems 15 and 16, without the aid of a calculator or tables, find the XY-coordinates of the point whose xy-coordinates are given for the indicated amount of rotation.

15. $(5, -5); \tan^{-1}\frac{4}{3}$

16. $(20, 10); \tan^{-1}\frac{1}{3}$

In Problems 17–22 use rotation of axes to eliminate the xy-term in the given equation. Identify and graph.

17. $x^2 + xy + y^2 = 4$

18. $2x^2 - 3xy - 2y^2 = 5$

19. $x^2 - 2xy + y^2 = 8x + 8y$

20. $3x^2 + 4xy = 16$

21. $4x^2 - 4xy + 7y^2 + 12x + 6y - 9 = 0$

22. $x^2 - xy + y^2 - 4x - 4y = 20$

23. Given $3x^2 + 2\sqrt{3}xy + y^2 + 2x - 2\sqrt{3}y = 0$.
  (a) By rotation of axes show that the graph of the equation is a parabola.
  (b) Find the XY-coordinates of the focus. Use this information to find the xy-coordinates of the focus.
  (c) Find an equation of the directrix in terms of the XY-coordinates. Use this information to find an equation of the directrix in terms of the xy-coordinates.

24. Given $13x^2 - 8xy + 7y^2 = 30$.
  (a) By rotation of axes show that the graph of the equation is an ellipse.
  (b) Find the XY-coordinates of the foci. Use this information to find the xy-coordinates of the foci.
  (c) Find the xy-coordinates of the vertices.

Miscellaneous Problems

25. (a) Use Figure 12.44 to show that $X = d \cos \phi$, $Y = d \sin \phi$ and $x = d \cos(\theta + \phi)$, $y = d \sin(\theta + \phi)$.

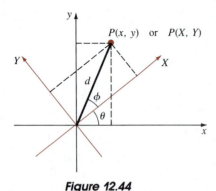

Figure 12.44

(b) Use the results in part **(a)** to derive the equations in (12.21).

26. Graph the equation $xy = 3x - 2y + 6$.

Except in degenerate cases, the graph of (12.23) is
(*i*)   a parabola if $b^2 - 4ac = 0$,
(*ii*) an ellipse if $b^2 - 4ac < 0$,
(*iii*) a hyperbola if $b^2 - 4ac > 0$.

The expression $b^2 - 4ac$ is called the **discriminant** of the quadratic equation. In Problems 27–32 determine the type of graph without actually graphing.

27. $x^2 - 3xy + y^2 = 5$

28. $2x^2 - 2xy + 2y^2 = 1$

29. $4x^2 - 4xy + y^2 - 6 = 0$

30. $x^2 + \sqrt{3}\,xy - \dfrac{1}{2}y^2 = 0$

31. $x^2 + xy + y^2 - x + 2y + 1 = 0$

32. $3x^2 + 2\sqrt{3}\,xy + y^2 - 2x + 2\sqrt{3}y - 4 = 0$

Chapter 12 Review Exercises

Answers to odd-numbered problems begin on page A-41.

In Problems 1–16 answer true or false.

1. For a parabola, the distance from the vertex to the focus is the same as the distance from the vertex to the directrix. _____

2. The directrix of $y^2 = 4px$, $p < 0$, is perpendicular to the positive *y*-axis. _____

3. The minor axis of an ellipse bisects the major axis. _____

4. For an ellipse whose equation is $x^2/a^2 + y^2/b^2 = 1$, the numbers *a*, *b*, and *c* are related by $a^2 = b^2 + c^2$. _____

5. For a hyperbola, the numbers *c* and *a* satisfy $c > a$. _____

6. The foci of $(x - h)^2/a^2 + (y - k)^2/b^2 = 1$ are $(h, k \pm c)$. _____

7. A beam of light originating at one focus of an ellipse and striking its mirrored surface will necessarily be reflected to the other focus. _____

8. The asymptotes of $x^2/a^2 - y^2/a^2 = 1$ are perpendicular. _____

9. The asymptotes of a hyperbola always pass through the origin. _____

10. The branches of a hyperbola are two parabolas. _____

11. If $(0, 4)$, $(3, 6)$, $(6, 4)$, and $(3, 2)$ are vertices of an ellipse, then its major axis is vertical. _____

12. If *P* is a point on the left branch of the hyperbola given by $x^2/a^2 - y^2/b^2 = 1$, then $d(P, F_1) - d(P, F_2) < 0$, where $F_1(-c, 0)$ and $F_2(c, 0)$ are the foci. _____

13. The directrix of a parabola must be either vertical or horizontal. _____

14. The *y*-intercepts of the graph of $x^2/a^2 - y^2/b^2 = 1$ are $\pm b$. _____

15. The point $(-2, 5)$ is on the ellipse $x^2/8 + y^2/50 = 1$. _____

16. The graphs of $y = x^2$ and $y^2 - x^2 = 1$ have at most two points in common. _____

In Problems 17–30, fill each blank with the required answer for the given equation.

17. $y = 2x^2$, focus _____

18. $\dfrac{x^2}{4} - \dfrac{y^2}{12} = 1$, foci _____

19. $4x^2 + 5(y - 2)^2 = 20$, center _____

20. $25y^2 - 4x^2 = 100$, asymptotes _____

21. $8(y + 3) = (x - 1)^2$, directrix _____

22. $\dfrac{(x + 1)^2}{36} + \dfrac{(y + 7)^2}{16} = 1$, vertices _____

23. $x^2 - 2y^2 = 18$, length of conjugate axis _____

24. $y = x^2 + 4x - 6$, vertex _____

25. $(x - 4)^2 - (y + 1)^2 = 4$, endpoints of transverse axis _____

26. $\dfrac{(x - 3)^2}{7} + \dfrac{(y + 3/2)^2}{8} = 1$, equation of line that contains major axis _____

27. $25x^2 + y^2 - 200x + 6y + 384 = 0$, center _____

28. $(x + 1)^2 + (y + 8)^2 = 100$, x-intercepts _____

29. $y^2 - (x - 2)^2 = 1$, y-intercepts _____

30. $y^2 - y + 3x = 3$, slope of tangent line at $(1, 1)$ _____

31. Show that an equation of the tangent line to the graph of an ellipse $x^2/a^2 + y^2/b^2 = 1$ at (x_1, y_1) is given by

$$\frac{xx_1}{a^2} + \frac{yy_1}{b^2} = 1$$

32. Show that an equation of the tangent line to the graph of a hyperbola $x^2/a^2 - y^2/b^2 = 1$ at (x_1, y_1) is given by

$$\frac{xx_1}{a^2} - \frac{yy_1}{b^2} = 1$$

33. A satellite revolves around the planet Jupiter in an elliptical orbit with the center of the planet at one focus. The length of the major axis of the orbit is 10^9 m and the length of the minor axis is 6×10^8 m. Find the minimum distance between the satellite and the center of Jupiter. What is the maximum distance?

34. Find an equation of the hyperbola whose asymptotes are $3y = 5x$ and $3y = -5x$ and that has vertices $(0, 10)$ and $(0, -10)$.

35. The distance between a point (x_0, y_0) and a line $ax + by + c = 0$ is given by $|ax_0 + by_0 + c|/\sqrt{a^2 + b^2}$. Use this result to find an equation of the directrix of a parabola whose vertex is at the origin and whose focus is at $(2, 2)$.

36. Use Definition 12.1 to find an equation of the parabola described in Problem 35.

13

Parametric Equations and Polar Coordinates

A rectangular or Cartesian equation is not the only, and often not the most convenient, way of describing a curve in the plane. In this chapter we shall consider two additional means by which a curve can be represented. One of these two approaches utilizes an entirely new kind of coordinate system.

13.1 Parametric Equations

Curvilinear Motion   The motion of a particle along a curve, in contrast to a straight line, is called **curvilinear motion**. In physics it is shown that the motion of a projectile in the xy-plane, such as a struck golf ball,* is governed by the fact that its acceleration in the x- and y-directions satisfies

$$a_x = 0, \qquad a_y = -g \tag{13.1}$$

where g is the acceleration of gravity and $a_x = d^2x/dt^2$, $a_y = d^2y/dt^2$. At $t = 0$ we take $x = 0$, $y = 0$ and the x- and y-components of the initial velocity v_0 to be

$$v_0 \cos \theta_0 \quad \text{and} \quad v_0 \sin \theta_0$$

respectively. See Figure 13.1. Taking two antiderivatives of (13.1), it follows from the initial conditions, that the x- and y-coordinates of the ball at time t are

$$x = (v_0 \cos \theta_0)t$$

$$y = -\frac{1}{2}gt^2 + (v_0 \sin \theta_0)t \tag{13.2}$$

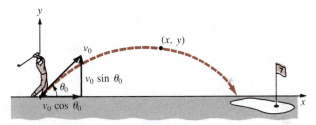

Figure 13.1

These equations, which give the ball's position in the xy-plane at a time t, are said to be **parametric equations**. The variable t is called a **parameter** and is restricted to an interval $0 \le t \le T$, where T is the time the ball hits the ground.

In general, a curve can be *defined* in terms of parametric equations.

DEFINITION 13.1   Plane Curve

A **plane curve** is a set C of ordered pairs $(f(t), g(t))$, where f and g are functions continuous on an interval I. The equations $x = f(t)$, $y = g(t)$, for t in I, are called **parametric equations** for C. The variable t is called a **parameter**.

The **graph** of a plane curve C is the set of all points (x, y) in the Cartesian plane corresponding to the ordered pairs $(f(t), g(t))$. Hereafter, we shall refer to a plane curve as a **curve**. Furthermore, for simplicity we shall not belabor the distinction between a *curve* and the *graph of a curve*.

---

*Assuming no slice or hook. The effects of air resistance are also ignored.

Example 1

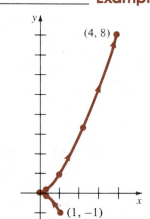

Figure 13.2

Graph the curve that has the parametric equations

$$x = t^2, \qquad y = t^3, \qquad -1 \le t \le 2$$

Solution   As shown in the accompanying table, for any choice of t in $[-1, 2]$, we obtain an ordered pair (x, y). By connecting the points with a curve, we obtain the graph in Figure 13.2.

t	-1	$-\frac{1}{2}$	0	$\frac{1}{2}$	1	$\frac{3}{2}$	2
x	1	$\frac{1}{4}$	0	$\frac{1}{4}$	1	$\frac{9}{4}$	4
y	-1	$-\frac{1}{8}$	0	$\frac{1}{8}$	1	$\frac{27}{8}$	8

In Example 1, if we think in terms of motion and t as time, then as t increases from -1 to 2 a particle at point P starts from $(1, -1)$, advances up the lower branch, passes to the upper branch, and finally stops at $(4, 8)$. In general, as we plot points corresponding to increasing values of the parameter, the curve C is traced in a certain direction. This direction is called the **orientation** of the curve. The arrowheads on the curve in Figure 13.2 indicate its orientation.

A parameter need have no relation to time. When the interval I is not specified, it is usually understood to be either $-\infty < t < \infty$ or the longest interval on which f and g are both continuous.

Example 2

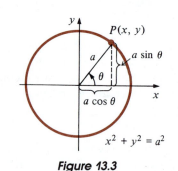

Figure 13.3

The circle $x^2 + y^2 = a^2$ with center at the origin and radius a can be parameterized in terms of a central angle θ. Using Figure 13.3 and trigonometry we see that the equations

$$x = a \cos \theta, \qquad y = a \sin \theta, \qquad 0 \le \theta \le 2\pi \qquad (13.3)$$

give every point P on the circle.

We note that the orientation of the circle in Example 2 is counterclockwise.

Example 3

In Example 2, the semicircle $x^2 + y^2 = a^2$, $0 \le y \le a$ is given parametrically by $x = a \cos \theta$, $y = a \sin \theta$, $0 \le \theta \le \pi$.

Eliminating the Parameter Given a set of parametric equations, we sometimes desire to **eliminate** or **clear the parameter** to obtain a Cartesian equation. To clear the parameter in (13.3), we simply square x and y:

$$x^2 + y^2 = a^2 \cos^2\theta + a^2 \sin^2\theta \quad \text{implies that} \quad x^2 + y^2 = a^2$$

since $\cos^2\theta + \sin^2\theta = 1$.

Example 4

(a) From the first equation in (13.2), $t = x/(v_0 \cos \theta_0)$ so that the second equation yields

$$y = -\frac{g}{2(v_0 \cos \theta_0)^2}x^2 + (\tan \theta_0)x$$

Thus, the trajectory of any projectile launched at an angle $0 < \theta_0 < \pi/2$ is necessarily a parabola.

(b) In Example 1, we can clear the parameter by solving the second equation for t in terms of y and substituting in the first equation. We find

$$t = y^{1/3} \quad \text{and so} \quad x = (y^{1/3})^2 = y^{2/3}$$

But for $-1 \le t \le 2$ we have correspondingly $-1 \le y \le 8$. Thus a Cartesian equation for the curve is given by $x = y^{2/3}$, $-1 \le y \le 8$.

A curve C can have more than one parameterization. For example, an examination of $x = t$, $y = 2t^2$, $-\infty < t < \infty$ and $x = t^3/4$, $y = t^6/8$, $-\infty < t < \infty$ reveals that both sets of equations represent $y = 2x^2$.* But, one has to be careful when working with parametric equations. Clearing the parameter in $x = t^2$, $y = 2t^4$, $-\infty < t < \infty$ would seem to yield the same parabola $y = 2x^2$. This is not the case because, for any value of t, $t^2 \ge 0$ and so $x \ge 0$. In other words, the last set is a parametric representation of only the right branch of the parabola: $y = x^2$, $x \ge 0$. You are encouraged to work Problems 17–22 in Exercises 13.1.

Example 5

Consider the curve C given parametrically by

$$x = \sin \theta, \quad y = \cos 2\theta, \quad 0 \le \theta \le \pi/2$$

Eliminate the parameter and obtain a Cartesian equation that has the same graph.

---

*Note that a point need not correspond to the same value of the parameter in each set. For example, $(1, 2)$ is obtained from $t = 1$ in the first set but from $t = \sqrt[3]{4}$ in the second set.

Solution   Using the double-angle formula for cos 2θ we can write

$$y = \cos^2\theta - \sin^2\theta$$
$$= (1 - \sin^2\theta) - \sin^2\theta$$
$$= 1 - 2\sin^2\theta$$

or $$y = 1 - 2x^2$$

Now the curve C described by the parametric equations does not consist of the entire parabola whose equation is $y = 1 - 2x^2$. See Figure 13.4(a). For $0 \le \theta \le \pi/2$ we have $0 \le \sin\theta \le 1$ and $-1 \le \cos 2\theta \le 1$. This means C is only that portion of the parabola for which the coordinates of a point $P(x, y)$ satisfy $0 \le x \le 1$ *and* $-1 \le y \le 1$. Thus C is the curve shown in color in Figure 13.4(b). A Cartesian equation of C is then $y = 1 - 2x^2$, $0 \le x \le 1$.

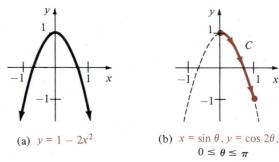

(a) $y = 1 - 2x^2$ (b) $x = \sin\theta$, $y = \cos 2\theta$,
                                $0 \le \theta \le \pi$

Figure 13.4

We can get the intercepts of a curve C without finding its Cartesian equation. For instance, in Example 5 we can find the x-intercept by solving $y = 0$. The equation $\cos 2\theta = 0$ yields $2\theta = \pi/2$ or $\theta = \pi/4$. The corresponding point at which C crosses the x-axis is $(1/\sqrt{2}, 0)$. Similarly, the y-intercept of C is found by solving $x = 0$.

Cycloidal Curves   **Cycloidal curves** were a popular topic of study by mathematicians in the seventeenth century. Suppose a point $P(x, y)$, marked on a circle of radius a, is at the origin when its diameter lies along the y-axis. As the circle rolls along the x-axis the point P traces out a curve C that is called a **cycloid**.* See Figure 13.6(a). Now suppose a circle of radius a rolls on the *inside* of a larger circle

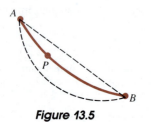

Figure 13.5

*Two problems were extensively studied in the seventeenth century by the Dutch physicist Christian Huygens (1629–1695) and the Swiss mathematicians Jakob (1654–1705) and Johann Bernoulli (1667–1748). Consider a flexible (frictionless) wire fixed at points A and B and a bead free to slide down the wire starting at P. See Figure 13.5. Is there a particular shape of the wire so that, regardless of where the bead starts, the time to slide down the wire to B will be the same? Also, what would the shape of the wire be so that the bead slides from P to B in the shortest time? The so-called *tautochrone* (same time) and *brachistochrone* (least time) were shown to be an inverted half arc of a cycloid. You may be interested in the excellent film *Cycloidal Curves or Tales from the Wankenburg Woods*, by Wards Natural Science Co., Monterey, Calif.

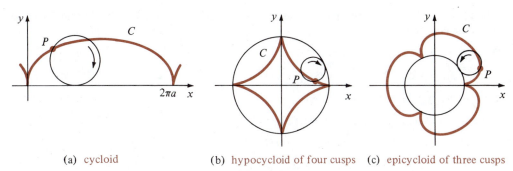

(a) cycloid (b) hypocycloid of four cusps (c) epicycloid of three cusps

Figure 13.6

of radius b. If a point $P(x, y)$, on the inside circle, starts from $(b, 0)$, the curve C traced out by P is called a **hypocycloid**. Specifically, if $b = 4a$, then C is a **hypocycloid of four cusps**. This is the curve illustrated in Figure 13.6(b). Finally, if a circle of radius a rolls on the *outside* of a circle of radius b, starting from $(b, 0)$, we obtain a curve C known as an **epicycloid**. The **epicycloid of three cusps** illustrated in Figure 13.6(c) results when $b = 3a$. It is left as exercises to obtain the parametric equations of the three curves shown in Figure 13.6.

Smooth Curves   A curve C, given parametrically by

$$x = f(t), \qquad y = g(t), \qquad a \le t \le b$$

is said to be **smooth** if f' and g' are continuous on $[a, b]$ and not simultaneously zero on (a, b). A curve C is said to be **piecewise smooth** if the interval $[a, b]$ can be partitioned into subintervals such that C is smooth on each subinterval. The curves in Examples 2, 3, and 5 are smooth; the curve in Example 1 is piecewise smooth.

Remark

A curve C described by a continuous function $y = f(x)$ can *always* be parameterized by letting $x = t$. A set of parametric equations for C is $x = t$, $y = f(t)$.

Exercises 13.1

Answers to odd-numbered problems begin on page A-41.

In Problems 1 and 2 fill in the table for the given set of parametric equations.

1. $x = 2t + 1$, $y = t^2 + t$

t	-3	-2	-1	0	1	2	3
x							
y							

2. $x = \cos t$, $y = \sin^2 t$

t	0	$\pi/6$	$\pi/4$	$\pi/3$	$\pi/2$	$5\pi/6$	$7\pi/4$
x							
y							

In Problems 3–10 graph the curve which has the given set of parametric equations.

3. $x = t - 1$, $y = 2t - 1$; $-1 \leq t \leq 5$

4. $x = 3t$, $y = t^2 - 1$; $-2 \leq t \leq 3$

5. $x = \sqrt{t}$, $y = 5 - t$; $t \geq 0$

6. $x = 3 + 2 \sin t$, $y = 4 + \sin t$; $-\pi/2 \leq t \leq \pi/2$

7. $x = 4 \cos t$, $y = 4 \sin t$; $-\pi/2 \leq t \leq \pi/2$

8. $x = t^3 + 1$, $y = t^2 - 1$; $-2 \leq t \leq 2$

9. $x = e^t$, $y = e^{3t}$; $0 \leq t \leq \ln 2$

10. $x = -e^t$, $y = e^{-t}$; $t \geq 0$

In Problems 11–16 eliminate the parameter from the given set of parametric equations and obtain a Cartesian equation that has the same graph.

11. $x = t^2$, $y = t^4 + 3t^2 - 1$

12. $x = t^3 + t + 4$, $y = -2t^3 - 2t$

13. $x = \cos 2\theta$, $y = \sin \theta$, $0 \leq \theta \leq 2\pi$

14. $x = e^t$, $y = \ln t$, $t > 0$

15. $x = t^3$, $y = 3 \ln t$, $t > 0$

16. $x = \tan t$, $y = \sec t$, $-\pi/2 < t < \pi/2$

In Problems 17–22 graphically show the difference between the given curves.

17. $y = x$ and $x = \sin t$, $y = \sin t$

18. $y = x^2$ and $x = -\sqrt{t}$, $y = t$, $t \geq 0$

19. $y = \dfrac{x^2}{4} - 1$ and $x = 2t$, $y = t^2 - 1$, $-1 \leq t \leq 2$

20. $y = -x^2$ and $x = e^t$, $y = -e^{2t}$, $t \geq 0$

21. $x^2 - y^2 = 1$ and $x = \cosh t$, $y = \sinh t$

22. $y = 2x - 2$ and $x = t^2 - 1$, $y = 2t^2 - 4$

In Problems 23–26 graphically show the difference between the given curves. Assume $a > 0$, $b > 0$.

23. $x = a \cos t$, $y = a \sin t$, $0 \leq t \leq \pi$
    $x = a \sin t$, $y = a \cos t$, $0 \leq t \leq \pi$

24. $x = a \cos t$, $y = b \sin t$, $a > b$, $\pi \leq t \leq 2\pi$
    $x = a \sin t$, $y = b \cos t$, $a > b$, $\pi \leq t \leq 2\pi$

25. $x = a \cos t$, $y = a \sin t$, $-\pi/2 \leq t \leq \pi/2$
    $x = a \cos 2t$, $y = a \sin 2t$, $-\pi/2 \leq t \leq \pi/2$

26. $x = a \cos \dfrac{t}{2}$, $y = a \sin \dfrac{t}{2}$, $0 \leq t \leq \pi$

    $x = a \cos\left(-\dfrac{t}{2}\right)$, $y = a \sin\left(-\dfrac{t}{2}\right)$, $-\pi \leq t \leq 0$

In Problems 27 and 28 graph the curve that has the given parametric equations.

27. $x = 1 + 2 \cosh t$, $y = 2 + 3 \sinh t$

28. $x = -3 + 3 \cos t$, $y = 5 + 5 \sin t$

29. Determine which of the following sets of parametric equations have the same graph as the Cartesian equation $xy = 1$.

(a) $x = \dfrac{1}{2t + 1}$, $y = 2t + 1$

(b) $x = t^{1/2}$, $y = t^{-1/2}$

(c) $x = \cos t$, $y = \sec t$

(d) $x = t^2 + 1$, $y = (t^2 + 1)^{-1}$

(e) $x = e^{-2t}$, $y = e^{2t}$

(f) $x = t^3$, $y = t^{-3}$

30. The ends of a rod of length L slide on horizontal and vertical tracks that coincide with the x- and y-axes, respectively.

(a) Parameterize the coordinates of the point P in Figure 13.7(a) in terms of ϕ. Show that P traces a circular path as ϕ varies from 0 to $\pi/2$.

(b) Show that if P is located as shown in Figure 13.7(b), then it traces an elliptical path as ϕ varies from 0 to $\pi/2$.

(c) Parameterize the coordinates of P in Figure 13.7(b) in terms of θ.

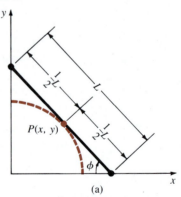

(a)

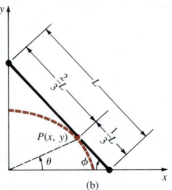

(b)

Figure 13.7

31. As shown in Figure 13.8, a piston is attached by means of a rod of length L to a circular crank mechanism of radius r. Parameterize the coordinates of the point P in terms of the angle ϕ.

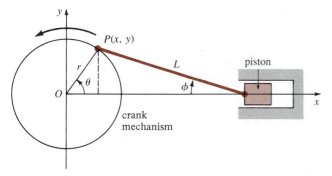

Figure 13.8

32. A point Q traces out a circular path of radius r and a point P moves in a manner as shown in Figure 13.9. If R is constant, find parametric equations of the path traced by P. This curve is called an **epitrochoid**. (Those knowledgeable about automobiles might recognize the curve traced by P as the shape of the rotor housing in the rotary or Wankel engine.)

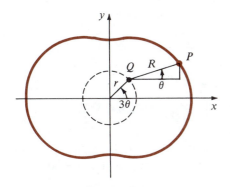

Figure 13.9

33. Consider a circle of radius a, which is tangent to the x-axis at the origin O. Let B be a point on a horizontal line through $(O, 2a)$ and let the line segment OB cut the circle at point A. As shown in Figure 13.10, the projection of AB on the vertical gives the line segment BP. Find parametric equations of the path traced by the point P as A varies around the circle. The curve is called the **Witch of Agnesi**.*

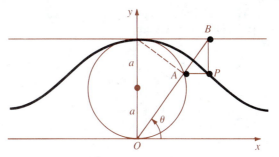

Figure 13.10

34. In Problem 33, eliminate the parameter and show that the curve has the Cartesian equation $y = 8a^3/(x^2 + 4a^2)$.

35. Use Figure 13.11 to show that parametric equations of a **cycloid** are given by

$$x = a(\theta - \sin \theta), \quad y = a(1 - \cos \theta)$$

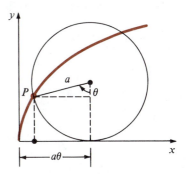

Figure 13.11

Maria Agnesi

Maria Gaetana Agnesi (1718–1799) Agnesi was an Italian mathematician and philosopher. The curve that bears her name was not discovered by her but was first studied by Pierre Fermat and Guido Grandi (1672–1742). Grandi called the curve "versoria," which is Latin for a certain kind of nautical rope. The curve "versoria" appeared in her two-volume text *Analytical Institutions*, which was published in 1748. This text on analytical geometry and calculus proved to be so popular that it was soon translated into English. The translator confused the word *versoria* with the Italian word *versiera*, which means "female goblin." In English "female goblin" became a "witch."

36. Use Figure 13.12 to show that parametric equations of a **hypocycloid** are

$$x = (b - a) \cos \theta + a \cos \frac{b - a}{a} \theta$$

$$y = (b - a) \sin \theta - a \sin \frac{b - a}{a} \theta$$

Use these equations to show that parametric equations of a **hypocycloid of four cusps** are

$$x = b \cos^3 \theta, \quad y = b \sin^3 \theta$$

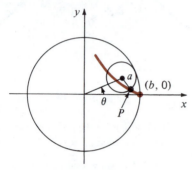

Figure 13.12

37. In Problem 36, eliminate the parameter and obtain a Cartesian equation for the hypocycloid of four cusps.

38. Use Figure 13.13 to show that parametric equations of an **epicycloid** are given by

$$x = (a + b) \cos \theta - a \cos \frac{a + b}{a} \theta$$

$$y = (a + b) \sin \theta - a \sin \frac{a + b}{a} \theta$$

Use these equations to show that parametric equations of an **epicycloid of three cusps** are

$$x = 4a \cos \theta - a \cos 4\theta, \quad y = 4a \sin \theta - a \sin 4\theta$$

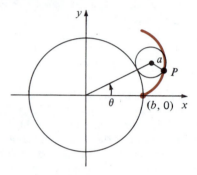

Figure 13.13

13.2 Slope of Tangent Lines, Arc Length

As with graphs of functions $y = f(x)$, we can often obtain useful information about a curve C defined parametrically by examining the derivative dy/dx.

Slope Let $x = f(t)$ and $y = g(t)$ be parametric equations of a smooth curve C. The **slope** of the tangent line at a point $P(x, y)$ on C is given by dy/dx. To calculate this derivative, we form

$$\Delta x = f(t + \Delta t) - f(t) \quad \text{and} \quad \Delta y = g(t + \Delta t) - g(t)$$

and

$$\frac{\Delta y}{\Delta x} = \frac{\Delta y / \Delta t}{\Delta x / \Delta t}$$

so that

$$\lim_{\Delta t \to 0} \frac{\Delta y}{\Delta x} = \frac{\lim_{\Delta t \to 0} \Delta y / \Delta t}{\lim_{\Delta t \to 0} \Delta x / \Delta t}$$

if the limit of the denominator is not zero. We summarize the result.

If $x = f(t)$ and $y = g(t)$ define a smooth curve C, then the slope of the tangent line at a point $P(x, y)$ on C is

$$\frac{dy}{dx} = \frac{dy/dt}{dx/dt} = \frac{g'(t)}{f'(t)} \qquad (13.4)$$

provided that $f'(t) \neq 0$.

Example 1

Find an equation of the tangent line to the curve $x = 1/(t^2 + 1)$, $y = t^3$ at the point corresponding to $t = -1$.

Solution   We first find the slope dy/dx of the tangent line. Since

$$\frac{dx}{dt} = -\frac{2t}{(t^2 + 1)^2}$$

$$\frac{dy}{dt} = 3t^2$$

it follows from (13.4) that $\dfrac{dy}{dx} = \dfrac{dy/dt}{dx/dt} = \dfrac{3t^2}{-2t/(t^2 + 1)^2} = -\dfrac{3}{2}t(t^2 + 1)^2$

Thus, at $t = -1$ we have

$$\left.\frac{dy}{dx}\right|_{t=-1} = -\frac{3}{2}(-1)(2)^2 = 6$$

By substituting $t = -1$ back into the original parametric equations we find the point of tangency to be $(\frac{1}{2}, -1)$. Hence, an equation of the tangent line is

$$y - (-1) = 6(x - \tfrac{1}{2}) \quad \text{or} \quad y = 6x - 4$$

Horizontal and Vertical Tangents   At a point (x, y) on a curve C at which $dy/dt = 0$ and $dx/dt \neq 0$, the tangent line is necessarily **horizontal** because $dy/dx = 0$ at that point. On the other hand, at a point at which $dx/dt = 0$ and $dy/dt \neq 0$, the tangent line is **vertical**. When both dy/dt and dx/dt are zero at a point, we can draw no immediate conclusion about the tangent line.

Example 2

Graph the curve that has the parametric equations

$$x = t^2 - 4, \quad y = t^3 - 3t$$

Solution

x-intercepts: $y = 0$ implies $t(t^2 - 3) = 0$ at $t = 0$, $t = -\sqrt{3}$, and $t = \sqrt{3}$.

y-intercepts: $x = 0$ implies $t^2 - 4 = 0$ at $t = -2$ and $t = 2$.

Horizontal tangents: $\dfrac{dy}{dt} = 3t^2 - 3$; $\dfrac{dy}{dt} = 0$ implies $3(t^2 - 1) = 0$ at $t = -1$ and $t = 1$. Note that $dx/dt \neq 0$ at $t = \pm 1$.

Vertical tangents: $\dfrac{dx}{dt} = 2t$; $\dfrac{dx}{dt} = 0$ implies $2t = 0$ at $t = 0$. Note that $dy/dt \neq 0$ at $t = 0$.

The points (x, y) on the curve corresponding to these values of the parameter are summarized in the following table.

t	-2	$-\sqrt{3}$	-1	0	1	$\sqrt{3}$	2
x	0	-1	-3	-4	-3	-1	0
y	-2	0	2	0	-2	0	2
	y-int	x-int	hor tan	ver tan, x-int	hor tan	x-int	y-int

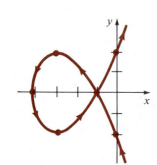

Figure 13.14

A curve plotted through these points, consistent with the orientation and tangent information, is illustrated in Figure 13.14.

The graph of a differentiable function $y = f(x)$ can have only one tangent line at a point on its graph. In contrast, since a curve defined parametrically may not be the graph of a function, it is possible that such a curve may have more than one tangent line at a point. In Example 2, for $t = -\sqrt{3}$ and $t = \sqrt{3}$ we get $(-1, 0)$. This means the curve intersects itself at that point. Now, since

$$\left.\frac{dy}{dx}\right|_{t=-\sqrt{3}} = -\sqrt{3} \quad \text{and} \quad \left.\frac{dy}{dx}\right|_{t=\sqrt{3}} = \sqrt{3}$$

we conclude that there are two tangent lines at $(-1, 0)$. This fact should be apparent in Figure 13.14.

Higher-order Derivatives   Higher-order derivatives can be found in exactly the same manner as dy/dx. Suppose (13.4) is written as

$$\frac{d}{dx}(\ \ ) = \frac{d(\ \ )/dt}{dx/dt} \tag{13.5}$$

If $y' = dy/dx$ is a differentiable function of t, it follows from (13.5) by replacing $(\ \ )$ by y' that

$$\frac{d^2y}{dx^2} = \frac{d}{dx}y' = \frac{dy'/dt}{dx/dt}$$

Similarly, if $y'' = d^2y/dx^2$ is a differentiable function of t, then the third derivative is

$$\frac{d^3y}{dx^3} = \frac{d}{dx}y'' = \frac{dy''/dt}{dx/dt}$$

and so on.

Example 3

Find d^3y/dx^3 for the curve given by $x = 4t + 6$, $y = t^2 + t - 2$.

Solution

$$\frac{dy}{dx} = \frac{dy/dt}{dx/dt} = \frac{2t+1}{4}$$

$$\frac{d^2y}{dx^2} = \frac{dy'/dt}{dx/dt} = \frac{2/4}{4} = \frac{1}{8}$$

$$\frac{d^3y}{dx^3} = \frac{dy''/dt}{dx/dt} = \frac{0}{4} = 0$$

Inspection of Example 3 shows that the curve has a horizontal tangent at $t = -\frac{1}{2}$ or $(4, -\frac{9}{4})$. Furthermore, since $d^2y/dx^2 > 0$ for all t, the graph of the curve is concave upward at every point.

Length of a Curve

In Section 6.5 we were able to find the length s of the graph of a smooth function $y = f(x)$ by means of a definite integral. We can now generalize the result given in (6.12) to curves defined parametrically. Suppose $x = f(t)$, $y = g(t)$, $a \le t \le b$ are parametric equations of a smooth curve C that does not intersect itself for $a < t < b$. If P is a partition of $[a, b]$ given by

$$a = t_0 < t_1 < t_2 < \cdots < t_{n-1} < t_n = b$$

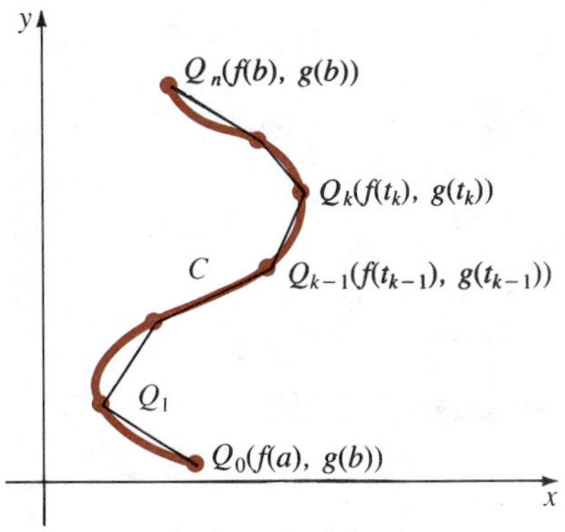

Figure 13.15

then, as shown in Figure 13.15, it seems reasonable that C can be approximated by a polygonal path through the points $Q_k(f(t_k), g(t_k))$, $k = 0, 1, \ldots, n$. Denoting the length of the line segment through Q_{k-1} and Q_k by $|Q_{k-1}Q_k|$, the approximate length of C is

$$\sum_{k=1}^{n} |Q_{k-1}Q_k| \tag{13.6}$$

where $|Q_{k-1}Q_k| = \sqrt{[f(t_k) - f(t_{k-1})]^2 + [g(t_k) - g(t_{k-1})]^2}$

Now, since f and g have continuous derivatives, the Mean Value Theorem (see Section 4.4) asserts that there exist numbers u_k^* and v_k^* in (t_{k-1}, t_k) such that

$$\begin{aligned} f(t_k) - f(t_{k-1}) &= f'(u_k^*)(t_k - t_{k-1}) \\ &= f'(u_k^*)\,\Delta t_k \end{aligned} \tag{13.7}$$

and

$$\begin{aligned} g(t_k) - g(t_{k-1}) &= g'(v_k^*)(t_k - t_{k-1}) \\ &= g'(v_k^*)\,\Delta t_k \end{aligned} \tag{13.8}$$

Substituting (13.7) and (13.8) in (13.6) and simplifying yields

$$\sum_{k=1}^{n} |Q_{k-1}Q_k| = \sum_{k=1}^{n} \sqrt{[f'(u_k^*)]^2 + [g'(v_k^*)]^2} \, \Delta t_k \qquad (13.9)$$

By taking $\|P\| \to 0$ in (13.9), we obtain a formula for the **length** s of a smooth curve:

$$s = \int_a^b \sqrt{[f'(t)]^2 + [g'(t)]^2} \, dt$$

$$= \int_a^b \sqrt{\left(\frac{dx}{dt}\right)^2 + \left(\frac{dy}{dt}\right)^2} \, dt \qquad (13.10)$$

Notice that the limit of the sum in (13.9) is not the usual definition of a definite integral since we are dealing with two numbers (u_k^* and v_k^*) rather than one in (t_{k-1}, t_k). Nevertheless, it *can* be shown rigorously that (13.10) results from (13.9) by taking $\|P\| \to 0$.

Example 4

Find the length of the curve given by $x = 4t$, $y = t^2$, $0 \le t \le 2$.

Solution   Since $f'(t) = 4$, $g'(t) = 2t$, (13.10) gives

$$s = \int_0^2 \sqrt{16 + 4t^2} \, dt$$

$$= 2 \int_0^2 \sqrt{4 + t^2} \, dt$$

Using the trigonometric substitution $t = 2 \tan \theta$, the last integral becomes

$$s = 8 \int_0^{\pi/4} \sec^3 \theta \, d\theta$$

Integration by parts leads to (see Example 3, Section 9.2),

$$s = [4 \sec \theta \tan \theta + 4 \ln|\sec \theta + \tan \theta|]_0^{\pi/4}$$

$$= 4\sqrt{2} + 4 \ln(\sqrt{2} + 1) \approx 9.1823$$

Exercises 13.2

Answers to odd-numbered problems begin on page A-41.

In Problems 1–6 find the slope of the tangent line at the point corresponding to the indicated value of the parameter.

1. $x = t^3 - t^2$, $y = t^2 + 5t$; $t = -1$

2. $x = 4/t$, $y = 2t^3 - t + 1$; $t = 2$

3. $x = \sqrt{t^2 + 1}$, $y = t^4$; $t = \sqrt{3}$

4. $x = e^{2t}$, $y = e^{-4t}$; $t = \ln 2$

5. $x = \cos^2\theta$, $y = \sin \theta$; $\theta = \pi/6$

6. $x = 2\theta - 2 \sin \theta$, $y = 2 - 2 \cos \theta$, $\theta = \pi/4$

In Problems 7 and 8 find an equation of the tangent line to the given curve at the point corresponding to the indicated value of the parameter.

7. $x = t^3 + 3t$, $y = 6t^2 + 1$; $t = -1$

8. $x = 2t + 4$, $y = t^2 + \ln t$; $t = 1$

In Problems 9 and 10 find an equation of the tangent line to the given curve at the indicated point.

9. $x = t^2 + t$, $y = t^2$; (2, 4)

10. $x = t^4 - 9$, $y = t^4 - t^2$; (0, 6)

11. What is the slope of the tangent line to the curve given by $x = 4 \sin 2t$, $y = 2 \cos t$, $0 \le t \le 2\pi$, at the point $(2\sqrt{3}, 1)$?

12. A curve C has parametric equations $x = t^2$, $y = t^3 + 1$. At what point on C is the tangent line given by $y + 3x - 5 = 0$?

13. A curve C has parametric equations $x = 2t - 5$, $y = t^2 - 4t + 3$. Find an equation of the tangent line to C that is parallel to the line $y = 3x + 1$.

14. Verify that the curve given by $x = \cos \theta - 2/\pi$, $y = \sin \theta - 2\theta/\pi$, $-\pi \le \theta \le \pi$, intersects itself. Find equations of tangent lines at the point of intersection.

In Problems 15–18 determine the points on the given curve at which the tangent line is either horizontal or vertical. Graph the curve.

15. $x = t^3 - t$, $y = t^2$

16. $x = \frac{1}{8}t^3 + 1$, $y = t^2 - 2t$

17. $x = t - 1$, $y = t^3 - 3t^2$

18. $x = \sin t$, $y = \cos 3t$, $0 \le t \le 2\pi$

In Problems 19–22 find dy/dx, d^2y/dx^2, and d^3y/dx^3.

19. $x = 3t^2$, $y = 6t^3$

20. $x = \cos t$, $y = \sin t$

21. $x = e^{-t}$, $y = e^{2t} + e^{3t}$

22. $x = \frac{1}{2}t^2 + t$, $y = \frac{1}{2}t^2 - t$

23. Use d^2y/dx^2 to determine the intervals of the parameter for which the curve in Problem 16 is concave upward and intervals for which it is concave downward.

24. Use d^2y/dx^2 to determine whether the curve given by $x = 2t + 5$, $y = 2t^3 + 6t^2 + 4t$ has any points of inflection.

In Problems 25–30 find the length of the given curve.

25. $x = \frac{5}{3}t^3 + 2$, $y = 4t^3 + 6$; $0 \le t \le 2$

26. $x = \frac{1}{3}t^3$, $y = \frac{1}{2}t^2$; $0 \le t \le \sqrt{3}$

27. $x = e^t \sin t$, $y = e^t \cos t$; $0 \le t \le \pi$

28. One arch of the cycloid:

$$x = a(\theta - \sin \theta), \quad y = a(1 - \cos \theta), \quad 0 \le \theta \le 2\pi$$

29. One arch of the hypocycloid of four cusps:

$$x = b \cos^3\theta, \quad y = b \sin^3\theta, \quad 0 \le \theta \le \pi/2$$

30. One arch of the epicycloid of three cusps:

$$x = 4a \cos \theta - a \cos 4\theta,$$
$$y = 4a \sin \theta - a \sin 4\theta, \quad 0 \le \theta \le 2\pi/3$$

Miscellaneous Problems

31. Use $dy/dx = (dy/dt)/(dx/dt)$ and

$$s = \int_{x_1}^{x_2} \sqrt{1 + [f'(x)]^2} \, dx$$

to derive (13.10).

32. Let C be a curve described by $y = F(x)$, where F is a continuous nonnegative function on $x_1 \le x \le x_2$. Show that if C is given parametrically by $x = f(t)$, $y = g(t)$, $a \le t \le b$, f' and g continuous, then the **area** under the graph of C is $\int_a^b g(t)f'(t) \, dt$.

33. Use Problem 32 to show that the area under one arch of the cycloid in Figure 13.5(a) is three times the area of the circle.

13.3 Polar Coordinate System

Up to now we have been using a rectangular coordinate system to specify a point P in the plane. We can regard this system as a grid of horizontal and vertical lines. The coordinates of P are determined by the intersection of two lines, one perpendicular to a horizontal reference line called the *x*-axis and the other perpendicular to a vertical reference line called the *y*-axis. As an alternative, in **polar coordinates**, a point *P* can be described by means of a grid of circles centered at a point *O*, called the **pole**, and straight lines or rays emanating from *O*. We take as a reference axis a horizontal half-line directed to the right of *O* and call it the **polar axis**. By specifying a directed distance *r* from *O* and an angle *θ*, measured in radians, whose initial side is the polar axis and whose terminal side is the ray *OP*, the coordinates of the point *P* are then (*r*, *θ*). See Figure 13.16.

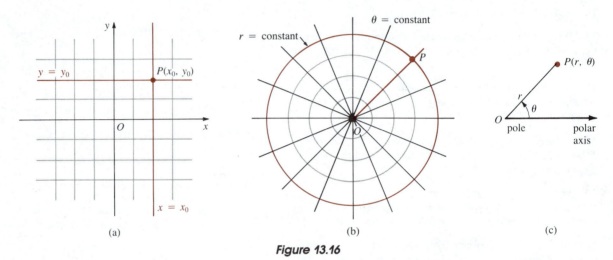

Figure 13.16

Conventions   In polar coordinates we adopt the following conventions.

(*i*)   Angles *θ* > 0 are measured counterclockwise from the polar axis, whereas angles *θ* < 0 are measured clockwise.

(*ii*) To graph a point (−*r*, *θ*), −*r* < 0, measure |*r*| units along the ray *θ* + *π*.

(*iii*) The coordinates of the pole *O* are (0, *θ*), *θ* any angle.

────────── **Example 1** ──────────────────────────────────

Graph the points whose polar coordinates are given.

(*a*) $\left(4, \dfrac{\pi}{6}\right)$ (*b*) $\left(2, -\dfrac{\pi}{4}\right)$ (*c*) $\left(-3, \dfrac{3\pi}{4}\right)$

Solution

(*a*)   Measure 4 units along the ray *π*/6. See Figure 13.17(a).

(*b*)   Measure 2 units along the ray −*π*/4. See Figure 13.17(b).

(c)   Measure 3 units along the ray $3\pi/4 + \pi = 7\pi/4$. Equivalently, we can measure 3 units along the ray $3\pi/4$ extended *backward* through the pole. Note carefully in Figure 13.17(c) that the point $(-3, 3\pi/4)$ is not in the same quadrant as the terminal side of the given angle.

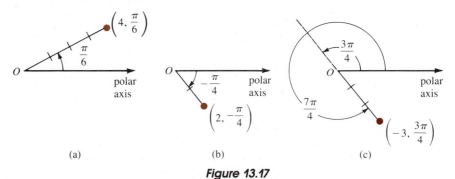

Figure 13.17

Unlike the rectangular coordinate system, the description of a point in polar coordinates is not unique. This is an immediate consequence of the fact that

$$(r, \theta) \quad \text{and} \quad (r, \theta + 2n\pi), \qquad n \text{ an integer}$$

are equivalent. To compound the problem, negative values of r can be used.

_____ **Example 2** _____

The following are some alternative representations of the point $(2, \pi/6)$:

$$\left(2, \frac{13\pi}{6}\right), \qquad \left(2, -\frac{11\pi}{6}\right), \qquad \left(-2, \frac{7\pi}{6}\right), \qquad \left(-2, -\frac{5\pi}{6}\right)$$

Conversion of Polar Coordinates to Rectangular Coordinates   By superimposing a rectangular coordinate system on a polar coordinate system, as shown in Figure 13.18, a polar description of a point can be converted to rectangular coordinates by using

$$x = r \cos \theta, \qquad y = r \sin \theta \qquad (13.11)$$

These values hold for any value of r.

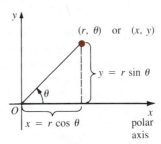

Figure 13.18

_____ **Example 3** _____

Convert $(2, \pi/6)$ in polar coordinates to rectangular coordinates.

Solution With $r = 2$, $\theta = \pi/6$, we have from (13.11)

$$x = 2 \cos\frac{\pi}{6} = 2\left(\frac{\sqrt{3}}{2}\right) = \sqrt{3}$$

$$y = 2 \sin\frac{\pi}{6} = 2\left(\frac{1}{2}\right) = 1$$

Thus, $(2, \pi/6)$ is equivalent to $(\sqrt{3}, 1)$ in rectangular coordinates.

Conversion of Rectangular Coordinates to Polar Coordinates

It should be evident from Figure 13.18 that x, y, r, and θ are also related by

$$r^2 = x^2 + y^2, \qquad \tan \theta = \frac{y}{x} \tag{13.12}$$

These latter equations are used to convert the rectangular coordinates (x, y) to polar coordinates (r, θ).

_____ **Example 4** _____

Convert $(-1, 1)$ in rectangular coordinates to polar coordinates.

Solution With $x = -1$, $y = 1$, we have from (13.12)

$$r^2 = 2 \quad \text{and} \quad \tan \theta = -1$$

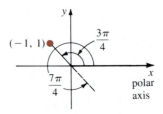

Figure 13.19

Now, $r = \pm\sqrt{2}$, and two, of many possible, angles that satisfy $\tan \theta = -1$ are $3\pi/4$ and $7\pi/4$. From Figure 13.19 we see that two representations for the given point are

$$\left(\sqrt{2}, \frac{3\pi}{4}\right) \quad \text{and} \quad \left(-\sqrt{2}, \frac{7\pi}{4}\right)$$

Caution: Note in Example 4 that $(-\sqrt{2}, 3\pi/4)$ and $(\sqrt{2}, 7\pi/4)$ are *not* polar representations of $(-1, 1)$.* In other words, we cannot pair just *any* angle θ and *any* value r that satisfy (13.12); these solutions must also be consistent with (13.11).

_____

*$(-\sqrt{2}, 3\pi/4)$ and $(\sqrt{2}, 7\pi/4)$ represent the rectangular coordinates $(1, -1)$.

By virtue of the equations in (13.11), a Cartesian equation can often be expressed as a polar equation $r = f(\theta)$.

Example 5

Find a polar equation that has the same graph as the circle $x^2 + y^2 = 4y$.

Solution Using $x = r \cos \theta$, $y = r \sin \theta$, we write

$$r^2 \cos^2\theta + r^2 \sin^2\theta = 4r \sin \theta$$
$$r^2(\cos^2\theta + \sin^2\theta) = 4r \sin \theta$$
$$r(r - 4 \sin \theta) = 0$$

The latter equation implies that

$$r = 0 \quad \text{or} \quad r = 4 \sin \theta$$

Since $r = 0$ determines *only* the pole O, we conclude that a polar equation of the circle is $r = 4 \sin \theta$. Observe that relative to this last equation, the coordinates of the pole can be taken to be $(0, \pi)$.

Example 6

Find a polar equation that has the same graph as the parabola $x^2 = 8(2 - y)$.

Solution

$$r^2 \cos^2\theta = 8(2 - r \sin \theta)$$
$$r^2(1 - \sin^2\theta) = 16 - 8r \sin \theta$$
$$r^2 = r^2 \sin^2\theta - 8r \sin \theta + 16$$
$$r^2 = (r \sin \theta - 4)^2$$
$$r = \pm(r \sin \theta - 4)$$

Solving for r gives

$$r = \frac{4}{1 + \sin \theta} \quad \text{or} \quad r = \frac{-4}{1 - \sin \theta} \tag{13.13}$$

Since replacement of (r, θ) by $(-r, \theta + \pi)$ in the second equation of (13.13) yields the first equation,* we may simply take the polar equation of the parabola to be $r = 4/(1 + \sin \theta)$.

---

*Remember that by convention (*ii*), (r, θ) and $(-r, \theta + \pi)$ represent the same point.

_____ **Example 7** _____

Find a Cartesian equation that has the same graph as the polar equation $r^2 = 9 \cos 2\theta$.

Solution   First, use the trigonometric identity for the cosine of a double angle:

$$r^2 = 9(\cos^2\theta - \sin^2\theta)$$

Then, from $r^2 = x^2 + y^2$, $\cos\theta = x/r$, $\sin\theta = y/r$, we have

$$x^2 + y^2 = 9\left(\frac{x^2}{x^2 + y^2} - \frac{y^2}{x^2 + y^2}\right)$$

or $$(x^2 + y^2)^2 = 9(x^2 - y^2)$$

The next section will be devoted to graphing polar equations.

_____ **Exercises 13.3** _____

Answers to odd-numbered problems begin on page A-42.

In Problems 1–4 graph the point whose polar coordinates are given.

1. $(2, \pi)$ **2.** $(-4, \pi/3)$

3. $(4, -3\pi/2)$ **4.** $(-5, -\pi/6)$

In Problems 5–10 find alternative polar coordinate representations of the given point that satisfy
(a) $r > 0$, $\theta < 0$;   **(b)** $r > 0$, $\theta > 2\pi$;
(c) $r < 0$, $\theta > 0$;   and   **(d)** $r < 0$, $\theta < 0$.

5. $(6, 3\pi/4)$ **6.** $(10, \pi/2)$

7. $(2, 2\pi/3)$ **8.** $(5, \pi/4)$

9. $(1, \pi/6)$ **10.** $(3, 7\pi/6)$

In Problems 11–16 find the rectangular coordinates of each point whose polar coordinates are given.

11. $(-1, 2\pi/3)$ **12.** $(1/2, 7\pi/4)$

13. $(-7, -\pi/3)$ **14.** $(\sqrt{3}, -11\pi/6)$

15. $(4, 5\pi/4)$ **16.** $(-5, \pi/2)$

In Problems 17–24 find polar coordinates that satisfy
(a) $r > 0$, $-\pi \le \theta \le \pi$ and **(b)** $r < 0$, $-\pi \le \theta \le \pi$ of each point whose rectangular coordinates are given.

17. $(-3, -3)$ **18.** $(1, 1)$

19. $(\sqrt{3}, -1)$ **20.** $(\sqrt{2}, \sqrt{6})$

21. $(0, -5)$ **22.** $(7, 0)$

23. $(-4, 3)$ **24.** $(1, 2)$

In Problems 25–34 find a polar equation that has the same graph as the given Cartesian equation.

25. $y = 5$ **26.** $x + 1 = 0$

27. $y = 7x$ **28.** $3x + 8y + 6 = 0$

29. $y^2 = -4x + 4$ **30.** $x^2 - 12y - 36 = 0$

31. $x^2 + y^2 = 36$ **32.** $x^2 - y^2 = 25$

33. $x^2 + y^2 + x = \sqrt{x^2 + y^2}$

34. $x^3 + y^3 - xy = 0$

35. Show that a polar equation that has the same graph as a Cartesian equation of a circle through the origin with center at $(a/2, 0)$ is $r = a \cos\theta$.

36. Show that a polar equation that has the same graph as $x^2 + y^2 = ay$ is $r = a \sin\theta$. What is the graph?

In Problems 37–48 find a Cartesian equation that has the same graph as the given polar equation.

37. $r = 2 \sec\theta$ **38.** $r \cos\theta = -4$

39. $r = 6 \sin 2\theta$ **40.** $2r = \tan\theta$

41. $r^2 = 4 \sin 2\theta$ **42.** $r^2 \cos 2\theta = 16$

43. $r + 5 \sin\theta = 0$ **44.** $r = 2 + \cos\theta$

45. $r = \dfrac{2}{1 + 3 \cos\theta}$ **46.** $r(4 - \sin\theta) = 10$

47. $r = \dfrac{5}{3 \cos\theta + 8 \sin\theta}$ **48.** $r = 3 + 3 \sec\theta$

Miscellaneous Problems

49. In Problems 43 and 44 of Section 12.2, the eccentricity of the ellipse $x^2/a^2 + y^2/b^2 = 1$ was defined to be $e = c/a$, where $c^2 = a^2 - b^2$. Show that a polar equation that has the same graph as the ellipse is $r^2(1 - e^2 \cos^2\theta) = b^2$.

50. Show that the distance $d(P_1, P_2)$ between two points whose polar coordinates are $P_1(r_1, \theta_1)$ and $P_2(r_2, \theta_2)$ is given by

$$d(P_1, P_2) = [r_1^2 + r_2^2 - 2r_1r_2\cos(\theta_2 - \theta_1)]^{1/2}$$

51. The two lever arms shown in Figure 13.20 rotate about a common hub so that $\theta_1 = 4t$ and $\theta_2 = 6t$, where t is measured in seconds. At what rate is the distance between the ends of the arms changing when $t = \pi/4$?

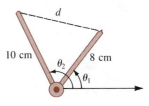

Figure 13.20

52. A line L, which is not through the pole, is completely determined by a point and a direction. Suppose $(p, \alpha), p > 0$, are polar coordinates of a point Q. Show that a polar equation of a line through Q and perpendicular to the line segment of length p, given in Figure 13.21, is $r \cos(\theta - \alpha) = p$.

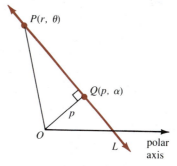

Figure 13.21

53. Show that a Cartesian equation for the line L in Problem 52 is $x \cos \alpha + y \sin \alpha = p$.

54. Use the result of Problem 53 to find the distance from the origin to the line $4x - 3y = 9$. (*Hint*: Divide the equation by $\sqrt{4^2 + 3^2} = 5$. Explain this.)

13.4 Graphs of Polar Equations

The graph of a polar equation $r = f(\theta)$ is the set of points P with *at least* one set of coordinates that satisfies the equation.

We begin our consideration of graphing polar equations with the polar analogue of the simple Cartesian equation $y = x$.

_____ **Example 1** _____

Graph $r = \theta$.

Solution   As $\theta \geq 0$ increases, r increases and the points (r, θ) wind around the pole in a counterclockwise manner. This is illustrated by the colored portion of the graph in Figure 13.22. The black portion of the graph is obtained by plotting points for $\theta < 0$.

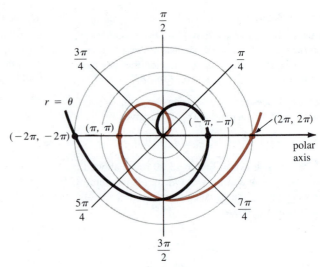

Figure 13.22

Many graphs in polar coordinates are given, for good reason, special names. The graph given in Example 1 is a special case of $r = a\theta$. A graph of this equation is called a **spiral of Archimedes**.

In addition to basic point plotting, *symmetry* can often be utilized to graph a polar equation.

Symmetry To facilitate graphing and discussion of graphs of polar equations, we shall, as shown in Figure 13.23, superimpose rectangular coordinates over the polar coordinate system. As shown in the figure, a polar graph can have three types of symmetry. A polar graph is

(*i*) **symmetric with respect to the y-axis** if, whenever (r, θ) is a point on the graph, $(r, \pi - \theta)$ is also a point on the graph;

(*ii*) **symmetric with respect to the x-axis** if, whenever (r, θ) is a point on the graph, $(r, -\theta)$ is also a point on the graph; and

(*iii*) **symmetric with respect to the origin** if, whenever (r, θ) is a point on the graph, $(-r, \theta)$ is also a point on the graph.

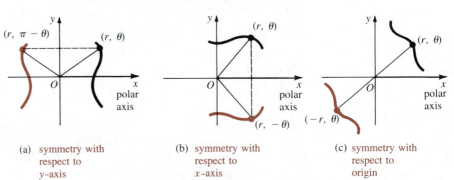

(a) symmetry with respect to y-axis

(b) symmetry with respect to x-axis

(c) symmetry with respect to origin

Figure 13.23

Hence, the graph of a polar equation is symmetric with respect to

(*i*)   the **y-axis** if replacing θ by $\pi - \theta$ results in an equivalent equation;

(*ii*)   the **x-axis** if replacing θ by $-\theta$ results in an equivalent equation;

(*iii*)   the **origin** if replacing r by $-r$ results in an equivalent equation.

Example 2

Since $\sin(\pi - \theta) = \sin \theta$, the graph of $r = 3 - 3 \sin \theta$ is symmetric with respect to the *y*-axis. Replacing, in turn, θ by $-\theta$ and r by $-r$ fail to give the original equation. Hence, no conclusion can be drawn regarding additional symmetries of the graph.

Example 3

Graph $r = 3 - 3 \sin \theta$.

Solution   By using the symmetry information from Example 2 and plotting the points that correspond to the data in the following table, we obtain the graph given in Figure 13.24.

θ	0	$\pi/6$	$\pi/3$	$\pi/2$	$2\pi/3$	$5\pi/6$	π	$7\pi/6$	$4\pi/3$	$3\pi/2$	$5\pi/3$	$11\pi/6$	2π
r	3	$\frac{3}{2}$	0.4	0	0.4	$\frac{3}{2}$	3	$\frac{9}{2}$	5.6	6	5.6	$\frac{9}{2}$	3

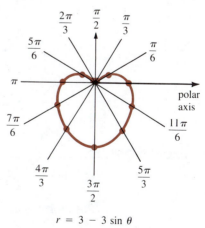

$r = 3 - 3 \sin \theta$

Figure 13.24

Cardioids   The equation in Example 3 is a member of a family of polar equations all having a "heart-shaped" graph that passes through the origin. A graph of any polar equation of the form

$$r = a \pm a \sin \theta \quad \text{or} \quad r = a \pm a \cos \theta$$

is called a **cardioid**. The only difference in the graphs of these four equations is their symmetry with respect to the y-axis ($r = a \pm a \sin \theta$) or the x-axis ($r = a \pm a \cos \theta$). See Figure 13.25.

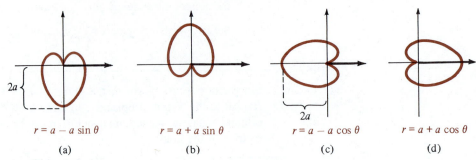

| $r = a - a \sin \theta$ | $r = a + a \sin \theta$ | $r = a - a \cos \theta$ | $r = a + a \cos \theta$ |
| (a) | (b) | (c) | (d) |

Figure 13.25

Limaçons   Cardioids are special cases of polar curves known as **limaçons**:

$$r = a \pm b \sin \theta \quad \text{or} \quad r = a \pm b \cos \theta$$

When $|a| > |b|$, a limaçon is similar to a cardioid but does not pass through the origin. When $|b| > |a|$, a limaçon has an interior loop. Both types of curves are illustrated in Figure 13.26.

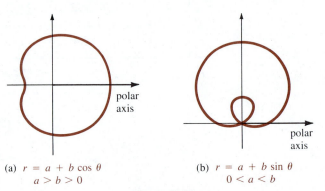

(a) $r = a + b \cos \theta$
    $a > b > 0$

(b) $r = a + b \sin \theta$
    $0 < a < b$

Figure 13.26

___

Example 4 _____

Graph $r = 2 \cos 2\theta$.

Solution   Since

$$\cos(-2\theta) = \cos 2\theta \quad \text{and} \quad \cos 2(\pi - \theta) = \cos 2\theta$$

we conclude that the graph is symmetric with respect to both the x- and y-axes. Although the given equation has period π,* a moment of reflection should

_____

*This does not mean that the graph is necessarily *complete* after graphing points for $0 \le \theta \le \pi$. Rather, it means the values of r repeat for $\pi \le \theta \le 2\pi$. In this case the graph is complete on $[0, 2\pi]$.

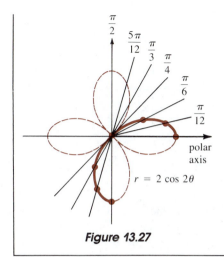

Figure 13.27

convince you that we need only consider $0 \leq \theta \leq \pi/2$. Using the data in the following table, we see that the dashed portion of the graph given in Figure 13.27 is that completed by symmetry. The graph is called a **rose curve with four petals**.

θ	0	$\pi/12$	$\pi/6$	$\pi/4$	$\pi/3$	$5\pi/12$	$\pi/2$
r	2	1.7	1	0	-1	-1.7	-2

Rose Curves   In general, if n is a positive integer, the graphs of

$$r = a \sin n\theta \quad \text{or} \quad r = a \cos n\theta, \qquad n \geq 2$$

are called **rose curves**. If n is odd, the number of loops or petals is n; if n is even, there are $2n$ petals. To graph a rose curve we can start by graphing one petal. To begin, we find an angle θ for which r is an extremum. This gives the center line of the petal. We then find corresponding values of θ for which the rose curve enters the pole ($r = 0$). To complete the graph we use the fact that the center lines of the petals are spaced $2\pi/n$ radians ($360/n$ degrees) apart. In Figure 13.28 we have drawn the graph of $r = a \sin 5\theta$, $a > 0$. The spacing between the center lines of the five petals is $2\pi/5$ radians ($72°$).

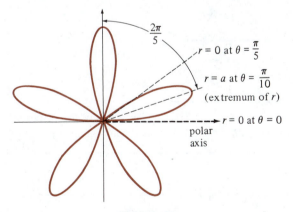

Figure 13.28

Circles   The graphs of $r = a \sin n\theta$ or $r = a \cos n\theta$ in the special case $n = 1$:

$$r = a \sin \theta \quad \text{or} \quad r = a \cos \theta$$

are **circles** passing through the origin with diameter $|a|$ and centers on the y-axis and x-axis, respectively. Figure 13.29 illustrates the graphs of $r = a \sin \theta$ and $r = a \cos \theta$ in the case $a > 0$.

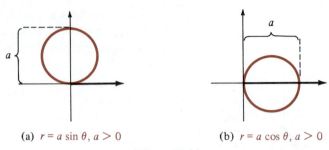

(a) $r = a \sin \theta, a > 0$ (b) $r = a \cos \theta, a > 0$

Figure 13.29

Slope of a Tangent to   Somewhat surprisingly, the slope of a tangent line to the graph of a polar equation
a Polar Graph $r = f(\theta)$ is *not* the derivative $dr/d\theta$. The slope of a tangent is still dy/dx. To
find this latter derivative, we use $r = f(\theta)$ and $x = r \cos \theta$, $y = r \sin \theta$ to write
parametric equations of the curve:

$$x = f(\theta)\cos \theta, \qquad y = f(\theta)\sin \theta$$

Then

$$\frac{dy}{dx} = \frac{dy/d\theta}{dx/d\theta}$$

_____ **Example 5** _____

Find the slope of the tangent to the graph of $r = 4 \sin 3\theta$ at $\theta = \pi/6$.

Solution   From the parametric equations

$$x = 4 \sin 3\theta \cos \theta, \qquad y = 4 \sin 3\theta \sin \theta$$

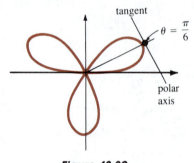

we find

$$\frac{dy}{dx} = \frac{dy/d\theta}{dx/d\theta}$$

$$= \frac{4 \sin 3\theta \cos \theta + 12 \cos 3\theta \sin \theta}{-4 \sin 3\theta \sin \theta + 12 \cos 3\theta \cos \theta}$$

$$\left. \frac{dy}{dx} \right|_{\theta = \pi/6} = -\sqrt{3}$$

The graph of the equation, which we recognize as a rose curve with three petals,
and the tangent line are illustrated in Figure 13.30.

Figure 13.30

_____ **Example 6** _____

Find the points on the graph of $r = 3 - 3 \sin \theta$ at which the tangent line is
horizontal and the points at which the tangent line is vertical.

Solution   Recall from Section 13.2 that a horizontal tangent occurs at a point
for which $dy/d\theta = 0$ and $dx/d\theta \neq 0$, whereas a vertical tangent occurs at a
point for which $dx/d\theta = 0$ and $dy/d\theta \neq 0$. Now, from the parametric equations

$$x = (3 - 3 \sin \theta)\cos \theta, \qquad y = (3 - 3 \sin \theta)\sin \theta$$

we get

$$\frac{dx}{d\theta} = (3 - 3 \sin \theta)(-\sin \theta) + \cos \theta(-3 \cos \theta)$$

$$= -3 \sin \theta + 3 \sin^2\theta - 3 \cos^2\theta$$

$$= -3 - 3 \sin \theta + 6 \sin^2\theta$$

$$= 3(2 \sin \theta + 1)(\sin \theta - 1)$$

$$\frac{dy}{d\theta} = (3 - 3 \sin \theta)\cos \theta + \sin \theta(-3 \cos \theta)$$

$$= 3 \cos \theta(1 - 2 \sin \theta)$$

From these derivatives we see that:

$$\frac{dy}{d\theta} = 0 \left(\frac{dx}{d\theta} \neq 0\right) \text{ at } \theta = \frac{\pi}{6}, \quad \theta = \frac{5\pi}{6}, \quad \text{and } \theta = \frac{3\pi}{2}$$

$$\frac{dx}{d\theta} = 0 \left(\frac{dy}{d\theta} \neq 0\right) \text{ at } \theta = \frac{7\pi}{6} \text{ and } \theta = \frac{11\pi}{6}$$

Thus, there are

Horizontal tangents at: $\left(\frac{3}{2}, \frac{\pi}{6}\right), \left(\frac{3}{2}, \frac{5\pi}{6}\right), \left(6, \frac{3\pi}{2}\right)$

Vertical tangents at: $\left(\frac{9}{2}, \frac{7\pi}{6}\right), \left(\frac{9}{2}, \frac{11\pi}{6}\right)$

These points, along with the tangent lines, are shown in Figure 13.31.

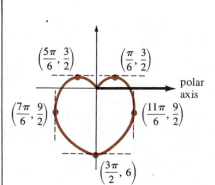

Figure 13.31

Note in Example 6 that $dy/d\theta = 0$ and $dx/d\theta = 0$ at $\theta = \pi/2$. We shall not draw any conclusion about the tangent line at the pole $(0, \pi/2)$.

Remarks

(*i*) This is a good opportunity to review the difference between necessary and sufficient conditions. The tests for symmetry of the graph of a polar equation are sufficient but not necessary. Recall, when a test for symmetry of the graph of a Cartesian equation fails, we can say definitely that the graph does not possess that particular symmetry. In rectangular coordinates, the tests for symmetry are necessary and sufficient. By way of contrast, if, say, the test for symmetry about the *x*-axis fails for a polar equation, the graph may still have that symmetry. Notice in Example 2 we stated "no conclusion" rather than "not symmetric" when a test failed. Because a point has many polar coordinates, it is possible to devise alternative tests. See Problems 51–54 in Exercises 13.4.

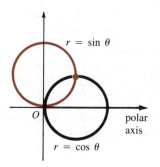

Figure 13.32

(*ii*) Problems also arise when determining where two polar curves intersect. For example, Figure 13.32 shows that the circles $r = \cos \theta$ and $r = \sin \theta$ have two points of intersection. But solving the two equations simultaneously: $\cos \theta = \sin \theta$ leads only to $(\sqrt{2}/2, \pi/4)$. The problem here is that the pole is $(0, \pi/2)$ on the first curve but is $(0, 0)$ on the second. This is analogous to the curves reaching the same point at different times. Furthermore, *a point can be on the graph of a polar equation even though its given coordinates do not satisfy the equation.* You should verify that $(2, \pi/2)$ is an alternative description of the point $(-2, 3\pi/2)$ on the graph of $r = 1 + 3 \sin \theta$. But, note that the coordinates of $(2, \pi/2)$ do not satisfy the equation.

——— Exercises 13.4 ———

Answers to odd-numbered problems begin on page A-42.

In Problems 1–24 graph the given polar equation.

1. $r = 6$

2. $r = -1$

3. $\theta = \pi/3$

4. $\theta = 5\pi/6$

5. $r = 2\theta, \theta \le 0$

6. $r = 3\theta, \theta \ge 0$

7. $r = 1 + \cos \theta$

8. $r = 5 - 5 \sin \theta$

9. $r = 2(1 + \sin \theta)$

10. $2r = 1 - \cos \theta$

11. $r = 3 + 2 \cos \theta$

12. $r = -1 + 2 \sin \theta$

13. $r = 2 + 4 \sin \theta$

14. $r = 4 - \cos \theta$

15. $r = 3 \cos 3\theta$

16. $r = 3 \sin 4\theta$

17. $r = \cos 5\theta$

18. $r = 2 \sin 9\theta$

19. $r = 6 \cos \theta$

20. $r = -2 \cos \theta$

21. $r = -3 \sin \theta$

22. $r = 5 \sin \theta$

23. $r = \tan \theta$

24. $r = \sec \theta$

In Problems 25–30 graph the **lemniscates**.

25. $r^2 = 4 \sin 2\theta$

26. $r^2 = 4 \cos 2\theta$

27. $r^2 = -25 \cos 2\theta$

28. $r^2 = -9 \sin 2\theta$

29. $r^2 = \sin \theta$

30. $r^2 = \cos \theta$

In Problems 31 and 32 graph the **spirals**.

31. $r = 2^\theta, \theta \ge 0$ (logarithmic)

32. $r\theta = \pi, \theta > 0$ (hyperbolic)

33. Graph the **cissoid of Diocles** $r = \sec \theta - \cos \theta$.

34. Graph the **conchoid of Nicomedes** $r = \csc \theta - 2$.

In Problems 35–38, find the slope of the tangent at the indicated value of θ.

35. $r = \theta; \theta = \pi/2$

36. $r = \sin \theta; \theta = \pi/6$

37. $r = 2 + 3 \cos \theta; \theta = \pi/3$

38. $r = 10 \sin 5\theta; \theta = \pi/4$

39. Find a Cartesian equation of the tangent line to the graph of $r = 1/(1 + \cos \theta)$ at $\theta = \pi/2$.

40. Find a polar equation of the tangent line to the graph of $r = 2 \cos 3\theta$ at $\theta = 2\pi/3$. (*Hint*: First find a Cartesian equation.)

In Problems 41 and 42 find the points on the graph of the given equation at which the tangent line is horizontal and the points at which the tangent line is vertical.

41. $r = 2 + 2 \cos \theta$

42. $r = \sin \theta$

In Problems 43–46, find the points of intersection of the graphs of the given pair of polar equations.

43. $r = 2$
$r = 4 \sin \theta$

44. $r = \sin \theta$
$r = \sin 2\theta$

45. $r = 1 - \cos \theta$
$r = 1 + \cos \theta$

46. $r = -\sin \theta$
$r = \cos \theta$

In Problems 47 and 48 use the fact that

$$r = f(\theta) \quad \text{and} \quad -r = f(\theta + \pi)$$

describe the same curve as an aid in finding the points of intersection of the given pair of polar equations.

47. $r = 3$
$r = 6 \sin 2\theta$

48. $r = \cos 2\theta$
$r = 1 + \cos \theta$

49. Graph the **bifolium** $r = 4 \sin \theta \cos^2\theta$ and the circle $r = \sin \theta$. Find all points of intersection of the graphs.

50. By means of carefully drawn graphs, verify that the cardioid $r = 1 + \cos \theta$ and the lemniscate $r^2 = 4 \cos \theta$ intersect at four points. Determine whether these points of intersection can be found by solving the equations simultaneously.

In Problems 51–54 identify the symmetries if the given pair of points are on the graph of $r = f(\theta)$.

51. $(r, \theta), (-r, \pi - \theta)$

52. $(r, \theta), (r, \theta + \pi)$

53. $(r, \theta), (-r, \theta + 2\pi)$

54. $(r, \theta), (-r, -\theta)$

In Problems 55 and 56, let $r = f(\theta)$ be a polar equation. Interpret the given result geometrically.

55. $f(-\theta) = f(\theta)$ (even function)

56. $f(-\theta) = -f(\theta)$ (odd function)

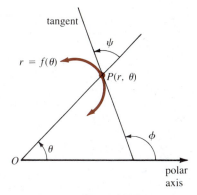

Miscellaneous Problems

57. Show that the tangent of the angle ψ, measured counterclockwise from the line OP to the tangent to the graph of $r = f(\theta)$ at $P(r, \theta)$, is given by

$$\tan \psi = \frac{r}{dr/d\theta}$$

Assume, as shown in Figure 13.33, that $\phi > \theta$. (*Hint*: How are $\tan \psi$ and $\tan(\phi - \theta)$ related? What is $\tan \phi$?)

Figure 13.33

58. Use Problem 57 to find the angle ψ for $r = 1/(1 + \cos \theta)$ at $\theta = \pi/3$.

13.5 Area in Polar Coordinates, Arc Length

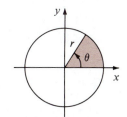

Figure 13.34

In the discussion that follows we shall use the fact that the area of a circular sector, shown in Figure 13.34, is given by

$$\text{area} = \tfrac{1}{2}r^2\theta$$

where θ is measured in radians.*

Area Suppose $r = f(\theta)$ is a nonnegative continuous function on $[\alpha, \beta]$, where $0 \le \alpha < \beta \le 2\pi$. To find the area A of the region shown in Figure 13.35(a) that is bounded by the graph of f and the rays $\theta = \alpha$ and $\theta = \beta$, we start by forming a partition P of $[\alpha, \beta]$:

$$\alpha = \theta_0 < \theta_1 < \theta_2 < \cdots < \theta_n = \beta$$

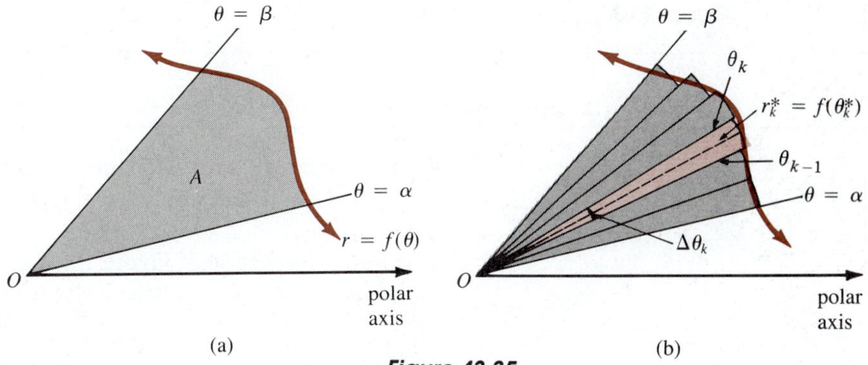

Figure 13.35

---

*Since the area of the circle is πr^2, the area of a sector determined by a central angle θ measured in radians is $(\theta/2\pi)(\pi r^2) = \tfrac{1}{2}r^2\theta$.

If θ_k^* is any number in the kth subinterval $[\theta_{k-1}, \theta_k]$, then the area of the *circular sector* of radius $r_k = f(\theta_k^*)$ indicated in Figure 13.35(b) is

$$\Delta A_k = \frac{1}{2}[f(\theta_k^*)]^2 \, \Delta \theta_k$$

where $\Delta \theta_k = \theta_k - \theta_{k-1}$ is its central angle. In turn, the Riemann sum

$$\sum_{k=1}^{n} \frac{1}{2}[f(\theta_k^*)]^2 \, \Delta \theta_k \tag{13.14}$$

gives an approximation to A. We then define A by the limit of (13.14) as the norm of the partition approaches zero:

$$A = \lim_{\|P\| \to 0} \sum_{k=1}^{n} \frac{1}{2}[f(\theta_k^*)]^2 \, \Delta \theta_k$$

$$= \int_{\alpha}^{\beta} \frac{1}{2}[f(\theta)]^2 \, d\theta$$

or simply
$$A = \frac{1}{2} \int_{\alpha}^{\beta} r^2 \, d\theta \tag{13.15}$$

Example 1

Find the area of the region that is bounded by the spiral $r = \theta$, $\theta \geq 0$, between the rays $\theta = 0$ and $\theta = 7\pi/4$.

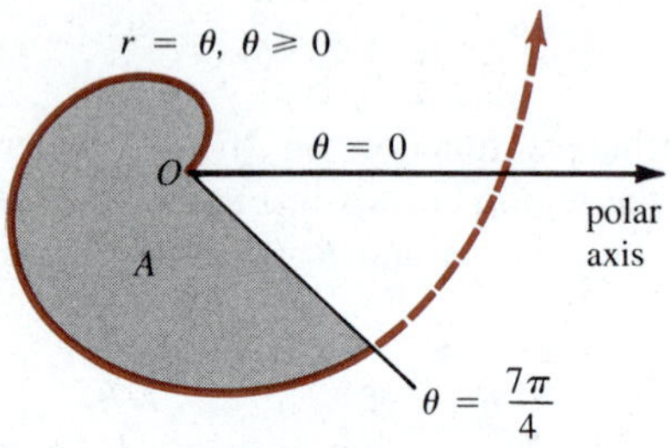

Solution   From (13.15) the area shown in Figure 13.36 is

$$A = \frac{1}{2} \int_{0}^{7\pi/4} \theta^2 \, d\theta$$

$$= \frac{1}{2} \frac{\theta^3}{3} \Bigg]_{0}^{7\pi/4}$$

$$= \frac{343}{384} \pi^3 \approx 27.70 \text{ square units}$$

$r = \theta,\ \theta \geq 0$

$\theta = 0$

polar axis

A

$\theta = \dfrac{7\pi}{4}$

Figure 13.36

Example 2

Find the area of the region that is common to the interiors of the cardioid $r = 2 - 2 \cos \theta$ and the limaçon $r = 2 + \cos \theta$.

Solution   Inspection of Figure 13.37 shows that we need two integrals. Solving the given equations simultaneously:

$$2 - 2 \cos \theta = 2 + \cos \theta \quad \text{or} \quad \cos \theta = 0$$

yields $\theta = \pi/2$ so that a point of intersection is $(2, \pi/2)$. By symmetry, it follows that

$$A = 2\left\{\frac{1}{2}\int_0^{\pi/2}(2 - 2\cos\theta)^2\,d\theta + \frac{1}{2}\int_{\pi/2}^{\pi}(2 + \cos\theta)^2\,d\theta\right\}$$

$$= 4\int_0^{\pi/2}(1 - 2\cos\theta + \cos^2\theta)\,d\theta + \int_{\pi/2}^{\pi}(4 + 4\cos\theta + \cos^2\theta)\,d\theta$$

$$= 4\int_0^{\pi/2}\left(1 - 2\cos\theta + \frac{1 + \cos 2\theta}{2}\right)d\theta + \int_{\pi/2}^{\pi}\left(4 + 4\cos\theta + \frac{1 + \cos 2\theta}{2}\right)d\theta$$

$$= 4\left[\frac{3}{2}\theta - 2\sin\theta + \frac{1}{4}\sin 2\theta\right]_0^{\pi/2} + \left[\frac{9}{2}\theta + 4\sin\theta + \frac{1}{4}\sin 2\theta\right]_{\pi/2}^{\pi}$$

$$= \frac{21}{4}\pi - 12 \approx 4.49 \text{ square units}$$

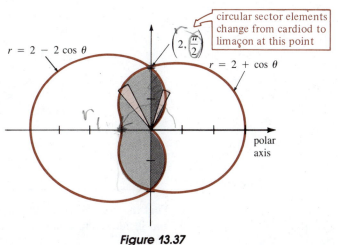

$r = 2 - 2\cos\theta$

$\left(2, \frac{\pi}{2}\right)$

circular sector elements change from cardiod to limaçon at this point

$r = 2 + \cos\theta$

polar axis

Figure 13.37

Example 3

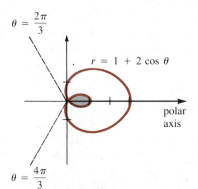

$\theta = \frac{2\pi}{3}$

$r = 1 + 2\cos\theta$

polar axis

$\theta = \frac{4\pi}{3}$

Figure 13.38

Find the area of the inner loop of the limaçon $r = 1 + 2\cos\theta$.

Solution As seen in Figure 13.38, the inner loop corresponds to $2\pi/3 \le \theta \le 4\pi/3$. Although $r \le 0$ for these values of θ, the fact that (13.15) utilizes $r^2 \ge 0$ enables us to write

$$A = \frac{1}{2}\int_{2\pi/3}^{4\pi/3}(1 + 2\cos\theta)^2\,d\theta$$

$$= \pi - \frac{3\sqrt{3}}{2} \approx 0.54 \text{ square unit}$$

Area Bounded by Two Graphs   The area A of the shaded region shown in Figure 13.39 can be found by subtracting areas. If f and g are continuous on $[\alpha, \beta]$ and $f(\theta) \ge g(\theta)$ on the interval, then the area bounded by the graphs of $r = f(\theta)$, $r = g(\theta)$, $\theta = \alpha$, and $\theta = \beta$ is

$$A = \frac{1}{2} \int_\alpha^\beta [f(\theta)]^2 \, d\theta - \frac{1}{2} \int_\alpha^\beta [g(\theta)]^2 \, d\theta$$

Written as a single integral, the area is given by

$$A = \frac{1}{2} \int_\alpha^\beta ([f(\theta)]^2 - [g(\theta)]^2) \, d\theta \qquad (13.16)$$

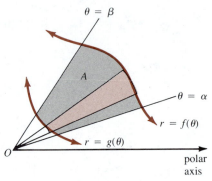

Figure 13.39

_____ **Example 4** _____

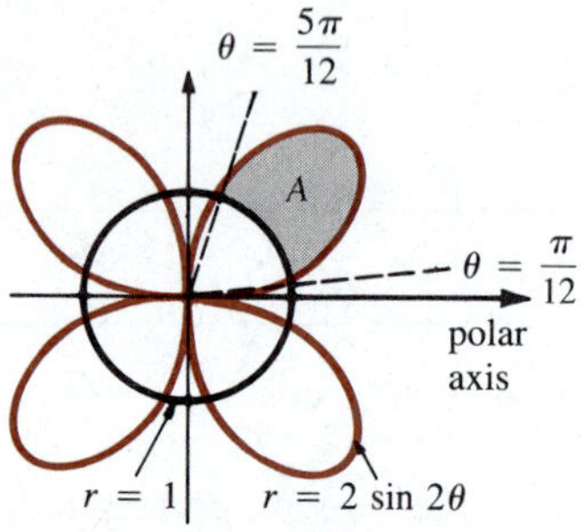

Find the area of the region in the first quadrant that is outside the circle $r = 1$ and inside the rose curve $r = 2 \sin 2\theta$.

Solution   Solving the two equations simultaneously:

$$1 = 2 \sin 2\theta \quad \text{or} \quad \sin 2\theta = \frac{1}{2}$$

implies that $2\theta = \pi/6$ and $2\theta = 5\pi/6$. Thus, two points of intersection in the first quadrant are $(1, \pi/12)$ and $(1, 5\pi/12)$. The area in question is shaded in Figure 13.40. From (13.16),

$$A = \frac{1}{2} \int_{\pi/12}^{5\pi/12} [(2 \sin 2\theta)^2 - 1^2] \, d\theta$$

$$= \frac{1}{2} \int_{\pi/12}^{5\pi/12} [4 \sin^2\theta - 1] \, d\theta$$

$$= \frac{1}{2} \int_{\pi/12}^{5\pi/12} \left[4\left(\frac{1 - \cos 4\theta}{2} \right) - 1 \right] d\theta$$

$$= \frac{1}{2} \left[\theta - \frac{1}{2} \sin 4\theta \right]_{\pi/12}^{5\pi/12}$$

$$= \frac{\pi}{6} + \frac{\sqrt{3}}{4} \approx 0.96 \text{ square unit}$$

Figure 13.40 (margin): $\theta = \frac{5\pi}{12}$, $\theta = \frac{\pi}{12}$, polar axis, A, $r = 1$, $r = 2 \sin 2\theta$

Figure 13.40

Arc Length for Polar Graphs We have seen that if $r = f(\theta)$ is the equation of a curve C in polar coordinates, then parametric equations of C are

$$x = f(\theta)\cos\,\theta, \qquad y = f(\theta)\sin\,\theta, \qquad \alpha \le \theta \le \beta$$

If f has a continuous derivative, then it is a straightforward matter to derive a formula for **arc length in polar coordinates**. Since

$$\frac{dx}{d\theta} = f'(\theta)\cos\,\theta - f(\theta)\sin\,\theta, \qquad \frac{dy}{dx} = f'(\theta)\sin\,\theta + f(\theta)\cos\,\theta$$

$$\left(\frac{dx}{d\theta}\right)^2 + \left(\frac{dy}{d\theta}\right)^2 = [f'(\theta)]^2 + [f(\theta)]^2$$

$$= \left(\frac{dr}{d\theta}\right)^2 + r^2$$

(13.10) of Section 13.2 becomes

$$s = \int_\alpha^\beta \sqrt{\left(\frac{dr}{d\theta}\right)^2 + r^2}\; d\theta \tag{13.17}$$

_____ **Example 5** _____

Find the length of the cardioid $r = 1 + \cos\,\theta$ for $0 \le \theta \le \pi$.

Solution The portion of the complete graph is shown in Figure 13.41. Now, $dr/d\theta = -\sin\,\theta$ so that

$$\left(\frac{dr}{d\theta}\right)^2 + r^2 = \sin^2\theta + (1 + 2\cos\,\theta + \cos^2\theta)$$

$$= 2 + 2\cos\,\theta$$

and

$$s = \sqrt{2}\int_0^\pi \sqrt{1 + \cos\,\theta}\; d\theta$$

To evaluate this integral, we employ the trigonometric identity $\cos^2(\theta/2) = (1 + \cos\,\theta)/2$:

$$s = 2\int_0^\pi \cos\frac{\theta}{2}\; d\theta$$

$$= 4\,\sin\frac{\theta}{2}\Bigg]_0^\pi = 4$$

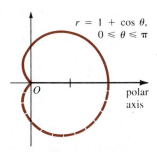

$r = 1 + \cos\,\theta,$
$0 \le \theta \le \pi$

O polar axis

Figure 13.41

It is left as an exercise for you to ponder why the length of the complete cardioid ($0 \le \theta \le 2\pi$) in Example 5 is *not* $s = 2\int_0^{2\pi}\cos(\theta/2)\,d\theta$.

Exercises 13.5

Answers to odd-numbered problems begin on page A-43.

In Problems 1–10 find the area of the region that is bounded by the graph of the given polar equation.

1. $r = 2 \sin \theta$

2. $r = 10 \cos \theta$

3. $r = 4 + 4 \cos \theta$

4. $r = 1 - \sin \theta$

5. $r = 3 + 2 \sin \theta$

6. $r = 2 + \cos \theta$

7. $r = 3 \sin 2\theta$

8. $r = \cos 4\theta$

9. $r = 2 \sin 3\theta$

10. $r = 4 \cos 5\theta$

11. Consider the lemniscate $r^2 = 9 \cos 2\theta$.

   (a) Explain why $\frac{1}{2} \int_0^{2\pi} 9 \cos 2\theta \, d\theta$ is not the area of the region bounded by the graph.

   (b) Using an appropriate integral, find the area of the region bounded by the graph.

12. Repeat Problem 11 for $r^2 = \sin 2\theta$.

In Problems 13–18 find the area of the region that is bounded by the graph of the given polar equation and the indicated rays.

13. $r = 2\theta, \ \theta \geq 0, \ \theta = 0, \ \theta = 3\pi/2$

14. $r\theta = \pi, \ \theta > 0, \ \theta = \pi/2, \ \theta = \pi$

15. $r = e^\theta, \ \theta = 0, \ \theta = \pi$

16. $r = 10e^{-\theta}, \ \theta = 1, \ \theta = 2$

17. $r = \tan \theta, \ \theta = 0, \ \theta = \pi/4$

18. $r \sin \theta = 5, \ \theta = \pi/6, \ \theta = \pi/3$

19. Find the area of the region that is outside the circle $r = 1$ and inside the rose curve $r = 2 \cos 3\theta$.

20. Find the area of the region that is common to the interiors of the circles $r = \cos \theta$ and $r = \sin \theta$.

21. Find the area of the region that is inside the circle $r = 5 \sin \theta$ and outside the limaçon $r = 3 - \sin \theta$.

22. Find the area of the region that is common to the interiors of the graphs of the equations in Problem 21.

23. Find the area of the region that is inside the cardioid $r = 4 - 4 \cos \theta$ and outside the circle $r = 6$.

24. Find the area of the region that is common to the interiors of the graphs of the equations in Problem 23.

25. Find the area of the region that is inside the limaçon $r = 1 + 2 \sin \theta$. (*Hint*: The area is *not*

$$\frac{1}{2} \int_0^{2\pi} (1 + 2 \sin \theta)^2 \, d\theta.)$$

26. For the polar equation in Problem 25, find the area that is outside the interior loop but inside the limaçon.

27. Find the area of the region that is outside the circle $r = 1$ and inside the lemniscate $r^2 = 4 \cos 2\theta$.

28. Find the area of the region that is common to the interiors of the circle $r = 4 \sin \theta$ and the cardioid $r = 1 + \sin \theta$.

In Problems 29–32 find the length of the curve for the indicated values of θ.

29. $r = e^{\theta/2}, \ 0 \leq \theta \leq 4$

30. $r = 2e^{-\theta}, \ 0 \leq \theta \leq \pi$

31. $r = \theta, \ 0 \leq \theta \leq 1$

32. $r = 3 - 3 \cos \theta, \ 0 \leq \theta \leq \pi$

Miscellaneous Problem

33. In polar coordinates the **angular momentum** of a moving particle of mass m is defined to be $L = mr^2 \, d\theta/dt$. Assume that the coordinates of the particle are (r_1, θ_1) and (r_2, θ_2) at times $t = a$ and $t = b$, $a < b$, respectively. If L is constant, show that the area A swept out by r is $A = L(b - a)/2m$.*

13.6 Conic Sections Revisited

In Chapter 12 we saw that the parabola, ellipse, and hyperbola are defined by three different properties. However, it is possible to state one general definition of a conic section that unifies these three previous definitions.

---

*When the sun is taken to be at the origin, this equation proves **Kepler's second law of planetary motion**: The radius vector joining a planet with the sun sweeps out equal areas in equal times. See page 629.

A **conic section** is a set of points P in the plane for which the ratio of the distance $d(P, F)$ from a fixed point F to the distance $d(P, Q)$ to a fixed line L is a constant.

The fixed line L is a **directrix** and the constant ratio $d(P, F)/d(P, Q)$ is called the **eccentricity** of the conic. The eccentricity is usually denoted by the letter e.*

Polar Equations of Conics

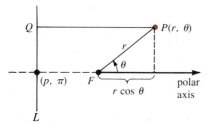

Figure 13.42

The equation

$$\frac{d(P, F)}{d(P, Q)} = e \quad \text{or} \quad d(P, F) = ed(P, Q) \tag{13.18}$$

is readily interpreted using polar coordinates. Suppose F is placed at the pole and L is p units ($p > 0$) to the left of F perpendicular to the extended polar axis. We see from Figure 13.42 that (13.18) is the same as

$$r = e(p + r \cos \theta) \quad \text{or} \quad r - er \cos \theta = ep \tag{13.19}$$

Solving for r yields

$$r = \frac{ep}{1 - e \cos \theta} \tag{13.20}$$

If the directrix is chosen to the right of F through $(p, 0)$, then the only change in (13.20) is that the negative sign in the denominator becomes a positive sign.

To see that Definition 13.2 yields the familiar equations of the conics, let us superimpose a rectangular coordinate system on the polar coordinate system with the origin at the pole and the positive x-axis coinciding with the polar axis. We then express (13.19) in rectangular coordinates and simplify:

$$\sqrt{x^2 + y^2} - ex = ep$$
$$x^2 + y^2 = e^2 x^2 + 2e^2 px + e^2 p^2$$
$$(1 - e^2)x^2 - 2e^2 px + y^2 = e^2 p^2 \tag{13.21}$$

When $e = 1$, (13.21) becomes

$$-2px + y^2 = p^2 \quad \text{or} \quad y^2 = 2p\left(x + \frac{p}{2}\right)$$

which is an equation of a parabola whose axis is the x-axis, whose vertex is at $(-p/2, 0)$, and whose focus is at the origin. By completing the square, it can be shown that the graph of (13.21) is an ellipse when $0 < e < 1$ and a hyperbola when $e > 1$. In each case F is a **focus** of the conic.

When the directrix is chosen parallel to the polar axis, then the equation of the conic is found to be either

$$r = \frac{ep}{1 + e \sin \theta} \quad \text{or} \quad r = \frac{ep}{1 - e \sin \theta}$$

*Do not confuse this with the number e of Chapter 8.

Specifically, $r = ep/(1 + e \sin \theta)$ describes a conic whose directrix is parallel to the polar axis and passes through $(p, \pi/2)$, whereas the directrix for $r = ep/(1 - e \sin \theta)$ passes through $(p, 3\pi/2)$.

We summarize the preceding discussion:

The polar equation

$$r = \frac{ep}{1 \pm e \cos \theta} \tag{13.22}$$

is a conic section with a focus at the origin and whose axis lies along the x-axis.

The polar equation

$$r = \frac{ep}{1 \pm e \sin \theta} \tag{13.23}$$

is a conic section with a focus at the origin and whose axis lies along the y-axis.

Equations (13.22) and (13.23) yield the various conics for the following choices of e:

a **parabola** if $e = 1$

an **ellipse** if $0 < e < 1$

a **hyperbola** if $e > 1$

You should compare (13.18) when $e = 1$ with Definition 12.1.

Example 1

Identify the conics

$$\text{(a)}\quad r = \frac{6}{1 + 2 \sin \theta} \qquad \text{(b)}\quad r = \frac{4}{3 - \cos \theta}$$

Solution

(a) Comparison with (13.23) enables us to make the identification $e = 2$. The conic section is a hyperbola.

(b) By dividing numerator and denominator by 3, the given equation becomes

$$r = \frac{\frac{4}{3}}{1 - \frac{1}{3} \cos \theta}$$

Thus, from (13.22) we see that $e = \frac{1}{3}$ and so the conic section is an ellipse.

Graphs   A rough **graph** of a conic defined by (13.22) or (13.23) can be obtained by knowing the orientation of its axis and by finding x- and y-intercepts and vertices. In the case of (13.22), the two vertices on the axis of an ellipse or a hyperbola occur at $\theta = 0$ and $\theta = \pi$. When $e = 1$ the denominator of (13.22) contains $1 - \cos\theta$ or $1 + \cos\theta$ and, consequently, the vertex of a parabola can occur at only one of the values: $\theta = 0$ or $\theta = \pi$. For (13.23) the vertices of an ellipse or a hyperbola occur at $\theta = \pi/2$ and $\theta = 3\pi/2$. One of the values $\theta = \pi/2$ or $\theta = 3\pi/2$ will yield the vertex of a parabola.

_____ **Example 2** _____

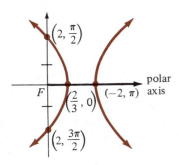

Figure 13.43

Graph $r = \dfrac{2}{1 + 2\cos\theta}$.

Solution   From (13.22) we see that $e = 2$ and so the equation describes a hyperbola whose transverse axis is horizontal (because of $\cos\theta$). In view of the foregoing discussion, we obtain:

$$\text{Vertices:} \quad (\tfrac{2}{3}, 0),\ (-2, \pi)$$

$$y\text{-intercepts:} \quad (2, \pi/2),\ (2, 3\pi/2)$$

The graph of the equation is given in Figure 13.43.

_____ **Example 3** _____

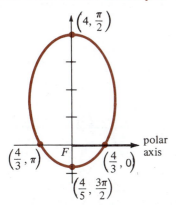

Figure 13.44

Graph $r = \dfrac{4}{3 - 2\sin\theta}$.

Solution   By writing $r = \dfrac{\frac{4}{3}}{1 - \frac{2}{3}\sin\theta}$

we see that $e = \frac{2}{3}$ and so the conic section is an ellipse whose major axis is vertical (because of $\sin\theta$). It follows that:

$$\text{Vertices:} \quad (4, \pi/2),\ (\tfrac{4}{5}, 3\pi/2)$$

$$x\text{-intercepts:} \quad (\tfrac{4}{3}, 0),\ (\tfrac{4}{3}, \pi)$$

The graph of the equation is given in Figure 13.44.

In Example 3, the rectangular coordinates of the vertices on the major axis are $(0, 4)$ and $(0, -\frac{4}{5})$. By averaging the y-coordinates of these points, we find that the center of the ellipse is $(0, \frac{8}{5})$. Now, since the distance from the center to the focus at the origin is $\frac{8}{5}$ we conclude that $c = \frac{8}{5}$. Hence, the other focus of the ellipse is at $(0, 4 - \frac{8}{5})$ or $(0, \frac{12}{5})$. Also, one-half of the length of the major axis yields the number $a = \frac{12}{5}$. Using $b^2 = a^2 - c^2$, we get $b = 4\sqrt{5}/5$ and so the vertices on the minor axis are $(-4\sqrt{5}/5, \frac{8}{5})$ and $(4\sqrt{5}/5, \frac{8}{5})$.

Example 4

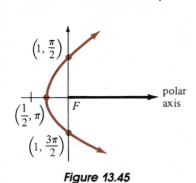

Figure 13.45

Graph $r = \dfrac{1}{1 - \cos \theta}$.

Solution   Inspection of the equation reveals that $e = 1$ and so the conic section is a parabola whose axis is horizontal (because of $\cos \theta$). Since r is undefined at $\theta = 0$, the vertex of the parabola occurs at $\theta = \pi$:

$$Vertex: \quad (\tfrac{1}{2}, \pi)$$

$$y\text{-}intercepts: \quad (1, \pi/2), (1, 3\pi/2)$$

The graph of the equation is given in Figure 13.45.

The shapes of ellipses and hyperbolas are related to the magnitudes of their eccentricities. When e is close to zero, the graph of an ellipse is nearly circular. An ellipse whose eccentricity is close to 1 is necessarily elongated. Similar remarks hold for the hyperbola. See Figure 13.46. A circle can be considered as a limiting case of (13.22) or (13.23) in which $e \to 0$ and $p \to \infty$.

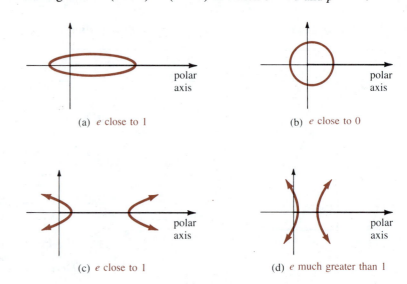

(a) *e* close to 1

(b) *e* close to 0

(c) *e* close to 1

(d) *e* much greater than 1

Figure 13.46

The orbits of seven of the nine planets have eccentricities less than 0.01 and, hence, are not far from circular. Mercury and Pluto are the exceptions, with orbits having eccentricities 0.21 and 0.25, respectively. Many asteroids and comets have highly eccentric orbits. The orbit of asteroid Hidalgo is one of the most eccentric, with $e = 0.66$. Of course, a notable case is the orbit of comet Halley with $e = 0.97$.

Orbits   The orbit of a satellite around the sun (earth or moon) is an ellipse with the sun (earth or moon) at one focus. Suppose an equation of the orbit is given by $r = ep/(1 - e \cos \theta)$, $0 < e < 1$, and that r_p is the value of r at **perihelion** (perigee or perilune) and that r_a is the value of r at **aphelion** (apogee or apolune).

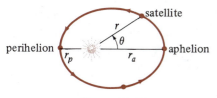

Figure 13.47

These are the points in the orbit, occurring on the *x*-axis, at which the satellite is closest and farthest, respectively, from the sun (earth or moon). See Figure 13.47. It is left as an exercise to show that the eccentricity of the orbit is related to r_p and r_a by

$$e = \frac{r_a - r_p}{r_a + r_p} \tag{13.24}$$

Example 5

Find a polar equation of the orbit of the planet Mercury around the sun if $r_p = 2.85 \times 10^7$ mi and $r_a = 4.36 \times 10^7$ mi.

Solution From (13.24) the eccentricity of Mercury's orbit is

$$e = \frac{(4.36 - 2.85) \times 10^7}{(4.36 + 2.85) \times 10^7} = 0.21$$

Hence we can write

$$r = \frac{0.21p}{1 - 0.21 \cos \theta}$$

To find *p* we note that aphelion occurs at $\theta = 0$:

$$4.36 \times 10^7 = \frac{0.21p}{1 - 0.21}$$

The last equation yields $0.21p = 3.44 \times 10^7$ or $p = 16.38 \times 10^7$. Hence a polar equation of the orbit is

$$r = \frac{3.44 \times 10^7}{1 - 0.21 \cos \theta}$$

Remark

In rectangular coordinates the eccentricity of an ellipse (hyperbola) is defined as the ratio of the distance between the foci to the length of the major axis (transverse axis). See Problems 43 and 44 of Exercises 12.2 and 12.3. You are asked to verify that $e = c/a$ in the case of $r = ep/(1 - e \cos \theta)$ for $0 < e < 1$. See Problem 27 of Exercises 13.6.

Exercises 13.6

Answers to odd-numbered problems begin on page A-43.

In Problems 1–10 determine the eccentricity, identify the conic section, and sketch its graph.

1. $r = \dfrac{2}{1 + \cos \theta}$

2. $r = \dfrac{2}{2 - \sin \theta}$

3. $r = \dfrac{15}{4 - \cos \theta}$

4. $r = \dfrac{5}{2 - 2 \sin \theta}$

5. $r = \dfrac{4}{1 + 2 \sin \theta}$

6. $r = \dfrac{12}{6 + 2 \sin \theta}$

7. $r = \dfrac{18}{3 + 6 \cos \theta}$ **8.** $r = \dfrac{6 \sec \theta}{\sec \theta - 1}$

9. $r = \dfrac{10}{5 + 4 \sin \theta}$ **10.** $r = \dfrac{2}{2 + 5 \cos \theta}$

11. Consider the ellipse whose equation is

$$r = \dfrac{6}{2 + \cos \theta}$$

Find the rectangular coordinates of the center, foci, and four vertices.

12. Consider the hyperbola whose equation is

$$r = \dfrac{8}{1 - 3 \sin \theta}$$

Find the rectangular coordinates of the center and foci. What is the length of the conjugate axis?

In Problems 13–16 find a polar equation of the conic section with a focus at the pole and with the indicated eccentricity and directrix.

13. $e = 1, r = -3 \sec \theta$ **14.** $e = \dfrac{2}{3}, r = -2 \csc \theta$

15. $e = \dfrac{5}{2}, r = 6 \csc \theta$ **16.** $e = \dfrac{1}{2}, r = 4 \sec \theta$

In Problems 17 and 18 find a polar equation of the parabola with focus at the pole and vertex at the given point.

17. $(\frac{5}{2}, \pi/2)$ **18.** $(3, \pi)$

19. Use the equation $r = ep/(1 - e \cos \theta)$ to derive the result given in (13.24).

20. A communications satellite is 12,000 km above the earth at its apogee. The eccentricity of its orbit is 0.2. Use (13.24) to find the perigee distance.

21. Determine a polar equation of the orbit of the satellite in Problem 20.

22. Find a polar equation of the orbit of the earth around the sun if $r_p = 1.47 \times 10^8$ km and $r_a = 1.52 \times 10^8$ km.

23. The eccentricity of the orbit of Halley's comet is 0.97 and the length of the major axis of its orbit is 3.34×10^9 mi. Find a polar equation of its orbit of the form $r = ep/(1 - e \cos \theta)$.

24. Use the equation obtained in Problem 23 to obtain r_p and r_a for the orbit of Halley's comet.

Miscellaneous Problems

25. (a)   Derive $r = ep/(1 + e \cos \theta)$.

    (b)   Derive $r = ep/(1 \pm e \sin \theta)$.

26. For the orientation of the directrix L shown in Figure 13.48, show that the polar equation of a conic section is $r = ep/[1 + e \cos(\theta + \phi)]$.

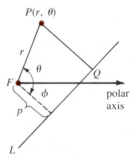

Figure 13.48

27. Consider the polar equation $r = ep/(1 - e \cos \theta)$, $0 < e < 1$.

    (a)   By completing the square, show that the graph of (13.21) is an ellipse and that F is a focus at the origin.

    (b)   Show that $e = c/a$.

28. Show that the area of the region that is bounded by the graph of the polar equation $r = ep/(1 + e \cos \theta)$, $0 < e < 1$, is $\pi e^2 p^2/(1 - e^2)^{3/2}$. (*Hint*: Use the substitution $u = \tan(\theta/2)$.)

_____ Chapter 13 Review Exercises _____

Answers to odd-numbered problems begin on page A-43.

In Problems 1–20 answer true or false.

1. If for all values of θ the points $(-r, \theta)$ and $(r, \theta + \pi)$ are on the graph of the polar equation $r = f(\theta)$, then the graph is symmetric with respect to the origin. _____

2. The graph of the curve $x = t^2, y = t^4 + 1$ is the same as the graph of $y = x^2 + 1$. _____

3. The graph of $r = 4 \cos \theta$ is a circle centered at $(2, 0)$. _____

4. $(3, \pi/6)$ and $(-3, -5\pi/6)$ are polar coordinates of the same point. _____

5. Rectangular coordinates of a point in the plane are unique. _____

6. The graph of the "rose" $r = 5 \sin 6\theta$ has six "petals." _____

7. The point $(4, 3\pi/2)$ is not on the graph of $r = 4 \cos 2\theta$ since its coordinates do not satisfy the equation. _____

8. The eccentricity of a parabola is $e = 1$. _____

9. The transverse axis of the hyperbola $r = 5/(2 + 3 \cos \theta)$ lies along the x-axis. _____

10. The graph of the ellipse $r = 90/(15 - \sin \theta)$ is nearly circular. _____

11. The rectangular coordinates of the point $(-\sqrt{2}, 5\pi/4)$ in polar coordinates are $(1, 1)$. _____

12. The graph of $r = -5 \sec \theta$ is a line. _____

13. The terminal side of the angle θ is always in the same quadrant as the point (r, θ). _____

14. The slope of the tangent to the graph of $r = e^\theta$ at $\theta = \pi/2$ is -1. _____

15. The graphs of the cardioids $r = 3 + 3 \cos \theta$ and $r = -3 + 3 \cos \theta$ are the same. _____

16. The area bounded by $r = \cos 2\theta$ is $2 \int_{-\pi/4}^{\pi/4} \cos^2 2\theta \, d\theta$. _____

17. The area bounded by $r = 2 \sin 3\theta$ is $6 \int_0^{\pi/3} \sin^2 3\theta \, d\theta$. _____

18. The area bounded by $r = 1 - 2 \cos \theta$ is $\frac{1}{2} \int_0^{2\pi} (1 - 2 \cos \theta)^2 \, d\theta$. _____

19. The area bounded by $r^2 = 36 \cos 2\theta$ is $18 \int_0^{2\pi} \cos 2\theta \, d\theta$. _____

20. The θ coordinate of a point of intersection of the graphs of $r = f(\theta)$ and $r = g(\theta)$ must satisfy the equation $f(\theta) = g(\theta)$. _____

21. Find an equation of the line that is normal to the graph of the curve $x = t - \sin t$, $y = 1 - \cos t$, $0 \le t \le 2\pi$ at $t = \pi/3$.

22. Find the length of the curve given in Problem 21.

23. Find the points on the graph of the curve $x = t^2 + 4$, $y = t^3 - 9t^2 + 2$ at which the tangent line is parallel to $6x + y = 8$.

24. Find the points on the graph of the curve $x = t^2 + 1$, $y = 2t$ at which the tangent line passes through $(1, 5)$.

25. Consider the Cartesian equation $y^2 = 4x^2(1 - x^2)$.

(a) Explain why it is necessary that $|x| \le 1$.

(b) If $x = \sin t$, the $|x| \le 1$. Find parametric equations that have the same graph as the given equation.

(c) Using parametric equations, find the points on the graph of the curve at which the tangent is horizontal.

(d) Graph the curve.

26. Find the area of the region that is outside the circle $r = 4 \cos \theta$ and inside the limaçon $r = 3 + \cos \theta$.

27. Find the area of the region that is common to the interiors of the circle $r = 3 \sin \theta$ and the cardioid $r = 1 + \sin \theta$.

28. In polar coordinates, sketch the region whose area A is described by $A = \int_0^{\pi/2} (25 - 25 \sin^2 \theta) \, d\theta$.

29. Find **(a)** a Cartesian equation, and **(b)** a polar equation, of the tangent line to the graph of $r = 2 \sin 2\theta$ at $\theta = \pi/4$.

30. Find a Cartesian equation that has the same graph as $r = \cos \theta + \sin \theta$.

31. Find a polar equation that has the same graph as $(x^2 + y^2 - 2x)^2 = 9(x^2 + y^2)$.

32. Determine the rectangular coordinates of the four vertices of the ellipse whose polar equation is $r = 2/(2 - \sin \theta)$.

33. Find a polar equation for the set of points that are equidistant from the pole and the line $r = -\sec \theta$.

34. Find a polar equation of the hyperbola with focus at the origin, vertices (in rectangular coordinates) $(0, -\frac{4}{3})$, $(0, -4)$, and eccentricity 2.

In Problems 35 and 36 (Figures 13.49 and 13.50) find the area of the shaded region. Each circle has radius $\frac{1}{2}$.

35.

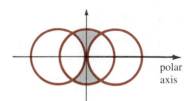

Figure 13.49

36.

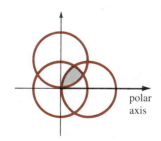

Figure 13.50

37. The **folium of Descartes**, shown in Figure 13.51, has the Cartesian equation $x^3 + y^3 = 3axy$. Use the substitution $y = tx$ to find parametric equations for the curve.

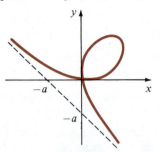

Figure 13.51

38. Use the parametric equations found in Problem 37 to find the points on the folium of Descartes where the tangent line is horizontal.

39. (a) Find a polar equation for the folium of Descartes in Problem 37.

(b) Use the polar equation to find the area bounded by the loop in the first quadrant. (*Hint:* Let $u = \tan \theta$.)

40. Use the parametric equations found in Problem 37 to show that the folium of Descartes has the asymptote $x + y + a = 0$. This is the dashed line in Figure 13.51. (*Hint:* Consider what happens to x, y, and $x + y$ as $t \to -1$.)

14

Vectors and 3-Space

Until now we have carried out most of our endeavors in calculus in the flatland of the two-dimensional Cartesian plane or 2-space. For the next several chapters, we shall be primarily interested in examining mathematical life in three dimensions or 3-space.

14.1 Vectors in 2-Space

Scalars and Vectors In science, mathematics, and engineering we distinguish two important quantities: **scalars** and **vectors**. A **scalar** is simply a real number or a quantity that has *magnitude*. For example, length, temperature, and blood pressure are represented by numbers such as 80 m, 20°C, and the systolic–diastolic ratio 120/80. A **vector**, on the other hand, is usually described as a quantity that has both *magnitude* and *direction*. Pictorially, a vector is represented by an arrow and is written either as a boldface symbol **v** or as either $\vec{v}$ or $\overrightarrow{AB}$. Examples of vector quantities shown in Figure 14.1 are weight **w**, velocity **v**, and the retarding force of friction $\mathbf{F}_f$.

(a)

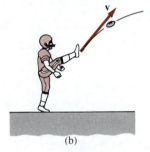
(b)

(c)

Figure 14.1

Notation and Terminology A vector whose initial point (or end) is A and whose terminal point (or tip) is B is written $\overrightarrow{AB}$. The magnitude of a vector is written $|\overrightarrow{AB}|$. Two vectors that have the same magnitude and same direction are said to be **equal**. Thus, in Figure 14.2, we have $\overrightarrow{AB} = \overrightarrow{CD}$. Vectors are said to be **free**, which means that a vector can be moved from one position to another provided its magnitude and direction are not changed. The **negative** of a vector $\overrightarrow{AB}$, written $-\overrightarrow{AB}$, is a vector that has the same magnitude of $\overrightarrow{AB}$ but is opposite in direction. If $k \neq 0$ is a scalar, the **scalar multiple** of a vector, $k\overrightarrow{AB}$, is a vector that is $|k|$ times as long as $\overrightarrow{AB}$ and that has a direction depending on the algebraic sign of k. When $k = 0$, we say $0\overrightarrow{AB} = \mathbf{0}$ is the **zero vector**.* Two vectors are **parallel** if and only if they are nonzero scalar multiples of each other. See Figure 14.3.

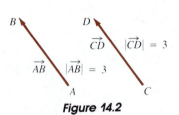

$\overrightarrow{CD}$ $|\overrightarrow{CD}| = 3$

$\overrightarrow{AB}$ $|\overrightarrow{AB}| = 3$

Figure 14.2

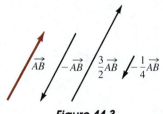

$\overrightarrow{AB}$ $-\overrightarrow{AB}$ $\dfrac{3}{2}\overrightarrow{AB}$ $-\dfrac{1}{4}\overrightarrow{AB}$

Figure 14.3

---

*The question of what is the direction of **0** is usually answered by saying that the zero vector can be assigned *any* direction. More to the point, **0** is needed in order to have a vector algebra.

Addition and Subtraction

Two vectors can be considered as having a common initial point, such as A in Figure 14.4(a). Thus, if nonparallel vectors $\overrightarrow{AB}$ and $\overrightarrow{AC}$ are the sides of a parallelogram in Figure 14.4(b), we say the vector that is the main diagonal, or $\overrightarrow{AD}$, is the **sum** or **resultant** of $\overrightarrow{AB}$ and $\overrightarrow{AC}$. We write

$$\overrightarrow{AD} = \overrightarrow{AB} + \overrightarrow{AC}$$

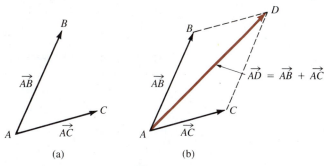

(a)   (b)

Figure 14.4

The **difference** of two vectors $\overrightarrow{AB}$ and $\overrightarrow{AC}$ is defined by

$$\overrightarrow{AB} - \overrightarrow{AC} = \overrightarrow{AB} + (-\overrightarrow{AC})$$

As seen in Figure 14.5(a), the difference $\overrightarrow{AB} - \overrightarrow{AC}$ can be interpreted as the main diagonal of the parallelogram with sides $\overrightarrow{AB}$ and $-\overrightarrow{AC}$. However, as shown in Figure 14.5(b), we can also interpret the same vector difference as the third side of a triangle with sides $\overrightarrow{AB}$ and $\overrightarrow{AC}$. In this second interpretation, observe that the vector difference $\overrightarrow{CB} = \overrightarrow{AB} - \overrightarrow{AC}$ points toward the terminal point of the vector *from* which we are subtracting the second vector. If $\overrightarrow{AB} = \overrightarrow{AC}$, then

$$\overrightarrow{AB} - \overrightarrow{AC} = \mathbf{0}$$

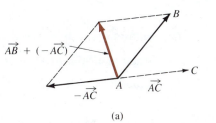

(a)   (b)

Figure 14.5

Vectors in 2-space

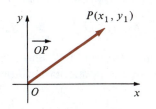

Figure 14.6

To describe a vector analytically, let us suppose for the remainder of this section that the vectors we are considering lie in a two-dimensional coordinate plane or **2-space**. The vector shown in Figure 14.6, whose initial point is the origin O and whose terminal point is $P(x_1, y_1)$, is called the **position vector** of the point P and is written

$$\overrightarrow{OP} = \langle x_1, y_1 \rangle$$

Components   In general, a vector **a** in 2-space is any ordered pair of real numbers,

$$\mathbf{a} = \langle a_1, a_2 \rangle$$

The numbers a_1 and a_2 are said to be the **components** of the vector **a**.

As we see in the first example, the vector **a** is not necessarily a position vector.

Example 1

The displacement between the points (x, y) and $(x + 4, y + 3)$ in Figure 14.7(a) is written $\langle 4, 3 \rangle$. As seen in Figure 14.7(b), the position vector of $\langle 4, 3 \rangle$ is the vector emanating from the origin and terminating at the point $P(4, 3)$.

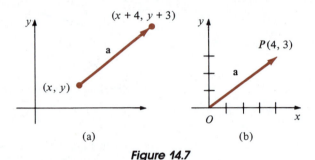

(a)   (b)

Figure 14.7

Addition and subtraction of vectors, multiplication of vectors by scalars, and so on, are defined in terms of their components.

> **DEFINITION 14.1**   Let $\mathbf{a} = \langle a_1, a_2 \rangle$ and $\mathbf{b} = \langle b_1, b_2 \rangle$ be vectors in 2-space.
>
> *(i)*   **Addition:** $\mathbf{a} + \mathbf{b} = \langle a_1 + b_1, a_2 + b_2 \rangle$ (14.1)
>
> *(ii)*   **Scalar multiplication:** $k\mathbf{a} = \langle ka_1, ka_2 \rangle$ (14.2)
>
> *(iii)*   **Equality:** $\mathbf{a} = \mathbf{b}$ if and only if $a_1 = b_1, a_2 = b_2$ (14.3)

Subtraction   Using (14.2) we define the **negative** of a vector **b** by

$$-\mathbf{b} = (-1)\mathbf{b} = \langle -b_1, -b_2 \rangle$$

We can then define the **subtraction**, or the difference, of two vectors as

$$\mathbf{a} - \mathbf{b} = \mathbf{a} + (-\mathbf{b}) = \langle a_1 - b_1, a_2 - b_2 \rangle \qquad (14.4)$$

In Figure 14.8(a), we see the sum of two vectors $\overrightarrow{OP_1}$ and $\overrightarrow{OP_2}$ illustrated. In Figure 14.8(b) the vector $\overrightarrow{P_1P_2}$, whose initial point is P_1 and terminal point is P_2, is the difference of position vectors

$$\overrightarrow{P_1P_2} = \overrightarrow{OP_2} - \overrightarrow{OP_1} = \langle x_2 - x_1, y_2 - y_1 \rangle$$

As shown in Figure 14.8(b), the vector $\overrightarrow{P_1P_2}$ can be drawn either starting from the terminal point of $\overrightarrow{OP_1}$ and ending at the terminal point of $\overrightarrow{OP_2}$, or as the position vector $\overrightarrow{OP}$ whose terminal point has coordinates $(x_2 - x_1, y_2 - y_1)$. Remember, $\overrightarrow{OP}$ and $\overrightarrow{P_1P_2}$ are considered equal since they have the same magnitude and direction.

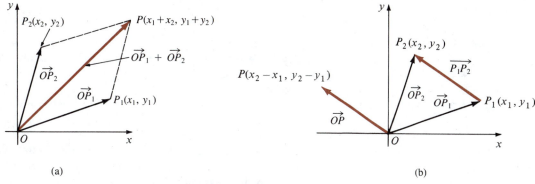

(a)

(b)

Figure 14.8

Example 2

If $\mathbf{a} = \langle 1, 4 \rangle$ and $\mathbf{b} = \langle -6, 3 \rangle$, find

(a) $\mathbf{a} + \mathbf{b}$ (b) $\mathbf{a} - \mathbf{b}$ (c) $2\mathbf{a} + 3\mathbf{b}$

Solution   We use (14.1), (14.2), and (14.4).

(a) $\mathbf{a} + \mathbf{b} = \langle 1 + (-6), 4 + 3 \rangle = \langle -5, 7 \rangle$

(b) $\mathbf{a} - \mathbf{b} = \langle 1 - (-6), 4 - 3 \rangle = \langle 7, 1 \rangle$

(c) $2\mathbf{a} + 3\mathbf{b} = \langle 2, 8 \rangle + \langle -18, 9 \rangle = \langle -16, 17 \rangle$

Properties   The component definition of a vector can be used to verify each of the following **properties**:

(i) $\mathbf{a} + \mathbf{b} = \mathbf{b} + \mathbf{a}$ (Commutative law)

(ii) $\mathbf{a} + (\mathbf{b} + \mathbf{c}) = (\mathbf{a} + \mathbf{b}) + \mathbf{c}$ (Associative law)

(iii) $\mathbf{a} + \mathbf{0} = \mathbf{a}$ (Additive identity)

(iv) $\mathbf{a} + (-\mathbf{a}) = \mathbf{0}$ (Additive inverse)

(v) $k(\mathbf{a} + \mathbf{b}) = k\mathbf{a} + k\mathbf{b}$, k a scalar

(vi) $(k_1 + k_2)\mathbf{a} = k_1\mathbf{a} + k_2\mathbf{a}$, k_1 and k_2 scalars

(vii) $(k_1)(k_2\mathbf{a}) = (k_1 k_2)\mathbf{a}$, k_1 and k_2 scalars

(viii) $1\mathbf{a} = \mathbf{a}$

(ix) $0\mathbf{a} = \mathbf{0}$

The **zero vector 0** in properties (*iii*), (*iv*), and (*ix*) is defined as

$$\mathbf{0} = \langle 0, \, 0 \rangle$$

Magnitude   The **magnitude**, **length**, or **norm** of a vector **a** is denoted by $|\mathbf{a}|$. Motivated by the Pythagorean Theorem and Figure 14.9, we define the magnitude of a vector

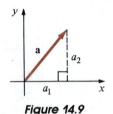

Figure 14.9

$$\mathbf{a} = \langle a_1, \, a_2 \rangle \quad \text{to be} \quad |\mathbf{a}| = \sqrt{a_1^2 + a_2^2}$$

Clearly, $|\mathbf{a}| \geq 0$ for any vector **a**, and $|\mathbf{a}| = 0$ if and only if $\mathbf{a} = \mathbf{0}$. For example, if $\mathbf{a} = \langle 6, \, -2 \rangle$ then

$$|\mathbf{a}| = \sqrt{6^2 + (-2)^2} = \sqrt{40} = 2\sqrt{10}$$

Unit Vectors   A vector that has magnitude 1 is called a **unit vector**. We can obtain a unit vector **u** in the same direction as a nonzero vector **a** by multiplying **a** by the reciprocal of its magnitude. The vector $\mathbf{u} = (1/|\mathbf{a}|)\mathbf{a}$ is a unit vector since

$$|\mathbf{u}| = \left| \frac{1}{|\mathbf{a}|}\mathbf{a} \right| = \frac{1}{|\mathbf{a}|}|\mathbf{a}| = 1$$

Example 3

Given $\mathbf{a} = \langle 2, \, -1 \rangle$. Form a unit vector

(*a*)   in the same direction as **a**

(*b*)   in the opposite direction of **a**

Solution   First, we find the magnitude:

$$|\mathbf{a}| = \sqrt{4 + (-1)^2} = \sqrt{5}$$

(*a*)   A unit vector in the same direction as **a** is

$$\mathbf{u} = \frac{1}{\sqrt{5}}\langle 2, \, -1 \rangle = \left\langle \frac{2}{\sqrt{5}}, \, \frac{-1}{\sqrt{5}} \right\rangle$$

(*b*)   A unit vector in the opposite direction of **a** is

$$-\mathbf{u} = \left\langle -\frac{2}{\sqrt{5}}, \, \frac{1}{\sqrt{5}} \right\rangle$$

*The **i**, **j** Vectors*   In view of (14.1) and (14.2), any vector $\mathbf{a} = \langle a_1, \, a_2 \rangle$ can be written as a sum:

$$\langle a_1, \, a_2 \rangle = \langle a_1, \, 0 \rangle + \langle 0, \, a_2 \rangle$$

$$= a_1\langle 1, \, 0 \rangle + a_2\langle 0, \, 1 \rangle \tag{14.5}$$

The unit vectors $\langle 1, 0 \rangle$ and $\langle 0, 1 \rangle$ are usually given the special symbols **i** and **j**. See Figure 14.10(a). Thus, if

$$\mathbf{i} = \langle 1, 0 \rangle \quad \text{and} \quad \mathbf{j} = \langle 0, 1 \rangle$$

then (14.5) becomes
$$\mathbf{a} = a_1\mathbf{i} + a_2\mathbf{j} \tag{14.6}$$

The vector sum $a_1\mathbf{i} + a_2\mathbf{j}$ is called a **linear combination** of **i** and **j**. The unit vectors **i** and **j** are said to form a **basis** for the system of two-dimensional vectors since any vector **a** can be written uniquely as a linear combination of **i** and **j**. If $\mathbf{a} = a_1\mathbf{i} + a_2\mathbf{j}$ is a position vector, then Figure 14.10(b) shows that **a** is the resultant of the vectors $a_1\mathbf{i}$ and $a_2\mathbf{j}$, which have the origin as a common initial point and which lie on the *x*- and *y*-axes, respectively. The scalar a_1 is called the **horizontal component** of **a** and the scalar a_2 is called the **vertical component** of **a**.

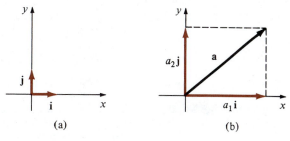

(a)   (b)

Figure 14.10

Example 4

(a) $\langle 4, 7 \rangle = 4\mathbf{i} + 7\mathbf{j}$

(b) $(2\mathbf{i} - 5\mathbf{j}) + (8\mathbf{i} + 13\mathbf{j}) = 10\mathbf{i} + 8\mathbf{j}$

(c) $|\mathbf{i} + \mathbf{j}| = \sqrt{2}$

(d) $10(3\mathbf{i} - \mathbf{j}) = 30\mathbf{i} - 10\mathbf{j}$

(e) $\mathbf{a} = 6\mathbf{i} + 4\mathbf{j}$ and $\mathbf{b} = 9\mathbf{i} + 6\mathbf{j}$ are parallel since **b** is a scalar multiple of **a**. We see $\mathbf{b} = \frac{3}{2}\mathbf{a}$.

Example 5

Let $\mathbf{a} = 4\mathbf{i} + 2\mathbf{j}$ and $\mathbf{b} = -2\mathbf{i} + 5\mathbf{j}$. Graph

(a) $\mathbf{a} + \mathbf{b}$ (b) $\mathbf{a} - \mathbf{b}$

Solution

(a) $\mathbf{a} + \mathbf{b} = 2\mathbf{i} + 7\mathbf{j}$

(b) $\mathbf{a} - \mathbf{b} = 6\mathbf{i} - 3\mathbf{j}$

The graphs of these two vectors in the *xy*-plane are given in Figure 14.11.

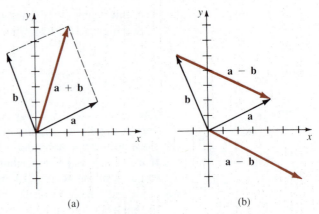

(a)   (b)

Figure 14.11

Exercises 14.1

Answers to odd-numbered problems begin on page A-44.

In Problems 1–8 find **(a)** 3**a**, **(b) a** + **b**, **(c) a** − **b**, **(d)** |**a** + **b**|, and **(e)** |**a** − **b**|.

1. **a** = 2**i** + 4**j**, **b** = −**i** + 4**j**

2. **a** = ⟨1, 1⟩, **b** = ⟨2, 3⟩

3. **a** = ⟨4, 0⟩, **b** = ⟨0, −5⟩

4. **a** = $\frac{1}{6}$**i** − $\frac{1}{6}$**j**, **b** = $\frac{1}{2}$**i** + $\frac{5}{6}$**j**

5. **a** = −3**i** + 2**j**, **b** = 7**j**

6. **a** = ⟨1, 3⟩, **b** = −5**a**

7. **a** = −**b**, **b** = 2**i** − 9**j**

8. **a** = ⟨7, 10⟩, **b** = ⟨1, 2⟩

In Problems 9–14 find **(a)** 4**a** − 2**b**, **(b)** −3**a** − 5**b**.

9. **a** = ⟨1, −3⟩, **b** = ⟨−1, 1⟩

10. **a** = **i** + **j**, **b** = 3**i** − 2**j**

11. **a** = **i** − **j**, **b** = −3**i** + 4**j**

12. **a** = ⟨2, 0⟩, **b** = ⟨0, −3⟩

13. **a** = ⟨4, 10⟩, **b** = −2⟨1, 3⟩

14. **a** = ⟨3, 1⟩ + ⟨−1, 2⟩, **b** = ⟨6, 5⟩ − ⟨1, 2⟩

In Problems 15–18 find the vector $\overrightarrow{P_1P_2}$. Graph $\overrightarrow{P_1P_2}$ and its corresponding position vector.

15. $P_1(3, 2)$, $P_2(5, 7)$

16. $P_1(-2, -1)$, $P_2(4, -5)$

17. $P_1(3, 3)$, $P_2(5, 5)$

18. $P_1(0, 3)$, $P_2(2, 0)$

19. Find the terminal point of the vector $\overrightarrow{P_1P_2}$ = 4**i** + 8**j** if its initial point is (−3, 10).

20. Find the initial point of the vector $\overrightarrow{P_1P_2}$ = ⟨−5, −1⟩ if its terminal point is (4, 7).

21. Determine which of the following vectors are parallel to **a** = 4**i** + 6**j**.

(a)   −4**i** − 6**j**

(b)   −**i** − $\frac{3}{2}$**j**

(c)   10**i** + 15**j**

(d)   2(**i** − **j**) − 3$\left(\frac{1}{2}\textbf{i} - \frac{5}{12}\textbf{j}\right)$

(e)   8**i** + 12**j**

(f)   (5**i** + **j**) − (7**i** + 4**j**)

22. Determine a scalar c so that $\mathbf{a} = 3\mathbf{i} + c\mathbf{j}$ and $\mathbf{b} = -\mathbf{i} + 9\mathbf{j}$ are parallel.

In Problems 23 and 24 find $\mathbf{a} + (\mathbf{b} + \mathbf{c})$ for the given vectors.

23. $\mathbf{a} = \langle 5, 1 \rangle$, $\mathbf{b} = \langle -2, 4 \rangle$, $\mathbf{c} = \langle 3, 10 \rangle$

24. $\mathbf{a} = \langle 1, 1 \rangle$, $\mathbf{b} = \langle 4, 3 \rangle$, $\mathbf{c} = \langle 0, -2 \rangle$

In Problems 25–28 find a unit vector **(a)** in the same direction as $\mathbf{a}$, and **(b)** in the opposite direction of $\mathbf{a}$.

25. $\mathbf{a} = \langle 2, 2 \rangle$ **26.** $\mathbf{a} = \langle -3, 4 \rangle$

27. $\mathbf{a} = \langle 0, -5 \rangle$ **28.** $\mathbf{a} = \langle 1, -\sqrt{3} \rangle$

In Problems 29 and 30, $\mathbf{a} = \langle 2, 8 \rangle$ and $\mathbf{b} = \langle 3, 4 \rangle$. Find a unit vector in the same direction as the given vector.

29. $\mathbf{a} + \mathbf{b}$ **30.** $2\mathbf{a} - 3\mathbf{b}$

In Problems 31 and 32 find a vector $\mathbf{b}$ that is parallel to the given vector and has the indicated magnitude.

31. $\mathbf{a} = 3\mathbf{i} + 7\mathbf{j}$, $|\mathbf{b}| = 2$

32. $\mathbf{a} = \dfrac{1}{2}\mathbf{i} - \dfrac{1}{2}\mathbf{j}$, $|\mathbf{b}| = 3$

33. Find a vector in the opposite direction of $\mathbf{a} = \langle 4, 10 \rangle$ but $\frac{3}{4}$ as long.

34. Given that $\mathbf{a} = \langle 1, 1 \rangle$ and $\mathbf{b} = \langle -1, 0 \rangle$ find a vector in the same direction as $\mathbf{a} + \mathbf{b}$ but 5 times as long.

In Problems 35 and 36 (Figures 14.12 and 14.13) use the given figure to draw the indicated vector.

35. $3\mathbf{b} - \mathbf{a}$ **36.** $\mathbf{a} + (\mathbf{b} + \mathbf{c})$

Figure 14.12

Figure 14.13

In Problems 37 and 38 (Figures 14.14 and 14.15) express the vector $\mathbf{x}$ in terms of vectors $\mathbf{a}$ and $\mathbf{b}$.

37.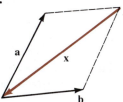

Figure 14.14

38.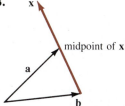

Figure 14.15

In Problems 39 and 40 (Figures 14.16 and 14.17) use the given figure to prove the given result.

39. $\mathbf{a} + \mathbf{b} + \mathbf{c} = 0$

Figure 14.16

40. $\mathbf{a} + \mathbf{b} + \mathbf{c} + \mathbf{d} = 0$

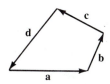

Figure 14.17

41. When walking, a person's foot strikes the ground with a force $\mathbf{F}$ at an angle θ from the vertical. In Figure 14.18, the vector $\mathbf{F}$ is resolved into vector components $\mathbf{F}_g$, which is parallel to the ground, and $\mathbf{F}_n$, which is perpendicular to the ground. In order that the foot does not slip, the force $\mathbf{F}_g$ must be offset by the opposing force $\mathbf{F}_f$ of friction; that is, $\mathbf{F}_f = -\mathbf{F}_g$.

    (a) Use the fact that $|\mathbf{F}_f| = \mu|\mathbf{F}_n|$, where μ is the coefficient of friction, to show that $\tan \theta = \mu$. The foot will not slip for angles less than or equal to θ.

    (b) Given that $\mu = 0.6$ for a rubber heel striking an asphalt sidewalk, find the "no-slip" angle.

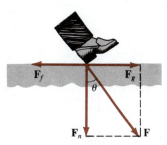

Figure 14.18

42. A 200-lb traffic light supported by two cables hangs in equilibrium. As shown in Figure 14.19(b)(page 688), let the weight of the light be represented by $\mathbf{w}$ and the forces in the two cables by $\mathbf{F}_1$ and $\mathbf{F}_2$. From Figure 14.19(c), we see that a condition of equilibrium is

$$\mathbf{w} + \mathbf{F}_1 + \mathbf{F}_2 = 0 \qquad (14.7)$$

(See Problem 39.) If

$$\mathbf{w} = -200\mathbf{j}$$
$$\mathbf{F}_1 = (|\mathbf{F}_1|\cos 20°)\mathbf{i} + (|\mathbf{F}_1|\sin 20°)\mathbf{j}$$
$$\mathbf{F}_2 = -(|\mathbf{F}_2|\cos 15°)\mathbf{i} + (|\mathbf{F}_2|\sin 15°)\mathbf{j}$$

use (14.7) to determine the magnitudes of $\mathbf{F}_1$ and $\mathbf{F}_2$. (*Hint:* Reread (*iii*) of Definition 14.1.)

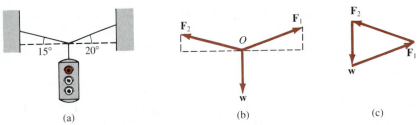

Figure 14.19

Miscellaneous Problems

43. Let P_1 and P_2 be points in 2-space and let M be the midpoint of the line segment between P_1 and P_2. Show that

$$\overrightarrow{OM} = \tfrac{1}{2}(\overrightarrow{OP_1} + \overrightarrow{OP_2})$$

44. Using vectors, show that the diagonals of a parallelogram bisect each other. (*Hint:* Let M be the midpoint of one diagonal and N the midpoint of the other.)

45. Using vectors, show that the line segment between the midpoints of two sides of a triangle is parallel to the third side and half as large.

46. Let P_1, P_2, and P_3 be distinct points such that $\mathbf{a} = \overrightarrow{P_1P_2}$, $\mathbf{b} = \overrightarrow{P_2P_3}$, and $\mathbf{a} + \mathbf{b} = \overrightarrow{P_1P_3}$.

(a)   What is the relation of $|\mathbf{a} + \mathbf{b}|$ to $|\mathbf{a}| + |\mathbf{b}|$?

(b)   Under what condition is $|\mathbf{a} + \mathbf{b}| = |\mathbf{a}| + |\mathbf{b}|$?

14.2   3-Space

14.2.1 Rectangular Coordinate System in Three Dimensions

In the plane, or 2-space, one way of describing the position of a point P is to assign to it coordinates relative to two mutually orthogonal axes called the x- and y-axes. If P is the point of intersection of the line $x = a$ (perpendicular to the x-axis) and the line $y = b$ (perpendicular to the y-axis), then the **ordered pair** (a, b) is said to be the **rectangular** or **Cartesian coordinates** of the point. See Figure 14.20.

In three dimensions, or **3-space**, a rectangular coordinate system is constructed using three mutually orthogonal axes. The point at which these axes intersect is called the **origin** O. These axes, shown in Figure 14.21(a), are labeled in accordance with the so-called **right-hand rule**: If the fingers of the right hand, pointing in the direction of the positive x-axis, are curled toward the positive y-axis, then the thumb will point in the direction of a new axis perpendicular to the plane of the x- and y-axes. This new axis is labeled the z-axis.* The dashed lines in Figure 14.21(a) represent the negative axes. Now, if

$$x = a, \qquad y = b, \qquad z = c$$

are planes perpendicular to the x-axis, y-axis, and z-axis, respectively, the point P at which these planes intersect can be represented by an **ordered triple** of numbers (a, b, c) said to be the **rectangular** or **Cartesian coordinates** of the point. The numbers a, b, and c are, in turn, called the **x-**, **y-**, and **z-coordinates** of $P(a, b, c)$. See Figure 14.21(b).

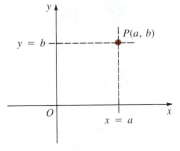

Figure 14.20

---

*If the x- and y-axes are interchanged in Figure 14.21(a), the coordinate system is said to be **left-handed**.

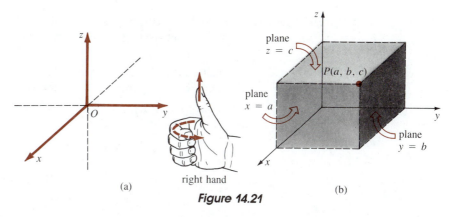

Figure 14.21

Octants   Each pair of coordinate axes determines a **coordinate plane**. As shown in Figure 14.22 the x- and y-axes determine the xy-plane, the x- and z-axes determine the xz-plane, and so on. The coordinate planes divide 3-space into eight parts known as **octants**. The octant in which all three coordinates of a point are *positive* is called the **first octant**. There is no agreement for naming the other seven octants.

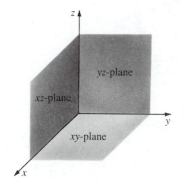

Figure 14.22

The following table summarizes the coordinates of a point on either a coordinate axis or in a coordinate plane. As seen in the table, we can also describe, say, the xy-plane by the simple equation $z = 0$. Similarly, the xz-plane is $y = 0$ and the yz-plane is $x = 0$.

Axes	Coordinates	Plane	Coordinates
x	$(a, 0, 0)$	xy	$(a, b, 0)$
y	$(0, b, 0)$	xz	$(a, 0, c)$
z	$(0, 0, c)$	yz	$(0, b, c)$

_____ **Example 1** _____

Graph the points $(4, 5, 6)$, $(3, -3, -1)$, $(-2, -2, 0)$.

Solution   Of the three points shown in Figure 14.23 only $(4, 5, 6)$ is in the first octant. The point $(-2, -2, 0)$ is in the xy-plane.

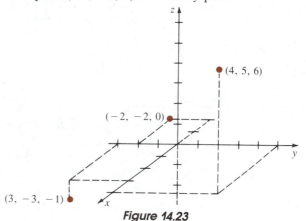

Figure 14.23

Distance Formula   To find the **distance** between two points $P_1(x_1, y_1, z_1)$ and $P_2(x_2, y_2, z_2)$ in 3-space, let us first consider their projections onto the *xy*-plane. As seen in Figure 14.24, the distance between $(x_1, y_1, 0)$ and $(x_2, y_2, 0)$ follows from the usual distance formula in the plane and is $\sqrt{(x_2 - x_1)^2 + (y_2 - y_1)^2}$. Hence, from the Pythagorean Theorem applied to the right triangle $P_1P_3P_2$, we have

$$[d(P_1, P_2)]^2 = [\sqrt{(x_2 - x_1)^2 + (y_2 - y_1)^2}]^2 + |z_2 - z_1|^2$$

or $$d(P_1, P_2) = \sqrt{(x_2 - x_1)^2 + (y_2 - y_1)^2 + (z_2 - z_1)^2} \qquad (14.7)$$

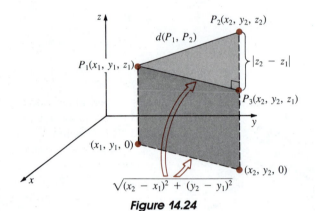

Figure 14.24

_____ **Example 2** _____

Find the distance between $(2, -3, 6)$ and $(-1, -7, 4)$.

Solution   From (14.7),

$$d = \sqrt{(2 - (-1))^2 + (-3 - (-7))^2 + (6 - 4)^2} = \sqrt{29}$$

Midpoint Formula   The formulas for finding the midpoint of a line segment between two points in 2-space (see (1.5), page 11) carry over in an analogous fashion to 3-space. If $P_1(x_1, y_1, z_1)$ and $P_2(x_2, y_2, z_2)$ are two distinct points, then the **coordinates of the midpoint** of the line segment between them are

$$\left(\frac{x_1 + x_2}{2}, \frac{y_1 + y_2}{2}, \frac{z_1 + z_2}{2} \right) \qquad (14.8)$$

_____ **Example 3** _____

Find the coordinates of the midpoint of the line segment between the two points in Example 2.

Solution   From (14.8) we obtain

$$\left(\frac{2 + (-1)}{2}, \frac{-3 + (-7)}{2}, \frac{6 + 4}{2} \right) \quad \text{or} \quad \left(\frac{1}{2}, -5, 5 \right)$$

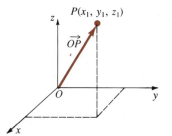

Figure 14.25

14.2.2 Vectors in 3-Space

A **vector a in 3-space** is any ordered triple of real numbers

$$\mathbf{a} = \langle a_1, a_2, a_3 \rangle$$

where a_1, a_2, and a_3 are the **components** of the vector. The **position vector** of a point $P(x_1, y_1, z_1)$ in space is the vector $\overrightarrow{OP} = \langle x_1, y_1, z_1 \rangle$ whose initial point is the origin O and whose terminal point is P. See Figure 14.25.

The component definitions of addition, subtraction, scalar multiplication, and so on, are natural generalizations of those given for vectors in 2-space.

DEFINITION 14.2   Let $\mathbf{a} = \langle a_1, a_2, a_3 \rangle$ and $\mathbf{b} = \langle b_1, b_2, b_3 \rangle$ be vectors in 3-space.

  (*i*)   Addition: $\mathbf{a} + \mathbf{b} = \langle a_1 + b_1, a_2 + b_2, a_3 + b_3 \rangle$

  (*ii*) Scalar multiplication: $k\mathbf{a} = \langle ka_1, ka_2, ka_3 \rangle$

  (*iii*) Equality: $\mathbf{a} = \mathbf{b}$ if and only if $a_1 = b_1, a_2 = b_2, a_3 = b_3$

  (*iv*) Negative: $-\mathbf{b} = (-1)\mathbf{b} = \langle -b_1, -b_2, -b_3 \rangle$

  (*v*)   Subtraction: $\mathbf{a} - \mathbf{b} = \mathbf{a} + (-\mathbf{b}) = \langle a_1 - b_1, a_2 - b_2, a_3 - b_3 \rangle$

  (*vi*) Zero vector: $\mathbf{0} = \langle 0, 0, 0 \rangle$

  (*vii*) Magnitude: $|\mathbf{a}| = \sqrt{a_1^2 + a_2^2 + a_3^2}$

If $\overrightarrow{OP_1}$ and $\overrightarrow{OP_2}$ are the position vectors of the points $P_1(x_1, y_1, z_1)$ and $P_2(x_2, y_2, z_2)$, then the vector $\overrightarrow{P_1P_2}$ is given by

$$\overrightarrow{P_1P_2} = \overrightarrow{OP_2} - \overrightarrow{OP_1} = \langle x_2 - x_1, y_2 - y_1, z_2 - z_1 \rangle \tag{14.9}$$

As in 2-space $\overrightarrow{P_1P_2}$ can be drawn either as a vector whose initial point is P_1 and whose terminal point is P_2 or as a position vector $\overrightarrow{OP}$ with terminal point $P(x_2 - x_1, y_2 - y_1, z_2 - z_1)$. See Figure 14.26.

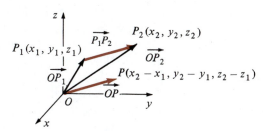

Figure 14.26

_____ **Example 4** _____

Find the vector $\overrightarrow{P_1P_2}$ if the points P_1 and P_2 are given by P_1 (4, 6, −2) and P_2 (1, 8, 3).

Solution If the position vectors of the points are $\overrightarrow{OP_1} = \langle 4, 6, -2 \rangle$ and $\overrightarrow{OP_2} = \langle 1, 8, 3 \rangle$, then from (14.9) we have

$$\overrightarrow{P_1P_2} = \overrightarrow{OP_2} - \overrightarrow{OP_1}$$
$$= \langle 1 - 4, 8 - 6, 3 - (-2) \rangle = \langle -3, 2, 5 \rangle$$

_____ **Example 5** _____

From part (*vii*) of Definition 14.2 we see that $\mathbf{a} = \left\langle -\dfrac{2}{7}, \dfrac{3}{7}, \dfrac{6}{7} \right\rangle$ is a unit vector since

$$|\mathbf{a}| = \sqrt{\left(-\frac{2}{7}\right)^2 + \left(\frac{3}{7}\right)^2 + \left(\frac{6}{7}\right)^2}$$
$$= \sqrt{\frac{4 + 9 + 36}{49}} = 1$$

The i, j, k Vectors We saw in the preceding section that the unit vectors $\mathbf{i} = \langle 1, 0 \rangle$ and $\mathbf{j} = \langle 0, 1 \rangle$ are a basis for the system of two-dimensional vectors in that any vector $\mathbf{a}$ in 2-space can be written as a linear combination of $\mathbf{i}$ and $\mathbf{j}$: $\mathbf{a} = a_1 \mathbf{i} + a_2 \mathbf{j}$. A basis for the system of three-dimensional vectors is given by the set of unit vectors:

$$\mathbf{i} = \langle 1, 0, 0 \rangle, \qquad \mathbf{j} = \langle 0, 1, 0 \rangle, \qquad \mathbf{k} = \langle 0, 0, 1 \rangle$$

Any vector $\mathbf{a} = \langle a_1, a_2, a_3 \rangle$ in 3-space can be expressed as a linear combination of $\mathbf{i}$, $\mathbf{j}$, and $\mathbf{k}$:

$$\langle a_1, a_2, a_3 \rangle = \langle a_1, 0, 0 \rangle + \langle 0, a_2, 0 \rangle + \langle 0, 0, a_3 \rangle$$
$$= a_1 \langle 1, 0, 0 \rangle + a_2 \langle 0, 1, 0 \rangle + a_3 \langle 0, 0, 1 \rangle$$

That is, $\mathbf{a} = a_1 \mathbf{i} + a_2 \mathbf{j} + a_3 \mathbf{k}$

The vectors $\mathbf{i}$, $\mathbf{j}$, and $\mathbf{k}$ are illustrated in Figure 14.27(a). In Figure 14.27(b) we see that a position vector $\mathbf{a} = a_1 \mathbf{i} + a_2 \mathbf{j} + a_3 \mathbf{k}$ is the resultant of the vectors $a_1 \mathbf{i}$, $a_2 \mathbf{j}$, and $a_3 \mathbf{k}$, which lie along the coordinate axes and have the origin as a common initial point.

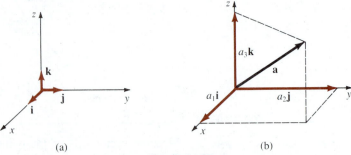

(a)   (b)

Figure 14.27

_____ **Example 6** _____

The vector

$$\mathbf{a} = \langle 7, -5, 13 \rangle \quad \text{is the same as} \quad \mathbf{a} = 7\mathbf{i} - 5\mathbf{j} + 13\mathbf{k}$$

When the third dimension is taken into consideration, any vector in the xy-plane is equivalently described as a three-dimensional vector that lies in the coordinate plane $z = 0$. Although the vectors $\langle a_1, a_2 \rangle$ and $\langle a_1, a_2, 0 \rangle$ are technically not equal, we shall ignore the distinction. That is why, for example, we denoted $\langle 1, 0 \rangle$ and $\langle 1, 0, 0 \rangle$ by the same symbol $\mathbf{i}$. But to avoid any possible confusion, hereafter we shall always consider a vector a three-dimensional vector, and the symbols $\mathbf{i}$ and $\mathbf{j}$ will only represent $\langle 1, 0, 0 \rangle$ and $\langle 0, 1, 0 \rangle$, respectively. Similarly, a vector in either the yz-plane or the xz-plane must have one zero component. In the yz-plane a vector

$$\mathbf{b} = \langle 0, b_2, b_3 \rangle \quad \text{is written} \quad \mathbf{b} = b_2\mathbf{j} + b_3\mathbf{k}$$

In the xz-plane a vector

$$\mathbf{c} = \langle c_1, 0, c_3 \rangle \quad \text{is the same as} \quad \mathbf{c} = c_1\mathbf{i} + c_3\mathbf{k}$$

_____ **Example 7** _____

(_a_)   The vector $\mathbf{a} = 5\mathbf{i} + 3\mathbf{k}$ is in the xz-coordinate plane.

(_b_) $|5\mathbf{i} + 3\mathbf{k}| = \sqrt{5^2 + 3^2} = \sqrt{34}$

_____ **Example 8** _____

If $\mathbf{a} = 3\mathbf{i} - 4\mathbf{j} + 8\mathbf{k}$ and $\mathbf{b} = \mathbf{i} - 4\mathbf{k}$, find $5\mathbf{a} - 2\mathbf{b}$.

Solution   We treat $\mathbf{b}$ as a three-dimensional vector and write, for emphasis, $\mathbf{b} = \mathbf{i} + 0\mathbf{j} - 4\mathbf{k}$. From

$$5\mathbf{a} = 15\mathbf{i} - 20\mathbf{j} + 40\mathbf{k} \quad \text{and} \quad 2\mathbf{b} = 2\mathbf{i} + 0\mathbf{j} - 8\mathbf{k}$$

we get $5\mathbf{a} - 2\mathbf{b} = (15\mathbf{i} - 20\mathbf{j} + 40\mathbf{k}) - (2\mathbf{i} + 0\mathbf{j} - 8\mathbf{k})$
$$= 13\mathbf{i} - 20\mathbf{j} + 48\mathbf{k}$$

_____ **Exercises 14.2** _____

Answers to odd-numbered problems begin on page A-44.

[14.2.1]

In Problems 1–6 graph the given point. Use the same coordinate axes.

1. $(1, 1, 5)$ **2.** $(0, 0, 4)$ **3.** $(3, 4, 0)$

4. $(6, 0, 0)$ **5.** $(6, -2, 0)$ **6.** $(5, -4, 3)$

In Problems 7–10 describe geometrically all points $P(x, y, z)$ that satisfy the given condition.

7. $z = 5$ **8.** $x = 1$

9. $x = 2, y = 3$ **10.** $x = 4, y = -1, z = 7$

11. Give the coordinates of the vertices of the rectangular parallelepiped whose sides are the coordinate planes and the planes $x = 2$, $y = 5$, $z = 8$.

12. In Figure 14.28, two vertices are shown of a rectangular parallelepiped having sides parallel to the coordinate planes. Find the coordinates of the remaining six vertices.

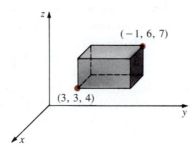

Figure 14.28

13. Consider the point $P(-2, 5, 4)$.

(a) If lines are drawn from P perpendicular to the coordinate planes, what are the coordinates of the point at the base of each perpendicular?

(b) If a line is drawn from P to the plane $z = -2$, what are the coordinates of the point at the base of the perpendicular?

(c) Find the point in the plane $x = 3$ that is closest to P.

14. Determine an equation of a plane parallel to a coordinate plane that contains the given pair of points.

(a) $(3, 4, -5)$, $(-2, 8, -5)$

(b) $(1, -1, 1)$, $(1, -1, -1)$

(c) $(-2, 1, 2)$, $(2, 4, 2)$

In Problems 15–20 describe the locus of points $P(x, y, z)$ that satisfy the given equation(s).

15. $xyz = 0$

16. $x^2 + y^2 + z^2 = 0$

17. $(x + 1)^2 + (y - 2)^2 + (z + 3)^2 = 0$

18 $(x - 2)(z - 8) = 0$

19. $z^2 - 25 = 0$

20. $x = y = z$

In Problems 21 and 22 find the distance between the given points.

21. $(3, -1, 2)$, $(6, 4, 8)$

22. $(-1, -3, 5)$, $(0, 4, 3)$

23. Find the distance from the point $(7, -3, -4)$ to (a) the yz-plane and (b) the x-axis.

24. Find the distance from the point $(-6, 2, -3)$ to (a) the xz-plane and (b) the origin.

In Problems 25–28 the given three points form a triangle. Determine which triangles are isosceles and which are right triangles.

25. $(0, 0, 0)$, $(3, 6, -6)$, $(2, 1, 2)$

26. $(0, 0, 0)$, $(1, 2, 4)$, $(3, 2, 2\sqrt{2})$

27. $(1, 2, 3)$, $(4, 1, 3)$, $(4, 6, 4)$

28. $(1, 1, -1)$, $(1, 1, 1)$, $(0, -1, 1)$

In Problems 29 and 30 use the distance formula to prove that the given points are collinear.

29. $P_1(1, 2, 0)$, $P_2(-2, -2, -3)$, $P_3(7, 10, 6)$

30. $P_1(2, 3, 2)$, $P_2(1, 4, 4)$, $P_3(5, 0, -4)$

In Problems 31 and 32 solve for the unknown.

31. $P_1(x, 2, 3)$, $P_2(2, 1, 1)$; $d(P_1, P_2) = \sqrt{21}$

32. $P_1(x, x, 1)$, $P_2(0, 3, 5)$; $d(P_1, P_2) = 5$

In Problems 33 and 34 find the coordinates of the midpoint of the line segment between the given points.

33. $\left(1, 3, \dfrac{1}{2}\right)$, $\left(7, -2, \dfrac{5}{2}\right)$

34. $(0, 5, -8)$, $(4, 1, -6)$

35. The coordinates of the midpoint of the line segment between $P_1(x_1, y_1, z_1)$ and $P_2(2, 3, 6)$ are $(-1, -4, 8)$. Find the coordinates of P_1.

36. Let P_3 be the midpoint of the line segment between $P_1(-3, 4, 1)$ and $P_2(-5, 8, 3)$. Find the coordinates of the midpoint of the line segment

(a) between P_1 and P_3

(b) between P_3 and P_2

[14.2.2]

In Problems 37–40 find the vector $\overrightarrow{P_1 P_2}$.

37. $P_1(3, 4, 5)$, $P_2(0, -2, 6)$

38. $P_1(-2, 4, 0)$, $P_2\left(6, \dfrac{3}{2}, 8\right)$

39. $P_1(0, -1, 0)$, $P_2(2, 0, 1)$

40. $P_1\left(\dfrac{1}{2}, \dfrac{3}{4}, 5\right)$, $P_2\left(-\dfrac{5}{2}, -\dfrac{9}{4}, 12\right)$

In Problems 41–48 $\mathbf{a} = \langle 1, -3, 2\rangle$, $\mathbf{b} = \langle -1, 1, 1\rangle$, and $\mathbf{c} = \langle 2, 6, 9\rangle$. Find the indicated vector or scalar.

41. $\mathbf{a} + (\mathbf{b} + \mathbf{c})$

42. $2\mathbf{a} - (\mathbf{b} - \mathbf{c})$

43. $\mathbf{b} + 2(\mathbf{a} - 3\mathbf{c})$

44. $4(\mathbf{a} + 2\mathbf{c}) - 6\mathbf{b}$

45. $|\mathbf{a} + \mathbf{c}|$ **46.** $|\mathbf{c}|\,|2\mathbf{b}|$

47. $\left|\dfrac{\mathbf{a}}{|\mathbf{a}|} + 5\dfrac{\mathbf{b}}{|\mathbf{b}|}\right|$ **48.** $|\mathbf{b}|\mathbf{a} + |\mathbf{a}|\mathbf{b}$

49. Find a unit vector in the opposite direction of $\mathbf{a} = \langle 10, -5, 10 \rangle$.

50. Find a unit vector in the same direction as $\mathbf{a} = \mathbf{i} - 3\mathbf{j} + 2\mathbf{k}$.

51. Find a vector $\mathbf{b}$ that is four times as long as $\mathbf{a} = \mathbf{i} - \mathbf{j} + \mathbf{k}$ in the same direction as $\mathbf{a}$.

52. Find a vector $\mathbf{b}$ for which $|\mathbf{b}| = \frac{1}{2}$ that is parallel to $\mathbf{a} = \langle -6, 3, -2 \rangle$ but has the opposite direction.

53. Using the vectors $\mathbf{a}$ and $\mathbf{b}$ shown in Figure 14.29, sketch the "average vector" $\frac{1}{2}(\mathbf{a} + \mathbf{b})$.

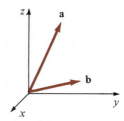

Figure 14.29

14.3 The Dot Product

In this and the following section, we shall consider two kinds of products between vectors that originated in the study of mechanics and electricity and magnetism. The first of these products yields a scalar.

DEFINITION 14.3   Dot Product of Two Vectors*

The **dot product** of two vectors $\mathbf{a}$ and $\mathbf{b}$ is the scalar

$$\mathbf{a} \cdot \mathbf{b} = |\mathbf{a}|\,|\mathbf{b}| \cos\theta \qquad (14.10)$$

where θ is the angle between the vectors such that $0 \le \theta \le \pi$.

Figure 14.30 illustrates the angle θ in three cases. If the vectors $\mathbf{a}$ and $\mathbf{b}$ are not parallel, then θ is the *smaller* of the two possible angles between them.

(a)

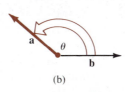
(b)

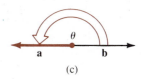
(c)

Figure 14.30

Example 1

From (14.10) we obtain

$$\mathbf{i} \cdot \mathbf{i} = 1, \qquad \mathbf{j} \cdot \mathbf{j} = 1, \qquad \mathbf{k} \cdot \mathbf{k} = 1 \qquad (14.11)$$

since $|\mathbf{i}| = |\mathbf{j}| = |\mathbf{k}| = 1$, and, in each case, $\cos\theta = 1$.

---

*Also known as the **inner product** and the **scalar product**.

Properties   The dot product possesses the following properties:

> (*i*) $\mathbf{a} \cdot \mathbf{b} = 0$ if $\mathbf{a} = \mathbf{0}$ or $\mathbf{b} = \mathbf{0}$
>
> (*ii*) $\mathbf{a} \cdot \mathbf{b} = \mathbf{b} \cdot \mathbf{a}$ (Commutative law)
>
> (*iii*) $\mathbf{a} \cdot (\mathbf{b} + \mathbf{c}) = \mathbf{a} \cdot \mathbf{b} + \mathbf{a} \cdot \mathbf{c}$ (Distributive law)
>
> (*iv*) $\mathbf{a} \cdot (k\mathbf{b}) = (k\mathbf{a}) \cdot \mathbf{b} = k(\mathbf{a} \cdot \mathbf{b})$, k a scalar
>
> (*v*) $\mathbf{a} \cdot \mathbf{a} \geq 0$
>
> (*vi*) $\mathbf{a} \cdot \mathbf{a} = |\mathbf{a}|^2$

Each of these properties, with the possible exception of (*iii*), should be apparent from (14.10). Notice that (*vi*) states that the magnitude of a vector

$$\mathbf{a} = a_1\mathbf{i} + a_2\mathbf{j} + a_3\mathbf{k}$$

can be written

$$|\mathbf{a}| = \sqrt{\mathbf{a} \cdot \mathbf{a}} = \sqrt{a_1^2 + a_2^2 + a_3^2}$$

Orthogonal Vectors   If $\mathbf{a}$ and $\mathbf{b}$ are nonzero vectors, then Definition 14.3 implies that

> (*i*) $\mathbf{a} \cdot \mathbf{b} > 0$ if and only if θ is acute,
>
> (*ii*) $\mathbf{a} \cdot \mathbf{b} < 0$ if and only if θ is obtuse, and
>
> (*iii*) $\mathbf{a} \cdot \mathbf{b} = 0$ if and only if $\cos \theta = 0$.

But, in the last case the only number in $[0, \pi]$ for which $\cos \theta = 0$ is $\theta = \pi/2$. When $\theta = \pi/2$, we say that the vectors are **perpendicular** or **orthogonal**. Thus, we are led to the following result.

> **THEOREM 14.1**   Two nonzero vectors $\mathbf{a}$ and $\mathbf{b}$ are orthogonal if and only if $\mathbf{a} \cdot \mathbf{b} = 0$.

Since $\mathbf{0} \cdot \mathbf{b} = 0$ for every vector $\mathbf{b}$, the zero vector is regarded to be orthogonal to every vector.

_____ **Example 2** _____

It follows immediately from Theorem 14.1 and the fact that the dot product is commutative that

$$\mathbf{i} \cdot \mathbf{j} = \mathbf{j} \cdot \mathbf{i} = 0$$
$$\mathbf{j} \cdot \mathbf{k} = \mathbf{k} \cdot \mathbf{j} = 0 \qquad\qquad (14.12)$$
$$\mathbf{k} \cdot \mathbf{i} = \mathbf{i} \cdot \mathbf{k} = 0$$

Alternative Definition of the
Dot Product

From the distributive property of the dot product we can write

$$\mathbf{a} \cdot \mathbf{b} = (a_1\mathbf{i} + a_2\mathbf{j} + a_3\mathbf{k}) \cdot (b_1\mathbf{i} + b_2\mathbf{j} + b_3\mathbf{k})$$

$$= a_1\mathbf{i} \cdot (b_1\mathbf{i} + b_2\mathbf{j} + b_3\mathbf{k}) + a_2\mathbf{j} \cdot (b_1\mathbf{i} + b_2\mathbf{j} + b_3\mathbf{k})$$

$$+ a_3\mathbf{k} \cdot (b_1\mathbf{i} + b_2\mathbf{j} + b_3\mathbf{k})$$

$$= a_1b_1(\mathbf{i} \cdot \mathbf{i}) + a_1b_2(\mathbf{i} \cdot \mathbf{j}) + a_1b_3(\mathbf{i} \cdot \mathbf{k})$$

$$+ a_2b_1(\mathbf{j} \cdot \mathbf{i}) + a_2b_2(\mathbf{j} \cdot \mathbf{j}) + a_2b_3(\mathbf{j} \cdot \mathbf{k})$$

$$+ a_3b_1(\mathbf{k} \cdot \mathbf{i}) + a_3b_2(\mathbf{k} \cdot \mathbf{j}) + a_3b_3(\mathbf{k} \cdot \mathbf{k}) \qquad (14.13)$$

Now, in view of (14.11) and (14.12), (14.13) simplifies to an alternative form of the dot product:

$$\mathbf{a} \cdot \mathbf{b} = a_1b_1 + a_2b_2 + a_3b_3 \qquad (14.14)$$

In other words, the dot product of two vectors is *the sum of the products of their corresponding components.**

___ **Example 3** _____

If $\mathbf{a} = 10\mathbf{i} + 2\mathbf{j} - 6\mathbf{k}$ and $\mathbf{b} = -\frac{1}{2}\mathbf{i} + 4\mathbf{j} - 3\mathbf{k}$, then

$$\mathbf{a} \cdot \mathbf{b} = (10)(-\tfrac{1}{2}) + (2)(4) + (-6)(-3) = 21$$

___ **Example 4** _____

If $\mathbf{a} = -3\mathbf{i} - \mathbf{j} + 4\mathbf{k}$ and $\mathbf{b} = 2\mathbf{i} + 14\mathbf{j} + 5\mathbf{k}$, then

$$\mathbf{a} \cdot \mathbf{b} = (-3)(2) + (-1)(14) + (4)(5) = 0$$

From Theorem 14.1, we conclude that $\mathbf{a}$ and $\mathbf{b}$ are orthogonal.

Angle between Two Vectors

By equating the two forms of the dot product, (14.10) and (14.14), we can determine **the angle between two vectors** from

$$\cos \theta = \frac{a_1b_1 + a_2b_2 + a_3b_3}{|\mathbf{a}||\mathbf{b}|} \qquad (14.15)$$

___ **Example 5** _____

Find the angle between $\mathbf{a} = 2\mathbf{i} + 3\mathbf{j} + \mathbf{k}$ and $\mathbf{b} = -\mathbf{i} + 5\mathbf{j} + \mathbf{k}$.

Solution

$$|\mathbf{a}| = \sqrt{14}, \qquad |\mathbf{b}| = \sqrt{27}, \qquad \mathbf{a} \cdot \mathbf{b} = 14$$

_____

*The result in (14.14) can also be obtained from the Law of Cosines. You are encouraged to work Problem 45 in Exercises 14.3.

Hence, (14.15) gives

$$\cos \theta = \frac{14}{\sqrt{14}\sqrt{27}} = \frac{\sqrt{42}}{9}$$

and so

$$\theta = \cos^{-1}\left(\frac{\sqrt{42}}{9}\right) \approx 0.77 \text{ radian}$$

or $\theta \approx 44.9°$.

Direction Cosines

For a nonzero vector $\mathbf{a} = a_1\mathbf{i} + a_2\mathbf{j} + a_3\mathbf{k}$ in 3-space, the angles α, β, and γ, between $\mathbf{a}$ and the unit vectors $\mathbf{i}$, $\mathbf{j}$, and $\mathbf{k}$, respectively, are called **direction angles of a**. See Figure 14.31. Now, by (14.15)

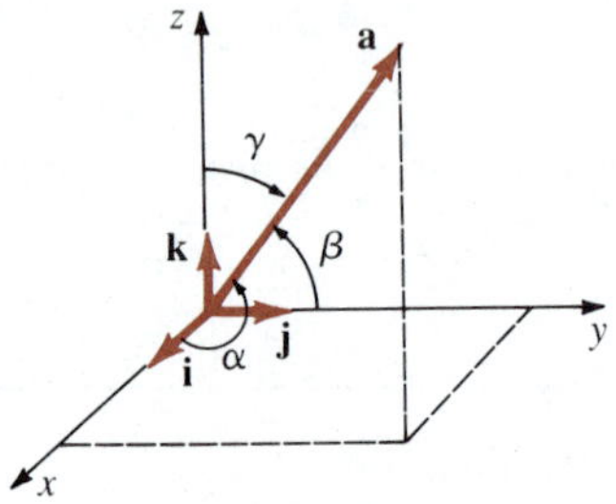

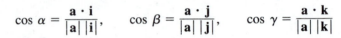

$$\cos \alpha = \frac{\mathbf{a} \cdot \mathbf{i}}{|\mathbf{a}|\,|\mathbf{i}|}, \qquad \cos \beta = \frac{\mathbf{a} \cdot \mathbf{j}}{|\mathbf{a}|\,|\mathbf{j}|}, \qquad \cos \gamma = \frac{\mathbf{a} \cdot \mathbf{k}}{|\mathbf{a}|\,|\mathbf{k}|}$$

which simplify to

$$\cos \alpha = \frac{a_1}{|\mathbf{a}|}, \qquad \cos \beta = \frac{a_2}{|\mathbf{a}|}, \qquad \cos \gamma = \frac{a_3}{|\mathbf{a}|}$$

We say that $\cos \alpha$, $\cos \beta$, $\cos \gamma$, are the **direction cosines** of $\mathbf{a}$. The direction cosines of a nonzero vector $\mathbf{a}$ are simply the components of the unit vector $(1/|\mathbf{a}|)\mathbf{a}$:

Figure 14.31

$$\frac{1}{|\mathbf{a}|}\mathbf{a} = \frac{a_1}{|\mathbf{a}|}\mathbf{i} + \frac{a_2}{|\mathbf{a}|}\mathbf{j} + \frac{a_3}{|\mathbf{a}|}\mathbf{k}$$

$$= (\cos \alpha)\mathbf{i} + (\cos \beta)\mathbf{j} + (\cos \gamma)\mathbf{k}$$

Since the magnitude of $(1/|\mathbf{a}|)\mathbf{a}$ is 1, it follows from the last equation that

$$\cos^2\alpha + \cos^2\beta + \cos^2\gamma = 1$$

Example 6

Find the direction cosines and direction angles of the vector $\mathbf{a} = 2\mathbf{i} + 5\mathbf{j} + 4\mathbf{k}$.

Solution From $|\mathbf{a}| = \sqrt{2^2 + 5^2 + 4^2} = \sqrt{45} = 3\sqrt{5}$, we see that the direction cosines are

$$\cos \alpha = \frac{2}{3\sqrt{5}}, \qquad \cos \beta = \frac{5}{3\sqrt{5}}, \qquad \cos \gamma = \frac{4}{3\sqrt{5}}$$

The direction angles are

$$\alpha = \cos^{-1}\left(\frac{2}{3\sqrt{5}}\right) \approx 1.27 \text{ radians} \quad \text{or} \quad \alpha \approx 72.7°$$

$$\beta = \cos^{-1}\left(\frac{5}{3\sqrt{5}}\right) \approx 0.73 \text{ radian} \quad \text{or} \quad \beta \approx 41.8°$$

$$\gamma = \cos^{-1}\left(\frac{4}{3\sqrt{5}}\right) \approx 0.93 \text{ radian} \quad \text{or} \quad \gamma \approx 53.4°$$

Observe in Example 6 that

$$\cos^2\alpha + \cos^2\beta + \cos^2\gamma = \frac{4}{45} + \frac{25}{45} + \frac{16}{45} = 1$$

Component of **a** *on* **b**   Using the distributive law and (14.12) enables us to express the components of a vector $\mathbf{a} = a_1\mathbf{i} + a_2\mathbf{j} + a_3\mathbf{k}$ in terms of the dot product:

$$a_1 = \mathbf{a} \cdot \mathbf{i}, \qquad a_2 = \mathbf{a} \cdot \mathbf{j}, \qquad a_3 = \mathbf{a} \cdot \mathbf{k} \qquad (14.16)$$

Symbolically, we write the components of **a** as

$$\text{comp}_\mathbf{i}\mathbf{a} = \mathbf{a} \cdot \mathbf{i}, \qquad \text{comp}_\mathbf{j}\mathbf{a} = \mathbf{a} \cdot \mathbf{j}, \qquad \text{comp}_\mathbf{k}\mathbf{a} = \mathbf{a} \cdot \mathbf{k} \qquad (14.17)$$

We shall now see that the procedure indicated in (14.17) carries over to finding the **component of a on an arbitrary vector b**. Note that in either of the two cases shown in Figure 14.32,

$$\text{comp}_\mathbf{b}\mathbf{a} = |\mathbf{a}| \cos\theta \qquad (14.18)$$

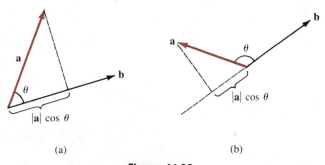

(a)   (b)

Figure 14.32

In Figure 14.32(b), $\text{comp}_\mathbf{b}\mathbf{a} < 0$ since $\pi/2 < \theta \le \pi$. Now, by writing (14.18) as

$$\text{comp}_\mathbf{b}\mathbf{a} = \frac{|\mathbf{a}|\,|\mathbf{b}|\cos\theta}{|\mathbf{b}|} = \frac{\mathbf{a} \cdot \mathbf{b}}{|\mathbf{b}|}$$

we see that

$$\text{comp}_\mathbf{b}\mathbf{a} = \mathbf{a} \cdot \left(\frac{1}{|\mathbf{b}|}\mathbf{b}\right) \qquad (14.19)$$

In other words,

> *to find the component of **a** on a vector **b**, we dot **a** with a unit vector in the direction of **b**.*

___ **Example 7** ___

Let $\mathbf{a} = 2\mathbf{i} + 3\mathbf{j} - 4\mathbf{k}$ and $\mathbf{b} = \mathbf{i} + \mathbf{j} + 2\mathbf{k}$. Find

(a) $\text{comp}_\mathbf{b}\mathbf{a}$ (b) $\text{comp}_\mathbf{a}\mathbf{b}$

Solution

(*a*)   We first form a unit vector in the direction of **b**:

$$|\mathbf{b}| = \sqrt{6}, \qquad \frac{1}{|\mathbf{b}|}\mathbf{b} = \frac{1}{\sqrt{6}}(\mathbf{i} + \mathbf{j} + 2\mathbf{k})$$

Then from (14.19) we have

$$\text{comp}_\mathbf{b}\mathbf{a} = (2\mathbf{i} + 3\mathbf{j} - 4\mathbf{k}) \cdot \frac{1}{\sqrt{6}}(\mathbf{i} + \mathbf{j} + 2\mathbf{k}) = -\frac{3}{\sqrt{6}}$$

(*b*)   By modifying (14.19) accordingly, we have

$$\text{comp}_\mathbf{a}\mathbf{b} = \mathbf{b} \cdot \left(\frac{1}{|\mathbf{a}|}\mathbf{a}\right)$$

Therefore,

$$|\mathbf{a}| = \sqrt{29}, \qquad \frac{1}{|\mathbf{a}|}\mathbf{a} = \frac{1}{\sqrt{29}}(2\mathbf{i} + 3\mathbf{j} - 4\mathbf{k})$$

and

$$\text{comp}_\mathbf{a}\mathbf{b} = (\mathbf{i} + \mathbf{j} + 2\mathbf{k}) \cdot \frac{1}{\sqrt{29}}(2\mathbf{i} + 3\mathbf{j} - 4\mathbf{k}) = -\frac{3}{\sqrt{29}}$$

Projection of a onto b   As illustrated in Figure 14.33(a), the projection of a vector **a** in any of the directions determined by **i**, **j**, **k** is simply the *vector* formed by multiplying the component of **a** in the specified direction with a unit vector in that direction; for example,

$$\text{proj}_\mathbf{i}\mathbf{a} = (\text{comp}_\mathbf{i}\mathbf{a})\mathbf{i} = (\mathbf{a} \cdot \mathbf{i})\mathbf{i} = a_1\mathbf{i}$$

and so on. Figure 14.33(b) shows the general case of the **projection of a onto b**:*

$$\text{proj}_\mathbf{b}\mathbf{a} = (\text{comp}_\mathbf{b}\mathbf{a})\left(\frac{1}{|\mathbf{b}|}\mathbf{b}\right) \qquad (14.20)$$

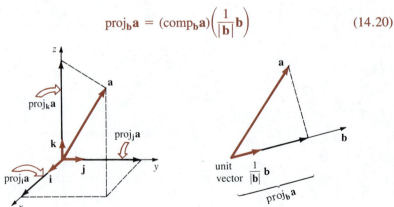

(a)   (b)

Figure 14.33

*Also called the **orthogonal projection of a onto b**.

_____ **Example 8** _____

Find the projection of $\mathbf{a} = 4\mathbf{i} + \mathbf{j}$ on the vector $\mathbf{b} = 2\mathbf{i} + 3\mathbf{j}$. Graph.

Solution   First, we find the component of $\mathbf{a}$ on $\mathbf{b}$. Since $|\mathbf{b}| = \sqrt{13}$, we find from (14.19),

$$\mathrm{comp}_\mathbf{b}\mathbf{a} = (4\mathbf{i} + \mathbf{j}) \cdot \frac{1}{\sqrt{13}}(2\mathbf{i} + 3\mathbf{j}) = \frac{11}{\sqrt{13}}$$

Thus, from (14.20),

$$\mathrm{proj}_\mathbf{b}\mathbf{a} = \left(\frac{11}{\sqrt{13}}\right)\left(\frac{1}{\sqrt{13}}\right)(2\mathbf{i} + 3\mathbf{j})$$

$$= \frac{22}{13}\mathbf{i} + \frac{33}{13}\mathbf{j}$$

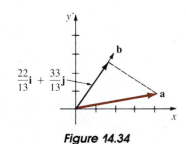

Figure 14.34

The graph of this vector is shown in Figure 14.34.

Projection of a onto b⊥   If $\mathbf{b} \neq \mathbf{0}$, any vector $\mathbf{a}$ can be projected onto $\mathbf{b}$ as well as onto a vector $\mathbf{b}^\perp$, of magnitude $|\mathbf{b}|$, that is orthogonal to $\mathbf{b}$. From Figure 14.35, we see that $\mathbf{a}$ can be written as the sum of two projections:

$$\mathrm{proj}_\mathbf{b}\mathbf{a} + \mathrm{proj}_{\mathbf{b}^\perp}\mathbf{a} = \mathbf{a} \tag{14.21}$$

Equation (14.21) enables us to define the projection of $\mathbf{a}$ onto $\mathbf{b}^\perp$:

$$\mathrm{proj}_{\mathbf{b}^\perp}\mathbf{a} = \mathbf{a} - \mathrm{proj}_\mathbf{b}\mathbf{a} \tag{14.22}$$

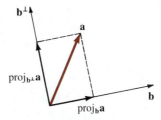

Figure 14.35

_____ **Example 9** _____

Let $\mathbf{a} = 3\mathbf{i} - \mathbf{j} + 5\mathbf{k}$ and $\mathbf{b} = 2\mathbf{i} + \mathbf{j} + 2\mathbf{k}$. Find

(a) $\mathrm{proj}_\mathbf{b}\mathbf{a}$ (b) $\mathrm{proj}_{\mathbf{b}^\perp}\mathbf{a}$

Solution   Since $|\mathbf{b}| = 3$, we have

$$\mathrm{comp}_\mathbf{b}\mathbf{a} = (3\mathbf{i} - \mathbf{j} + 5\mathbf{k}) \cdot \frac{1}{3}(2\mathbf{i} + \mathbf{j} + 2\mathbf{k})$$

$$= \frac{15}{3} = 5$$

(a) $\text{proj}_b\mathbf{a} = (5)\left(\dfrac{1}{3}\right)(2\mathbf{i} + \mathbf{j} + 2\mathbf{k})$

$$= \frac{10}{3}\mathbf{i} + \frac{5}{3}\mathbf{j} + \frac{10}{3}\mathbf{k}$$

(b) $\text{proj}_{b^\perp}\mathbf{a} = \mathbf{a} - \text{proj}_b\mathbf{a}$

$$= (3\mathbf{i} - \mathbf{j} + 5\mathbf{k}) - \left(\frac{10}{3}\mathbf{i} + \frac{5}{3}\mathbf{j} + \frac{10}{3}\mathbf{k}\right)$$

$$= -\frac{1}{3}\mathbf{i} - \frac{8}{3}\mathbf{j} + \frac{5}{3}\mathbf{k}$$

Physical Interpretation of the Dot Product

In Section 6.9 we saw that when a constant force of magnitude F moves an object a distance d in the same direction of the force, the work done is simply

$$W = Fd \tag{14.23}$$

However, if a constant force $\mathbf{F}$ applied to a body acts at an angle θ to the direction of motion, then the work done by $\mathbf{F}$ is defined to be the product of the component of $\mathbf{F}$ in the direction of the displacement and the distance $|\mathbf{d}|$ that the body moves:

$$W = (|\mathbf{F}|\cos\theta)|\mathbf{d}| = |\mathbf{F}||\mathbf{d}|\cos\theta$$

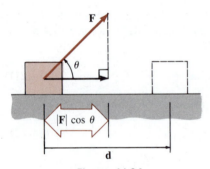

See Figure 14.36. It follows from Definition 14.3 that if $\mathbf{F}$ causes a displacement $\mathbf{d}$ of a body, then the work done is

$$W = \mathbf{F} \cdot \mathbf{d} \tag{14.24}$$

Note that (14.24) reduces to (14.23) when $\theta = 0$.

Figure 14.36

Example 10

Find the work done by a constant force $\mathbf{F} = 2\mathbf{i} + 4\mathbf{j}$ if its point of application to a block moves from $P_1(1, 1)$ to $P_2(4, 6)$. Assume that $|\mathbf{F}|$ is measured in newtons and $|\mathbf{d}|$ is measured in meters.

Solution   The displacement of the block is given by

$$\mathbf{d} = \overrightarrow{P_1P_2} = \overrightarrow{OP_2} - \overrightarrow{OP_1}$$
$$= 3\mathbf{i} + 5\mathbf{j}$$

It follows from (14.24) that the work done is

$$W = (2\mathbf{i} + 4\mathbf{j}) \cdot (3\mathbf{i} + 5\mathbf{j}) = 26 \text{ nt-m}$$

Exercises 14.3

Answers to odd-numbered problems begin on page A-44.

In Problems 1 and 2 find $\mathbf{a} \cdot \mathbf{b}$ if the smaller angle between $\mathbf{a}$ and $\mathbf{b}$ is as given.

1. $|\mathbf{a}| = 10, |\mathbf{b}| = 5, \theta = \dfrac{\pi}{4}$

2. $|\mathbf{a}| = 6, |\mathbf{b}| = 12, \theta = \dfrac{\pi}{6}$

In Problems 3–14, $\mathbf{a} = \langle 2, -3, 4 \rangle$, $\mathbf{b} = \langle -1, 2, 5 \rangle$, and $\mathbf{c} = \langle 3, 6, -1 \rangle$. Find the indicated number or vector.

3. $\mathbf{a} \cdot \mathbf{b}$

4. $\mathbf{b} \cdot \mathbf{c}$

5. $\mathbf{a} \cdot \mathbf{c}$

6. $\mathbf{a} \cdot (\mathbf{b} + \mathbf{c})$

7. $\mathbf{a} \cdot (4\mathbf{b})$

8. $\mathbf{b} \cdot (\mathbf{a} - \mathbf{c})$

9. $\mathbf{a} \cdot \mathbf{a}$

10. $(2\mathbf{b}) \cdot (3\mathbf{c})$

11. $\mathbf{a} \cdot (\mathbf{a} + \mathbf{b} + \mathbf{c})$

12. $(2\mathbf{a}) \cdot (\mathbf{a} - 2\mathbf{b})$

13. $\left(\dfrac{\mathbf{a} \cdot \mathbf{b}}{\mathbf{b} \cdot \mathbf{b}} \right) \mathbf{b}$

14. $(\mathbf{c} \cdot \mathbf{b})\mathbf{a}$

15. Determine which pairs of the following vectors are orthogonal:

    (a) $\langle 2, 0, 1 \rangle$

    (b) $3\mathbf{i} + 2\mathbf{j} - \mathbf{k}$

    (c) $2\mathbf{i} - \mathbf{j} - \mathbf{k}$

    (d) $\mathbf{i} - 4\mathbf{j} + 6\mathbf{k}$

    (e) $\langle 1, -1, 1 \rangle$

    (f) $\langle -4, 3, 8 \rangle$

16. Determine a scalar c so that the given vectors are orthogonal.

    (a) $\mathbf{a} = 2\mathbf{i} - c\mathbf{j} + 3\mathbf{k}$
       $\mathbf{b} = 3\mathbf{i} + 2\mathbf{j} + 4\mathbf{k}$

    (b) $\mathbf{a} = \langle c, \frac{1}{2}, c \rangle$
       $\mathbf{b} = \langle -3, 4, c \rangle$

17. Find a vector $\mathbf{v} = \langle x_1, y_1, 1 \rangle$ that is orthogonal to both $\mathbf{a} = \langle 3, 1, -1 \rangle$ and $\mathbf{b} = \langle -3, 2, 2 \rangle$.

18. Determine a scalar c so that the angle between $\mathbf{a} = \mathbf{i} + c\mathbf{j}$ and $\mathbf{b} = \mathbf{i} + \mathbf{j}$ is 45°.

In Problems 19–22 find the angle θ between the given vectors.

19. $\mathbf{a} = 3\mathbf{i} - \mathbf{k}, \mathbf{b} = 2\mathbf{i} + 2\mathbf{k}$

20. $\mathbf{a} = 2\mathbf{i} + \mathbf{j}, \mathbf{b} = -3\mathbf{i} - 4\mathbf{j}$

21. $\mathbf{a} = \langle 2, 4, 0 \rangle, \mathbf{b} = \langle -1, -1, 4 \rangle$

22. $\mathbf{a} = \langle \frac{1}{2}, \frac{1}{2}, \frac{3}{2} \rangle, \mathbf{b} = \langle 2, -4, 6 \rangle$

In Problems 23–26 find the direction cosines and direction angles of the given vector.

23. $\mathbf{a} = \mathbf{i} + 2\mathbf{j} + 3\mathbf{k}$

24. $\mathbf{a} = 6\mathbf{i} + 6\mathbf{j} - 3\mathbf{k}$

25. $\mathbf{a} = \langle 1, 0, -\sqrt{3} \rangle$

26. $\mathbf{a} = \langle 5, 7, 2 \rangle$

27. An airplane is 4 km high, 5 km south, and 7 km east of an airport. See Figure 14.37. Find the direction angles of the plane.

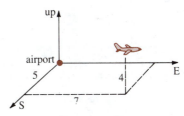

Figure 14.37

28. Determine a unit vector whose direction angles, relative to the three coordinate axes, are equal.

In Problems 29–32, $\mathbf{a} = \langle 1, -1, 3 \rangle$ and $\mathbf{b} = \langle 2, 6, 3 \rangle$. Find the indicated number.

29. $\text{comp}_{\mathbf{b}}\mathbf{a}$

30. $\text{comp}_{\mathbf{a}}\mathbf{b}$

31. $\text{comp}_{\mathbf{a}}(\mathbf{b} - \mathbf{a})$

32. $\text{comp}_{2\mathbf{b}}(\mathbf{a} + \mathbf{b})$

In Problems 33 and 34 find the component of the given vector in the direction from the origin to the indicated point.

33. $\mathbf{a} = 4\mathbf{i} + 6\mathbf{j}; P(3, 10)$

34. $\mathbf{a} = \langle 2, 1, -1 \rangle, P(1, -1, 1)$

In Problems 35–38 find (a) proj$_b$a, and (b) proj$_{b^\perp}$a.

35. $\mathbf{a} = -5\mathbf{i} + 5\mathbf{j}$, $\mathbf{b} = -3\mathbf{i} + 4\mathbf{j}$

36. $\mathbf{a} = 4\mathbf{i} + 2\mathbf{i}$, $\mathbf{b} = -3\mathbf{i} + \mathbf{j}$

37. $\mathbf{a} = -\mathbf{i} - 2\mathbf{j} + 7\mathbf{k}$, $\mathbf{b} = 6\mathbf{i} - 3\mathbf{j} - 2\mathbf{k}$

38. $\mathbf{a} = \langle 1, 1, 1 \rangle$, $\mathbf{b} = \langle -2, 2, -1 \rangle$

In Problems 39 and 40 $\mathbf{a} = 4\mathbf{i} + 3\mathbf{j}$ and $\mathbf{b} = -\mathbf{i} + \mathbf{j}$. Find the indicated vector.

39. proj$_{(a+b)}$a
40. proj$_{(a-b)^\perp}$b

41. A sled is pulled horizontally over ice by a rope attached to its front. A 20-lb force acting at an angle of 60° with the horizontal moves the sled 100 ft. Find the work done.

42. Find the work done if the point at which the constant force $\mathbf{F} = 4\mathbf{i} + 3\mathbf{j} + 5\mathbf{k}$ is applied to an object moves from $P_1(3, 1, -2)$ to $P_2(2, 4, 6)$. Assume $|\mathbf{F}|$ is measured in newtons and $|\mathbf{d}|$ is measured in meters.

43. A block with weight **w** is pulled along a frictionless horizontal surface by a constant force **F** of magnitude 30 nt in the direction given by a vector **d**. See Figure 14.38. Assume $|\mathbf{d}|$ is measured in meters.

   (a) What is the work done by the weight **w**?

   (b) What is the work done by the force **F** if $\mathbf{d} = 4\mathbf{i} + 3\mathbf{j}$?

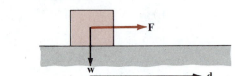

Figure 14.38

44. A constant force **F** of magnitude 3 lb is applied to the block shown in Figure 14.39. **F** has the same direction as the vector $\mathbf{a} = 3\mathbf{i} + 4\mathbf{j}$. Find the work done in the direction of motion if the block moves from $P_1(3, 1)$ to $P_2(9, 3)$. Assume distance is measured in feet.

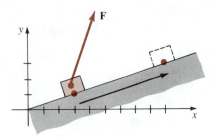

Figure 14.39

Miscellaneous Problems

45. Use Figure 14.40 with $\mathbf{a} = a_1\mathbf{i} + a_2\mathbf{j} + a_3\mathbf{k}$ and $\mathbf{b} = b_1\mathbf{i} + b_2\mathbf{j} + b_3\mathbf{k}$ and the Law of Cosines to prove (14.14).

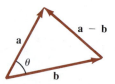

Figure 14.40

46. Use the dot product to prove the **Cauchy–Schwarz inequality**: $|\mathbf{a} \cdot \mathbf{b}| \leq |\mathbf{a}||\mathbf{b}|$.

47. Use the dot product to prove the triangle inequality, $|\mathbf{a} + \mathbf{b}| \leq |\mathbf{a}| + |\mathbf{b}|$. (*Hint:* Consider property (*vi*).)

48. Prove that the vector $\mathbf{n} = a\mathbf{i} + b\mathbf{j}$ is perpendicular to the line whose equation is $ax + by + c = 0$. (*Hint:* Let $P_1(x_1, y_1)$ and $P_2(x_2, y_2)$ be distinct points on the line.)

49. Use the result of Problem 47 and Figure 14.41 to show that the distance d from a point $P_1(x_1, y_1)$ to a line $ax + by + c = 0$ is $d = |ax_1 + by_1 + c|/\sqrt{a^2 + b^2}$.

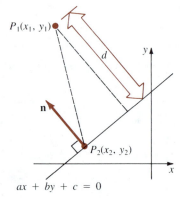

Figure 14.41

50. Prove that if two nonzero vectors **a** and **b** are orthogonal, then their direction cosines satisfy

$$\cos \alpha_1 \cos \alpha_2 + \cos \beta_1 \cos \beta_2 + \cos \gamma_1 \cos \gamma_2 = 0$$

14.4 The Cross Product

Review

Since knowledge of determinants of order 2 and order 3 is important to the discussion that follows, we recall the following facts:

(a) $\begin{vmatrix} a_1 & a_2 \\ b_1 & b_2 \end{vmatrix} = a_1 b_2 - a_2 b_1$

(b) $\begin{vmatrix} a_1 & a_2 & a_3 \\ b_1 & b_2 & b_3 \\ c_1 & c_2 & c_3 \end{vmatrix} = a_1 \begin{vmatrix} b_2 & b_3 \\ c_2 & c_3 \end{vmatrix} - a_2 \begin{vmatrix} b_1 & b_3 \\ c_1 & c_3 \end{vmatrix} + a_3 \begin{vmatrix} b_1 & b_2 \\ c_1 & c_2 \end{vmatrix}$

This is called **expanding the determinant by cofactors** of the first row.

(c) When two rows of a determinant are interchanged, the resulting determinant is the negative of the original.

See Appendix I for a review of the properties of determinants.

In contrast to the dot product, which is a number or a scalar, the next special product of two vectors is a vector.

DEFINITION 14.4   Cross Product of Two Vectors*

The **cross product** of two vectors **a** and **b** is the vector

$$\mathbf{a} \times \mathbf{b} = (|\mathbf{a}||\mathbf{b}| \sin \theta)\mathbf{n} \qquad (14.25)$$

where θ is the angle between the vectors such that $0 \le \theta \le \pi$ and **n** is a unit vector perpendicular to the plane of **a** and **b** with direction given by the right-hand rule.

As seen in Figure 14.42(a)(page 706), if the fingers of the right hand point along the vector **a** and then curl toward the vector **b**, the thumb will give the direction of **n** and hence **a** × **b**. In Figure 14.42(b) the right-hand rule shows the direction of **b** × **a**.

---

*The cross product is also called the **vector product**.

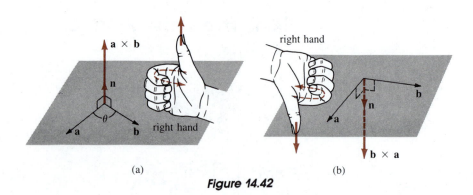

Figure 14.42

Example 1

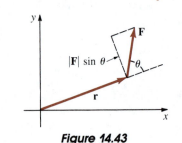

Figure 14.43

In physics a force **F** acting at the end of a position vector **r**, as shown in Figure 14.43, is said to produce a **torque** τ defined by $\tau = \mathbf{r} \times \mathbf{F}$. For example, if $|\mathbf{F}| = 20$ nt, $|\mathbf{r}| = 3.5$ m, $\theta = 30°$, then from (14.25) $|\tau| = (3.5)(20)\sin 30° = 35$ nt · m. If **F** and **r** are in the plane of the page, the right-hand rule implies that the direction of τ is outward from, and perpendicular to, the page (toward the reader).

Properties   The cross product has the following **properties**.

> (*i*) $\mathbf{a} \times \mathbf{b} = \mathbf{0}$ if $\mathbf{a} = \mathbf{0}$ or $\mathbf{b} = \mathbf{0}$
>
> (*ii*) $\mathbf{a} \times \mathbf{b} = -\mathbf{b} \times \mathbf{a}$
>
> (*iii*) $\mathbf{a} \times (\mathbf{b} + \mathbf{c}) = (\mathbf{a} \times \mathbf{b}) + (\mathbf{a} \times \mathbf{c})$
>
> (*iv*) $(\mathbf{a} + \mathbf{b}) \times \mathbf{c} = (\mathbf{a} \times \mathbf{c}) + (\mathbf{b} \times \mathbf{c})$ (Distributive laws)
>
> (*v*) $\mathbf{a} \times (k\mathbf{b}) = (k\mathbf{a}) \times \mathbf{b} = k(\mathbf{a} \times \mathbf{b})$, k a scalar
>
> (*vi*) $\mathbf{a} \times \mathbf{a} = \mathbf{0}$
>
> (*vii*) $\mathbf{a} \cdot (\mathbf{a} \times \mathbf{b}) = 0$
>
> (*viii*) $\mathbf{b} \cdot (\mathbf{a} \times \mathbf{b}) = 0$

Property (*vi*) follows from (14.25) because $\theta = 0$. Properties (*vii*) and (*viii*) are simply statements of the fact that $\mathbf{a} \times \mathbf{b}$ is perpendicular to the plane containing **a** and **b**. Property (*ii*) should be intuitively clear from Figure 14.42.

Parallel Vectors   When the angle between two nonzero vectors is either $\theta = 0$ or $\theta = \pi$, then $\sin \theta = 0$ and so we must have $\mathbf{a} \times \mathbf{b} = \mathbf{0}$. This is stated formally in the next theorem.

_____ **Example 2** _____

(*a*)   From property (*vi*) we have

$$\mathbf{i} \times \mathbf{i} = \mathbf{0}, \qquad \mathbf{j} \times \mathbf{j} = \mathbf{0}, \qquad \mathbf{k} \times \mathbf{k} = \mathbf{0} \qquad (14.26)$$

(*b*)   If $\mathbf{a} = 2\mathbf{i} + \mathbf{j} - \mathbf{k}$ and $\mathbf{b} = -6\mathbf{i} - 3\mathbf{j} + 3\mathbf{k} = -3\mathbf{a}$, then **a** and **b** are parallel. Hence, from Theorem 14.2 $\mathbf{a} \times \mathbf{b} = \mathbf{0}$. Note that this result also follows by combining properties (*v*) and (*vi*).

From (14.25) if $\mathbf{a} = \mathbf{i}$, $\mathbf{b} = \mathbf{j}$, then

$$\mathbf{i} \times \mathbf{j} = \left(|\mathbf{i}||\mathbf{j}| \sin \frac{\pi}{2} \right)\mathbf{n} = \mathbf{n} \qquad (14.27)$$

But, since a unit vector perpendicular to the plane that contains **i** and **j** with the direction given by the right-hand rule is **k**, it follows from (14.27) that $\mathbf{n} = \mathbf{k}$. In other words, $\mathbf{i} \times \mathbf{j} = \mathbf{k}$.

_____ **Example 3** _____

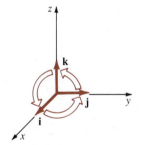

Figure 14.44

The cross products of any pair of vectors in the set **i**, **j**, **k** can be obtained by the circular mnemonic

that is,

$$
\left.
\begin{array}{l}
\mathbf{i} \times \mathbf{j} = \mathbf{k} \\
\mathbf{j} \times \mathbf{k} = \mathbf{i} \\
\mathbf{k} \times \mathbf{i} = \mathbf{j}
\end{array}
\right\}
\quad \text{and from property (\textit{ii})} \quad
\left\{
\begin{array}{l}
\mathbf{j} \times \mathbf{i} = -\mathbf{k} \\
\mathbf{k} \times \mathbf{j} = -\mathbf{i} \\
\mathbf{i} \times \mathbf{k} = -\mathbf{j}
\end{array}
\right.
\qquad (14.28)
$$

See Figure 14.44.

Alternative Definition of the Cross Product   As we did for the dot product, we can use the distributive law (*iii*) to arrive at an alternative form of the cross product.

$$
\begin{aligned}
\mathbf{a} \times \mathbf{b} &= (a_1\mathbf{i} + a_2\mathbf{j} + a_3\mathbf{k}) \times (b_1\mathbf{i} + b_2\mathbf{j} + b_3\mathbf{k}) \\
&= a_1\mathbf{i} \times (b_1\mathbf{i} + b_2\mathbf{j} + b_3\mathbf{k}) + a_2\mathbf{j} \times (b_1\mathbf{i} + b_2\mathbf{j} + b_3\mathbf{k}) \\
&\quad + a_3\mathbf{k} \times (b_1\mathbf{i} + b_2\mathbf{j} + b_3\mathbf{k}) \\
&= a_1b_1(\mathbf{i} \times \mathbf{i}) + a_1b_2(\mathbf{i} \times \mathbf{j}) + a_1b_3(\mathbf{i} \times \mathbf{k}) \\
&\quad + a_2b_1(\mathbf{j} \times \mathbf{i}) + a_2b_2(\mathbf{j} \times \mathbf{j}) + a_2b_3(\mathbf{j} \times \mathbf{k}) \\
&\quad + a_3b_1(\mathbf{k} \times \mathbf{i}) + a_3b_2(\mathbf{k} \times \mathbf{j}) + a_3b_3(\mathbf{k} \times \mathbf{k}) \qquad (14.29)
\end{aligned}
$$

Using the results in (14.27) and (14.28), (14.29) simplifies to

$$\mathbf{a} \times \mathbf{b} = (a_2 b_3 - a_3 b_2)\mathbf{i} - (a_1 b_3 - a_3 b_1)\mathbf{j} + (a_1 b_2 - a_2 b_1)\mathbf{k} \quad (14.30)$$

By taking a quick glance at (*a*) of the introductory review, we note that the components of the vector in (14.30) can written as determinants of order 2:

$$\mathbf{a} \times \mathbf{b} = \begin{vmatrix} a_2 & a_3 \\ b_2 & b_3 \end{vmatrix}\mathbf{i} - \begin{vmatrix} a_1 & a_3 \\ b_1 & b_3 \end{vmatrix}\mathbf{j} + \begin{vmatrix} a_1 & a_2 \\ b_1 & b_2 \end{vmatrix}\mathbf{k} \quad (14.31)$$

In turn, inspection of (*b*) of the review suggests that (14.31) can be written as

$$\mathbf{a} \times \mathbf{b} = \begin{vmatrix} \mathbf{i} & \mathbf{j} & \mathbf{k} \\ a_1 & a_2 & a_3 \\ b_1 & b_2 & b_3 \end{vmatrix} \quad (14.32)$$

The expression on the right-hand side of the equality in (14.32) is not an actual determinant since its entries are not all scalars; (14.32) is simply a way of remembering the complicated expression in (14.31).

Example 4

Let $\mathbf{a} = 4\mathbf{i} - 2\mathbf{j} + 5\mathbf{k}$ and $\mathbf{b} = 3\mathbf{i} + \mathbf{j} - \mathbf{k}$. Find $\mathbf{a} \times \mathbf{b}$.

Solution   From (14.32) we have

$$\mathbf{a} \times \mathbf{b} = \begin{vmatrix} \mathbf{i} & \mathbf{j} & \mathbf{k} \\ 4 & -2 & 5 \\ 3 & 1 & -1 \end{vmatrix}$$

$$= \begin{vmatrix} -2 & 5 \\ 1 & -1 \end{vmatrix}\mathbf{i} - \begin{vmatrix} 4 & 5 \\ 3 & -1 \end{vmatrix}\mathbf{j} + \begin{vmatrix} 4 & -2 \\ 3 & 1 \end{vmatrix}\mathbf{k}$$

$$= -3\mathbf{i} + 19\mathbf{j} + 10\mathbf{k}$$

The form of the cross product given in (14.31) enables us to prove some of the properties (*i*)–(*viii*). For example, to prove (*ii*) we write

$$\mathbf{a} \times \mathbf{b} = \begin{vmatrix} a_2 & a_3 \\ b_2 & b_3 \end{vmatrix}\mathbf{i} - \begin{vmatrix} a_1 & a_3 \\ b_1 & b_3 \end{vmatrix}\mathbf{j} + \begin{vmatrix} a_1 & a_2 \\ b_1 & b_2 \end{vmatrix}\mathbf{k}$$

$$= -\begin{vmatrix} b_2 & b_3 \\ a_2 & a_3 \end{vmatrix}\mathbf{i} + \begin{vmatrix} b_1 & b_3 \\ a_1 & a_3 \end{vmatrix}\mathbf{j} - \begin{vmatrix} b_1 & b_2 \\ a_1 & a_2 \end{vmatrix}\mathbf{k} \text{ [from (\textit{c}) of the review]}$$

$$= -\left(\begin{vmatrix} b_2 & b_3 \\ a_2 & a_3 \end{vmatrix}\mathbf{i} - \begin{vmatrix} b_1 & b_3 \\ a_1 & a_3 \end{vmatrix}\mathbf{j} + \begin{vmatrix} b_1 & b_2 \\ a_1 & a_2 \end{vmatrix}\mathbf{k} \right) = -\mathbf{b} \times \mathbf{a}$$

We leave the proof of property (*iii*) as an exercise.

Special Products The so-called **triple scalar product** of vectors **a**, **b**, and **c** is

$$\mathbf{a} \cdot (\mathbf{b} \times \mathbf{c})$$

Now,

$$\mathbf{a} \cdot (\mathbf{b} \times \mathbf{c}) = (a_1\mathbf{i} + a_2\mathbf{j} + a_3\mathbf{k}) \cdot \left[\begin{vmatrix} b_2 & b_3 \\ c_2 & c_3 \end{vmatrix}\mathbf{i} - \begin{vmatrix} b_1 & b_3 \\ c_1 & c_3 \end{vmatrix}\mathbf{j} + \begin{vmatrix} b_1 & b_2 \\ c_1 & c_2 \end{vmatrix}\mathbf{k} \right]$$

$$= a_1\begin{vmatrix} b_2 & b_3 \\ c_2 & c_3 \end{vmatrix} - a_2\begin{vmatrix} b_1 & b_3 \\ c_1 & c_3 \end{vmatrix} + a_3\begin{vmatrix} b_1 & b_2 \\ c_1 & c_2 \end{vmatrix}$$

Thus, we see

$$\mathbf{a} \cdot (\mathbf{b} \times \mathbf{c}) = \begin{vmatrix} a_1 & a_2 & a_3 \\ b_1 & b_2 & b_3 \\ c_1 & c_2 & c_3 \end{vmatrix} \tag{14.33}$$

Furthermore, from the properties of determinants, we also have

$$\mathbf{a} \cdot (\mathbf{b} \times \mathbf{c}) = (\mathbf{a} \times \mathbf{b}) \cdot \mathbf{c}$$

The **triple vector product** of three vectors **a**, **b**, and **c** is

$$\mathbf{a} \times (\mathbf{b} \times \mathbf{c})$$

It is left as an exercise to show that

$$\mathbf{a} \times (\mathbf{b} \times \mathbf{c}) = (\mathbf{a} \cdot \mathbf{c})\mathbf{b} - (\mathbf{a} \cdot \mathbf{b})\mathbf{c} \tag{14.34}$$

Areas and Volume Two nonzero and nonparallel vectors **a** and **b** can be considered to be the sides of a parallelogram. The **area** A **of a parallelogram** is

$$A = (\text{base}) \cdot (\text{altitude})$$

From Figure 14.45(a), we see

$$A = |\mathbf{b}|(|\mathbf{a}| \sin \theta) = |\mathbf{a}||\mathbf{b}| \sin \theta$$

or $$A = |\mathbf{a} \times \mathbf{b}| \tag{14.35}$$

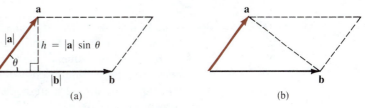

Figure 14.45

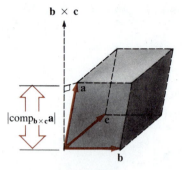

Figure 14.46

Likewise from Figure 14.45(b), we see that the **area of a triangle** with sides **a** and **b** is

$$A = \frac{1}{2}|\mathbf{a} \times \mathbf{b}| \tag{14.36}$$

Similarly, if the vectors **a**, **b**, and **c** do not lie in the same plane, then the **volume of the parallelepiped** with edges **a**, **b**, and **c** shown in Figure 14.46 is

$$V = (\text{area of base})(\text{height})$$
$$= |\mathbf{b} \times \mathbf{c}||\text{comp}_{\mathbf{b}\times\mathbf{c}}\mathbf{a}|$$
$$= |\mathbf{b} \times \mathbf{c}|\left|\mathbf{a} \cdot \left(\frac{1}{|\mathbf{b} \times \mathbf{c}|}\mathbf{b} \times \mathbf{c}\right)\right|$$

or $$V = |\mathbf{a} \cdot (\mathbf{b} \times \mathbf{c})| \tag{14.37}$$

Example 5

Find the area of the triangle determined by the points $P_1(1, 1, 1)$, $P_2(2, 3, 4)$, and $P_3(3, 0, -1)$.

Solution   The vectors $\overrightarrow{P_1P_2}$ and $\overrightarrow{P_2P_3}$ can be taken as two sides of the triangle. Since

$$\overrightarrow{P_1P_2} = \mathbf{i} + 2\mathbf{j} + 3\mathbf{k} \quad \text{and} \quad \overrightarrow{P_2P_3} = \mathbf{i} - 3\mathbf{j} - 5\mathbf{k}$$

we have

$$\overrightarrow{P_1P_2} \times \overrightarrow{P_2P_3} = \begin{vmatrix} \mathbf{i} & \mathbf{j} & \mathbf{k} \\ 1 & 2 & 3 \\ 1 & -3 & -5 \end{vmatrix} = \begin{vmatrix} 2 & 3 \\ -3 & -5 \end{vmatrix}\mathbf{i} - \begin{vmatrix} 1 & 3 \\ 1 & -5 \end{vmatrix}\mathbf{j} + \begin{vmatrix} 1 & 2 \\ 1 & -3 \end{vmatrix}\mathbf{k}$$

$$= -\mathbf{i} + 8\mathbf{j} - 5\mathbf{k}$$

From (14.36) we see that the area is

$$A = \frac{1}{2}|-\mathbf{i} + 8\mathbf{j} - 5\mathbf{k}| = \frac{3}{2}\sqrt{10} \text{ square units}$$

Coplanar Vectors   Vectors that lie in the same plane are said to be **coplanar**. We have just seen that if the vectors **a**, **b**, and **c** are not coplanar, then necessarily $\mathbf{a} \cdot (\mathbf{b} \times \mathbf{c}) \neq 0$ since the volume of a parallelepiped with edges **a**, **b**, and **c** has nonzero volume. Equivalently stated, this means that if $\mathbf{a} \cdot (\mathbf{b} \times \mathbf{c}) = 0$, then the vectors **a**, **b**, and **c** are coplanar. Since the converse of this last statement is also true, we have

$$\mathbf{a} \cdot (\mathbf{b} \times \mathbf{c}) = 0 \text{ *if and only if* } \mathbf{a}, \mathbf{b}, \text{ *and* } \mathbf{c} \text{ *are coplanar.*}$$

Remark

When working with vectors, one should be careful not to mix the symbols $\cdot$ and $\times$ with the symbols for ordinary multiplication, and to be especially careful in the use, or lack of use, of parentheses. For example, expressions such as

$$\mathbf{a} \times \mathbf{b} \times \mathbf{c}, \qquad \mathbf{a} \cdot \mathbf{b} \times \mathbf{c}, \qquad \mathbf{a} \cdot \mathbf{b} \cdot \mathbf{c}, \qquad \mathbf{a} \cdot \mathbf{bc}$$

are not meaningful or well-defined.

_____ Exercises 14.4 _____

Answers to odd-numbered problems begin on page A-44.

In Problems 1–10 find $\mathbf{a} \times \mathbf{b}$.

1. $\mathbf{a} = \mathbf{i} - \mathbf{j}, \mathbf{b} = 3\mathbf{j} + 5\mathbf{k}$

2. $\mathbf{a} = 2\mathbf{i} + \mathbf{j}, \mathbf{b} = 4\mathbf{i} - \mathbf{k}$

3. $\mathbf{a} = \langle 1, -3, 1 \rangle, \mathbf{b} = \langle 2, 0, 4 \rangle$

4. $\mathbf{a} = \langle 1, 1, 1 \rangle, \mathbf{b} = \langle -5, 2, 3 \rangle$

5. $\mathbf{a} = 2\mathbf{i} - \mathbf{j} + 2\mathbf{k}, \mathbf{b} = -\mathbf{i} + 3\mathbf{j} - \mathbf{k}$

6. $\mathbf{a} = 4\mathbf{i} + \mathbf{j} - 5\mathbf{k}, \mathbf{b} = 2\mathbf{i} + 3\mathbf{j} - \mathbf{k}$

7. $\mathbf{a} = \langle \frac{1}{2}, 0, \frac{1}{2} \rangle, \mathbf{b} = \langle 4, 6, 0 \rangle$

8. $\mathbf{a} = \langle 0, 5, 0 \rangle, \mathbf{b} = \langle 2, -3, 4 \rangle$

9. $\mathbf{a} = \langle 2, 2, -4 \rangle, \mathbf{b} = \langle -3, -3, 6 \rangle$

10. $\mathbf{a} = \langle 8, 1, -6 \rangle, \mathbf{b} = \langle 1, -2, 10 \rangle$

In Problems 11 and 12 find $\overrightarrow{P_1P_2} \times \overrightarrow{P_1P_3}$.

11. $P_1(2, 1, 3), P_2(0, 3, -1), P_3(-1, 2, 4)$

12. $P_1(0, 0, 1), P_2(0, 1, 2), P_3(1, 2, 3)$

In Problems 13 and 14 find a vector that is perpendicular to both $\mathbf{a}$ and $\mathbf{b}$.

13. $\mathbf{a} = 2\mathbf{i} + 7\mathbf{j} - 4\mathbf{k}, \mathbf{b} = \mathbf{i} + \mathbf{j} - \mathbf{k}$

14. $\mathbf{a} = \langle -1, -2, 4 \rangle, \mathbf{b} = \langle 4, -1, 0 \rangle$

In Problems 15 and 16 verify that $\mathbf{a} \cdot (\mathbf{a} \times \mathbf{b}) = 0$ and $\mathbf{b} \cdot (\mathbf{a} \times \mathbf{b}) = 0$.

15. $\mathbf{a} = \langle 5, -2, 1 \rangle, \mathbf{b} = \langle 2, 0, -7 \rangle$

16. $\mathbf{a} = \frac{1}{2}\mathbf{i} - \frac{1}{4}\mathbf{j}, \mathbf{b} = 2\mathbf{i} - 2\mathbf{j} + 6\mathbf{k}$

In Problems 17 and 18 **(a)** calculate $\mathbf{b} \times \mathbf{c}$ followed by $\mathbf{a} \times (\mathbf{b} \times \mathbf{c})$. **(b)** Verify the results in part **(a)** by (14.34) of this section.

17. $\mathbf{a} = \mathbf{i} - \mathbf{j} + 2\mathbf{k}$
   $\mathbf{b} = 2\mathbf{i} + \mathbf{j} + \mathbf{k}$
   $\mathbf{c} = 3\mathbf{i} + \mathbf{j} + \mathbf{k}$

18. $\mathbf{a} = 3\mathbf{i} - 4\mathbf{k}$
   $\mathbf{b} = \mathbf{i} + 2\mathbf{j} - \mathbf{k}$
   $\mathbf{c} = -\mathbf{i} + 5\mathbf{j} + 8\mathbf{k}$

In Problems 19–36 find the indicated number or vector *without* using (14.32), (14.33), or (14.34).

19. $(2\mathbf{i}) \times \mathbf{j}$

20. $\mathbf{i} \times (-3\mathbf{k})$

21. $\mathbf{k} \times (2\mathbf{i} - \mathbf{j})$

22. $\mathbf{i} \times (\mathbf{j} \times \mathbf{k})$

23. $[(2\mathbf{k}) \times (3\mathbf{j})] \times (4\mathbf{j})$

24. $(2\mathbf{i} - \mathbf{j} + 5\mathbf{k}) \times \mathbf{i}$

25. $(\mathbf{i} + \mathbf{j}) \times (\mathbf{i} + 5\mathbf{k})$

26. $\mathbf{i} \times \mathbf{k} - 2(\mathbf{j} \times \mathbf{i})$

27. $\mathbf{k} \cdot (\mathbf{j} \times \mathbf{k})$

28. $\mathbf{i} \cdot [\mathbf{j} \times (-\mathbf{k})]$

29. $|4\mathbf{j} - 5(\mathbf{i} \times \mathbf{j})|$

30. $(\mathbf{i} \times \mathbf{j}) \cdot (3\mathbf{j} \times \mathbf{i})$

31. $\mathbf{i} \times (\mathbf{i} \times \mathbf{j})$

32. $(\mathbf{i} \times \mathbf{j}) \times \mathbf{i}$

33. $(\mathbf{i} \times \mathbf{i}) \times \mathbf{j}$

34. $(\mathbf{i} \cdot \mathbf{i})(\mathbf{i} \times \mathbf{j})$

35. $2\mathbf{j} \cdot [\mathbf{i} \times (\mathbf{j} - 3\mathbf{k})]$

36. $(\mathbf{i} \times \mathbf{k}) \times (\mathbf{j} \times \mathbf{i})$

In Problems 37–44, $\mathbf{a} \times \mathbf{b} = 4\mathbf{i} - 3\mathbf{j} + 6\mathbf{k}$ and $\mathbf{c} = 2\mathbf{i} + 4\mathbf{j} - \mathbf{k}$. Find the indicated number or vector.

37. $\mathbf{a} \times (3\mathbf{b})$

38. $\mathbf{b} \times \mathbf{a}$

39. $(-\mathbf{a}) \times \mathbf{b}$

40. $|\mathbf{a} \times \mathbf{b}|$

41. $(\mathbf{a} \times \mathbf{b}) \times \mathbf{c}$

42. $(\mathbf{a} \times \mathbf{b}) \cdot \mathbf{c}$

43. $\mathbf{a} \cdot (\mathbf{b} \times \mathbf{c})$

44. $(4\mathbf{a}) \cdot (\mathbf{b} \times \mathbf{c})$

In Problems 45–48 find the area of the triangle determined by the given points.

45. $P_1(1, 1, 1), P_2(1, 2, 1), P_3(1, 1, 2)$

46. $P_1(0, 0, 0), P_2(0, 1, 2), P_3(2, 2, 0)$

47. $P_1(1, 2, 4), P_2(1, -1, 3), P_3(-1, -1, 2)$

48. $P_1(1, 0, 3), P_2(0, 0, 6), P_3(2, 4, 5)$

In Problems 49 and 50 find the volume of the parallelepiped for which the given vectors are three edges.

49. $\mathbf{a} = \mathbf{i} + \mathbf{j}, \mathbf{b} = -\mathbf{i} + 4\mathbf{j}, \mathbf{c} = 2\mathbf{i} + 2\mathbf{j} + 2\mathbf{k}$

50. $\mathbf{a} = 3\mathbf{i} + \mathbf{j} + \mathbf{k}, \mathbf{b} = \mathbf{i} + 4\mathbf{j} + \mathbf{k}, \mathbf{c} = \mathbf{i} + \mathbf{j} + 5\mathbf{k}$

51. Determine whether the vectors $\mathbf{a} = 4\mathbf{i} + 6\mathbf{j}$, $\mathbf{b} = -2\mathbf{i} + 6\mathbf{j} - 6\mathbf{k}$, and $\mathbf{c} = \frac{5}{2}\mathbf{i} + 3\mathbf{j} + \frac{1}{2}\mathbf{k}$ are coplanar.

52. Determine whether the four points $P_1(1, \ 1, \ -2)$, $P_2(4, 0, -3)$, $P_3(1, -5, 10)$, and $P_4(-7, 2, 4)$ lie in the same plane.

53. As shown in Figure 14.47, the vector $\mathbf{a}$ lies in the xy-plane and the vector $\mathbf{b}$ lies along the positive z-axis. Their magnitudes are $|\mathbf{a}| = 6.4$ and $|\mathbf{b}| = 5$.

(a) Use Definition 14.4 to find $|\mathbf{a} \times \mathbf{b}|$.

(b) Use the right-hand rule to find the direction of $\mathbf{a} \times \mathbf{b}$.

(c) Use part (b) to express $\mathbf{a} \times \mathbf{b}$ in terms of the unit vectors $\mathbf{i}$, $\mathbf{j}$, $\mathbf{k}$.

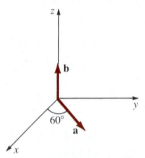

Figure 14.47

54. Two vectors $\mathbf{a}$ and $\mathbf{b}$ lie in the xz-plane so that the angle between them is $120°$. If $|\mathbf{a}| = \sqrt{27}$ and $|\mathbf{b}| = 8$, find all possible values of $\mathbf{a} \times \mathbf{b}$.

Miscellaneous Problems

55. Use (14.31) to prove property (*iii*) of the cross product.

56. Prove $\mathbf{a} \times (\mathbf{b} \times \mathbf{c}) = (\mathbf{a} \cdot \mathbf{c})\mathbf{b} - (\mathbf{a} \cdot \mathbf{b})\mathbf{c}$.

57. Prove or disprove $\mathbf{a} \times (\mathbf{b} \times \mathbf{c}) = (\mathbf{a} \times \mathbf{b}) \times \mathbf{c}$.

58. Prove $\mathbf{a} \cdot (\mathbf{b} \times \mathbf{c}) = (\mathbf{a} \times \mathbf{b}) \cdot \mathbf{c}$.

59. Prove $\mathbf{a} \times (\mathbf{b} \times \mathbf{c}) + \mathbf{b} \times (\mathbf{c} \times \mathbf{a}) + \mathbf{c} \times (\mathbf{a} \times \mathbf{b}) = \mathbf{0}$.

60. Prove **Lagrange's* identity**

$$|\mathbf{a} \times \mathbf{b}|^2 = |\mathbf{a}|^2|\mathbf{b}|^2 - (\mathbf{a} \cdot \mathbf{b})^2$$

61. Does $\mathbf{a} \times \mathbf{b} = \mathbf{a} \times \mathbf{c}$ imply that $\mathbf{b} = \mathbf{c}$?

62. Show that $(\mathbf{a} + \mathbf{b}) \times (\mathbf{a} - \mathbf{b}) = 2\mathbf{b} \times \mathbf{a}$.

14.5 Lines in 3-Space

Vector Equation   As in the plane, any two distinct points in 3-space determine only one line between them. To find an equation of the line through $P_1(x_1, \ y_1, \ z_1)$ and $P_2(x_2, \ y_2, \ z_2)$ let us assume that $P(x, \ y, z)$ is *any* point on the line. Referring to Figure 14.48, if $\mathbf{r} = \overrightarrow{OP}$, $\mathbf{r}_1 = \overrightarrow{OP_1}$, and $\mathbf{r}_2 = \overrightarrow{OP_2}$, we see that vector $\mathbf{a} = \mathbf{r}_2 - \mathbf{r}_1$ is parallel to vector $\mathbf{r} - \mathbf{r}_2$. Thus,

$$\mathbf{r} - \mathbf{r}_2 = t(\mathbf{r}_2 - \mathbf{r}_1) \qquad (14.38)$$

If we write

$$\mathbf{a} = \mathbf{r}_2 - \mathbf{r}_1 = \langle x_2 - x_1, y_2 - y_1, z_2 - z_1 \rangle = \langle a_1, a_2, a_3 \rangle$$

then (14.38) implies that a **vector equation** for the line $\mathscr{L}_a$ is

$$\mathbf{r} = \mathbf{r}_2 + t\mathbf{a} \qquad (14.39)$$

The vector $\mathbf{a}$ is called a **direction vector** of the line.

Note: Since $\mathbf{r} - \mathbf{r}_1$ is also parallel to $\mathscr{L}_a$, an alternative vector equation for the line is $\mathbf{r} = \mathbf{r}_1 + t\mathbf{a}$. Indeed $\mathbf{r} = \mathbf{r}_1 + t(-\mathbf{a})$ and $\mathbf{r} = \mathbf{r}_1 + t(k\mathbf{a})$, k a nonzero scalar, are also equations for $\mathscr{L}_a$.

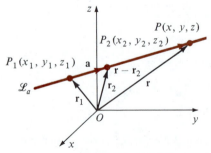

Figure 14.48

---

*Joseph Louis Lagrange (1738–1813), a French mathematician, was called by Napoleon Bonaparte a "lofty pyramid of . . . mathematical sciences."

Example 1

Find a vector equation for the line through $(2, -1, 8)$ and $(5, 6, -3)$.

Solution Define

$$\mathbf{a} = \langle 2 - 5, -1 - 6, 8 - (-3) \rangle = \langle -3, -7, 11 \rangle$$

The following are three possible vector equations for the line:

$$\langle x, y, z \rangle = \langle 2, -1, 8 \rangle + t\langle -3, -7, 11 \rangle \tag{14.40}$$
$$\langle x, y, z \rangle = \langle 5, 6, -3 \rangle + t\langle -3, -7, 11 \rangle \tag{14.41}$$
$$\langle x, y, z \rangle = \langle 5, 6, -3 \rangle + t\langle 3, 7, -11 \rangle \tag{14.42}$$

Parametric Equations By writing (14.39) as

$$\langle x, y, z \rangle = \langle x_2 + t(x_2 - x_1), y_2 + t(y_2 - y_1), z_2 + t(z_2 - z_1) \rangle$$
$$= \langle x_2 + a_1 t, y_2 + a_2 t, z_2 + a_3 t \rangle$$

and equating components, we obtain

$$x = x_2 + a_1 t, \qquad y = y_2 + a_2 t, \qquad z = z_2 + a_3 t \tag{14.43}$$

The equations in (14.43) are called **parametric equations** for the line through P_1 and P_2.

Example 2

Find parametric equations for the line in Example 1.

Solution From (14.40) it follows that

$$x = 2 - 3t, \qquad y = -1 - 7t, \qquad z = 8 + 11t \tag{14.44}$$

An alternative set of parametric equations can be obtained from (14.42):

$$x = 5 + 3t, \qquad y = 6 + 7t, \qquad z = -3 - 11t \tag{14.45}$$

Note in Example 2 that the value $t = 0$ in (14.44) gives $(2, -1, 8)$, whereas in (14.45), $t = -1$ must be used to obtain the same point.

Example 3

Find a vector $\mathbf{a}$ that is parallel to the line $\mathscr{L}_a$ whose parametric equations are

$$x = 4 + 9t, \qquad y = -14 + 5t, \qquad z = 1 - 3t$$

Solution The coefficients (or a nonzero constant multiple of the coefficients) of the parameter in each equation are the components of a vector that is parallel to the line. Thus, $\mathbf{a} = 9\mathbf{i} + 5\mathbf{j} - 3\mathbf{k}$ is parallel to $\mathscr{L}_a$ and, hence, is a direction vector of the line.

Symmetric Equations   From (14.43) observe that we can clear the parameter by writing

$$t = \frac{x - x_2}{a_1} = \frac{y - y_2}{a_2} = \frac{z - z_2}{a_3}$$

provided that the three numbers a_1, a_2, and a_3 are nonzero. The resulting equations

$$\frac{x - x_2}{a_1} = \frac{y - y_2}{a_2} = \frac{z - z_2}{a_3} \qquad (14.46)$$

are said to be **symmetric equations** for the line through P_1 and P_2.

___ **Example 4** _____

Find symmetric equations for the line through $(4, 10, -6)$ and $(7, 9, 2)$.

Solution   Define $a_1 = 7 - 4 = 3$, $a_2 = 9 - 10 = -1$, and $a_3 = 2 - (-6) = 8$. It follows from (14.46) that symmetric equations for the line are

$$\frac{x - 7}{3} = \frac{y - 9}{-1} = \frac{z - 2}{8}$$

If one of the numbers a_1, a_2, or a_3 is zero in (14.43) we use the remaining two equations to eliminate the parameter t. For example, if $a_1 = 0$, $a_2 \neq 0$, $a_3 \neq 0$, then (14.43) yields

$$x = x_2 \quad \text{and} \quad t = \frac{y - y_2}{a_2} = \frac{z - z_2}{a_3}$$

In this case,

$$x = x_2, \qquad \frac{y - y_2}{a_2} = \frac{z - z_2}{a_3}$$

are symmetric equations for the line.

___ **Example 5** _____

Find symmetric equations for the line through $(5, 3, 1)$ and $(2, 1, 1)$.

Solution   Define $a_1 = 5 - 2 = 3$, $a_2 = 3 - 1 = 2$, and $a_3 = 1 - 1 = 0$. From the preceding discussion it follows that symmetric equations for the line are

$$\frac{x - 5}{3} = \frac{y - 3}{2}, \qquad z = 1$$

In other words, the symmetric equations describe a line in the plane $z = 1$.

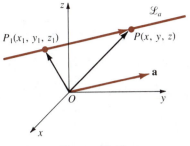

Figure 14.49

A line in space is also determined by specifying a point $P_1(x_1, y_1, z_1)$ and a nonzero direction vector **a**. Through the point P_1 there passes only one line $\mathcal{L}_a$ parallel to the given vector. If $P(x, y, z)$ is a point on the line $\mathcal{L}_a$, shown in Figure 14.49, then, as before,

$$\overrightarrow{OP} - \overrightarrow{OP_1} = t\mathbf{a}$$

or

$$\mathbf{r} = \mathbf{r}_1 + t\mathbf{a}$$

_____ **Example 6** _____

Write vector, parametric, and symmetric equations for the line through $(4, 6, -3)$ and parallel to $\mathbf{a} = 5\mathbf{i} - 10\mathbf{j} + 2\mathbf{k}$.

Solution With $a_1 = 5$, $a_2 = -10$, and $a_3 = 2$ we have immediately,

$$\textit{Vector:}\quad \langle x, y, z \rangle = \langle 4, 6, -3 \rangle + t\langle 5, -10, 2 \rangle$$

$$\textit{Parametric:}\quad x = 4 + 5t, \qquad y = 6 - 10t, \qquad z = -3 + 2t$$

$$\textit{Symmetric:}\quad \frac{x - 4}{5} = \frac{y - 6}{-10} = \frac{z + 3}{2}$$

Orthogonal and Parallel Lines

Let **a** and **b** be direction vectors for lines $\mathcal{L}_a$ and $\mathcal{L}_b$, respectively.

> **DEFINITION 14.5**
>
> (i) $\mathcal{L}_a$ and $\mathcal{L}_b$ are **orthogonal** if $\mathbf{a} \cdot \mathbf{b} = 0$, and
>
> (ii) $\mathcal{L}_a$ and $\mathcal{L}_b$ are **parallel** if $\mathbf{a} = k\mathbf{b}$ for some nonzero scalar k.

_____ **Example 7** _____

The lines

$$\mathcal{L}_a: x = 4 - 2t, \qquad y = 1 + 4t, \qquad z = 3 + 10t$$

$$\mathcal{L}_b: x = s, \qquad y = 6 - 2s, \qquad z = \frac{1}{2} - 5s$$

are parallel since $\mathbf{a} = -2\mathbf{b}$ (or $\mathbf{b} = -\frac{1}{2}\mathbf{a}$), where

$$\mathbf{a} = -2\mathbf{i} + 4\mathbf{j} + 10\mathbf{k} \quad \text{and} \quad \mathbf{b} = \mathbf{i} - 2\mathbf{j} - 5\mathbf{k}$$

_____ **Example 8** _____

Determine whether the lines

$$\mathcal{L}_a\text{: } x = -6 - t, \qquad y = 20 + 3t, \qquad z = 1 + 2t$$
$$\mathcal{L}_b\text{: } x = 5 + 2s, \qquad y = -9 - 4s, \qquad z = 1 + 7s$$

are orthogonal.

Solution   Reading off the coefficients of the parameters, we see that

$$\mathbf{a} = -\mathbf{i} + 3\mathbf{j} + 2\mathbf{k} \quad \text{and} \quad \mathbf{b} = 2\mathbf{i} - 4\mathbf{j} + 7\mathbf{k}$$

are the direction vectors for $\mathcal{L}_a$ and $\mathcal{L}_b$, respectively. Since

$$\mathbf{a} \cdot \mathbf{b} = -2 - 12 + 14 = 0$$

the lines are orthogonal.

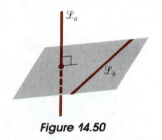

Figure 14.50

Notice that (*i*) of Definition 14.5 does not demand that the two lines intersect in order to be orthogonal. Figure 14.50 shows two orthogonal lines that do not intersect. In other words, $\mathcal{L}_a$ can be perpendicular to the plane containing $\mathcal{L}_b$.

_____ **Example 9** _____

Determine whether the lines $\mathcal{L}_a$ and $\mathcal{L}_b$ in Example 8 intersect.

Solution   Since a point (x, y, z) of intersection is common to both lines, we must have

$$\left.\begin{aligned} -6 - t &= 5 + 2s \\ 20 + 3t &= -9 - 4s \\ 1 + 2t &= 1 + 7s \end{aligned}\right\} \quad \text{or} \quad \left\{\begin{aligned} 2s + t &= -11 \\ 4s + 3t &= -29 \\ -7s + 2t &= 0 \end{aligned}\right. \qquad (14.47)$$

We now solve any *two* of the equations simultaneously and use the remaining equation as a check. Choosing the first and third, we find from

$$2s + t = -11$$
$$-7s + 2t = 0$$

that $s = -2$ and $t = -7$. Substitution of these values in the second equation yields the identity $-8 - 21 = -29$. Thus, $\mathcal{L}_a$ and $\mathcal{L}_b$ intersect. To find the point of intersection, we use, say, $s = -2$:

$$x = 5 + 2(-2), \qquad y = -9 - 4(-2), \qquad z = 1 + 7(-2)$$

or $(1, -1, -13)$.

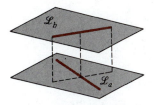

Figure 14.51

In Example 9, had the remaining equation not been satisfied when the values $s = -2$ and $t = -7$ were substituted, then the three equations would not be satisfied simultaneously and so the lines would not intersect.

Two lines $\mathscr{L}_a$ and $\mathscr{L}_b$ in 3-space that do not intersect and are not parallel are called **skew lines**. As shown in Figure 14.51, skew lines lie in parallel planes.

Exercises 14.5

Answers to odd-numbered problems begin on page A-44.

In Problems 1–6 find a vector equation for the line through the given points.

1. $(1, 2, 1)$, $(3, 5, -2)$

2. $(0, 4, 5)$, $(-2, 6, 3)$

3. $\left(\frac{1}{2}, -\frac{1}{2}, 1\right)$, $\left(-\frac{3}{2}, \frac{5}{2}, -\frac{1}{2}\right)$

4. $(10, 2, -10)$, $(5, -3, 5)$

5. $(1, 1, -1)$, $(-4, 1, -1)$

6. $(3, 2, 1)$, $\left(\frac{5}{2}, 1, -2\right)$

In Problems 7–12 find parametric equations for the line through the given points.

7. $(2, 3, 5)$, $(6, -1, 8)$

8. $(2, 0, 0)$, $(0, 4, 9)$

9. $(1, 0, 0)$, $(3, -2, -7)$

10. $(0, 0, 5)$, $(-2, 4, 0)$

11. $\left(4, \frac{1}{2}, \frac{1}{3}\right)$, $\left(-6, -\frac{1}{4}, \frac{1}{6}\right)$

12. $(-3, 7, 9)$, $(4, -8, -1)$

In Problems 13–18 find symmetric equations for the line through the given points.

13. $(1, 4, -9)$, $(10, 14, -2)$

14. $\left(\frac{2}{3}, 0, -\frac{1}{4}\right)$, $\left(1, 3, \frac{1}{4}\right)$

15. $(4, 2, 1)$, $(-7, 2, 5)$

16. $(-5, -2, -4)$, $(1, 1, 2)$

17. $(5, 10, -2)$, $(5, 1, -14)$

18. $\left(\frac{5}{6}, -\frac{1}{4}, \frac{1}{5}\right)$, $\left(\frac{1}{3}, \frac{3}{8}, -\frac{1}{10}\right)$

In Problems 19–22 find parametric and symmetric equations for the line through the given point parallel to the given vector.

19. $(4, 6, -7)$, $\mathbf{a} = \left\langle 3, \frac{1}{2}, -\frac{3}{2}\right\rangle$

20. $(1, 8, -2)$, $\mathbf{a} = -7\mathbf{i} - 8\mathbf{j}$

21. $(0, 0, 0)$, $\mathbf{a} = 5\mathbf{i} + 9\mathbf{j} + 4\mathbf{k}$

22. $(0, -3, 10)$, $\mathbf{a} = \langle 12, -5, -6\rangle$

23. Find parametric equations for the line through $(6, 4, -2)$ that is parallel to the line $x/2 = (1 - y)/3 = (z - 5)/6$.

24. Find symmetric equations for the line through $(4, -11, -7)$ that is parallel to the line $x = 2 + 5t$, $y = -1 + t/3$, $z = 9 - 2t$.

25. Find parametric equations for the line through $(2, -2, 15)$ that is parallel to the xz-plane and the xy-plane.

26. Find parametric equations for the line through $(1, 2, 8)$ that is **(a)** parallel to the y-axis, **(b)** perpendicular to the xy-plane.

27. Show that the lines given by $\mathbf{r} = t\langle 1, 1, 1\rangle$ and $\mathbf{r} = \langle 6, 6, 6\rangle + t\langle -3, -3, -3\rangle$ are the same.

28. Determine which of the following lines are orthogonal and which are parallel:

(a) $\mathbf{r} = \langle 1, 0, 2\rangle + t\langle 9, -12, 6\rangle$

(b) $x = 1 + 9t$, $y = 12t$, $z = 2 - 6t$

(c) $x = 2t$, $y = -3t$, $z = 4t$

(d) $x = 5 + t$, $y = 4t$, $z = 3 + \frac{5}{2}t$

(e) $x = 1 + t$, $y = \frac{3}{2}t$, $z = 2 - \frac{3}{2}t$

(f) $\frac{x + 1}{-3} = \frac{y + 6}{4} = \frac{z - 3}{-2}$

In Problems 29 and 30 determine the points of intersection of the given line and the three coordinate planes.

29. $x = 4 - 2t, \ y = 1 + 2t, \ z = 9 + 3t$

30. $\dfrac{x-1}{2} = \dfrac{y+2}{3} = \dfrac{z-4}{2}$

In Problems 31–34 determine whether the given lines intersect. If so, find the point of intersection.

31. $x = 4 + t, \ y = 5 + t, \ z = -1 + 2t$
$x = 6 + 2s, \ y = 11 + 4s, \ z = -3 + s$

32. $x = 1 + t, \ y = 2 - t, \ z = 3t$
$x = 2 - s, \ y = 1 + s, \ z = 6s$

33. $x = 2 - t, \ y = 3 + t, \ z = 1 + t$
$x = 4 + s, \ y = 1 + s, \ z = 1 - s$

34. $x = 3 - t, \ y = 2 + t, \ z = 8 + 2t$
$x = 2 + 2s, \ y = -2 + 3s, \ z = -2 + 8s$

The angle between two lines $\mathcal{L}_a$ and $\mathcal{L}_b$ is the angle between their direction vectors **a** and **b**. In Problems 35 and 36 find the angle between the given lines.

35. $x = 4 - t, \ y = 3 + 2t, \ z = -2t;$
$x = 5 + 2s, \ y = 1 + 3s, \ z = 5 - 6s$

36. $\dfrac{x-1}{2} = \dfrac{y+5}{7} = \dfrac{z-1}{-1}; \dfrac{x+3}{-2} = y - 9 = \dfrac{z}{4}$

In Problems 37 and 38 the given lines lie in the same plane. Find parametric equations for the line through the indicated point that is perpendicular to this plane.

37. $x = 3 + t, \ y = -2 + t, \ z = 9 + t$
$x = 1 - 2s, \ y = 5 + s, \ z = -2 - 5s; \ (4, 1, 6)$

38. $\dfrac{x-1}{3} = \dfrac{y+1}{2} = \dfrac{z}{4}$
$\dfrac{x+4}{6} = \dfrac{y-6}{4} = \dfrac{z-10}{8}; \ (1, -1, 0)$

In Problems 39 and 40 show that the given lines are skew.

39. $x = -3 + t, \ y = 7 + 3t, \ z = 5 + 2t$
$x = 4 + s, \ y = 8 - 2s, \ z = 10 - 4s$

40. $x = 6 + 2t, \ y = 6t, \ z = -8 + 10t$
$x = 7 + 8s, \ y = 4 - 4s, \ z = 3 - 24s$

41. Suppose $\mathcal{L}_a$ and $\mathcal{L}_b$ are skew lines. Let P_1 and P_2 be points on line $\mathcal{L}_a$ and let P_3 and P_4 be points on line $\mathcal{L}_b$. Use the vector $\overrightarrow{P_1 P_3}$, shown in Figure 14.52, to show that the shortest distance d between $\mathcal{L}_a$ and $\mathcal{L}_b$ is

$$d = \frac{|\overrightarrow{P_1 P_3} \cdot (\overrightarrow{P_1 P_2} \times \overrightarrow{P_3 P_4})|}{|\overrightarrow{P_1 P_2} \times \overrightarrow{P_3 P_4}|}$$

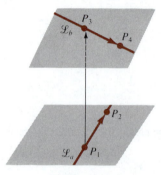

Figure 14.52

14.6 Planes

Vector Equation Figure 14.53(a) illustrates the fact that through a given point $P_1(x_1, y_1, z_1)$ there pass an infinite number of planes. However, as shown in Figure 14.53(b), if a point P_1 and a vector **n** are specified, there is only *one* plane $\mathcal{P}$ containing P_1 with **n normal**, or perpendicular, to the plane. Moreover, if $P(x, y, z)$ is any point on $\mathcal{P}$, and $\mathbf{r} = \overrightarrow{OP}, \ \mathbf{r} = \overrightarrow{OP_1}$, then, as shown in Figure 14.53(c), $\mathbf{r} - \mathbf{r}_1$ is in the plane. It follows that a **vector equation** of the plane is

$$\mathbf{n} \cdot (\mathbf{r} - \mathbf{r}_1) = 0 \qquad (14.48)$$

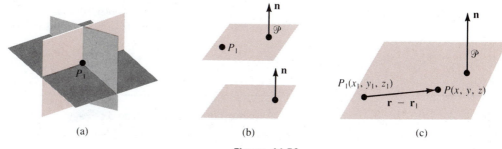

Figure 14.53

Cartesian Equation   Specifically, if the normal vector is $\mathbf{n} = a\mathbf{i} + b\mathbf{j} + c\mathbf{k}$, then (14.48) yields a **Cartesian equation** of the plane containing $P_1(x_1, y_1, z_1)$:

$$a(x - x_1) + b(y - y_1) + c(z - z_1) = 0 \qquad (14.49)$$

Example 1 _____

Find an equation of the plane which contains the point $(4, -1, 3)$ and is perpendicular to the vector $\mathbf{n} = 2\mathbf{i} + 8\mathbf{j} - 5\mathbf{k}$.

Solution   It follows immediately from (14.49) that

$$2(x - 4) + 8(y + 1) - 5(z - 3) = 0 \quad \text{or} \quad 2x + 8y - 5z + 15 = 0$$

The equation (14.49) can always be written as $ax + by + cz + d = 0$ by identifying $d = -ax_1 - by_1 - cz_1$. Conversely, we shall now prove that any linear equation

$$ax + by + cz + d = 0, \qquad a, b, c \text{ not all zero} \qquad (14.50)$$

is a plane.

> **THEOREM 14.3**   The graph of any equation $ax + by + cz + d = 0$, a, b, c not all zero, is a plane with the normal vector $\mathbf{n} = a\mathbf{i} + b\mathbf{j} + c\mathbf{k}$.

Proof   Suppose $x_0, y_0,$ and z_0 are numbers that satisfy the given equation. Then, $ax_0 + by_0 + cz_0 + d = 0$ implies that $d = -ax_0 - by_0 - cz_0$. Replacing this latter value of d in the original equation gives, after simplifying,

$$a(x - x_0) + b(y - y_0) + c(z - z_0) = 0$$

or, in terms of vectors,

$$[a\mathbf{i} + b\mathbf{j} + c\mathbf{k}] \cdot [(x - x_0)\mathbf{i} + (y - y_0)\mathbf{j} + (z - z_0)\mathbf{k}] = 0$$

This last equation implies that $a\mathbf{i} + b\mathbf{j} + c\mathbf{k}$ is normal to the plane containing the point (x_0, y_0, z_0) and the vector

$$(x - x_0)\mathbf{i} + (y - y_0)\mathbf{j} + (z - z_0)\mathbf{k}$$ ∎

_____ **Example 2** _____

A vector normal to the plane $3x - 4y + 10z - 8 = 0$ is $\mathbf{n} = 3\mathbf{i} - 4\mathbf{j} + 10\mathbf{k}$.

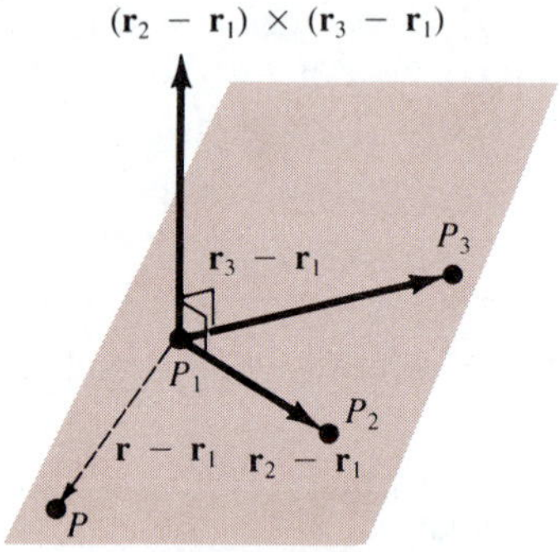

Of course, a nonzero scalar multiple of a normal vector is still perpendicular to the plane.

Three noncollinear points P_1, P_2, and P_3 also determine a plane.* To obtain an equation of the plane, we need only form two vectors between two pairs of points. As shown in Figure 14.54, their cross product is a vector normal to the plane containing these vectors. If $P(x, y, z)$ represents any point on the plane, and $\mathbf{r} = \overrightarrow{OP}$, $\mathbf{r}_1 = \overrightarrow{OP_1}$, $\mathbf{r}_2 = \overrightarrow{OP_2}$, $\mathbf{r}_3 = \overrightarrow{OP_3}$, then $\mathbf{r} - \mathbf{r}_1$ (or, for that matter, $\mathbf{r} - \mathbf{r}_2$ or $\mathbf{r} - \mathbf{r}_3$) is in the plane. Hence,

$$(\mathbf{r} - \mathbf{r}_1) \cdot [(\mathbf{r}_2 - \mathbf{r}_1) \times (\mathbf{r}_3 - \mathbf{r}_1)] = 0 \qquad (14.51)$$

is a vector equation of the plane. You are urged not to memorize the last formula. The procedure is the same as (14.48) with the exception that the vector normal to the plane is obtained by means of the cross product.

Figure 14.54

_____ **Example 3** _____

Find an equation of the plane that contains $(1, 0, -1)$, $(3, 1, 4)$, and $(2, -2, 0)$.

Solution We need three vectors. Pairing the points on the left as shown yields the vectors on the right. The order in which we subtract is irrelevant.

$$\left.\begin{array}{r}(1, 0, -1) \\ (3, 1, 4) \\ (2, -2, 0) \\ (x, y, z)\end{array}\right\} \quad \begin{array}{l}\mathbf{u} = 2\mathbf{i} + \mathbf{j} + 5\mathbf{k} \\ \mathbf{v} = \mathbf{i} + 3\mathbf{j} + 4\mathbf{k} \\ \mathbf{w} = (x - 2)\mathbf{i} + (y + 2)\mathbf{j} + z\mathbf{k}\end{array}$$

Now,
$$\mathbf{u} \times \mathbf{v} = \begin{vmatrix} \mathbf{i} & \mathbf{j} & \mathbf{k} \\ 2 & 1 & 5 \\ 1 & 3 & 4 \end{vmatrix} = -11\mathbf{i} - 3\mathbf{j} + 5\mathbf{k}$$

From (14.51), we see that a vector equation of the plane is then $\mathbf{w} \cdot (\mathbf{u} \times \mathbf{v}) = 0$. The latter equation yields

$$-11(x - 2) - 3(y + 2) + 5z = 0 \quad \text{or} \quad -11x - 3y + 5z + 16 = 0$$

---

*If you have ever sat at a four-legged table that rocks, you might consider replacing it with a three-legged table.

***Orthogonal and
Parallel Planes***
Figure 14.55 illustrates the plausibility of the following definition about **orthogonal** and **parallel** planes.

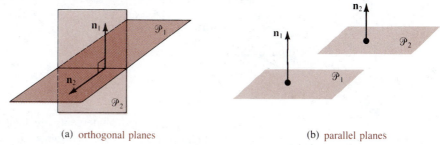

(a) orthogonal planes   (b) parallel planes

Figure 14.55

> **DEFINITION 14.6**   Let $\mathbf{n}_1$ be a normal vector to plane $\mathscr{P}_1$ and let $\mathbf{n}_2$ be a normal vector to plane $\mathscr{P}_2$. Then
>
> (*i*) $\mathscr{P}_1$ and $\mathscr{P}_2$ are **orthogonal** if $\mathbf{n}_1 \cdot \mathbf{n}_2 = 0$.
>
> (*ii*) $\mathscr{P}_1$ and $\mathscr{P}_2$ are **parallel** if $\mathbf{n}_1 = k\mathbf{n}_2$ for some nonzero scalar k.

Example 4

The planes given by

$$\mathscr{P}_1: \quad 2x - 4y + 8z = 7$$
$$\mathscr{P}_2: \quad x - 2y + 4z = 0$$
$$\mathscr{P}_3: -3x + 6y - 12z = 1$$

are parallel since their respective normal vectors

$$\mathbf{n}_1 = 2\mathbf{i} - 4\mathbf{j} + 8\mathbf{k}$$
$$\mathbf{n}_2 = \mathbf{i} - 2\mathbf{j} + 4\mathbf{k} = \tfrac{1}{2}\mathbf{n}_1$$
$$\mathbf{n}_3 = -3\mathbf{i} + 6\mathbf{j} - 12\mathbf{k} = -\tfrac{3}{2}\mathbf{n}_1$$

are parallel.

Graphs
The graph of (14.50) with one or even two variables missing is still a plane. For example, we saw in Section 14.2 that the graphs of

$$x = x_0, \qquad y = y_0, \qquad z = z_0$$

where x_0, y_0, z_0 are constants, are planes perpendicular to the x-, y-, and z-axes, respectively. In general, to graph a plane, we should try to find

      (*i*)   the x-, y-, and z-intercepts, and if necessary,

      (*ii*)   the trace of the plane in each coordinate plane.

A **trace** of a plane in a coordinate plane is the line of intersection of the plane with a coordinate plane.

Example 5

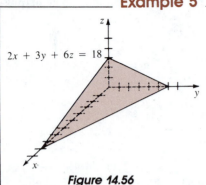

$2x + 3y + 6z = 18$

Figure 14.56

Graph the equation $2x + 3y + 6z = 18$.

Solution   Setting:

$$y = 0, z = 0 \qquad \text{gives } x = 9$$
$$x = 0, z = 0 \qquad \text{gives } y = 6$$
$$x = 0, y = 0 \qquad \text{gives } z = 3$$

The x-, y-, and z-intercepts are then 9, 6, and 3, respectively. As shown in Figure 14.56, we use the points $(9, 0, 0)$, $(0, 6, 0)$, and $(0, 0, 3)$ to draw the graph of the plane in the first octant.

Example 6

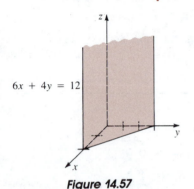

$6x + 4y = 12$

Figure 14.57

Graph the equation $6x + 4y = 12$.

Solution   In two dimensions the graph of the equation is a line with x-intercept 2 and y-intercept 3. However, in three dimensions this line is the trace of a plane in the xy-coordinate plane. Since z is not specified, it can be any real number. In other words, (x, y, z) is a point on the plane provided that x and y are related by the given equation. As shown in Figure 14.57, the graph is a plane parallel to the z-axis.

Example 7

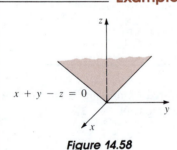

$x + y - z = 0$

Figure 14.58

Graph the equation $x + y - z = 0$.

Solution   First observe that the plane passes through the origin $(0, 0, 0)$. Now, the trace of the plane in the xz-plane ($y = 0$) is $z = x$, whereas its trace in the yz-plane ($x = 0$) is $z = y$. Drawing these two lines leads to the graph given in Figure 14.58.

Two planes $\mathcal{P}_1$ and $\mathcal{P}_2$ that are not parallel must intersect in a line $\mathcal{L}$. See Figure 14.59. Example 8 illustrates one way of finding parametric equations for the line of intersection. In Example 9 we see how to find a point of intersection (x_0, y_0, z_0) of a plane $\mathcal{P}$ and a line $\mathcal{L}$. See Figure 14.60.

Figure 14.59

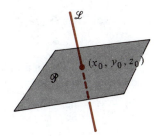

Figure 14.60

Example 8

Find parametric equations for the line of intersection of

$$2x - 3y + 4z = 1$$
$$x - y - z = 5$$

Solution   In a system of two equations and three unknowns, we choose one variable arbitrarily, say, $z = t$ and solve for x and y from

$$2x - 3y = 1 - 4t$$
$$x - y = 5 + t$$

Proceeding, we find

$$x = 14 + 7t, \qquad y = 9 + 6t, \qquad z = t$$

These are parametric equations for the line of intersection of the given planes.

Example 9

Find the point of intersection of the plane $3x - 2y + z = -5$ and the line $x = 1 + t, y = -2 + 2t, z = 4t$.

Solution   If (x_0, y_0, z_0) denotes the point of intersection, then we must have

$$3x_0 - 2y_0 + z_0 = -5$$

and $\qquad x_0 = 1 + t_0, \qquad y = -2 + 2t_0, \qquad z = 4t_0$

for some number t_0. Substituting the latter equations into the equation of the plane gives

$$3(1 + t_0) - 2(-2 + 2t_0) + 4t_0 = -5 \quad \text{or} \quad t_0 = -4$$

From the parametric equations for the line we then obtain $x_0 = -3$, $y_0 = -10$, $z_0 = -16$. The point of intersection is $(-3, -10, -16)$.

Exercises 14.6

Answers to odd-numbered problems begin on page A-45.

In Problems 1–6 find an equation of the plane that contains the given point and is perpendicular to the indicated vector.

1. $(5, 1, 3)$; $2\mathbf{i} - 3\mathbf{j} + 4\mathbf{k}$

2. $(1, 2, 5)$; $4\mathbf{i} - 2\mathbf{j}$

3. $(6, 10, -7)$; $-5\mathbf{i} + 3\mathbf{k}$

4. $(0, 0, 0)$; $6\mathbf{i} - \mathbf{j} + 3\mathbf{k}$

5. $(\frac{1}{2}, \frac{3}{4}, -\frac{1}{2})$; $6\mathbf{i} + 8\mathbf{j} - 4\mathbf{k}$

6. $(-1, 1, 0)$; $-\mathbf{i} + \mathbf{j} - \mathbf{k}$

In Problems 7–12 find, if possible, an equation of a plane that contains the given points.

7. $(3, 5, 2)$, $(2, 3, 1)$, $(-1, -1, 4)$

8. $(0, 1, 0)$, $(0, 1, 1)$, $(1, 3, -1)$

9. $(0, 0, 0)$, $(1, 1, 1)$, $(3, 2, -1)$

10. $(0, 0, 3)$, $(0, -1, 0)$, $(0, 0, 6)$

11. $(1, 2, -1)$, $(4, 3, 1)$, $(7, 4, 3)$

12. $(2, 1, 2)$, $(4, 1, 0)$, $(5, 0, -5)$

In Problems 13–22 find an equation of the plane that satisfies the given conditions.

13. Contains $(2, 3, -5)$ and is parallel to $x + y - 4z = 1$

14. Contains the origin and is parallel to $5x - y + z = 6$

15. Contains $(3, 6, 12)$ and is parallel to the xy-plane

16. Contains $(-7, -5, 18)$ and is perpendicular to the y-axis

17. Contains the lines $x = 1 + 3t$, $y = 1 - t$, $z = 2 + t$; $x = 4 + 4s$, $y = 2s$, $z = 3 + s$

18. Contains the lines $\dfrac{x - 1}{2} = \dfrac{y + 1}{-1} = \dfrac{z - 5}{6}$; $\mathbf{r} = \langle 1, -1, 5 \rangle + t\langle 1, 1, -3 \rangle$

19. Contains the parallel lines $x = 1 + t$, $y = 1 + 2t$, $z = 3 + t$; $x = 3 + s$, $y = 2s$, $z = -2 + s$

20. Contains the point $(4, 0, -6)$ and the line $x = 3t$, $y = 2t$, $z = -2t$

21. Contains $(2, 4, 8)$ and is perpendicular to the line $x = 10 - 3t$, $y = 5 + t$, $z = 6 - \frac{1}{2}t$

22. Contains $(1, 1, 1)$ and is perpendicular to the line through $(2, 6, -3)$, $(1, 0, -2)$

23. Determine which of the following planes are orthogonal and which are parallel:

(a) $2x - y + 3z = 1$

(b) $x + 2y + 2z = 9$

(c) $x + y - \dfrac{3}{2}z = 2$

(d) $-5x + 2y + 4z = 0$

(e) $-8x - 8y + 12z = 1$

(f) $-2x + y - 3z = 5$

24. Find parametric equations for the line that contains $(-4, 1, 7)$ and is perpendicular to the plane $-7x + 2y + 3z = 1$.

25. Determine which of the following planes are perpendicular to the line $x = 4 - 6t$, $y = 1 + 9t$, $z = 2 + 3t$:

(a) $4x + y + 2x = 1$

(b) $2x - 3y + z = 4$

(c) $10x - 15y - 5z = 2$

(d) $-4x + 6y + 2z = 9$

26. Determine which of the following planes are parallel to the line $(1 - x)/2 = (y + 2)/4 = z - 5$:

(a) $x - y + 3z = 1$

(b) $6x - 3y = 1$

(c) $x - 2y + 5z = 0$

(d) $-2x + y - 2z = 7$

In Problems 27–30 find parametric equations for the line of intersection of the given planes.

27. $5x - 4y - 9z = 8$
    $x + 4y + 3z = 4$

28. $x + 2y - z = 2$
    $3x - y + 2z = 1$

29. $4x - 2y - z = 1$
    $x + y + 2z = 1$

30. $2x - 5y + z = 0$
    $y = 0$

In Problems 31–34 find the point of intersection of the given plane and line.

31. $2x - 3y + 2z = -7$; $x = 1 + 2t$, $y = 2 - t$, $z = -3t$

32. $x + y + 4z = 12$; $x = 3 - 2t$, $y = 1 + 6t$,
$z = 2 - \dfrac{1}{2}t$

33. $x + y - z = 8$; $x = 1$, $y = 2$, $z = 1 + t$

34. $x - 3y + 2z = 0$; $x = 4 + t$, $y = 2 + t$, $z = 1 + 5t$

In Problems 35 and 36 find parametric equations for the line through the indicated point that is parallel to the given planes.

35. $x + y - 4z = 2$
$2x - y + z = 10$; $(5, 6, -12)$

36. $2x \quad + z = 0$
$-x + 3y + z = 1$; $(-3, 5, -1)$

In Problems 37 and 38 find an equation of the plane that contains the given line and that is orthogonal to the indicated plane.

37. $x = 4 + 3t$, $y = -t$, $z = 1 + 5t$; $x + y + z = 7$

38. $\dfrac{2 - x}{3} = \dfrac{y + 2}{5} = \dfrac{z - 8}{2}$; $2x - 4y - z + 16 = 0$

In Problems 39–43 graph the given equation.

39. $5x + 2y + z = 10$

40. $3x + 2z = 9$

41. $-y - 3z + 6 = 0$

42. $3x + 4y - 2z - 12 = 0$

43. $-x + 2y + z = 4$

Miscellaneous Problems

44. Let $P_1(x_1, y_1, z_1)$ be a point on the plane $ax + by + cz + d = 0$ and let **n** be a normal vector to the plane. See Figure

14.61. Show that if $P_2(x_2, y_2, z_2)$ is any point not on the plane, then **the distance** D **from a point to a plane is given by**

$$D = \frac{|ax_2 + by_2 + cz_2 + d|}{\sqrt{a^2 + b^2 + c^2}}$$

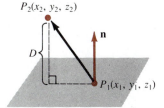

Figure 14.61

In Problems 45 and 46 find the distance from the point to the plane.

45. $(2, 1, 4)$; $x - 3y + z - 6 = 0$

46. $(4, 5, 4)$; $2x + 6y + 3z = -6$

As shown in Figure 14.62, **the angle between two planes** is defined to be the acute angle between their normal vectors. In Problems 47 and 48 find the angles between the given planes.

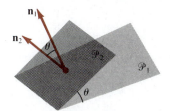

Figure 14.62

47. $x - 3y + 2z = 14$
$-x + y + z = 10$

48. $2x + 6y + 3z = 13$
$4x - 2y + 4z = -7$

14.7 Cylinders and Spheres

14.7.1 Cylinders

In 2-space the graph of $x^2 + y^2 = 1$ is a circle centered at the origin. However, in 3-space we can interpret the graph of the set

$$\{(x, y, z) \mid x^2 + y^2 = 1, z \text{ arbitrary}\}$$

as a **surface** that is the right circular *cylinder* shown in Figure 14.63(b). We

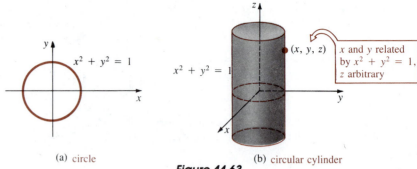

(a) circle   (b) circular cylinder

Figure 14.63

have already seen in Example 6 of Section 14.6 that the graph of an equation $ax + by + c = 0$ is a line in 2-space but a plane in 3-space.

Example 1

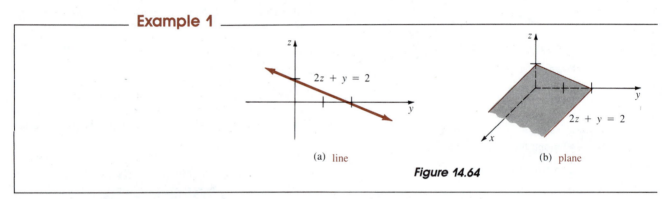

(a) line   (b) plane

Figure 14.64

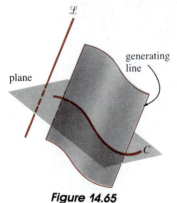

Figure 14.65

Figure 14.66

The surface illustrated in Example 1 is also called a cylinder. We use the term *cylinder* in a more general sense than that of a right circular cylinder. If C is a curve in a plane and $\mathscr{L}$ is a line not parallel to the plane, then the set of all points (x, y, z) generated by a moving line traversing C parallel to $\mathscr{L}$ is called a **cylinder**. The curve C is called the **directrix** of the cylinder. See Figure 14.65.

Thus, an equation of a curve in a coordinate plane, when considered in three dimensions, is an equation of a cylinder perpendicular to that coordinate plane.

If the graphs of $f(x, y) = c_1$, $g(y, z) = c_2$, $h(x, z) = c_3$ are *curves* in the 2-space of their respective coordinate planes, then their graphs in 3-space are *surfaces* called *cylinders*. A cylinder is generated by a moving line that traverses the curve parallel to the coordinate axis, which is represented by the variable missing in its equation.

Figure 14.66 shows a curve C defined by $f(x, y) = c_1$ in the xy-plane and a sequence of lines called **rulings** that represent various positions of a generating line that is traversing C while moving parallel to the z-axis.

In the next four examples, we compare the graph of an equation in a coordinate plane with its interpretation as a cylinder in 3-space (Figures 14.67–14.70). As in Figure 14.64(b), we shall show only a portion of the cylinder.

Example 2

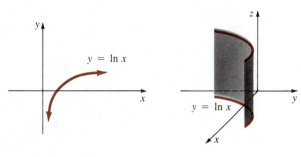

(a) logarithmic curve   (b) logarithmic cylinder

Figure 14.67

Example 3

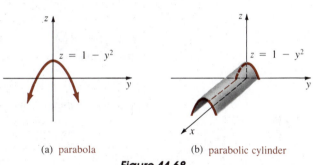

(a) parabola   (b) parabolic cylinder

Figure 14.68

Example 4

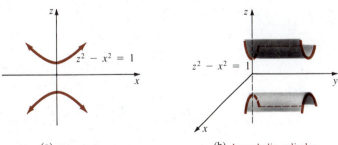

(a) hyperbola   (b) hyperbolic cylinder

Figure 14.69

_____ **Example 5** _____

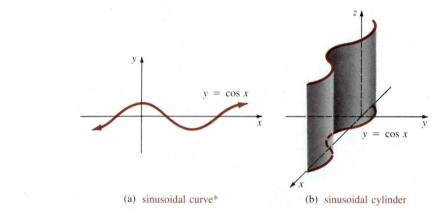

(a) sinusoidal curve*　　　　　　(b) sinusoidal cylinder

Figure 14.70

14.7.2 Spheres

Like a circle, a **sphere** can be defined by means of the distance formula.

> **DEFINITION 14.7**　　　A **sphere** is the set of all points P in 3-space that are equidistant from a fixed point called the **center**.

If r denotes the fixed distance, or **radius** of the sphere, and if the center is $P_1(a, b, c)$, then a point $P(x, y, z)$ is on the sphere if and only if $[d(P_1, P)]^2 = r^2$, or

$$(x - a)^2 + (y - b)^2 + (z - c)^2 = r^2 \qquad (14.52)$$

_____ **Example 6** _____

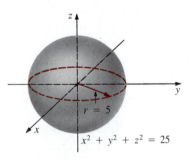

$r = 5$

$x^2 + y^2 + z^2 = 25$

Figure 14.71

Graph $x^2 + y^2 + z^2 = 25$.

Solution　We identify $a = 0$, $b = 0$, $c = 0$, and $r^2 = 25$ in (14.52), and so the graph of $x^2 + y^2 + z^2 = 25$ is a sphere of radius 5 whose center is at the origin. The graph of the equation is given in Figure 14.71.

_____

*The graph of cos x is the graph of the *sine* function shifted $\pi/2$ radians to the left.

Trace of a Surface In general, a **trace of a surface** in any plane is the curve formed by the intersection of the surface and the plane. Note that in Figure 14.71 the trace of the sphere in the xy-plane is the dashed circle $x^2 + y^2 = 25$. In the xz- and yz-planes, the traces of the sphere are the circles $x^2 + z^2 = 25$ and $y^2 + z^2 = 25$, respectively.

Example 7

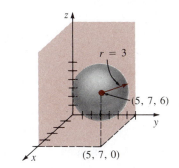

$r = 3$

$(5, 7, 6)$

$(5, 7, 0)$

Figure 14.72

Graph $(x - 5)^2 + (y - 7)^2 + (z - 6)^2 = 9$.

Solution In this case we identify $a = 5$, $b = 7$, $c = 6$, and $r^2 = 9$. From (14.52) we see that the graph of $(x - 5)^2 + (y - 7)^2 + (z - 6)^2 = 9$ is a sphere whose center is at (5, 7, 6) and radius 3. Its graph lies entirely in the first octant and is shown in Figure 14.72.

Example 8

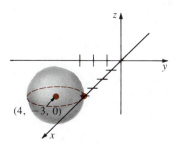

$(4, -3, 0)$

Figure 14.73

Find an equation of the sphere whose center is $(4, -3, 0)$ that is tangent to the xz-plane.

Solution The perpendicular distance from the given point to the xz-plane, and hence the radius of the sphere, is the absolute value of the y-coordinate, $|-3| = 3$. Thus, an equation of the sphere is $(x - 4)^2 + (y + 3)^2 + z^2 = 3^2$. See Figure 14.73.

Example 9

Find the center and radius of the sphere whose equation is

$$16x^2 + 16y^2 + 16z^2 - 16x + 8y - 32z + 16 = 0$$

Solution Dividing by 16 and completing the square in x, y, and z yields

$$\left(x - \frac{1}{2}\right)^2 + \left(y + \frac{1}{4}\right)^2 + (z - 1)^2 = 5$$

The center and radius of the sphere are $(\frac{1}{2}, -\frac{1}{4}, 1)$ and $\sqrt{5}$, respectively.

_____ **Exercises 14.7** _____

Answers to odd-numbered problems begin on page A-45.

[14.7.1]

In Problems 1–12 sketch the graph of the given cylinder.

1. $y = x^2$

2. $x^2 + z^2 = 25$

3. $y^2 + z^2 = 9$

4. $z = y^2$

5. $z = e^{-x}$

6. $z = 1 - e^y$

7. $y^2 - x^2 = 4$

8. $z = \cosh y$

9. $z = \sin x$

10. $y = \dfrac{1}{x^2}$

11. $yz = 1$

12. $z = x^3 - 3x$

[14.7.2]

In Problems 13–16 sketch the graph of the given equation.

13. $x^2 + y^2 + z^2 = 9$

14. $x^2 + y^2 + (z - 3)^2 = 16$

15. $(x - 1)^2 + (y - 1)^2 + (z - 1)^2 = 1$

16. $(x + 3)^2 + (y + 4)^2 + (z - 5)^2 = 4$

In Problems 17–20 find the center and radius of the sphere with the given equation.

17. $x^2 + y^2 + z^2 + 8x - 6y - 4z - 7 = 0$

18. $4x^2 + 4y^2 + 4z^2 + 4x - 12z + 9 = 0$

19. $x^2 + y^2 + z^2 - 16z = 0$

20. $x^2 + y^2 + z^2 - x + y = 0$

In Problems 21–28 find an equation of a sphere that satisfies the given conditions.

21. Center $(-1, 4, 6)$; radius $\sqrt{3}$

22. Center $(0, -3, 0)$; diameter $\frac{5}{2}$

23. Center $(1, 1, 4)$; tangent to the xy-plane

24. Center $(5, 2, -2)$; tangent to the yz-plane

25. Center on the positive y-axis; radius 2; tangent to $x^2 + y^2 + z^2 = 36$

26. Center on the line $x = 2t$, $y = 3t$, $z = 6t$, $t > 0$, at a distance 21 units from the origin; radius 5

27. Diameter has endpoints $(0, -4, 7)$ and $(2, 12, -3)$

28. Center $(-3, 1, 2)$; passing through the origin

14.8 Quadric Surfaces and Surfaces of Revolution

The equation of the sphere given in (14.52) is just a particular case of the second-degree equation

$$Ax^2 + By^2 + Cz^2 + Dx + Ey + Fz + G = 0 \qquad (14.53)$$

When A, B, and C are not all zero, the graph of an equation of form (14.53), describing a real locus, is said to be a **quadric surface**. For example, both the elliptical cylinder $x^2/4 + y^2/9 = 1$ and the parabolic cylinder $z = y^2$ are quadric surfaces. We conclude this section by considering six additional quadric surfaces.

Ellipsoid   The graph of any equation of the form

$$\frac{x^2}{a^2} + \frac{y^2}{b^2} + \frac{z^2}{c^2} = 1, \qquad a > 0, b > 0, c > 0 \qquad (14.54)$$

is called an **ellipsoid**. For $|y_0| < b$, the equation

$$\frac{x^2}{a^2} + \frac{z^2}{c^2} = 1 - \frac{y_0^2}{b^2}$$

represents a family of ellipses (or circles if $a = c$) parallel to the xz-plane that are formed by slicing the surface by planes $y = y_0$. By choosing, in turn, $x = x_0$, $z = z_0$, we would find that slices of the surface are ellipses (or circles) parallel to the yz- and xy-planes, respectively. Figure 14.74 summarizes the traces in the coordinate planes and gives a typical graph.

Coordinate plane	Trace
xy ($z = 0$)	ellipse: $\dfrac{x^2}{a^2} + \dfrac{y^2}{b^2} = 1$
xz ($y = 0$)	ellipse: $\dfrac{x^2}{a^2} + \dfrac{z^2}{c^2} = 1$
yz ($x = 0$)	ellipse: $\dfrac{y^2}{b^2} + \dfrac{z^2}{c^2} = 1$

(a)

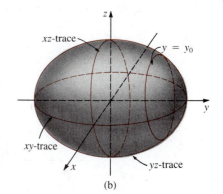

(b)

Figure 14.74

Hyperboloid of One Sheet

The graph of an equation of the form

$$\frac{x^2}{a^2} + \frac{y^2}{b^2} - \frac{z^2}{c^2} = 1, \qquad a > 0,\ b > 0,\ c > 0 \tag{14.55}$$

is called a **hyperboloid of one sheet**. In this case, a plane $z = z_0$, parallel to the xy-plane, slices the surface into elliptical (or circular if $a = b$) cross sections. The equations of these ellipses are

$$\frac{x^2}{a^2} + \frac{y^2}{b^2} = 1 + \frac{z_0^2}{c^2}$$

The smallest ellipse, $z_0 = 0$, corresponds to the trace in the xy-plane. A summary of the traces and a typical graph of (14.55) are given in Figure 14.75.

Coordinate plane	Trace
xy ($z = 0$)	ellipse: $\dfrac{x^2}{a^2} + \dfrac{y^2}{b^2} = 1$
xz ($y = 0$)	hyperbola: $\dfrac{x^2}{a^2} - \dfrac{z^2}{c^2} = 1$
yz ($x = 0$)	hyperbola: $\dfrac{y^2}{b^2} - \dfrac{z^2}{c^2} = 1$

(a)

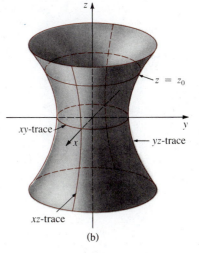

(b)

Figure 14.75

Hyperboloid of Two Sheets   As seen in Figure 14.76(b), a graph of

$$-\frac{x^2}{a^2} + \frac{y^2}{b^2} - \frac{z^2}{c^2} = 1, \qquad a > 0, b > 0, c > 0 \qquad (14.56)$$

is appropriately called a **hyperboloid of two sheets**.

Coordinate plane	Trace
xy ($z = 0$)	hyperbola: $-\dfrac{x^2}{a^2} + \dfrac{y^2}{b^2} = 1$
xz ($y = 0$)	no locus
yz ($x = 0$)	hyperbola: $\dfrac{y^2}{b^2} - \dfrac{z^2}{c^2} = 1$

(a)

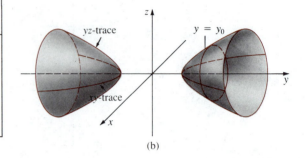

(b)

Figure 14.76

For $|y_0| > b$ the equation $\dfrac{x^2}{a^2} + \dfrac{z^2}{c^2} = \dfrac{y_0^2}{b^2} - 1$ describes the elliptical curve of intersection of the surface with the plane $y = y_0$.

Paraboloid   The graph of an equation of the form

$$\frac{x^2}{a^2} + \frac{y^2}{b^2} = cz \qquad (14.57)$$

is called a **paraboloid**. In Figure 14.77(b) we see that for $c > 0$, planes $z = z_0 > 0$, parallel to the xy-plane, slice the surface in ellipses whose equations are

$$\frac{x^2}{a^2} + \frac{y^2}{b^2} = cz_0$$

Coordinate plane	Trace
xy ($z = 0$)	point: $(0, 0)$
xz ($y = 0$)	parabola: $\dfrac{x^2}{a^2} = cz$
yz ($x = 0$)	parabola: $\dfrac{y^2}{b^2} = cz$

(a)

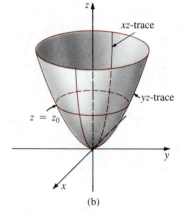

(b)

Figure 14.77

Cone The graph of an equation of the form

$$\frac{x^2}{a^2} + \frac{y^2}{b^2} = \frac{z^2}{c^2}, \qquad a > 0, b > 0, c > 0 \tag{14.58}$$

is called an **elliptical** (or circular if $a = b$) **cone**. For arbitrary z_0, planes parallel to the xy-plane slice the surface in ellipses whose equations are

$$\frac{x^2}{a^2} + \frac{y^2}{b^2} = \frac{z_0^2}{c^2}$$

A typical graph of (14.58) is shown in Figure 14.78(b).

Coordinate plane	Trace
$xy\ (z = 0)$	point: $(0, 0)$
$xz\ (y = 0)$	lines: $z = \pm\dfrac{c}{a}x$
$yz\ (x = 0)$	lines: $z = \pm\dfrac{c}{b}y$

(a)

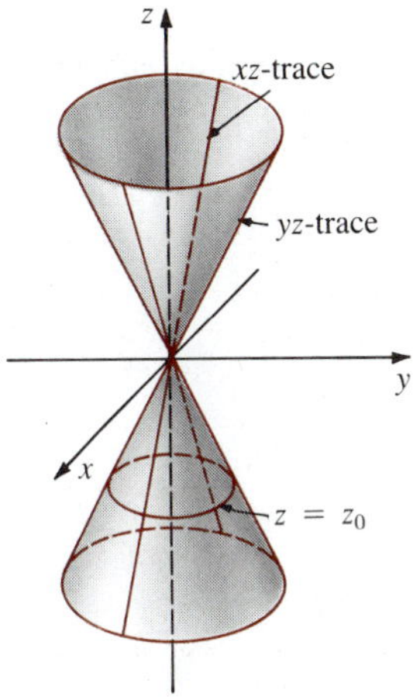

(b)

Figure 14.78

Hyperbolic Paraboloid The last quadric surface we shall consider, known as a **hyperbolic paraboloid**, is the graph of any equation of the form

$$\frac{y^2}{a^2} - \frac{x^2}{b^2} = cz, \qquad a > 0, b > 0 \tag{14.59}$$

Note that for $c > 0$, planes $z = z_0$, parallel to the xy-plane, cut the surface in hyperbolas whose equations are

$$\frac{y^2}{a^2} - \frac{x^2}{b^2} = cz_0$$

The characteristic saddle shape of a hyperbolic paraboloid is shown in Figure 14.79(b).

Coordinate plane	Trace
$xy\ (z = 0)$	lines: $y = \pm\dfrac{a}{b}x$
$xz\ (y = 0)$	parabola: $-\dfrac{x^2}{b^2} = cz$
$yz\ (x = 0)$	parabola: $\dfrac{y^2}{a^2} = cz$

(a)

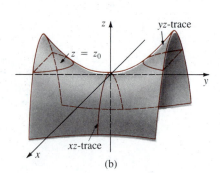

(b)

Figure 14.79

Variation of the Equations

Interchanging the position of the variables in equations (14.54)–(14.59) does not change the basic nature of a surface, but *does* change the surface's orientation in space. For example, graphs of

$$\frac{x^2}{a^2} - \frac{y^2}{b^2} + \frac{z^2}{c^2} = 1 \quad \text{and} \quad -\frac{x^2}{a^2} + \frac{y^2}{b^2} + \frac{z^2}{c^2} = 1 \qquad (14.60)$$

are still hyperboloids of one sheet. Similarly, the two minus signs in (14.56) that characterize hyperboloids of two sheets can occur anywhere in the equation. Similarly,

$$\frac{x^2}{a^2} + \frac{z^2}{b^2} = cy \quad \text{and} \quad \frac{y^2}{a^2} + \frac{z^2}{b^2} = cx \qquad (14.61)$$

are paraboloids. Graphs of equations of the form

$$\frac{x^2}{a^2} - \frac{z^2}{b^2} = cy \quad \text{and} \quad \frac{y^2}{a^2} - \frac{z^2}{b^2} = cx \qquad (14.62)$$

are hyperbolic paraboloids.

Example 1

Identify (*a*) $y = x^2 + z^2$ and (*b*) $y = x^2 - z^2$. Compare the graphs.

Solution   From the first equations in (14.61) and (14.62) with $a = 1$, $b = 1$, and $c = 1$, we identify the graph of (*a*) as a paraboloid and the graph of (*b*) as a hyperbolic paraboloid. In the case of equation (*a*), a plane $y = y_0$, $y_0 > 0$ slices the surface in circles whose equations are $y_0 = x^2 + z^2$. On the other hand, a plane $y = y_0$ slices the graph of equation (*b*) in hyperbolas $y_0 = x^2 - z^2$. The graphs are compared in Figure 14.80.

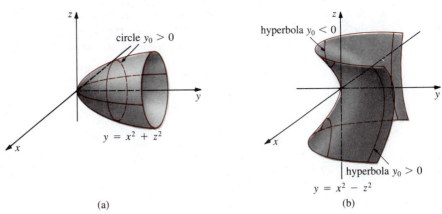

Figure 14.80

(a)   (b)

Example 2

Identify (*a*) $2x^2 - 4y^2 + z^2 = 0$ and (*b*) $-2x^2 + 4y^2 + z^2 = -36$.

Solution

(*a*) From

$$\frac{x^2}{2} + \frac{z^2}{4} = y^2$$

we identify the graph as a cone.

(*b*) From

$$\frac{x^2}{18} - \frac{y^2}{9} - \frac{z^2}{36} = 1$$

we see the graph as a hyperboloid of two sheets.

Origin at (h, k, l)   When the origin is translated to (h, k, l), the equations of the quadric surfaces become

$$\frac{(x - h)^2}{a^2} + \frac{(y - k)^2}{b^2} + \frac{(z - l)^2}{c^2} = 1$$

$$\frac{(x - h)^2}{a^2} + \frac{(y - k)^2}{b^2} - \frac{(z - l)^2}{c^2} = 1$$

and so on.

Example 3

Graph $z = 4 - x^2 - y^2$.

Solution   By writing the equation as

$$-(z - 4) = x^2 + y^2$$

we recognize the equation of a paraboloid. The minus sign in front of the term on the left side of the equality indicates that the graph of the paraboloid opens downward from (0, 0, 4). See Figure 14.81.

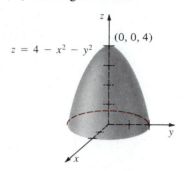

$$z = 4 - x^2 - y^2$$

(0, 0, 4)

Figure 14.81

Surfaces of Revolution

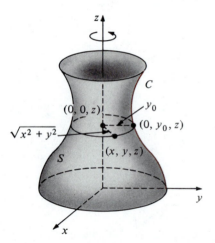

Figure 14.82

In Section 6.6 we saw that a surface S could be generated by revolving a plane curve C about an axis. In the discussion that follows we shall find equations of **surfaces of revolution** when C is a curve in a coordinate plane and the axis of revolution is a coordinate axis.

For the sake of discussion, let us suppose that $f(y, z) = 0$ is an equation of a curve C in the yz-plane and that C is revolved about the z-axis generating a surface S. Let us also suppose for the moment that the y- and z-coordinates of points on C are nonnegative. If (x, y, z) denotes a general point on S that results from revolving the point $(0, y_0, z)$ on C, then we see from Figure 14.82 that the distance from (x, y, z) to $(0, 0, z)$ is the same as the distance from $(0, y_0, z)$ to $(0, 0, z)$; that is, $y_0 = \sqrt{x^2 + y^2}$. From the fact that $f(y_0, z) = 0$ we arrive at an equation for S:

$$f(\sqrt{x^2 + y^2}, z) = 0 \qquad (14.63)$$

A curve in a coordinate plane can, of course, be revolved about each coordinate axis. If the curve C in the yz-plane defined by $f(y, z) = 0$ is now revolved about the y-axis, it can be shown that an equation of the resulting surface of revolution is

$$f(y, \sqrt{x^2 + z^2}) = 0 \qquad (14.64)$$

Finally, we note that if there are points $(0, y, z)$ on C for which the y- or z-coordinates are negative, then we replace $\sqrt{x^2 + y^2}$ in (14.63) by $\pm\sqrt{x^2 + y^2}$ and $\sqrt{x^2 + z^2}$ in (14.64) by $\pm\sqrt{x^2 + z^2}$.

Equations of surfaces of revolution generated when a curve in the xy- or xz-plane is revolved about a coordinate axis are analogous to (14.63) and (14.64). As the following table shows, an equation of a surface generated by revolving a curve in a coordinate plane about the

$$
\left.\begin{array}{l}
x\text{-axis} \\
y\text{-axis} \\
z\text{-axis}
\end{array}\right\} \text{ involves the term }
\left\{\begin{array}{l}
\sqrt{y^2 + z^2} \\
\sqrt{x^2 + z^2} \\
\sqrt{x^2 + y^2}
\end{array}\right.
$$

Equation of curve C	Axis of revolution	Equation of surface S
$f(x, y) = 0$	x-axis y-axis	$f(x, \pm\sqrt{y^2 + z^2}) = 0$ $f(\pm\sqrt{x^2 + z^2}, y) = 0$
$f(x, z) = 0$	x-axis z-axis	$f(x, \pm\sqrt{y^2 + z^2}) = 0$ $f(\pm\sqrt{x^2 + y^2}, z) = 0$
$f(y, z) = 0$	y-axis z-axis	$f(y, \pm\sqrt{x^2 + z^2}) = 0$ $f(\pm\sqrt{x^2 + y^2}, z) = 0$

Example 4

(a) In Example 1, the equation $y = x^2 + z^2$ can be written as

$$y = (\pm\sqrt{x^2 + z^2})^2$$

Hence, from the preceding table we see that the surface is generated by revolving either the parabola $y = z^2$ or the parabola $y = x^2$ about the y-axis. The surface shown in Figure 14.80(a) is called a **paraboloid of revolution**.

(b) In Example 3, the equation $-(z - 4) = x^2 + y^2$ can be written as

$$-(z - 4) = (\pm\sqrt{x^2 + y^2})^2$$

The surface is also a paraboloid of revolution. In this case the surface is generated by revolving either the parabola $-(z - 4) = y^2$ or the parabola $-(z - 4) = x^2$ about the z-axis.

Example 5

The graph of $4x^2 + y^2 = 16$ is revolved about the x-axis. Find an equation of the surface of revolution.

Solution The given equation has the form $f(x, y) = 0$. Since the axis of revolution is the x-axis, we see from the table that an equation of the surface of revolution can be found by replacing y by $\pm\sqrt{y^2 + z^2}$. It follows that

$$4x^2 + (\pm\sqrt{y^2 + z^2})^2 = 16 \quad \text{or} \quad 4x^2 + y^2 + z^2 = 16$$

The surface is called an **ellipsoid of revolution**.

Example 6

The graph of $z = y$, $y \geq 0$ is revolved about the z-axis. Find an equation of the surface of revolution.

Solution Since there are no points on the graph of $z = y$, $y \geq 0$, with a negative y-coordinate we obtain an equation for the surface of revolution by substituting $\sqrt{x^2 + y^2}$ for y:

$$z = \sqrt{x^2 + y^2} \tag{14.65}$$

Observe that (14.65) is not the same as $z^2 = x^2 + y^2$. The latter equation is equivalent to $z = \pm\sqrt{x^2 + y^2}$, which is an equation of the surface obtained by revolving the graph of $z = y$ about the z-axis. In other words, the equation $z = \sqrt{x^2 + y^2}$ describes only the upper nappe of the cone whose equation is $z^2 = x^2 + y^2$. See Figure 14.83.

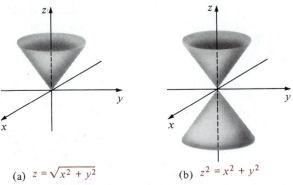

(a) $z = \sqrt{x^2 + y^2}$ (b) $z^2 = x^2 + y^2$

Figure 14.83

Exercises 14.8

Answers to odd-numbered problems begin on page A-46.

In Problems 1–14 identify and graph the quadric surface.

1. $x^2 + y^2 = z$

2. $-x^2 + y^2 = z^2$

3. $9x^2 + 36y^2 + 4z^2 = 36$

4. $x^2 + y^2 - z^2 = -4$

5. $36x^2 - y^2 + 9z^2 = 144$

6. $4x^2 + 4y^2 + z^2 = 100$

7. $y^2 + 5z^2 = x^2$

8. $-9x^2 + 16y^2 = 144z$

9. $y = 4x^2 - z^2$

10. $9z + x^2 + y^2 = 0$

11. $x^2 - y^2 - z^2 = 4$

12. $-x^2 + 9y^2 + z^2 = 9$

13. $y^2 + 4z^2 = x$

14. $x^2 + y^2 - z^2 = 1$

In Problems 15–18 graph the quadric surface.

15. $z = 3 + x^2 + y^2$

16. $y + x^2 + 4z^2 = 4$

17. $(x - 4)^2 + (y - 6)^2 - z^2 = 1$

18. $5x^2 + (y - 5)^2 + 5z^2 = 25$

In Problems 19–22 the given equation is an equation of a surface of revolution obtained by revolving a curve C in a coordinate plane about a coordinate axis. Find an equation for C and identify the axis of revolution.

19. $x^2 + y^2 + z^2 = 1$

20. $-9x^2 + 4y^2 + 4z^2 = 36$

21. $y = e^{x^2 + z^2}$

22. $x^2 + y^2 = \sin^2 z$

In Problems 23–30 the graph of the given equation is revolved about the indicated axis. Find an equation of the surface of revolution.

23. $y = 2x$; y-axis **24.** $y = \sqrt{z}$; y-axis

25. $z = 9 - x^2$, $x \geq 0$; x-axis

26. $z = 1 + y^2$, $y \geq 0$; z-axis

27. $x^2 - z^2 = 4$; x-axis

28. $3x^2 + 4z^2 = 12$; z-axis

29. $z = \ln y$; z-axis

30. $xy = 1$; x-axis

31. Which of the surfaces in Problems 1–14 are surfaces of revolution? Identify the axis of revolution.

32. Sketch a graph of the equation in Problem 22 for $0 \leq z \leq 2\pi$.

───── Chapter 14 Review Exercises ─────

Answers to odd-numbered problems begin on page A-46.

In Problems 1–10 answer true or false.

1. The vectors $\langle -4, -6, 10 \rangle$ and $\langle -10, -15, 25 \rangle$ are parallel. _____

2. In 3-space any three distinct points determine a plane. _____

3. The line $x = 1 + 5t$, $y = 1 - 2t$, $z = 4 + t$ and the plane $2x + 3y - 4z = 1$ are perpendicular. _____

4. Nonzero vectors $\mathbf{a}$ and $\mathbf{b}$ are parallel if $\mathbf{a} \times \mathbf{b} = \mathbf{0}$. _____

5. If $\mathbf{a} \cdot \mathbf{b} < 0$, the angle between $\mathbf{a}$ and $\mathbf{b}$ is obtuse. _____

6. If $\mathbf{a}$ is a unit vector, then $\mathbf{a} \cdot \mathbf{a} = 1$. _____

7. The cross product of two vectors is not commutative. _____

8. The terminal point of the vector $\mathbf{a} - \mathbf{b}$ is at the terminal point of $\mathbf{a}$. _____

9. $(\mathbf{a} \times \mathbf{b}) \cdot \mathbf{c} = \mathbf{a} \cdot (\mathbf{b} \times \mathbf{c})$ _____

10. If $\mathbf{a}$, $\mathbf{b}$, $\mathbf{c}$, and $\mathbf{d}$ are nonzero coplanar vectors, then $(\mathbf{a} \times \mathbf{b}) \times (\mathbf{c} \times \mathbf{d}) = \mathbf{0}$. _____

In Problems 11–30 fill in the blanks.

11. The resultant of $3\mathbf{i} + 4\mathbf{j} + 5\mathbf{k}$ and $6\mathbf{i} - 2\mathbf{j} - 3\mathbf{k}$ is _____.

12. If $\mathbf{a} \cdot \mathbf{b} = 0$, the nonzero vectors $\mathbf{a}$ and $\mathbf{b}$ are _____.

13. $(-\mathbf{k}) \times (5\mathbf{j}) =$ _____

14. $\mathbf{i} \cdot (\mathbf{i} \times \mathbf{j}) =$ _____

15. $|-12\mathbf{i} + 4\mathbf{j} + 6\mathbf{k}| =$ _____

16. $\begin{vmatrix} \mathbf{i} & \mathbf{j} & \mathbf{k} \\ 2 & 1 & 5 \\ 0 & 4 & -1 \end{vmatrix} =$ _____

17. A vector that is normal to the plane $-6x + y - 7z + 10 = 0$ is _____.

18. The trace of the surface $x^2 - y^2 + z^2 = 4$ in the xz-plane is _____.

19. The point of intersection of the line $x - 1 = (y + 2)/3 = (z + 1)/2$ and the plane $x + 2y - z = 13$ is _____.

20. A unit vector that has the opposite direction of $\mathbf{a} = 4\mathbf{i} + 3\mathbf{j} - 5\mathbf{k}$ is _____.

21. If $\overrightarrow{P_1P_2} = \langle 3, 5, -4 \rangle$ and P_1 has coordinates $(2, 1, 7)$, then the coordinates of P_2 are _____.

22. The midpoint of the line segment between $P_1(4, 3, 10)$ and $P_2(6, -2, -5)$ has coordinates _____.

23. If $|\mathbf{a}| = 7.2$, $|\mathbf{b}| = 10$, and the angle between $\mathbf{a}$ and $\mathbf{b}$ is 135°, then $\mathbf{a} \cdot \mathbf{b} =$ _____.

24. If $\mathbf{a} = \langle 3, 1, 0 \rangle$, $\mathbf{b} = \langle -1, 2, 1 \rangle$, and $\mathbf{c} = \langle 0, -2, 2 \rangle$, then $\mathbf{a} \cdot (2\mathbf{b} + 4\mathbf{c}) =$ _____.

25. The x-, y-, and z-intercepts of the plane $2x - 3y + 4z = 24$ are, respectively, _____.

26. The angle θ between the vectors $\mathbf{a} = \mathbf{i} + \mathbf{j}$ and $\mathbf{b} = \mathbf{i} - \mathbf{k}$ is _____.

27. The area of a triangle with two sides given by $\mathbf{a} = \langle 1, 3, -1 \rangle$ and $\mathbf{b} = \langle 2, -1, 2 \rangle$ is _____.

28. An equation of a sphere with center $(-5, 7, -9)$ and radius $\sqrt{6}$ is _____.

29. The distance from the plane $y = -5$ to the point $(4, -3, 1)$ is _____.

30. The vectors $\langle 1, 3, c \rangle$ and $\langle -2, -6, 5 \rangle$ are parallel for $c =$ _____ and orthogonal for $c =$ _____.

31. Find a unit vector that is perpendicular to both $\mathbf{a} = \mathbf{i} + \mathbf{j}$ and $\mathbf{b} = \mathbf{i} - 2\mathbf{i} + \mathbf{k}$.

32. Find the direction cosines and direction angles of the vector $\mathbf{a} = \frac{1}{2}\mathbf{i} + \frac{1}{2}\mathbf{j} - \frac{1}{4}\mathbf{k}$.

In Problems 33–36 let $\mathbf{a} = \langle 1, 2, -2 \rangle$ and $\mathbf{b} = \langle 4, 3, 0 \rangle$. Find the indicated number or vector.

33. $\text{comp}_{\mathbf{b}}\mathbf{a}$

34. $\text{proj}_{\mathbf{a}}\mathbf{b}$

35. $\text{proj}_{\mathbf{a}^\perp}\mathbf{b}$

36. $\text{proj}_{\mathbf{b}^\perp}(\mathbf{a} - \mathbf{b})$

In Problems 37–42 identify the surface whose equation is given.

37. $x^2 + 4y^2 = 16$

38. $y + 2x^2 + 4z^2 = 0$

39. $x^2 + 4y^2 - z^2 = -9$

40. $x^2 + y^2 + z^2 = 10z$

41. $9z - x^2 + y^2 = 0$

42. $2x - 3y = 6$

43. Find an equation of the surface of revolution obtained by revolving the graph of $x^2 - y^2 = 1$ about the y-axis. About the x-axis. Identify the surface in each case.

44. A surface of revolution has an equation $y = 1 + \sqrt{x^2 + z^2}$. Find an equation of a curve C in a coordinate plane that, when revolved about a coordinate axis, generates the surface.

45. Let $\mathbf{r}$ be the position vector of a variable point $P(x, y, z)$ in space and let $\mathbf{a}$ be a constant vector. Determine the surface described by the following equations:

(a) $(\mathbf{r} - \mathbf{a}) \cdot \mathbf{r} = 0$

(b) $(\mathbf{r} - \mathbf{a}) \cdot \mathbf{a} = 0$

46. Use the dot product to determine whether the points $(4, 2, -2)$, $(2, 4, -3)$, and $(6, 7, -5)$ are vertices of a right triangle.

47. Find symmetric equations for the line through the point $(7, 3, -5)$ that is parallel to $(x - 3)/4 = (y + 4)/(-2) = (z - 9)/6$.

48. Find parametric equations for the line through the point $(5, -9, 3)$ that is perpendicular to the plane $8x + 3y - 4z = 13$.

49. Show that the lines $x = 1 - 2t$, $y = 3t$, $z = 1 + t$ and $x = 1 + 2s$, $y = -4 + s$, $z = -1 + s$ intersect orthogonally.

50. Find an equation of the plane containing the points $(0, 0, 0)$, $(2, 3, 1)$, $(1, 0, 2)$.

51. Find an equation of the plane containing the lines $x = t$, $y = 4t$, $z = -2t$, and $x = 1 + t$, $y = 1 + 4t$, $z = 3 - 2t$.

52. Find an equation of the plane containing $(1, 7, -1)$ that is perpendicular to the line of intersection of $-x + y - 8z = 4$ and $3x - y + 2z = 0$.

53. A constant force of 10 nt in the direction of $\mathbf{a} = \mathbf{i} + \mathbf{j}$ moves a block on a frictionless surface from $P_1(4, 1, 0)$ to $P_2(7, 4, 0)$. Suppose distance is measured in meters. Find the work done.

54. In Problem 53, find the work done in moving the block between the same points if another constant force of 50 nt in the direction of $\mathbf{b} = \mathbf{i}$ acts simultaneously with the original force.

55. Water rushing from a fire hose exerts a horizontal force $\mathbf{F}_1$ of magnitude 200 lb. See Figure 14.84. What is the magnitude of the force $\mathbf{F}_3$ that a firefighter must exert to hold the hose at an angle of 45° from the horizontal?

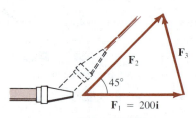

Figure 14.84

56. A uniform ball of weight 50 lb is supported by two frictionless planes as shown in Figure 14.85. Let the force exerted by the supporting plane $\mathscr{P}_1$ on the ball be $\mathbf{F}_1$ and the force exerted by the plane $\mathscr{P}_2$ be $\mathbf{F}_2$. Since the ball is held in equilibrium, we must have $\mathbf{w} + \mathbf{F}_1 + \mathbf{F}_2 = \mathbf{0}$, where $\mathbf{w} = -50\mathbf{j}$. Find the magnitudes of the forces $\mathbf{F}_1$ and $\mathbf{F}_2$. (*Hint:* Assume the forces $\mathbf{F}_1$ and $\mathbf{F}_2$ are normal to the planes $\mathscr{P}_1$ and $\mathscr{P}_2$, respectively, and act along lines through the center C of the ball. Place the origin of a two-dimensional coordinate system at C.)

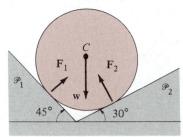

Figure 14.85

15

Vector-Valued Functions

Curves in the plane as well as in 3-space can be described by means of parametric equations. Using the functions in a set of parametric equations as components, we construct a vector function that gives vectorially the position of a point on the curve. In this chapter we consider the calculus and applications of these vector functions.

15.1 Vector Functions

15.1.1 Definition of a Vector Function

Recall, a curve C in the xy-plane can be parameterized by

$$x = f(t), \qquad y = g(t), \qquad a \leq t \leq b$$

It is often convenient in science and engineering to introduce a vector $\mathbf{r}$ with the functions f and g as components,

$$\mathbf{r}(t) = \langle f(t), g(t) \rangle = f(t)\mathbf{i} + g(t)\mathbf{j}$$

We say that $\mathbf{r}$ is a **vector function**. Similarly, a **space-curve** is parameterized by three equations

$$x = f(t), \qquad y = g(t), \qquad z = h(t), \qquad a \leq t \leq b \qquad (15.1)$$

Correspondingly, a vector function is given by

$$\mathbf{r}(t) = \langle f(t), g(t), h(t) \rangle = f(t)\mathbf{i} + g(t)\mathbf{j} + h(t)\mathbf{k}$$

As shown in Figure 15.1, for a given number t_0, the vector $\mathbf{r}(t_0)$ is the *position vector* of a point P on the curve C. In other words, as t varies, we can envision the curve C being traced out by the moving arrowhead of $\mathbf{r}(t)$.

We have already seen an example of parametric equations as well as the vector function of a space curve in Section 14.5, when we discussed the line in 3-space.

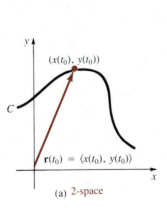

(a) 2-space

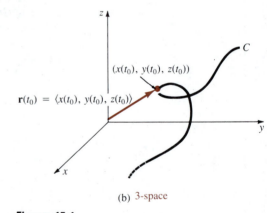

(b) 3-space

Figure 15.1

Example 1 _____

Graph the curve traced by the vector function

$$\mathbf{r}(t) = 2 \cos t\,\mathbf{i} + 2 \sin t\,\mathbf{j} + t\mathbf{k}, \qquad t \geq 0$$

Solution   The parametric equations of the curve are $x = 2 \cos t$, $y = 2 \sin t$, $z = t$. By eliminating the parameter t from the first two equations, we see that a point on the curve lies on the circular cylinder $x^2 + y^2 = 4$. As seen in Figure

15.2 and the accompanying table, as the value of t increases, the curve winds upward in a spiral or a *circular helix*.

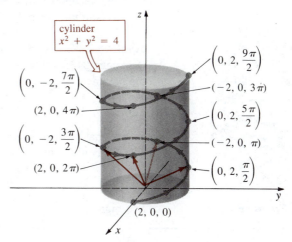

Figure 15.2

t	0	$\pi/2$	π	$3\pi/2$	2π	$5\pi/2$	3π	$7\pi/2$	4π	$9\pi/2$
x	2	0	-2	0	2	0	-2	0	2	0
y	0	2	0	-2	0	2	0	-2	0	2
z	0	$\pi/2$	π	$3\pi/2$	2π	$5\pi/2$	3π	$7\pi/2$	4π	$9\pi/2$

The curve in Example 1 is a special case of the vector function

$$\mathbf{r}(t) = a \cos t\mathbf{i} + b \sin t\mathbf{j} + ct\mathbf{k}, \qquad a > 0, b > 0, c > 0$$

which describes an *elliptical helix*. When $a = b$ the helix is circular. The *pitch* of a helix is defined to be the number $2\pi c$. Problems 9 and 10 in Exercises 15.1 illustrate two other kinds of helixes.

Example 2

Graph the curve traced by the vector function

$$\mathbf{r}(t) = 2 \cos t\mathbf{i} + 2 \sin t\mathbf{j} + 3\mathbf{k}$$

Solution   The parametric equations of this curve are $x = 2 \cos t$, $y = 2 \sin t$, $z = 3$. As in Example 1, we see that a point on the curve must also lie on the cylinder $x^2 + y^2 = 4$. However, since the z-coordinate of any point has the constant value $z = 3$ the vector function $\mathbf{r}(t)$ traces out a circle 3 units above the xy-plane. See Figure 15.3.

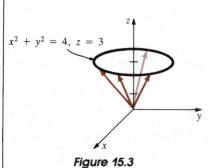

Figure 15.3

Example 3

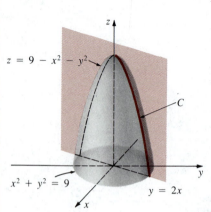

$z = 9 - x^2 - y^2$

C

$x^2 + y^2 = 9$

$y = 2x$

Figure 15.4

Find the vector function that describes the curve C of intersection of the plane $y = 2x$ and the paraboloid $z = 9 - x^2 - y^2$.

Solution   We first parameterize the curve C of intersection by letting $x = t$. (See page 643.) It follows that $y = 2t$ and $z = 9 - t^2 - (2t)^2 = 9 - 5t^2$. From the parametric equations

$$x = t, \qquad y = 2t, \qquad z = 9 - 5t^2$$

we see that a vector function describing the trace of the paraboloid in the plane $y = 2x$ is given by

$$\mathbf{r}(t) = t\mathbf{i} + 2t\mathbf{j} + (9 - 5t^2)\mathbf{k}$$

See Figure 15.4.

15.1.2 Calculus of Vector Functions

Limits and Continuity   The fundamental notion of **limit** of a vector function $\mathbf{r}(t) = \langle f(t), g(t), h(t) \rangle$ is defined in terms of the limits of the component functions.

> **DEFINITION 15.1**   If $\lim_{t \to a} f(t)$, $\lim_{t \to a} g(t)$, and $\lim_{t \to a} h(t)$ exist, then
>
> $$\lim_{t \to a} \mathbf{r}(t) = \left\langle \lim_{t \to a} f(t), \ \lim_{t \to a} g(t), \ \lim_{t \to a} h(t) \right\rangle$$

As an immediate consequence of Definition 15.1, we have the following result.

> **THEOREM 15.1**   If $\lim_{t \to a} \mathbf{r}_1(t) = \mathbf{L}_1$ and $\lim_{t \to a} \mathbf{r}_2(t) = \mathbf{L}_2$, then
>
> (i) $\lim_{t \to a} c\mathbf{r}_1(t) = c\mathbf{L}_1$, c a scalar
>
> (ii) $\lim_{t \to a} [\mathbf{r}_1(t) + \mathbf{r}_2(t)] = \mathbf{L}_1 + \mathbf{L}_2$
>
> (iii) $\lim_{t \to a} \mathbf{r}_1(t) \cdot \mathbf{r}_2(t) = \mathbf{L}_1 \cdot \mathbf{L}_2$

> **DEFINITION 15.2**   A vector function $\mathbf{r}$ is said to be **continuous at a number** a if
>
> (i) $\mathbf{r}(a)$ is defined,
>
> (ii) $\lim_{t \to a} \mathbf{r}(t)$ exists, and
>
> (iii) $\lim_{t \to a} \mathbf{r}(t) = \mathbf{r}(a)$.

Equivalently, $\mathbf{r}(t)$ is continuous at a number a if and only if the component functions f, g, and h are continuous there.

Derivatives of Vector Functions

> **DEFINITION 15.3**   The **derivative** of a vector function $\mathbf{r}$ is
>
> $$\mathbf{r}'(t) = \lim_{\Delta t \to 0} \frac{1}{\Delta t}[\mathbf{r}(t + \Delta t) - \mathbf{r}(t)] \qquad (15.2)$$
>
> for all t for which the limit exists.

The derivative of $\mathbf{r}$ is also written $d\mathbf{r}/dt$. The next theorem shows that on a practical level the derivative of a vector function is obtained by simply differentiating its component functions.

> **THEOREM 15.2**   If $\mathbf{r}(t) = \langle f(t), g(t), h(t) \rangle$, where f, g, and h are differentiable, then
>
> $$\mathbf{r}'(t) = \langle f'(t), g'(t), h'(t) \rangle$$

Proof   From (15.2) we have

$$\mathbf{r}'(t) = \lim_{\Delta t \to 0} \frac{1}{\Delta t}[\langle f(t + \Delta t), g(t + \Delta t), h(t + \Delta t) \rangle - \langle f(t), g(t), h(t) \rangle]$$

$$= \lim_{\Delta t \to 0} \left\langle \frac{f(t + \Delta t) - f(t)}{\Delta t}, \frac{g(t + \Delta t) - g(t)}{\Delta t}, \frac{h(t + \Delta t) - h(t)}{\Delta t} \right\rangle$$

Taking the limit of each component yields the desired result. ∎

Smooth Curves   When the component functions of a vector function $\mathbf{r}$ have continuous first derivatives and $\mathbf{r}'(t) \neq \mathbf{0}$ for all t in the open interval (a, b), then $\mathbf{r}$ is said to be a **smooth function** and the curve C traced by $\mathbf{r}$ is called a **smooth curve**.

Geometric Interpretation of r'(t)   If the vector $\mathbf{r}'(t)$ is not $\mathbf{0}$ at a point P, then it may be drawn *tangent to the curve* at P. As seen in parts (a) and (b) of Figure 15.5, the vectors

$$\Delta\mathbf{r} = \mathbf{r}(t + \Delta t) - \mathbf{r}(t) \quad \text{and} \quad \frac{\Delta\mathbf{r}}{\Delta t} = \frac{1}{\Delta t}[\mathbf{r}(t + \Delta t) - \mathbf{r}(t)]$$

are parallel. Assuming $\lim_{\Delta t \to 0} \Delta\mathbf{r}/\Delta t$ exists, it seems reasonable to conclude that as $\Delta t \to 0$, $\mathbf{r}(t)$ and $\mathbf{r}(t + \Delta t)$ become close, and, as a consequence, the limiting position of the vector $\Delta\mathbf{r}/\Delta t$ is the tangent line at P. Indeed, the tangent line at P is *defined* as that line through P parallel to $\mathbf{r}'(t)$.

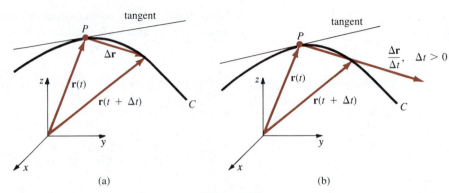

Figure 15.5

Example 4

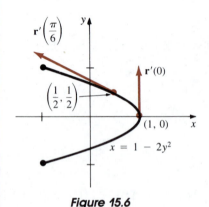

Figure 15.6

Graph the curve C that is traced by a point P whose position is given by $\mathbf{r}(t) = \cos 2t\,\mathbf{i} + \sin t\,\mathbf{j}$, $0 \le t \le 2\pi$. Graph $\mathbf{r}'(0)$ and $\mathbf{r}'(\pi/6)$.

Solution   By clearing the parameter from the parametric equations $x = \cos 2t$, $y = \sin t$, $0 \le t \le 2\pi$, we find that C is the parabola $x = 1 - 2y^2$, $-1 \le x \le 1$. From

$$\mathbf{r}'(t) = -2 \sin 2t\,\mathbf{i} + \cos t\,\mathbf{j}$$

we find $\mathbf{r}'(0) = \mathbf{j}$ and $\mathbf{r}'\left(\dfrac{\pi}{6}\right) = -\sqrt{3}\,\mathbf{i} + \dfrac{\sqrt{3}}{2}\,\mathbf{j}$

In Figure 15.6 these vectors are drawn tangent to the curve C at $(1, 0)$ and $(\frac{1}{2}, \frac{1}{2})$, respectively.

Example 5

Find parametric equations of the tangent line to the graph of the curve C whose parametric equations are

$$x = t^2, \qquad y = t^2 - t, \qquad z = -7t$$

at $t = 3$.

Solution   The vector function that gives the position of a point P on the curve is given by

$$\mathbf{r}(t) = t^2\mathbf{i} + (t^2 - t)\mathbf{j} - 7t\mathbf{k}$$

Now, $\mathbf{r}'(t) = 2t\mathbf{i} + (2t - 1)\mathbf{j} - 7\mathbf{k}$

and $\mathbf{r}'(3) = 6\mathbf{i} + 5\mathbf{j} - 7\mathbf{k}$

which is tangent to C at the point whose position vector is

$$\mathbf{r}(3) = 9\mathbf{i} + 6\mathbf{j} - 21\mathbf{k}$$

that is, $P(9, 6, -21)$. Using the components of $\mathbf{r}'(3)$, we see that parametric equations of the tangent line are

$$x = 9 + 6t, \qquad y = 6 + 5t, \qquad z = -21 - 7t$$

Higher-order Derivatives   Higher-order derivatives of a vector function are also obtained by differentiating its components. In the case of the **second derivative**, we have

$$\mathbf{r}''(t) = \langle f''(t), g''(t), h''(t) \rangle = f''(t)\mathbf{i} + g''(t)\mathbf{j} + h''(t)\mathbf{k}$$

Example 6

If $\qquad\qquad\qquad\qquad \mathbf{r}(t) = (t^3 - 2t^2)\mathbf{i} + 4t\mathbf{j} + e^{-t}\mathbf{k}$

then $\qquad\qquad\qquad \mathbf{r}'(t) = (3t^2 - 4t)\mathbf{i} + 4\mathbf{j} - e^{-t}\mathbf{k}$

and $\qquad\qquad\qquad \mathbf{r}''(t) = (6t - 4)\mathbf{i} + e^{-t}\mathbf{k}$

THEOREM 15.3   **Chain Rule**

If $\mathbf{r}$ is a differentiable vector function and $s = u(t)$ is a differentiable scalar function, then the derivative of $\mathbf{r}(s)$ with respect to t is

$$\frac{d\mathbf{r}}{dt} = \frac{d\mathbf{r}}{ds}\frac{ds}{dt} = \mathbf{r}'(s)\, u'(t)$$

Example 7

If $\mathbf{r}(s) = \cos 2s\mathbf{i} + \sin 2s\mathbf{j} + e^{-3s}\mathbf{k}$, where $s = t^4$, then

$$\frac{d\mathbf{r}}{dt} = [-2\sin 2s\mathbf{i} + 2\cos 2s\mathbf{j} - 3e^{-3s}\mathbf{k}]4t^3$$

$$= -8t^3\sin(2t^4)\mathbf{i} + 8t^3\cos(2t^4)\mathbf{j} - 12t^3e^{-3t^4}\mathbf{k}$$

Details of the proof of the next theorem are left as exercises.

THEOREM 15.4   Let $\mathbf{r}_1$ and $\mathbf{r}_2$ be differentiable vector functions and $u(t)$ a differentiable scalar function.

$$(i)\quad \frac{d}{dt}[\mathbf{r}_1(t) + \mathbf{r}_2(t)] = \mathbf{r}_1'(t) + \mathbf{r}_2'(t)$$

$$(ii)\quad \frac{d}{dt}[u(t)\mathbf{r}_1(t)] = u(t)\mathbf{r}_1'(t) + u'(t)\mathbf{r}_1(t)$$

$$(iii)\quad \frac{d}{dt}[\mathbf{r}_1(t) \cdot \mathbf{r}_2(t)] = \mathbf{r}_1(t) \cdot \mathbf{r}_2'(t) + \mathbf{r}_1'(t) \cdot \mathbf{r}_2(t)$$

$$(iv)\quad \frac{d}{dt}[\mathbf{r}_1(t) \times \mathbf{r}_2(t)] = \mathbf{r}_1(t) \times \mathbf{r}_2'(t) + \mathbf{r}_1'(t) \times \mathbf{r}_2(t)$$

Note: Since the cross product of two vectors is not commutative, the order in which $\mathbf{r}_1$ and $\mathbf{r}_2$ appear in part (*iv*) of Theorem 15.4 must be strictly observed.

Integrals of Vector Functions

If f, g, and h are integrable, then the indefinite and definite integrals of a vector function $\mathbf{r}(t) = f(t)\mathbf{i} + g(t)\mathbf{j} + h(t)\mathbf{k}$ are defined, respectively, by

$$\int \mathbf{r}(t)\, dt = \left[\int f(t)\, dt\right]\mathbf{i} + \left[\int g(t)\, dt\right]\mathbf{j} + \left[\int h(t)\, dt\right]\mathbf{k}$$

$$\int_a^b \mathbf{r}(t)\, dt = \left[\int_a^b f(t)\, dt\right]\mathbf{i} + \left[\int_a^b g(t)\, dt\right]\mathbf{j} + \left[\int_a^b h(t)\, dt\right]\mathbf{k}$$

The indefinite integral of $\mathbf{r}$ is another vector $\mathbf{R} + \mathbf{c}$ such that $\mathbf{R}'(t) = \mathbf{r}(t)$.

Example 8

If
$$\mathbf{r}(t) = 6t^2\mathbf{i} + 4e^{-2t}\mathbf{j} + 8\cos 4t\mathbf{k}$$

then
$$\int \mathbf{r}(t)\, dt = \left[\int 6t^2\, dt\right]\mathbf{i} + \left[\int 4e^{-2t}\, dt\right]\mathbf{j} + \left[\int 8\cos 4t\, dt\right]\mathbf{k}$$

$$= [2t^3 + c_1]\mathbf{i} + [-2e^{-2t} + c_2]\mathbf{j} + [2\sin 4t + c_3]\mathbf{k}$$

$$= 2t^3\mathbf{i} - 2e^{-2t}\mathbf{j} + 2\sin 4t\mathbf{k} + \mathbf{c}$$

where $\mathbf{c} = c_1\mathbf{i} + c_2\mathbf{j} + c_3\mathbf{k}$.

Length of a Space Curve

If $\mathbf{r}(t) = f(t)\mathbf{i} + g(t)\mathbf{j} + h(t)\mathbf{k}$ is a smooth function, then, in a manner similar to that given in Section 13.1, it can be shown that the **length** of the smooth curve traced by $\mathbf{r}$ is given by

$$s = \int_a^b \sqrt{[f'(t)]^2 + [g'(t)]^2 + [h'(t)]^2}\, dt = \int_a^b |\mathbf{r}'(t)|\, dt \qquad (15.3)$$

Arc Length as a Parameter

A curve in the plane or in space can be parameterized in terms of the arc length s.

Example 9

Consider the helix of Example 1. Since $|\mathbf{r}'(t)| = \sqrt{5}$, it follows from (15.3) that the length of the curve starting at $\mathbf{r}(0)$ to an arbitrary point $\mathbf{r}(t)$ is

$$s = \int_0^t \sqrt{5}\, du = \sqrt{5}t$$

where we have used u as a dummy variable of integration. Using $t = s/\sqrt{5}$, we obtain a vector equation of the helix as a function of arc length:

$$\mathbf{r}(s) = 2\cos\frac{s}{\sqrt{5}}\mathbf{i} + 2\sin\frac{s}{\sqrt{5}}\mathbf{j} + \frac{s}{\sqrt{5}}\mathbf{k} \qquad (15.4)$$

Parametric equations of the helix are then

$$f(s) = 2 \cos(s/\sqrt{5}), \qquad g(s) = 2 \sin(s/\sqrt{5}), \qquad h(s) = s/\sqrt{5}$$

The derivative of a vector function $\mathbf{r}(t)$ with respect to the parameter t is a tangent vector to the curve traced by $\mathbf{r}$. However, if the curve is parameterized in terms of arc length s, then

$$\mathbf{r}'(s) \text{ is a unit tangent vector}$$

To see this, let a curve be described by $\mathbf{r}(s)$, where s is arc length. From (15.3) the length of the curve from $\mathbf{r}(0)$ to $\mathbf{r}(s)$ is

$$s = \int_0^s |\mathbf{r}'(u)| \, du$$

Differentiation of this last equation with respect to s then yields $|\mathbf{r}'(s)| = 1$.

Exercises 15.1

Answers to odd-numbered problems begin on page A-46.

[15.1.1]

In Problems 1–10 graph the curve traced by the given vector function.

1. $\mathbf{r}(t) = 2 \sin t\mathbf{i} + 4 \cos t\mathbf{j} + t\mathbf{k}, \ t \geq 0$

2. $\mathbf{r}(t) = \cos t\mathbf{i} + t\mathbf{j} + \sin t\mathbf{k}, \ t \geq 0$

3. $\mathbf{r}(t) = t\mathbf{i} + 2t\mathbf{j} + \cos t\mathbf{k}, \ t \geq 0$

4. $\mathbf{r}(t) = 4\mathbf{i} + 2 \cos t\mathbf{j} + 3 \sin t\mathbf{k}$

5. $\mathbf{r}(t) = \langle e^t, e^{2t} \rangle$

6. $\mathbf{r}(t) = \cosh t\mathbf{i} + 3 \sinh t\mathbf{j}$

7. $\mathbf{r}(t) = \langle \sqrt{2} \sin t, \ \sqrt{2} \sin t, \ 2 \cos t \rangle, \ 0 \leq t \leq \pi/2$

8. $\mathbf{r}(t) = t\mathbf{i} + t^3\mathbf{j} + t\mathbf{k}$

9. $\mathbf{r}(t) = e^t \cos t\mathbf{i} + e^t \sin t\mathbf{j} + e^t\mathbf{k}$

10. $\mathbf{r}(t) = \langle t \cos t, t \sin t, t^2 \rangle$

In Problems 11–14 find the vector function that describes the curve C of intersection between the given surfaces. Sketch the curve C. Use the indicated parameter.

11. $z = x^2 + y^2, \ y = x; \ x = t$

12. $x^2 + y^2 - z^2 = 1, \ y = 2x; \ x = t$

13. $x^2 + y^2 = 9, \ z = 9 - x^2; \ x = 3 \cos t$

14. $z = x^2 + y^2, \ z = 1; \ x = \sin t$

[15.1.2]

15. Given that $\mathbf{r}(t) = \dfrac{\sin 2t}{t}\mathbf{i} + (t - 2)^5\mathbf{j} + t \ln t\mathbf{k}$ find $\lim_{t \to 0^+} \mathbf{r}(t)$.

16. Given that $\lim_{t \to a} \mathbf{r}_1(t) = \mathbf{i} - 2\mathbf{j} + \mathbf{k}$ and $\lim_{t \to a} \mathbf{r}_2(t) = 2\mathbf{i} + 5\mathbf{j} + 7\mathbf{k}$, find

    (a) $\displaystyle\lim_{t \to a} [-4\mathbf{r}_1(t) + 3\mathbf{r}_2(t)]$

    (b) $\displaystyle\lim_{t \to a} \mathbf{r}_1(t) \cdot \mathbf{r}_2(t)$

In Problems 17–20 find $\mathbf{r}'(t)$ and $\mathbf{r}''(t)$ for the given vector function.

17. $\mathbf{r}(t) = \ln t\mathbf{i} + \dfrac{1}{t}\mathbf{j}, \ t > 0$

18. $\mathbf{r}(t) = \langle t \cos t - \sin t, \ t + \cos t \rangle$

19. $\mathbf{r}(t) = \langle te^{2t}, \ t^3, \ 4t^2 - t \rangle$

20. $\mathbf{r}(t) = t^2\mathbf{i} + t^3\mathbf{j} + \tan^{-1}t\mathbf{k}$

In Problems 21–24 graph the curve C that is described by $\mathbf{r}$ and graph $\mathbf{r}'$ at the indicated value of t.

21. $\mathbf{r}(t) = 2 \cos t\mathbf{i} + 6 \sin t\mathbf{j}; \ t = \pi/6$

22. $\mathbf{r}(t) = t^3\mathbf{i} + t^2\mathbf{j}; \ t = -1$

23. $\mathbf{r}(t) = 2\mathbf{i} + t\mathbf{j} + \dfrac{4}{1 + t^2}\mathbf{k}; \ t = 1$

24. $\mathbf{r}(t) = 3 \cos t\mathbf{i} + 3 \sin t\mathbf{j} + 2t\mathbf{k}; \ t = \pi/4$

In Problems 25 and 26 find parametric equations of the tangent line to the given curve at the indicated value of t.

25. $x = t, \ y = \dfrac{1}{2}t^2, \ z = \dfrac{1}{3}t^3; \ t = 2$

26. $x = t^3 - t, \ y = \dfrac{6t}{t + 1}, \ z = (2t + 1)^2; \ t = 1$

In Problems 27–32 find the indicated derivative. Assume that all vector functions are differentiable.

27. $\dfrac{d}{dt}[\mathbf{r}(t) \times \mathbf{r}'(t)]$

28. $\dfrac{d}{dt}[\mathbf{r}(t) \cdot (t\mathbf{r}(t))]$

29. $\dfrac{d}{dt}[\mathbf{r}(t) \cdot (\mathbf{r}'(t) \times \mathbf{r}''(t))]$

30. $\dfrac{d}{dt}[\mathbf{r}_1(t) \times (\mathbf{r}_2(t) \times \mathbf{r}_3(t))]$

31. $\dfrac{d}{dt}\left[\mathbf{r}_1(2t) + \mathbf{r}_2\left(\dfrac{1}{t}\right)\right]$

32. $\dfrac{d}{dt}[t^3\mathbf{r}(t^2)]$

In Problems 33–36 evaluate the given integral.

33. $\displaystyle\int_{-1}^{2} (t\mathbf{i} + 3t^2\mathbf{j} + 4t^3\mathbf{k}) \, dt$

34. $\displaystyle\int_{0}^{4} (\sqrt{2t + 1}\mathbf{i} - \sqrt{t}\mathbf{j} + \sin \pi t\mathbf{k}) \, dt$

35. $\displaystyle\int (te^t\mathbf{i} - e^{-2t}\mathbf{j} + te^{t^2}\mathbf{k}) \, dt$

36. $\displaystyle\int \dfrac{1}{1 + t^2}(\mathbf{i} + t\mathbf{j} + t^2\mathbf{k}) \, dt$

In Problems 37–40 find a vector function $\mathbf{r}$ that satisfies the indicated conditions.

37. $\mathbf{r}'(t) = 6\mathbf{i} + 6t\mathbf{j} + 3t^2\mathbf{k}; \ \mathbf{r}(0) = \mathbf{i} - 2\mathbf{j} + \mathbf{k}$

38. $\mathbf{r}'(t) = t \sin t^2\mathbf{i} - \cos 2t\mathbf{j}; \ \mathbf{r}(0) = \dfrac{3}{2}\mathbf{i}$

39. $\mathbf{r}''(t) = 12t\mathbf{i} - 3t^{-1/2}\mathbf{j} + 2\mathbf{k}; \ \mathbf{r}'(1) = \mathbf{j}, \ \mathbf{r}(1) = 2\mathbf{i} - \mathbf{k}$

40. $\mathbf{r}''(t) = \sec^2 t\mathbf{i} + \cos t\mathbf{j} - \sin t\mathbf{k}; \ \mathbf{r}'(0) = \mathbf{i} + \mathbf{j} + \mathbf{k},$ $\mathbf{r}(0) = -\mathbf{j} + 5\mathbf{k}$

In Problems 41–44 find the length of the curve traced by the given vector function on the indicated interval.

41. $\mathbf{r}(t) = a \cos t\mathbf{j} + a \sin t\mathbf{j} + ct\mathbf{k}; \ 0 \leq t \leq 2\pi$

42. $\mathbf{r}(t) = t\mathbf{i} + t \cos t\mathbf{j} + t \sin t\mathbf{k}; \ 0 \leq t \leq \pi$

43. $\mathbf{r}(t) = e^t \cos 2t\mathbf{i} + e^t \sin 2t\mathbf{j} + e^t\mathbf{k}; \ 0 \leq t \leq 3\pi$

44. $\mathbf{r}(t) = 3t\mathbf{i} + \sqrt{3}t^2\mathbf{j} + \dfrac{2}{3}t^3\mathbf{k}; \ 0 \leq t \leq 1$

45. Express the vector equation of a circle $\mathbf{r}(t) = a \cos t\mathbf{i} + a \sin t\mathbf{j}$ as a function of arc length s. Verify that $\mathbf{r}'(s)$ is a unit vector.

46. If $\mathbf{r}(s)$ is the vector function given in (15.4), verify that $\mathbf{r}'(s)$ is a unit vector.

47. If m is the mass of a particle, Newton's second law of motion can be written in vector form:

$$\mathbf{F} = m\mathbf{a} = \dfrac{d}{dt}(m\mathbf{v}) = \dfrac{d\mathbf{p}}{dt}$$

where $\mathbf{p} = m\mathbf{v}$ is called **linear momentum**. The **angular momentum** of the particle with respect to the origin is defined to be $\mathbf{L} = \mathbf{r} \times \mathbf{p}$, where $\mathbf{r}$ is the position vector. If the torque of the particle about the origin is $\boldsymbol{\tau} = \mathbf{r} \times \mathbf{F} = \mathbf{r} \times d\mathbf{p}/dt$, show that $\boldsymbol{\tau}$ is the time rate of change of angular momentum.

48. The velocity of a particle in a moving fluid is described by means of a **velocity field** $\mathbf{v} = v_1\mathbf{i} + v_2\mathbf{j} + v_3\mathbf{k}$, where the components $v_i, \ i = 1, 2, 3$ are functions of x, y, z, and time t. If $\mathbf{r}(t)$ is the vector function giving the position of a particle at any time, then $d\mathbf{r}/dt = \mathbf{v}$. Given that $\mathbf{v} = 6t^2x\mathbf{i} - 4ty^2\mathbf{j} + 2t(z + 1)\mathbf{k}$, find $\mathbf{r}(t)$. (*Hint*: Use separation of variables.)

49. Suppose $\mathbf{r}$ is a differentiable vector function for which $|\mathbf{r}(t)| = c$ for all t. Show that the tangent vector $\mathbf{r}'(t)$ is perpendicular to the position vector $\mathbf{r}(t)$ for all t.

50. In Problem 49 describe geometrically the kind of curve C for which $|\mathbf{r}(t)| = c$.

Miscellaneous Problems

51. Prove Theorem 15.4(*ii*).

52. Prove Theorem 15.4(*iii*).

53. Prove Theorem 15.4(*iv*).

54. If $\mathbf{v}$ is a constant vector and $\mathbf{r}$ is integrable on $[a, b]$, prove that $\int_a^b \mathbf{v} \cdot \mathbf{r}(t) \, dt = \mathbf{v} \cdot \int_a^b \mathbf{r}(t) \, dt$.

15.2 Motion on a Curve; Velocity and Acceleration

Suppose a moving body or particle traces a path C and that its position on the path is given by the vector function

$$\mathbf{r}(t) = f(t)\mathbf{i} + g(t)\mathbf{j} + h(t)\mathbf{k}$$

where t represents time. If f, g, and h have second derivatives, then the vectors

$$\mathbf{v}(t) = \mathbf{r}'(t) = f'(t)\mathbf{i} + g'(t)\mathbf{j} + h'(t)\mathbf{k}$$
$$\mathbf{a}(t) = \mathbf{r}''(t) = f''(t)\mathbf{i} + g''(t)\mathbf{j} + h''(t)\mathbf{k}$$

are called the **velocity** and **acceleration** of the particle, respectively. The scalar function $|\mathbf{v}(t)|$ is the **speed** of the particle. Since

$$|\mathbf{v}(t)| = \left|\frac{d\mathbf{r}}{dt}\right| = \sqrt{\left(\frac{dx}{dt}\right)^2 + \left(\frac{dy}{dt}\right)^2 + \left(\frac{dz}{dt}\right)^2}$$

speed is related to arc length s by $s'(t) = |\mathbf{v}(t)|$. In other words, arc length is given by

$$s = \int_{t_0}^{t_1} |\mathbf{v}(t)|\, dt$$

It also follows from the discussion of Section 15.1 that if $P(x_1, y_1, z_1)$ is the position of the particle on C at time t_1, then we may draw

$$\mathbf{v}(t_1) \text{ tangent to } C \text{ at } P$$

Similar remarks hold for curves traced by the vector function

$$\mathbf{r}(t) = f(t)\mathbf{i} + g(t)\mathbf{j}$$

___ **Example 1** ___

The position of a moving particle is given by

$$\mathbf{r}(t) = t^2\mathbf{i} + t\mathbf{j} + \frac{5}{2}t\mathbf{k}$$

Graph the path and the vectors $\mathbf{v}(2)$ and $\mathbf{a}(2)$.

Solution Since $x = t^2$, $y = t$, the path of the particle is above the parabola $x = y^2$. When $t = 2$ the position vector $\mathbf{r}(2) = 4\mathbf{i} + 2\mathbf{j} + 5\mathbf{k}$ indicates that the particle is at the point $P(4, 2, 5)$. Now,

$$\mathbf{v}(t) = \mathbf{r}'(t) = 2t\mathbf{i} + \mathbf{j} + \frac{5}{2}\mathbf{k}$$
$$\mathbf{a}(t) = \mathbf{r}''(t) = 2\mathbf{i}$$

so that $\mathbf{v}(2) = 4\mathbf{i} + \mathbf{j} + \dfrac{5}{2}\mathbf{k}$ and $\mathbf{a}(2) = 2\mathbf{i}$

These vectors are shown in Figure 15.7.

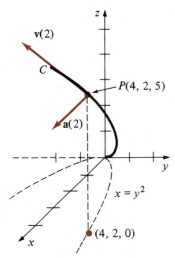

Figure 15.7

If a particle moves with a constant speed c, then its acceleration vector is perpendicular to the velocity vector $\mathbf{v}$. To see this, note that

$$|\mathbf{v}|^2 = c^2 \quad \text{or} \quad \mathbf{v} \cdot \mathbf{v} = c^2$$

We differentiate both sides with respect to t, and obtain, with the aid of Theorem 15.4(iii),

$$\frac{d}{dt}(\mathbf{v} \cdot \mathbf{v}) = \mathbf{v} \cdot \frac{d\mathbf{v}}{dt} + \frac{d\mathbf{v}}{dt} \cdot \mathbf{v} = 2\mathbf{v} \cdot \frac{d\mathbf{v}}{dt} = 0$$

Thus $\dfrac{d\mathbf{v}}{dt} \cdot \mathbf{v} = 0 \quad \text{or} \quad \mathbf{a}(t) \cdot \mathbf{v}(t) = 0 \qquad$ for all t

Example 2

Suppose the vector function in Example 2 of Section 15.1 represents the position of a particle moving in a circular orbit. Graph the velocity and acceleration vector at $t = \pi/4$.

Solution   Recall that

$$\mathbf{r}(t) = 2 \cos t\,\mathbf{i} + 2 \sin t\,\mathbf{j} + 3\mathbf{k}$$

is the position vector of a particle moving in a circular orbit of radius 2 in the plane $z = 3$. When $t = \pi/4$ the particle is at the point $P(\sqrt{2}, \sqrt{2}, 3)$. Now,

$$\mathbf{v}(t) = \mathbf{r}'(t) = -2 \sin t\,\mathbf{i} + 2 \cos t\,\mathbf{j}$$
$$\mathbf{a}(t) = \mathbf{r}''(t) = -2 \cos t\,\mathbf{i} - 2 \sin t\,\mathbf{j}$$

Since the speed is $|\mathbf{v}(t)| = 2$ for all time t, it follows from the discussion preceding this example that $\mathbf{a}(t)$ is perpendicular to $\mathbf{v}(t)$. (Verify this.) As shown in Figure 15.8, the vectors

$$\mathbf{v}\left(\frac{\pi}{4}\right) = -2 \sin \frac{\pi}{4}\mathbf{i} + 2 \cos \frac{\pi}{4}\mathbf{j} = -\sqrt{2}\mathbf{i} + \sqrt{2}\mathbf{j}$$

$$\mathbf{a}\left(\frac{\pi}{4}\right) = -2 \cos \frac{\pi}{4}\mathbf{i} - 2 \sin \frac{\pi}{4}\mathbf{j} = -\sqrt{2}\mathbf{i} - \sqrt{2}\mathbf{j}$$

are drawn at the point P. The vector $\mathbf{v}(\pi/4)$ is tangent to the circular path and is $\mathbf{a}(\pi/4)$ points along a radius toward the center of the circle.

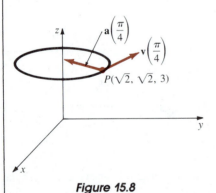

Figure 15.8

Centripetal Acceleration   For circular motion in the plane, described by $\mathbf{r}(t) = r_0 \cos \omega t\,\mathbf{i} + r_0 \sin \omega t\,\mathbf{j}$, r_0 and ω constants, it is evident that $\mathbf{r}'' = -\omega^2\mathbf{r}$. This means that the acceleration vector $\mathbf{a}(t) = \mathbf{r}''(t)$ points in the direction opposite to that of the position vector $\mathbf{r}(t)$. We then say $\mathbf{a}(t)$ is **centripetal acceleration**. See Figure 15.9. If $v = |\mathbf{v}(t)|$ and $a = |\mathbf{a}(t)|$, we leave it as an exercise to show that $a = v^2/r_0$.

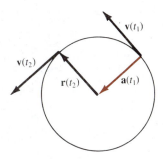

Figure 15.9

Curvilinear Motion in the Plane

Many important applications of vector functions occur in describing curvilinear motion in a plane. For example, planetary and projectile motion take place in a plane.

In analyzing the motion of short-range ballistic* projectiles, we begin with the acceleration of gravity written in vector form

$$\mathbf{a}(t) = -g\mathbf{j}$$

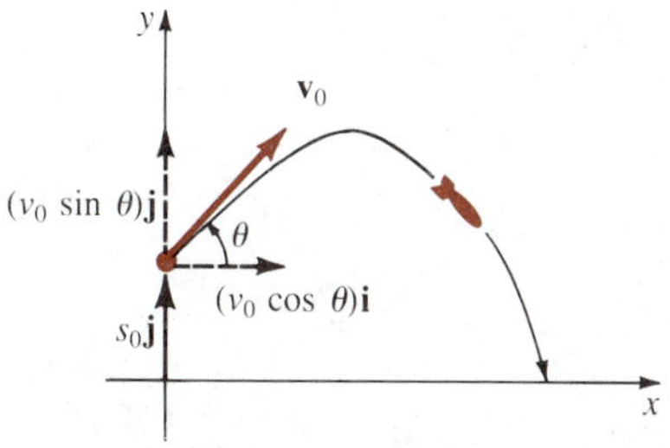

If, as shown in Figure 15.10, a projectile is launched with an initial velocity $\mathbf{v}_0 = v_0 \cos \theta \mathbf{i} + v_0 \sin \theta \mathbf{j}$ from an initial height $\mathbf{s}_0 = s_0\mathbf{j}$, then

$$\mathbf{v}(t) = \int (-g\mathbf{j})\, dt = -gt\mathbf{j} + \mathbf{c}_1$$

where $\mathbf{v}(0) = \mathbf{v}_0$ implies that $\mathbf{c}_1 = \mathbf{v}_0$. Therefore,

$$\mathbf{v}(t) = (-gt + v_0 \sin \theta)\mathbf{j} + (v_0 \cos \theta)\mathbf{i}$$

Integrating again and using $\mathbf{r}(0) = \mathbf{s}_0$ yields

$$\mathbf{r}(t) = \left[-\frac{1}{2}gt^2 + (v_0 \sin \theta)t + s_0 \right]\mathbf{j} + (v_0 \cos \theta)t\mathbf{i}$$

Hence, parametric equations for the trajectory of the projectile are

$$x(t) = (v_0 \cos \theta)t, \qquad y(t) = -\frac{1}{2}gt^2 + (v_0 \sin \theta)t + s_0 \qquad (15.5)$$

We are naturally interested in finding the maximum height H and the range R attained by a projectile. As shown in Figure 15.11, these quantities are the maximum values of $y(t)$ and $x(t)$, respectively.

Figure 15.10

---

*The projectile is shot or hurled rather than self-propelled. In the analysis of *long-range* ballistic motion, the curvature of the earth must be taken into consideration.

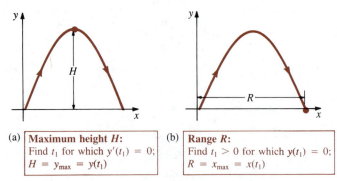

(a) | **Maximum height H:**
Find t_1 for which $y'(t_1) = 0$;
$H = y_{max} = y(t_1)$

(b) | **Range R:**
Find $t_1 > 0$ for which $y(t_1) = 0$;
$R = x_{max} = x(t_1)$

Figure 15.11

___ **Example 3** ___

A shell is fired from ground level with an initial speed of 768 ft/sec at an angle of elevation of 30°. Find

(*a*)   the vector function and parametric equations of the shell's trajectory,

(*b*)   the maximum altitude attained,

(*c*)   the range of the shell, and

(*d*)   the speed at impact.

Solution

(*a*)   Initially, we have $\mathbf{s}_0 = \mathbf{0}$ and

$$\mathbf{v}_0 = (768 \cos 30°)\mathbf{i} + (768 \sin 30°)\mathbf{j} = 384\sqrt{3}\mathbf{i} + 384\mathbf{j} \quad (15.6)$$

Integrating $\mathbf{a}(t) = -32\mathbf{j}$ and using (15.6) gives

$$\mathbf{v}(t) = (-32t + 384)\mathbf{j} + (384\sqrt{3})\mathbf{i} \quad (15.7)$$

Integrating again gives

$$\mathbf{r}(t) = (-16t^2 + 384t)\mathbf{j} + (384\sqrt{3}t)\mathbf{i}$$

Hence, the parametric equations of the shell's trajectory are

$$x(t) = 384\sqrt{3}t, \qquad y(t) = -16t^2 + 384t \quad (15.8)$$

(*b*)   From (15.8) we see that $dy/dt = 0$ when

$$-32t + 384 = 0 \quad \text{or} \quad t = 12$$

Thus, the maximum height H attained by the shell is

$$H = y(12) = -16(12)^2 + 384(12) = 2304 \text{ ft}$$

(c)   From (15.8) we see that $y(t) = 0$ when

$$-16t(t - 24) = 0 \quad \text{or} \quad t = 0, \quad t = 24$$

The range R is then

$$R = x(24) = 384\sqrt{3}(24) \approx 15,963 \text{ ft}$$

(d)   From (15.7) we obtain the impact speed of the shell,

$$|\mathbf{v}(24)| = \sqrt{(-384)^2 + (384\sqrt{3})^2} = 768 \text{ ft/sec}$$

Remark

We have seen that the rate of change of arc length ds/dt is the same as the speed $|\mathbf{v}(t)| = |\mathbf{r}'(t)|$. However, as we shall see in the next section, it does *not* follow that the *scalar acceleration* d^2s/dt^2 is the same as $|\mathbf{a}(t)| = |\mathbf{r}''(t)|$. See Problem 20 in Exercises 15.2.

_____ Exercises 15.2 _____

Answers to odd-numbered problems begin on page A-47.

In Problems 1–8 $\mathbf{r}(t)$ is the position vector of a moving particle. Graph the curve and the velocity and acceleration vectors at the indicated time. Find the speed at that time.

1. $\mathbf{r}(t) = t^2\mathbf{i} + \dfrac{1}{4}t^4\mathbf{j}; \ t = 1$

2. $\mathbf{r}(t) = t^2\mathbf{i} + \dfrac{1}{t^2}\mathbf{j}; \ t = 1$

3. $\mathbf{r}(t) = -\cosh 2t\mathbf{i} + \sinh 2t\mathbf{j}; \ t = 0$

4. $\mathbf{r}(t) = 2\cos t\mathbf{i} + (1 + \sin t)\mathbf{j}; \ t = \pi/3$

5. $\mathbf{r}(t) = 2\mathbf{i} + (t - 1)^2\mathbf{j} + t\mathbf{k}; \ t = 2$

6. $\mathbf{r}(t) = t\mathbf{i} + t\mathbf{j} + t^3\mathbf{k}; \ t = 2$

7. $\mathbf{r}(t) = t\mathbf{i} + t^2\mathbf{j} + t^3\mathbf{k}; \ t = 1$

8. $\mathbf{r}(t) = t\mathbf{i} + t^3\mathbf{j} + t\mathbf{k}; \ t = 1$

9. Suppose $\mathbf{r}(t) = t^2\mathbf{i} + (t^3 - 2t)\mathbf{j} + (t^2 - 5t)\mathbf{k}$ is the position vector of a moving particle. At what points does the particle pass through the xy-plane? What are its velocity and acceleration at these points?

10. Suppose a particle moves in space so that $\mathbf{a}(t) = \mathbf{0}$ for all time t. Describe its path.

11. A shell is fired from ground level with an initial speed of 480 ft/sec at an angle of elevation of 30°. Find

(a)   a vector function and parametric equations of the shell's trajectory,

(b)   the maximum altitude attained,

(c)   the range of the shell,

(d)   the speed at impact.

12. Rework Problem 11 if the shell is fired with the same initial speed and the same angle of elevation but from a cliff 1600 ft high.

13. A used car is pushed off an 81-ft-high sheer seaside cliff with a speed of 4 ft/sec. Find the speed at which the car hits the water.

14. A small projectile is launched with an initial speed of 98 m/sec. Find the possible angles of elevation so that its range is 490 m.

15. A football quarterback throws a 100-yard "bomb" at an angle of 45° from the horizontal. What is the initial speed of the football at the point of release?

16. A quarterback throws a football with the same initial speed at an angle of 60° from the horizontal and then at an angle of 30° from the horizontal. Show that the range of the football is the same in each case. Generalize this result for any release angle $0 < \theta < \pi/2$.

17. A projectile is fired from a cannon directly at a target that is dropped from rest simultaneously as the cannon is fired. Show that the projectile will strike the target in midair. See Figure 15.12. (*Hint:* Assume that the origin is at the muzzle of the cannon and that the angle of elevation is θ. If $\mathbf{r}_p$ and $\mathbf{r}_t$ are position vectors of the projectile and target, respectively, is there a time at which $\mathbf{r}_p = \mathbf{r}_t$?)

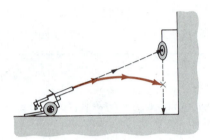

Figure 15.12

18. In army field maneuvers sturdy equipment and supply packs are simply dropped from planes that fly horizontally at a slow speed and a low altitude. A supply plane flies horizontally over a target at an altitude of 1024 ft at a constant speed of 180 mph. Use (15.5) to determine the horizontal distance a supply pack travels relative to the point from which it was dropped. At what line-of-sight angle α should the supply pack be released in order to hit the target indicated in Figure 15.13?

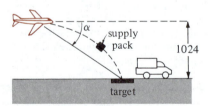

Figure 15.13

19. Suppose that $\mathbf{r}(t) = r_0 \cos \omega t \mathbf{i} + r_0 \sin \omega t \mathbf{j}$ is the position vector of an object that is moving in a circle of radius r_0 in the xy-plane. If $|\mathbf{v}(t)| = v$, show that the magnitude of the centripetal acceleration is $a = |\mathbf{a}(t)| = v^2/r_0$.

20. The motion of a particle in space is described by

$$\mathbf{r}(t) = b \cos t \mathbf{i} + b \sin t \mathbf{j} + ct\mathbf{k}, \qquad t \geq 0$$

(a) Compute $|\mathbf{v}(t)|$.

(b) Compute $s = \int_0^t |\mathbf{v}(t)| \, dt$ and verify that ds/dt is the same as the result of part **(a)**.

(c) Verify that $d^2s/dt^2 \neq |\mathbf{a}(t)|$.

21. The **effective weight** w_e of a body of mass m at the equator of the earth is defined by $w_e = mg - ma$, where a is the magnitude of the centripetal acceleration given in Problem 19. Determine the effective weight of a 192-lb person if the radius of the earth is 4000 mi, $g = 32$ ft/sec^2, and $v = 1530$ ft/sec.

22. Consider a bicyclist riding on a flat circular track of radius r_0. If m is the combined mass of the rider and bicycle, fill in the blanks in Figure 15.14. (*Hint:* Use Problem 19 and force = mass × acceleration. Assume that the positive directions are upward and to the left.) The resultant vector $\mathbf{U}$ gives the direction the bicyclist must be tipped to avoid falling. Find the angle ϕ from the vertical at which the bicyclist must be tipped if her speed is 44 ft/sec and the radius of the track is 60 ft.

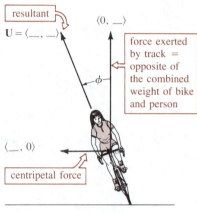

Figure 15.14

Miscellaneous Problems

23. Use the results given in (15.5) to prove that the trajectory of a ballistic projectile is parabolic.

24. A projectile is launched with an initial speed v_0 from ground level at an angle of elevation θ. Use (15.5) to show that the maximum height and range of the projectile are

$$H = \frac{v_0^2 \sin^2\theta}{2g} \quad \text{and} \quad R = \frac{v_0^2 \sin 2\theta}{g}$$

respectively.

15.3 Components of Acceleration; Curvature

Unit Tangent and Principal Normal

Let C be a space curve traced by $\mathbf{r}(t) = f(t)\mathbf{i} + g(t)\mathbf{j} + h(t)\mathbf{k}$, where f, g, and h have second derivatives. If $|\mathbf{r}'(t)| \neq 0$ at a point P on C, we define a **unit tangent** at P by

$$\mathbf{T} = \frac{\mathbf{r}'(t)}{|\mathbf{r}'(t)|} \qquad (15.9)$$

Now, velocity of a moving particle on C is $\mathbf{v}(t) = \mathbf{r}'(t)$, whereas its speed is $ds/dt = v = |\mathbf{v}(t)|$. Thus, (15.9) implies that

$$\mathbf{v}(t) = v\mathbf{T} \qquad (15.10)$$

Differentiating (15.10) with respect to t gives acceleration

$$\mathbf{a}(t) = v\frac{d\mathbf{T}}{dt} + \frac{dv}{dt}\mathbf{T} \qquad (15.11)$$

Furthermore, with the help of Theorem 15.4(*iii*), it follows from differentiation of $\mathbf{T} \cdot \mathbf{T} = 1$ that $\mathbf{T} \cdot d\mathbf{T}/dt = 0$. Hence, $\mathbf{T}$ and $d\mathbf{T}/dt$ are orthogonal at P. If $|d\mathbf{T}/dt| \neq 0$, the vector

$$\mathbf{N} = \frac{d\mathbf{T}/dt}{|d\mathbf{T}/dt|} \qquad (15.12)$$

is a unit normal to the curve C at P with direction given by $d\mathbf{T}/dt$. The vector $\mathbf{N}$ is also called the **principal normal**. If we define $\kappa = |d\mathbf{T}/dt| / ds/dt$, then $d\mathbf{T}/dt = \kappa v\mathbf{N}$ and so (15.11) becomes

$$\mathbf{a}(t) = \kappa v^2\mathbf{N} + \frac{dv}{dt}\mathbf{T} \qquad (15.13)$$

The scalar function κ is called the **curvature** of the curve C at P.*

Tangential and Normal Components of Acceleration

By writing (15.13) as

$$\mathbf{a}(t) = a_N\mathbf{N} + a_T\mathbf{T} \qquad (15.14)$$

we see that the acceleration vector $\mathbf{a}$ is the resultant of two orthogonal vectors $a_N\mathbf{N}$ and $a_T\mathbf{T}$. See Figure 15.15. The scalar functions $a_T = dv/dt$ and $a_N = \kappa v^2$ are called the **tangential** and **normal components of the acceleration**, respectively. Note that the tangential component of the acceleration results from a change in the *magnitude* of the velocity $\mathbf{v}$, whereas the normal component of the acceleration results from a change in *direction* of $\mathbf{v}$.

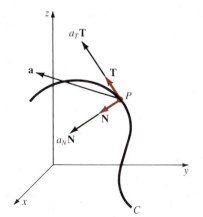

Figure 15.15

---

*Recall from Section 15.1 that a space curve can be parameterized in terms of arc length s; in this case the Chain Rule gives

$$\kappa = \left| \frac{d\mathbf{T}}{dt} \frac{dt}{ds} \right| = \left| \frac{d\mathbf{T}}{ds} \right|$$

In other words, the curvature is the magnitude of the rate of change of the unit vector $\mathbf{T}$ with respect to arc length.

The Binormal   A third unit vector defined by

$$\mathbf{B} = \mathbf{T} \times \mathbf{N}$$

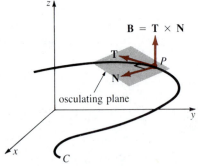

Figure 15.16

is called the **binormal**. The three unit vectors $\mathbf{T}$, $\mathbf{N}$, and $\mathbf{B}$ form a right-handed set of mutually orthogonal vectors called the **moving trihedral**. The plane of $\mathbf{T}$ and $\mathbf{N}$ is called the **osculating plane**;* the plane of $\mathbf{N}$ and $\mathbf{B}$ is said to be the **normal plane**, whereas the plane of $\mathbf{T}$ and $\mathbf{B}$ is the **rectifying plane**. See Figure 15.16.

Example 1

The position of a moving particle is given by $\mathbf{r}(t) = 2 \cos t\mathbf{i} + 2 \sin t\mathbf{j} + 3t\mathbf{k}$. Find the vectors $\mathbf{T}$, $\mathbf{N}$, and $\mathbf{B}$. Find the curvature.

Solution   Since $\mathbf{r}'(t) = -2 \sin t\mathbf{i} + 2 \cos t\mathbf{j} + 3\mathbf{k}$, $|\mathbf{r}'(t)| = \sqrt{13}$, and so from (15.9), we see that a unit tangent is

$$\mathbf{T} = -\frac{2}{\sqrt{13}}\sin t\mathbf{i} + \frac{2}{\sqrt{13}}\cos t\mathbf{j} + \frac{3}{\sqrt{13}}\mathbf{k}$$

Next, we have

$$\frac{d\mathbf{T}}{dt} = -\frac{2}{\sqrt{13}}\cos t\mathbf{i} - \frac{2}{\sqrt{13}}\sin t\mathbf{j} \quad \text{and} \quad \left|\frac{d\mathbf{T}}{dt}\right| = \frac{2}{\sqrt{13}}$$

Hence, (15.12) gives the principal normal

$$\mathbf{N} = -\cos t\mathbf{i} - \sin t\mathbf{j}$$

Now, the binormal is

$$\mathbf{B} = \mathbf{T} \times \mathbf{N} = \begin{vmatrix} \mathbf{i} & \mathbf{j} & \mathbf{k} \\ -\dfrac{2}{\sqrt{13}}\sin t & \dfrac{2}{\sqrt{13}}\cos t & \dfrac{3}{\sqrt{13}} \\ -\cos t & -\sin t & 0 \end{vmatrix}$$

$$= \frac{3}{\sqrt{13}}\sin t\mathbf{i} - \frac{3}{\sqrt{13}}\cos t\mathbf{j} + \frac{2}{\sqrt{13}}\mathbf{k}$$

Finally, the curvature is given by

$$\kappa = \frac{2/\sqrt{13}}{\sqrt{13}} = \frac{2}{13}$$

---

*Literally, this means the "kissing" plane.

By dotting, and in turn crossing, the vector $\mathbf{v} = v\mathbf{T}$ with (15.14), it is possible to obtain explicit formulas for the tangential and normal components of the acceleration and the curvature. Observe that

$$\mathbf{v} \cdot \mathbf{a} = a_N(\underbrace{v\mathbf{T} \cdot \mathbf{N}}_{0}) + a_T(\underbrace{v\mathbf{T} \cdot \mathbf{T}}_{1}) = a_T v$$

yields the tangential component of acceleration

$$a_T = \frac{dv}{dt} = \frac{\mathbf{v} \cdot \mathbf{a}}{|\mathbf{v}|} = \frac{\mathbf{r}'(t) \cdot \mathbf{r}''(t)}{|\mathbf{r}'(t)|} \tag{15.15}$$

On the other hand,

$$\mathbf{v} \times \mathbf{a} = a_N(\underbrace{v\mathbf{T} \times \mathbf{N}}_{\mathbf{B}}) + a_T(\underbrace{v\mathbf{T} \times \mathbf{T}}_{0}) = a_N v\mathbf{B}$$

Since $|\mathbf{B}| = 1$, it follows that the normal component of acceleration is

$$a_N = \kappa v^2 = \frac{|\mathbf{v} \times \mathbf{a}|}{|\mathbf{v}|} = \frac{|\mathbf{r}'(t) \times \mathbf{r}''(t)|}{|\mathbf{r}'(t)|} \tag{15.16}$$

Solving (15.16) for the curvature gives

$$\kappa = \frac{|\mathbf{v} \times \mathbf{a}|}{|\mathbf{v}|^3} = \frac{|\mathbf{r}'(t) \times \mathbf{r}''(t)|}{|\mathbf{r}'(t)|^3} \tag{15.17}$$

Example 2

The curve traced by $\mathbf{r}(t) = t\mathbf{i} + \frac{1}{2}t^2\mathbf{j} + \frac{1}{3}t^3\mathbf{k}$ is said to be a "twisted cube." If $\mathbf{r}(t)$ is the position vector of a moving particle, find the tangential and normal components of the acceleration at any t. Find the curvature.

Solution
$$\mathbf{v}(t) = \mathbf{r}'(t) = \mathbf{i} + t\mathbf{j} + t^2\mathbf{k}$$
$$\mathbf{a}(t) = \mathbf{r}''(t) = \mathbf{j} + 2t\mathbf{k}$$

Since $\mathbf{v} \cdot \mathbf{a} = t + 2t^3$ and $|\mathbf{v}| = \sqrt{1 + t^2 + t^4}$, it follows from (15.15) that

$$a_T = \frac{dv}{dt} = \frac{t + 2t^3}{\sqrt{1 + t^2 + t^4}}$$

Now,
$$\mathbf{v} \times \mathbf{a} = \begin{vmatrix} \mathbf{i} & \mathbf{j} & \mathbf{k} \\ 1 & t & t^2 \\ 0 & 1 & 2t \end{vmatrix} = t^2\mathbf{i} - 2t\mathbf{j} + \mathbf{k}$$

and $|\mathbf{v} \times \mathbf{a}| = \sqrt{t^4 + 4t^2 + 1}$. Thus, (15.16) gives

$$a_N = \kappa v^2 = \frac{\sqrt{t^4 + 4t^2 + 1}}{\sqrt{1 + t^2 + t^4}} = \sqrt{\frac{t^4 + 4t^2 + 1}{t^4 + t^2 + 1}}$$

From (15.17) we find that the curvature of the twisted cube is given by

$$\kappa = \frac{(t^4 + 4t^2 + 1)^{1/2}}{(1 + t^2 + t^4)^{3/2}}$$

Radius of Curvature

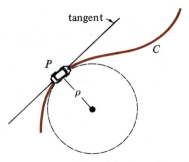

tangent

P

C

ρ

Figure 15.17

The reciprocal of the curvature, $\rho = 1/\kappa$, is called the **radius of curvature**. The radius of curvature at a point P on a curve C is the radius of a circle that "fits" the curve there better than any other circle. The circle at P is called the **circle of curvature** and its center is the **center of curvature**. The circle of curvature has the same tangent line at P as the curve C, and its center lies on the concave side of C. For example, a car moving on a curved track, as shown in Figure 15.17, can, at any instant, be thought to be moving on a circle of radius ρ. Hence, the normal component of its acceleration $a_N = \kappa v^2$ must be the same as the magnitude of its centripetal acceleration $a = v^2/\rho$. Therefore, $\kappa = 1/\rho$ and $\rho = 1/\kappa$. Knowing the radius of curvature, it is possible to determine the speed v at which a car can negotiate a banked curve without skidding. (This is essentially the idea in Problem 22 of Exercises 15.2.)

Remark

By writing (15.11) as

$$\mathbf{a}(t) = \frac{ds}{dt}\frac{d\mathbf{T}}{dt} + \frac{d^2s}{dt^2}\mathbf{T}$$

we note that the so-called scalar acceleration d^2s/dt^2, referred to in the last remark, is now seen to be the tangential component of the acceleration a_T.

Exercises 15.3

Answers to odd-numbered problems begin on page A-47.

In Problems 1 and 2, for the given position function, find the unit tangent.

1. $\mathbf{r}(t) = (t \cos t - \sin t)\mathbf{i} + (t \sin t + \cos t)\mathbf{j} + t^2\mathbf{k}$, $t > 0$

2. $\mathbf{r}(t) = e^t \cos t\mathbf{i} + e^t \sin t\mathbf{j} + \sqrt{2}e^t\mathbf{k}$

3. Use the procedure outlined in Example 1 to find $\mathbf{T}$, $\mathbf{N}$, $\mathbf{B}$, and κ for motion on a general circular helix that is described by
$$\mathbf{r}(t) = a \cos t\mathbf{i} + a \sin t\mathbf{j} + ct\mathbf{k}$$

4. Use the procedure outlined in Example 1 to show on the "twisted cube" of Example 2 that at $t = 1$:

$$\mathbf{T} = \frac{1}{\sqrt{3}}(\mathbf{i} + \mathbf{j} + \mathbf{k}), \qquad \mathbf{N} = -\frac{1}{\sqrt{2}}(\mathbf{i} - \mathbf{k}),$$

$$\mathbf{B} = -\frac{1}{\sqrt{6}}(-\mathbf{i} + 2\mathbf{j} - \mathbf{k}), \quad \text{and} \quad \kappa = \frac{\sqrt{2}}{3}$$

In Problems 5 and 6 find an equation of the osculating plane to the given space curve at the point that corresponds to the indicated value of t.

5. The circular helix of Example 1; $t = \pi/4$

6. The "twisted cube" of Example 2; $t = 1$

In Problems 7–16 $\mathbf{r}(t)$ is the position vector of a moving particle. Find the tangential and normal components of the acceleration at any t.

7. $\mathbf{r}(t) = \mathbf{i} + t\mathbf{j} + t^2\mathbf{k}$

8. $\mathbf{r}(t) = 3 \cos t\mathbf{i} + 2 \sin t\mathbf{j} + t\mathbf{k}$

9. $\mathbf{r}(t) = t^2\mathbf{i} + (t^2 - 1)\mathbf{j} + 2t^2\mathbf{k}$

10. $\mathbf{r}(t) = t^2\mathbf{i} - t^3\mathbf{j} + t^4\mathbf{k}$

11. $\mathbf{r}(t) = 2t\mathbf{i} + t^2\mathbf{j}$

12. $\mathbf{r}(t) = \tan^{-1}t\,\mathbf{i} + \dfrac{1}{2}\ln(1 + t^2)\mathbf{j}$

13. $\mathbf{r}(t) = 5\cos t\,\mathbf{i} + 5\sin t\,\mathbf{j}$

14. $\mathbf{r}(t) = \cosh t\,\mathbf{i} + \sinh t\,\mathbf{j}$

15. $\mathbf{r}(t) = e^{-t}(\mathbf{i} + \mathbf{j} + \mathbf{k})$

16. $\mathbf{r}(t) = t\,\mathbf{i} + (2t - 1)\mathbf{j} + (4t + 2)\mathbf{k}$

17. Find the curvature of an elliptical helix that is described by $\mathbf{r}(t) = a\cos t\,\mathbf{i} + b\sin t\,\mathbf{j} + ct\,\mathbf{k}$, $a > 0$, $b > 0$, $c > 0$.

18. **(a)** Find the curvature of an elliptical orbit that is described by $\mathbf{r}(t) = a\cos t\,\mathbf{i} + b\sin t\,\mathbf{j} + c\mathbf{k}$, $a > 0$, $b > 0$, $c > 0$.

   (b) Show that when $a = b$, the curvature of a circular orbit is the constant $\kappa = 1/a$.

19. Show that the curvature of a straight line is the constant $\kappa = 0$. (*Hint*: Use (14.39) of Section 14.5.)

20. Find the curvature of the cycloid that is described by

$$\mathbf{r}(t) = a(t - \sin t)\mathbf{i} + a(1 - \cos t)\mathbf{j}, \qquad a > 0$$

at $t = \pi$.

21. Let C be a plane curve traced by $\mathbf{r}(t) = f(t)\mathbf{i} + g(t)\mathbf{j}$, where f and g have second derivatives. Show that the curvature at a point is given by

$$\kappa = \frac{|f'(t)g''(t) - g'(t)f''(t)|}{([f'(t)]^2 + [g'(t)]^2)^{3/2}}$$

22. Show that if $y = F(x)$, the formula for κ in Problem 21 reduces to

$$\kappa = \frac{|F''(x)|}{[1 + (F'(x))^2]^{3/2}}$$

In Problems 23 and 24 use the result of Problem 22 to find the curvature and radius of curvature of the curve at the indicated points. Decide at which point the curve is "sharper."

23. $y = x^2$; $(0, 0)$, $(1, 1)$

24. $y = x^3$; $(-1, -1)$, $(\frac{1}{2}, \frac{1}{8})$

25. Discuss the curvature near a point of inflection of $y = F(x)$.

26. Show that $|\mathbf{a}(t)|^2 = a_N^2 + a_T^2$.

Chapter 15 Review Exercises

Answers to odd-numbered problems begin on page A-47.

In Problems 1–10 answer true or false.

1. A particle whose position vector is $\mathbf{r}(t) = \cos t\,\mathbf{i} + \cos t\,\mathbf{j} + \sqrt{2}\sin t\,\mathbf{k}$ moves with constant speed. _____

2. The path of a moving particle whose position vector is $\mathbf{r}(t) = (t^2 + 1)\mathbf{i} + 4\mathbf{j} + t^4\mathbf{k}$ lies in a plane. _____

3. The binormal vector is perpendicular to the osculating plane. _____

4. If $\mathbf{r}(t)$ is the position vector of a moving particle, then the velocity vector $\mathbf{v}(t) = \mathbf{r}'(t)$ and the acceleration vector $\mathbf{a}(t) = \mathbf{r}''(t)$ are orthogonal. _____

5. If s is arc length of a curve C, then the magnitude of the velocity of a particle moving on C is ds/dt. _____

6. If s is arc length of a curve C, then the magnitude of the acceleration of a particle on C is d^2s/dt^2. _____

7. If the binormal is defined by $\mathbf{B} = \mathbf{T} \times \mathbf{N}$, then the principal normal is $\mathbf{N} = \mathbf{B} \times \mathbf{T}$. _____

8. If $\lim_{t \to a}\mathbf{r}_1(t) = 2\mathbf{i} + \mathbf{j}$ and $\lim_{t \to a}\mathbf{r}_2(t) = -\mathbf{i} + 2\mathbf{j}$, then $\lim_{t \to a}\mathbf{r}_1(t) \cdot \mathbf{r}_2(t) = 0$. _____

9. $\displaystyle\int_a^b [\mathbf{r}_1(t) \cdot \mathbf{r}_2(t)]\,dt = \left[\int_a^b \mathbf{r}_1(t)\,dt\right] \cdot \left[\int_a^b \mathbf{r}_2(t)\,dt\right]$ _____

10. If $\mathbf{r}(t)$ is differentiable, then $\dfrac{d}{dt}|\mathbf{r}(t)|^2 = 2\mathbf{r}(t) \cdot \dfrac{d\mathbf{r}}{dt}$. _____

11. Find the length of the curve that is traced by the vector function $\mathbf{r}(t) = \sin t\,\mathbf{i} + (1 - \cos t)\mathbf{j} + t\mathbf{k}$ on the interval $0 \le t \le \pi$.

12. The position vector of a moving particle is given by $\mathbf{r}(t) = 5t\,\mathbf{i} + (1 + t)\mathbf{j} + 7t\mathbf{k}$. Given that the particle starts at the point corresponding to $t = 0$, find the distance the particle travels to the point corresponding to $t = 3$. At what point will the particle have traveled $80\sqrt{3}$ units along the curve?

13. Find parametric equations for the tangent line to the curve that is traced by

$$\mathbf{r}(t) = -3t^2\mathbf{i} + 4\sqrt{t + 1}\,\mathbf{j} + (t - 2)\mathbf{k}$$

at $t = 3$.

14. Show that the curve traced by $\mathbf{r}(t) = t \cos t\mathbf{i} + t \sin t\mathbf{j} + t\mathbf{k}$ lies on the surface of a cone. Sketch the curve.

15. Sketch the curve traced by $\mathbf{r}(t) = \cosh t\mathbf{i} + \sinh t\mathbf{j} + t\mathbf{k}$.

16. Given that $\mathbf{r}_1(t) = t^2\mathbf{i} + 2t\mathbf{j} + t^3\mathbf{k}$ and $\mathbf{r}_2(t) = -t\mathbf{i} + t^2\mathbf{j} + (t^2 + 1)\mathbf{k}$, calculate $(d/dt)[\mathbf{r}_1(t) \times \mathbf{r}_2(t)]$ in two different ways.

17. Given that $\mathbf{r}_1(t) = \cos t\mathbf{i} - \sin t\mathbf{j} + 4t^3\mathbf{k}$ and $\mathbf{r}_2(t) = t^2\mathbf{i} + \sin t\mathbf{j} + e^{2t}\mathbf{k}$, calculate $(d/dt)[\mathbf{r}_1(t) \cdot \mathbf{r}_2(t)]$ in two different ways.

18. Given that $\mathbf{r}_1$, $\mathbf{r}_2$, and $\mathbf{r}_3$ are differentiable, find $(d/dt)[\mathbf{r}_1(t) \cdot (\mathbf{r}_2(t) \times \mathbf{r}_3(t))]$.

19. A particle of mass m is acted on by a continuous force of magnitude 2, which is directed parallel to the positive y-axis. If the particle starts with an initial velocity $\mathbf{v}(0) = \mathbf{i} + \mathbf{j} + \mathbf{k}$ from $(1, 1, 0)$, find the position vector of the particle and parametric equations of its path. (*Hint*: $\mathbf{F} = m\mathbf{a}$.)

20. The position vector of a moving particle is $\mathbf{r}(t) = t\mathbf{i} + (1 - t^3)\mathbf{j}$.

  (a) Sketch the path of the particle.

  (b) Sketch the velocity and acceleration vectors at $t = 1$.

  (c) Find the speed at $t = 1$.

21. Find the velocity and acceleration of a particle whose position vector is $\mathbf{r}(t) = 6t\mathbf{i} + t\mathbf{j} + t^2\mathbf{k}$ as it passes through the plane $-x + y + z = -4$.

22. The velocity of a moving particle is $\mathbf{v}(t) = -10t\mathbf{i} + (3t^2 - 4t)\mathbf{j} + \mathbf{k}$. If the particle starts at $t = 0$ at $(1, 2, 3)$, what is its position at $t = 2$?

23. The acceleration of a moving particle is $\mathbf{a}(t) = \sqrt{2} \sin t\mathbf{i} + \sqrt{2} \cos t\mathbf{j}$. Given that the velocity and position of the particle at $t = \pi/4$ are $\mathbf{v}(\pi/4) = -\mathbf{i} + \mathbf{j} + \mathbf{k}$ and $\mathbf{r}(\pi/4) = \mathbf{i} + 2\mathbf{j} + (\pi/4)\mathbf{k}$, respectively, what was the position of the particle at $t = 3\pi/4$?

24. Given that

$$\mathbf{r}(t) = \frac{1}{2}t^2\mathbf{i} + \frac{1}{3}t^3\mathbf{j} - \frac{1}{2}t^2\mathbf{k}$$

is the position vector of a moving particle, find the tangential and normal components of the acceleration at any t. Find the curvature.

25. Suppose that the vector function of Problem 15 is the position vector of a moving particle. Find the vectors $\mathbf{T}$, $\mathbf{N}$, and $\mathbf{B}$ at $t = 1$. Find the curvature at that point.

16

Differential Calculus of Functions of Several Variables

Up to this point in our study of calculus, we have only considered functions of a single variable. Previously considered concepts such as limits, continuity, derivatives, tangents, maxima and minima, integrals, and so on, extend to functions of several variables as well. The present chapter is devoted primarily to the differential calculus of such multivariable functions.

16.1   Functions of Two or More Variables

Recall that a function of one variable $y = f(x)$ is a *rule of correspondence* that assigns to an element x of a subset of the real numbers R, called the *domain* of f, one and only one real number y. The set $\{y \mid y = f(x)\}$ is called the *range* of f. You are probably already aware of the existence of functions of two or more variables.

Example 1

(a) $A = xy$, area of a rectangle

(b) $V = \pi r^2 h$, volume of a circular cylinder

(c) $V = \dfrac{\pi}{3} r^2 h$, volume of a cone

(d) $P = 2x + 2y$, perimeter of a rectangle

Example 2

(a)   The pressure P exerted by an enclosed ideal gas is a function of its temperature T and volume V,

$$P = k\left(\frac{T}{V}\right), \qquad k \text{ a constant}$$

(b)   The area S of the surface of a human body is a function of its weight w and its height h,

$$S = 0.1091\, w^{0.425} h^{0.725}$$

Functions of Two Variables   The formal definition of a function of two variables follows.

> **DEFINITION 16.1   Function of Two Variables**
>
> A **function of two variables** is a rule of correspondence that assigns to each ordered pair of real numbers (x, y) of a subset of the xy-plane one and only one number z in the set R of real numbers.

The set of ordered pairs (x, y) is called the **domain** of the function and the set of corresponding values of z is called the **range**. A function of two variables is usually written $z = f(x, y)$ and read "f of x, y." The variables x and y are called the **independent variables** of the function and z is called the **dependent variable**.

Graphs   The **graph** of a function $z = f(x, y)$ is a *surface* in 3-space. See Figure 16.1.

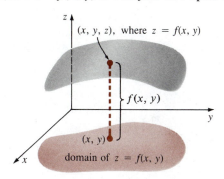

Figure 16.1

_____ **Example 3** _____

An equation of a plane $ax + by + cz = d$, $c \neq 0$ describes a function by writing

$$z = -\frac{a}{c}x - \frac{b}{c}y + \frac{d}{c} \quad \text{or} \quad f(x, y) = -\frac{a}{c}x - \frac{b}{c}y + \frac{d}{c}$$

Since z is defined for any choice of x and y, the domain of the function consists of the entire xy-plane.

_____ **Example 4** _____

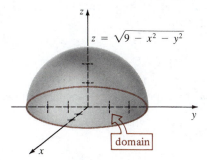

Figure 16.2

The graph of the function

$$f(x, y) = \sqrt{9 - x^2 - y^2}$$

is the hemisphere* shown in Figure 16.2. The domain of the function is the set of ordered pairs (x, y) that satisfies

$$9 - x^2 - y^2 \geq 0 \quad \text{or} \quad x^2 + y^2 \leq 9$$

That is, the domain of f consists of the interior and boundary of the circle $x^2 + y^2 = 9$. Inspection of Figure 16.2 shows that the range of the function is defined by $0 \leq z \leq 3$.

_____ **Example 5** _____

From the discussion of quadric surfaces in Section 14.7 you should recognize the graph of the function $f(x, y) = x^2 + 9y^2$ as a paraboloid. Since f is defined for every ordered pair of real numbers, its domain is the entire xy-plane. From the fact that $x^2 \geq 0$ and $y^2 \geq 0$ we see that the range of f is given by $z \geq 0$.

---

*Verify this by replacing the symbol $f(x, y)$ by z and squaring both sides of the equation.

Example 6

Given that $f(x, y) = 4 + \sqrt{x^2 - y^2}$, find

(a) $f(1, 0), f(5, 3), f(4, -2)$

(b)   Sketch the domain of the function.

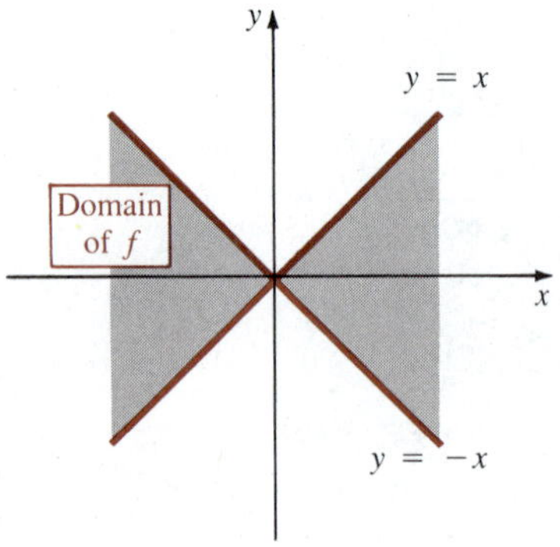

Solution

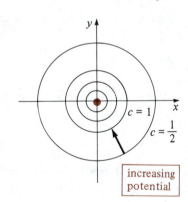

Figure 16.3

(a)
$$f(1, 0) = 4 + \sqrt{1 - 0} = 5$$
$$f(5, 3) = 4 + \sqrt{25 - 9} = 4 + \sqrt{16} = 8$$
$$f(4, -2) = 4 + \sqrt{16 - (-2)^2} = 4 + \sqrt{20} = 4 + 2\sqrt{5}$$

(b)   The domain of f consists of all ordered pairs (x, y) for which $x^2 - y^2 \geq 0$. As shown in Figure 16.3, the domain consists of all points on the lines $y = x$ and $y = -x$ and in the shaded region between them.

In science, one often encounters the words **isothermal**, **equipotential**, or **isobaric**. These terms apply to lines or curves on which either temperature, potential, or barometric pressure is *constant*.

Example 7

The electrostatic potential at a point $P(x, y)$ in the plane due to a unit point charge at the origin is given by $U = 1/\sqrt{x^2 + y^2}$. If the potential is a constant, say $U = c$, where c is a positive constant, then

$$\frac{1}{\sqrt{x^2 + y^2}} = c \quad \text{or} \quad x^2 + y^2 = \frac{1}{c^2}$$

Thus, as shown in Figure 16.4, the curves of equipotential are concentric circles surrounding the charge.

Figure 16.4

Note that in Figure 16.4, we can get a feeling of the behavior of the function U, specifically where it is increasing (or decreasing), by observing the direction of increasing c.

Level Curves   In general, if a function of two variables is given by $z = f(x, y)$, then the curves defined by $f(x, y) = c$, for suitable c, are called the **level curves** of f. The word *level* arises from the fact that we can interpret $f(x, y) = c$ as the projection onto the xy-plane of the curve of intersection, or trace, of $z = f(x, y)$ and the (horizontal or level) plane $z = c$. See Figure 16.5.

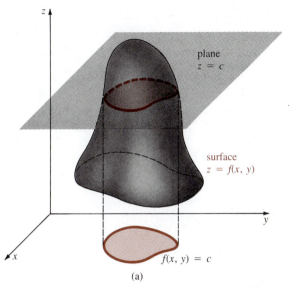

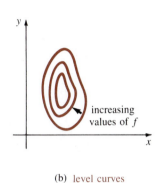

(a)

(b) level curves

Figure 16.5

Example 8

The level curves of the function $f(x, y) = y^2 - x^2$ are defined by $y^2 - x^2 = c$. As shown in Figure 16.6, when $c > 0$ or $c < 0$, a member of this family of curves is a hyperbola. For $c = 0$, we obtain the lines $y = x$ and $y = -x$.

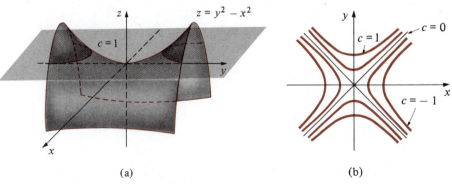

(a)

(b)

Figure 16.6

In the plane, the curves $f(x, y) = c$ are also called **contour lines** of a surface. In Figure 16.7, we see that a contour map illustrates the various segments of a hill that have a given altitude. This is the idea of the contours in Figure 16.8, which show the thickness of volcanic ash surrounding the volcano El Chichon. El Chichon, in the state of Chiapas, Mexico, erupted on March 28 and April 4, 1982.

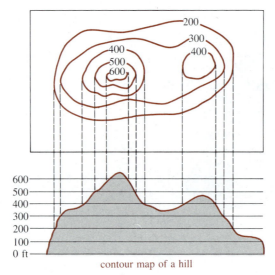

contour map of a hill

Figure 16.7

Bay of Campeche

El Chichon

GUATEMALA

MEXICO

thickness (in mm) of rain-
compacted volcanic ash
surrounding El Chichon*

Figure 16.8

Computer Graphics   In many instances the task of graphing a function of two variables is formidable. The use of **computer graphics** has become widespread in analyzing complicated surfaces in 3-space. See Figures 16.9–16.16 (pages 770–771). Note the computer-generated level curves given in Figures 16.9 and 16.10. In Figure 16.15 the function is

$$
f(x, y) = 40 \exp\left[-\left(\frac{x-30}{20}\right)^2 - \left(\frac{y-35}{20}\right)^2\right] + 20 \exp\left[-\left(\frac{x-75}{10}\right)^2 - \left(\frac{y-15}{10}\right)^2\right]
$$

$$
+ 15 \exp\left[-\left(\frac{x-90}{10}\right)^2 - \left(\frac{y-55}{10}\right)^2\right] + 25 \exp\left[-\left(\frac{x-120}{12}\right)^2 - \left(\frac{y-20}{12}\right)^2\right]
$$

In Figure 16.16 the function is defined by

$$
f(x, y) = 2 + \sum_{k=1}^{6} (-1)^{k+1}[(x - \alpha_k)^2 + (y - \beta_k)^2]^{-1/2}
$$

where

k	1	2	3	4	5	6
α_k	12	12	15	15	18	18
β_k	102.5	107.5	107.5	102.5	102.5	107.5

The spikes in this latter graph correspond to the points (α_k, β_k), $k = 1, 2, \ldots, 6$ at which the function is not defined.

---

*Adapted, with permission, from *National Geographic Magazine*.

Functions of Three or More Variables

The definitions of functions of three or more variables are simply generalizations of Definition 16.1. For example, a **function of three variables** is a rule of correspondence that assigns to each ordered triple of real numbers (x, y, z) in a subset of 3-space, one and only one number w in the set R of real numbers.

Example 9

$$F(x, y, z) = \frac{2x + 3y + z}{4 - x^2 - y^2 - z^2}$$

is an example of a function of three variables. Its domain is the set of points (x, y, z) that satisfies $x^2 + y^2 + z^2 \neq 4$; that is, the domain of F is all of 3-space *except* the points on the surface of a sphere of radius 2 centered at the origin.

Example 10

(a) $V = xyz$, volume of a rectangular box

(b)   Poiseuille's law states that the discharge rate, or rate of flow, of a viscous fluid (such as blood) through a tube (such as an artery) is

$$Q = k\frac{R^4}{L}(p_1 - p_2), \qquad k \text{ a constant}$$

where R is the radius of the tube, L its length, and p_1 and p_2 pressures at the end of the tube. This is an example of a function of *four* variables.

Note: Since it would take four dimensions, we cannot graph a function of three variables.

Level Surfaces

For a function of three variables, $w = F(x, y, z)$, the surfaces defined by $F(x, y, z) = c$, for suitable values of c, are called **level surfaces**.*

Example 11

The level surfaces of $F(x, y, z) = 2x - 3y + 4z$ are parallel planes defined by $2x - 3y + 4z = c$.

---

*An unfortunate, but standard, choice of words since level surfaces are not usually level.

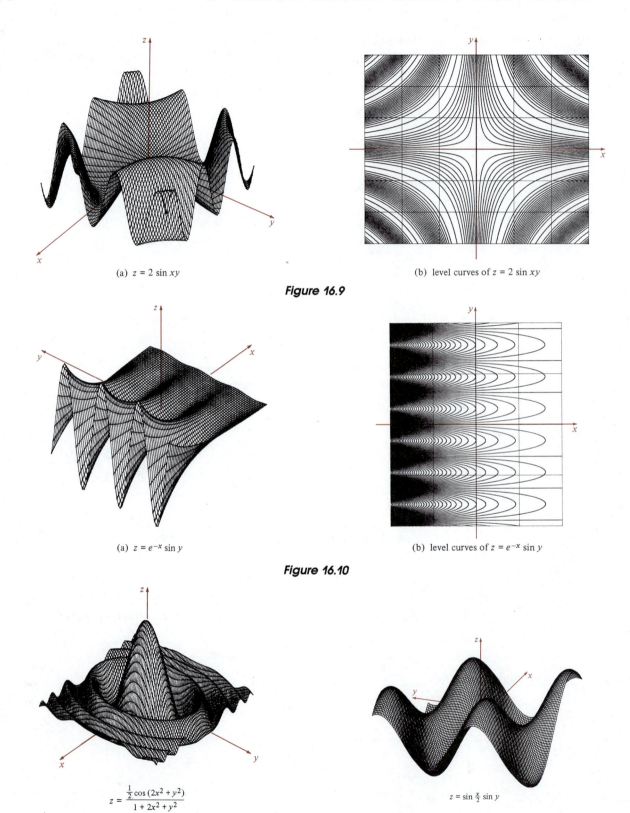

(a) $z = 2 \sin xy$

(b) level curves of $z = 2 \sin xy$

Figure 16.9

(a) $z = e^{-x} \sin y$

(b) level curves of $z = e^{-x} \sin y$

Figure 16.10

$$z = \frac{\frac{1}{2} \cos (2x^2 + y^2)}{1 + 2x^2 + y^2}$$

Figure 16.11

$z = \sin \frac{x}{2} \sin y$

Figure 16.12

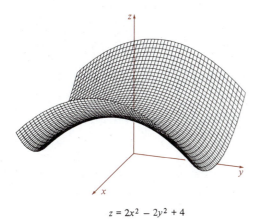

$$z = 2x^2 - 2y^2 + 4$$

Figure 16.13

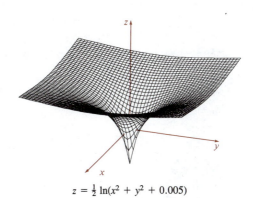

$$z = \tfrac{1}{2} \ln(x^2 + y^2 + 0.005)$$

Figure 16.14

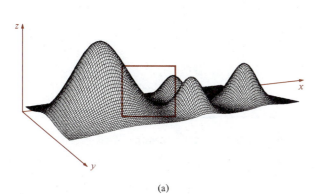

(a)

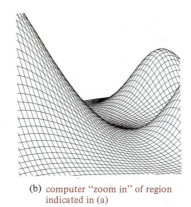

(b) computer "zoom in" of region
indicated in (a)

Figure 16.15

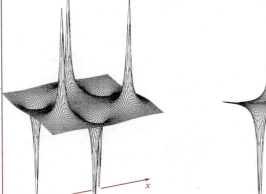

(a) top view of surface

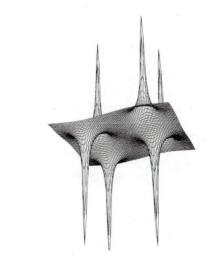

(b) side view of surface

(c) bottom view of surface

Figure 16.16

Example 12

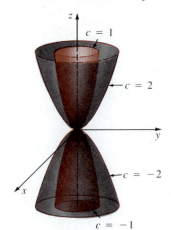

$c = 1$

$c = 2$

z

y

$c = -2$

x

$c = -1$

Figure 16.17

Describe the level surfaces of the function $F(x, y, z) = (x^2 + y^2)/z$.

Solution   For $c \neq 0$ the level surfaces are given by

$$\frac{x^2 + y^2}{z} = c$$

or

$$x^2 + y^2 = cz$$

A few members of this family of paraboloids are shown in Figure 16.17.

Exercises 16.1

Answers to odd-numbered problems begin on page A-48.

In Problems 1–10 find the domain of the given functions.

1. $f(x, y) = \dfrac{xy}{x^2 + y^2}$

2. $f(x, y) = (x^2 - 9y^2)^{-2}$

3. $f(x, y) = \dfrac{y^2}{y + x^2}$

4. $f(x, y) = x^2 - y^2\sqrt{4 + y}$

5. $f(s, t) = s^3 - 2t^2 + 8st$

6. $f(u, v) = \dfrac{u}{\ln(u^2 + v^2)}$

7. $g(r, s) = e^{2r}\sqrt{s^2 - 1}$

8. $g(\theta, \phi) = \dfrac{\tan \theta + \tan \phi}{1 - \tan \theta \tan \phi}$

9. $H(u, v, w) = \sqrt{u^2 + v^2 + w^2 - 16}$

10. $F(x, y, z) = \dfrac{\sqrt{25 - x^2 - y^2}}{z - 5}$

In Problems 11–18 match the given function with the figure (Figures 16.18–16.25) that most closely resembles its domain.

11. $f(x, y) = \sqrt{y - x^2}$

12. $f(x, y) = \ln(x - y^2)$

13. $f(x, y) = \sqrt{x} + \sqrt{y - x}$

14. $f(x, y) = \sqrt{\dfrac{x}{y} - 1}$

15. $f(x, y) = \sqrt{xy}$

16. $f(x, y) = \sin^{-1}(xy)$

17. $f(x, y) = \dfrac{x^4 + y^4}{xy}$

18. $f(x, y) = \dfrac{\sqrt{x^2 + y^2 - 1}}{y - x}$

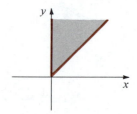

Figure 16.18

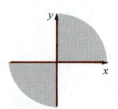

Figure 16.19

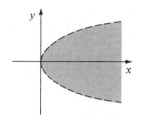

Figure 16.20

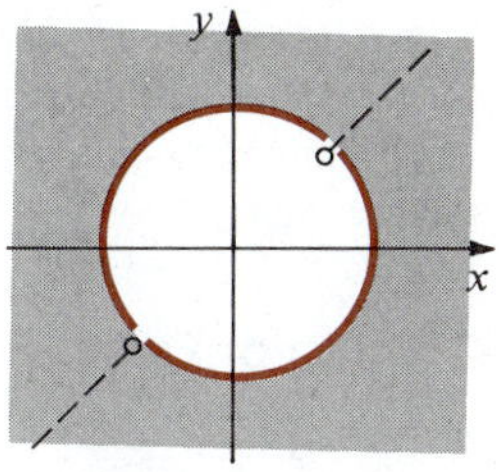

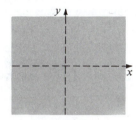

Figure 16.21

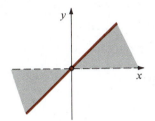

Figure 16.22

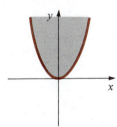

Figure 16.23

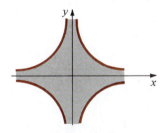

Figure 16.24

Figure 16.25

In Problems 27–30 evaluate the given function at the indicated points.

27. $f(x, y) = \int_x^y (2t - 1) \, dt$; $(2, 4)$, $(-1, 1)$

28. $f(x, y) = \ln\dfrac{x^2}{x^2 + y^2}$; $(3, 0)$, $(5, -5)$

29. $F(x, y, z) = (x + 2y + 3z)^2$; $(-1, 1, -1)$, $(2, 3, -2)$

30. $F(x, y, z) = \dfrac{1}{x^2} + \dfrac{1}{y^2} + \dfrac{1}{z^2}$; $(\sqrt{3}, \sqrt{2}, \sqrt{6})$, $\left(\dfrac{1}{4}, \dfrac{1}{5}, \dfrac{1}{3}\right)$

In Problems 31–36 discuss the graph of the given function.

31. $z = x$ **32.** $z = y^2$

33. $z = \sqrt{x^2 + y^2}$ **34.** $z = \sqrt{1 + x^2 + y^2}$

35. $z = \sqrt{9 - x^2 - 3y^2}$

36. $z = -\sqrt{16 - x^2 - y^2}$

In Problems 37–42 sketch some of the level curves associated with the given function.

37. $f(x, y) = x + 2y$ **38.** $f(x, y) = y^2 - x$

39. $f(x, y) = \sqrt{x^2 - y^2 - 1}$

40. $f(x, y) = \sqrt{36 - 4x^2 - 9y^2}$

41. $f(x, y) = e^{y - x^2}$

42. $f(x, y) = \tan^{-1}(y - x)$

In Problems 43–46 describe the level surfaces but do not graph.

43. $F(x, y, z) = \dfrac{x^2}{9} + \dfrac{z^2}{4}$

44. $F(x, y, z) = x^2 + y^2 + z^2$

45. $F(x, y, z) = x^2 + 3y^2 + 6z^2$

46. $F(x, y, z) = 4y - 2z + 1$

47. Graph some of the level surfaces associated with $F(x, y, z) = x^2 + y^2 - z^2$ for $c = 0$, $c > 0$, and $c < 0$.

48. Given that

$$F(x, y, z) = \frac{x^2}{16} + \frac{y^2}{4} + \frac{z^2}{9}$$

find the x-, y-, and z-intercepts of the level surface that passes through $(-4, 2, -3)$.

In Problems 19–22 sketch the domain of the given function.

19. $f(x, y) = \sqrt{x} - \sqrt{y}$

20. $f(x, y) = \sqrt{(1 - x^2)(y^2 - 4)}$

21. $f(x, y) = \sqrt{\ln(y - x + 1)}$

22. $f(x, y) = e^{\sqrt{xy+1}}$

In Problems 23–26 find the range of the given function.

23. $f(x, y) = 10 + x^2 + 2y^2$

24. $f(x, y) = x + y$

25. $F(x, y, z) = \sin(x + 2y + 3z)$

26. $F(x, y, z) = 7 - e^{xyz}$

49. The temperature, pressure, and volume of an enclosed ideal gas are related by $T = 0.01\,PV$, where T, P, and V are measured in degrees Kelvin, atmospheres, and liters, respectively. Sketch the isotherms $T = 300°K$, $400°K$, and $600°K$.

50. Express the height of a rectangular box with a square bottom as a function of the volume and the length of one side of the box.

51. A soda pop can is constructed with a tin lateral side and with an aluminum top and bottom. Given that the cost of the top is 1.8 cents per square unit, 1 cent per square unit for the bottom, and 2.3 cents per square unit for the side, determine the cost function $C(r, h)$, where r is the radius of the can and h is its height.

52. A closed rectangular box is to be constructed from 500 cm² of cardboard. Express the volume V as a function of the length x and width y.

53. As shown in Figure 16.26, a conical cap rests on top of a circular cylinder. If the height of the cap is two-thirds the height of the cylinder, express the volume of the solid as a function of the indicated variables.

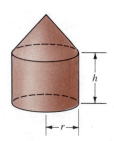

Figure 16.26

54. An obliquely cut cylinder (such as used in tissue samples) is shown in Figure 16.27. Express the thickness t of the cut as a function of x, y, and z.

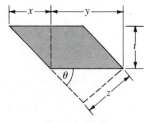

Figure 16.27

55. In medicine, formulas for surface area (see Example 2) are often used to calibrate drug doses since it is assumed that a drug dose D and surface area S are directly proportional. The following simple function can be used to obtain a quick estimate of the body surface area of a human: $S = 2ht$, where h is the

height (in cm) and t is the maximum thigh circumference (in cm). Estimate the surface area of a 156 cm-tall-person with a maximum thigh circumference of 50 cm. Estimate your own surface area.

Calculator Problems

56. The power required for a hummingbird to hover is given by

$$P = \sqrt{w^3/2\rho A}$$

where w is its weight, ρ is the density of air, and A is the area swept out by its moving wings. Find P for a hummingbird whose mass is 2.5×10^{-3} kg, $A = 2.2 \times 10^{-3}$ m², and $\rho = 1.2$ kg/m³. (*Hint:* Use $w = mg$.)

57. During his investigation of the winter of 1941 in the Antarctic, Dr. Paul A. Siple devised the following function for defining the **wind chill factor**:

$$H(v, T) = (10\sqrt{v} - v + 10.5)(33 - T)$$

where H is measured in kcal/m²hr, v is wind velocity in m/sec, and T is temperature in °C. An example of this index is: 1000 = very cold, 1200 = bitterly cold, and 1400 = exposed flesh freezes. Determine the wind chill factor at $-6.67°C(20°F)$ with a wind velocity of 20m/sec (45 mph).

58. In Problem 57, the wind chill factor is often expressed as an equivalent temperature. Use the table at the top of page 775 to find the wind chill factors.

 (a) $H(45, 20)$ **(b)** $H(20, -5)$

 (c) $H(15, -25)$ **(d)** $H(40, -5)$

59. The total energy consumption C in calories per hour of a person with a metabolic rate r, weight w, and height h is given by $C = 0.2rw^{0.425}h^{0.725}$. Find the energy consumption of a jogger who weighs 80 kg, is 1.6 m tall, and whose metabolic rate is 600 cal/m²hr.

60. In the clean and jerk competition, a weight lifter in the heavyweight class who weighs 110 kg makes a lift of 210 kg. In the flyweight class, a person who weighs 50 kg lifts 130 kg. How can these feats of strength be compared? In an overall competition, who would be judged the superior lifter? Several different formulas for handicapping lifts have been proposed. Let w_l denote the weight lifted (in kg) and w_b be the body weight of the lifter (in kg).

 (a) Used in ABC's Superstars competition:
$h = w_l - w_b$

 (b) Austin formula: $h = w_l/w_b^{3/4}$

 (c) Classical formula: $h = w_l/w_b^{2/3}$

 (d) O'Carroll formula: $h = w_l/(w_b - 35)^{1/3}$

Use these formulas to determine whether the heavyweight or the flyweight is the superior lifter.

Wind v (mph)	Temperature T(°F)															
	45	40	35	30	25	20	15	10	5	0	−5	−10	−15	−20	−25	−30
4	45	40	35	30	25	20	15	10	5	0	−5	−10	−15	−20	−25	−30
5	43	37	32	27	22	16	11	6	0	−5	−10	−15	−21	−26	−31	−36
10	34	28	22	16	10	3	−3	−9	−15	−22	−27	−34	−40	−46	−52	−58
15	29	23	16	9	2	−5	−11	−18	−25	−31	−38	−45	−51	−58	−65	−72
20	26	19	12	4	−3	−10	−17	−24	−31	−39	−46	−53	−60	−67	−74	−81
25	23	16	8	1	−7	−15	−22	−29	−36	−44	−51	−59	−66	−74	−81	−88
30	21	13	6	−2	−10	−18	−25	−33	−41	−49	−56	−64	−71	−79	−86	−93
35	20	12	4	−4	−12	−20	−27	−35	−43	−52	−58	−67	−74	−82	−89	−97
40	19	11	3	−5	−13	−21	−29	−37	−45	−53	−60	−69	−76	−84	−92	−100
45	18	10	2	−6	−14	−22	−30	−38	−46	−54	−62	−70	−78	−85	−93	−102

16.2 Limits and Continuity

16.2.1 An Informal Discussion

For functions of one variable, in many instances, we were able to make a judgment about the existence of $\lim_{x \to a} f(x)$ from the graph of $y = f(x)$. Also, we utilized the fact that $\lim_{x \to a} f(x)$ exists if and only if $\lim_{x \to a^-} f(x)$ and $\lim_{x \to a^+} f(x)$ exist and are equal to the same number L. In which case, $\lim_{x \to a} f(x) = L$.

The situation is more demanding when we consider limits of functions of two variables. To analyze a limit by sketching a graph of $z = f(x, y)$ is not convenient nor even routinely possible for most functions. Intuitively, f has a limit at a point (a, b) if the functional values $f(x, y)$ are approaching a number L as (x, y) approaches (a, b). We write $f(x, y) \to L$ as $(x, y) \to (a, b)$, or

$$\lim_{(x,y) \to (a,b)} f(x, y) = L$$

To be more precise, f has a limit L at a point (a, b) if the points in space $(x, y, f(x, y))$ can be made arbitrarily close to (a, b, L) whenever (x, y) is close enough to (a, b).

The notation of (x, y) "approaching" a point (a, b) is not as simple as in functions of one variable where $x \to a$ means that x can approach a only from the left and from the right. In the xy-plane, there are, of course, an infinite number of ways of approaching a point (a, b). As shown in Figure 16.28, in order that $\lim_{(x,y) \to (a,b)} f(x, y)$ exist, we now require that f approach the same number L along *every* possible curve or **path** through (a, b). Put in a negative way:

> If $f(x, y)$ does *not* approach the same number L for two different paths to (a, b), then $\lim_{(x,y) \to (a,b)} f(x, y)$ does not exist.　(16.1)

In the discussion of $\lim_{(x,y) \to (a,b)} f(x, y)$ that follows we shall assume that the function f is defined at every point (x, y) in the interior of a circle centered at (a, b) but not necessarily *at* (a, b) itself.

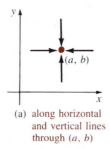
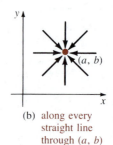
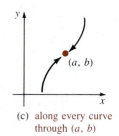

(a) along horizontal
and vertical lines
through (a, b)

(b) along every
straight line
through (a, b)

(c) along every curve
through (a, b)

Figure 16.28

___ **Example 1** _____

Show that $\lim\limits_{(x,y)\to(0,0)} \dfrac{x^2 - 3y^2}{x^2 + 2y^2}$ does not exist.

Solution   The function $f(x, y) = (x^2 - 3y^2)/(x^2 + 2y^2)$ is defined everywhere except at $(0, 0)$. Two ways of approaching $(0, 0)$ are along the x-axis ($y = 0$) and along the y-axis ($x = 0$). We have

$$\text{on } y = 0, \qquad \lim\limits_{(x,0)\to(0,0)} f(x, 0) = \lim\limits_{(x,0)\to(0,0)} \frac{x^2 - 0}{x^2 + 0} = 1$$

$$\text{on } x = 0, \qquad \lim\limits_{(0,y)\to(0,0)} f(0, y) = \lim\limits_{(0,y)\to(0,0)} \frac{0 - 3y^2}{0 + 2y^2} = -\frac{3}{2}$$

In view of (16.1) we conclude that the limit does not exist.

___ **Example 2** _____

Show that $\lim\limits_{(x,y)\to(0,0)} \dfrac{xy}{x^2 + y^2}$ does not exist.

Solution   In this case the limits along the x- and y-axes are the same:

$$\lim\limits_{(x,0)\to(0,0)} f(x, 0) = \lim\limits_{(0,y)\to(0,0)} f(0, y) = 0$$

However, this does *not* mean $\lim_{(x,y)\to(0,0)} f(x, y)$ exists since we have not examined every path to $(0, 0)$. We now try any line through the origin given by $y = mx$:

$$\lim\limits_{(x,y)\to(0,0)} f(x, y) = \lim\limits_{(x,y)\to(0,0)} \frac{mx^2}{x^2 + m^2 x^2} = \frac{m}{1 + m^2}$$

Since $\lim_{(x,y)\to(0,0)} f(x, y)$ depends on the slope of the line on which we approach the origin, we conclude that the limit does not exist. For example,

$$\text{on } y = x, \qquad \lim\limits_{(x,y)\to(0,0)} f(x, y) = \frac{1}{2}$$

$$\text{on } y = 2x, \qquad \lim\limits_{(x,y)\to(0,0)} f(x, y) = \frac{2}{5}$$

Example 3

Show that $\displaystyle\lim_{(x,y)\to(0,0)} \frac{x^3y}{x^6 + y^2}$ does not exist.

Solution Let $f(x, y) = x^3y/(x^6 + y^2)$. You are encouraged to show that along the x- and y-axes, along any line $y = mx$, $m \neq 0$ through $(0, 0)$, and along any parabola $y = kx^2$, $k \neq 0$ through $(0, 0)$, $\lim_{(x,y)\to(0, 0)} f(x, y) = 0$. Although this is an infinite number of paths to the origin, the limit *still* does not exist since on $y = x^3$:

$$\lim_{(x,y)\to(0,0)} f(x, y) = \lim_{(x,y)\to(0,0)} \frac{x^6}{x^6 + x^6} = \frac{1}{2}$$

We state the next two results without proof. Theorem 16.2 is the analogue, for functions of two variables, of Theorem 2.5.

THEOREM 16.1

$(i) \qquad \displaystyle\lim_{(x,y)\to(a,b)} x = a$

$(ii) \qquad \displaystyle\lim_{(x,y)\to(a,b)} y = b$

THEOREM 16.2

If $\displaystyle\lim_{(x,y)\to(a,b)} f(x, y) = L_1$ and $\displaystyle\lim_{(x,y)\to(a,b)} g(x, y) = L_2$, then

$(i) \quad \displaystyle\lim_{(x,y)\to(a,b)} [f(x, y) + g(x, y)] = \lim_{(x,y)\to(a,b)} f(x, y) + \lim_{(x,y)\to(a,b)} g(x, y) = L_1 + L_2$

$(ii) \quad \displaystyle\lim_{(x,y)\to(a,b)} f(x, y) \cdot g(x, y) = \lim_{(x,y)\to(a,b)} f(x, y) \cdot \lim_{(x,y)\to(a,b)} g(x, y) = L_1 L_2$

$(iii) \quad \displaystyle\lim_{(x,y)\to(a,b)} \frac{f(x, y)}{g(x, y)} = \frac{\displaystyle\lim_{(x,y)\to(a,b)} f(x, y)}{\displaystyle\lim_{(x,y)\to(a,b)} g(x, y)} = \frac{L_1}{L_2}, \qquad L_2 \neq 0$

Example 4

Evaluate $\displaystyle\lim_{(x,y)\to(2,3)} (x + y^2)$

Solution From Theorem 16.1 we first note that

$$\lim_{(x,y)\to(2,3)} x = 2 \quad \text{and} \quad \lim_{(x,y)\to(2,3)} y = 3$$

Then from parts (*i*) and (*ii*) of Theorem 16.2 it follows that

$$\lim_{(x,y)\to(2,3)} (x + y^2) = \lim_{(x,y)\to(2,3)} x + \lim_{(x,y)\to(2,3)} y^2$$

$$= \lim_{(x,y)\to(2,3)} x + \left(\lim_{(x,y)\to(2,3)} y\right)\left(\lim_{(x,y)\to(2,3)} y\right)$$

$$= 2 + 3 \cdot 3 = 11$$

Use of Polar Coordinates

In some cases polar coordinates can be used to evaluate a limit of the form $\lim_{(x,y)\to(0,\,0)} f(x,\,y)$. If $x = r \cos \theta$, $y = r \sin \theta$, $r^2 = x^2 + y^2$, then $(x,\,y) \to (0,\,0)$ if and only if $r \to 0$.

Example 5

Evaluate

$$\lim_{(x,y)\to(0,0)} \frac{xy^2}{x^2 + y^2}$$

Solution   If $x = r \cos \theta$, $y = r \sin \theta$, then

$$\frac{xy^2}{x^2 + y^2} = \frac{r^3 \cos \theta \sin^2\theta}{r^2}$$

$$= r \cos \theta \sin^2\theta$$

Since $\lim_{r\to 0} r \cos \theta \sin^2\theta = 0$, we conclude that $\lim_{(x,y)\to(0,\,0)} xy^2/(x^2 + y^2) = 0$.

Continuity

A function $z = f(x,\,y)$ is **continuous** at $(a,\,b)$ if $f(a,\,b)$ is defined, $\lim_{(x,y)\to(a,b)} f(x,\,y)$ exists, and the limit is the same as $f(a,\,b)$; that is, f is continuous at $(a,\,b)$ if

$$\lim_{(x,y)\to(a,b)} f(x,\,y) = f(a,\,b)$$

If f is not continuous at $(a,\,b)$, it is said to be **discontinuous**. A function $z = f(x,\,y)$ is **continuous on a region *R*** of the xy-plane if f is continuous at every point in R. The **sum** and **product** of two continuous functions is continuous; the **quotient** of two continuous functions is continuous, except at points where the denominator is zero. Also, if g is a function of two variables continuous at $(a,\,b)$ and if F is a function of one variable continuous at $g(a,\,b)$, then the **composition** $f(x,\,y) = F(g(x,\,y))$ is continuous at $(a,\,b)$. Lastly, the **graph** of a continuous function is a surface with no breaks. In Figure 16.29 we see a computer-generated graph of $f(x,\,y) = 1/(9x^2 + y^2)$. As the figure shows, the function f is discontinuous at $(0,\,0)$.

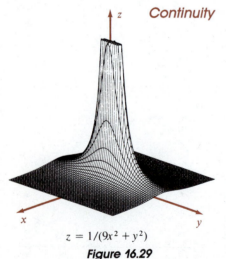

$z = 1/(9x^2 + y^2)$

Figure 16.29

Polynomial and Rational Functions

A **polynomial function** of two variables consists of the sum of powers $x^m y^n$, where m and n are nonnegative integers. The quotient of two polynomial functions is called a **rational function**. Polynomial functions are continuous throughout the entire xy-plane and rational functions are continuous except at points where the denominator is zero.

___ **Example 6** _____

$f(x, y) = 3x^4y^2 - x^3y + xy^3$ is a polynomial function of two variables.

___ **Example 7** _____

The rational function $f(x, y) = xy/(y - x)$ is continuous except at points on the line $y = x$.

___ **Example 8** _____

Evaluate $\displaystyle\lim_{(x,y)\to(1,4)} \frac{x + 2y}{x^2 + y}$.

Solution   Since the rational function $f(x, y) = (x + 2y)/(x^2 + y)$ is continuous at $(1, 4)$, the limit is $f(1, 4) = \frac{9}{5}$.

___ **Example 9** _____

(*i*)   The function $f(x, y) = \dfrac{x^4 - y^4}{x^2 + y^2}$ is discontinuous at $(0, 0)$ since $f(0, 0)$ is not defined.

(*ii*)   The function f defined by

$$f(y, x) = \begin{cases} \dfrac{x^4 - y^4}{x^2 + y^2}, & (x, y) \neq (0, 0) \\ 0, & (x, y) = (0, 0) \end{cases}$$

is continuous at $(0, 0)$ since $f(0, 0) = 0$ and

$$\lim_{(x,y)\to(0,0)} \frac{x^4 - y^4}{x^2 + y^2} = \lim_{(x,y)\to(0,0)} \frac{(x^2 + y^2)(x^2 - y^2)}{x^2 + y^2}$$

$$= \lim_{(x,y)\to(0,0)} (x^2 - y^2)$$

$$= 0^2 - 0^2 = 0$$

We see that $\lim_{(x,y)\to(0, 0)} f(x, y) = f(0, 0)$.

___ **Example 10** _____

Since $F(x) = e^x$ is continuous for all real numbers, the composite function $f(x, y) = e^{x/(y+2x)}$ is continuous, except at points on the line $y = -2x$.

In Figures 16.30–16.35 we have used a computer to illustrate six functions that are discontinuous at points on a curve.

$$f(x, y) = \frac{4}{6 - x^2 - y^2}$$

discontinuous on $x^2 + y^2 = 6$

Figure 16.30

$$f(x, y) = \ln|x^2 + y^2 - 4|$$

discontinuous on $x^2 + y^2 = 4$

Figure 16.31

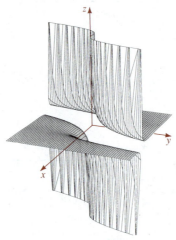

$$f(x, y) = \frac{1}{2y - x^3}$$

discontinuous on $y = \frac{1}{2}x^3$

Figure 16.32

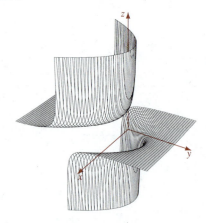

$$f(x, y) = \frac{-4}{4y + x^2 - 2}$$

discontinuous on $y = -\frac{1}{4}x^2 + \frac{1}{2}$

Figure 16.33

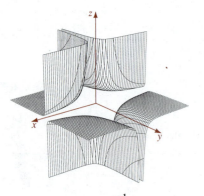

$$f(x, y) = \frac{-1}{x^2 y}$$

discontinuous on $x = 0$ and $y = 0$

Figure 16.34

$$f(x, y) = 3 - \frac{1}{2|\sin x + \cos y|}$$

discontinuous on $y = -x + (4n - 1)\dfrac{\pi}{2}$ and $y = x + (4n + 1)\dfrac{\pi}{2}$

Figure 16.35

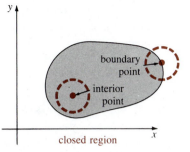

Figure 16.36

closed region

Terminology   At this point we need to introduce some terminology that will be used in the following definition as well as in the next chapter. If R is some region of the xy-plane, then a point (a, b) is said to be an **interior point** of R if there is *some* circle centered at (a, b) that contains only points of R. In contrast, we say (a, b) is a **boundary point** of R if every circle centered at (a, b) contains both points in R and points not in R. The region R is said to be **open** if it contains no boundary points and **closed** if it contains all boundary points. See Figure 16.36

[O] 16.2.2 The ε–δ Definition of a Limit

As in Section 2.5, the formal definition of a limit of a function $z = f(x, y)$ at a point (a, b) is given in terms of ε–δ.

DEFINITION 16.2   Let f be a function of two variables that is defined at every point (x, y) in the interior of a circle centered at (a, b), except possibly at (a, b). Then

$$\lim_{(x,y)\to(a,b)} f(x, y) = L$$

means that for every $\varepsilon > 0$, there exists a number $\delta > 0$ such that

$$|f(x, y) - L| < \varepsilon \quad \text{whenever} \quad 0 < \sqrt{(x - a)^2 + (y - b)^2} < \delta$$

As illustrated in Figure 16.37, when f has a limit at (a, b), for a given $\varepsilon > 0$, regardless how small, we can find a circle of radius δ centered at (a, b) so that $L - \varepsilon < f(x, y) < L + \varepsilon$ for every interior point $(x, y) \neq (a, b)$ of the circle. As mentioned previously, the values of f are close to L whenever (x, y) is close enough to (a, b). The concept of "close enough" is defined by the number δ.

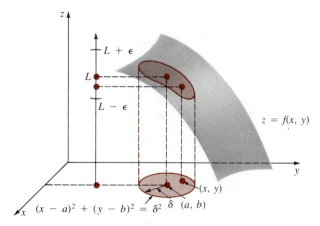

Figure 16.37

Remark

The definitions of limit and continuity for functions of three or more variables are natural extensions of those just considered. For example, $w = F(x, y, z)$ is continuous at (a, b, c) if

$$\lim_{(x,y,z)\to(a,b,c)} F(x, y, z) = F(a, b, c)$$

The polynomial function in three variables $F(x, y, z) = xy^2z^3$ is continuous throughout 3-space. The rational function

$$F(x, y, z) = \frac{xy^2}{x^2 + y^2 + (z - 1)^2}$$

is continuous except at the point $(0, 0, 1)$. The function

$$F(x, y, z) = \frac{x + 3y}{2x + 5y + z}$$

is continuous except at points on the plane $2x + 5y + z = 0$.

Exercises 16.2

Answers to odd-numbered problems begin on page A-48.

[16.2.1]

In Problems 1–30 evaluate the given limit, if it exists.

1. $\lim_{(x,y)\to(5,-1)} (x^2 + y^2)$

2. $\lim_{(x,y)\to(2,1)} \dfrac{x^2 - y}{x - y}$

3. $\lim_{(x,y)\to(0,0)} \dfrac{5x^2 + y^2}{x^2 + y^2}$

4. $\lim_{(x,y)\to(1,2)} \dfrac{4x^2 + y^2}{16x^4 + y^4}$

5. $\lim_{(x,y)\to(1,1)} \dfrac{4 - x^2 - y^2}{x^2 + y^2}$

6. $\lim_{(x,y)\to(0,0)} \dfrac{2x^2 - y}{x^2 + 2y^2}$

7. $\lim_{(x,y)\to(0,0)} \dfrac{x^2y}{x^4 + y^2}$

8. $\lim_{(x,y)\to(0,0)} \dfrac{6xy^2}{x^2 + y^4}$

9. $\lim_{(x,y)\to(1,2)} x^3y^2(x + y)^3$

10. $\lim_{(x,y)\to(2,3)} \dfrac{xy}{x^2 - y^2}$

11. $\lim_{(x,y)\to(0,0)} \dfrac{e^{xy}}{x + y + 1}$

12. $\lim_{(x,y)\to(0,0)} \dfrac{\sin xy}{x^2 + y^2}$

13. $\lim_{(x,y)\to(2,2)} \dfrac{xy}{x^3 + y^2}$

14. $\lim_{(x,y)\to(\pi, \pi/4)} \cos(3x + y)$

15. $\lim_{(x,y)\to(0,0)} \dfrac{x^2 - 3y + 1}{x + 5y - 3}$

16. $\lim_{(x,y)\to(0,0)} \dfrac{x^2y^2}{x^4 + 5y^4}$

17. $\lim_{(x,y)\to(4,3)} xy^2\left(\dfrac{x + 2y}{x - y}\right)$

18. $\lim_{(x,y)\to(1,0)} \dfrac{x^2y}{x^3 + y^3}$

19. $\lim_{(x,y)\to(1,1)} \dfrac{xy - x - y + 1}{x^2 + y^2 - 2x - 2y + 2}$

20. $\lim_{(x,y)\to(0,3)} \dfrac{xy - 3y}{x^2 + y^2 - 6y + 9}$

21. $\lim_{(x,y)\to(0,0)} \dfrac{x^3y + xy^3 - 3x^2 - 3y^2}{x^2 + y^2}$

22. $\lim_{(x,y)\to(-2,2)} \dfrac{y^3 + 2x^3}{x + 5xy^2}$

23. $\lim_{(x,y)\to(1,1)} \ln(2x^2 - y^2)$

24. $\lim_{(x,y)\to(1,2)} \dfrac{\sin^{-1}(x/y)}{\cos^{-1}(x - y)}$

25. $\lim_{(x,y)\to(0,0)} \dfrac{(x^2 - y^2)^2}{x^2 + y^2}$

26. $\lim_{(x,y)\to(0,0)} \dfrac{\sin(3x^2 + 3y^2)}{x^2 + y^2}$

27. $\displaystyle\lim_{(x,y)\to(0,0)} \frac{6xy}{\sqrt{x^2+y^2}}$

28. $\displaystyle\lim_{(x,y)\to(0,0)} \frac{x^2-y^2}{\sqrt{x^2+y^2}}$

29. $\displaystyle\lim_{(x,y)\to(0,0)} \frac{x^3}{x^2+y^2}$

30. $\displaystyle\lim_{(x,y)\to(0,0)} \frac{x^3+y^3}{x^2+y^2}$

In Problems 31–34 determine where the given function is continuous.

31. $f(x,y) = \sqrt{x}\,\cos\sqrt{x+y}$

32. $f(x,y) = y^2 e^{1/xy}$

33. $f(x,y) = \tan\dfrac{x}{y}$

34. $f(x,y) = \ln(4x^2 + 9y^2 + 36)$

In Problems 35 and 36 determine whether the given function is continuous on the indicated regions.

35. $f(x,y) = \begin{cases} x+y, & x \ge 2 \\ 0, & x < 2 \end{cases}$

    (a) $x^2+y^2 < 1$

    (b) $x \ge 0$

    (c) $y > x$

36. $f(x,y) = \dfrac{xy}{\sqrt{x^2+y^2-25}}$

    (a) $y \ge 3$

    (b) $|x| + |y| < 1$

    (c) $(x-2)^2 + y^2 < 1$

37. Determine whether the function f defined by

$$f(x,y) = \begin{cases} \dfrac{6x^2y^3}{(x^2+y^2)^2}, & (x,y) \ne (0,0) \\ 0, & (x,y) = (0,0) \end{cases}$$

is continuous at $(0,0)$.

38. Show that

$$f(x,y) = \begin{cases} \dfrac{xy}{2x^2+2y^2}, & (x,y) \ne (0,0) \\ 0, & (x,y) = (0,0) \end{cases}$$

is continuous in each variable separately at $(0,0)$; that is, that $f(x,0)$ and $f(0,y)$ are continuous at $x=0$ and $y=0$, respectively. Show, however, that f is not continuous at $(0,0)$.

[16.2.2]

In Problems 39 and 40 use Definition 16.2 to prove the given result; that is, find δ for any arbitrary $\varepsilon > 0$. (*Hint:* Use polar coordinates.)

39. $\displaystyle\lim_{(x,y)\to(0,0)} \frac{3xy^2}{x^2+y^2} = 0$

40. $\displaystyle\lim_{(x,y)\to(0,0)} \frac{x^2y^2}{x^2+y^2} = 0$

41. Use Definition 16.2 to prove that

$$\lim_{(x,y)\to(a,b)} y = b$$

16.3 Partial Differentiation

The derivative of a function of **one variable** $y = f(x)$ is given by

$$\frac{dy}{dx} = \lim_{\Delta x \to 0} \frac{f(x+\Delta x) - f(x)}{\Delta x}$$

In exactly the same manner, we can define a derivative of a function of **two variables** $z = f(x, y)$ with respect to *each* variable.

DEFINITION 16.3   Partial Derivatives

If $z = f(x, y)$, then the **partial derivative with respect to x** is

$$\frac{\partial z}{\partial x} = \lim_{\Delta x \to 0} \frac{f(x+\Delta x,\, y) - f(x,\, y)}{\Delta x} \qquad (16.2)$$

and the **partial derivative with respect to y** is

$$\frac{\partial z}{\partial y} = \lim_{\Delta y \to 0} \frac{f(x, \, y + \Delta y) - f(x, \, y)}{\Delta y} \qquad (16.3)$$

provided each limit exists.

In (16.2) the variable y does not change in the limiting process; that is, y is held fixed. Similarly, in (16.3) the variable x is held fixed. The two partial derivatives (16.2) and (16.3) then represent the *rates of change* of f with respect to x and y, respectively. On a practical level:

To compute $\partial z/\partial x$, use the laws of ordinary differentiation while treating y as a constant.

To compute $\partial z/\partial y$, use the laws of ordinary differentiation while treating x as a constant.

Example 1

If $z = 4x^3y^2 - 4x^2 + y^6 + 1$, find

(a) $\dfrac{\partial z}{\partial x}$ (b) $\dfrac{\partial z}{\partial y}$

Solution

(a)   We hold y fixed and treat constants in the usual manner. Thus,

$$\frac{\partial z}{\partial x} = 12x^2y^2 - 8x$$

(b)   By treating x as a constant, we obtain
$$\frac{\partial z}{\partial y} = 8x^3y + 6y^5$$

Alternative Symbols   The partial derivatives $\partial z/\partial x$ and $\partial z/\partial y$ are often represented by alternative symbols. If $z = f(x, \, y)$, then

$$\frac{\partial z}{\partial x} = \frac{\partial f}{\partial x} = z_x = f_x \quad \text{and} \quad \frac{\partial z}{\partial y} = \frac{\partial f}{\partial y} = z_y = f_y$$

A symbol such as $\partial/\partial x$ denotes the *operation* of taking a partial derivative, in this case with respect to x; for example,

$$\frac{\partial}{\partial x}(x^2 - y^2) = 2x$$

Example 2

If $f(x, y) = x^5 y^{10} \cos(xy^2)$, find f_y.

Solution   When x is held fixed, observe that

$$f(x, y) = x^5 y^{10} \cos(xy^2)$$

product of two
functions of y

Hence, by the Product Rule,

$$f_y(x, y) = x^5 y^{10}[-2xy \sin(xy^2)] + 10x^5 y^9 \cos(xy^2)$$

Example 3

The function $S = 0.1091 w^{0.425} h^{0.725}$ relates the surface area (in square feet) of a person's body as a function of weight w (in pounds) and height h (in inches). Find $\partial S / \partial w$ when $w = 150$ and $h = 72$. Interpret.

Solution $$\frac{\partial S}{\partial w} = (0.1091)(0.425) w^{-0.575} h^{0.725}$$

$$\left. \frac{\partial S}{\partial w} \right|_{(150,72)} = (0.1091)(0.425)(150)^{-0.575}(72)^{0.725} \approx 0.058$$

The partial derivative $\partial S / \partial w$ is the rate at which the surface area of a person of fixed height, such as an adult, changes with respect to weight. Since the units for the derivative are sq ft/lb, we see that a gain of 1 lb, while h is fixed at 72, results in an increase in the area of the skin by approximately $0.058 \approx 1/17$ sq ft.

Geometric Interpretation   As seen in Figure 16.32(a)(page 788), when y is constant, say $y = b$, the trace of the surface $z = f(x, y)$ in the plane $y = b$ is a curve C. If we define the slope of the secant through the indicated points P and R as

$$\frac{f(a + \Delta x, b) - f(a, b)}{\Delta x}$$

we have

$$\left. \frac{\partial z}{\partial x} \right|_{(a,b)} = \lim_{\Delta x \to 0} \frac{f(a + \Delta x, b) - f(a, b)}{\Delta x}$$

In other words, we can interpret $\partial z / \partial x$ as the slope of the tangent at any point (for which the limit exists) on a curve C of intersection between the surface $z = f(x, y)$ and a plane $y = $ constant. In turn, an inspection of Figure 16.38(b) reveals that $\partial z / \partial y$ is the slope of the tangent at a point on a curve of intersection between the surface $z = f(x, y)$ and a plane $x = $ constant.

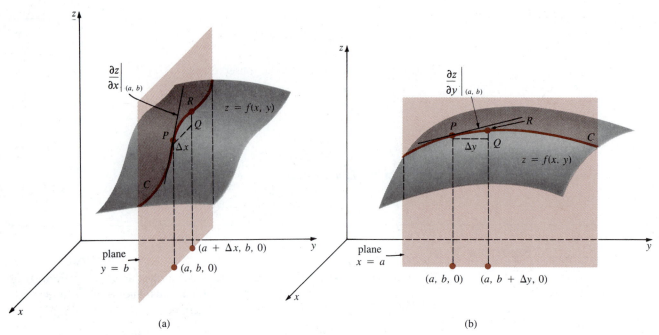

Figure 16.38

Example 4

For $z = 9 - x^2 - y^2$, find the slope of the tangent line at $(2, 1, 4)$ in (*a*) the plane $x = 2$, and (*b*) the plane $y = 1$.

Solution

(*a*)   By specifying the plane $x = 2$, we are holding all values of x constant. Hence, we compute the partial derivative with respect to y:

$$\frac{\partial z}{\partial y} = -2y$$

At $(2, 1, 4)$ the slope is

$$\left.\frac{\partial z}{\partial y}\right|_{(2,1)} = -2$$

(*b*)   In the plane $y = 1$, y is constant and so we find the partial derivative with respect to x:

$$\frac{\partial z}{\partial x} = -2x$$

At $(2, 1, 4)$ the slope is

$$\left.\frac{\partial z}{\partial x}\right|_{(2,1)} = -4$$

See Figure 16.39.

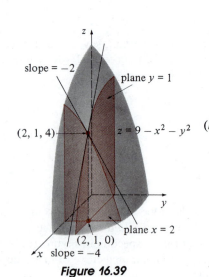

Figure 16.39

Functions of Three or More Variables

The rates of change of a function of three variables $w = F(x, y, z)$ in the x, y, and z directions are $\partial w/\partial x$, $\partial w/\partial y$, and $\partial w/\partial z$, respectively. To compute, say, $\partial w/\partial x$, we differentiate with respect to x in the usual manner while holding *both* y and z constant. In this manner we extend the process of partial differentiation to functions of any number of variables.

Example 5

If $w = \dfrac{x^2 - z^2}{y^2 + z^2}$, find $\dfrac{\partial w}{\partial z}$.

Solution   We use the Quotient Rule while holding x and y constant:

$$\frac{\partial w}{\partial z} = \frac{(y^2 + z^2)(-2z) - (x^2 - z^2)2z}{(y^2 + z^2)^2}$$

$$= -\frac{2z(x^2 + y^2)}{(y^2 + z^2)^2}$$

Example 6

If $F(x, y, t) = e^{-3\pi t} \cos 4x \sin 6y$, then the partial derivatives with respect to x, y, and t are, in turn,

$$F_x(x, y, t) = -4e^{-3\pi t} \sin 4x \sin 6y$$
$$F_y(x, y, t) = 6e^{-3\pi t} \cos 4x \cos 6y$$
$$F_t(x, y, t) = -3\pi e^{-3\pi t} \cos 4x \sin 6y$$

Higher-order and Mixed Derivatives

For a function of two variables $z = f(x, y)$, the partial derivatives $\partial z/\partial x$ and $\partial z/\partial y$ are themselves functions of x and y. Consequently, we can compute **second**, and higher, **partial derivatives**. Indeed, we can find the partial derivative of $\partial z/\partial x$ with respect to y, and the partial derivative of $\partial z/\partial y$ with respect to x. The latter types of partial derivatives are called **mixed partial derivatives**. In summary, for $z = f(x, y)$:

Second-order partial derivatives:

$$\frac{\partial^2 z}{\partial x^2} = \frac{\partial}{\partial x}\left(\frac{\partial z}{\partial x}\right) \quad \text{and} \quad \frac{\partial^2 z}{\partial y^2} = \frac{\partial}{\partial y}\left(\frac{\partial z}{\partial y}\right)$$

Third-order partial derivatives:

$$\frac{\partial^3 z}{\partial x^3} = \frac{\partial}{\partial x}\left(\frac{\partial^2 z}{\partial x^2}\right) \quad \text{and} \quad \frac{\partial^3 z}{\partial y^3} = \frac{\partial}{\partial y}\left(\frac{\partial^2 z}{\partial y^2}\right)$$

Mixed second-order partial derivatives:

$$\frac{\partial^2 z}{\partial x \, \partial y} = \frac{\partial}{\partial x}\left(\frac{\partial z}{\partial y}\right) \quad \text{and} \quad \frac{\partial^2 z}{\partial y \, \partial x} = \frac{\partial}{\partial y}\left(\frac{\partial z}{\partial x}\right)$$

Higher-order partial derivatives for functions of three or more variables are defined in a similar manner.

Alternative Symbols   The second- and third-order partial derivatives are denoted by f_{xx}, f_{yy}, f_{xxx}, and so on. The subscript notation for mixed second partial derivatives is f_{xy} or f_{yx}. Note that

$$f_{xy} = (f_x)_y = \frac{\partial}{\partial y}\left(\frac{\partial z}{\partial x}\right) = \frac{\partial^2 z}{\partial y\, \partial x} \quad \text{and} \quad f_{yx} = \frac{\partial^2 z}{\partial x\, \partial y}$$

Although we shall not prove it, if a function f has continuous second partial derivatives, then the order in which a mixed second partial derivative is done is irrelevant; that is,

$$f_{xy} = f_{yx} \tag{16.4}$$

___
Example 7
___

If
$$z = x^2 y^2 - y^3 + 3x^4 + 5$$

find
$$\frac{\partial^2 z}{\partial x^2}, \frac{\partial^3 z}{\partial x^3}, \frac{\partial^2 z}{\partial y^2}, \frac{\partial^3 z}{\partial y^3}, \quad \text{and} \quad \frac{\partial^2 z}{\partial x\, \partial y}$$

Solution
$$\frac{\partial z}{\partial x} = 2xy^2 + 12x^3, \qquad \frac{\partial z}{\partial y} = 2x^2 y - 3y^2,$$

$$\frac{\partial^2 z}{\partial x^2} = 2y^2 + 36x^2, \qquad \frac{\partial^2 z}{\partial y^2} = 2x^2 - 6y,$$

$$\frac{\partial^3 z}{\partial x^3} = 72x, \qquad \frac{\partial^3 z}{\partial y^3} = -6,$$

$$\frac{\partial^2 z}{\partial x\, \partial y} = \frac{\partial}{\partial x}\,(2x^2 y - 3y^2) = 4xy$$

You should also verify that $\dfrac{\partial^2 z}{\partial y\, \partial x} = 4xy$.

___
Example 8
___

If $F(x, y, z) = \sqrt{x^2 + y^4 + z^6}$, find F_{yzz}.

Solution F_{yzz} is a mixed third-order partial derivative. First we find the partial derivative with respect to y by the Power Rule for Functions:

$$F_y = \frac{1}{2}(x^2 + y^4 + z^6)^{-1/2} 4y^3$$

$$= 2y^3(x^2 + y^4 + z^6)^{-1/2}$$

Then,

$$F_{yz} = (F_y)_z = (2y^3)\left(-\frac{1}{2}\right)(x^2 + y^4 + z^6)^{-3/2} 6z^5$$

$$= -6y^3 z^5 (x^2 + y^4 + z^6)^{-3/2}$$

Finally, by the Product Rule,

$$F_{yzz} = (F_{yz})_z = -6y^3z^5\left(-\frac{3}{2}\right)(x^2 + y^4 + z^6)^{-5/2}(6z^5)$$

$$-30y^3z^4(x^2 + y^4 + z^6)^{-3/2}$$

$$= y^3z^4(x^2 + y^4 + z^6)^{-5/2}(24z^6 - 30x^2 - 30y^4)$$

Remark

If $w = F(x, y, z)$ has continuous partial derivatives of any order, then analogous to (16.4) we have

$$F_{xyz} = F_{yzx} = F_{zyx}$$

$$F_{xxy} = F_{yxx} = F_{xyx}$$

and so on.

Exercises 16.3

Answers to odd-numbered problems begin on page A-48.

In Problems 1–4 use Definition 16.3 to compute $\partial z/\partial x$ and $\partial z/\partial y$ for the given function.

1. $z = 7x + 8y^2$

2. $z = xy$

3. $z = 3x^2y + 4xy^2$

4. $z = \dfrac{x}{x + y}$

In Problems 5–24 find the first partial derivatives of the given function.

5. $z = x^2 - xy^2 + 4y^5$

6. $z = -x^3 + 6x^2y^3 + 5y^2$

7. $z = 5x^4y^3 - x^2y^6 + 6x^5 - 4y$

8. $z = \tan(x^3y^2)$

9. $z = \dfrac{4\sqrt{x}}{3y^2 + 1}$

10. $z = 4x^3 - 5x^2 + 8x$

11. $z = (x^3 - y^2)^{-1}$

12. $z = (-x^4 + 7y^2 + 3y)^6$

13. $z = \cos^2 5x + \sin^2 5y$

14. $z = e^{x^2 \tan^{-1} y^2}$

15. $f(x, y) = xe^{x^3y}$

16. $f(\theta, \phi) = \phi^2 \sin\dfrac{\theta}{\phi}$

17. $f(x, y) = \dfrac{3x - y}{x + 2y}$

18. $f(x, y) = \dfrac{xy}{(x^2 - y^2)^2}$

19. $g(u, v) = \ln(4u^2 + 5v^3)$

20. $h(r, s) = \dfrac{\sqrt{r}}{s} - \dfrac{\sqrt{s}}{r}$

21. $w = 2\sqrt{xy} - ye^{y/z}$

22. $w = xy \ln xz$

23. $F(u, v, x, t) = u^2w^2 - uv^3 + vw \cos(ut^2) + (2x^2t)^4$

24. $G(p, q, r, s) = (p^2q^3)^{r^4s^5}$

In Problems 25 and 26 suppose $z = 4x^3y^4$.

25. Find the slope of the tangent line at $(1, -1, 4)$ in the plane $x = 1$.

26. Find the slope of the tangent line at $(1, -1, 4)$ in the plane $y = -1$.

In Problems 27 and 28 suppose $f(x, y) = 18xy/(x + y)$.

27. Find parametric equations for the tangent line at $(-1, 4, -24)$ in the plane $x = -1$.

28. Find symmetric equations for the tangent line at $(-1, 4, -24)$ in the plane $y = 4$.

In Problems 29 and 30 suppose $z = \sqrt{9 - x^2 - y^2}$.

29. At what rate is z changing with respect to x in the plane $y = 2$ at the point $(2, 2, 1)$?

30. At what rate is z changing with respect to y in the plane $x = \sqrt{2}$ at the point $(\sqrt{2}, \sqrt{3}, 2)$?

In Problems 31–38 find the indicated partial derivative.

31. $z = e^{xy}; \dfrac{\partial^2 z}{\partial x^2}$ **32.** $z = x^4 y^{-2}; \dfrac{\partial^3 z}{\partial y^3}$

33. $f(x, y) = 5x^2 y^2 - 2xy^3; f_{xy}$

34. $f(p, q) = \ln \dfrac{p + q}{q^2}; f_{qp}$

35. $w = u^2 v^3 t^3; w_{tuv}$ **36.** $w = \dfrac{\cos(u^2 v)}{t^3}; w_{vvt}$

37. $F(r, \theta) = e^{r^2} \cos \theta; F_{r\theta r}$ **38.** $H(s, t) = \dfrac{s + t}{s - t}; H_{tts}$

In Problems 39 and 40 verify that $\dfrac{\partial^2 z}{\partial x\, \partial y} = \dfrac{\partial^2 z}{\partial y\, \partial x}$.

39. $z = x^6 - 5x^4 y^3 + 4xy^2$ **40.** $z = \tan^{-1}(2xy)$

In Problems 41 and 42 verify that the indicated partial derivatives are equal.

41. $w = u^3 v^4 - 4u^2 v^2 t^3 + v^2 t; w_{uvt}, w_{tvu}, w_{vut}$

42. $F(\eta, \xi, \tau) = (\eta^3 + \xi^2 + \tau)^2; F_{\eta\xi\eta}, F_{\xi\eta\eta}, F_{\eta\eta\xi}$

In Problems 43–46 suppose the given equation defines z as a function of the remaining two variables. Use implicit differentiation to find the first partial derivatives.

43. $x^2 + y^2 + z^2 = 25$ **44.** $z^2 = x^2 + y^2 z$

45. $z^2 + u^2 v^3 - uvz = 0$ **46.** $se^z - e^{st} + 4s^3 t = z$

47. The area A of a parallelogram with base x and height $y \sin \theta$ is $A = xy \sin \theta$. Find all first partial derivatives.

48. The volume of the frustum of a cone shown in Figure 16.40 is $V = (\pi/3)h(r^2 + rR + R^2)$. Find all first partial derivatives.

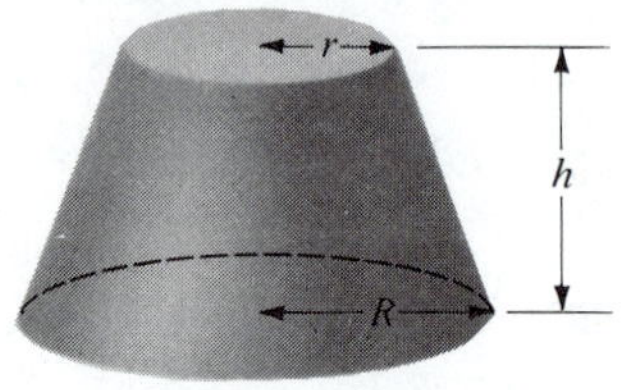

Figure 16.40

In Problems 49 and 50 verify that the given function satisfies **Laplace's equation**

$$\frac{\partial^2 z}{\partial x^2} + \frac{\partial^2 z}{\partial y^2} = 0$$

49. $z = \ln(x^2 + y^2)$ **50.** $z = e^{x^2 - y^2} \cos 2xy$

In Problems 51 and 52 verify that the given function satisfies the **wave equation**

$$a^2 \frac{\partial^2 u}{\partial x^2} = \frac{\partial^2 u}{\partial t^2}$$

51. $u = \cos at \sin x$

52. $u = \cos(x + at) + \sin(x - at)$

53. The molecular concentration $C(x, t)$ of a liquid is given by $C(x, t) = t^{-1/2} e^{-x^2/kt}$. Verify that this function satisfies the **diffusion equation**

$$\frac{k}{4} \frac{\partial^2 C}{\partial x^2} = \frac{\partial C}{\partial t}$$

54. The electrostatic potential at a point $P(x, y)$ in the plane due to a unit point at the origin is given by $U = 1/\sqrt{x^2 + y^2}$. At $(3, 4)$ find the rate of change of U in the direction of **(a)** the x-axis, **(b)** the y-axis.

55. The vertical displacement of a long string fastened at the origin but falling under its own weight is given by

$$u(x, t) = \begin{cases} -\dfrac{g}{2a^2}(2axt - x^2), & 0 \le x \le at \\ -\dfrac{1}{2}gt^2, & x > at \end{cases}$$

See Figure 16.41.

(a) Find $\partial u/\partial t$. Interpret.

(b) Find $\partial u/\partial x$ for $x > at$. Interpret.

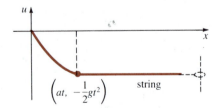

Figure 16.41

56. The pressure P exerted by an enclosed ideal gas is given by $P = k(T/V)$, where k is a constant, T is temperature, and V is volume. Find

(a) the rate of change of P with respect to V,

(b) the rate of change of V with respect to T, and

(c) the rate of change of T with respect to P.

57. The temperature T in a flat rectangular plate is given by $T(x, y) = xy(2 - x)(2 - y)$, $0 \leq x \leq 2$, $0 \leq y \leq 2$. At $(1, \frac{1}{2})$, find the rate of change of T

(a) with respect to x, and

(b) with respect to y.

Calculator Problem

58. For the skin-area function

$$S = 0.1091 w^{0.425} h^{0.725}$$

discussed in Example 3, find $\partial S / \partial h$ at $w = 60$, $h = 36$. If a girl grows in height from 36 to 37 inches, while holding her weight fixed at 60 lb, what is the approximate increase in the area of skin?

Miscellaneous Problems

59. State a limit definition which is analogous to Definition 16.3 for

(a) $\dfrac{\partial^2 z}{\partial x^2}$ (b) $\dfrac{\partial^2 z}{\partial y^2}$ (c) $\dfrac{\partial^2 z}{\partial x\, \partial y}$

60. Find a function $z = f(x, y)$ such that

$$\frac{\partial z}{\partial x} = 2xy^3 + 2y + \frac{1}{x} \quad \text{and} \quad \frac{\partial z}{\partial y} = 3x^2 y^2 + 2x + 1$$

61. Suppose the function $w = F(x, y, z)$ has continuous third-order partial derivatives. How many different third-order partial derivatives are there?

16.4 The Total Differential

Increment of the Dependent Variable

The notion of the differentiability of a function of any number of independent variables depends on the **increment** of the dependent variable. Recall that for a function of *one variable* $y = f(x)$,

$$\Delta y = f(x + \Delta x) - f(x)$$

Analogously, for a function of *two variables* $z = f(x, y)$, we define

$$\Delta z = f(x + \Delta x, y + \Delta y) - f(x, y) \tag{16.5}$$

Figure 16.42 shows that Δz gives the amount of **change** in the function as (x, y) changes to $(x + \Delta x, y + \Delta y)$.

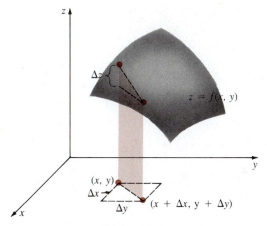

Figure 16.42

_____ **Example 1** _____

Find Δz for $z = x^2 - xy$. What is the change in the function from $(1, 1)$ to $(1.2, 0.7)$?

Solution   From (16.5),

$$\Delta z = [(x + \Delta x)^2 - (x + \Delta x)(y + \Delta y)] - (x^2 - xy)$$
$$= (2x - y)\, \Delta x - x\, \Delta y + (\Delta x)^2 - \Delta x\, \Delta y \qquad (16.6)$$

With $x = 1$, $y = 1$, $\Delta x = 0.2$, $\Delta y = -0.3$,

$$\Delta z = (1)(0.2) - (1)(-0.3) + (0.2)^2 - (0.2)(-0.3) = 0.6$$

A Fundamental Increment Formula

A brief reinspection of the increment Δz in (16.6) shows that in the first two terms the coefficients of Δx and Δy are $\partial z/\partial x$ and $\partial z/\partial y$, respectively. The following important theorem shows that this is no accident.

> **THEOREM 16.3**   Let $z = f(x, y)$ have continuous partial derivatives $f_x(x, y)$ and $f_y(x, y)$ in a rectangular region that is defined by $a < x < b$, $c < y < d$. If (x, y) is any point in this region, then there exist ε_1 and ε_2, which are functions of Δx and Δy such that
>
> $$\Delta z = f_x(x, y)\, \Delta x + f_y(x, y)\, \Delta y + \varepsilon_1\, \Delta x + \varepsilon_2\, \Delta y \qquad (16.7)$$
>
> where $\varepsilon_1 \to 0$ and $\varepsilon_2 \to 0$ when $\Delta x \to 0$ and $\Delta y \to 0$.

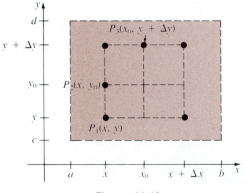

Figure 16.43

Proof   By adding and subtracting $f(x, y + \Delta y)$ in (16.5), we have

$$\Delta z = [f(x + \Delta x, y + \Delta y) - f(x, y + \Delta y)]$$
$$+ [f(x, y + \Delta y) - f(x, y)]$$

Applying the Mean Value Theorem to each set of brackets then gives

$$\Delta z = f_x(x_0, y + \Delta y)\, \Delta x + f_y(x, y_0)\, \Delta y \qquad (16.8)$$

where, as shown in Figure 16.43, $x < x_0 < x + \Delta x$ and $y < y_0 < \Delta y$. Now, define

$$\varepsilon_1 = f_x(x_0, y + \Delta y) - f_x(x, y)$$
$$\varepsilon_2 = f_y(x, y_0) - f_y(x, y) \qquad (16.9)$$

As $\Delta x \to 0$ and $\Delta y \to 0$, then, as shown in the figure, $P_2 \to P_1$ and $P_3 \to P_1$. Since f_x and f_y are assumed continuous in the region, we have

$$\lim_{(\Delta x, \Delta y) \to (0,0)} \varepsilon_1 = 0 \quad \text{and} \quad \lim_{(\Delta x, \Delta y) \to (0,0)} \varepsilon_2 = 0$$

Solving (16.9) for $f_x(x_0, y + \Delta y)$ and $f_y(x, y_0)$ and substituting in (16.8) gives the result

$$\Delta z = f_x(x, y)\, \Delta x + f_y(x, y)\, \Delta y + \varepsilon_1\, \Delta x + \varepsilon_2\, \Delta y \qquad \blacksquare$$

_____ **Example 2** _____

In Example 1 we can take $\varepsilon_1 = \Delta x$ and $\varepsilon_2 = -\Delta x$.

Total Differential for   Inspired by (16.7) we now define the **total differential**, or simply **differential**,
$z = f(x, y)$ of a function $z = f(x, y)$.

> **DEFINITION 16.4**   **Differentials**
>
> Let $z = f(x, y)$ be a function for which the first partials f_x and f_y exist. Then,
>
> (*i*)   The **differentials** of the independent variables are
>
> $$dx = \Delta x, \qquad dy = \Delta y$$
>
> (*ii*)   The **total differential** of the function is
>
> $$dz = f_x(x, y)\, dx + f_y(x, y)\, dy$$
>
> $$= \frac{\partial z}{\partial x}\, dx + \frac{\partial z}{\partial y}\, dy$$

_____ **Example 3** _____

If $z = x^2 - xy$, then

$$\frac{\partial z}{\partial x} = 2x - y, \qquad \frac{\partial z}{\partial y} = -x$$

so that $$dz = (2x - y)\, dx - x\, dy$$

It follows immediately from Theorem 16.3 that when f_x and f_y are continuous and when Δx and Δy are small, then dz is an approximation for Δz:

$$dz \approx \Delta z$$

_____ **Example 4** _____

The change in the function in Example 1 can be approximated using the differential in Example 3,

$$dz = (1)(0.2) - (1)(-0.3) = 0.5$$

Example 5

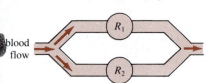

Figure 16.44

The human cardiovascular system is similar to electrical series and parallel circuits. For example, when blood flows through two resistances in parallel, as shown in Figure 16.44, then the equivalent resistance R of the network is

$$\frac{1}{R} = \frac{1}{R_1} + \frac{1}{R_2} \quad \text{or} \quad R = \frac{R_1 R_2}{R_1 + R_2}$$

If the percentage errors in measuring R_1 and R_2 are $\pm 0.2\%$ and $\pm 0.6\%$, respectively, find the approximate maximum percentage error in R.

Solution   We have $\Delta R_1 = \pm 0.002 R_1$ and $\Delta R_2 = \pm 0.006 R_2$. Now,

$$dR = \frac{R_2^2}{(R_1 + R_2)^2}\, dR_1 + \frac{R_1^2}{(R_1 + R_2)^2}\, dR_2$$

and so

$$|\Delta R| \approx |dR| \le \left| \frac{R_2^2}{(R_1 + R_2)^2}(\pm 0.002 R_1) \right| + \left| \frac{R_1^2}{(R_1 + R_2)^2}(\pm 0.006 R_2) \right|$$

$$= R \left[\frac{0.002 R_2}{R_1 + R_2} + \frac{0.006 R_1}{R_1 + R_2} \right]$$

$$\le R \left[\frac{0.006 R_2}{R_1 + R_2} + \frac{0.006 R_1}{R_1 + R_2} \right] = (0.006) R$$

Thus, the maximum relative error is given by the approximation $|dR|/R \approx 0.006$; therefore, the maximum percentage error is approximately 0.6%.

Differentiability of
$z = f(x, y)$

Although we have considered partial derivatives and the differential of a function $z = f(x, y)$, we have not yet defined **differentiability** of f. The latter concept is defined in terms of the fundamental increment formula (16.7).

> **DEFINITION 16.5**   The function $z = f(x, y)$ is said to be **differentiable** at (x, y) if Δz can be written
>
> $$\Delta z = f_x(x, y)\, \Delta x + f_y(x, y)\, \Delta y + \varepsilon_1\, \Delta x + \varepsilon_2\, \Delta y$$
>
> where $\displaystyle \lim_{(\Delta x, \Delta y) \to (0,0)} \varepsilon_1 = \lim_{(\Delta x, \Delta y) \to (0,0)} \varepsilon_2 = 0$

> **THEOREM 16.4**   If $f_x(x, y)$ and $f_y(x, y)$ are continuous at every point (x, y) in a rectangular region that is defined by $a < x < b$, $c < y < d$, then $z = f(x, y)$ is differentiable over the region.

Not surprisingly,

differentiability of $z = f(x, y)$ implies continuity

> **THEOREM 16.5**   If $z = f(x, y)$ is differentiable at (x, y), then f is continuous at (x, y).

Total Differential for
w = F(x, y, z)

Definition 16.4 generalizes to functions of three or more variables. Specifically, if $w = F(x, y, z)$, then its total differential is given by

$$dw = \frac{\partial w}{\partial x}\, dx + \frac{\partial w}{\partial y}\, dy + \frac{\partial w}{\partial z}\, dz$$

_____ **Example 6** _____

If $w = x^2 + 2y^3 + 3z^4$, then

$$\frac{\partial w}{\partial x} = 2x, \quad \frac{\partial w}{\partial y} = 6y^2, \quad \frac{\partial w}{\partial z} = 12x^3$$

and so $dw = 2x\, dx + 6y^2\, dy + 12z^3\, dz$

Remarks

(i)   Since $dy \approx \Delta y$ when $f'(x)$ exists and Δx is small, it seems reasonable to expect $dz = f_x(x, y)\, \Delta x + f_y(x, y)\, \Delta y$ to give a good approximation for Δz for small Δx and Δy. But life is not so simple for functions of several variables; there are functions for which dz and Δz are not close even when Δx and Δy are small. The guarantee $dz \approx \Delta z$ for small increments in the independent variables comes from the continuity and not simply from the existence of $f_x(x, y)$ and $f_y(x, y)$.

(ii)   When you work Problems 5–8 in Exercises 16.4 you will discover that the functions ε_1 and ε_2 introduced in (16.7) of Theorem 16.3 are not unique.

_____ **Exercises 16.4** _____

Answers to odd-numbered problems begin on page A-49.

In Problems 1–4 compare the values of Δz and dz for the given function as (x, y) varies from the first to the second point.

1. $z = 3x + 4y + 8$; (2, 4), (2.2, 3.9)

2. $z = 2x^2y + 5y$; (0, 0), (0.2, −0.1)

3. $z = (x + y)^2$; (3, 1), (3.1, 0.8)

4. $z = x^2 + x^2y^2 + 2$; (1, 1), (0.9, 1.1)

In Problems 5–8 find functions ε_1 and ε_2 from Δz as defined in (16.7) of Theorem 16.3.

5. $z = 5x^2 + 3y - xy$

6. $z = 10y^2 + 3x - x^2$

7. $z = x^2y^2$

8. $z = x^3 - y^3$

In Problems 9–20 find the total differential of the given function.

9. $z = x^2 \sin 4y$

10. $z = xe^{x^2 - y^2}$

11. $z = \sqrt{2x^2 - 4y^3}$

12. $z = (5x^3y + 4y^5)^3$

13. $f(s, t) = \dfrac{2s - t}{s + 3t}$

14. $g(r, \theta) = r^2 \cos 3\theta$

15. $w = x^2y^4z^{-5}$

16. $w = e^{-z^2} \cos(x^2 + y^4)$

17. $F(r, s, t) = r^3 + s^{-2} - 4t^{1/2}$

18. $G(\rho, \theta, \phi) = \rho \sin \phi \cos \theta$

19. $w = \ln\left(\dfrac{uv}{st}\right)$

20. $w = \sqrt{u^2 + s^2t^2 - v^2}$

21. When blood flows through three resistances R_1, R_2, R_3 in parallel, the equivalent resistance R of the network is

$$\frac{1}{R} = \frac{1}{R_1} + \frac{1}{R_2} + \frac{1}{R_3}$$

Given that the percentage error in measuring each resistance is $\pm 0.9\%$, find the approximate maximum percentage error in R.

22. The pressure P of an enclosed ideal gas is given by $P = k(T/V)$, where V is volume, T is temperature, and k is a constant. Given that the percentage errors in measuring T and V are at most 0.6% and 0.8%, respectively, find the approximate maximum percentage error in P.

23. The tension T in the string of the yo-yo shown in Figure 16.45 is

$$T = mg\frac{R}{2r^2 + R^2}$$

where mg is its constant weight. Find the approximate change in the tension if R and r are increased from 4 cm and 0.8 cm to 4.1 cm and 0.9 cm, respectively. Does the tension increase or decrease?

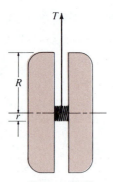

Figure 16.45

24. Find the approximate increase in the volume of a right circular cylinder if its height is increased from 10 to 10.5 cm and its radius is increased from 5 to 5.3 cm. What is the approximate new volume?

25. If the length, width, and height of a closed rectangular box are increased by 2%, 5%, and 8%, respectively, what is the approximate percentage increase in volume?

26. In Problem 25 if the original length, width, and height were 3 ft, 1 ft, and 2 ft, respectively, what is the approximate increase in surface area? What is the approximate new surface area?

27. The function $S = 0.1091w^{0.425}h^{0.725}$ gives the surface area of a person's body in terms of weight w and height h. If the error in the measurement of w is at most 3% and the error in the measurement of h is at most 5%, what is the approximate maximum percentage error in the measurement of S?

28. The impedance Z of the series circuit shown in Figure 16.46 is $Z = \sqrt{R^2 + X^2}$, where R is resistance, $X = 1000L - 1/(1000C)$ is net reactance, L is inductance, and C is capacitance. If the values of R, L, and C given in the figure are increased to 425 ohms, 0.45 henry, and 11.1×10^{-5} farad, respectively, what is the approximate change in the impedance of the circuit? What is the approximate new impedance?

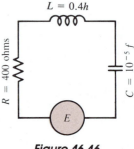

Figure 16.46

In Problems 29 and 30 use the concept of the differential to find an approximation to the given expression.

29. $\sqrt{102} + \sqrt[4]{80}$

30. $\sqrt[4]{\dfrac{35}{63}}$

16.5   Exact Differentials

Often it is important to be able to tell at a glance when a differential expression

$$P(x, y) \, dx + Q(x, y) \, dy \qquad (16.10)$$

is the total differential of a function $z = f(x, y)$.

> **DEFINITION 16.6   Exact Differential**
>
> A differential expression (16.10) is said to be an **exact differential** if there exists a function f such that
>
> $$df = P(x, y) \, dx + Q(x, y) \, dy$$

_____ **Example 1** _____

$x^2 y^3 \, dx + x^3 y^2 \, dy$ is an exact differential since it is the total differential of $f(x, y) = \frac{1}{3} x^3 y^3$. Verify this.

The following result provides a test for determining when (16.10) is exact.

> **THEOREM 16.6   Let P and Q** be continuous and have continuous first partial derivatives in a rectangular region of the xy-plane. Then, (16.10) is an exact differential if and only if
>
> $$\frac{\partial P}{\partial y} = \frac{\partial Q}{\partial x} \qquad (16.11)$$
>
> for all (x, y) in the region.

Proof   If (16.10) is an exact differential, there exists a function f such that

$$df = P(x, y) \, dx + Q(x, y) \, dy = \frac{\partial f}{\partial x} dx + \frac{\partial f}{\partial y} dy$$

Since $\partial f / \partial x = P$ and $\partial f / \partial y = Q$, we have

$$\frac{\partial^2 f}{\partial y \, \partial x} = \frac{\partial P}{\partial y} \quad \text{and} \quad \frac{\partial^2 f}{\partial x \, \partial y} = \frac{\partial Q}{\partial x}$$

By continuity, the mixed partials are equal and thus we obtain the results in (16.11). Conversely, let us now assume $\partial P / \partial y = \partial Q / \partial x$. We wish to find a function f such that $\partial f / \partial x = P$ and $\partial f / \partial y = Q$. It turns out that

$$f(x, y) = \int_{x_0}^{x} P(x, y_0) \, dx + \int_{y_0}^{y} Q(x, y) \, dy \qquad (16.12)$$

where (x_0, y_0) is a fixed point in the region, is the function that we seek. In the second integral in (16.12), the variable x is held fixed.

Now,
$$\frac{\partial f}{\partial x} = \frac{\partial}{\partial x} \int_{x_0}^{x} P(x, y_0) \, dx + \frac{\partial}{\partial x} \int_{y_0}^{y} Q(x, y) \, dy$$

$$= P(x, y_0) + \int_{y_0}^{y} \frac{\partial Q}{\partial x} \, dy$$
$$\downarrow$$
$$= P(x, y_0) + \int_{y_0}^{y} \frac{\partial P}{\partial y} \, dy \qquad \text{(by assumption)}$$

$$= P(x, y_0) + P(x, y) - P(x, y_0)$$

$$= P(x, y)$$

Verifying that $\partial f / \partial y = Q$ is left as an exercise.　　　　■

Example 2

Determine whether the following differential expressions are exact:

(a)　$(2y^2 - 2y) \, dx + (2xy - x) \, dy$

(b)　$2xy \, dx + (x^2 - 1) \, dy$

Solution

(a)
$$P = 2y^2 - 2y \qquad\qquad Q = 2xy - x$$
$$\frac{\partial P}{\partial y} = 4y - 2 \qquad\qquad \frac{\partial Q}{\partial x} = 2y - 1$$

Since $\partial P / \partial y \neq \partial Q / \partial x$, the expression is not exact. In other words, there is no function f whose total differential is $(2y^2 - 2y) \, dx + (2xy - x) \, dy$.

(b)
$$P = 2xy \qquad\qquad Q = x^2 - 1$$
$$\frac{\partial P}{\partial y} = 2x \qquad\qquad \frac{\partial Q}{\partial x} = 2x$$

Since $\partial P / \partial y = \partial Q / \partial x$, the expression is exact.

Exact Differential Equations　　The **first-order differential equation**

$$P(x, y) \, dx + Q(x, y) \, dy = 0 \qquad\qquad (16.13)$$

is also said to be **exact** when $P \, dx + Q \, dy$ is an exact differential. When (16.13) is exact, it is equivalent to $df(x, y) = 0$. Hence, a family of **solutions** of the equation is given implicitly by $f(x, y) = C$.

Example 3

Solve $2xy \, dx + (x^2 - 1) \, dy = 0$.

Solution　It was proved in Example 2, part (b) that the equation is exact. Consequently, there exists a function f such that

$$\overbrace{\frac{\partial f}{\partial x} = 2xy}^{P} \quad \text{and} \quad \overbrace{\frac{\partial f}{\partial y} = x^2 - 1}^{Q}$$

Integrating the first of these equations* with respect to x gives

$$f = x^2 y + g(y)$$

where $g(y)$ is the "constant" of integration. Taking the partial derivative of this last expression with respect to y and setting the result equal to Q gives

$$\frac{\partial f}{\partial y} = x^2 + g'(y) = x^2 - 1$$

It follows that $g'(y) = -1$, $g(y) = -y$ and so $f = x^2 y - y$. A family of solutions of the equation is then $x^2 y - y = C$.

___

Example 4

Solve
$$(\cos x \sin x - xy^2)\,dx + (y - x^2 y)\,dy = 0$$

subject to $y(0) = 2$.

Solution   The equation is exact since

$$\frac{\partial P}{\partial y} = -2xy = \frac{\partial Q}{\partial x}$$

Now, $\overbrace{\frac{\partial f}{\partial y} = y - x^2 y}^{Q}$ implies that $f = \frac{1}{2}y^2 - \frac{1}{2}x^2 y^2 + h(x)$

We then take the partial derivative of this expression with respect to x and equate the result equal to P:

$$\frac{\partial f}{\partial x} = -xy^2 + h'(x) = \cos x \sin x - xy^2$$

$$h'(x) = \cos x \sin x$$

$$h(x) = \int \sin x (\cos x\,dx)$$

$$= \frac{1}{2}\sin^2 x$$

From $f = \frac{1}{2}(y^2 - x^2 y^2 + \sin^2 x)$, we find a family of solutions $y^2 - x^2 y^2 + \sin^2 x = C_1$, where we have relabeled $2C$ as C_1. Finally, substituting the given conditions $x = 0$, $y = 2$ in the last equation, immediately gives $C_1 = 4$. Consequently, a solution to the problem is $y^2 - x^2 y^2 + \sin^2 x = 4$.

___

*We could integrate either of these equations.

——— Exercises 16.5 ———

Answers to odd-numbered problems begin on page A-49.

In Problems 1–24 determine whether the given differential equation is exact. If exact, solve.

1. $(2x + 4) \, dx + (3y - 1) \, dy = 0$

2. $(2x + y) \, dx - (x + 6y) \, dy = 0$

3. $(5x + 4y) \, dx + (4x - 8y^3) \, dy = 0$

4. $(\sin y - y \sin x) \, dx + (\cos x + x \cos y - y) \, dy = 0$

5. $(2y^2 x - 3) \, dx + (2yx^2 + 4) \, dy = 0$

6. $\left(2y - \dfrac{1}{x} + \cos 3x\right)\dfrac{dy}{dx} + \dfrac{y}{x^2} - 4x^3 + 3y \sin 3x = 0$

7. $(x + y)(x - y) \, dx + x(x - 2y) \, dy = 0$

8. $\left(1 + \ln x + \dfrac{y}{x}\right) dx = (1 - \ln x) \, dy$

9. $(y^3 - y^2 \sin x - x) \, dx + (3xy^2 + 2y \cos x) \, dy = 0$

10. $(x^3 + y^3) \, dx + 3xy^2 \, dy = 0$

11. $(y \ln y - e^{-xy}) \, dx + \left(\dfrac{1}{y} + x \ln y\right) dy = 0$

12. $\dfrac{2x}{y} \, dx - \dfrac{x^2}{y^2} \, dy = 0$ **13.** $x\dfrac{dy}{dx} = 2xe^x - y + 6x^2$

14. $(3x^2 y + e^y) \, dx + (x^3 + xe^y - 2y) \, dy = 0$

15. $\left(1 - \dfrac{3}{x} + y\right) dx + \left(1 - \dfrac{3}{y} + x\right) dy = 0$

16. $(e^y + 2xy \cosh x)y' + xy^2 \sinh x + y^2 \cosh x = 0$

17. $\left(x^2 y^3 - \dfrac{1}{1 + 9x^2}\right) dx + x^3 y^2 \, dy = 0$

18. $(5y - 2x)y' - 2y = 0$

19. $(\tan x - \sin x \sin y) \, dx + \cos x \cos y \, dy = 0$

20. $(3x \cos 3x + \sin 3x - 3) \, dx + (2y + 5) \, dy = 0$

21. $(1 - 2x^2 - 2y)\dfrac{dy}{dx} = 4x^3 + 4xy$

22. $(2y \sin x \cos x - y + 2y^2 e^{xy^2}) \, dx$
$= (x - \sin^2 x - 4xye^{xy^2}) \, dy$

23. $(4x^3 y - 15x^2 - y) \, dx + (x^4 + 3y^2 - x) \, dy = 0$

24. $\left(\dfrac{1}{x} + \dfrac{1}{x^2} - \dfrac{y}{x^2 + y^2}\right) dx + \left(ye^y + \dfrac{x}{x^2 + y^2}\right) dy = 0$

In Problems 25–28 solve the given differential equation subject to the indicated condition.

25. $(x + y)^2 \, dx + (2xy + x^2 - 1) \, dy = 0, \ y(1) = 1$

26. $(e^x + y) \, dx + (2 + x + ye^y) \, dx = 0, \ y(0) = 1$

27. $(4y + 2x - 5) \, dx + (6y + 4x - 1) \, dy = 0,$
$y(-1) = 2$

28. $(y^2 \cos x - 3x^2 y - 2x) \, dx$
$+ (2y \sin x - x^3 + \ln y) \, dy = 0, \ y(0) = e$

Miscellaneous Problems

29. Determine a function $P(x, y)$ such that

$$P(x, y) \, dx + \left(xe^{xy} + 2xy + \dfrac{1}{x}\right) dy = 0$$

is an exact differential equation.

30. Determine a function $Q(x, y)$ such that

$$\left(y^{1/2}x^{-1/2} + \dfrac{x}{x^2 + y}\right) dx + Q(x, y) \, dy = 0$$

is an exact differential equation.

31. Find a constant k so that

$$(y^3 + kxy^4 - 2x) \, dx + (3xy^2 + 20x^2 y^3) \, dy = 0$$

is an exact differential equation.

32. If f is the function defined in (16.12), verify that $\partial f / \partial y = Q$.

33. Solve $\dfrac{dy}{dx} = \dfrac{5y - 2x}{-5x + 3y^2}$ subject to $y(2) = -3$.

16.6   The Chain Rule

The Chain Rule for functions of one variable states that if $y = f(u)$ is a differentiable function of u, and $u = g(x)$ is a differentiable function of x, then the derivative of the composite function is

$$\frac{dy}{dx} = \frac{dy}{du}\frac{du}{dx} \tag{16.14}$$

For a composite function of two variables $z = f(u, v)$, where $u = g(x, y)$ and $v = h(x, y)$, we would naturally expect *two* formulas analogous to (16.14) since we can compute both $\partial z/\partial x$ and $\partial z/\partial y$. The **Chain Rule** for functions of two variables is summarized as follows.

THEOREM 16.7   **Chain Rule**

If $z = f(u, v)$ is differentiable and $u = g(x, y)$ and $v = h(x, y)$ have continuous first partial derivatives, then

$$\frac{\partial z}{\partial x} = \frac{\partial z}{\partial u}\frac{\partial u}{\partial x} + \frac{\partial z}{\partial v}\frac{\partial v}{\partial x} \tag{16.15}$$

$$\frac{\partial z}{\partial y} = \frac{\partial z}{\partial u}\frac{\partial u}{\partial y} + \frac{\partial z}{\partial v}\frac{\partial v}{\partial y}$$

Proof   We prove the second of these results. If $\Delta x = 0$, then

$$\Delta z = f(g(x, y + \Delta y), h(x, y + \Delta y)) - f(g(x, y), h(x, y))$$

Now, if

$$\Delta u = g(x, y + \Delta y) - g(x, y), \qquad \Delta v = h(x, y + \Delta y) - h(x, y)$$

then $g(x, y + \Delta y) = u + \Delta u, \qquad h(x, y + \Delta y) = v + \Delta v$

Hence, Δz can be rewritten as

$$\Delta z = f(u + \Delta u, v + \Delta v) - f(u, v)$$

Since f is differentiable, it follows from the increment formula (16.7) of Section 16.4 that

$$\Delta z = \frac{\partial z}{\partial u}\Delta u + \frac{\partial z}{\partial v}\Delta v + \varepsilon_1\,\Delta u + \varepsilon_2\,\Delta v$$

where, recall, $\varepsilon_1(\Delta u, \Delta v)$ and $\varepsilon_2(\Delta u, \Delta v)$ are functions of Δu and Δv with property

$$\lim_{(\Delta u, \Delta v)\to(0,0)} \varepsilon_1 = 0, \qquad \lim_{(\Delta u, \Delta v)\to(0,0)} \varepsilon_2 = 0$$

Since ε_1 and ε_2 are not uniquely defined functions, a pair of functions can always be found for which $\varepsilon_1(0, 0) = 0$, $\varepsilon_2(0, 0) = 0$. Hence, ε_1 and ε_2 are continuous at $(0, 0)$. Therefore,

$$\frac{\Delta z}{\Delta y} = \frac{\partial z}{\partial u}\frac{\Delta u}{\Delta y} + \frac{\partial z}{\partial v}\frac{\Delta v}{\Delta y} + \varepsilon_1\frac{\Delta u}{\Delta y} + \varepsilon_2\frac{\Delta v}{\Delta y}$$

Now, taking the limit of the last line as $\Delta y \to 0$ gives

$$\frac{\partial z}{\partial y} = \frac{\partial z}{\partial u}\frac{\partial u}{\partial y} + \frac{\partial z}{\partial v}\frac{\partial v}{\partial y} + 0 \cdot \frac{\partial u}{\partial y} + 0 \cdot \frac{\partial v}{\partial y}$$

$$= \frac{\partial z}{\partial u}\frac{\partial u}{\partial y} + \frac{\partial z}{\partial v}\frac{\partial v}{\partial y}$$

since Δu and Δv both approach zero as $\Delta y \to 0$.   ■

Example 1

If $z = u^2 - v^3$ and $u = e^{2x-3y}$, $v = \sin(x^2 - y^2)$, find $\partial z/\partial x$ and $\partial z/\partial y$.

Solution   Since $\partial z/\partial u = 2u$ and $\partial z/\partial v = -3v^2$, it follows from (16.15) that

$$\frac{\partial z}{\partial x} = 2u(2e^{2x-3y}) - 3v^2[2x\cos(x^2 - y^2)]$$

$$= 4ue^{2x-3y} - 6xv^2\cos(x^2 - y^2) \qquad (16.16)$$

$$\frac{\partial z}{\partial y} = 2u(-3e^{2x-3y}) - 3v^2[(-2y)\cos(x^2 - y^2)]$$

$$= -6ue^{2x-3y} + 6yv^2\cos(x^2 - y^2) \qquad (16.17)$$

Of course, in Example 1, we could substitute the expressions for u and v in the original function and then find the partial derivatives directly. In the same manner, the answers (16.16) and (16.17) can be expressed in terms of x and y.

Special Case   If $z = f(u, v)$ is differentiable and $u = g(t)$ and $v = h(t)$ are differentiable functions of a single variable t, then Theorem 16.7 implies that the ordinary derivative dz/dt is

$$\frac{dz}{dt} = \frac{\partial z}{\partial u}\frac{du}{dt} + \frac{\partial z}{\partial v}\frac{dv}{dt} \qquad (16.18)$$

Generalizations   The results given in (16.15) and (16.18) immediately generalize to any number of variables. If $z = f(u_1, u_2, \ldots, u_n)$ and each of the variables $u_1, u_2, u_3, \ldots, u_n$ are functions of $x_1, x_2, \ldots, x_k$, then under the same assumptions as in Theorem 16.7, we have

$$\frac{\partial z}{\partial x_i} = \frac{\partial z}{\partial u_1}\frac{\partial u_1}{\partial x_i} + \frac{\partial z}{\partial u_2}\frac{\partial u_2}{\partial x_i} + \cdots + \frac{\partial z}{\partial u_n}\frac{\partial u_n}{\partial x_i} \qquad (16.19)$$

where $i = 1, 2, \ldots, k$. Similarly, if the $u_i, i = 1, \ldots, n$ are differentiable functions of a single variable t, then

$$\frac{dz}{dt} = \frac{\partial z}{\partial u_1}\frac{du_1}{dt} + \frac{\partial z}{\partial u_2}\frac{du_2}{dt} + \cdots + \frac{\partial z}{\partial u_n}\frac{du_n}{dt} \qquad (16.20)$$

Tree Diagrams   The results in (16.15) can be memorized in terms of a **tree diagram**. The dots in the first diagram that follows indicate the fact that z depends on u and v; u and v depend, in turn, on x and y. To compute $\partial z / \partial x$, for example, we read left to right and follow the *two* colored polygonal paths leading from z to x, multiply the partial derivatives on each path, and then add the products. The result given in (16.18) is represented by the second tree diagram.

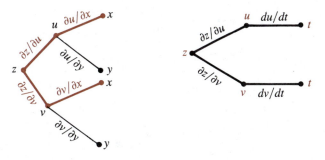

We use tree diagrams in the next two examples to illustrate special cases of (16.19) and (16.20).

Example 2

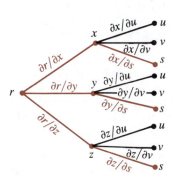

If $r = x^2 + y^5 z^3$ and $x = uve^{2s}$, $y = u^2 - v^2 s$, $z = \sin(uvs^2)$, find $\partial r / \partial s$.

Solution   From the colored paths of the accompanying tree diagram we obtain

$$\frac{\partial r}{\partial s} = \frac{\partial r}{\partial x}\frac{\partial x}{\partial s} + \frac{\partial r}{\partial y}\frac{\partial y}{\partial s} + \frac{\partial r}{\partial z}\frac{\partial z}{\partial s}$$

$$= 2x(2uve^{2s}) + 5y^4 z^3(-v^2) + 3y^5 z^2(2uvs\,\cos(uvs^2))$$

Example 3

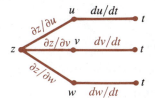

If $z = u^2 v^3 w^4$ and $u = t^2$, $v = 5t - 8$, $w = t^3 + t$, find dz/dt.

Solution   In this case the tree diagram indicates that

$$\frac{dz}{dt} = \frac{\partial z}{\partial u}\frac{du}{dt} + \frac{\partial z}{\partial v}\frac{dv}{dt} + \frac{\partial z}{\partial w}\frac{dw}{dt}$$

$$= 2uv^3 w^4(2t) + 3u^2 v^2 w^4(5) + 4u^2 v^3 w^3(3t^2 + 1)$$

Alternative Solution   Differentiate $z = t^4(5t - 8)^3(t^3 + t)^4$ by the Product Rule.

Exercises 16.6

Answers to odd-numbered problems begin on page A-49.

In Problems 1–10 find the indicated partial derivatives.

1. $z = e^{uv^2}$; $u = x^3$, $v = x - y^2$; $\dfrac{\partial z}{\partial x}, \dfrac{\partial z}{\partial y}$

2. $z = u^2 \cos 4v$; $u = x^2y^3$, $v = x^3 + y^3$; $\dfrac{\partial z}{\partial x}, \dfrac{\partial z}{\partial y}$

3. $z = 4x - 5y^2$; $x = u^4 - 8v^3$, $y = (2u - v)^2$; $\dfrac{\partial z}{\partial u}, \dfrac{\partial z}{\partial v}$

4. $z = \dfrac{x - y}{x + y}$; $x = \dfrac{u}{v}$, $y = \dfrac{v^2}{u}$; $\dfrac{\partial z}{\partial u}, \dfrac{\partial z}{\partial v}$

5. $w = (u^2 + v^2)^{3/2}$; $u = e^{-t} \sin \theta$, $v = e^{-t} \cos \theta$; $\dfrac{\partial w}{\partial t}, \dfrac{\partial w}{\partial \theta}$

6. $w = \tan^{-1}\sqrt{uv}$; $u = r^2 - s^2$, $v = r^2s^2$; $\dfrac{\partial w}{\partial r}, \dfrac{\partial w}{\partial s}$

7. $R = rs^2t^4$; $r = ue^{v^2}$, $s = ve^{-u^2}$, $t = e^{u^2v^2}$; $\dfrac{\partial R}{\partial u}, \dfrac{\partial R}{\partial v}$

8. $Q = \ln(pqr)$; $p = t^2 \sin^{-1}x$, $q = \dfrac{x}{t^2}$, $r = \tan^{-1}\dfrac{x}{t}$;
$\dfrac{\partial Q}{\partial x}, \dfrac{\partial Q}{\partial t}$

9. $w = \sqrt{x^2 + y^2}$; $x = \ln(rs + tu)$, $y = \dfrac{t}{u} \cosh rs$;
$\dfrac{\partial w}{\partial t}, \dfrac{\partial w}{\partial r}, \dfrac{\partial w}{\partial u}$

10. $s = p^2 + q^2 - r^2 + 4t$; $p = \phi e^{3\theta}$, $q = \cos(\phi + \theta)$,
$r = \phi\theta^2$, $t = 2\phi + 8\theta$; $\dfrac{\partial s}{\partial \phi}, \dfrac{\partial s}{\partial \theta}$

In Problems 11–16 find the indicated derivative.

11. $z = \ln(u^2 + v^2)$; $u = t^2$, $v = t^{-2}$; $\dfrac{dz}{dt}$

12. $z = u^3v - uv^4$; $u = e^{-5t}$, $v = \sec 5t$; $\dfrac{dz}{dt}$

13. $w = \cos(3u + 4v)$; $u = 2t + \dfrac{\pi}{2}$, $v = -t - \dfrac{\pi}{4}$; $\dfrac{dw}{dt}\bigg|_{t=\pi}$

14. $w = e^{xy}$; $x = \dfrac{4}{2t + 1}$, $y = 3t + 5$; $\dfrac{dw}{dt}\bigg|_{t=0}$

15. $p = \dfrac{r}{2s + t}$; $r = u^2$, $s = \dfrac{1}{u^2}$, $t = \sqrt{u}$; $\dfrac{dp}{du}$

16. $r = \dfrac{xy^2}{z^3}$; $x = \cos s$, $y = \sin s$, $z = \tan s$; $\dfrac{dr}{ds}$

17. If F and G have second partial derivatives, show that $u(x, t) = F(x + at) + G(x - at)$ satisfies the **wave equation**

$$a^2 \frac{\partial^2 u}{\partial x^2} = \frac{\partial^2 u}{\partial t^2}$$

18. Let $\eta = x + at$ and $\xi = x - at$. Show that the wave equation in Problem 17 becomes

$$\frac{\partial^2 u}{\partial \eta \, \partial \xi} = 0$$

where $u = f(\eta, \xi)$.

19. If $u = f(x, y)$ and $x = r \cos \theta$, $y = r \sin \theta$, show that **Laplace's equation** $\partial^2u/\partial x^2 + \partial^2u/\partial y^2 = 0$ becomes

$$\frac{\partial^2 u}{\partial r^2} + \frac{1}{r}\frac{\partial u}{\partial r} + \frac{1}{r^2}\frac{\partial^2 u}{\partial \theta^2} = 0$$

20. If $z = f(u)$ is a differentiable function of one variable and $u = g(x, y)$ possesses first partial derivatives, what are $\partial z/\partial x$ and $\partial z/\partial y$?

21. Use the result of Problem 20 to show that for any differentiable function f, $z = f(y/x)$ satisfies the equation $x\,\partial z/\partial x + y\,\partial z/\partial y = 0$.

22. The voltage across a conductor is increasing at a rate of 2 volts/min and the resistance is decreasing at a rate of 1 ohm/min. Use $I = E/R$ and the Chain Rule to find the rate at which the current passing through the conductor is changing when $R = 50$ ohms and $E = 60$ volts.

23. The side labeled x of the triangle in Figure 16.47 increases at a rate of 0.3 cm/sec, the side labeled y increases at a rate of 0.5 cm/sec, and the included angle θ increases at a rate of 0.1 rad/sec. Use the Chain Rule to find the rate at which the area of the triangle is changing at the instant $x = 10$ cm, $y = 8$ cm, and $\theta = \pi/6$.

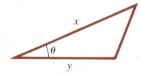

Figure 16.47

24. Van der Waals'* equation of state for the real gas CO_2 is

$$P = \frac{0.08T}{V - 0.0427} - \frac{3.6}{V^2}$$

If dT/dt and dV/dt are rates at which the temperature and volume change, respectively, use the Chain Rule to find dP/dt.

---

*J. D. Van der Waals (1837–1923), a Dutch physicist.

Calculator Problems

25. A very young child grows at a rate of 2 in/yr and gains weight at a rate of 4.2 lb/yr. Use $S = 0.1091w^{0.425}h^{0.725}$ and the Chain Rule to find the rate at which the surface area of the child is changing when it weighs 25 lb and is 29 in. tall.

26. A particle moves in 3-space so that its coordinates at any time are $x = 4 \cos t$, $y = 4 \sin t$, $z = 5t$, $t \geq 0$. Use the Chain Rule to find the rate at which its distance

$$w = \sqrt{x^2 + y^2 + z^2}$$

from the origin is changing at $t = 5\pi/2$ sec.

Miscellaneous Problems

27. If $z = f(x, y)$ is a differentiable function and y is a differentiable function of x, show that

$$\frac{dz}{dx} = \frac{\partial f}{\partial x} + \frac{\partial f}{\partial y}\frac{dy}{dx}$$

28. If $f(x, y) = 0$ defines y implicitly as a differentiable function of x, show that

$$\frac{dy}{dx} = -\frac{f_x(x, y)}{f_y(x, y)}$$

In Problems 29–32 find dy/dx by two methods: **(a)** implicit differentiation, **(b)** the result of Problem 28.

29. $x^3 - 2x^2y^2 + y = 1$ **30.** $x + 2y^2 = e^y$

31. $y = \sin xy$ **32.** $(x + y)^{2/3} = xy$

33. If $F(x, y, z) = 0$ defines z implicitly as a differentiable function of x and y, show that

$$\frac{\partial z}{\partial x} = -\frac{F_x(x, y, z)}{F_z(x, y, z)}$$

34. Repeat Problem 33 to show that

$$\frac{\partial z}{\partial y} = -\frac{F_y(x, y, z)}{F_z(x, y, z)}$$

In Problems 35–38 use the results of Problems 33 and 34 to find $\partial z/\partial x$ and $\partial z/\partial y$.

35. $x^2 + y^2 - z^2 = 1$ **36.** $x^{2/3} + y^{2/3} + z^{2/3} = a^{2/3}$

37. $xy^2z^3 + x^2 - y^2 = 5z^2$ **38.** $z = \ln(xyz)$

39. The equation of state for a thermodynamic system is $F(P, V, T) = 0$, where P, V, and T are pressure, volume, and temperature, respectively. If the equation defines V as a function of P and T, and also defines T as a function of V and P, use, in turn, the results of Problems 33 and 34 to show that

$$\frac{\partial V}{\partial T} = -\frac{\dfrac{\partial F}{\partial T}}{\dfrac{\partial F}{\partial V}} = \frac{1}{\dfrac{\partial T}{\partial V}}$$

40. A function f is said to be **homogeneous of degree n** if $f(\lambda x, \lambda y) = \lambda^n f(x, y)$. If f has first partial derivatives, show that

$$x\frac{\partial f}{\partial x} + y\frac{\partial f}{\partial y} = nf$$

16.7 The Directional Derivative

The Gradient of a Function

In this and the next section, it is convenient to introduce a new vector based on partial differentiation. When the **vector differential operator**

$$\nabla = \mathbf{i}\frac{\partial}{\partial x} + \mathbf{j}\frac{\partial}{\partial y} \quad \text{or} \quad \nabla = \mathbf{i}\frac{\partial}{\partial x} + \mathbf{j}\frac{\partial}{\partial y} + \mathbf{k}\frac{\partial}{\partial z}$$

is applied to a differentiable function $z = f(x, y)$ or $w = F(x, y, z)$, we say that the vectors

$$\nabla f(x, y) = \frac{\partial f}{\partial x}\mathbf{i} + \frac{\partial f}{\partial y}\mathbf{j} \tag{16.21}$$

$$\nabla F(x, y, z) = \frac{\partial F}{\partial x}\mathbf{i} + \frac{\partial F}{\partial y}\mathbf{j} + \frac{\partial F}{\partial z}\mathbf{k} \tag{16.22}$$

are the **gradients** of the respective functions. The symbol ∇, an inverted capital Greek delta, is called "del" or "nabla." The vector ∇f is usually read "grad f."

___Example 1___

Compute $\nabla f(x, y)$ for $f(x, y) = 5y - x^3y^2$.

Solution   From (16.21),

$$\nabla f(x, y) = \frac{\partial}{\partial x}(5y - x^3y^2)\mathbf{i} + \frac{\partial}{\partial y}(5y - x^3y^2)\mathbf{j}$$

$$= -3x^2y^2\mathbf{i} + (5 - 2x^3y)\mathbf{j}$$

___Example 2___

If $F(x, y, z) = xy^2 + 3x^2 - z^3$, find $\nabla F(x, y, z)$ at $(2, -1, 4)$.

Solution   From (16.22),

$$\nabla F(x, y, z) = (y^2 + 6x)\mathbf{i} + 2xy\mathbf{j} - 3z^2\mathbf{k}$$

and so $\nabla F(2, -1, 4) = 13\mathbf{i} - 4\mathbf{j} - 48\mathbf{k}$

A Generalization of Partial Differentiation

The partial derivatives $\partial z/\partial x$ and $\partial z/\partial y$ give the slope of a tangent to the trace, or curve of intersection, of the surface given by $z = f(x, y)$ and vertical planes which are, respectively, parallel to the x- and y-coordinate axes. Equivalently, $\partial z/\partial x$ is the rate of change of the function in the direction given by the vector $\mathbf{i}$, and $\partial z/\partial y$ is the rate of change of $z = f(x, y)$ in the $\mathbf{j}$-direction. There is no reason to confine our attention to just two directions; we can find the rate of change of a differentiable function in *any* direction. See Figure 16.48.

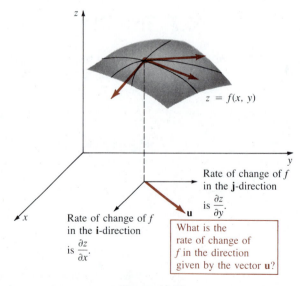

Figure 16.48

Suppose $\mathbf{u} = \cos \theta \mathbf{i} + \sin \theta \mathbf{j}$ is a unit vector in the xy-plane that makes an angle θ with the positive x-axis and that is parallel to the vector $\mathbf{v}$ from $(x, y, 0)$ to $(x + \Delta x, y + \Delta y, 0)$. If $h = \sqrt{(\Delta x)^2 + (\Delta y)^2} > 0$, then $\mathbf{v} = h\mathbf{u}$. Furthermore, let the plane perpendicular to the xy-plane that contains these points slice the surface $z = f(x, y)$ in a curve C. We ask: What is the slope of the tangent line to C at a point P with coordinates $(x, y, f(x, y))$ in the direction given by $\mathbf{v}$? See Figure 16.49.

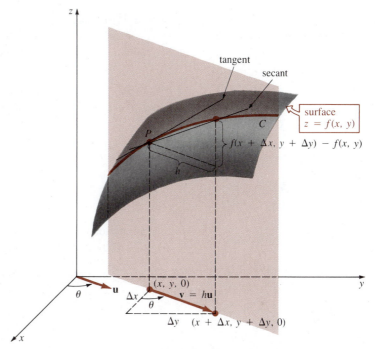

Figure 16.49

From Figure 16.43, we see that $\Delta x = h \cos \theta$ and $\Delta y = h \sin \theta$ so that the slope of the indicated secant line is

$$\frac{f(x + \Delta x, y + \Delta y) - f(x, y)}{h} = \frac{f(x + h \cos \theta, y + h \sin \theta) - f(x, y)}{h} \qquad (16.23)$$

We expect the slope of the tangent at P to be the limit of (16.23) as $h \to 0$. This slope is the rate of change of f at P in the direction specified by the unit vector $\mathbf{u}$. This leads us to the following definition.

DEFINITION 16.7   **Directional Derivative**

The **directional derivative** of $z = f(x, y)$ in the direction of a unit vector $\mathbf{u} = \cos \theta \mathbf{i} + \sin \theta \mathbf{j}$ is

$$D_{\mathbf{u}} f(x, y) = \lim_{h \to 0} \frac{f(x + h \cos \theta, y + h \sin \theta) - f(x, y)}{h} \qquad (16.24)$$

provided that the limit exists.

Observe that (16.24) is truly a generalization of (16.2) and (16.3) of Section 16.3 since:

$$\theta = 0 \quad \text{implies that} \quad D_{\mathbf{i}} f(x, y) = \lim_{h \to 0} \frac{f(x + h, y) - f(x, y)}{h} = \frac{\partial z}{\partial x}$$

$$\text{and} \quad \theta = \frac{\pi}{2} \quad \text{implies that} \quad D_{\mathbf{j}} f(x, y) = \lim_{h \to 0} \frac{f(x, y + h) - f(x, y)}{h} = \frac{\partial z}{\partial y}$$

Method for Computing the Directional Derivative

While (16.24) could be used to find $D_{\mathbf{u}} f(x, y)$ for a given function, as usual we seek a more efficient procedure. The next theorem shows how the concept of the gradient of a function plays a key role in computing a directional derivative.

> **THEOREM 16.8**   If $z = f(x, y)$ is a differentiable function of x and y and $\mathbf{u} = \cos\theta\mathbf{i} + \sin\theta\mathbf{j}$, then
>
> $$D_{\mathbf{u}} f(x, y) = \nabla f(x, y) \cdot \mathbf{u} \qquad (16.25)$$

Proof   Let x, y, and θ be fixed so that

$$g(t) = f(x + t \cos\theta, y + t \sin\theta)$$

is a function of one variable. We wish to compare the value of $g'(0)$, which is found by two different methods. First, by the definition of a derivative

$$g'(0) = \lim_{h \to 0} \frac{g(0 + h) - g(0)}{h}$$

$$= \lim_{h \to 0} \frac{f(x + h \cos\theta, y + h \sin\theta) - f(x, y)}{h}$$

$$= D_{\mathbf{u}} f(x, y) \qquad (16.26)$$

Secondly, by the Chain Rule,

$$g'(t) = f_1(x + t \cos\theta, y + t \sin\theta) \frac{d}{dt}(x + t \cos\theta)$$

$$+ f_2(x + t \cos\theta, y + t \sin\theta) \frac{d}{dt}(x + t \sin\theta)$$

$$= f_1(x + t \cos\theta, y + t \sin\theta)\cos\theta$$
$$+ f_2(x + t \cos\theta, y + t \sin\theta)\sin\theta \qquad (16.27)$$

Here the subscripts 1 and 2 refer to the partial derivatives of $f(x + t \cos\theta, y + t \sin\theta)$ with respect to $x + t \cos\theta$ and $y + t \sin\theta$, respectively. When $t = 0$, we note that $x + t \cos\theta$ and $y + t \sin\theta$ are simply x and y and therefore (16.27) becomes

$$g'(0) = f_x(x, y)\cos\theta + f_y(x, y)\sin\theta \qquad (16.28)$$

Comparing (16.28) with (16.26) then yields

$$D_{\mathbf{u}}f(x,\ y) = f_x(x,\ y)\cos\ \theta + f_y(x,\ y)\sin\ \theta$$
$$= [f_x(x,\ y)\mathbf{i} + f_y(x,\ y)\mathbf{j}] \cdot (\cos\ \theta\mathbf{i} + \sin\ \theta\mathbf{j})$$
$$= \nabla f(x,\ y) \cdot \mathbf{u} \qquad\qquad\blacksquare$$

Example 3

Find the directional derivative of $f(x,\ y) = 2x^2y^3 + 6xy$ at $(1,\ 1)$ in the direction of a unit vector whose angle with the positive x-axis is $\pi/6$.

Solution   Since

$$\frac{\partial f}{\partial x} = 4xy^3 + 6y, \qquad \frac{\partial f}{\partial y} = 6x^2y^2 + 6x$$

we have

$$\nabla f(x,\ y) = (4xy^3 + 6y)\mathbf{i} + (6x^2y^2 + 6x)\mathbf{j} \quad\text{and}\quad \nabla f(1,\ 1) = 10\mathbf{i} + 12\mathbf{j}$$

Now, at $\theta = \pi/6$, $\mathbf{u} = \cos\ \theta\mathbf{i} + \sin\ \theta\mathbf{j}$ becomes

$$\mathbf{u} = \frac{\sqrt{3}}{2}\mathbf{i} + \frac{1}{2}\mathbf{j}$$

Therefore,

$$D_{\mathbf{u}}f(1,\ 1) = \nabla f(1,\ 1) \cdot \mathbf{u} = (10\mathbf{i} + 12\mathbf{j}) \cdot \left(\frac{\sqrt{3}}{2}\mathbf{i} + \frac{1}{2}\mathbf{j}\right) = 5\sqrt{3} + 6$$

Example 4

Consider the plane that is perpendicular to the xy-plane and passes through the points $P(2,\ 1)$ and $Q(3,\ 2)$. What is the slope of the tangent line to the curve of intersection of this plane with the surface $f(x,\ y) = 4x^2 + y^2$ at $(2,\ 1,\ 17)$ in the direction of Q?

Solution   We want $D_{\mathbf{u}}f(2,\ 1)$ in the direction given by the vector $\overrightarrow{PQ} = \mathbf{i} + \mathbf{j}$. But since $\overrightarrow{PQ}$ is not a unit vector, we form $\mathbf{u} = (1/\sqrt{2})\mathbf{i} + (1/\sqrt{2})\mathbf{j}$. Now,

$$\nabla f(x,\ y) = 8x\mathbf{i} + 2y\mathbf{j} \quad\text{and}\quad \nabla f(2,\ 1) = 16\mathbf{i} + 2\mathbf{j}$$

Therefore, the required slope is

$$D_{\mathbf{u}}f(2,\ 1) = (16\mathbf{i} + 2\mathbf{j}) \cdot \left(\frac{1}{\sqrt{2}}\mathbf{i} + \frac{1}{\sqrt{2}}\mathbf{j}\right) = 9\sqrt{2}$$

Functions of Three Variables   For a function $w = F(x, y, z)$ the directional derivative is defined by

$$D_{\mathbf{u}}f(x, y, z) = \lim_{h \to 0} \frac{F(x + h \cos \alpha, \; y + h \cos \beta, \; z + h \cos \gamma) - F(x, y, z)}{h}$$

where α, β, and γ are the direction angles of the vector $\mathbf{u}$ measured relative to the positive x-, y-, and z-axes, respectively.* But in the same manner as before, we can show that

$$D_{\mathbf{u}}F(x, y, z) = \nabla F(x, y, z) \cdot \mathbf{u} \qquad (16.29)$$

Notice, since $\mathbf{u}$ is a unit vector, it follows from (14.19) of Section 14.3 that

$$D_{\mathbf{u}}f(x, y) = \text{comp}_{\mathbf{u}} \nabla f(x, y) \quad \text{and} \quad D_{\mathbf{u}}F(x, y, z) = \text{comp}_{\mathbf{u}} \nabla F(x, y, z)$$

In addition, (16.29) reveals that

$$D_{\mathbf{k}}F(x, y, z) = \frac{\partial w}{\partial z}$$

_____ **Example 5** _____

Find the directional derivative of $F(x, y, z) = xy^2 - 4x^2y + z^2$ at $(1, -1, 2)$ in the direction of $6\mathbf{i} + 2\mathbf{j} + 3\mathbf{k}$.

Solution   We have

$$\frac{\partial F}{\partial x} = y^2 - 8xy, \qquad \frac{\partial F}{\partial y} = 2xy - 4x^2, \qquad \frac{\partial F}{\partial z} = 2z$$

so that
$$\nabla F(x, y, z) = (y^2 - 8xy)\mathbf{i} + (2xy - 4x^2)\mathbf{j} + 2z\mathbf{k}$$
$$\nabla F(1, -1, 2) = 9\mathbf{i} - 6\mathbf{j} + 4\mathbf{k}$$

Since
$$|6\mathbf{i} + 2\mathbf{j} + 3\mathbf{k}| = 7, \quad \text{then} \quad \mathbf{u} = \frac{6}{7}\mathbf{i} + \frac{2}{7}\mathbf{j} + \frac{3}{7}\mathbf{k}$$

is a unit vector in the indicated direction. It follows from (16.29) that

$$D_{\mathbf{u}}F(1, -1, 2) = (9\mathbf{i} - 6\mathbf{j} + 4\mathbf{k}) \cdot \left(\frac{6}{7}\mathbf{i} + \frac{2}{7}\mathbf{j} + \frac{3}{7}\mathbf{k}\right) = \frac{54}{7}$$

_____

*Note that the numerator of (16.24) can be written

$$f(x + h \cos \alpha, \; y + h \cos \beta) - f(x, y)$$

where $\beta = (\pi/2) - \alpha$.

**Maximum Value of the
Directional Derivative**

Let f represent a function of either two or three variables. Since (16.25) and (16.29) express the directional derivative as a dot product, we see from Definition 14.2 that

$$D_{\mathbf{u}}f = |\nabla f|\,|\mathbf{u}|\cos\theta$$
$$= |\nabla f|\cos\theta \qquad (|\mathbf{u}| = 1)$$

Because $0 \le \theta \le \pi$, we have $-1 \le \cos\theta \le 1$ and, consequently,

$$-|\nabla f| \le D_{\mathbf{u}}f \le |\nabla f|$$

In other words,

> the maximum value of the directional derivative is $|\nabla f|$ and occurs when $\mathbf{u}$ has the same direction as ∇f (when $\cos\theta = 1$), (16.30)

and

> the minimum value of the directional derivative is $-|\nabla f|$ and occurs when $\mathbf{u}$ and ∇f have opposite directions (when $\cos\theta = -1$). (16.31)

_____ **Example 6** _____

In Example 5 the maximum value of the directional derivative at F at $(1, -1, 2)$ is $|\nabla F(1, -1, 2)| = \sqrt{133}$. The minimum value of $D_{\mathbf{u}}F(1, -1, 2)$ is then $-\sqrt{133}$.

**Gradient Points in Direction of
Most Rapid Increase of f**

Put yet another way, (16.30) and (16.31) state:

> the gradient vector ∇f points in the direction in which f increases most rapidly; whereas $-\nabla f$ points in the direction of the most rapid decrease of f.

_____ **Example 7** _____

Each year in Los Angeles there is a bicycle race up to the top of a hill by a road known to be steepest in the city. To understand why a bicyclist, with a modicum of sanity, will zig-zag up the road, let us suppose the graph of $f(x, y) = 4 - \frac{2}{3}\sqrt{x^2 + y^2}, 0 \le z \le 4$, shown in Figure 16.50(a) is a mathematical model of the hill. The gradient of f is

$$\nabla f(x, y) = \frac{2}{3}\left[\frac{-x}{\sqrt{x^2 + y^2}}\mathbf{i} + \frac{-y}{\sqrt{x^2 + y^2}}\mathbf{j}\right] = \frac{2/3}{\sqrt{x^2 + y^2}}\mathbf{r}$$

where $\mathbf{r} = -x\mathbf{i} - y\mathbf{j}$ is a vector pointing to the center of the circular base.

Thus, the steepest ascent up the hill is a straight road whose projection in the xy-plane is a radius of the circular base. Since $D_{\mathbf{u}}f = \text{comp}_{\mathbf{u}}\nabla f$, a bicyclist will zig-zag, or seek a direction $\mathbf{u}$ other than ∇f, in order to reduce this component.

road

(a)

(b)

Figure 16.50

___ **Example 8** ___

The temperature in a rectangular box is approximated by

$$T(x, y, z) = xyz(1 - x)(2 - y)(3 - z), \qquad 0 \leq x \leq 1, 0 \leq y \leq 2, 0 \leq z \leq 3$$

If a mosquito is located at $(\frac{1}{2}, 1, 1)$, in which direction should it fly to cool off as rapidly as possible?

Solution   The gradient of T is

$$\nabla T(x, y, z) = yz(2 - y)(3 - z)(1 - 2x)\mathbf{i} + xz(1 - x)(3 - z)(2 - 2y)\mathbf{j}$$
$$+ xy(1 - x)(2 - y)(3 - 2z)\mathbf{k}$$

Therefore,
$$\nabla T\left(\frac{1}{2}, 1, 1\right) = \frac{1}{4}\mathbf{k}$$

To cool off most rapidly, the mosquito should fly in the direction of $-\frac{1}{4}\mathbf{k}$; that is, it should dive for the floor of the box, where the temperature is $T(x, y, 0) = 0$.

___ **Exercises 16.7** ___

Answers to odd-numbered problems begin on page A-49.

In Problems 1–4 compute the gradient for the given function.

1. $f(x, y) = x^2 - x^3y^2 + y^4$

2. $f(x, y) = y - e^{-2x^2y}$

3. $F(x, y, z) = \dfrac{xy^2}{z^3}$

4. $F(x, y, z) = xy \cos yz$

In Problems 5–8 find the gradient of the given function at the indicated point.

5. $f(x, y) = x^2 - 4y^2$; $(2, 4)$

6. $f(x, y) = \sqrt{x^3y - y^4}$; $(3, 2)$

7. $F(x, y, z) = x^2z^2 \sin 4y$; $(-2, \pi/3, 1)$

8. $F(x, y, z) = \ln(x^2 + y^2 + z^2)$; $(-4, 3, 5)$

In Problems 9 and 10 use Definition 16.7 to find $D_{\mathbf{u}}f(x, y)$ given that $\mathbf{u}$ makes the indicated angle with the positive x-axis.

9. $f(x, y) = x^2 + y^2$; $\theta = 30°$

10. $f(x, y) = 3x - y^2$; $\theta = 45°$

In Problems 11–20 find the directional derivative of the given function at the given point in the indicated direction.

11. $f(x, y) = 5x^3y^6$; $(-1, 1)$, $\theta = \pi/6$

12. $f(x, y) = 4x + xy^2 - 5y$; $(3, -1)$, $\theta = \pi/4$

13. $f(x, y) = \tan^{-1}\dfrac{y}{x}$; $(2, -2)$, $\mathbf{i} - 3\mathbf{j}$

14. $f(x, y) = \dfrac{xy}{x + y}$; $(2, -1)$, $6\mathbf{i} + 8\mathbf{j}$

15. $f(x, y) = (xy + 1)^2$; $(3, 2)$, in the direction of $(5, 3)$

16. $f(x, y) = x^2 \tan y$; $\left(\dfrac{1}{2}, \dfrac{\pi}{3}\right)$, in the direction of the negative x-axis

17. $F(x, y, z) = x^2y^2(2z + 1)^2$; $(1, -1, 1)$, $\langle 0, 3, 3\rangle$

18. $F(x, y, z) = \dfrac{x^2 - y^2}{z^2}$; $(2, 4, -1)$, $\mathbf{i} - 2\mathbf{j} + \mathbf{k}$

19. $F(x, y, z) = \sqrt{x^2y + 2y^2z}$; $(-2, 2, 1)$, in the direction of the negative z-axis

20. $F(x, y, z) = 2x - y^2 + z^2$; $(4, -4, 2)$, in the direction of the origin

In Problems 21 and 22 consider the plane through the points P and Q that is perpendicular to the xy-plane. Find the slope of the tangent at the indicated point to the curve of intersection of this plane and the graph of the given function in the direction of Q.

21. $f(x, y) = (x - y)^2$; $P(4, 2)$, $Q(0, 1)$; $(4, 2, 4)$

22. $f(x, y) = x^3 - 5xy + y^2$; $P(1, 1)$, $Q(-1, 6)$; $(1, 1, -3)$

In Problems 23–26 find a vector that gives the direction in which the given function increases most rapidly at the indicated point. Find the maximum rate.

23. $f(x, y) = e^{2x} \sin y$; $(0, \pi/4)$

24. $f(x, y) = xye^{x-y}$; $(5, 5)$

25. $F(x, y, z) = x^2 + 4xz + 2yz^2$; $(1, 2, -1)$

26. $F(x, y, z) = xyz$; $(3, 1, -5)$

In Problems 27–30 find a vector that gives the direction in which the given function decreases most rapidly at the indicated point. Find the minimum rate.

27. $f(x, y) = \tan(x^2 + y^2)$; $(\sqrt{\pi/6}, \sqrt{\pi/6})$

28. $f(x, y) = x^3 - y^3$; $(2, -2)$

29. $F(x, y, z) = \sqrt{xz}e^y$; $(16, 0, 9)$

30. $F(x, y, z) = \ln\dfrac{xy}{z}$; $\left(\dfrac{1}{2}, \dfrac{1}{6}, \dfrac{1}{3}\right)$

31. Find the directional derivative(s) of $f(x, y) = x + y^2$ at $(3, 4)$ in the direction of a tangent vector to the graph of $2x^2 + y^2 = 9$ at $(2, 1)$.

32. If $f(x, y) = x^2 + xy + y^2 - x$, find all points where $D_{\mathbf{u}}f(x, y)$ in the direction of $\mathbf{u} = (1/\sqrt{2})(\mathbf{i} + \mathbf{j})$ is zero.

33. Suppose $\nabla f(a, b) = 4\mathbf{i} + 3\mathbf{j}$. Find a unit vector $\mathbf{u}$ so that
(a) $D_{\mathbf{u}}f(a, b) = 0$
(b) $D_{\mathbf{u}}f(a, b)$ is a maximum
(c) $D_{\mathbf{u}}f(a, b)$ is a minimum

34. Suppose $D_{\mathbf{u}}f(a, b) = 6$. What is the value of $D_{-\mathbf{u}}f(a, b)$?

35. (a) If $f(x, y) = x^3 - 3x^2y^2 + y^3$, find the directional derivative of f at a point (x, y) in the direction of $\mathbf{u} = (1/\sqrt{10})(3\mathbf{i} + \mathbf{j})$.
(b) If $F(x, y) = D_{\mathbf{u}}f(x, y)$ of part (a), find $D_{\mathbf{u}}F(x, y)$.

36. Consider the gravitational potential

$$U(x, y) = \frac{-Gm}{\sqrt{x^2 + y^2}}$$

where G and m are constants. Show that U increases or decreases most rapidly along a line through the origin.

37. If $f(x, y) = x^3 - 12x + y^2 - 10y$, find all points at which $|\nabla f| = 0$.

38. Suppose
$$D_{\mathbf{u}}f(a, b) = 7, \qquad D_{\mathbf{v}}f(a, b) = 3,$$

$$\mathbf{u} = \frac{5}{13}\mathbf{i} - \frac{12}{13}\mathbf{j}, \quad \text{and} \quad \mathbf{v} = \frac{5}{13}\mathbf{i} + \frac{12}{13}\mathbf{j}$$

Find $\nabla f(a, b)$.

39. Consider the rectangular plate shown in Figure 16.45. The temperature at a point (x, y) on the plate is given by $T(x, y) = 5 + 2x^2 + y^2$. Determine the direction an insect should take, starting at $(4, 2)$, in order to cool off as rapidly as possible.

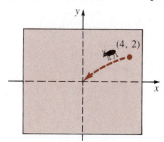

Figure 16.51

40. In Problem 39, observe that $(0, 0)$ is the coolest point of the plate. Find the path the cold-seeking insect, starting at $(4, 2)$, will take to the origin. If $\langle x(t), y(t)\rangle$ is the vector equation of the path, then use the fact that $-\nabla T(x, y) = \langle x'(t), y'(t)\rangle$. Why is this? (*Hint:* Remember separation of variables?)

41. The temperature at a point (x, y) on a rectangular metal plate is given by $T(x, y) = 100 - 2x^2 - y^2$. Find the path a heat-seeking particle will take, starting at $(3, 4)$, as it moves in the direction in which the temperature increases most rapidly.

42. The temperature T at a point (x, y, z) in space is inversely proportional to the square of the distance from (x, y, z) to the origin. It is known that $T(0, 0, 1) = 500$. Find the rate of change of T at $(2, 3, 3)$ in the direction of $(3, 1, 1)$. In which direction from $(2, 3, 3)$ does the temperature T increase most rapidly? At $(2, 3, 3)$ what is the maximum rate of change of T?

Miscellaneous Problems

43. Find a function f such that

$$\nabla f = (3x^2 + y^3 + ye^{xy})\mathbf{i} + (-2y^2 + 3xy^2 + xe^{xy})\mathbf{j}$$

44. Let f_x, f_y, f_{xy}, f_{yx} be continuous and $\mathbf{u}$ and $\mathbf{v}$ be unit vectors. Show that $D_{\mathbf{u}}D_{\mathbf{v}}f = D_{\mathbf{v}}D_{\mathbf{u}}f$.

In Problems 45–48 assume that f and g are differentiable functions of two variables. Prove the given identity.

45. $\nabla(cf) = c\,\nabla f$

46. $\nabla(f + g) = \nabla f + \nabla g$

47. $\nabla(fg) = f\,\nabla g + g\,\nabla f$

48. $\nabla\left(\dfrac{f}{g}\right) = \dfrac{g\,\nabla f - f\,\nabla g}{g^2}$

49. If $F(x, y, z) = f_1(x, y, z)\mathbf{i} + f_2(x, y, z)\mathbf{j} + f_3(x, y, z)\mathbf{k}$ and

$$\nabla = \mathbf{i}\frac{\partial}{\partial x} + \mathbf{j}\frac{\partial}{\partial y} + \mathbf{k}\frac{\partial}{\partial z}$$

show that

$$\nabla \times F = \left(\frac{\partial f_3}{\partial y} - \frac{\partial f_2}{\partial z}\right)\mathbf{i} + \left(\frac{\partial f_1}{\partial z} - \frac{\partial f_3}{\partial x}\right)\mathbf{j}$$
$$+ \left(\frac{\partial f_2}{\partial x} - \frac{\partial f_1}{\partial y}\right)\mathbf{k}$$

16.8 The Tangent Plane

Geometric Interpretation of the Gradient—Functions of Two Variables

Suppose $f(x, y) = c$ is the *level curve* of the differentiable function $z = f(x, y)$ that passes through a specified point $P(x_0, y_0)$; that is, $f(x_0, y_0) = c$. If this level curve is parameterized by the differentiable functions

$$x = g(t), \quad y = h(t) \quad \text{such that} \quad x_0 = g(t_0), \quad y_0 = h(t_0)$$

then the derivative of $f(g(t), h(t)) = c$ with respect to t is

$$\frac{\partial f}{\partial x}\frac{dx}{dt} + \frac{\partial f}{\partial y}\frac{dy}{dt} = 0 \tag{16.32}$$

By introducing the vectors

$$\nabla f(x, y) = \frac{\partial f}{\partial x}\mathbf{i} + \frac{\partial f}{\partial y}\mathbf{j} \quad \text{and} \quad \mathbf{r}'(t) = \frac{dx}{dt}\mathbf{i} + \frac{dy}{dt}\mathbf{j}$$

(16.32) becomes $\nabla f \cdot \mathbf{r}' = 0$. Specifically, at $t = t_0$, we have

$$\nabla f(x_0, y_0) \cdot \mathbf{r}'(t_0) = 0 \tag{16.33}$$

Thus, if $\mathbf{r}'(t_0) \neq 0$, the vector $\nabla f(x_0, y_0)$ is orthogonal to the tangent vector $\mathbf{r}'(t_0)$ at $P(x_0, y_0)$. We interpret this to mean

∇f is perpendicular to the level curve at P.

See Figure 16.52.

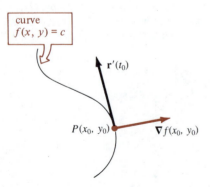

Figure 16.52

Example 1

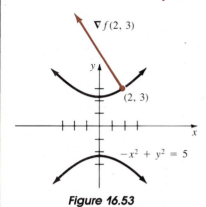

Figure 16.53

Find the level curve of $f(x, y) = -x^2 + y^2$ passing through $(2, 3)$. Graph the gradient at the point.

Solution   Since $f(2, 3) = -4 + 9 = 5$, the level curve is the hyperbola $-x^2 + y^2 = 5$. Now,

$$\nabla f(x, y) = -2x\mathbf{i} + 2y\mathbf{j}$$

and

$$\nabla f(2, 3) = -4\mathbf{i} + 6\mathbf{j}$$

Figure 16.53 shows the level curve and $\nabla f(2, 3)$.

Geometric Interpretation of the Gradient—Functions of Three Variables

Proceeding as before, let $F(x, y, z) = c$ be the *level surface* of a differentiable function $w = F(x, y, z)$ that passes through $P(x_0, y_0, z_0)$. If the differentiable functions

$$x = f(t), \qquad y = g(t), \qquad z = h(t)$$

are the parametric equations of a curve C on the surface for which $x_0 = f(t_0)$, $y_0 = g(t_0)$, $z_0 = h(t_0)$, then the derivative of $F(f(t), g(t), h(t)) = 0$ implies that

$$\frac{\partial F}{\partial x}\frac{dx}{dt} + \frac{\partial F}{\partial y}\frac{dy}{dt} + \frac{\partial F}{\partial z}\frac{dz}{dt} = 0$$

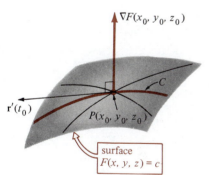

Figure 16.54

or

$$\left(\frac{\partial F}{\partial x}\mathbf{i} + \frac{\partial F}{\partial y}\mathbf{j} + \frac{\partial F}{\partial z}\mathbf{k}\right) \cdot \left(\frac{dx}{dt}\mathbf{i} + \frac{dy}{dt}\mathbf{j} + \frac{dz}{dt}\mathbf{k}\right) = 0 \qquad (16.34)$$

In particular, at $t = t_0$, (16.34) is

$$\nabla F(x_0, y_0, z_0) \cdot \mathbf{r}'(t_0) = 0 \qquad (16.35)$$

Thus, when $\mathbf{r}'(t_0) \neq 0$, the vector $\nabla F(x_0, y_0, z_0)$ is orthogonal to the tangent vector $\mathbf{r}'(t_0)$. Since this argument holds for any differentiable curve through $P(x_0, y_0, z_0)$ on the surface, we conclude that

∇F **is perpendicular (normal) to the level surface at P.**

See Figure 16.54.

Example 2

Find the level surface of $F(x, y, z) = x^2 + y^2 + z^2$ passing through $(1, 1, 1)$. Graph the gradient at the point.

Solution   Since $F(1, 1, 1) = 3$, the level surface passing through $(1, 1, 1)$ is sphere $x^2 + y^2 + z^2 = 3$. The gradient of the function is

$$\nabla F(x, y, z) = 2x\mathbf{i} + 2y\mathbf{j} + 2z\mathbf{k}$$

and so, at the given point,

$$\nabla F(1, 1, 1) = 2\mathbf{i} + 2\mathbf{j} + 2\mathbf{k}$$

The level surface and $\nabla F(1, 1, 1)$ are illustrated in Figure 16.55.

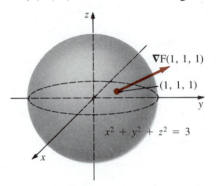

Figure 16.55

Tangent Plane   In the earlier chapters of this text, we found equations of tangent lines to graphs of functions. In 3-space we can now solve the analogous problem of finding equations of **tangent planes** to surfaces. If $F(x, y, z) = c$ is the equation of a surface, we define the tangent plane at $P(x_0, y_0, z_0)$ as the plane through P with normal vector $\nabla F(x_0, y_0, z_0)$ provided that this gradient is not zero. Thus, if (x, y, z) and (x_0, y_0, z_0) are points on the tangent plane and $\mathbf{r}$ and $\mathbf{r}_0$ are their respective position vectors, a vector equation of the tangent plane is

$$\nabla F(x_0, y_0, z_0) \cdot (\mathbf{r} - \mathbf{r}_0) = 0$$

or

$$F_x(x_0, y_0, z_0)(x - x_0) + F_y(x_0, y_0, z_0)(y - y_0) + F_z(x_0, y_0, z_0)(z - z_0) = 0 \qquad (16.36)$$

See Figure 16.56.

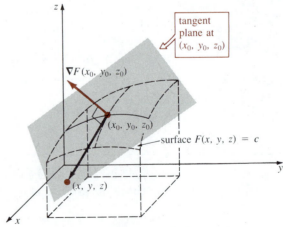

Figure 16.56

_____ **Example 3** _____

Find an equation of the tangent plane to the graph of $x^2 - 4y^2 + z^2 = 16$ at $(2, 1, 4)$.

Solution   By defining $F(x, y, z) = x^2 - 4y^2 + z^2$, the given surface is the level surface $F(x, y, z) = F(2, 1, 4) = 16$ passing through $(2, 1, 4)$. Now,

$$\nabla F(x, y, z) = 2x\mathbf{i} - 8y\mathbf{j} + 2z\mathbf{k} \quad \text{and} \quad \nabla F(2, 1, 4) = 4\mathbf{i} - 8\mathbf{j} + 8\mathbf{k}$$

It follows from (16.36) that an equation of the tangent plane is

$$4(x - 2) - 8(y - 1) + 8(z - 4) = 0 \quad \text{or} \quad x - 2y + 2z = 8$$

Surfaces Given by
$z = f(x, y)$ For a surface given explicitly by $z = f(x, y)$, we define $F(x, y, z) = f(x, y) - z$ or $F(x, y, z) = z - f(x, y)$. Thus, a point (x_0, y_0, z_0) is on the graph of $z = f(x, y)$ if and only if it is also on the level surface $F(x, y, z) = 0$. This follows from $F(x_0, y_0, z_0) = f(x_0, y_0) - z_0 = 0$.

_____ **Example 4** _____

Find an equation of the tangent plane to the graph of $z = 4x + y^2$ at $(-1, 3, 5)$.

Solution   Define $F(x, y, z) = 4x + y^2 - z$ so that the level surface of F passing through the given point is $F(x, y, z) = F(-1, 3, 5)$ or $F(x, y, z) = 0$. Now,

$$\nabla F(x, y, z) = 4\mathbf{i} + 2y\mathbf{j} - \mathbf{k} \quad \text{and} \quad \nabla F(-1, 3, 5) = 4\mathbf{i} + 6\mathbf{j} - \mathbf{k}$$

Thus, from (16.36) the desired equation is

$$4(x + 1) + 6(y - 3) - (z - 5) = 0 \quad \text{or} \quad 4x + 6y - z = 9$$

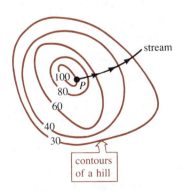

contours
of a hill

Figure 16.57

Remark

Water flowing down a hill chooses a path in the direction of greatest change in altitude. Figure 16.57 shows the contours, or level curves, of a hill. As shown in the figure, a stream starting at point P will take a path that is perpendicular to the contours. After reading Sections 16.7 and 16.8 you should be able to explain why.

Exercises 16.8

Answers to odd-numbered problems begin on page A-49.

In Problems 1–12 sketch the level curve or surface passing through the indicated point. Sketch the gradient at the point.

1. $f(x, y) = x - 2y;$ (6, 1)

2. $f(x, y) = \dfrac{y + 2x}{x};$ (1, 3)

3. $f(x, y) = y - x^2;$ (2, 5)

4. $f(x, y) = x^2 + y^2;$ (−1, 3)

5. $f(x, y) = \dfrac{x^2}{4} + \dfrac{y^2}{9};$ (−2, −3)

6. $f(x, y) = \dfrac{y^2}{x};$ (2, 2)

7. $f(x, y) = (x - 1)^2 - y^2;$ (1, 1)

8. $f(x, y) = \dfrac{y - 1}{\sin x};$ $\left(\dfrac{\pi}{6}, \dfrac{3}{2}\right)$

9. $F(x, y, z) = y + z;$ (3, 1, 1)

10. $F(x, y, z) = x^2 + y^2 - z;$ (1, 1, 3)

11. $F(x, y, z) = \sqrt{x^2 + y^2 + z^2};$ (3, 4, 0)

12. $F(x, y, z) = x^2 - y^2 + z;$ (0, −1, 1)

In Problems 13 and 14 find the points on the given surface at which the gradient is parallel to the indicated vector.

13. $z = x^2 + y^2;$ $4\mathbf{i} + \mathbf{j} + \dfrac{1}{2}\mathbf{k}$

14. $x^3 + y^2 + z = 15;$ $27\mathbf{i} + 8\mathbf{j} + \mathbf{k}$

In Problems 15–24 find an equation of the tangent plane to the graph of the given equation at the indicated point.

15. $x^2 + y^2 + z^2 = 9;$ (−2, 2, 1)

16. $5x^2 - y^2 + 4z^2 = 8;$ (2, 4, 1)

17. $x^2 - y^2 - 3z^2 = 5;$ (6, 2, 3)

18. $xy + yz + zx = 7;$ (1, −3, −5)

19. $z = 25 - x^2 - y^2;$ (3, −4, 0)

20. $xz = 6;$ (2, 0, 3)

21. $z = \cos(2x + y);$ $\left(\dfrac{\pi}{2}, \dfrac{\pi}{4}, -\dfrac{1}{\sqrt{2}}\right)$

22. $x^2y^3 + 6z = 10;$ (2, 1, 1)

23. $z = \ln(x^2 + y^2);$ $\left(\dfrac{1}{\sqrt{2}}, \dfrac{1}{\sqrt{2}}, 0\right)$

24. $z = 8e^{-2y} \sin 4x;$ $\left(\dfrac{\pi}{24}, 0, 4\right)$

In Problems 25 and 26 find the points on the given surface at which the tangent plane is parallel to the indicated plane.

25. $x^2 + y^2 + z^2 = 7;$ $2x + 4y + 6z = 1$

26. $x^2 - 2y^2 - 3z^2 = 33;$ $8x + 4y + 6z = 5$

27. Find points on the surface $x^2 + 4x + y^2 + z^2 - 2z = 11$ at which the tangent plane is horizontal.

28. Find points on the surface $x^2 + 3y^2 + 4z^2 - 2xy = 16$ at which the tangent plane is parallel to **(a)** the xz-plane, **(b)** the yz-plane, and **(c)** the xy-plane.

Miscellaneous Problems

In Problems 29 and 30 show that the second equation is an equation of the tangent plane to the graph of the first equation at (x_0, y_0, z_0).

29. $\dfrac{x^2}{a^2} + \dfrac{y^2}{b^2} + \dfrac{z^2}{c^2} = 1;$ $\dfrac{xx_0}{a^2} + \dfrac{yy_0}{b^2} + \dfrac{zz_0}{c^2} = 1$

30. $\dfrac{x^2}{a^2} - \dfrac{y^2}{b^2} + \dfrac{z^2}{c^2} = 1;$ $\dfrac{xx_0}{a^2} - \dfrac{yy_0}{b^2} + \dfrac{zz_0}{c^2} = 1$

31. Show that every tangent plane to the graph of $z^2 = x^2 + y^2$ passes through the origin.

32. Show that the sum of the x-, y-, and z-intercepts of every tangent plane to the graph of $\sqrt{x} + \sqrt{y} + \sqrt{z} = \sqrt{a}$, $a > 0$, is the number a.

The line perpendicular to the tangent plane to a surface at a point P is said to be a **normal line** at P. In Problems 33 and 34, find parametric equations for the normal line at the indicated point. In Problems 35 and 36, find symmetric equations for the normal line.

33. $x^2 + 2y^2 + z^2 = 4;$ (1, −1, 1)

34. $z = 2x^2 - 4y^2;$ (3, −2, 2)

35. $z = 4x^2 + 9y^2 + 1;$ $\left(\dfrac{1}{2}, \dfrac{1}{3}, 3\right)$

36. $x^2 + y^2 - z^2 = 0;$ (3, 4, 5)

37. Show that every normal line to the graph $x^2 + y^2 + z^2 = a^2$ passes through the origin.

38. Two surfaces are said to be **orthogonal** at a point P of intersection if their normal lines at P are orthogonal. Prove that the surfaces given by $F(x, y, z) = 0$ and $G(x, y, z) = 0$ are orthogonal at P if and only if $F_x G_x + F_y G_y + F_z G_z = 0$.

In Problems 39 and 40 use the result of Problem 38 to show that the given surfaces are orthogonal at a point of intersection.

39. $x^2 + y^2 + z^2 = 25$; $-x^2 + y^2 + z^2 = 0$

40. $x^2 - y^2 + z^2 = 4$; $z = 1/xy^2$

16.9 Extrema for Functions of Two Variables

If $z = f(x, y)$ is continuous on a closed rectangular region R whose sides are parallel to the x- and y-coordinate axes, then, analogous to Theorem 4.1, f has both an **absolute maximum** and an **absolute minimum** on the region. That is, there are points (c, d) and (a, b) in R so that

$$f(c, d) \leq f(x, y) \leq f(a, b)$$

for every point (x, y) in R.

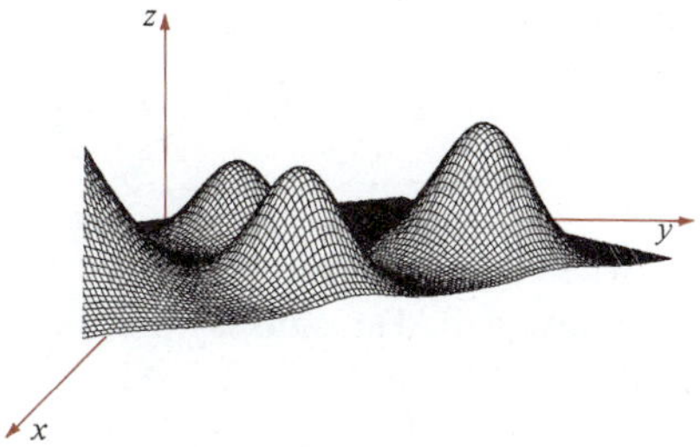

A function f of two variables can, of course, have **relative**, or **local**, **extrema**. We say that $z = f(x, y)$ has a **relative maximum** at (a, b) if there is some rectangular region containing (a, b) such that $f(x, y) \leq f(a, b)$ for all (x, y) in the region. Similarly, $z = f(x, y)$ has a **relative minimum** at (a, b) if there is some rectangular region containing (a, b) such that $f(x, y) \geq f(a, b)$ for all (x, y) in the region. Figure 16.58 shows a function with several relative maxima.

Suppose for the sake of illustration that (a, b) is an interior point of rectangular region R at which f has a maximum and, furthermore, suppose that f has first and second partial derivatives. As seen in Figure 16.59, on the curve C_1 of intersection of the surface and the plane $x = a$, we must have

$$f_y(a, b) = 0 \quad \text{and} \quad f_{yy}(a, b) \leq 0*$$

Figure 16.58

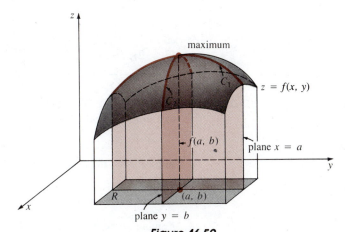

Figure 16.59

---

*Do you remember the relationship between concavity and the second derivative?

Similarly, on the curve C_2, which is the trace of the surface in the plane $y = b$, we have

$$f_x(a, b) = 0 \quad \text{and} \quad f_{xx}(a, b) \leq 0$$

In contrast, if $z = f(x, y)$ has a minimum at (a, b) we expect

$$f_x(a, b) = 0, \qquad f_y(a, b) = 0 \quad \text{and} \quad f_{xx}(a, b) \geq 0, \qquad f_{yy}(a, b) \geq 0$$

Critical Points   The foregoing discussion suggests that to find an extremum of $z = f(x, y)$, we must solve $f_x(x, y) = 0$ and $f_y(x, y) = 0$ *simultaneously*.* This leads to the following definition.

> **DEFINITION 16.8   Critical Points**
>
> If $z = f(x, y)$ has first partial derivatives, then the solutions of
>
> $$f_x(x, y) = 0 \quad \text{and} \quad f_y(x, y) = 0$$
>
> are called **critical points**.

The critical points correspond to points where f could *possibly* have a relative extremum. In some texts critical points are also called **stationary points**.

_____ **Example 1** _____

Find all critical points for $f(x, y) = x^3 + y^3 - 27x - 12y$.

Solution   The first partial derivatives are

$$f_x(x, y) = 3x^2 - 27 \quad \text{and} \quad f_y(x, y) = 3y^2 - 12.$$

Hence, $f_x(x, y) = 0$ and $f_y(x, y) = 0$ imply that

$$x^2 = 9 \quad \text{and} \quad y^2 = 4$$

and so $x = \pm 3$, $y = \pm 2$. Thus, there are four critical points $(3, 2)$, $(-3, 2)$, $(3, -2)$, and $(-3, -2)$.

Second Partials Test   The following theorem gives sufficient conditions for ascertaining relative extrema. The proof of the theorem will not be given.

_____

*Alternatively, the equation of a tangent plane to the graph of $z = f(x, y)$ at any point (a, b) is

$$z - f(a, b) = (x - a)f_x(a, b) + (y - b)f_y(a, b)$$

Arguing as we did for tangent lines, we look for horizontal tangent planes. At a point (a, b), where the tangent plane is horizontal, its equation must be $z = f(a, b)$, and so $f_x(a, b) = 0$ and $f_y(a, b) = 0$.

THEOREM 16.9   **Second Partials Test for Relative Extrema**

Let (a, b) be a critical point of $z = f(x, y)$ and suppose f_{xx}, f_{yy}, and f_{xy} are continuous in a rectangular region containing (a, b). Let $D(x, y) = f_{xx}(x, y)f_{yy}(x, y) - [f_{xy}(x, y)]^2$.

(*i*)   If $D(a, b) > 0$ and $f_{xx}(a, b) > 0$, then $f(a, b)$ is a relative minimum.

(*ii*)   If $D(a, b) > 0$ and $f_{xx}(a, b) < 0$, then $f(a, b)$ is a relative maximum.

(*iii*)   If $D(a, b) < 0$, then $f(a, b)$ is not an extremum.

(*iv*)   If $D(a, b) = 0$, no conclusion can be drawn concerning a relative extremum.

Example 2

Find the extrema for $f(x, y) = 4x^2 + 2y^2 - 2xy - 10y - 2x$.

Solution   The first partial derivatives are

$$f_x(x, y) = 8x - 2y - 2 \quad \text{and} \quad f_y(x, y) = 4y - 2x - 10$$

Solving the simultaneous equations

$$8x - 2y = 2 \quad \text{and} \quad -2x + 4y = 10$$

yields the single critical point $(1, 3)$. Now,

$$f_{xx}(x, y) = 8, \qquad f_{yy}(x, y) = 4, \qquad f_{xy}(x, y) = -2$$

and so $D(x, y) = (8)(4) - (-2)^2 = 28$. Since $D(1, 3) > 0$ and $f_{xx}(1, 3) > 0$, it follows from (*i*) of Theorem 16.9 that $f(1, 3) = -16$ is a relative minimum.

Example 3

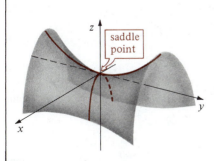

Figure 16.60

The graph of $f(x, y) = y^2 - x^2$ is the hyperbolic paraboloid given in Figure 16.60. From $f_x(x, y) = -2x$ and $f_y(x, y) = 2y$ we see that $(0, 0)$ is a critical point and that $f(0, 0) = 0$ is the only possible extremum of the function. But from

$$f(0, y) = y^2 \geq 0 \quad \text{and} \quad f(x, 0) = -x^2 \leq 0$$

we see that, in a neighborhood of $(0, 0)$, the points along the y-axis correspond to functional values that are greater than or equal to $f(0, 0) = 0$ and the points along the x-axis correspond to functional values that are less than or equal to $f(0, 0) = 0$. Hence, $f(0, 0) = 0$ is not an extremum. We say that $(0, 0)$ is a **saddle point** of the function.
   You should verify that $D(0, 0) < 0$.

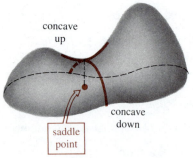

concave up

concave down

saddle point

Figure 16.61

In general, the critical point (a, b) in case (iii) of Theorem 16.9 is called a **saddle point**. When $D(a, b) < 0$ for a critical point (a, b), then the graph of the function behaves essentially like the saddle-shaped hyperbolic paraboloid in a neighborhood of (a, b). See Figure 16.61.

Example 4

Find the extrema for $f(x, y) = 4xy - x^2 - y^2 - 14x + 4y + 10$.

Solution From $f_x(x, y) = 4y - 2x - 14$ and $f_y(x, y) = 4x - 2y + 4$, we find the solution of the system

$$4y - 2x - 14 = 0 \quad \text{and} \quad 4x - 2y + 4 = 0$$

to be $(1, 4)$. Now, $f_{xx}(x, y) = -2$, $f_{yy}(x, y) = -2$, and $f_{xy}(x, y) = 4$. By themselves the equations

$$f_{xx}(0, 0) = -2 \quad \text{and} \quad f_{yy}(0, 0) = -2$$

would seem to imply that $f(1, 4) = 11$ is a relative maximum. However, $D(1, 4) = (-2)(-2) - (4)^2 < 0$ indicates that $f(1, 4)$ is not *any* extremum and $(1, 4)$ is a saddle point.

Example 5

Find the extrema of $f(x, y) = x^3 + y^3 - 3x^2 - 3y^2 - 9x$.

Solution From

$$f_x(x, y) = 3x^2 - 6x - 9 = 3(x - 3)(x + 1)$$

and

$$f_y(x, y) = 3y^2 - 6y = 3y(y - 2)$$

and the equations

$$(x - 3)(x + 1) = 0 \quad \text{and} \quad y(y - 2) = 0$$

we find four critical points $(3, 0)$, $(3, 2)$, $(-1, 0)$, $(-1, 2)$. Since

$$f_{xx} = 6x - 6, \qquad f_{yy} = 6y - 6, \quad \text{and} \quad f_{xy} = 0$$

we find $D(x, y) = 36(x - 1)(y - 1)$. The second partial derivatives test is summarized in the accompanying table.

Critical Point (a, b)	$D(a, b)$	$f_{xx}(a, b)$	$f(a, b)$	Conclusion
$(3, 0)$	negative	positive	-27	no extremum
$(3, 2)$	positive	positive	-31	rel. min.
$(-1, 0)$	positive	negative	5	rel. max.
$(-1, 2)$	negative	negative	1	no extremum

Remarks

(*i*)   You may have noticed a slight difference between the definition of a critical point and that of a critical value. Recall that a number c in the domain of $y = f(x)$ is a critical value if $f'(c) = 0$ or $f'(c)$ does not exist. Similarly, $z = f(x, y)$ can have an extremum at a point (a, b), where f_x and f_y do not exist, but Theorem 16.9 is certainly not applicable at such a point. See Problems 34 and 35 in Exercises 16.9. For functions of two variables, there is no convenient first partial derivative test to fall back on.

(*ii*)   The method of solution for the system $f_x(x, y) = 0, f_y(x, y) = 0$ will not always be obvious especially when f_x and f_y are not linear. Do not be afraid to exercise your algebraic skills in the problems that follow.

Exercises 16.9

Answers to odd-numbered problems begin on page A-50.

In Problems 1–20 find any relative extrema of the given function.

1. $f(x, y) = x^2 + y^2 + 5$

2. $f(x, y) = 4x^2 + 8y^2$

3. $f(x, y) = -x^2 - y^2 + 8x + 6y$

4. $f(x, y) = 3x^2 + 2y^2 - 6x + 8y$

5. $f(x, y) = 5x^2 + 5y^2 + 20x - 10y + 40$

6. $f(x, y) = -4x^2 - 2y^2 - 8x + 12y + 5$

7. $f(x, y) = 4x^3 + y^3 - 12x - 3y$

8. $f(x, y) = -x^3 + 2y^3 + 27x - 24y + 3$

9. $f(x, y) = 2x^2 + 4y^2 - 2xy - 10x - 2y + 2$

10. $f(x, y) = 5x^2 + 5y^2 + 5xy - 10x - 5y + 18$

11. $f(x, y) = (2x - 5)(y - 4)$

12. $f(x, y) = (x + 5)(2y + 6)$

13. $f(x, y) = -2x^3 - 2y^3 + 6xy + 10$

14. $f(x, y) = x^3 + y^3 - 6xy + 27$

15. $f(x, y) = xy - \dfrac{2}{x} - \dfrac{4}{y} + 8$

16. $f(x, y) = -3x^2y - 3xy^2 + 36xy$

17. $f(x, y) = xe^x \sin y$

18. $f(x, y) = e^{y^2 - 3y + x^2 + 4x}$

19. $f(x, y) = \sin x + \sin y$

20. $f(x, y) = \sin xy$

21. Find three positive numbers whose sum is 21 such that their product P is a maximum. (*Hint:* Express P as a function of only two variables.)

22. Find the dimensions of a rectangular box with volume of one cubic foot that has a minimal surface area S.

23. Find the point on the plane $x + 2y + z = 1$ closest to the origin. (*Hint:* Consider the square of the distance.)

24. Find the least distance between the point $(2, 3, 1)$ and the plane $x + y + z = 1$.

25. Find all points on the surface $xyz = 8$ that are closest to the origin. Find the least distance.

26. Find the shortest distance between the lines

$$\mathcal{L}_1: x = t, \; y = 4 - 2t, \; z = 1 + t$$
$$\mathcal{L}_2: x = 3 + 2s, \; y = 6 + 2s, \; z = 8 - 2s$$

At what points on the lines does the minimum occur?

27. Find the maximum volume of a rectangular box with sides parallel to the coordinate planes that can be inscribed in the ellipsoid

$$\frac{x^2}{a^2} + \frac{y^2}{b^2} + \frac{z^2}{c^2} = 1, \qquad a > 0, b > 0, c > 0$$

28. The volume of an ellipsoid

$$\frac{x^2}{a^2} + \frac{y^2}{b^2} + \frac{z^2}{c^2} = 1, \qquad a > 0, b > 0, c > 0$$

is $V = 4\pi abc/3$. Show that the ellipsoid of greatest volume that satisfies $a + b + c = $ constant, is a sphere.

29. A closed rectangular box is to be made so that its volume is 60 ft^3. The costs of the material for the top and bottom are 10 cents per square foot and 20 cents per square foot, respectively. The cost of the sides is 2 cents per square foot. Determine the cost function $C(x, y)$, where x and y are the length and width of the box, respectively. Find the dimensions of the box that will give a minimum cost.

30. A revenue function is $R(x, y) = x(100 - 6x) + y(192 - 4y)$, where x and y denote the number of items of two commodities sold. Given that the corresponding cost function is

$$C(x, y) = 2x^2 + 2y^2 + 4xy - 8x + 20$$

find the maximum profit. (*Hint:* profit = revenue − cost.)

31. In statistics one often wants to fit an approximating curve to a set of data points (x_i, y_i), $i = 1, \ldots, n$. A measure of the "goodness of fit" is the sum of the squares S of the distances between ordinates on the approximating curve and the ordinates of the data points. For a linear approximating function $y = ax + b$, called a regression line, we have

$$S = \sum_{i=1}^{n} [(ax_i + b) - y_i]^2$$

See Figure 16.62. Of course, it is desired that S be a minimum. Show that $\partial S/\partial a = 0$ and $\partial S/\partial b = 0$ when

$$a = \frac{n \sum_{i=1}^{n} x_i y_i - \sum_{i=1}^{n} x_i \sum_{i=1}^{n} y_i}{n \sum_{i=1}^{n} x_i^2 - \left(\sum_{i=1}^{n} x_i\right)^2}$$

and

$$b = \frac{\sum_{i=1}^{n} x_i^2 \sum_{i=1}^{n} y_i - \sum_{i=1}^{n} x_i y_i \sum_{i=1}^{n} x_i}{n \sum_{i=1}^{n} x_i^2 - \left(\sum_{i=1}^{n} x_i\right)^2}$$

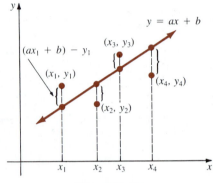

Figure 16.62

32. A 24-in.-wide piece of tin is bent into a trough whose cross section is an isosceles trapezoid. See Figure 16.63. Find x and θ so that the cross-sectional area is a maximum. What is the maximum area?

Figure 16.63

33. The pentagon shown in Figure 16.64, formed by an isosceles triangle surmounted on a rectangle, has a fixed perimeter P. Find x, y, and θ so that the area of the pentagon is a maximum.

Figure 16.64

Miscellaneous Problems

In Problems 34–37 show that the given function has an absolute extremum but that Theorem 16.9 is not applicable.

34. $f(x, y) = \sqrt{x^2 + y^2}$

35. $f(x, y) = 16 - x^{2/3} - y^{2/3}$

36. $f(x, y) = 1 - x^4 y^2$

37. $f(x, y) = 5x^2 + y^4 - 8$

A function f that is continuous on a closed bounded set S has both an absolute maximum and an absolute minimum in S. In Problems 38–41 find the absolute extrema of the given function over the set of points defined by $x^2 + y^2 \leq 1$. (*Hint:* Examine the function in the interior *and* on the boundary of S. Parameterize the boundary.)

38. $f(x, y) = -x^2 - 3y^2 + 4y + 1$

39. $f(x, y) = x + \sqrt{3}y$

40. $f(x, y) = xy$

41. $f(x, y) = x^2 + xy + y^2$

42. The function $f(x, y) = \sin xy$ is continuous on the closed rectangular region defined by $0 \leq x \leq \pi, 0 \leq y \leq 1$. Find the critical points in the region. Find the points where f has an absolute extremum. Graph the function on the rectangular region.

16.10 Lagrange Multipliers

In Problems 21–28 of Exercises 16.9, you were asked to find the maximum or minimum of a function subject to a given side condition or **constraint**. The side condition was used to eliminate one of the variables in the function so that the second partial derivatives test was applicable. In the present discussion, we examine another procedure for determining the so-called **constrained extrema** of a function.

Example 1

Determine geometrically whether $f(x, y) = 9 - x^2 - y^2$ subject to $x + y = 3$ has an extremum.

Solution As seen in Figure 16.65(b), the graph of $x + y = 3$ is a plane that intersects the paraboloid given by $f(x, y) = 9 - x^2 - y^2$. It appears from the

x	y	$f(x, y)$
0.5	2.5	2.5
1	2	4
1.25	1.75	4.375
1.5	1.5	4.5
1.75	1.25	4.375
2	1	4
2.5	0.5	2.5

(a)

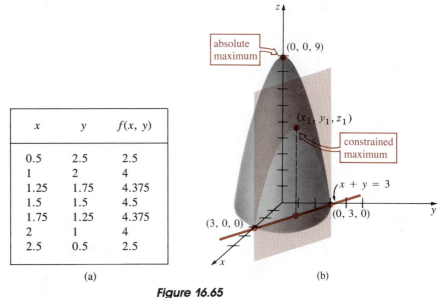

(b)

Figure 16.65

figure that the function has a *constrained maximum* for some x_1 and y_1 satisfying $0 < x_1 < 3$, $0 < y_1 < 3$ and $x_1 + y_1 = 3$. The accompanying table would also seem to indicate that this new maximum is $f(1.5, 1.5) = 4.5$. Note that we cannot use numbers such as $x = 1.7$ and $y = 2.4$ since these values do not satisfy the constraint $x + y = 3$.

Alternatively, we can analyze Example 1 by means of level curves. As shown in Figure 16.66, increasing values of f correspond to increasing c in the level curves $9 - x^2 - y^2 = c$. The maximum value of f (that is, c) subject to the constraint occurs where the level curve $c = \frac{9}{2}$ intersects, or more precisely is tangent to, the line $x + y = 3$. By solving $x^2 + y^2 = \frac{9}{2}$ and $x + y = 3$ simultaneously, we find the point of tangency is $(\frac{3}{2}, \frac{3}{2})$.

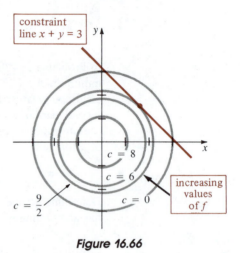

constraint line $x + y = 3$

$c = 8$

$c = 6$

$c = \dfrac{9}{2}$

$c = 0$

increasing values of f

Figure 16.66

Functions of Two Variables   To generalize the foregoing discussion, suppose we wish to:

find extrema of the function $z = f(x, y)$ subject to a constraint given by $g(x, y) = 0$.

It seems plausible from Figure 16.67 that to find, say, a constrained maximum of f, we need only find the highest level curve $f(x, y) = c$ that is tangent to the

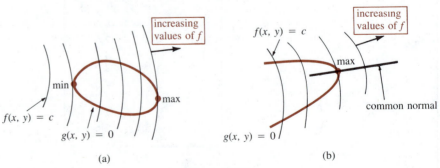

increasing values of f

min

max

$f(x, y) = c$

$g(x, y) = 0$

(a)

$f(x, y) = c$

increasing values of f

max

common normal

$g(x, y) = 0$

(b)

Figure 16.67

graph of the constraint equation $g(x, y) = 0$. Now, recall that the gradients ∇f and ∇g are perpendicular to the curves $f(x, y) = c$ and $g(x, y) = 0$, respectively. Hence if $\nabla g \neq \mathbf{0}$ at a point P of tangency of the curves, then ∇f and ∇g are parallel at P; that is, they lie along a common normal. Therefore, for some nonzero scalar λ (the lowercase Greek letter lambda), we must have $\nabla f = \lambda \nabla g$. We state this result in a formal fashion.

> **THEOREM 16.10**   Suppose f has an extremum at a point (x_0, y_0) on the graph of the constraint equation $g(x, y) = 0$. If f and g have continuous first partial derivatives and $\nabla g(x_0, y_0) \neq \mathbf{0}$, then there is a real number λ such that $\nabla f(x_0, y_0) = \lambda \nabla g(x_0, y_0)$.

Method of Lagrange Multipliers

The real number λ for which $\nabla f = \lambda \nabla g$, or, after equating components,

$$f_x(x, y) = \lambda g_x(x, y), \qquad f_y(x, y) = \lambda g_y(x, y)$$

is called a **Lagrange multiplier**. If f has a constrained extremum at (x_0, y_0), then we have just seen there is a number λ such that

$$
\begin{aligned}
f_x(x_0, y_0) &= \lambda g_x(x_0, y_0) \\
f_y(x_0, y_0) &= \lambda g_y(x_0, y_0) \\
g(x_0, y_0) &= 0
\end{aligned}
\tag{16.37}
$$

The equations in (16.37) suggest the following procedure, known as the **method of Lagrange multipliers**, for finding constrained extrema.

To find the extrema of $z = f(x, y)$ subject to the constraint $g(x, y) = 0$, solve the system of equations:

$$
\begin{aligned}
f_x(x, y) &= \lambda g_x(x, y) \\
f_y(x, y) &= \lambda g_y(x, y) \\
g(x, y) &= 0
\end{aligned}
\tag{16.38}
$$

Among the solutions (x, y, λ) of the system will be the points (x_i, y_i), where f has an extremum. When f has a maximum (or minimum), it will be the largest (or smallest) number in the list of functional values $f(x_i, y_i)$.

Example 2

Use the method of Lagrange multipliers to find the maximum of $f(x, y) = 9 - x^2 - y^2$ subject to $x + y = 3$.

Solution   By defining $g(x, y) = x + y - 3$ and using $f_x = -2x, f_y = -2y$, $g_x = 1, g_y = 1$, the system in (16.38) is

$$
\begin{aligned}
-2x &= \lambda \\
-2y &= \lambda \\
x + y - 3 &= 0
\end{aligned}
$$

Equating the first and second equations gives $-2x = -2y$ or $x = y$. Substituting this result into the third equation yields $2y - 3 = 0$ or $y = \frac{3}{2}$. Thus, $x = y = \frac{3}{2}$ and the constrained maximum is $f(\frac{3}{2}, \frac{3}{2}) = \frac{9}{2}$.

Example 3

Find the extrema of $f(x, y) = y^2 - 4x$ subject to $x^2 + y^2 = 9$.

Solution   If we define $g(x, y) = x^2 + y^2 - 9$, then $f_x = -4$, $f_y = 2y$, $g_x = 2x$, and $g_y = 2y$. Therefore, (16.38) becomes

$$-4 = 2x\lambda$$
$$2y = 2y\lambda$$
$$x^2 + y^2 - 9 = 0$$

From the second of these equations, $y(\lambda - 1) = 0$, we see that either $y = 0$ or $\lambda = 1$. If $y = 0$, the third equation in the system gives $x^2 = 9$ or $x = \pm 3$. Hence, $(-3, 0)$ and $(3, 0)$ are two points at which f might possibly have an extremum. Now if $\lambda = 1$, the first equation yields $x = -2$. Substituting this value into $x^2 + y^2 - 9 = 0$ gives $y^2 = 5$ or $y = \pm\sqrt{5}$. Two more points are $(-2, -\sqrt{5})$ and $(-2, \sqrt{5})$. For the list of functional values

$$f(-3, 0) = 12, \quad f(3, 0) = -12, \quad f(-2, -\sqrt{5}) = 13, \quad f(-2, \sqrt{5}) = 13$$

we conclude that f has a constrained minimum of -12 at $(3, 0)$ and a constrained maximum of 13 at $(-2, -\sqrt{5})$ and at $(-2, \sqrt{5})$.

Figure 16.68 shows the graph of the constraint equation $x^2 + y^2 = 9$ and some of the level curves $y^2 - 4x = c$.

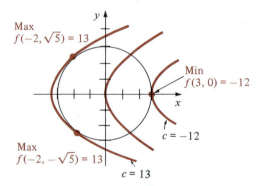

Figure 16.68

In applying the method of Lagrange multipliers we really are not very interested in finding the values of λ that satisfy the system (16.38). Notice in Example 1 that we did not even bother to find λ. In Example 3, we used the value $\lambda = 1$ to help find $x = -2$, but after that we ignored it.

_____ **Example 4** _____

A closed right circular cylinder will have a volume of 1000 ft³. The top and bottom of the cylinder are made of metal that costs 2 dollars per square foot. The lateral side is wrapped in metal costing 2.5 dollars per square foot. Find the minimum cost of construction.

Solution   The cost function is

$$C(r, h) = 2(2\pi r^2) + 2.5(2\pi rh)$$

↑   ↑

| cost of bottom and top | | cost of side |

$$= 4\pi r^2 + 5\pi rh$$

Now, from the constraint $1000 = \pi r^2 h$, we can identify $g(r, h) = \pi r^2 h - 1000$, and so $C_r = 8\pi r + 5\pi h$, $C_h = 5\pi r$, $g_r = 2\pi rh$, and $g_h = \pi r^2$. We must then solve the system

$$8\pi r + 5\pi h = 2\pi rh\lambda$$
$$5\pi r = \pi r^2\lambda$$
$$\pi r^2 h - 1000 = 0$$

Multiplying the first equation by r and the second equation by $2h$ and subtracting yields

$$8\pi r^2 - 5\pi rh = 0 \quad \text{or} \quad \pi r(8r - 5h) = 0$$

Since $r = 0$ does not satisfy the constraint equation, we take $r = \frac{5}{8}h$. The constraint then gives

$$h^3 = \frac{1000 \cdot 64}{25\pi} \quad \text{or} \quad h = \frac{40}{\sqrt[3]{25\pi}}$$

Thus, $r = 25/\sqrt[3]{25\pi}$. The constrained minimum cost is

$$C\left(\frac{25}{\sqrt[3]{25\pi}}, \frac{40}{\sqrt[3]{25\pi}}\right) = 4\pi\left(\frac{25}{\sqrt[3]{25\pi}}\right)^2 + 5\pi\left(\frac{25}{\sqrt[3]{25\pi}}\right)\left(\frac{40}{\sqrt[3]{25\pi}}\right)$$
$$= 300\sqrt[3]{25\pi} \approx \$1284.75$$

Functions of Three Variables   To find the extrema of a function of three variables $w = F(x, y, z)$ subject to the constraint $g(x, y, z) = 0$, we solve a system of four equations:

$$F_x(x, y, z) = \lambda g_x(x, y, z)$$
$$F_y(x, y, z) = \lambda g_y(x, y, z) \quad\quad (16.39)$$
$$F_z(x, y, z) = \lambda g_z(x, y, z)$$
$$g(x, y, z) = 0$$

Example 5

Find the extrema of $F(x, y, z) = x^2 + y^2 + z^2$ subject to $2x - 2y - z = 5$.

Solution   With $g(x, y, z) = 2x - 2y - z - 5$, the system (16.39) is

$$2x = 2\lambda$$
$$2y = -2\lambda$$
$$2z = -\lambda$$
$$2x - 2y - z - 5 = 0$$

Using $\lambda = x = -y = -2z$, the last equation gives $x = \frac{10}{9}$ and so $y = -\frac{10}{9}$, $z = -\frac{5}{9}$. Thus, a constrained extremum is

$$F\left(\frac{10}{9}, -\frac{10}{9}, -\frac{5}{9}\right) = \frac{225}{81}$$

Two Constraints   In order to optimize a function $w = F(x, y, z)$ subject to *two* constraints, $g(x, y, z) = 0$ and $h(x, y, z) = 0$, we must introduce a second Lagrange multiplier μ (the lowercase Greek letter mu) and solve the system

$$F_x(x, y, z) = \lambda g_x(x, y, z) + \mu h_x(x, y, z)$$
$$F_y(x, y, z) = \lambda g_y(x, y, z) + \mu h_y(x, y, z)$$
$$F_z(x, y, z) = \lambda g_z(x, y, z) + \mu h_z(x, y, z) \qquad (16.40)$$
$$g(x, y, z) = 0$$
$$h(x, y, z) = 0$$

Example 6

Find the point on the curve C of intersection of the sphere $x^2 + y^2 + z^2 = 9$ and the plane $x - y + 3z = 6$ that is farthest from the xy-plane. Find the point on C that is closest to the xy-plane.

Solution   Figure 16.69 suggests that there does exist two such points P_1 and P_2 with nonnegative z-coordinates. Thus the distance from these points to the xy-plane is simply $F(x, y, z) = z$. If we take $g(x, y, z) = x^2 + y^2 + z^2 - 9$ and $h(x, y, z) = x - y + 3z - 6$, then the system (16.40) is

$$0 = 2x\lambda + \mu$$
$$0 = 2y\lambda - \mu$$
$$1 = 2z\lambda + 3\mu$$
$$x^2 + y^2 + z^2 - 9 = 0$$
$$x - y + 3z - 6 = 0$$

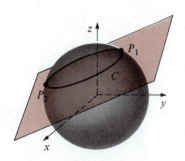

Figure 16.69

Adding the first and second equations gives $2\lambda(y + x) = 0$. If $\lambda = 0$, then the first equation implies $\mu = 0$, but the third equation leads to a contradiction. By taking $y = -x$, the last two equations become

$$x^2 + x^2 + z^2 - 9 = 0 \qquad \text{or} \qquad 2x^2 + z^2 = 9$$
$$x + x + 3z - 6 = 0 \qquad \qquad 2x + 3z = 6$$

Solving this system simultaneously leads to

$$x = \frac{6}{11} + \frac{9}{22}\sqrt{14}, \qquad z = \frac{18}{11} - \frac{3}{11}\sqrt{14}$$

and

$$x = \frac{6}{11} - \frac{9}{22}\sqrt{14}, \qquad z = \frac{18}{11} + \frac{3}{11}\sqrt{14}$$

Thus, the points on C that are farthest and closest to the xy-plane are, respectively,

$$P_1\left(\frac{6}{11} - \frac{9}{22}\sqrt{14}, \; -\frac{6}{11} + \frac{9}{22}\sqrt{14}, \; \frac{18}{11} + \frac{3}{11}\sqrt{14}\right) \quad \text{and} \quad P_2\left(\frac{6}{11} + \frac{9}{22}\sqrt{14}, \; -\frac{6}{11} - \frac{9}{22}\sqrt{14}, \; \frac{18}{11} - \frac{3}{11}\sqrt{14}\right)$$

The approximate coordinates of P_1 and P_2 are $(-0.99, \; 0.99, \; 2.66)$ and $(2.08, \; -2.08, \; 0.62)$.

Remark

Notice the use of the word *extremum* in Example 5. The method of Lagrange multipliers does not have a built-in indicator that flashes $\boxed{\text{MAX}}$ or $\boxed{\text{MIN}}$ when a single extremum is found. In addition to the graphical procedure discussed at the beginning of this section, another way of convincing oneself as to the nature of the extremum is to compare it with values obtained by calculating the given function at other points that satisfy the constraint equation. Indeed, in this manner we find that $225/81$ of Example 5 is actually a constrained *minimum* of the function.

─────── **Exercises 16.10** ───────

Answers to odd-numbered problems begin on page A-50.

In Problems 1 and 2 sketch the graphs of the level curves of the given function f and the indicated constraint equation. Determine whether f has a constrained extremum.

1. $f(x, y) = x + 3y$, subject to $x^2 + y^2 = 1$

2. $f(x, y) = xy$, subject to $\frac{1}{2}x + y = 1$, $x \geq 0$, $y \geq 0$

In Problems 3–20 use the method of Lagrange multipliers to find the constrained extrema of the given function.

3. Problem 1

4. Problem 2

5. $f(x, y) = xy$, subject to $x^2 + y^2 = 2$

6. $f(x, y) = x^2 + y^2$, subject to $2x + y = 5$

7. $f(x, y) = 3x^2 + 3y^2 + 5$, subject to $x - y = 1$

8. $f(x, y) = 4x^2 + 2y^2 + 10$, subject to $4x^2 + y^2 = 4$

9. $f(x, y) = x^2 + y^2$, subject to $x^4 + y^4 = 1$

10. $f(x, y) = 8x^2 - 8xy + 2y^2$, subject to $x^2 + y^2 = 10$

11. $f(x, y) = x^3y$, subject to $\sqrt{x} + \sqrt{y} = 1$

12. $f(x, y) = xy^2$, subject to $x^2 + y^2 = 27$

13. $F(x, y, z) = x + 2y + z$, subject to $x^2 + y^2 + z^2 = 30$

14. $F(x, y, z) = x^2 + y^2 + z^2$, subject to $x + 2y + 3z = 4$

15. $F(x, y, z) = xyz$, subject to $x^2 + y^2/4 + z^2/9 = 1$, $x > 0, y > 0, z > 0$

16. $F(x, y, z) = xyz + 5$, subject to $x^3 + y^3 + z^3 = 24$

17. $F(x, y, z) = x^3 + y^3 + z^3$, subject to $x + y + z = 1$, $x > 0, y > 0, z > 0$

18. $F(x, y, z) = 4x^2y^2z^2$, subject to $x^2 + y^2 + z^2 = 9$, $x > 0, y > 0, z > 0$

19. $F(x, y, z) = x^2 + y^2 + z^2$, subject to $2x + y + z = 1$, $-x + 2y - 3z = 4$

20. $F(x, y, z) = x^2 + y^2 + z^2$, subject to $4x + z = 7$, $z^2 = x^2 + y^2$

21. Give a geometric interpretation of the extrema in Problem 9.

22. Give a geometric interpretation of the extrema in Problem 14.

23. Give a geometric interpretation of the extremum in Problem 19.

24. Give a geometric interpretation of the extremum in Problem 20.

25. Find the maximum area of a right triangle whose perimeter is 4.

26. Find the dimensions of an open rectangular box with maximum volume if its surface area is 75 cm³. What are the dimensions if the box is closed?

27. A right-cylindrical tank is surmounted by a conical cap as shown in Figure 16.70. The radius of the tank is 3 m and its total surface area is 81π m². Find heights x and y so that the volume of the tank is a maximum. (*Hint:* The surface area of the cone is $3\pi\sqrt{9 + y^2}$.)

28. In business a utility index U is a function that gives a measure of satisfaction obtained from the purchasing of variable amounts, x and y, of two commodities that are purchased on a regular basis. If $U(x, y) = x^{1/3}y^{2/3}$ is a utility index, find its extrema subject to $x + 6y = 18$.

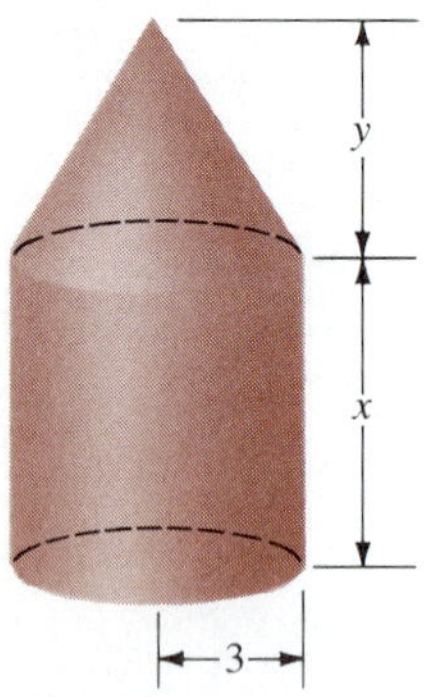

Figure 16.70

29. The Haber–Bosch process* produces ammonia by a direct union of nitrogen and hydrogen under conditions of constant pressure P and constant temperature:

$$N_2 + 3H_2 \overset{\text{catalyst}}{\rightleftharpoons} 2NH_3$$

The partial pressures x, y, and z of hydrogen, nitrogen, and ammonia satisfy $x + y + z = P$ and the equilibrium law $z^2/xy^3 = k$, where k is a constant. The maximum amount of ammonia occurs when the maximum partial pressure of ammonia is obtained. Find the maximum value of z.

30. Find the point $P(x, y)$, $x > 0$, $y > 0$, on the surface $xy^2 = 1$, which is closest to the origin. Show that the line segment from the origin to P is perpendicular to the tangent at P.

31. Find the maximum value of $F(x, y, z) = \sqrt[3]{xyz}$ on the plane $x + y + z = k$.

32. Use the result of Problem 31 to prove the inequality

$$\sqrt[3]{xyz} \le \frac{x + y + z}{3}$$

33. Find the point on the curve C of intersection of the cylinder $x^2 + z^2 = 1$ and the plane $x + y + 2z = 4$ that is farthest from the xz-plane. Find the point on C that is closest to the xz-plane.

Miscellaneous Problem

34. Express the first three equations of (16.40) as a gradient.

---

*Fritz Haber (1868–1934) was a German chemist. For inventing this process, Haber won the Nobel Prize in chemistry in 1918. Carl Bosch (1874–1940) was Haber's brother-in-law and a chemical engineer who made this process practical on a large scale. Bosch won the Nobel Prize in chemistry in 1931. During World War I the German government used the Haber–Bosch process to produce large quantities of fertilizers and explosives. Haber was subsequently expelled from Germany by Adolf Hitler.

Chapter 16 Review Exercises

Answers to odd-numbered problems begin on page A-50.

In Problems 1–10 answer true or false.

1. If $\lim_{(x,y)\to(a,b)} f(x, y)$ has the same value for an infinite number of approaches to (a, b), then the limit exists. _____

2. The domain of the functions

$$f(x, y) = \sqrt{\ln(x^2 + y^2 - 16)}$$
and
$$g(x, y) = \ln(x^2 + y^2 - 16)$$

are the same. _____

3. $f(x, y) = \begin{cases} \dfrac{1 - \cos(x^2 + y^2)}{x^2 + y^2}, & (x, y) \neq (0, 0) \\ 0, & (x, y) = (0, 0) \end{cases}$

is continuous at $(0, 0)$. _____

4. The function $f(x, y) = x^2 + 2xy + y^3$ is continuous everywhere. _____

5. If $\partial z/\partial x = 0$, then $z = $ constant. _____

6. If $\nabla f = 0$, then $f = $ constant. _____

7. ∇z is perpendicular to the graph of $z = f(x, y)$. _____

8. ∇f points in the direction in which f increases most rapidly. _____

9. If f has continuous second partial derivatives, then $f_{xy} = f_{yx}$. _____

10. If $f_x(x, y) = 0$ and $f_y(x, y) = 0$ at (a, b), then $f(a, b)$ is a relative extremum. _____

In Problems 11–22 fill in the blanks.

11. $\lim\limits_{(x,y)\to(1,1)} \dfrac{3x^2 + xy^2 - 3xy - 2y^3}{5x^2 - y^2} = $ _____

12. $f(x, y) = \dfrac{xy^2 + 1}{x - y + 1}$ is continuous except for points _____.

13. For $f(x, y) = 3x^2 + y^2$ the level curve that passes through $(2, -4)$ is _____.

14. If $p = g(\eta, \xi)$, $q = h(\eta, \xi)$, then $\dfrac{\partial}{\partial \xi} T(p, q)$ = _____.

15. If $r = g(w)$, $s = h(w)$, then $\dfrac{d}{dw} F(r, s) = $ _____.

16. If s is the distance that a body falls in time t, then the acceleration of gravity g can be obtained from $g = 2s/t^2$. Small

errors Δs and Δt in the measurements of s and t, respectively, will result in an approximate error in g of _____.

17. $\dfrac{\partial^4 f}{\partial x\, \partial z\, \partial y^2}$ in subscript notation is _____.

18. f_{xyy} in ∂ notation is _____.

19. If $f(x, y) = \displaystyle\int_x^y F(t)\, dt$, then $\dfrac{\partial f}{\partial x} = $ _____.

20. At (x_0, y_0, z_0) the function $F(x, y, z) = x + y + z$ increases most rapidly in the direction of _____.

21. If $F(x, y, z) = f(x, y)g(y)h(z)$, then $F_{xyz} = $ _____.

22. If $z = f(x, y)$ has continuous partial derivatives of any order, list all possible fourth-order partial derivatives. _____.

In Problems 23–30 compute the indicated derivative.

23. $z = ye^{-x^3y}$; z_y

24. $z = \ln(\cos(uv))$; z_u

25. $f(r, \theta) = \sqrt{r^3 + \theta^2}$; $f_{r\theta}$

26. $f(x, y) = (2x + xy^2)^2$; $\dfrac{\partial^2 f}{\partial x^2}$

27. $z = \cosh(x^2y^3)$; $\dfrac{\partial^2 z}{\partial y^2}$

28. $z = (e^{x^2} + e^{-y^2})^2$; $\dfrac{\partial^3 z}{\partial x^2\, \partial y}$

29. $F(s, t, v) = s^3 t^5 v^{-4}$; F_{stv}

30. $w = \dfrac{xy}{z} + \dfrac{xz}{y} + \dfrac{yz}{x}$; $\dfrac{\partial^4 w}{\partial x\, \partial y^2\, \partial z}$

In Problems 31 and 32 find the gradient of the given function at the indicated point.

31. $f(x, y) = \tan^{-1}\dfrac{y}{x}$; $(1, -1)$

32. $F(x, y, z) = \dfrac{x^2 - 3y^3}{z^4}$; $(1, 2, 1)$

In Problems 33 and 34 find the directional derivative of the given function in the indicated direction.

33. $f(x, y) = x^2y - y^2x$; $D_{\mathbf{u}}f$ in the direction of $2\mathbf{i} + 6\mathbf{j}$

34. $F(x, y, z) = \ln(x^2 + y^2 + z^2)$; $D_{\mathbf{u}}F$ in the direction of $-2\mathbf{i} + \mathbf{j} + 2\mathbf{k}$

In Problems 35 and 36 sketch the domain of the given function.

35. $f(x, y) = \sqrt{1 - (x + y)^2}$

36. $f(x, y) = \dfrac{1}{\ln(y - x)}$

In Problems 37 and 38 find Δz for the given function.

37. $z = 2xy - y^2$

38. $z = x^2 - 4y^2 + 7x - 9y + 10$

In Problems 39 and 40 find the total differential of the given function.

39. $z = \dfrac{x - 2y}{4x + 3y}$

40. $A = 2xy + 2yz + 2zx$

In Problems 41 and 42 determine whether the given equation is exact. If exact, solve.

41. $2x \cos y^3 \, dx = (3x^2 y^2 \sin y^3 + 1) \, dy$

42. $(3x^2 + 2y^3) \, dx + y^2(6x + 1) \, dy = 0$

43. Find symmetric equations of the tangent line at $(-\sqrt{5}, 1, 3)$ to the trace of $z = \sqrt{x^2 + 4y^2}$ in the plane $x = -\sqrt{5}$.

44. Find the slope of the tangent line at $(2, 3, 10)$ to the curve of intersection of the surface $z = xy + x^2$ and the vertical plane that passes through $P(2, 3)$ and $Q(4, 5)$ in the direction of Q.

45. Consider the function $f(x, y) = x^2 y^4$. At $(1, 1)$ what is:

(a) The rate of change of f in the direction of $\mathbf{i}$?

(b) The rate of change of f in the direction of $\mathbf{i} - \mathbf{j}$?

(c) The rate of change of f in the direction of $\mathbf{j}$?

46. Let $w = \sqrt{x^2 + y^2 + z^2}$.

(a) If $x = 3 \sin 2t$, $y = 4 \cos 2t$, $z = 5t^3$, find $\dfrac{dw}{dt}$.

(b) If $x = 3 \sin 2\dfrac{t}{r}$, $y = 4 \cos 2\dfrac{r}{t}$, $z = 5t^3 r^3$, find $\dfrac{\partial w}{\partial t}$.

47. Find an equation of the tangent plane to the graph of $z = \sin xy$ at $\left(\dfrac{1}{2}, \dfrac{2\pi}{3}, \dfrac{\sqrt{3}}{2}\right)$.

48. Determine whether there are any points on the surface $z^2 + xy - 2x - y^2 = 1$ at which the tangent plane is parallel to $z = 2$.

49. Find an equation of the tangent plane to the cylinder $x^2 + y^2 = 25$ at $(3, 4, 6)$.

50. At what point is the directional derivative of $f(x, y) = x^3 + 3xy + y^3 - 3x^2$ in the direction of $\mathbf{i} + \mathbf{j}$ a minimum?

51. Find the dimensions of the rectangular box of greatest volume that is bounded in the first octant by the coordinate planes and the plane $x + 2y + z = 6$. See Figure 16.71.

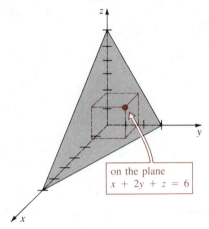

on the plane
$x + 2y + z = 6$

Figure 16.71

52. The percentage error in measuring the height of a right circular cone is $\pm 2\%$ and the radius is measured with an error of $\pm 1.5\%$. What is the approximate maximum percentage error in measuring the volume?

53. The velocity of the conical pendulum shown in Figure 16.72 is given by $v = r\sqrt{g/y}$, where $g = 980$ cm/sec^2. If r decreases from 20 cm to 19 cm and y increases from 25 cm to 26 cm, what is the approximate change in the velocity of the pendulum?

Figure 16.72

54. Find the directional derivative of $f(x, y) = x^2 + y^2$ at $(3, 4)$ in the direction of (a) $\nabla f(1, -2)$, (b) $\nabla f(3, 4)$.

55. The so-called steady-state temperatures inside a circle of radius R are given by **Poisson's formula***

$$U(r, \theta) = \dfrac{1}{2\pi} \int_{-\pi}^{\pi} \dfrac{R^2 - r^2}{R^2 - 2rR \cos(\theta - \phi) + r^2} f(\phi) \, d\phi$$

*Siméon Denis Poisson (1781–1840), a French mathematician and physicist.

By formally differentiating under the integral sign, show that U satisfies

$$r^2 U_{rr} + r U_r + U_{\theta\theta} = 0$$

56. The **Cobb–Douglas production function** $z = f(x, y)$ is defined by $z = Ax^\alpha y^\beta$, where A, α, and β are constants. The value of z is called the efficient output for inputs x and y. Show that

$$f_x = \frac{\alpha z}{x}, \qquad f_y = \frac{\beta z}{y}, \qquad f_{xx} = \frac{\alpha(\alpha-1)z}{x^2},$$

$$f_{yy} = \frac{\beta(\beta-1)z}{y^2}, \quad \text{and} \quad f_{xy} = f_{yx} = \frac{\alpha\beta z}{xy}$$

In Problems 57–60 suppose that $f_x(a, b) = 0$, $f_y(a, b) = 0$. If the given higher-order partial derivatives are evaluated at (a, b), determine, if possible, whether $f(a, b)$ is a relative extremum.

57. $f_{xx} = 4, f_{yy} = 6, f_{xy} = 5$

58. $f_{xx} = 2, f_{yy} = 7, f_{xy} = 0$

59. $f_{xx} = -5, f_{yy} = -9, f_{xy} = 6$

60. $f_{xx} = -2, f_{yy} = -8, f_{xy} = 4$

61. Express the area A of a right triangle as a function of the length L of its hypotenuse and one of its acute angles θ.

62. In Figure 16.73 express the height h of the mountain as a function of angles θ and ϕ.

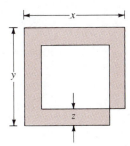

Figure 16.73

63. A rectangular walkway shown in Figure 16.74 has a uniform width z. Express the area A of the walkway in terms of x, y, and z.

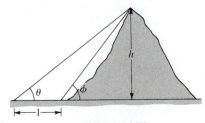

Figure 16.74

64. An open box made of plastic has the shape of a rectangular parallelepiped. The outer dimensions of the box are given in Figure 16.75. If the plastic is $\frac{1}{2}$ cm thick, find the approximate volume of the plastic.

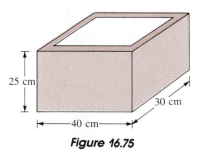

25 cm
30 cm
40 cm

Figure 16.75

65. A rectangular box, shown in Figure 16.76, is inscribed in the cone $z = 4 - \sqrt{x^2 + y^2}$, $0 \le z \le 4$. Express the volume V of the box in terms of x and y.

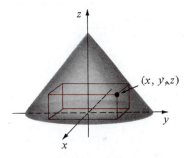

(x, y, z)

Figure 16.76

66. The rectangular box shown in Figure 16.77 has a cover and twelve compartments. The box is made out of heavy plastic that costs 1.5¢ per square inch. Find a function giving the cost C of construction of the box.

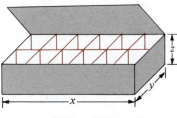

Figure 16.77

17

Multiple Integrals

We conclude our study of the calculus of multivariable functions with the definitions and applications of the two-dimensional and three-dimensional definite integrals. These integrals are more commonly called the **double integral** and the **triple integral**, respectively.

17.1 The Double Integral

Recall from Section 5.5 that the definition of the definite integral of a function of a single variable is given by a limit of a sum:

$$\int_a^b f(x)\ dx = \lim_{\|P\|\to 0} \sum_{k=1}^{n} f(x_k^*)\ \Delta x_k$$

The five steps leading to this definition are outlined in the left-hand column of the following table. The analogous steps, which lead to the concept of a *two-dimensional definite integral* known simply as a **double integral** of a function *f* of two variables, are given in the right-hand column.

$y = f(x)$	$z = f(x, y)$	
1. Let *f* be defined on a closed interval [*a*, *b*].	Let *f* be defined in a closed and bounded region *R*.	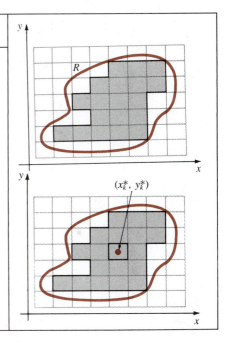
2. Form a partition *P* of the interval [*a*, *b*] into *n* subintervals of lengths Δx_k.	By means of a grid of vertical and horizontal lines parallel to the coordinate axes, form a partition *P* of *R* into *n* rectangular subregions R_k of areas ΔA_k that lie entirely in *R*.	
3. Let $\|P\|$ be the norm of the partition or the length of the longest subinterval.	Let $\|P\|$ be the norm of the partition or the length of the longest diagonal of the R_k.	
4. Choose a number x_k^* in each subinterval $[x_{k-1}, x_k]$.	Choose a point (x_k^*, y_k^*) in each subregion R_k.	
5. Form the sum	Form the sum	
$$\sum_{k=1}^{n} f(x_k^*)\ \Delta x_k$$	$$\sum_{k=1}^{n} f(x_k^*, y_k^*)\ \Delta A_k$$	

Thus, we have the following definition.

DEFINITION 17.1   The Double Integral

Let *f* be a function of two variables defined on a closed region *R*. Then the **double integral of *f* over *R*** is given by

$$\iint_R f(x, y)\ dA = \lim_{\|P\|\to 0} \sum_{k=1}^{n} f(x_k^*, y_k^*)\ \Delta A_k \qquad (17.1)$$

Integrability   If the limit in (17.1) exists, we say *f* is **integrable** over *R* and that *R* is the **region of integration**. When *f* is continuous on *R*, then *f* is necessarily integrable over *R*. For a partition *P* of *R* into subregions R_k with (x_k^*, y_k^*) in R_k, a sum of the form $\sum_{k=1}^{n} f(x_k^*, y_k^*)\ \Delta A_k$ is called a **Riemann sum**. The partition described in step 2, where the R_k lie entirely with R, is called an **inner partition** of R. The collection of shaded rectangles in the accompanying two figures illustrate an inner partition.

Example 1

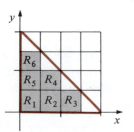

Figure 17.1

Consider the region R in the first quadrant bounded by the graphs of $x + y = 2$, $y = 0$, and $x = 0$. Approximate the double integral $\iint_R (6x + 2y + 3)\, dA$ using a Riemann sum, the R_k shown in Figure 17.1, and (x_k^*, y_k^*), the geometric center of each R_k.

Solution   From 17.1 we see that $\Delta A_k = \frac{1}{2} \cdot \frac{1}{2} = \frac{1}{4}$, $k = 1, 2, \ldots, 6$ and the (x_k^*, y_k^*), $k = 1, 2, \ldots, 6$, are in turn $(\frac{1}{4}, \frac{1}{4})$, $(\frac{3}{4}, \frac{1}{4})$, $(\frac{5}{4}, \frac{1}{4})$, $(\frac{3}{4}, \frac{3}{4})$, $(\frac{1}{4}, \frac{3}{4})$, $(\frac{1}{4}, \frac{5}{4})$. Hence the Riemann sum is

$$\sum_{k=1}^{n} f(x_k^*, y_k^*)\, \Delta A_k = f(\tfrac{1}{4}, \tfrac{1}{4})\tfrac{1}{4} + f(\tfrac{3}{4}, \tfrac{1}{4})\tfrac{1}{4} + f(\tfrac{5}{4}, \tfrac{1}{4})\tfrac{1}{4} + f(\tfrac{3}{4}, \tfrac{3}{4})\tfrac{1}{4} + f(\tfrac{1}{4}, \tfrac{3}{4})\tfrac{1}{4} + f(\tfrac{1}{4}, \tfrac{5}{4})\tfrac{1}{4}$$

$$= \tfrac{5}{4} + \tfrac{8}{4} + \tfrac{11}{4} + \tfrac{9}{4} + \tfrac{6}{4} + \tfrac{7}{4} = 11.5$$

Area   When $f(x, y) = 1$ on R, then $\lim_{\|P\| \to 0} \sum_{k=1}^{n} \Delta A_k$ will simply give the **area** A of the region; that is,

$$A = \iint_R dA \tag{17.2}$$

Volume   If $f(x, y) \geq 0$ on R, then, as shown in Figure 17.2, the product $f(x_k^*, y_k^*)\, \Delta A_k$ can be interpreted as the volume of a rectangular prism of height $f(x_k^*, y_k^*)$ and

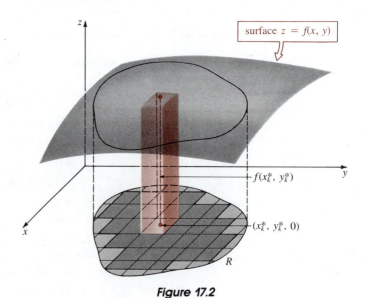

Figure 17.2

base of area ΔA_k. The summation of volumes $\sum_{k=1}^{n} f(x_k^*, y_k^*) \, \Delta A_k$ is an approximation to the **volume** V of the solid *above* the region R and *below* the surface $z = f(x, y)$. The limit of this sum as $\|P\| \to 0$, if it exists, will give the exact volume of this solid; that is, if f is nonnegative on R, then

$$V = \iint_R f(x, y) \, dA \tag{17.3}$$

Properties   The following properties of the double integral are analogous to those of the definite integral given in Theorem 5.7.

THEOREM 17.1   Let f and g be functions of two variables that are integrable over a region R. Then

(*i*) $\displaystyle\iint_R kf(x, y) \, dA = k \iint_R f(x, y) \, dA$, where k is any constant

(*ii*) $\displaystyle\iint_R [f(x, y) \pm g(x, y)] \, dA = \iint_R f(x, y) \, dA \pm \iint_R g(x, y) \, dA$

(*iii*) $\displaystyle\iint_R f(x, y) \, dA = \iint_{R_1} f(x, y) \, dA + \iint_{R_2} f(x, y) \, dA$, where R_1 and R_2 are subregions of R that do not overlap and $R = R_1 \cup R_2$.

Part (*iii*) of Theorem 17.1 is the two-dimensional equivalent of $\int_a^b f(x) \, dx = \int_a^c f(x) \, dx + \int_c^b f(x) \, dx$. Figure 17.3 illustrates the division of a region into subregions R_1 and R_2 for which $R = R_1 \cup R_2$. R_1 and R_2 can have no points in common except possibly on their common border. Furthermore, Theorem 17.1(*iii*) extends to any finite number of nonoverlapping subregions whose union is R.

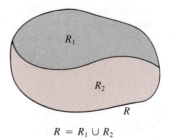

$$R = R_1 \cup R_2$$
Figure 17.3

Remark

Of course not every double integral gives volume. For the surface $z = f(x, y)$ shown in Figure 17.4, $\iint_R f(x, y) \, dA$ is a real number but it is not volume since f is not nonnegative on R.

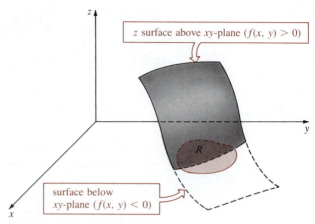

Figure 17.4

Exercises 17.1

Answers to odd-numbered problems begin on page A-51.

1. Consider the region R in the first quadrant that is bounded by the graphs of $x^2 + y^2 = 16$, $y = 0$, and $x = 0$. Approximate the double integral $\iint_R (x + 3y + 1)\, dA$ using a Riemann sum and the R_k shown in Figure 17.5. Choose (x_k^*, y_k^*) at the geometric center of each R_k.

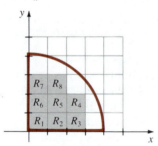

Figure 17.5

2. Consider the region R in the first quadrant bounded by the graphs of $x + y = 1$, $x + y = 3$, $y = 0$, and $x = 0$. Approximate the double integral $\iint_R (2x + 4y)\, dA$ using a Riemann sum and the R_k shown in Figure 17.6. Choose the (x_k^*, y_k^*) at the upper right-hand corner of each R_k.

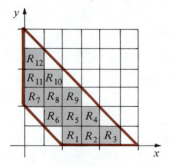

Figure 17.6

3. Consider the rectangular region R shown in Figure 17.7. Approximate the double integral $\iint_R (x + y)\, dA$ using a Riemann sum and the R_k shown in the figure. Choose the (x_k^*, y_k^*) at **(a)** the geometric center of each R_k, **(b)** at the upper left-hand corner of each R_k.

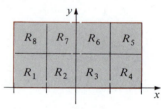

Figure 17.7

4. Consider the region R bounded by the graphs of $y = x^2$ and $y = 4$. Place a rectangular grid over R corresponding to the lines $x = -2$, $x = -\frac{3}{2}$, $x = -1$, . . . , $x = 2$, and $y = 0$, $y = \frac{1}{2}$, $y = 1$, . . . , $y = 4$. Approximate the double integral $\iint_R xy\, dA$ using a Riemann sum, where the (x_k^*, y_k^*) are chosen at the lower right-hand corner of each complete rectangular R_k in R.

In Problems 5–8 (Figures 17.8–17.11) evaluate $\iint_R 10\, dA$ over the given region R.

5.

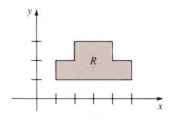

Figure 17.8

6.

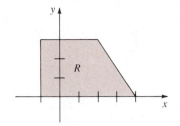

Figure 17.9

7.

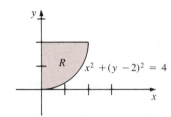

Figure 17.10

8.

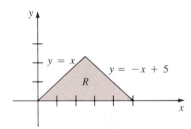

Figure 17.11

9. Consider the region R bounded by the graph of $(x - 3)^2 + y^2 = 9$. Does the double integral $\iint_R (x + 5y) \, dA$ represent a volume? Explain.

10. Consider the region R in the second quadrant that is bounded by the graphs of $-2x + y = 6$, $x = 0$, and $y = 0$. Does the double integral $\iint_R (x^2 + y^2) \, dA$ represent a volume? Explain.

In Problems 11–16 suppose that

$$\iint_R x \, dA = 3, \qquad \iint_R y \, dA = 7$$

and the area of R is 8. Evaluate the given double integral.

11. $\displaystyle\iint_R 10 \, dA$ **12.** $\displaystyle\iint_R -5x \, dA$

13. $\displaystyle\iint_R (2x + 4y) \, dA$ **14.** $\displaystyle\iint_R (x - y) \, dA$

15. $\displaystyle\iint_R (3x + 7y + 1) \, dA$

16. $\displaystyle\iint_R y^2 \, dA - \iint_R (2 + y)^2 \, dA$

In Problems 17 and 18 let R_1 and R_2 be nonoverlapping regions for which $R = R_1 \cup R_2$.

17. If $\iint_{R_1} f(x, y) \, dA = 4$ and $\iint_{R_2} f(x, y) \, dA = 14$, what is the value of $\iint_R f(x, y) \, dA$?

18. Suppose $\iint_R f(x, y) \, dA = 25$ and $\iint_{R_1} f(x, y) \, dA = 30$. What is the value of $\iint_{R_2} f(x, y) \, dA$?

17.2 Iterated Integrals

Partial Integration

Analogous to the process of partial differentiation we can define **partial integration**.

If $F(x, y)$ is a function such that $F_y(x, y) = f(x, y)$, then the **partial integral of f with respect to y** is

$$\int_{g_1(x)}^{g_2(x)} f(x, y) \, dy = F(x, y) \Big]_{g_1(x)}^{g_2(x)} = F(x, g_2(x)) - F(x, g_1(x))$$

Similarly, if $G(x, y)$ is a function such that $G_x(x, y) = f(x, y)$, then the **partial integral of f with respect to x** is

$$\int_{h_1(y)}^{h_2(y)} f(x, y) \, dx = G(x, y) \Big]_{h_1(y)}^{h_2(y)} = G(h_2(y), y) - G(h_1(y), y)$$

In other words to evaluate $\int_{g_1(x)}^{g_2(x)} f(x, y) \, dy$, we hold x fixed, whereas in $\int_{h_1(y)}^{h_2(y)} f(x, y) \, dx$ we hold y fixed.

Example 1

Evaluate (*a*) $\int_{1}^{2}\left(6xy^2 - 4\dfrac{x}{y}\right)dy,$ (*b*) $\int_{-1}^{3}\left(6xy^2 - 4\dfrac{x}{y}\right)dx.$

Solution

(*a*) $\int_{1}^{2}\left(6xy^2 - 4\dfrac{x}{y}\right)dy = \left[2xy^3 - 4x\ln|y|\right]_{1}^{2}$
  (with annotations: *x* fixed)

$= (16x - 4x\ln 2) - (2x - 4x\ln 1)$

$= 14x - 4x\ln 2$

(*b*) $\int_{-1}^{3}\left(6xy^2 - 4\dfrac{x}{y}\right)dx = \left[3x^2y^2 - 2\dfrac{x^2}{y}\right]_{-1}^{3}$
  (with annotations: *y* fixed)

$= \left(27y^2 - \dfrac{18}{y}\right) - \left(3y^2 - \dfrac{2}{y}\right)$

$= 24y^2 - \dfrac{16}{y}$

In Example 1 you should note that

$$\frac{\partial}{\partial y}[2xy^3 - 4x\ln|y|] = 6xy^2 - 4\frac{x}{y}$$

and

$$\frac{\partial}{\partial x}\left[3x^2y^2 - 2\frac{x^2}{y}\right] = 6xy^2 - 4\frac{x}{y}$$

Example 2

Evaluate $\int_{x^2}^{x}\sin xy\,dy.$

Solution By treating x as a constant, we obtain

$$\int_{x^2}^{x}\sin xy\,dy = -\frac{\cos xy}{x}\Big]_{x^2}^{x} = -\frac{\cos x^2}{x} + \frac{\cos x^3}{x}$$

Regions of Type I and II The region shown in Figure 17.12(a),

$$R: a \le x \le b, \qquad g_1(x) \le y \le g_2(x)$$

where the boundary functions g_1 and g_2 are continuous, is called a **region of Type I**. In Figure 17.12(b), the region

$$R: c \le y \le d, \qquad h_1(y) \le x \le h_2(y)$$

where h_1 and h_2 are continuous, is called a **region of Type II**.

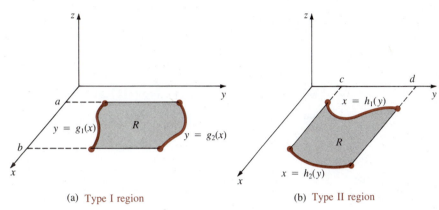

(a) Type I region   (b) Type II region

Figure 17.12

Iterated Integrals   Since the partial integral $\int_{g_1(x)}^{g_2(x)} f(x,\ y)\ dy$ is a function of x alone, we may in turn integrate the resulting function now with respect to x. If f is continuous on a region of Type I, we define an **iterated integral** of f over the region by

$$\int_a^b \int_{g_1(x)}^{g_2(x)} f(x,\ y)\ dy\ dx = \int_a^b \left[\int_{g_1(x)}^{g_2(x)} f(x,\ y)\ dy \right] dx \qquad (17.4)$$

The basic idea in (17.4) is to carry out *successive integrations*. The partial integral gives a function of x, which is then integrated in the usual manner from $x = a$ to $x = b$. The end result of both integrations will be a real number. In a similar manner, we define an iterated integral of a continuous function f over a region of Type II by

$$\int_c^d \int_{h_1(y)}^{h_2(y)} f(x,\ y)\ dx\ dy = \int_c^d \left[\int_{h_1(y)}^{h_2(y)} f(x,\ y)\ dx \right] dy \qquad (17.5)$$

Example 3

Evaluate the iterated integral of $f(x,\ y) = 2xy$ over the region shown in Figure 17.13.

Solution   The region is of Type I and so by (17.4) we have

$$\int_{-1}^2 \int_x^{x^2+1} 2xy\ dy\ dx = \int_{-1}^2 \left[\int_x^{x^2+1} 2xy\ dy \right] dx$$

$$= \int_{-1}^2 \left[xy^2 \right]_x^{x^2+1} dx$$

$$= \int_{-1}^2 [x(x^2 + 1)^2 - x^3]\ dx$$

$$= \left[\frac{1}{6}(x^2 + 1)^3 - \frac{x^4}{4} \right]_{-1}^2 = \frac{63}{4}$$

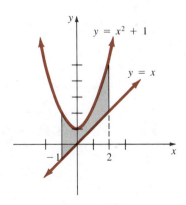

Figure 17.13

Example 4

Evaluate $\int_0^4 \int_y^{2y} (8x + e^y)\, dx\, dy$.

Solution From (17.5) we see that

$$\int_0^4 \int_y^{2y} (8x + e^y)\, dx\, dy = \int_0^4 \left[\int_y^{2y} (8x + e^y)\, dx \right] dy$$

$$= \int_0^4 \left[4x^2 + xe^y \right]_y^{2y} dy$$

$$= \int_0^4 [(16y^2 + 2ye^y) - (4y^2 + ye^y)]\, dy$$

$$= \int_0^4 (12y^2 + ye^y)\, dy$$

$$= \left[4y^3 + ye^y - e^y \right]_0^4 \quad \boxed{\text{Integration by parts}}$$

$$= 257 + 3e^4 \approx 420.79$$

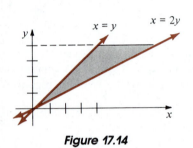

Figure 17.14

Comparing the iterated integral to (17.5), we see that the region of integration is a Type II region. See Figure 17.14.

Example 5

Evaluate $\int_{-1}^3 \int_1^2 \left(6xy^2 - 4\dfrac{x}{y} \right) dy\, dx$.

Solution From the result in (a) of Example 1, we have

$$\int_{-1}^3 \int_1^2 \left(6xy^2 - 4\frac{x}{y} \right) dy\, dx = \int_{-1}^3 \left[\int_1^2 \left(6xy^2 - 4\frac{x}{y} \right) dy \right] dx$$

$$= \int_{-1}^3 (14x - 4x \ln 2)\, dx$$

$$= \left[7x^2 - 2x^2 \ln 2 \right]_{-1}^3$$

$$= 56 - 16 \ln 2 \approx 44.91$$

Inspection of Figure 17.15 should convince you that a rectangular region R defined by $a \le x \le b$, $c \le y \le d$ is simultaneously a Type I and a Type II region. If f is continuous on R, it can be proved that

$$\int_a^b \int_c^d f(x, y)\, dy\, dx = \int_c^d \int_a^b f(x, y)\, dx\, dy$$

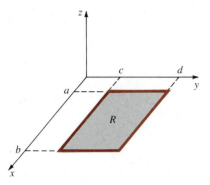

Figure 17.15

You should verify that

$$\int_1^2 \int_{-1}^3 \left(6xy^2 - 4\frac{x}{y}\right) dx\, dy$$

gives the same result as the given iterated integral in Example 5.

Remark

A rectangular region is not the only region that can be both Type I and Type II. If f is continuous on a region R that is simultaneously a Type I and a Type II region, the two iterated integrals of f over R are equal. See Problems 39 and 40.

Exercises 17.2

Answers to odd-numbered problems begin on page A-51.

In Problems 1–8 evaluate the given partial integral.

1. $\displaystyle\int_{-1}^3 (6xy - 5e^y)\, dx$

2. $\displaystyle\int_1^2 \tan xy\, dy$

3. $\displaystyle\int_1^{3x} x^3 e^{xy}\, dy$

4. $\displaystyle\int_{\sqrt{y}}^{y^3} (8x^3 y - 4xy^2)\, dx$

5. $\displaystyle\int_0^{2x} \frac{xy}{x^2 + y^2}\, dy$

6. $\displaystyle\int_{x^3}^x e^{2y/x}\, dy$

7. $\displaystyle\int_{\tan y}^{\sec y} (2x + \cos y)\, dx$

8. $\displaystyle\int_{\sqrt{y}}^1 y \ln x\, dx$

In Problems 9–30 evaluate the given iterated integral.

9. $\displaystyle\int_1^2 \int_{-x}^{x^2} (8x - 10y + 2)\, dy\, dx$

10. $\displaystyle\int_{-1}^1 \int_0^y (x + y)^2\, dx\, dy$

11. $\displaystyle\int_0^{\sqrt{2}} \int_{-\sqrt{2-y^2}}^{\sqrt{2-y^2}} (2x - y)\, dx\, dy$

12. $\displaystyle\int_0^{\pi/4} \int_0^{\cos x} (1 + 4y \tan^2 x)\, dy\, dx$

13. $\displaystyle\int_0^\pi \int_y^{3y} \cos(2x + y)\, dx\, dy$

14. $\displaystyle\int_1^2 \int_0^{\sqrt{x}} 2y \sin \pi x^2\, dy\, dx$

15. $\displaystyle\int_1^{\ln 3} \int_0^x 6e^{x+2y}\, dy\, dx$

16. $\displaystyle\int_0^1 \int_0^{2y} e^{-y^2}\, dx\, dy$

17. $\displaystyle\int_0^3 \int_{x+1}^{2x+1} \frac{1}{\sqrt{y-x}}\, dy\, dx$

18. $\displaystyle\int_0^1 \int_0^y x(y^2 - x^2)^{3/2}\, dx\, dy$

19. $\displaystyle\int_1^9 \int_0^x \frac{1}{x^2 + y^2}\, dy\, dx$

20. $\displaystyle\int_0^{1/2} \int_0^y \frac{1}{\sqrt{1 - x^2}}\, dx\, dy$

21. $\displaystyle\int_1^e \int_1^y \frac{y}{x}\, dx\, dy$

22. $\displaystyle\int_1^4 \int_1^{\sqrt{x}} 2y e^{-x}\, dy\, dx$

23. $\displaystyle\int_0^6 \int_0^{\sqrt{25-y^2}/2} \frac{1}{\sqrt{(25 - y^2) - x^2}}\, dx\, dy$

24. $\displaystyle\int_0^2 \int_{y^2}^{\sqrt{20-y^2}} y\, dx\, dy$

25. $\displaystyle\int_{\pi/2}^\pi \int_{\cos y}^0 e^x \sin y\, dx\, dy$

26. $\displaystyle\int_0^1 \int_0^{y^{1/3}} 6x^2 \ln(y + 1)\, dx\, dy$

27. $\displaystyle\int_\pi^{2\pi} \int_0^x (\cos x - \sin y)\, dy\, dx$

28. $\displaystyle\int_1^3 \int_0^{1/x} \frac{1}{x+1}\, dy\, dx$

29. $\displaystyle\int_{\pi/12}^{5\pi/12} \int_1^{\sqrt{2\,\sin\,2\theta}} r\, dr\, d\theta$

30. $\displaystyle\int_0^{\pi/3} \int_{3\,\cos\,\theta}^{1+\cos\,\theta} r\, dr\, d\theta$

In Problems 31–34 sketch the region of integration for the given iterated integral.

31. $\displaystyle\int_0^2 \int_1^{2x+1} f(x,\, y)\, dy\, dx$

32. $\displaystyle\int_1^4 \int_{-\sqrt{y}}^{\sqrt{y}} f(x,\, y)\, dx\, dy$

33. $\displaystyle\int_{-1}^3 \int_0^{\sqrt{16-y^2}} f(x,\, y)\, dx\, dy$

34. $\displaystyle\int_{-1}^2 \int_{-x^2}^{x^2+1} f(x,\, y)\, dy\, dx$

In Problems 35–38 verify the given equality.

35. $\displaystyle\int_{-1}^2 \int_0^3 x^2\, dy\, dx = \int_0^3 \int_{-1}^2 x^2\, dx\, dy$

36. $\displaystyle\int_{-2}^2 \int_2^4 (2x+4y)\, dx\, dy = \int_2^4 \int_{-2}^2 (2x+4y)\, dy\, dx$

37. $\displaystyle\int_1^3 \int_0^\pi (3x^2 y - 4\,\sin\, y)\, dy\, dx$

$\displaystyle = \int_0^\pi \int_1^3 (3x^2 y - 4\,\sin\, y)\, dx\, dy$

38. $\displaystyle\int_0^1 \int_0^2 \left(\frac{8y}{x+1} - \frac{2x}{y^2+1} \right) dx\, dy$

$\displaystyle = \int_0^2 \int_0^1 \left(\frac{8y}{x+1} - \frac{2x}{y^2+1} \right) dy\, dx$

In Problems 39 and 40 the region given in the figure (Figures 17.16 and 17.17) is both Type I and Type II. Verify that the iterated integrals are equal.

39.

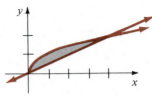

Figure 17.16

Type I: $\frac{1}{2}x \leq y \leq \sqrt{x},\ 0 \leq x \leq 4$

Type II: $y^2 \leq x \leq 2y,\ 0 \leq y \leq 2$

$\displaystyle\int_0^4 \int_{\frac{1}{2}x}^{\sqrt{x}} x^2 y\, dy\, dx = \int_0^2 \int_{y^2}^{2y} x^2 y\, dx\, dy$

40.

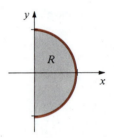

Figure 17.17

Type I: $-\sqrt{1-x^2} \leq y \leq \sqrt{1-x^2},\ 0 \leq x \leq 1$

Type II: $0 \leq x \leq \sqrt{1-y^2},\ -1 \leq y \leq 1$

$\displaystyle\int_0^1 \int_{-\sqrt{1-x^2}}^{\sqrt{1-x^2}} 2x\, dy\, dx = \int_{-1}^1 \int_0^{\sqrt{1-y^2}} 2x\, dx\, dy$

Miscellaneous Problem

41. If f and g are integrable, prove that

$\displaystyle\int_c^d \int_a^b f(x)g(y)\, dx\, dy = \left(\int_a^b f(x)\, dx \right)\left(\int_c^d g(y)\, dy \right)$

17.3 Evaluation of Double Integrals

The iterated integrals of the preceding section provide the means for evaluating a double integral $\iint_R f(x,\, y)\, dA$ over a region of Type I or Type II or a region that can be expressed as a union of a finite number of these regions.

THEOREM 17.2   Let f be continuous on a region R.

(i)   If R is of Type I, then

$$\iint\limits_{R} f(x,\ y)\ dA = \int_{a}^{b} \int_{g_1(x)}^{g_2(x)} f(x,\ y)\ dy\ dx \qquad (17.6)$$

(ii)   If R is of Type II, then

$$\iint\limits_{R} f(x,\ y)\ dA = \int_{c}^{d} \int_{h_1(y)}^{h_2(y)} f(x,\ y)\ dx\ dy \qquad (17.7)$$

Theorem 17.2 is the double integral analogue of Theorem 5.11, the Fundamental Theorem of Calculus. While Theorem 17.2 is difficult to prove, we can get some intuitive feeling for its significance by considering volumes. Let R be a Type I region and $z = f(x,\ y)$ be continuous and nonnegative on R. The area A of the vertical plane, as shown in Figure 17.18, is the area under the trace of the surface $z = f(x,\ y)$ in the plane $x = $ constant and hence is given by the partial integral

$$A(x) = \int_{g_1(x)}^{g_2(x)} f(x,\ y)\ dy$$

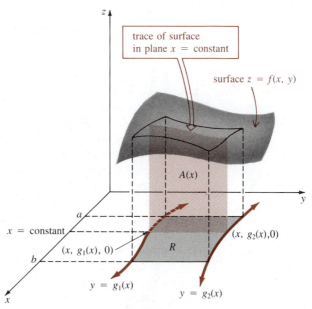

Figure 17.18

By summing all these areas from $x = a$ to $x = b$ we obtain the volume V of the solid above R and below the surface:

$$V = \int_{a}^{b} A(x)\ dx$$

$$= \int_{a}^{b} \int_{g_1(x)}^{g_2(x)} f(x,\ y)\ dy\ dx$$

But, as we have already seen in (17.3) of Section 17.1, this volume is also given by the double integral

$$V = \iint_R f(x, \ y) \ dA$$

Example 1

Evaluate the double integral $\iint_R e^{x+3y} \ dA$ over the region bounded by the graphs of $y = 1$, $y = 2$, $y = x$, and $y = -x + 5$.

Solution As seen in Figure 17.19 the region is of Type II; hence, by (17.7) we integrate first with respect to x from the left boundary $x = y$ to the right boundary $x = 5 - y$:

$$\iint_R e^{x+3y} \ dA = \int_1^2 \int_y^{5-y} e^{x+3y} \ dx \ dy$$

$$= \int_1^2 e^{x+3y} \Big]_y^{5-y} \ dy$$

$$= \int_1^2 (e^{5+2y} - e^{4y}) \ dy$$

$$= \left[\frac{1}{2} e^{5+2y} - \frac{1}{4} e^{4y} \right]_1^2$$

$$= \frac{1}{2} e^9 - \frac{1}{4} e^8 - \frac{1}{2} e^7 + \frac{1}{4} e^4 \approx 2771.64$$

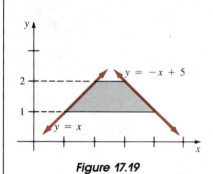

Figure 17.19

As an aid in reducing a double integral to an iterated integral with correct limits of integration, it is useful to visualize, as just suggested in the foregoing discussion, the double integral as a double summation process.* Over a Type I region the iterated integral $\int_a^b \int_{g_1(x)}^{g_2(x)} f(x, \ y) \ dy \ dx$ is first a summation in the y-direction. Pictorially, this is indicated by the vertical arrow in Figure 17.20(a); the typical rectangle in the arrow has area $dy \ dx$. The dy placed before the dx signifies that the "volumes" $f(x, \ y) \ dy \ dx$ of prisms built up on the rectangles are summed vertically with respect to y from the lower boundary curve g_1 to the upper boundary curve g_2. The dx following the dy signifies that the result of each vertical summation is then summed horizontally with respect to x from left ($x = a$) to right ($x = b$). Similar remarks hold for double integrals over regions of Type II. See Figure 17.20(b). Recall from (17.2) of Section 17.1 that when $f(x, \ y) = 1$, the double integral $A = \iint_R \ dA$ gives the area of the region. Thus, Figure 17.20(a) shows that $\int_a^b \int_{g_1(x)}^{g_2(x)} \ dy \ dx$ adds the rectangular *areas* vertically and then horizontally, whereas Figure 17.20(b) shows that $\int_c^d \int_{h_1(y)}^{h_2(y)} \ dx \ dy$ adds the rectangular areas horizontally and then vertically.

*Although we shall not pursue the details, the double integral can be defined in terms of a limit of a double sum such as

$$\sum_i \sum_j f(x_i^*, \ y_j^*) \ \Delta y_j \ \Delta x_i \quad \text{or} \quad \sum_j \sum_i f(x_i^*, \ y_j^*) \ \Delta x_i \ \Delta y_i$$

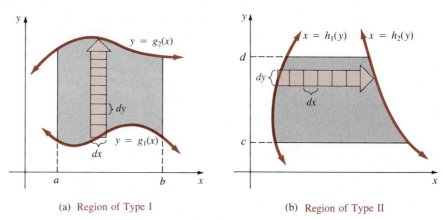

(a) Region of Type I   (b) Region of Type II

Figure 17.20

Example 2

Use a double integral to find the area of the region bounded by the graphs of $y = x^2$ and $y = 8 - x^2$.

Solution   The graphs and their points of intersection are given in Figure 17.21. Since R is evidently of Type I, we have from (17.6)

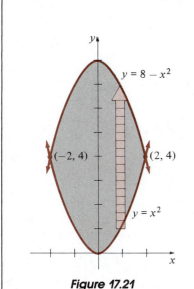

Figure 17.21

$$A = \iint_R dA$$

$$= \int_{-2}^{2} \int_{x^2}^{8-x^2} dy \, dx$$

$$= \int_{-2}^{2} [(8 - x^2) - x^2] \, dx$$

$$= \int_{-2}^{2} (8 - 2x^2) \, dx$$

$$= \left[8x - \frac{2}{3}x^3 \right]_{-2}^{2} = \frac{64}{3} \text{ square units}$$

You should recognize

$$A = \iint_R dA = \int_a^b \int_{g_1(x)}^{g_2(x)} dy \, dx = \int_a^b [g_2(x) - g_1(x)] \, dx$$

as the formula, discussed in Section 6.1, for the area bounded by two graphs.

Example 3

Use a double integral to find the volume V of the solid in the first octant that is bounded by the coordinate planes and the graphs of $x^2 + y^2 = 1$ and $z = 3 - x - y$.

Solution From Figure 17.22(a) we see that the volume is given by $V = \iint_R (3 - x - y)\, dA$. Since Figure 17.22(b) shows that R is of Type I, we have from (17.6)

$$V = \int_0^1 \int_0^{\sqrt{1-x^2}} (3 - x - y)\, dy\, dx$$

$$= \int_0^1 \left[3y - xy - \frac{y^2}{2} \right]_0^{\sqrt{1-x^2}} dx$$

$$= \int_0^1 \left(3\sqrt{1 - x^2} - x\sqrt{1 - x^2} - \frac{1}{2} + \frac{1}{2}x^2 \right) dx$$

$$\boxed{\text{Trig substitution}}$$

$$= \left[3 \sin^{-1} x + \frac{3}{2} x\sqrt{1 - x^2} + \frac{1}{3}(1 - x^2)^{3/2} - \frac{1}{2}x + \frac{1}{6}x^3 \right]_0^1$$

$$= \frac{3\pi}{2} - \frac{3}{2} \approx 4.05 \text{ cubic units}$$

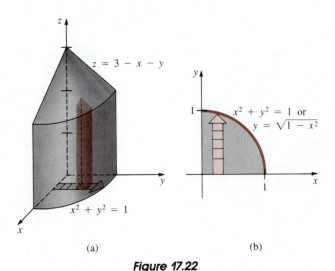

(a)   (b)

Figure 17.22

The reduction of a double integral to either of the iterated integrals (17.6) or (17.7) depends on (a) the type of region, and (b) the function itself. The next two examples illustrate each case.

Example 4

Evaluate $\iint_R (x + y)\, dA$ on the region bounded by the graphs of $x = y^2$ and $y = \frac{1}{2}x - \frac{3}{2}$.

Solution The region, which is shown in Figure 17.23(a), can be written as the union $R = R_1 \cup R_2$ of two Type I regions. By solving $y^2 = 2y + 3$ we find that the points of intersection of the two graphs are $(1, -1)$ and $(9, 3)$.

Thus, from (17.6) and Theorem 17.1(*iii*), we have

$$\iint_R (x + y)\, dA = \iint_{R_1} (x + y)\, dA + \iint_{R_2} (x + y)\, dA$$

$$= \int_0^1 \int_{-\sqrt{x}}^{\sqrt{x}} (x + y)\, dy\, dx + \int_1^9 \int_{x/2-3/2}^{\sqrt{x}} (x + y)\, dy\, dx$$

$$= \int_0^1 \left[xy + \frac{y^2}{2} \right]_{-\sqrt{x}}^{\sqrt{x}} dx + \int_1^9 \left[xy + \frac{y^2}{2} \right]_{x/2-3/2}^{\sqrt{x}} dx$$

$$= \int_0^1 2x^{3/2}\, dx + \int_1^9 \left(x^{3/2} + \frac{11}{4} x - \frac{5}{8} x^2 - \frac{9}{8} \right) dx$$

$$= \frac{4}{5} x^{5/2} \Big]_0^1 + \left[\frac{2}{5} x^{5/2} + \frac{11}{8} x^2 - \frac{5}{24} x^3 - \frac{9}{8} x \right]_1^9 \approx 46.93$$

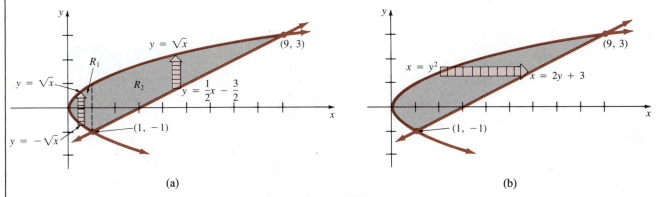

(a)   (b)

Figure 17.23

Alternative Solution   By interpreting the region as a single region of Type II, we see from Figure 17.23(b) that

$$\iint_R (x + y)\, dA = \int_{-1}^3 \int_{y^2}^{2y+3} (x + y)\, dx\, dy$$

$$= \int_{-1}^3 \left[\frac{x^2}{2} + xy \right]_{y^2}^{2y+3} dy$$

$$= \int_{-1}^3 \left(-\frac{y^4}{2} - y^3 + 4y^2 + 9y + \frac{9}{2} \right) dy$$

$$= \left[-\frac{y^5}{10} - \frac{y^4}{4} + \frac{4}{3} y^3 + \frac{9}{2} y^2 + \frac{9}{2} y \right]_{-1}^3 \approx 46.93$$

Note that the answer in Example 4 does not represent the volume of the solid above R and below the plane $z = x + y$. Why not?

Reversing the Order of   As the last example illustrates, a problem may become easier when the order of
Integration   integration is **changed** or **reversed**. Also, some iterated integrals that may be
impossible to evaluate using one order of integration can, perhaps, be evaluated
using the reverse order of integration.

Example 5

Evaluate $\iint_R xe^{y^2}\, dA$ over the region R in the first quadrant bounded by the
graphs of $y = x^2$, $x = 0$, $y = 4$.

Solution   When viewed as a region of Type I, we have from Figure 17.24(a),
$0 \le x \le 2$, $x^2 \le y \le 4$, and so

$$\iint_R xe^{y^2}\, dA = \int_0^2 \int_{x^2}^4 xe^{y^2}\, dy\, dx$$

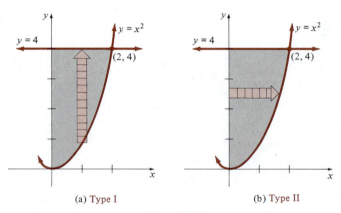

(a) Type I   (b) Type II

Figure 17.24

The difficulty here is that the partial integral $\int_{x^2}^4 xe^{y^2}\, dy$ cannot be evaluated
since e^{y^2} has no elementary antiderivative with respect to y. However, as we see
in Figure 17.24(b), we can interpret the same region as a Type II region defined
by $0 \le y \le 4$, $0 \le x \le \sqrt{y}$. Hence, from (17.7)

$$\iint_R xe^{y^2}\, dA = \int_0^4 \int_0^{\sqrt{y}} xe^{y^2}\, dx\, dy$$

$$= \int_0^4 \frac{x^2}{2} e^{y^2}\bigg]_0^{\sqrt{y}} dy$$

$$= \int_0^4 \frac{1}{2} ye^{y^2}\, dy$$

$$= \frac{1}{4} e^{y^2}\bigg]_0^4 = \frac{1}{4}(e^{16} - 1)$$

Remarks

(*i*)   You are encouraged to take advantage of symmetries to minimize your work when finding areas and volumes by double integration. In the case of volumes, make sure *both* the region and the surface over the region possess corresponding symmetries. See Problem 19 in Exercises 17.3

(*ii*)   Before attempting to evaluate a double integral, *always* try to sketch an accurate picture of the region R of integration.

Exercises 17.3

Answers to odd-numbered problems begin on page A-51.

In Problems 1–10 evaluate the double integral over the region R that is bounded by the graphs of the given equations. Choose the most convenient order of integration.

1. $\iint\limits_{R} x^3 y^2 \, dA$, $y = x$, $y = 0$, $x = 1$

2. $\iint\limits_{R} (x + 1) \, dA$, $y = x$, $x + y = 4$, $x = 0$

3. $\iint\limits_{R} (2x + 4y + 1) \, dA$, $y = x^2$, $y = x^3$

4. $\iint\limits_{R} xe^y \, dA$, R the same as in Problem 1

5. $\iint\limits_{R} 2xy \, dA$, $y = x^3$, $y = 8$, $x = 0$

6. $\iint\limits_{R} \dfrac{x}{\sqrt{y}} \, dA$, $y = x^2 + 1$, $y = 3 - x^2$

7. $\iint\limits_{R} \dfrac{y}{1 + xy} \, dA$, $y = 0$, $y = 1$, $x = 0$, $x = 1$

8. $\iint\limits_{R} \sin \dfrac{\pi x}{y} \, dA$, $x = y^2$, $x = 0$, $y = 1$, $y = 2$

9. $\iint\limits_{R} \sqrt{x^2 + 1} \, dA$, $x = y$, $x = -y$, $x = \sqrt{3}$

10. $\iint\limits_{R} x \, dA$, $y = \tan^{-1}x$, $y = 0$, $x = 1$

In Problems 11 and 12 (Figures 17.25 and 17.26) evaluate $\iint_R (x + y) \, dA$, where R is the given region.

11.

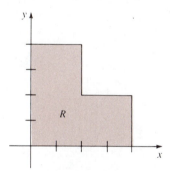

Figure 17.25

12.

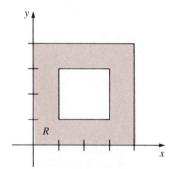

Figure 17.26

In Problems 13–18 use a double integral to find the area of the region R that is bounded by the graphs of the given equations.

13. $y = -x$, $y = 2x - x^2$

14. $x = y^2$, $x = 2 - y^2$

15. $y = e^x$, $y = \ln x$, $x = 1$, $x = 4$

16. $\sqrt{x} + \sqrt{y} = 2$, $x + y = 4$

17. $y = -2x + 3$, $y = x^3$, $x = -2$

18. $y = -x^2 + 3x$, $y = -2x + 4$, $y = 0$, $0 \le x \le 2$

19. Consider the solid bounded by the graphs of $x^2 + y^2 = 4$, $z = 4 - y$ and $z = 0$ shown in Figure 17.27. Choose and evaluate the correct integral representing the volume V of the solid.

(a) $4 \int_0^2 \int_0^{\sqrt{4-x^2}} (4 - y)\, dy\, dx$

(b) $2 \int_{-2}^2 \int_0^{\sqrt{4-x^2}} (4 - y)\, dy\, dx$

(c) $2 \int_{-2}^2 \int_0^{\sqrt{4-y^2}} (4 - y)\, dx\, dy$

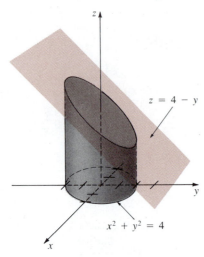

Figure 17.27

20. Consider the solid bounded by the graphs of $x^2 + y^2 = 4$ and $y^2 + z^2 = 4$. An eighth of the solid is shown in Figure 17.28. Choose and evaluate the correct integral representing the volume V of the solid.

(a) $4 \int_{-2}^2 \int_{-\sqrt{4-x^2}}^{\sqrt{4-x^2}} (4 - y^2)^{1/2}\, dy\, dx$

(b) $8 \int_0^2 \int_0^{\sqrt{4-y^2}} (4 - y^2)^{1/2}\, dx\, dy$

(c) $8 \int_0^2 \int_0^{\sqrt{4-x^2}} (4 - x^2)^{1/2}\, dy\, dx$

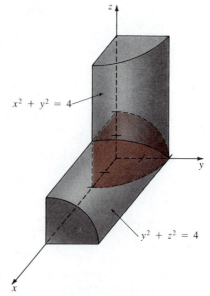

Figure 17.28

In Problems 21–30 find the volume of the solid bounded by the graphs of the given equations.

21. $2x + y + z = 6$, $x = 0$, $y = 0$, $z = 0$, first octant

22. $z = 4 - y^2$, $x = 3$, $x = 0$, $y = 0$, $z = 0$, first octant

23. $x^2 + y^2 = 4$, $x - y + 2z = 4$, $x = 0$, $y = 0$, $z = 0$, first octant

24. $y = x^2$, $y + z = 3$, $z = 0$

25. $z = 1 + x^2 + y^2$, $3x + y = 3$, $x = 0$, $y = 0$, $z = 0$, first octant

26. $z = x + y$, $x^2 + y^2 = 9$, $x = 0$, $y = 0$, $z = 0$, first octant

27. $yz = 6$, $x = 0$, $x = 5$, $y = 1$, $y = 6$, $z = 0$

28. $z = 4 - x^2 - \frac{1}{4}y^2$, $z = 0$

29. $z = 4 - y^2$, $x^2 + y^2 = 2x$, $z = 0$

30. $z = 1 - x^2$, $z = 1 - y^2$, $x = 0$, $y = 0$, $z = 0$, first octant

If $f_2(x, y) \ge f_1(x, y)$ for all (x, y) in a region R, then the volume of the solid bounded by the two surfaces over R is $V = \iint_R [f_2(x, y) - f_1(x, y)]\, dA$. In Problems 31–34 use this result to find the volume bounded by the graphs of the given equations.

31. $x + 2y + z = 4$, $z = x + y$, $x = 0$, $y = 0$, first octant

32. $z = x^2 + y^2$, $z = 9$

33. $z = x^2$, $z = -x + 2$, $x = 0$, $y = 0$, $y = 5$, first octant

34. $2z = 4 - x^2 - y^2$, $z = 2 - y$

In Problems 35–40 reverse the order of integration.

35. $\displaystyle\int_0^2 \int_0^{y^2} f(x, y)\, dx\, dy$

36. $\displaystyle\int_{-5}^5 \int_0^{\sqrt{25-y^2}} f(x, y)\, dx\, dy$

37. $\displaystyle\int_0^3 \int_1^{e^x} f(x, y)\, dy\, dx$

38. $\displaystyle\int_0^2 \int_{y/2}^{3-y} f(x, y)\, dx\, dy$

39. $\displaystyle\int_0^1 \int_0^{\sqrt[3]{x}} f(x, y)\, dy\, dx + \int_1^2 \int_0^{2-x} f(x, y)\, dy\, dx$

40. $\displaystyle\int_0^1 \int_0^{\sqrt{y}} f(x, y)\, dx\, dy + \int_1^2 \int_0^{\sqrt{2-y}} f(x, y)\, dx\, dy$

In Problems 41–46 evaluate the given iterated integral by reversing the order of integration.

41. $\displaystyle\int_0^1 \int_x^1 x^2\sqrt{1 + y^4}\, dy\, dx$

42. $\displaystyle\int_0^1 \int_{2y}^2 e^{-y/x}\, dx\, dy$

43. $\displaystyle\int_0^2 \int_{y^2}^4 \cos \sqrt{x^3}\, dx\, dy$

44. $\displaystyle\int_{-1}^1 \int_{-\sqrt{1-x^2}}^{\sqrt{1-x^2}} x\sqrt{1 - x^2 - y^2}\, dy\, dx$

45. $\displaystyle\int_0^1 \int_x^1 \frac{1}{1 + y^4}\, dy\, dx$

46. $\displaystyle\int_0^4 \int_{\sqrt{y}}^2 \sqrt{x^3 + 1}\, dx\, dy$

17.4 Center of Mass and Moments

Laminas with Variable Density—Center of Mass In Section 6.12 we saw that if ρ is a constant density (mass per unit area), then the mass of the lamina coinciding with a region bounded by the graphs of $y = f(x)$, the x-axis, and the lines $x = a$ and $x = b$ is

$$m = \lim_{\|P\|\to 0} \sum_{k=1}^n \rho f(x_k^*)\, \Delta x_k = \int_a^b \rho f(x)\, dx \qquad (17.8)$$

If a lamina corresponding to a region R has a *variable density* $\rho(x, y)$, where ρ is nonnegative and continuous on R, then analogous to (17.8) we define its **mass** m by the double integral

$$m = \lim_{\|P\|\to 0} \sum_{k=1}^n \rho(x_k^*, y_k^*)\, \Delta A_k = \iint_R \rho(x, y)\, dA \qquad (17.9)$$

As in Section 6.12, we define the coordinates of the **center of mass** of the lamina by

$$\bar{x} = \frac{M_y}{m}, \qquad \bar{y} = \frac{M_x}{m} \qquad (17.10)$$

where $\displaystyle M_y = \iint_R x\rho(x, y)\, dA$ and $\displaystyle M_x = \iint_R y\rho(x, y)\, dA \qquad (17.11)$

are the **moments** of the lamina about the y- and x-axes, respectively. The center of mass is the point where we consider all the mass of the lamina to be concentrated. If $\rho(x, y) = $ a constant, the center of mass is called the **centroid** of the lamina.

Example 1

A lamina has the shape of the region in the first quadrant that is bounded by the graphs of $y = \sin x$, $y = \cos x$, between $x = 0$ and $x = \pi/4$. Find its center of mass if the density is $\rho(x, y) = y$.

Solution   From Figure 17.29 we see that

$$m = \iint_R y \, dA$$

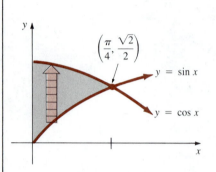

$$= \int_0^{\pi/4} \int_{\sin x}^{\cos x} y \, dy \, dx$$

$$= \int_0^{\pi/4} \frac{y^2}{2}\Big]_{\sin x}^{\cos x} dx$$

$$= \frac{1}{2} \int_0^{\pi/4} (\cos^2 x - \sin^2 x) \, dx$$

$$= \frac{1}{2} \int_0^{\pi/4} \cos 2x \, dx = \frac{1}{4}\sin 2x\Big]_0^{\pi/4} = \frac{1}{4}$$

Figure 17.29

> Double-angle formula

Now,

$$M_y = \iint_R xy \, dA$$

$$= \int_0^{\pi/4} \int_{\sin x}^{\cos x} xy \, dy \, dx$$

$$= \int_0^{\pi/4} \frac{1}{2}xy^2\Big]_{\sin x}^{\cos x} dx$$

$$= \frac{1}{2} \int_0^{\pi/4} x \cos 2x \, dx$$

> Integration by parts

$$= \left[\frac{1}{4}x \sin 2x + \frac{1}{8}\cos 2x\right]_0^{\pi/4} = \frac{\pi - 2}{16}$$

Similarly,

$$M_x = \iint_R y^2 \, dA$$

$$= \int_0^{\pi/4} \int_{\sin x}^{\cos x} y^2 \, dy \, dx$$

$$= \frac{1}{3} \int_0^{\pi/4} (\cos^3 x - \sin^3 x) \, dx$$

$$= \frac{1}{3} \int_0^{\pi/4} [\cos x(1 - \sin^2 x) - \sin x(1 - \cos^2 x)] \, dx$$

$$= \frac{1}{3}\left[\sin x - \frac{1}{3}\sin^3 x + \cos x - \frac{1}{3}\cos^3 x\right]_0^{\pi/4} = \frac{5\sqrt{2} - 4}{18}$$

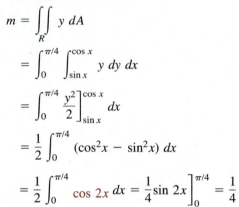

Hence, from (17.10)

$$\bar{x} = \frac{M_y}{m} = \frac{(\pi - 2)/16}{1/4} = \frac{\pi - 2}{4} \approx 0.29$$

$$\bar{y} = \frac{M_x}{m} = \frac{(5\sqrt{2} - 4)/18}{1/4} = \frac{10\sqrt{2} - 8}{9} \approx 0.68$$

Thus, the center of mass has the approximate coordinates (0.29, 0.68).

Example 2

A lamina has the shape of the region bounded by the graph of $x^2/4 + y^2/16 = 1$, $0 \le y \le 4$, and $y = 0$. Find its center of mass if the density is $\rho(x, y) = |x| y$.

Solution From Figure 17.30 we see that the region is symmetric with respect to the y-axis. Furthermore, since $\rho(-x, y) = \rho(x, y)$, ρ is symmetric about this axis. Since the y-coordinate of the center of mass must lie on the axis of symmetry, we have $\bar{x} = 0$. Utilizing symmetry, we have

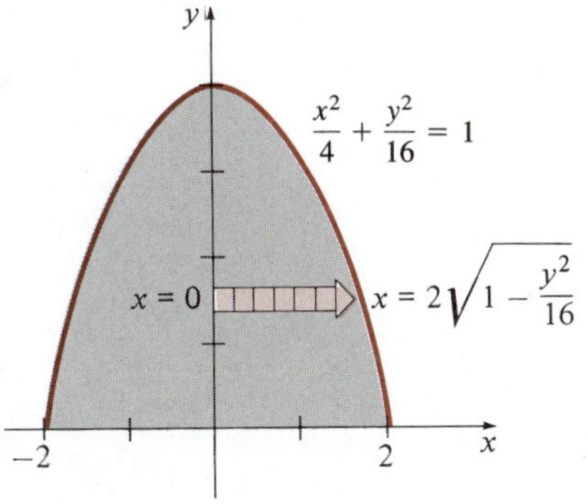

$$m = \iint\limits_R |x| y \, dA$$

$$= 2 \int_0^4 \int_0^{2\sqrt{1-y^2/16}} xy \, dx \, dy$$

$$= \int_0^4 x^2 y \Big]_0^{2\sqrt{1-y^2/16}} dy$$

$$= 4 \int_0^4 \left(y - \frac{1}{16} y^3 \right) dy$$

$$= 4 \left[\frac{y^2}{2} - \frac{1}{64} y^4 \right]_0^4 = 16$$

$$\frac{x^2}{4} + \frac{y^2}{16} = 1$$

$$x = 0 \qquad x = 2\sqrt{1 - \frac{y^2}{16}}$$

Figure 17.30

Similarly,

$$M_x = \iint\limits_R |x| y^2 \, dA$$

$$= 2 \int_0^4 \int_0^{2\sqrt{1-y^2/16}} xy^2 \, dx \, dy = \frac{512}{15}$$

From (17.10)

$$\bar{y} = \frac{512/15}{16} = \frac{32}{15}$$

The coordinates of the center of mass are then $(0, \frac{32}{15})$.

Moments of Inertia The integrals M_x and M_y in (17.11) are also called the **first moments** of a lamina about the x-axis and y-axis, respectively. The so-called **second moments** of a lamina or **moments of inertia**, about the x- and y-axes are, in turn, defined by the double integrals

$$I_x = \iint\limits_R y^2\rho(x,\ y)\ dA \quad \text{and} \quad I_y = \iint\limits_R x^2\rho(x,\ y)\ dA \qquad (17.12)$$

A moment of inertia is the rotational equivalent of mass. For translational motion, kinetic energy is given by $K = \frac{1}{2}mv^2$, where m is mass and v is linear speed. The kinetic energy of a particle of mass m rotating at a distance r from an axis is $K = \frac{1}{2}mv^2 = \frac{1}{2}m(r\omega)^2 = \frac{1}{2}(mr^2)\omega^2 = \frac{1}{2}I\omega^2$, where $I = mr^2$ is its moment of inertia about the axis of rotation and ω is angular speed.

_____ **Example 3** _____

Find the moment of inertia about the y-axis of the thin homogeneous disk of mass m shown in Figure 17.31.

Solution Since the disk is homogeneous, its density is the constant $\rho(x,\ y) = m/\pi r^2$. Hence, from (17.12)

$$I_y = \iint\limits_R x^2\left(\frac{m}{\pi r^2}\right) dA$$

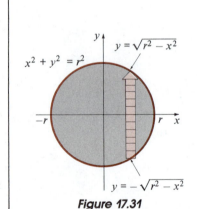

$$= \frac{m}{\pi r^2} \int_{-r}^{r} \int_{-\sqrt{r^2-x^2}}^{\sqrt{r^2-x^2}} x^2\ dy\ dx$$

$$= \frac{2m}{\pi r^2} \int_{-r}^{r} x^2\sqrt{r^2 - x^2}\ dx$$

$\boxed{\text{Trig substitution}}$

$$= \frac{2mr^2}{\pi} \int_{-\pi/2}^{\pi/2} \sin^2\theta \cos^2\theta\ d\theta$$

$$= \frac{mr^2}{2\pi} \int_{-\pi/2}^{\pi/2} \sin^2 2\theta\ d\theta$$

$$= \frac{mr^2}{4\pi} \int_{-\pi/2}^{\pi/2} (1 - \cos 4\theta)\ d\theta = \frac{1}{4}mr^2$$

Figure 17.31

Radius of Gyration The **radius of gyration** of a lamina of mass m and the moment of inertia I about an axis is defined by

$$R_g = \sqrt{\frac{I}{m}} \qquad (17.13)$$

Since (17.13) implies that $I = mR_g^2$, the radius of gyration is interpreted as the radial distance the lamina, considered as a point mass, can rotate about the axis without changing the rotational inertia of the body. In Example 3 the radius of gyration is $R_g = \sqrt{I_y/m} = \sqrt{(mr^2/4)/m} = r/2$.

Remark

Do not conclude from Example 2 that the center of mass must always lie on an axis of symmetry of a lamina. Bear in mind that the density function $\rho(x, y)$ must also be symmetric with respect to that axis.

──────── **Exercises 17.4** ────────────────────────────

Answers to odd-numbered problems begin on page A-51.

In Problems 1–10 find the center of mass of the lamina that has the given shape and density.

1. $x = 0$, $x = 4$, $y = 0$, $y = 3$; $\rho(x, y) = xy$

2. $x = 0$, $y = 0$, $2x + y = 4$; $\rho(x, y) = x^2$

3. $y = x$, $x + y = 6$, $y = 0$; $\rho(x, y) = 2y$

4. $y = |x|$, $y = 3$; $\rho(x, y) = x^2 + y^2$

5. $y = x^2$, $x = 1$, $y = 0$; $\rho(x, y) = x + y$

6. $x = y^2$, $x = 4$; $\rho(x, y) = y + 5$

7. $y = 1 - x^2$, $y = 0$; density at a point P directly proportional to the distance from the x-axis

8. $y = \sin x$, $0 \le x \le \pi$, $y = 0$; density at a point P directly proportional to the distance from the y-axis

9. $y = e^x$, $x = 0$, $x = 1$, $y = 0$; $\rho(x, y) = y^3$

10. $y = \sqrt{9 - x^2}$, $y = 0$; $\rho(x, y) = x^2$

In Problems 11–14 find the moment of inertia about the x-axis of the lamina that has the given shape and density.

11. $x = y - y^2$, $x = 0$; $\rho(x, y) = 2x$

12. $y = x^2$, $y = \sqrt{x}$; $\rho(x, y) = x^2$

13. $y = \cos x$, $-\dfrac{\pi}{2} \le x \le \dfrac{\pi}{2}$, $y = 0$; $\rho(x, y) = k$(constant)

14. $y = \sqrt{4 - x^2}$, $x = 0$, $y = 0$, first quadrant; $\rho(x, y) = y$

In Problems 15–18 find the moment of inertia about the y-axis of the lamina that has the given shape and density.

15. $y = x^2$, $x = 0$, $y = 4$, first quadrant; $\rho(x, y) = y$

16. $y = x^2$, $y = \sqrt{x}$; $\rho(x, y) = x^2$

17. $y = x$, $y = 0$, $y = 1$, $x = 3$; $\rho(x, y) = 4x + 3y$

18. Same R and density as in Problem 7.

In Problems 19 and 20 find the radius of gyration about the indicated axis of the lamina that has the given shape and density.

19. $x = \sqrt{a^2 - y^2}$, $x = 0$; $\rho(x, y) = x$; y-axis

20. $x + y = a$, $a > 0$, $x = 0$, $y = 0$; $\rho(x, y) = k$(constant); x-axis

The **polar moment of inertia** of a lamina with respect to the origin is defined to be

$$I_0 = \iint\limits_{R} (x^2 + y^2)\rho(x, y)\, dA = I_x + I_y$$

In Problems 21–24 find the polar moment of inertia of the lamina that has the given shape and density.

21. $x + y = a$, $a > 0$, $x = 0$, $y = 0$; $\rho(x, y) = k$(constant)

22. $y = x^2$, $y = \sqrt{x}$; $\rho(x, y) = x^2$ (*Hint*: See Problems 12 and 16.)

23. $x = y^2 + 2$, $x = 6 - y^2$; density at a point P inversely proportional to the square of the distance from the origin

24. $y = x$, $y = 0$, $y = 3$, $x = 4$; $\rho(x, y) = k$(constant)

25. Find the radius of gyration in Problem 21.

26. Show that the polar moment of inertia about the center of a thin homogeneous rectangular plate of mass m, width w, and length l is $I_0 = m(l^2 + w^2)/12$.

17.5 Double Integrals in Polar Coordinates

Suppose R is a region bounded by the graphs of the polar equations $r = g_1(\theta)$, $r = g_2(\theta)$, and the rays $\theta = \alpha$, $\theta = \beta$, and f is a function of r and θ that is continuous on R. In order to define the double integral of f over R, we use rays

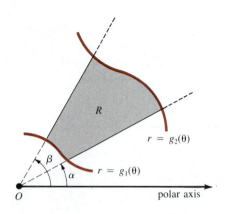

(a) region R

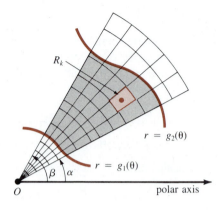

(b) subregion R_k

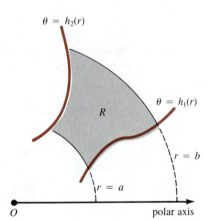

(c) enlargement of R_k

Figure 17.32

and concentric circles to partition the region into a grid of "polar rectangles" or subregions R_k. See Figures 17.32(a) and (b). The area ΔA_k of typical subregion R_k, shown in Figure 17.32(c), is the difference of areas of two circular sectors:

$$\Delta A_k = \frac{1}{2} r_{k+1}^2 \, \Delta \theta_k - \frac{1}{2} r_k^2 \, \Delta \theta_k$$

$$= \frac{1}{2}(r_{k+1}^2 - r_k^2) \, \Delta \theta_k$$

$$= \frac{1}{2}(r_{k+1} + r_k)(r_{k+1} - r_k) \, \Delta \theta_k$$

$$= r_k^* \, \Delta r_k \, \Delta \theta_k$$

where $\Delta r_k = r_{k+1} - r_k$ and r_k^* denotes the average radius $(r_{k+1} + r_k)/2$. By choosing (r_k^*, θ_k^*) on each R_k, the double integral of f over R is

$$\lim_{\|P\| \to 0} \sum_{k=1}^{n} f(r_k^*, \theta_k^*) r_k^* \, \Delta r_k \, \Delta \theta_k = \iint\limits_{R} f(r, \theta) \, dA$$

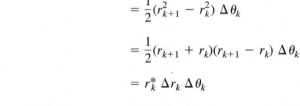

Figure 17.33

The double integral is then evaluated by means of the iterated integral,

$$\iint\limits_{R} f(r, \theta) \, dA = \int_{\alpha}^{\beta} \int_{g_1(\theta)}^{g_2(\theta)} f(r, \theta) \, r \, dr \, d\theta \qquad (17.14)$$

On the other hand, if the region R is as given in Figure 17.33, the double integral of f over R is then

$$\iint\limits_{R} f(r, \theta) \, dA = \int_{a}^{b} \int_{h_1(r)}^{h_2(r)} f(r, \theta) \, r \, d\theta \, dr \qquad (17.15)$$

_____ **Example 1** _____

Find the center of mass of the lamina that corresponds to the region bounded by one leaf of the rose $r = 2 \sin 2\theta$ in the first quadrant if the density at a point P in the lamina is directly proportional to the distance from the pole.

Solution By varying θ from 0 to $\pi/2$, we obtain the graph in Figure 17.34. Now, $d(0, P) = |r|$. Hence, $\rho(r, \theta) = k|r|$, where k is a constant of proportionality. From (17.9) of Section 17.4, we have

$$m = \iint\limits_{R} k|r|\, dA$$

$$= k \int_{0}^{\pi/2} \int_{0}^{2 \sin 2\theta} (r)r\, dr\, d\theta$$

$$= k \int_{0}^{\pi/2} \frac{r^3}{3} \Big]_{0}^{2 \sin 2\theta} d\theta$$

$$= \frac{8}{3}k \int_{0}^{\pi/2} \sin^3 2\theta\, d\theta$$

$$= \frac{8}{3}k \int_{0}^{\pi/2} (1 - \cos^2 2\theta)\sin 2\theta\, d\theta$$

$$= \frac{8}{3}k \left[-\frac{1}{2}\cos 2\theta + \frac{1}{6}\cos^3 2\theta \right]_{0}^{\pi/2} = \frac{16}{9}k$$

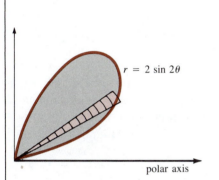

$r = 2 \sin 2\theta$

polar axis

Figure 17.34

Since $x = r \cos \theta$, we can write

$$M_y = k \iint\limits_{R} x|r|\, dA$$

as

$$M_y = \int_{0}^{\pi/2} \int_{0}^{2 \sin 2\theta} r^3 \cos \theta\, dr\, d\theta$$

$$= k \int_{0}^{\pi/2} \frac{r^4}{4} \cos \theta \Big]_{0}^{2 \sin 2\theta} d\theta$$

$$= 4k \int_{0}^{\pi/2} \sin^4 2\theta \cos \theta\, d\theta$$

$$\downarrow$$

$$= 4k \int_{0}^{\pi/2} 16 \sin^4 \theta \cos^4 \theta \cos \theta\, d\theta \qquad \boxed{\text{Double-angle formula}}$$

$$= 64k \int_{0}^{\pi/2} \sin^4 \theta \cos^5 \theta\, d\theta$$

$$= 64k \int_{0}^{\pi/2} \sin^4 \theta (1 - \sin^2 \theta)^2 \cos \theta\, d\theta$$

$$= 64k \int_{0}^{\pi/2} (\sin^4 \theta - 2 \sin^6 \theta + \sin^8 \theta)\cos \theta\, d\theta$$

$$= 64k \left[\frac{1}{5}\sin^5 \theta - \frac{2}{7}\sin^7 \theta + \frac{1}{9}\sin^9 \theta \right]_{0}^{\pi/2} = \frac{512}{315}k$$

Similarly, by using $x = r \sin \theta$, we find*

$$M_x = k \int_0^{\pi/2} \int_0^{2 \sin 2\theta} r^2 \sin \theta \, dr \, d\theta = \frac{512}{315} k$$

Here the rectangular coordinates of the center of mass are

$$\bar{x} = \bar{y} = \frac{512k/315}{16k/9} = \frac{32}{35}$$

_____ **Example 2** _____

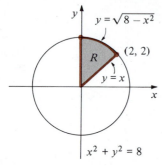

Figure 17.35

Use polar coordinates to evaluate

$$\int_0^2 \int_x^{\sqrt{8-x^2}} \frac{1}{5 + x^2 + y^2} \, dy \, dx$$

Solution   From $x \le y \le \sqrt{8 - x^2}$, $0 \le x \le 2$, we have sketched the region R of integration in Figure 17.35. Since $x^2 + y^2 = r^2$, the polar description of the circle $x^2 + y^2 = 8$ is $r = \sqrt{8}$. Hence, in polar coordinates, the region of R is given by $0 \le r \le \sqrt{8}$, $\pi/4 \le \theta \le \pi/2$. Using $1/(5 + x^2 + y^2) = 1/(5 + r^2)$ the original integral becomes

$$\int_0^2 \int_x^{\sqrt{8-x^2}} \frac{1}{5 + x^2 + y^2} \, dy \, dx = \int_{\pi/4}^{\pi/2} \int_0^{\sqrt{8}} \frac{1}{5 + r^2} r \, dr \, d\theta$$

$$= \frac{1}{2} \int_{\pi/4}^{\pi/2} \int_0^{\sqrt{8}} \frac{2r \, dr}{5 + r^2} \, d\theta$$

$$= \frac{1}{2} \int_{\pi/4}^{\pi/2} \ln(5 + r^2) \Big]_0^{\sqrt{8}} \, d\theta$$

$$= \frac{1}{2} (\ln 13 - \ln 5) \int_{\pi/4}^{\pi/2} d\theta$$

$$= \frac{1}{2} (\ln 13 - \ln 5) \left(\frac{\pi}{2} - \frac{\pi}{4} \right) = \frac{\pi}{8} \ln \frac{13}{5}$$

_____ **Example 3** _____

Find the volume of the solid that is under the hemisphere $z = \sqrt{1 - x^2 - y^2}$ and above the region bounded by the graph of the circle $r = \sin \theta$.

Solution   From Figure 17.36, we see that

$$V = \iint_R \sqrt{1 - x^2 - y^2} \, dA$$

_____

*We could have argued to the fact that $M_x = M_y$ and hence $\bar{x} = \bar{y}$ from the fact that the lamina and the density function are symmetric about the ray $\theta = \pi/4$.

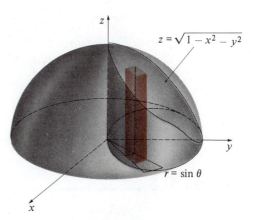

$z = \sqrt{1 - x^2 - y^2}$

$r = \sin\theta$

Figure 17.36

In polar coordinates $x = r\cos\theta$, $y = r\sin\theta$, and so the equation of the hemisphere becomes $z = \sqrt{1 - x^2 - y^2} = \sqrt{1 - r^2}$. Using symmetry, we then have

$$V = \iint_R \sqrt{1 - r^2}\, dA$$

$$= 2\int_0^{\pi/2} \int_0^{\sin\theta} (1 - r^2)^{1/2} r\, dr\, d\theta$$

$$= 2\int_0^{\pi/2} \left[-\frac{1}{3}(1 - r^2)^{3/2} \right]_0^{\sin\theta} d\theta$$

$$= \frac{2}{3}\int_0^{\pi/2} [1 - (1 - \sin^2\theta)^{3/2}]\, d\theta$$

$$= \frac{2}{3}\int_0^{\pi/2} [1 - (\cos^2\theta)^{3/2}]\, d\theta$$

$$= \frac{2}{3}\int_0^{\pi/2} [1 - \cos^3\theta]\, d\theta$$

$$= \frac{2}{3}\int_0^{\pi/2} [1 - (1 - \sin^2\theta)\cos\theta]\, d\theta$$

$$= \frac{2}{3}\left[\theta - \sin\theta + \frac{1}{3}\sin^3\theta \right]_0^{\pi/2} = \frac{\pi}{3} - \frac{4}{9} \approx 0.60 \text{ cubic unit}$$

Area   Note that in (17.14) if $f(r, \theta) = 1$, then the **area** of the region R in Figure 17.32(a) is given by

$$A = \iint_R dA = \int_\alpha^\beta \int_{g_1(\theta)}^{g_2(\theta)} r\, dr\, d\theta$$

The same observation holds for (17.15) and Figure 17.33 when $f(r, \theta) = 1$.

Remark

You are invited to reexamine Example 3. The graph of the circle $r = \sin \theta$ is obtained by varying θ from 0 to π. However, carry out the iterated integration

$$V = \int_0^\pi \int_0^{\sin \theta} (1 - r^2)^{1/2} r \, dr \, d\theta$$

and see if you obtain the *incorrect* answer $\pi/3$. What goes wrong?

——————— Exercises 17.5 ———————————————

Answers to odd-numbered problems begin on page A-51.

In Problems 1–4 use a double integral in polar coordinates to find the area of the region bounded by the graphs of the given polar equations.

1. $r = 3 + 3 \sin \theta$

2. $r = 2 + \cos \theta$

3. $r = 2 \sin \theta$, $r = 1$, common area

4. $r = 8 \sin 4\theta$, one petal

In Problems 5–10 find the volume of the solid bounded by the graphs of the given equations.

5. One petal of $r = 5 \cos 3\theta$, $z = 0$, $z = 4$

6. $x^2 + y^2 = 4$, $z = \sqrt{9 - x^2 - y^2}$, $z = 0$

7. Between $x^2 + y^2 = 1$ and $x^2 + y^2 = 9$, $z = \sqrt{16 - x^2 - y^2}$, $z = 0$

8. $z = \sqrt{x^2 + y^2}$, $x^2 + y^2 = 25$, $z = 0$

9. $r = 1 + \cos \theta$, $z = y$, $z = 0$, first octant

10. $r = \cos \theta$, $z = 2 + x^2 + y^2$, $z = 0$

In Problems 11–16 find the center of mass of the lamina that has the given shape and density.

11. $r = 1$, $r = 3$, $x = 0$, $y = 0$, first quadrant; $\rho(r, \theta) = k$(constant)

12. $r = \cos \theta$; density at a point P directly proportional to the distance from the pole

13. $y = \sqrt{3}x$, $y = 0$, $x = 3$; $\rho(r, \theta) = r^2$

14. $r = 4 \cos 2\theta$, petal on the polar axis; $\rho(r, \theta) = k$(constant)

15. Outside $r = 2$ and inside $r = 2 + 2 \cos \theta$, $y = 0$, first quadrant; density at a point P inversely proportional to the distance from the pole

16. $r = 2 + 2 \cos \theta$, $y = 0$, first and second quadrants; $\rho(r, \theta) = k$(constant)

In Problems 17–20 find the indicated moment of inertia of the lamina that has the given shape and density.

17. $r = a$; $\rho(r, \theta) = k$(constant); I_x

18. $r = a$; $\rho(r, \theta) = \dfrac{1}{1 + r^4}$; I_x

19. Outside $r = a$ and inside $r = 2a \cos \theta$; density at a point P inversely proportional to the cube of the distance from the pole; I_y

20. Outside $r = 1$ and inside $r = 2 \sin 2\theta$, first quadrant; $\rho(r, \theta) = \sec^2\theta$; I_y

In Problems 21–24 find the **polar moment of inertia** $I_0 = \iint_R r^2 \rho(r, \theta) \, dA = I_x + I_y$ of the lamina that has the given shape and density.

21. $r = a$; $\rho(r, \theta) = k$(constant) (*Hint*: Use Problem 17 and the fact that $I_x = I_y$.)

22. $r = \theta$, $0 \le \theta \le \pi$, $y = 0$; density at a point P proportional to the distance from the pole

23. $r\theta = 1$, $\frac{1}{3} \le \theta \le 1$, $r = 1$, $r = 3$, $y = 0$; density at a point P inversely proportional to the distance from the pole (*Hint*: Integrate first with respect to θ.)

24. $r = 2a \cos \theta$; $\rho(r, \theta) = k$(constant)

In Problems 25–32 evaluate the given iterated integral by changing to polar coordinates.

25. $\displaystyle\int_{-3}^{3} \int_0^{\sqrt{9 - x^2}} \sqrt{x^2 + y^2} \, dy \, dx$

26. $\displaystyle\int_0^{\sqrt{2}/2} \int_y^{\sqrt{1 - y^2}} \frac{y^2}{\sqrt{x^2 + y^2}} \, dx \, dy$

27. $\displaystyle\int_0^1 \int_0^{\sqrt{1 - y^2}} e^{x^2 + y^2} \, dx \, dy$

28. $\displaystyle\int_{-\sqrt{\pi}}^{\sqrt{\pi}}\int_{0}^{\sqrt{\pi-x^2}}\sin(x^2+y^2)\,dy\,dx$

29. $\displaystyle\int_{0}^{1}\int_{\sqrt{1-x^2}}^{\sqrt{4-x^2}}\frac{x^2}{x^2+y^2}\,dy\,dx+\int_{1}^{2}\int_{0}^{\sqrt{4-x^2}}\frac{x^2}{x^2+y^2}\,dy\,dx$

30. $\displaystyle\int_{0}^{1}\int_{0}^{\sqrt{2y-y^2}}(1-x^2-y^2)\,dx\,dy$

31. $\displaystyle\int_{-5}^{5}\int_{0}^{\sqrt{25-x^2}}(4x+3y)\,dy\,dx$

32. $\displaystyle\int_{0}^{1}\int_{0}^{\sqrt{1-y^2}}\frac{1}{1+\sqrt{x^2+y^2}}\,dx\,dy$

33. Evaluate $\iint_R (x+y)\,dA$ over the region shown in Figure 17.37.

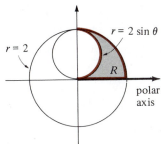

Figure 17.37

Miscellaneous Problems

34. The improper integral $\int_0^\infty e^{-x^2}\,dx$ is important in the theory of probability, statistics, and other areas of applied mathematics. If I denotes the integral, then

$$I=\int_{0}^{\infty}e^{-x^2}\,dx\quad\text{and}\quad I=\int_{0}^{\infty}e^{-y^2}\,dy$$

In view of Problem 41 of Exercises 17.2 we have

$$I^2=\left(\int_{0}^{\infty}e^{-x^2}\,dx\right)\left(\int_{0}^{\infty}e^{-y^2}\,dy\right)$$
$$=\int_{0}^{\infty}\int_{0}^{\infty}e^{-(x^2+y^2)}\,dx\,dy$$

Use polar coordinates to evaluate the last integral. Find the value of I.

17.6 Surface Area

In the plane we saw that the length of an arc of the graph of $y=f(x)$ from $x=a$ to $x=b$ was given by

$$s=\int_{a}^{b}\sqrt{1+\left(\frac{dy}{dx}\right)^2}\,dx \tag{17.16}$$

The problem in three dimensions, which is the counterpart of the arc length problem, is to find the area $A(S)$ of that portion of the surface S given by a function $z=f(x,y)$ having continuous first partial derivatives on a closed region R in the xy-plane.

Suppose, as shown in Figure 17.38(a), that an inner partition P of R is formed using lines parallel to the x- and y-axes. P then consists of n rectangular elements R_k of area $\Delta A_k=\Delta x_k\,\Delta y_k$ that lie entirely within R. Let $(x_k,y_k,0)$ denote any point in an R_k. As we see in Figure 17.38(a), by projecting the sides of R_k upward, we determine two quantities: a portion S_k of the surface and a portion T_k of a tangent plane at $(x_k,y_k,f(x_k,y_k))$. It seems reasonable to assume that when R_k is small, the area ΔT_k of T_k is approximately the same as the area ΔS_k of S_k.

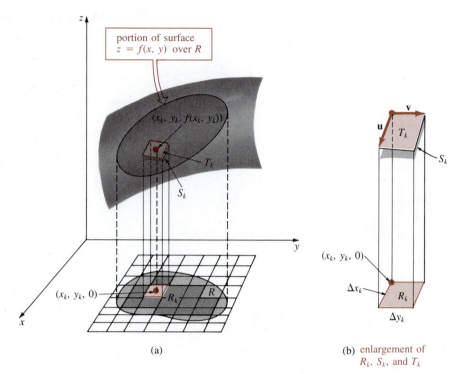

(a)

(b) enlargement of
R_k, S_k, and T_k

Figure 17.38

To find the area of T_k let us choose $(x_k, y_k, 0)$ at a corner of R_k as shown in Figure 17.38(b). The indicated vectors **u** and **v**, which form two sides of T_k, are given by

$$\mathbf{u} = \Delta x_k \mathbf{i} + f_x(x_k, y_k)\, \Delta x_k \mathbf{k}$$

$$\mathbf{v} = \Delta y_k \mathbf{j} + f_y(x_k, y_k)\, \Delta y_k \mathbf{k}$$

where $f_x(x_k, y_k)$ and $f_y(x_k, y_k)$ are the slopes of the lines containing **u** and **v**, respectively. Now from (14.35) of Section 14.4 we know that

$$\Delta T_k = |\mathbf{u} \times \mathbf{v}|$$

where

$$\mathbf{u} \times \mathbf{v} = \begin{vmatrix} \mathbf{i} & \mathbf{j} & \mathbf{k} \\ \Delta x_k & 0 & f_x(x_k, y_k)\, \Delta x_k \\ 0 & \Delta y_k & f_y(x_k, y_k)\, \Delta y_k \end{vmatrix}$$

$$= [-f_x(x_k, y_k)\mathbf{i} - f_y(x_k, y_k)\mathbf{j} + \mathbf{k}]\, \Delta x_k\, \Delta y_k$$

In other words,

$$\Delta T_k = \sqrt{[f_x(x_k, y_k)]^2 + [f_y(x_k, y_k)]^2 + 1}\; \Delta x_k\, \Delta y_k$$

Consequently, the area A is approximately

$$\sum_{k=1}^{n} \sqrt{1 + [f_x(x_k, y_k)]^2 + [f_y(x_k, y_k)]^2}\; \Delta x_k\, \Delta y_k$$

Taking the limit of the foregoing sum as $\|P\| \to 0$ leads us to the next definition.

> **DEFINITION 17.2**   **Surface Area**
>
> Let f be a function for which the first partial derivatives f_x and f_y are continuous on a closed region R. Then the **area of the surface over R** is given by
>
> $$A(S) = \iint_R \sqrt{1 + [f_x(x, y)]^2 + [f_y(x, y)]^2} \, dA \qquad (17.17)$$

Note: One could have almost guessed the form of (17.17) by naturally extending the one-variable structure of (17.16) to two variables.

Example 1

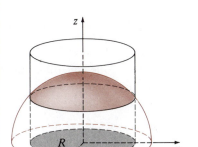

Figure 17.39

Find the surface area of that portion of the sphere $x^2 + y^2 + z^2 = a^2$ that is above the xy-plane and within the cylinder $x^2 + y^2 = b^2$, $0 < b < a$.

Solution   If we define $z = f(x, y)$ by $f(x, y) = \sqrt{a^2 - x^2 - y^2}$, then

$$f_x(x, y) = \frac{-x}{\sqrt{a^2 - x^2 - y^2}} \quad \text{and} \quad f_y(x, y) = \frac{-y}{\sqrt{a^2 - x^2 - y^2}}$$

and so

$$1 + [f_x(x, y)]^2 + [f_y(x, y)]^2 = \frac{a^2}{a^2 - x^2 - y^2}$$

Hence, (17.17) is

$$A(S) = \iint_R \frac{a}{\sqrt{a^2 - x^2 - y^2}} \, dA$$

where R is indicated in Figure 17.39. To evaluate this double integral, we change to polar coordinates:

$$A(S) = a \int_0^{2\pi} \int_0^b (a^2 - r^2)^{-1/2} r \, dr \, d\theta$$

$$= a \int_0^{2\pi} \left[-(a^2 - r^2)^{1/2} \right]_0^b d\theta$$

$$= a(a - \sqrt{a^2 - b^2}) \int_0^{2\pi} d\theta$$

$$= 2\pi a(a - \sqrt{a^2 - b^2}) \text{ square units}$$

Example 2

Find the surface area of the portions of the sphere $x^2 + y^2 + z^2 = 4$ that are within the cylinder $(x - 1)^2 + y^2 = 1$.

Solution   The surface area, which is above and below the xy-plane, is shaded in Figure 17.40. As in Example 1, (17.17) simplifies to

$$A(S) = 2 \iint_R \frac{2}{\sqrt{4 - x^2 - y^2}} \, dA$$

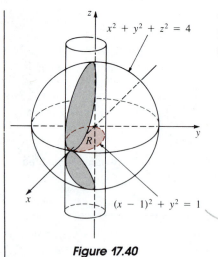

$x^2 + y^2 + z^2 = 4$

$(x - 1)^2 + y^2 = 1$

Figure 17.40

where R is the region bounded by the graph of $(x - 1)^2 + y^2 = 1$. The extra factor of 2 comes from using symmetry. Now, in polar coordinates the boundary of R is simply $r = 2 \cos \theta$. Thus,

$$A(S) = 4 \int_0^\pi \int_0^{2 \cos \theta} (4 - r^2)^{-1/2} r \, dr \, d\theta$$

$$= 8(\pi - 2)$$

$$\approx 9.13 \text{ square units}$$

Differential of Surface Area The function $dS = \sqrt{1 + [f_x(x, y)]^2 + [f_y(x, y)]^2} \, dA$ is called the **differential of the surface area**. We will use this function in Section 18.3.

--- **Exercises 17.6** ---

Answers to odd-numbered problems begin on page A-51.

1. Find the surface area of that portion of the plane $2x + 3y + 4z = 12$ that is bounded by the coordinate planes in the first octant.

2. Find the surface area of that portion of the plane $2x + 3y + 4z = 12$ that is above the region in the first quadrant bounded by the graph $r = \sin 2\theta$.

3. Find the surface area of that portion of the cylinder $x^2 + z^2 = 16$ that is above the region in the first quadrant bounded on the graphs of $x = 0$, $x = 2$, $y = 0$, $y = 5$.

4. Find the surface area of that portion of the paraboloid $z = x^2 + y^2$ that is below the plane $z = 2$.

5. Find the surface area of that portion of the paraboloid $z = 4 - x^2 - y^2$ that is above the xy-plane.

6. Find the surface area of the portions of the sphere $x^2 + y^2 + z^2 = 2$ that are within the cone $z^2 = x^2 + y^2$.

7. Find the surface area of that portion of the sphere $x^2 + y^2 + z^2 = 25$ that is above the region in the first quadrant bounded by the graphs of $x = 0$, $y = 0$, $4x^2 + y^2 = 25$. (*Hint:* Integrate first with respect to x.)

8. Find the surface area of that portion of the graph of $z = x^2 - y^2$ that is in the first octant within the cylinder $x^2 + y^2 = 4$.

9. Find the surface area of the portions of the sphere $x^2 + y^2 + z^2 = a^2$ that are within the cylinder $x^2 + y^2 = ay$.

10. Find the surface area of the portions of the cone $z^2 = \frac{1}{4}(x^2 + y^2)$ that are within the cylinder $(x - 1)^2 + y^2 = 1$.

11. Find the surface area of the portions of the cylinder $y^2 + z^2 = a^2$ that are within the cylinder $x^2 + y^2 = a^2$. (*Hint:* See Figure 17.28.)

12. Use the result given in Example 1 to prove that the surface area of a sphere of radius a is $4\pi a^2$. (*Hint:* Consider a limit as $b \to a$.)

13. Find the surface area of that portion of the sphere $x^2 + y^2 + z^2 = a^2$ that is bounded between $y = c_1$ and $y = c_2$, $0 < c_1 < c_2 < a$. (*Hint:* Use polar coordinates in the xz-plane.)

14. Show that the area found in Problem 13 is the same as the surface area of the cylinder $x^2 + z^2 = a^2$ between $y = c_1$ and $y = c_2$.

17.7 The Triple Integral

The steps leading to the definition of the *three-dimensional definite integral*, or **triple integral**, $\iiint_D F(x, y, z) \, dV$ are quite similar to those of the double integral.

$$w = F(x, y, z)$$

1. Let F be defined over a closed and bounded region D of space.

2. By means of a three-dimensional grid of vertical and horizontal planes parallel to the coordinate planes, form a partition P of D into n subregions (boxes) D_k of volumes ΔV_k that lie entirely in D.

3. Let $\|P\|$ be the norm of the partition or the length of the longest diagonal of the D_k.

4. Choose a point (x_k^*, y_k^*, z_k^*) in each subregion D_k.

5. Form the sum

$$\sum_{k=1}^{n} F(x_k^*, y_k^*, z_k^*) \, \Delta V_k$$

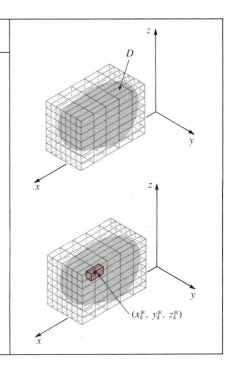

A sum of the form $\sum_{k=1}^{n} F(x_k^*, y_k^*, z_k^*) \, \Delta V_k$, where (x_k^*, y_k^*, z_k^*) is an arbitrary point within each D_k and ΔV_k denotes the volume of each D_k, is called a **Riemann sum**. The type of partition used in step 2, where all the D_k lie completely within D is called an **inner partition** of D.

DEFINITION 17.3   The Triple Integral

Let F be a function of three variables defined over a closed region D of space. Then the **triple integral of F over D** is given by

$$\iiint_D F(x, y, z) \, dV = \lim_{\|P\| \to 0} \sum_{k=1}^{n} F(x_k^*, y_k^*, z_k^*) \, \Delta V_k \qquad (17.18)$$

As in our previous discussions on the integral, when F is continuous over D then the limit in (17.18) exists; that is, F is **integrable** over D.

If the region D is bounded above by the graph of $z = f_2(x, y)$ and bounded below by the graph of $z = f_1(x, y)$, then it can be shown that the triple
integral (17.18) can be expressed as a double integral of the partial integral
$\int_{f_1(x, y)}^{f_2(x, y)} F(x, y, z)\, dz$. That is,

$$\iiint_D F(x, y, z)\, dV = \iint_R \left[\int_{f_1(x, y)}^{f_2(x, y)} F(x, y, z)\, dz \right] dA$$

where R is the orthogonal projection of D onto the xy-plane. In particular, if R
is a Type I region, then, as shown in Figure 17.41, the triple integral of F over
D can be written as an iterated integral:

$$\iiint_D F(x, y, z)\, dV = \int_a^b \int_{g_1(x)}^{g_2(x)} \int_{f_1(x, y)}^{f_2(x, y)} F(x, y, z)\, dz\, dy\, dx \qquad (17.19)$$

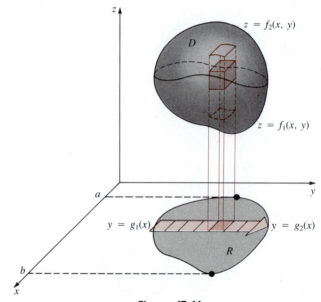

Figure 17.41

To evaluate the iterated integral in (17.19), we begin by evaluating the partial
integral

$$\int_{f_1(x, y)}^{f_2(x, y)} F(x, y, z)\, dz$$

in which *both* x and y are held fixed.

In a double integral there are only two possible orders of integration, *dy*
dx and *dx dy*. The triple integral in (17.19) illustrates one of *six* possible orders
of integration:

dz dy dx	*dz dx dy*	*dy dx dz*
dx dy dz	*dx dz dy*	*dy dz dx*

The last two differentials tell us the coordinate plane in which the region R is situated. For example, the iterated integral corresponding to the order of integration $dx\ dz\ dy$ must have the form

$$\int_c^d \int_{k_1(y)}^{k_2(y)} \int_{h_1(y,z)}^{h_2(y,z)} F(x,\ y,\ z)\ dx\ dz\ dy$$

The geometric interpretation of this integral and the region R of integration in the yz-plane are shown in Figure 17.42.

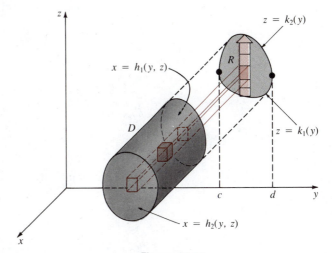

Figure 17.42

Applications A list of some of the standard applications of the triple integral follows.

Volume

If $F(x,\ y,\ z) = 1$, then the **volume** of the solid D is

$$V = \iiint\limits_D dV$$

Mass

If $\rho(x,\ y,\ z)$ is density, then the **mass** of the solid D is given by

$$m = \iiint\limits_D \rho(x,\ y,\ z)\ dV$$

First Moments

The **first moments** of the solid about the coordinate planes indicated by the subscripts are given by

$$M_{xy} = \iiint_D z\rho(x, y, z)\, dV, \qquad M_{xz} = \iiint_D y\rho(x, y, z)\, dV$$

$$M_{yz} = \iiint_D x\rho(x, y, z)\, dV$$

Center of Mass

The coordinates of the **center of mass** of D are given by

$$\bar{x} = \frac{M_{yz}}{m}, \qquad \bar{y} = \frac{M_{xz}}{m}, \qquad \bar{z} = \frac{M_{xy}}{m}$$

Centroid

If $\rho(x, y, z) = $ a constant, the center of mass is called the **centroid** of the solid.

Second Moments

The **second moments**, or **moments of inertia** of D about the coordinate axes indicated by the subscripts, are given by

$$I_x = \iiint_D (y^2 + z^2)\rho(x, y, z)\, dV, \qquad I_y = \iiint_D (x^2 + z^2)\rho(x, y, z)\, dV,$$

$$I_z = \iiint_D (x^2 + y^2)\rho(x, y, z)\, dV$$

Radius of Gyration

As in Section 17.4, if I is a moment of inertia of the solid about a given axis, then the **radius of gyration** is

$$R_g = \sqrt{\frac{I}{m}}$$

Example 1

Find the volume of the solid in the first octant bounded by the graphs of $z = 1 - y^2$, $y = 2x$, and $x = 3$.

Solution   As indicated in Figure 17.43(a), the first integration with respect to z will be from 0 to $1 - y^2$. Furthermore, from Figure 17.43(b) we see that the projection of the solid D in the xy-plane is a region of Type II. Hence, we next integrate, with respect to x, from $y/2$ to 3. The last integration is with respect to y from 0 to 1. Thus,

$$V = \iiint\limits_{D} dV$$

$$= \int_{0}^{1} \int_{y/2}^{3} \int_{0}^{1-y^2} dz\ dx\ dy$$

$$= \int_{0}^{1} \int_{y/2}^{3} (1 - y^2)\ dx\ dy$$

$$= \int_{0}^{1} \left[x - xy^2 \right]_{y/2}^{3} dy$$

$$= \int_{0}^{1} \left(3 - 3y^2 - \frac{1}{2}y + \frac{1}{2}y^3 \right) dy$$

$$= \left[3y - y^3 - \frac{1}{4}y^2 + \frac{1}{8}y^4 \right]_{0}^{1} = \frac{15}{8} \text{ cubic units}$$

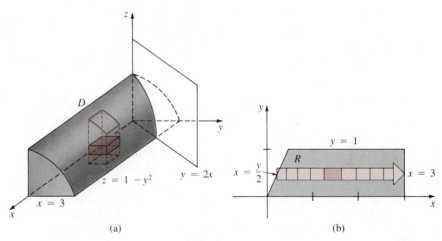

(a)   (b)

Figure 17.43

You should observe that the volume in the foregoing example could have just as easily been obtained by means of a double integral.

_____ **Example 2** _____

Find the triple integral that gives the volume of the solid that has the shape determined by the cone $x = \sqrt{y^2 + z^2}$ and the paraboloid $x = 6 - y^2 - z^2$.

Solution   By substituting $y^2 + z^2 = x^2$ in $y^2 + z^2 = 6 - x$, we find that $x^2 = 6 - x$ or $(x + 3)(x - 2) = 0$. Thus, the two surfaces intersect in the plane $x = 2$. The projection onto the yz-plane of the curve of intersection is $y^2 + z^2 = 4$. Using symmetry and referring to Figures 17.44(a) and (b), we see that

$$V = \iiint_D dV$$

$$= 4 \int_0^2 \int_0^{\sqrt{4-y^2}} \int_{\sqrt{y^2+z^2}}^{6-y^2-z^2} dx \, dz \, dy$$

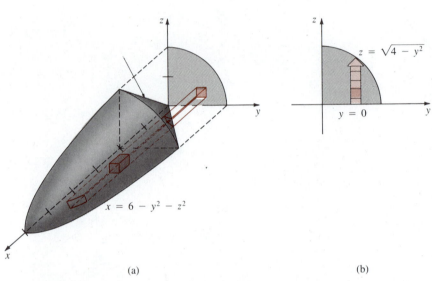

(a)   (b)

Figure 17.44

While evaluation of this integral is straightforward, it is admittedly "messy." We shall return to this integral in the next section after we have examined triple integrals in other coordinate systems.

Example 3

A solid has the shape determined by the graphs of the cylinder $|x| + |y| = 1$* and the planes $z = 2$ and $z = 4$. Find its center of mass if the density is given by $\rho(x, y, z) = kz$, k a constant.

Solution   The solid and its orthogonal projection onto a region R of Type I in the xy-plane are shown in Figure 17.45(a). Since the density function $\rho(x, y, z) = kz$ is symmetric over R, we conclude that the center of mass is on the z-axis; that is, we need only compute m and M_{xy}. From symmetry and Figure 17.45(b) it follows that

---

*This is equivalent to four lines: $x + y = 1$, $x > 0$, $y > 0$; $x - y = 1$, $x > 0$, $y < 0$; and so on.

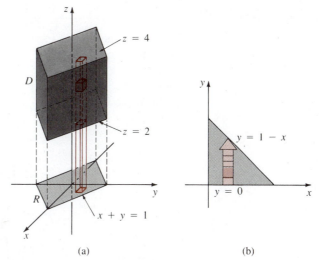

(a)   (b)

Figure 17.45

$$m = 4 \int_0^1 \int_0^{1-x} \int_2^4 kz \; dz \; dy \; dx$$

$$= 4k \int_0^1 \int_0^{1-x} \frac{z^2}{2} \Big]_2^4 \; dy \; dx$$

$$= 24k \int_0^1 \int_0^{1-x} dy \; dx$$

$$= 24k \int_0^1 (1 - x) \; dx$$

$$= 24k \left[x - \frac{x^2}{2} \right]_0^1 = 12k$$

and

$$M_{xy} = 4 \int_0^1 \int_0^{1-x} \int_2^4 kz^2 \; dz \; dy \; dx$$

$$= 4k \int_0^1 \int_0^{1-x} \frac{z^3}{3} \Big]_2^4 \; dy \; dx$$

$$= \frac{224}{3}k \int_0^1 \int_0^{1-x} dy \; dx = \frac{112}{3}k$$

Hence,

$$\bar{z} = \frac{M_{xy}}{m} = \frac{112k/3}{12k} = \frac{28}{9}$$

The coordinates of the center of mass are then $(0, 0, \frac{28}{9})$.

Example 4

Find the moment of inertia of the solid in Example 3 about the z-axis. Find the radius of gyration.

Solution $I_z = \iiint_D (x^2 + y^2) kz \, dV$. Using symmetry again, we can write this triple integral as

$$I_z = 4k \int_0^1 \int_0^{1-x} \int_2^4 (x^2 + y^2)z \, dz \, dy \, dx$$

$$= 4k \int_0^1 \int_0^{1-x} (x^2 + y^2)\frac{z^2}{2}\Big]_2^4 \, dy \, dx$$

$$= 24k \int_0^1 \int_0^{1-x} (x^2 + y^2) \, dy \, dx$$

$$= 24k \int_0^1 \left[x^2y + \frac{y^3}{3} \right]_0^{1-x} dx$$

$$= 24k \int_0^1 \left[x^2 - x^3 + \frac{1}{3}(1 - x)^3 \right] dx$$

$$= 24k \left[\frac{x^3}{3} - \frac{x^4}{4} - \frac{1}{12}(1 - x)^4 \right]_0^1 = 4k$$

Since $m = 12k$, it follows that the radius of gyration is

$$R_g = \sqrt{\frac{I_z}{m}} = \sqrt{\frac{4k}{12k}} = \frac{\sqrt{3}}{3}$$

Example 5

Change the order of integration in

$$\int_0^6 \int_0^{4-2x/3} \int_0^{3-x/2-3y/4} F(x, y, z) \, dz \, dy \, dx$$

to $dy \, dx \, dz$.

Solution As seen in Figure 17.46(a), the region D is the solid in the first octant bounded by the three coordinate planes and the plane $2x + 3y + 4z = 12$. Referring to Figure 17.46(b) and the included table, we conclude that

$$\int_0^6 \int_0^{4-2x/3} \int_0^{3-x/2-3y/4} F(x, y, z) \, dz \, dy \, dx = \int_0^3 \int_0^{6-2z} \int_0^{4-2x/3-4z/3} F(x, y, z) \, dy \, dx \, dz$$

Order of Integration	First Integration		Second Integration		Third Integration	
$dz \, dy \, dx$	0 to	$3 - x/2 - 3y/4$	0 to	$4 - 2x/3$	0 to	6
$dy \, dx \, dz$	0 to	$4 - 2x/3 - 4z/3$	0 to	$6 - 2z$	0 to	3

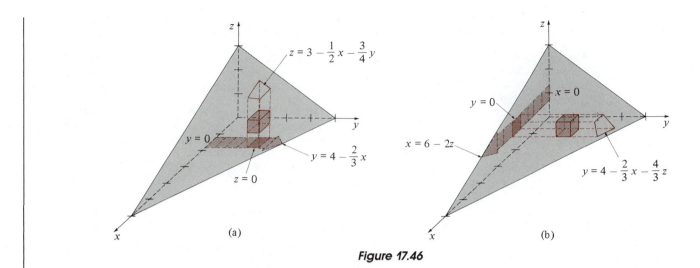

$$z = 3 - \frac{1}{2}x - \frac{3}{4}y$$

$y = 0$

$y = 4 - \frac{2}{3}x$

$z = 0$

(a)

$y = 0$

$x = 0$

$x = 6 - 2z$

$y = 4 - \frac{2}{3}x - \frac{4}{3}z$

(b)

Figure 17.46

Exercises 17.7

Answers to odd-numbered problems begin on page A-51.

In Problems 1–8 evaluate the given iterated integral.

1. $\displaystyle\int_2^4 \int_{-2}^2 \int_{-1}^1 (x + y + z)\, dx\, dy\, dz$

2. $\displaystyle\int_1^3 \int_1^x \int_2^{xy} 24xy\, dz\, dy\, dx$

3. $\displaystyle\int_0^6 \int_0^{6-x} \int_0^{6-x-z} dy\, dz\, dx$

4. $\displaystyle\int_0^1 \int_0^{1-x} \int_0^{\sqrt{y}} 4x^2 z^3\, dz\, dy\, dx$

5. $\displaystyle\int_0^{\pi/2} \int_0^{y^2} \int_0^y \cos\!\left(\frac{x}{y}\right) dz\, dx\, dy$

6. $\displaystyle\int_0^{\sqrt{2}} \int_{\sqrt{y}}^2 \int_0^{e x^2} x\, dz\, dx\, dy$

7. $\displaystyle\int_0^1 \int_0^1 \int_0^{2-x^2-y^2} xy e^z\, dz\, dx\, dy$

8. $\displaystyle\int_0^4 \int_0^{1/2} \int_0^{x^2} \frac{1}{\sqrt{x^2 - y^2}}\, dy\, dx\, dz$

9. Evaluate $\iiint_D z\, dV$, where D is the region in the first octant bounded by the graphs of $y = x$, $y = x - 2$, $y = 1$, $y = 3$, $z = 0$, and $z = 5$.

10. Evaluate $\iiint_D (x^2 + y^2)\, dV$, where D is the region bounded by the graphs of $y = x^2$, $z = 4 - y$, and $z = 0$.

In Problems 11 and 12 change the indicated order of integration to each of the other five orders.

11. $\displaystyle\int_0^2 \int_0^{4-2y} \int_{x+2y}^4 F(x, y, z)\, dz\, dx\, dy$

12. $\displaystyle\int_0^2 \int_0^{\sqrt{36-9x^2}/2} \int_1^3 F(x, y, z)\, dz\, dy\, dx$

In Problems 13 and 14 consider the solid given in the figure (Figures 17.47 and 17.48). Set up, but do not evaluate, the integrals, giving the volume V of the solid using the indicated orders of integration.

13.

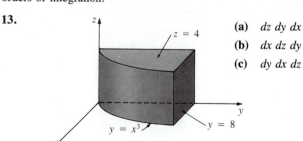

$z = 4$

$y = x^3$

$y = 8$

(a) $dz\, dy\, dx$
(b) $dx\, dz\, dy$
(c) $dy\, dx\, dz$

Figure 17.47

14.

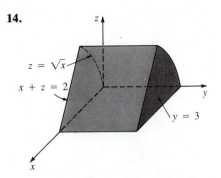

$z = \sqrt{x}$

$x + z = 2$

$y = 3$

(a) $dx\,dz\,dy$

(b) $dy\,dx\,dz$

(c) $dz\,dx\,dy$

(*Hint*: This will require two integrals.)

Figure 17.48

In Problems 15–20 sketch the region D whose volume V is given by the iterated integral.

15. $\displaystyle\int_0^4\int_0^3\int_0^{2-2z/3} dx\,dz\,dy$

16. $\displaystyle 4\int_0^3\int_0^{\sqrt{9-y^2}}\int_4^{\sqrt{25-x^2-y^2}} dz\,dx\,dy$

17. $\displaystyle\int_{-1}^1\int_{-\sqrt{1-x^2}}^{\sqrt{1-x^2}}\int_0^5 dz\,dy\,dx$

18. $\displaystyle\int_0^2\int_0^{\sqrt{4-x^2}}\int_{x^2+y^2}^4 dz\,dy\,dx$

19. $\displaystyle\int_0^2\int_0^{2-y}\int_{-\sqrt{y}}^{\sqrt{y}} dx\,dz\,dy$

20. $\displaystyle\int_1^3\int_0^{1/x}\int_0^3 dy\,dz\,dx$

In Problems 21–24 find the volume of the solid bounded by the graphs of the given equations.

21. $x = y^2,\ 4 - x = y^2,\ z = 0,\ z = 3$

22. $x^2 + y^2 = 4,\ z = x + y$, the coordinate planes, first octant

23. $y = x^2 + z^2,\ y = 8 - x^2 - z^2$

24. $x = 2,\ y = x,\ y = 0,\ z = x^2 + y^2,\ z = 0$

25. Find the center of mass of the solid given in Figure 17.47 if the density at a point P is directly proportional to the distance from the xy-plane.

26. Find the centroid of the solid in Figure 17.48, if the density is constant.

27. Find the center of mass of the solid bounded by the graphs of $x^2 + z^2 = 4$, $y = 0$, and $y = 3$ if the density at a point P is directly proportional to the distance from the xz-plane.

28. Find the center of mass of the solid bounded by the graphs of $y = x^2$, $y = x$, $z = y + 2$, and $z = 0$ if the density at a point P is directly proportional to the distance from the xy-plane.

In Problems 29 and 30 set up, but do not evaluate, the iterated integrals giving the mass of the solid that has the given shape and density.

29. $x^2 + y^2 = 1,\ z + y = 8,\ z - 2y = 2$; $\rho(x, y, z) = x + y + 4$

30. $x^2 + y^2 - z^2 = 1,\ z = -1,\ z = 2$; $\rho(x, y, z) = z^2$ (*Hint:* Do not use $dz\,dy\,dx$.)

31. Find the moment of inertia of the solid in Figure 17.47 about the y-axis if the density is as given in Problem 25. Find the radius of gyration.

32. Find the moment of inertia of the solid in Figure 17.48 about the x-axis if the density is constant. Find the radius of gyration.

33. Find the moment of inertia about the z-axis of the solid in the first octant that is bounded by the coordinate planes and the graph of $x + y + z = 1$ if the density is constant.

34. Find the moment of inertia about the y-axis of the solid bounded by the graphs of $z = y$, $z = 4 - y$, $z = 1$, $z = 0$, $x = 2$, and $x = 0$ if the density at a point P is directly proportional to the distance from the yz-plane.

In Problems 35 and 36 set up, but do not evaluate, the iterated integral giving the indicated moment of inertia of the solid having the given shape and density.

35. $z = \sqrt{x^2 + y^2},\ z = 5$; density at a point P directly proportional to the distance from the origin; I_z

36. $x^2 + z^2 = 1,\ y^2 + z^2 = 1$; density at a point P directly proportional to the distance from the yz-plane; I_y

17.8 Triple Integrals in Other Coordinate Systems

Depending on the geometry of a region in 3-space, the evaluation of a triple integral over that region may be made easier by utilizing a new coordinate system.

17.8.1 Cylindrical Coordinates

The **cylindrical coordinate system** combines the polar description of a point in the plane with the rectangular description of the z-component of a point in space. As seen in Figure 17.49(a), the cylindrical coordinates of a point P are denoted by the ordered triple (r, θ, z). The word *cylindrical* arises from the fact that a point P in space is determined by the intersection of the planes $z = $ constant, $\theta = $ constant, with a cylinder $r = $ constant. See Figure 17.49(b).

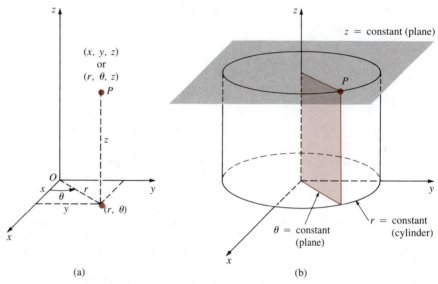

(a)   (b)

Figure 17.49

Conversion of Cylindrical Coordinates to Rectangular Coordinates

From Figure 17.49(a) we also see that the rectangular coordinates (x, y, z) of a point can be obtained from the cylindrical coordinates (r, θ, z) by means of

$$x = r \cos \theta, \qquad y = r \sin \theta, \qquad z = z \qquad (17.20)$$

_____ **Example 1** _____

Convert $(8, \pi/3, 7)$ in cylindrical coordinates to rectangular coordinates.

Solution   From (17.20),

$$x = 8 \cos \frac{\pi}{3} = 8 \left(\frac{1}{2}\right) = 4$$

$$y = 8 \sin \frac{\pi}{3} = 8 \left(\frac{\sqrt{3}}{2}\right) = 4\sqrt{3}$$

$$z = 7$$

Thus, $(8, \pi/3, 7)$ is equivalent to $(4, 4\sqrt{3}, 7)$ in rectangular coordinates.

Conversion of Rectangular Coordinates to Cylindrical Coordinates

To express rectangular coordinates (x, y, z) as cylindrical coordinates, we use

$$r^2 = x^2 + y^2, \qquad \tan \theta = \frac{y}{x}, \qquad z = z \qquad (17.21)$$

Example 2

Convert $(-\sqrt{2}, \sqrt{2}, 1)$ in rectangular coordinates to cylindrical coordinates.

Solution From (17.21) we see that

$$r^2 = (-\sqrt{2})^2 + (\sqrt{2})^2 = 4$$

$$\tan \theta = \frac{\sqrt{2}}{-\sqrt{2}} = -1$$

$$z = 1$$

If we take $r = 2$, then, consistent with the fact that $x < 0$ and $y > 0$, we take $\theta = 3\pi/4$.* Consequently, $(-\sqrt{2}, \sqrt{2}, 1)$ is equivalent to $(2, 3\pi/4, 1)$ in cylindrical coordinates. See Figure 17.50.

$(-\sqrt{2}, \sqrt{2}, 1)$
or
$(2, 3\pi/4, 1)$

$(-\sqrt{2}, \sqrt{2}, 0)$

$z = 1$

$r = 2$

$\theta = \frac{3\pi}{4}$

Figure 17.50

Triple Integrals in Cylindrical Coordinates

Recall from Section 17.5 that the area of a "polar rectangle" is $\Delta A = r^* \, \Delta r \, \Delta \theta$, where r^* is the average radius. From Figure 17.51(a) we see that the volume of a "cylindrical wedge" is simply

$$\Delta V = (\text{area of base}) \cdot (\text{height}) = r^* \, \Delta r \, \Delta \theta \, \Delta z$$

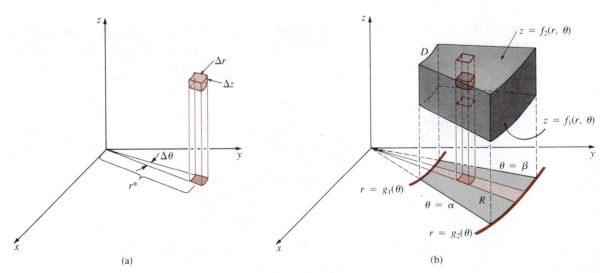

(a)

(b)

Figure 17.51

---

*If we use $\theta = \tan^{-1}(-1) = -\pi/4$, then we can use $r = -2$. Notice that the combinations $r = 2$, $\theta = -\pi/4$ and $r = -2$, $\theta = 3\pi/4$ are inconsistent.

Thus, if $F(r, \theta, z)$ is a continuous function over the region D, as shown in Figure 17.51(b), then the triple integral of F over D is given by

$$\iiint\limits_{D} F(r, \theta, z) \; dV = \iint\limits_{R} \left[\int_{f_1(r, \theta)}^{f_2(r, \theta)} F(r, \theta, z) \; dz \right] dA$$

$$= \int_{\alpha}^{\beta} \int_{g_1(\theta)}^{g_2(\theta)} \int_{f_1(r, \theta)}^{f_2(r, \theta)} F(r, \theta, z) r \; dz \; dr \; d\theta$$

Example 3

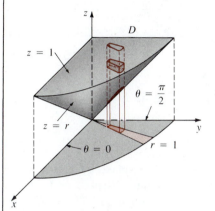

Figure 17.52

A solid in the first octant has the shape determined by the graph of the cone $z = \sqrt{x^2 + y^2}$ and the planes $z = 1$, $x = 0$, and $y = 0$. Find the center of mass if the density is given by $\rho(r, \theta, z) = r$.

Solution   In view of (17.21), the equation of the cone is $z = r$. Hence, we see from Figure 17.52 that

$$m = \iiint\limits_{D} r \; dV$$

$$= \int_{0}^{\pi/2} \int_{0}^{1} \int_{r}^{1} r(r \; dz \; dr \; d\theta)$$

$$= \int_{0}^{\pi/2} \int_{0}^{1} r^2 z \bigg]_{r}^{1} dr \; d\theta$$

$$= \int_{0}^{\pi/2} \int_{0}^{1} (r^2 - r^3) \; dr \; d\theta = \frac{\pi}{24}$$

$$M_{xy} = \iiint\limits_{D} zr \; dV$$

$$= \int_{0}^{\pi/2} \int_{0}^{1} \int_{r}^{1} zr^2 \; dz \; dr \; d\theta$$

$$= \int_{0}^{\pi/2} \int_{0}^{1} \frac{z^2}{2} r^2 \bigg]_{r}^{1} dr \; d\theta$$

$$= \frac{1}{2} \int_{0}^{\pi/2} \int_{0}^{1} (r^2 - r^4) \; dr \; d\theta = \frac{\pi}{30}$$

$$M_{xz} = \iiint\limits_{D} r^2 \sin \theta \; dV \qquad (y = r \sin \theta)$$

$$= \int_{0}^{\pi/2} \int_{0}^{1} \int_{r}^{1} r^3 \sin \theta \; dz \; dr \; d\theta$$

$$= \int_{0}^{\pi/2} \int_{0}^{1} r^3 z \sin \theta \bigg]_{r}^{1} dr \; d\theta$$

$$= \int_{0}^{\pi/2} \int_{0}^{1} (r^3 - r^4)\sin \theta \; dr \; d\theta = \frac{1}{20}$$

$$M_{yz} = \iiint\limits_{D} r^2 \cos \theta \; dV \qquad (x = r \cos \theta)$$

$$= \int_{0}^{\pi/2} \int_{0}^{1} \int_{r}^{1} r^3 \cos \theta \; dz \; dr \; d\theta = \frac{1}{20}$$

Hence,

$$\bar{x} = \frac{M_{yz}}{m} = \frac{1/20}{\pi/24} = \frac{6}{5\pi} \approx 0.38$$

$$\bar{y} = \frac{M_{xz}}{m} = \frac{1/20}{\pi/24} = \frac{6}{5\pi} \approx 0.38$$

$$\bar{z} = \frac{M_{xy}}{m} = \frac{\pi/30}{\pi/24} = \frac{4}{5} = 0.8$$

The center of mass has the approximate coordinates $(0.38, 0.38, 0.8)$.

Example 4

Evaluate the volume integral

$$V = 4 \int_0^2 \int_0^{\sqrt{4-y^2}} \int_{\sqrt{y^2+z^2}}^{6-y^2-x^2} dx \, dz \, dy$$

of Example 2 of Section 17.7.

Solution If we introduce polar coordinates in the yz-plane by $y = r \cos \theta$, $z = r \sin \theta$, then the cylindrical coordinates of a point in 3-space are (r, θ, x). The polar analogue of Figure 17.44(b) is given in Figure 17.53. Now, since $y^2 + z^2 = r^2$, we have

$$x = \sqrt{y^2 + z^2} = r \quad \text{and} \quad x = 6 - y^2 - z^2 = 6 - r^2$$

Hence, the integral becomes

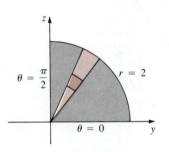

Figure 17.53

$$V = 4 \int_0^{\pi/2} \int_0^2 \int_r^{6-r^2} r \, dx \, dr \, d\theta$$

$$= 4 \int_0^{\pi/2} \int_0^2 rx \Big]_r^{6-r^2} dr \, d\theta$$

$$= 4 \int_0^{\pi/2} \int_0^2 (6r - r^3 - r^2) \, dr \, d\theta$$

$$= 4 \int_0^{\pi/2} \left[3r^2 - \frac{r^4}{4} - \frac{r^3}{3} \right]_0^2 d\theta$$

$$= \frac{64}{3} \int_0^{\pi/2} d\theta = \frac{32\pi}{3} \text{ cubic units}$$

17.8.2 Spherical Coordinates

As seen in Figure 17.54(a), the **spherical coordinates** of a point P are given by the ordered triple (ρ, ϕ, θ), where ρ is the distance from the origin to P, ϕ is the angle between the positive z-axis and the vector $\overrightarrow{OP}$, and θ is the angle measured from the positive x-axis to the vector projection $\overrightarrow{OQ}$ of $\overrightarrow{OP}$.* Figure 17.54(b) shows that a point P in space is determined by the intersection of a

*θ is the same angle as in polar and cylindrical coordinates.

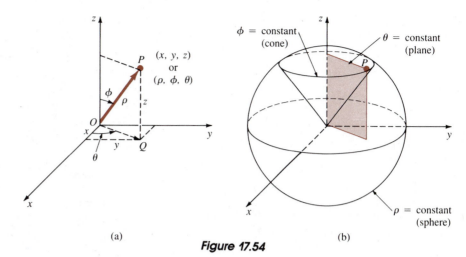

Figure 17.54

cone ϕ = constant, a plane θ = constant, and a sphere ρ = constant; whence arises the name "spherical" coordinates.

Conversion of Spherical Coordinates to Rectangular and Cylindrical Coordinates

To transform from spherical coordinates (ρ, ϕ, θ) to rectangular coordinates (x, y, z), we observe from Figure 17.54(a) that

$$x = |\overrightarrow{OQ}|\cos\theta, \qquad y = |\overrightarrow{OQ}|\sin\theta, \qquad z = |\overrightarrow{OP}|\cos\phi$$

Since $|\overrightarrow{OQ}| = \rho\sin\phi$ and $|\overrightarrow{OP}| = \rho$, the foregoing equations become

$$x = \rho\sin\phi\cos\theta, \qquad y = \rho\sin\phi\sin\theta, \qquad z = \rho\cos\phi \quad (17.22)$$

It is customary to take $\rho \geq 0$ and $0 \leq \phi \leq \pi$. Also, since $|\overrightarrow{OQ}| = \rho\sin\phi = r$, the formulas

$$r = \rho\sin\phi, \qquad \theta = \theta, \qquad z = \rho\cos\phi \quad (17.23)$$

enable us to transform from spherical coordinates (ρ, ϕ, θ) to cylindrical coordinates (r, θ, z).

Example 5

Convert $(6, \pi/4, \pi/3)$ in spherical coordinates to (a) rectangular coordinates, (b) cylindrical coordinates.

Solution

(a) Identifying $\rho = 6$, $\phi = \pi/4$, and $\theta = \pi/3$, it follows from (17.22) that the rectangular coordinates of the point are given by

$$x = 6\sin\frac{\pi}{4}\cos\frac{\pi}{3} = 6\left(\frac{\sqrt{2}}{2}\right)\left(\frac{1}{2}\right) = \frac{3\sqrt{2}}{2}$$

$$y = 6\sin\frac{\pi}{4}\sin\frac{\pi}{3} = 6\left(\frac{\sqrt{2}}{2}\right)\left(\frac{\sqrt{3}}{2}\right) = \frac{3\sqrt{6}}{2}$$

$$z = 6\cos\frac{\pi}{4} = 6\left(\frac{\sqrt{2}}{2}\right) = 3\sqrt{2}$$

(b)   From (17.23) we obtain

$$r = 6 \sin \frac{\pi}{4} = 6\left(\frac{\sqrt{2}}{2}\right) = 3\sqrt{2}$$

$$\theta = \frac{\pi}{3}$$

$$z = 6 \cos \frac{\pi}{4} = 6\left(\frac{\sqrt{2}}{2}\right) = 3\sqrt{2}$$

Thus, the cylindrical coordinates of the point are $(3\sqrt{2}, \pi/3, 3\sqrt{2})$.

Conversion of Rectangular Coordinates to Spherical Coordinates

To transform from rectangular coordinates to spherical coordinates, we use

$$\rho^2 = x^2 + y^2 + z^2, \qquad \tan \theta = \frac{y}{x}, \qquad \cos \phi = \frac{z}{\sqrt{x^2 + y^2 + z^2}} \qquad (17.24)$$

Triple Integrals in Spherical Coordinates

As seen in Figure 17.55, the volume of a "spherical wedge" is given by the approximation

$$\Delta V \approx \rho^2 \sin \phi \, \Delta \rho \, \Delta \phi \, \Delta \theta$$

Thus, in a triple integral of a continuous spherical-coordinate function $F(\rho, \phi, \theta)$, the differential of volume dV is given by

$$dV = \rho^2 \sin \phi \, d\rho \, d\phi \, d\theta$$

A typical triple integral in spherical coordinates has the form

$$\iiint_D F(\rho, \phi, \theta) \, dV = \int_\alpha^\beta \int_{g_1(\theta)}^{g_2(\theta)} \int_{f_1(\phi, \theta)}^{f_2(\phi, \theta)} F(\rho, \phi, \theta)\rho^2 \sin \phi \, d\rho \, d\phi \, d\theta$$

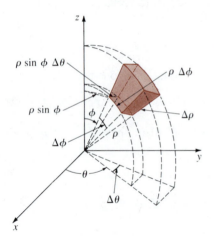

Figure 17.55

_____ **Example 6** _____

Use spherical coordinates to find the volume of the solid in Example 3.

Solution   Using (17.22)

$$z = 1 \quad \text{becomes} \quad \rho = \sec \phi$$

and $\qquad\qquad z = \sqrt{x^2 + y^2} \quad \text{becomes} \quad \phi = \pi/4$

As indicated in Figure 17.56,

$$V = \iiint_D dV$$

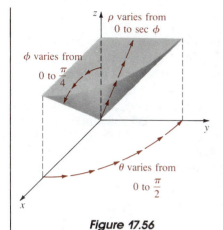

Figure 17.56

written as an iterated integral, is

$$V = \int_0^{\pi/2} \int_0^{\pi/4} \int_0^{\sec \phi} \rho^2 \sin \phi \, d\rho \, d\phi \, d\theta$$

Hence,

$$V = \int_0^{\pi/2} \int_0^{\pi/4} \frac{\rho^3}{3}\bigg]_0^{\sec \phi} \sin \phi \, d\phi \, d\theta$$

$$= \frac{1}{3} \int_0^{\pi/2} \int_0^{\pi/4} \sec^3\phi \sin \phi \, d\phi \, d\theta$$

$$= \frac{1}{3} \int_0^{\pi/2} \int_0^{\pi/4} \tan \phi \sec^2\phi \, d\phi \, d\theta$$

$$= \frac{1}{3} \int_0^{\pi/2} \frac{1}{2} \tan^2\phi \bigg]_0^{\pi/4} d\theta$$

$$= \frac{1}{6} \int_0^{\pi/2} d\theta = \frac{\pi}{12} \text{ cubic units}$$

Example 7

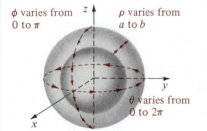

Figure 17.57

Find the moment of inertia about the z-axis of the homogeneous solid bounded between the spheres

$$x^2 + y^2 + z^2 = a^2 \quad \text{and} \quad x^2 + y^2 + z^2 = b^2, \qquad a < b$$

Solution   If $\delta(\rho, \phi, \theta) = k$ is the density,* then

$$I_z = \iiint_D (x^2 + y^2)k \, dV$$

From (17.22) we find $x^2 + y^2 = \rho^2 \sin^2 \phi$, and from the first equation in (17.23), we see that the equations of the spheres are simply $\rho = a$ and $\rho = b$. See Figure 17.57. Consequently, in spherical coordinates the foregoing integral becomes

$$I_z = k \int_0^{2\pi} \int_0^{\pi} \int_a^{b} \rho^2 \sin^2\phi (\rho^2 \sin \phi \, d\rho \, d\phi \, d\theta)$$

$$= k \int_0^{2\pi} \int_0^{\pi} \int_a^{b} \rho^4 \sin^3\phi \, d\rho \, d\phi \, d\theta$$

$$= k \int_0^{2\pi} \int_0^{\pi} \frac{\rho^5}{5} \sin^3\phi \bigg]_a^{b} d\phi \, d\theta$$

$$= \frac{k}{5}(b^5 - a^5) \int_0^{2\pi} \int_0^{\pi} (1 - \cos^2\phi)\sin \phi \, d\phi \, d\theta$$

$$= \frac{k}{5}(b^5 - a^5) \int_0^{2\pi} \left[-\cos \phi + \frac{1}{3}\cos^3\phi \right]_0^{\pi} d\theta$$

$$= \frac{4k}{15}(b^5 - a^5) \int_0^{2\pi} d\theta = \frac{8\pi k}{15}(b^5 - a^5)$$

---

*We must use a different symbol to denote density to avoid confusion with the symbol ρ of spherical coordinates.

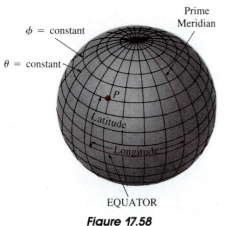

Prime
Meridian

Figure 17.58

Remark

Spherical coordinates are used in navigation. If we think of the earth as a sphere of fixed radius centered at the origin, then a point P can be located by specifying two angles θ and ϕ. As shown in Figure 17.58, when ϕ is held constant the resulting curve is called a **parallel**. Fixed values of θ result in curves called **great circles**. A half of one of these great circles joining the north and south poles is called a **meridian**. The intersection of a parallel and a meridian gives the position of a point P. If $0° \leq \phi \leq 180°$ and $-180° \leq \theta \leq 180°$, the angles $90° - \phi$ and θ are said to be the **latitude** and **longitude** of P, respectively. The **prime meridian** corresponds to a longitude of $0°$. The latitude of the equator is $0°$; the latitudes of the north and south poles are, in turn, $+90°$ (or $90°$ North) and $-90°$ (or $90°$ South).

Exercises 17.8

Answers to odd-numbered problems begin on page A-52.

[17.8.1]

In Problems 1–6 convert the point given in cylindrical coordinates to rectangular coordinates.

1. $\left(10, \dfrac{3\pi}{4}, 5\right)$ **2.** $\left(2, \dfrac{5\pi}{6}, -3\right)$

3. $\left(\sqrt{3}, \dfrac{\pi}{3}, -4\right)$ **4.** $\left(4, \dfrac{7\pi}{4}, 0\right)$

5. $\left(5, \dfrac{\pi}{2}, 1\right)$ **6.** $\left(10, \dfrac{5\pi}{3}, 2\right)$

In Problems 7–12 convert the point given in rectangular coordinates to cylindrical coordinates.

7. $(1, -1, -9)$ **8.** $(2\sqrt{3}, 2, 17)$

9. $(-\sqrt{2}, \sqrt{6}, 2)$ **10.** $(1, 2, 7)$

11. $(0, -4, 0)$ **12.** $(\sqrt{7}, -\sqrt{7}, 3)$

In Problems 13–16 convert the given equation to cylindrical coordinates.

13. $x^2 + y^2 + z^2 = 25$ **14.** $x + y - z = 1$

15. $x^2 + y^2 - z^2 = 1$ **16.** $x^2 + z^2 = 16$

In Problems 17–20 convert the given equation to rectangular coordinates.

17. $z = r^2$ **18.** $z = 2r \sin \theta$

19. $r = 5 \sec \theta$ **20.** $\theta = \pi/6$

In the remaining problems use triple integrals and cylindrical coordinates.

In Problems 21–24 find the volume of the solid that is bounded by the graphs of the given equations.

21. $x^2 + y^2 = 4$, $x^2 + y^2 + z^2 = 16$, $z = 0$

22. $z = 10 - x^2 - y^2$, $z = 1$

23. $z = x^2 + y^2$, $x^2 + y^2 = 25$, $z = 0$

24. $y = x^2 + z^2$, $2y = x^2 + z^2 + 4$

25. Find the centroid of the homogeneous solid that is bounded by the hemisphere $z = \sqrt{a^2 - x^2 - y^2}$ and the plane $z = 0$.

26. Find the center of mass of the solid that is bounded by the graphs of $y^2 + z^2 = 16$, $x = 0$, and $x = 5$ if the density at a point P is directly proportional to distance from the yz-plane.

27. Find the moment of inertia about the z-axis of the solid that is bounded above by the hemisphere $z = \sqrt{9 - x^2 - y^2}$ and below by the plane $z = 2$ if the density at a point P is inversely proportional to the square of the distance from the z-axis.

28. Find the moment of inertia about the x-axis of the solid that is bounded by the cone $z = \sqrt{x^2 + y^2}$ and the plane $z = 1$ if the density at a point P is directly proportional to the distance from the z-axis.

[17.8.2]

In Problems 29–34 convert the point given in spherical coordinates to **(a)** rectangular coordinates, and **(b)** cylindrical coordinates.

29. $\left(\dfrac{2}{3}, \dfrac{\pi}{2}, \dfrac{\pi}{6}\right)$

30. $\left(5, \dfrac{5\pi}{4}, \dfrac{2\pi}{3}\right)$

31. $\left(8, \dfrac{\pi}{4}, \dfrac{3\pi}{4}\right)$

32. $\left(\dfrac{1}{3}, \dfrac{5\pi}{3}, \dfrac{\pi}{6}\right)$

33. $\left(4, \dfrac{3\pi}{4}, 0\right)$

34. $\left(1, \dfrac{11\pi}{6}, \pi\right)$

In Problems 35–40 convert the points given in rectangular coordinates to spherical coordinates.

35. $(-5, -5, 0)$

36. $(1, -\sqrt{3}, 1)$

37. $\left(\dfrac{\sqrt{3}}{2}, \dfrac{1}{2}, 1\right)$

38. $\left(-\dfrac{\sqrt{3}}{2}, 0, -\dfrac{1}{2}\right)$

39. $(3, -3, 3\sqrt{2})$

40. $(1, 1, -\sqrt{6})$

In Problems 41–44 convert the given equation to spherical coordinates.

41. $x^2 + y^2 + z^2 = 64$

42. $x^2 + y^2 + z^2 = 4z$

43. $z^2 = 3x^2 + 3y^2$

44. $-x^2 - y^2 + z^2 = 1$

In Problems 45–48 convert the given equation to rectangular coordinates.

45. $\rho = 10$

46. $\phi = \pi/3$

47. $\rho = 2 \sec \phi$

48. $\rho \sin^2 \phi = \cos \phi$

In the remaining problems use triple integrals and spherical coordinates.

In Problems 49–52 find the volume of the solid that is bounded by the graphs of the given equations.

49. $z = \sqrt{x^2 + y^2}$, $x^2 + y^2 + z^2 = 9$

50. $x^2 + y^2 + z^2 = 4$, $y = x$, $y = \sqrt{3}x$, $z = 0$, first octant

51. $z^2 = 3x^2 + 3y^2$, $x = 0$, $y = 0$, $z = 2$, first octant

52. Inside $x^2 + y^2 + z^2 = 1$ and outside $z^2 = x^2 + y^2$

53. Find the centroid of the homogeneous solid that is bounded by the cone $z = \sqrt{x^2 + y^2}$ and the sphere $x^2 + y^2 + z^2 = 2z$.

54. Find the center of mass of the solid that is bounded by the hemisphere $z = \sqrt{1 - x^2 - y^2}$ and the plane $z = 0$ if the density at a point P is directly proportional to distance from the xy-plane.

55. Find the mass of the solid that is bounded above by the hemisphere $z = \sqrt{25 - x^2 - y^2}$ and below by the plane $z = 4$ if the density at a point P is inversely proportional to the distance from the origin. (*Hint:* Express the upper ϕ limit of integration as an inverse cosine.)

56. Find the moment of inertia about the z-axis of the solid that is bounded by the sphere $x^2 + y^2 + z^2 = a^2$ if the density at a point P is directly proportional to the distance from the origin.

Chapter 17 Review Exercises

Answers to odd-numbered problems begin on page A-52.

In Problems 1–4 answer true or false.

1. $\displaystyle\int_{-2}^{3} \int_{1}^{5} e^{x^2 - y} \, dx \, dy = \int_{1}^{5} \int_{-2}^{3} e^{x^2 - y} \, dy \, dx$ _____

2. If I is the partial integral $\displaystyle\int_{g_1(x)}^{g_2(x)} f(x, y) \, dy$, then $\partial I / \partial y = 0$. _____

3. For every continuous function f,

$$\int_{-1}^{1} \int_{x^2}^{1} f(x, y) \, dy \, dx = 2 \int_{0}^{1} \int_{x^2}^{1} f(x, y) \, dy \, dx$$ _____

4. The center of mass of a lamina possessing symmetry always lies on the axis of symmetry. _____

In Problems 5–16 fill in the blanks.

5. $\displaystyle\int_{y^2+1}^{5} \left(8y^3 - \dfrac{5y}{x}\right) dx =$ _____

6. If R_1 and R_2 are nonoverlapping regions such that $R = R_1 \cup R_2$, $\iint_R f(x, y) \, dA = 10$, and $\iint_{R_2} f(x, y) \, dA = -6$, then $\iint_{R_1} f(x, y) \, dA =$ _____.

7. $\displaystyle\int_{-a}^{a} \int_{-a}^{a} dx \, dy$ gives the area of a _____.

8. The region bounded by the graphs of $9x^2 + y^2 = 36$, $y = -2$, $y = 5$ is of Type _____.

9. $\displaystyle\int_{2}^{4} f_y(x, y) \, dy =$ _____

10. If $\rho(x, y, z)$ is density, then the iterated integral giving the mass of the solid bounded by the ellipsoid $x^2/a^2 + y^2/b^2 + z^2/c^2 = 1$ is _____ .

11. $\int_0^2 \int_{y^2}^{2y} f(x, y)\, dx\, dy = \int_-^- \int_-^- f(x, y)\, dy\, dx$

12. The rectangular coordinates of the point $\left(6, \dfrac{5\pi}{3}, \dfrac{5\pi}{6}\right)$ given in spherical coordinates are _____ .

13. The cylindrical coordinates of the point $\left(2, \dfrac{\pi}{4}, \dfrac{2\pi}{3}\right)$ given in spherical coordinates are _____ .

14. The region R bounded by the graphs of $y = 4 - x^2$ and $y = 0$ is both Type I and Type II. Interpreted as a Type II region, $\iint_R f(x, y)\, dA = \int_-^- \int_-^- f(x, y) \underline{\quad}\ \underline{\quad}$.

15. The equation of the paraboloid $z = x^2 + y^2$ in cylindrical coordinates is _____ , whereas in spherical coordinates the equation is _____ .

16. In cylindrical *and* spherical coordinates the equation of the plane $y = x$ is _____ .

In Problems 17–28 evaluate the given integral.

17. $\int_{y^3}^y y^2 \sin xy\, dx$

18. $\int_{1/x}^{e^x} \dfrac{x}{y^2}\, dy$

19. $\int_0^2 \int_0^{2x} ye^{y-x}\, dy\, dx$

20. $\int_0^4 \int_x^4 \dfrac{1}{16 + x^2}\, dy\, dx$

21. $\int_0^1 \int_x^{\sqrt{x}} \dfrac{\sin y}{y}\, dy\, dx$

22. $\int_e^{e^2} \int_0^{1/x} \ln x\, dy\, dx$

23. $\int_0^5 \int_0^{\pi/2} \int_0^{\cos \theta} 3r^2\, dr\, d\theta\, dz$

24. $\int_{\pi/4}^{\pi/2} \int_0^{\sin z} \int_0^{\ln x} e^y\, dy\, dx\, dz$

25. $\iint_R 5\, dA$, where R is bounded by the circle

$x^2 + y^2 = 64$

26. $\iint_R dA$, where R is bounded by the cardioid

$r = 1 + \cos \theta$

27. $\iint_R (2x + y)\, dA$, where R is bounded by the graphs of

$y = \tfrac{1}{2}x, x = y^2 + 1, y = 0$

28. $\iiint_D x\, dV$, where D is bounded by the planes

$z = x + y, z = 6 - x - y, x = 0, y = 0$

29. Using rectangular coordinates, express as an iterated integral:

$$\iint_R \dfrac{1}{x^2 + y^2}\, dA$$

where R is the region in the first quadrant that is bounded by the graphs of $x^2 + y^2 = 1, x^2 + y^2 = 9, x = 0$ and $y = x$. Do not evaluate.

30. Evaluate the double integral in Problem 29 using polar coordinates.

In Problems 31 and 32 sketch the region of integration.

31. $\int_{-2}^2 \int_{-x^2}^{x^2} f(x, y)\, dy\, dx$

32. $\int_{-1}^1 \int_{-1}^1 \int_0^{x^2+y^2} F(x, y, z)\, dz\, dx\, dy$

33. Reverse the order of integration and evaluate

$$\int_0^1 \int_y^{\sqrt[3]{y}} \cos x^2\, dx\, dy$$

34. Consider $\iiint_D F(x, y, z)\, dV$, where D is the region in the first octant bounded by the planes $z = 8 - 2x - y, z = 4, x = 0, y = 0$. Express the triple integral as six different iterated integrals.

In Problems 35 and 36 use an appropriate coordinate system to evaluate the given integral.

35. $\int_0^2 \int_{1/2}^1 \int_0^{\sqrt{x-x^2}} (4z + 1)\, dy\, dx\, dz$

36. $\int_0^1 \int_0^{\sqrt{1-x^2}} \int_{-\sqrt{1-x^2-y^2}}^{\sqrt{1-x^2-y^2}} (x^2 + y^2 + z^2)^4\, dz\, dy\, dx$

37. Find the surface area of that portion of the graph of $z = xy$ within the cylinder $x^2 + y^2 = 1$.

38. Use a double integral to find the volume of the solid shown in Figure 17.59.

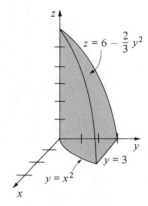

Figure 17.59

39. Express the volume of the solid shown in Figure 17.60 as one or more iterated integrals using the order of integration **(a)** $dy\, dx$, **(b)** $dx\, dy$. Choose either part **(a)** or part **(b)** to find the volume.

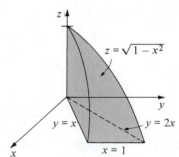

Figure 17.60

40. A lamina has the shape of the region in the first quadrant bounded by the graphs of $y = x^2$ and $y = x^3$. Find the center of mass if the density at a point P is directly proportional to the square of the distance from the origin.

41. Find the moment of inertia of the lamina described in Problem 40 about the y-axis.

42. Find the volume of the sphere $x^2 + y^2 + z^2 = a^2$ using a triple integral in **(a)** rectangular coordinates, **(b)** cylindrical coordinates, and **(c)** spherical coordinates.

43. Find the volume of the solid that is bounded between the cones $z = \sqrt{x^2 + y^2}$, $z = \sqrt{9x^2 + 9y^2}$, and the plane $z = 3$.

44. Find the volume of the solid shown in Figure 17.61.

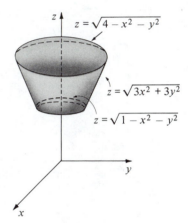

Figure 17.61

18

Vector Integral Calculus

So far we have studied three kinds of integrals, the definite integral, the double integral, and the triple integral. In this present chapter we will introduce two new kinds of integrals, **line** and **surface integrals**. The development of these concepts depends heavily on vector methods.

18.1 Line Integrals

The notion of integration of a function defined over an interval can be general-ized to integration of a function *defined along a curve*. To this end we need to introduce some terminology about curves. Suppose C is a curve parameterized by $x = f(t)$, $y = g(t)$, $a \leq t \leq b$, and A and B are the points $(f(a), g(a))$ and $(f(b), g(b))$, respectively. We say that

(*i*) C is a **smooth curve** if f' and g' are continuous on $[a, b]$ and not simul-taneously zero on (a, b).

(*ii*) C is **piecewise smooth** if it can be expressed as a union of a finite number of smooth curves.

(*iii*) C is a **closed curve** if $A = B$.

(*iv*) C is a **simple closed curve** if $A = B$ and the curve does not cross itself.

(*v*) If C is not a closed curve, then the **positive direction** on C is the direction corresponding to increasing values of t.

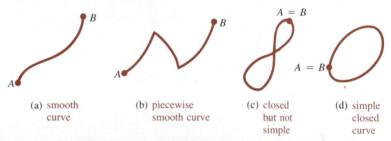

(a) smooth curve   (b) piecewise smooth curve   (c) closed but not simple   (d) simple closed curve

Figure 18.1

Figure 18.1 illustrates each type of curve defined in (*i*)–(*iv*). The region R, *interior* to a simple closed curve, is said to be **simply connected**. A simply connected region does not have any "holes" in it. See Figure 18.2.

(a) R simply connected   (b) R not simply connected

Figure 18.2

The following five steps lead to the definitions of three **line integrals*** in the plane.

---

*Another unfortunate choice of names. **Curve integrals** would be more appropriate.

$$z = G(x, y)$$

1. Let G be defined in some region that contains the smooth curve C defined by $x = f(t)$, $y = g(t)$, $a \leq t \leq b$.

2. Divide C into n subarcs of lengths Δs_k according to the partition $a = t_0 < t_1 < t_2 < \cdots < t_n = b$ of $[a, b]$. Let the projection of each subarc onto the x- and y-axes have lengths Δx_k and Δy_k, respectively.

3. Let $\|P\|$ be the **norm** of the subdivision or the length of the longest subarc.

4. Choose a point (x_k^*, y_k^*) on each subarc.

5. Form the sums

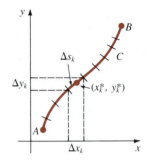

$$\sum_{k=1}^{n} G(x_k^*, y_k^*) \, \Delta x_k, \qquad \sum_{k=1}^{n} G(x_k^*, y_k^*) \, \Delta y_k, \qquad \sum_{k=1}^{n} G(x_k^*, y_k^*) \, \Delta s_k$$

DEFINITION 18.1   Line Integrals

Let G be a function of two variables x and y defined on a region of the plane containing a smooth curve C. Then

(*i*) the **line integral of G along C from A to B with respect to x** is

$$\int_C G(x, y) \, dx = \lim_{\|P\| \to 0} \sum_{k=1}^{n} G(x_k^*, y_k^*) \, \Delta x_k$$

(*ii*) the **line integral of G along C from A to B with respect to y** is

$$\int_C G(x, y) \, dy = \lim_{\|P\| \to 0} \sum_{k=1}^{n} G(x_k^*, y_k^*) \, \Delta y_k$$

(*iii*) the **line integral of G along C from A to B with respect to arc length** is

$$\int_C G(x, y) \, ds = \lim_{\|P\| \to 0} \sum_{k=1}^{n} G(x_k^*, y_k^*) \, \Delta s_k$$

It can be proved that if $G(x, y)$ is continuous on C, then the integrals defined in (*i*), (*ii*), and (*iii*) exist. We shall assume continuity of G as a matter of course.

Method of Evaluation   The line integrals in Definition 18.1 can be evaluated in two ways, depending on whether the curve C is defined parametrically or defined by an explicit function. In either case the basic idea is to convert the line integral to a definite integral in a single variable. If C is a smooth curve parameterized by $x = f(t)$, $y = g(t)$, $a \leq t \leq b$, then we have immediately

$$\int_C G(x, y)\, dx = \int_a^b G(f(t),\, g(t))f'(t)\, dt \qquad (18.1)$$

$$\int_C G(x, y)\, dy = \int_a^b G(f(t),\, g(t))g'(t)\, dt \qquad (18.2)$$

Furthermore, using (6.14) of Section 6.5 and the given parameterization we find $ds = \sqrt{[f'(t)]^2 + [g'(t)]^2}\, dt$. Hence,

$$\int_C G(x, y)\, ds = \int_a^b G(f(t),\, g(t))\sqrt{[f'(t)]^2 + [g'(t)]^2}\, dt \qquad (18.3)$$

Alternatively, if the curve C is defined by an explicit function $y = f(x)$, $a \leq x \leq b$, we can use x as a parameter. Employing $dy = f'(x)\, dx$ and $ds = \sqrt{1 + [f'(x)]^2}\, dx$, the foregoing line integrals become, in turn,

$$\int_C G(x, y)\, dx = \int_a^b G(x, f(x))\, dx \qquad (18.4)$$

$$\int_C G(x, y)\, dy = \int_a^b G(x, f(x))f'(x)\, dx \qquad (18.5)$$

$$\int_C G(x, y)\, ds = \int_a^b G(x, f(x))\sqrt{1 + [f'(x)]^2}\, dx \qquad (18.6)$$

A line integral along a *piecewise smooth* curve C is defined as the *sum* of the integrals over the various smooth curves whose union comprises C. For example, if C is composed of smooth curves C_1 and C_2, then

$$\int_C G(x, y)\, ds = \int_{C_1} G(x, y)\, ds + \int_{C_2} G(x, y)\, ds$$

Example 1

Evaluate (a)$\int_C xy^2\, dx$,   (b) $\int_C xy^2\, dy$,   (c) $\int_C xy^2\, ds$ on the quarter-circle C defined by $x = 4\cos t$, $y = 4\sin t$, $0 \leq t \leq \pi/2$. See Figure 18.3.

Solution
(a)   From (18.1),

$$\int_C xy^2\, dx = \int_0^{\pi/2} (4\cos t)(16\sin^2 t)(-4\sin t\, dt)$$

$$= -256 \int_0^{\pi/2} \sin^3 t\, \cos t\, dt$$

$$= -256 \left[\frac{1}{4}\sin^4 t\right]_0^{\pi/2} = -64$$

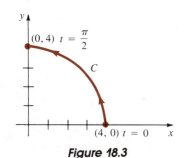

Figure 18.3

(*b*)   From (18.2),

$$\int_C xy^2 \, dy = \int_0^{\pi/2} (4 \cos t)(16 \sin^2 t)(4 \cos t \, dt)$$

$$= 256 \int_0^{\pi/2} \sin^2 t \, \cos^2 t \, dt$$

$\boxed{\text{Trig identities}}$

$$= 256 \int_0^{\pi/2} \frac{1}{4} \sin^2 2t \, dt$$

$$= 64 \int_0^{\pi/2} \frac{1}{2}(1 - \cos 4t) \, dt$$

$$= 32 \left[t - \frac{1}{4} \sin 4t \right]_0^{\pi/2} = 16\pi$$

(*c*)   From (18.3),

$$\int_C xy^2 \, ds = \int_0^{\pi/2} (4 \cos t)(16 \sin^2 t)\sqrt{16(\cos^2 t + \sin^2 t)} \, dt$$

$$= 256 \int_0^{\pi/2} \sin^2 t \, \cos t \, dt$$

$$= 256 \left[\frac{1}{3} \sin^3 t \right]_0^{\pi/2} = \frac{256}{3}$$

Notation   In many applications, line integrals appear as a sum

$$\int_C P(x, \, y) \, dx + \int_C Q(x, \, y) \, dy$$

It is common practice to write this sum as one integral without parentheses as

$$\int_C P(x, \, y) \, dx + Q(x, \, y) \, dy \quad \text{or simply} \quad \int_C P \, dx + Q \, dy \qquad (18.7)$$

A line integral along a *closed* curve *C* is very often denoted by

$$\oint_C P \, dx + Q \, dy$$

Example 2

Evaluate $\int_C xy \, dx + x^2 \, dy$, where *C* is given by $y = x^3, \, -1 \leq x \leq 2$.

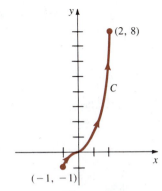

Figure 18.4

Solution   The curve C is illustrated in Figure 18.4. Since $y = x^3$, we have $dy = 3x^2\ dx$. Using x as the parameter, it follows from (18.4) and (18.5) that

$$\int_C xy\ dx + x^2\ dy = \int_{-1}^{2} \overbrace{x(x^3)}^{y}\ dx + x^2\overbrace{(3x^2\ dx)}^{dy}$$

$$= \int_{-1}^{2} 4x^4\ dx$$

$$= \frac{4}{5}x^5\Big]_{-1}^{2} = \frac{132}{5}$$

_____ **Example 3** _____

Evaluate $\oint_C x\ dx$, where C is the circle $x = \cos t$, $y = \sin t$, $0 \le t \le 2\pi$.

Solution   From (18.1),

$$\oint_C x\ dx = \int_0^{2\pi} \cos t(-\sin t\ dt) = \frac{1}{2}\cos^2 t\Big]_0^{2\pi} = \frac{1}{2}[1 - 1] = 0$$

_____ **Example 4** _____

Evaluate $\oint_C y^2\ dx - x^2\ dy$ on the closed curve C that is shown in Figure 18.5(a).

Solution   Since C is piecewise smooth, we express the integral as a sum of integrals. Symbolically, we write

$$\oint_C = \int_{C_1} + \int_{C_2} + \int_{C_3}$$

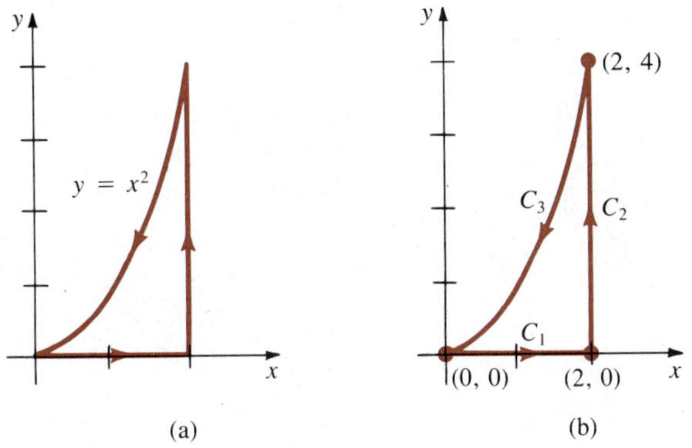

(a)

(b)

Figure 18.5

where C_1, C_2, and C_3 are the curves shown in Figure 18.5(b). On C_1, we use x as a parameter. Since $y = 0$, $dy = 0$; therefore,

$$\int_{C_1} y^2 \, dx - x^2 \, dy = \int_0^2 0 \, dx - x^2(0) = 0$$

On C_2, we use y as a parameter. From $x = 2$, $dx = 0$, we have

$$\int_{C_2} y^2 \, dx - x^2 \, dy = \int_0^4 y^2(0) - 4 \, dy$$

$$= -\int_0^4 4 \, dy = -4y \Big]_0^4 = -16$$

Finally, on C_3, we again use x as a parameter. From $y = x^2$, we get $dy = 2x \, dx$ and so

$$\int_{C_3} y^2 \, dx - x^2 \, dy = \int_2^0 x^4 \, dx - x^2(2x \, dx)$$

$$= \int_2^0 (x^4 - 2x^3) \, dx$$

$$= \left(\frac{1}{5}x^5 - \frac{1}{2}x^4 \right) \Big]_2^0 = \frac{8}{5}$$

Hence, $$\oint_C y^2 \, dx - x^2 \, dy = 0 - 16 + \frac{8}{5} = -\frac{72}{5}$$

It is important to note that

a line integral is independent of the parameterization of the curve C provided C is given the same orientation by all sets of parametric equations defining the curve.

See Problem 37 in Exercises 18.1. Also, recall for definite integrals that

$$\int_b^a f(x) \, dx = -\int_a^b f(x) \, dx$$

Line integrals possess a similar property. Suppose, as shown in Figure 18.6, that $-C$ denotes the curve having the opposite orientation of C. Then it can be shown that

$$\int_{-C} P \, dx + Q \, dy = -\int_C P \, dx + Q \, dy$$

or equivalently,

$$\int_{-C} P \, dx + Q \, dy + \int_C P \, dx + Q \, dy = 0 \tag{18.8}$$

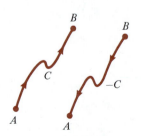

Figure 18.6

For example, in part (*a*) of Example 1, $\int_{-C} xy^2 \, dx = 64$.

Line Integrals in Space   Line integrals of a function G of three variables, $\int_C G(x, y, z)\, dx$, $\int_C G(x, y, z)\, dy$, and $\int_C G(x, y, z)\, ds$ are defined in a manner analogous to Definition 18.1. However, to that list we add a fourth **line integral along a space curve C with respect to z:**

$$\int_C G(x, y, z)\, dz = \lim_{\|P\| \to 0} \sum_{k=1}^{n} G(x_k^*, y_k^*, z_k^*)\, \Delta z_k \qquad (18.9)$$

Method of Evaluation   If C is a smooth curve in 3-space defined by the parametric equations

$$x = f(t), \qquad y = g(t), \qquad z = h(t), \qquad a \le t \le b$$

then the integral in (18.9) can be evaluated by using

$$\int_C G(x, y, z)\, dz = \int_a^b G(f(t), g(t), h(t)) h'(t)\, dt$$

The integrals $\int_C G(x, y, z)\, dx$ and $\int_C G(x, y, z)\, dy$ are evaluated in a similar fashion. The line integral with respect to arc length is

$$\int_C G(x, y, z)\, ds = \int_a^b G(f(t), g(t), h(t)) \sqrt{[f'(t)]^2 + [g'(t)]^2 + [h'(t)]^2}\, dt$$

As in (18.7), in 3-space we are often concerned with line integrals in the form of a sum:

$$\int_C P(x, y, z)\, dx + Q(x, y, z)\, dy + R(x, y, z)\, dz$$

_____ **Example 5** _____

Evaluate $\int_C y\, dx + x\, dy + z\, dz$, where C is the helix $x = 2 \cos t$, $y = 2 \sin t$, $z = t$, $0 \le t \le 2\pi$.

Solution   Substituting the expressions for x, y, and z along with $dx = -2 \sin t\, dt$, $dy = 2 \cos t\, dt$, $dz = dt$, it follows that

$$\int_C y\, dx + x\, dy + z\, dz = \int_0^{2\pi} \underbrace{-4 \sin^2 t\, dt + 4 \cos^2 t\, dt}_{} + t\, dt$$

$$\boxed{\text{Double-angle formula}}$$

$$= \int_0^{2\pi} (4 \cos 2t + t)\, dt$$

$$= \left(2 \sin 2t + \frac{t^2}{2} \right) \Big]_0^{2\pi} = 2\pi^2$$

Vector Fields   Vector functions such as

$$\mathbf{F}(x,\ y) = P(x,\ y)\mathbf{i} + Q(x,\ y)\mathbf{j}$$

and $$\mathbf{F}(x,\ y,\ z) = P(x,\ y,\ z)\mathbf{i} + Q(x,\ y,\ z)\mathbf{j} + R(x,\ y,\ z)\mathbf{k}$$

are also called **vector fields**. For example, the motion of a wind or a fluid can be described by a *velocity field* in that a vector can be assigned at each point representing the velocity of a particle at the point. See Figures 18.7(a) and 18.7(b). The concept of a *force field* plays an important role in mechanics, electricity, and magnetism. See Figures 18.7(c) and 18.7(d).

(a) airflow around an airplane wing; $|\mathbf{v}_a| > |\mathbf{v}_b|$

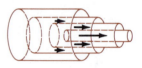

(b) laminar flow of blood in an artery; cylindrical layers of blood flow faster near the center of the artery

(c) inverse square force field; magnitude of the attractive force is large near the particle

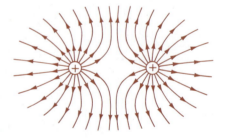

(d) lines of force around two equal positive charges

Figure 18.7

Work   In Chapter 6 we saw that the work done in moving an object from $x = a$ to $x = b$ by a force $F(x)$, which varies in magnitude but not in direction, is given by the integral

$$W = \int_a^b F(x)\ dx \qquad\qquad (18.10)$$

In general, a force field $\mathbf{F}(x,\ y) = P(x,\ y)\mathbf{i} + Q(x,\ y)\mathbf{j}$ acting at each point on a smooth curve $C: x = f(t),\ y = g(t),\ a \le t \le b$, varies in both magnitude and direction. See Figure 18.8(a). If A and B are the points $(f(a),\ g(a))$ and

$(f(b),\ g(b))$, respectively, we ask: What is the work done by $\mathbf{F}$ as its point of application moves along C from A to B? To answer this question, suppose C is divided into n subarcs of lengths Δs_k. On each subarc $\mathbf{F}(x_k^*,\ y_k^*)$ is a constant force. If, as shown in Figure 18.8(b), the length of the vector $\Delta\mathbf{r}_k = (x_k - x_{k-1})\mathbf{i} + (y_k - y_{k-1})\mathbf{j} = \Delta x_k\mathbf{i} + \Delta y_k\mathbf{j}$ is an approximation to the length of the kth subarc, then the approximate work done by $\mathbf{F}$ over the subarc is

$$(|\mathbf{F}(x_k^*,\ y_k^*)|\ \cos\ \theta)|\Delta\mathbf{r}_k| = \mathbf{F}(x_k^*,\ y_k^*)\cdot\Delta\mathbf{r}_k$$
$$= P(x_k^*,\ y_k^*)\ \Delta x_k + Q(x_k^*,\ y_k^*)\ \Delta y_k$$

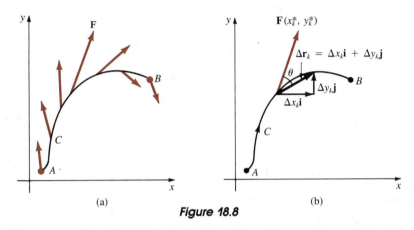

(a)   (b)

Figure 18.8

By summing these elements of work and passing to the limit, it is natural to define the **work done by F along C** as the line integral

$$W = \int_C P(x,\ y)\ dx + Q(x,\ y)\ dy \qquad (18.11)$$

Vector Form   If C is a curve traced by a vector function $\mathbf{r}(t) = f(t)\mathbf{i} + g(t)\mathbf{j}$, then $d\mathbf{r}/dt = f'(t)\mathbf{i} + g'(t)\mathbf{j}$ prompts us to define $d\mathbf{r} = [f'(t)\mathbf{i} + g'(t)\mathbf{j}]\ dt = dx\ \mathbf{i} + dy\ \mathbf{j}$. Thus, (18.11) can be written in **vector form**

$$W = \int_C \mathbf{F}\cdot d\mathbf{r} \qquad (18.12)$$

where $\mathbf{F} = P\mathbf{i} + Q\mathbf{j}$. Now, since

$$\frac{d\mathbf{r}}{dt} = \frac{d\mathbf{r}}{ds}\frac{ds}{dt}$$

we let $d\mathbf{r} = \mathbf{T}\ ds$, where $\mathbf{T} = d\mathbf{r}/ds$ is a unit tangent* to C. Hence,

$$W = \int_C \mathbf{F}\cdot d\mathbf{r} = \int_C \mathbf{F}\cdot\mathbf{T}\ ds = \int_C \text{comp}_{\mathbf{T}}\mathbf{F}\ ds \qquad (18.13)$$

---

*See Section 15.1.

In other words

> *the work done by a force* **F** *along a curve C is due entirely to the tangential component of* **F**.

The definition of work given in (18.12) and (18.13) extends to a force $\mathbf{F}(x, y, z) = P(x, y, z)\mathbf{i} + Q(x, y, z)\mathbf{j} + R(x, y, z)\mathbf{k}$ acting along a space curve C. In this case $d\mathbf{r} = dx\,\mathbf{i} + dy\,\mathbf{j} + dz\,\mathbf{k}$.

_____ Example 6 _____

Find the work done by (*a*) $\mathbf{F} = x\mathbf{i} + y\mathbf{j}$ and (*b*) $\mathbf{F} = \frac{3}{4}\mathbf{i} + \frac{1}{2}\mathbf{j}$ along the curve C traced by $\mathbf{r}(t) = \cos t\mathbf{i} + \sin t\mathbf{j}$ from $t = 0$ to $t = \pi$.

Solution

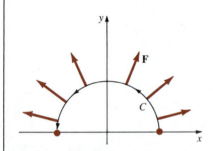

Figure 18.9

(*a*)   The vector function $\mathbf{r}(t)$ gives the parametric equations $x = \cos t$, $y = \sin t$, $0 \leq t \leq \pi$, which we recognize as a half-circle. As seen in Figure 18.9, the force field **F** is perpendicular to C at every point. (Why is this?) Since the tangential components of **F** are zero, the work done along C is zero. To see this we use (18.12):

$$W = \int_C \mathbf{F} \cdot d\mathbf{r} = \int_C (x\mathbf{i} + y\mathbf{j}) \cdot d\mathbf{r}$$

$$= \int_0^\pi (\cos t\mathbf{i} + \sin t\mathbf{j}) \cdot (-\sin t\mathbf{i} + \cos t\mathbf{j})\, dt$$

$$= \int_0^\pi (-\cos t \sin t + \sin t \cos t)\, dt = 0$$

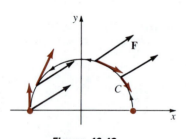

Figure 18.10

(*b*)   In Figure 18.10 the vectors in color are the projections of **F** on the unit tangent vectors. The work done by **F** is

$$W = \int_C \mathbf{F} \cdot d\mathbf{r} = \int_C \left(\frac{3}{4}\mathbf{i} + \frac{1}{2}\mathbf{j}\right) \cdot d\mathbf{r}$$

$$= \int_0^\pi \left(\frac{3}{4}\mathbf{i} + \frac{1}{2}\mathbf{j}\right) \cdot (-\sin t\mathbf{i} + \cos t\mathbf{j})\, dt$$

$$= \int_0^\pi \left(-\frac{3}{4}\sin t + \frac{1}{2}\cos t\right) dt$$

$$= \left(\frac{3}{4}\cos t + \frac{1}{2}\sin t\right)\Bigg]_0^\pi = -\frac{3}{2}$$

The units of work depend on the units of $|\mathbf{F}|$ and on the units of distance.

Circulation   A line integral of a vector field **F** around a simple closed curve C is said to be the **circulation** of **F** around C; that is,

$$\text{circulation} = \oint_C \mathbf{F} \cdot d\mathbf{r} = \oint_C \mathbf{F} \cdot \mathbf{T} \, ds$$

In particular, if **F** is the velocity field of a fluid, then the circulation is a measure of the amount by which the fluid tends to turn the curve C in a counterclockwise direction. For example, if **F** is perpendicular to **T** for every (x, y) on C, then $\int_C \mathbf{F} \cdot \mathbf{T} \, ds = 0$ and the curve does not move at all. On the other hand, $\int_C \mathbf{F} \cdot \mathbf{T} \, ds > 0$ and $\int_C \mathbf{F} \cdot \mathbf{T} \, ds < 0$ means that the fluid tends to rotate C in the counterclockwise and clockwise directions, respectively. See Figure 18.11.

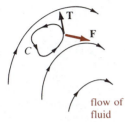

flow of fluid

Figure 18.11

Exercises 18.1

Answers to odd-numbered problems begin on page A-52.

In Problems 1–4 evaluate $\int_C G(x, y) \, dx$, $\int_C G(x, y) \, dy$, and $\int_C G(x, y) \, ds$ on the indicated curve C.

1. $G(x, y) = 2xy$; $x = 5 \cos t$, $y = 5 \sin t$, $0 \le t \le \pi/4$

2. $G(x, y) = x^3 + 2xy^2 + 2x$; $x = 2t$, $y = t^2$, $0 \le t \le 1$

3. $G(x, y) = 3x^2 + 6y^2$; $y = 2x + 1$, $-1 \le x \le 0$

4. $G(x, y) = x^2/y^3$; $2y = 3x^{2/3}$, $1 \le x \le 8$

In Problems 5 and 6 evaluate $\int_C G(x, y, z) \, dx$, $\int_C G(x, y, z) \, dy$, $\int_C G(x, y, z) \, dz$, and $\int_C G(x, y, z) \, ds$ on the indicated curve C.

5. $G(x, y, z) = z$; $x = \cos t$, $y = \sin t$, $z = t$, $0 \le t \le \pi/2$

6. $G(x, y, z) = 4xyz$; $x = \frac{1}{3}t^3$, $y = t^2$, $z = 2t$, $0 \le t \le 1$

In Problems 7–10 (Figures 18.12 and 18.13 for Problems 9 and 10) evaluate $\int_C (2x + y) \, dx + xy \, dy$ on the given curve C between $(-1, 2)$ and $(2, 5)$.

7. $y = x + 3$

8. $y = x^2 + 1$

9.

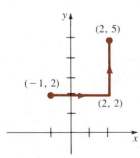

Figure 18.12

10.

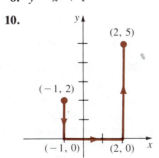

Figure 18.13

In Problems 11–14 evaluate $\int_C y \, dx + x \, dy$ on the given curve C between $(0, 0)$ and $(1, 1)$.

11. $y = x^2$

12. $y = x$

13. C consists of the line segments from $(0, 0)$ to $(0, 1)$ and from $(0, 1)$ to $(1, 1)$.

14. C consists of the line segments from $(0, 0)$ to $(1, 0)$ and from $(1, 0)$ to $(1, 1)$.

15. Evaluate $\int_C (6x^2 + 2y^2) \, dx + 4xy \, dy$, where C is given by $x = \sqrt{t}$, $y = t$, $4 \le t \le 9$.

16. Evaluate $\int_C -y^2 \, dx + xy \, dy$, where C is given by $x = 2t$, $y = t^3$, $0 \le t \le 2$.

17. Evaluate $\int_C 2x^3y \, dx + (3x + y) \, dy$, where C is given by $x = y^2$ from $(1, -1)$ to $(1, 1)$.

18. Evaluate $\int_C 4x \, dx + 2y \, dy$, where C is given by $x = y^3 + 1$ from $(0, -1)$ to $(9, 2)$.

In Problems 19 and 20 (Figures 18.14 and 18.15) evaluate $\oint_C (x^2 + y^2) \, dx - 2xy \, dy$ on the given closed curve C.

19.

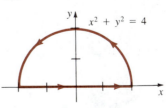

Figure 18.14

20.

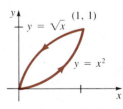

Figure 18.15

In Problems 21 and 22 (Figures 18.16 and 18.17) evaluate $\oint_C x^2 y^3 \, dx - xy^2 \, dy$ on the given closed curve C.

21.

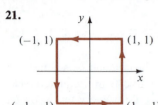

22.

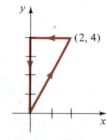

Figure 18.16   **Figure 18.17**

23. Evaluate $\oint_C (x^2 - y^2) \, ds$, where C is given by

$$x = 5 \cos t, \quad y = 5 \sin t, \quad 0 \le t \le 2\pi$$

24. Evaluate $\int_{-C} y \, dx - x \, dy$, where C is given by

$$x = 2 \cos t, \quad y = 3 \sin t, \quad 0 \le t \le \pi$$

In Problems 25–28 (Figures 18.18 and 18.19 for Problems 27 and 28) evaluate $\int_C y \, dx + z \, dy + x \, dz$ on the given curve C between $(0, 0, 0)$ and $(6, 8, 5)$.

25. C consists of the line segments from $(0, 0, 0)$ to $(2, 3, 4)$ and from $(2, 3, 4)$ to $(6, 8, 5)$.

26. $x = 3t, \ y = t^3, \ z = \dfrac{5}{4}t^2, \ 0 \le t \le 2$

27.

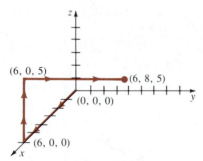

Figure 18.18

28.

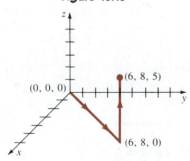

Figure 18.19

In Problems 29 and 30 evaluate $\int_C \mathbf{F} \cdot d\mathbf{r}$.

29. $\mathbf{F}(x, y) = y^3 \mathbf{i} - x^2 y \mathbf{j}; \ \mathbf{r}(t) = e^{-2t}\mathbf{i} + e^t \mathbf{j}, \ 0 \le t \le \ln 2$

30. $\mathbf{F}(x, y, z) = e^x \mathbf{i} + xe^{xy} \mathbf{j} + xye^{xyz} \mathbf{k}; \ \mathbf{r}(t) = t\mathbf{i} + t^2 \mathbf{j} + t^3 \mathbf{k}, \ 0 \le t \le 1$

31. Find the work done by the force $\mathbf{F}(x, y) = y\mathbf{i} + x\mathbf{j}$ acting along $y = \ln x$ from $(1, 0)$ to $(e, 1)$.

32. Find the work done by the force $\mathbf{F}(x, y) = 2xy\mathbf{i} + 4y^2\mathbf{j}$ acting along the piecewise smooth curve consisting of the line segments from $(-2, 2)$ to $(0, 0)$ and from $(0, 0)$ to $(2, 3)$.

33. Find the work done by the force $\mathbf{F}(x, y) = (x + 2y)\mathbf{i} + (6y - 2x)\mathbf{j}$ acting counterclockwise once around the triangle with vertices $(1, 1)$, $(3, 1)$, and $(3, 2)$.

34. Find the work done by the force $\mathbf{F}(x, y, z) = yz\mathbf{i} + xz\mathbf{j} + xy\mathbf{k}$ acting along the curve given by $\mathbf{r}(t) = t^3\mathbf{i} + t^2\mathbf{j} + t\mathbf{k}$ from $t = 1$ to $t = 3$.

35. Find the work done by a constant force $\mathbf{F}(x, y) = a\mathbf{i} + b\mathbf{j}$ acting counterclockwise once around the circle $x^2 + y^2 = 9$.

36. In an inverse square force field $\mathbf{F} = c\mathbf{r}/|\mathbf{r}|^3$, where c is a constant and $\mathbf{r} = x\mathbf{i} + y\mathbf{j} + z\mathbf{k}$.* Find the work done in moving a particle along the line from $(1, 1, 1)$ to $(3, 3, 3)$.

37. Verify that the line integral $\int_C y^2 \, dx + xy \, dy$ has the same value of C for each of the following parameterizations:

$$
\begin{array}{llll}
C: & x = 2t + 1, & y = 4t + 2, & 0 \le t \le 1 \\
C: & x = t^2, & y = 2t^2, & 1 \le t \le \sqrt{3} \\
C: & x = \ln t, & y = 2 \ln t, & e \le t \le e^3
\end{array}
$$

38. Consider the three curves between $(0, 0)$ and $(2, 4)$:

$$
\begin{array}{llll}
C_1: & x = t, & y = 2t, & 0 \le t \le 2 \\
C_2: & x = t, & y = t^2, & 0 \le t \le 2 \\
C_3: & x = 2t - 4, & y = 4t - 8, & 2 \le t \le 3
\end{array}
$$

Show that $\int_{C_1} xy \, ds = \int_{C_3} xy \, ds$ but $\int_{C_1} xy \, ds \ne \int_{C_2} xy \, ds$. Explain.

Miscellaneous Problems

39. Assume a smooth curve C is described by the vector function $\mathbf{r}(t)$ for $a \le t \le b$. Let acceleration, velocity, and speed be given by $\mathbf{a} = d\mathbf{v}/dt$, $\mathbf{v} = d\mathbf{r}/dt$, and $v = |\mathbf{v}|$, respectively. Using Newton's second law $\mathbf{F} = m\mathbf{a}$, show that, in the absence of friction, the work done by $\mathbf{F}$ in moving a particle of constant

---

*Note that the magnitude of $\mathbf{F}$ is inversely proportional to $|\mathbf{r}|^2$.

mass m from point A at $t = a$ to point B at $t = b$ is the same as the change in kinetic energy:

$$K(B) - K(A) = \frac{1}{2}m[v(b)]^2 - \frac{1}{2}m[v(a)]^2$$

$$\left(\textit{Hint: } \text{Consider } \frac{d}{dt}v^2 = \frac{d}{dt}\, \mathbf{v} \cdot \mathbf{v}.\right)$$

40. If $\rho(x, y)$ is density of a wire (mass per unit length), then $m = \int_C \rho(x, y)\, ds$ is the mass of the wire. Find the mass of a wire having the shape of the semicircle $x = 1 + \cos t$,

$y = \sin t,\ 0 \le t \le \pi$ if the density at a point P is directly proportional to distance from the y-axis.

41. The coordinates of the center of mass of a wire with variable density are given by $\bar{x} = M_y/m,\ \bar{y} = M_x/m$, where

$$m = \int_C \rho(x, y)\, ds, \qquad M_x = \int_C y\rho(x, y)\, ds,$$

and $$M_y = \int_C x\rho(x, y)\, ds$$

Find the center of mass of the wire in Problem 40.

18.2 Line Integrals Independent of the Path

The concept of an exact differential plays an important role in the discussion that follows. A review of Section 16.5 is recommended.

The value of a line integral $\int_C P\, dx + Q\, dy$ generally depends on the curve or **path** between two points A and B. However, there are exceptions. A line integral whose value is the same for *every* curve connecting A and B is said to be **independent of the path**.

Example 1

The integral $\int_C y\, dx + x\, dy$ has the same value on each path C between $(0, 0)$ and $(1, 1)$ shown in Figure 18.20. You may recall from Problems 11–14 of Exercises 18.1 that on these paths

$$\int_C y\, dx + x\, dy = 1$$

In Example 2 we shall prove that the given integral is independent of the path.

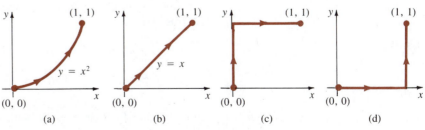

Figure 18.20

A Fundamental Theorem   The next theorem establishes an important relationship between the seemingly disparate notions of path independence and exact differentials. In addition, it provides a means of evaluating path-independent line integrals in a manner analogous to the Fundamental Theorem of Calculus:

$$\int_a^b f'(x)\,dx = f(b) - f(a)$$

For this reason the result is sometimes called the **Fundamental Theorem for line integrals**.

THEOREM 18.1   Suppose there exists a function $\phi(x, y)$ such that $d\phi = P\,dx + Q\,dy$; that is, $P\,dx + Q\,dy$ is an exact differential. Then $\int_C P\,dx + Q\,dy$ depends only on the endpoints A and B of the path C, and

$$\int_C P\,dx + Q\,dy = \phi(B) - \phi(A)$$

Proof   Let C be a smooth path defined parametrically by $x = f(t)$, $y = g(t)$, $a \le t \le b$, and let the coordinates of A and B be $(f(a), g(a))$ and $(f(b), g(b))$, respectively. Then, by the Chain Rule

$$\int_C P\,dx + Q\,dy = \int_a^b \left(\frac{\partial \phi}{\partial x} \frac{dx}{dt} + \frac{\partial \phi}{\partial y} \frac{dy}{dt} \right) dt$$

$$= \int_a^b \frac{d\phi}{dt}\,dt$$

$$= \phi(f(t), g(t)) \Big]_a^b$$

$$= \phi(f(b), g(b)) - \phi(f(a), g(a))$$

$$= \phi(B) - \phi(A) \qquad \blacksquare$$

Two observations are in order. The theorem is also valid for piecewise smooth curves, but the foregoing proof would have to be modified by considering each smooth arc of the curve C. Furthermore, the converse of the theorem is also true. Hence,

$$\int_C P\,dx + Q\,dy \text{ is independent of the path if and only if}$$
$$P\,dx + Q\,dy \text{ is an exact differential} \tag{18.14}$$

Notation   A line integral $\int_C P\,dx + Q\,dy$, which is independent of the path between the endpoints A and B, is often written

$$\int_A^B P\,dx + Q\,dy$$

Example 2

In Example 1 note that $d(xy) = y\,dx + x\,dy$. That is, $y\,dx + x\,dy$ is an exact differential. Hence $\int_C y\,dx + x\,dy$ is independent of the path between any two points A and B. Specifically, if A and B are, respectively, $(0, 0)$ and $(1, 1)$, we then have, from Theorem 18.1,

$$\int_{(0,0)}^{(1,1)} y\,dx + x\,dy = \int_{(0,0)}^{(1,1)} d(xy) = xy \Big]_{(0,0)}^{(1,1)} = 1$$

Test for Path Independence   In view of the statement given in (18.14), the same criteria given in Theorem 16.6 for determining an exact differential become the criteria, or test, for path independence.

> **THEOREM 18.2**   Let P and Q be continuous and have continuous first partial derivatives in a simply connected region. Then $\int_C P\,dx + Q\,dy$ is independent of the path C if and only if
>
> $$\frac{\partial P}{\partial y} = \frac{\partial Q}{\partial x}$$
>
> for all (x, y) in the region.

Example 3

Show that the integral $\int_C (x^2 - 2y^3)\,dx + (x + 5y)\,dy$ is not independent of the path C.

Solution   From $P = x^2 - 2y^3$ and $Q = x + 5y$, we find

$$\frac{\partial P}{\partial y} = -6y^2 \quad \text{and} \quad \frac{\partial Q}{\partial x} = 1$$

Since $\partial P/\partial y \neq \partial Q/\partial x$, it follows from Theorem 18.2 that the integral is not independent of the path.

Example 4

Show that $\int_C (y^2 - 6xy + 6)\,dx + (2xy - 3x^2)\,dy$ is independent of any path C between $(-1, 0)$ and $(3, 4)$. Evaluate.

Solution   Identifying $P = y^2 - 6xy + 6$ and $Q = 2xy - 3x^2$ yields

$$\frac{\partial P}{\partial y} = 2y - 6x \quad \text{and} \quad \frac{\partial Q}{\partial x} = 2y - 6x$$

Since $\partial P/\partial y = \partial Q/\partial x$, the integral is independent of the path and so there exists a function ϕ such that $\partial\phi/\partial x = y^2 - 6xy + 6$ and $\partial\phi/\partial y = 2xy - 3x^2$. Employing the partial integration method of Section 16.5, we find

$$\phi = xy^2 - 3x^2y + 6x$$

In other words,

$$d(xy^2 - 3x^2y + 6x) = (y^2 - 6xy + 6)\, dx + (2xy - 3x^2)\, dy$$

Hence,

$$\int_{(-1,0)}^{(3,4)} (y^2 - 6xy + 6)\, dx + (2xy - 3x^2)\, dy = \int_{(-1,0)}^{(3,4)} d(xy^2 - 3x^2y + 6x)$$

$$= (xy^2 - 3x^2y + 6x)\Big]_{(-1,0)}^{(3,4)}$$

$$= (48 - 108 + 18) - (-6)$$

$$= -36$$

Alternative Solution   Since the integral is independent of the path, we can integrate along any convenient curve connecting the given points. In particular, $y = x + 1$ is such a curve. Using x as a parameter then gives

$$\int_C (y^2 - 6xy + 6)\, dx + (2xy - 3x^2)\, dy$$

$$= \int_{-1}^{3} [(x + 1)^2 - 6x(x + 1) + 6]\, dx + [2x(x + 1) - 3x^2]\, dx$$

$$= \int_{-1}^{3} (-6x^2 - 2x + 7)\, dx = -36$$

Conservative Vector Fields   If $\int_C P\, dx + Q\, dy$ is independent of the path C, we know there exists a function ϕ such that

$$d\phi = \frac{\partial \phi}{\partial x}\, dx + \frac{\partial \phi}{\partial y}\, dy = P\, dx + Q\, dy$$

$$= (P\mathbf{i} + Q\mathbf{j}) \cdot (dx\, \mathbf{i} + dy\, \mathbf{j}) = \mathbf{F} \cdot d\mathbf{r}$$

where $\mathbf{F} = P\mathbf{i} + Q\mathbf{j}$ is a vector field and $P = \partial\phi/\partial x$, $Q = \partial\phi/\partial y$. In other words, the vector field $\mathbf{F}$ is a gradient of the function ϕ. Since $\mathbf{F} = \nabla\phi$, $\mathbf{F}$ is sometimes said to be a **gradient field** and the function ϕ is then said to be a **potential function** for $\mathbf{F}$. In a gradient force field $\mathbf{F}$, the work done by the force upon a particle moving from position A to position B is the same for all paths between these points. Moreover, the work done by the force along a closed path is *zero*. See Problem 25 in Exercises 18.2. For this reason, such a force field is also said to be **conservative**. In a conservative field $\mathbf{F}$ the *law of conservation of mechanical energy* holds: For a particle moving along a path in a conservative field,

kinetic energy + potential energy = constant

See Problem 27. In a simply connected domain the hypotheses of Theorem 18.2 imply that a force field $\mathbf{F}(x, y) = P(x, y)\mathbf{i} + Q(x, y)\mathbf{j}$ is a gradient field (that is, conservative) if and only if

$$\frac{\partial P}{\partial y} = \frac{\partial Q}{\partial x} \tag{18.15}$$

_____ **Example 5** _____

Show that the vector field $\mathbf{F} = (y^2 + 5)\mathbf{i} + (2xy - 8)\mathbf{j}$ is a gradient field. Find a potential function for $\mathbf{F}$.

Solution By identifying $P = y^2 + 5$ and $Q = 2xy - 8$, we see from (18.15) that

$$\frac{\partial P}{\partial y} = \frac{\partial Q}{\partial x} = 2y$$

Hence, $\mathbf{F}$ is a gradient field and so there exists a potential function ϕ satisfying

$$\frac{\partial \phi}{\partial x} = y^2 + 5 \quad \text{and} \quad \frac{\partial \phi}{\partial y} = 2xy - 8$$

Proceeding as in Section 16.5, we find that $\phi = xy^2 - 8y + 5x$.

Check
$$\nabla \phi = \frac{\partial \phi}{\partial x}\mathbf{i} + \frac{\partial \phi}{\partial y}\mathbf{j}$$
$$= (y^2 + 5)\mathbf{i} + (2xy - 8)\mathbf{j}$$

Remarks

(*i*) The test for path independence of line integrals in 3-space is given in the exercises.

(*ii*) A frictional force such as air resistance is *nonconservative*. Nonconservative forces are *dissipative* in that their action reduces kinetic energy without a corresponding increase in potential energy. In other words, if the work done $\int_C \mathbf{F} \cdot d\mathbf{r}$ depends on the path, then $\mathbf{F}$ is nonconservative.

_____ **Exercises 18.2** _____

Answers to odd-numbered problems begin on page A-52.

In Problems 1–10 show that the given integral is independent of the path. Evaluate in two ways: **(a)** Find a function ϕ such that $d\phi = P \, dx + Q \, dy$, and **(b)** integrate along any convenient path between the points.

1. $\int_{(0,0)}^{(2,2)} x^2 \, dx + y^2 \, dy$

2. $\int_{(1,1)}^{(2,4)} 2xy \, dx + x^2 \, dy$

3. $\int_{(1,0)}^{(3,2)} (x + 2y) \, dx + (2x - y) \, dy$

4. $\int_{(0,0)}^{(\pi/2,0)} \cos x \cos y \, dx + (1 - \sin x \sin y) \, dy$

5. $\int_{(4,1)}^{(4,4)} \frac{-y \, dx + x \, dy}{y^2}$ on any path not crossing the x-axis

6. $\int_{(1,0)}^{(3,4)} \frac{x \, dx + y \, dy}{\sqrt{x^2 + y^2}}$ on any path not through the origin

7. $\int_{(1,2)}^{(3,6)} (2y^2 x - 3) \, dx + (2yx^2 + 4) \, dy$

8. $\int_{(-1,1)}^{(0,0)} (5x + 4y) \, dx + (4x - 8y^3) \, dy$

9. $\int_{(0,0)}^{(2,8)} (y^3 + 3x^2 y) \, dx + (x^3 + 3y^2 x + 1) \, dy$

10. $\int_{(-2,0)}^{(1,0)} (2x - y \sin xy - 5y^4) \, dx - (20xy^3 + x \sin xy) \, dy$

In Problems 11–16 determine if the given vector field is a gradient field. If so, find the potential function ϕ for $\mathbf{F}$.

11. $\mathbf{F}(x, y) = (4x^3y^3 + 3)\mathbf{i} + (3x^4y^2 + 1)\mathbf{j}$

12. $\mathbf{F}(x, y) = 2xy^3\mathbf{i} + 3y^2(x^2 + 1)\mathbf{j}$

13. $\mathbf{F}(x, y) = y^2 \cos xy^2\mathbf{i} - 2xy \sin xy^2\mathbf{j}$

14. $\mathbf{F}(x, y) = (x^2 + y^2 + 1)^{-2}(x\mathbf{i} + y\mathbf{j})$

15. $\mathbf{F}(x, y) = (x^3 + y)\mathbf{i} + (x + y^3)\mathbf{j}$

16. $\mathbf{F}(x, y) = 2e^{2y}\mathbf{i} + xe^{2y}\mathbf{j}$

In Problems 17 and 18 (Figures 18.21 and 18.22) find the work done by the force $\mathbf{F}(x, y) = (2x + e^{-y})\mathbf{i} + (4y - xe^{-y})\mathbf{j}$ along the indicated curve.

17.   **18.**

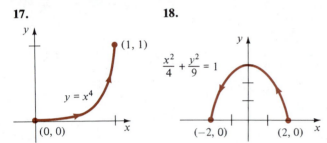

Figure 18.21   **Figure 18.22**

If $\partial P/\partial y$, $\partial P/\partial z$, $\partial Q/\partial x$, $\partial Q/\partial z$, $\partial R/\partial x$, $\partial R/\partial y$ are continuous in a rectangular region in 3-space, then $\int_C P\, dx + Q\, dy + R\, dz$ is independent of the path if and only if

$$\frac{\partial P}{\partial y} = \frac{\partial Q}{\partial x}, \qquad \frac{\partial Q}{\partial z} = \frac{\partial R}{\partial y}, \qquad \frac{\partial R}{\partial x} = \frac{\partial P}{\partial z}. \qquad (18.16)$$

In Problems 19–22 use this test to show that the given integral is independent of the path. Evaluate.

19. $\int_{(1,1,1)}^{(2,4,8)} yz\, dx + xz\, dy + xy\, dz$

20. $\int_{(0,0,0)}^{(1,1,1)} 2x\, dx + 3y^2\, dy + 4z^3\, dz$

21. $\int_{(1,1,\ln 3)}^{(2,2,\ln 3)} e^{2z}\, dx + 3y^2\, dy + 2xe^{2z}\, dz$

22. $\int_{(-2,3,1)}^{(0,0,0)} 2xz\, dx + 2yz\, dy + (x^2 + y^2)\, dz$

23. The inverse square law of gravitational attraction between two masses m_1 and m_2 is given by $\mathbf{F} = -Gm_1m_2\mathbf{r}/|\mathbf{r}|^3$, where $\mathbf{r} = x\mathbf{i} + y\mathbf{j} + z\mathbf{k}$. Use (18.16) to show that $\mathbf{F}$ is conservative. Find a potential function for $\mathbf{F}$.

24. Find the work done by the force $\mathbf{F}(x, y, z) = 8xy^3z\mathbf{i} + 12x^2y^2z\mathbf{j} + 4x^2y^3\mathbf{k}$ acting along the helix $\mathbf{r}(t) = 2 \cos t\mathbf{i} + 2 \sin t\mathbf{j} + t\mathbf{k}$ from $(2, 0, 0)$ to $(1, \sqrt{3}, \pi/3)$. From $(2, 0, 0)$ to $(0, 2, \pi/2)$. (*Hint:* Use (18.16) to show that $\mathbf{F}$ is conservative.)

25. If $\mathbf{F}$ is a conservative force field, show that the work done along any simple closed path is zero.

26. A particle in the plane is attracted to the origin with a force $\mathbf{F} = |\mathbf{r}|^n\mathbf{r}$, where n is a positive integer and $\mathbf{r} = x\mathbf{i} + y\mathbf{j}$ is the position vector of the particle. Show that $\mathbf{F}$ is conservative. Find the work done in moving the particle between (x_1, y_1) and (x_2, y_2).

Miscellaneous Problems

27. Suppose $\mathbf{F}$ is a conservative force field with potential function ϕ. In physics the function $p = -\phi$ is called **potential energy**. Since $\mathbf{F} = -\nabla p$, Newton's second law becomes

$$m\mathbf{r}'' = -\nabla p \quad \text{or} \quad m\frac{d\mathbf{v}}{dt} + \nabla p = \mathbf{0}$$

By integrating $m\dfrac{d\mathbf{v}}{dt} \cdot \dfrac{d\mathbf{r}}{dt} + \nabla p \cdot \dfrac{d\mathbf{r}}{dt} = 0$ with respect to t, derive the law of conservation of mechanical energy: $\frac{1}{2}mv^2 + p = \text{constant}$. (*Hint:* See Problem 39 in Exercises 18.1.)

28. Suppose C is a smooth curve between points A (at $t = a$) and B (at $t = b$) and that p is potential energy, defined in Problem 27. If $\mathbf{F}$ is a conservative force field and $K = \frac{1}{2}mv^2$ is kinetic energy, show that $p(B) + K(B) = p(A) + K(A)$.

18.3 Surface Integrals

The last kind of integral that we shall consider in this text is called a **surface integral** and involves a function G of three variables defined on a surface S. The five steps preparatory to the definition of this integral are similar to combinations of the steps leading to the line integral, with respect to arc length, and the steps leading to the double integral.

$$w = G(x, y, z)$$

1. Let G be defined in a region of 3-space that contains a surface S, which is the graph of a function $z = f(x, y)$. Let the projection R of the surface onto the xy-plane be either a Type I or a Type II region.

2. Divide the surface into n pieces of areas ΔS_k corresponding to a partition P of R into n rectangles R_k of areas ΔA_k.

3. Let $\|P\|$ be the **norm** of the partition or the length of the longest diagonal of the R_k.

4. Choose a point (x_k^*, y_k^*, z_k^*) on each element of surface area.

5. Form the sum

$$\sum_{k=1}^{n} G(x_k^*, y_k^*, z_k^*) \, \Delta S_k$$

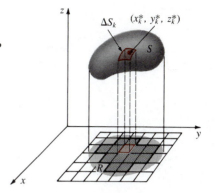

DEFINITION 18.2   Surface Integral

Let G be a function of three variables defined over a region of space containing surface S. Then the **surface integral of F over S** is given by

$$\iint_S G(x, y, z) \, dS = \lim_{\|P\| \to 0} \sum_{k=1}^{n} G(x_k^*, y_k^*, z_k^*) \, \Delta S_k \quad (18.17)$$

Method of Evaluation   Recall from Section 17.6 that if $z = f(x, y)$ is the equation of a surface S, then the differential of the surface is

$$dS = \sqrt{1 + [f_x(x, y)]^2 + [f_y(x, y)]^2} \, dA$$

Thus, if G, f, f_x, and f_y are continuous throughout a region containing S, we can evaluate (18.17) by means of a double integral.

$$\iint_S G(x, y, z) \, dS = \iint_R G(x, y, f(x, y))\sqrt{1 + [f_x(x, y)]^2 + [f_y(x, y)]^2} \, dA \quad (18.18)$$

Note that when $G = 1$, (18.18) reduces to the formula for surface area.

Projection of S into Other Planes   If $y = g(x, z)$ is the equation of a surface S that projects onto a region R of the xz-plane, then

$$\iint_S G(x, y, z) \, dS = \iint_R G(x, g(x, z), z)\sqrt{1 + [g_z(x, z)]^2 + [g_z(x, z)]^2} \, dA \quad (18.19)$$

Similarly, if $x = h(y, z)$ is the equation of a surface S that projects onto the yz-plane, then the analogue of (18.18) is

$$\iint_S G(x, y, z) \, dS = \iint_R G(h(y, z), y, z)\sqrt{1 + [h_y(y, z)]^2 + [h_z(y, z)]^2} \, dA \quad (18.20)$$

Mass of a Surface   Suppose $\rho(x, y, z)$ represents the density of surface at any point, or mass per unit surface area, then the **mass** m of the surface is

$$m = \iint_S \rho(x, y, z)\, dS \qquad (18.21)$$

_____ **Example 1** _____

Find the mass of the surface of the paraboloid $z = 1 + x^2 + y^2$ in the first octant for $1 \le z \le 5$ if the density at a point P on the surface is directly proportional to distance from the xy-plane.

Solution   The surface in question and its projection onto the xy-plane are shown in Figure 18.23. Now, since $\rho(x, y, z) = kz$ and $z = 1 + x^2 + y^2$, (18.21) and (18.18) give

$$m = \iint_S kz\, dS = k \iint_R (1 + x^2 + y^2)\sqrt{1 + 4x^2 + 4y^2}\, dA$$

By changing to polar coordinates, we obtain

$$m = k \int_0^{\pi/2} \int_0^2 (1 + r^2)\sqrt{1 + 4r^2}\, r\, dr\, d\theta$$

$$= k \int_0^{\pi/2} \int_0^2 [r(1 + 4r^2)^{1/2} + r^3(1 + 4r^2)^{1/2}]\, dr\, d\theta$$

⎡ Integration by parts ⎤

$$= k \int_0^{\pi/2} \left[\frac{1}{12}(1 + 4r^2)^{3/2} + \frac{1}{12}r^2(1 + 4r^2)^{3/2} - \frac{1}{120}(1 + 4r^2)^{5/2}\right]_0^2 d\theta$$

$$= \frac{k\pi}{2}\left[\frac{5(17)^{3/2}}{12} - \frac{17^{5/2}}{120} - \frac{3}{40}\right] \approx 19.2k$$

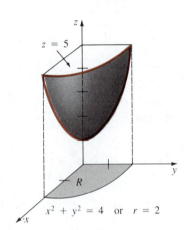

$z = 5$

R

$x^2 + y^2 = 4 \quad \text{or} \quad r = 2$

Figure 18.23

_____ **Example 2** _____

Evaluate $\iint_S xz^2\, dS$, where S is that portion of the cylinder $y = 2x^2 + 1$ in the first octant bounded by $x = 0$, $x = 2$, $z = 4$, and $z = 8$.

Solution   We shall use (18.19) with $g(x, z) = 2x^2 + 1$ and R the rectangular region in the xz-plane shown in Figure 18.24. Since $g_x(x, z) = 4x$ and $g_z(x, z) = 0$ it follows that

$$\iint_S xz^2\, dS = \int_0^2 \int_4^8 xz^2\sqrt{1 + 16x^2}\, dz\, dx$$

$$= \int_0^2 \frac{z^3}{3}x\sqrt{1 + 16x^2}\Big]_4^8 dx$$

$$= \frac{448}{3}\int_0^2 x(1 + 16x^2)^{1/2}\, dx = \frac{28}{9}(1 + 16x^2)^{3/2}\Big]_0^2$$

$$= \frac{28}{9}[65^{3/2} - 1] \approx 1627.3$$

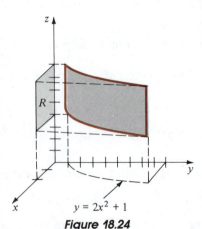

R

$y = 2x^2 + 1$

Figure 18.24

Integrals of Vector Fields   If $\mathbf{F}(x, y, z) = P(x, y, z)\mathbf{i} + Q(x, y, z)\mathbf{j} + R(x, y, z)\mathbf{k}$ is the velocity field of fluid, then, as shown in Figure 18.25(b), the volume of the fluid flowing through an element of surface area dS per unit time is approximated by

$$(\text{height}) \cdot (\text{area of base}) = (\text{comp}_{\mathbf{n}}\mathbf{F})\, dS = (\mathbf{F} \cdot \mathbf{n})\, dS$$

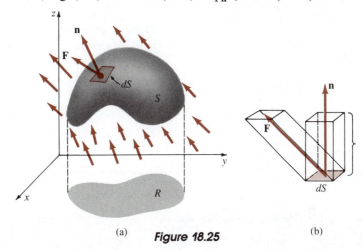

(a)   **Figure 18.25**   (b)

where $\mathbf{n}$ is a unit normal to the surface. The total volume of a fluid passing through S per unit time is called the **flux of F through** S and is given by

$$\text{flux} = \iint_S (\mathbf{F} \cdot \mathbf{n})\, dS \qquad (18.22)$$

In the case of a closed surface S, if $\mathbf{n}$ is the outer (inner) normal, then (18.22) gives the volume of fluid flowing out (in) through S per unit time.

Example 3 _____

Let $\mathbf{F}(x, y, z) = z\mathbf{j} + z\mathbf{k}$ represent the flow of a liquid. Find the flux of $\mathbf{F}$ through the surface S given by that part of the plane $z = 6 - 3x - 2y$ in the first octant.

Solution   The vector field and the surface are illustrated in Figure 18.26. If the plane is defined by $g(x, y, z) = 3x + 2y + z - 6 = 0$, then a unit normal is

$$\mathbf{n} = \frac{\nabla g}{|\nabla g|} = \frac{3}{\sqrt{14}}\mathbf{i} + \frac{2}{\sqrt{14}}\mathbf{j} + \frac{1}{\sqrt{14}}\mathbf{k}$$

Hence

$$\text{flux} = \iint_S (\mathbf{F} \cdot \mathbf{n})\, dS = \frac{1}{\sqrt{14}} \iint_S 3z\, dS$$

With R the projection of the surface onto the xy-plane we find from (18.18) that

$$\text{flux} = \frac{1}{\sqrt{14}} \iint_R 3(6 - 3x - 2y)(\sqrt{14}\, dA)$$

$$= 3 \int_0^2 \int_0^{3-3x/2} (6 - 3x - 2y)\, dy\, dx = 18$$

Figure 18.26

$3x + 2y = 6$

Remarks

(*i*) Depending on the nature of the vector field, the integral in (18.22) can represent other kinds of flux. For example, (18.22) could also give electric flux, magnetic flux, flux of heat, and so on.

(*ii*) There is an ambiguity in Example 3 since there are always two normals to a surface: an "upper" and a "lower" normal. In the problems that follow we shall always take the "upper normal"—that is, the normal with a positive **k** component. For a closed surface we shall take **n** to be an outer normal.

Exercises 18.3

Answers to odd-numbered problems begin on page A-52.

In Problems 1–10 evaluate $\iint_S G(x, y, z)\, dS$.

1. $G(x, y, z) = x$; S the portion of the cylinder $z = 2 - x^2$ in the first octant bounded by $x = 0$, $y = 0$, $y = 4$, $z = 0$

2. $G(x, y, z) = xy(9 - 4z)$; same surface as in Problem 1

3. $G(x, y, z) = xz^3$; S the cone $z = \sqrt{x^2 + y^2}$ inside the cylinder $x^2 + y^2 = 1$

4. $G(x, y, z) = x + y + z$; S the cone $z = \sqrt{x^2 + y^2}$ between $z = 1$ and $z = 4$

5. $G(x, y, z) = (x^2 + y^2)z$; S that portion of the sphere $x^2 + y^2 + z^2 = 36$ in the first octant

6. $G(x, y, z) = z^2$; S that portion of the plane $z = x + 1$ within the cylinder $y = 1 - x^2$, $0 \le y \le 1$

7. $G(x, y, z) = xy$; S that portion of the paraboloid $2z = 4 - x^2 - y^2$ within $0 \le x \le 1$, $0 \le y \le 1$

8. $G(x, y, z) = 2z$; S that portion of the paraboloid $2z = 1 + x^2 + y^2$ in the first octant bounded by $x = 0$, $y = \sqrt{3}x$, $z = 1$

9. $G(x, y, z) = 24\sqrt{yz}$; S that portion of the cylinder $y = x^2$ in the first octant bounded by $y = 0$, $y = 4$, $z = 0$, $z = 3$

10. $G(x, y, z) = (1 + 4y^2 + 4z^2)^{1/2}$; S that portion of the paraboloid $x = 4 - y^2 - z^2$ in the first octant outside the cylinder $y^2 + z^2 = 1$

In Problems 11 and 12 (Figures 18.27 and 18.28) evaluate $\iint_S (3z^2 + 4yz)\, dS$, where S is that portion of the plane $x + 2y + 3z = 6$ in the first octant. Use the projection of S onto the coordinate plane indicated in the given figure.

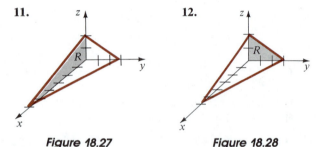

Figure 18.27   **Figure 18.28**

In Problems 13 and 14 find the mass of the given surface with the indicated density function.

13. S that portion of the plane $x + y + z = 1$ in the first octant; density at a point P directly proportional to the square of the distance from the yz-plane

14. S the hemisphere $z = \sqrt{4 - x^2 - y^2}$; $\rho(x, y, z) = |xy|$

In Problems 15–18 let **F** be a vector field. Find the flux of **F** through the given surface. Use the upper normal **n** to S.

15. $\mathbf{F} = x\mathbf{i} + 2z\mathbf{j} + y\mathbf{k}$; S that portion of the cylinder $y^2 + z^2 = 4$ in the first octant bounded by $x = 0$, $x = 3$, $y = 0$, $z = 0$

16. $\mathbf{F} = z\mathbf{k}$; S that part of the paraboloid $z = 5 - x^2 - y^2$ inside the cylinder $x^2 + y^2 = 4$

17. $\mathbf{F} = x\mathbf{i} + y\mathbf{j} + z\mathbf{k}$; same surface S as in Problem 16

18. $\mathbf{F} = -x^3y\mathbf{i} + yz^3\mathbf{j} + xy^3\mathbf{k}$; S that portion of the plane $z = x + 3$ in the first octant within the cylinder $x^2 + y^2 = 2x$

19. Let $T(x, y, z) = x^2 + y^2 + z^2$ represent temperature and let the "flow" of heat be given by the vector field $\mathbf{F} = -\nabla T$.

Find the flux of heat out of the sphere $x^2 + y^2 + z^2 = a^2$. (*Hint:* The surface area of a sphere of radius a is $4\pi a^2$.)

20. Find the flux of $\mathbf{F} = x\mathbf{i} + y\mathbf{j} + z\mathbf{k}$ out of the unit cube $0 \le x \le 1, 0 \le y \le 1, 0 \le z \le 1$. See Figure 18.29. Use the fact that the flux out of the cube is the sum of the fluxes out of the sides.

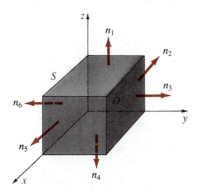

Figure 18.29

21. Coulomb's law states that the electric field $\mathbf{E}$ due to a point charge q at the origin is given by $\mathbf{E} = kq\mathbf{r}/|\mathbf{r}|^3$, where k is a constant and $\mathbf{r} = x\mathbf{i} + y\mathbf{j} + z\mathbf{k}$. Determine the flux out of a sphere $x^2 + y^2 + z^2 = a^2$.

22. If $\sigma(x, y, z)$ is charge density in an electrostatic field, then the total charge on a surface S is $Q = \iint_S \sigma(x, y, z) \, dS$.

Find the total charge on that part of the hemisphere $z = \sqrt{16 - x^2 - y^2}$ that is inside the cylinder $x^2 + y^2 = 9$ if the charge density at a point P on the surface is directly proportional to distance from the xy-plane.

Miscellaneous Problems

23. Let $z = f(x, y)$ be the equation of a surface S and $\mathbf{F}$ be the vector field $\mathbf{F}(x, y, z) = P(x, y, z)\mathbf{i} + Q(x, y, z)\mathbf{j} + R(x, y, z)\mathbf{k}$. Show that

$$\iint_S (\mathbf{F} \cdot \mathbf{n}) \, dS =$$

$$\iint_R \left[-P(x, y, z)\frac{\partial z}{\partial x} - Q(x, y, z)\frac{\partial z}{\partial y} + R(x, y, z) \right] dA$$

24. The coordinates of the centroid of a surface are given by

$$\bar{x} = \iint_S x \, dS/A(S), \qquad \bar{y} = \iint_S y \, dS/A(S),$$

$$\bar{z} = \iint_S z \, dS/A(S)$$

where $A(S)$ is the area of the surface. Find the centroid of that portion of the plane $2x + 3y + z = 6$ in the first octant.

25. Use the information in Problem 24 to find the centroid of the hemisphere $z = \sqrt{a^2 - x^2 - y^2}$.

18.4 Divergence and Curl

We have seen that if a vector force field $\mathbf{F}$ is conservative, it can be written as the gradient of a potential function ϕ,

$$\mathbf{F} = \nabla\phi = \frac{\partial\phi}{\partial x}\mathbf{i} + \frac{\partial\phi}{\partial y}\mathbf{j} + \frac{\partial\phi}{\partial z}\mathbf{k}$$

The del operator

$$\nabla = \mathbf{i}\frac{\partial}{\partial x} + \mathbf{j}\frac{\partial}{\partial y} + \mathbf{k}\frac{\partial}{\partial z}$$

used in the gradient can be combined with a vector field $\mathbf{F}(x, y, z) = P(x, y, z)\mathbf{i} + Q(x, y, z)\mathbf{j} + R(x, y, z)\mathbf{k}$ in two additional ways. We will assume hereafter that P, Q, and R have continuous partial derivatives.

DEFINITION 18.3   The **curl** of a vector field $\mathbf{F} = P\mathbf{i} + Q\mathbf{j} + R\mathbf{k}$ is the vector

$$\operatorname{curl} \mathbf{F} = \left(\frac{\partial R}{\partial y} - \frac{\partial Q}{\partial z}\right)\mathbf{i} + \left(\frac{\partial P}{\partial z} - \frac{\partial R}{\partial x}\right)\mathbf{j} + \left(\frac{\partial Q}{\partial x} - \frac{\partial P}{\partial y}\right)\mathbf{k}$$

In practice, curl $\mathbf{F}$ can be computed from the cross product of the del operator and the vector $\mathbf{F}$:

$$\operatorname{curl} \mathbf{F} = \nabla \times \mathbf{F} = \begin{vmatrix} \mathbf{i} & \mathbf{j} & \mathbf{k} \\ \dfrac{\partial}{\partial x} & \dfrac{\partial}{\partial y} & \dfrac{\partial}{\partial z} \\ P & Q & R \end{vmatrix} \qquad (18.23)$$

DEFINITION 18.4   The **divergence** of a vector field $\mathbf{F} = P\mathbf{i} + Q\mathbf{j} + R\mathbf{k}$ is the scalar function

$$\operatorname{div} \mathbf{F} = \frac{\partial P}{\partial x} + \frac{\partial Q}{\partial y} + \frac{\partial R}{\partial z}$$

Observe that div $\mathbf{F}$ can also be written in terms of the del operator as:

$$\operatorname{div} \mathbf{F} = \nabla \cdot \mathbf{F} = \frac{\partial}{\partial x}P(x, y, z) + \frac{\partial}{\partial y}Q(x, y, z) + \frac{\partial}{\partial z}R(x, y, z) \quad (18.24)$$

Example 1

If $\mathbf{F} = (x^2 y^3 - z^4)\mathbf{i} + 4x^5 y^2 z\mathbf{j} - y^4 z^6\mathbf{k}$, find

(a)   curl $\mathbf{F}$ (b)   div $\mathbf{F}$

Solution

(a)   From (18.23),

$$\operatorname{curl} \mathbf{F} = \nabla \times \mathbf{F} = \begin{vmatrix} \mathbf{i} & \mathbf{j} & \mathbf{k} \\ \dfrac{\partial}{\partial x} & \dfrac{\partial}{\partial y} & \dfrac{\partial}{\partial z} \\ x^2 y^3 - z^4 & 4x^5 y^2 z & -y^4 z^6 \end{vmatrix}$$
$$= (-4y^3 z^6 - 4x^5 y^2)\mathbf{i} + 4z^3\mathbf{j} + (20x^4 y^2 z - 3x^2 y^2)\mathbf{k}$$

(b)   From (18.24),

$$\operatorname{div} \mathbf{F} = \nabla \cdot \mathbf{F} = \frac{\partial}{\partial x}(x^2 y^3 - z^4) + \frac{\partial}{\partial y}(4x^5 y^2 z) + \frac{\partial}{\partial z}(-y^4 z^6)$$
$$= 2xy^3 + 8x^5 yz - 6y^4 z^5$$

We ask you to prove the following two important properties. If f is a *scalar function* with continuous second partial derivatives, then

$$\text{curl}(\text{grad } f) = \nabla \times \nabla f = \mathbf{0} \qquad (18.25)$$

Also, if $\mathbf{F}$ is a *vector field* having continuous second partial derivatives, then

$$\text{div}(\text{curl } \mathbf{F}) = \nabla \cdot (\nabla \times \mathbf{F}) = 0 \qquad (18.26)$$

See Problems 19 and 20 in Exercises 18.4.

Physical Interpretations

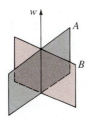

Figure 18.30

The word *curl* was first introduced by Maxwell* in his studies of electromagnetic fields. However, the curl is easily understood in connection with the flow of fluids. If a paddle device, such as shown in Figure 18.30, is inserted in a flowing fluid, then the curl of the velocity field $\mathbf{F}$ is a measure of the tendency of the fluid to turn the device about its vertical axis w. If curl $\mathbf{F} = \mathbf{0}$, then the flow of the fluid is said to be **irrotational**, which means that it is free of vertices or whirlpools that would cause the paddle to rotate.[†] In Figure 18.31 the axis w of the paddle points straight out of the page. Note from Figure 18.31(b) that "irrotational" does *not* mean that the fluid does not rotate. See Problem 28 in Exercises 18.4 for another interpretation of the curl.

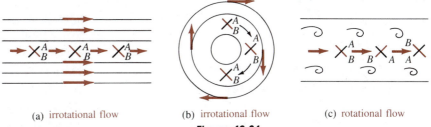

    (a) irrotational flow   (b) irrotational flow   (c) rotational flow

Figure 18.31

The divergence of a vector field can also be interpreted in the context of fluid flow. A measure of the rate of change of the density of the fluid at a point is simply div $\mathbf{F}$. In other words, div $\mathbf{F}$ is a measure of the fluid's compressibility. If $\nabla \cdot \mathbf{F} = 0$, the fluid is said to be **incompressible**. In electromagnetic theory, if $\nabla \cdot \mathbf{F} = 0$, the vector field $\mathbf{F}$ is said to be **solenoidal**.

Exercises 18.4

Answers to odd-numbered problems begin on page A-52.

In Problems 1–6 find the curl and the divergence of the given vector field.

1. $\mathbf{F}(x, y, z) = 3x^2 y \mathbf{i} + 2xz^3 \mathbf{j} + y^4 \mathbf{k}$

2. $\mathbf{F}(x, y, z) = 5y^3 \mathbf{i} + \left(\frac{1}{2}x^3 y^2 - xy\right)\mathbf{j} - (x^3 yz - xz)\mathbf{k}$

3. $\mathbf{F}(x, y, z) = xe^{-z}\mathbf{i} + 4yz^2\mathbf{j} + 3ye^{-z}\mathbf{k}$

---

*James Clerk Maxwell (1831–1879), a Scottish physicist.

[†]In science texts the word *rotation* is sometimes used instead of *curl*. The symbol curl $\mathbf{F}$ is then replaced by rot $\mathbf{F}$.

4. $F(x, y, z) = yz \ln x\mathbf{i} + (2x - 3yz)\mathbf{j} + xy^2z^3\mathbf{k}$

5. $F(x, y, z) = xye^x\mathbf{i} - x^3yze^z\mathbf{j} + xy^2e^y\mathbf{k}$

6. $F(x, y, z) = x^2 \sin yz\mathbf{i} + z \cos xz^3\mathbf{j} + ye^{5xy}\mathbf{k}$

In Problems 7–14 let **a** be a constant vector and $\mathbf{r} = x\mathbf{i} + y\mathbf{j} + z\mathbf{k}$. Verify the given identity.

7. div $\mathbf{r} = 3$

8. curl $\mathbf{r} = \mathbf{0}$

9. $(\mathbf{a} \times \nabla) \times \mathbf{r} = -2\mathbf{a}$

10. $\nabla \times (\mathbf{a} \times \mathbf{r}) = 2\mathbf{a}$

11. $\nabla \cdot (\mathbf{a} \times \mathbf{r}) = 0$

12. $\mathbf{a} \times (\nabla \times \mathbf{r}) = \mathbf{0}$

13. $\nabla \times [(\mathbf{r} \cdot \mathbf{r})\mathbf{a}] = 2(\mathbf{r} \times \mathbf{a})$

14. $\nabla \cdot [(\mathbf{r} \cdot \mathbf{r})\mathbf{a}] = 2(\mathbf{r} \cdot \mathbf{a})$

In Problems 15–22 verify the given identity. Assume continuity of all partial derivatives.

15. $\nabla \cdot (\mathbf{F} + \mathbf{G}) = \nabla \cdot \mathbf{F} + \nabla \cdot \mathbf{G}$

16. $\nabla \times (\mathbf{F} + \mathbf{G}) = \nabla \times \mathbf{F} + \nabla \times \mathbf{G}$

17. $\nabla \cdot (f\mathbf{F}) = f(\nabla \cdot \mathbf{F}) + \mathbf{F} \cdot \nabla f$

18. $\nabla \times (f\mathbf{F}) = f(\nabla \times \mathbf{F}) + (\nabla f) \times \mathbf{F}$

19. curl(grad f) = $\mathbf{0}$

20. div(curl $\mathbf{F}$) = 0

21. div($\mathbf{F} \times \mathbf{G}$) = $\mathbf{G} \cdot$ curl $\mathbf{F} - \mathbf{F} \cdot$ curl $\mathbf{G}$

22. curl(curl $\mathbf{F}$ + grad f) = curl(curl $\mathbf{F}$)

23. Show that

$$\nabla \cdot \nabla f = \frac{\partial^2 f}{\partial x^2} + \frac{\partial^2 f}{\partial y^2} + \frac{\partial^2 f}{\partial z^2}$$

This is known as the **Laplacian** and is also written $\nabla^2 f$.

24. Show that $\nabla \cdot (f \nabla f) = f \nabla^2 f + |\nabla f|^2$ where $\nabla^2 f$ is the Laplacian defined in Problem 23. (*Hint:* See Problem 17.)

25. Any scalar function f for which $\nabla^2 f = 0$ is said to be **harmonic**. Verify that $f(x, y, z) = (x^2 + y^2 + z^2)^{-1/2}$ is harmonic except at the origin. $\nabla^2 f = 0$ is called **Laplace's equation**.

26. Verify that

$$f(x, y) = \arctan\left(\frac{2y}{x^2 + y^2 - 1}\right), \qquad x^2 + y^2 \neq 1$$

satisfies Laplace's equation in two variables

$$\nabla^2 f = \frac{\partial^2 f}{\partial x^2} + \frac{\partial^2 f}{\partial y^2} = 0$$

27. Let $\mathbf{r} = x\mathbf{i} + y\mathbf{j} + z\mathbf{k}$ be the position vector of a mass m_1 and let the mass m_2 be located at the origin. If the force of gravitational attraction is

$$F = -\frac{Gm_1m_2}{|\mathbf{r}|^3}\mathbf{r}$$

verify that curl $\mathbf{F} = \mathbf{0}$ and div $\mathbf{F} = 0$.

28. Suppose a body rotates with a constant angular velocity $\boldsymbol{\omega}$ about an axis. If $\mathbf{r}$ is the position vector of a point P on the body measured from the origin, then the linear velocity vector $\mathbf{v}$ of rotation is $\mathbf{v} = \boldsymbol{\omega} \times \mathbf{r}$. See Figure 18.32. If $\mathbf{r} = x\mathbf{i} + y\mathbf{j} + z\mathbf{k}$ and $\boldsymbol{\omega} = \omega_1\mathbf{i} + \omega_2\mathbf{j} + \omega_3\mathbf{k}$, show that $\boldsymbol{\omega} = \frac{1}{2}$ curl $\mathbf{v}$.

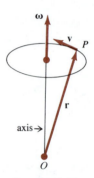

Figure 18.32

In Problems 29 and 30 assume that f and g have continuous second partial derivatives. Show that the given vector field is solenoidal. (*Hint:* See Problem 21.)

29. $\mathbf{F} = \nabla f \times \nabla g$

30. $\mathbf{F} = \nabla f \times (f \nabla g)$

31. If $\mathbf{F} = y^3\mathbf{i} + x^3\mathbf{j} + z^3\mathbf{k}$, find the flux of $\nabla \times \mathbf{F}$ through that portion of the ellipsoid $x^2 + y^2 + 4z^2 = 4$ in the first octant that is bounded by $y = 0$, $y = x$, $z = 0$.

32. The velocity vector field for the two-dimensional flow of an ideal fluid around a cylinder is given by

$$V(x, y) = A\left[\left(1 - \frac{x^2 - y^2}{(x^2 + y^2)^2}\right)\mathbf{i} - \frac{2xy}{(x^2 + y^2)^2}\mathbf{j}\right]$$

for some positive constant A. See Figure 18.33.

(a) Show that when the point (x, y) is far from the origin, $V(x, y) \approx A\mathbf{i}$.

(b) Show that V is irrotational.

(c) Show that V is incompressible.

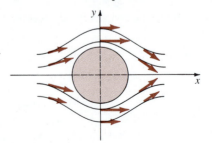

Figure 18.33

18.5 Green's Theorem

One of the most important theorems in vector integral calculus relates a line integral around a piecewise smooth *simple closed curve* with a double integral over the region bounded by the curve.

Line Integrals along Simple Closed Curves

We say the **positive direction** around a simple closed curve C is that direction a point on the curve must move, or the direction a person must walk on C, in order to keep the region R bounded by C to the left. See Figure 18.34(a). Roughly, as shown in Figure 18.34(b) and 18.34(c), the *positive* and *negative* directions correspond to the *counterclockwise* and *clockwise directions*, respectively. Line integrals on simple closed curves are written

$$\oint_C P(x, y)\, dx + Q(x, y)\, dy, \quad \oint_C P(x, y)\, dx + Q(x, y)\, dy, \quad \oint_C F(x, y)\, ds$$

and so on. The symbols $\oint_C$ and $\oint_C$ refer, in turn, to integrations in the positive and negative directions.

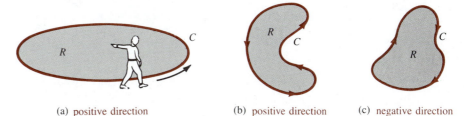

(a) positive direction   (b) positive direction   (c) negative direction

Figure 18.34

THEOREM 18.3   Green's Theorem in the Plane*

Suppose that C is a piecewise smooth simple closed curve bounding a region R. If P, Q, $\partial P/\partial y$, and $\partial Q/\partial x$ are continuous on R, then

$$\oint_C P\, dx + Q\, dy = \iint_R \left(\frac{\partial Q}{\partial x} - \frac{\partial P}{\partial y} \right) dA \qquad (18.27)$$

Partial Proof   We shall prove (18.27) only for a region R that is simultaneously of Type I and Type II:

$$R: g_1(x) \le y \le g_2(x), \qquad a \le x \le b$$
$$R: h_1(y) \le x \le h_2(y), \qquad c \le y \le d$$

---

*Named after George Green (1793–1841), an English mathematician and physicist. The words *in the plane* suggest that the theorem generalizes to 3-space. It does—read on.

Using Figure 18.35(a), we have

$$-\iint_R \frac{\partial P}{\partial y} \, dA = -\int_a^b \int_{g_1(x)}^{g_2(x)} \frac{\partial P}{\partial y} \, dy \, dx$$

$$= -\int_a^b [P(x, g_2(x)) - P(x, g_1(x))] \, dx$$

$$= \int_a^b P(x, g_1(x)) \, dx + \int_b^a P(x, g_2(x)) \, dx$$

$$= \oint_C P(x, y) \, dx \tag{18.28}$$

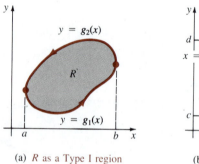
(a) R as a Type I region

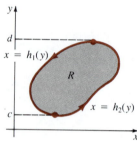
(b) R as a Type II region

Figure 18.35

Similarly, from Figure 18.35(b)

$$\iint_R \frac{\partial Q}{\partial x} \, dA = \int_c^d \int_{h_1(y)}^{h_2(y)} \frac{\partial Q}{\partial x} \, dx \, dy$$

$$= \int_c^d [Q(h_2(y), y) - Q(h_1(y), y)] \, dy$$

$$= \int_c^d Q(h_2(y), y) \, dy + \int_d^c Q(h_1(y), y) \, dy$$

$$= \oint_C Q(x, y) \, dy \tag{18.29}$$

Adding the results in (18.28) and (18.29) yields (18.27). ∎

Figure 18.36

Although the foregoing proof is not valid, the theorem is applicable to more complicated regions, such as those shown in Figure 18.36. The proof consists of decomposing R into a finite number of subregions to which (18.27) can be applied and then adding the results.

_____ **Example 1** _____

Evaluate $\oint_C (x^2 - y^2)\, dx + (2y - x)\, dy$, where C consists of the boundary of the region in the first quadrant that is bounded by the graphs of $y = x^2$ and $y = x^3$.

Solution If $P(x, y) = x^2 - y^2$ and $Q(x, y) = 2y - x$, then $\partial P/\partial y = -2y$ and $\partial Q/\partial x = -1$. From (18.27) and Figure 18.37, we have

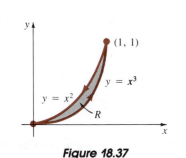

$$\oint_C (x^2 - y^2)\, dx + (2y - x)\, dy = \iint_R (-1 + 2y)\, dA$$

$$= \int_0^1 \int_{x^3}^{x^2} (-1 + 2y)\, dy\, dx$$

$$= \int_0^1 \left(-y + y^2\right)\Big]_{x^3}^{x^2}\, dx$$

$$= \int_0^1 (-x^6 + x^4 + x^3 - x^2)\, dx = -\frac{11}{420}$$

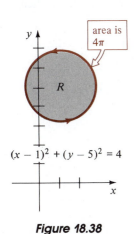

Figure 18.37

We note that the line integral in Example 1 could have been evaluated in a straightforward manner using the variable x as a parameter. However, in the next example, ponder the problem of evaluating the given line integral in the usual manner.

_____ **Example 2** _____

Evaluate $\oint_C (x^5 + 3y)\, dx + (2x - e^{y^3})\, dy$, where C is the circle $(x - 1)^2 + (y - 5)^2 = 4$.

Solution Identifying $P(x, y) = x^5 + 3y$ and $Q(x, y) = 2x - e^{y^3}$, we have $\partial P/\partial y = 3$ and $\partial Q/\partial x = 2$. Hence, (18.27) gives

$$\oint_C (x^5 + 3y)\, dx + (2x - e^{y^3})\, dy = \iint_R [2 - 3]\, dA$$

$$= -\iint_R dA$$

area is 4π

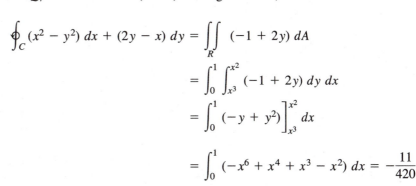

Figure 18.38

Now the double integral $\iint_R dA$ gives the area of the region R bounded by the circle of radius 2 shown in Figure 18.38. Since the area of the circle is $\pi 2^2 = 4\pi$ it follows that

$$\oint_C (x^5 + 3y)\, dx + (2x - e^{y^3})\, dy = -4\pi$$

Example 3

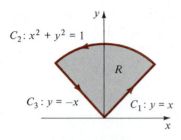

$C_2: x^2 + y^2 = 1$

$C_3: y = -x$ $C_1: y = x$

Figure 18.39

Find the work done by the force $\mathbf{F} = (-16y + \sin x^2)\mathbf{i} + (4e^y + 3x^2)\mathbf{j}$ acting along the simple closed curve C shown in Figure 18.39.

Solution   From (18.12) of Section 18.1 the work done by $\mathbf{F}$ is given by

$$W = \oint_C \mathbf{F} \cdot d\mathbf{r}$$

$$= \oint_C (-16y + \sin x^2)\, dx + (4e^y + 3x^2)\, dy$$

and so by Green's Theorem

$$W = \iint_R (6x + 16)\, dA$$

In view of the region R the last integral is best handled in polar coordinates. Since R is defined by $0 \le r \le 1$, $\pi/4 \le \theta \le 3\pi/4$,

$$W = \int_{\pi/4}^{3\pi/4} \int_0^1 (6r \cos \theta + 16)r\, dr\, d\theta$$

$$= \int_{\pi/4}^{3\pi/4} (2r^3 \cos \theta + 8r^2) \Big]_0^1 d\theta$$

$$= \int_{\pi/4}^{3\pi/4} (2 \cos \theta + 8)\, d\theta = 4\pi$$

Example 4

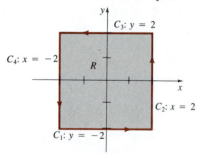

$C_3: y = 2$

$C_4: x = -2$ R

$C_2: x = 2$

$C_1: y = -2$

Figure 18.40

Let C be the closed curve consisting of the four straight line segments C_1, C_2, C_3, and C_4 shown in Figure 18.40. Green's Theorem is *not* applicable to the line integral

$$\oint_C \frac{x}{x^2 + y^2}\, dx + \frac{y}{x^2 + y^2}\, dy$$

since P, Q, $\partial P/\partial y$, and $\partial Q/\partial x$ are not continuous at the origin.

Example 5

Evaluate $\oint_C \dfrac{x}{x^2 + y^2}\, dx + \dfrac{y}{x^2 + y^2}\, dy$, where C is the closed curve consisting of the seven parts shown in Figure 18.41.

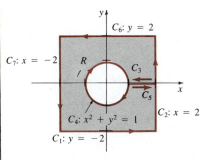

Figure 18.41

Solution Because

$$P(x, y) = \frac{x}{x^2 + y^2}, \qquad Q(x, y) = \frac{y}{x^2 + y^2},$$

$$\frac{\partial P}{\partial y} = -\frac{2xy}{(x^2 + y^2)^2}, \qquad \frac{\partial Q}{\partial x} = -\frac{2xy}{(x^2 + y^2)^2}$$

are continuous on the region R and bounded by C, we can apply Green's Theorem:

$$\oint_C \frac{x}{x^2 + y^2}\, dx + \frac{y}{x^2 + y^2}\, dy = \iint_R \left[-\frac{2xy}{(x^2 + y^2)^2} - \left(-\frac{2xy}{(x^2 + y^2)^2} \right) \right] dA = 0$$

Exercises 18.5

Answers to odd-numbered problems begin on page A-52.

In Problems 1 and 2 verify Green's Theorem by evaluating both integrals.

1. $\oint_C (x - y)\, dx + xy\, dy = \iint_R (y + 1)\, dA$, where C is the triangle with vertices $(0, 0)$, $(1, 0)$, $(1, 3)$

2. $\oint_C -y^2\, dx + x^2\, dy = \iint_R (2x + 2y)\, dA$, where C is the circle $x = 3 \cos t$, $y = 3 \sin t$, $0 \le t \le 2\pi$

In Problems 3–12 use Green's Theorem to evaluate the given integral.

3. $\oint_C 2y\, dx + 5x\, dy$, where C is the circle $(x - 1)^2 + (y + 3)^2 = 25$

4. $\oint_C (x + y^2)\, dx + (2x^2 - y)\, dy$, where C is the boundary of the region determined by the graphs of $y = x^2$, $y = 4$

5. $\oint_C (x^4 - 2y^3)\, dx + (2x^3 - y^4)\, dy$, where C is the circle $x^2 + y^2 = 4$

6. $\oint_C (x - 3y)\, dx + (4x + y)\, dy$, where C is the rectangle with vertices $(-2, 0)$, $(3, 0)$, $(3, 2)$, $(-2, 2)$

7. $\oint_C 2xy\, dx + 3xy^2\, dy$, where C is the triangle with vertices $(1, 2)$, $(2, 2)$, $(2, 4)$

8. $\oint_C e^{2x} \sin 2y\, dx + e^{2x} \cos 2y\, dy$, where C is the ellipse $9(x - 1)^2 + 4(y - 3)^2 = 36$

9. $\oint_C xy\, dx + x^2\, dy$, where C is the boundary of the region determined by the graphs of $x = 0$, $x^2 + y^2 = 1$, $x \ge 0$

10. $\oint_C e^{x^2}\, dx + 2 \tan^{-1}x\, dy$, where C is the triangle with vertices $(0, 0)$, $(0, 1)$, $(-1, 1)$

11. $\oint_C \frac{1}{3}y^3\, dx + (xy + xy^2)\, dy$, where C is the boundary of the region in the first quadrant determined by the graphs of $y = 0$, $x = y^2$, $x = 1 - y^2$

12. $\oint_C xy^2\, dx + 3 \cos y\, dy$, where C is the boundary of the region in the first quadrant determined by the graphs of $y = x^2$, $y = x^3$

In Problems 13 and 14 evaluate the given integral on any piecewise smooth simple closed curve C.

13. $\oint_C ay\, dx + bx\, dy$

14. $\oint_C P(x)\, dx + Q(y)\, dy$

In Problems 15 and 16 let R be the region bounded by a piecewise smooth simple closed curve C. Prove the given result.

15. $\oint_C x\, dy = -\oint_C y\, dx = \text{area of } R$

16. $\frac{1}{2} \oint_C - y\, dx + x\, dy = \text{area of } R$

In Problems 17 and 18 use the results of Problems 15 and 16 to find the area of the region bounded by the given closed curve.

17. The hypocycloid $x = a \cos^3 t$, $y = a \sin^3 t$, $a > 0$, $0 \le t \le 2\pi$

18. The ellipse $x = a \cos t$, $y = b \sin t$, $a > 0$, $b > 0$, $0 \le t \le 2\pi$

19. (a) Show that

$$\int_C -y\,dx + x\,dy = x_1 y_2 - x_2 y_1$$

where C is the line segment from the point (x_1, y_1) to (x_2, y_2).

(b) Use part **(a)** and Problem 16 to show that the area A of a polygon with vertices (x_1, y_1), (x_2, y_2), . . . , (x_n, y_n), labeled counterclockwise, is

$$A = \frac{1}{2}(x_1 y_2 - x_2 y_1) + \frac{1}{2}(x_2 y_3 - x_3 y_2)$$

$$+ \frac{1}{2}(x_{n-1} y_n - x_n y_{n-1}) + \frac{1}{2}(x_n y_1 - x_1 y_n)$$

20. Use part **(b)** of Problem 19 to find the area of the quadrilateral with vertices $(-1, 3)$, $(1, 1)$, $(4, 2)$, and $(3, 5)$.

21. Evaluate $\oint_C (\cos x^2 - y)\,dx + \sqrt{y^3 + 1}\;dy$ on the closed curve shown in Figure 18.42. (*Hint:* See Problem 18.)

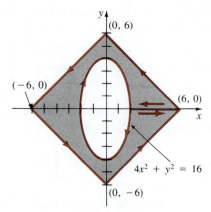

Figure 18.42

22. Consider the integral

$$\oint_{C_1} \frac{-y^3\,dx + xy^2\,dy}{(x^2 + y^2)^2}$$

where C_1 is the ellipse $x^2 + 4y^2 = 4$.

(a) Explain why an application of Green's Theorem gives the *incorrect* result: $\oint_{C_1} = 0$.

(b) If C_2 is the circle $x^2 + y^2 = 9$, use the curve in Figure 18.43 and Green's Theorem to show that $\oint_{C_1} = \oint_{C_2}$. (*Hint:* See (18.8) of Section 18.1.)

(c) Find the value of the integral in part **(a)**. (*Hint:* Use part **(b)**.)

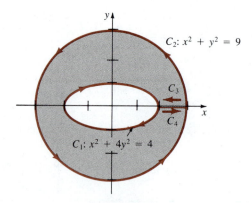

Figure 18.43

In Problems 23 and 24 use Green's Theorem to find the work done by the given force $\mathbf{F}$ around the closed curve in Figure 18.44.

23. $\mathbf{F} = (x - y)\mathbf{i} + (x + y)\mathbf{j}$

24. $\mathbf{F} = -xy^2\mathbf{i} + x^2 y\mathbf{j}$

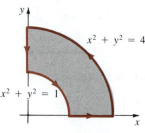

Figure 18.44

Miscellaneous Problems

25. Let P and Q be continuous and have continuous first partial derivatives in a simply connected region of the xy-plane. If $\int_A^B P\,dx + Q\,dy$ is independent of the path, show that $\oint_C P\,dx + Q\,dy = 0$ on every piecewise smooth simple closed curve C in the region.

26. Let R be a region bounded by a piecewise smooth simple closed curve C. Show that the coordinates of the centroid of the region are given by

$$\bar{x} = \frac{1}{2A}\oint_C x^2\,dy, \qquad \bar{y} = -\frac{1}{2A}\oint_C y^2\,dx$$

27. Find the work done by the force $\mathbf{F} = -y\mathbf{i} + x\mathbf{j}$ acting along the cardioid $r = 1 + \cos\theta$.

18.6 Stokes' Theorem and the Divergence Theorem

Vector Form of Green's Theorem

If $\mathbf{F}(x, y) = P(x, y)\mathbf{i} + Q(x, y)\mathbf{j}$ is a two-dimensional vector field, then

$$\text{curl } \mathbf{F} = \nabla \times \mathbf{F} = \begin{vmatrix} \mathbf{i} & \mathbf{j} & \mathbf{k} \\ \dfrac{\partial}{\partial x} & \dfrac{\partial}{\partial y} & \dfrac{\partial}{\partial z} \\ P & Q & 0 \end{vmatrix} = \left(\dfrac{\partial Q}{\partial x} - \dfrac{\partial P}{\partial y} \right) \mathbf{k} \qquad (18.30)$$

From (18.12) and (18.13) of Section 18.1, Green's Theorem (18.27) can be written in vector notation as

$$\oint_C \mathbf{F} \cdot d\mathbf{r} = \oint_C \mathbf{F} \cdot \mathbf{T} \, ds = \iint_R (\text{curl } \mathbf{F}) \cdot \mathbf{k} \, dA \qquad (18.31)$$

That is, the line integral of the tangential component of $\mathbf{F}$ is the double integral of the normal component of curl $\mathbf{F}$.

Green's Theorem in 3-space

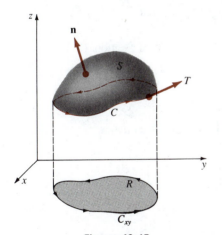

Figure 18.45

The vector form of Green's theorem given in (18.31) generalizes from a simple closed curve in the plane to a simple closed curve in 3-space. Suppose $z = f(x, y)$ is a continuous function having continuous first partial derivatives whose graph is a surface S over a region R on the xy-plane. Let C be a simple closed curve forming the boundary of S and let its projection C_{xy} onto the xy-plane form the boundary of R. The positive direction on C is defined by the positive direction on C_{xy}. Furthermore, let $\mathbf{T}$ be a unit tangent to C and let $\mathbf{n}$ be a unit *upper* normal to S. See Figure 18.45. The three-dimensional form of Green's Theorem, which we now give, is called **Stokes' Theorem.***

THEOREM 18.4   **Stokes' Theorem**

Let $\mathbf{F}(x, y, z) = P(x, y, z)\mathbf{i} + Q(x, y, z)\mathbf{j} + R(x, y, z)\mathbf{k}$ be a vector field for which P, Q, and R are continuous and have continuous first partial derivatives in a region containing a surface S. If C is the boundary of S traversed in the positive direction, then

$$\oint_C \mathbf{F} \cdot d\mathbf{r} = \oint_C (\mathbf{F} \cdot \mathbf{T}) \, ds = \iint_S (\text{curl } \mathbf{F} \cdot \mathbf{n}) \, dS \qquad (18.32)$$

Theorem 18.4 is not the full statement of Stokes' Theorem. It is limited to surfaces such as those shown in Figure 18.45, on which we can take $\mathbf{n}$ as an upper normal. The complete statement and proof of Theorem 18.4 for two-sided surfaces, or surfaces with an *orientation*, can be found in most advanced calculus texts.

---

*George G. Stokes (1819–1903), an Irish mathematician and physicist.

_____ **Example 1** _____

Let S be the part of the cylinder $z = 1 - x^2$ for $0 \le x \le 1$, $-2 \le y \le 2$. Verify Stokes' Theorem if $\mathbf{F} = xy\mathbf{i} + yz\mathbf{j} + xz\mathbf{k}$.

Solution   The surface S, the curve C (which is composed of the union of C_1, C_2, C_3, and C_4), and the region R are shown in Figure 18.46.

The Surface Integral: From $\mathbf{F} = xy\mathbf{i} + yz\mathbf{j} + xz\mathbf{k}$, we find

$$\text{curl } \mathbf{F} = \begin{vmatrix} \mathbf{i} & \mathbf{j} & \mathbf{k} \\ \dfrac{\partial}{\partial x} & \dfrac{\partial}{\partial y} & \dfrac{\partial}{\partial z} \\ xy & yz & xz \end{vmatrix} = -y\mathbf{i} - z\mathbf{j} - x\mathbf{k}$$

Now, if $g(x, y, z) = z + x^2 - 1 = 0$ defines the cylinder, the upper normal is

$$\mathbf{n} = \frac{\nabla g}{|\nabla g|} = \frac{2x\mathbf{i} + \mathbf{k}}{\sqrt{4x^2 + 1}}$$

Therefore,

$$\iint_S (\text{curl } \mathbf{F} \cdot \mathbf{n}) \, dS = \iint_S \frac{-2xy - x}{\sqrt{4x^2 + 1}} \, dS$$

To evaluate the latter surface integral, we use (18.18) of Section 18.3:

$$\iint_S \frac{-2xy - x}{\sqrt{4x^2 + 1}} \, dS = \iint_R (-2xy - x) \, dA$$

$$= \int_0^1 \int_{-2}^2 (-2xy - x) \, dy \, dx$$

$$= \int_0^1 \left[-xy^2 - xy \right]_{-2}^2 dx$$

$$= \int_0^1 (-4x) \, dx = -2 \tag{18.33}$$

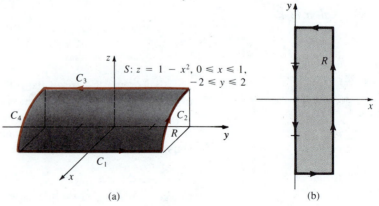

(a)   (b)

Figure 18.46

The Line Integral: We write $\oint_C = \int_{C_1} + \int_{C_2} + \int_{C_3} + \int_{C_4}$. On C_1: $x = 1$, $z = 0$, $dx = 0$, $dz = 0$, so

$$\int_{C_1} y(0) + y(0)\, dy + 0 = 0$$

On C_2: $y = 2$, $z = 1 - x^2$, $dy = 0$, $dz = -2x\, dx$, so

$$\int_{C_2} 2x\, dx + 2(1 - x^2)0 + x(1 - x^2)(-2x\, dx)$$

$$= \int_1^0 (2x - 2x^2 + 2x^4)\, dx = -\frac{11}{15}$$

On C_3: $x = 0$, $z = 1$, $dx = 0$, $dz = 0$, so

$$\int_{C_3} 0 + y\, dy + 0 = \int_2^{-2} y\, dy = 0$$

On C_4: $y = -2$, $z = 1 - x^2$, $dy = 0$, $dz = -2x\, dx$, so

$$\int_{C_4} -2x\, dx - 2(1 - x^2)0 + x(1 - x^2)(-2x\, dx)$$

$$= \int_0^1 (-2x - 2x^2 + 2x^4)\, dx = -\frac{19}{15}$$

Hence,

$$\oint_C xy\, dx + yz\, dy + xz\, dz = 0 - \frac{11}{15} + 0 - \frac{19}{15} = -2$$

which, of course, agrees with (18.33).

___

Example 2 _____

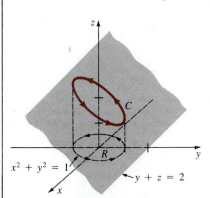

Figure 18.47

Evaluate $\oint_C z\, dx + x\, dy + y\, dz$, where C is the trace of the cylinder $x^2 + y^2 = 1$ in the plane $y + z = 2$. See Figure 18.47.

Solution   If $\mathbf{F} = z\mathbf{i} + x\mathbf{j} + y\mathbf{k}$, then

$$\text{curl } \mathbf{F} = \begin{vmatrix} \mathbf{i} & \mathbf{j} & \mathbf{k} \\ \dfrac{\partial}{\partial x} & \dfrac{\partial}{\partial y} & \dfrac{\partial}{\partial z} \\ z & x & y \end{vmatrix} = \mathbf{i} + \mathbf{j} + \mathbf{k}$$

Moreover, if $g(x, y, z) = y + z - 2 = 0$ defines the plane, then the upper normal is

$$\mathbf{n} = \frac{\nabla g}{|\nabla g|} = \frac{1}{\sqrt{2}}\mathbf{j} + \frac{1}{\sqrt{2}}\mathbf{k}$$

Hence, from (18.32)

$$\oint_C \mathbf{F} \cdot d\mathbf{r} = \iint_S \left[(\mathbf{i} + \mathbf{j} + \mathbf{k}) \cdot \left(\frac{1}{\sqrt{2}} \mathbf{j} + \frac{1}{\sqrt{2}} \mathbf{k} \right) \right] dS$$

$$= \sqrt{2} \iint_S dS$$

$$= \sqrt{2} \iint_R \sqrt{2} \, dA = 2\pi$$

Note that if $\mathbf{F}$ is the gradient of a scalar function, then, in view of (18.26) of Section 18.4, (18.32) implies that the circulation $\oint_C \mathbf{F} \cdot d\mathbf{r}$ is zero. Conversely, it can be shown that if the circulation is zero for every simple closed curve, then $\mathbf{F}$ is the gradient of a scalar function. In other words, $\mathbf{F}$ is irrotational if and only if $\mathbf{F} = \nabla\phi$, where ϕ is the potential for $\mathbf{F}$. Equivalently, this gives a test for a conservative vector field.

$\mathbf{F}$ is a conservative vector field if and only if curl $\mathbf{F} = \mathbf{0}$.

Divergence Theorem Let $\mathbf{F} = P\mathbf{i} + Q\mathbf{j}$ and let $\mathbf{T} = (dx/ds)\mathbf{i} + (dy/ds)\mathbf{j}$ be a *unit tangent* to a simple closed plane curve C. In (18.31) we saw that $\oint_C (\mathbf{F} \cdot \mathbf{T}) \, ds$ can be evaluated by a double integral involving curl $\mathbf{F}$. Similarly, if $\mathbf{n} = (dy/ds)\mathbf{i} - (dx/ds)\mathbf{j}$ is a *unit normal* to C (check $\mathbf{T} \cdot \mathbf{n}$), then $\oint_C (\mathbf{F} \cdot \mathbf{n}) \, ds$ can be expressed in terms of a double integral of div $\mathbf{F}$. From Green's Theorem,

$$\oint_C (\mathbf{F} \cdot \mathbf{n}) \, ds = \oint_C P \, dy - Q \, dx$$

$$= \iint_R \left[\frac{\partial P}{\partial x} - \left(-\frac{\partial Q}{\partial y} \right) \right] dA$$

$$= \iint_R \left[\frac{\partial P}{\partial x} + \frac{\partial Q}{\partial y} \right] dA$$

That is,

$$\oint_C (\mathbf{F} \cdot \mathbf{n}) \, ds = \iint_R \text{div } \mathbf{F} \, dA \qquad (18.34)$$

The result in (18.34) is a special case of the **Divergence** or **Gauss' Theorem**.* (See the biographical sketch on page 927.) The following is a generalization of (18.34) to 3-space.

THEOREM 18.5   Divergence Theorem

Let $\mathbf{F}(x, y, z) = P(x, y, z)\mathbf{i} + Q(x, y, z)\mathbf{j} + R(x, y, z)\mathbf{k}$ be a vector field for which P, Q, and R are continuous and have continuous first partial

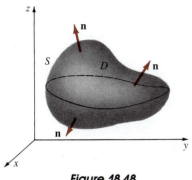

Figure 18.48

derivatives in a region that contains a surface S, which forms the boundary of a closed and bounded region D in space. Then

$$\iint_S (\mathbf{F} \cdot \mathbf{n}) \, dS = \iiint_D \text{div } \mathbf{F} \, dV \qquad (18.35)$$

where $\mathbf{n}$ is an outer normal to S.

A typical surface S, satisfying the hypothesis of Theorem 18.5 along with various positions of n, is shown in Figure 18.48. For the proof of the theorem, refer to an advanced text.

Example 3

Let D be the region bounded by the hemisphere $x^2 + y^2 + (z - 1)^2 = 9$, $1 \le z \le 4$, and the plane $z = 1$. Verify the Divergence Theorem if $\mathbf{F} = x\mathbf{i} + y\mathbf{j} + (z - 1)\mathbf{k}$.

Solution The closed region is shown in Figure 18.49.

Karl Friedrich Gauss

*Karl Friedrich Gauss (1777–1855)** Gauss was the first of a new breed of precise and demanding mathematicians—the "rigorists." We have already seen in earlier biographical sketches that Augustin Louis Cauchy and Karl Wilhelm Weierstrass were two mathematicians who followed in his footsteps. Gauss was a child prodigy in mathematics. As an adult, he often remarked that he could calculate or "reckon" before he could talk. However, as a college student Gauss was torn between two loves: philology and mathematics. But inspired by some original mathematical achievements as a teenager, and encouraged by the mathematician Wolfgang Bolyai, the choice was not too difficult. At the age of 20 Gauss settled on a career in mathematics. At the age of 22 he completed a book on number theory, *Disquisitiones Arithmeticae*. Published in 1801, this text was recognized as a masterpiece, and even today remains a classic in its field. Gauss' doctoral dissertation of 1799 also remains a memorable document. Using the theory of functions of a complex variable, he was the first to prove the so-called fundamental theorem of algebra: Every polynomial equation has at least one root.

Although Gauss was certainly recognized and respected as an outstanding mathematician during his lifetime, the full extent of his genius was not realized until the publication of his scientific diary in 1898, 44 years after his death. Much to the chagrin of some nineteenth-century mathematicians, the diary revealed that Gauss had foreseen, sometimes by decades, many of their discoveries or, perhaps more accurately, rediscoveries. Oblivious to fame, his mathematical researches were often pursued, like a child playing on a beach, simply for pleasure and self-satisfaction and not for the instruction that could be given to others through publication.

On any list of "Greatest Mathematicians Who Ever Lived," Karl Friedrich Gauss must surely rank near or at the top. For his profound impact on so many branches of mathematics, Gauss is sometimes referred to as "the prince of mathematicians."

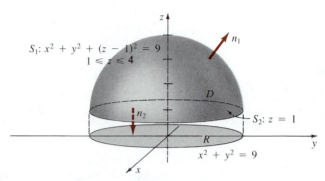

Figure 18.49

The Triple Integral: Since $\mathbf{F} = x\mathbf{i} + y\mathbf{j} + (z - 1)\mathbf{k}$, we see div $\mathbf{F} = 3$. Hence,

$$\iiint_D \operatorname{div} \mathbf{F}\, dV = \iiint_D 3\, dV$$

$$= 3 \iiint_D dV$$

$$= 3\left[\frac{2}{3}\pi 3^3\right] = 54\pi \qquad (18.36)$$

In the last calculation, we used the fact that $\iiint_D dV$ gives the volume of the hemisphere.

The Surface Integral: We write $\iint_S = \iint_{S_1} + \iint_{S_2}$, where S_1 is the hemisphere and S_2 is the plane $z = 1$. If S_1 is a level surface of $g(x, y, z) = x^2 + y^2 + (z - 1)^2$, then a unit outer normal is

$$\mathbf{n} = \frac{\nabla g}{|\nabla g|} = \frac{x\mathbf{i} + y\mathbf{j} + (z - 1)\mathbf{k}}{\sqrt{x^2 + y^2 + (z - 1)^2}} = \frac{x}{3}\mathbf{i} + \frac{y}{3}\mathbf{j} + \frac{z - 1}{3}\mathbf{k}$$

Now
$$\mathbf{F} \cdot \mathbf{n} = \frac{x^2}{3} + \frac{y^2}{3} + \frac{(z - 1)^2}{3} = 3$$

and so from (18.18) of Section 18.3,

$$\iint_{S_1} (\mathbf{F} \cdot \mathbf{n})\, dS = \iint_R (3)\left(\frac{3}{\sqrt{9 - x^2 - y^2}}\, dA\right) \qquad \boxed{\text{Polar coordinates}}$$

$$= 9 \int_0^{2\pi} \int_0^3 (9 - r^2)^{-1/2} r\, dr\, d\theta = 54\pi$$

On S_2, we take

$$\mathbf{n} = -\mathbf{k} \quad \text{so that} \quad \mathbf{F} \cdot \mathbf{n} = -z + 1$$

But, since $z = 1$, $\iint_{S_2} (-z + 1)\, dS = 0$. Hence, we see that

$$\iint_S (\mathbf{F} \cdot \mathbf{n})\, dS = 54\pi + 0 = 54\pi$$

agrees with (18.36).

Example 4

If $\mathbf{F} = xy\mathbf{i} + y^2z\mathbf{j} + z^3\mathbf{k}$, evaluate $\iint_S (\mathbf{F} \cdot \mathbf{n})\, dS$, where S is the unit cube defined by $0 \le x \le 1$, $0 \le y \le 1$, $0 \le z \le 1$.

Solution   See Figure 18.29 and Problem 20 of Exercises 18.3. Rather than evaluate six surface integrals, we apply the Divergence Theorem. Since div $\mathbf{F} = \nabla \cdot \mathbf{F} = y + 2yz + 3z^2$, we have from (18.35),

$$\iint_S (\mathbf{F} \cdot \mathbf{n})\, dS = \iiint_D (y + 2yz + 3z^2)\, dV$$

$$= \int_0^1 \int_0^1 \int_0^1 (y + 2yz + 3z^2)\, dx\, dy\, dz$$

$$= \int_0^1 \int_0^1 (y + 2yz + 3z^2)\, dy\, dz$$

$$= \int_0^1 \left(\frac{y^2}{2} + y^2z + 3yz^2 \right) \Big]_0^1 dz$$

$$= \int_0^1 \left(\frac{1}{2} + z + 3z^2 \right) dz$$

$$= \left(\frac{1}{2}z + \frac{z^2}{2} + z^3 \right) \Big]_0^1 = 2$$

Exercises 18.6

In Problems 1 and 2 verify Stokes' Theorem.

1. $\mathbf{F} = 5y\mathbf{i} - 5x\mathbf{j} + 3\mathbf{k}$; S that portion of the plane $z = 1$ within the cylinder $x^2 + y^2 = 4$

2. $\mathbf{F} = 2z\mathbf{i} - 3x\mathbf{j} + 4y\mathbf{k}$; S that portion of the paraboloid $z = 16 - x^2 - y^2$ for $z \ge 0$

3. Use Stokes' Theorem to evaluate $\oint_C (2z + x)\, dx + (y - z)\, dy + (x + y)\, dz$, where C is the triangle with vertices $(1, 0, 0)$, $(0, 1, 0)$, $(0, 0, 1)$.

4. If $\mathbf{F} = z^2y \cos xy\mathbf{i} + z^2x(1 + \cos xy)\mathbf{j} + 2z \sin xy\mathbf{k}$, use Stokes' Theorem to evaluate $\oint_C \mathbf{F} \cdot d\mathbf{r}$, where C is the boundary of the plane $z = 1 - y$ shown in Figure 18.50.

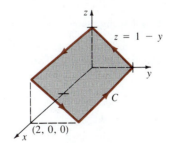

Figure 18.50

5. Use Stokes' Theorem to evaluate $\oint_C xy\,dx + 2yz\,dy + xz\,dz$, where C is given in Problem 4.

6. Use Stokes' Theorem to evaluate $\oint_C (x + 2z)\,dx + (3x + y)\,dy + (2y - z)\,dz$, where C is the curve of intersection of the plane $x + 2y + z = 4$ with the coordinate planes.

7. Use Stokes' Theorem to evaluate the circulation $\oint_C \mathbf{F} \cdot d\mathbf{r}$, where $\mathbf{F} = y^3\mathbf{i} - x^3\mathbf{j} + z^3\mathbf{k}$ and C is the trace of the
cylinder $x^2 + y^2 = 1$ in the plane $x + y + z = 1$.

8. If $\mathbf{F} = y\mathbf{i} + (y - x)\mathbf{j} + z^2\mathbf{k}$, use Stokes' theorem to evaluate $\iint_S (\text{curl } \mathbf{F} \cdot \mathbf{n})\,dS$, where S is that portion of the sphere $x^2 + y^2 + (z - 4)^2 = 25$ for $z \geq 0$.

9. Let D be the region bounded by the unit cube defined by $0 \leq x \leq 1,\ 0 \leq y \leq 1,\ 0 \leq z \leq 1$. Verify the Divergence Theorem if $\mathbf{F} = xy\mathbf{i} + yz\mathbf{j} + xz\mathbf{k}$.

10. If $\mathbf{F} = x^2\mathbf{i} + 2yz\mathbf{j} + 4z^3\mathbf{k}$, use the Divergence Theorem to evaluate $\iint_S (\mathbf{F} \cdot \mathbf{n})\,dS$, where S is the parallelepiped defined by $0 \leq x \leq 1,\ 0 \leq y \leq 2,\ 0 \leq z \leq 3$.

11. If $\mathbf{F} = x^3\mathbf{i} + y^3\mathbf{j} + z^3\mathbf{k}$, use the Divergence Theorem to evaluate $\iint_S (\mathbf{F} \cdot \mathbf{n})\,dS$, where S is the sphere $x^2 + y^2 + z^2 = a^2$.

12. If $\mathbf{F} = (x\mathbf{i} + y\mathbf{j} + z\mathbf{k})/(x^2 + y^2 + z^2)$, use the Divergence Theorem to evaluate $\iint_S (\mathbf{F} \cdot \mathbf{n})\,dS$, where S is the surface of the region bounded by the concentric spheres $x^2 + y^2 + z^2 = a^2,\ x^2 + y^2 + z^2 = b^2,\ b > a$.

13. If $\mathbf{F} = y^3\mathbf{i} + x^3\mathbf{j} + z^3\mathbf{k}$, use the Divergence Theorem to evaluate $\iint_S (\mathbf{F} \cdot \mathbf{n})\,dS$, where S is the surface of the region bounded by $z = \sqrt{4 - x^2 - y^2},\ x^2 + y^2 = 3,\ z = 0$.

14. If $\mathbf{F} = (x^2 + \sin y)\mathbf{i} + z^2\mathbf{j} + xy^3\mathbf{k}$, use the Divergence Theorem to evaluate $\iint_S (\mathbf{F} \cdot \mathbf{n})\,dS$, where S is the surface of the region bounded by $y = x^2,\ z = 9 - y,\ z = 0$.

Miscellaneous Problems

In Problems 15–19 assume that S forms the boundary of a closed region D.

15. If $\mathbf{a}$ is a constant vector, show that $\iint_S (\mathbf{a} \cdot \mathbf{n})\,dS = 0$.

16. If $\mathbf{F} = P\mathbf{i} + Q\mathbf{j} + R\mathbf{k}$ and P, Q, and R have continuous second partial derivatives, prove that $\iint_S (\text{curl } \mathbf{F} \cdot \mathbf{n})\,dS = 0$.

17. If f and g are scalar functions with continuous second partial derivatives, prove that*

$$\iint_S (f\,\nabla g) \cdot \mathbf{n}\,dS = \iiint_D (f\,\nabla^2 g + \nabla f \cdot \nabla g)\,dV$$

18. If f and g are scalar functions with continuous second partial derivatives, prove that

$$\iint_S (f\,\nabla g - g\,\nabla f) \cdot \mathbf{n}\,dS = \iiint_D (f\,\nabla^2 g - g\,\nabla^2 f)\,dV$$

19. If f is a scalar function with continuous first partial derivatives, prove that

$$\iint_S f\mathbf{n}\,dS = \iiint_D \nabla f\,dV$$

(*Hint:* Use (18.35) on $f\mathbf{a}$, where $\mathbf{a}$ is a constant vector, and Problem 17 of Exercises 18.4.)

20. The buoyancy force on a floating object is $\mathbf{B} = -\iint_S p\mathbf{n}\,dS$, where p is the fluid pressure. The pressure p is related to the density of the fluid $\rho(x, y, z)$ by a law of hydrostatics: $\nabla p = \rho(x, y, z)\mathbf{g}$, where $\mathbf{g}$ is the constant acceleration of gravity. If the weight of the object is $\mathbf{W} = m\mathbf{g}$, use the result of Problem 19 to prove Archimedes' principle, $\mathbf{B} + \mathbf{W} = \mathbf{0}$. See Figure 18.51.

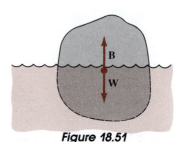

Figure 18.51

21. Use Stokes' Theorem to evaluate $\oint_C z\,dx + x\,dy + y\,dz$, where C is the curve of intersection of the plane $x + y + z = 0$ and the sphere $x^2 + y^2 + z^2 = 1$. (*Hint:* Review Section 12.4. Recall that the area of an ellipse $x^2/a^2 + y^2/b^2 = 1$ is πab.)

22. Evaluate the line integral in Problem 7 by direct methods.

---

*The results in Problems 17 and 18 are called *Green's identities*.

Chapter 18 Review Exercises

Answers to odd-numbered problems begin on page A-53.

In Problems 1–12 answer true or false. Where appropriate, assume continuity of P, Q, and their first partial derivatives.

1. The integral $\int_C (x^2 + y^2)\, dx + 2xy\, dy$, where C is given by $y = x^3$ from $(0, 0)$ to $(1, 1)$ has the same value on the curve $y = x^6$ from $(0, 0)$ to $(1, 1)$. _____

2. The value of the integral $\int_C 2xy\, dx - x^2\, dy$ between two points A and B depends on the path C. _____

3. If C_1 and C_2 are two smooth curves such that $\int_{C_1} P\, dx + Q\, dy = \int_{C_2} P\, dx + Q\, dy$, then $\int_C P\, dx + Q\, dy$ is independent of the path. _____

4. If the work $\int_C \mathbf{F} \cdot d\mathbf{r}$ depends on the curve C, then $\mathbf{F}$ is nonconservative. _____

5. If $\partial P/\partial x = \partial Q/\partial y$, then $\int_C P\, dx + Q\, dy$ is independent of the path. _____

6. In a conservative force field $\mathbf{F}$, the work done by $\mathbf{F}$ around a simple closed curve is zero. _____

7. Assuming continuity of all partial derivatives, $\nabla \times \nabla f = \mathbf{0}$. _____

8. The surface integral of the normal component of the curl of a conservative vector field $\mathbf{F}$ over a surface S is equal to zero. _____

9. Work done by a force $\mathbf{F}$ along a curve C is due entirely to the tangential component of $\mathbf{F}$. _____

10. For a two-dimensional vector field $\mathbf{F}$ in the plane $z = 0$, Stokes' Theorem is the same as Green's Theorem. _____

11. If $\mathbf{F}$ is a conservative force field, then the sum of the potential and kinetic energies of an object is constant. _____

12. If $\int_C P\, dx + Q\, dy$ is independent of the path, then $\mathbf{F} = P\mathbf{i} + Q\mathbf{j}$ is the gradient of some function ϕ. _____

In Problems 13–20 fill in the blanks.

13. If $\phi = \dfrac{1}{\sqrt{x^2 + y^2}}$ is a potential function for a conservative force field $\mathbf{F}$, then $\mathbf{F} =$ _____.

14. If $\mathbf{F} = f(x)\mathbf{i} + g(y)\mathbf{j} + h(z)\mathbf{k}$, then curl $\mathbf{F} =$ _____.

In Problems 15–18, $\mathbf{F} = x^2 y\mathbf{i} + xy^2\mathbf{j} + 2xyz\mathbf{k}$.

15. $\nabla \cdot \mathbf{F} =$ _____

16. $\nabla \times \mathbf{F} =$ _____

17. $\nabla \cdot (\nabla \times \mathbf{F}) =$ _____

18. $\nabla(\nabla \cdot \mathbf{F}) =$ _____

19. If C is the ellipse $2(x - 10)^2 + 9(y + 13)^2 = 3$, then $\oint_C (y - 7e^{x^3})\, dx + (x + \ln \sqrt{y})\, dy =$ _____.

20. If $\mathbf{F}$ is a velocity field of a fluid for which curl $\mathbf{F} = \mathbf{0}$, then $\mathbf{F}$ is said to be _____.

21. Evaluate $\displaystyle\int_C \frac{z^2}{x^2 + y^2}\, ds$, where C is given by

$$x = \cos 2t,\ y = \sin 2t,\ z = 2t,\ \pi \le t \le 2\pi$$

22. Evaluate $\int_C (xy + 4x)\, ds$, where C is given by $2x + y = 2$ from $(1, 0)$ to $(0, 2)$.

23. Evaluate $\int_C 3x^2 y^2\, dx + (2x^3 y - 3y^2)\, dy$, where C is given by $y = 5x^4 + 7x^2 - 14x$ from $(0, 0)$ to $(1, -2)$.

24. Show that $\displaystyle\oint_C \frac{-y\, dx + x\, dy}{x^2 + y^2} = 2\pi$, where C is the circle $x^2 + y^2 = a^2$.

25. Evaluate $\int_C y \sin \pi z\, dx + x^2 e^y\, dy + 3xyz\, dz$, where C is given by $x = t,\ y = t^2,\ z = t^3$ from $(0, 0, 0)$ to $(1, 1, 1)$.

26. If $\mathbf{F} = 4y\mathbf{i} + 6x\mathbf{j}$ and C is given by $x^2 + y^2 = 1$, evaluate $\oint_C \mathbf{F} \cdot d\mathbf{r}$ in two different ways.

27. Find the work done by the force $\mathbf{F} = x \sin y\mathbf{i} + y \sin x\mathbf{j}$ acting along the line segments from $(0, 0)$ to $(\pi/2, 0)$ and from $(\pi/2, 0)$ to $(\pi/2, \pi)$.

28. Find the work done by $\mathbf{F} = \dfrac{2}{x^2 + y^2}\mathbf{i} + \dfrac{1}{x^2 + y^2}\mathbf{j}$ from $(-\frac{1}{2}, \frac{1}{2})$ to $(1, \sqrt{3})$ acting on the path shown in Figure 18.52.

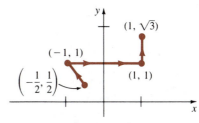

Figure 18.52

29. Evaluate $\iint_S (z/xy)\, dS$, where S is that portion of the cylinder $z = x^2$ in the first octant that is bounded by $y = 1$, $y = 3$, $z = 1$, $z = 4$.

30. If $\mathbf{F} = \mathbf{i} + 2\mathbf{j} + 3\mathbf{k}$, find the flux of $\mathbf{F}$ through the square $0 \le x \le 1, 0 \le y \le 1, z = 2$.

31. If $\mathbf{F} = c\nabla(1/r)$, where c is constant and $|\mathbf{r}| = r$, $\mathbf{r} = x\mathbf{i} + y\mathbf{j} + z\mathbf{k}$, find the flux of $\mathbf{F}$ through the sphere $x^2 + y^2 + z^2 = a^2$.

32. Explain why the Divergence Theorem is not applicable in Problem 31.

33. Find the flux of $\mathbf{F} = c\nabla(1/r)$ through any surface S that forms the boundary of a closed bounded region of space not containing the origin.

34. If $\mathbf{F} = 6x\mathbf{i} + 7z\mathbf{j} + 8y\mathbf{k}$, use Stokes' Theorem to evaluate $\iint_S (\text{curl } \mathbf{F} \cdot \mathbf{n}) \, dS$, where S is that portion of the paraboloid $z = 9 - x^2 - y^2$ within the cylinder $x^2 + y^2 = 4$.

35. Use Stokes' Theorem to evaluate $\oint_C -2y \, dx + 3x \, dy + 10z \, dz$, where C is the circle $(x - 1)^2 + (y - 3)^2 = 25$, $z = 3$.

36. Find the work $\oint_C \mathbf{F} \cdot d\mathbf{r}$ done by the force $\mathbf{F} = x^2\mathbf{i} + y^2\mathbf{j} + z^2\mathbf{k}$ around the curve C that is formed by the intersection of the plane $z = 2 - y$ and the sphere $x^2 + y^2 + z^2 = 4z$.

37. If $\mathbf{F} = x\mathbf{i} + y\mathbf{j} + z\mathbf{k}$, use the Divergence Theorem to evaluate $\iint_S (\mathbf{F} \cdot \mathbf{n}) \, dS$, where S is the surface of the region bounded by $x^2 + y^2 = 1$, $z = 0$, $z = 1$.

38. Repeat Problem 37 for $\mathbf{F} = \frac{1}{3}x^3\mathbf{i} + \frac{1}{3}y^3\mathbf{j} + \frac{1}{3}z^3\mathbf{k}$.

39. If $\mathbf{F} = (x^2 - e^y \tan^{-1}z)\mathbf{i} + (x + y)^2\mathbf{j} - (2yz + x^{10})\mathbf{k}$, use the Divergence Theorem to evaluate $\iint_S (\mathbf{F} \cdot \mathbf{n}) \, dS$, where S is the surface of the region in the first octant bounded by $z = 1 - x^2$, $z = 0$, $z = 2 - y$, $y = 0$.

40. Suppose $\mathbf{F} = x\mathbf{i} + y\mathbf{j} + (z^2 + 1)\mathbf{k}$ and S is the surface of the region bounded by $x^2 + y^2 = a^2$, $z = 0$, $z = c$. Evaluate $\iint_S (\mathbf{F} \cdot \mathbf{n}) \, dS$ without the aid of the Divergence Theorem. (*Hint:* The lateral surface area of the cylinder is $2\pi ac$.)

19

Differential Equations

We have already touched upon the notion of a first-order differential equation $F(x, y, y') = 0$. In Section 8.7 recall that we examined separable equations and in Section 16.5 we considered exact equations. In this further, and admittedly brief, discussion of differential equations, we shall focus our attention on two additional kinds of first-order equations and on an important class of second-order equations, $F(x, y, y', y'') = 0$. The basic idea in the study of differential equations is to *solve* them; that is, find suitable differentiable functions, defined explicitly or implicitly, which, when substituted in the equation, reduces it to an identity.

19.1 Basic Definitions and Terminology

An equation containing the derivatives or differentials of one or more dependent variables, with respect to one or more independent variables, is said to be a **differential equation**. Differential equations are classified according to *type*, *order*, and *linearity*.

Classification by Type   If an equation contains only ordinary derivatives of one or more dependent variables, with respect to a single independent variable, it is then said to be an **ordinary differential equation**.

___

Example 1

The equations

$$\frac{dy}{dx} - 5y = 1$$

$$(x + y)\, dx - 4y\, dy = 0$$

$$\frac{du}{dx} - \frac{dv}{dx} = x$$

$$\frac{d^2y}{dx^2} - 2\frac{dy}{dx} + 6y = 0$$

are ordinary differential equations.

___

An equation involving the partial derivatives of one or more dependent variables of two or more independent variables is called a **partial differential equation**.

___

Example 2

The equations

$$\frac{\partial u}{\partial y} = -\frac{\partial v}{\partial x}$$

$$x\frac{\partial u}{\partial x} + y\frac{\partial u}{\partial y} = u$$

$$\frac{\partial^2 u}{\partial x\, \partial y} = x + y$$

$$a^2\frac{\partial^2 u}{\partial x^2} = \frac{\partial^2 u}{\partial t^2} - 2k\frac{\partial u}{\partial t}$$

are partial differential equations.

___

Classification by Order The order of the highest derivative in a differential equation is called the **order of the equation**.

Example 3

(a) The equation $\dfrac{d^2y}{dx^2} + 5 \left(\dfrac{dy}{dx}\right)^3 - 4y = x$ is a second-order ordinary differential equation.

(b) Since the differential equation

$$x^2\, dy + y\, dx = 0$$

can be put into the form

$$x^2 \frac{dy}{dx} + y = 0$$

by dividing by the differential dx, it is an example of a first-order ordinary differential equation.

(c) The equation

$$c^2 \frac{\partial^4 u}{\partial x^4} + \frac{\partial^2 u}{\partial t^2} = 0$$

is a fourth-order partial differential equation.

Although partial differential equations are very important, their study demands a good foundation in the theory of ordinary differential equations. Consequently, in this chapter we shall confine our attention to ordinary differential equations.

A general *n*th-order, ordinary differential equation is often represented by the symbolism

$$F\left(x,\ y,\ \frac{dy}{dx},\ \ldots,\ \frac{d^n y}{dx^n}\right) = 0 \tag{19.1}$$

The following is a special case of (19.1).

Classification as Linear or Nonlinear A differential equation is said to be **linear** if it has the form

$$a_n(x) \frac{d^n y}{dx^n} + a_{n-1}(x) \frac{d^{n-1} y}{dx^{n-1}} + \cdots + a_1(x) \frac{dy}{dx} + a_0(x) y = g(x)$$

It should be observed that linear differential equations are characterized by two properties: (a) the dependent variable y and all its derivatives are of the first degree; that is, the power of each term involving y is 1; and (b) each coefficient depends only on the independent variable x. An equation that is not linear is said to be **nonlinear**.

Example 4

The equations

$$x\,dy + y\,dx = 0$$
$$y'' - 2y' + y = 0$$

and
$$x^3\frac{d^3y}{dx^3} - x^2\frac{d^2y}{dx^2} + 3x\frac{dy}{dx} + 5y = e^x$$

are linear first-, second-, and third-order ordinary differential equations, respectively. On the other hand,

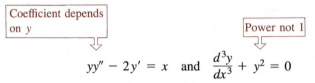

Coefficient depends on y　　　　　Power not 1

$$yy'' - 2y' = x \quad \text{and} \quad \frac{d^3y}{dx^3} + y^2 = 0$$

are nonlinear second- and third-order ordinary differential equations, respectively.

Solutions　　As mentioned before, our goal is to solve, or find solutions of, differential equations.

> **DEFINITION 19.1**　　Any function f defined on some interval, which, when substituted into a differential equation, reduces the equation to an identity, is said to be a **solution** of the equation.

Example 5

The function $y = e^{x^2}$ is a solution of

$$\frac{dy}{dx} - 2xy = 0 \qquad (19.2)$$

Since
$$\frac{dy}{dx} = 2xe^{x^2}$$

we see that
$$\frac{dy}{dx} - 2xy = 2xe^{x^2} - 2x(e^{x^2}) = 0$$

Example 6

The first-order differential equations

$$\left(\frac{dy}{dx}\right)^2 + 1 = 0 \quad \text{and} \quad (y')^2 + y^2 + 4 = 0$$

possess no real solutions. Why?

Explicit and Implicit Solutions

Solutions of differential equations are further distinguished as either **explicit** or **implicit solutions**. For example, we have already seen that $y = e^{x^2}$ is an explicit solution of equation (19.2). A relation $G(x, y) = 0$ is said to define a solution of (19.1) implicitly on an interval I provided it defines one or more solutions on I.

Example 7

$y = xe^x$ is an explicit solution of $y'' - 2y' + y = 0$. To see this, we compute

$$y' = xe^x + e^x \quad \text{and} \quad y'' = xe^x + 2e^x$$

Observe that $y'' - 2y' + y = (xe^x + 2e^x) - 2(xe^x + e^x) + xe^x = 0$

Example 8

For $-2 < x < 2$ the relation $x^2 + y^2 - 4 = 0$ is an implicit solution of the differential equation

$$\frac{dy}{dx} = -\frac{x}{y}$$

By implicit differentiation it follows that

$$\frac{d}{dx}(x^2) + \frac{d}{dx}(y^2) - \frac{d}{dx}(4) = 0$$

$$2x + 2y\frac{dy}{dx} = 0 \quad \text{or} \quad \frac{dy}{dx} = -\frac{x}{y}$$

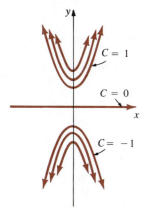

Figure 19.1

You should become accustomed to the fact that a given differential equation will usually possess an infinite number of solutions. By direct substitution we can show that any function in the one-parameter family $y = Ce^{x^2}$, where C is any arbitrary constant, also satisfies equation (19.2). As indicated in Figure 19.1, $y = 0$, obtained by setting $C = 0$, is also a solution of the equation. In Example 7 we saw that $y = xe^x$ is a solution of $y'' - 2y' + y = 0$; tracing back through the work reveals $y = Cxe^x$ represents a family of solutions. By again choosing $C = 0$ in the family of solutions $y = Cxe^x$ we see $y = 0$ is also a solution of $y'' - 2y' + y = 0$. The constant function $y = 0$, $-\infty < x < \infty$, that satisfies a given differential equation is often referred to as a **trivial solution**.

Further Terminology

The study of differential equations is similar to integral calculus. When evaluating an antiderivative or indefinite integral we utilize a single constant of integration. When solving a first-order differential equation, we have already seen that we obtain a one-parameter family of functions $G(x, y, C) = 0$ such that each member of the family is a solution of the differential equation. Naturally, we expect an **n-parameter family of solutions** $G(x, y, C_1, \ldots, C_n) = 0$ when solving an nth-order differential equation $F(x, y, y', \ldots, y^{(n)}) = 0$.

A solution of a differential equation that is free of arbitrary parameters is called a **particular solution**. One way of obtaining a particular solution is to choose specific values of the parameters in a family of solutions. For example, it is readily seen that $y = Ce^x$ is a one-parameter family of solutions of the simple first-order equation $y' = y$. For $C = 0, -2$, and 5, we get the particular solutions $y = 0$, $y = -2e^x$, and $y = 5e^x$, respectively.

Sometimes a differential equation possesses a solution that cannot be obtained by specializing the parameters in a family of solutions. Such a solution is called a **singular solution**. The concept of a singular solution is left for an advanced course.

If *every* solution of an nth-order differential equation can be obtained from $G(x, y, C_1, \ldots, C_n) = 0$ by appropriate choices of the C_i, $i = 1, 2, \ldots, n$, we then say that the n-parameter family is the **general** or **complete** solution of the equation.

Exercises 19.1

Answers to odd-numbered problems begin on page A-53.

In Problems 1–10 state whether the given differential equation is linear or nonlinear. Give the order of each equation.

1. $(1 - x)y'' - 4xy' + 5y = \cos x$

2. $x\dfrac{d^3y}{dx^3} - 2\left(\dfrac{dy}{dx}\right)^4 + y = 0$

3. $yy' + 2y = 1 + x^2$

4. $x^2\, dy + (y - xy - xe^x)\, dx = 0$

5. $x^3y^{(4)} - x^2y'' + 4xy' - 3y = 0$

6. $\dfrac{d^2y}{dx^2} + 9y = \sin y$

7. $\dfrac{dy}{dx} = \sqrt{1 + \left(\dfrac{d^2y}{dx^2}\right)^2}$

8. $\dfrac{d^2r}{dt^2} = -\dfrac{k}{r^2}$

9. $(\sin x)y''' - (\cos x)y' = 2$

10. $(1 - y^2)\, dx + x\, dy = 0$

In Problems 11–34 verify that the indicated function is a solution of the given differential equation. Where appropriate, C_1 and C_2 denote constants.

11. $2y' + y = 0$; $y = e^{-x/2}$

12. $y' + 4y = 32$; $y = 8$

13. $\dfrac{dy}{dx} - 2y = e^{3x}$; $y = e^{3x} + 10e^{2x}$

14. $\dfrac{dy}{dt} + 20y = 24$; $y = \dfrac{6}{5} - \dfrac{6}{5}e^{-20t}$

15. $y' = 25 + y^2$; $y = 5 \tan 5x$

16. $\dfrac{dy}{dx} = \sqrt{\dfrac{y}{x}}$; $y = (\sqrt{x} + C_1)^2$, $x > 0$, $C_1 > 0$

17. $y' + y = \sin x$; $y = \dfrac{1}{2}\sin x - \dfrac{1}{2}\cos x + 10e^{-x}$

18. $2xy\, dx + (x^2 + 2y)\, dy = 0$; $x^2y + y^2 = C_1$

19. $x^2\, dy + 2xy\, dx = 0$; $y = -\dfrac{1}{x^2}$

20. $(y')^3 + xy' = y$; $y = x + 1$

21. $y = 2xy' + y(y')^2$; $y^2 = C_1\left(x + \dfrac{1}{4}C_1\right)$

22. $y' = 2\sqrt{|y|}$; $y = x|x|$

23. $y' - \dfrac{1}{x}y = 1$; $y = x \ln x$, $x > 0$

24. $\dfrac{dP}{dt} = P(a - bP)$; $P = \dfrac{aC_1e^{at}}{1 + bC_1e^{at}}$

25. $\dfrac{dX}{dt} = (2 - X)(1 - X)$; $\ln \dfrac{2 - X}{1 - X} = t$

26. $y' + 2xy = 1$; $y = e^{-x^2} \int_0^x e^{t^2} dt + C_1 e^{-x^2}$

27. $(x^2 + y^2) dx + (x^2 - xy) dy = 0$; $C_1(x + y)^2 = xe^{y/x}$

28. $y'' + y' - 12y = 0$; $y = C_1 e^{3x} + C_2 e^{-4x}$

29. $y'' - 6y' + 13y = 0$; $y = e^{3x} \cos 2x$

30. $\dfrac{d^2y}{dx^2} - 4\dfrac{dy}{dx} + 4y = 0$; $y = e^{2x} + xe^{2x}$

31. $y'' = y$; $y = \cosh x + \sinh x$

32. $y'' + 25y = 0$; $y = C_1 \cos 5x$

33. $y'' + (y')^2 = 0$; $y \ln|x + C_1| + C_2$

34. $y'' + y = \tan x$; $y = -\cos x \ln(\sec x + \tan x)$

In Problems 35 and 36 find values of m so that $y = e^{mx}$ is a solution of the given differential equation.

35. $y'' - 5y' + 6y = 0$ **36.** $y'' + 10y' + 25y = 0$

In Problems 37 and 38 find values of m so that $y = x^m$ is a solution of the given differential equation.

37. $xy'' - 3y' = 0$ **38.** $x^2y'' + 6xy' + 4y = 0$

19.2 Homogeneous First-order Differential Equations

Note: Before working through this section you are encouraged to review Section 8.7 on separable first-order differential equations.

Suppose a first-order differential equation $F(x, y, y') = 0$ has the differential form

$$P(x, y) \, dx + Q(x, y) \, dy = 0$$

and the property that

$$P(tx, ty) = t^n P(x, y) \quad \text{and} \quad Q(tx, ty) = t^n Q(x, y)$$

We then say the equation has **homogeneous coefficients**, or is a **homogeneous equation**. The important point in the subsequent discussion is the fact that a homogeneous differential equation *can always be reduced to a separable equation* through an appropriate algebraic substitution. Before pursuing the method of solution for this type of differential equation, let us closely examine the nature of homogeneous functions.

DEFINITION 19.2   Homogeneous Function

If $f(tx, ty) = t^n f(x, y)$, for some real number n, then $f(x, y)$ is said to be a **homogeneous function** of degree n.

Example 1

(a) $f(x, y) = x - 3\sqrt{xy} + 5y$
$$f(tx, ty) = (tx) - 3\sqrt{(tx)(ty)} + 5(ty)$$
$$= tx - 3\sqrt{t^2xy} + 5ty$$
$$= t[x - 3\sqrt{xy} + 5y] = tf(x, y)$$

The function is homogeneous of degree 1.

(b) $f(x, y) = \sqrt{x^3 + y^3}$

$f(tx, ty) = \sqrt{t^3 x^3 + t^3 y^3}$

$= t^{3/2}\sqrt{x^3 + y^3} = t^{3/2}f(x, y)$

The function is homogeneous of degree $\frac{3}{2}$.

(c) $f(x, y) = x^2 + y^2 + 1$

$f(tx, ty) = t^2 x^2 + t^2 y^2 + 1 \neq t^2 f(x, y)$

since $t^2 f(x, y) = t^2 x^2 + t^2 y^2 + t^2$. The function is not homogeneous.

(d) $f(x, y) = \dfrac{x}{2y} + 4$

$f(tx, ty) = \dfrac{tx}{2ty} + 4$

$= \dfrac{x}{2y} + 4 = t^0 f(x, y)$

The function is homogeneous of degree 0.

As parts (c) and (d) of Example 1 show, a constant added to a function destroys homogeneity, *unless* the function is homogeneous of degree 0. Also, in many instances a homogeneous function can be recognized by examining the total degree of each term.

Example 2

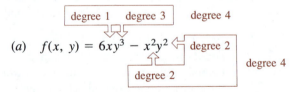

(a) $f(x, y) = 6xy^3 - x^2 y^2$

The function is homogeneous of degree 4.

(b) $f(x, y) = x^2 - y$

The function is not homogeneous.

The Method of Solution   An equation of the form $P(x, y)\,dx + Q(x, y)\,dy = 0$, where P and Q have the same degree of homogeneity can be reduced to separable variables by *either* the substitution $y = ux$ or $x = vy$, where u and v are new dependent variables. In particular, if we choose $y = ux$, then $dy = u\,dx + x\,du$. Hence the differential equation becomes

$$P(x, ux)\,dx + Q(x, ux)[u\,dx + x\,du] = 0$$

Now, by homogeneity of P and Q we can write

$$x^n P(1, u) \, dx + x^n Q(1, u)[u \, dx + x \, du] = 0$$

or

$$[P(1, u) + uQ(1, u)] \, dx + xQ(1, u) \, du = 0$$

which gives

$$\frac{dx}{x} + \frac{Q(1, u) \, du}{P(1, u) + uQ(1, u)} = 0$$

We hasten to point out that the preceding formula should *not* be memorized; rather, *the procedure should be worked through each time*. The proof that the substitution $x = vy$ also leads to a separable equation is left as an exercise.

Example 3

Solve

$$(x^2 + y^2) \, dx + (x^2 - xy) \, dy = 0$$

Solution Both $P(x, y)$ and $Q(x, y)$ are homogeneous of degree 2. If we let $y = ux$, it follows that

$$(x^2 + u^2 x^2) \, dx + (x^2 - ux^2)[u \, dx + x \, du] = 0$$
$$x^2(1 + u) \, dx + x^3(1 - u) \, du = 0$$
$$\frac{1 - u}{1 + u} \, du + \frac{dx}{x} = 0$$
$$\left[-1 + \frac{2}{1 + u} \right] du + \frac{dx}{x} = 0$$

Integrating then yields

$$-u + 2 \ln |1 + u| + \ln |x| + \ln |C| = 0^*$$
$$-\frac{y}{x} + 2 \ln \left| 1 + \frac{y}{x} \right| + \ln |x| + \ln |C| = 0$$

Using the properties of logarithms, this solution can be written in the alternative form

$$C(x + y)^2 = xe^{y/x}$$

Although the substitution $y = ux$ can be used for every homogeneous equation, in practice we use the alternative substitution $x = vy$ whenever the function $P(x, y)$ is simpler than $Q(x, y)$. Also, it could happen that after using one substitution we may encounter integrals that are difficult or impossible to evaluate; switching substitutions may result in an easier problem.

---

*If most of the integrations result in logarithms, a judicious choice of the constant of integration is $\ln |C|$ rather than C.

_____ **Example 4** _____

Solve $\qquad\qquad\qquad\qquad 2x^3y\,dx + (x^4 + y^4)\,dy = 0$

Solution Each coefficient is a homogeneous function of degree 4. Since the coefficient of dx is slightly simpler than the coefficient of dy, we try $x = vy$. It follows that

$$2v^3y^4[v\,dy + y\,dv] + (v^4y^4 + y^4)\,dy = 0$$

$$2v^3y^5\,dv + y^4(3v^4 + 1)\,dy = 0$$

$$\frac{2v^3\,dv}{3v^4 + 1} + \frac{dy}{y} = 0$$

$$\frac{1}{6}\ln(3v^4 + 1) + \ln|y| = \ln|C_1|$$

or $\qquad\qquad\qquad\qquad 3x^4y^2 + y^6 = C$

Had $y = ux$ been used then

$$\frac{dx}{x} + \frac{u^4 + 1}{u^5 + 3u}\,du = 0$$

It is worth a minute of your time to reflect on how to evaluate the integral of the second term in the last equation.

_____ **Exercises 19.2** _____

Answers to odd-numbered problems begin on page A-53.

In Problems 1–20 solve the given differential equation by using an appropriate substitution.

1. $(x - y)\,dx + x\,dy = 0$

2. $(x + y)\,dx + x\,dy = 0$

3. $x\,dx + (y - 2x)\,dy = 0$

4. $y\,dx = 2(x + y)\,dy$

5. $(y^2 + yx)\,dx - x^2\,dy = 0$

6. $(y^2 + yx)\,dx + x^2\,dy = 0$

7. $\dfrac{dy}{dx} = \dfrac{y - x}{y + x}$

8. $\dfrac{dy}{dx} = \dfrac{x + 3y}{3x + y}$

9. $-y\,dx + (x + \sqrt{xy})\,dy = 0$

10. $x\dfrac{dy}{dx} - y = \sqrt{x^2 + y^2}$

11. $2x^2y\,dx = (3x^3 + y^3)\,dy$

12. $(x^4 + y^4)\,dx - 2x^3y\,dy = 0$

13. $\dfrac{dy}{dx} = \dfrac{y}{x} + \dfrac{x}{y}$

14. $\dfrac{dy}{dx} = \dfrac{y}{x} + \dfrac{x^2}{y^2} + 1$

15. $y\dfrac{dx}{dy} = x + 4ye^{-2x/y}$

16. $(x^2e^{-y/x} + y^2)\,dx = xy\,dy$

17. $\left(y + x\cot\dfrac{y}{x}\right)dx - x\,dy = 0$

18. $\dfrac{dy}{dx} = \dfrac{y}{x}\ln\dfrac{y}{x}$

19. $(x^2 + xy - y^2)\,dx + xy\,dy = 0$

20. $(x^2 + xy + 3y^2)\,dx - (x^2 + 2xy)\,dy = 0$

21. Show that the substitution $x = vy$ reduces a homogeneous equation to one with separate variables.

22. If $f(x, y)$ is a homogeneous function of degree n, show that

$$x\frac{\partial f}{\partial x} + y\frac{\partial f}{\partial y} = nf$$

19.3 Linear First-order Differential Equations

In Section 19.1 we defined the general form of a linear differential equation of order n to be

$$a_n(x)\frac{d^n y}{dx^n} + a_{n-1}(x)\frac{d^{n-1} y}{dx^{n-1}} + \cdots + a_1(x)\frac{dy}{dx} + a_0(x)y = g(x)$$

Now when $n = 1$, we obtain the **linear first-order equation**

$$a_1(x)\frac{dy}{dx} + a_0(x)y = g(x) \tag{19.3}$$

Example 1

By direct comparison with (19.3), we see that

$$x\frac{dy}{dx} - 4y = x^6 e^x$$

is a linear first-order differential equation. You should be able to convince yourself that this first-order equation is not separable, exact, or homogeneous.

By dividing (19.3) by $a_1(x)$, we obtain the more useful form

$$\frac{dy}{dx} + P(x)y = f(x) \tag{19.4}$$

We seek solutions of (19.4) on an interval I for which P and f are continuous.

An Integrating Factor   Let us suppose equation (19.4) is written in the differential form

$$dy + [P(x)y - f(x)]\,dx = 0 \tag{19.5}$$

Linear equations possess the pleasant property that a function $\mu(x)$ can always be found such that the multiple of (19.5)

$$\mu(x)\,dy + \mu(x)[P(x)y - f(x)]\,dx = 0 \tag{19.6}$$

is an exact differential equation. By Theorem 16.6 we know that the left side of equation (19.6) will be an exact differential if

$$\frac{\partial}{\partial x}\,\mu(x) = \frac{\partial}{\partial y}\,\mu(x)[P(x)y - f(x)] \tag{19.7}$$

or

$$\frac{d\mu}{dx} = \mu P(x)$$

This is a separable equation from which we can determine $\mu(x)$. We have

$$\frac{d\mu}{\mu} = P(x)\ dx$$

$$\ln|\mu| = \int P(x)\ dx \tag{19.8}$$

so that

$$\mu(x) = e^{\int P(x)\ dx} \tag{19.9}$$

The function $\mu(x)$ defined in (19.9) is called an **integrating factor** for the linear equation. Note that we need not use a constant of integration in (19.8) since (19.6) is unaffected by a constant multiple. Also, $\mu(x) \neq 0$ for every x in I and is continuous and differentiable.

It is interesting to observe that equation (19.5) is still an exact differential equation even when $f(x) = 0$. In fact, $f(x)$ plays no part in determining $\mu(x)$ since we see from (19.7) that $\partial/\partial y\ (\mu(x)f(x)) = 0$. Thus both

$$e^{\int P(x)\ dx}\ dy + e^{\int P(x)\ dx}[P(x)y - f(x)]\ dx$$

and

$$e^{\int P(x)\ dx}\ dy + e^{\int P(x)\ dx}P(x)y\ dx$$

are exact differentials. We now write (19.6) in the form

$$e^{\int P(x)\ dx}\ dy + e^{\int P(x)\ dx}P(x)y\ dx = e^{\int P(x)\ dx}f(x)\ dx$$

and recognize that we can write the equation as

$$d[e^{\int P(x)\ dx}y] = e^{\int P(x)\ dx}f(x)\ dx$$

Integrating the last equation gives

$$e^{\int P(x)\ dx}y = \int e^{\int P(x)\ dx}f(x)\ dx + C$$

or

$$y = e^{-\int P(x)\ dx}\int e^{\int P(x)\ dx}f(x)\ dx + Ce^{-\int P(x)\ dx} \tag{19.10}$$

In other words, if (19.4) has a solution it must be of form (19.10). Conversely, it is a straightforward matter to verify that (19.10) constitutes a one-parameter family of solutions of equation (19.4). However, no attempt should be made to memorize formula (19.10). The procedure should be followed each time, so for convenience we summarize the results.

The Method of Solution   To solve a linear first-order differential equation first put it into the form (19.4); that is, make the coefficient of y' unity. Then multiply the entire equation by the integrating factor

$$e^{\int P(x)\ dx}$$

The left side of

$$e^{\int P(x)\ dx}\frac{dy}{dx} + P(x)e^{\int P(x)\ dx}y = e^{\int P(x)\ dx}f(x)$$

is the derivative of the product of the integrating factor and the dependent variable: $e^{\int P(x)\,dx}y$. Write the equation in the form

$$\frac{d}{dx}\left[e^{\int P(x)\,dx}y\right] = e^{\int P(x)\,dx}f(x)$$

and, finally, integrate both sides.

Example 2

Solve
$$x\frac{dy}{dx} - 4y = x^6 e^x$$

Solution   Write the equation as

$$\frac{dy}{dx} - \frac{4}{x}y = x^5 e^x \tag{19.11}$$

and determine the integrating factor

$$e^{-4\int dx/x} = e^{-4\ln|x|} = e^{\ln x^{-4}} = x^{-4}$$

Here we have used the basic identity $b^{\log_b N} = N$. Now multiply (19.11) by this term

$$x^{-4}\frac{dy}{dx} - 4x^{-5}y = xe^x \tag{19.12}$$

and obtain*
$$\frac{d}{dx}[x^{-4}y] = xe^x \tag{19.13}$$

It follows from integration by parts that

$$x^{-4}y = xe^x - e^x + C$$

or
$$y = x^5 e^x - x^4 e^x + Cx^4$$

Example 3

Solve
$$\frac{dy}{dx} - 3y = 0$$

Solution   The equation is already in form (19.4). Hence the integrating factor is

$$e^{\int(-3)\,dx} = e^{-3x}$$

---

*You should perform the indicated differentiations a few times in order to be convinced that all equations, such as (19.11), (19.12), and (19.13), are formally equivalent.

and thus
$$e^{-3x}\frac{dy}{dx} - 3e^{-3x}y = 0$$

$$\frac{d}{dx}[e^{-3x}y] = 0$$

$$e^{-3x}y = C$$

and so
$$y = Ce^{3x}$$

Notice that the differential equation in Example 3 can also be solved by separation of variables.

The General Solution   If it is assumed that $P(x)$ and $f(x)$ are continuous on an interval I and x_0 is any point on the interval, then it can be proved that there exists only one solution of the **initial-value problem**:

Solve:
$$\frac{dy}{dx} + P(x)y = f(x)$$

Subject to:
$$y(x_0) = y_0$$

(19.14)

Indeed, *every* solution of the equation on the interval I is of form (19.10). Thus, obtaining the solution of (19.14) is a simple matter of finding an appropriate value of C in (19.10). As a consequence we are justified in calling (19.10) the **general solution** of the differential equation.

Example 4

Find the general solution of

$$(x^2 + 9)\frac{dy}{dx} + xy = 0$$

Solution   We write

$$\frac{dy}{dx} + \frac{x}{x^2 + 9}y = 0$$

The function $P(x) = x/(x^2 + 9)$ is continuous on $(-\infty, \infty)$. Now the integrating factor for the equation is

$$e^{\int x\, dx/(x^2 + 9)} = e^{(1/2)\int 2x\, dx/(x^2 + 9)} = e^{(1/2)\ln(x^2 + 9)} = \sqrt{x^2 + 9}$$

so that
$$\sqrt{x^2 + 9}\,\frac{dy}{dx} + \frac{x}{\sqrt{x^2 + 9}}y = 0$$

$$\frac{d}{dx}[\sqrt{x^2 + 9}\,y] = 0$$

$$\sqrt{x^2 + 9}\,y = C$$

Hence, the general solution is

$$y = \frac{C}{\sqrt{x^2 + 9}}$$

Example 5

Solve $\qquad\qquad \dfrac{dy}{dx} + 2xy = x \qquad\qquad$ subject to $y(0) = -3$.

Solution   The functions $P(x) = 2x$ and $f(x) = x$ are continuous on $(-\infty, \infty)$. The integrating factor is

$$e^{2\int x\,dx} = e^{x^2}$$

so that $\qquad\qquad e^{x^2}\dfrac{dy}{dx} + 2xe^{x^2}y = xe^{x^2}$

$$\frac{d}{dx}\,[e^{x^2}y] = xe^{x^2}$$

$$e^{x^2}y = \int xe^{x^2}\,dx$$

$$= \frac{1}{2}\int e^{x^2}(2x\,dx)$$

$$= \frac{1}{2}e^{x^2} + C$$

Thus, the general solution of the differential equation is

$$y = \frac{1}{2} + Ce^{-x^2}$$

The condition $y(0) = -3$ gives $C = -\frac{7}{2}$, and hence the solution of the initial-value problem on the interval is

$$y = \frac{1}{2} - \frac{7}{2}e^{-x^2}$$

See Figure 19.2.

$C = 0$
$C = 1$
$C = -1$
$y = \frac{1}{2}$
$C = -\frac{7}{2}$
$(0, -3)$

Figure 19.2

Example 6

Solve $\qquad\qquad \dfrac{dy}{dx} = \dfrac{1}{x + y^2}$

Solution   Comparison with (19.3) clearly shows that the differential equation is not linear in the variable y. However, the reciprocal

$$\frac{dx}{dy} = x + y^2 \quad\text{or}\quad \frac{dx}{dy} - x = y^2$$

is recognized as *linear in* x. In this case the integrating factor is $e^{-\int dy} = e^{-y}$.

Therefore, it follows that

$$\frac{d}{dy}[e^{-y}x] = y^2 e^{-y}$$

$$e^{-y}x = \int y^2 e^{-y}\, dy$$

Integration by parts

$$= -y^2 e^{-y} - 2ye^{-y} - 2e^{-y} + C$$

$$x = -y^2 - 2y - 2 + Ce^y.$$

Remark

Formula (19.10), representing the general solution of (19.4), actually consists of the sum of two solutions. We define

$$y = y_c + y_p \tag{19.15}$$

where $\quad y_c = Ce^{-\int P(x)\,dx}\quad$ and $\quad y_p = e^{-\int P(x)\,dx}\int e^{\int P(x)\,dx} f(x)\, dx$

The function y_c is readily shown to be the general solution of $y' + P(x)y = 0$, whereas y_p is a solution of $y' + P(x)y = f(x)$. As we shall see in Section 19.5 the additivity property of solution (19.15) to form a general solution is an intrinsic property of linear equations.

Exercises 19.3

Answers to odd-numbered problems begin on page A-53.

In Problems 1–22 find the general solution of the given differential equation.

1. $\dfrac{dy}{dx} = 4y$

2. $\dfrac{dy}{dx} + 2y = 0$

3. $2\dfrac{dy}{dx} + 10y = 1$

4. $x\dfrac{dy}{dx} + 2y = 3$

5. $\dfrac{dy}{dx} + y = e^{3x}$

6. $\dfrac{dy}{dx} = y + e^x$

7. $y' + 3x^2 y = x^2$

8. $y' + 2xy = x^3$

9. $x^2 y' + xy = 1$

10. $y' = 2y + x^2 + 5$

11. $(x + 4y^2)\, dy + 2y\, dx = 0$

12. $\dfrac{dx}{dy} = x + y$

13. $x\, dy = (x \sin x - y)\, dx$

14. $(1 + x^2)\, dy + (xy + x^3 + x)\, dx = 0$

15. $(1 + e^x)\dfrac{dy}{dx} + e^x y = 0$

16. $(1 - x^3)\dfrac{dy}{dx} = 3x^2 y$

17. $\cos x\dfrac{dy}{dx} + y \sin x = 1$

18. $\dfrac{dy}{dx} + y \cot x = 2 \cos x$

19. $x\dfrac{dy}{dx} + 4y = x^3 - x$

20. $(1 + x)y' - xy = x + x^2$

21. $\dfrac{dr}{d\theta} + r \sec \theta = \cos \theta$

22. $\dfrac{dP}{dt} + 2tP = P + 4t - 2$

In Problems 23–30 solve the given differential equation subject to the indicated initial condition.

23. $\dfrac{dy}{dx} + 5y = 20, \quad y(0) = 2$

24. $y' = 2y + x(e^{3x} - e^{2x}), \quad y(0) = 2$

25. $(x + 1)\dfrac{dy}{dx} + y = \ln x, \quad y(1) = 10$

26. $xy' + y = e^x, \quad y(1) = 2$

27. $x(x - 2)y' + 2y = 0, \quad y(3) = 6$

28. $\sin x \dfrac{dy}{dx} + (\cos x)y = 0, \quad y\left(-\dfrac{\pi}{2}\right) = 1$

29. $\dfrac{dy}{dx} = \dfrac{y}{y - x}, \quad y(5) = 2$

30. $\cos^2 x \dfrac{dy}{dx} + y = 1, \quad y(0) = -3$

31. In Problem 27 of Exercises 8.7 we saw that the differential equation for the current $i(t)$ in a simple L-R series circuit is $L(di/dt) + Ri = E$, where L and R are constants called inductance and resistance, respectively, and E is the impressed voltage. Solve the differential equation under the assumption that $E = E_0 \sin \omega t$ and $i(0) = i_0$.

32. Suppose a cell is suspended in a solution containing a solute of constant concentration C_s. Suppose further that the cell has constant volume V and that the area of its permeable membrane is the constant A. By Fick's* law, the rate of change of its mass m is directly proportional to the area A and the difference $C_s - C(t)$, where $C(t)$ is the concentration of the solute inside the cell at any time t. Find $C(t)$ if $m = VC(t)$ and $C(0) = C_0$. See Figure 19.3.

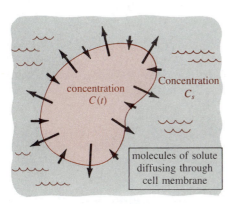

Figure 19.3

33. Initially, 50 lb of salt is dissolved in a large tank holding 300 gallons of water. A brine solution is pumped into the tank at a rate of 3 gallons per minute, and a well-stirred solution is pumped out at a rate of 2 gallons per minute. When the concentration of the solution entering is 2 lb per gallon, the differential equation for the amount A of salt (in lb) in the tank

---

*Adolf Fick (1829–1901), a German physiologist.

at any time is

$$\frac{dA}{dt} = (3)(2) - 2\left(\frac{A}{300 + t}\right)$$

Determine the amount of salt in the tank at any time. See Figure 19.4.

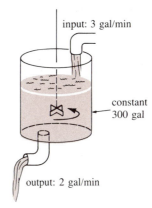

input: 3 gal/min

constant 300 gal

output: 2 gal/min

Figure 19.4

34. A heart pacemaker, shown in Figure 19.5, consists of a battery, a capacitor, and the heart as a resistor. When the switch S is at P the capacitor charges; when S is at Q the capacitor discharges, sending an electrical stimulus to the heart. During this time the voltage E applied to the heart is given by

$$\frac{dE}{dt} = -\frac{1}{RC}E, \quad t_1 < t < t_2$$

where R and C are constants. Determine $E(t)$ if $E(t_1) = E_0$. (Of course, the opening and closing of the switch is periodic in time to simulate the natural heartbeat.)

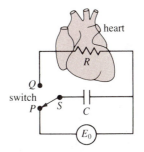

heart

R

Q
switch
P S C

E_0

Figure 19.5

Miscellaneous Problems

35. Find a continuous solution of the initial-value problem:

$$\frac{dy}{dx} + y = f(x) \quad \text{where } f(x) = \begin{cases} 1, & 0 \le x \le 1 \\ 0, & x > 1 \end{cases}$$

subject to $y(0) = 0$. (*Hint:* Solve the problem in two parts.)

36. The differential equation

$$\frac{dy}{dx} + P(x)y = f(x)y^n$$

where n is any real number, is called **Bernoulli's equation.*** Show, for $n \neq 0$ and $n \neq 1$, that the substitution $w = y^{1-n}$ leads to the linear equation

$$\frac{dw}{dx} + (1 - n)P(x)w = (1 - n)f(x)$$

In Problems 37 and 38 solve the given Bernoulli equation.

37. $\dfrac{dy}{dx} + \dfrac{1}{x}y = xy^2$

38. $\dfrac{dy}{dx} = y(xy^3 - 1)$

19.4 Linear Second-order Differential Equations

A linear nth-order differential equation

$$a_n(x)\frac{d^n y}{dx^n} + a_{n-1}(x)\frac{d^{n-1}y}{dx^{n-1}} + \cdots + a_1(x)\frac{dy}{dx} + a_0(x)y = g(x)$$

is said to be **nonhomogeneous** if $g(x) \neq 0$ for some x. If $g(x) = 0$ for every x, then the differential equation is said to be **homogeneous.**[†] In this and the following section we shall only be concerned with finding solutions of linear *second-order* equations with real *constant* coefficients:

$$ay'' + by' + cy = g(x)$$

We begin by considering the homogeneous equation

$$ay'' + by' + cy = 0 \qquad (19.16)$$

THEOREM 19.1   Superposition Principle

Let y_1 and y_2 be solutions of the homogeneous second-order differential equation (19.16). Then the linear combination

$$y = C_1 y_1(x) + C_2 y_2(x)$$

where C_1 and C_2 are arbitrary constants, is also a solution of the equation.

Proof   If $y = C_1 y_1(x) + C_2 y_2(x)$, then

$$a[C_1 y_1'' + C_2 y_2''] + b[C_1 y_1' + C_2 y_2'] + c[C_1 y_1 + C_2 y_2]$$

$$= C_1 \underbrace{[ay_1'' + by_1' + cy_1]}_{\text{zero}} + C_2 \underbrace{[ay_2'' + by_2' + cy_2]}_{\text{zero}}$$

$$= C_1 \cdot 0 + C_2 \cdot 0 = 0$$

∎

---

*James Bernoulli (1654–1705), a Swiss mathematician.

[†]In this context the word *homogeneous* does not refer to coefficients that are homogeneous functions.

Corollaries   (*a*) A constant multiple $y = C_1 y_1(x)$ of a solution $y_1(x)$ of (19.16) is also a solution. (*b*) Equation (19.16) always possesses the trivial solution $y = 0$.

> **THEOREM 19.2**   Let y_1 and y_2 be solutions of the homogeneous second-order differential equation (19.16) such that neither function is a constant multiple of the other. Then every solution of (19.16) can be obtained from the linear combination
>
> $$y = C_1 y_1(x) + C_2 y_2(x) \qquad (19.17)$$

General Solution   In Theorem 19.2 we say (19.17) is the **general solution** of the differential equation. In other words, any solution of (19.16) can be obtained from (19.17) by specializing the constants C_1 and C_2. Also, two solutions of (19.16) such that neither is a constant multiple of the other are said to be **linearly independent**. The general solution of (19.16) consists of a linear combination of linearly independent solutions.

Example 1

You should verify that $y_1 = 0$ and $y_2 = e^{2x}$ are both solutions of the equation $y'' + 2y' - 8y = 0$. Although y_2 is not a constant multiple of y_1, y_1 and y_2 are not linearly independent since $y_1 = 0 \cdot y_2$.

The surprising fact about equation (19.16) is that *all* solutions are either exponential functions or are constructed out of exponential functions.

The Auxiliary Equation   If we try a solution of the form $y = e^{mx}$, then $y' = me^{mx}$ and $y'' = m^2 e^{mx}$ so that (19.16) becomes

$$am^2 e^{mx} + bme^{mx} + ce^{mx} = 0 \quad \text{or} \quad e^{mx}[am^2 + bm + c] = 0$$

Because e^{mx} is never zero for real values of x, it is apparent that the only way that this exponential function can satisfy the differential equation is to choose m so that it is a root of the quadratic equation

$$am^2 + bm + c = 0$$

This latter equation is called the **auxiliary equation** or **characteristic equation** of the differential equation (19.16). We shall consider three cases, namely, the solutions corresponding to real distinct roots, real but equal roots, and lastly, complex conjugate roots.

Case I   Under the assumption that the auxiliary equation of (19.16) has two unequal real roots m_1 and m_2, we find two solutions

$$y_1 = e^{m_1 x} \quad \text{and} \quad y_2 = e^{m_2 x}$$

Since y_1 and y_2 are linearly independent, it follows that the general solution of the equation is

$$y = C_1 e^{m_1 x} + C_2 e^{m_2 x} \qquad (19.18)$$

Example 2

Solve $$2y'' - 5y' - 3y = 0$$

Solution   We first solve the auxiliary equation:

$$2m^2 - 5m - 3 = 0 \quad \text{or} \quad (2m + 1)(m - 3) = 0$$

$$m_1 = -\frac{1}{2}, \qquad m_2 = 3$$

Hence, by (19.18) the general solution is

$$y = C_1 e^{-x/2} + C_2 e^{3x}$$

Case II   When $m_1 = m_2$ we necessarily obtain only one exponential solution $y_1 = e^{m_1 x}$. However, it is a straightforward matter of substitution into (19.16) to show that $y = u(x)e^{m_1 x}$ is also a solution whenever $u(x) = x$. See Problem 37 in Exercises 19.4. That is, $y_1 = e^{m_1 x}$ and $y_2 = xe^{m_1 x}$ are linearly independent solutions. Hence, the general solution is

$$y = C_1 e^{m_1 x} + C_2 x e^{m_1 x} \qquad (19.19)$$

Example 3

Solve $$y'' - 10y' + 25y = 0$$

Solution   From the auxiliary equation $m^2 - 10m + 25 = 0$ or $(m - 5)^2 = 0$ we see that $m_1 = m_2 = 5$. Thus, by (19.19) the general solution is

$$y = C_1 e^{5x} + C_2 x e^{5x}$$

Complex Numbers   The last case deals with complex numbers.* Recall from algebra that a number of the form $z = \alpha + i\beta$, where α and β are real numbers and $i^2 = -1$ (sometimes written $i = \sqrt{-1}$), is called a **complex number**. The complex number $\bar{z} = \alpha - i\beta$ is called the **complex conjugate** of z. Now, from the quadratic formula the roots of $am^2 + bm + c = 0$ can be written

$$m_1 = \frac{-b + \sqrt{b^2 - 4ac}}{2a}, \qquad m_2 = \frac{-b - \sqrt{b^2 - 4ac}}{2a}$$

When $b^2 - 4ac < 0$, m_1 and m_2 are complex conjugates.

Case III   If m_1 and m_2 are complex, then we can write

$$m_1 = \alpha + i\beta \quad \text{and} \quad m_2 = \alpha - i\beta$$

---

*See Appendix I.

where α and β are real and $i^2 = -1$. Formally there is no difference between this case and case I, and hence the general solution is

$$y = c_1 e^{(\alpha + i\beta)x} + c_2 e^{(\alpha - i\beta)x} \qquad (19.20)$$

However, in practice we would prefer to work with real functions instead of complex exponentials. Now, we can rewrite (19.20) in a more practical form by using **Euler's formula***

$$e^{i\theta} = \cos \theta + i \sin \theta$$

where θ is any real number. From this result we can write

$$e^{i\beta x} = \cos \beta x + i \sin \beta x \quad \text{and} \quad e^{-i\beta x} = \cos \beta x - i \sin \beta x$$

where we have used $\cos(-\beta x) = \cos \beta x$ and $\sin(-\beta x) = -\sin \beta x$. Thus, (19.20) becomes

$$
\begin{aligned}
y &= e^{\alpha x}[c_1 e^{i\beta x} + c_2 e^{-i\beta x}] \\
&= e^{\alpha x}[c_1 \{\cos \beta x + i \sin \beta x\} + c_2 \{\cos \beta x - i \sin \beta x\}] \\
&= e^{\alpha x}[(c_1 + c_2)\cos \beta x + (c_1 i - c_2 i)\sin \beta x]
\end{aligned}
$$

Since $e^{\alpha x} \cos \beta x$ and $e^{\alpha x} \sin \beta x$ are themselves linearly independent solutions of the given differential equation, we can simply relabel $c_1 + c_2$ as C_1 and $c_1 i - c_2 i$ as C_2 and use the superposition principle to write the general solution

$$
\begin{aligned}
y &= C_1 e^{\alpha x} \cos \beta x + C_2 e^{\alpha x} \sin \beta x \\
&= e^{\alpha x}(C_1 \cos \beta x + C_2 \sin \beta x) \qquad (19.21)
\end{aligned}
$$

Example 4

Solve
$$y'' + y' + y = 0$$

Solution From the quadratic formula we find that the auxiliary equation $m^2 + m + 1 = 0$ has the complex roots

$$m_1 = -\frac{1}{2} + \frac{\sqrt{3}}{2}i \quad \text{and} \quad m_2 = -\frac{1}{2} - \frac{\sqrt{3}}{2}i$$

Identifying $\alpha = -\frac{1}{2}$ and $\beta = \sqrt{3}/2$, we see from (19.21) that the general solution of the equation is

$$y = e^{-x/2}\left(C_1 \cos \frac{\sqrt{3}}{2}x + C_2 \sin \frac{\sqrt{3}}{2}x\right)$$

---

*You were asked to prove this in Problem 41 of Exercises 11.8.

___ **Example 5** ___

The two equations

$$y'' + k^2 y = 0 \quad \text{and} \quad y'' - k^2 y = 0$$

are frequently encountered in the study of applied mathematics. For the former equation the auxiliary equation $m^2 + k^2 = 0$ has complex roots $m_1 = ki$ and $m_2 = -ki$. It follows from (19.21) that its general solution is

$$y = C_1 \cos kx + C_2 \sin kx$$

The latter equation has the auxiliary equation $m^2 - k^2 = 0$ with real roots $m_1 = k$ and $m_2 = -k$, so that its general solution is

$$y = C_1 e^{kx} + C_2 e^{-kx} \tag{19.22}$$

Notice that if we choose $C_1 = C_2 = \frac{1}{2}$ in (19.22), then,

$$y = \frac{e^{kx} + e^{-kx}}{2} = \cosh kx$$

is also a solution. Furthermore, when $C_1 = \frac{1}{2}$, $C_2 = -\frac{1}{2}$, then (19.22) becomes

$$y = \frac{e^{kx} - e^{-kx}}{2} = \sinh kx$$

Since $\cosh kx$ and $\sinh kx$ are linearly independent, we obtain an alternative form of the general solution for $y'' - k^2 y = 0$:

$$y = C_1 \cosh kx + C_2 \sinh kx$$

Initial-Value Problem   The problem,

> *Solve:* $ay'' + by' + cy = g(x)$
>
> *Subject to:* $y(x_0) = y_0, \qquad y'(x_0) = y_0'$

where y_0 and y_0' are arbitrary constants is called an **initial-value problem**. A solution of the problem is a function whose graph passes through (x_0, y_0) such that the slope of the tangent to the curve at that point is y_0'. The next example illustrates an initial-value problem for a homogeneous equation.

___ **Example 6** ___

Solve $y'' - 4y' + 13y = 0$ subject to $y(0) = -1, \; y'(0) = 2.$

Solution   The roots of the auxiliary equation

$$m^2 - 4m + 13 = 0$$

are $\qquad\qquad m_1 = 2 + 3i \quad$ and $\quad m_2 = 2 - 3i$

so that $\qquad\qquad y = e^{2x}(C_1 \cos 3x + C_2 \sin 3x)$

The condition $y(0) = -1$ implies that

$$-1 = e^0(C_1 \cos 0 + .C_2 \sin 0) = C_1$$

from which we can write

$$y = e^{2x}(-\cos 3x + C_2 \sin 3x)$$

Differentiating this latter expression and using the second condition gives

$$y' = e^{2x}(3 \sin 3x + 3C_2 \cos 3x) + 2e^{2x}(-\cos 3x + C_2 \sin 3x)$$
$$2 = 3C_2 - 2$$

so that $C_2 = \frac{4}{3}$. Hence,

$$y = e^{2x}\left(-\cos 3x + \frac{4}{3} \sin 3x\right)$$

Exercises 19.4

Answers to odd-numbered problems begin on page A-53.

In Problems 1–20 find the general solution of the given differential equation.

1. $3y'' - y' = 0$

2. $2y'' + 5y' = 0$

3. $y'' - 16y = 0$

4. $y'' - 8y = 0$

5. $y'' + 9y = 0$

6. $4y'' + y = 0$

7. $y'' - 3y' + 2y = 0$

8. $y'' - y' - 6y = 0$

9. $\dfrac{d^2y}{dx^2} + 8\dfrac{dy}{dx} + 16y = 0$

10. $\dfrac{d^2y}{dx^2} - 10\dfrac{dy}{dx} + 25y = 0$

11. $y'' + 3y' - 5y = 0$

12. $y'' + 4y' - y = 0$

13. $12y'' - 5y' - 2y = 0$

14. $8y'' + 2y' - y = 0$

15. $y'' - 4y' + 5y = 0$

16. $2y'' - 3y' + 4y = 0$

17. $3y'' + 2y' + y = 0$

18. $2y'' + 2y' + y = 0$

19. $9y'' + 6y' + y = 0$

20. $15y'' - 16y' - 7y = 0$

In Problems 21–30 solve the given differential equation subject to the indicated initial conditions.

21. $y'' + 16y = 0, \quad y(0) = 2, \quad y'(0) = -2$

22. $y'' - y = 0, \quad y(0) = y'(0) = 1$

23. $y'' + 6y' + 5y = 0, \quad y(0) = 0, \quad y'(0) = 3$

24. $y'' - 8y' + 17y = 0, \quad y(0) = 4, \quad y'(0) = -1$

25. $2y'' - 2y' + y = 0, \quad y(0) = -1, \quad y'(0) = 0$

26. $y'' - 2y' + y = 0, \quad y(0) = 5, \quad y'(0) = 10$

27. $y'' + y' + 2y = 0, \quad y(0) = y'(0) = 0$

28. $4y'' - 4y' - 3y = 0, \quad y(0) = 1, \quad y'(0) = 5$

29. $y'' - 3y' + 2y = 0, \quad y(1) = 0, \, y'(1) = 1$

30. $y'' + y = 0, \quad y(\pi/3) = 0, \quad y'(\pi/3) = 2$

31. The roots of an auxiliary equation are $m_1 = 4$ and $m_2 = -5$. What is the corresponding differential equation?

32. The roots of an auxiliary equation are $m_1 = 3 + i$ and $m_2 = 3 - i$. What is the corresponding differential equation?

Miscellaneous Problems

In Problems 33 and 34 use the assumed solution $y = e^{mx}$ to find the auxiliary equation, roots, and general solution of the given third-order differential equation.

33. $y''' - 4y'' - 5y' = 0$

34. $y''' + 3y'' - 4y' - 12y = 0$

In Problems 35 and 36 find the general solution of the given third-order differential equation if it is known that y_1 is a solution.

35. $y''' - 9y'' + 25y' - 17y = 0;$ $y_1 = e^x$

36. $y''' + 6y'' + y' - 34y = 0;$ $y_1 = e^{-4x} \cos x$

37. **(a)** In case II show that if $m_1 = m_2$, the differential equation (19.16) must be of the form $y'' - 2m_1 y' + m_1^2 y = 0$.

(b) Show that the substitution $y = u(x)e^{m_1 x}$ in the differential equation in part **(a)** leads to $u(x) = x$.

19.5 Nonhomogeneous Linear Second-order Differential Equations

Particular Solutions   We now turn our attention to finding the general solution of the nonhomogeneous equation

$$ay'' + by' + cy = g(x) \qquad (19.23)$$

where a, b, and c are constants and g is continuous. Any function y_p, free of arbitrary parameters that satisfies (19.23) is said to be a **particular solution** of the equation.

_____ **Example 1** _____

$y_p = x^3 - x$ is a particular solution of

$$y'' - y' + 6y = 6x^3 - 3x^2 + 1$$

since $y_p' = 3x^2 - 1$, $y_p'' = 6x$, and

$$y'' - y' + 6y = 6x - (3x^2 - 1) + 6(x^3 - x)$$
$$= 6x^3 - 3x^2 + 1$$

The General Solution   The procedure for solving the nonhomogeneous equation (19.23) consists of two steps:

(*i*)   Solve the associated homogeneous equation

$$ay'' + by' + cy = 0$$

(*ii*)   and then find *any* particular solution of the nonhomogeneous equation (19.23).

The sum of the solutions in parts (*i*) and (*ii*) constitutes the **general solution** of (19.23).

> **THEOREM 19.3**   Let y_p be a given solution of the nonhomogeneous equation (19.23) and let
>
> $$y_c = C_1 y_1(x) + C_2 y_2(x)$$
>
> be the general solution of the associated homogeneous equation. Then the **general solution** of the nonhomogeneous equation is
>
> $$y = C_1 y_1(x) + C_2 y_2(x) + y_p(x)$$
> $$= y_c(x) + y_p(x) \qquad (19.24)$$

The proof that (19.24) is a solution of (19.23) is left as an exercise.

Complementary Function

In Theorem 19.3 the linear combination $y_c(x) = C_1 y_1(x) + C_2 y_2(x)$ is called the **complementary function** of equation (19.23). In other words, the general solution of a nonhomogeneous equation is

$$y = complementary\ function + any\ particular\ solution$$

Variation of Parameters

One of the most popular ways of determining a particular solution y_p of the linear second-order differential equation (19.23) is called the method of **variation of parameters**. To apply this method we shall put (19.23) into the form

$$y'' + Py' + Qy = f(x) \qquad (19.25)$$

by dividing through by a.

Suppose y_1 and y_2 are linearly independent solutions of the associated homogeneous form of (19.25). That is,

$$y_1'' + Py_1' + Qy_1 = 0 \quad \text{and} \quad y_2'' + Py_2' + Qy_2 = 0$$

Now we ask: Can two functions u_1 and u_2 be found so that

$$y_p = u_1(x)y_1(x) + u_2(x)y_2(x) \qquad (19.26)$$

is a particular solution of (19.23)? Notice our assumption for y_p is the same as $y_c = C_1 y_1 + C_2 y_2$, but we have replaced C_1 and C_2 by the "variable parameters" u_1 and u_2. Using the product rule to differentiate y_p gives

$$y_p' = u_1 y_1' + y_1 u_1' + u_2 y_2' + y_2 u_2' \qquad (19.27)$$

If we make the further demand that u_1 and u_2 be functions for which

$$y_1 u_1' + y_2 u_2' = 0 \qquad (19.28)$$

then (19.27) becomes

$$y_p' = u_1 y_1' + u_2 y_2'$$

Continuing, we find

$$y_p'' = u_1 y_1'' + y_1' u_1' + u_2 y_2'' + y_2' u_2'$$

and hence

$$y_p'' + Py_p' + Qy_p = u_1 y_1'' + y_1' u_1' + u_2 y_2'' + y_2' u_2'$$
$$+ Pu_1 y_1' + Pu_2 y_2' + Qu_1 y_1 + Qu_2 y_2$$
$$= u_1 \underbrace{[y_1'' + Py_1' + Qy_1]}_{\text{zero}} + u_2 \underbrace{[y_2'' + Py_2' + Qy_2]}_{\text{zero}}$$
$$+ y_1' u_1' + y_2' u_2' = f(x)$$

In other words, u_1 and u_2 must be functions that also satisfy the condition

$$y_1' u_1' + y_2' u_2' = f(x) \tag{19.29}$$

Equations (19.28) and (19.29) constitute a linear system of equations for the derivatives u_1' and u_2'. That is, we must solve

$$y_1 u_1' + y_2 u_2' = 0$$
$$y_1' u_1' + y_2' u_2' = f(x)$$

By Cramer's rule* we obtain

$$u_1' = \frac{\begin{vmatrix} 0 & y_2 \\ f(x) & y_2' \end{vmatrix}}{\begin{vmatrix} y_1 & y_2 \\ y_1' & y_2' \end{vmatrix}} \quad \text{and} \quad u_2' = \frac{\begin{vmatrix} y_1 & 0 \\ y_1' & f(x) \end{vmatrix}}{\begin{vmatrix} y_1 & y_2 \\ y_1' & y_2' \end{vmatrix}} \tag{19.30}$$

The determinant $\begin{vmatrix} y_1 & y_2 \\ y_1' & y_2' \end{vmatrix}$ is called the **Wronskian**† and is usually denoted by W.

Example 2 _____

Solve $4y'' + 36y = \csc 3x$

Solution First write the equation in the form

$$y'' + 9y = \frac{1}{4} \csc 3x$$

Since the roots of the auxiliary equation $m^2 + 9 = 0$ are $m_1 = 3i$ and $m_2 = -3i$,

$$y_c = C_1 \cos 3x + C_2 \sin 3x.$$

$$W = \begin{vmatrix} \cos 3x & \sin 3x \\ -3 \sin 3x & 3 \cos 3x \end{vmatrix} = 3$$

---

*See Appendix I.

†Named after the Polish philosopher-mathematician, Joseph M. H. Wronski (1778–1853).

From (19.30) we then find

$$u_1' = -\frac{(\sin 3x)(\frac{1}{4} \csc 3x)}{3} = -\frac{1}{12}$$

and
$$u_2' = \frac{(\cos 3x)(\frac{1}{4} \csc 3x)}{3} = \frac{1}{12} \frac{\cos 3x}{\sin 3x}$$

Integrating u_1' and u_2' yields

$$u_1 = -\frac{1}{12}x \quad \text{and} \quad u_2 = \frac{1}{36} \ln |\sin 3x|$$

Hence, from (19.26)

$$y_p = -\frac{1}{12}x \cos 3x + \frac{1}{36}(\sin 3x)\ln |\sin 3x|$$

Therefore, the general solution is

$$y = y_c + y_p$$
$$= C_1 \cos 3x + C_2 \sin 3x - \frac{1}{12}x \cos 3x + \frac{1}{36}(\sin 3x)\ln |\sin 3x|$$

Constants of Integration When computing the indefinite integrals of u_1' and u_2', we need not introduce any constants. This is because

$$y = y_c + y_p$$
$$= C_1 y_1 + C_2 y_2 + (u_1 + a_1)y_1 + (u_2 + b_1)y_2$$
$$= (C_1 + a_1)y_1 + (C_2 + b_1)y_2 + u_1 y_1 + u_2 y_2$$
$$= c_1 y_1 + c_2 y_2 + u_1 y_1 + u_2 y_2$$

where $c_1 = C_1 + a_1$ and $c_2 = C_2 + b_1$.

Undetermined Coefficients When $g(x)$ consists of

    (*i*)   a constant k, or

    (*ii*) a polynomial in x, or

    (*iii*) an exponential function $e^{\alpha x}$, or

    (*iv*) $\sin \beta x$, $\cos \beta x$,

or finite sums and products of these functions, it is possible to find a particular solution of (19.23) by the **method of undetermined coefficients**. In the special case* where the n distinct functions $f_k(x)$ appearing in $g(x)$ *or* in its derivatives

---

*See the author's *A First Course in Differential Equations* for further details of this method.

do not appear in the complementary function y_c, a particular solution y_p can be found consisting of a linear combination:

$$y = \sum_{k=1}^{n} A_k f_k(x)$$

We find the coefficients A_k by substituting this expression into the given differential equation.

___ **Example 3** _____

Solve $\dfrac{d^2y}{dx^2} + 3\dfrac{dy}{dx} + 2y = 4x^2$

Solution   The complementary function is

$$y_c = C_1 e^{-x} + C_2 e^{-2x}$$

Now, since

$$g(x) = 4x^2, \qquad g'(x) = 8x, \qquad g''(x) = 8 \cdot 1$$

$$\underbrace{\qquad}_{f_1(x)} \qquad \underbrace{\qquad}_{f_2(x)} \qquad \underbrace{\qquad}_{f_3(x)}$$

we seek a particular solution having the basic structure

$$y_p = Ax^2 + Bx + C \tag{19.31}$$

Differentiating (19.31) and substituting into the given equation gives

$$y_p'' + 3y_p' + 2y_p = 2A + 3B + 2C + (6A + 2B)x + 2Ax^2$$
$$= 4x^2$$

By equating coefficients in the last identity we obtain the system of equations

$$2A + 3B + 2C = 0$$
$$6A + 2B \qquad = 0$$
$$2A \qquad\qquad = 4$$

Solving gives $A = 2$, $B = -6$, and $C = 7$. Thus, $y_p = 2x^2 - 6x + 7$ and the general solution of the differential equation is $y = y_c + y_p$, or

$$y = C_1 e^{-x} + C_2 e^{-2x} + 2x^2 - 6x + 7$$

___ **Example 4** _____

Solve $y'' + 2y' + 2y = -10xe^x + 5\sin x$

Solution   The roots of the auxiliary equation $m^2 + 2m + 2 = 0$ are $m_1 = -1 + i$ and $m_2 = -1 - i$ and so

$$y_c = e^{-x}(C_1 \cos x + C_2 \sin x)$$

In this case

$$g(x) = -10xe^x + 5 \sin x, \qquad g'(x) = 10xe^x - 10e^x + 5 \cos x$$

$$\underset{f_1(x)}{\uparrow} \quad \underset{f_2(x)}{\uparrow} \qquad\qquad \underset{f_3(x)}{\uparrow} \quad \underset{f_4(x)}{\uparrow}$$

indicates that a particular solution can be found of the form

$$y_p = Axe^x + Be^x + C \sin x + D \cos x$$

Substituting y_p in the differential equation and simplifying yields

$$y_p'' + 2y_p' + 2y_p = 5Axe^x + (4A + 5B)e^x + (C - 2D)\sin x$$
$$+ (2C + D)\cos x$$
$$= -10xe^x + 5 \sin x$$

and
$$5A \quad\quad\quad = -10$$
$$4A + 5B = 0$$
$$C - 2D = 5$$
$$2C + \ D = 0$$

We find $A = -2$, $B = \frac{8}{5}$, $C = 1$, and $D = -2$. Hence,

$$y_p = -2xe^x + \frac{8}{5}e^x + \sin x - 2 \cos x$$

The general solution is then

$$y = e^{-x}(C_1 \cos x + C_2 \sin x) - 2xe^x + \frac{8}{5}e^x + \sin x - 2 \cos x$$

Exercises 19.5

Answers to odd-numbered problems begin on page A-53.

In Problems 1–20 solve the given differential equation by variation of parameters.

1. $y'' + y = \sec x$

2. $y'' + y = \tan x$

3. $y'' + y = \sin x$

4. $y'' + y = \sec x \tan x$

5. $y'' + y = \cos^2 x$

6. $y'' + y = \sec^2 x$

7. $y'' - y = \cosh x$

8. $y'' - y = \sinh 2x$

9. $y'' - 4y = e^{2x}/x$

10. $y'' - 9y = 9x/e^{3x}$

11. $y'' + 3y' + 2y = 1/(1 + e^x)$

12. $y'' - 3y' + 2y = e^{3x}/(1 + e^x)$

13. $y'' + 3y' + 2y = \sin e^x$

14. $y'' - 2y' + y = e^x \arctan x$

15. $y'' - 2y' + y = e^x/(1 + x^2)$

16. $y'' - 2y' + 2y = e^x \sec x$

17. $y'' + 2y' + y = e^{-x} \ln x$

18. $y'' + 10y' + 25y = e^{-10x}/x^2$

19. $4y'' - 4y' + y = 8e^{-x} + x$

20. $4y'' - 4y' + y = e^{x/2}\sqrt{1 - x^2}$

In Problems 21 and 22 solve the given differential equation by variation of parameters subject to the initial conditions $y(0) = 1$ and $y'(0) = 0$.

21. $y'' - y = xe^x$

22. $2y'' + y' - y = x + 1$

In Problems 23–32 solve the given differential equation by undetermined coefficients.

23. $y'' - 9y = 54$

24. $2y'' - 7y' + 5y = -29$

25. $y'' + 4y' + 4y = 2x + 6$

26. $y'' - 2y' + y = x^3 + 4x$

27. $y'' + 25y = 6 \sin x$

28. $y'' - 4y = 7e^{4x}$

29. $y'' - 2y' - 3y = 4e^{2x} + 2x^3$

30. $y'' + y' + y = x^2e^x + 3$

31. $y'' - 8y' + 25y = e^{3x} - 6 \cos 2x$

32. $y'' - 5y' + 4y = 2 \sinh 3x$

In Problems 33 and 34 solve the given differential equation by undetermined coefficients subject to the initial conditions $y(0) = 1$ and $y'(0) = 0$.

33. $y'' - 64y = 16$

34. $y'' + 5y' - 6y = 10e^{2x}$

Miscellaneous Problems

35. Given that $y_1 = x$ and $y_2 = x \ln x$ are linearly independent solutions of $x^2y'' - xy' + y = 0$, use variation of parameters to solve $x^2y'' - xy' + y = 4x \ln x$.

36. Given that $y_1 = x^2$ and $y_2 = x^3$ are linearly independent solutions of $x^2y'' - 4xy' + 6y = 0$, use variation of parameters to solve $x^2y'' - 4xy' + 6y = 1/x$.

37. Prove that (19.24) is a solution of (19.23).

19.6   Power Series Solutions

Some homogeneous linear second-order differential equations with *variable coefficients* can be solved by use of power series. The procedure consists of assuming a solution of the form

$$y = c_0 + c_1x + c_2x^2 + c_3x^3 + \cdots = \sum_{n=0}^{\infty} c_nx^n$$

differentiating

$$\frac{dy}{dx} = c_1 + 2c_2x + 3c_3x^2 + \cdots = \sum_{n=1}^{\infty} nc_nx^{n-1}$$

$$\frac{d^2y}{dx^2} = 2c_2 + 6c_3x + \cdots = \sum_{n=2}^{\infty} n(n-1)c_nx^{n-2}$$

and substituting the results into the differential equation with the expectation of determining a **recurrence relation** that will yield the coefficients c_n. The power series will then represent a solution of the differential equation on some interval of convergence.

_____ **Example 1** _____

Solve
$$y'' - 2xy = 0$$

Solution   If we assume a solution $y = \sum_{n=0}^{\infty} c_nx^n$, then, as just shown,

$$y'' = \sum_{n=2}^{\infty} n(n-1)c_nx^{n-2}$$

Hence, we have

$$y'' - 2xy = \sum_{n=2}^{\infty} n(n-1)c_n x^{n-2} - \sum_{n=0}^{\infty} 2c_n x^{n+1}$$

$$= 2 \cdot 1 c_2 x^0 + \underbrace{\sum_{n=3}^{\infty} n(n-1)c_n x^{n-2} - \sum_{n=0}^{\infty} 2c_n x^{n+1}}_{\text{both series start with } x^1},$$

Letting $k = n - 2$ in the first series and $k = n + 1$ in the second yields

$$y'' - 2xy = 2c_2 + \sum_{k=1}^{\infty} (k+2)(k+1)c_{k+2} x^k - \sum_{k=1}^{\infty} 2c_{k-1} x^k$$

$$= 2c_2 + \sum_{k=1}^{\infty} [(k+2)(k+1)c_{k+2} - 2c_{k-1}]x^k = 0$$

We must then have

$$2c_2 = 0 \quad \text{and} \quad (k+2)(k+1)c_{k+2} - 2c_{k-1} = 0$$

The last expression written as

$$c_{k+2} = \frac{2c_{k-1}}{(k+2)(k+1)}, \qquad k = 1, 2, 3 \ldots$$

is called a *recurrence relation* for the c_n. Iteration of this formula gives

$$c_3 = \frac{2c_0}{3 \cdot 2}$$

$$c_4 = \frac{2c_1}{4 \cdot 3}$$

$$c_5 = \frac{2c_2}{5 \cdot 4} = 0$$

$$c_6 = \frac{2c_3}{6 \cdot 5} = \frac{2^2}{6 \cdot 5 \cdot 3 \cdot 2}c_0$$

$$c_7 = \frac{2c_4}{7 \cdot 6} = \frac{2^2}{7 \cdot 6 \cdot 4 \cdot 3}c_1$$

$$c_8 = \frac{2c_5}{8 \cdot 7} = 0$$

$$c_9 = \frac{2c_6}{9 \cdot 8} = \frac{2^3}{9 \cdot 8 \cdot 6 \cdot 5 \cdot 3 \cdot 2}c_0$$

$$c_{10} = \frac{2c_7}{10 \cdot 9} = \frac{2^3}{10 \cdot 9 \cdot 7 \cdot 6 \cdot 4 \cdot 3}c_1$$

$$c_{11} = \frac{2c_8}{11 \cdot 10} = 0$$

and so on. It should be apparent that both c_0 and c_1 are arbitrary. Now

$$y = c_0 + c_1x + c_2x^2 + c_3x^3 + c_4x^4 + c_5x^5 + c_6x^6 + c_7x^7 + c_8x^8$$
$$+ c_9x^9 + c_{10}x^{10} + c_{11}x^{11} + \cdots$$

$$= c_0 + c_1x + 0 + \frac{2}{3 \cdot 2}c_0x^3 + \frac{2}{4 \cdot 3}c_1x^4 + 0 + \frac{2^2}{6 \cdot 5 \cdot 3 \cdot 2}c_0x^6$$

$$+ \frac{2^2}{7 \cdot 6 \cdot 4 \cdot 3}c_1x^7 + 0 + \frac{2^3}{9 \cdot 8 \cdot 6 \cdot 5 \cdot 3 \cdot 2}c_0x^9$$

$$+ \frac{2^3}{10 \cdot 9 \cdot 7 \cdot 6 \cdot 4 \cdot 3}c_1x^{10} + 0 + \cdots$$

$$= c_0\left[1 + \frac{2}{3 \cdot 2}x^3 + \frac{2^2}{6 \cdot 5 \cdot 3 \cdot 2}x^6 + \frac{2^3}{9 \cdot 8 \cdot 6 \cdot 5 \cdot 3 \cdot 2}x^9 + \cdots\right]$$

$$+ c_1\left[x + \frac{2}{4 \cdot 3}x^4 + \frac{2^2}{7 \cdot 6 \cdot 4 \cdot 3}x^7 + \frac{2^3}{10 \cdot 9 \cdot 7 \cdot 6 \cdot 4 \cdot 3}x^{10} + \cdots\right]$$

Although the pattern of the coefficients in the preceding example should be clear, it is sometimes useful to write the solutions in terms of summation notation. By using the properties of the factorial we can write

$$y_1(x) = c_0\left[1 + \sum_{k=1}^{\infty} \frac{2^k[1 \cdot 4 \cdot 7 \cdots (3k-2)]}{(3k)!}x^{3k}\right]$$

and

$$y_2(x) = c_1\left[x + \sum_{k=1}^{\infty} \frac{2^k[2 \cdot 5 \cdot 8 \cdots (3k-1)]}{(3k+1)!}x^{3k+1}\right]$$

In this form the ratio test can be used to show that each series converges for $|x| < \infty$.

Example 2

Solve
$$(x^2 + 1)y'' + xy' - y = 0$$

Solution   The assumption $y = \sum_{n=0}^{\infty} c_nx^n$ leads to

$$(x^2 + 1)\sum_{n=2}^{\infty} n(n-1)c_nx^{n-2} + x\sum_{n=1}^{\infty} nc_nx^{n-1} - \sum_{n=0}^{\infty} c_nx^n$$

$$= \sum_{n=2}^{\infty} n(n-1)c_nx^n + \sum_{n=2}^{\infty} n(n-1)c_nx^{n-2} + \sum_{n=1}^{\infty} nc_nx^n - \sum_{n=0}^{\infty} c_nx^n$$

$$= 2c_2x^0 - c_0x^0 + 6c_3x + c_1x - c_1x + \underbrace{\sum_{n=2}^{\infty} n(n-1)c_nx^n}_{k=n}$$

$$+ \sum_{n=4}^{\infty} n(n-1)c_n x^{n-2} + \sum_{n=2}^{\infty} nc_n x^n - \sum_{n=2}^{\infty} c_n x^n$$

$$\underbrace{\quad\quad\quad\quad\quad}_{k = n - 2} \quad \underbrace{\quad\quad}_{k = n} \quad \underbrace{\quad\quad}_{k = n}$$

$$= 2c_2 - c_0 + 6c_3 x$$

$$+ \sum_{k=2}^{\infty} [k(k-1)c_k + (k+2)(k+1)c_{k+2} + kc_k - c_k]x^k$$

$$= 2c_2 - c_0 + 6c_3 x$$

$$+ \sum_{k=2}^{\infty} [(k+1)(k-1)c_k + (k+2)(k+1)c_{k+2}]x^k = 0$$

Thus, we must have

$$2c_2 - c_0 = 0$$

$$c_3 = 0$$

$$(k+1)(k-1)c_k + (k+2)(k+1)c_{k+2} = 0$$

or, after dividing by $k + 1$,

$$c_2 = \frac{1}{2}c_0$$

$$c_3 = 0$$

$$c_{k+2} = -\frac{k-1}{k+2}c_k, \qquad k = 2, 3, 4 \dots$$

Iteration of the last formula gives

$$c_4 = -\frac{1}{4}c_2 = -\frac{1}{2 \cdot 4}c_0 = -\frac{1}{2^2 2!}c_0$$

$$c_5 = -\frac{2}{5}c_3 = 0$$

$$c_6 = -\frac{3}{6}c_4 = \frac{3}{2 \cdot 4 \cdot 6}c_0 = \frac{1 \cdot 3}{2^3 3!}c_0$$

$$c_7 = -\frac{4}{7}c_5 = 0$$

$$c_8 = -\frac{5}{8}c_6 = -\frac{3 \cdot 5}{2 \cdot 4 \cdot 6 \cdot 8}c_0 = -\frac{1 \cdot 3 \cdot 5}{2^4 4!}c_0$$

$$c_9 = -\frac{6}{9}c_7 = 0$$

$$c_{10} = -\frac{7}{10}c_8 = \frac{3 \cdot 5 \cdot 7}{2 \cdot 4 \cdot 6 \cdot 8 \cdot 10}c_0 = \frac{1 \cdot 3 \cdot 5 \cdot 7}{2^5 5!}c_0$$

and so on. Therefore,

$$y = c_0 + c_1 x + c_2 x^2 + c_3 x^3 + c_4 x^4 + c_5 x^5 + c_6 x^6 + c_7 x^7$$
$$+ c_8 x^8 + \cdots$$

$$= c_1 x + c_0 \left[1 + \frac{1}{2} x^2 - \frac{1}{2^2 2!} x^4 \right.$$

$$\left. + \frac{1 \cdot 3}{2^3 3!} x^6 - \frac{1 \cdot 3 \cdot 5}{2^4 4!} x^8 + \frac{1 \cdot 3 \cdot 5 \cdot 7}{2^5 5!} x^{10} + \cdots \right]$$

The solutions are

$$y_1(x) = c_0 \left[1 + \frac{1}{2} x^2 + \sum_{n=2}^{\infty} (-1)^{n-1} \frac{1 \cdot 3 \cdot 5 \cdots (2n-3)}{2^n n!} x^{2n} \right]$$

$$y_2(x) = c_1 x$$

Exercises 19.6

Answers to odd-numbered problems begin on page A-54.

In Problems 1–18 find power series solutions of the given differential equation.

1. $y'' + y = 0$

2. $y'' - y = 0$

3. $y'' = y'$

4. $2y'' + y' = 0$

5. $y'' = xy$

6. $y'' + x^2 y = 0$

7. $y'' - 2xy' + y = 0$

8. $y'' - xy' + 2y = 0$

9. $y'' + x^2 y' + xy = 0$

10. $y'' + 2xy' + 2y = 0$

11. $(x - 1)y'' + y' = 0$

12. $(x + 2)y'' + xy' - y = 0$

13. $(x^2 - 1)y'' + 4xy' + 2y = 0$

14. $(x^2 + 1)y'' - 6y = 0$

15. $(x^2 + 2)y'' + 3xy' - y = 0$

16. $(x^2 - 1)y'' + xy' - y = 0$

17. $y'' - (x + 1)y' - y = 0$

18. $y'' - xy' - (x + 2)y = 0$

In Problems 19 and 20 use the power series method to solve the given differential equation subject to the indicated initial conditions.

19. $(x - 1)y'' - xy' + y = 0$; $y(0) = -2$, $y'(0) = 6$

20. $y'' - 2xy' + 8y = 0$; $y(0) = 3$, $y'(0) = 0$

19.7 Vibrational Models

Hooke's Law Suppose, as in Figure 19.6(b), a mass m_1 is attached to a flexible spring suspended from a rigid support. When m_1 is replaced with a different mass m_2, the amount of stretch, or elongation of the spring, will of course be different.

By Hooke's law, the spring itself exerts a restoring force F opposite to the direction of elongation and proportional to the amount of elongation s. Simply stated, $F = ks$, where k is a constant of proportionality. Although masses with different weights stretch a spring by different amounts, the spring is essentially characterized by the number k. For example, if a mass weighing 10 lb stretches a spring $\frac{1}{2}$ ft, then

$$10 = k\left(\frac{1}{2}\right) \quad \text{implies} \quad k = 20 \text{ lb/ft}$$

Necessarily then, a mass weighing 8 lb stretches the same spring $\frac{2}{5}$ ft.

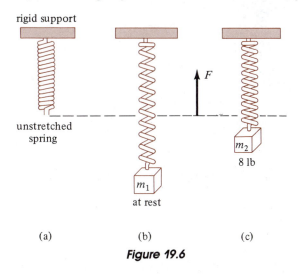

(a)   (b)   (c)

Figure 19.6

Newton's Second Law   After a mass m is attached to a spring it will stretch the spring by an amount s and attain a position of equilibrium at which its weight W is balanced by the restoring force ks. Recall that weight is defined by $W = mg$, where the mass m is measured in slugs, kilograms, or grams and $g = 32$ ft/sec^2, 9.8 m/sec^2, or 980 cm/sec^2, respectively. As indicated in Figure 19.7(b), the condition of equilibrium is $mg = ks$ or $mg - ks = 0$. If the mass is now displaced by an amount x from its equilibrium position and released, the net force F in this dynamic case is given by **Newton's second law of motion**, $F = ma$, where a

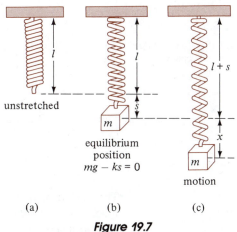

(a)   (b)   (c)

Figure 19.7

is the acceleration d^2x/dt^2. Assuming that there are no retarding forces acting on the system and assuming that the mass vibrates free of other external influencing forces—**free motion**—we can then equate F to the resultant force of the weight and the restoring force:

$$m\frac{d^2x}{dt^2} = -k(s + x) + mg$$

$$= -kx + \underbrace{mg - ks}_{\text{zero}} = -kx \qquad (19.32)$$

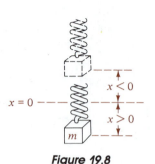

Figure 19.8

The negative sign in (19.32) indicates that the restoring force of the spring acts opposite to the direction of motion. Furthermore, we shall adopt the convention that displacements measured *below* the equilibrium position are positive. See Figure 19.8.

Free Undamped Motion   By dividing (19.32) by the mass m we obtain the second-order differential equation

$$\frac{d^2x}{dt^2} + \frac{k}{m}x = 0 \quad \text{or} \quad \frac{d^2x}{dt^2} + \omega^2x = 0 \qquad (19.33)$$

where $\omega^2 = k/m$. Equation (19.33) is said to describe **simple harmonic motion**, or **free undamped motion**. There are two obvious initial conditions associated with this differential equation:

$$x(0) = \alpha, \qquad \left.\frac{dx}{dt}\right|_{t=0} = \beta \qquad (19.34)$$

representing the amount of initial displacement and the initial velocity, respectively. For example, if $\alpha > 0$, $\beta < 0$, the mass would start from a point *below* the equilibrium position with an imparted *upward* velocity. If $\alpha < 0$, $\beta = 0$, the mass would be released from *rest* from a point $|\alpha|$ units *above* the equilibrium position, and so on.

The Solution and the Equation   To solve (19.33) we note that the solutions of the auxiliary equation
of Motion $m^2 + \omega^2 = 0$ are the complex numbers

$$m_1 = \omega i, \qquad m_2 = -\omega i$$

Thus, from (19.21) of Section 19.4, we find the general solution of the equation to be

$$x(t) = C_1 \cos \omega t + C_2 \sin \omega t \qquad (19.35)$$

The **period** of free vibrations described by (19.35) is $T = 2\pi/\omega$ and the **frequency** is $f = 1/T = \omega/2\pi$. Finally, when the initial conditions (19.34) are used to determine the constants C_1 and C_2 in (19.35) we say that the resulting particular solution is the **equation of motion** of the mass.

Example 1

Solve and interpret the initial-value problem

$$\frac{d^2x}{dt^2} + 16x = 0$$

$$x(0) = 10, \qquad \frac{dx}{dt}\bigg|_{t=0} = 0$$

Solution The problem is equivalent to pulling a mass on a spring down 10 units below the equilibrium position, holding it until $t = 0$, and then releasing it from rest. Applying the initial conditions to the solution

$$x(t) = C_1 \cos 4t + C_2 \sin 4t$$

gives
$$x(0) = 10 = C_1 \cdot 1 + C_2 \cdot 0$$

so that $C_1 = 10$, and hence

$$x(t) = 10 \cos 4t + C_2 \sin 4t$$

$$\frac{dx}{dt} = -40 \sin 4t + 4C_2 \cos 4t$$

$$\frac{dx}{dt}\bigg|_{t=0} = 0 = 4C_2 \cdot 1$$

The latter equation implies that $C_2 = 0$, so therefore the equation of motion is $x(t) = 10 \cos 4t$.

    The solution clearly shows that once the system is set in motion, it stays in motion with the mass bouncing back and forth 10 units on either side of the equilibrium position $x = 0$. As shown in Figure 19.9(b), the period of oscillation is $2\pi/4 = \pi/2$ sec.

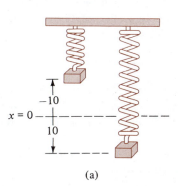

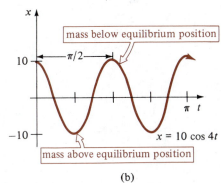

(a)   (b)

Figure 19.9

Free Damped Motion   The discussion of simple harmonic motion in Example 1 is somewhat unrealistic. Unless the mass is suspended in a perfect vacuum, there will be at least a resisting force due to the surrounding medium. For example, as Figure 19.10 shows, the mass m could be suspended in a viscous medium or connected to a dashpot

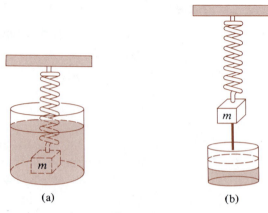

Figure 19.10

damping device. In the study of mechanics, damping forces acting on a body are considered to be proportional to a power of the instantaneous velocity. In particular, we shall assume that this force is given by a constant multiple of dx/dt. Thus, when no other external forces are impressed on the system it follows from Newton's second law that

$$m\frac{d^2x}{dt^2} = -kx - \beta\frac{dx}{dt} \tag{19.36}$$

where β is a positive *damping constant* and the negative sign is a consequence of the fact that the damping force acts in a direction opposite to the motion. Dividing (19.36) by the mass m, the differential equation of **free damped motion** then is

$$\frac{d^2x}{dt^2} + \frac{\beta}{m}\frac{dx}{dt} + \frac{k}{m}x = 0 \quad \text{or} \quad \frac{d^2x}{dt^2} + 2\lambda\frac{dx}{dt} + \omega^2 x = 0 \tag{19.37}$$

where $2\lambda = \beta/m$ and $\omega^2 = k/m$. The symbol 2λ is used only for algebraic convenience since the auxiliary equation is $m^2 + 2\lambda m + \omega^2 = 0$ and the corresponding roots are then

$$m_1 = -\lambda + \sqrt{\lambda^2 - \omega^2}, \qquad m_2 = -\lambda - \sqrt{\lambda^2 - \omega^2}$$

We can now distinguish three possible cases depending on the algebraic sign of $\lambda^2 - \omega^2$. Since each solution will contain the *damping factor* $e^{-\lambda t}$, $\lambda > 0$, the displacements of the mass will become negligible for large time.

Case I $\lambda^2 - \omega^2 > 0$. In this situation the system is said to be **overdamped** since the damping coefficient β is large when compared to the spring constant k. The corresponding solution of (19.37) is

$$x(t) = C_1 e^{m_1 t} + C_2 e^{m_2 t}$$

or

$$x(t) = e^{-\lambda t}(C_1 e^{\sqrt{\lambda^2 - \omega^2}\,t} + C_2 e^{-\sqrt{\lambda^2 - \omega^2}\,t}) \tag{19.38}$$

This equation represents a smooth and nonoscillatory motion. Figure 19.11 shows two possible graphs of $x(t)$.

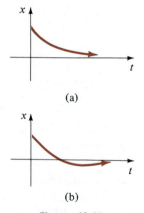

Figure 19.11

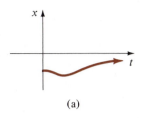

(a)

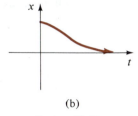

(b)

Figure 19.12

Figure 19.13

Case II $\lambda^2 - \omega^2 = 0$. The system is said to be **critically damped** since any slight decrease in the damping force would result in oscillatory motion. The general solution of (19.37) is

$$x(t) = C_1 e^{m_1 t} + C_2 t e^{m_1 t}$$

or $$x(t) = e^{-\lambda t}(C_1 + C_2 t) \qquad (19.39)$$

Some graphs of typical motion are given in Figure 19.12. Notice that the motion is quite similar to that of an overdamped system. It is also apparent from (19.39) that the mass can pass through the equilibrium position at most one time.

Case III $\lambda^2 - \omega^2 < 0$. In this case the system is said to be **underdamped** since the damping coefficient is small compared to the spring constant. The roots m_1 and m_2 are now complex:

$$m_1 = -\lambda + \sqrt{\omega^2 - \lambda^2}\,i, \qquad m_2 = -\lambda - \sqrt{\omega^2 - \lambda^2}\,i$$

and so the general solution of equation (19.37) is

$$x(t) = e^{-\lambda t}(C_1 \cos \sqrt{\omega^2 - \lambda^2}\,t + C_2 \sin \sqrt{\omega^2 - \lambda^2}\,t) \qquad (19.40)$$

As indicated in Figure 19.13 the motion described by (19.40) is oscillatory, but because of the coefficient $e^{-\lambda t}$ we see that the amplitudes of vibration $\to 0$ as $t \to \infty$.

_____ **Example 2** _____

An 8-lb weight stretches a spring 2 ft. Assuming a damping force numerically equal to two times the instantaneous velocity acts on the system, determine the equation of motion if the weight is released from the equilibrium position with an upward velocity of 3 ft/sec.

Solution From Hooke's law we have

$$8 = k(2), \qquad k = 4 \text{ lb/ft}$$

and from $m = W/g$,

$$m = \frac{8}{32} = \frac{1}{4} \text{ slug}$$

Thus the differential equation of motion is

$$\frac{1}{4}\frac{d^2 x}{dt^2} = -4x - 2\frac{dx}{dt} \quad \text{or} \quad \frac{d^2 x}{dt^2} + 8\frac{dx}{dt} + 16x = 0$$

The initial conditions are

$$x(0) = 0, \qquad \left.\frac{dx}{dt}\right|_{t=0} = -3$$

Now the auxiliary equation of the differential equation is

$$m^2 + 8m + 16 = (m + 4)^2 = 0$$

so that $m_1 = m_2 = -4$. Hence the system is critically damped and

$$x(t) = C_1 e^{-4t} + C_2 t e^{-4t}$$

The initial condition $x(0) = 0$ immediately demands that $C_1 = 0$, whereas using $x'(0) = -3$ gives $C_2 = -3$. Thus the equation of motion is

$$x(t) = -3t e^{-4t}$$

To graph $x(t)$ we find the time at which $x'(t) = 0$:

$$x'(t) = -3(-4t e^{-4t} + e^{-4t})$$
$$= -3 e^{-4t}(1 - 4t)$$

Clearly $x'(t) = 0$ when $t = \frac{1}{4}$. The corresponding displacement is

$$x\left(\frac{1}{4}\right) = -3\left(\frac{1}{4}\right) e^{-1} = -0.276 \text{ ft}$$

As shown in Figure 19.14 we interpret this value to mean that the weight reaches a maximum height of 0.276 ft above the equilibrium position.

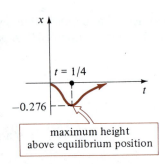

maximum height
above equilibrium position

Figure 19.14

_____ **Example 3** _____

A 16-lb weight is attached to a 5-ft long spring. At equilibrium the spring measures 8.2 ft. If the weight is pushed up and released from rest at a point 2 ft above the equilibrium position, find the displacements $x(t)$ if it is further known that the surrounding medium offers a resistance numerically equal to the instantaneous velocity.

Solution   The elongation of the spring after the weight is attached is $8.2 - 5 = 3.2$ ft so it follows from Hooke's law that

$$16 = k(3.2), \qquad k = 5 \text{ lb/ft}$$

In addition,

$$m = \frac{16}{32} = \frac{1}{2} \text{ slug}$$

so that the differential equation is given by

$$\frac{1}{2}\frac{d^2x}{dt^2} = -5x - \frac{dx}{dt} \quad \text{or} \quad \frac{d^2x}{dt^2} + 2\frac{dx}{dt} + 10x = 0$$

This latter equation is solved subject to the conditions

$$x(0) = -2, \qquad \left.\frac{dx}{dt}\right|_{t=0} = 0$$

Proceeding, we find that the roots of $m^2 + 2m + 10 = 0$ are $m_1 = -1 + 3i$ and $m_2 = -1 - 3i$, which then implies the system is underdamped and

$$x(t) = e^{-t}(C_1 \cos 3t + C_2 \sin 3t)$$

Now

$$x(0) = -2 = C_1$$

$$x(t) = e^{-t}(-2 \cos 3t + C_2 \sin 3t)$$

$$x'(t) = e^{-t}(6 \sin 3t + 3C_2 \cos 3t) - e^{-t}(-2 \cos 3t + C_2 \sin 3t)$$

$$x'(0) = 0 = 3C_2 + 2$$

which gives $C_2 = -\frac{2}{3}$. Thus we finally obtain

$$x(t) = e^{-t}\left(-2 \cos 3t - \frac{2}{3} \sin 3t\right)$$

Forced Motion   Suppose we now take into consideration an external force $f(t)$ acting on a vibrating mass on a spring. For example, $f(t)$ could represent a driving force causing an oscillatory vertical motion of the support of the spring (Figure 19.15). The inclusion of $f(t)$ in the formulation of Newton's second law gives

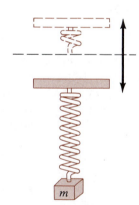

Figure 19.15

$$m\frac{d^2x}{dt^2} = -kx - \beta\frac{dx}{dt} + f(t)$$

$$\frac{d^2x}{dt^2} + \frac{\beta}{m}\frac{dx}{dt} + \frac{k}{m}x = \frac{f(t)}{m} \quad \text{or} \quad \frac{d^2x}{dt^2} + 2\lambda\frac{dx}{dt} + \omega^2 x = F(t)$$

where $F(t) = f(t)/m$ and, as before, $2\lambda = \beta/m$, $\omega^2 = k/m$. To solve the latter nonhomogeneous equation, we can employ either the method of undetermined coefficients or variation of parameters.

The next example illustrates undamped forced motion.

Example 4

Solve the initial-value problem

$$\frac{d^2x}{dt^2} + \omega^2 x = F_0 \sin \gamma t, \qquad F_0 = \text{constant}$$

$$x(0) = 0, \qquad \left.\frac{dx}{dt}\right|_{t=0} = 0$$

Solution   The complementary function is $x_c(t) = C_1 \cos \omega t + C_2 \sin \omega t$.

To obtain a particular solution we use the method of undetermined coefficients and assume $x_p(t) = A \cos \gamma t + B \sin \gamma t$, so that

$$x_p' = -A\gamma \sin \gamma t + B\gamma \cos \gamma t$$

$$x_p'' = -A\gamma^2 \cos \gamma t - B\gamma^2 \sin \gamma t$$

$$x_p'' + \omega^2 x_p = A(\omega^2 - \gamma^2) \cos \gamma t + B(\omega^2 - \gamma^2) \sin \gamma t$$

$$= F_0 \sin \gamma t$$

It follows that $A = 0$ and $B = \dfrac{F_0}{\omega^2 - \gamma^2}$, $\gamma \neq \omega$

Therefore, $x_p(t) = \dfrac{F_0}{\omega^2 - \gamma^2} \sin \gamma t$

Applying the given initial conditions to the general solution

$$x(t) = C_1 \cos \omega t + C_2 \sin \omega t + \frac{F_0}{\omega^2 - \gamma^2} \sin \gamma t$$

yields $C_1 = 0$ and $C_2 = -\gamma F_0 / \omega(\omega^2 - \gamma^2)$. Thus the solution is

$$x(t) = \frac{F_0}{\omega(\omega^2 - \gamma^2)}(-\gamma \sin \omega t + \omega \sin \gamma t), \qquad \gamma \neq \omega \qquad (19.41)$$

Pure Resonance   Although (19.41) is not defined for $\gamma = \omega$, it is interesting to observe that its limiting value as $\gamma \to \omega$ can be obtained by applying L'Hôpital's Rule. This limiting process is analogous to "tuning in" the frequency of the driving force $(\gamma/2\pi)$ to the frequency of free vibrations $(\omega/2\pi)$. Intuitively, we expect that over a length of time we should be able to substantially increase the amplitudes of vibration. For $\gamma = \omega$ we define the solution to be

$$x(t) = \lim_{\gamma \to \omega} F_0 \frac{-\gamma \sin \omega t + \omega \sin \gamma t}{\omega(\omega^2 - \gamma^2)}$$

$$= F_0 \lim_{\gamma \to \omega} \frac{\dfrac{d}{d\gamma}[-\gamma \sin \omega t + \omega \sin \gamma t]}{\dfrac{d}{d\gamma}[\omega^3 - \omega\gamma^2]}$$

$$= F_0 \lim_{\gamma \to \omega} \frac{-\sin \omega t + \omega t \cos \gamma t}{-2\omega\gamma}$$

$$= \frac{F_0}{2\omega^2}[\sin \omega t - \omega t \cos \omega t]$$

$$= \frac{F_0}{2\omega^2} \sin \omega t - \frac{F_0}{2\omega}t \cos \omega t \qquad (19.42)$$

As suspected, when $t \to \infty$ the displacements become large, in fact, $|x(t)| \to \infty$. The phenomenon we have just described is known as **pure resonance**. The graph in Figure 19.16 displays typical motion in this case.

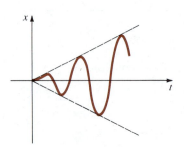

Figure 19.16

In conclusion, it should be noted that there is no actual need to use a limiting process on (19.41) to obtain the solution for $\gamma = \omega$. Alternatively, (19.42) follows by solving the initial-value problem

$$\frac{d^2x}{dt^2} + \omega^2 x = F_0 \sin \omega t$$

$$x(0) = 0, \qquad \frac{dx}{dt}\bigg|_{t=0} = 0$$

directly by conventional methods.

Remark

If a mechanical system were actually described by a function such as (19.42) of this section, it would necessarily fail; large oscillations of a weight on a spring would eventually force the spring beyond its elastic limit. One might argue too that the resonating model presented (Figure 19.16) is completely unrealistic since it ignores the retarding effects of ever-present damping forces. While it is true that pure resonance cannot occur when the smallest amount of damping is taken into consideration, nevertheless, large and equally destructive amplitudes of vibration (though bounded as $t \to \infty$) could take place.

If you have ever looked out a window while in flight, you have probably observed that wings on an airplane are not perfectly rigid. A reasonable amount of flutter is not only tolerated but necessary to prevent the wing from snapping like a piece of peppermint stick candy. In late 1959 and early 1960 two commercial plane crashes occurred, with a then relatively new model of prop-jet, that illustrate the destructive effects of large mechanical oscillations.

The unusual aspect of these crashes was that they both happened while the planes were in midflight. Barring mid-air collisions, the safest period during any flight is when the plane has attained its cruising altitude. It is well known that a plane is most vulnerable to an accident when it is least maneuverable, namely, either during takeoff or landing. So having two planes simply fall out of the sky was at least an embarrassment to the aircraft industry as well as a thoroughly puzzling problem to aerodynamic engineers. In crashes of this sort, a structural failure of some kind is immediately suspected. After a subsequent and massive technical investigation, the problem was eventually traced in each case to an outboard engine and engine housing. Roughly, it was determined that when each plane surpassed a critical speed of approximately 400 mph, a propeller and engine began to wobble, causing a gyroscopic force, which could not be quelled or damped by the engine housing. This external vibrational force was then transferred to the already oscillating wing. This in itself need not have been destructively dangerous since the aircraft wings are designed to withstand the stress of unusual and excessive forces. (In fact, the particular wing in question was so incredibly strong that test engineers and pilots, who were deliberately trying, failed to snap a wing under every conceivable flight condition.) Unfortunately, after a short period of time during which the engine wobbled rapidly, the frequency of the impressed force actually slowed to a point where it approached and finally coincided with the maximum frequency of wing flutter (around 3 cycles per second). The resulting resonance situation finally accomplished what the test engineers could not do; namely, the amplitudes of wing flutter became large enough to snap the wing. See Figure 19.17. The problem

(a) normal flutter

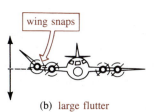

(b) large flutter

Figure 19.17

was solved in two steps. All models of this particular plane were required to fly at speeds substantially below 400 mph until each plane could be modified by considerably strengthening (or stiffening) the engine housings. A strengthened engine housing was shown to be able to impart a damping effect capable of preventing the critical resonance phenomenon even in the unlikely event of a subsequent engine wobble.*

You may be aware that soldiers usually do not march in step across bridges. The reason for breaking stride is simply to avoid any possibility of resonance occurring between the natural vibrations inherent in the bridge's structure and the frequency of the external force of a multitude of feet stomping in unison on the bridge.

Bridges are good examples of vibrating mechanical systems, which are constantly being subjected to external forces, either from people walking, cars and trucks driving on them, water pushing against their foundations, or wind blowing against their superstructures. On November 7, 1940, the Tacoma Narrows Bridge at Puget Sound in the state of Washington collapsed. However, the crash came as no surprise since "Galloping Gertie," as the bridge was called by local residents, was famous for a vertical undulating motion of its roadway, which gave many motorists a very exciting crossing. On November 7, only four months after its grand opening, the amplitudes of these undulations became so large that the bridge failed and a substantial portion was sent splashing into the water below. In the investigation that followed, it was found that a poorly designed superstructure caused the wind blowing across it to vortex in a periodic manner. When the frequency of this periodic force approached the natural frequency of the bridge, large upheavals of the road resulted. In a word, the bridge was another victim of the destructive effect of mechanical resonance. Since this disaster developed over a matter of months, there was sufficient opportunity to record on film the strange and frightening phenomenon of a bucking and heaving bridge and its ultimate collapse.[†] See Figure 19.18.

Acoustic vibrations can be as destructive as large mechanical vibrations. In recent television commercials, jazz singers have inflicted destruction on the lowly wine glass. See Figure 19.19. Sounds from organs and piccolos have been known to crack windows.

As the horns blew, the people began to shout. When they heard the signal horn, they raised a tremendous shout. The wall collapsed. . . . (*Joshua* 6:20)

Did the power of acoustic resonance cause the walls of Jericho to tumble down? This is the conjecture of some contemporary scholars.

The phenomenon of resonance is not always destructive. It is resonance of an electrical circuit that enables a radio to be tuned to a specific station.

Figure 19.18

© 1987 Memtek Products

Figure 19.19

---

*For a fascinating nontechnical account of the investigation see Robert J. Serling, *Loud and Clear* (New York: Dell, 1970), Chapter 5.

[†]National Committee for Fluid Mechanics Films, Educational Services, Inc., Watertown, Mass. See also *American Society of Civil Engineers: Proceedings*, "Failure of the Tacoma Narrows Bridge," Vol. 69, pp. 1555–86, Dec. 1943.

Exercises 19.7

Answers to odd-numbered problems begin on page A-54.

In Problems 1 and 2 state in words a possible physical interpretation of the given initial-value problem.

1. $\dfrac{4}{32}x'' + 3x = 0$

$x(0) = -3, \qquad x'(0) = -2$

2. $\dfrac{1}{16}x'' + 4x = 0$

$x(0) = 0.7, \qquad x'(0) = 0$

3. An 8-lb weight attached to a spring exhibits simple harmonic motion. Determine the equation of motion if the spring constant is 1 lb/ft and if the weight is released 6 in. below the equilibrium position with a downward velocity of $\frac{3}{2}$ ft/sec.

4. A 24-lb weight attached to a spring exhibits simple harmonic motion. When placed on the spring, the weight stretched the spring 4 in. Find the equation of motion if the weight is released from rest from a point 3 in above the equilibrium position.

5. A force of 400 newtons stretches a spring 2 m. A mass of 50 kg attached to the spring exhibits simple harmonic motion. Find the equation of motion if the mass is released from the equilibrium position with an upward velocity of 10 m/sec.

6. A mass weighing 2 lb attached to a spring exhibits simple harmonic motion. At $t = 0$ the mass is released from a point 8 in below the equilibrium position with an upward velocity of $\frac{4}{3}$ ft/sec. If the spring constant is $k = 4$ lb/ft, find the equation of motion.

In Problems 7 and 8 state in words a possible physical interpretation of the given initial-value problem.

7. $\dfrac{1}{16}x'' + 2x' + x = 0$

$x(0) = 0, \qquad \left. \dfrac{dx}{dt} \right|_{t=0} = -1.5$

8. $\dfrac{16}{32}x'' + x' + 2x = 0$

$x(0) = -2, \qquad x'(0) = 1$

9. A 4-lb weight attached to a spring exhibits free damped motion. The spring constant is 2 lb/ft and the medium offers a resistance to the motion of the weight numerically equal to the instantaneous velocity. If the weight is released from a point 1 ft above the equilibrium position with a downward velocity of 8 ft/sec, determine the time that the weight passes through the equilibrium position. Find the time for which the weight attains its maximum displacement from the equilibrium position. What is the position of the weight at this instant?

10. A mass of 40 g stretches a spring 10 cm. A damping device imparts a resistance to motion numerically equal to 560 times the instantaneous velocity. Find the equation of free motion if the mass is released from the equilibrium position with a downward velocity of 2 cm/sec.

11. After a 10-lb weight is attached to a 5-ft spring, the spring measures 7 ft long. The 10-lb weight is removed and replaced with an 8-lb weight and the entire system is placed in a medium offering a resistance numerically equal to the instantaneous velocity. Find the equation of motion if the weight is released $\frac{1}{2}$ ft below the equilibrium position with a downward velocity of 1 ft/sec.

12. A 24-lb weight stretches a spring 4 ft. The subsequent motion takes place in a medium offering a resistance numerically equal to β ($\beta > 0$) times the instantaneous velocity. If the weight starts from the equilibrium position with an upward velocity of 2 ft/sec, show that if $\beta > 3\sqrt{2}$, the equation of motion is

$$x(t) = \frac{-3}{\sqrt{\beta^2 - 18}} e^{-2\beta t/3} \sinh \frac{2}{3} \sqrt{\beta^2 - 18}\, t$$

13. A 10-lb weight attached to a spring stretches it 2 ft. The system is then set in motion in a medium, which offers a resistance numerically equal to $\beta(\beta > 0)$ times the instantaneous velocity. Determine the values of β so that the motion is **(a)** overdamped, **(b)** critically damped, and **(c)** underdamped.

14. A mass of 1 slug when attached to a spring stretches it 2 ft and then comes to rest in the equilibrium position. Starting at $t = 0$ an external force equal to $f(t) = 8 \sin 4t$ is applied to the system. Find the equation of motion if the surrounding medium offers a damping force numerically equal to 8 times the instantaneous velocity.

15. Solve Problem 14 when $f(t) = e^{-t} \sin 4t$. Analyze the displacements for $t \to \infty$.

16. Solve and interpret the initial-value problem:

$$\frac{d^2x}{dt^2} + 9x = 5 \sin 3t$$

$$x(0) = 2, \qquad \left. \frac{dx}{dt} \right|_{t=0} = 0$$

Chapter 19 Review Exercises

Answers to odd-numbered problems begin on page A-54.

In Problems 1–10 answer true or false.

1. The function $f(x, y) = x^2/y^2 + 9$ is homogeneous of degree 0. _____

2. The first-order equation $x^2y' + 2xy = y^2$ is homogeneous. _____

3. Every separable first-order differential equation is homogeneous. _____

4. The first-order equation $x^{10}y' = e^{9x}y + x^{20}$ is linear. _____

5. An integrating factor for $y' + 2y = 0$ is e^{2x}. _____

6. If y_1 is a solution of $ay'' + by' + cy = 0$, then C_1y_1 is also a solution for every real number C_1. _____

7. The general solution of $y'' + y = 0$ is $y = C_1 \cos x + C_2 \sin x$. _____

8. $y_1 = e^{-x}$ and $y_2 = 0$ are linearly independent solutions of $y'' + y' = 0$. _____

9. Undamped and unforced motion of a mass on a spring is called simple harmonic motion. _____

10. Pure resonance cannot occur when damping is present. _____

In Problems 11 and 12 solve the given differential equation.

11. $y\,dx + x\,dy = 0$ **12.** $xy\dfrac{dy}{dx} = 3y^2 + x^2$

In Problems 13–16 find the general solution of the given differential equation.

13. $y' - 10y = 30$

14. $xy' + (1 + x)y = e^{-x} \sin 2x$

15. $x^2y' + x(x + 2)y = e^x$

16. $(2x + y)y' = 1$

In Problems 17 and 18 solve the given initial-value problem.

17. $(x^2 + 4)\dfrac{dy}{dx} = 2x - 8xy$, $y(0) = -1$

18. $y\dfrac{dx}{dy} - x = 2y^2$, $y(1) = 5$

In Problems 19–22 find the general solution of the given differential equation.

19. $y'' - 2y' - 2y = 0$ **20.** $y'' - 8y = 0$

21. $y'' - 3y' - 10y = 0$

22. $4y'' + 20y' + 25y = 0$

In Problems 23 and 24 solve the given initial-value problem.

23. $y'' + 36y = 0$, $y\left(\dfrac{\pi}{2}\right) = 24$, $y'\left(\dfrac{\pi}{2}\right) = -18$

24. $y'' + 4y' + 4y = 0$, $y(0) = 0$, $y'(0) = 0$

In Problems 25 and 26 find power series solutions of the given differential equation.

25. $y'' + xy = 0$ **26.** $(x - 1)y'' + 3y = 0$

In Problems 27 and 28 solve each differential equation by the method of variation of parameters.

27. $y'' - 2y' + 2y = e^x \tan x$

28. $y'' - y = 2e^x/(e^x + e^{-x})$

29. Solve $y'' + y = \sec^3 x$ subject to $y(0) = 1$, $y'(0) = \frac{1}{2}$.

30. Without solving, determine the form of a particular solution of

$$y'' + 6y' + 9y = 5x^2 - 6x + 4x^2e^{2x} + 3e^{5x} - 2e^x \sin 3x$$

31. Use the method of undetermined coefficients to solve $y'' - y' - 12y = (x + 1)e^{2x}$.

32. The vertical motion of a weight attached to a spring is described by the initial-value problem

$$\frac{1}{4}\frac{d^2x}{dt^2} + \frac{dx}{dt} + x = 0$$

$$x(0) = 4, \qquad x'(0) = 2$$

Determine the maximum vertical displacement.

33. A spring with constant $k = 2$ is suspended in a liquid that offers a damping force numerically equal to 4 times the instantaneous velocity. If a mass m is suspended from the spring, determine the values of m for which the subsequent free motion is nonoscillatory.

34. Find a particular solution for $\dfrac{d^2x}{dt^2} + 2\lambda\dfrac{dx}{dt} + \omega^2 x = A$, where A is a constant force.

35. A 4-lb weight is suspended from a spring whose constant is 3 lb/ft. The entire system is immersed in a fluid offering a damping force numerically equal to the instantaneous velocity. Beginning at $t = 0$, an external force equal to $f(t) = e^{-t}$ is impressed on the system. Determine the equation of motion if the weight is released from rest at a point 2 ft below the equilibrium position.

Appendices

Greek Alphabet

α	A	Alpha		ν	N	Nu
β	B	Beta		ξ	Ξ	Xi
γ	Γ	Gamma		o	O	Omicron
δ	Δ	Delta		π	Π	Pi
ε	E	Epsilon		ρ	P	Rho
ζ	Z	Zeta		σ	Σ	Sigma
η	H	Eta		τ	T	Tau
θ	Θ	Theta		υ	Υ	Upsilon
ι	I	Iota		ϕ	Φ	Phi
κ	K	Kappa		χ	X	Chi
λ	Λ	Lambda		ψ	Ψ	Psi
μ	M	Mu		ω	Ω	Omega

Appendix I Review of Basic Mathematics

Sets A set A is a collection of objects called the **elements** of A. The symbolism, $a \in A$, is used to denote that a is an element of a set A. A set B is a **subset** of a set A, written $B \subset A$, if every element of B is also an element of A. Sets A and B are equal, $A = B$, if they contain the same elements. The **union** $A \cup B$ of two sets A and B is the set containing the elements that are either in A or in B. The **intersection** $A \cap B$ is the set of elements that are elements in A and in B. The intersection of sets A and B is the set of elements that are *common* to both A and B. The set $\varnothing$ with no elements is called the **empty set**. If $A \cap B = \varnothing$, then A and B are said to be **disjoint** sets.

Laws of Exponents If n is a positive integer,

$$x^{-n} = \frac{1}{x^n}, \qquad x \neq 0$$

Also,
$$x^0 = 1, \qquad x \neq 0$$

If m and n are integers, then

$$(i) \quad x^m x^n = x^{m+n}$$

$$(ii) \quad \frac{x^m}{x^n} = x^{m-n}, \qquad x \neq 0$$

$$(iii) \quad (x^m)^n = x^{mn}$$

$$(iv) \quad (xy)^n = x^n y^n$$

$$(v) \quad \left(\frac{x}{y}\right)^n = \frac{x^n}{y^n}, \qquad y \neq 0$$

Rational Exponents and Radicals If m and n are positive integers,

$$x^{m/n} = (x^m)^{1/n} = (x^{1/n})^m$$
$$x^{1/n} = \sqrt[n]{x}$$

provided all expressions represent real numbers. The laws of exponents hold for rational numbers.

────────────── **Example 1** ──────────────

$$16^{3/4} = (16^{1/4})^3 = 2^3 = 8$$

Binomial Theorem If n is a positive integer, the **Binomial Theorem** is

$$(a + b)^n = a^n + \frac{n}{1!}a^{n-1}b + \frac{n(n-1)}{2!}a^{n-2}b^2 + \cdots$$

$$+ \frac{n(n-1)\cdots(n-r+1)}{r!}a^{n-r}b^r + \cdots + b^n$$

where $r! = 1 \cdot 2 \cdot 3 \cdots (r-1)r$. For example,

$$(a + b)^2 = a^2 + 2ab + b^2$$
$$(a + b)^3 = a^3 + 3a^2b + 3ab^2 + b^3$$

Example 2

Expand $(2x + 4)^3$.

Solution   Identifying $a = 2x$ and $b = 4$, the Binomial Theorem gives

$$(2x + 4)^3 = (2x)^3 + 3(2x)^2 4 + 3(2x)(4)^2 + (4)^3$$
$$= 8x^3 + 48x^2 + 96x + 64$$

Special Factors   **Difference of two squares:** $X^2 - Y^2 = (X + Y)(X - Y)$

Difference of two cubes: $X^3 - Y^3 = (X - Y)(X^2 + XY + Y^2)$

Sum of two cubes: $X^3 + Y^3 = (X + Y)(X^2 - XY + Y^2)$

Example 3

(a) $x^3 - 1 = (x - 1)(x^2 + x + 1)$

(b) $x^3 + 8 = (x + 2)(x^2 - 2x + 4)$

Completing the Square   To **complete the square** for a quadratic expression $x^2 + bx$, we add and subtract the square of one-half the coefficient of x:

$$x^2 + bx = x^2 + bx + \underbrace{\left(\frac{b}{2}\right)^2 - \left(\frac{b}{2}\right)^2}_{\text{zero}}$$

$$= \left(x^2 + bx + \frac{b^2}{4}\right) - \frac{b^2}{4}$$

$$= \left(x + \frac{b}{2}\right)^2 - \frac{b^2}{4}$$

Example 4

Complete the square for $x^2 + 16x$.

Solution   We have $b/2 = \frac{16}{2} = 8$ and $(b/2)^2 = 64$. Therefore,

$$x^2 + 16x = x^2 + 16x + 64 - 64$$
$$= (x + 8)^2 - 64$$

Note: For a quadratic expression $ax^2 + bx$, $a \neq 1$, we first write

$$ax^2 + bx = a\left(x^2 + \frac{b}{a}x\right)$$

and then add and subtract $(b/2a)^2$ *inside* the parentheses.

Equations The equation $ax + b = 0$ is called a **first-degree** or **linear equation**. A linear equation has the unique solution $x = -b/a$, $a \neq 0$. A **second-degree** or **quadratic equation** is $ax^2 + bx + c = 0$, $a \neq 0$. Two methods for finding the solutions of a quadratic equation are factoring and the quadratic formula. Recall, the latter is given by

$$x = \frac{-b \pm \sqrt{b^2 - 4ac}}{2a}$$

When $b^2 - 4ac > 0$ the equation has two distinct real solutions. For example, we see from the quadratic formula that the solutions of $x^2 + x - 1 = 0$ are the irrational numbers $(-1 - \sqrt{5})/2$ and $(-1 + \sqrt{5})/2$. When $b^2 - 4ac = 0$ the solutions of a quadratic equation are **equal** (or have *multiplicity 2*.) When $b^2 - 4ac < 0$ the solutions of a quadratic equation are **complex numbers**.

Complex Numbers A **complex number** is any expression of the form

$$z = a + bi \quad \text{where} \quad i^2 = -1$$

The real numbers a and b are called the **real** and **imaginary** parts of z, respectively. In practice the symbol i is written as $i = \sqrt{-1}$. The number $\bar{z} = a - bi$ is called the **complex conjugate** of z.

Example 5

If $z_1 = 4 + 5i$ and $z_2 = 3 - 2i$ are complex numbers, then their complex conjugates are, respectively,

$$\bar{z}_1 = 4 - 5i \quad \text{and} \quad \bar{z}_2 = 3 - (-2)i = 3 + 2i$$

Sum, Difference, and Product The **sum**, **difference**, and **product** of two complex numbers $z_1 = a_1 + b_1 i$ and $z_2 = a_2 + b_2 i$ are defined as follows.

> (i) $z_1 + z_2 = (a_1 + a_2) + (b_1 + b_2)i$
>
> (ii) $z_1 - z_2 = (a_1 - a_2) + (b_1 - b_2)i$
>
> (iii) $z_1 z_2 = (a_1 a_2 - b_1 b_2) + (a_1 b_2 + b_1 a_2)i$

In other words, to add or subtract two complex numbers we simply add or subtract the corresponding real and imaginary parts. To multiply two complex numbers we use the distributive law and the fact that $i^2 = -1$.

Example 6

If $z_1 = 4 + 5i$ and $z_2 = 3 - 2i$, then

$$z_1 + z_2 = (4 + 3) + (5 - 2)i = 7 + 3i$$
$$z_1 - z_2 = (4 - 3) + (5 - (-2))i = 1 + 7i$$
$$z_1 z_2 = (4 + 5i)(3 - 2i)$$
$$= (4 + 5i)3 + (4 + 5i)(-2i)$$
$$= 12 + 15i - 8i - 10i^2$$
$$= (12 + 10) + (15 - 8)i = 22 + 7i$$

The product of a complex number $z = a + bi$ and its complex conjugate $\bar{z} = a - bi$ is the *real number*

$$z\bar{z} = a^2 + b^2$$

Quotient   The **quotient** of two complex numbers is found by multiplying numerator and denominator of the expression by the complex conjugate of the denominator.

Example 7

If z_1 and z_2 are the complex numbers in Example 6, then

$$\frac{z_1}{z_2} = \frac{4 + 5i}{3 - 2i} = \frac{(4 + 5i)(3 + 2i)}{(3 - 2i)(3 + 2i)}$$

$$= \frac{2 + 23i}{3^2 + 2^2} = \frac{2}{13} + \frac{23}{13}i$$

If $a > 0$, then the square root of the negative number $-a$ is defined by

$$\sqrt{-a} = \sqrt{a}\sqrt{-1} = \sqrt{a}\, i$$

Example 8

$$\sqrt{-25} = \sqrt{25}\sqrt{-1} = 5i$$

Example 9

By the quadratic formula, the solutions of

$$x^2 - 4x + 20 = 0$$

are

$$x = \frac{4 \pm \sqrt{-64}}{2} = \frac{4 \pm \sqrt{64}i}{2}$$

or

$$x_1 = 2 + 4i, \qquad x_2 = 2 - 4i$$

As illustrated in the last example, the complex solutions of a quadratic equation with real coefficients are complex conjugates.

Every real number can also be considered a complex number by simply taking $b = 0$ in $z = a + bi$. The relationship between real and complex numbers is shown in the following diagram.

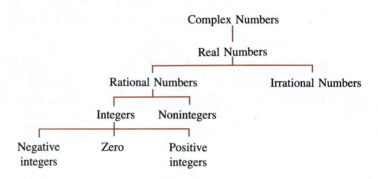

Determinants A **determinant of order 2** is the number

$$\begin{vmatrix} a_1 & a_2 \\ b_1 & b_2 \end{vmatrix} = a_1 b_2 - a_2 b_1$$

Example 10

$$\begin{vmatrix} 4 & -3 \\ 2 & 9 \end{vmatrix} = 4 \cdot 9 - (-3)(2) = 42$$

A **determinant of order 3** can be evaluated by **expanding the determinant into cofactors of the first row:**

$$\begin{vmatrix} a_1 & a_2 & a_3 \\ b_1 & b_2 & b_3 \\ c_1 & c_2 & c_3 \end{vmatrix} = a_1 \begin{vmatrix} b_2 & b_3 \\ c_2 & c_3 \end{vmatrix} - a_2 \begin{vmatrix} b_1 & b_3 \\ c_1 & c_2 \end{vmatrix} + a_3 \begin{vmatrix} b_1 & b_2 \\ c_1 & c_2 \end{vmatrix}$$

Example 11

$$\begin{vmatrix} 8 & 5 & 4 \\ 2 & 4 & 6 \\ -1 & 2 & 3 \end{vmatrix} = 8 \begin{vmatrix} 4 & 6 \\ 2 & 3 \end{vmatrix} - 5 \begin{vmatrix} 2 & 6 \\ -1 & 3 \end{vmatrix} + 4 \begin{vmatrix} 2 & 4 \\ -1 & 2 \end{vmatrix}$$

$$= 8(0) - 5(12) + 4(8) = -28$$

Cofactors In general, the **cofactor** of an entry in the ith row and jth column of a determinant is $(-1)^{i+j}$ times that determinant formed by deleting the ith row and jth column. Thus, the cofactors of a_1, a_2, and a_3 are, respectively,

$$\begin{vmatrix} b_2 & b_3 \\ c_2 & c_3 \end{vmatrix}, \qquad -\begin{vmatrix} b_1 & b_3 \\ c_1 & c_3 \end{vmatrix}, \quad \text{and} \quad \begin{vmatrix} b_1 & b_2 \\ c_1 & c_2 \end{vmatrix}$$

The cofactor of, say, b_3 is $-\begin{vmatrix} a_1 & a_2 \\ c_1 & c_2 \end{vmatrix}$,

Some Properties of Determinants

(*i*)   If every entry in a row (or column) is zero, then the value of the determinant is zero.

(*ii*)   If two rows (or columns) are equal, then the value of the determinant is zero.

(*iii*)   Interchanging two rows (or columns) results in the negative of the value of the original determinant.

(*iv*)   A determinant can be expanded by the cofactors of any row (or column).

Example 12

(*a*)   From (*i*), $\begin{vmatrix} 1 & 6 & 0 \\ 3 & 2 & 0 \\ 4 & 5 & 0 \end{vmatrix} = 0.$

(*b*)   From (*ii*), $\begin{vmatrix} 1 & 6 & 6 \\ 3 & 2 & 2 \\ 4 & 5 & 5 \end{vmatrix} = 0.$

(*c*)   Since $\begin{vmatrix} 2 & 3 \\ 4 & 5 \end{vmatrix} = -2$, it follows from (*iii*) that $\begin{vmatrix} 4 & 5 \\ 2 & 3 \end{vmatrix} = 2.$

Example 13

Expand the determinant $\begin{vmatrix} 3 & 4 & 3 \\ 2 & 0 & 0 \\ 7 & 1 & 2 \end{vmatrix}.$

Solution   Taking advantage of the zeros in the second row, we expand the determinant by the cofactors of the second row:

$$\begin{vmatrix} 3 & 4 & 3 \\ 2 & 0 & 0 \\ 7 & 1 & 2 \end{vmatrix} = -2\begin{vmatrix} 4 & 3 \\ 1 & 2 \end{vmatrix} + 0\begin{vmatrix} 3 & 3 \\ 7 & 2 \end{vmatrix} - 0\begin{vmatrix} 3 & 4 \\ 7 & 1 \end{vmatrix}$$

$$= -2(8 - 3) = -10$$

Cramer's Rule   Systems of linear equations can be solved by means of determinants. The solution of the system

$$a_1x + b_1y = c_1$$
$$a_2x + b_2y = c_2$$

is

$$x = \frac{\begin{vmatrix} c_1 & b_1 \\ c_2 & b_2 \end{vmatrix}}{\begin{vmatrix} a_1 & b_1 \\ a_2 & b_2 \end{vmatrix}}, \qquad y = \frac{\begin{vmatrix} a_1 & c_1 \\ a_2 & c_2 \end{vmatrix}}{\begin{vmatrix} a_1 & b_1 \\ a_2 & b_2 \end{vmatrix}}$$

provided that the determinant of the coefficients $\begin{vmatrix} a_1 & b_1 \\ a_2 & b_2 \end{vmatrix}$ is not zero. This last condition also guarantees that there is only one solution. Similarly, the system of three equations in three unknowns

$$a_1 x + b_1 y + c_1 z = d_1$$
$$a_2 x + b_2 y + c_2 z = d_2$$
$$a_3 x + b_3 y + c_3 z = d_3$$

has the unique solution

$$x = \frac{\begin{vmatrix} d_1 & b_1 & c_1 \\ d_2 & b_2 & c_2 \\ d_3 & b_3 & c_3 \end{vmatrix}}{\begin{vmatrix} a_1 & b_1 & c_1 \\ a_2 & b_2 & c_2 \\ a_3 & b_3 & c_3 \end{vmatrix}}, \qquad y = \frac{\begin{vmatrix} a_1 & d_1 & c_1 \\ a_2 & d_2 & c_2 \\ a_3 & d_3 & c_3 \end{vmatrix}}{\begin{vmatrix} a_1 & b_1 & c_1 \\ a_2 & b_2 & c_2 \\ a_3 & b_3 & c_3 \end{vmatrix}}, \qquad z = \frac{\begin{vmatrix} a_1 & b_1 & d_1 \\ a_2 & b_2 & d_2 \\ a_3 & b_3 & d_3 \end{vmatrix}}{\begin{vmatrix} a_1 & b_1 & c_1 \\ a_2 & b_2 & c_2 \\ a_3 & b_3 & c_3 \end{vmatrix}}$$

provided that the determinant of the coefficients is not zero. The foregoing procedure illustrates **Cramer's Rule.***

Example 14

Solve

$$x + 2y + z = 1$$
$$x \qquad + 3z = 4$$
$$4x + 2y + z = 5$$

Solution   Since $\begin{vmatrix} 1 & 2 & 1 \\ 1 & 0 & 3 \\ 4 & 2 & 1 \end{vmatrix} = 18$, the system can be solved by Cramer's Rule:

$$x = \frac{\begin{vmatrix} 1 & 2 & 1 \\ 4 & 0 & 3 \\ 5 & 2 & 1 \end{vmatrix}}{\begin{vmatrix} 1 & 2 & 1 \\ 1 & 0 & 3 \\ 4 & 2 & 1 \end{vmatrix}} = \frac{24}{18} = \frac{4}{3}, \qquad y = \frac{\begin{vmatrix} 1 & 1 & 1 \\ 1 & 4 & 3 \\ 4 & 5 & 1 \end{vmatrix}}{\begin{vmatrix} 1 & 2 & 1 \\ 1 & 0 & 3 \\ 4 & 2 & 1 \end{vmatrix}} = -\frac{11}{18}, \qquad z = \frac{\begin{vmatrix} 1 & 2 & 1 \\ 1 & 0 & 4 \\ 4 & 2 & 5 \end{vmatrix}}{\begin{vmatrix} 1 & 2 & 1 \\ 1 & 0 & 3 \\ 4 & 2 & 1 \end{vmatrix}} = \frac{16}{18} = \frac{8}{9}$$

---

*Named after the Swiss mathematician Gabriel Cramer (1704–1752).

Similar Triangles

The lengths of corresponding sides in similar triangles are proportional. Thus, if the figure represents two similar triangles, the ratios of the lengths of the corresponding sides are equal:

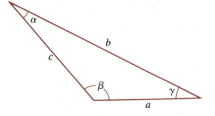

$$\frac{a}{a'} = \frac{b}{b'} = \frac{c}{c'}$$

Law of Sines

For the triangle shown, if we know (*a*) the length of one side and two angles, or (*b*) the lengths of two sides and the angle opposite one of them, then the remaining parts of the triangle can be found from the **law of sines**:

$$\frac{\sin \alpha}{a} = \frac{\sin \beta}{b} = \frac{\sin \gamma}{c}$$

Law of Cosines

Referring to the accompanying triangle, if we know (*a*) the lengths of three sides, or (*b*) the length of two sides and the angle between them, then the remaining parts of the triangle can be found from the **law of cosines**:

$$a^2 = b^2 + c^2 - 2bc \cos \alpha$$
$$b^2 = a^2 + c^2 - 2ac \cos \beta$$
$$c^2 = a^2 + b^2 - 2ab \cos \gamma$$

———— Appendix I Exercises ————

Answers to odd-numbered problems begin on page A-55.

In Problems 1–4 evaluate the given expression.

1. $8^{2/3}$

2. $16^{-5/4}$

3. $\sqrt{(-5)^2}$

4. $\sqrt[3]{8x^3 y^9}$

In Problems 5–8 use the Binomial Theorem to expand the given expression.

5. $(3r - 4s)^2$

6. $(x + y - 1)^2$

7. $(2x - y)^3$

8. $(t^2 + r^3)^3$

In Problems 9–12 factor the given expression.

9. $4x^2 y^2 - 9$

10. $x^4 - y^4$

11. $64a^3 - b^3$

12. $x^6 - 1$

In Problems 13–16 complete the square for the given expression.

13. $x^2 - 24x$

14. $x^2 + 3x$

15. $2x^2 + 12x$

16. $-x^2 + 10x$

In Problems 17–26 solve the given equation.

17. $2x - 3 = 6x + 9$

18. $3(x - 2) = -4(x - 1)$

19. $4x^2 = 9$

20. $36x^2 - 4 = 0$

21. $5x^2 = 2x$

22. $3x^2 + x = 0$

23. $x^2 + 2x - 15 = 0$

24. $x^2 + 6x + 9 = 0$

25. $2x^2 + 2x - 1 = 0$

26. $-x^2 + 3x + 2 = 0$

In Problems 27–36 perform the indicated operation. Write the answer in the form $a + bi$.

27. $(6 + 5i) - (-8 + 2i)$

28. $3i(2 + i) + 4(1 - 2i)$

29. $(4 + 6i)(-2 + 5i)$

30. $(1 - i)(1 + i)(3 + 7i)$

31. $\dfrac{1 + 3i}{3 - 4i}$

32. $\dfrac{i}{1 + i}$

33. $i\dfrac{2+i}{1-i}$

34. $\dfrac{1}{i(2-i)}$

35. i^5

36. i^{-8}

37. For the determinant $\begin{vmatrix} -5 & 6 & 0 \\ 3 & 4 & 2 \\ -2 & 1 & 5 \end{vmatrix}$ find **(a)** the value of the

cofactor of the entry in the third row and second column, and **(b)** the value of the cofactor of the entry in the second row and second column.

38. If $\begin{vmatrix} a_1 & b_1 & c_1 \\ a_2 & b_2 & c_2 \\ a_3 & b_3 & c_3 \end{vmatrix} = 3$, determine the values of the

following:

(a) $\begin{vmatrix} a_3 & b_3 & c_3 \\ a_2 & b_2 & c_2 \\ a_1 & b_1 & c_1 \end{vmatrix}$

(b) $\begin{vmatrix} b_1 & c_1 & a_1 \\ b_2 & c_2 & a_2 \\ b_3 & c_3 & a_3 \end{vmatrix}$

(c) $\begin{vmatrix} a_1 & b_1 & c_1 \\ a_1 & b_1 & c_1 \\ a_2 & b_2 & c_2 \end{vmatrix}$

(d) $\begin{vmatrix} a_1 & b_1 & 0 \\ a_2 & b_2 & 0 \\ a_3 & b_3 & 0 \end{vmatrix}$

In Problems 39 and 40 expand the given determinant by cofactors other than those of the first row.

39. $\begin{vmatrix} 2 & -3 & 1 \\ 0 & 2 & -4 \\ 0 & 6 & 5 \end{vmatrix}$

40. $\begin{vmatrix} 4 & 3 & -1 \\ 7 & 0 & -2 \\ -1 & 5 & 8 \end{vmatrix}$

In Problems 41 and 42 find the values of λ satisfying the given equation.

41. $\begin{vmatrix} 3-\lambda & 5 \\ 1 & -1-\lambda \end{vmatrix} = 0$

42. $\begin{vmatrix} 1-\lambda & -4 & 4 \\ 0 & 3-\lambda & -2 \\ -2 & 0 & -3-\lambda \end{vmatrix} = 0$

In Problems 43–46 use Cramer's Rule to solve the given system of equations.

43. $-3x + y = 3$
$\quad\ \ 2x - 4y = -6$

44. $x + y = 4$
$\quad 2x - y = 2$

45. $x - 2y - 3z = 3$
$\quad x + y - z = 5$
$\quad 3x + 2y \quad\ \ = -4$

46. $x - y + 6z = -2$
$\quad -x + 2y + 4z = 9$
$\quad 2x + 3y - z = \frac{1}{2}$

In Problems 47 and 48 determine x and y.

47.

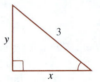

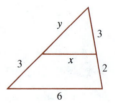

48.

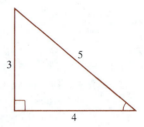

In Problems 49 and 50 find the indicated unknowns in the given triangle.

49.

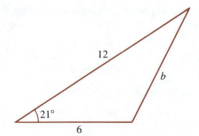

50.

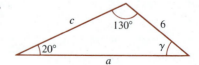

Appendix II Some ε–δ Proofs

Proof of Theorem 2.5(i) Let $\varepsilon > 0$ be given. To prove (i) we must find a $\delta > 0$ so that

$$|f(x) + g(x) - L_1 - L_2| < \varepsilon \quad \text{whenever} \quad 0 < |x - a| < \delta$$

Since $\lim_{x \to a} f(x) = L_1$ and $\lim_{x \to a} g(x) = L_2$, we know there exist numbers $\delta_1 > 0$ and $\delta_2 > 0$ for which

$$|f(x) - L_1| < \frac{\varepsilon}{2} \quad \text{whenever} \quad 0 < |x - a| < \delta_1 \qquad \text{(II.1)}$$

$$|g(x) - L_2| < \frac{\varepsilon}{2} \quad \text{whenever} \quad 0 < |x - a| < \delta_2 \qquad \text{(II.2)}$$

Now, if we choose δ the smaller of δ_1 and δ_2, then (II.1) and (II.2) *both* hold and so

$$
\begin{aligned}
|f(x) + g(x) - L_1 - L_2| &= |f(x) - L_1 + g(x) - L_2| \\
&\leq |f(x) - L_1| + |g(x) - L_2| \\
&< \frac{\varepsilon}{2} + \frac{\varepsilon}{2} = \varepsilon
\end{aligned}
$$

whenever $0 < |x - a| < \delta$. ∎

Proof of Theorem 2.5(ii) Consider

$$
\begin{aligned}
|f(x)g(x) - L_1 L_2| &= |f(x)g(x) - f(x)L_2 + f(x)L_2 - L_1 L_2| \\
&\leq |f(x)g(x) - f(x)L_2| + |f(x)L_2 - L_1 L_2| \\
&= |f(x)|\,|g(x) - L_2| + |L_2|\,|f(x) - L_1| \\
&\leq |f(x)|\,|g(x) - L_2| + (1 + |L_2|)|f(x) - L_1| \qquad \text{(II.3)}
\end{aligned}
$$

Since $\lim_{x \to a} f(x) = L_1$ and $\lim_{x \to a} g(x) = L_2$, we know there exist numbers $\delta_1 > 0$, $\delta_2 > 0$, $\delta_3 > 0$ such that $|f(x) - L_1| < 1$ or

$$|f(x)| < 1 + |L_1| \quad \text{whenever} \quad 0 < |x - a| < \delta_1 \qquad \text{(II.4)}$$

$$|f(x) - L_1| < \frac{\varepsilon/2}{1 + |L_2|} \quad \text{whenever} \quad 0 < |x - a| < \delta_2 \qquad \text{(II.5)}$$

$$|g(x) - L_2| < \frac{\varepsilon/2}{1 + |L_1|} \quad \text{whenever} \quad 0 < |x - a| < \delta_3 \qquad \text{(II.6)}$$

Hence, if we choose δ the smaller of δ_1, δ_2, and δ_3, we have from (II.3), (II.4), (II.5), and (II.6),

$$
\begin{aligned}
|f(x)g(x) - L_1 L_2| &< (1 + |L_1|) \cdot \frac{\varepsilon/2}{1 + |L_1|} \\
&\quad + (1 + |L_2|) \cdot \frac{\varepsilon/2}{1 + |L_2|} = \varepsilon
\end{aligned}
$$
∎

Proof of Theorem 2.5(*iii*)   We shall first prove that

$$\lim_{x \to a} \frac{1}{g(x)} = \frac{1}{L_2}, \quad L_2 \neq 0$$

Consider

$$\left| \frac{1}{g(x)} - \frac{1}{L_2} \right| = \frac{|g(x) - L_2|}{|L_2| \, |g(x)|} \tag{II.7}$$

Since $\lim_{x \to a} g(x) = L_2$, there exists a $\delta_1 > 0$ such that

$$|g(x) - L_2| < \frac{|L_2|}{2} \quad \text{whenever} \quad 0 < |x - a| < \delta_1$$

For these values of x,

$$|L_2| = |g(x) - (g(x) - L_2)| \leq |g(x)| + |g(x) - L_2| < |g(x)| + \frac{|L_2|}{2}$$

gives

$$|g(x)| > \frac{|L_2|}{2} \quad \text{and} \quad \frac{1}{|g(x)|} < \frac{2}{|L_2|}$$

Thus from (II.7)

$$\left| \frac{1}{g(x)} - \frac{1}{L_2} \right| < \frac{2}{|L_2|^2} |g(x) - L_2| \tag{II.8}$$

Now for $\varepsilon > 0$ there exists a $\delta_2 > 0$ such that

$$|g(x) - L_2| < \frac{|L_2|^2}{2} \varepsilon \quad \text{whenever} \quad 0 < |x - a| < \delta_2$$

By choosing δ the smaller of δ_1 and δ_2, it follows from (II.8) that

$$\left| \frac{1}{g(x)} - \frac{1}{L_2} \right| < \varepsilon \quad \text{whenever} \quad 0 < |x - a| < \delta$$

We conclude the proof using Theorem 2.5(*ii*):

$$\lim_{x \to a} \frac{f(x)}{g(x)} = \lim_{x \to a} \frac{1}{g(x)} \cdot f(x) = \lim_{x \to a} \frac{1}{g(x)} \cdot \lim_{x \to a} f(x) = \frac{L_1}{L_2} \quad \blacksquare$$

———— Appendix II Exercises ————

1. Use Definition 2.2 to prove Theorem 2.2.

2. Use Definition 2.2 to prove Theorem 2.3.

3. If $\lim_{x \to a} f(x) = L_1$ and $\lim_{x \to a} g(x) = L_2$, use Definition 2.2 to prove $\lim_{x \to a} [f(x) - g(x)] = L_1 - L_2$.

4. Use Definition 2.2 to prove the Squeeze Theorem. (*Hint:* For $\varepsilon > 0$ there exist numbers $\delta_1 > 0$ and $\delta_2 > 0$ such that $L - \varepsilon < f(x) < L + \varepsilon$ whenever $0 < |x - a| < \delta_1$ and $L - \varepsilon < h(x) < L + \varepsilon$ whenever $0 < |x - a| < \delta_2$. Let δ be the smaller of δ_1 and δ_2.)

Appendix III

Proof of Taylor's Theorem   Let x be a fixed number in $(a - r, a + r)$ and let the difference between $f(x)$ and the nth degree Taylor polynomial of f at a be denoted by

$$R_n(x) = f(x) - P_n(x)$$

For any t in the interval $[a, x]$ we define

$$F(t) = f(x) - f(t) - \frac{f'(t)}{1!}(x - t) - \frac{f''(t)}{2!}(x - t)^2 - \cdots$$

$$- \frac{f^{(n)}(t)}{n!}(x - t)^n - \frac{R_n(x)}{(x - a)^{n+1}}(x - t)^{n+1} \qquad \text{(III.1)}$$

With x held constant we differentiate F with respect to t using the Product and Power Rules:

$$F'(t) = -f'(t) + \left[f'(t) - \frac{f''(t)}{1!}(x - t) \right]$$

$$+ \left[\frac{f''(t)}{1!}(x - t) - \frac{f'''(t)}{2!}(x - t)^2 \right] + \cdots$$

$$+ \left[\frac{f^{(n)}(t)}{(n - 1)!}(x - t)^{n-1} - \frac{f^{(n+1)}(t)}{n!}(x - t)^n \right]$$

$$+ \frac{R_n(x)(n + 1)}{(x - a)^{n+1}}(x - t)^n$$

for all t in (a, x). Since the last sum telescopes, we obtain

$$F'(t) = -\frac{f^{(n+1)}(t)}{n!}(x - t)^n + \frac{R_n(x)(n + 1)}{(x - a)^{n+1}}(x - t)^n \qquad \text{(III.2)}$$

Now it is evident from (III.1) that F is continuous on $[a, x]$ and that

$$F(x) = f(x) - f(x) - 0 - \cdots - 0 = 0$$

Furthermore,

$$F(a) = f(x) - P_n(x) - R_n(x) = 0$$

Thus $F(t)$ satisfies the hypotheses of Rolle's Theorem (Theorem 4.4) and so there exists a number c between a and x for which $F'(c) = 0$. From (III.2) we obtain

$$R_n(x) = \frac{f^{(n+1)}(c)}{(n + 1)!}(x - a)^{n+1} \qquad \blacksquare$$

Appendix IV Tables

Table I   Trigonometric Functions

Degrees	Radians	sin	tan	cot	cos		
0	0.000	0.000	0.000		1.000	1.571	90
1	0.017	0.017	0.017	57.29	1.000	1.553	89
2	0.035	0.035	0.035	28.64	0.999	1.536	88
3	0.052	0.052	0.052	19.081	0.999	1.518	87
4	0.070	0.070	0.070	14.301	0.998	1.501	86
5	0.087	0.087	0.087	11.430	0.996	1.484	85
6	0.105	0.105	0.105	9.514	0.995	1.466	84
7	0.122	0.122	0.123	8.144	0.993	1.449	83
8	0.140	0.139	0.141	7.115	0.990	1.431	82
9	0.157	0.156	0.158	6.314	.0988	1.414	81
10	0.175	0.174	0.176	5.671	0.985	1.396	80
11	0.192	0.191	0.194	5.145	0.982	1.379	79
12	0.209	0.208	0.213	4.705	0.978	1.361	78
13	0.227	0.225	0.231	4.331	0.974	1.344	77
14	0.224	0.242	0.249	4.011	0.970	1.326	76
15	0.262	0.259	0.268	3.732	0.966	1.309	75
16	0.279	0.276	0.287	3.487	0.961	1.292	74
17	0.297	0.292	0.306	3.271	0.956	1.274	73
18	0.314	0.309	0.325	3.078	0.951	1.257	72
19	0.332	0.326	0.344	2.904	0.946	1.239	71
20	0.349	0.342	0.364	2.747	0.940	1.222	70
21	0.367	0.358	0.384	2.605	0.934	1.204	69
22	0.384	0.375	0.404	2.475	0.927	1.187	68
23	0.401	0.391	0.424	2.356	0.921	1.169	67
24	0.419	0.407	0.445	2.246	0.914	1.152	66
25	0.436	0.423	0.466	2.144	0.906	1.134	65
26	0.454	0.438	0.488	2.050	0.899	1.117	64
27	0.471	0.454	0.510	1.963	0.891	1.100	63
28	0.489	0.469	0.532	1.881	0.883	1.082	62
29	0.506	0.485	0.554	1.804	0.875	1.065	61
30	0.524	0.500	0.577	1.732	0.866	1.047	60
31	0.541	0.515	0.601	1.664	0.857	1.030	59
32	0.559	0.530	0.625	1.600	0.848	1.012	58
33	0.576	0.545	0.649	1.540	0.839	0.995	57
34	0.593	0.559	0.675	1.483	0.829	0.977	56
35	0.611	0.574	0.700	1.428	0.819	0.960	55
36	0.628	0.588	0.727	1.376	0.809	0.942	54
37	0.646	0.602	0.754	1.327	0.799	0.925	53
38	0.663	0.616	0.781	1.280	0.788	0.908	52
39	0.681	0.629	0.810	1.235	0.777	0.890	51
40	0.698	0.643	0.839	1.192	0.766	0.873	50
41	0.716	0.656	0.869	1.150	0.755	0.855	49
42	0.733	0.669	0.900	1.111	0.743	0.838	48
43	0.750	0.682	0.933	1.072	0.731	0.820	47
44	0.768	0.695	0.966	1.036	0.719	0.803	46
45	0.785	0.707	1.000	1.000	0.707	0.785	45
		cos	cot	tan	sin	Radians	Degrees

Table II   Exponential Functions

x	e^x	e^{-x}	x	e^x	e^{-x}
0.00	1.0000	1.0000	2.50	12.182	0.0821
0.05	1.0513	0.9512	2.60	13.464	0.0743
0.10	1.1052	0.9048	2.70	14.880	0.0672
0.15	1.1618	0.8607	2.80	16.445	0.0608
0.20	1.2214	0.8187	2.90	18.174	0.0550
0.25	1.2840	0.7788	3.00	20.086	0.0498
0.30	1.3499	0.7408	3.10	22.198	0.0450
0.35	1.4191	0.7047	3.20	24.533	0.0408
0.40	1.4918	0.6703	3.30	27.113	0.0369
0.45	1.5683	0.6376	3.40	29.964	0.0334
0.50	1.6487	0.6065	3.50	33.115	0.0302
0.55	1.7333	0.5769	3.60	36.598	0.0273
0.60	1.8221	0.5488	3.70	40.447	0.0247
0.65	1.9155	0.5220	3.80	44.701	0.0224
0.70	2.0138	0.4966	3.90	49.402	0.0202
0.75	2.1170	0.4724	4.00	54.598	0.0183
0.80	2.2255	0.4493	4.10	60.340	0.0166
0.85	2.3396	0.4274	4.20	66.686	0.0150
0.90	2.4596	0.4066	4.30	73.700	0.0136
0.95	2.5857	0.3867	4.40	81.451	0.0123
1.00	2.7183	0.3679	4.50	90.017	0.0111
1.10	3.0042	0.3329	4.60	99.484	0.0101
1.20	3.3201	0.3012	4.70	109.95	0.0091
1.30	3.6693	0.2725	4.80	121.51	0.0082
1.40	4.0552	0.2466	4.90	134.29	0.0074
1.50	4.4817	0.2231	5.00	148.41	0.0067
1.60	4.9530	0.2019	6.00	403.43	0.0025
1.70	5.4739	0.1827	7.00	1096.6	0.0009
1.80	6.0496	0.1653	8.00	2981.0	0.0003
1.90	6.6859	0.1496	9.00	8103.1	0.0001
2.00	7.3891	0.1353	10.00	22026.0	0.00005
2.10	8.1662	0.1225			
2.20	9.0250	0.1108			
2.30	9.9742	0.1003			
2.40	11.0232	0.0907			

Table III   Natural Logarithms

n	0.0	0.1	0.2	0.3	0.4	0.5	0.6	0.7	0.8	0.9
0*		7.697	8.391	8.796	9.084	9.307	9.489	9.643	9.777	9.895
1	0.000	0.095	0.182	0.262	0.336	0.405	0.470	0.531	0.588	0.642
2	0.693	0.742	0.788	0.833	0.875	0.916	0.956	0.993	1.030	1.065
3	1.099	1.131	1.163	1.194	1.224	1.253	1.281	1.308	1.335	1.361
4	1.386	1.411	1.435	1.459	1.482	1.504	1.526	1.548	1.569	1.589
5	1.609	1.629	1.649	1.668	1.686	1.705	1.723	1.740	1.758	1.775
6	1.792	1.808	1.825	1.841	1.856	1.872	1.887	1.902	1.917	1.932
7	1.946	1.960	1.974	1.988	2.001	2.015	2.028	2.041	2.054	2.067
8	2.079	2.092	2.104	2.116	2.128	2.140	2.152	2.163	2.175	2.186
9	2.197	2.208	2.219	2.230	2.241	2.251	2.262	2.272	2.282	2.293
10	2.303	2.313	2.322	2.332	2.342	2.351	2.361	2.370	2.380	2.389

*Subtract 10 if $n < 1$; for example, $\ln 0.3 \approx 8.796 - 10 = -1.204$.

ANSWERS TO ODD-NUMBERED PROBLEMS

Exercises 1.1, Page 7

1. $[-4, 20)$ **3.** $(-\infty, -2)$ **5.** $\frac{3}{2} < x < 6$
7. $x \geq 20$ **9.** $[-1, \infty)$ **11.** $[3, 5]$
13. $(-\infty, -3)$ **15.** $(\frac{9}{4}, \infty)$ **17.** $(-\infty, 7]$
19. $[-2, 5)$ **21.** $[-1, 2]$ **23.** $(1, 9)$
25. $[-5, \frac{2}{3}]$ **27.** $(-\infty, \infty)$ **29.** $-4 + a$
31. $a + 10$ **33.** -9 or 9 **35.** $-\frac{19}{5}$ or 5
37. $(-4, 4)$ **39.** $[0, 1]$ **41.** $(-5, -1)$
43. $(-\infty, -6) \cup (6, \infty)$ **45.** $(-\infty, -1) \cup (6, \infty)$
47. No. The solution is $(-\infty, 0) \cup (\frac{1}{4}, \infty)$.
49. $|x - 9| < 4$; $(5, 13)$
51. $|x - 2| > 2$; $(-\infty, 0) \cup (4, \infty)$
53. $4 \leq x \leq 14$ **55.** $<$ **57.** $>$
59. $|x - 4| < 4$ **61.** $|x - \frac{1}{2}| < \frac{7}{2}$

Exercises 1.2, Page 18

1. Fourth quadrant **3.** Third quadrant
5. Second quadrant **7.** $2\sqrt{5}$ **9.** 3
11. Not a right triangle **13.** Collinear
15. $x = 4$ or $x = -2$ **17.** $x^2 = 4y$
19. $(6, 2)$ **21.** $(5, 5)$ **23.** $(3, 0), (5, -6)$
25. No symmetry;

27. Symmetry with respect to x-axis;

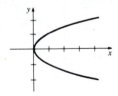

29. Symmetry with respect to y-axis;

31. Symmetry with respect to x- and y-axes, and the origin;

33. **35.**

37. $(x - 4)^2 + (y + 6)^2 = 64$
39. $(x - 3)^2 + (y + 4)^2 = 25$
41. $(x - 1)^2 + (y - 1)^2 = 17$
43. Circle, $C(-4, 3)$, $r = 5$
45. Circle, $C(3, -1)$, $r = \sqrt{28/3}$
47. Equation does not describe a circle but rather the single point $(6, -4)$.

49. 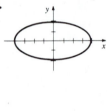 **51.**

53. $\dfrac{(x - 2)^2}{4^2} + \dfrac{(y + 5)^2}{2^2} = 1$, $C(2, -5)$
55. The graph consists of the x- and y-axes.
57.

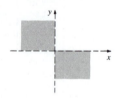

59. $(x + 9)^2 + (y - 3)^2 = 100$
$(x - 9)^2 + (y + 3)^2 = 100$

Exercises 1.3, Page 25

1. $-\frac{3}{2}$ **3.** -3 **5.** $P_2(7, 3)$
7. $y = -x + 3$ **9.** $y = 3x$ **11.** $y = x + 8$
13. $y = -\frac{3}{2}$ **15.** $y = -2x + 4$
17. $y = 4x$ **19.** $x = -3x + 9$
21. $y = -x + 7$ **23.** $y = -x + 6$
25. $2, -\frac{3}{2}, 3$ **27.** 0, none, $\frac{5}{2}$

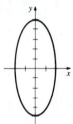

29. $-\frac{5}{3}, 3, 5$

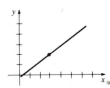

31. $\frac{3}{8}, \frac{10}{3}, -\frac{5}{4}$

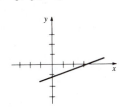

33.

35. $-\frac{2}{5}$ **37.** $-\frac{3}{4}$ **39.** $-\frac{3}{7}$
41. Collinear   **43.** $(63/16, 31/8)$
49. $x/5 + y/2 = 1$ **51.** $18\sqrt{13}/13$
53. $17\sqrt{37}/37$

Exercises 1.4, Page 38

1. $12, 0, 3 - \sqrt{3}, a^2 + a$ **3.** 6
5. $3a^2 + 3ah + h^2$ **7.** $[-1, \infty)$
9. $(0, \infty)$ **11.** $[-5, 5]$
13. The set of real numbers except 4
15. The set of real numbers except 0 and $\frac{1}{2}$
17. $[0, \infty)$ **19.** $(-\infty, 3)$ **21.** $[1, \infty)$
23. $(-\infty, 4]$ **25.** A function
27. Not a function   **29.** $P = 4\sqrt{A}$
31. $A = (\sqrt{3}/4)s^2$ **33.** $V = A^{3/2}$
35. $d = 20\sqrt{13t^2 + 8t + 4}$ **37.** $V = (\pi/27)h^3$
39. $A = (-1 + \sqrt{5})x^2/2$
41. $1, 3, 3.9, 1 + 2h + h^2$ **43.** $5, -4$
45. Even   **47.** Neither even nor odd
49. Odd   **51.** Odd

53.

55.

57.

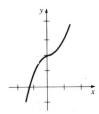

59.

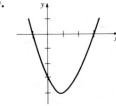

61.

63.

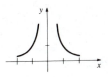

65.

67.

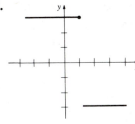

69.

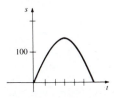

71. $-2, 4; -8$ **73.** $-3, 1, 3, 6; -54$
75. $-\frac{2}{5}, \frac{2}{5}; -\frac{4}{3}$ **77.** $[1, \infty), [-1, \infty)$
79. 123.9 long tons   **81.** $T_f = \frac{9}{5}T_c + 32$
83. $t = 0$ and $t = 6$

85. 50 g   **89.** $\Delta x = 5, \Delta y = 1$

Exercises 1.5, Page 45

1. $-2x + 13; 6x - 3; -8x^2 - 4x + 40;$
$(2x + 5)/(-4x + 8), x \neq 2$
3. $3x^2 + 4x^3; 3x^2 - 4x^3; 12x^5; 3/(4x), x \neq 0$
5. $\dfrac{x^2 + x + 1}{x(x + 1)}; \dfrac{x^2 - x - 1}{x(x + 1)}; \dfrac{1}{x + 1}; \dfrac{x^2}{x + 1},$
$x \neq 0$ and $x \neq -1$
7. $2x^2 + 5x - 7; -x + 1; x^4 + 5x^3 - x^2 - 17x + 12;$
$(x + 3)/(x + 4), x \neq 1$ and $x \neq -4$
9. $5x^3 - 3x^2; -3x^3 + 5x^2; 4x^6 - 4x^4;$
$(x + 1)/(4x - 4), x \neq 0$ and $x \neq 1$
11. $3x + 16; 3x + 4$
13. $4x^2 + 1; 16x^2 + 8x + 1$
15. $(3x + 3)/x; 3/(3 + x)$
17. $-1/x; -(x + 5)/(2x + 1)$ **19.** $x; x$
21. $4x^6; 128x^9; 1/(4x^9); 1/(16x^9)$
23. $4/x^4; x^4/8; 8/x^4; 2/x^4$
25. $-2; 2; -24; 24$
27. $10; 353/64; 1153; 19$
29. $[0, 5]; [0, 5); [-5, 5]$
31. $x \neq 0; x \neq 0, x \neq \pm 1; x \neq \pm 1$
33. $f(x) = 2x^2 - x, g(x) = x^2$
35. $f(x) = x^2 + 6\sqrt{x}, g(x) = x - 1$
37. $36x^2 - 36x + 15$
39. $f(x) = \sqrt{x}, g(x) = x + 2, h(x) = x^3$

41. $g(x) = -2x + 9$

43.

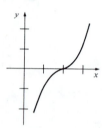

45.

47.

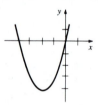

49.

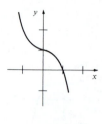

51. $y = |x - \frac{3}{2}|$ **53.** $y = -(x - 2)^2 + 3$

55. $y = (x + 2)^2 - 4$;

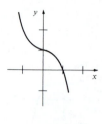

57. False

59. $f + g$ and $f - g$ are odd, fg and f/g are even.

61.

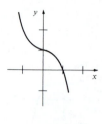

63. $y = 2 - 3U(x - 2) + U(x - 3)$

65. $(f + g)(x) = \begin{cases} x + x^2, & x \le -1 \\ 2x - 5, & -1 < x < 0 \\ 2x - 3, & x \ge 0 \end{cases}$

$(fg)(x) = \begin{cases} x^3, & x \le -1 \\ x^2 - 5x, & -1 < x < 0 \\ x^2 - 3x - 10, & x \ge 0 \end{cases}$

Exercises 1.6, Page 54

1. $9°$ **3.** $330°$ **5.** $-240°$ **7.** $7\pi/6$

9. $5\pi/3$ **11.** $-5\pi/6$ **13.** $-\frac{1}{2}$

15. $-\sqrt{3}/2$ **17.** $2/\sqrt{3}$ **19.** $-1/\sqrt{3}$

21. $1/\sqrt{3}$ **23.** 0

25. $\tan t = -2$

$\cot t = -\frac{1}{2}$

$\sec t = \sqrt{5}$

$\csc t = -\sqrt{5}/2$

27. $\cos t = \sqrt{15}/4$

$\tan t = -1/\sqrt{15}$

$\cot t = -\sqrt{15}$

$\sec t = 4/\sqrt{15}$

$\csc t = -4$

29. $\sin t = -1/\sqrt{2}$

$\cos t = -1/\sqrt{2}$

$\tan t = 1$

$\cot t = 1$

$\sec t = -\sqrt{2}$

31. $\pi/4$, $\pi/2$, $3\pi/2$, $7\pi/4$ **33.** $(\sqrt{6} - \sqrt{2})/4$

35. $-\sqrt{2 - \sqrt{2}}/2$

43. For all values of t, $|\sin t| \le 1$ and $|\cos t| \le 1$

45.

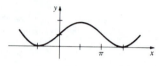

47.

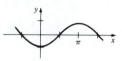

49.

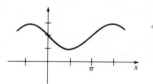

51.

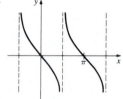

53.

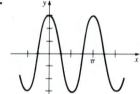

55.

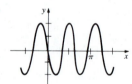

57.

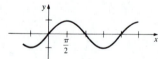

59. $y = 2 + \sin 4x$ **61** $y = \sin(x - \pi/6)$

63. $(1 + 4x)\cos x$; $(1 + 4x)/\cos x$; $1 + 4\cos x$; $\cos(1 + 4x)$

65. $\sin x \cos x$; $\tan x$; $\sin(\cos x)$; $\cos(\sin x)$

67. $\sqrt{x} + \sqrt{x} \cos 2x$; $(1 + \cos 2x)/\sqrt{x}$; $1 + \cos(2\sqrt{x})$; $\sqrt{1 + \cos 2x}$

69. $t = 9$ and $t = 21$; $80°$ occurs at $t = 15$; $60°$ occurs at $t = 3$

71. Even   **73.** Odd   **75.** Even

79.

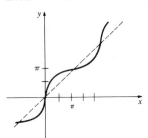

81.

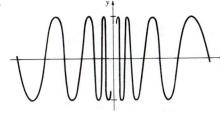

Chapter 1 Review Exercises, Page 56

1. True   **3.** True   **5.** True   **7.** False

9. False   **11.** True   **13.** False

15. True   **17.** $-a$ **19.** $-12, 9$

21. -27 **23.** Fourth   **25.** $(0, 10]$

27. -7 and 5 **29.** $4x - 24, 8x - 56$

31. $50°$ **33.** $y = 11$ **35.** $y = -2x - 2$

37. $(x - h)^2 + (y - h)^2 = 2h^2$

39. x-intercepts: $-6\sqrt{2}, 6\sqrt{2}$
y-intercepts: $-2, 4$
center: $(0, 1)$

41. $V = 6l^3, V = 2w^3/9, V = 3h^3/4$ **43.** $34,000

45. First; $\tan t = \frac{3}{4}$, $\cot t = \frac{4}{3}$, $\sec t = \frac{5}{4}$, $\csc t = \frac{5}{3}$;
$\frac{24}{25}, \frac{7}{25}$

47. $\pi/3$ and $-\pi/3$;   **49.**
$\pi/3$ and $2\pi/3$

51.

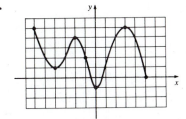

53. $f(-4) = 3, f(-3) = 0$
$f(-2) = -2, f(-1) = 0$
$f(0) = 2.5, f(1) = 2$
$f(1.5) = 0.5, f(2) = 0$
$f(3.5) = 3, f(4) = 4$

55. $A = (\frac{16}{3})x\sqrt{9 - x^2}$

Exercises 2.1, Page 66

1. 8   **3.** Does not exist   **5.** 2

7. Does not exist   **9.** 3   **11.** 0

13. Does not exist; 2; does not exist

15. 3; 3; 3   **17.** 0; 2; does not exist; 5

19. Does not exist; 0; 0; 1   **21.** Does not exist

23. The correct statement should be $\lim_{x \to 0^+} \sqrt{x} = 0$.

25. Since $\lim_{x \to 0^-} \sqrt[3]{x} = \lim_{x \to 0^+} \sqrt[3]{x} = 0$, the given
statement is correct.

27. $-\frac{1}{4}$ **29.** $\frac{1}{3}$ **31.** $\frac{1}{4}$ **33.** 5

Exercises 2.2, Page 75

1. 17   **3.** -12 **5.** 4   **7.** 14

9. $\frac{28}{9}$ **11.** $\sqrt{7}$ **13.** Does not exist

15. 3   **17.** 1   **19.** $-\frac{3}{2}$ **21.** $-\frac{1}{8}$

23. 3   **25.** Does not exist   **27.** 2

29. $\frac{128}{3}$ **31.** Does not exist   **33.** -2

35. $a^2 - 2ab + b^2$ **37.** $2x$ **39.** $-1/x^2$

41. $\frac{1}{2}$ **43.** $\frac{1}{5}$ **45.** 12   **49.** 3

Exercises 2.3, Page 83

1.

VA: $x = -3$
HA: $y = 0$

3.

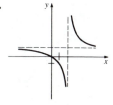

VA: $x = 2$
HA: $y = 1$

5.

VA: $x = 2$
HA: $y = 0$

7.

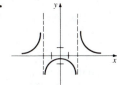

VA: $x = -2$ and $x = 2$
HA: $y = 0$

9.

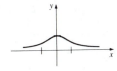

VA: None
HA: $y = 0$

11.

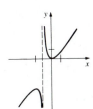

VA: $x = -1$
HA: None

13.

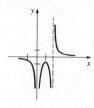

15.

VA: $x = 0$ and $x = 2$ VA: $x = 1$
HA: $y = 0$ HA: $y = 1$

17. 1; 3; does not exist; 2; 1; 3   **19.** 2; $-\infty$; 0; 2
21. $-\infty$; $-\frac{3}{2}$; ∞; 0   **23.** $-\infty$ **25.** ∞
27. ∞ **29.** ∞ **31.** $\frac{1}{4}$ **33.** 5
35. 0   **37.** $-\frac{1}{4}$ **39.** $\frac{5}{2}$ **41.** 4
43. $-2/\sqrt{3}$ **45.** $1/\sqrt{2}$ **47.** 0
49. -1
51.

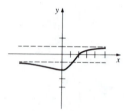

53.

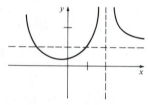

55.

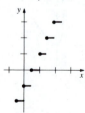

57. $m \to \infty$ **59.** 2   **61.** 1
63. ∞; approximately 2.72; 1   **65.** 0
67. ∞ **69.** ∞

Exercises 2.4, Page 91

1. None   **3.** 3 and 6
5. $n\pi/2$, $n = 0, \pm 1, \pm 2, \ldots$ **7.** 2   **9.** 0
11. Continuous; continuous
13. Continuous; continuous
15. Not continuous; not continuous
17. Continuous; not continuous
19. Not continuous; not continuous
21. Not continuous; continuous   **23.** $m = 4$
25. $m = 1$, $n = 3$
27.

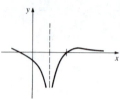

Discontinuous at $n/2$, where n is any integer
29. $\sqrt{3}/2$ **31.** 0   **33.** 1   **35.** 1

37. $(-3, \infty)$ **39.** $c = 4$
41. $c = -\sqrt{2}$, $c = 0$, $c = \sqrt{2}$
55. 0, $\pm\pi/2$, $\pm 3\pi/2$, $\pm 2\pi, \ldots$
57. Define $f(9) = 6$.

Exercises 2.5, Page 100

1. Choose $\delta = \varepsilon$.   **3.** Choose $\delta = \varepsilon$.
5. Choose $\delta = \varepsilon$.   **7.** Choose $\delta = \varepsilon/3$.
9. Choose $\delta = 2\varepsilon$.   **11.** Choose $\delta = \varepsilon$.
13. Choose $\delta = \varepsilon/8$.   **15.** Choose $\delta = \sqrt{\varepsilon}$.
17. Choose $\delta = \varepsilon^2/5$.   **19.** Choose $\delta = \varepsilon/2$.
25. Since $2 < x < 4$ is equivalent to $|x - 3| < 1$,
choose δ to be the minimum of 1 and $\varepsilon/7$.
27. Choose $\delta = \sqrt{a\varepsilon}$.   **29.** Choose $N = 7/4\varepsilon$.
31. Choose $N = -30/\varepsilon$.

Chapter 2 Review Exercises, Page 101

1. True   **3.** False   **5.** True   **7.** False
9. False   **11.** True   **13.** False
15. True   **17.** True   **19.** 4   **21.** $-\frac{1}{5}$
23. 3^- **25.** $-\infty$ **27.** -2 **29.** 10
31. $\frac{1}{6}$ **33.** 26   **35.** $\frac{1}{2}$
37. Does not exist   **39.** 4   **41.** $2a$
43.

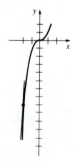

f is continuous at every real number.
45. $[-2, 1)$ and $(1, 2]$ **47.** Yes
49. Vertical asymptotes: $x = -\frac{5}{2}$, $x = \frac{5}{2}$, $x = -9$;
horizontal asymptote: $y = \frac{1}{2}$ **51.** 1; no

Exercises 3.1, Page 113

1. -4.5;   **3.** 7;

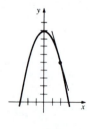

5. $(3\sqrt{3} - 6)/\pi$;

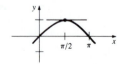

7. 2   **9.** 6   **11.** 8   **13.** 3   **15.** −9
17. $-2/x_0^3$; $\frac{1}{4}$; −54   **19.** $y = -10x + 26$
21. $y = 6x$ **23.** $y = -2x + 1$ **25.** 1
27. Possible horizontal tangents at $x = 1$ and $x = 3$
29. Possible horizontal tangents at $x = -2$ and $x = 2$;
possible vertical tangents at $x = 5$; tangent does not exist
at $x = 6$
31. The line is not tangent to the graph at the point.

35. $(\sqrt{3\,\Delta x + 1} - 1)/\Delta x$;

0.2649	1.3246
0.1402	1.4018
0.0149	1.4889
−0.3675	1.8377
−0.1633	1.6334
−0.0151	1.5114

; 1.5

37. 58 mph   **39.** 3.8 hr   **41.** −14
43. −4.9 m/sec; 5 sec; −49 m/sec

Exercises 3.2, Page 123

1. 0   **3.** −3   **5.** $6x$ **7.** $8x - 1$
9. $2x + 2$ **11.** $3x^2 + 1$
13. $-3x^2 + 30x - 1$ **15.** $-1/x^2$
17. $-1/(x - 1)^2$ **19.** $-1/x^2 - 2/x^3$
21. $y = -x - 4$ **23.** $y = 2x - 2$
25. $(-4, -6)$
27. $1/(2\sqrt{x})$; $(0, \infty)$; By Theorem 2.8

$f'_+(0) = \lim\limits_{\Delta x \to 0^+} \dfrac{1}{\sqrt{\Delta x}}$ does not exist.

29. $(2x + 1)^{-1/2}$; $(-\frac{1}{2}, \infty)$
31. $\frac{1}{3}(x - 4)^{-2/3}$; $(-\infty, 4) \cup (4, \infty)$
35. $f'_-(2) = -1$, whereas $f'_+(2) = 2$
37. f is differentiable at $x = 0$ and $f'(0) = 0$
39. The graph of a horizontal line $f(x) = k$ has zero
slope ($f'(x) = 0$).
41.

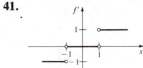

43.

47. $\frac{3}{2}a^{1/2}$

Exercises 3.3, Page 130

1. $9x^8$ **3.** 0   **5.** $14x - 4$
7. $x^4 - 12x^3 + 18x$ **9.** $20x^4 - 20x^3 - 18x^2$
11. $2x + 2$ **13.** $192u^2$ **15.** $18z^2 + 2az$

17. $-12\beta^3 + 21\beta^2 - 10\beta$ **19.** $6x^5 + 40x^3 + 50x$
21. $y = 6x + 3$ **23.** $y = 20x - 56$
25. $(4, -11)$ **27.** $(3, -25), (-1, 7)$
29. $y = \frac{1}{4}x - \frac{7}{2}$ **31.** $x = 4$ **33.** $(2, 8)$
35. $(\frac{1}{4}, -\frac{3}{16})$ **37.** $(-1, -4), (-5, 20)$
39. $y = \frac{1}{3}x + \frac{2}{27}$, $y = \frac{1}{3}x - \frac{2}{27}$
41. On graph of f, $(-\frac{3}{2}, \frac{3}{4})$; on graph of g, $(-\frac{3}{2}, -\frac{1}{2})$
43. $b = 8$, $c = -9$ **45.** $S = 4\pi r^2$
47. −15 nt

Exercises 3.4, Page 136

1. $-1/x^2$ **3.** $-10(5x)^{-3}$ **5.** $12x - 2x^{-3}$
7. $5x^4 - 9x^2 + 4x - 28$

9. $8 + \dfrac{8}{x^3} + \dfrac{3}{x^4}$ **11.** $\dfrac{-20x}{(x^2 + 1)^2}$

13. $\dfrac{-17}{(2x - 5)^2}$ **15.** $72x - 12$

17. $\dfrac{t^2 + 2t}{(2t^2 + t + 1)^2}$ **19.** $18z^2 + 22z + 6$

21. $\dfrac{6x^2 + 8x - 3}{(3x + 2)^2}$ **23.** $\dfrac{4x^3 + 9x^2 + 1}{(3x^3 + x)^2}$

25. $-\dfrac{1}{u^2} - \dfrac{2}{u^3} - \dfrac{3}{u^4} - \dfrac{4}{u^5}$

27. $\dfrac{2x^3 + 8x^2 - 6x - 8}{(x + 3)^2}$

29. $\dfrac{-6}{(3x + 1)^3}$ **31.** $6, x \neq 0$

33. $y = -16x + 12$ **35.** $y = -20x - 12$
37. $(0, 24), (-\sqrt{5}, -1), (\sqrt{5}, -1)$
39. $(-1, \frac{1}{2}), (0, 0), (1, \frac{1}{2})$ **41.** $(-\frac{1}{4}, -16), (\frac{1}{4}, 16)$
43. $(3, \frac{3}{2}), (-5, \frac{1}{2})$ **45.** −6   **47.** $\frac{11}{5}$
49. $x < 0$ **51.** $x > -\frac{9}{4}$ **53.** $-16km_1m_2$
57. $2f(x)f'(x)$; $3[f(x)]^2f'(x)$; $n[f(x)]^{n-1}f'(x)$

Exercises 3.5, Page 143

1. $\frac{3}{2}$ **3.** 0   **5.** 1   **7.** 4   **9.** 0
11. 36   **13.** $\frac{1}{2}$ **15.** Does not exist
17. 3   **19.** $\frac{3}{7}$ **21.** 0   **23.** 4   **25.** $\frac{1}{2}$
29.

0.0157	0.0167	0.0173	0.0174	0.0175	0.0175

;

0.0175; the limit appears to be $\pi/180$
31. $2x + \sin x$ **33.** $7 \cos x - \sec^2 x$
35. $x \cos x + \sin x$ **37.** $\cos 2x$
39. $x^2 \sec x \tan x + 2x \sec x + \sec^2 x$ **41.** $\cos x$

43. $-\dfrac{\csc^2 x + x \csc^2 x + \cot x}{(x + 1)^2}$

45. $\dfrac{4x \tan x - 2x^2 \sec^2 x + 2x}{(1 + 2 \tan x)^2}$ **47.** $\dfrac{1}{1 + \cos \theta}$

49. $2 \sin u \cos u$ **51.** 0

53. $x^2 \sin x \sec^2 x + x^2 \sin x + 2x \sin x \tan x$

55. $(\pi/2, \pi/2)$ **57.** $3\sqrt{3}x + 6y = \sqrt{3}\pi + 3$

59. $6\sqrt{3}x - 9\sqrt{3}y = \sqrt{3}\pi - 18$

61. $12x - 6y = 16\pi + 3\sqrt{3}$

63. $y = x - 2\pi$ **69.** $2 \cos 2x$

Exercises 3.6, Page 150

1. $-150(-5x)^{29}$ **3.** $200(2x^2 + x)^{199}(4x + 1)$

5. $-4(x^3 - 2x^2 + 7)^{-5}(3x^2 - 4x)$

7. $-10(3x - 1)^4(-2x + 9)^4 + 12(-2x + 9)^5(3x - 1)^3$

9. $3 \sin^2 x \cos x$ **11.** $\dfrac{8x(x^2 - 1)}{(x^2 + 1)^3}$

13. $10[x + (x^2 - 4)^3]^9[1 + 6x(x^2 - 4)^2]$

15. $4(t^{-1} + t^{-2} + t^{-3})^{-5}(t^{-2} + 2t^{-3} + 3t^{-4})$

17. $-3(2 + u \sin u)^{-4}(u \cos u + \sin u)$

19. $\dfrac{(2v - 5)^3(-6v^2 + 45v - 5)}{(v + 1)^9}$

21. $\pi \cos(\pi x + 1)$ **23.** $8 \sin 4x \cos 4x$

25. $-3x^5 \sin x^3 + 3x^2 \cos x^3$

27. $\left(\sec^2 \dfrac{1}{x}\right)\left(-\dfrac{1}{x^2}\right)$

29. $\dfrac{5 \cos 6\theta \cos 5\theta + 6 \sin 5\theta \sin 6\theta}{\cos^2 6\theta}$

31. $5(\sec 4x + \tan 2x)^4(4 \sec 4x \tan 4x + 2 \sec^2 2x)$

33. $\cos(\sin x)\cos x$

35. $24x \sin^2(4x^2 - 1)\cos(4x^2 - 1)$

37. -54 **39.** -7 **41.** $y = -8x - 3$

43. $12x - 2y = 3\pi + 2$

45. $(-\sqrt{3}/3, -3\sqrt{3}/16), (\sqrt{3}/3, 3\sqrt{3}/16)$; no

47. $\theta = (2n + 1)\pi/4, n = 0, 1, 2, \ldots$

49. $5/8\pi$ in./min   **51.** $(3u^2 + 2)(9x^8 + 8x)$

53. $3F'(3x)$ **55.** $-10/(-10x + 7)$

59. $f'(g(h(x)))g'(h(x))h'(x)$

Exercises 3.7, Page 154

1. -2 **3.** 32 **5.** $60x^{-4}$

7. $6x + 16 - 40x^{-6}$

9. $(3x - 4)(180x^2 - 192x + 32)$

11. $-14(t + 2)^{-3}$ **13.** $-100 \cos 10x$

15. $-x \sin x + 2 \cos x$

17. $\dfrac{4 \cos^2\theta + 6 \cos \theta + 8 \sin^2\theta}{(3 + 2 \cos \theta)^3}$

19. $\sec^3 x + \tan^2 x \sec x$ **21.** $1440x^2 + 120x$

23. $-\pi^3 \cos \pi x$ **25.** $n!$ **27.** $(-4, 48)$

29. $(1, \infty); (-\infty, 1)$ **31.** $y = -7x$ **33.** $\frac{1}{18}$

35. -32

43. $2; \kappa \to 0$ as $x \to \infty$ implies that the graph becomes close to linear for very large values of x.

Exercises 3.8, Page 159

1. $1/(2y - 2)$ **3.** $(2x - y^2)/(2xy)$

5. $(y + 1)/(2y - x)$

7. $(4x - 3x^2y^2)/(2x^3y - 2y)$

9. $[x^2 - 4x(x^2 + y^2)^5]/[y^2 + 4y(x^2 + y^2)^5]$

11. $(2x^4y^4 + 3y^{10} - 6x^9y)/(6xy^9 - 3x^{10})$

13. $(1 - x)/(y + 4)$ **15.** $3/[2y(x + 2)^2]$

17. $[\cos(x + y) - y]/[x - \cos(x + y)]$

19. $\cos y \cot y$ **21.** $(\cos 2\theta)/r$ **23.** $-\frac{2}{5}$

25. $-\frac{1}{3}$ and $-\frac{2}{3}$ **27.** $-8x + 3y = 22$

29. $2x - 4y = 2 - \pi$ **31.** $(1, 2), (-1, -2)$

33. $(-\sqrt{5}, 2\sqrt{5}), (\sqrt{5}, -2\sqrt{5})$

35. $(3 - x)/(y + 4); (x - 3)^2 + (y + 4)^2 = -2$ does not describe a real locus of points

37. $(y^3 - 2x^2)/y^5$ **39.** $-(\sin y)/(1 - \cos y)^3$

41. $-2/(y - x)^3$

43. $y' = (1 - 3x^2y)/x^3$ is equivalent to $y' = -2x^{-3} - 3x^{-4}$

45. $y = 1 - \sqrt{x - 2}$

47. $y = \begin{cases} \sqrt{4 - x^2}, & -2 \le x < 0 \\ -\sqrt{4 - x^2}, & 0 \le x \le 2 \end{cases}$

49. $-(x/y) \, dx/dt$ **51.** $\pi/4$

Exercises 3.9, Page 164

1. $15x^{1/2}$ **3.** $-\frac{4}{3}x^{-7/3}$ **5.** $\frac{1}{2}x^{-1/2} - \frac{1}{2}x^{-3/2}$

7. $\frac{4}{3}(x^3 + x)(x^2 - 4)^{-1/3} + 2x(x^2 - 4)^{2/3}$

9. $\dfrac{33}{2} \dfrac{(x - 9)^{1/2}}{(x + 2)^{5/2}}$ **11.** $\frac{1}{2}x^{-1/2}\cos \sqrt{x}$

13. $1 + x(x^2 + 1)^{-1/2}$

15. $\frac{1}{3}[(t^2 - 1)(t^3 + 4t)]^{-2/3}(5t^4 + 9t^2 - 4)$

17. $-\frac{1}{2}(\theta + \sin \theta)^{-3/2}(1 + \cos \theta)$

19. $\frac{8}{5}s^3(s^4 + 1)^{-3/5}$ **21.** $-\frac{1}{4}x^{-3/2}$

23. $-\frac{2}{9}x^{-4/3} - \frac{4}{3}x^{-5/3} - 10x^{-8/3}$

25. $-2.4(\sin \theta)^{2.4} + 3.36(\sin \theta)^{0.4} \cos^2\theta$

27. $x - 12y = -16$ **29.** $7x - 8y = 3$

31. $-(y + 2\sqrt{xy})/(x + 2\sqrt{xy})$

33. $y^{-1/2}(x^2 - 1)^{3/2}(6x^2 - 1)$

35. $8\sqrt{3}x + 8y = 4\sqrt{3}, 8\sqrt{3}x - 8y = 4\sqrt{3}$

37. Vertical tangent at $(4, 0)$

39. No vertical tangents

41. Vertical tangent at $(0, 1)$ **43.** $\pi/8$

45. $-ky^{-2}(2k/y - 2k/R + v_0^2)^{-1/2}$

49. $(0.22W^{-1/3})/3$

Exercises 3.10, Page 169

1. $dy = 0$ **3.** $dy = -(2x)^{-3/2} \, dx$

5. $dy = 16x^3(x^4 - 1)^{-2/3} \, dx$

7. $dy = \dfrac{4x}{(x^2 + 1)^2} \, dx$ **9.** $dy = -x \sin x \, dx$

11. $\Delta y = 2x \, \Delta x + (\Delta x)^2, dy = 2x \, dx$

13. $\Delta y = 2(x + 1) \, \Delta x + (\Delta x)^2, dy = 2(x + 1) \, dx$

15. $\Delta y = -\dfrac{\Delta x}{x(x + \Delta x)}, dy = -\dfrac{1}{x^2} \, dx$

17. $\Delta y = \cos x \sin \Delta x + \sin x(\cos \Delta x - 1)$,
$dy = \cos x\, dx$

19.

25	20	5
11.25	10	1.25
2.05	2	0.05
0.2005	0.2	0.0005

21. 6.0833 **23.** 16 **25.** 0.325

27. 0.4 **29.** $\dfrac{1}{2} + \dfrac{\sqrt{3}\pi}{120} \approx 0.5453$

31. 9π; 8π

33. Exact volume is $\Delta V = \frac{4}{3}\pi(3r^2t + 3rt^2 + t^3)$, approximate volume is $dV = 4\pi r^2 t$, where $t = \Delta r$; $(0.1024)\pi$ in^3

35. ± 6 cm^2; ± 0.06; $\pm 6\%$ **39.** 2048 ft; 160 ft

41. 2.8957, 0.1532, 15.3246% **43.** 0.01 sec

Exercises 3.11, Page 176

1. One real root **3.** No real roots **5.** 3.1623

7. 1.5874 **9.** 0.6823 **11.** $-1.1414, 1.1414$

13. 0, 0.8767 **15.** 2.4981 **17.** 1.6560

19. 0.7297

Chapter 3 Review Exercises, Page 177

1. True **3.** False **5.** True **7.** False

9. False **11.** True **13.** True

15. False **17.** 0 **19.** 3; 27

21. $(x^3)^2 \cdot 3x^2$ **23.** $\sin x$ **25.** -3

27. 0 **29.** 18

31. $x = \sqrt{n\pi}$, $n = 0, 1, 2, \ldots$

33. $y = 6x - 9$, $y = -6x - 9$

35. 0.2448 **37.** $2y^{1/3}/3x^{5/3}$

39. $-(16x \sin 4x + 4 \sin 4x + 4 \cos 4x)/(4x + 1)^2$

41. $2 - 2x^{-3} + 2x$

43. $10(t + \sqrt{t^2 + 1})^9(1 + t(t^2 + 1)^{-1/2})$

45. $0.08x^{-0.9}$

47. $x^2(x^4 + 16)^{1/4}(x^3 + 8)^{-2/3} + x^3(x^4 + 16)^{-3/4}(x^3 + 8)^{1/3}$

49. $\dfrac{2 + x^{1/3}}{3x^{1/3}(1 + x^{1/3})^2}$ **51.** $\frac{405}{8}(3x)^{-1/2}$

53. $2 + 6t^{-4}$ **55.** $\pi/3, 2\pi/3$

57. $(-16 \sin 4x)F'(\sin 4x) + (16 \cos^2 4x)F''(\sin 4x)$

59. 1.6751 **61.** 0.0915

Exercises 4.1, Page 184

1. $-1, -2, 2, 8$; 19, 18, 18, 8

3. $18, -23, 23, 18$; 6, 1, 1, -6

5. $-\frac{15}{4}, 17, 17, -128$; 0, 2, 2, -2

7. $1, 1 - \pi, \pi - 1, 0$; $\frac{1}{2}, 1, 1, \pi^2$

9. 6 and -6; 8 and -8 **11.** $6\sqrt{2}, -6\sqrt{2}$; 15

13. $(-\infty, -3), (0, 3)$ **15.** $v(t) = 2t$, $a(t) = 2$

17. $v(t) = 2t - 4$, $a(t) = 2$

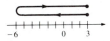

19. $v(t) = 6t^2 - 12t$, $a(t) = 12t - 12$

21. $v(t) = 12t^3 - 24t^2$, $a(t) = 36t^2 - 48t$

23. $v(t) = 1 - 2t^{-1/2}$, $a(t) = t^{-3/2}$

25.

positive	negative
zero	zero
positive	positive
positive	negative
negative	negative
negative	positive

27. $v > 0$ on $(-\infty, \frac{3}{2})$, $v < 0$ on $(\frac{3}{2}, \infty)$; 36 ft

29. $64\sqrt{2}$ ft/sec; 16 ft/sec^2

31. $-8\sqrt{\pi}$ ft/sec; the y-coordinate is decreasing

Exercises 4.2, Page 191

1. $\dfrac{dV}{dt} = 3x^2\dfrac{dx}{dt}$ **3.** $8\sqrt{3}$ cm^2/hr

5. $\frac{4}{3}$ in./min

7. $s \cos \theta \dfrac{d\theta}{dt} + \sin \theta \dfrac{ds}{dt} = \dfrac{dx}{dt}$

9. 24 cm/min **11.** -6 or 6 **13.** $\frac{4}{9}$ cm^2/hr

15. $\frac{35}{4}$ lb/in.^{2}/sec **17.** $5/(32\pi)$m/min

19. $\sqrt{3}/10$ ft/min

21. $-1/(4\pi)$ ft/min; $-1/(12\pi)$ ft/min

23. $-\sqrt{2}/2$ ft/min **25.** -360 mi/hr

27. $8\pi/9$ km/min **29.** $\pi/12$ km/sec

33. 17 knots **35.** $\dfrac{dh}{dt} = \dfrac{16}{\pi(12 - h)^2}$

Exercises 4.3, Page 201

1. $\frac{3}{2}$ **3.** $-1, 6$ **5.** $2, \frac{3}{4}$ **7.** 1

9. $\frac{3}{4}$ **11.** $-2, -\frac{11}{7}, 1$

13. $2n\pi$, $n = 0, \pm 1, \pm 2, \ldots$

15. Abs. max. $f(3) = 9$;
abs. min. $f(1) = 5$

17. Abs. max. $f(8) = 4$;
abs. min. $f(0) = 0$

19. Abs. max. $f(0) = 2$;
abs. min. $f(-3) = -79$

21. Abs. max. $f(3) = 8$;
abs. min. $f(-4) = -125$

23. Abs. max. $f(2) = 16$;
abs. min. $f(0) = f(1) = 0$

25. Abs. max. $f\left(\dfrac{\pi}{6}\right) = f\left(\dfrac{5\pi}{6}\right) = \dfrac{3}{2}$;

abs. min. $f\left(\dfrac{\pi}{2}\right) = f\left(\dfrac{3\pi}{2}\right) = -3$

27. Endpoint abs. max. $f(3) = 3$;
rel. max. $f(0) = 0$;
abs. min. $f(-1) = f(1) = -1$

29. c_1, c_3, c_4, c_{10}; $c_2, c_5, c_6, c_7, c_8, c_9$;
abs. min. $f(c_7)$, endpoint abs. max. $f(b)$;
rel. maxima $f(c_3), f(c_5), f(c_9)$;
rel. minima $f(c_2), f(c_4), f(c_7), f(c_{10})$

31.

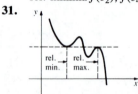

35. $s(t) \geq 0$ only for $0 \leq t \leq 20$;
$s(10) = 1600$

37. n an even positive integer

39. $f(-a)$ is a rel. min.

Exercises 4.4, Page 208

1. $c = 0$ **3.** $f(-3) = 0$ but $f(-2) \neq f(-3)$

5. $c = -\dfrac{2}{3}$ **7.** $c = -\pi/2, \pi/2,$ or $3\pi/2$

9. f is not differentiable on the interval.

11. $f(a) \neq 0$ and $f(b) = 0$ so $f(a) \neq f(b)$

13. $c = 3$ **15.** $c = \sqrt{13}$

17. f is not continuous on the interval.

19. $c = \dfrac{9}{4}$ **21.** $c = 1 - \sqrt{6}$

23. f is not continuous on $[a, b]$.

25. Increasing on $[0, \infty)$, decreasing on $(-\infty, 0]$

27. Increasing on $[-3, \infty)$, decreasing on $(-\infty, -3]$

29. Increasing on $(-\infty, 0]$ and $[2, \infty)$, decreasing on $[0, 2]$

31. Increasing on $[3, \infty)$, decreasing on $(-\infty, 0]$ and $[0, 3]$

33. Decreasing on $(-\infty, 0]$ and $[0, \infty)$

35. Increasing on $(-\infty, -1]$ and $[1, \infty)$, decreasing on $[-1, 0)$ and $(0, 1]$

37. Increasing on $[-2, 2]$, decreasing on $(-2\sqrt{2}, -2]$ and $[2, 2\sqrt{2})$

39. Increasing on $(-\infty, 0]$, decreasing on $[0, \infty)$

41. Increasing on $(-\infty, 1]$ and $[3, \infty)$, decreasing on $[1, 3]$

43. Increasing on $\left[-\dfrac{\pi}{2} + 2n\pi, \dfrac{\pi}{2} + 2n\pi\right]$,

decreasing on $\left[\dfrac{\pi}{2} + 2n\pi, \dfrac{3\pi}{2} + 2n\pi\right]$, where n is an integer

45. Since the average speed on the time interval is 60 mph, the Mean Value Theorem implies there must be some time in the interval at which the speed is exactly equal to 60 mph. This exceeds the legal speed limit.

51. $c \approx 0.3451$ radian

Exercises 4.5, Page 215

1. Rel. max. $f(1) = 2$

3. Rel. max. $f(-1) = 2$;
rel. min. $f(1) = -2$

5. Rel. max. $f(\tfrac{2}{3}) = \tfrac{32}{27}$;
rel. min. $f(2) = 0$

7. Rel. min. $f(-1) = -3$

9. Rel. min. $f(0) = 0$

11. Rel. max. $f(0) = 0$;
rel. min. $f(1) = -1$

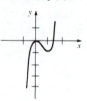

13. Rel. max. $f(-3) = -6$;
rel. min. $f(1) = 2$

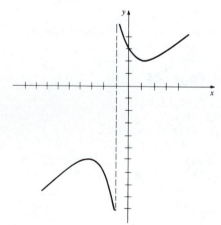

15. Rel. max. $f(\sqrt{3}) = 2/3\sqrt{3}$;
rel. min. $f(-\sqrt{3}) = -2/3\sqrt{3}$

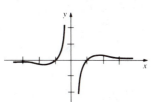

17. Rel. max. $f(0) = 10$ **19.** Rel. min. $f(4) = 0$

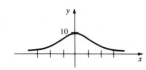

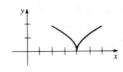

21. Rel. max. $f(\sqrt{2}/2) = \frac{1}{2}$;
rel. min. $f(-\sqrt{2}/2) = -\frac{1}{2}$

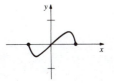

23. Rel. max. $f(-8) = 16$;
rel. min. $f(8) = -16$

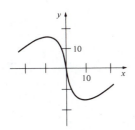

25. Rel. max. $f(0) = 0$;
rel. min. $f(-2) = -12(2)^{2/3}$;
rel. min. $f(2) = -12(2)^{2/3}$

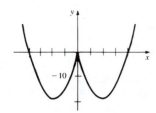

27. 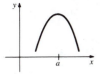   **29.**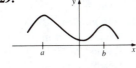

31.

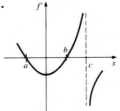

33.

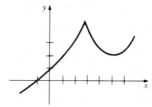

35.

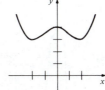

37. Rel. min. $f'(-2) = -13$
39. $(n\pi, \pi/2 + n\pi)$, $(\pi/2 + n\pi, \pi + n\pi)$;
$n\pi/2$, $n = 0, \pm1, \pm2, \ldots$;
rel. max. $f(-\pi/2) = f(\pi/2) = \cdots = 1$;
rel. min. $f(0) = f(\pi) = \cdots = 0$

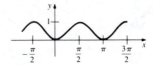

41. $f(x) = -\frac{1}{2}x^2 + 2x + 4$ **43.** Yes

Exercises 4.6, Page 223

1. Concave downward on $(-\infty, \infty)$
3. Concave upward on $(-\infty, 2)$,
concave downward on $(2, \infty)$
5. Concave upward on $(-\infty, 2)$ and on $(4, \infty)$,
concave downward on $(2, 4)$
7. Concave upward on $(-\infty, 0)$,
concave downward on $(0, \infty)$
9. Concave upward on $(0, \infty)$,
concave downward on $(-\infty, 0)$
11. Concave upward on $(-\infty, -1)$ and $(1, \infty)$,
concave downward on $(-1, 1)$
13. f' decreasing on $(-\infty, -1]$,
f' increasing on $[-1, 0]$,
f' decreasing on $[0, 1]$,
f' increasing on $[1, \infty)$

17. $(-\sqrt{2}, -21 - \sqrt{2}), (\sqrt{2}, -21 + \sqrt{2})$

19. $(0, 0), (\pm\pi, 0), (\pm 2\pi, 0), \dots$

21. $(0, 0), (\pm\pi, \pm\pi), (\pm 2\pi, \pm 2\pi), \dots$

23. Rel. max. $f(\frac{5}{2}) = 0$

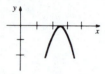

25. Point of inflection $(-1, 0)$

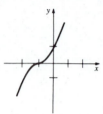

27. Rel. max. $f(-1) = 4$;
rel. min. $f(1) = -4$;
points of inflection $(0, 0)$,
$(-\sqrt{2}/2, 7\sqrt{2}/4), (\sqrt{2}/2, -7\sqrt{2}/4)$

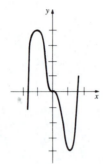

29. Rel. max. $f(\sqrt{2}) = \sqrt{2}/4$;
rel. min. $f(-\sqrt{2}) = -\sqrt{2}/4$;
points of inflection $(0, 0)$,
$(-\sqrt{6}, -\sqrt{6}/8), (\sqrt{6}, \sqrt{6}/8)$

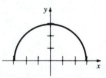

31. Rel. max. $f(0) = 3$

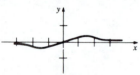

33. Rel. min. $f(-\frac{1}{4}) = -\dfrac{3}{4^{4/3}}$

points of inflection $(0, 0)$, $\left(\dfrac{1}{2}, \dfrac{3}{2^{4/3}}\right)$

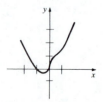

35. Rel. max. $f(2\pi/3) = f(4\pi/3) = 1$;
rel. min. $f(\pi/3) = f(\pi) = f(5\pi/3) = -1$;
points of inflection $(\pi/6, 0), (\pi/2, 0), (5\pi/6, 0)$,
$(7\pi/6, 0), (9\pi/6, 0), (11\pi/6, 0)$

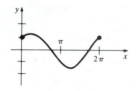

37. Rel. max. $f(\pi/4) = \sqrt{2}$;
rel. min. $f(5\pi/4) = -\sqrt{2}$;
points of inflection $(3\pi/4, 0), (7\pi/4, 0)$

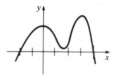

39. Rel. max. $f(\pi/4) = \frac{1}{2}$

41. Rel. min. $f(\pi) = 0$

43.

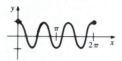

45.

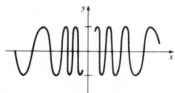

47. $f(x) = -\frac{1}{6}x^3 + \frac{1}{2}x^2 + \frac{2}{3}x$

49.

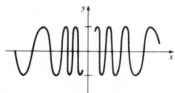

53. $(c, f(c))$ is a point of inflection.

Exercises 4.7, Page 231

1. 30, 30   **3.** $\frac{1}{2}$ **5.** $\frac{1}{3}$ and $\frac{2}{3}$

7. $(2, 2\sqrt{3}), (2, -2\sqrt{3}); (0, 0)$ **9.** $(\frac{4}{3}, -\frac{128}{27})$

11. 750 ft by 750 ft   **13.** 2000 m by 1000 m

15. Base: $4\sqrt{2}$; height: 16   **17.** Base: $\frac{3}{2}$; height: 1

19. Base: 40 cm by 40 cm; height: 20 cm

21. Base: $\frac{80}{3}$ cm by $\frac{80}{3}$ cm; height: $\frac{20}{3}$ cm;
maximum volume: 128,000/27 cm³

23. Radius: $2/3\pi$ m; height: 2/3 m

25. Radius: $2R/3$; height: $H/3$

27. Height: $\frac{15}{2}$ cm; width: 15 cm

29. Radius: $\sqrt[3]{16/\pi}$; height: $2\sqrt[3]{16/\pi}$

31. $\sqrt{\frac{2}{3}}R$; $(2\sqrt{3}/27)\pi R^3$

33. Radius of circular portion: $10/(4 + \pi)$; width: $20/(4 + \pi)$; height of rectangular portion: $10/(4 + \pi)$
35. The wire should not be cut at all. Bend the entire portion into a circle of radius $1/(2\pi)$ m.
37. Length of cross section: $\sqrt{3}d/3$; width of cross section: $\sqrt{6}d/3$
39. Radius of cylinder (and hemisphere): $\sqrt[3]{9}$; length of cylinder: $2\sqrt[3]{9}$
41. 35　　**43.** $-\frac{1}{8}$
45. $y = h/2$; maximum distance: h
47. $\frac{50}{11}$ m from I_1　　**49.** $16\sqrt{2}$ ft
51. Minimum costs occur when $x = 4$.
55. Let y denote the difference between Young's rule and Cowling's rule. Then y_{max} occurs at $t = 12(\sqrt{2} - 1) \approx 5$ yr. At this age $y_{max} \approx 0.04\, D_a$.

Exercises 4.8, Page 240

1. 40,000; 16,100; 60　　**3.** 60; 59
5. 183; 180; 186
7. $R''(x) < 0$ for all $x > 0$　　**11.** $\frac{2}{3}$, inelastic
13. $\frac{9}{8}$, elastic　　**15.** $\frac{7}{16}$, inelastic
17. $\frac{18}{7}$, decreases by approximately 15.4%
19. 4.47

Chapter 4 Review Exercises, Page 240

1. False　　**3.** False　　**5.** True　　**7.** False
9. True
11. f need not be differentiable at c.
13. Abs. max. $f(-3) = 348$; abs. min. $f(4) = -86$
15. Abs. max. $f(3) = \frac{9}{7}$; abs. min. $f(0) = 0$
17.

19. Maximum velocity is $v(2) = 12$, maximum speed is $|v(-1)| = |v(5)| = 15$

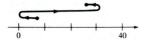

23. Rel. max. $f(-3) = 81$; rel. min. $f(2) = -44$

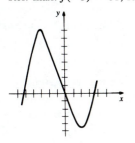

25. Rel. max. $f(0) = 2$; rel. min. $f(1) = 0$

27. Rel. min. $f(0) = 0$;
points of inflection $(-3, 27)$, $(-1, 11)$
29. Point of inflection $(3, 10)$　　**31.** $(a + b + c)/3$
33. $\frac{5}{4}$ ft/sec　　**37.** 2
41. Fly to point 17.75 km from the nest.
43. A 10-ft section of fence should be attached to the 40-ft-long side of the house; the other three sections should be 50 ft long.
45. $x = 195$, $y = 390$; 57,037.5 ft²
47. 10 ft from the 10-ft mast
53. Approximate dimensions: $11.83 \times 8.66 \times 3.17$; approximate maximum volume: 324.76 in.³
55. 36,300; 13,056; 236; 600; 224; 400

Exercises 5.1, Page 251

1. $3x + C$　　**3.** $\frac{1}{6}x^6 + C$　　**5.** $\frac{3}{2}x^{2/3} + C$
7. $t - \frac{25}{12}t^{0.48} + C$　　**9.** $x^3 + x^2 - x + C$
11. $\frac{16}{3}x^3 + 4x^2 + x + C$　　**13.** $\frac{1}{3}x^3 - 4x + C$
15. $-r^{-1} + 5r^{-2} + C$
17. $-\frac{1}{2}x^{-2} + \frac{1}{3}x^{-3} - \frac{1}{4}x^{-4} + C$
19. $16w^4 - 16w^3 + 6w^2 - w + C$
21. $f(x) = x^2 - x + 1$
23. $f'(x) = x^2 + C_1, f(x) = \frac{1}{3}x^3 + C_1x + C_2$
25. $f(x) = x^4 + x^2 - 3x + 2$
33. $x^2 - 4x + 5$　　**35.** $y = \frac{\omega^2}{2g}x^2$

Exercises 5.2, Page 259

1. $-\frac{1}{6}(1 - 4x)^{3/2} + C$　　**3.** $-\frac{1}{10}(5x + 1)^{-2} + C$
5. $-\frac{5}{32}(3 - 4x)^{8/5} + C$　　**7.** $\frac{1}{7}(x - 1)^7 + C$
9. $\frac{2}{3}(x^2 + 4)^{3/2} + C$　　**11.** $\frac{3}{4}(z^2 + 9)^{2/3} + C$
13. $\frac{1}{40}(4x^2 - 16x + 7)^5 + C$
15. $\frac{1}{2}(x^3 + 3x - 16)^{2/3} + C$
17. $\frac{1}{3}\left(3 - \frac{2}{v}\right)^{3/2} + C$　　**19.** $-\frac{9}{4}(1 - \sqrt[3]{x})^{4/3} + C$
21. $-\frac{1}{4}\cos 4x + C$　　**23.** $\frac{1}{2}\sin x^2 + C$
25. $\frac{1}{5}\sin(5x + 1) + C$　　**27.** $-2\cos\theta + C$
29. $-2\csc\sqrt{x} + C$　　**31.** $\frac{1}{18}\sin^6 3x + C$
33. $\frac{1}{6}\tan^3 2x + C$　　**35.** $\frac{1}{7}\tan 7x - x + C$
37. $-2\cot x - \csc x + C$
39. $\frac{1}{2}x - \frac{1}{4}\sin 2x + C$
41. $-\frac{1}{2}\cot(z^2 + 2z) + C$
43. $\frac{3}{2}x + \sin 2x + \frac{1}{8}\sin 4x + C$
45. $f(x) = x + \cos x - \pi$

49. $\sin x - \frac{1}{3}\sin^3 x + C$

51. $\frac{2}{3}(t + 2)^{3/2} - 4(t + 2)^{1/2} + C$

Exercises 5.3, Page 265

1. $3 + 6 + 9 + 12 + 15$ **3.** $\frac{2}{1} + \frac{2^2}{2} + \frac{2^3}{3} + \frac{2^4}{4}$

5. $-\frac{1}{7} + \frac{1}{9} - \frac{1}{11} + \frac{1}{13} - \frac{1}{15} + \frac{1}{17} - \frac{1}{19} + \frac{1}{21} - \frac{1}{23} + \frac{1}{25}$

7. $(2^2 - 4) + (3^2 - 6) + (4^2 - 8) + (5^2 - 10)$

9. $-1 + 1 - 1 + 1 - 1$ **11.** $\sum_{k=1}^{6} (2k)^2$

13. $\sum_{k=1}^{7} (2k + 1)$ **15.** $\sum_{k=0}^{12} (3k + 1)$

17. $\sum_{k=1}^{5} (-1)^{k+1}\frac{1}{k}$ **19.** $\sum_{k=1}^{8} 6$

21. 420 **23.** 65 **25.** 109 **27.** 3069

29. $101{,}262$ **31.** $70{,}940$ **33.** 20

35. $-f(0) + f(n); f(k) = \sqrt{k}$ and $f(k) = -1/(k + 1)$

37. Yes, let $j = k + 3$.

Exercises 5.4, Page 273

1. 12 **3.** 18 **5.** 28 **7.** $\frac{8}{3}$

9. $\frac{4}{3}$ **11.** $\frac{16}{3}$ **13.** $\frac{1}{4}$ **15.** 4

17. $\frac{32}{5}$ **19.** 5 **21.** $\frac{77}{60}; \frac{25}{12}$

23. $\lim_{n\to\infty} \sum_{k=1}^{n} f\left(a + (k - 1)\frac{b - a}{n}\right)\frac{b - a}{n};$

$\lim_{n\to\infty} \sum_{k=1}^{n} f\left(a + (2k - 1)\frac{b - a}{2n}\right)\frac{b - a}{n}$

25. 9 **27.** $\frac{1}{6}$ **29.**

$y = \sqrt{4 - x^2}$

31. $\frac{2}{3}$

Exercises 5.5, Page 281

1. $\frac{19}{2}; \frac{5}{2}$ **3.** $\frac{33}{2}; 1$ **5.** $\frac{189}{256}; \frac{3}{4}$

7. $(3 - \sqrt{2})\pi/4; \pi$ **9.** 5 **11.** 24

13. -4 **15.** $\frac{2}{3}$ **17.** $\frac{5}{6}$ **19.** $-\frac{3}{4}$

25. $\frac{2}{3}$

27. $\int_{-2}^{4} \sqrt{9 + x^2}\,dx$ **29.** $\int_{0}^{2} (x + 1)\,dx$

31. 1.1564135

Exercises 5.6, Page 285

1. 12 **3.** 10 **5.** $-\frac{3}{2}$ **7.** -20

9. 56 **11.** -20 **13.** 24 **15.** 0

17. 20 **19.** -56 **21.** 64 **23.** 20

25. 8 **27.** 0 **29.** 3 **31.** $\frac{1}{2}$

33. $(\pi + 2)/4$ **35.** 0 **37.** 36 **39.** 10

43. $x^3 \le x^2$ for all x in $[0, 1]$.

45. The function $f(x) = (x^3 + 1)^{1/2}$ is increasing on the interval $[0, 1]$. Therefore, $f(0) \le f(x) \le f(1)$ or $1 \le f(x) \le 2^{1/2}$. The result follows from the comparison property with $b - a = 1$, $m = 1$, and $M = 2^{1/2}$.

Exercises 5.7, Page 297

1. $3x^2 - 2x$ **3.** 4 **5.** 12 **7.** 46

9. 1 **11.** $\frac{2}{3}$ **13.** 0 **15.** $-\frac{2}{3}$

17. $-\frac{28}{3}$ **19.** $\frac{8}{3}$ **21.** $\frac{9}{2}$ **23.** 21

25. $\frac{128}{3}$ **27.** 1 **29.** $\frac{65}{4}$ **31.** $\sqrt{6} - \sqrt{3}$

33. $\frac{1}{2}$ **35.** 1 **37.** $\frac{2}{3}$

39. $\dfrac{4\pi + 6}{(\pi + 3)(\pi + 2)}$ **41.** 0 **43.** 0

45. 5 **47.** 22 **49.** 4 **51.** $\frac{19}{6}$

53. 9 **55.** $\frac{38}{3}$ **57.** $\frac{116}{15}$ **59.** 3

61. $\int_{a}^{x^2} f(t)\,dt; 2xf(x^2)$ **63.** $-6\sqrt{24x + 5}$

65. 28

67. The limit of the sum is the same as $\int_{0}^{\pi} \sin x\,dx$.

69. $f(x) = x^{-2}$ is not continuous on $[-1, 1]$.

Exercises 5.8, Page 308

1. $22, 22.5$ **3.** 1.8667 **5.** 1.1484

7. 0.4228 **9.** 0.4900 **11.** $n \ge 11$

13. 1.11 **15.** 7 **17.** $\frac{26}{3}, 8.6611$

19. 1.6222 **21.** 0.7854 **23.** 0.4339

25. 11.1053

27. For Simpson's Rule: $n \ge 26$; for Trapezoidal Rule: $n \ge 366$ **29.** 130

Chapter 5 Review Exercises, Page 309

1. False **3.** True **5.** True **7.** False

9. False **11.** True **13.** True **15.** True

17. True **19.** True **21.** $f(x)$ **23.** $\sqrt{6}$

25. 3720 **27.** 250 **29.** Of the same length

31. -6 **33.** $\frac{1}{505}(5t + 1)^{101} + C$ **35.** $\frac{1}{2}$

37. 0 **39.** $-\frac{1}{56}\cot^7 8x + C$ **41.** -6

43. $\frac{11}{2}$ **45.** 156 lb; in approximately 20 min

47. $E_4 = |\frac{8}{3} - 2.75| \approx 0.0833$

49. Trapezoidal Rule: 0.3160; Simpson's Rule: 0.3099

Exercises 6.1, Page 320

1. $\frac{4}{3}$ **3.** $\frac{81}{4}$ **5.** $\frac{9}{2}$ **7.** $\frac{11}{2}$

9. $\frac{11}{4}$ **11.** $\frac{11}{6}$ **13.** 2

15. $\frac{3}{4}(2^{4/3} + 3^{4/3})$ **17.** 4   **19.** 2π

21. $\frac{27}{2}$ **23.** $\frac{32}{3}$ **25.** $\frac{81}{4}$ **27.** 4

29. $\frac{10}{3}$ **31.** $\frac{64}{3}$ **33.** $\frac{128}{5}$ **35.** $\frac{118}{3}$

37. $8\sqrt{2}$ **39.** $\frac{9}{2}$ **41.** $\frac{8}{3}$ **43.** 8

45. $2\sqrt{2} - 2$ **47.** $4\sqrt{3} - 4\pi/3$

49. $\frac{19}{4}$ **51.** $\frac{5}{2}$ **53.** $\pi a^2/4$ **55.** $\frac{52}{3}$

Exercises 6.2, Page 324

1. $(5625)\sqrt{3}/16$ **3.** $\frac{1024}{3}$ **5.** 128

7. $10\pi/3$ **9.** 9   **11.** $2a^3/3$

Exercises 6.3, Page 330

1. $\pi/2$ **3.** $4\pi/5$ **5.** $\pi/6$

7. $1296\pi/5$ **9.** $\pi/2$ **11.** $8\pi/3$

13. $32\pi/5$ **15.** 32π **17.** 8π

19. $256\pi/15$ **21.** $3\pi/5$ **23.** 36π

25. $500\pi/3$ **27.** $16\pi/105$ **29.** π^2

31. $(4\pi - \pi^2)/4$ **33.** $3\pi^2/8$ **37.** $4ab^2\pi/3$

39. $74\pi/15$ **41.** 6.4

Exercises 6.4, Page 336

1. $4\pi/5$ **3.** $\pi/6$ **5.** $8\pi/15$

7. $250\pi/3$ **9.** $36\sqrt{3}\pi/5$ **11.** $3\pi/2$

13. 16π **15.** $8\pi/5$ **17.** $21\pi/10$

19. $\pi/6$ **21.** $248\pi/15$ **23.** 4π

25. $625\pi/6$ **27.** $45\pi/2$ **29.** $(\pi^2 - 2\pi)/2$

33. $4a^2b\pi/3$ **35.** $43\pi/2$

Exercises 6.5, Page 340

1. $2\sqrt{2}$ **3.** $\frac{1}{27}[13^{3/2} - 8] \approx 1.44$

5. 45   **7.** $\frac{10}{3}$ **9.** $\frac{4685}{288} \approx 16.27$

11. $\int_{-1}^{3} \sqrt{1 + 4x^2}\, dx$ **13.** $\int_{0}^{\pi} \sqrt{1 + \cos^2 x}\, dx$

15. $\frac{1}{27}[40^{3/2} - 8] \approx 9.07$

17. The integrand of $\int_{0}^{1} x^{-1/3}\, dx$ is discontinuous on

$[0, 1]$; 6   **19.** $\pi/2$ **21.** 1.4804

Exercises 6.6, Page 344

1. $208\pi/3$ **3.** $(\pi/27)[10^{3/2} - 1] \approx 3.56$

5. $(\pi/6)[37^{3/2} - 1] \approx 117.32$ **7.** $100\sqrt{5}\pi$

9. $253\pi/20$ **11.** $2\pi r(b - a)$ **13.** $12\pi a^2/5$

19. $(2\pi/3)\int_{1}^{8} (4x^{-1/3} - x^{1/3})\sqrt{9x^{2/3} + 4}\, dx$

Exercises 6.7, Page 349

1. -4 **3.** $\frac{34}{3}$ **5.** 3   **7.** 0   **9.** 2

11. $\frac{61}{9}$ **13.** 24   **15.** $\frac{1}{12}$ **17.** 0

19. $3\sqrt{3}/\pi$ **21.** $-1 + \dfrac{2\sqrt{3}}{3} \approx 0.15$

23. 12   **25.** $103°$ **29.** $2kt_1/3$

31. $[f(x + h) - f(x)]/h$

35. The car travels at a constant rate of 60 mph for 40 miles and at a constant rate of 40 mph for the next 30 miles; 51.4; 49.4

Exercises 6.8, Page 353

1. $s(t) = 6t - 7$ **3.** $s(t) = \frac{1}{3}t^3 - 2t^2 + 15$

5. $s(t) = -\frac{5}{2}\sin(4t + \pi/6) + \frac{5}{2}$

7. $v(t) = -5t + 9$,

$s(t) = -\frac{5}{2}t^2 + 9t - \frac{9}{2}$

9. $v(t) = t^3 - 2t^2 + 5t - 3$,

$s(t) = \frac{1}{4}t^4 - \frac{2}{3}t^3 + \frac{5}{2}t^2 - 3t + 10$

11. $v(t) = \frac{21}{4}t^{4/3} - t - 26$,

$s(t) = \frac{63}{28}t^{7/3} - \frac{1}{2}t^2 - 26t - 48$

13. $\frac{11}{225}$ km   **15.** 256 ft   **17.** 30.62 m

19. 400 ft; 6 sec   **21.** -80 ft/sec

23. 17 cm   **25.** 34 cm   **27.** 24 cm

Exercises 6.9, Page 361

1. 3300 ft-lb   **3.** 222,750 ft-lb   **5.** $\frac{2}{5}$ ft

7. 10 nt-m; 27.5 nt-m   **9.** 7.5 ft-lb; 37.5 ft-lb

11. 453.1×10^8 joules   **13.** 127,030.9 ft-lb

15. 45,741.6 ft-lb   **17.** 57,408 ft-lb

19. 900,000 ft-lb   **21.** 24,960 ft-lb

23. $2,700,000 - 100x$; 2.65×10^9 ft-lb

27. $v_0 = \sqrt{2km_1/R}$; 11,183.1 m/sec; 5086.4 m/sec

29. 126 nt-m

Exercises 6.10, Page 366

1. 196,000 nt/m², 4,900,000π nt; 196,000 nt/m², 784,000π nt; 196,000 nt/m², 19,600,000π nt

3. 59,904 lb; 29,952 lb   **5.** 121.59 lb

7. 1280 lb   **9.** 3660.8 lb   **11.** 13,977.6 lb

13. 9984π lb   **15.** 5990.4 lb

17. $30,420\sqrt{17}$ lb   **19.** $4,992,000\sqrt{2}$ lb

Exercises 6.11, Page 371

1. $\bar{x} = -\frac{2}{7}$ **3.** $\bar{x} = -\frac{13}{30}$ **5.** $\bar{x} = 1$

7. $\bar{x} = \frac{115}{36}$ **9.** $\bar{x} = \frac{4}{7}$ **11.** $\bar{x} = \frac{19}{5}$

13. $\bar{x} = \frac{11}{10}$ **15.** $\bar{x} = \frac{15}{2}$

17. $\bar{x} = 3$ since $\rho(x)$ is symmetric about the line $x = 3$.

19. Put the origin of an xy-coordinate system at the midpoint of the line containing m_1 and m_2 with the x-axis along this line. Then the center of mass has coordinates $(0, 2\sqrt{3}/5)$.

Exercises 6.12, Page 378

1. $\bar{x} = -\frac{2}{7}, \bar{y} = \frac{17}{7}$ **3.** $\bar{x} = \frac{17}{11}, \bar{y} = -\frac{20}{11}$

5. $\bar{x} = \frac{10}{9}, \bar{y} = \frac{28}{9}$ **7.** $\bar{x} = \frac{3}{4}, \bar{y} = \frac{3}{10}$

9. $\bar{x} = \frac{12}{5}, \bar{y} = \frac{54}{7}$ **11.** $\bar{x} = \frac{93}{35}, \bar{y} = \frac{45}{56}$

13. $\bar{x} = \frac{1}{2}, \bar{y} = \frac{8}{5}$ **15.** $\bar{x} = \frac{16}{35}, \bar{y} = \frac{16}{35}$

17. $\bar{x} = \frac{3}{2}, \bar{y} = \frac{121}{540}$ **19.** $\bar{x} = -\frac{7}{10}, \bar{y} = \frac{7}{8}$

21. $\bar{x} = 0, \bar{y} = 2$ **23.** $\bar{x} = \frac{5}{3}, \bar{y} = \frac{2}{3}$

25. $\bar{x} = 0, \bar{y} = (\pi + 8)/8$

27. $\bar{x} = 0, \bar{y} = 20/3\pi$ **31.** $\bar{x} = \frac{23}{10}, \bar{y} = \frac{17}{10}$

Exercises 6.13, Page 383

1. 0.11 liters/second $\approx$ 6.76 liters/minute
3. 1.85 milligrams/liter   **5.** 94%
7. \$49,600   **9.** 320 dollars/year
11. CS = \$625; PS = \$625
13. CS = \$12.50; PS = \$2.33
15. CS = \$45; PS = \$31.50

19. $\dfrac{1-k}{2-k} \dfrac{b^{2-k} - a^{2-k}}{b^{1-k} - a^{1-k}}$

Chapter 6 Review Exercises, Page 384

1. False   **3.** True   **5.** True   **7.** True
9. True   **11.** False   **13.** Joules
15. 2500 ft-lb   **17.** 6 square units

19. $\displaystyle\int_a^b f(x)\, dx$

21. $\displaystyle\int_a^b f(x)\, dx - \int_b^c f(x)\, dx + \int_c^d f(x)\, dx$

23. $\displaystyle\int_a^b 2\, dx$

25. $-\displaystyle\int_a^b 2f(x)\, dx + \int_b^c 2f(x)\, dx$

27. $-\displaystyle\int_a^0 \frac{1}{2}x\, dx + \int_0^{2b} \left(b - \frac{1}{2}x\right) dx$

29. $\bar{x} = \dfrac{\displaystyle\int_0^2 x(f(x) - g(x))\, dx}{\displaystyle\int_0^2 (f(x) - g(x))\, dx},$

$\bar{y} = \dfrac{\dfrac{1}{2}\displaystyle\int_0^2 ([f(x)]^2 - [g(x)]^2)\, dx}{\displaystyle\int_0^2 (f(x) - g(x))\, dx}$

31. $2\pi \displaystyle\int_0^2 x(f(x) - g(x))\, dx$

33. $\displaystyle\int_0^2 [f(x) - g(x)]^2\, dx$ **35.** $\frac{256}{45}$

37. 37.5 joules   **39.** 624,000 ft-lb
41. 2040 ft-lb   **43.** 691,612.8 ft-lb
45. $(40^{3/2} - 8)/27$ **47.** 17,066.7 nt

Exercises 7.1, Page 395

1.
Not one-to-one

3.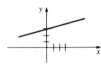
$f^{-1}(x) = 3x - 9$

5.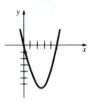
Not one-to-one

7.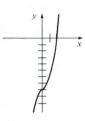
$f^{-1}(x) = \sqrt[3]{x + 8}$

9.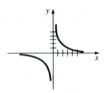
$f^{-1}(x) = 4/x$

11.
Not one-to-one

13. $f^{-1}(x) = x - 5$ **15.** $f^{-1}(x) = \sqrt[3]{\dfrac{x - 7}{3}}$

17. $f^{-1}(x) = \frac{1}{2}x^3 + 2$

19. $f^{-1}(x) = \sqrt[3]{(1 - x)/8x}$

21. $f^{-1}(x) = \dfrac{2 - x}{1 - x}$

23. $f'(x) > 0$ for all x shows that f is increasing on $(-\infty, \infty)$. The result follows from Theorem 7.2.
25. Since f is a polynomial function, it is continuous. Now $f(-2)$ is a relative maximum and $f(1)$ is a relative minimum. Since these are the only extrema, any value of y chosen between $f(1)$ and $f(-2)$ must correspond to at least two values of x. Thus, f is not one-to-one.
27. Domain: $[0, \infty)$
Range: $[-2, \infty)$
29. Domain: All real numbers except 0
Range: All real numbers except -3
31. Domain: $[0, \infty)$
Range: $[5, \infty)$

33.

35.

37. $\frac{2}{3}$ **39.** $(5, 3)$; $y = \frac{1}{10}x + \frac{5}{2}$

41. $(8, 1)$; $y = \frac{1}{60}x + \frac{13}{15}$

43. $f'(x) = 1/(x + 1) > 0$ for $x > -1$. Hence by Theorem 7.2, f^{-1} exists and so f is one-to-one; 2

45. $f^{-1}(x) = 1/(x - 2)$;
$(f^{-1})'(x) = -1/(x - 2)^2$

47. $F(x) = (5 - 2x)^2$, $x \geq 5/2$;
$$F^{-1}(x) = \frac{5 - \sqrt{x}}{2}$$

49. $F(x) = x^2 + 2x + 4$, $x \geq -1$;
$F^{-1}(x) = -1 + \sqrt{x - 3}$

51. $\frac{51}{4}$ **55.** No

Exercises 7.2, Page 402

1. 0   **3.** $3\pi/4$ **5.** $\pi/4$ **7.** $3\pi/4$

9. $-\pi/3$ **11.** $7\pi/6$ **13.** $\sqrt{3}/2$

15. $-\pi/4$ **17.** $\frac{4}{5}$ **19.** 2   **21.** $\sqrt{2}$

23. $\pi/3$ **25.** π **27.** $-1/\sqrt{3}$ **29.** -1

31. $4\sqrt{2}/9$ **33.** $\sqrt{3}(2 + \sqrt{10})/9$

35. $13/\sqrt{170}$ **37.** $\sqrt{1 - x^2}$ **39.** $\sqrt{x^2 - 1}$

41. $2x/(x^2 + 1)$ **45.** 3

47. $\cos t = \sqrt{5}/5$, $\tan t = -2$, $\cot t = -\frac{1}{2}$,
$\sec t = \sqrt{5}$, $\csc t = -\sqrt{5}/2$

49. 5 is not in the interval $(-\pi/2, \pi/2)$.

51. 0.28 radian (16.2°); 114.26 ft/sec

53. 0.20 radian or 11.31° south of west

Exercises 7.3, Page 411

1. $\dfrac{5}{\sqrt{1 - (5x - 1)^2}}$ **3.** $-8/(4 + x^2)$

5. $\dfrac{1}{1 + x} + x^{-1/2} \tan^{-1}\sqrt{x}$

7. $\dfrac{2(\cos^{-1}2x + \sin^{-1}2x)}{\sqrt{1 - 4x^2}(\cos^{-1}2x)^2}$

9. $\dfrac{-2x}{(1 + x^4)(\tan^{-1}x^2)^2}$ **11.** $\dfrac{2 - x}{\sqrt{1 - x^2}} + \cos^{-1}x$

13. $3\left(x^2 - 9 \tan^{-1}\dfrac{x}{3}\right)^2\left(2x - \dfrac{27}{9 + x^2}\right)$

15. $1/(t^2 + 1)$ **17.** $-4 \sin 4x/|\sin 4x|$

19. $\dfrac{2x \sec^2(\sin^{-1}x^2)}{\sqrt{1 - x^4}}$

21. $2x(1 + y^2)/(1 - 2y - 2y^3)$

23. $\frac{1}{5} \tan^{-1}5x + C$ **25.** $\pi/12$

27. $-2(1 - x^2)^{1/2} - 3 \sin^{-1}x + C$

29. $-x + 2 \tan^{-1}x + C$

31. $-\frac{4}{3}(2 - 3t^2)^{1/2} + C$

33. $\dfrac{\sqrt{10}}{10} \tan^{-1}\dfrac{\sqrt{10}}{5}x + C$ **35.** $\dfrac{2}{3}\left(\dfrac{\pi}{4}\right)^{3/2}$

37. $\sec^{-1}(x + 1) + C$

39. $\frac{1}{2}x^2 + \frac{1}{2} \tan^{-1}\left(\dfrac{x + 2}{2}\right) + C$

41. $\frac{1}{4} \tan^{-1}(2t^2 + 1) + C$ **43.** $\pi/48$

45. $\sqrt{3}/3$

47. $y = ((2 + \pi)/4)x - \frac{1}{2}$

49. Rel. max. $f(\sqrt{2}/2) = -\sqrt{2} + 3 \tan^{-1}(\sqrt{2}/2)$;
rel. min. $f(-\sqrt{2}/2) = \sqrt{2} + 3 \tan^{-1}(-\sqrt{2}/2)$

51.

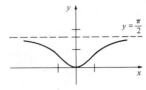

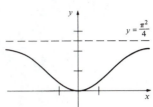

53. $\sin^{-1}(\frac{3}{5}) - \sin^{-1}(\frac{1}{5})$

55. $\dfrac{\pi}{4}[\sec^{-1}(4) - \sec^{-1}(\frac{25}{16})]$

57. $\sqrt{2}/120$ rad/sec   **59.** $10\sqrt{3}$ ft

61. 5.5730   **63.** 0.4389   **69.** f is constant.

Chapter 7 Review Exercises, Page 413

1. True   **3.** True   **5.** True   **7.** False

9. True   **11.** True   **13.** True   **15.** True

17. False   **19.** False

21. $f^{-1}(x) = \sqrt[3]{\dfrac{x - 8}{x}}$; $(f^{-1})'(x) = \dfrac{8}{3x^2}\left(\dfrac{x - 8}{x}\right)^{-2/3}$

23. $-\dfrac{3|x|}{x^2\sqrt{x^2 - 9}}$ **25.** $\dfrac{(\cot^{-1}x)^{-2}}{1 + x^2}$

27. $-4x^2/\sqrt{1 - x^2}$

29. $\dfrac{(4x \cos x^2)(\tan^{-1}(\sin x^2))}{1 + \sin^2x^2}$

31. $\sqrt{1 - y^2}/(\sqrt{1 - y^2} - 1)$

33. $\frac{1}{2} \sin^{-1}\frac{2}{3}x + C$ **35.** $\pi/6$

37. $\sin^{-1}\left(\dfrac{x + 3}{4}\right) + C$ **39.** $1/\pi$ **41.** $\pi/18$

43. $\frac{5}{13}$

45.

47. $\pi/6 - \tan^{-1}(\frac{1}{2})$

Exercises 8.1, Page 423

1. $x > -1$ **3.** $x \neq \pm 1$ **5.** No

7. Yes   **9.** No   **11.** $10/x$ **13.** $1/2x$

15. $(4x^3 + 6x)/(x^4 + 3x^2 + 1)$

17. $3x + 6x \ln x$ **19.** $(1 - \ln x)/x^2$
21. $1/x(x + 1)$ **23.** $\tan x$ **25.** $-1/x(\ln x)^2$
27. $(1 + \ln x)/x \ln x$
29. $1/[(x \ln x)\ln(\ln x)]$
31. $12(t^2 + 1)/t(3t^2 + 6)$
33. $(x^2 + 6x + 7)/(x + 1)(x + 2)(x + 3)$
35. $y/x(2y^2 - 1)$ **37.** $y(1 - x)/x(2y^2 + 1)$
39. $(2x - x^2y - y^3)/(x^3 + xy^2 - 2y)$
41. $y = x - 1$ **43.** $y = 4x - 8$
45. $(\frac{1}{4}, -\ln 2)$
47.

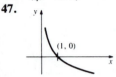

49.

51. x- and y-intercepts are at the origin; symmetry with respect to the y-axis; no asymptotes; $f(0) = 0$ is a relative minimum; concave downward on $(-\infty, -1)$ and $(1, \infty)$, concave upward on $(-1, 1)$

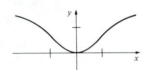

55.

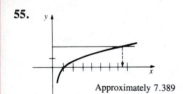

Approximately 7.389

Exercises 8.2, Page 430

1. $-e^{-x}$ **3.** $\frac{1}{2}x^{-1/2}e^{\sqrt{x}}$
5. $-e^{-2x}(2x + 1)/x^2$
7. $-\frac{5}{2}(1 + e^{-5x})^{-1/2}e^{-5x}$
9. $(4x^3 + 2xe^{x^2})/(x^4 + e^{x^2})$
11. $-4/(e^x - e^{-x})^2$ **13.** 1
15. $2xe^{3x}/(x^2 + 1) + 3e^{3x}\ln(x^2 + 1)$
17. $-4e^{(x+2)/(x-2)}/(x - 2)^2$ **19.** $2xe^{x^2}e^{e^{x^2}}$
21. $9e^{9x}$ **23.** $\frac{1}{3}t^{-2/3}e^{t^{1/3}} + \frac{1}{3}e^{t/3}$
25. $(2x + 1)^2 e^{-(1-x)^4}[4(2x + 1)(1 - x)^3 + 6]$
27. $(x^2e^x + e^{2x})/(x + e^x)^2$
29. $2e^{2x}/(1 + e^{4x})$ **31.** $e^{x+y}/(1 - e^{x+y})$
33. $(-ye^{xy} \sin e^{xy})/(1 + xe^{xy} \sin e^{xy})$
35. $(-y^2 + ye^{x/y})/(2y^3 + xe^{x/y})$
37. $-64e^{-4x}$ **39.** $e^x/(e^x + 1)^2$

41. $y = ex$ **43.** $-\frac{1}{2}$
45.

47.

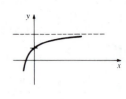

49.

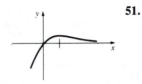

51.

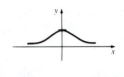

$f(1) = e^{-1}$ is an abs. max. $f(0) = 1$ is an abs. max.

53.

$f(e^{-1}) = -e^{-1}$ is an abs. min.

55. Since $f' < 0$ for all $x > 0$, f is decreasing on $(0, \infty)$. By Theorem 7.2, f has an inverse.

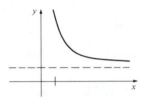

Range of $f = \{y \mid y > 1\}$; $f^{-1}(x) = 2/(\ln x)$; domain of $f^{-1} = \{x \mid x > 1\}$

57.

59. $y = 5e^x$

$$\frac{dy}{dx} = \begin{cases} e^x, & x > 0 \\ -e^{-x}, & x < 0 \end{cases}; \quad \text{no}$$

65. $P'(t) > 0$ for all t; $(0, 1)$ is a point of inflection;

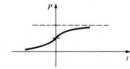

67. 88.22 cm; 9.95 cm/yr; $\frac{1}{4}$ yr

69.

$f(n)$
2.7048138
2.7169238
2.7181459
2.7182546
2.7182818

71. x-intercept: 1.3863 (approximately); y-intercept: -12

Exercises 8.3, Page 438

1. $\frac{1}{3}\ln|x| + C$ **3.** $\frac{1}{2}\ln|2x - 1| + C$

5. $\frac{1}{2}\ln(x^2 + 1) + C$ **7.** $\frac{1}{2}\ln|x^2 + 4x - 3| + C$

9. $x - \ln|x + 1| + C$

11. $\frac{1}{2}x^2 + 4x + \ln|x + 2| + C$

13. $\ln|\ln x| + C$ **15.** $\frac{1}{2}(\ln x)^2 + C$

17. $\frac{3}{4}(1 + \ln x)^{4/3} + C$ **19.** $\ln|1 + \sin t| + C$

21. $\frac{1}{10}e^{10x} + C$

23. $-\frac{1}{6}e^{-2x^3} + C$

25. $2e^{\sqrt{x}} + C$ **27.** $-\frac{1}{4}e^{-2x^2+4} + C$

29. $x + 3e^x + \frac{3}{2}e^{2x} + \frac{1}{3}e^{3x} + C$

31. $-e^{-t} + t + C$ **33.** $\frac{1}{3}e^{3x} + 2e^{-x} - \frac{1}{5}e^{-5x} + C$

35. $\ln(e^x + e^{-x}) + C$ **37.** $e^x - \ln(e^x + 1) + C$

39. $\frac{1}{2}e^{x/2} + C$ **41.** $-\frac{1}{5}\ln|\cos 5x| + C$

43. $\theta + 2\ln|\sec\theta + \tan\theta| + \tan\theta + C$

45. $\frac{1}{2}\ln 9$ **47.** 0   **49.** $-2 + \ln 4$

51. $2\ln 8$ **53.** $\frac{40}{9} - 2\ln 3$ **55.** $\frac{1}{2}(1 - e^{-4})$

57. $e - e^{-1} - 2$ is not area; area is $e + e^{-1} - 2$.

59. $\frac{\pi}{2}(e^4 - 1)$ **61.** $\pi(1 - e^{-1})$ **63.** $\frac{3}{4}$

65. $15\pi/16 + \pi\ln 2$

67. $C_1 x^4$ where C_1 is an arbitrary constant; $C_1(x + 1)^{-2}e^x$ where C_1 is an arbitrary constant

69. 0.744; 6; 58   **75.** 1

79. $\frac{x^2}{2}\ln x - \frac{x^2}{4} + C$

Exercises 8.4, Page 448

1. $e^{\pi\ln 2}$; 8.8   **3.** $e^{e\ln 10}$; 522

5. $e^{-\sqrt{5}\ln 7}$; 0.01   **7.** $4^x\ln 4$

9. $-2(10^{-2x})\ln 10$ **11.** $2x20^{x^2}\ln 20$

13. 0   **15.** $x^2 2^x\ln 2 + 2x2^x$

17. $\sqrt{5}x^{\sqrt{5}-1}$ **19.** $(\sin 2)x^{(\sin 2)-1}$

21. $(t^4 + 3t^2)^{(\ln 4)-1}(4t^3 + 6t)\ln 4$

23. $4^x(\ln 4 + e^x\ln 4 - e^x)/(1 + e^x)^2$

25. $\frac{1}{2}5^{x/2}\ln 5$ **27.** $-(\ln x)^{-1-1/e}/ex$

29. $1/(x\ln 4)$

31. $\ln x/(x\ln 10) + (\log_{10}x)/x$

33. $6x^\pi/(6x - 4)\ln 3 + \pi x^{\pi-1}\log_3|6x - 4|$

35. $3x^{3+1}/x\ln 3 + x^2 3x^{3+1}(\ln 3)(\log_3 x^3)$

37. $9^x(\ln 9)^2$ **39.** $8e(e - 1)(e - 2)(2x)^{e-3}$

41. $y/(2^y\ln 2 - x)$ **43.** $1/(x + xe^y)\ln 10$

45. $7^x/\ln 7 + C$ **47.** $3(10^{x/3})/\ln 10 + C$

49. $(10^{-2} - 10^{-3})/\ln 10$

51. $e^x + x^{e+1}/(e + 1) + (e^e)x + C$

53. $(1 + 2^t)^{21}/(21\ln 2) + C$

55. $(2^e - 2^{e-1})/\ln 2$

57. $\tan\theta + 2^{1+\tan\theta}/\ln 2 + 2^{-1+2\tan\theta}/\ln 2 + C$

59. $[\ln(1 + 5^x)]/\ln 5 + C$ **61.** $64\ln 4$

63. $(-1/\ln 4, -4^{-1/\ln 4}/\ln 4)$

65. $225/(16\ln 2)$

67. Approximately 251 times as strong; the San Fernando Valley earthquake was approximately 5 times as strong

69. 7.4; base

Exercises 8.5, Page 451

1.

$(e^{\Delta x} - 1)/\Delta x$
1.0517092
1.0050167
1.0005000
1.0000500
0.9999500

3.

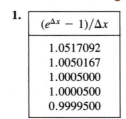

$2^{\sqrt{3}} \approx 3.4$

Exercises 8.6, Page 454

1. $\frac{1}{2}\sqrt{\dfrac{(2x + 1)(3x + 2)}{4x + 3}}$

$\cdot\left[\dfrac{2}{2x + 1} + \dfrac{3}{3x + 2} - \dfrac{4}{4x + 3}\right]$

3. $\dfrac{(x^3 - 1)^5(x^4 + 3x^3)^4}{(7x + 5)^9}$

$\cdot\left[\dfrac{15x^2}{x^3 - 1} + \dfrac{16x^3 + 36x^2}{x^4 + 3x^3} - \dfrac{63}{7x + 5}\right]$

5. $\dfrac{x^{2/3}}{\sqrt{(x^2 + 2x + 3)(2x^4 + x^2 + 1)}}$

$\cdot\left[\dfrac{2}{3x} - \dfrac{x + 1}{x^2 + 2x + 3} - \dfrac{4x^3 + x}{2x^4 + x^2 + 1}\right]$

7. $(x^2 + 4)^{2x}\left[\dfrac{4x^2}{x^2 + 4} + 2\ln(x^2 + 4)\right]$

9. $x^x 2^x [1 + \ln 2 + \ln x]$

11. $x(x - 1)^x \left[\dfrac{1}{x} + \dfrac{x}{x - 1} + \ln(x - 1) \right]$

13. $(x^6 + x^4)^{\ln x} \left[\dfrac{(6x^2 + 4)\ln x}{x^3 + x} + \dfrac{\ln(x^6 + x^4)}{x} \right]$

15. $(\ln x)^x \left[\dfrac{1}{\ln x} + \ln(\ln x) \right]$

17. $x^{\cos x} \left[\dfrac{\cos x}{x} - (\sin x)\ln x \right]$

19. $x^{e^x} \left[\dfrac{e^x}{x} + e^x \ln x \right]$ **21.** $y = 3x - 2$

23. Rel. min. $f(e^{-1}) = e^{-3e^{-1}}$

25. $x^x(1 + \ln x)\sec^2 x^x$

27. $2x^{2x} \left[2(1 + \ln x)^2 + \dfrac{1}{x} \right]$

29. $vu^{v-1} \dfrac{du}{dx} + (u^v \ln u) \dfrac{dv}{dx}$

Exercises 8.7, Page 459

1. $y = \frac{1}{2} \sin 2x + C$

3. $y = -1/x + C$ **5.** $y^{-2} = 2x^{-1} + C$

7. $y = Cx - 1$

9. $-3e^{-2y} = 2e^{3x} + C$

11. $(2y + 3)^{-1} = (8x + 10)^{-1} + C$

13. $4\cos y = 2x + \sin 2x + C$

15. $y = x + 5\ln \left| \dfrac{y + 3}{x + 4} \right| + C$

17. $P(t) = P_0 e^{kt}$; approximately 7.9 years; 10 years

19. $A(t) = 100e^{-0.0051t}$; 88.5 milligrams

21. Approximately 24,180 years

23. \$6665.45; 12; \$6651.82

25. Approximately 64.5 degrees

27. $i(t) = \frac{6}{5} - \frac{6}{5}e^{-20t}$

29. $v(t) = \dfrac{mg}{k} + \left(v_0 - \dfrac{mg}{k} \right) e^{-kt/m}$;

  $v \to mg/k$ as $t \to \infty$;

  $s(t) = \dfrac{mg}{k} t - \dfrac{m}{k} \left(v_0 - \dfrac{mg}{k} \right) e^{-kt/m}$

  $+ \dfrac{m}{k} \left(v_0 - \dfrac{mg}{k} \right) + s_0$

31. $2h^{1/2} = -\dfrac{t}{25} + 4\sqrt{5}$; $100\sqrt{5}$ sec

33. $v^2 = 2k/y - 2k/R + v_0^2$

Exercises 8.8, Page 469

1. $\sinh x = -\frac{1}{2}$, $\cosh x = \sqrt{5}/2$, $\tanh x = -\sqrt{5}/5$, $\coth x = -\sqrt{5}$, $\text{sech } x = 2\sqrt{5}/5$, $\text{csch } x = -2$

19. $10\sinh 10x$ **21.** $\frac{1}{2}x^{-1/2}\,\text{sech}^2\sqrt{x}$

23. $-6(3x - 1)\text{sech}(3x - 1)^2 \tanh(3x - 1)^2$

25. $-3\sinh 3x\,\text{csch}^2(\cosh 3x)$

27. $3\sinh 2x \sinh 3x + 2\cosh 2x \cosh 3x$

29. $2x^2 \sinh x^2 + \cosh x^2$

31. $3\sinh^2 x \cosh x$

33. $\frac{2}{3}(x - \cosh x)^{-1/3}(1 - \sinh x)$

35. $4\tanh 4x$ **37.** $(e^x + 1)/(1 + \cosh x)^2$

39. $e^{\sinh t} \cosh t$

41. $\dfrac{1 + \sinh 2t - 2\sqrt{1 - t^2}\,\sin^{-1}t \cosh 2t}{\sqrt{1 - t^2}(1 + \sinh 2t)^2}$

43. $y = 3x$ **45.** $\frac{1}{8}\cosh 8x + C$

47. $\frac{1}{5}\sinh(5x - 4) + C$ **49.** $\frac{1}{3}\tanh x^3 + C$

51. $-3\,\text{csch}\sqrt[3]{x} + C$ **53.** $\frac{1}{3}\cosh^3 x + C$

55. $\frac{1}{5}\ln(7 + \cosh 5x) + C$

57. $-\frac{1}{3}e^{-\cosh 3x} + C$ **59.** $\frac{1}{7}\sinh^7 x + C$

61. $\sinh e^x + C$ **63.** $2\sinh 1$

65. $\sinh 3 + \sinh 1 - 4$ **67.** $\pi(\cosh 3 - 1)$

69. $(\pi/2)(b - a) + (\pi/4)(\sinh 2b - \sinh 2a)$

71. $(x^2 + 1)/2x$, $x > 0$

73. $-2\,\text{sech}^2 x \tanh x$

Exercises 8.9, Page 477

1. $3/\sqrt{9x^2 + 1}$ **3.** $-2x/[1 - (1 - x^2)^2]$

5. $\sec x$ **7.** $3x^3/\sqrt{x^6 + 1} + \sinh^{-1}x^3$

9. $-1/x^2\sqrt{1 - x^2} - (\text{sech}^{-1}x)/x^2$

11. $-1/(x\sqrt{1 - x^2}\,\text{sech}^{-1}\,x)$

13. $3(\cosh^{-1}6x)^{-1/2}/\sqrt{36x^2 - 1}$

15. $(x + 1)/\sqrt{x^2 - 1}$ **17.** $\sinh x/\sqrt{\cosh^2 x + 1}$

19. $\coth^{-1}x$ **21.** $\frac{1}{2}\sinh^{-1}2x + C$

23. $\begin{cases} \frac{1}{2}\tanh^{-1}2x + C, & |x| < \frac{1}{2} \\ \frac{1}{2}\coth^{-1}2x + C, & |x| > \frac{1}{2} \end{cases}$

  or $\frac{1}{4}\ln \left| \dfrac{1 + 2x}{1 - 2x} \right| + C$, $2x \neq 1$

25. $\frac{1}{3}\cosh^{-1}\frac{3}{4}x + C$, $3x > 4$

27. $\frac{1}{2}\sinh^{-1}x^2 + C$

29. $\begin{cases} -\tanh^{-1}(x + 1) + C, & |x + 1| < 1 \\ -\coth^{-1}(x + 1) + C, & |x + 1| > 1 \end{cases}$

  or $\frac{1}{2}\ln \left| \dfrac{x + 2}{-x} \right| + C$, $x + 1 \neq 1$

31. $\sinh^{-1}(x + 2) + C$ **33.** $\cosh^{-1}e^x + C$

35. $\sinh^{-1}(\sin \theta) + C$ **37.** $(t^2 + 16)^{1/2} + C$

39. $x - \ln \left| \dfrac{1 + x}{1 - x} \right| + C$

41. Approximately 2.6998

43. Approximately 0.9730

45. $\pi/6$; $\pi(\sqrt{3} - 1)$; $\pi \ln\left(\dfrac{2 + \sqrt{3}}{6 - 3\sqrt{3}} \right)$

47. $y = (T/w)\cosh(wx/T) + 1 - T/w$

57. $\frac{1}{2}\ln|1 + x| - \frac{1}{2}\ln|1 - x| + C = \frac{1}{2}\ln \left| \dfrac{1 + x}{1 - x} \right| + C$

Chapter 8 Review Exercises, Page 479

1. True   **3.** False   **5.** False   **7.** True
9. False   **11.** False   **13.** True
15. False   **17.** True   **19.** True   **21.** 3
23. $a - b$ **25.** 2
27. Logarithmic differentiation   **29.** Catenary
31. $1/x + 2/(4x - 1)$
33. $\cos(e^x \ln x)(e^x/x + e^x \ln x)$
35. $e^x + 2e^{2x}$ **37.** $7x^6 + 7^x \ln 7$

39. $(x^2 + e^{2x})^{\ln x}\left[\dfrac{(2x + 2e^{2x})\ln x}{x^2 + e^{2x}} + \dfrac{\ln(x^2 + e^{2x})}{x}\right]$

41. $(e^x - y^2)/(2xy + e^y)$ **43.** $3x^2 e^{x^3}\cosh e^{x^3}$
45. $\sinh(\cosh x)\sinh x$
47. $1/(\sqrt{1 - x^2}\sqrt{(\sin^{-1}x)^2 + 1})$

49. $e^{x\cosh^{-1}x}\left[\dfrac{x^2}{\sqrt{x^2 - 1}} + x \cosh^{-1}x + 1\right]$

51. $-\frac{1}{10}\ln|1 - 5x^2| + C$ **53.** $\frac{1}{2}\ln|x(x + 2)| + C$

55. $\dfrac{1}{8} + \dfrac{6}{\ln 7} + 7^\pi$ **57.** $\frac{1}{4}\ln|\sin 4x| + C$

59. $\frac{1}{3}(e^{-6} - e^{-15})$ **51.** $(24)^t/\ln 24 + C$
63. $-\cosh(1/x) + C$ **65.** $-\frac{1}{9}(\sinh 3x)^{-3} + C$

67. $\begin{cases} \frac{1}{2}\tanh^{-1}\left(\dfrac{x - 7}{2}\right) + C, & |x - 7| < 2 \\[2mm] \frac{1}{2}\coth^{-1}\left(\dfrac{x - 7}{2}\right) + C, & |x - 7| > 2 \end{cases}$

69. $\frac{1}{3}\sinh^{-1}3$
71.

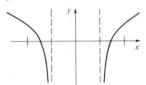

73. $1225\pi/72$ **75.** $y = C\cos x$

77. $y = \dfrac{2(Ce^{4x} - 1)}{Ce^{4x} + 1}$

79. Approximately 55,800 years

Exercises 9.1, Page 485

1. $\frac{1}{5}(x + 1)^5 - \frac{1}{4}(x + 1)^4 + C$
3. $\frac{4}{5}(x - 5)^{5/2} + \frac{22}{3}(x - 5)^{3/2} + C$
5. $\frac{2}{3}(x - 1)^{3/2} + 2(x - 1)^{1/2} + C$
7. $\frac{2}{9}(3x - 4)^{1/2} - \frac{26}{9}(3x - 4)^{-1/2} + C$
9. $2\sqrt{x} - 2\tan^{-1}\sqrt{x} + C$
11. $t - 8\sqrt{t} + 8\ln(\sqrt{t} + 1) + C$
13. $\frac{3}{10}(x^2 + 1)^{5/3} - \frac{3}{4}(x^2 + 1)^{2/3} + C$
15. $-(x - 1)^{-1} - (x - 1)^{-2} - \frac{1}{3}(x - 1)^{-3} + C$
17. $2x^{1/2} + 3x^{1/3} + 6x^{1/6} + 6\ln|x^{1/6} - 1| + C$
19. $2\sqrt{e^x - 1} - 2\tan^{-1}\sqrt{e^x - 1} + C$
21. $-\frac{4}{3}(1 - \sqrt{v})^{3/2} + \frac{4}{5}(1 - \sqrt{v})^{5/2} + C$
23. $\frac{4}{3}(1 + \sqrt{t})^{3/2} + C$

25. $\ln|x^2 + 2x + 5| + \frac{5}{2}\tan^{-1}\left(\dfrac{x + 1}{2}\right) + C$

27. $-2\sqrt{16 - 6x - x^2} - \sin^{-1}\left(\dfrac{x + 3}{5}\right) + C$

29. $\frac{506}{375}$ **31.** $6 + 20\ln\frac{11}{14}$ **33.** $\frac{177}{2}$
35. $\frac{1}{1326}$ **37.** $3 + 3\ln\frac{2}{3}$ **39.** $\frac{1}{168}$
41. $-\frac{3}{2} + 3\ln 2$ **43.** $32\pi/3 - 4\pi\ln 3$
45. $\frac{232}{15}$

Exercises 9.2, Page 492

1. $\frac{2}{3}x(x + 3)^{3/2} - \frac{4}{15}(x + 3)^{5/2} + C$
3. $x\ln 4x - x + C$
5. $(x^2/2)\ln 2x - x^2/4 + C$
7. $-x^{-1}\ln x - x^{-1} + C$
9. $t(\ln t)^2 - 2t\ln t + 2t + C$ **11.** $\frac{3}{2}\ln 3$
13. $\pi/4 - \frac{1}{2}\ln 2$ **15.** $x\sin^{-1}x + (1 - x^2)^{1/2} + C$
17. $\frac{1}{3}xe^{3x} - \frac{1}{9}e^{3x} + C$ **19.** $\frac{1}{2}e^{x^2}(x^2 - 1) + C$
21. $-12e^{-2} + 8e^{-1}$ **23.** $\frac{1}{8}t\sin 8t + \frac{1}{64}\cos 8t + C$
25. $-x^2\cos x + 2x\sin x + 2\cos x + C$
27. $\frac{1}{17}e^x[\sin 4x - 4\cos 4x] + C$
29. $\theta\sec\theta - \ln|\sec\theta + \tan\theta| + C$
31. $\frac{1}{3}\cos x\cos 2x + \frac{2}{3}\sin x\sin 2x + C$
33. $\frac{1}{3}x^2(x^2 + 4)^{3/2} - \frac{2}{15}(x^2 + 4)^{5/2} + C$
35. $\frac{1}{2}x[\sin(\ln x) - \cos(\ln x)] + C$
37. $-\frac{1}{2}\csc x\cot x + \frac{1}{2}\ln|\csc x - \cot x| + C$
39. $\frac{1}{4}x^2 - \frac{1}{4}x\sin 2x - \frac{1}{8}\cos 2x + C$
41. $x(\sin^{-1}x)^2 + 2\sqrt{1 - x^2}\sin^{-1}x - 2x + C$

43. $-\dfrac{x^2 e^x}{x + 2} + xe^x - e^x + C$

45. $5\tan^{-1}2 - 2$
47. $-2\sqrt{x + 2}\cos\sqrt{x + 2} + 2\sin\sqrt{x + 2} + C$
49. $\frac{1}{2}(\sin^{-1}x)^2 + C$ **51.** $3\ln 3 + e^{-1}$
53. $5\pi(\ln 5)^2 - 10\pi\ln 5 + 8\pi$ **55.** $2\pi^2$
61. $-\frac{1}{3}\sin^2x\cos x - \frac{2}{3}\cos x + C$

Exercises 9.3, Page 499

1. $\frac{2}{3}(\sin x)^{3/2} + C$ **3.** $\sin x - \frac{1}{3}\sin^3x + C$
5. $\frac{1}{4}\sin^4x - \frac{1}{6}\sin^6x + C$ **7.** $25\sqrt{2}/168$
9. $\frac{1}{60}$ **11.** $\frac{3}{8}t - \frac{1}{4}\sin 2t + \frac{1}{32}\sin 4t + C$
13. $\frac{1}{16}x - \frac{1}{64}\sin 4x + \frac{1}{48}\sin^32x + C$
15. $\frac{3}{128}x - \frac{1}{128}\sin 4x + \frac{1}{1024}\sin 8x + C$
17. $\sqrt{3} - \pi/3$ **19.** $\frac{1}{12}\tan^62t + \frac{1}{8}\tan^42t + C$
21. $\tan x + \frac{1}{3}\tan^3x + C$
23. $-\frac{1}{11}\cot^{11}x - \frac{1}{13}\cot^{13}x + C$
25. $\frac{1}{4}\tan^4x - \frac{1}{2}\tan^2x - \ln|\cos x| + C$
27. $\frac{2}{3}\sec^{3/2}x + 2\sec^{-1/2}x + C$
29. $\frac{1}{4}\sec^3x\tan x - \frac{1}{8}\sec x\tan x$
$-\frac{1}{8}\ln|\sec x + \tan x| + C$

31. $\ln|\sin x| + \frac{1}{2}\cos^2 x + C$

33. $\frac{1}{7}\tan^{-7}(1-t) + \frac{1}{5}\tan^{-5}(1-t) + C$

35. $-\frac{1}{2}\cot^2 t - \ln|\sin t| + C$

37. $\pi^2/12 + \sqrt{3}\pi/16$

39. $-\frac{1}{6}\cos 3x + \frac{1}{2}\cos x + C$

41. $\frac{1}{4}\sin 2x - \frac{1}{12}\sin 6x + C$ **43.** $\frac{5}{12}$

Exercises 9.4, Page 506

1. $-\sin^{-1}x - \sqrt{1-x^2}/x + C$

3. $\ln|x/6 + \sqrt{x^2-36}/6| + C$

5. $\frac{1}{2}x\sqrt{x^2+4} + 2\ln|\sqrt{x^2+4}/2 + x/2| + C$

7. $-\frac{1}{3}(1-x^2)^{3/2} + \frac{1}{5}(1-x^2)^{5/2} + C$

9. $-x/4\sqrt{x^2-4} + C$ **11.** $\frac{1}{3}(x^2+7)^{3/2} + C$

13. $\sin^{-1}(x/5) + C$

15. $\frac{1}{4}\ln|4/x - \sqrt{16-x^2}/x| + C$

17. $\ln|\sqrt{x^2+1}/x - 1/x| + C$

19. $-(1-x^2)^{3/2}/3x^3 + C$

21. $x/\sqrt{9-x^2} - \sin^{-1}(x/3) + C$

23. $\frac{1}{2}\tan^{-1}x + x/[2(1+x^2)] + C$

25. $x/[16\sqrt{4+x^2}] - x^3/[48(4+x^2)^{3/2}] + C$

27. $\ln|\sqrt{x^2+2x+10}/3 + (x+1)/3| + C$

29. $\frac{1}{16}\tan^{-1}[(x+3)/2]$
$+ (x+3)/[8(x^2+6x+13)] + C$

31. $-(5x+1)/[9\sqrt{5-4x-x^2}] + C$

33. $\ln|x^2+4x+13| + C$

35. $x - 4\tan^{-1}(x/4) + C$

37. $\sqrt{3} + 2\pi/3$ **39.** $\sqrt{2}/50$

41. $2\sqrt{3} - \frac{172}{81}$ **43.** $\sec^{-1}(e^x) + C$

45. $(\sqrt{3}/3)\ln\left(\dfrac{\sqrt{2}-1}{2-\sqrt{3}}\right)$

49. $(\sqrt{3}\pi/9)(\sqrt{3}-1-\pi/12)$

51. $12\sqrt{2}\pi - 4\pi\ln(\sqrt{2}+1)$

53. $2 - \sqrt{2} - \ln(\sqrt{6} - \sqrt{3})$

55. $\frac{1}{3}x^3\sin^{-1}x + \frac{1}{3}\sqrt{1-x^2} - \frac{1}{9}(1-x^2)^{3/2} + C$

57. $-\sqrt{9+x^2}/9x + C$

Exercises 9.5, Page 515

1. $-\frac{1}{2}\ln|x| + \frac{1}{2}\ln|x-2| + C$

3. $-2\ln|x| + \frac{5}{2}\ln|2x-1| + C$

5. $-\frac{1}{6}\ln|x+3| + \frac{1}{6}\ln|x-3| + C$

7. $\frac{5}{8}\ln|x-4| + \frac{3}{8}\ln|x+4| + C$

9. $-\frac{1}{2}\ln|x+3| + \frac{1}{2}\ln|x+1| + C$

11. $-\frac{1}{6}\ln|2x+1| + \frac{2}{3}\ln|x+2| + C$

13. $6\ln|x| - \frac{7}{2}\ln|x+1| - \frac{3}{2}\ln|x-1| + C$

15. $\frac{1}{2}\ln|x+1| - \ln|x+2| + \frac{1}{2}\ln|x+3| + C$

17. $-2\ln|t| - t^{-1} + 6\ln|t-1| + C$

19. $\ln|x| - \ln|x+1| + (x+1)^{-1} + C$

21. $-\frac{1}{3}(x-3)^{-3} + C$

23. $-2(x+1)^{-1} + \frac{3}{2}(x+1)^{-2} + C$

25. $-\frac{1}{2}(x^2-1)^{-1} + C$

27. $-\frac{1}{32}\ln|x+1| - \frac{1}{16}(x+1)^{-1} + \frac{1}{32}\ln|x+5|$
$- \frac{1}{16}(x+5)^{-1} + C$

29. $-\frac{19}{16}\ln|x| - \frac{19}{8}x^{-1} + \frac{11}{8}x^{-2} - \frac{3}{2}x^{-3}$
$+ \frac{35}{16}\ln|x+2| + C$

31. $\frac{1}{3}x^3 - x^2 + 6x - 10\ln|x+1| - 8(x+1)^{-1} + C$

33. $-\frac{1}{2}\ln 3$ **35.** $2\ln\frac{5}{3} - \frac{14}{15}$

37. $\ln|1+\sin x| - \ln|2+\sin x| + C$

39. $\frac{1}{9}\ln\left|\dfrac{e^t-2}{e^t+1}\right| + \dfrac{1}{3(e^t+1)} + C$

41. $\frac{1}{4}\ln(1+\sqrt{1-x^2}) + \frac{1}{4}(1+\sqrt{1-x^2})^{-1}$
$- \frac{1}{4}\ln|1-\sqrt{1-x^2}| - \frac{1}{4}(1-\sqrt{1-x^2})^{-1} + C$

43. $\frac{1}{4}\ln\frac{15}{7}$ **45.** $8\pi\ln\frac{2}{3} + 11\pi/3$

47. $8\pi\ln 2 - 4\pi$

49. $\frac{1}{3}\tan^{-1}x - \frac{1}{6}\tan^{-1}(x/2) + C$

51. $\frac{1}{2}\ln|x^2+2x+5| - \frac{1}{2}\ln|x^2+6x+10|$
$- 2\tan^{-1}(x+3) + C$

53. $-\ln|x| + \frac{1}{2}\ln(x^2+1) + \tan^{-1}x + C$

55. $\frac{1}{6}\ln|x-3| - \frac{1}{12}\ln(x^2+9) + \frac{1}{2}\tan^{-1}(x/3) + C$

57. $\frac{1}{2}(x+1)^{-1} + \frac{1}{2}\tan^{-1}x + C$

59. $-\frac{1}{4}\ln|t+1| + \frac{1}{4}\ln|t-1| - \frac{1}{2}\tan^{-1}t + C$

61. $\frac{1}{16}\tan^{-1}(x/2) + (x-8)/[8(x^2+4)] + C$

63. $5\ln|x+1| - \ln|x^2+2x+2|$
$- 7\tan^{-1}(x+1) + C$

65. $\frac{1}{3}\ln|x-1| - \frac{1}{6}\ln|x^2+x+1|$
$- (\sqrt{3}/3)\tan^{-1}[(2x+1)/\sqrt{3}] + C$

67. $-x^{-1} - \frac{3}{2}\tan^{-1}x - x/[2(x^2+1)] + C$

69. $\frac{1}{2}\ln(x^2+4) - \frac{11}{16}\tan^{-1}(x/2)$
$+ (12+5x)/[8(x^2+4)] + C$

71. $-\frac{1}{4}(4x^2+5)^{-1} + C$

73. $\frac{3}{4}\ln|x| - \frac{3}{8}\ln|x^2+2x+2| - 3\tan^{-1}(x+1)$
$- (9x+8)/[4(x^2+2x+2)] + C$

75. $\frac{1}{6}\ln\frac{8}{3} + (\sqrt{2}/6)\tan^{-1}(\sqrt{2}/2)$ **77.** 0

79. $3(x+1)^{1/3} + \ln|(x+1)^{1/3} - 1|$
$- \frac{1}{2}\ln|(x+1)^{2/3} + (x+1)^{1/3} + 1|$
$- \sqrt{3}\tan^{-1}\{[2(x+1)^{1/3} + 1]/\sqrt{3}\} + C$

81. $\ln 9 - \frac{1}{2}\ln 17$ **83.** $\pi^2/2 - \pi$

Exercises 9.6, Page 518

1. $\ln\left|1 + \tan\left(\dfrac{x}{2}\right)\right| + C$

3. $(2\sqrt{3}/3)\tan^{-1}\left[\left(\dfrac{\sqrt{3}}{3}\right)\tan\left(\dfrac{x}{2}\right)\right] + C$

5. $-\tan(x/2) + 2\tan[\tan^{-1}(x/2)] + C$

7. $\frac{1}{2}\ln|\tan(x/2)| - \frac{1}{4}\tan^2(x/2) + C$

9. $-\tan(x/2) + \ln\left|\dfrac{1+\tan(x/2)}{1-\tan(x/2)}\right| + C$

11. $\dfrac{1}{\sqrt{3}}\ln\left|\dfrac{\sqrt{3}+2-\tan(x/2)}{\sqrt{3}-2+\tan(x/2)}\right| + C$

13. $2/(1+\sqrt{3})$ **15.** $\frac{1}{2}\ln\frac{5}{3}$

Exercises 9.7, Page 519

1. $(\pi/8)[3\sqrt{2} - \ln(\sqrt{2} + 1)]$
3. $\pi/4 - \frac{1}{2}\ln 2$ 5. $\frac{1}{5}\ln\frac{4}{9}$
7. $s(t) = \frac{1}{2} - \frac{1}{2}e^{-t}\sin t - \frac{1}{2}e^{-t}\cos t$
9. Approximately 31.8 ft-lb
11. Approximately 49.01 lb
13. $\bar{x} = 3\sqrt{5}/5$ 15. $\bar{x} = 1$, $\bar{y} = \pi/8$
17. $(1/\ln(2 + \sqrt{3}), \pi/[6\ln(2 + \sqrt{3})])$
19. $aP_0/[bP_0 + (a - bP_0)e^{-at}]$
21. $x = 10\ln(10/y + \sqrt{100 - y^2}/y) - \sqrt{100 - y^2}$

Exercises 9.8, Page 522

1. $-\sqrt{9 + x^2}/9x + C$
3. $\sqrt{x^2 - 5} - \sqrt{5}\cos^{-1}(\sqrt{5}/x) + C$
5. $1/(16 + 20x) - \frac{1}{16}\ln|(4 + 5x)/x| + C$
7. $\frac{1}{7}t^2(1 + 2t)^{3/2} - \frac{1}{105}(3t - 1)(1 + 2t)^{3/2} + C$
9. $-(3 - x) + 9/(3 - x) + 6\ln|3 - x| + C$
11. $\frac{1}{4}\tan^4\theta - \frac{1}{2}\tan^2\theta - \ln|\cos\theta| + C$
13. $x\ln(x^2 + 16) - 2x + 8\tan^{-1}(x/4) + C$
15. $-\frac{1}{2}\tan[(\pi/4) - x] + C$
17. $-\sqrt{2x - x^2} + \cos^{-1}(1 - x) + C$
19. $63\pi/512$ 21. $\frac{3}{13}(e^{2\pi} + 1)$

Chapter 9 Review Exercises, Page 523

1. True   3. True   5. True   7. False
9. False   11. $2\sqrt{x} - 18\ln(\sqrt{x} + 9) + C$
13. $(x^2 + 4)^{1/2} + C$
15. $\frac{3}{256}\tan^{-1}(x/2) + x/[32(x^2 + 4)] + x/[32(x^2 + 4)^2]$ $- x^3/[128(x^2 + 4)^2] + C$
17. $x - 4x^{-1} + C$
19. $\frac{1}{2}\ln(x^2 + 4) - \frac{5}{2}\tan^{-1}(x/2) + C$
21. $(\ln x)^{10}/10 + C$
23. $(t^2/2)\sin^{-1}t - \frac{1}{4}\sin^{-1}t + \frac{1}{4}t\sqrt{1 - t^2} + C$
25. $\frac{1}{5}(x + 1)^5 - \frac{3}{4}(x + 1)^4 + C$
27. $x\ln(x^2 + 4) - 2x + 4\tan^{-1}(x/2) + C$
29. $-\frac{2}{125}\ln|x| - \frac{1}{25}x^{-1} + \frac{2}{125}\ln|x + 5|$ $- \frac{1}{25}(x + 5)^{-1} + C$
31. $-\frac{1}{12}\ln|x + 3| - \frac{1}{2}(x + 3)^{-1} + \frac{1}{12}\ln|x - 3| + C$
33. $\tan t - t + C$ 35. $\frac{1}{13}\tan^{13}x + \frac{1}{11}\tan^{11}x + C$
37. $y\sin y + \cos y + C$ 39. $\sin t - \frac{1}{5}\sin^5t + C$
41. $\frac{1}{6}(1 + e^w)^6 + C$
43. $-\frac{1}{8}\csc^2 4x - \frac{1}{4}\ln|\sin 4x| + C$
45. $\frac{1}{4}$ 47. $\sec x - \tan x + x + C$
49. $\frac{5}{2}\ln 2 - \frac{3}{2}\ln 3$
51. $(e^x/10)[\cos 3x + 3\sin 3x] + C$
53. $\frac{1}{2}t\cos(\ln t) + \frac{1}{2}t\sin(\ln t) + C$
55. $2\sqrt{x}\sin\sqrt{x} + 2\cos\sqrt{x} + C$
57. $-\frac{2}{3}\cos^3 x + C$
59. $\frac{1}{2}(x + 1)\sqrt{x^2 + 2x + 5}$ $+ 2\ln|\sqrt{x^2 + 2x + 5}/2 + (x + 1)/2| + C$
61. $\frac{1}{7}\sec^7 x - \frac{2}{5}\sec^5 x + \frac{1}{3}\sec^3 x + C$

63. $\frac{1}{4}t^4 - \frac{1}{2}t^2 + \frac{1}{2}\ln(1 + t^2) + C$
65. $\frac{5}{2}\ln(x^2 + 1) + \tan^{-1}x - \frac{1}{2}(x^2 + 1)^{-1} + C$
67. $\frac{1}{4}x^2 - \frac{1}{4}x\sin 2x - \frac{1}{8}\cos 2x + C$
69. $\sin x\, e^{\sin x} - e^{\sin x} + C$ 71. $\sqrt{6} - 2$
73. $t\sinh^{-1}t - (t^2 + 1)^{1/2} + C$
75. $\ln\frac{3}{2}$ 77. $\frac{1}{39}\tan^{13}3u + \frac{1}{45}\tan^{15}3u + C$
79. $3\tan x + \sec x + C$
81. $\frac{1}{2}x^2(1 + \ln x)^2 - \frac{1}{2}x^2(1 + \ln x) + \frac{1}{4}x^2 + C$
83. $e^{e^x} + C$
85. $\ln\left|\tan\left(\dfrac{x}{2}\right)\right| + \dfrac{1}{2}\cot^2\left(\dfrac{x}{2}\right) + C$
87. $\frac{1}{5}\sin^{-1}(5x + 2) + C$
89. $\sin x\ln|\sin x| - \sin x + C$

Exercises 10.1, Page 535

1. 0   3. 2   5. $\frac{1}{3}$ 7. 10
9. -6 11. $\frac{1}{2}$ 13. $\frac{7}{5}$ 15. $\frac{1}{6}$
17. Does not exist   19. $\frac{1}{2}$ 21. 0
23. 0   25. 5   27. Does not exist
29. -2 31. $-\frac{1}{8}$ 33. -1
35. Does not exist   37. $\frac{1}{9}$ 39. 3
45. $-\frac{1}{2}$ 47. 1   49. 1   51. 0
53. -4 55. $\frac{1}{4}$ 57. 1   59. e^3
61. 1   63. $\frac{1}{4}$ 65. 0   67. $\frac{1}{5}$
69. 0   71. 1   73. $\frac{1}{24}$
75. Does not exist   77. $e^{-1/3}$ 79. 1

81.

0.1	13,781
0.01	1.6×10^{43}

83. Does not exist; 0
85. 1 if $n = 1$; does not exist if $n > 1$

Exercises 10.2, Page 544

1. $\frac{1}{81}$ 3. Diverges   5. $\frac{1}{2}e^6$
7. Diverges   9. $\frac{1}{2}$ 11. 0   13. $-\frac{1}{18}$
15. $3e^{-2}$ 17. 1   19. $\pi/2$ 21. $\frac{1}{2}$
23. 4   25. $\ln 2$ 27. $\frac{1}{4}\ln\frac{7}{3}$ 29. $\frac{1}{21}$
31. $\frac{1}{6}$
33. The integral $\displaystyle\int_1^\infty (1/x)\,dx$ diverges;

$\displaystyle\pi\int_1^\infty (1/x^2)\,dx = \pi$

35. 2.86×10^{10} joules   39. $1/s$, $s > 0$
41. $1/(s - 1)$, $s > 1$ 43. $1/(s^2 + 1)$, $s > 0$
45. Diverges for $k \le 1$; converges for $k > 1$
47. Diverges for $k \ge 0$; converges for $k < 0$
49. Since $\sin^2 x \le 1$, use $g(x) = 1/x^2$; the integral

$\displaystyle\int_1^\infty (1/x^2)\,dx$ converges

51. On $[0, \infty)$, $1/(x + e^x) < 1/e^x = e^{-x}$;

use $g(x) = e^{-x}$; the integral $\int_0^\infty e^{-x}\,dx$ converges

55. $c = \beta^{n+1}/n!$ **57.** Diverges   **59.** 100
61. $2\sqrt{2}$ **63.** Diverges   **65.** 6   **67.** $-\frac{1}{4}$
69. Diverges   **71.** Diverges   **73.** $-\frac{4}{3}$
75. $\pi/4$ **77.** $\pi/2$ **79.** 4
81. $2\sqrt{3}$; the integral $\pi \int_{-2}^1 dx/(x + 2)$ diverges

83. $\frac{1}{2}\ln 2$
85. Diverges for $k \geq 1$; converges for $k < 1$
87. $\pi/2$ **89.** π
91. It is common practice to define $f(0) = 1$. Hence it is not an improper integral.

Chapter 10 Review Exercises, Page 547

1. False   **3.** False   **5.** False   **7.** True
9. True   **11.** True   **13.** True   **15.** False
17. False   **19.** False   **21.** 0
23. $8\sqrt{3}\pi/9$ **25.** Does not exist   **27.** $\frac{4}{25}$
29. -2 **31.** 1   **33.** e^{-1} **35.** Does not exist   **37.** $\frac{3}{2}\sqrt[3]{9}$ **39.** 0   **41.** Diverges
43. 0   **45.** Diverges   **47.** $2 - 2e^{-1}$
49. $\frac{2}{3}$ **51.** 2π; the areas are infinite   **55.** $\frac{1}{2}$
57. 5 yr

Exercises 11.1, Page 556

1. $\frac{1}{3}, \frac{1}{5}, \frac{1}{7}, \frac{1}{9}, \ldots$ **3.** $-1, \frac{1}{2}, -\frac{1}{3}, \frac{1}{4}, \ldots$
5. $3, -3, 3, -3, \ldots$
7. $10, 100, 1000, 10{,}000, \ldots$
9. $1, 1 + \frac{1}{2}, 1 + \frac{1}{2} + \frac{1}{3}, 1 + \frac{1}{2} + \frac{1}{3} + \frac{1}{4}, \ldots$
11. $2, 4, 12, 48, \ldots$ **13.** Choose $N > 1/\varepsilon$.
15. Choose $N > (1 - \varepsilon)/\varepsilon$.   **17.** Choose $N > -(\ln \varepsilon)/\ln 10$.   **19.** 0   **21.** 0   **23.** $\frac{1}{2}$
25. Sequence diverges   **27.** Sequence diverges
29. $\frac{1}{4}$ **31.** 0   **33.** 0
35. Sequence diverges   **37.** 0   **39.** $\frac{5}{7}$
41. 1   **43.** 6   **45.** 1   **47.** 1
49. $\ln \frac{4}{3}$ **51.** 0   **53.** $\{2n/(2n - 1)\}$
55. $\{(-1)^{n+1}(2n + 1)\}$ **57.** $\left\{\dfrac{2}{3^{n-1}}\right\}$
59. $\frac{40}{9}$ ft; $15(\frac{2}{3})^n$
61. $15, 18, 18.6, 18.72, 18.74, 18.75, \ldots$
63. After first year: $1 + r$;
after second year: $(1 + r) + r(1 + r) = (1 + r)^2$;
after third year: $(1 + r)^2 + r(1 + r)^2 = (1 + r)^3$
65. Use $a_n = 0$ and $c_n = 1/4^n$.
67. Use $a_n = 4$ and $c_n = 4 + 1/n$.
69. $-\frac{1}{2}, -\frac{1}{4}, -\frac{1}{8}, -\frac{1}{16}, \ldots$ **71.** $3, 1, \frac{1}{3}, \frac{1}{3}, \ldots$

73.

34	55	89	144	233
21	34	55	89	144
55	89	144	233	377

77. $1 + \dfrac{1}{2 + \dfrac{1}{2 + \dfrac{1}{2 + \frac{1}{2}}}}$, $1 + \dfrac{1}{2 + \dfrac{1}{2 + \dfrac{1}{2 + \dfrac{1}{2 + \frac{1}{2}}}}}$;

$1.5, 1.4, 1.4167, 1.4138, 1.4143, \ldots$; the sequence appears to converge to $\sqrt{2}$.
79. 3

Exercises 11.2, Page 562

1. Increasing   **3.** Not monotonic
5. Increasing   **7.** Nonincreasing
9. Increasing   **11.** Not monotonic
13. Bounded and increasing
15. Bounded and increasing
17. Bounded and decreasing
19. Bounded and decreasing
21. Bounded and increasing
23. Bounded and decreasing   **25.** $\{(-1)^n\}$
27. The sequence is increasing. It is also bounded since

$$0 \leq \int_1^n e^{-t^2}\,dt \leq \int_1^n e^{-t}\,dt < e^{-1}.$$

Exercises 11.3, Page 570

1. $3 + \frac{5}{2} + \frac{7}{3} + \frac{9}{4} + \cdots$
3. $\frac{1}{2} - \frac{1}{6} + \frac{1}{12} - \frac{1}{20} + \cdots$
5. $1 + 2 + \frac{3}{2} + \frac{4}{6} + \cdots$
7. $2 + \frac{8}{3} + \frac{48}{15} + \frac{384}{105} + \cdots$
9. $-\frac{1}{7} + \frac{1}{9} - \frac{1}{11} + \frac{1}{13} - \cdots$
11. Converges to $\frac{15}{4}$ **13.** Converges to $\frac{2}{3}$
15. Diverges   **17.** Converges to 9000
19. Diverges   **21.** $\frac{2}{9}$ **23.** $\frac{61}{99}$ **25.** $\frac{1313}{999}$
27. 1   **29.** $\frac{1}{2}$ **31.** $\frac{17}{6}$
33. By Theorem 11.7   **35.** By Theorem 11.7
37. By Theorem 11.7
39. The series is a constant multiple of the divergent harmonic series and, hence, divergent by Theorem 11.8.
41. By Theorem 11.10   **43.** $|x| < 2$
45. $-2 < x < 0$ **47.** 75 ft
49. 1000   **51.** 18.75 milligrams
61. The argument is invalid since
$1 + 2 + 4 + 8 + \cdots$ is a divergent series.

Exercises 11.4, Page 580

1. Converges   **3.** Diverges
5. Diverges   **7.** Converges
9. Converges   **11.** Diverges
13. Converges   **15.** Converges
17. Converges   **19.** Converges
21. Converges   **23.** Converges
25. Diverges   **27.** Converges
29. Converges   **31.** Diverges

33. Converges   **35.** Converges
37. Converges   **39.** Diverges
43. Converges   **45.** Diverges
47. Converges   **49.** Converges
51. Diverges   **53.** Converges
55. Converges   **57.** Converges
59. Diverges   **61.** Diverges
63. $0 \le p \le 1$ **65.** p any real number

Exercises 11.5, Page 587

1. Converges   **3.** Diverges
5. Converges   **7.** Converges
9. Converges   **11.** Converges
13. Diverges   **15.** Conditionally convergent
17. Absolutely convergent
19. Absolutely convergent
21. Absolutely convergent   **23.** Diverges
25. Conditionally convergent   **27.** Diverges
29. Conditionally convergent
31. Absolutely convergent   **33.** Diverges
35. 0.84147   **37.** 5   **39.** 0.9492
41. Less than $\frac{1}{101} \approx 0.009901$

43. The series contains mixed algebraic signs but is not an alternating series; converges
45. The series is not an alternating series; converges
47. $a_{k+1} \le a_k$ is not satisfied for k sufficiently large; diverges
49. Diverges; converges; converges; diverges
51. Adding zeros to the series for $S/2$ does not affect its value.

Exercises 11.6, Page 592

1. $(-1, 1]$ **3.** $[-\frac{1}{2}, \frac{1}{2})$ **5.** $[2, 4]$
7. $(-5, 15)$ **9.** $\{0\}$ **11.** $[0, \frac{2}{3}]$
13. $[-1, 1)$ **15.** $(-16, 2)$ **17.** $(-\frac{75}{32}, \frac{75}{32})$
19. $[\frac{2}{3}, \frac{4}{3}]$ **21.** $(-\infty, \infty)$ **23.** $(-3, 3)$
25. $(-\infty, \infty)$ **27.** $(-\frac{15}{4}, -\frac{9}{4})$ **29.** 4
31. $x > 1$ or $x < -1$ **33.** $x < -\frac{1}{2}$
35. $(-2, 2)$ **37.** $x < 0$
39. $0 \le x < \pi/3, 2\pi/3 < x < 4\pi/3, 5\pi/3 < x \le 2\pi$

Exercises 11.7, Page 596

1. $\sum_{k=0}^{\infty} x^k; (-1, 1)$ **3.** $\sum_{k=0}^{\infty} \frac{(-1)^k 3^k}{5^{k+1}} x^k; \left(-\frac{5}{3}, \frac{5}{3}\right)$

5. $\sum_{k=2}^{\infty} \frac{k(k-1)}{2} x^{k-2}; (-1, 1)$

7. $\ln 4 + \sum_{k=1}^{\infty} \frac{(-1)^{k+1}}{k4^k} x^k; (-4, 4]$

9. $\sum_{k=0}^{\infty} (-1)^k x^{2k}; (-1, 1)$

11. $\sum_{k=0}^{\infty} \frac{(-1)^k}{2k+1} x^{2k+1}; [-1, 1]$

13. $\sum_{k=0}^{\infty} \frac{(-1)^k}{(k+1)(2k+3)} x^{2k+3}; [-1, 1]$
15. $(-3, 3]$ **17.** 0.0953   **19.** 0.4854
21. 0.0088   **25.** $\sum_{k=0}^{\infty} \frac{(-1)^k}{k!} x^k$ **27.** 0.3678
29. 0.1973   **31.** Set $x = 1$.

Exercises 11.8, Page 604

1. $\sum_{k=0}^{\infty} \frac{x^k}{2^{k+1}}$ **3.** $\sum_{k=0}^{\infty} \frac{(-1)^k}{k+1} x^{k+1}$

5. $\sum_{k=0}^{\infty} \frac{(-1)^k}{(2k+1)!} x^{2k+1}$ **7.** $\sum_{k=0}^{\infty} \frac{x^k}{k!}$

9. $\sum_{k=0}^{\infty} \frac{x^{2k+1}}{(2k+1)!}$ **15.** $\sum_{k=0}^{\infty} \frac{(-1)^k}{5^{k+1}} (x-4)^k$

17. $(\sqrt{2}/2) + (\sqrt{2}/2)(x - \pi/4)$
$- (\sqrt{2}/2 \cdot 2!)(x - \pi/4)^2$
$- (\sqrt{2}/2 \cdot 3!)(x - \pi/4)^3 + \cdots$
19. $\frac{1}{2} - (\sqrt{3}/2)(x - \pi/3) - (1/2 \cdot 2!)(x - \pi/3)^2$
$+ (\sqrt{3}/2 \cdot 3!)(x - \pi/3)^3 + \cdots$

21. $\ln 2 + \sum_{k=1}^{\infty} \frac{(-1)^{k+1}}{k2^k} (x-2)^k$

23. $\sum_{k=0}^{\infty} \frac{e}{k!} (x-1)^k$

25. $x + \frac{1}{3} x^3 + \frac{2}{15} x^5 + \frac{17}{315} x^7 + \cdots$

27. $\sum_{k=0}^{\infty} \frac{(-1)^k}{k!} x^{2k}$ **29.** $\sum_{k=0}^{\infty} \frac{(-1)^k}{(2k)!} x^{2k+1}$

31. $- \sum_{k=0}^{\infty} \frac{x^{k+1}}{k+1}$

33. $1 + x^2 + \frac{2}{3} x^4 + \frac{17}{45} x^6 + \cdots$
35. 0.71934; four decimal places
37. 1.34983; four decimal places
39. 0.3103   **43.** $0 + 0 + 0 + \cdots$
45. Equate $\sec x = (R + y)/R$ with $1 + x^2/2$ and solve for y. From $L = Rx$, $y = Rx^2/2 = (Rx)^2/2R$ becomes $y = L^2/2R$. For $L = 1$ mi, $y = 7.92$ in.

Exercises 11.9, Page 607

1. $1 + \frac{1}{3} x - \frac{2}{3^2 2!} x^2 + \frac{1 \cdot 2 \cdot 5}{3^3 3!} x^3 + \cdots; 1$

3. $3 - \frac{3}{2 \cdot 9} x - \frac{3}{2^2 2! 9^2} x^2 - \frac{3(1 \cdot 3)}{2^3 3! 9^3} x^3 + \cdots; 9$

5. $1 - \frac{1}{2} x^2 + \frac{1 \cdot 3}{2^2 2!} x^4 - \frac{1 \cdot 3 \cdot 5}{2^3 3!} x^6 + \cdots; 1$

7. $8 + \frac{8 \cdot 3}{2 \cdot 4} x + \frac{8 \cdot 3 \cdot 1}{2^2 2! 4^2} x^2 - \frac{8 \cdot 3 \cdot 1}{2^3 3! 4^3} x^3 + \cdots; 4$

9. $\frac{1}{4} x - \frac{2}{4 \cdot 2} x^2 + \frac{2 \cdot 3}{4 \cdot 2^2 2!} x^3 - \frac{2 \cdot 3 \cdot 4}{4 \cdot 2^3 3!} x^4 + \cdots; 2$

11. $|S_2 - S| < a_3 = \frac{x^2}{9}$

13. $x + \sum_{k=1}^{\infty} \frac{1 \cdot 3 \cdot 5 \cdots (2k-1)}{2^k k! (2k+1)} x^{2k+1}$

15. Expand the integrand of $2\int_0^{l/2}\sqrt{1+(64d^2/l^4)x^2}\,dx$

in a binomial series and integrate term-by-term.

17. $P_0(x)=1$, $P_1(x)=x$, $P_2(x)=\frac{1}{2}(3x^2-1)$

19. $\sqrt{2}+\dfrac{\sqrt{2}}{2^2}(x-1)-\dfrac{\sqrt{2}}{2^42!}(x-1)^2$

$+\dfrac{\sqrt{2}(1\cdot 3)}{2^63!}(x-1)^3+\cdots$

Chapter 11 Review Exercises, Page 608

1. False **3.** True **5.** True **7.** False
9. True **11.** False **13.** True
15. False **17.** False **19.** False
21. False **23.** True
25. True **27.** $20,9,\frac{4}{5},16$ **29.** 12
31. $-\pi/4$

33. 4 **35.** Converges **37.** Converges
39. Converges **41.** Diverges **43.** Diverges
45. $\frac{61,004}{201}$ **47.** $[-\frac{1}{3},\frac{1}{3}]$ **49.** $\{-5\}$

51. $\frac{4}{3}$ **53.** $1-\dfrac{1}{3}x^5+\dfrac{4}{3^22!}x^{10}+\cdots$

55. $x-\dfrac{2^2}{3!}x^3+\dfrac{2^4}{5!}x^5+\cdots$

57. $\displaystyle\sum_{k=0}^{\infty}\dfrac{(-1)^{k+1}}{(2k+1)!}(x-\pi/2)^{2k+1}$

Exercises 12.1, Page 616

1. $(0,0)$; $(0,-\frac{1}{8})$; $y=\frac{1}{8}$
3. $(0,0)$; $(3,0)$; $x=-3$
5. $(0,0)$; $(-\frac{5}{2},0)$; $x=\frac{5}{2}$
7. $(1,0)$; $(1,1)$; $y=-1$
9. $(4,3)$; $(\frac{9}{2},3)$; $x=\frac{7}{2}$
11. $(4,-3)$; $(4,-2)$; $y=-4$
13. $(-2,2)$; $(-2,\frac{9}{4})$; $y=\frac{7}{4}$
15. $(-3,3)$; $(-2,3)$; $x=-4$
17. $(-2,-\frac{5}{8})$; $(-2,-\frac{7}{24})$; $y=-\frac{23}{24}$
19. $(\frac{1}{8},-\frac{5}{2})$; $(-\frac{3}{8},-\frac{5}{2})$; $x=\frac{5}{8}$
21. $y=4x^2$ **23.** $(y-1)^2=-12(x-1)$
25. $(y-4)^2=-12(x-1)$
27. $(y-4)^2=10(x+\frac{5}{2})$ **29.** $y=x^2+2$
31. $y=(\frac{2}{9})x^2-2$; $y=(\frac{2}{9})(x+2)^2-2$
33. $-x+8y=8$ **37.** $y=\frac{3}{500}x^2$; 5.4 m
43. $(-b/2a,(4ac-b^2)/4a)$,
$(-b/2a,(1+4ac-b^2)/4a)$
45. $x^2+y^2+2xy+2x-14y+1=0$

Exercises 12.2, Page 621

1. $(\pm4,0)$, $(0,\pm5)$; $(0,\pm3)$
3. $(\pm1,0)$, $(0,\pm\sqrt{10})$; $(0,\pm3)$

5. $(\pm4,0)$, $(0,\pm2\sqrt{2})$; $(\pm2\sqrt{2},0)$
7. $(\pm\sqrt{7},0)$, $(0,\pm2)$; $(\pm\sqrt{3},0)$
9. $(-4,3)$, $(6,3)$, $(1,-3)$, $(1,9)$; $(1,3-\sqrt{11})$, $(1,3+\sqrt{11})$
11. $(-6,2)$, $(10,2)$, $(2,-4)$, $(2,8)$; $(2-2\sqrt{7},2)$, $(2+2\sqrt{7},2)$
13. $(-\frac{1}{2}-\sqrt{2},1)$, $(-\frac{1}{2}+\sqrt{2},1)$, $(-\frac{1}{2},-1)$, $(-\frac{1}{2},3)$; $(-\frac{1}{2},1-\sqrt{2})$, $(-\frac{1}{2},1+\sqrt{2})$
15. $(-\sqrt{3},3)$, $(\sqrt{3},3)$, $(0,0)$, $(0,6)$; $(0,3-\sqrt{6})$, $(0,3+\sqrt{6})$
17. $(-6,-1)$, $(4,-1)$, $(-1,-4)$, $(-1,2)$; $(-5,-1)$, $(3,-1)$

19. $\left(4-\dfrac{\sqrt{5}}{5},2\right)$, $\left(4+\dfrac{\sqrt{5}}{5},2\right)$, $(4,1)$, $(4,3)$;
$\left(4,2-\dfrac{2\sqrt{5}}{5}\right)$, $\left(4,2+\dfrac{2\sqrt{5}}{5}\right)$

21. $x^2/25+y^2/4=1$ **23.** $x^2/12+y^2/16=1$
25. $(x+1)^2/16+(y+2)^2/25=1$
27. $4x^2/31+4y^2/3=1$ **29.** $x^2/8+y^2/2=1$
31. $(x-4)^2/64+(y+2)^2/100=1$ **33.** $-\frac{18}{5},\frac{18}{5}$
37. $(4,1)$; $(2,1)$, $(6,1)$, $(4,4)$, $(4,-2)$;
$(4,1-\sqrt{5})$, $(4,1+\sqrt{5})$ **39.** $20\sqrt{2}$ ft
43. $\frac{3}{5}$ **45.** $x^2/16+3y^2/32=1$ **47.** 0.97
51. 25.5296

Exercises 12.3, Page 629

1. $(\pm4,0)$; $(\pm5,0)$; $y=\pm3x/4$
3. $(0,\pm3)$; $(0,\pm3\sqrt{2})$; $y=\pm x$
5. $(\pm5,0)$; $(\pm\sqrt{29},0)$; $y=\pm2x/5$
7. $(0,\pm4)$, $(0,\pm5)$; $y=\pm4x/3$
9. $(0,2)$, $(2,2)$; $(1-\sqrt{2},2)$, $(1+\sqrt{2},2)$;
$y=x+1$, $y=-x+3$
11. $(4,-7)$, $(4,5)$; $(4,-1-2\sqrt{10})$,
$(4,-1+2\sqrt{10})$; $y=3x-13$, $y=-3x+11$
13. $(-\frac{3}{2},-\frac{4}{3})$, $(\frac{5}{2},-\frac{4}{3})$; $(\frac{1}{2}-2\sqrt{2},-\frac{4}{3})$,
$(\frac{1}{2}+2\sqrt{2},-\frac{4}{3})$; $6x-6y=11$, $6x+6y=-5$
15. $(-2,3)$, $(-2,7)$; $(-2,5-\sqrt{29})$,
$(-2,5+\sqrt{29})$; $-2x+5y=29$, $2x+5y=21$
17. $(2-\sqrt{3},1)$, $(2+\sqrt{3},1)$; $(2-\sqrt{6},1)$,
$(2+\sqrt{6},1)$; $y=x-1$, $y=-x+3$
19. $(-9,-1)$, $(1,-1)$; $(-4-5\sqrt{2},-1)$,
$(-4+5\sqrt{2},-1)$; $y=x+3$, $y=-x-5$
21. $x^2/9-y^2/16=1$
23. $(x-2)^2/25-(y+7)^2/25=1$
25. $(y-2)^2/9-(x-1)^2/7=1$
27. $y^2/16-x^2/2=1$
29. $4x^2/9-4y^2/27=1$
31. $(x+1)^2/16-(y-2)^2/20=1$
33. $(2\sqrt{2},\sqrt{2})$, $(-2\sqrt{2},-\sqrt{2})$
35. $(-2\sqrt{26},-8)$, $(2\sqrt{26},-8)$ **37.** 12π
41. $3,6,9,13,16,17,19$ **43.** 3
45. $y^2/36-x^2/64=1$

Exercises 12.4, Page 635

1. $(2, -1)$ **3.** $(-\frac{3}{2}, \frac{7}{2})$

5. The graph of $y = |x|$ is translated one unit to the right and up four units.

7. The graph of $y = \sin x$ has been translated $\pi/2$ units to the left.

9. The graph of $x^2 - y^2 = 4$ has been translated to the left one unit and up one unit.

11. $(4\sqrt{2}, -2\sqrt{2})$

13. $(-(1 + \sqrt{3})/2, (\sqrt{3} - 1)/2)$

15. $(-1, -7)$

17. Ellipse rotated 45°, $3X^2 + Y^2 = 8$

19. Parabola rotated 45°, $Y^2 = 4\sqrt{2}X$

21. Ellipse rotated by $\theta \approx 26.6°$, $(X + \sqrt{5})^2/8 + Y^2/3 = 1$

23. $Y = X^2$; XY: $(0, \frac{1}{4})$; xy: $(-\frac{1}{8}, \sqrt{3}/8)$; $Y = -\frac{1}{4}$, $2x - 2\sqrt{3}y = 1$

27. Hyperbola   **29.** Parabola   **31.** Ellipse

Chapter 12 Review Exercises, Page 636

1. True   **3.** True   **5.** True   **7.** True

9. False   **11.** False   **13.** False

15. True   **17.** $(0, \frac{1}{8})$ **19.** $(0, 2)$

21. $y = -5$ **23.** 6   **25.** $(2, -1), (6, -1)$

27. $(4, -3)$ **29.** $-\sqrt{5}, \sqrt{5}$

33. 10^8 m; 9×10^8 m   **35.** $x + y + 4 = 0$

Exercises 13.1, Page 643

1.

x	-5	-3	-1	1	3	5	7
y	6	2	0	0	2	6	12

3.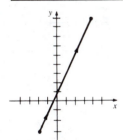

5.

7.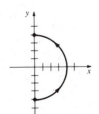

9.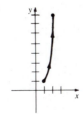

11. $y = x^2 + 3x - 1$, $x \geq 0$

13. $x = 1 - 2y^2$, $-1 \leq x \leq 1$

15. $y = \ln x$, $x > 0$

17.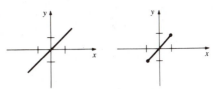

19.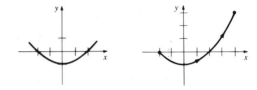

21.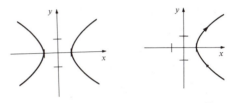

23.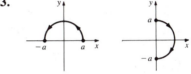

25.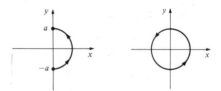

27.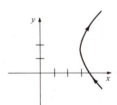

29. **(a), (f)**

31. $x = \pm\sqrt{r^2 - L^2 \sin^2\phi}$, $y = L \sin \phi$

33. $x = 2a \cot \theta$, $y = 2a \sin^2\theta$

37. $x^{2/3} + y^{2/3} = b^{2/3}$

Exercises 13.2, Page 650

1. $\frac{3}{5}$ **3.** 24   **5.** -1 **7.** $y = -2x - 1$

9. $4x - 3y = -4$ **11.** $\sqrt{3}/4$

13. $y = 3x - 7$

15. Hor. tan. at $(0, 0)$;
ver. tan. at $(2/3\sqrt{3}, \frac{1}{3})$, $(-2/3\sqrt{3}, \frac{1}{3})$

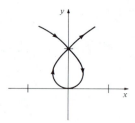

17. Hor. tan. at $(-1, 0)$ and $(1, -4)$;
no ver. tan.

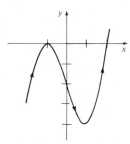

19. $3t$, $1/(2t)$, $-1/(12t^3)$
21. $-2e^{3t} - 3e^{4t}$, $6e^{4t} + 12e^{5t}$, $-24e^{5t}$, $-60e^{6t}$
23. Concave upward for $0 < t < 2$,
concave downward for $t < 0$ and $t > 2$
25. $\frac{104}{3}$ **27.** $\sqrt{2}(e^\pi - 1)$ **29.** $3b/2$

Exercises 13.3, Page 656

1.   **3.**

5. $(6, -5\pi/4)$; $(6, 11\pi/4)$; $(-6, 7\pi/4)$; $(-6, -\pi/4)$
7. $(2, -4\pi/3)$; $(2, 8\pi/3)$; $(-2, 5\pi/3)$; $(-2, -\pi/3)$
9. $(1, -11\pi/6)$; $(1, 13\pi/6)$; $(-1, 7\pi/6)$; $(-1, -5\pi/6)$
11. $(\frac{1}{2}, -\sqrt{3}/2)$ **13.** $(-\frac{7}{2}, 7\sqrt{3}/2)$
15. $(-2\sqrt{2}, -2\sqrt{2})$
17. $(3\sqrt{2}, -3\pi/4)$; $(-3\sqrt{2}, \pi/4)$
19. $(2, -\pi/6)$; $(-2, 5\pi/6)$
21. $(5, -\pi/2)$; $(-5, \pi/2)$
23. $(5, 2.4981)$; $(-5, -0.6435)$
25. $r = 5\csc\theta$ **27.** $\theta = \tan^{-1}7$
29. $r = 2/(1 + \cos\theta)$ **31.** $r = 6$
33. $r = 1 - \cos\theta$ **37.** $x = 2$
39. $(x^2 + y^2)^3 = 144x^2y^2$ **41.** $(x^2 + y^2)^2 = 8xy$
43. $x^2 + y^2 + 5y = 0$
45. $8x^2 - 12x - y^2 + 4 = 0$
47. $3x + 8y = 5$ **51.** $80/\sqrt{41}$ cm/sec

Exercises 13.4, Page 664

1.    **3.**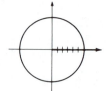

5. 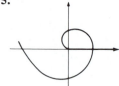   **7.**

9. 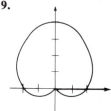   **11.**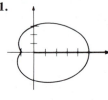

13. 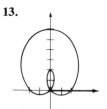   **15.**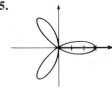

17. 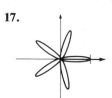   **19.**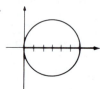

21. 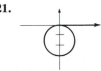   **23.**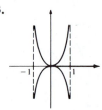

25. 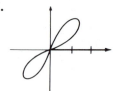   **27.**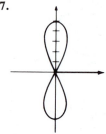

29.

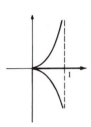

31.

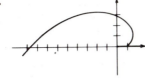

33.

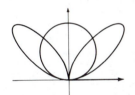

35. $-2/\pi$ **37.** $\sqrt{3}/15$ **39.** $y = -x + 1$

41. Hor. tan. at $(3, \pi/3)$ and $(3, 5\pi/3)$;
ver. tan. at $(4, 0)$, $(1, 2\pi/3)$, and $(1, 4\pi/3)$

43. $(2, \pi/6)$, $(2, 5\pi/6)$

45. $(1, \pi/2)$, $(1, 3\pi/2)$, origin

47. $(3, \pi/12)$, $(3, 5\pi/12)$, $(3, 13\pi/12)$,
$(3, 17\pi/12)$, $(3, -\pi/12)$, $(3, -5\pi/12)$,
$(3, -13\pi/12)$, $(3, -17\pi/12)$

49. $(0, 0)$, $(\sqrt{3}/2, \pi/3)$, $(\sqrt{3}/2, 2\pi/3)$

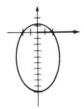

51. Symmetry with respect to the x-axis

53. Symmetry with respect to the origin

55. Symmetry with respect to the x-axis

Exercises 13.5, Page 670

1. π **3.** 24π **5.** 11π

7. $9\pi/2$ **9.** π

11. $r^2 = 9 \cos 2\theta$ is not nonnegative on $[0, 2\pi]$; 9

13. $9\pi^3/4$ **15.** $\frac{1}{4}(e^{2\pi} - 1)$ **17.** $\frac{1}{8}(4 - \pi)$

19. $\frac{1}{6}(2\pi + 3\sqrt{3})$ **21.** $\pi + 6\sqrt{3}$

23. $18\sqrt{3} - 4\pi$ **25.** $\frac{1}{2}(4\pi + 3\sqrt{3})$

27. $\sqrt{15} - \cos^{-1}(\frac{1}{4})$ **29.** $\sqrt{5}(e^2 - 1)$

31. $\frac{1}{2}[\sqrt{2} + \ln(\sqrt{2} + 1)]$

33. $A = \dfrac{1}{2} \displaystyle\int_{\theta_1}^{\theta_2} r^2 \, d\theta = \dfrac{1}{2} \displaystyle\int_{a}^{b} r^2 \, \dfrac{d\theta}{dt} \, dt$

$\qquad = \dfrac{1}{2} \displaystyle\int_{a}^{b} (L/m) \, dt = L(b - a)/2m$

Exercises 13.6, Page 675

1. $e = 1$, parabola

3. $e = \frac{1}{4}$, ellipse

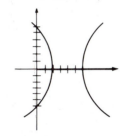

5. $e = 2$, hyperbola

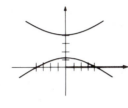

7. $e = 2$, hyperbola

9. $e = \frac{4}{5}$, ellipse

11. Center: $(-2, 0)$; foci: $(0, 0)$, $(-4, 0)$;
vertices: $(-6, 0)$, $(2, 0)$, $(-2, -2\sqrt{3})$, $(-2, 2\sqrt{3})$

13. $r = 3/(1 - \cos\theta)$

15. $r = 30/(2 + 5\sin\theta)$

17. $r = 5/(1 + \sin\theta)$

21. $r = 9600/(1 - 0.2\cos\theta)$

23. $r = (8.7 \times 10^8)/(1 - 0.97\cos\theta)$

Chapter 13 Review Exercises, Page 676

1. False   **3.** True   **5.** True   **7.** False

9. True   **11.** True   **13.** False   **15.** True

17. True   **19.** False   **21.** $3x + 3\sqrt{3}y = \pi$

23. $(8, -26)$

25. $y^2 < 0$ for $|x| > 1$; $x = \sin t$, $y = \sin 2t$;
$(\sqrt{2}/2, 1)$, $(\sqrt{2}/2, -1)$, $(-\sqrt{2}/2, 1)$, $(-\sqrt{2}/2, -1)$;

27. $5\pi/4$

29. $x + y = 2\sqrt{2}$; $r = 2\sqrt{2}/(\cos\theta + \sin\theta)$

31. $r = 3 + 2\cos\theta$ **33.** $r = \dfrac{1}{1 - \cos\theta}$

35. $\sqrt{3} - \pi/3$

37. $x = 3at/(1 + t^3)$, $y = 3at^2/(1 + t^3)$

39. $r = \dfrac{3a \cos \theta \sin \theta}{\cos^3\theta + \sin^3\theta}$; $3a^2/2$

Exercises 14.1, Page 686

1. $6\mathbf{i} + 12\mathbf{j}$; $\mathbf{i} + 8\mathbf{j}$; $3\mathbf{i}$; $\sqrt{65}$; 3

3. $\langle 12, 0\rangle$; $\langle 4, -5\rangle$; $\langle 4, 5\rangle$; $\sqrt{41}$; $\sqrt{41}$

5. $-9\mathbf{i} + 6\mathbf{j}$; $-3\mathbf{i} + 9\mathbf{j}$; $-3\mathbf{i} - 5\mathbf{j}$; $3\sqrt{10}$; $\sqrt{34}$

7. $-6\mathbf{i} + 27\mathbf{j}$; $\mathbf{0}$; $-4\mathbf{i} + 18\mathbf{j}$; 0; $2\sqrt{85}$

9. $\langle 6, -14\rangle$; $\langle 2, 4\rangle$ **11.** $10\mathbf{i} - 12\mathbf{j}$; $12\mathbf{i} - 17\mathbf{j}$

13. $\langle 20, 52\rangle$; $\langle -2, 0\rangle$

15. $2\mathbf{i} + 5\mathbf{j}$ **17.** $2\mathbf{i} + 2\mathbf{j}$

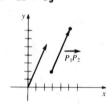

19. $(1, 18)$ **21.** (b), (c), (e), (f)   **23.** $\langle 6, 15\rangle$

25. $\left\langle \dfrac{1}{\sqrt{2}}, \dfrac{1}{\sqrt{2}}\right\rangle$; $\left\langle -\dfrac{1}{\sqrt{2}}, -\dfrac{1}{\sqrt{2}}\right\rangle$

27. $\langle 0, -1\rangle$; $\langle 0, 1\rangle$

29. $\left\langle \dfrac{5}{13}, \dfrac{12}{13}\right\rangle$ **31.** $\dfrac{6}{\sqrt{58}}\mathbf{i} + \dfrac{14}{\sqrt{58}}\mathbf{j}$

33. $\left\langle -3, -\dfrac{15}{2}\right\rangle$ **35.**

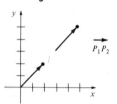

37. $-(\mathbf{a} + \mathbf{b})$ **41.** Approximately $31°$

Exercises 14.2, Page 693

1.–5.

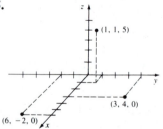

7. The set $\{(x, y, 5) \mid x, y \text{ real numbers}\}$ is a plane perpendicular to the z-axis, 5 units above the xy-plane.

9. The set $\{(2, 3, z) \mid z \text{ a real number}\}$ is a line perpendicular to the xy-plane at $(2, 3, 0)$.

11. $(0, 0, 0)$, $(2, 0, 0)$, $(2, 5, 0)$, $(0, 5, 0)$, $(0, 0, 8)$, $(2, 0, 8)$, $(2, 5, 8)$, $(0, 5, 8)$

13. $(-2, 5, 0)$, $(-2, 0, 4)$, $(0, 5, 4)$; $(-2, 5, -2)$; $(3, 5, 4)$

15. The union of the coordinate planes

17. The point $(-1, 2, -3)$

19. The union of the planes $z = -5$ and $z = 5$

21. $\sqrt{70}$ **23.** 3; 5 **25.** Right triangle

27. Isosceles

29. $d(P_1, P_2) + d(P_1, P_3) = d(P_2, P_3)$

31. 6 or -2 **33.** $\left(4, \dfrac{1}{2}, \dfrac{3}{2}\right)$

35. $P_1(-4, -11, 10)$ **37.** $\langle -3, -6, 1\rangle$

39. $\langle 2, 1, 1\rangle$ **41.** $\langle 2, 4, 12\rangle$

43. $\langle -11, -41, -49\rangle$ **45.** $\sqrt{139}$ **47.** 6

49. $\left\langle -\dfrac{2}{5}, \dfrac{1}{3}, -\dfrac{2}{5}\right\rangle$

51. $4\mathbf{i} - 12\mathbf{j} + 8\mathbf{k}$

53.

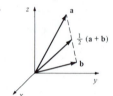

Exercises 14.3, Page 703

1. $25\sqrt{2}$ **3.** 12 **5.** -16 **7.** 48

9. 29 **11.** 25 **13.** $\left\langle -\dfrac{2}{5}, \dfrac{4}{5}, 2\right\rangle$

15. (a) and (f), (c) and (d), (b) and (e)

17. $\left\langle \dfrac{4}{9}, -\dfrac{1}{3}, 1\right\rangle$ **19.** 1.11 radians or $63.43°$

21. 1.89 radians or $108.43°$

23. $\cos\alpha = 1/\sqrt{14}$, $\cos\beta = 2/\sqrt{14}$, $\cos\gamma = 3/\sqrt{14}$; $\alpha = 74.5°$, $\beta = 57.69°$, $\gamma = 36.7°$

25. $\cos\alpha = \dfrac{1}{2}$, $\cos\beta = 0$, $\cos\gamma = -\sqrt{3}/2$; $\alpha = 60°$, $\beta = 90°$, $\gamma = 150°$

27. $\alpha = 58.19°$, $\beta = 42.45°$, $\gamma = 65.06°$

29. $\dfrac{5}{7}$ **31.** $-6\sqrt{11}/11$ **33.** $72\sqrt{109}/109$

35. $\langle -\dfrac{21}{5}, \dfrac{28}{5}\rangle$; $\langle -\dfrac{4}{5}, -\dfrac{3}{5}\rangle$

37. $\langle -\dfrac{12}{7}, \dfrac{6}{7}, \dfrac{4}{7}\rangle$; $\langle \dfrac{5}{7}, -\dfrac{20}{7}, \dfrac{45}{7}\rangle$ **39.** $\langle \dfrac{72}{25}, \dfrac{96}{25}\rangle$

41. 1000 ft-lb   **43.** 0; 150 nt-m

Exercises 14.4, Page 711

1. $-5\mathbf{i} - 5\mathbf{j} + 3\mathbf{k}$ **3.** $\langle -12, -2, 6\rangle$

5. $-5\mathbf{i} + 5\mathbf{k}$ **7.** $\langle -3, 2, 3\rangle$ **9.** $\mathbf{0}$

11. $6\mathbf{i} + 14\mathbf{j} + 4\mathbf{k}$ **13.** $-3\mathbf{i} - 2\mathbf{j} - 5\mathbf{k}$

17. $-\mathbf{i} + \mathbf{j} + \mathbf{k}$ **19.** $2\mathbf{k}$ **21.** $\mathbf{i} + 2\mathbf{j}$

23. $-24\mathbf{k}$ **25.** $5\mathbf{i} - 5\mathbf{j} - \mathbf{k}$ **27.** $\mathbf{0}$

29. $\sqrt{41}$ **31.** $-\mathbf{j}$ **33.** $\mathbf{0}$ **35.** 6

37. $12\mathbf{i} - 9\mathbf{j} + 18\mathbf{k}$ **39.** $-4\mathbf{i} + 3\mathbf{j} - 6\mathbf{k}$

41. $-21\mathbf{i} + 16\mathbf{j} + 22\mathbf{k}$ **43.** -10 **45.** $\dfrac{1}{2}$

47. $\dfrac{7}{2}$ **49.** 10 **51.** Coplanar

53. 32; in the xy-plane, $30°$ from the positive x-axis in the direction of the negative y-axis; $16\sqrt{3}\mathbf{i} - 16\mathbf{j}$

61. No

Exercises 14.5, Page 717

1. $\langle x, y, z\rangle = \langle 1, 2, 1\rangle + t\langle 2, 3, -3\rangle$

3. $\langle x, y, z\rangle = \langle \dfrac{1}{2}, -\dfrac{1}{2}, 1\rangle + t\langle -2, 3, -\dfrac{3}{2}\rangle$

5. $\langle x, y, z \rangle = \langle 1, 1, -1 \rangle + t\langle 5, 0, 0 \rangle$

7. $x = 2 + 4t, \ y = 3 - 4t, \ z = 5 + 3t$

9. $x = 1 + 2t, \ y = -2t, \ z = -7t$

11. $x = 4 + 10t, \ y = \frac{1}{2} + \frac{3}{4}t, \ z = \frac{1}{3} + \frac{1}{6}t$

13. $\dfrac{x - 1}{9} = \dfrac{y - 4}{10} = \dfrac{z + 9}{7}$

15. $\dfrac{x + 7}{11} = \dfrac{z - 5}{-4}, \ y = 2$

17. $x = 5, \ \dfrac{y - 10}{9} = \dfrac{z + 2}{12}$

19. $x = 4 + 3t, \ y = 6 + \frac{1}{2}t, \ z = -7 - \frac{3}{2}t;$

$\dfrac{x - 4}{3} = 2y - 12 = \dfrac{2z + 14}{-3}$

21. $x = 5t, \ y = 9t, \ z = 4t; \ \dfrac{x}{5} = \dfrac{y}{9} = \dfrac{z}{4}$

23. $x = 6 + 2t, \ y = 4 - 3t, \ z = -2 + 6t$

25. $x = 2 + t, \ y = -2, \ z = 15$

27. Both lines pass through the origin and have parallel direction vectors.

29. $(0, 5, 15), \ (5, 0, \frac{15}{2}), \ (10, -5, 0)$

31. $(2, 3, -5)$ **33.** Lines do not intersect.

35. $40.37°$

37. $x = 4 - 6t, \ y = 1 + 3t, \ z = 6 + 3t$

39. The lines are not parallel and do not intersect.

Exercises 14.6, Page 724

1. $2x - 3y + 4z = 19$ **3.** $5x - 3z = 51$

5. $6x + 8y - 4z = 11$ **7.** $5x - 3y + z = 2$

9. $3x - 4y + z = 0$

11. The points are collinear.

13. $x + y - 4z = 25$ **15.** $z = 12$

17. $-3x + y + 10z = 18$

19. $9x - 7y + 5z = 17$

21. $6x - 2y + z = 12$

23. Orthogonal: (a) and (d), (b) and (c), (d) and (f), (b) and (e); parallel: (a) and (f), (c) and (e)

25. (c) and (d)

27. $x = 2 + t, \ y = \frac{1}{2} - t, \ z = t$

29. $x = \frac{1}{2} - \frac{1}{2}t, \ y = \frac{1}{2} - \frac{3}{2}t, \ z = t$

31. $(-5, 5, 9)$ **33.** $(1, 2, -5)$

35. $x = 5 + t, \ y = 6 + 3t, \ z = -12 + t$

37. $3x - y - 2z = 10$

39.

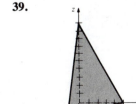

41.

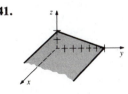

43.

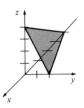

45. $3/\sqrt{11}$ **47.** $107.98°$

Exercises 14.7, Page 730

1.

3.

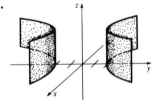

5.

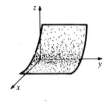

7.

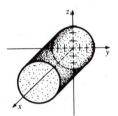

9.

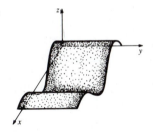

11.

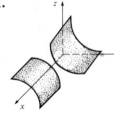

13.

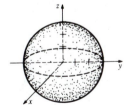

15.

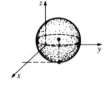

17. Center $(-4, 3, 2)$; radius 6

19. Center $(0, 0, 8)$; radius 8

21. $(x + 1)^2 + (y - 4)^2 + (z - 6)^2 = 3$

23. $(x - 1)^2 + (y - 1)^2 + (z - 4)^2 = 16$

25. $x^2 + (y - 4)^2 + z^2 = 4$ or
$x^2 + (y - 8)^2 + z^2 = 4$
27. $(x - 1)^2 + (y - 4)^2 + (z - 2)^2 = 90$

Exercises 14.8, Page 738

1. Paraboloid

3. Ellipsoid

5. Hyperboloid of one sheet

7. Elliptical cone

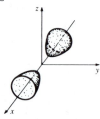

9. Hyperbolic paraboloid

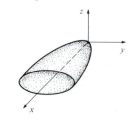

11. Hyperboloid of two sheets

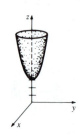

13. Ellipsoid

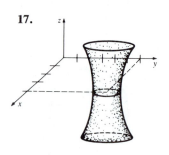

15.

17.

19. One possibility is $y^2 + z^2 = 1$; z-axis
21. One possibility is $y = e^{x^2}$; y-axis
23. $y^2 = 4(x^2 + z^2)$
25. $y^2 + z^2 = (9 - x^2)^2$, $x \geq 0$
27. $x^2 - y^2 - z^2 = 4$

29. $z = \ln \sqrt{x^2 + y^2}$
31. 1, z-axis; 2, y-axis; 4, z-axis; 6, z-axis; 10, z-axis; 11, x-axis; 14, z-axis

Chapter 14 Review Exercises, Page 739

1. True   **3.** False   **5.** True   **7.** True
9. True   **11.** $9\mathbf{i} + 2\mathbf{j} + 2\mathbf{k}$ **13.** $5\mathbf{i}$
15. 14   **17.** $-6\mathbf{i} + \mathbf{j} - 7\mathbf{k}$ **19.** $(4, 7, 5)$
21. $(5, 6, 3)$ **23.** $-36\sqrt{2}$
25. 12, -8, and 6   **27.** $3\sqrt{10}/2$
29. 2 units   **31.** $(\mathbf{i} - \mathbf{j} - 3\mathbf{k})/\sqrt{11}$
33. 2   **35.** $\frac{26}{9}\mathbf{i} + \frac{7}{9}\mathbf{j} + \frac{20}{9}\mathbf{k}$
37. Elliptic cylinder
39. Hyperboloid of two sheets
41. Hyperbolic paraboloid
43. $x^2 - y^2 + z^2 = 1$, hyperboloid of one sheet;
$x^2 - y^2 - z^2 = 1$, hyperboloid of two sheets
45. Sphere; plane   **47.** $\dfrac{x - 7}{4} = \dfrac{y - 3}{-2} = \dfrac{z + 5}{6}$
49. The direction vectors are orthogonal and the point of intersection is $(3, -3, 0)$.
51. $14x - 5y - 3z = 0$ **53.** $30\sqrt{2}$ nt-m
55. Approximately 153 lb

Exercises 15.1, Page 749

1.

3.

5.

7.

9.

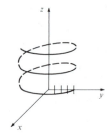

11.

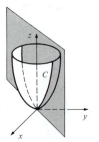

$$\mathbf{r}(t) = t\mathbf{i} + t\mathbf{j} + 2t^2\mathbf{k}$$

13.

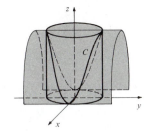

$\mathbf{r}(t) = 3 \cos t\mathbf{i} + 3 \sin t\mathbf{j} + 9 \sin^2 t\mathbf{k}$

15. $2\mathbf{i} - 32\mathbf{j}$

17. $(1/t)\mathbf{i} - (1/t^2)\mathbf{j}$; $-(1/t^2)\mathbf{i} + (2/t^3)\mathbf{j}$

19. $\langle e^{2t}(2t + 1), 3t^2, 8t - 1 \rangle$; $\langle 4e^{2t}(t + 1), 6t, 8 \rangle$

21. 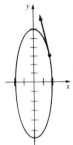   **23.**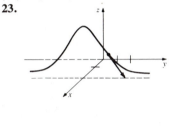

25. $x = 2 + t, \ y = 2 + 2t, \ z = \frac{8}{3} + 4t$

27. $\mathbf{r}(t) \times \mathbf{r}''(t)$ **29.** $\mathbf{r}(t) \cdot [\mathbf{r}'(t) \times \mathbf{r}'''(t)]$

31. $2\mathbf{r}_1'(2t) - (1/t^2)\mathbf{r}_2'(1/t)$ **33.** $\frac{3}{2}\mathbf{i} + 9\mathbf{j} + 15\mathbf{k}$

35. $e^t(t - 1)\mathbf{i} + \frac{1}{2}e^{-2t}\mathbf{j} + \frac{1}{2}e^{t^2}\mathbf{k} + \mathbf{c}$

37. $(6t + 1)\mathbf{i} + (3t^2 - 2)\mathbf{j} + (t^3 + 1)\mathbf{k}$

39. $(2t^3 - 6t + 6)\mathbf{i} + (7t - 4t^{3/2} - 3)\mathbf{j} + (t^2 - 2t)\mathbf{k}$

41. $2\sqrt{a^2 + c^2}\,\pi$ **43.** $\sqrt{6}(e^{3\pi} - 1)$

45. $a \cos(s/a)\mathbf{i} + a \sin(s/a)\mathbf{j}$

49. Differentiate $\mathbf{r}(t) \cdot \mathbf{r}(t) = c^2$.

Exercises 15.2, Page 755

1. Speed is $\sqrt{5}$.

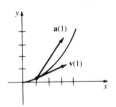

3. Speed is 2.   **5.** Speed is $\sqrt{5}$.

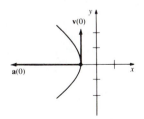   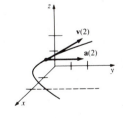

7. Speed is $\sqrt{14}$.

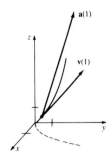

9. $(0, 0, 0)$ and $(25, 115, 0)$;
$\mathbf{v}(0) = -2\mathbf{j} - 5\mathbf{k}, \ \mathbf{a}(0) = 2\mathbf{i} + 2\mathbf{k}$,
$\mathbf{v}(5) = 10\mathbf{i} + 73\mathbf{j} + 5\mathbf{k}, \ \mathbf{a}(5) = 2\mathbf{i} + 30\mathbf{j} + 2\mathbf{k}$

11. $\mathbf{r}(t) = (-16t^2 + 240t)\mathbf{j} + 240\sqrt{3}t\mathbf{i}$ and
$x(t) = 240\sqrt{3}t, \ y(t) = -16t^2 + 240t$;
900 ft; approximately 6235 ft; 480 ft/sec

13. 72.11 ft/sec   **15.** 97.98 ft/sec

17. Assume that (x_0, y_0) are the coordinates of the center of the target at $t = 0$. Then $\mathbf{r}_p = \mathbf{r}_t$ when $t = x_0/(v_0 \cos \theta)$ $= y_0/(v_0 \sin \theta)$. This implies $\tan \theta = y_0/x_0$. In other words, aim directly at the target at $t = 0$.

21. 191.33 lb

23. $y = -\dfrac{g}{2v_0^2 \cos^2\theta}x^2 + (\tan \theta)x + s_0$ is an equation of a parabola.

Exercises 15.3, Page 760

1. $\mathbf{T} = (\sqrt{5}/5)(-\sin t\mathbf{i} + \cos t\mathbf{j} + 2\mathbf{k})$

3. $\mathbf{T} = (a^2 + c^2)^{-1/2}(-a \sin t\mathbf{i} + a \cos t\mathbf{j} + c\mathbf{k})$,
$\mathbf{N} = -\cos t\mathbf{i} - \sin t\mathbf{j}$,
$\mathbf{B} = (a^2 + c^2)^{-1/2}(c \sin t\mathbf{i} - c \cos t\mathbf{j} + a\mathbf{k})$,
$\kappa = a/(a^2 + c^2)$

5. $3\sqrt{2}x - 3\sqrt{2}y + 4z = 3\pi$

7. $4t/\sqrt{1 + 4t^2}, \ 2/\sqrt{1 + 4t^2}$

9. $2\sqrt{6}, 0, t > 0$ **11.** $2t/\sqrt{1 + t^2}, \ 2/\sqrt{1 + t^2}$

13. $0, 5$ **15.** $-\sqrt{3}e^{-t}, 0$

17. $\kappa = \dfrac{\sqrt{b^2c^2 \sin^2t + a^2c^2 \cos^2t + a^2b^2}}{(a^2 \sin^2t + b^2 \cos^2t + c^2)^{3/2}}$

23. $\kappa = 2, \rho = \frac{1}{2}; \kappa = 2/\sqrt{125} \approx 0.18$,
$\rho = \sqrt{125}/2 \approx 5.59$; the curve is sharper at $(0, 0)$

25. κ is close to zero.

Chapter 15 Review Exercises, Page 761

1. True   **3.** True   **5.** True
7. True   **9.** False   **11.** $\sqrt{2}\pi$
13. $x = -27 - 18t, \ y = 8 + t, \ z = 1 + t$
15.

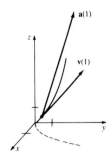

17. $-t^2 \sin t + 2t \cos t - 2 \sin t \cos t$
$+ 8t^3e^{2t} + 12t^2e^{2t}$

19. $(t + 1)\mathbf{i} + \left(\dfrac{1}{m}t^2 + t + 1\right)\mathbf{j} + t\mathbf{k}$;

$x = t + 1,\ y = \dfrac{1}{m}t^2 + t + 1,\ z = t$

21. $\mathbf{v}(1) = 6\mathbf{i} + \mathbf{j} + 2\mathbf{k},\ \mathbf{v}(4) = 6\mathbf{i} + \mathbf{j} + 8\mathbf{k}$,
$\mathbf{a}(t) = 2\mathbf{k}$ for all t **23.** $\mathbf{i} + 4\mathbf{j} + (3\pi/4)\mathbf{k}$
25. $\mathbf{T} = (\tanh 1\mathbf{i} + \mathbf{j} + \operatorname{sech} 1\mathbf{k})/\sqrt{2}$
$\mathbf{N} = \operatorname{sech} 1\mathbf{i} - \tanh 1\mathbf{k}$
$\mathbf{B} = (-\tanh 1\mathbf{i} + \mathbf{j} - \operatorname{sech} 1\mathbf{k})/\sqrt{2}$
$\kappa = \frac{1}{2} \operatorname{sech}^2 1$

Exercises 16.1, Page 772

1. $\{(x, y) \mid (x, y) \neq (0, 0)\}$
3. $\{(x, y) \mid y \neq x^2\}$
5. $\{(s, t) \mid s, t \text{ any real numbers}\}$
7. $\{(r, s) \mid |s| \geq 1\}$
9. $\{(u, v, w) \mid u^2 + v^2 + w^2 \geq 16\}$
11. Figure 16.25   **13.** Figure 16.18
15. Figure 16.19   **17.** Figure 16.23
19.   **21.**

   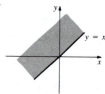

23. $\{z \mid z \geq 10\}$ **25.** $\{w \mid |w| \leq 1\}$
27. $10, -2$ **29.** $4, 4$
31. A plane through the origin perpendicular to the xz-plane   **33.** Upper half of a cone
35. Upper half of an ellipsoid
37.   **39.**

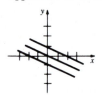   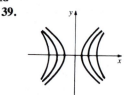

41.

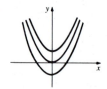

43. Elliptical cylinders   **45.** Ellipsoids
47.

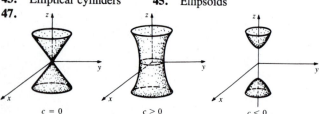

$c = 0$ $c > 0$ $c < 0$

49.

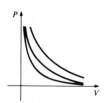

51. $C(r, h) = 2.8\pi r^2 + 4.6\pi rh$
53. $V(r, h) = \frac{11}{9}\pi r^2 h$ **55.** 15,600 cm^2
57. 1397.2   **59.** 1086.4 cal/hr

Exercises 16.2, Page 782

1. 26   **3.** Does not exist   **5.** 1
7. Does not exist   **9.** 108   **11.** 1   **13.** $\frac{1}{3}$
15. $-\frac{1}{3}$ **17.** 360   **19.** Does not exist
21. -3 **23.** 0   **25.** 0   **27.** 0
29. 0
31. $\{(x, y) \mid x \geq 0 \text{ and } y \geq -x\}$
33. $\{(x, y) \mid x/y \neq (2n + 1)\pi/2,$
$n = 0, \pm 1, \pm 2, \ldots, y \neq 0\}$
35. Continuous; not continuous; not continuous
37. f is continuous at $(0, 0)$
39. Choose $\delta = \varepsilon/3$.   **41.** Choose $\delta = \varepsilon$.

Exercises 16.3, Page 789

1. $\partial z/\partial x = 7,\ \partial z/\partial y = 16y$
3. $\partial z/\partial x = 6xy + 4y^2,\ \partial z/\partial y = 3x^2 + 8xy$
5. $\partial z/\partial x = 2x - y^2,\ \partial z/\partial y = -2xy + 20y^4$
7. $\partial z/\partial x = 20x^3y^3 - 2xy^6 + 30x^4,$
$\partial z/\partial y = 15x^4y^2 - 6x^2y^5 - 4$
9. $\partial z/\partial x = 2x^{-1/2}/(3y^2 + 1),$
$\partial z/\partial y = -24\sqrt{x}y/(3y^2 + 1)^2$
11. $\partial z/\partial x = -3x^2(x^3 - y^2)^{-2},$
$\partial z/\partial y = 2y(x^3 - y^2)^{-2}$
13. $\partial z/\partial x = -10 \cos 5x \sin 5x,$
$\partial z/\partial y = 10 \sin 5y \cos 5y$
15. $f_x = e^{x^3y}(3x^3y + 1), f_y = x^4e^{x^3y}$
17. $f_x = 7y/(x + 2y)^2, f_y = -7x/(x + 2y)^2$
19. $g_u = 8u/(4u^2 + 5v^3),\ g_v = 15v^2/(4u^2 + 5v^3)$
21. $w_x = x^{-1/2}y,\ \ w_y = 2\sqrt{x} - (y/z)e^{y/z} - e^{y/z},$
$w_z = (y^2/z^2)e^{y/z}$
23. $F_u = 2uw^2 - v^3 - vwt^2 \sin(ut^2),$
$F_v = -3uv^2 + w \cos(ut^2),\ F_x = 128x^7t^4,$
$F_t = -2uvwt \sin(ut^2) + 64x^8t^3$
25. -16 **27.** $x = -1, y = 4 + t, z = -24 + 2t$
29. -2 **31.** y^2e^{xy} **33.** $20xy - 6y^2$
35. $18uv^2t^2$ **37.** $-e^{r^2}(4r^2 + 2)\sin \theta$
39. $-60x^3y^2 + 8y$
43. $\partial z/\partial x = -x/z,\ \partial z/\partial y = -y/z$
45. $\partial z/\partial u = (vz - 2uv^3)/(2z - uv),$
$\partial z/\partial v = (uz - 3u^2v^2)/(2z - uv)$
47. $A_x = y \sin \theta,\ A_y = x \sin \theta,\ A_\theta = xy \cos \theta$

55. $\dfrac{\partial u}{\partial t} = \begin{cases} -gx/a, & 0 \le x \le at \\ -gt, & x > at \end{cases}$; for $x > at$ the

motion is that of a freely falling body;

$\dfrac{\partial u}{\partial x} = 0$ for $x > at$; the string is horizontal

57. 0; 1

59. $\dfrac{\partial^2 z}{\partial x^2} = \lim_{\Delta x \to 0} \dfrac{f_x(x + \Delta x, y) - f_x(x, y)}{\Delta x}$,

$\dfrac{\partial^2 z}{\partial y^2} = \lim_{\Delta y \to 0} \dfrac{f_y(x, y + \Delta y) - f_y(x, y)}{\Delta y}$,

$\dfrac{\partial^2 z}{\partial x \, \partial y} = \lim_{\Delta x \to 0} \dfrac{f_y(x + \Delta x, y) - f_y(x, y)}{\Delta x}$

61. 10

Exercises 16.4, Page 795

1. $\Delta z = 0.2,\ dz = 0.2$
3. $\Delta z = -0.79,\ dz = -0.8$
5. $\varepsilon_1 = 5\,\Delta x,\ \varepsilon_2 = -\Delta x$
7. $\varepsilon_1 = y^2\,\Delta x + 4xy\,\Delta y + 2y\,\Delta x\,\Delta y$,
$\varepsilon_2 = x^2\,\Delta y + 2x\,\Delta x\,\Delta y + (\Delta x)^2\,\Delta y$
9. $dz = 2x \sin 4y\,dx + 4x^2 \cos 4y\,dy$
11. $dz = 2x(2x^2 - 4y^3)^{-1/2}\,dx$
$- 6y^2(2x^2 - 4y^3)^{-1/2}\,dy$
13. $df = 7t\,ds/(s + 3t)^2 - 7s\,dt/(s + 3t)^2$
15. $dw = 2xy^4z^{-5}\,dx + 4x^2y^3z^{-5}\,dy - 5x^2y^4z^{-6}\,dz$
17. $dF = 3r^2\,dr - 2s^{-3}\,ds - 2t^{-1/2}\,dt$
19. $dw = du/u + dv/v - ds/s - dt/t$
21. 0.9%   **23.** $-mg(0.009)$, decreases
25. 15%   **27.** 4.9%   **29.** 13.0907

Exercises 16.5, Page 800

1. $x^2 + 4x + \frac{3}{2}y^2 - y = C$
3. $\frac{5}{2}x^2 + 4xy - 2y^4 = C$
5. $x^2y^2 - 3x + 4y = C$ **7.** Not exact
9. $xy^3 + y^2 \cos x - \frac{1}{2}x^2 = C$ **11.** Not exact
13. $xy - 2xe^x + 2e^x - 2x^3 = C$
15. $x + y + xy - 3 \ln|xy| = C$
17. $x^3y^3 - \tan^{-1}(3x) = C$
19. $-\ln|\cos x| + \cos x \sin y = C$
21. $y - 2x^2y - y^2 - x^4 = C$
23. $x^4y - 5x^3 - xy + y^3 = C$
25. $xy^2 + x^2y - y + \frac{1}{3}x^3 = \frac{4}{3}$
27. $4xy + x^2 - 5x + 3y^2 - y = 8$
29. $P(x, y) = ye^{xy} + y^2 - y/x^2 + h(x)$
31. $k = 10$ **33.** $x^2 - 5xy + y^3 = 7$

Exercises 16.6, Page 804

1. $\partial z/\partial x = 3x^2v^2e^{uv^2} + 2uve^{uv^2},\ \partial z/\partial y = -4yuve^{uv^2}$
3. $\partial z/\partial u = 16u^3 - 40y(2u - v)$,
$\partial z/\partial v = -96v^2 + 20y(2u - v)$
5. $\partial w/\partial t = -3u(u^2 + v^2)^{1/2}e^{-t} \sin\theta$
$-3v(u^2 + v^2)^{1/2}e^{-t}\cos\theta$,
$\partial w/\partial\theta = 3u(u^2 + v^2)^{1/2}e^{-t}\cos\theta$
$- 3v(u^2 + v^2)^{1/2}e^{-t}\sin\theta$

7. $\partial R/\partial u = s^2t^4e^{v^2} - 4rst^4uve^{-u^2} + 8rs^2t^3uv^2e^{u^2v^2}$,
$\partial R/\partial v = 2s^2t^4uve^{v^2} + 2rst^4e^{-u^2} + 8rs^2t^3u^2ve^{u^2v^2}$

9. $\dfrac{\partial w}{\partial t} = \dfrac{xu}{(x^2 + y^2)^{1/2}(rs + tu)} + \dfrac{y \cosh rs}{u(x^2 + y^2)^{1/2}}$,

$\dfrac{\partial w}{\partial r} = \dfrac{xs}{(x^2 + y^2)^{1/2}(rs + tu)} + \dfrac{sty \sinh rs}{u(x^2 + y^2)^{1/2}}$,

$\dfrac{\partial w}{\partial u} = \dfrac{xt}{(x^2 + y^2)^{1/2}(rs + tu)} - \dfrac{ty \cosh rs}{u^2(x^2 + y^2)^{1/2}}$

11. $dz/dt = (4ut - 4vt^{-3})/(u^2 + v^2)$
13. $dw/dt|_{t=\pi} = -2$
15. $dp/du = 2u/(2s + t) + 4r/[u^3(2s + t)^2]$
$- r/[2u^{1/2}(2s + t)^2]$
23. 5.31 cm^2/sec   **25.** 0.5976 in^2/yr
29. $dy/dx = (4xy^2 - 3x^2)/(1 - 4x^2y)$
31. $dy/dx = y \cos xy/(1 - x \cos xy)$
35. $\partial z/\partial x = x/z,\ \partial z/\partial y = y/z$
37. $\partial z/\partial x = (2x + y^2z^3)/(10z - 3xy^2z^2)$,
$\partial z/\partial y = (2xyz^3 - 2y)/(10z - 3xy^2z^2)$

Exercises 16.7, Page 812

1. $(2x - 3x^2y^2)\mathbf{i} + (-2x^3y + 4y^3)\mathbf{j}$
3. $(y^2/z^3)\mathbf{i} + (2xy/z^3)\mathbf{j} - (3xy^2/z^4)\mathbf{k}$
5. $4\mathbf{i} - 32\mathbf{j}$ **7.** $2\sqrt{3}\mathbf{i} - 8\mathbf{j} - 4\sqrt{3}\mathbf{k}$
9. $\sqrt{3}x + y$ **11.** $\frac{15}{2}(\sqrt{3} - 2)$
13. $-1/2\sqrt{10}$ **15.** $98/\sqrt{5}$
17. $-3\sqrt{2}$ **19.** -1 **21.** $-12/\sqrt{17}$
23. $\sqrt{2}\mathbf{i} + (\sqrt{2}/2)\mathbf{j},\ \sqrt{5/2}$
25. $-2\mathbf{i} + 2\mathbf{j} - 4\mathbf{k},\ 2\sqrt{6}$
27. $-8\sqrt{\pi/6}\mathbf{i} - 8\sqrt{\pi/6}\mathbf{j},\ -8\sqrt{\pi/3}$
29. $-\frac{3}{8}\mathbf{i} - 12\mathbf{j} - \frac{2}{3}\mathbf{k},\ -\sqrt{83281}/24$
31. $\pm 31/\sqrt{17}$
33. $\mathbf{u} = \frac{3}{5}\mathbf{i} - \frac{4}{5}\mathbf{j};\ \mathbf{u} = \frac{4}{5}\mathbf{i} + \frac{3}{5}\mathbf{j};\ \mathbf{u} = -\frac{4}{5}\mathbf{i} - \frac{3}{5}\mathbf{j}$
35. $D_\mathbf{u}f = (9x^2 + 3y^2 - 18xy^2 - 6x^2y)/\sqrt{10}$;
$D_\mathbf{u}F = (-6x^2 - 54y^2 + 54x + 6y - 72xy)/10$
37. $(2, 5), (-2, 5)$ **39.** $-16\mathbf{i} - 4\mathbf{j}$
41. $x = 3e^{-4t},\ y = 4e^{-2t}$
43. One possible function is

$$f(x, y) = x^3 - \frac{2}{3}y^3 + xy^3 + e^{xy}$$

Exercises 16.8, Page 818

1.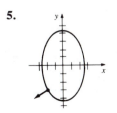

3.

5.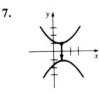

7.

9.

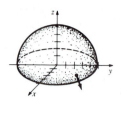

11.

13. $(-4, -1, 17)$ **15.** $-2x + 2y + z = 9$

17. $6x - 2y - 9z = 5$ **19.** $6x - 8y + z = 50$

21. $2x + y - \sqrt{2}z = (4 + 5\pi)/4$

23. $\sqrt{2}x + \sqrt{2}y - z = 2$

25. $(1/\sqrt{2}, \sqrt{2}, 3/\sqrt{2}), (-1/\sqrt{2}, -\sqrt{2}, -3/\sqrt{2})$

27. $(-2, 0, 5), (-2, 0, -3)$

33. $x = 1 + 2t, y = -1 - 4t, z = 1 + 2t$

35. $(x - \frac{1}{2})/4 = (y - \frac{1}{3})/6 = -(z - 3)$

Exercises 16.9, Page 823

1. Rel. min. $f(0, 0) = 5$

3. Rel. max. $f(4, 3) = 25$

5. Rel. min. $f(-2, 1) = 15$

7. Rel. max. $f(-1, -1) = 10$;
rel. min. $f(1, 1) = -10$

9. Rel. min. $f(3, 1) = -14$ **11.** No extrema

13. Rel. max. $f(1, 1) = 12$

15. Rel. min. $f(-1, -2) = 14$

17. Rel. max. $f(-1, (2n + 1)\pi/2) = e^{-1}$, n odd;
rel. min. $f(-1, (2n + 1)\pi/2) = -e^{-1}$, n even

19. Rel. max. $f((2m + 1)\pi/2, (2n + 1)\pi/2) = 2$,
m and n even;
rel. min. $f((2m + 1)\pi/2, (2n + 1)\pi/2) = -2$,
m and n odd

21. $x = 7, y = 7, z = 7$ **23.** $(\frac{1}{6}, \frac{1}{3}, \frac{1}{6})$

25. $(2, 2, 2), (2, -2, -2), (-2, 2, -2), (-2, -2, 2)$;
at these points the least distance is $2\sqrt{3}$

27. $8\sqrt{3}abc/9$ **29.** $x = 2, y = 2, z = 15$

33. $\theta = 30°, x = P/(4 + 2\sqrt{3}), y = P(\sqrt{3} - 1)/2\sqrt{3}$

35. Abs. max. $f(0, 0) = 16$

37. Abs. min. $f(0, 0) = -8$

39. Abs. max. $f(\frac{1}{2}, \sqrt{3}/2) = 2$;
abs. min. $f(-\frac{1}{2}, -\sqrt{3}/2) = -2$

41. Abs. max. $f(\sqrt{2}/2, \sqrt{2}/2) = f(-\sqrt{2}/2, -\sqrt{2}/2) = \frac{3}{2}$
abs. min. $f(0, 0) = 0$

Exercises 16.10, Page 831

1.

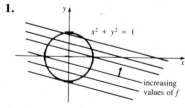

f appears to have a constrained maximum and a constrained minimum.

3. Max. $f(1/\sqrt{10}, 3/\sqrt{10}) = \sqrt{10}$;
min. $f(-1/\sqrt{10}, -3/\sqrt{10}) = -\sqrt{10}$

5. Max. $f(1, 1) = f(-1, -1) = 1$;
min. $f(1, -1) = f(-1, 1) = -1$

7. Min. $f(\frac{1}{2}, -\frac{1}{2}) = \frac{13}{2}$

9. Max. $f(1/\sqrt[4]{2}, 1/\sqrt[4]{2}) = f(-1/\sqrt[4]{2}, -1/\sqrt[4]{2})$
$= f(1/\sqrt[4]{2}, -1/\sqrt[4]{2})$
$= f(-1/\sqrt[4]{2}, 1/\sqrt[4]{2}) = \sqrt{2}$
min. $f(0, 1) = f(0, -1) = f(1, 0) = f(-1, 0) = 1$

11. Max. $f(\frac{9}{16}, \frac{1}{16}) = \frac{729}{65,536}$;
min. $f(0, 1) = f(1, 0) = 0$

13. Max. $F(\sqrt{5}, 2\sqrt{5}, \sqrt{5}) = 6\sqrt{5}$;
min. $F(-\sqrt{5}, -2\sqrt{5}, -\sqrt{5}) = -6\sqrt{5}$

15. Max. $F(1/\sqrt{3}, 2/\sqrt{3}, \sqrt{3}) = 2/\sqrt{3}$

17. Min. $F(\frac{1}{3}, \frac{1}{3}, \frac{1}{3}) = \frac{1}{9}$

19. Min. $F(\frac{1}{3}, \frac{16}{15}, -\frac{11}{15}) = \frac{402}{225}$

21. f is the square of the distance from a point (x, y) on the graph of $x^4 + y^4 = 1$ to the origin. The points $(\pm 1, 0)$ and $(0, \pm 1)$ are closest to the origin, whereas the points $(\pm 1/\sqrt[4]{2}, \pm 1/\sqrt[4]{2})$ are farthest from the origin.

23. F is the square of the distance from a point (x, y) on the line of intersection of the two given planes. The point $(\frac{1}{3}, \frac{16}{15}, -\frac{11}{15})$ is closest to the origin.

25. Max. $A(4/(2 + \sqrt{2}), 4/(2 + \sqrt{2})) = 4/(3 + 2\sqrt{2})$

27. Max. $V(12 - 9/2\sqrt{5}, 6/\sqrt{5}) = (9\pi/2)(24 - \sqrt{5})$

29. Max. $z = P + 4(2 - \sqrt{4 + P\sqrt{27k}}/\sqrt{27k}$

31. Max. $F(k/3, k/3, k/3) = k/3$

33. $(-1/\sqrt{5}, 4 + \sqrt{5}, -2/\sqrt{5})$ is the point farthest from the xz-plane, whereas $(1/\sqrt{5}, 4 - \sqrt{5}, 2/\sqrt{5})$ is the point closest to the xz-plane.

Chapter 16 Review Exercises, Page 833

1. False   **3.** True   **5.** False   **7.** False

9. True   **11.** $-\frac{1}{4}$

13. The ellipse $3x^2 + y^2 = 28$.

15. $\dfrac{\partial F}{\partial r}g'(w) + \dfrac{\partial F}{\partial s}h'(w)$ **17.** f_{yyzx} **19.** $-F(x)$

21. $f_x(x, y)g'(y)h'(z) + f_{xy}(x, y)g(y)h'(z)$

23. $e^{-x^3y}(-x^3y + 1)$ **25.** $-\frac{3}{2}r^2\theta(r^3 + \theta^2)^{-3/2}$

27. $6x^2y \sinh(x^2y^3) + 9x^4y^4 \cosh(x^2y^3)$

29. $-60s^2t^4v^{-5}$ **31.** $\frac{1}{2}\mathbf{i} + \frac{1}{2}\mathbf{j}$

33. $(6x^2 - 2y^2 - 8xy)/\sqrt{40}$

35.

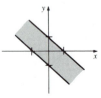

37. $2x \Delta y + 2y \Delta x + 2 \Delta x \Delta y - 2y \Delta y - (\Delta y)^2$

39. $dz = 11y \, dx/(4x + 3y)^2 - 11x \, dy/(4x + 3y)^2$

41. $x^2 \cos y^3 - y = C$

43. $x = -\sqrt{5}, z - 3 = \frac{4}{3}(y - 1)$
45. $2; -2/\sqrt{2}; 4$
47. $4\pi x + 3y - 12z = 4\pi - 6\sqrt{3}$
49. $3x + 4y = 25$ **51.** $x = 2, y = 1, z = 2$
53. -8.77 cm/sec **57.** Not an extremum
59. Relative maximum   **61.** $A = \frac{1}{2}L^2 \cos\theta \sin\theta$
63. $A = 2xz + 2yz - 5z^2$
65. $V = 16xy - 4xy\sqrt{x^2 + y^2}$

Exercises 17.1, Page 840

1. 52 **3.** $8; 8$ **5.** 60 **7.** 10π
9. The integrand $f(x, y) = x + 5y$ is not nonnegative over the region.
11. 80 **13.** 34 **15.** 66 **17.** 18

Exercises 17.2, Page 845

1. $24y - 20e^y$ **3.** $x^2 e^{3x^2} - x^2 e^x$ **5.** $\frac{x}{2}\ln 5$
7. $2 - \sin y$ **9.** 37 **11.** $-4\sqrt{2}/3$
13. $-\frac{4}{21}$ **15.** $18 - e^3 + 3e$ **17.** $\frac{10}{3}$
19. $(\pi/4)\ln 9$ **21.** $\frac{1}{4}e^2 + \frac{1}{4}$ **23.** π
25. e^{-1} **27.** $2 - \pi$ **29.** $(3\sqrt{3} - \pi)/6$
31. 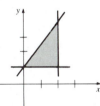   **33.**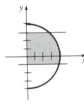

35. Both integrals equal 9.
37. Both integrals equal $13\pi^2 - 16$.
39. Both integrals equal $\frac{32}{5}$.

Exercises 17.3, Page 853

1. $\frac{1}{21}$ **3.** $\frac{25}{84}$ **5.** 96 **7.** $2\ln 2 - 1$
9. $\frac{14}{3}$ **11.** 40 **13.** $\frac{9}{2}$
15. $e^4 - e + 3 - 4\ln 4$ **17.** $\frac{63}{4}$
19. c. 16π **21.** 18 **23.** 2π **25.** 4
27. $30\ln 6$ **29.** $15\pi/4$ **31.** $\frac{16}{9}$ **33.** $\frac{35}{6}$
35. $\int_0^4 \int_{\sqrt{x}}^2 f(x, y)\, dy\, dx$
37. $\int_1^{e^3} \int_{\ln y}^3 f(x, y)\, dx\, dy$
39. $\int_0^1 \int_{y^3}^{2-y} f(x, y)\, dx\, dy$ **41.** $(2^{3/2} - 1)/18$
43. $\frac{2}{3}\sin 8$ **45.** $\pi/8$

Exercises 17.4, Page 859

1. $\bar{x} = \frac{8}{3}, \bar{y} = 2$ **3.** $\bar{x} = 3, \bar{y} = \frac{3}{2}$
5. $\bar{x} = \frac{17}{21}, \bar{y} = \frac{55}{147}$ **7.** $\bar{x} = 0, \bar{y} = \frac{4}{7}$

9. $\bar{x} = (3e^4 + 1)/[4(e^4 - 1)]$,
$\bar{y} = 16(e^5 - 1)/[25(e^4 - 1)]$
11. $\frac{1}{105}$ **13.** $4k/9$ **15.** $\frac{256}{21}$ **17.** $\frac{941}{10}$
19. $a\sqrt{10}/5$ **21.** $ka^4/6$ **23.** $16\sqrt{2}k/3$
25. $a\sqrt{3}/3$

Exercises 17.5, Page 864

1. $27\pi/2$ **3.** $(4\pi - 3\sqrt{3})/6$ **5.** $25\pi/3$
7. $(2\pi/3)(15^{3/2} - 7^{3/2})$ **9.** $\frac{5}{4}$
11. $\bar{x} = 13/3\pi, \bar{y} = 13/3\pi$
13. $\bar{x} = \frac{12}{5}, \bar{y} = 3\sqrt{3}/2$
15. $\bar{x} = (4 + 3\pi)/6, \bar{y} = \frac{4}{3}$ **17.** $\pi a^4 k/4$
19. $(ka/12)(15\sqrt{3} - 4\pi)$ **21.** $\pi a^4 k/2$
23. $4k$ **25.** 9π **27.** $(\pi/4)(e - 1)$
29. $3\pi/8$ **31.** 250 **33.** 2

Exercises 17.6, Page 868

1. $3\sqrt{29}$ **3.** $10\pi/3$ **5.** $(\pi/6)(17^{3/2} - 1)$
7. $25\pi/6$ **9.** $2a^2(\pi - 2)$ **11.** $8a^2$
13. $2\pi a(c_2 - c_1)$

Exercises 17.7, Page 877

1. 48 **3.** 36 **5.** $\pi - 2$ **7.** $\frac{1}{4}e^2 - \frac{1}{2}e$
9. 50
11. $\int_0^4 \int_0^{2 - (x/2)} \int_{x+2y}^4 F(x, y, z)\, dz\, dy\, dx$;
$\int_0^2 \int_{2y}^4 \int_0^{z - 2y} F(x, y, z)\, dx\, dz\, dy$;
$\int_0^4 \int_0^{z/2} \int_0^{z - 2y} F(x, y, z)\, dx\, dy\, dz$;
$\int_0^4 \int_x^4 \int_0^{(z - x)/2} F(x, y, z)\, dy\, dz\, dx$;
$\int_0^4 \int_0^z \int_0^{(z - x)/2} F(x, y, z)\, dy\, dx\, dz$
13. $\int_0^2 \int_{x^3}^8 \int_0^4 dz\, dy\, dx$;
$\int_0^8 \int_0^4 \int_0^{\sqrt[3]{y}} dx\, dz\, dy$;
$\int_0^4 \int_0^2 \int_{x^3}^8 dy\, dx\, dz$
15. 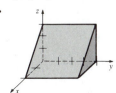   **17.**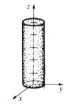

19.

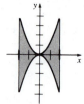

21. $16\sqrt{2}$ **23.** 16π

25. $\bar{x} = \frac{4}{5}, \bar{y} = \frac{32}{7}, \bar{z} = \frac{8}{3}$

27. $\bar{x} = 0, \bar{y} = 2, \bar{z} = 0$

29. $\int_{-1}^{1} \int_{-\sqrt{1-x^2}}^{\sqrt{1-x^2}} \int_{2y+2}^{8-y} (x + y + 4)\, dz\, dy\, dx$

31. $2560k/3; \sqrt{80/9}$ **33.** $k/30$

35. $k \int_{-5}^{5} \int_{-\sqrt{25-x^2}}^{\sqrt{25-x^2}} \int_{\sqrt{x^2+y^2}}^{5} \cdot (x^2 + y^2) \cdot \sqrt{x^2 + y^2 + z^2}\, dz\, dy\, dx$

Exercises 17.8, Page 886

1. $(-10/\sqrt{2}, 10/\sqrt{2}, 5)$ **3.** $(\sqrt{3}/2, \frac{3}{2}, -4)$

5. $(0, 5, 1)$ **7.** $(\sqrt{2}, -\pi/4, -9)$

9. $(2\sqrt{2}, 2\pi/3, 2)$ **11.** $(4, -\pi/2, 0)$

13. $r^2 + z^2 = 25$ **15.** $r^2 - z^2 = 1$

17. $z = x^2 + y^2$ **19.** $x = 5$

21. $(2\pi/3)(64 - 12^{3/2})$ **23.** $625\pi/2$

25. $(0, 0, 3a/8)$ **27.** $8\pi k/3$

29. $(\sqrt{3}/3, \frac{1}{3}, 0); (\frac{2}{3}, \pi/6, 0)$

31. $(-4, 4, 4\sqrt{2}); (4\sqrt{2}, 3\pi/4, 4\sqrt{2})$

33. $(2\sqrt{2}, 0, -2\sqrt{2}); (2\sqrt{2}, 0, -2\sqrt{2})$

35. $(5\sqrt{2}, \pi/2, 5\pi/4)$ **37.** $(\sqrt{2}, \pi/4, \pi/6)$

39. $(6, \pi/4, -\pi/4)$ **41.** $\rho = 8$

43. $\phi = \pi/6$ **45.** $x^2 + y^2 + z^2 = 100$

47. $z = 2$ **49.** $9\pi(2 - \sqrt{2})$

51. $2\pi/9$ **53.** $(0, 0, \frac{7}{6})$ **55.** πk

Chapter 17 Review Exercises, Page 887

1. True   **3.** False

5. $32y^3 - 8y^5 + 5y \ln(y^2 + 1) - 5y \ln 5$

7. Square region   **9.** $f(x, 4) - f(x, 2)$

11. $\int_{0}^{4} \int_{x/2}^{\sqrt{x}} f(x, y)\, dy\, dx$ **13.** $(\sqrt{2}, 2\pi/3, \sqrt{2})$

15. $z = r^2, \rho = \csc\phi \cot\phi$

17. $-y \cos y^2 + y \cos y^4$

19. $e^2 - e^{-2} + 4$ **21.** $1 - \sin 1$ **23.** $\frac{10}{3}$

25. 320π **27.** $\frac{37}{60}$

29. $\int_{0}^{1/\sqrt{2}} \int_{\sqrt{1-x^2}}^{\sqrt{9-x^2}} \frac{1}{x^2 + y^2}\, dy\, dx$

$+ \int_{1/\sqrt{2}}^{3/\sqrt{2}} \int_{x}^{\sqrt{9-x^2}} \frac{1}{x^2 + y^2}\, dy\, dx$

31.

33. $(1 - \cos 1)/2$ **35.** $5\pi/8$

37. $(2\pi/3)(2^{3/2} - 1)$

39. $\int_{0}^{1} \int_{x}^{2x} \sqrt{1 - x^2}\, dy\, dx;$

$\int_{0}^{1} \int_{y/2}^{y} \sqrt{1 - x^2}\, dx\, dy + \int_{1}^{2} \int_{y/2}^{1} \sqrt{1 - x^2}\, dx\, dy; \frac{1}{3}$

41. $41k/1512$ **43.** 8π

Exercises 18.1, Page 901

1. $-125/3\sqrt{2}; -250(\sqrt{2} - 4)/12; \frac{125}{2}$

3. $3; 6; 3\sqrt{5}$

5. $-1; (\pi - 2)/2; \pi^2/8; \sqrt{2}\pi^2/8$ **7.** 21

9. 30 **11.** 1 **13.** 1 **15.** 460

17. $\frac{26}{9}$ **19.** $-\frac{64}{3}$ **21.** $-\frac{8}{3}$ **23.** 0

25. $\frac{123}{2}$ **27.** 70 **29.** $-\frac{19}{8}$ **31.** e

33. -4 **35.** 0

37. On each curve the line integral has the value $\frac{208}{3}$.

41. $\bar{x} = \frac{3}{2}, \bar{y} = 2/\pi$

Exercises 18.2, Page 907

1. $\frac{16}{3}$ **3.** 14 **5.** 3 **7.** 330

9. 1096 **11.** $\phi = x^4y^3 + 3x + y$

13. Not a gradient field

15. $\phi = \frac{1}{4}x^4 + xy + \frac{1}{4}y^4$ **17.** $3 + e^{-1}$

19. 63 **21.** 16 **23.** $\phi = (Gm_1m_2)/|\mathbf{r}|$

Exercises 18.3, Page 912

1. $\frac{26}{3}$ **3.** 0 **5.** 972π

7. $(3^{5/2} - 2^{7/2} + 1)/15$ **9.** $9(17^{3/2} - 1)$

11. 36 **13.** $k\sqrt{3}/12$ **15.** 18 **17.** 28π

19. $-8\pi a^3$ **21.** $4\pi kq$ **25.** $(0, 0, a/2)$

Exercises 18.4, Page 915

1. $(4y^3 - 6xz^2)\mathbf{i} + (2z^3 - 3x^2)\mathbf{k}; 6xy$

3. $(3e^{-z} - 8yz)\mathbf{i} - xe^{-z}\mathbf{j}; e^{-z} + 4z^2 - 3ye^{-z}$

5. $(xy^2e^y + 2xye^y + x^3yze^z + x^3ye^z)\mathbf{i} - y^2e^y\mathbf{j}$
$+ (-3x^2yze^z - xe^x)\mathbf{k}; xye^x + ye^x - x^3ze^z$

31. 6

Exercises 18.5, Page 921

1. 3 **3.** 75π **5.** 48π **7.** $\frac{56}{3}$

9. $\frac{2}{3}$ **11.** $\frac{1}{8}$

13. $(b - a) \times$ (area of the region bounded by C)
17. $3a^2\pi/8$ **21.** $72 - 8\pi$ **23.** $3\pi/2$
27. 3π

Exercises 18.6, Page 929

1. -40π **3.** $\frac{3}{2}$ **5.** -3 **7.** $-3\pi/2$
9. $\frac{3}{2}$ **11.** $12a^5\pi/5$ **13.** $62\pi/5$
21. $\sqrt{3}\pi$

Chapter 18 Review Exercises, Page 931

1. True   **3.** False   **5.** True   **7.** True
9. True   **11.** True

13. $\nabla\phi = -\dfrac{x}{(x^2 + y^2)^{3/2}}\mathbf{i} - \dfrac{y}{(x^2 + y^2)^{3/2}}\mathbf{j}$

15. $6xy$ **17.** 0   **19.** 0   **21.** $56\sqrt{2}\pi^3/3$
23. 12   **25.** $2 + 2/3\pi$ **27.** $\pi^2/2$
29. $(\ln 3)(17^{3/2} - 5^{3/2})/12$ **31.** $-4\pi c$ **33.** 0
35. 125π **37.** 3π **39.** $\frac{5}{3}$

Exercises 19.1, Page 938

1. Linear, second-order
3. Nonlinear, first-order
5. Linear, fourth-order
7. Nonlinear, second-order
9. Linear, third-order
35. $m_1 = 2$, $m_2 = 3$;
corresponding solutions are $y_1 = e^{2x}$ and $y_2 = e^{3x}$
37. $m_1 = 0$, $m_2 = 4$;
corresponding solutions are $y = 1$ and $y = x^4$

Exercises 19.2, Page 942

1. $x \ln|x| + y = Cx$
3. $(x - y)\ln|x - y| = y + C(x - y)$
5. $x + y \ln|x| = Cy$
7. $\ln(x^2 + y^2) + 2 \tan^{-1}(y/x) = C$
9. $\ln|y| = 2\sqrt{x/y} + C$
11. $y^9 = C(x^3 + y^3)^2$
13. $(y/x)^2 = 2 \ln|x| + C$
15. $e^{2x/y} = 8 \ln|y| + C$
17. $x \cos(y/x) = C$ **19.** $y + x = Cx^2 e^{y/x}$

Exercises 19.3, Page 948

1. $y = Ce^{4x}$ **3.** $y = \frac{1}{10} + Ce^{-5x}$
5. $y = \frac{1}{4}e^{3x} + Ce^{-x}$ **7.** $y = \frac{1}{3} + Ce^{-x^3}$
9. $y = x^{-1}\ln x + Cx^{-1}$
11. $x = -\frac{4}{5}y^2 + Cy^{-1/2}$
13. $y = -\cos x + (\sin x)/x + C/x$
15. $y = C/(e^x + 1)$
17. $y = \sin x + C \cos x$
19. $y = \frac{1}{7}x^3 - \frac{1}{5}x + Cx^{-4}$
21. $(\sec \theta + \tan \theta)r = \theta - \cos \theta + C$

23. $y = 4 - 2e^{-5x}$
25. $(x + 1)y = x \ln x - x + 21$
27. $y = 2x/(x - 2)$ **29.** $x = \frac{1}{2}y + 8/y$
31. $i(t) = \dfrac{E_0 L}{R^2 + \omega^2 L^2}\left[\dfrac{R}{L}\sin \omega t - \omega \cos \omega t\right]$
$+ \left[i_0 + \dfrac{\omega E_0 L}{R^2 + \omega^2 L^2}\right]e^{-Rt/L}$
33. $A(t) = 600 + 2t - (4.95 \times 10^7)(300 + t)^{-2}$
35. $y = \begin{cases} 1 - e^{-x}, & 0 \le x \le 1 \\ (e - 1)e^{-x}, & x > 1 \end{cases}$
37. $y = 1/(Cx - x^2)$

Exercises 19.4, Page 955

1. $y = C_1 + C_2 e^{x/3}$
3. $y = C_1 e^{-4x} + C_2 e^{4x}$
5. $y = C_1 \cos 3x + C_2 \sin 3x$
7. $y = C_1 e^x + C_2 e^{2x}$
9. $y = C_1 e^{-4x} + C_2 x e^{-4x}$
11. $y = C_1 e^{(-3 + \sqrt{29})x/2} + C_2 e^{(-3 - \sqrt{29})x/2}$
13. $y = C_1 e^{2x/3} + C_2 e^{-x/4}$
15. $y = e^{2x}(C_1 \cos x + C_2 \sin x)$
17. $y = e^{-x/3}\left(C_1 \cos \dfrac{\sqrt{2}}{3}x + C_2 \sin \dfrac{\sqrt{2}}{3}x\right)$
19. $y = C_1 e^{-x/3} + C_2 x e^{-x/3}$
21. $y = 2 \cos 4x - \frac{1}{2}\sin 4x$
23. $y = -\frac{3}{4}e^{-5x} + \frac{3}{4}e^{-x}$
25. $y = -e^{x/2}\cos(x/2) + e^{x/2}\sin(x/2)$
27. $y = 0$ **29.** $y = e^{2(x-1)} - e^{x-1}$
31. $y'' + y' - 20y = 0$
33. $y = C_1 + C_2 e^{-x} + C_3 e^{5x}$
35. $y = C_1 e^x + C_2 e^{4x}\cos x + C_3 e^{4x}\sin x$
37. The auxiliary equation must be $(m - m_1)^2$
$= m^2 - 2m_1 m + m_1^2 = 0$. Thus, the differential equation
is $y'' - 2m_1 y' + m_1^2 y = 0$.

Exercises 19.5, Page 961

1. $y = C_1 \cos x + C_2 \sin x + x \sin x$
$+ \cos x \ln|\cos x|$
3. $y = C_1 \cos x + C_2 \sin x + \frac{1}{2}\sin x$
$- \frac{1}{2}x \cos x$
$= C_1 \cos x + C_3 \sin x - \frac{1}{2}x \cos x$
5. $y = C_1 \cos x + C_2 \sin x + \frac{1}{2} - \frac{1}{6}\cos 2x$
7. $y = C_1 e^x + C_2 e^{-x} + \frac{1}{4}x e^x - \frac{1}{4}x e^{-x}$
$= C_1 e^x + C_2 e^{-x} + \frac{1}{2}x \sinh x$
9. $y = C_1 e^{2x} + C_2 e^{-2x}$
$+ \dfrac{1}{4}\left(e^{2x}\ln|x| - e^{-2x}\displaystyle\int_{x_0}^{x}\dfrac{e^{4t}}{t}dt\right)$
$x_0 > 0$
11. $y = C_1 e^{-x} + C_2 e^{-2x} + (e^{-x} + e^{-2x})$
$\times \ln(1 + e^x)$
13. $y = C_1 e^{-2x} + C_2 e^{-x} - e^{-2x}\sin e^x$

15. $y = C_1 e^x + C_2 x e^x - \frac{1}{2} e^x \ln(1 + x^2)$
$\qquad + x e^x \tan^{-1} x$

17. $y = C_1 e^{-x} + C_2 x e^{-x} + \frac{1}{2} x^2 e^{-x} \ln x$
$\qquad - \frac{3}{4} x^2 e^{-x}$

19. $y = C_1 e^{x/2} + C_2 x e^{x/2} + \frac{8}{9} e^{-x} + x + 4$

21. $y = \frac{3}{8} e^{-x} + \frac{5}{8} e^x + \frac{1}{4} x^2 e^x - \frac{1}{4} x e^x$

23. $y = C_1 e^{-3x} + C_2 e^{3x} - 6$

25. $y = C_1 e^{-2x} + C_2 x e^{-2x} + \frac{1}{2} x + 1$

27. $y = C_1 \cos 5x + C_2 \sin 5x + \frac{1}{4} \sin x$

29. $y = C_1 e^{-x} + C_2 e^{3x} - \frac{4}{3} e^{2x} - \frac{2}{3} x^3 + \frac{4}{3} x^2$
$\qquad - \frac{28}{9} x + \frac{80}{27}$

31. $y = e^{4x}(C_1 \cos 3x + C_2 \sin 3x) + \frac{1}{10} e^{3x}$
$\qquad - \frac{126}{697} \cos 2x + \frac{96}{697} \sin 2x$

33. $y = \frac{5}{8} e^{-8x} + \frac{5}{8} e^{8x} - \frac{1}{4}$

35. $y = C_1 x + C_2 x \ln x + \frac{2}{3} x (\ln x)^3$

Exercises 19.6, Page 966

1. $y_1(x) = c_0 \sum_{n=0}^{\infty} \frac{(-1)^n}{(2n)!} x^{2n};$

$\qquad y_2(x) = c_1 \sum_{n=0}^{\infty} \frac{(-1)^n}{(2n+1)!} x^{2n+1}$

3. $y_1(x) = c_0$

$\qquad y_2(x) = c_1 \sum_{n=1}^{\infty} \frac{x^n}{n!}$

5. $y_1(x) = c_0 \left[1 + \frac{1}{3 \cdot 2} x^3 + \frac{1}{6 \cdot 5 \cdot 3 \cdot 2} x^6 \right.$

$\qquad \left. + \frac{1}{9 \cdot 8 \cdot 6 \cdot 5 \cdot 3 \cdot 2} x^9 + \cdots \right]$

$\qquad y_2(x) = c_1 \left[x + \frac{1}{4 \cdot 3} x^4 + \frac{1}{7 \cdot 6 \cdot 4 \cdot 3} x^7 \right.$

$\qquad \left. + \frac{1}{10 \cdot 9 \cdot 7 \cdot 6 \cdot 4 \cdot 3} x^{10} + \cdots \right]$

7. $y_1(x) = c_0 \left[1 - \frac{1}{2} x^2 - \frac{3}{4!} x^4 - \frac{21}{6!} x^6 - \cdots \right]$

$\qquad y_2(x) = c_1 \left[x + \frac{1}{3!} x^3 + \frac{5}{5!} x^5 + \frac{45}{7!} x^7 + \cdots \right]$

9. $y_1(x) = c_0 \left[1 - \frac{1}{3!} x^3 + \frac{4^2}{6!} x^6 - \frac{7^2 \cdot 4^2}{9!} x^9 + \cdots \right]$

$\qquad y_2(x) = c_1 \left[x - \frac{2^2}{4!} x^4 + \frac{5^2 \cdot 2^2}{7!} x^7 \right.$

$\qquad \left. - \frac{8^2 \cdot 5^2 \cdot 2^2}{10!} x^{10} + \cdots \right]$

11. $y_1(x) = c_0; \; y_2(x) = c_1 \sum_{n=1}^{\infty} \frac{1}{n} x^n$

13. $y_1(x) = c_0 \sum_{n=0}^{\infty} x^{2n}; \; y_2(x) = c_1 \sum_{n=0}^{\infty} x^{2n+1}$

15. $y_1(x) = c_0 \left[1 + \frac{1}{4} x^2 - \frac{7}{4 \cdot 4!} x^4 \right.$

$\qquad \left. + \frac{23 \cdot 7}{8 \cdot 6!} x^6 - \cdots \right]$

$\qquad y_2(x) = c_1 \left[x - \frac{1}{6} x^3 + \frac{14}{2 \cdot 5!} x^5 \right.$

$\qquad \left. - \frac{34 \cdot 14}{4 \cdot 7!} x^7 - \cdots \right]$

17. $y_1(x) = c_0 [1 + \frac{1}{2} x^2 + \frac{1}{6} x^3 + \frac{1}{6} x^4 + \cdots]$

$\qquad y_2(x) = c_1 [x + \frac{1}{2} x^2 + \frac{1}{2} x^3 + \frac{1}{4} x^4 + \cdots]$

19. $y = 6x - 2[1 + \frac{1}{2!} x^2 + \frac{1}{3!} x^3 + \frac{1}{4!} x^4 + \cdots]$

Exercises 19.7, Page 977

1. A weight of 4 lb ($\frac{1}{8}$ slug), attached to a spring, is released from a point 3 units above the equilibrium position with an initial velocity of 2 ft/sec. The spring constant is 3 lb/ft.

3. $x(t) = \frac{1}{2} \cos 2t + \frac{3}{4} \sin 2t$

5. $x(t) = -5 \sin 2t$

7. A weight of 2 lb ($\frac{1}{16}$ slug) is attached to a spring whose constant is 1 lb/ft. The system is damped with a resisting force numerically equal to 2 times the instantaneous velocity. The weight starts from the equilibrium position with an upward velocity of 1.5 ft/sec.

9. $\frac{1}{4}$ sec, $\frac{1}{2}$ sec, $x(\frac{1}{2}) = e^{-2} \approx 0.14$

11. $x(t) = e^{-2t}(\frac{1}{2} \cos 4t + \frac{1}{2} \sin 4t)$

13. Overdamped for $\beta > \frac{5}{2}$; critically damped for $\beta = \frac{5}{2}$; underdamped for $0 < \beta < \frac{5}{2}$

15. $x(t) = \frac{1}{625} e^{-4t}(24 + 100t)$
$\qquad - \frac{1}{625} e^{-t}(24 \cos 4t + 7 \sin 4t);$
$\qquad x(t) \to 0$ as $t \to \infty$

Chapter 19 Review Exercises, Page 978

1. True　　**3.** False　　**5.** True
7. True　　**9.** True　　**11.** $xy = C$

13. $y = -3 + C e^{10x}$

15. $y = \frac{1}{2x^2} e^x + \frac{1}{x^2} e^{-x}$

17. $y = \frac{1}{4} - 320(x^2 + 4)^{-4}$

19. $y = C_1 e^{(1-\sqrt{3})x} + C_2 e^{(1+\sqrt{3})x}$

21. $y = C_1 e^{-2x} + C_2 e^{5x}$

23. $y = -24 \cos 6x + 3 \sin 6x$

25. $y_1(x) = c_0 \left[1 - \frac{1}{3 \cdot 2} x^3 + \frac{1}{6 \cdot 5 \cdot 3 \cdot 2} x^6 \right.$

$\qquad \left. - \frac{1}{9 \cdot 8 \cdot 6 \cdot 5 \cdot 3 \cdot 2} x^9 + \cdots \right]$

$\qquad y_2(x) = c_1 \left[x - \frac{1}{4 \cdot 3} x^4 + \frac{1}{7 \cdot 6 \cdot 4 \cdot 3} x^7 \right.$

$\qquad \left. - \frac{1}{10 \cdot 9 \cdot 7 \cdot 6 \cdot 4 \cdot 3} x^{10} + \cdots \right]$

27. $y = e^x(C_1 \cos x + C_2 \sin x)$
$\qquad - e^x \cos x \ln |\sec x + \tan x|$

29. $y = \frac{1}{2} \cos x + \frac{1}{2} \sin x + \frac{1}{2} \sec x$

31. $y = C_1 e^{-3x} + C_2 e^{4x} - \frac{1}{10} x e^{2x} - \frac{13}{100} e^{2x}$

33. $0 < m \leq 2$

35. $x(t) = e^{-4t}\left(\dfrac{26}{17}\cos 2\sqrt{2}t + \dfrac{28\sqrt{2}}{17}\sin 2\sqrt{2}t\right)$
$\qquad + \dfrac{8}{17}e^{-t}$

Appendix I, Page A-9

1. 4 **3.** 5 **5.** $9r^2 - 24rs + 16s^2$

7. $8x^3 - 12x^2y + 6xy^2 - y^3$

9. $(2xy - 3)(2xy + 3)$

11. $(4a - b)(16a^2 + 4ab + b^2)$

13. $(x - 12)^2 - 144$ **15.** $2(x + 3)^2 - 18$

17. -3 **19.** $-\dfrac{3}{2}, \dfrac{3}{2}$ **21.** $0, \dfrac{2}{5}$

23. $-5, 3$ **25.** $-\dfrac{1}{2} - \dfrac{1}{2}\sqrt{3}, -\dfrac{1}{2} + \dfrac{1}{2}\sqrt{3}$

27. $14 + 3i$ **29.** $-38 + 8i$ **31.** $-\dfrac{9}{25} + \dfrac{13}{25}i$

33. $-\dfrac{3}{2} + \dfrac{1}{2}i$ **35.** i **37.** $10; -25$

39. 68 **41.** $-2, 4$ **43.** $x = -\dfrac{3}{5}, y = \dfrac{6}{5}$

45. $x = -4, y = 4, z = -5$

47. $x = \dfrac{12}{5}, y = \dfrac{9}{5}$ **49.** $b = 6.75$

Index